Biomechanics and Mechanobiology of Bone

Biomechanics of Living Organs

Biomechanics and Mechanobiology of Bone

From Model to Patient

Edited by

Luis Cardoso

Jean-Francois Ganghoffer

Mitchell B. Schaffler

ELSEVIER

ACADEMIC PRESS
An imprint of Elsevier

Academic Press is an imprint of Elsevier
125 London Wall, London EC2Y 5AS, United Kingdom
50 Hampshire Street, 5th Floor, Cambridge, MA 02139, United States

ISBN: 978-0-443-26739-0

For information on all Academic Press publications visit our website at https://www.elsevier.com/books-and-journals

Publisher: Mara Conner
Acquisitions Editor: Sonnini Yura
Editorial Project Manager: Pratishtha Gupta
Production Project Manager: Bharatwaj Varatharajan
Cover Designer: Raman Kumar

Typeset by TNQ Tech

Contents

Contributors

A. H. Rezwanuddin Ahmed
The City College of New York, Biomedical Engineering, New York, NY, United States

Matthew Allen
Indiana University School of Medicine, Department of Anatomy, Cell Biology, and Physiology, Indianapolis, IN, United States; Richard L. Roudebush Veterans' Administration Medical Center, Indianapolis, IN, United States

Peter Augat
Paracelsus Medical University, Institute of Biomechanics, Salzburg, Austria; Berufsgenossenschaftliche Unfallklinik Murnau, Institute of Biomechanics, Murnau, Germany

Subrata Bandhu Ghosh
Manipal University Jaipur, Engineered Biomedical Materials Research and Innovation Centre, Department of Mechanical Engineering, Jaipur, Rajastan, India

Sanchita Bandyopadhyah-Ghosh
Manipal University Jaipur, Engineered Biomedical Materials Research and Innovation Centre, Department of Mechanical Engineering, Jaipur, Rajastan, India

Emily D. Berestesky
Vanderbilt University, Department of Biomedical Engineering, Nashville, TN, United States

Eda Biricik
The City College of New York, Department of Biomedical Engineering, New York, NY, United States

Steven K. Boyd
University of Calgary, Department of Radiology, Department of Biomedical Engineering, Schulich School of Engineering, McCaig Institute for Bone and Joint Health, Calgary, AB, Canada

David Burr
Indiana University School of Medicine, Department of Anatomy, Cell Biology, and Physiology, Indiana Center for Musculoskeletal Health, Indianapolis, IN, United States

José Luis Calvo-Gallego
Departmento de Ingeniería Mecánica y Fabricación, Universidad de Sevilla, Seville, Spain; Instituto de Biomedicina de Sevilla (IBIS), Sevilla, Spain

Luis Cardoso
Department of Biomedical Engineering, City University of New York, New York, NY, United States

Alessandra Carriero
The City College of New York, Department of Biomedical Engineering, New York, NY, United States

Natalia M. Castoldi
Faculty of Mechanical Engineering, Technische Hochschule Nürnberg Georg Simon Ohm, Nuremberg, Germany

Cyrille Confavreux
Université Lyon 1, INSERM, LYOS UMR 1033, Lyon, France; Hôpital Lyon Sud, Hospices Civils de Lyon, Centre Expert des Métastases Osseuses (CEMOS), Lyon, France

Niovi Dollas
The City College of New York, Department of Biomedical Engineering, New York, NY, United States

Eve Donnelly
Cornell University, Department of Materials Science and Engineering, Ithaca, NY, United States; Hospital for Special Surgery, Skeletal Health and Orthopedic Research Program, New York, NY, United States

Hélène Follet
Université Lyon 1, INSERM, LYOS UMR 1033, Lyon, France

Susannah P. Fritton
The City College of New York, Department of Biomedical Engineering, New York, NY, United States

Jean-François Ganghoffer
Universite de Lorraine, LEM3 - CNRS - Arts et Métiers Paris Tech, Nancy, France

Marc Gardegaront
Université Eiffel, Université Lyon 1, LBMC UMR_T9406, Lyon, France

Rémy Gauthier
INSA Lyon, Université Lyon 1, CNRS, MATEIS, UMR5510, Villeurbanne, France

Ibrahim Goda
Universite de Lorraine, LEM3 - CNRS - Arts et Métiers Paris Tech, Nancy, France

Thomas Grenier
INSA Lyon, Université Lyon 1, CNRS, INSERM, CREATIS, UMR 5220, UMR S1294, Villeurbanne, France

Tanmay Gupta
University of Toronto, Department of Mechanical and Industrial Engineering, Toronto, ON, Canada

Kishore Hazarika
Manipal University Jaipur, Engineered Biomedical Materials Research and Innovation Centre, Department of Mechanical Engineering, Jaipur, Rajastan, India

Muyin Huang
Cornell University, Meinig School of Biomedical Engineering, Ithaca, NY, United States

Oran Kennedy
Royal College of Surgeons in Ireland, Department of Anatomy and Regenerative Medicine, Dublin, Ireland

Halima Kerdjoudj
Université de Reims Champagne Ardenne, UR BIOS, Reims, France

Aurélie Levillain
Université Eiffel, Université Lyon 1, LBMC UMR_T9406, Lyon, France

Robert Majeska
The City College of New York, Department of Biomedical Engineering, New York, NY, United States

Javier Martínez-Reina
Departmento de Ingeniería Mecánica y Fabricación, Universidad de Sevilla, Seville, Spain; Instituto de Biomedicina de Sevilla (IBIS), Sevilla, Spain

Bryn E. Matheson
University of Calgary, Department of Radiology, Department of Biomedical Engineering, Schulich School of Engineering, McCaig Institute for Bone and Joint Health, Calgary, AB, Canada

Corinne Metzger
Indiana University School of Medicine, Department of Anatomy, Cell Biology & Physiology, Indianapolis, IN, United States

David Mitton
Université Eiffel, Université Lyon 1, LBMC UMR_T9406, Lyon, France

Monzur Murshed
McGill University, Department of Medicine, and Faculty of Dental Medicine and Oral Health Sciences, Montreal, QC, Canada; Shriners Hospital for Children, Montreal, QC, Canada

Michał Nowak
Poznan University of Technology, Department of Virtual Engineering, Poznan, Poland

Jeffry S. Nyman
Vanderbilt University Medical Center, Department of Orthopaedic Surgery, Nashville, TN, United States; Vanderbilt University, Department of Biomedical Engineering, Nashville, TN, United States; United States Department of Veteran Affairs, Tennessee Valley Healthcare System, Nashville, TN, United States

Jean-Baptiste Pialat
INSA Lyon, Université Lyon 1, CNRS, INSERM, CREATIS, UMR 5220, UMR S1294, Villeurbanne, France; Hôpital Lyon Sud, Hospices Civils de Lyon, Centre Expert des Métastases Osseuses (CEMOS), Lyon, France; Hôpital Lyon Sud, Hospices Civils de Lyon, Service de radiologie, Lyon, France

Peter Pivonka
School of Mechanical, Medical and Process Engineering, Queensland University of Technology, Brisbane, QLD, Australia

Jan Polak
Poznan University of Technology, Department of Virtual Engineering, Poznan, Poland; Systems Research Institute of the Polish Academy of Sciences, Warszawa, Poland

Alexander Robling
Indiana University School of Medicine, Department of Anatomy, Cell Biology, and Physiology, Indiana Center for Musculoskeletal Health, Indianapolis, IN, United States

Emile Saillard
INSA Lyon, Université Lyon 1, CNRS, INSERM, CREATIS, UMR 5220, UMR S1294, Villeurbanne, France; Université Lyon 1, INSERM, LYOS UMR 1033, Lyon, France

Mohini Sain
University of Toronto, Department of Mechanical and Industrial Engineering, Toronto, ON, Canada

Vittorio Sansalone
Univ Paris Est Creteil, Univ Gustave Eiffel, CNRS, UMR 8208, MSME, Créteil, France

Mitchell Schaffler
The City College of New York, Department of Biomedical Engineering, New York, NY, United States

Jessica Schiavi-Tritz
University of Lorraine, CNRS, LRGP, Vandoeuvre-les-Nancy, France

Jan Sokołowski
Systems Research Institute of the Polish Academy of Sciences, Warszawa, Poland

Piyush Uniyal
KU Leuven, Department of Mechanical Engineering, Leuven, Belgium

Marjolein van der Meulen
Cornell University, Meinig School of Biomedical Engineering, Ithaca, NY, United States; Cornell University, Sibley School of Mechanical & Aerospace Engineering, Ithaca, NY, United States; Hospital for Special Surgery, New York, NY, United States

G. Harry van Lenthe
KU Leuven, Department of Mechanical Engineering, Leuven, Belgium

Deepak Vashishth
Rensselaer Polytechnic Institute - Icahn School of Medicine at Mount Sinai Center for Engineering and Precision Medicine, New York, NY, United States; Rensselaer Polytechnic Institute, Department of Biomedical Engineering, New York, NY, United States

Abhijit Vyas
Manipal University Jaipur, Engineered Biomedical Materials Research and Innovation Centre, Department of Mechanical Engineering, Jaipur, Rajastan, India; University Centre for Research and Development (UCRD), Chandigarh University, Mohali, Punjab, India

Joseph Wallace
Purdue University, Weldon School of Biomedical Engineering, Indiana Center for Musculoskeletal Research, Indianapolis, IN, United States

Matthias Walle
University of Calgary, Department of Radiology, Department of Biomedical Engineering, Schulich School of Engineering, McCaig Institute for Bone and Joint Health, Calgary, AB, Canada

Bowen Wang
Rensselaer Polytechnic Institute - Icahn School of Medicine at Mount Sinai Center for Engineering and Precision Medicine, New York, NY, United States

Sihong Wang
The City College of New York, Biomedical Engineering, New York, NY, United States

Quinton Wright
Cornell University, Department of Materials Science and Engineering, Ithaca, NY, United States

Elizabeth A. Zimmermann
McGill University, Faculty of Dental Medicine and Oral Health Sciences, Montreal, QC, Canada

Antoni Żochowski
Systems Research Institute of the Polish Academy of Sciences, Warszawa, Poland

Philippe K. Zysset
ARTORG Centre for Biomedical Engineering Research, University of Bern, Switzerland

Editors' biographies

Luis Cardoso is a Professor of Biomedical Engineering at the City College of New York (CCNY). He is a former postdoctoral fellow on the Bone Loss team at the NASA National Space Biomedical Research Institute (NSBRI), and he is currently the director of the multiscale and functional imaging laboratory at CCNY. A major focus of Dr. Cardoso's research is the biomechanics of the musculoskeletal system, using experimental, analytical, and numerical approaches to investigate the structure-function relationship of calcified tissues. Dr. Cardoso's laboratory has particular interests in ultrasound wave propagation in porous media, imaging of bone microarchitecture, and the characterization of bone mechanical competence using the Fabric Tensor. Dr. Cardoso is the author of 90 peer-reviewed papers in international journals, more than 150 peer-reviewed abstracts, 4 book chapters, and 6 patents. He has served on multiple review panels at the National Institutes of Health, the National Science Foundation, the National Aeronautics and Space Administration, and other international funding agencies, as well as a manuscript peer reviewer for several international journals. Dr. Cardoso has been an invited lecturer at several international venues, including the Gordon Conference and the Benjamin Franklin Medal in Biomedical Engineering Symposium. He has received several distinctions, including the New Investigator Recognition Award from the Orthopaedic Research Society in the USA, and was elected a Fellow of the American Institute for Medical and Biological Engineering (AIMBE).

Jean-François Ganghoffer was a CNRS fellow from 1992 to 2000, before being hired as Full Professor of mechanics at the Université de Lorraine (Nancy, France), where he is currently active. He is the head of the biomechanics team of LEM3 (http://www.lem3.univ-lorraine.fr), the Laboratoire d'Etude des Microstructures et de Mécanique des Matériaux. He is the leader of a national Federation of Laboratories (IMOA, Ingénierie Mécanobiologie ostéoarticulaire, FR CNRS 2003) in the field of osteoarticular mechanobiology. He has been the Director of a GDR research group on fibrous media. His activities cover bone biomechanics, homogenization methods, mechanics of fibrous media, the multiphysical behavior of architected materials, and topology optimization methods. He has developed different types of bone remodeling models over the past decade, and he has been more recently investigating piezoelectric and flexoelectric effects in bone. He is the author of about 300 scientific papers in international journals. He has been invited to many conferences and universities for both research and teaching activities (ETHZ, Georgia Tech, Metz, TU Delft, TU Eindhoven, UBC, and Vancouver). He is collaborating with many universities worldwide (Brazil, Germany, Italy, Bulgaria, Poland, Canada, and the US), and he has been invited to deliver lectures at many international conferences. He has published two books, including the monograph *Multiscale Biomechanics* in Wiley in 2018. He is a frequent reviewer for over 20 international journals. He is the 2025 laureate of the ERC Advanced Grant BONEREPAIR devoted to microrepair in bone using electromechanical stimulations. He was elected as a Fellow of the European Academy of Sciences in 2026. He received several distinctions, including the Beltrami Prize in 2017, the Grand Prix de la Société Industrielle de l'Est in 2005, and the scientific award of excellence by the French Ministry of Research for more than 10 years.

Mitchell B. Schaffler is a Distinguished Professor and Wallace H. Coulter Professor of Biomedical Engineering and Director of the Bone and Joint Laboratory at the City College of New York (CCNY), as well as Adjunct Professor of Orthpedic Surgery at the Icahn School of Medicine at Mount Sinai. His research aims to understand how skeletal tissues (bone, ligament, tendon, and cartilage) develop, maintain, and repair themselves in order to meet mechanical demands throughout life. In particular, he focuses on the cellular and integrative processes that control the architectural features of bone and tendon, and that govern how they respond to physical challenges normally and in aging and in diseases such as osteoporosis, genetic defects, and diabetes. He has longstanding research efforts focused on fatigue and repair in bone and tendon, with specific emphasis on discovering how living cells in these tissues detect and repair wear and tear damage before it accumulates to the point of mechanical failure. Dr. Schaffler's laboratory also examines how osteocytes (the tissue-resident bone cells) influence mechanical function, both directly by modulating local matrix composition, and indirectly by controlling local bone remodeling activities. In related studies, Dr. Schaffler's laboratory also examines how osteocytes function as mechanical sensors that allow bone to perceive and react to mechanical loading. He has published more than 150 full-length scientific papers in international journals and more than 400 transaction papers. He has been elected a Fellow of numerous prestigious societies, including the American Institute for Medical and Biological Engineering, the American Society for Bone and Mineral Research, the Orthopaedic Research Society, and the American Association of Anatomists. He received the AcroMed Award for Outstanding Research from the North American Spine Society and the Gideon Rodan Esteemed Award from the American Society for Bone and Mineral Research and has been an Editorial Board member of the *Journal of Orthopaedic Research*, *BONE*, and *The Anatomical Record*.

Foreword

It is a privilege to introduce *Biomechanics and Mechanobiology of Bone: From Model to Patient*. The title itself signals where the field has arrived. Bone research is no longer divided into separate conversations about tissue mechanics, cellular signaling, imaging, modeling, and clinical care. The urgent questions that now define the field require all of these perspectives at once: How does living tissue sense and interpret mechanical loading? How do matrix composition, architecture, hydration, and damage determine strength? How can mechanobiological knowledge be translated into better ways to predict fracture, treat disease, and restore function?

Throughout my own career in bone biomechanics, I have seen how much progress becomes possible when careful experimentation, rigorous mechanics, and an appreciation of cell biology are brought together. For many years, bone has served as one of the clearest examples of a living material that is both engineered and self-engineering. Its architecture follows mechanical need, but that adaptation is executed by cells embedded in a complex mineralized matrix. As students, many of us first encountered this principle through Wolff's law and through the elegant correspondence between trabecular patterns and load paths. Yet the modern field has moved far beyond description. We now ask how osteocytes and other bone cells perceive mechanical stimuli; how fluid flow, matrix strain, and biochemical signaling are coupled; how tissue-level mechanical properties emerge from microstructure and composition; and how computational models can predict the behavior of an individual bone or patient.

This volume, edited by Luis Cardoso, Jean-François Ganghoffer, and Mitchell Schaffler, provides a timely and much-needed synthesis of this expanded landscape. The editors are especially well-positioned to bring together the biological, mechanical, and computational dimensions of the field. Their organization of the book reflects the way rigorous bone research is now conducted: beginning with fundamental structure and physiology, proceeding through material and multiphysical characterization, using models to test mechanisms and predictions, and finally confronting the clinical and translational questions that give the work its human importance.

The opening part grounds the reader in bone structure, growth, adaptation, repair, matrix mineralization, cell function, mechanosensing, mechanotransduction, and the essential distinction between bone quantity and bone quality. This foundation is important because mechanics without biology can become an exercise in abstraction, while biology without mechanics can miss the physical cues that shape skeletal form and function. The chapters in this section remind us that bone strength begins not at the scale of a whole femur or vertebra, but in a coordinated system of cells, matrix, mineral, and remodeling.

The second part turns to bone as a hierarchical, hydrated, multiphase material. Poromechanics, fabric, volume fraction, experimental testing, water, whole-bone strength, fracture toughness, damage, and fatigue are treated not as isolated topics, but as connected determinants of skeletal performance. This integration is essential.

The ability of bone to resist fracture depends not only on mass or density, but also on architecture, anisotropy, tissue quality, crack initiation and propagation, and the effects of repeated loading over time. These chapters provide the mechanical vocabulary and experimental perspective necessary for understanding both normal skeletal function and disease-related fragility.

The modeling section is one of the book's particularly valuable features. Contemporary bone biomechanics depends increasingly on quantitative frameworks capable of linking scales and testing hypotheses that cannot be addressed by experiment alone. The chapters on adaptation models, topology optimization, microstructural analysis, osteocyte fluid flow, and in silico trials of osteoporosis treatments demonstrate how far the field has progressed. At the same time, they remind us that computational power is valuable only when paired with careful assumptions, well-defined parameters, and meaningful validation. The future of patient-specific skeletal assessment depends on this combination of mechanistic insight and rigorous modeling.

The final part moves toward clinical translation. Fracture pathology and management, quantitative computed tomography, reproducibility in numerical fracture prediction, biomaterials, additive manufacturing, and emerging frontiers are not peripheral to the field; they are central to its promise. Bone biomechanics and mechanobiology must ultimately improve the care of patients with osteoporosis, traumatic injury, skeletal defects, impaired repair, and age-related fragility. The inclusion of both technological advances and practical challenges gives this volume a welcome balance between ambition and realism.

I expect this book to serve several audiences well. Graduate students and postdoctoral researchers will find a structured map of a complex field. Engineers and physical scientists will gain a deeper appreciation of living tissue as an adaptive biological system. Biologists and clinicians will find the mechanical and computational concepts needed to interpret skeletal adaptation, strength, failure, and repair. Established investigators will value the breadth of expertise assembled here and the clarity with which the chapters connect methods to biological and clinical questions.

A successful reference in a rapidly advancing field must do more than summarize established knowledge. It must help readers ask better questions. This volume does that. It makes clear that the next advances in skeletal health will come from investigators who can move fluently across scales, methods, and disciplines—from molecular regulation and cellular mechanosensing to tissue mechanics, imaging, simulation, biomaterials, and patient care.

I congratulate the editors and contributing authors on producing a rigorous, timely, and forward-looking book. I am confident that *Biomechanics and Mechanobiology of Bone*: *From Model to Patient* will become an important resource for those who seek to understand bone as a living mechanical system and to translate that understanding into better science, better technologies, and better care.

X. Edward Guo, PhD
Stanley Dicker Professor of Biomedical Engineering
Professor of Medical Sciences (in Medicine)
Columbia University, New York, USA

Preface

The skeleton is a dynamic organ system shaped by a continuous interplay between physical forces and biological responses. To engineers, biologists, and clinicians alike, bone represents a remarkable multiscale system governed by complex material behavior and living feedback mechanisms. Far from being an inert structural material, bone continuously senses mechanical stimuli, adapts its architecture, and modulates its material properties in response to the physical demands of its environment. Understanding how bone detects mechanical strain, redistributes stress, remodels its microarchitecture, and ultimately fails under extreme or repetitive loading is central to modern biomedical engineering and clinical orthopedics.

This book explores the evolving scientific landscape at the interface between mechanics and biology by integrating fundamental physiology, advanced experimental mechanics, computational modeling, and clinical applications. It was conceived in response to the growing need for a comprehensive resource that captures the breadth of contemporary skeletal research. Leading international experts have contributed their specialized knowledge to create a multidisciplinary reference that bridges foundational cellular biology with rigorous mechanical and multiphysical frameworks. Together, these perspectives provide the conceptual and technical vocabulary required to advance the next generation of diagnostics, therapeutics, and regenerative strategies.

The evolution of a discipline

The idea that bone adapts to mechanical loading is not new. In the nineteenth century, Julius Wolff and his contemporaries observed the striking correspondence between trabecular architecture and principal stress trajectories. However, Wolff's Law remained primarily descriptive, and for more than a century the mechanisms linking physical activity to new bone formation remained poorly understood.

A major conceptual advance emerged in the mid-to-late twentieth century through the work of Harold Frost. Through the development of new experimental techniques, Frost established the foundations of modern bone histomorphometry. His work clarified that the physiological processes governing changes in bone shape and architecture—commonly associated with bone modeling—differ fundamentally from the remodeling processes responsible for maintaining tissue integrity and microarchitecture throughout life. Each process follows distinct biological and biomechanical principles.

Frost subsequently proposed that bone adaptation could be understood as a closed-loop mechanical feedback system. In this framework, the "mechanostat" functions as a regulatory mechanism capable of sensing whether mechanical strains fall below or exceed physiological thresholds and of signaling bone cells to remove or form tissue where necessary. This concept transformed the understanding of skeletal adaptation from a largely descriptive phenomenon into a quantitatively interpretable biological system.

The beginning of the 21st century marked a major biological revolution in skeletal research. The identification of the osteocyte as the principal mechanosensor, the discovery of the Wnt/β-catenin signaling pathway, and the recognition of strain amplification mediated by interstitial fluid flow within the lacunar-canalicular network profoundly reshaped the field. Scientific attention progressively shifted from asking whether bone adapts to mechanical loading to understanding precisely how mechanical forces alter cellular signaling, gene expression, and tissue organization. Bone mechanobiology entered an integrated era in which continuum mechanics, molecular biology, imaging, and computational modeling became deeply interconnected disciplines.

Rationale and philosophy of this book

Numerous outstanding texts address skeletal biology, while others provide rigorous treatments of continuum mechanics. Among these, Cowin's *Bone Mechanics Handbook* has long served as a foundational reference for graduate students and experts in bone biomechanics because of its successful integration of biological and mechanical perspectives without compromising the depth of either field. However, substantial advances in skeletal mechanobiology, experimental characterization, imaging, and computational modeling have emerged since the publication of its last edition in 2001. This volume therefore aims to provide an updated and integrative perspective on bone biomechanics and mechanotransduction.

The philosophy of this book is rooted in a structured and progressive framework that guides the reader from biological foundations to advanced mechanical characterization, predictive computational modeling, and ultimately clinical translation. The text is organized into four interconnected parts that reflect the contemporary workflow of skeletal research.

Part 1: Bone physiology, adaptation, and disease

The first part establishes the biological and physiological foundations of skeletal function. It examines bone structure, growth, and repair mechanisms, followed by the molecular regulation and hierarchical organization of matrix mineralization. Particular attention is devoted to the roles of resident bone cells and to the mechanisms of mechanosensing and mechanotransduction that govern skeletal adaptation. This section concludes by examining how these biological processes collectively determine bone quantity, quality, and susceptibility to disease.

Part 2: Mechanical and multiphysical properties of bone and constitutive laws

The second part shifts the focus toward engineering mechanics and physical principles. Bone is analyzed through the frameworks of poromechanics, constitutive behavior, and structure–property relationships across multiple length scales. The chapters present rigorous approaches to experimental bone mechanics and emphasize the critical role of water in determining whole-bone mechanical behavior. This section further investigates the physical determinants of skeletal strength through detailed analyses of fracture toughness, progressive damage accumulation, and cyclic fatigue.

Part 3: Biomechanics and mechanobiology modeling

The third part explores computational and mathematical approaches for simulating skeletal adaptation and mechanobiological processes. It traces the evolution of theoretical models from classical interpretations of Wolff's Law to contemporary multiscale paradigms. Advanced numerical frameworks are introduced, including topology optimization, microstructural analysis, and boundary-value formulations describing interstitial fluid flow around osteocytes. Particular emphasis is placed on the growing role of in silico clinical trials and predictive simulation platforms that integrate biochemical signaling pathways with pharmacokinetic-pharmacodynamic (PK–PD) models for evaluating osteoporosis therapies and other treatment strategies.

Part 4: Bone diseases, treatments, challenges, and future trends

The final part bridges theoretical models and laboratory discoveries with clinical realities. It examines the pathology and treatment of skeletal diseases and fractures while addressing the technical challenges associated with quantitative imaging and patient-specific assessment using computed tomography. This section also discusses important translational challenges, including reproducibility in fracture prediction models and the integration of computational tools into clinical workflows. Finally, it considers the global health and economic burden of skeletal disorders and highlights future directions for therapeutic and diagnostic innovation.

Intended audience

This work is intended for an interdisciplinary readership spanning biomedical engineering, orthopedics, biophysics, materials science, and computational mechanics. It is designed both as a primary textbook for graduate students and doctoral candidates and as a reference resource for established researchers and clinicians. Life scientists will find the poromechanical and computational frameworks clearly contextualized within biological systems, while mechanical and computational engineers will gain a deeper appreciation of the cellular and physiological processes that distinguish living skeletal tissue from conventional inert materials.

Acknowledgments

An interdisciplinary undertaking of this scope is possible only through the collaboration of an exceptional international community. We extend our deepest gratitude to the contributing authors—representing institutions across Australia, Austria,

Belgium, Canada, France, India, Ireland, Poland, Spain, Switzerland, and the United States—for generously sharing their expertise, perspectives, and scientific contributions.

We also thank the peer reviewers whose careful evaluations ensured that both the mechanical analyses and biological interpretations presented in this volume meet the highest standards of scientific rigor. We are deeply grateful to the editorial and production teams for their guidance in transforming a complex manuscript into a cohesive and accessible work. Finally, we thank our families for their unwavering encouragement and support throughout this project.

It is our sincere hope that this book will inspire the next generation of researchers and innovators. Despite remarkable progress, bone mechanobiology remains far from fully understood. Many questions remain unresolved within the microscopic architecture of the skeleton, awaiting new advances at the intersection of mechanics, biology, and medicine.

Luis Cardoso
Jean-François Ganghoffer
Mitchell B. Schaffler

Note of the series editors

We are very pleased to announce that the *Biomechanics of Living Organs* series, published by Elsevier, is enriched with this new exciting volume devoted entirely to the Mechanobiology of Bone. This series, which we have been coordinating since 2017, already accounted for the following eight volumes:

- *Hyperelastic Constitutive Laws for Finite Element Modeling of Living Organs (2017),*
- *Biomechanics of Coronary Atherosclerotic Plaque (2020),*
- *Biomechanics of Female Reproductive System, Breast and Pelvic Floor (2023),*
- *Reduced order Models for the Biomechanics of Living Organs (2023),*
- *Biomechanics of the Aorta (2024),*
- *Model Validation and Uncertainty Quantification in Biomechanics*
    - *Part I: From Soft Biological Tissue to Blood Flow (2026),*
    - Part II: Sources and Methods of *Uncertainty and Variability Analysis (2026)*, and
- *Biomechanics of the Heart (2026).*

Four additional volumes devoted to the *Biomechanics of Orofacial Organs, Machine Learning in Cardiovascular Biomechanics, Biomechanics of the Human Neuromusculoskeletal System*, and *Experimental Characterization of Soft Biological Tissues*, are currently in preparation. We are infinitely grateful to all the volume editors and chapter contributors, experts in their fields and internationally recognized leaders, for accepting our invitation to publish a book in our Elsevier series. On behalf of the international biomechanics community and as series editors, we would like to express our sincere thanks. Without you, your hard work, motivation, and support, such a project for the successful development of the *Biomechanics of Living Organs* series would not have been possible.

The current new volume of our series, *Biomechanics and Mechanobiology of Bone: From Model to Patient,* presents (i) a survey on bone physiology, adaptation, and disease, including cellular aspects at the root of bone mechanotransduction and leading to bone growth and remodeling, and management of diseases, (ii) the mechanical, multiphysical, multiscale properties of bone and constitutive laws including bone viscoelasticity, poroelasticity, bone damage, fatigue and fracture, measurement techniques of bone properties, (iii) biosubstitutes and bone implants from the point of view of design, properties, computational modeling, and choice of biomaterials, and (iv) the most recent biomechanics and mechanobiology modeling approaches of bone, including biological models, multiphysical approaches, surface and volumetric growth models, and bone fluid flow models. This book aims to bridge the gap between modeling approaches and medical treatments for bone diseases and pathologies. It is therefore also intended for clinicians to help them adapt treatment based on predictive computational models to the specific conditions of each patient. This book offers students, researchers, and clinicians the opportunity to acquire, in a single monograph written by experts, in-depth knowledge of bone physiology, the relationship between structure and function, and biomechanical aspects by studying the wide variety of topics covered.

We are pleased to offer our scientific community this long-awaited volume, which serves as a natural follow-up to Stephen C. Cowin's 2001 book *Bone Mechanics Handbook* and provides a detailed examination of key topics such as multiphysical phenomena and their modeling, the design of bone implants, and recent research on surface and volumetric growth.

Thank you, Luis, Jean-François, and Mitchell, for offering this fascinating book to our scientific community!

Jacques Ohayon, PhD

Yohan Payan, PhD

Editors of the *Biomechanics of Living Organs* Series, Elsevier

May 15, 2026

PART

1

Bone physiology, mechanobiology, and adaptation

CHAPTER

Bone structure, growth, adaptation and repair

1

Corinne Metzger[1] and Matthew Allen[2, 3]

[1]*Indiana University School of Medicine, Department of Anatomy, Cell Biology & Physiology, Indianapolis, IN, United States;* [2]*Indiana University School of Medicine, Department of Anatomy, Cell Biology, and Physiology, Indianapolis, IN, United States;* [3]*Richard L. Roudebush Veterans, Administration Medical Center, Indianapolis, IN, United States*

1. Introduction

The skeleton is the foundational structural support element of the human body. It serves essential life functions, including protecting internal organs and facilitating weight-bearing activity. These functions are facilitated through intricate physiological processes that balance altering structure/properties while allowing continued weight-bearing activities. The dynamic processes of modeling and remodeling are most active during the early years of life while the skeleton is growing, yet the skeleton continues to renew and repair throughout the lifespan. To accomplish these goals, a combination of mineralized and nonmineralized bone/cartilage is produced and removed through coordinated processes in time and space. Tissue development and adaptation is driven by mechanical and endocrine signals, and supported by vascular tissue. This chapter is focused on overviewing these processes, highlighting fundamental principles and providing clinically relevant examples that illustrate function/dysfunction.

2. Bone composition

Bone is a classic example of a composite material. It includes minerals, organic material (predominantly collagen), and water. These components are arranged in a hierarchical fashion, from the nanoscale to macroscale. This results in a material and structure that are both rigid and tough. Such contrasting properties are essential to the function of bone. Being rigid/stiff, driven by the mineral component, enables the bone to support the body, protect vital organs, and provide the structure necessary for locomotion. Yet a structure that has high stiffness can be brittle, which can lead to catastrophic failure. The integration of organic material and water offset the mineral-driven brittleness, allowing the bone to absorb and dissipate energy to control internal damage.

Biomechanics and Mechanobiology of Bone. https://doi.org/10.1016/B978-0-443-26739-0.00011-9

3. Bone organization—Macro/Microscale

At the macroscale, bone is classified as cortical or cancellous (Fig. 1.1) (Burr, 2002). This classification is predominantly driven by the structural location of the bone, with cortical bone being predominantly the outer shell of bones and cancellous bone being internal (ends of long bones) and throughout other bones such as vertebrae. The main difference between the two types of bone is their porosity. Cancellous bone is more void (marrow) than bone matrix and cortical bone is more bone matrix than void.

3.1 Cortical bone

Cortical bone provides structural support. Healthy cortical bone has minimal porosity, typically <5%. The fundamental structural component of cortical bone is the osteon. Primary osteons are formed during development; secondary osteons are formed through remodeling. Central canals within osteons house the vascular supply for cortical bone.

3.2 Cancellous bone

Cancellous bone provides structural integrity but is often considered to have a more metabolic function. Structurally, it helps direct loads to the cortical bone. This can be appreciated by examining the trabecular architecture in complex loaded regions like the proximal femur. Trabecular bone is highly porous, it's almost universally more void space than bone matrix. A typical region of healthy cancellous bone will have

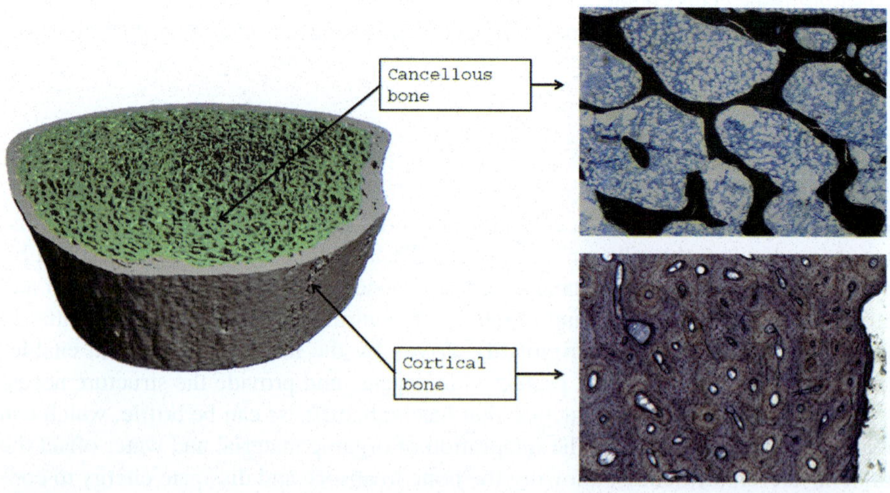

FIGURE 1.1

Macroscopically, bone exists as either porous cancellous bone or dense cortical bone.

~30% bone volume per unit of total volume (so porosity of 70%). The matrix of cancellous bone is arranged in struts, called trabecule. These are structurally described as being in the shape of rods or plates (although in reality, there is a continuum between these two contrasting shapes). Bone marrow exists within the void of the cancellous bone struts. The bone matrix within trabecule is organized as "hemi-osteons"—invoking the same sort of organization as cortical bone but without central canals (as the vascular supply exists within the adjacent marrow).

3.3 Blood supply and innervation

Bone is highly vascular. Cancellous bone receives its blood supply through vessels and sinusoids within the adjacent marrow. Cortical bone presents more unique vascular challenges due to its low porosity, which it overcomes through the central canal channels within osteons. The vascular network of most bones begins with one or more principle nutrient arteries, which pierce through the cortical bone through a nutrient foramen and then bifurcate into branches that supply the marrow and osteonal vessels. The periosteum is also highly vascularized, and likely contributes to nutrient delivery to some of the outer regions of the cortex. The bone is innervated by primary afferent nerves and autonomic fibers from the sympathetic nervous system, the latter most often associated physically and physiologically with the vascular system. The distribution of nerves differs between the mineralized matrix, the marrow and the periosteum, the latter being the most richly innervated.

4. Bone organization—Nanoscale

4.1 Collagen

Collagen represents the vast majority of the extracellular matrix of bone, specifically type I collagen (90% of total collagen in both with some Type III and V). Collagen molecules are produced by osteoblasts and, following exocytosis, undergo enzymatic cleavage of the propetpide regions. This produces N- and C-telopeptide fragments. These are traditional biomarkers of bone formation. Collagen aggregation occurs to form microfibrils and then fibers. In this process, collagen crosslinking occurs through both enzymatic and non-enzymatic processes. These crosslinks provide stability to the fiber and may have roles in biological processes.

4.2 Noncollagenous extracellular matrix

These components represent a relatively small portion of the extracellular matrix but serve important physiological functions. Most descriptions group these components into classes. Proteoglycans, defined as a core protein with glyscoaminoglycan sidechains, are a group of molecules that regulate mineralization, collagen fiber formation/growth, and cell function. More commonly studied members of this class are heparan sulfate, biglycan, and fibromodulin. Glycoproteins, a protein molecule with branched chains formed through glycosylation, that play prominent roles in bone include alkaline phosphatase and fibronectin. These mainly influence bone formation, with bone-specific alkaline phosphatase levels considered a useful biomarker for bone formation/mineralization. Phosphoproteins, a protein with phosphate groups attached, with prominent skeletal impact include bone saliaprotein, dentin-matrix-protein 1, and osteopontin. The components all have notable influences on bone formation and mineralization. Additionally, osteocalcin is a noncollagenous matrix protein that enhances calcium binding and regulates mineral deposition—it is often used as a biomarker of bone formation.

4.3 Bone mineral

Bone mineral, hydroxyapatite, nucleates within the gaps of the collagen fiber network. The initial deposition of mineral is predominantly calcium phosphate and calcium carbonate; as the mineral matures the carbonate level goes down. The mineral crystals congregate and coalesce into aggregates over time. Incorporation of the mineral into the bone matrix occurs in two phases, a relatively rapid primary mineralization where the crystals are deposited followed by a more prolonged secondary mineralization where the crystals grow and mature.

4.4 Water

Water exists in various bone compartments. Free water (also called pore water) can freely move through the vascular/lacunar-canalicular system. It is likely important in cell nourishment and signal transmission among cells. Bound water, on collagen, mineral, or between the two, is divided into three categories. Loosely bound water exists at the collagen-mineral interface and is thought to be important in load transfer and sliding between the two divergent materials. As the name implies, this category of water is thought to be labile. Tightly bound water exists within the collagen matrix, serving as a stabilizer of the collagen. Structural water is found within the mineralized crystals and is a component of the mineral aggregation.

5. Bone development

Bones are formed through one of two processes: intramembranous or endochondral ossification (Allen & Burr, 2019, pp. 85–100). Intramembranous ossification is the process of formation within most bones in the skull, endochondral ossification is the process for most long bones. The key difference between the two processes is that the intramembranous process occurs through formation of only bone tissue while the endochondral process occurs through the formation of a cartilage template that is eventually remodeled into bone tissue.

5.1 Endochondral ossification

Most bones in the skeleton are formed through endochondral ossification (Allen & Burr, 2019, pp. 85–100). They are originally formed as cartilaginous templates, comprised mainly of hyaline cartilage except for some select regions that are fibrocartilage (such as regions between vertebrae). The process to transform the cartilage template, termed an anlage, is highly coordinated in space and time (Fig. 1.2). Through a mechanism that is not well understood, cells within the perichondrium (a fibrous covering on the cartilage template) differentiate into osteoblasts. These cells produce osteoid (which is unmineralized bone matrix) that then mineralizes to form a thin bone shell, termed a "bone collar" on the cartilaginous anlage. The bone collar typically forms near the mid-diaphysis of the bone, extending proximally and distally over time. The formation of the mineralized collar around the bone alters the dynamics of nutrient flow to the non-vascularized cartilage

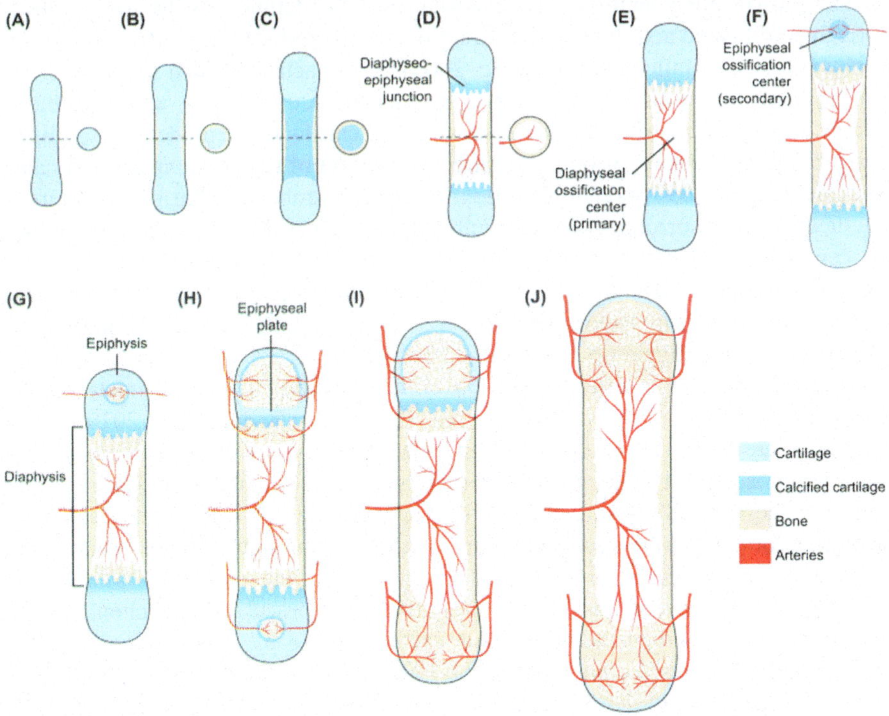

FIGURE 1.2

The process of endochondral ossification begins with a cartilage template that transforms into bone through a series of stages involving the coordinated activity of chondrocytes, osteoblasts, and osteoclasts.

- Reproduced with permission from Allen, M. R. & Burr, D. B. (2019). Bone growth, modeling, and remodeling (pp. 85–100). Elsevier BV. https://doi.org/10.1016/b978-0-12-813259-3.00005-1

anlage– effectively reducing isolating chondrocytes within the central region from freely receiving nutrients through diffusion. The bone collar is eventually pierced by vasculature, which penetrates the central portion of the cartilaginous analge. This forms the primary ossification center, where the central chondrocytes and cartilaginous tissue are removed and replaced by marrow cells and bone tissue.

The primary ossification center grows proximally and distally over time, forming an enlarged bone marrow space and a defined cortical shell. Meanwhile, the original cartilaginous analge continues to grow longitudinally (see the section below on Longitudinal growth), through a combination of appositional and interstitial growth of the matrix. Secondary ossification centers eventually form when vessels pierce the bone collar near each end of the bone. This produces segmented regions, called epiphyses, metaphyses and diaphysis. The interface between the epiphysis and metaphysis is called the growth plate. As the name implies, the growth plates represent the regions that permit the continued growth of the skeleton that is otherwise mineralized.

The growth plate facilitates longitudinal growth (detailed in the next section) through coordinated activities of the chondrocytes to produce new matrix and hypertrophy (Fig. 1.3) (Ballock and O'keefe, 2003). These activities lead to the compaction of the surrounding bone matrix, restricting nutrient flow and leading to calcification of the matrix. Over time the calcified cartilage is remodeled into mineralized bone. This occurs through the direct formation of new bone matrix on calcified cartilage surfaces (modeling), often resulting in trabecule that have centers of calcified cartilage surrounded by mineralized bone. Over time, through remodeling, the central core of calcified cartilage is removed and replaced with mineralized bone. Growth plate dynamics are regulated by various factors, including sex hormones and circulating factors like IGF-1, which alter the speed of longitudinal growth and eventually cause growth to cease.

Osteochondrosis is a term for multiple clinical diseases that are characterized by joint pain and bone deformity (McCoy et al., 2013; West & Jaramillo, 2019). These conditions are the result of disrupted endochondral ossification. There are various osteochondroses, such as Legg-Calve-Perthes and Osgood-Schlater, which are named based on the skeletal site affected. A key pathogenesis of osteochondrosis is the disruption of the vascular supply. This can occur through various mechanisms including trauma, genetics, or various other factors. The manifestation of osteochondrosis illustrates the essential nature of vasculature in the process of endochondral ossification. The bone within the ossification center affected by reduced flow becomes necrotic. This disrupts the normal temporal sequence and slows growth. If not corrected, the disruption of development can be permanent, leading to altered longitudinal growth or altered bone shape (both of which can alter bone mechanics and gait).

Genetic conditions such as Prader–Willi syndrome, Turner syndrome, and Noonan syndrome can cause short stature. These conditions are caused by different genetic defects, but they all share poor linear growth as a common feature, driven by disruptions in the normal process of growth plate dynamics. In the same way, conditions such as Beckwith-Wiedemann syndrome, Sotos syndrome,

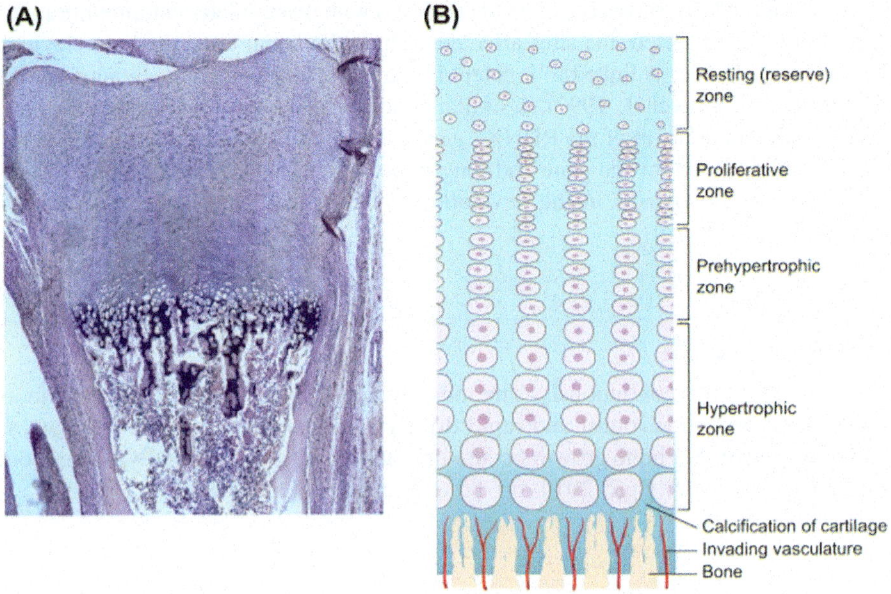

FIGURE 1.3

The epiphyseal growth plate is classified into five zones (A and B) based on the different histologic appearance of the cells and matrix: Resting (reserve) zone; proliferative zone; hypertrophic zone; zone of calcified cartilage; and zone of ossification. As long bone growth occurs, cells progress through the different zones. In the low power photomicrograph (A), the individual delineations between zones are not obvious, although the overall structure of the region can be appreciated. (B) The schematic more clearly shows the cell/tissue morphology of the zones within the growth plate.

- Reproduced with permission from Allen, M. R. & Burr, D. B. (2019). Bone growth, modeling, and remodeling (pp. 85–100). Elsevier BV. https://doi.org/10.1016/b978-0-12-813259-3.00005-1

Simpson-Golabi-Behmel syndrome, and Weaver syndrome can result in tall stature due to accentuated/prolonged growth plate activity.

5.2 Intramembranous ossification

A few bones in the skeleton, most of those in the skull along with select others such as the clavicle, are formed through direct deposition of the bone matrix without a cartilage template (Allen & Burr, 2019, pp. 85–100). The process begins with condensation of mesenchymal tissue. Cells within the condensed mass differentiate into osteoblasts that begin to produce unmineralized bone matrix. The initial matrix is unorganized (woven)—but is eventually remodeled into lamellar bone. Some bones formed through intramembranous ossification develop marrow cavities while others do not. Although most prominent during embryonic development, intramembranous ossification occurs during fracture healing.

Cleidocranial dysostosis is a clinical condition with dysregulated intramembranous ossification. As expected, the sites affected include the facial bones and the clavicle. This genetic disorder is linked to a defect in the RUNX2 gene, an essential driver of bone formation (Otto et al., 1997). A similar condition can be recapitulated in animals, where complete removal of the RUNX2 gene results in no bone formation, while animals heterogeneous for the gene had long bone formation (via endochondral ossification) yet severe defects in the development of the skull and clavicle (Mundlos et al., 1997).

6. Bone growth

The skeleton grows continuously. The most dynamic growth occurs early in life as longitudinal growth is robust and rapid driven by dynamics at the growth plate as described above. Yet bone growth occurs in other ways, mainly radially, throughout the lifespan (Allen & Burr, 2019, pp. 85–100; Carter et al., 1996).

6.1 Longitudinal growth

The bone anlage, formed as a starting point for endochondral ossification, experiences some longitudinal growth while still a cartilage template. This occurs through two matrix producing processes. Appositional growth occurs when chondrocytes/chondroblasts on the outer surface of the anlage produce matrix directly on the surface. This can extend the anlage both radially and longitudinally. Chondrocytes embedded in the matrix can also produce matrix, a process known as interstitial growth, where the new matrix effectively grows from within to extend the anlage in any/all directions. Once the bone collar forms early in endochondral ossification, appositional formation can only occur on those surfaces not covered by the bone collar and interstitial growth can only extend the bone longitudinally (as it's restricted from expanding radially by the collar).

As the bone collar forms on the diaphyseal shaft and primary/secondary centers of ossification form, the growth plate comes into existence and becomes the focal point for longitudinal growth. The growth plate has five distinct "zones" that are morphologically distinct (Fig. 1.3) although they exist as a continuum with cells progressing through each region during the growth process.

6.1.1 Resting zone

This region is nearest to the joint space at each end of the bone. It is comprised of chondrocytes embedded in a hyaline matrix. Although the region is often referred to as resting zone, the chondrocytes do produce matrix and thus contribute to growth both appositionally and interstitially. The chondroblasts that produce matrix on the outer surface of bone template arise from the perichondrium that is adjacent to the surface. This is a cellular/fibrous connective tissue that lays adjacent to the cartilage surface. As the cells produce matrix, some of them become embedded and become chondrocytes. Embedded chondrocytes also give rise to additional chondrocytes through cell division. These dividing cells form what are known as isogenous groups, clusters of cells within a single lacunar space. Eventually, these cells produce matrix and isolate themselves from each other in their own lacunae.

6.1.2 Proliferative zone

This is the site of active chondrocyte mitosis. It is a histologically distinct site as the cells are flattened and stacked, resembling a stack of coins or blueberry pancakes. The divided cells each produce matrix. Major regulators of cell proliferation in this zone include growth hormone, insulin-like growth factors, Indian hedgehog proteins, bone morphogenic proteins and several members of the Wnt-B-catenin signaling pathway.

6.1.3 Hypertrophic zone

The cells that were produced in the zone of proliferation grow in size through hypertrophic processes. This region produces growth as the cells increase their volume/surface area as much as 10-fold. The major factor promoting chondrocyte hypertrophy is thyroxine. Cells within this region transition from producing only Type II collagen to also producing Type X collagen. Type X collagen fibers have advantages for a stiffer matrix yet disadvantages in that they limit diffusion. This limits the nutrient supply in and out of the chondrocytes. Type X collagen fibers are also essential for signaling of vascular invasion of the matrix (described in the next region).

6.1.4 Calcified cartilage zone

The limitations of diffusion caused by Type X collagen and the compression of the matrix caused by hypertrophy eventually cause matrix calcification/mineralization. Calcification completely shuts off nutrient exchange and eventually, the chondrocytes die. Chondrocyte apoptosis plays an active role in matrix mineralization (through release of vesicles). These signals eventually stimulate vascular invasion and recruit chondroclasts, specialized cells that resorb calcified cartilage.

6.1.5 Ossification zone

In this region mineralized bone matrix is produced. Osteoblasts are recruited to the area and produce new bone matrix directly on surfaces of calcified cartilage. The new bone is typically woven (unorganized) although over time it is remodeled to produce lamellar bone. Longitudinal growth continues until cell division ceases. This is caused by chondrocyte senescence within the resting zone. Once division ceases, the growth plate calcifies. Estrogen plays a key role in growth cessation.

6.2 Radial growth

Long bones expand radially through appositional bone formation. This process is driven by osteoblasts that differentiate from precursors in the periosteum, a multilayer connective tissue membrane that covers most outer bone surfaces. The periosteum inner cambium layer is comprised of cells while the outer fibrous layer is mainly connective tissue. Periosteum thickness decreases with age, mainly due to fewer cells within the cambium layer.

The signals that drive appositional bone formation and outward expansion are likely multi-factorial but a key stimulus is mechanical loading (Robling et al., 2006). Numerous studies over the years have solidified the fact that increasing the mechanical load on the bone stimulates periosteal apposition and bone expansion. This is a finely tuned process, both in space and magnitude. Bone expansion occurs in proportion to the strain imparted on the bone. Expansion also occurs non-uniformly, as illustrated by experiments that show direct positive relationships between regions of periosteal bone formation and the regions of high strain.

Sex steroids exert a notable influence on a bones' radial growth across the lifespan (Seeman, 2003). During puberty, the rise in estrogen in women suppresses periosteal bone expansion while the rise in testosterone in men stimulates expansion (Fig. 1.4). Later in life, the loss of estrogen in women removes the inhibition in periosteal apposition, renewing radial bone growth. Modulation of radial growth through sex steroids, both in pre-clinical and clinical studies or conditions support the direct effects of these hormones on this process.

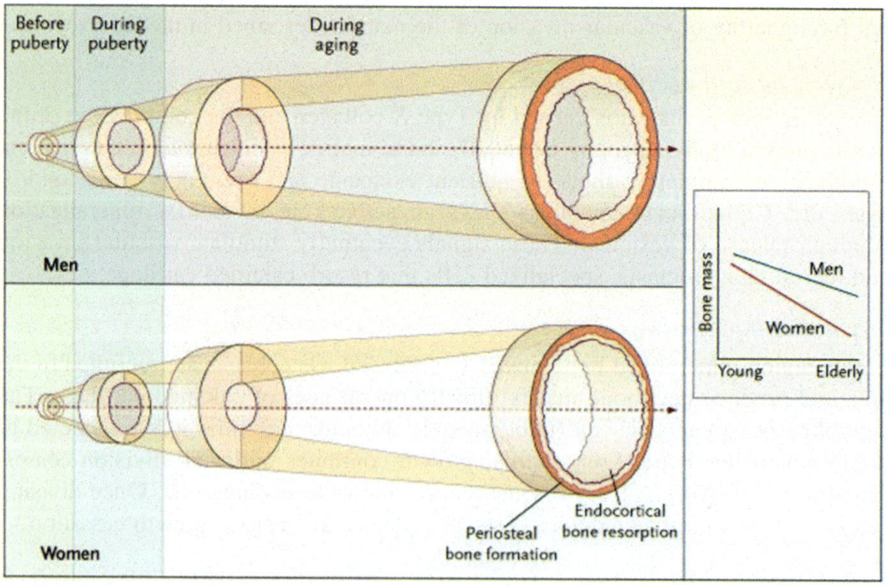

FIGURE 1.4

Differences between men and women in periosteal apposition and net bone loss during aging.

- Reproduced with permission from Seeman, Ego. (2003). Periosteal bone formation—A Neglected determinant of bone strength. New England Journal of Medicine, 349*(4), 320–323. https://doi.org/10.1056/nejmp038101*

Radial bone growth is important in the overall scope of bone health due to its' influence on mechanical properties. Engineering principles dictate that the cross-sectional moment of inertia, the distribution of material relative to a central axis, has important mechanical implications for a structure (Martin et al., 2015, pp. 355–422). Estimates suggest only a third of bone loss from the endocortical surface needs to be added to the periosteal surface to maintain similar mechanical integrity. The addition of periosteal bone can offset small increases in cortical porosity.

7. Bone modeling and remodeling

Bone adaptively responds to both internal and external stimuli through two processes—modeling and remodeling. Both processes work through osteoclast and osteoblast activity on the bone surface to resorb, form, or shape bone; however, there are distinct differences between these two processes.

7.1 Bone modeling

Bone modeling is the independent actions of osteoblast or osteoclasts on the surface of the bone. Modeling can occur on the trabecular, endosteal, or periosteal surfaces of bone (Fig. 1.5). An activation signal (i.e., mechanical loading) leads to either modeling formation or modeling resorption, but these two actions do not occur consecutively on the same bone surface. Modeling most often occurs during childhood, but can occur throughout the lifespan. For example, during longitudinal growth of the tibia, formation and resorptive modeling occur through the metaphyseal region to maintain the shape of the bone. Similarly, during radial growth of a long bone, formation modeling works on the periosteal surface to expand the bone while resorption modeling occurs on the endosteal surface to maintain the proper thickness of the cortical bone.

Modeling also often occurs in response to increased or altered mechanical forces on the bone. This effect is often most prominently seen before puberty; for example, when comparing bone mass between the dominant and non-dominant arm of tennis and squash players, those who began rigorous training prior to puberty had a greater differential in bone mass in their dominant play arm versus non-dominant than those who began after puberty (Kannus et al., 1995). Similarly, altered mechanical loading on bone can stimulate modeling. For example, in children that underwent fibular harvesting to correct an upper limb segment defect, subtle morphological differences in the shape of the tibia occurred over time due to slightly different loading patterns (Taddei et al., 2009).

A prominent clinical example of bone modeling is in correcting bones that are misshapen due to mineralization defects. Rickets, a condition resulting from vitamin D deficiency, results in bones that are undermineralized and, when it occurs during growth, can cause significant bowing of the leg bones. Once the underlying condition is corrected, the mineralized bone effectively moves through space (slowly) to

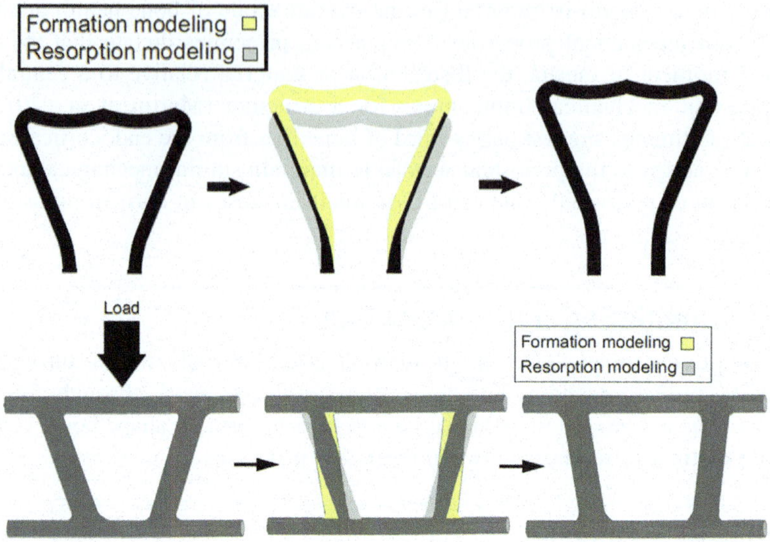

FIGURE 1.5

(a) As the bone grows longitudinally, there is coordinated modeling activity on the metaphyseal bone surfaces that serve to preserve the bone shape. (b) Modeling occurs on trabecular surfaces to normalize loading-induced strains. Areas of high strain undergo formation modeling, while regions of low strain are resorbed. These activities reorient individual trabecular structures in the direction of the principal loads.

- Reproduced with permission from Allen, M. R. & Burr, D.B. (2019). Bone growth, modeling, and remodeling (pp. 85–100). Elsevier BV. https://doi.org/10.1016/b978-0-12-813259-3.00005-1

eventually produce a straight bone (Frost, 1987). This occurs by formation modeling on some surfaces and resorption modeling on the others (Fig. 1.6).

7.2 Bone remodeling

In contrast to bone modeling, bone remodeling is a highly coordinated temporal and spatial event that includes osteoclasts and osteoblasts working in concert with osteocytes, marrow cells, blood vessels, and other local cell types to replace or renew. Importantly, bone remodeling involves bone resorption and bone formation occurring at the same bone site over time; contrasted to modeling which has surface-independent actions of these processes. Remodeling occurs on all bone surfaces - trabecular, endosteal, periosteal, as well as within the cortical shell (termed intracortical remodeling). Remodeling occurs throughout life and the net effect is maintenance of bone mass or a slight decrease in bone mass. The latter is heightened with aging or conditions that disrupt the normal balance of osteoblast and osteoclast activity. The remodeling process is typically described as having stages (Fig. 1.7).

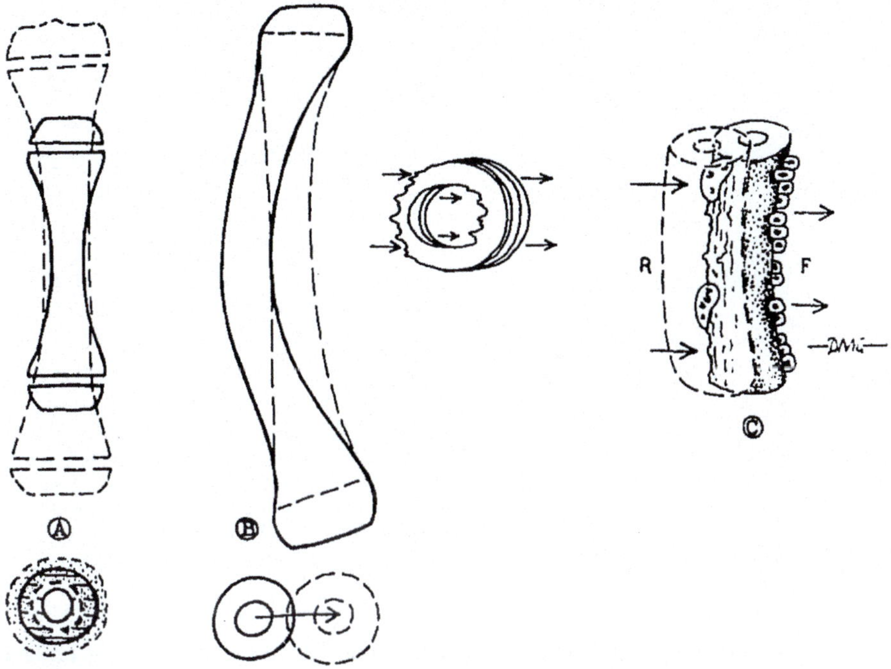

FIGURE 1.6

Bone modeling features (A) during normal bone growth, bone resorption and formation drifts move periosteal and cortical-endosteal surfaces through tissue space in patterned and stereotyped ways to maintain the overall bone shape in the face of great changes in size, as shown in longitudinal and cross-sections (called bone macromodeling). The term suggests that a micromodeling activity also occurs (B) when a child's fracture heals with a deformity, as shown here, resorption and formation drifts move the inner and outer surfaces of the shaft back to their normal location relative to the ends of the bone. (C) A resorption drift on the left (R) and a formation drift on the right (F).

- *Reproduced with permission from Frost, H. M. (1987). Osteogenesis Irnperfecta.* Clinical Orthopedics and Related Research, 216*(NA). https://doi.org/10.1097/00003086-198703000-00043*

7.2.1 Activation

Various signals can activate a remodeling cycle. Mineral homeostasis is a general signal. When serum calcium levels dip below an internal setpoint, parathyroid hormone is released from the parathyroid glands. In bone, parathyroid hormone triggers osteoclasts to resorb bone in an attempt to rapidly release calcium. This site of activation for this type of remodeling is thought to be random (often called stochastic remodeling) (Burr, 2002). Alternatively, there is remodeling that is more targeted, directed to occur at a specific site. Some signals for targeted activation of remodeling include microdamage, hypomineralization of bone, high collage crosslinks, osteocyte apoptosis, and other osteocyte signals like the release of receptor activator of

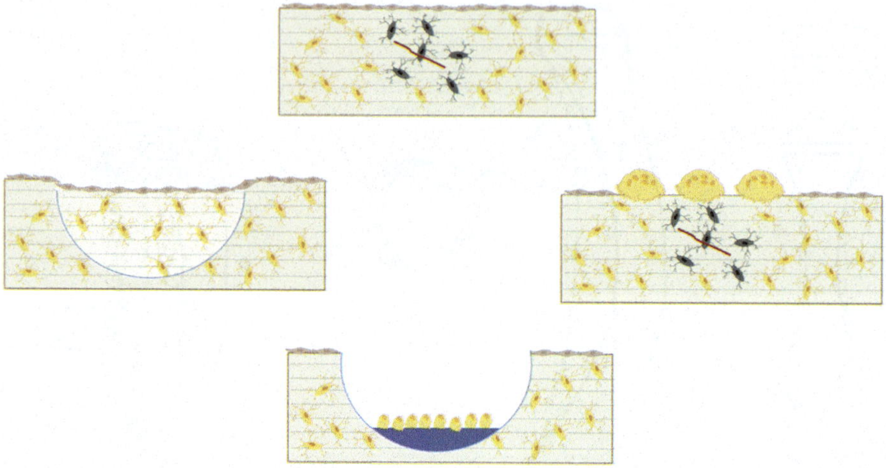

FIGURE 1.7

The remodeling cycle involves multiple stages that move in sequence from activation, to resorption, to formation and finally to quiescence. At any one time, remodeling cycles throughout the body are at various stages.

nuclear factor-κB ligand (RANKL). Targeted remodeling allows the bone to replace old or mechanically deficient bone in a precise way. Elegant studies using micro-damage as a signal for activation have shown that osteocytes near the damage produce pro-osteoclast signals while those more distant produce anti-apoptotic signals (Kennedy et al., 2012). This suggests a sort of "bullseye" for the remodeling unit to target.

7.2.2 Resorption

Following activation, bone resorption is initiated by osteoclasts recruited to the area by the activation signal. In humans, this process takes days to weeks. This suggests that stochastic remodeling is unlikely to result in a completely new remodeling event, and more likely the regulation of serum calcium is driven by modifications of existing remodeling units, or through processes independent of remodeling (such as liberation of calcium from osteocyte lacunar walls). At the start of the remodeling process on bone surfaces, the layer of bone lining cells retract to allow osteoclasts to interact with the bone mineral surface. Some evidence suggests bone lining cells may, rather than retract, elevate from the bone surface to form a canopy over the remodeling unit (Andersen et al., 2009). Once the process begins, additional osteoclasts may be recruited to the area, or existing osteoclasts may fuse. Recent data also support osteoclast fission, the role of this in resorption is unclear. Bone resorption cavities can have notable variation in size, in length and depth (on surfaces) and in length and diameter (intracortical remodeling).

7.2.3 Reversal

Following the exit of osteoclasts from the resorption surface, a brief period lasting only a few days in humans occurs where mononuclear "reversal cells" play a role in the remodeling cycle. There is very little known about these cells and their function. It is postulated that these cells do the final clean-up of the resorption pit, clearing away remaining collagen fragments and then lay an initial matrix that represents the reversal/cement line. This morphologically distinct entity allows clear demarcation of sites that have undergone remodeling, delineating where resorption ended and formation began. It is generally accepted that the reversal line has a different mineral content than the surrounding bone, although whether it is hypermineralized or hypomineralized is debated (Schaffler et al., 1987; Skedros et al., 2005).

7.2.4 Formation

Osteoblasts are recruited to the remodeling site and begin to lay down osteoid, the non-mineralized organic matrix of bone. Osteoid mineralizes in two phases, primary mineralization takes a few days and accounts for well over 50% of the final mineralization level, while secondary mineralization take much longer (months/years) (Fuchs et al., 2008). Many of the osteoblasts working on the bone surface undergo apoptosis, but some are buried in the osteoid matrix and eventually differentiate into osteocytes. The process of matrix formation within a remodeling unit can take up to 3 months in humans, with mineralization continuing for up to a year. Therefore, this step is the longest of the remodeling cycle.

7.2.5 Quiescence

Once formation is complete, the remodeling unit is covered by a single layer of bone lining cells and the surface is considered "resting".

Bone remodeling is often described as "coupled", meaning that the two processes, formation and resorption, are linked to each other. The signals that couple bone resorption to bone formation within the remodeling cycle have led to many studies to determine the underlying mechanisms. It is now clear there is not only one nor even a few coupling factors, but many derived from many cell sources. Coupling factors come from all bone cells (osteoblast, osteoclast, and osteocyte) as well as immune cells and other cells within the bone marrow. Therefore, the process of coupling is complex and influenced by multiple cells releasing many different factors that work in concert to accomplish the remodeling cycle (Sims & Martin, 2014, 2020). The fact that so many factors seem to play a role supports the concept of the remodeling canopy, where the bone lining cells help isolate the remodeling unit from the surrounding area and promote a higher concentration of local factors. Although canopies have been described in humans, there are still questions about the extent of their role in various aspects of bone physiology (Andersen et al., 2009; Wesseling-Perry, 2014).

Each bone remodeling unit has a bone balance, the relative amounts of bone resorbed and within a given unit (Fig. 1.8). Remodeling units in trabecular bone are typically considered to be either neutral or have a slightly negative balance.

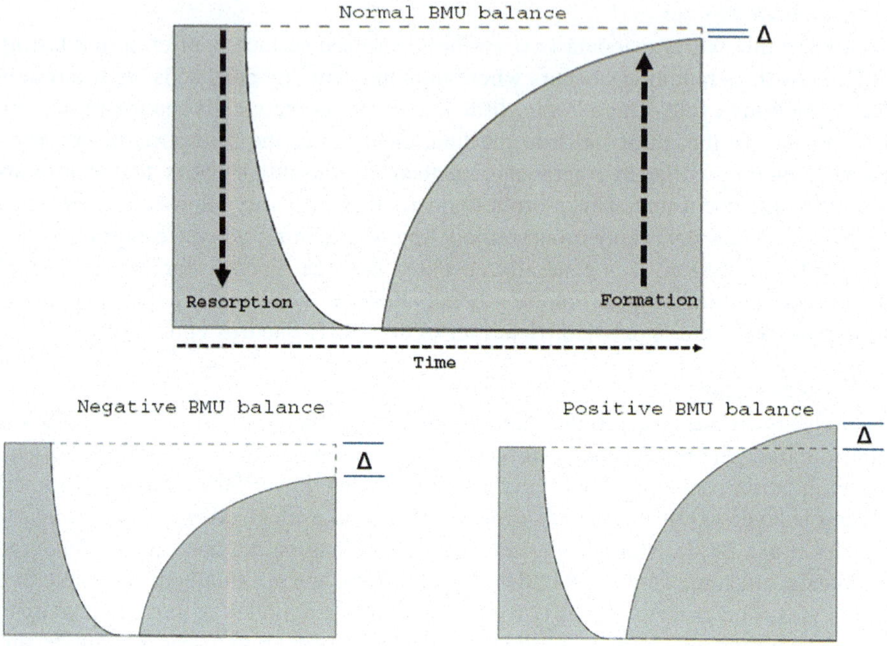

FIGURE 1.8

The amount of bone formed relative to the amount of bone resorbed by an individual bone remodeling unit is referred to as bone (or BMU) balance. In normal situations, BMU balance is slightly negative. In certain diseases, such as postmenopausal osteoporosis, BMU balance becomes even more negative, resulting in significant bone loss at each BMU. Pharmacological treatment with intermittent teriparatide/rhPTH has been shown to produce a positive bone balance at the BMU level.

- Reproduced with permission from Allen, M. R. & Burr, D. B. (2019). Bone growth, modeling, and remodeling (pp. 85–100). Elsevier BV. https://doi.org/10.1016/b978-0-12-813259-3.00005-1

Remodeling units in cortical bone are almost always in negative bone balance as each unit leaves a central canal. With aging, remodeling units, even in trabecular bone, tend to have a net negative balance due to age-related declines in osteoblast activity. Postmenopausal conditions accentuate the negative bone balance at each remodeling site, and combined with the increased rate of remodeling activity, accounts for the accelerated rate of bone loss during these years.

Several conditions highlight the impact of the bone remodeling cycle on overall bone health. For example, osteopetrosis is a rare heritable class of conditions characterized by a lack of osteoclasts and/or dysfunctional osteoclasts. In this condition, bone mass is high, but bones are brittle and fracture risk is elevated due to no mechanism to remove old bone. The ends of the bones are often mis-shapen due to the lack of osteoclasts that normally model bone into the correct shape. In some forms of renal osteodystrophy caused by chronic kidney disease, bone turnover is high (often caused by high circulating levels of parathyroid hormone) with increased bone resorption and bone formation resulting in bone loss and high fracture rates. As mentioned above, age-related bone loss is also caused in part by alterations in the bone remodeling cycle with maintained or increased bone resorption and decreased bone formation. Most approved therapeutics to protect bone work by altering the bone remodeling cycle. Antiresorptive therapeutics either inhibit the formation or function of osteoclasts leading to a suppression of bone resorption and, therefore, overall bone turnover. Anabolic therapies target osteoblasts to increase bone turnover and/or osteoblastic function to enhance bone formation. Understanding how a condition impacts bone turnover is important to determine which class of therapeutic options will best protect bone.

8. Fracture healing

When the force on a particular bone or part of a bone is greater than the strength of the bone, a fracture can occur. Fractures occur from a single traumatic event, like a fall, and due to repetitive forces in the case of stress fractures. Additionally, osteoporotic fractures can occur without significant trauma due to lower bone mass and altered bone quality. Once a fracture occurs, healing takes place either through primary or secondary methods.

Primary fracture healing occurs through a pure remodeling process. This is most notable when the bones are placed back in their original configuration and secured in a way that they do not move. Typically this would include the use of hardware (plates/screws). In primary bone healing, only bone remodeling occurs. Secondary osteons tunnel across the fracture line, remodeling the bone until there is no longer a crack.

The more common type of fracture healing is called secondary healing (Li et al., 2019, pp. 235–253). This occurs where bones are stabilized, but some movement is still occurring, such as with casting. In secondary healing, there is an initial callous that forms to provide initial stability to the site, and then over time this callous is remodeled into a healed bone. This process is often described as occurring in four stages (Fig. 1.9).

8.1 Hematoma formation

Bones have a rich blood supply which is disrupted following a fracture. Immediately post-fracture a clot forms which is the first stage in healing. A host of various cells arrive at the scene including platelets, macrophages, and neutrophils which in turn secrete cytokines. This pro-inflammatory response and the subsequent involvement of immune cells and cytokines is critical in fracture healing. Conditions with dysregulated immune responses like type 2 diabetes mellitus and autoimmune diseases often have longer fracture healing times and greater risk of complications believed to be caused by the altered immune response.

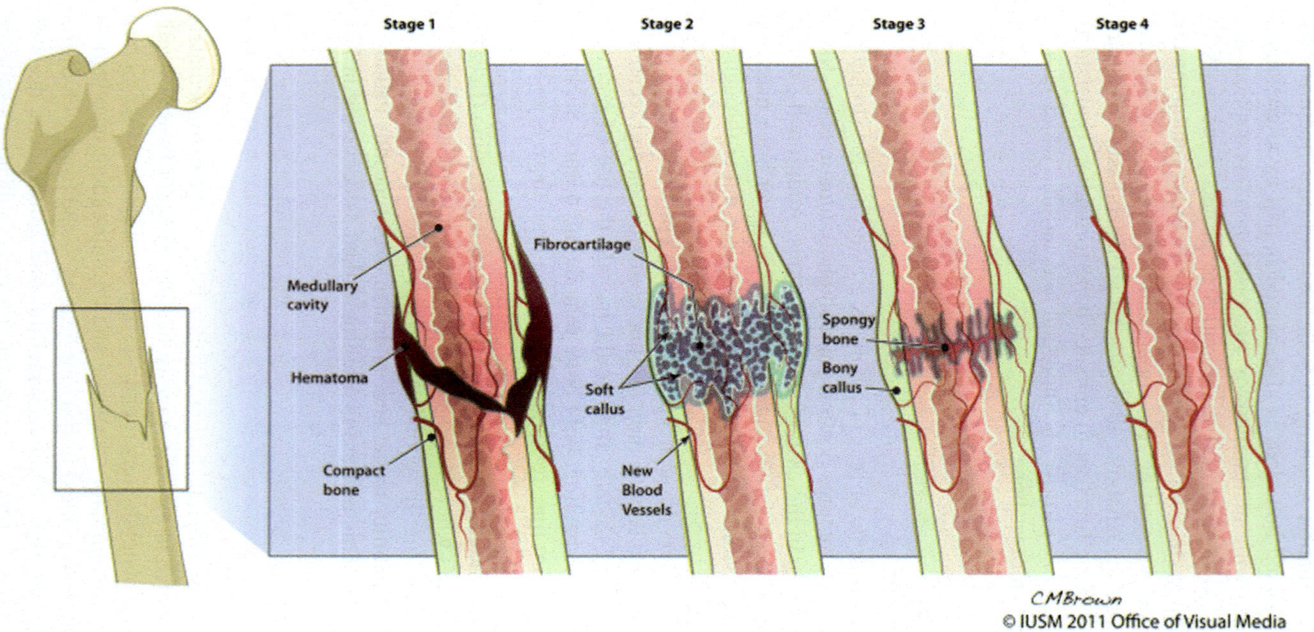

FIGURE 1.9

Stages of fracture repair. Stage 1: Following fracture or osteotomy, blood supply is disrupted and a blood clot (hematoma) forms. Stage 2: Progenitor cells in the periosteum and marrow differentiate into osteoblasts to facilitate intramembranous bone formation where an intact blood supply is preserved. In the fracture space where the tissue is hypoxic, the progenitor cells undergo chondrogenesis. The chondrocytes hypertrophy and their matrix becomes calcified, leading to chondrocyte apoptosis and invasion of the matrix by periosteal and marrow blood vessels. These vessels are accompanied by perivascular mesenchymal stem cells that differentiate into osteoblasts. Stage 3: The osteoblasts form woven bone on the calcified matrix. Stage 4: The remodeling process proceeds with osteoclasts and osteoblasts facilitating the conversion of woven bone into lamellar bone and eventually recreating the appropriate anatomical shape.

- Reproduced with permission from Li, J., Kacena, M. A., & Stocum, D. L. (2019). Fracture healing (pp. 235–253). Elsevier BV. https://doi.org/10.1016/b978-0-12-813259-3.00012-9

8.2 Fibrocartilaginous callus formation

Within roughly 2 weeks, a collagen-rich fibrocartilaginous tissue forms bridging the gap between the fracture ends. This serves the primary purpose of stabilizing the site. The extent of matrix formed is dependent on the extent of motion the site experiences during this phase. More motion will lead to higher amounts of fibrocartilage tissue compared to a site with lower amount of motion. Angiogenesis, the formation of new blood vessels, and proper blood supply remains important throughout these steps for oxygen and nutrient delivery as well as supplying a pool of progenitor cells.

8.3 Bony callus formation

Fibroblasts, likely derived from the endosteum and periosteum, differentiate into osteoblasts and the fibrocartilaginous callus undergoes some combination of endochondral ossification and/or intramembranous ossification. Woven bone, the initial bone formed at the site, serves the purpose of creating stability at the fracture site. Compared to lamellar bone, woven bone is disorganized, but rate of apposition of bone formation is 2–4 times faster than lamellar bone allowing for much faster stabilization of the site. This bony callus does not have the same structure as the rest of the bone and the strength of the woven bone is lower than lamellar bone. Yet, because a significant amount of woven bone is produced, the overall mechanical integrity of the site can be close to normal.

8.4 Remodeling

Over the months to years following formation of a bony callous, the bone undergoes remodeling to replace the woven bone with lamellar bone eventually leading to restored bone structure and strength. This stage represents the longest stage in the fracture healing process although adequate stability and strength is achieved prior to the complete remodeling of the site.

Research over the decades has shown the importance of the periosteum in fracture healing. The periosteum is a significant source of chondrocytes as well as an important factor in cortical bone blood perfusion. In pre-clinical models, removal of the periosteum resulted in significantly less chondrogenesis (Ozaki et al., 2000) and lower cortical bone perfusion (Kowalski et al., 1996). In the absence of the periosteum, non-union fractures are more common; however, skeletal muscle can provide the needed progenitors and support when the periosteum is damaged or insufficient (Liu et al., 2010; Rahmati et al., 2024). The surrounding skeletal muscle plays a critical role in the stages of fracture healing contributing the needed mesenchymal stem cells necessary for the development of the immediate callus as well as the following steps (Julien et al., 2021). Furthermore, it is likely that cross-talk between the bone and muscle contributes to the fracture healing process in multiple different ways (Rahmati et al., 2024). Therefore, bone repair does not occur in isolation and surrounding tissue and bone-muscle crosstalk enable the fracture healing process.

Additionally, restoration of proper blood flow and vascular supply to the fracture site is critical for proper healing. Multiple factors involved in angiogenesis have been shown to play key roles in fracture healing including vascular endothelial growth factor (VEGF). Poor vascularization at the site of injury is closely associated with non-union fracture (Stegen et al., 2015); however, recent research has demonstrated that focusing on blood flow alone for fracture healing overlooks the complexity of the multitude of factors involved in fracture healing (Menger et al., 2022). Therefore, fracture healing is a process involving many tissues and cell types working in concert to restore the bone to its original (or close to original) form and function.

References

Allen, M. R., & Burr, D. B. (2019). *Bone growth, modeling, and remodeling*. Elsevier BV. https://doi.org/10.1016/b978-0-12-813259-3.00005-1

Andersen, T. L., Sondergaard, T. E., Skorzynska, K. E., Dagnaes-Hansen, F., Plesner, T. L., Hauge, E. M., Plesner, T., & Delaisse, J. M. (2009). A physical mechanism for coupling bone resorption and formation in adult human bone. *American Journal Of Pathology, 174* (1), 239–247. https://doi.org/10.2353/ajpath.2009.080627

Ballock, R. T., & O'keefe, R. J. (2003). The biology of the growth plate. *Journal of Bone and Joint Surgery American Volume, 85*(4), 715–726. https://doi.org/10.2106/00004623-200304000-00021

Burr, D. B. (2002). Targeted and nontargeted remodeling. *Bone, 30*(1), 2–4. https://doi.org/10.1016/s8756-3282(01)00619-6

Carter, D. R., Van der Meulen, M. C. H., & Beaupré, G. S. (1996). Mechanical factors in bone growth and development. *Bone, 18*(1), S5. https://doi.org/10.1016/8756-3282(95)00373-8

Frost, H. M. (1987). Osteogenesis irnperfecta. *Clinical Orthopaedics and Related Research, 216*, 280–297. https://doi.org/10.1097/00003086-198703000-00043

Fuchs, R. K., Allen, M. R., Ruppel, M. E., Diab, T., Phipps, R. J., Miller, L. M., & Burr, D. B. (2008). In situ examination of the time-course for secondary mineralization of Haversian bone using synchrotron Fourier transform infrared microspectroscopy. *Matrix Biology, 27* (1), 34–41. https://doi.org/10.1016/j.matbio.2007.07.006

Julien, A., Kanagalingam, A., Martínez-Sarrà, E., Megret, J., Marine, L., Ménager, M., Relaix, F., & Colnot, C. (2021). Direct contribution of skeletal muscle mesenchymal progenitors to bone repair. *Nature Communications, 12*(1). https://doi.org/10.1038/s41467-021-22842-5

Kannus, P., Haapasalo, H., Sankelo, M., Sievanen, H., Pasanen, M., Heinonen, A., Oja, P., & Vuori, I. (1995). Effect of starting age of physical activity on bone mass in the dominant arm of tennis and squash players. *Annals of Internal Medicine, 123*(1), 27–31. https://doi.org/10.7326/0003-4819-123-1-199507010-00003

Kennedy, O. D., Herman, B. C., Laudier, D. M., Majeska, R. J., Sun, H. B., & Schaffler, M. B. (2012). Activation of resorption in fatigue-loaded bone involves both apoptosis and active pro-osteoclastogenic signaling by distinct osteocyte populations. *Bone, 50*(5), 1115–1122. https://doi.org/10.1016/j.bone.2012.01.025

Kowalski, M. J., Schemitsch, E. H., Kregor, P. J., Senft, D., & Swiontkowski, M. F. (1996). Effect of periosteal stripping on cortical bone perfusion: A laser doppler study in sheep. *Calcified Tissue International, 59*(1), 24–26. https://doi.org/10.1007/s002239900080

Li, J., Kacena, M. A., & Stocum, D. L. (2019). *Fracture healing*. Elsevier BV. https://doi.org/10.1016/b978-0-12-813259-3.00012-9

Liu, R., Schindeler, A., & Little, D. G. (2010). The potential role of muscle in bone repair. *Journal of Musculoskeletal and Neuronal Interactions, 10*(1), 71–76.

Martin, R. B., Burr, D. B., Sharkey, N. A., & Fyhrie, D. P. (2015). *Mechanical properties of bone*. Springer Science and Business Media LLC. https://doi.org/10.1007/978-1-4939-3002-9_7

McCoy, A. M., Toth, F., Dolvik, N. I., Ekman, S., Ellermann, J., Olstad, K., Ytrehus, B., & Carlson, C. S. (2013). Articular osteochondrosis: A comparison of naturally-occurring human and animal disease. *Osteoarthritis and Cartilage, 21*(11), 1638–1647. https://doi.org/10.1016/j.joca.2013.08.011

Menger, M. M., Laschke, M. W., Nussler, A. K., Menger, M. D., & Histing, T. (2022). The vascularization paradox of non-union formation. *Angiogenesis, 25*(3), 279–290. https://doi.org/10.1007/s10456-022-09832-x

Mundlos, S., Otto, F., Mundlos, C., Mulliken, J. B., S Aylsworth, A., Albright, S., Lindhout, D., Cole, W. G., Henn, W., M Knoll, J. H., J Owen, M., Mertelsmann, R., Zabel, B. U., & Olsen, B. R. (1997). Mutations involving the transcription factor CBFA1 cause cleidocranial dysplasia. *Cell, 89*(5), 773–779. https://doi.org/10.1016/s0092-8674(00)80260-3

Otto, F., Thornell, A. P., Crompton, T., Denzel, A., Gilmour, K. C., Rosewell, I. R., Stamp, G. W. H., Beddington, R. S. P., Mundlos, S., Olsen, B. R., Selby, P. B., & Owen, M. J. (1997). Cbfa1, a candidate gene for cleidocranial dysplasia syndrome, is essential for osteoblast differentiation and bone development. *Cell, 89*(5), 765–771. https://doi.org/10.1016/S0092-8674(00)80259-7

Ozaki, A., Tsunoda, M., Kinoshita, S., & Saura, R. (2000). Role of fracture hematoma and periosteum during fracture healing in rats: Interaction of fracture hematoma and the periosteum in the initial step of the healing process. *Journal of Orthopaedic Science, 5*(1), 64–70. https://doi.org/10.1007/s007760050010

Rahmati, M., Haffner, M., Lee, M. A., Leach, J. K., & Saiz, A. M. (2024). The critical impact of traumatic muscle loss on fracture healing: Basic science and clinical aspects. *Journal of Orthopaedic Research, 42*(2), 249–258. https://doi.org/10.1002/jor.25746

Robling, A. G., Castillo, A. B., & Turner, C. H. (2006). Biomechanical and molecular regulation of bone remodeling. *Annual Review of Biomedical Engineering, 8*, 455–498. https://doi.org/10.1146/annurev.bioeng.8.061505.095721

Schaffler, M. B., Burr, D. B., & Frederickson, R. G. (1987). Morphology of the osteonal cement line in human bone. *The Anatomical Record, 217*(3), 223–228. https://doi.org/10.1002/ar.1092170302

Seeman, E. (2003). Periosteal bone formation — A neglected determinant of bone strength. *New England Journal of Medicine, 349*(4), 320–323. https://doi.org/10.1056/nejmp038101

Sims, N. A., & Martin, T. J. (2014). Coupling the activities of bone formation and resorption: A multitude of signals within the basic multicellular unit. *BoneKEy Reports, 3*. https://doi.org/10.1038/bonekey.2013.215

Sims, N. A., & Martin, T. J. (2020). Osteoclasts provide coupling signals to osteoblast lineage cells through multiple mechanisms. *Annual Review of Physiology, 82*, 507–529. http://www.annualreviews.org/journal/physiol. https://doi.org/10.1146/annurev-physiol-021119-034425.

Skedros, J. G., Holmes, J. L., Vajda, E. G., & Bloebaum, R. D. (2005). Cement lines of secondary osteons in human bone are not mineral-deficient: New data in a historical perspective. *The Anatomical Record Part A Discoveries in Molecular, Cellular, and Evolutionary Biology, 286*(1), 781–803. https://doi.org/10.1002/ar.a.20214

Stegen, S., van Gastel, N., & Carmeliet, G. (2015). Bringing new life to damaged bone: The importance of angiogenesis in bone repair and regeneration. *Bone, 70*, 19–27. https://doi.org/10.1016/j.bone.2014.09.017

Taddei, F., Balestri, M., Rimondi, E., Viceconti, M., & Manfrini, M. (2009). Tibia adaptation after fibula harvesting: An in vivo quantitative study. *Clinical Orthopaedics and Related Research, 467*(8), 2149–2158. https://doi.org/10.1007/s11999-009-0776-z

Wesseling-Perry, K. (2014). The BRC canopy: An important player in bone remodeling. *American Journal Of Pathology, 184*(4), 924–926. https://doi.org/10.1016/j.ajpath.2014.01.004

West, E. Y., & Jaramillo, D. (2019). Imaging of osteochondrosis. *Pediatric Radiology, 49*(12), 1610–1616. https://doi.org/10.1007/s00247-019-04556-5

CHAPTER

2

Bone matrix mineralization: Molecular regulation, hierarchical structure, and emerging paradigms

Monzur Murshed[1,2]

[1]*McGill University, Department of Medicine, and Faculty of Dental Medicine and Oral Health Sciences, Montreal, QC, Canada;* [2]*Shriners Hospital for Children, Montreal, QC, Canada*

1. Introduction

Calcium phosphate mineral deposition within the collagen-rich extracellular matrix (ECM) in bones results in a unique composite material that carries tremendous load-bearing capacity with sufficient flexibility and toughness to support body weight. Mineralized skeleton facilitates rapid movement and mastication and also forms a protective covering for the soft tissues of many vital organs. At the same time, mineralized bones serve as a dynamic reservoir of calcium (Ca^{2+}) and phosphate (Pi), which are essential for many body functions, including muscle contraction, neuronal excitability, cellular signaling, and other physiological processes (Blaine et al., 2015). The skeletal system's ability to store and release these minerals forms an integral part of systemic mineral homeostasis, meeting physiological demands for these minerals even when dietary intake is insufficient.

From an evolutionary perspective, biomineralization emerged in the late Precambrian period, when rising mineral ion concentrations in seawater enabled marine organisms to develop calcified exoskeletons (Knoll, 2003). Early skeletons utilized both calcium carbonate and calcium phosphate minerals. Vertebrates ultimately evolved to carry internal skeletons mineralized primarily with calcium phosphate rather than calcium carbonate. A likely advantage of calcium phosphate is its greater stability under acidic conditions. During intense short burst physical activity (e.g., acute "fight or flight" responses that can involve anaerobic glycolysis), vertebrates produce lactic acid, leading to transient acidosis that could more readily demineralize a calcium carbonate–based skeleton than one mineralized with calcium phosphate (Ruben & Bennett, 1987). This vulnerability may have driven the evolutionary preference for calcium phosphate (apatite) as the skeletal mineral, offering greater resilience under physiological stress.

Biomechanics and Mechanobiology of Bone. https://doi.org/10.1016/B978-0-443-26739-0.00012-0

Bone mineralization is not merely a physico-chemical precipitation of salts, but a highly regulated biological process. Osteoblasts (bone-forming cells) actively direct this process by synthesizing the mineral scaffolding collagen-rich ECM and maintaining a complex balance of mineral ions and inhibitory molecules within the bone microenvironment (Murshed et al., 2005). There is strong evidence that these cells actively participate in the initial mineral nucleation process by releasing amorphous calcium phosphate (ACP)-filled vesicles (Mahamid et al., 2011). In the following sections, we explore the mechanism of bone mineralization and its determinants (Fig. 2.1) in detail.

2. Historical perspective

Our understanding of bone mineralization has dramatically changed over the past two and a half centuries. In the late eighteenth century, the Swedish chemist Carl Wilhelm Scheele first identified a phosphoric salt in bone, recognizing that bone

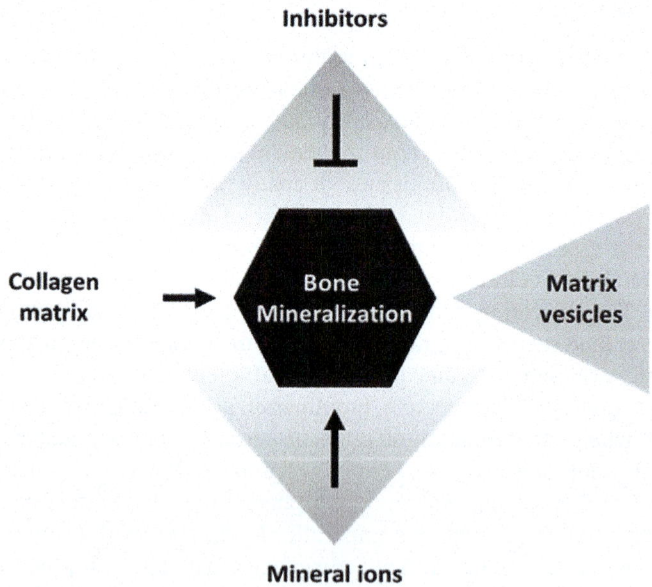

FIGURE 2.1 Key determinants regulating bone mineralization.

Bone extracellular matrix mineralization is governed by three key factors: The presence of a suitable collagen scaffold, systemic availability of calcium and phosphate ions, and the regulation of extracellular mineralization inhibitors within the bone matrix. Mineral deposition largely occurs through a collagen-guided process, facilitated by the absence or removal of inhibitors. In addition, intracellular enzymes contribute to the formation and release of specialized nanosized matrix vesicles and promote mineral nucleation within their protected microenvironment.

is largely composed of a calcium phosphate mineral (West, 2014). By the mid-20th century, studies using various analytical techniques had determined that bone mineral is a form of hydroxyapatite [$Ca_{10}(PO_4)_6(OH)_2$], similar to geological apatite (Termine & Posner, 1966; Weiner & Wagner, 1998).

Initially, many researchers believed that bone mineralization was a relatively passive process in which calcium and phosphate from body fluids simply crystallized onto the collagenous matrix once their concentrations exceeded a supersaturation threshold. However, even as the chemical nature of bone mineral was being elucidated, evidence for biological control of mineralization was emerging. In 1923, Robert Robison discovered that mineralizing cartilage and bone tissues contain high levels of an enzyme, alkaline phosphatase, and he postulated that this enzyme's role was to liberate inorganic phosphate (Pi) needed for mineral formation (Whyte, 2020). By 1932, Robison further hypothesized that an unknown "inhibitory factor" must exist to prevent unwanted calcification (Whyte, 2020).

In 1962, Harold Fleisch and colleagues provided the first conclusive evidence supporting Robison's hypothesis by identifying inorganic pyrophosphate (PPi) in plasma and demonstrating that PPi potently inhibits calcium-phosphate crystal growth (Fleisch & Bisaz, 1962). This finding provided a biochemical explanation for Robison's hypothetical inhibitor and revealed that the body actively produces such inhibitory molecules not only to prevent abnormal mineralization in soft tissues, but also to regulate the process of ECM mineralization in hard tissues. Around the same time, electron microscopy studies by Sheldon and Robinson showed mineral crystals forming in discrete locations within the collagen fibrils of developing bone (Sheldon & Robinson, 1957). In their images, minerals appeared within collagen fibrils near the calcification front, implying a structural guidance provided by the protein scaffold for crystal deposition and growth. Collectively, these observations established the notion that ECM mineralization is a genetically regulated process, and that cells in the mineralizing ECM produce both mineralization-promoting and inhibitory molecules. By confirming the active cellular involvement, this notion seriously challenged the initial assumption that bone ECM is mineralized by a passive process through spontaneous precipitation of calcium phosphate salts.

By the late 20th century, the concept of passive mineral precipitation in bone had been largely replaced by a model proposing the cell-mediated active regulation of ECM mineralization. Like many other branches of biology, new developments in gene manipulation techniques, culminating in the generation of numerous genetically modified mouse models, profoundly changed our understanding of basic biological pathways regulating bone mineralization and the pathophysiology of many associated diseases. This was further facilitated by the identification of new mouse models showing ECM mineralization abnormalities caused by spontaneous genetic mutations.

An explosion of genetic studies in the 1990s and 2000s identified numerous genes that, when mutated, lead to either deficient or excessive mineralization, underscoring that specific molecular pathways govern this process. For example, genetic experiments showing the effects of the *PHEX* gene mutation in X-linked hypophosphatemic rickets (Eicher et al., 1976; The HYP Consortium, 1995), and the *ENPP1* gene mutation in generalized arterial calcification of infancy (Nitschke & Rutsch, 2012; Okawa et al., 1998) demonstrated genetic control over phosphate regulation and PPi production, respectively. Likewise, knockout mouse models, e.g., mice lacking the vitamin D receptor or tissue-nonspecific alkaline phosphatase, reproduced human metabolic bone diseases and established the principle that the balance of mineral promoters and inhibitors is genetically programmed (Demay, 2013; Fedde et al., 1999).

More recently, advanced imaging techniques have provided direct visualization of the orchestrated process of bone mineralization. Cryo-electron microscopy and tomography of bone have captured intracellular mineral-containing vesicles in osteoblasts and initial mineral nuclei forming extracellularly within the unmineralized collagen matrix, or osteoid, in near-native (hydrated) states (Mahamid et al., 2011). Such observations confirm that bone mineralization is initiated and guided by cellular activities rather than occurring haphazardly as a passive process.

In summary, over two centuries we have moved from viewing bone mineralization as a simple chemical precipitation to recognizing it as a complex, regulated process involving specialized cellular structures, enzymes, inhibitors, and signaling pathways. This historical journey sets the stage for examining the detailed mechanisms by which bone ECM mineralization is regulated both locally and systemically.

3. Composition of bone mineral: Hydroxyapatite and amorphous precursors

The mineral phase of mature bone is predominantly composed of apatitic calcium phosphate crystals (Termine & Posner, 1966). However, these crystals are not uniform, ideal hydroxyapatite; rather, they are carbonated apatite with substantial carbonate and other ionic substitutions within the crystal lattice. This was established by multiple analytical methods, including chemical and infrared analyses and X-ray diffraction (Termine & Posner, 1966; Weiner & Wagner, 1998). High-resolution transmission electron microscopy (TEM) images have shown that bone apatite crystals are nanometer-scale units (Robinson, 1952) with dimensions on the order of tens of nanometers in length and only a few nanometers in thickness, as confirmed by later studies. These crystallites are typically plate-shaped and are organized within collagen fibrils with their long axes generally aligned with the collagen fibril axis in bone (Weiner & Traub, 1986). The small size and large surface area of bone crystals, along with their ionic substitutions, contribute to the relatively high solubility and surface reactivity of bone mineral (Rey et al., 2009). The presence of a relatively large interfacial area between the mineral and the organic matrix (collagen) has important implications for the mechanical properties of bone as a composite material.

One longstanding debate in bone biology has been how bone mineral initially forms: Does it precipitate directly as a crystalline apatite, or is there a precursor phase? In the 1960s and onward, in vitro mixing experiments were conducted with concentrated calcium and phosphate solutions. These experiments demonstrated the rapid formation of an amorphous calcium phosphate precipitate, which slowly transformed over a day or two into a poorly crystalline apatite (Posner & Betts, 1975). This suggested that an amorphous calcium phosphate (ACP) phase could be a transient precursor that eventually crystallizes into apatite. Indeed, amorphous mineral phases do exist in certain invertebrate skeletons and pathologic calcifications (Ben-tov et al., 2016; Gourgas et al., 2018; LeGeros, 2001), which raised the possibility that transient ACP is also present in vertebrate bones. Supporting this, Bonar et al. reported the presence of an amorphous mineral phase in developing chick bone (Bonar et al., 1983). However, a follow-up study suggested that the observed ACP might, in fact, be poorly crystalline apatite or reflect sample preparation artifacts (Grynpas et al., 1984). The existence of ACP in bone remained controversial for many years.

More recently, thanks to sophisticated analytical techniques, a consensus is emerging that while the fully mature mineral in bone is crystalline apatite, an initial disordered phase is likely part of the normal mineralization process. In a landmark study, Mahamid et al. used cryo-electron microscopy and tomography to demonstrate that newly forming fin bones in zebrafish are initially mineralized with nanospheres of ACP, which later crystallize into apatite and integrate with the collagen matrix (Mahamid et al., 2010).

Direct imaging of osteoblasts using cryo-electron tomography has provided even more compelling evidence for an ACP-based mineralization process. In murine bone, Mahamid et al. observed electron-dense granules within osteoblast vesicles that exhibited diffraction patterns consistent with an amorphous phase (Mahamid et al., 2011). Notably, these mineral granules contained lower calcium relative to phosphate than synthetic ACP, suggesting that osteoblasts may first concentrate phosphate in a labile, amorphous form before recruiting calcium to induce mineral maturation (Mahamid et al., 2011). This stepwise regulation likely minimizes premature or ectopic mineral precipitation and enables fine control over the initiation of mineral deposition. These findings were verified in an avian system by Kerschnitzki et al., who visualized the deposition of highly disordered calcium phosphate minerals inside intracellular vesicles as well as in the ECM in embryonic chick bone, supporting a conserved mechanism across species (Kerschnitzki et al., 2016).

Together, these studies have shifted the paradigm of bone mineralization by suggesting that bone mineral forms through a two-phase process: first, the controlled deposition of ACP, and second, a maturation phase in which the ACP transitions into carbonated apatite that is structurally and chemically integrated with the collagen matrix. While it is evident that such transitions occur rapidly, it remains unclear whether this two-phase mineralization process occurs at the same rate across all sites of new bone formation, and how this process is influenced by the age of the organism across various vertebrate species.

4. Hierarchical nanostructure of mineralized bone: Mineral–matrix integration

Mineralized bone is a strong yet flexible material because it combines two key components: a hard mineral phase, mainly carbonated apatite, and a hydrated protein matrix made mostly of type I collagen (Weiner & Wagner, 1998). The mineral gives bone stiffness and strength, while collagen provides toughness and flexibility. These two components are not simply added together; rather they are organized hierarchically at multiple structural levels. Tiny mineral crystals, only a few nanometers thick, are deposited within and around the collagen fibrils, closely integrated with the organic matrix. This close integration allows stress and force to be transferred between the two components, helping bone absorb impact and resist fracture.

Overall, bone has a hierarchical structure that comprises up to twelve nested levels of organization. At each level, smaller structural units are arranged in specific patterns and then assembled into larger units. Importantly, many of these levels show helical, twisted, braided, or rotated arrangements. These features are seen from the triple-helical structure of collagen and the subtle curvature of hydroxyapatite crystallites to the plywood-like rotation of mineralized collagen bundles around Haversian canals in cortical bones (Buss et al., 2021; Wagermaier et al., 2006). At higher levels of organization, these repeating rotated and twisted assemblies may also contribute to spiral-like patterns along the long axis of bone. Thus, bone can be viewed as a series of hierarchically nested spring-like structures. This architecture helps distribute mechanical energy during loading, reduces local stress concentration, and lowers the risk of sudden structural failure (Buss et al., 2021).

In short, bone's strength comes not just from what it is made of, but from how those parts are arranged. The intricate assembly of mineral and collagen, from the molecular scale to the whole organ, allows bone to be strong, light, and durable.

5. Importance of mineral homeostasis in bone Mineralization

Ca^{2+} and Pi are the core mineral components of the vertebrate skeleton. These ions must be present in supersaturated concentrations in the extracellular matrix (ECM) for nucleation and propagation of mineral crystals to occur. The critical ion product ($Ca^{2+} \times$ Pi) must exceed a thermodynamic threshold to drive calcium phosphate precipitation, initially as ACP that can subsequently mature into apatitic mineral (Termine & Eanes, 1972). A deficiency of these ions, especially during growth, impairs this process, leading to the accumulation of unmineralized collagenous matrix (osteoid), as seen in conditions such as rickets in children or osteomalacia in adults (Sabbagh et al., 2009; Xue et al., 2005).

Numerous clinical and experimental models have confirmed the essential role for Ca^{2+} and Pi in bone mineralization (Xue et al., 2005). In vitamin D deficiency, for instance, the reduced intestinal absorption of both Ca^{2+} and Pi leads to their systemic depletion and consequently to mineralization defects (Christakos et al., 2016). Similarly, systemic reduction of Pi caused by renal phosphate wasting in X-linked hypophosphatemia (XLH) results in defective bone ECM mineralization (Sabbagh et al., 2009). Therefore, maintaining optimal levels of these two ions is essential for healthy bone formation and remodeling.

The concentration of Ca^{2+} and Pi in the ECM is directly influenced by their levels in the blood. While apatite crystals form locally within the ECM, their ionic precursors are supplied primarily through systemic absorption from dietary intake and renal reabsorption. The matrix becomes mineralization-competent only when the local ionic milieu is favorable, which is itself a reflection of systemic homeostasis.

5.1 Hormonal regulation of calcium and phosphate availability

Three major cell types in bone, osteoblasts, osteocytes, and osteoclasts regulate Ca^{2+} and Pi homeostasis through both direct and indirect mechanisms. Direct regulation is primarily mediated by osteocytes, which secrete fibroblast growth factor 23 (FGF23), a key phosphaturic hormone. FGF23 acts on the kidneys to reduce phosphate reabsorption and suppress the synthesis of 1,25-dihydroxyvitamin D_3 (calcitriol), the active form of vitamin D (Shimada et al., 2004). This results in decreased renal Pi retention and reduced intestinal phosphate absorption, preventing hyperphosphatemia and maintaining phosphate homeostasis (Razzaque, 2009). However, in pathological states such as XLH, excessive FGF23 activity leads to chronic phosphate wasting, impaired bone mineralization, and the development of rickets or osteomalacia (Carpenter et al., 2011).

Bone cells also participate in the homeostasis of Ca^{2+} and Pi indirectly as targets of extraskeletal hormones. Parathyroid hormone (PTH), secreted in response to low serum Ca^{2+}, binds receptors on osteoblasts and osteocytes, stimulating the expression of receptor activator of nuclear factor kappa-B ligand (RANKL) (Boyle et al., 2003). RANKL, via its receptor RANK, promotes osteoclast differentiation and bone resorption, thereby mobilizing Ca^{2+} and Pi from the mineralized bone matrix into circulation. PTH simultaneously promotes renal absorption of Ca^{2+}, but inhibits renal phosphate reabsorption, causing phosphaturia. Through these combined actions, PTH raises serum Ca^{2+} levels but contributes to Pi loss, especially when chronically elevated (Schneider et al., 1975).

Calcitriol, primarily synthesized in the kidneys under the stimulatory control of PTH and suppressed by FGF23, enhances intestinal absorption of both Ca^{2+} and Pi, thus increasing systemic mineral availability (van Driel & van Leeuwen, 2017). It supports bone mineralization by ensuring adequate substrate delivery for mineral precipitation and maturation. In addition, calcitriol acts as a negative regulator of PTH synthesis and release, and modulates bone remodeling by promoting osteoblast differentiation and matrix protein synthesis, while indirectly supporting bone resorption under certain conditions (van Driel & van Leeuwen, 2017).

5.2 Hormonal pathologies affecting bone mineralization

The essential roles of PTH, calcitriol, and FGF23 in bone mineralization are well established through evidence from both human genetic disorders and corresponding genetic models. These hormonal pathways regulate the systemic availability of Ca^{2+} and Pi, the two critical components of bone mineral, and thereby support skeletal mineral deposition. Below is a brief overview of how pathologies arising from impairments in these key hormonal systems affect bone mineralization.

PTH Signaling Pathway: Although PTH is a key regulator of Ca^{2+} and Pi homeostasis, its deficiency does not typically result in classical osteomalacia. In humans, primary hypoparathyroidism caused by autoimmune mechanisms or loss-of-function mutations in the *GCM2* gene (causing agenesis of the parathyroid gland) leads to hypocalcemia and hyperphosphatemia, yet bone mineralization is generally preserved (Mannstadt et al., 2017). This apparent paradox is recapitulated in both $Pth^{-/-}$ and $Gcm2^{-/-}$ mice, where severe hypocalcemia does not impair matrix mineralization. The preservation of mineralization is attributed to marked hyperphosphatemia resulting from the loss of PTH-mediated inhibition of renal phosphate reabsorption (Günther et al., 2000; Miao et al., 2004). The elevated phosphate levels compensate for low Ca^{2+} by maintaining a sufficient Ca^{2+} x Pi ion product to permit hydroxyapatite formation.

In these models, calcitriol levels are reduced due to the absence of PTH stimulation of 1α-hydroxylase, leading to diminished intestinal absorption of Ca^{2+} and Pi. However, Pi remains elevated due to persistent renal reabsorption. FGF23 levels are also low, likely because calcitriol-dependent stimulation of FGF23 expression is impaired, despite elevated Pi levels (Agoro & White, 2023). Together, this hormonal profile - low Ca^{2+}, high Pi, and low calcitriol - creates a systemic environment that paradoxically remains permissive for bone mineral deposition, despite the absence of PTH.

Calcitriol (vitamin D) Pathway: In humans, mutations in *CYP27B1*, encoding renal 1α-hydroxylase, cause vitamin D–dependent rickets type 1A (VDDR1A) (Fu et al., 1997). Similarly, mutations in *CYP2R1* encoding the upstream enzyme, hepatic 25-hydroxylase, cause vitamin D–dependent rickets type 1B (VDDR1B) (Cheng et al., 2004). These mutations impair vitamin D activation, resulting in hypocalcemia, hypophosphatemia, and defective bone mineralization. In contrast, vitamin D–dependent rickets type 2A (VDDR2A), caused by mutations in the vitamin D receptor (VDR), leads to end-organ resistance to calcitriol (Malloy et al., 1999). Despite elevated hormone levels, the intestinal uptake of Ca^{2+} and Pi is impaired, leading to rickets and osteoid accumulation. These pathologies are faithfully modeled in mice. $Cyp27b1^{-/-}$ mice show hypocalcemia, hypophosphatemia, growth plate mineralization defects, and rickets, which are reversible by calcitriol therapy (Dardenne et al., 2001). $Vdr^{-/-}$ mice similarly exhibit impaired mineralization, secondary hyperparathyroidism, and skeletal deformities, reinforcing the essential function of both calcitriol production and signaling in bone mineral deposition (Demay, 2013).

FGF23 Signaling Pathway: Overproduction or impaired degradation of FGF23 leads to hypophosphatemia and defective bone mineralization (Fukumoto, 2021). In humans, XLH caused by mutations in PHEX results in elevated FGF23, renal phosphate wasting, reduced calcitriol levels, and rickets despite normal calcium levels (Francis et al., 1995). Other genetic conditions, including autosomal dominant hypophosphatemic rickets (ADHR) due to FGF23 gain-of-function mutations (White et al., 2001), and autosomal recessive forms (ARHR1/ARHR2) caused by mutations in DMP1 and ENPP1 (Michigami, 2019), also result in FGF23-mediated phosphate wasting and impaired skeletal mineralization. Notably, *ENPP1* mutations can also cause ectopic vascular calcification from PPi deficiency, alongside elevated FGF23 levels and skeletal hypomineralization, reflecting the dual regulatory role of ENPP1 in preventing ECM mineralization and phosphate balance (see below) (Lorenz-Depiereux et al., 2010). Additionally, tumor-induced osteomalacia (TIO) serves as an acquired model of FGF23 excess, with similarly impaired renal Pi reabsorption and defective bone mineralization (Shimada et al., 2001).

Mouse models provide robust mechanistic confirmation of the above pathologies. *Hyp* mice, bearing a *Phex* mutation, display high *Fgf23* expression, hypophosphatemia, and severe mineralization defects, closely recapitulating human XLH (Eicher et al., 1976). *Fgf23*-overexpressing mice exhibit low serum phosphate, suppressed calcitriol, and rickets, even in the absence of *Phex* mutation, confirming the pathogenic role of elevated FGF23 (Bai et al., 2004). Additionally, *Dmp1*$^{-/-}$ and *Enpp1*$^{-/-}$ mice develop hypophosphatemic rickets due to increased *Fgf23* expression (Feng et al., 2006; Mackenzie et al., 2012). Conversely, *Fgf23*$^{-/-}$ and *Klotho*$^{-/-}$ mice (lack FGF23 co-receptor), which cannot produce FGF23 or respond to FGF23 signaling, develop hyperphosphatemia and ectopic soft tissue calcification, but also show disorganized bone matrix and growth plate abnormalities, suggesting that loss of FGF23 signaling can disrupt normal skeletal mineralization patterns (Murali et al., 2016; Razzaque et al., 2006).

Collectively, these human diseases and corresponding mouse models provide strong evidence that PTH, calcitriol, and FGF23 signaling pathways are indispensable for bone mineralization. Each hormone regulates key steps in the systemic availability and local utilization of Ca^{2+} and Pi, and their disruption can lead to rickets or osteomalacia. The distinct skeletal phenotypes that emerge from each hormonal defect underscore their non-redundant, but complementary roles in mineral homeostasis. Understanding these pathways offers critical insights into the pathophysiology of mineralization disorders and provides the foundation for developing targeted therapies to restore bone health in affected individuals.

5.3 Phosphate as the dominant regulator of mineralization

While Ca^{2+} is indispensable for bone mineralization, Pi availability more consistently determines whether mineralization will proceed. Across numerous experimental models and clinical conditions, hypophosphatemia results in more profound skeletal defects than hypocalcemia. This is partly because calcium levels are under tighter homeostatic control and can be maintained even under conditions of dietary scarcity, while phosphate regulation is more dynamic and tolerates a wider range of fluctuations (Murshed, 2018). Evidence that physiologically Pi homeostasis is a dominant driver of bone mineralization comes from the observation that PTH deficiency leads to severe hypocalcemia but no osteomalacia, largely because of high Pi levels (Günther et al., 2000; Miao et al., 2004). Phosphate's role extends beyond its structural contribution; it directly influences the Pi/PPi ratio that governs the permissiveness of the ECM for mineralization (see below).

5.4 Magnesium and citrate as additional critical ions

In addition to Ca^{2+} and phosphate, magnesium and citrate modulate the formation, maturation, and structural organization of bone apatite. Magnesium acts as a cofactor for enzymes like alkaline phosphatase and plays a dual role in bone mineralization, facilitating crystal nucleation at early stages while preventing excessive crystal growth (Zhang et al., 2019). Citrate, an organic molecule derived from citric acid, binds to hydroxyapatite crystal surfaces, helping to stabilize nanocrystal size and spacing within the collagen matrix (Dirckx et al., 2022). As demonstrated by Dirckx et al., citrate transporter SLC13A5 deficiency in mice alters citrate handling in mineralized tissues and is associated with abnormal bone mineral organization and reduced mechanical strength, despite the absence of a primary systemic Ca^{2+} or Pi deficiency. Together, these ions ensure controlled and structurally sound mineral deposition during bone formation.

6. Collagen scaffold guiding mineral deposition in bone

An essential prerequisite for bone mineralization is the presence of an organized collagenous matrix. Type I collagen is the predominant protein of bone ECM, comprising roughly 90% of the organic matrix (Fratzl & Weinkamer, 2007). The type I collagen molecule is a triple helix formed by two α1(I) chains and one α2 (I) chain (encoded by *COL1A1* and *COL1A2* genes) that assemble into a rope-like helix about 300 nm in length (Fratzl & Weinkamer, 2007). These tropocollagen molecules pack in a staggered array to form collagen fibrils, giving rise to a periodic structure comprising a "hole zone" of roughly 40 nm (with an intermolecular spacing of ~1.5 nm) and a ~27 nm overlap zone (Stock, 2015). Collagen fibrils in bone further aggregate into larger fibers in a hierarchical fashion. This highly ordered scaffold not only provides tensile strength but also creates specific micro-environments that foster mineral nucleation (Fratzl & Weinkamer, 2007; Stock, 2015; Xu et al., 2015).

Bone mineralization occurs within a pre-existing collagen-rich organic matrix. During development, osteoblasts first deposit osteoid, after which mineralization begins, underscoring the essential role of the collagenous scaffold in organizing mineral deposition (Boskey & Posner, 1984). In intramembranous ossification (e.g., flat bones of the skull), osteoblasts secrete osteoid and then mineralize it directly. In endochondral ossification (e.g., long bones), a mineralized cartilage matrix is first formed by chondrocytes and subsequently replaced by osteoid produced by osteoblasts, which mineralize it at the cartilage–bone interface (Boskey & Posner, 1984). These observations indicate that simply having high concentrations of Ca^{2+} and Pi ions is not sufficient for bone minerals to form; rather, the presence of a collagen scaffold is a permissive factor for mineral deposition.

The role of collagen as the scaffold for apatite was first clearly demonstrated by ultrastructural studies in the 1950s (Sheldon & Robinson, 1957). Using electron microscopy on newborn mouse bone, Sheldon and Robinson showed that initial mineral crystals appear within collagen fibrils near the calcification front. Later studies further supported mineral localization at two specific locations in relation to collagen fibrils: (1) intrafibrillar sites, i.e., within the ~40 nm gap zones along each fibril, and (2) interfibrillar sites, i.e., in the spaces between adjacent collagen fibrils (Stock, 2015; Weiner & Wagner, 1998). These are precisely the locations where collagen's packing creates tiny voids or channels. The intrafibrillar gap zones are periodic voids inside each fibril, while interfibrillar spaces are the pores between fibrils; both are on the order of a few nanometers to tens of nanometers in size (Weiner & Wagner, 1998). Initial mineral crystals were observed to form in these zones, suggesting that collagen's architecture directs where mineral can nucleate.

The idea of matrix-directed mineralization is supported by both in vivo and in vitro studies. Urist's classic experiments showed that implanted demineralized bone matrix could undergo remineralization in vivo, whereas subsequent cell-free studies demonstrated that demineralized collagenous matrices can serve as templates for mineral deposition when exposed to a solution containing sufficient Ca^{2+} and Pi (Eskandari et al., 2006; Urist, 1965). Furthermore, mouse models with impaired collagen matrix production, such as ATF4-deficient mice, exhibit reduced mineralized bone formation (Yang et al., 2004). In contrast, in osteogenesis imperfecta, abnormal type I collagen structure and matrix organization are often associated with excess mineral accumulation within the organic matrix (Eyre & Weis, 2013; Roschger et al., 2008). Collectively, these examples establish the collagen matrix not only as a prerequisite for bone mineralization but also as a key regulator of the location, extent, and organization of mineral deposition.

How exactly does the collagen matrix facilitate mineral deposition? The mechanism is not yet fully resolved, but several hypotheses exist. One hypothesis emphasizes chemical interactions: charged amino acid residues on collagen (especially acidic residues like aspartate and glutamate) might bind calcium ions and contribute to mineral nucleation (Landis & Silver, 2008). However, an alternative and widely discussed theory is that collagen's primary role is biophysical confinement of mineral nucleation, rather than specific chemistry. According to the "size-exclusion" hypothesis, the narrow intermolecular channels (~1.5–2 nm wide) between collagen molecules within a fibril permit the entry of small ions (Ca^{2+} and Pi), but may restrict the entry of larger noncollagenous matrix proteins that have the capacity to inhibit mineralization (Toroian et al., 2007). A complementary model proposed by Kim et al. suggests that the confined nanoscale geometry of collagen gap zones may constrain the mobility and spatial arrangement of Ca^{2+} and Pi, favor ordered ionic organization, and reduce the energetic barrier for apatite nucleation, thereby creating a thermodynamically favorable environment for mineral formation within the fibril (Kim et al., 2018). Both mechanisms may work together to promote mineral crystal nucleation and growth within the collagen scaffold.

In essence, the nanoscale intrafibrillar spaces create microenvironments that promote mineral nucleation once the Ca^{2+} and Pi ion activity product exceeds the critical threshold for precipitation. Once apatite crystals form in these protected microenvironments, collagen's periodic structure may template the initial orientation of apatite, aligning crystals along the fibril's axis and thus integrating the mineral with the organic matrix (Nudelman et al., 2010; Weiner & Wagner, 1998). This model of matrix-mediated bone mineralization suggests that type I collagen acts as a selective physical barrier that restricts the access of inhibitory proteins, and through its spatial geometry, serves as a scaffold defining where mineral can form (Nudelman et al., 2010; Weiner & Wagner, 1998). Although large protein inhibitors may be restricted from entering the narrow intrafibrillar spaces of collagen, small diffusible inhibitors such as PPi that escape ALPL-mediated hydrolysis may still access these compartments. Thus, the collagen scaffold, local ALPL activity, and residual PPi may together create a microenvironment that suppresses uncontrolled precipitation while supporting spatially restricted and directional mineral propagation within the fibrils.

7. Mineralization inhibitors as regulators of skeletal mineralization

Although Ca^{2+} and Pi concentrations in extracellular fluids are generally in equilibrium with blood, mineralization is selectively restricted to skeletal and dental tissues. This raises a critical biological question: why do soft tissues like skin, blood vessels, and some of the cartilaginous tissues remain non-mineralized despite similar ionic environments and a high collagen content? The answer lies in the presence and localized regulation of potent endogenous inhibitors of mineralization, notably PPi, matrix Gla protein (MGP), fetuin-A, osteopontin (OPN), and ASARM peptides (Murshed, 2018). These molecules act through overlapping, but also distinct mechanisms to prevent indiscriminate hydroxyapatite deposition while allowing tightly controlled skeletal mineralization.

Inorganic pyrophosphate (PPi): PPi is a potent physicochemical inhibitor of hydroxyapatite crystal formation. It binds to nascent mineral nuclei and prevents further incorporation of Ca^{2+} and Pi ions, thereby blocking crystal growth. Extracellular PPi is generated through two sequential steps: (1) cellular export of ATP by ANK, a transmembrane transporter, and (2) extracellular cleavage of ATP by

ENPP1 (ectonucleotide pyrophosphatase/phosphodiesterase 1) to generate PPi. ANK was previously thought to transport PPi directly. However, recent findings by Szeri et al. reveal that ANK primarily acts as an ATP transporter (Szeri et al., 2022). This revision of the theory explaining extracellular PPi generation positions ANK and ENPP1 as a functional axis essential for PPi homeostasis in the ECM.

The importance of PPi in preventing soft tissue calcification is illustrated by human disorders such as generalized arterial calcification of infancy (GACI) and craniometaphyseal dysplasia, caused by mutations in ENPP1 and ANKH, respectively. Corresponding mouse models (*ttw/ttw* and *ank/ank*) show ectopic mineralization of joints, spinal tissues, and arteries (Ho et al., 2000; Koshizuka et al., 2001). Interestingly, both genes are highly expressed in osteoblasts, suggesting that bone tissue inherently possesses the machinery for mineralization inhibition (Murshed et al., 2005).

The apparent paradox, how mineralization proceeds in bone despite the presence of PPi, is resolved by the activity of tissue-nonspecific alkaline phosphatase (TNAP/ALPL). ALPL hydrolyzes PPi into two Pi ions, thus removing the inhibitor and supplying additional Pi favoring mineralization (Hessle et al., 2002). The Pi/PPi ratio is now recognized as a key determinant of matrix mineralization. This was demonstrated in *ank/ank;Hyp* compound-mutant mice, where reduced serum Pi levels attenuate the ectopic mineralization seen in *ank/ank* mice, underscoring the need for a threshold Pi/PPi ratio to initiate apatite formation (Murshed et al., 2005).

In humans, loss-of-function mutations in ALPL lead to hypophosphatasia, a spectrum of diseases characterized by elevated PPi levels and defective bone mineralization (Whyte, 2020). $Alpl^{-/-}$ mice show severe osteomalacia and perinatal lethality, and reintroduction of enzymatically active ALPL corrects these defects, highlighting the critical requirement for PPi clearance in the mineralizing ECM (Millan et al., 2009).

Matrix Gla Protein (MGP): MGP is a vitamin K–dependent, calcium-binding protein that inhibits extracellular mineralization (Cancela et al., 2021). It is highly expressed in vascular smooth muscle cells and growth plate chondrocytes, with relatively low expression in osteoblasts (Luo et al., 1995, 1997). It has been proposed that MGP binds Ca^{2+} and nascent crystals to prevent apatite nucleation and propagation (O'Young et al., 2011).

In humans, homozygous mutations in MGP cause Keutel syndrome, characterized by ectopic mineralization of cartilaginous structures, tracheobronchial calcification, midface hypoplasia, and, in some cases, arterial calcification (Cancela et al., 2021). $Mgp^{-/-}$ mice phenocopy these features but show more severe vascular calcification. Despite the absence of overt defects in bone mineralization in *Mgp*-null mice, cartilaginous skeletal tissues, including the growth plates and nasal septum undergo severe ectopic calcification, leading to shortening of endochondral bones and impaired midfacial growth (Luo et al., 1997; Marulanda et al., 2017). The ectopic mineral in the arterial walls of MGP-deficient mice contains amorphous calcium phosphate (ACP), suggesting that MGP blocks the earliest steps of mineral nucleation (Bak et al., 2024; Gourgas et al., 2018).

Importantly, osteoblast-specific overexpression of MGP causes moderate osteomalacia affecting both intramembranous and endochondral bones (Murshed et al., 2004), indicating that suppression of MGP expression is necessary for proper mineralization in bone. This supports the notion that MGP acts as a highly effective mineralization inhibitor whose activity must be spatially and temporally restricted. While the precise mechanisms underlying MGP's anti-mineralization function are still being elucidated, both in vitro and in vivo studies suggest a critical role of its phosphorylated serine residues in this function (Parashar et al., 2021; Parashar et al., 2022). Recently, dominant variants of MGP replacing its Cys19 residue have been shown to cause spondyloepiphyseal dysplasia, MGP type, in humans (Gourgas et al., 2023). This finding suggests that mutant MGP can perturb skeletal development through mechanisms distinct from the established anti-mineralization function of native MGP.

Osteopontin (OPN): OPN, also known as secreted phosphoprotein 1 (SPP1), is a highly phosphorylated, acidic matrix protein that binds to hydroxyapatite crystals and inhibits further growth (Si et al., 2020). It is thought to inhibit mineralization partly through its interactions with growing mineral crystals and steric hindrance of their growth, making it an effective regulator of mineralization at sites of both physiological and pathological calcification.

Although OPN-null ($Spp1^{-/-}$) mice show largely normal skeletal mineralization, they are more susceptible to ectopic soft tissue calcification, particularly under stress conditions such as high-phosphate diets or aortic valve implantation (Paloian et al., 2016; Steitz et al., 2002). Moreover, in $Mgp^{-/-}$;$Spp1^{-/-}$ double knockout mice, vascular calcification is even more pronounced than in *Mgp* deficiency alone, demonstrating additive or synergistic effects of these inhibitors (Speer et al., 2002). OPN is frequently upregulated in calcified lesions and is considered a compensatory inhibitor in pathological conditions. In soft tissues, its role appears to be protective, limiting the extent of calcification once it has initiated.

Recent studies using *Hyp* mice, a model for human XLH, have shown that $Spp1^{-/-}$;*Hyp* compound mutant mice exhibit modestly improved skeletal mineralization compared with *Hyp* mice alone, indicating that OPN contributes to osteomalacia under phosphate-wasting conditions (Hoac et al., 2020). When challenged with a high-phosphate diet, these compound mutants demonstrate improved rescue of trabecular and cortical bone mineralization (Hoac et al., 2020). However, periosteocytic lesions, hypomineralized areas surrounding osteocytes in *Hyp* mice, remain uncorrected, suggesting that OPN does not play a major role in this pathology (Hoac et al., 2020). These findings underscore the compartment-specific functions of OPN as a modulator of bone mineralization.

ASARM Peptides: ASARM (acidic serine- and aspartate-rich motif) peptides are small phosphopeptides derived from the proteolytic cleavage of SIBLING family proteins, including MEPE, osteopontin, and DMP1 (Rowe, 2012). These peptides bind strongly to hydroxyapatite and inhibit crystal growth by occupying active growth sites on the mineral surface (Rowe, 2012).

In healthy bone, ASARM peptides are tightly regulated through proteolytic degradation by PHEX, a Zn^{2+}-dependent endopeptidase (Addison et al., 2008). In

XLH, caused by inactivating mutations in PHEX, ASARM peptides accumulate in the ECM (Addison et al., 2008; Rowe, 2012). This contributes to the osteomalacia seen in XLH patients and *Hyp* mice, independently of systemic phosphate levels. The inhibitory action of ASARM peptides further worsens the osteomalacia caused by the phosphate-wasting effect of elevated FGF23 in this condition.

In summary, the process of bone mineralization is governed by a fine balance between pro-mineralization factors and inhibitory molecules that prevent pathological calcification while modulating mineral deposition in bone. Inhibitors such as PPi, MGP, OPN, and ASARM peptides act at different levels and sites to ensure that mineral is deposited only in appropriate locations and amounts. Recent findings underscore the importance of precise control over the generation and degradation of these inhibitors. The Pi/PPi ratio, shaped by the actions of ANKH, ENPP1, and ALPL, plays a central role in this regulation. Similarly, the spatially restricted expression of MGP, PHEX-mediated clearance of ASARM peptides, and the regulated induction of OPN expression all contribute to the specificity of skeletal mineralization. Collectively, these regulatory systems ensure that soft tissue ECMs, which are permissive to mineral deposition when Ca^{2+} and Pi levels reach the threshold for hydroxyapatite precipitation, remain protected from inappropriate calcification, while allowing regulated mineralization in hard tissues.

8. Matrix vesicles in bone mineralization: From controversy to consensus

Matrix vesicles (MVs) are nanosized membrane-bound bodies shed by mineralizing cells that are thought to provide a protected microenvironment for mineral nucleation (Anderson, 1969, 2003). According to the theory of MV-mediated ECM mineralization, MVs are the initial sites of mineral deposition where specific enzyme, transporter, and channel activities increase Ca^{2+} and Pi ions leading to their precipitation as ACP, which subsequently matures to HA crystals (Anderson, 1969, 2003). These crystals eventually breach the MV membrane to form mineralized nodules, which attach to adjacent collagen fibrils and propagate matrix mineralization. This theory has been supported by multiple ultrastructural studies (Hasegawa et al., 2017). It has been suggested that in mineralized tissues such as dentin, cementum, growth plate cartilage, and bone, this process corresponds to the initial or primary mineralization phase, setting the stage for more robust secondary mineralization aided by ALPL activity (Anderson et al., 2004; Murshed, 2018).

8.1 Molecular mechanisms of MV-mediated mineralization

MV membranes are laden with specific enzymes and transporters that orchestrate ion accumulation. Key membrane-associated proteins include ALPL and ENPP1 as well as sodium-dependent phosphate cotransporter Pit-1 (SLC20A1) and the ankylosis protein ANK (Cui et al., 2016; Hasegawa et al., 2017; Roberts et al.,

2007). Numerous annexins (A2, A5, A6) coat the MV membrane, where they bind phosphatidylserine (PS) and form Ca^{2+}-permeable channels. Phosphatidylserine, which is highly enriched on the inner leaflet of MVs, chelates Ca^{2+} to form stable Ca^{2+}–PS complexes that nucleate calcium phosphate within the vesicle (Cui et al., 2016; Hasegawa et al., 2017; Roberts et al., 2007). These coordinated transport functions enable rapid loading of Ca^{2+} and Pi: ALPL hydrolyzes PPi to Pi outside the vesicles, Pit-1 imports Pi into the vesicle lumen, and annexins facilitate Ca^{2+} influx, creating a favorable microenvironment for mineral precipitation (Cui et al., 2016; Hasegawa et al., 2017; Roberts et al., 2007).

It has been suggested that inside MVs, enzymes such as neutral sphingomyelinase 2 (SMPD3/nSMase2) and ENPP6 produce phosphorylcholine in the MV lumen: SMPD3 cleaves sphingomyelin to ceramide and phosphorylcholine, while ENPP6 (glycerophosphocholine phosphodiesterase) converts lysophosphatidylcholine into phosphorylcholine (Murshed, 2018; Stewart et al., 2018). PHOSPHO1, a downstream phosphatase highly expressed in MVs, hydrolyzes phosphorylcholine and phosphoethanolamine from other enzymatic reactions to generate free Pi (Roberts et al., 2007). The dual mechanisms of intravesicular Pi accumulation through PHOSPHO1 enzymatic activity and Pi import via Pit-1 and other transporters facilitate de novo HA nucleation within MVs (Roberts et al., 2007; Stewart et al., 2018).

8.2 Biological evidence from genetic models

Genetically, multiple loss-of-function models illustrate the role of various proteins in MV-mediated mineralization. For example, *Phospho1*$^{-/-}$ mice exhibit defective skeletal mineralization: their bones show excess unmineralized osteoid and patchy osteomalacia, despite apparently normal HA crystals. Crucially, the chondrocyte- and osteoblast-derived MVs in *Phospho1*-null mice have greatly reduced mineral content (Yadav et al., 2016). Combining *Phospho1* ablation with deletion of the phosphate cotransporter Pit1 (*Col2a1*-driven PiT1 knockout) greatly worsens skeletal defects: double-null MVs (~80%) remain unmineralized, far more than *Phospho1*$^{-/-}$ MVs alone (Yadav et al., 2016). When PHOSPHO1 is deleted together with *Alpl*$^{-/-}$, mineralization fails entirely, producing a complete mineralization arrest (Yadav et al., 2011). These compound knockouts demonstrate that PHOSPHO1 (intravesicular Pi) and ALPL (extravesicular Pi control) are non-redundant drivers of MV nucleation (Yadav et al., 2011). This genetic synergy supports the idea that both intravesicular phosphate production and transporter-mediated phosphate influx are required for mineral nucleation inside MVs.

Genetic studies show the importance of SMPD3 (nSMase2) in bone mineralization in vivo. The *fragilitas ossium* (*fro*) mouse carries a loss-of-function *Smpd3* mutation and phenocopies a non-collagenous form of osteogenesis imperfecta: *fro/fro* mice display severe skeletal dysplasia and poor bone mineralization (Khavandgar et al., 2011; Li et al., 2016). Importantly, *fro* mutation also impairs tooth mineralization: *fro/fro* dentin shows delayed mineralization of mantle dentin and subsequent enamel formation (Khavandgar et al., 2013). Transgenic expression of *Smpd3* in osteoblasts and odontoblasts in *fro/fro* mice rescues the bone and tooth mineralization defects, confirming local requirements for SM hydrolysis in ECM mineralization. This and other emerging data indicate that sphingomyelin breakdown yielding phosphorylcholine is needed to supply PHOSPHO1 with its substrate and sustain MV-mediated mineral nucleation in both bone and dentin. Apart from generating phosphorylcholine inside MVs, SMPD3 is also likely to be involved in MV release from the plasma membrane of cells in mineralizing tissues (Chen et al., 2016; Cui et al., 2016).

Enpp6$^{-/-}$ mice provide further mechanistic insight on MV-mediated mineralization. These mice show transient osteomalacia in early growth: young mutants show lower bone volume fraction and hypomineralized osteoid compared to wild-type, though the defect largely normalizes by 3 months (Dillon et al., 2021). Ultrastructurally, the mutant trabeculae show many ossification foci that fail to propagate (Dillon et al., 2021). This phenotype suggests that ENPP6-derived phosphorylcholine is particularly important during rapid bone growth, supplying PHOSPHO1 with its substrate for initial mineral seeding. It also suggests that SMPD3 and ENPP6 may cooperate at the early stages of bone mineralization involving MVs.

8.3 Controversies associated with the MV theory

Although recent genetic models and advanced imaging techniques have provided compelling evidence for the involvement of MVs in bone mineralization, this theory has remained controversial since its inception. A key concern continues to center on the universality and exclusivity of the MV-mediated mechanism. While MVs are readily observed during the early stages of endochondral ossification and in rapidly forming woven bone, the majority of mineral in mature lamellar bone appears to be deposited in direct association with the collagen matrix rather than within vesicles. High-resolution imaging of mature bone has generally failed to detect MVs at mineral nucleation sites within the collagen scaffold. This might be attributed to the hierarchical organization of type I collagen, which forms nanoscale compartments that are accessible to Ca^{2+} and Pi ions but too small to accommodate MVs. It is likely that collagen-mediated mineralization predominates in mature bone, and that MVs are not required for initiating mineral deposition within collagen fibrils. Supporting this view, in vitro studies have demonstrated that demineralized collagen can undergo re-mineralization in the absence of cells or vesicles (Eskandari et al., 2006; Soicher et al., 2013), further challenging the notion that MVs are indispensable for crystal nucleation.

Another concern remains regarding the early ultrastructural studies claiming widespread involvement of MVs in bone mineralization. These studies might have overestimated MV prevalence due to technical limitations, including artifacts introduced by dehydration and fixation. However, more recent cryo-electron microscopy studies of native frozen-hydrated bone have substantially reduced this concern by showing membrane-bound, calcium phosphate-loaded vesicles in osteoblast-lineage cells during early bone mineralization (Mahamid et al., 2011).

Finally, genetic ablation of key MV-associated enzymes, such as PHOSPHO1, SMPD3, and ENPP6 results in delayed or incomplete mineralization rather than a complete absence of mineral deposition, suggesting the existence of redundant or compensatory pathways that can sustain mineralization in the absence of vesicle-based mechanisms. While the roles of these enzymes in bone mineralization are well established in mouse models, their specific contribution to MV-mediated mineralization in human bone remains to be conclusively demonstrated.

Despite these controversies, current evidence supports the view that MVs play an important role during the early stages of skeletal mineralization, particularly under conditions that require rapid deposition of mineral, such as during embryonic development or fracture healing. While MVs may not be the dominant mechanism in mature lamellar bone, their role as an initiating or auxiliary mechanism in specific physiological contexts remains significant.

9. Concluding remarks

Bone mineralization is a highly coordinated, multi-step process that integrates systemic mineral homeostasis, local matrix architecture, and cell-mediated regulatory mechanisms. Far from being a passive precipitation of calcium and phosphate ions, it involves a dynamic interplay between collagen scaffolding, mineralization-promoting enzymes, inhibitory molecules, and specialized cell-secreted vesicular bodies, such as MVs (Fig. 2.2). Emerging evidence highlights that both amorphous precursors and hierarchical collagen structures influence the orderly deposition and integration of apatite crystals. While some controversies remain—particularly regarding the universality of MV-mediated nucleation—genetic models and advanced imaging technologies continue to refine our understanding of this complex process. Together, these insights not only shed light on the fundamental biology of skeletal mineralization but also offer possible therapeutic avenues for correcting mineralization disorders in bone and soft tissues.

AI Disclosure

During the preparation of this work the author(s) used ChatGPT (OpenAI, San Francisco, CA, USA) in order to assist with minor corrections of grammar and spelling, and occasionally to verify factual information. After using this tool/service, the author(s) reviewed and edited the content as needed and take(s) full responsibility for the content of the publication. The author gratefully acknowledges Natalie Reznikov (McGill University) for her assistance with the revision of Section 4. Julia Sebag (McGill University) contributed to the initial planning of this chapter. Acknowledgement The author gratefully acknowledges Natalie Reznikov (McGill University) for her assistance with the revision of Section 4. Julia Sebag (McGill University) contributed to the initial planning of this chapter.

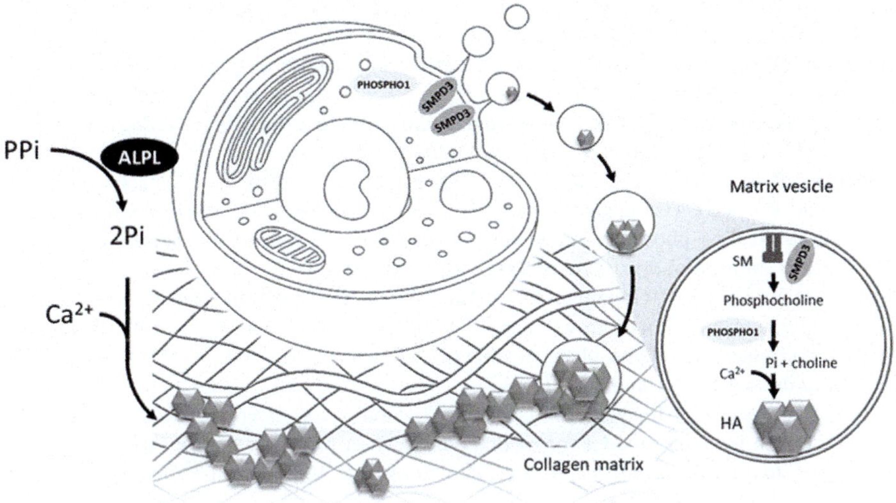

FIGURE 2.2 Model of bone mineralization.

Bone mineralization depends on serum calcium (Ca^{2+}), inorganic phosphate (Pi), and the collagen-rich extracellular matrix (ECM) that provides a scaffold for mineral deposition. Alkaline phosphatase (ALPL), an ectoenzyme anchored to the osteoblast membrane, hydrolyzes inorganic pyrophosphate (PPi), a potent inhibitor of mineralization, thereby generating Pi. This dual action promotes mineral deposition by shifting the local Pi/PPi ratio in favor of calcium phosphate mineral formation. The hierarchical organization of collagen fibrils and fibers creates intra- and interfibrillar nanoscale gaps that are permeable to Ca^{2+} and Pi but exclude larger inhibitory proteins, explaining why mineral deposits form both within and between collagen fibrils. In addition to this collagen-directed process, matrix vesicle (MV)–mediated mineralization provides an auxiliary mechanism. These nanoscale vesicles bud from mineralizing cells and contain enzymes that promote intravesicular mineralization. For example, SMPD3, ENPP6 and phospholipases cleave sphingomyelin (SM) or other phospholipids to generate phosphorylcholine, which is further hydrolyzed by PHOSPHO1 to release Pi. Accumulation of Pi inside the vesicles enables its precipitation with Ca^{2+}, giving rise to nascent hydroxyapatite (HA) crystals that eventually contribute to ECM mineralization. SMPD3 is also likely to be directly involved in MV release from the plasma membrane of cells in mineralizing tissues

References

Addison, W. N., Nakano, Y., Loisel, T., Crine, P., & McKee, M. D. (2008). MEPE-ASARM peptides control extracellular matrix mineralization by binding to hydroxyapatite: An inhibition regulated by PHEX cleavage of ASARM. *Journal of Bone and Mineral Research, 23*(10), 1638–1649. https://doi.org/10.1359/jbmr.080601

Agoro, R., & White, K. E. (2023). Regulation of FGF23 production and phosphate metabolism by bone–kidney interactions. *Nature Reviews Nephrology, 19*(3), 185–193. https://doi.org/10.1038/s41581-022-00665-x

Anderson, H. C. (1969). Vesicles associated with calcification in the matrix of epiphyseal cartilage. *The Journal of Cell Biology, 41*(1), 59–72. https://doi.org/10.1083/jcb.41.1.59

Anderson, H. C. (2003). Matrix vesicles and calcification. *Current Rheumatology Reports, 5*(3), 222–226. https://doi.org/10.1007/s11926-003-0071-z

Anderson, H. C., Sipe, J. B., Hessle, L., Dhamyamraju, R., Atti, E., Camacho, N. P., & Millán, J. L. (2004). Impaired calcification around matrix vesicles of growth plate and bone in alkaline phosphatase-deficient mice. *American Journal Of Pathology, 164*(3), 841–847. https://doi.org/10.1016/S0002-9440(10)63172-0

Bai, X., Miao, D., Li, J., Goltzman, D., & Karaplis, A. C. (2004). Transgenic mice overexpressing human fibroblast growth factor 23 (R176Q) delineate a putative role for parathyroid hormone in renal phosphate wasting disorders. *Endocrinology, 145*(11), 5269–5279. https://doi.org/10.1210/en.2004-0233

Bak, K., Parashar, A., Allgayer, R., Marulanda, J., Gourgas, O., Cerruti, M., & Murshed, M. (2024). An inducible model for medial calcification based on matrix Gla protein deficiency. *Journal of Structural Biology, 216*(4), Article 108144. https://doi.org/10.1016/j.jsb.2024.108144

Bentov, S., Aflalo, E. D., Tynyakov, J., Glazer, L., & Sagi, A. (2016). Calcium phosphate mineralization is widely applied in crustacean mandibles. *Scientific Reports, 6*(1). https://doi.org/10.1038/srep22118

Blaine, J., Chonchol, M., & Levi, M. (2015). Renal control of calcium, phosphate, and magnesium homeostasis. *Clinical Journal of the American Society of Nephrology, 10*(7), 1257–1272. https://doi.org/10.2215/CJN.09750913

Bonar, L. C., Roufosse, A. H., Sabine, W. K., Grynpas, M. D., & Glimcher, M. J. (1983). X-ray diffraction studies of the crystallinity of bone mineral in newly synthesized and density fractionated bone. *Calcified Tissue International, 35*(1), 202–209. https://doi.org/10.1007/bf02405032

Boskey, A. L., & Posner, A. S. (1984). Bone structure, composition, and mineralization. *Orthopedic Clinics of North America, 15*(4), 597–612.

Boyle, W. J., Simonet, W. S., & Lacey, D. L. (2003). Osteoclast differentiation and activation. *Nature, 423*(6937), 337–342. https://doi.org/10.1038/nature01658

Buss, D. J., Kröger, R., McKee, M. D., & Reznikov, N. (2021). Hierarchical organization of bone in three dimensions: A twist of twists. *Journal of Structural Biology, 6*(100057), 1–10.

Cancela, M. L., Laizé, V., Conceição, N., Kempf, H., & Murshed, M. (2021). Keutel syndrome, a review of 50 years of literature. *Frontiers in Cell and Developmental Biology, 9*. https://doi.org/10.3389/fcell.2021.642136

Carpenter, T. O., Imel, E. A., Holm, I. A., Jan De Beur, S. M., & Insogna, K. L. (2011). A clinician's guide to X-linked hypophosphatemia. *Journal of Bone and Mineral Research, 26*(7), 1381–1388. https://doi.org/10.1002/jbmr.340

Chen, Q., Bei, J. J., Liu, C., Feng, S. B., Zhao, W. B., Zhou, Z., Yu, Z. P., Du, X. J., & Hu, H. Y. (2016). HMGB1 induces secretion of matrix vesicles by macrophages to enhance ectopic mineralization. *PLoS One, 11*(5), 1–18.

Cheng, J. B., Levine, M. A., Bell, N. H., Mangelsdorf, D. J., & Russell, D. W. (2004). Genetic evidence that the human CYP2R1 enzyme is a key vitamin D 25-hydroxylase.

Proceedings of the National Academy of Sciences of the United States of America, 101(20), 7711–7715.

Christakos, S., Dhawan, P., Verstuyf, A., Verlinden, L., & Carmeliet, G. (2016). Vitamin D: Metabolism, molecular mechanism of action, and pleiotropic effects. *Physiological Reviews, 96*(1), 365–408. https://doi.org/10.1152/physrev.00014.2015

Cui, L., Houston, D. A., Farquharson, C., & MacRae, V. E. (2016). Characterisation of matrix vesicles in skeletal and soft tissue mineralisation. *Bone, 87*, 147–158. https://doi.org/10.1016/j.bone.2016.04.007

Dardenne, O., Prud'homme, J., Arabian, A., Glorieux, F. H., & St-Arnaud, R. (2001). Targeted inactivation of the 25-hydroxyvitamin D3-1α-hydroxylase gene (CYP27B1) creates an animal model of pseudovitamin D-deficiency rickets. *Endocrinology, 142*(7), 3135–3141. https://doi.org/10.1210/endo.142.7.8281

Demay, M. B. (2013). Physiological insights from the vitamin D receptor knockout mouse. *Calcified Tissue International, 92*(2), 99–105. https://doi.org/10.1007/s00223-012-9633-2

Dillon, S., Suchacki, K., Hsu, S.-N., Stephen, L. A., Wang, R., Cawthorn, W. P., Stewart, A. J., Nudelman, F., Morton, N. M., & Farquharson, C. (2021). Ablation of Enpp6 results in transient bone hypomineralization. *JBMR Plus, 5*(2). https://doi.org/10.1002/jbm4.10439

Dirckx, N., Zhang, Q., Chu, E. Y., Tower, R. J., Li, Z., Guo, S., Yuan, S., Khare, P. A., Zhang, C., Verardo, A., Alejandro, L. O., Park, A., Faugere, M. C., Helfand, S. L., Somerman, M. J., Riddle, R. C., de Cabo, R., Le, A., Schmidt-Rohr, K., & Clemens, T. L. (2022). A specialized metabolic pathway partitions citrate in hydroxyapatite to impact mineralization of bones and teeth. *Proceedings of the National Academy of Sciences of the United States of America, 119*(45), 1–12.

Eicher, E. M., Southard, J. L., Scriver, C. R., & Glorieux, F. H. (1976). Hypophosphatemia: Mouse model for human familial hypophosphatemic (vitamin D resistant) rickets. *Proceedings of the National Academy of Sciences of the United States of America, 73*(12), 4667–4671. https://doi.org/10.1073/pnas.73.12.4667

Eskandari, M. M., Eskandari, H. G., Aktas, M., & Atik, U. (2006). In vitro re-mineralization of demineralized bone matrix in human serum. *Clinical Chemistry and Laboratory Medicine, 44*(1), 54–58. https://doi.org/10.1515/CCLM.2006.011

Eyre, D. R., & Weis, M. A. (2013). Bone collagen: new clues to its mineralization mechanism from recessive osteogenesis imperfecta. *Calcified Tissue International, 93*(4), 338–347.

Fedde, K. N., Blair, L., Silverstein, J., Coburn, S. P., Ryan, L. M., Weinstein, R. S., Waymire, K., Narisawa, S., Millán, J. L., MacGregor, G. R., & Whyte, M. P. (1999). Alkaline phosphatase knock-out mice recapitulate the metabolic and skeletal defects of infantile hypophosphatasia. *Journal of Bone and Mineral Research, 14*(12), 2015–2026. https://doi.org/10.1359/jbmr.1999.14.12.2015

Feng, J. Q., Ward, L. M., Liu, S., Lu, Y., Xie, Y., Yuan, B., Yu, X., Rauch, F., Davis, S. I., Zhang, S., Rios, H., Drezner, M. K., Quarles, L. D., Bonewald, L. F., & White, K. E. (2006). Loss of DMP1 causes rickets and osteomalacia and identifies a role for osteocytes in mineral metabolism. *Nature Genetics, 38*(11), 1–17.

Fleisch, H., & Bisaz, S. (1962). Mechanism of calcification: Inhibitory role of pyrophosphate. *Nature, 195*, 911–911.

Francis, F., Hennig, S., Korn, B., Reinhardt, R., de Jong, P., Poustka, A., Lehrach, H., N Rowe, P. S., Goulding, J. N., Summerfield, T., Mountford, R., Read, A. P., Popowska, E., Pronicka, E., Davies, K. E., O'Riordan, J. L. H., Econs, M. J., Nesbitt, T., Drezner, M. K., … Meitinger, T. (1995). A gene (PEX) with homologies to endopeptidases is mutated in patients with X–linked hypophosphatemic rickets. *Nature Genetics, 11*(2), 130–136. https://doi.org/10.1038/ng1095-130

Fratzl, P., & Weinkamer, R. (2007). Nature's hierarchical materials. *Progress in Materials Science, 52*(8), 1263–1334. https://doi.org/10.1016/j.pmatsci.2007.06.001

Fu, G. K., Lin, D., Zhang, M. Y. H., Bikle, D. D., Shackleton, C. H. L., Miller, W. L., & Portale, A. A. (1997). Cloning of human 25-hydroxyvitamin D-1α-hydroxylase and mutations causing vitamin D-dependent rickets type 1. *Molecular Endocrinology, 11*(13), 1961–1970. https://doi.org/10.1210/mend.11.13.0035

Fukumoto, S. (2021). FGF23-related hypophosphatemic rickets/osteomalacia: Diagnosis and new treatment. *Journal of Molecular Endocrinology, 66*(2), R57. https://doi.org/10.1530/jme-20-0089

Gourgas, O., Lemire, G., Eaton, A. J., Alshahrani, S., Duker, A. L., Li, J., Carroll, R. S., Mackenzie, S., Nikkel, S. M., Bober, M. B., Boycott, K. M., Murshed, M., & Care4Rare Canada Consortium. (2023). Specific heterozygous variants in MGP lead to endoplasmic reticulum stress and cause spondyloepiphyseal dysplasia. *Nature Communications, 14* (7054), 1–20.

Gourgas, O., Marulanda, J., Zhang, P., Murshed, M., & Cerruti, M. (2018). Multidisciplinary approach to understand medial arterial calcification. *Arteriosclerosis, Thrombosis, and Vascular Biology, 38*(2), 363–372.

Grynpas, M. D., Bonar, L. C., & Glimcher, M. J. (1984). Failure to detect an amorphous calcium-phosphate solid phase in bone mineral: A radial distribution function study. *Calcified Tissue International, 36*(1), 291–301. https://doi.org/10.1007/BF02405333

Günther, T., Chen, Z.-F., Kim, J., Priemel, M., Rueger, J. M., Amling, M., Moseley, J. M., John Martin, T., Anderson, D. J., & Karsenty, G. (2000). Genetic ablation of parathyroid glands reveals another source of parathyroid hormone. *Nature, 406*(6792), 199–203. https://doi.org/10.1038/35018111

Hasegawa, T., Yamamoto, T., Tsuchiya, E., Hongo, H., Tsuboi, K., Kudo, A., Abe, M., Yoshida, T., Nagai, T., Khadiza, N., Yokoyama, A., Oda, K., Ozawa, H., Freitas, P. H. L.de, Li, M., & Amizuka, N. (2017). Ultrastructural and biochemical aspects of matrix vesicle-mediated mineralization. *Japanese Dental Science Review, 53*(2), 34–45. https://doi.org/10.1016/j.jdsr.2016.09.002

Hessle, L., Johnson, K. A., Clarke Anderson, H., Narisawa, S., Sali, A., Goding, J. W., Terkeltaub, R., & Luis Millán, J. (2002). Tissue-nonspecific alkaline phosphatase and plasma cell membrane glycoprotein-1 are central antagonistic regulators of bone mineralization. *Proceedings of the National Academy of Sciences, 99*(14), 9445–9449. https://doi.org/10.1073/pnas.142063399

Ho, A. M., Johnson, M. D., & Kingsley, D. M. (2000). Role of the mouse ank gene in control of tissue calcification and arthritis. *Science, 289*(5477), 265–270. https://doi.org/10.1126/science.289.5477.265

Hoac, B., Østergaard, M., Wittig, N. K., Boukpessi, T., Buss, D. J., Chaussain, C., Birkedal, H., Murshed, M., & McKee, M. D. (2020). Genetic ablation of osteopontin in osteomalacic Hyp mice partially rescues the deficient mineralization without correcting hypophosphatemia. *Journal of Bone and Mineral Research, 35*(10), 2032–2048. https://doi.org/10.1002/jbmr.4101

Kerschnitzki, M., Akiva, A., Ben Shoham, A., Asscher, Y., Wagermaier, W., Fratzl, P., Addadi, L., & Weiner, S. (2016). Bone mineralization pathways during the rapid growth of embryonic chicken long bones. *The Journal of Structural Biology, 195*(1), 82–92.

Khavandgar, Z., Alebrahim, S., Eimar, H., Tamimi, F., McKee, M. D., & Murshed, M. (2013). Local regulation of tooth mineralization by sphingomyelin phosphodiesterase 3. *Journal of Dental Research, 92*(4), 358–364. https://doi.org/10.1177/0022034513478429

Khavandgar, Z., Poirier, C., Clarke, C. J., Li, J., Wang, N., McKee, M. D., Hannun, Y. A., & Murshed, M. (2011). A cell-autonomous requirement for neutral sphingomyelinase 2 in bone mineralization. *Journal of Cell Biology, 194*(2), 277–289. https://doi.org/10.1083/jcb.201102051

Kim, D., Lee, B., Thomopoulos, S., & Jun, Y.-S. (2018). The role of confined collagen geometry in decreasing nucleation energy barriers to intrafibrillar mineralization. *Nature Communications, 9*(1). https://doi.org/10.1038/s41467-018-03041-1

Knoll, A. H. (2003). Biomineralization and evolutionary history. *Reviews in Mineralogy and Geochemistry, 54*(1), 329–356. https://doi.org/10.2113/0540329

Koshizuka, Y., Ikegawa, S., Sano, M., Nakamura, K., & Nakamura, Y. (2001). Isolation of novel mouse genes associated with ectopic ossification by differential display method using <i>ttw</i>, a mouse model for ectopic ossification. *Cytogenetic and Genome Research, 94*(3–4), 163–168. https://doi.org/10.1159/000048809

Landis, W. J., & Silver, F. H. (2008). Mineral deposition in the extracellular matrices of vertebrate tissues: Identification of possible apatite nucleation sites on type I collagen. *Cells Tissues Organs, 189*(1–4), 20–24. https://doi.org/10.1159/000151454

LeGeros, R. Z. (2001). Formation and transformation of calcium phosphates: Relevance to vascular calcification. *Zeitschrift fur Kardiologie, 90*(3), 116–124. https://doi.org/10.1007/s003920170032

Li, J., Manickam, G., Ray, S., Oh, C.-do, Yasuda, H., Moffatt, P., & Murshed, M. (2016). Smpd3 expression in both chondrocytes and osteoblasts is required for normal endochondral bone development. *Molecular and Cellular Biology, 36*(17), 2282–2299. https://doi.org/10.1128/mcb.01077-15

Lorenz-Depiereux, B., Schnabel, D., Tiosano, D., Häusler, G., & Strom, T. M. (2010). Loss-of-Function ENPP1 mutations cause both generalized arterial calcification of infancy and autosomal-recessive hypophosphatemic rickets. *The American Journal of Human Genetics, 86*(2), 267–272. https://doi.org/10.1016/j.ajhg.2010.01.006

Luo, G., D'Souza, R., Hogue, D., & Karsenty, G. D. R. (1995). The matrix Gla protein gene is a marker of the chondrogenesis cell lineage during mouse development. *Journal of Bone and Mineral Research, 10*(2), 325–334. https://doi.org/10.1002/jbmr.5650100221

Luo, G., Ducy, P., McKee, M. D., Pinero, G. J., Loyer, E., Behringer, R. R., & Karsenty, G. (1997). Spontaneous calcification of arteries and cartilage in mice lacking matrix GLA protein. *Nature, 386*(6620), 78–81. https://doi.org/10.1038/386078a0

Mackenzie, N. C., Zhu, D., Milne, E. M., van 't Hof, R., Martin, A., Darryl, Q. L., Millán, J. L., Farquharson, C., & MacRae, V. E. (2012). Altered bone development and an increase in FGF-23 expression in Enpp1(-/-) mice. *PLoS One, 7*(2), 1–12.

Mahamid, J., Aichmayer, B., Shimoni, E., Ziblat, R., Li, C., Siegel, S., Paris, O., Fratzl, P., Weiner, S., & Addadi, L. (2010). Mapping amorphous calcium phosphate transformation into crystalline mineral from the cell to the bone in zebrafish fin rays. *Proceedings of the National Academy of Sciences, 107*(14), 6316–6321. https://doi.org/10.1073/pnas.0914218107

Mahamid, J., Sharir, A., Gur, D., Zelzer, E., Addadi, L., & Weiner, S. (2011). Bone mineralization proceeds through intracellular calcium phosphate loaded vesicles: A cryo-electron microscopy study. *Journal of Structural Biology, 174*(3), 527–535. https://doi.org/10.1016/j.jsb.2011.03.014

Malloy, P. J., Pike, J. W., & Feldman, D. (1999). The vitamin D receptor and the syndrome of hereditary 1,25- dihydroxyvitamin D-resistant rickets. *Endocrine Reviews, 20*(2), 156–188. https://doi.org/10.1210/er.20.2.156

Mannstadt, M., Bilezikian, J. P., Thakker, R. V., Hannan, F. M., Clarke, B. L., Rejnmark, L., Mitchell, D. M., Vokes, T. J., Winer, K. K., & Shoback, D. M. (2017). Hypoparathyroidism. *Nature Reviews Disease Primers, 3*(1), 1–20.

Marulanda, J., Eimar, H., McKee, M. D., Berkvens, M., Nelea, V., Roman, H., Borrás, T., Tamimi, F., Ferron, M., & Murshed, M. (2017). Matrix Gla protein deficiency impairs nasal septum growth, causing midface hypoplasia. *Journal of Biological Chemistry, 292*(27), 11400–11412. https://doi.org/10.1074/jbc.m116.769802

Miao, D., He, B., Lanske, B., Bai, X.-Y., Tong, X.-K., Hendy, G. N., Goltzman, D., & Karaplis, A. C. (2004). Skeletal abnormalities in Pth-Null mice are influenced by dietary calcium. *Endocrinology, 145*(4), 2046–2053. https://doi.org/10.1210/en.2003-1097

Michigami, T. (2019). Skeletal mineralization: Mechanisms and diseases. *Annals of Pediatric Endocrinology & Metabolism, 24*(4), 213–219. https://doi.org/10.6065/apem.2019.24.4.213

Millan, J. L., Lemire, I., Narisawa, S., Loisel, T. P., Boileau, G., Leonard, P., McKee, M. D., Crine, P., & Whyte, M. P. (2009). Enzyme replacement therapy (ERT) for murine hypophosphatasia. *Bone, 44*, Article S52. https://doi.org/10.1016/j.bone.2009.01.137

Murali, S. K., Roschger, P., Zeitz, U., Klaushofer, K., Andrukhova, O., & Erben, R. G. (2016). FGF23 regulates bone mineralization in a 1,25(OH)2D3 and klotho-independent manner. *Journal of Bone and Mineral Research, 31*(1), 129–142. https://doi.org/10.1002/jbmr.2606

Murshed. (2018). Mechanism of bone mineralization. *Cold Spring Harbor Perspectives in Medicine, 8*(12).

Murshed, M., Harmey, D., Luis Millán, J., McKee, M. D., & Karsenty, G. (2005). Unique coexpression in osteoblasts of broadly expressed genes accounts for the spatial restriction of ECM mineralization to bone. *Genes & Development, 19*(9), 1093–1104.

Murshed, M., Schinke, T., McKee, M. D., & Karsenty, G. (2004). Extracellular matrix mineralization is regulated locally; different roles of two gla-containing proteins. *The Journal of Cell Biology, 165*(5), 625–630. https://doi.org/10.1083/jcb.200402046

Nitschke, Y., & Rutsch, F. (2012). Generalized arterial calcification of infancy and pseudoxanthoma elasticum: Two sides of the same coin. *Frontiers in Genetics, 3*(302), 1–3.

Nudelman, F., Pieterse, K., George, A., Bomans, P. H. H., Friedrich, H., Brylka, L. J., Hilbers, P. A. J., de With, G., & Sommerdijk, N. A. J. M. (2010). The role of collagen in bone apatite formation in the presence of hydroxyapatite nucleation inhibitors. *Nature Materials, 9*(12), 1004–1009. https://doi.org/10.1038/nmat2875

Okawa, A., Nakamura, I., Goto, S., Moriya, H., Nakamura, Y., & Ikegawa, S. (1998). Mutation in Npps in a mouse model of ossification of the posterior longitudinal ligament of the spine. *Nature Genetics, 19*, 271–273.

O'Young, J., Liao, Y., Xiao, Y., Jalkanen, J., Lajoie, G., Karttunen, M., Goldberg, H. A., & Hunter, G. K. (2011). Matrix Gla protein inhibits ectopic calcification by a direct interaction with hydroxyapatite crystals. *Journal of the American Chemical Society, 133*(45), 18406–18412.

Paloian, N. J., Leaf, E. M., & Giachelli, C. M. (2016). Osteopontin protects against high phosphate-induced nephrocalcinosis and vascular calcification. *Kidney International, 89* (5), 1027–1036.

Parashar, A., Bak, K., & Murshed, M. (2022). Prevention of arterial elastocalcinosis: Differential roles of the conserved glutamic acid and serine residues of matrix Gla protein. *Arteriosclerosis, Thrombosis, and Vascular Biology, 42*(6), Article E155. https://doi.org/10.1161/ATVBAHA.122.317518

Parashar, A., Gourgas, O., Lau, K., Li, J., Muiznieks, L., Sharpe, S., Davis, E., Cerruti, M., & Murshed, M. (2021). Elastin calcification in in vitro models and its prevention by MGP's N-terminal peptide. *Journal of Structural Biology, 213*(1), Article 107637. https://doi.org/10.1016/j.jsb.2020.107637

Posner, A. S., & Betts, F. (1975). Synthetic amorphous calcium phosphate and its relation to bone mineral structure. *Accounts of Chemical Research, 8*(8), 273–281. https://doi.org/10.1021/ar50092a003

Razzaque, M. S. (2009). The FGF23-Klotho axis: Endocrine regulation of phosphate homeostasis. *Nature Reviews Endocrinology, 5*(11), 611–619. https://doi.org/10.1038/nrendo.2009.196

Razzaque, M. S., Sitara, D., Taguchi, T., St-Arnaud, R., & Lanske, B. (2006). Premature aging-like phenotype in fibroblast growth factor 23 null mice is a vitamin D-mediated process. *Federation of American Societies for Experimental Biology Journal, 20*(6), 720–722. https://doi.org/10.1096/fj.05-5432fjeUnitedStates

Rey, C., Combes, C., Drouet, C., & Glimcher, M. J. (2009). Bone mineral: Update on chemical composition and structure. *Osteoporosis International, 20*(6), 1013–1021. https://doi.org/10.1007/s00198-009-0860-y

Roberts, S., Narisawa, S., Harmey, D., Luis Millán, J., & Farquharson, C. (2007). Functional involvement of PHOSPHO1 in matrix vesicle–mediated skeletal mineralization. *Journal of Bone and Mineral Research, 22*(4), 617–627. https://doi.org/10.1359/jbmr.070108

Robinson, R. A. (1952). An electron-microscopic study of the crystalline inorganic component of bone and its relationship to the organic matrix. *The Journal of bone and joint surgery. American, 34*(2), 389–435. https://doi.org/10.2106/00004623-195234020-00013

Roschger, P., Fratzl-Zelman, N., Misof, B. M., Glorieux, F. H., Klaushofer, K., & Rauch, F. (2008). Evidence that abnormal high bone mineralization in growing children with osteogenesis imperfecta is not associated with specific collagen mutations. *Calcified Tissue International, 82*(4), 263–270.

Rowe, P. S. N. (2012). Regulation of bone−renal mineral and energy metabolism: The PHEX, FGF23, DMP1, MEPE ASARM pathway. *Critical Reviews™ in Eukaryotic Gene Expression, 22*(1), 61–86. https://doi.org/10.1615/critreveukargeneexpr.v22.i1.50

Ruben, J. A., & Bennett, A. A. (1987). The evolution of bone. *Evolution, 41*(6), 1187–1197. https://doi.org/10.1111/j.1558-5646.1987.tb02460.x

Sabbagh, Y., O'Brien, S. P., Song, W., Boulanger, J. H., Stockmann, A., Arbeeny, C., & Schiavi, S. C. (2009). Intestinal Npt2b plays a major role in phosphate absorption and homeostasis. *Journal of the American Society of Nephrology, 20*(11), 2348–2358. https://doi.org/10.1681/asn.2009050559

Schneider, E. G., Goldsmith, R. S., Arnaud, C. D., & Knox, F. G. (1975). Role of parathyroid hormone in the phosphaturia of extracellular fluid volume expansion. *Kidney International, 7*(5), 317–324. https://doi.org/10.1038/ki.1975.44

Sheldon, H., & Robinson, R. A. (1957). Electron microscope studies of crystal-collagen relationships in bone. IV. The occurrence of crystals within collagen fibrils. *The Journal of Biophysical and Biochemical Cytology, 3*(6), 1011–1016.

Shimada, T., Hasegawa, H., Yamazaki, Y., Muto, T., Hino, R., Takeuchi, Y., Fujita, T., Nakahara, K., Fukumoto, S., & Yamashita, T. (2004). FGF-23 is a potent regulator of vitamin D metabolism and phosphate homeostasis. *Journal of Bone and Mineral Research, 19*(3), 429–435. https://doi.org/10.1359/jbmr.0301264

Shimada, T., Mizutani, S., Muto, T., Hino, R., Yoneya, T., Takeda, S., Takeuchi, Y., Fujita, T., Fukumoto, S., & Yamashita, T. (2001). Cloning and characterization of FGF23 as a causative factor of tumor-induced osteomalacia. *Proceedings of the National Academy of Sciences of the United States of America, 98*(11), 6500–6505.

Si, J., Wang, C., Zhang, D., Wang, B., Hou, W., & Zhou, Y. (2020). Osteopontin in bone metabolism and bone diseases. *Medical Science Monitor, 26*. https://doi.org/10.12659/msm.919159

Soicher, M. A., Christiansen, B. A., Stover, S. M., Leach, J. K., & Fyhrie, D. P. (2013). Remineralization of demineralized bone matrix (DBM) via alternating solution immersion (ASI). *Journal of the Mechanical Behavior of Biomedical Materials, 26*, 109–118. https://doi.org/10.1016/j.jmbbm.2013.05.007

Speer, M. Y., McKee, M. D., Guldberg, R. E., Liaw, L., Yang, H. Y., Tung, E., Karsenty, G, & Giachelli, C. M. (2002). Inactivation of the osteopontin gene enhances vascular calcification of matrix Gla protein-deficient mice: evidence for osteopontin as an inducible inhibitor of vascular calcification in vivo. *Journal of Experimental Medicine, 196*(8), 1047–1055.

Steitz, S. A., Speer, M. Y., McKee, M. D., Liaw, L., Almeida, M., Yang, H., & Giachelli, C. M. (2002). Osteopontin inhibits mineral deposition and promotes regression of ectopic calcification. *The American Journal of Pathology, 161*(6), 2035–2046.

Stewart, A. J., Leong, D. T. K., & Farquharson, C. (2018). PLA2 and ENPP6 may act in concert to generate phosphocholine from the matrix vesicle membrane during skeletal mineralization. *FASEB Journal, 32*(1), 20–25. https://doi.org/10.1096/fj.201700521R

Stock, S. R. (2015). The Mineral-Collagen Interface in Bone. *Calcified Tissue International, 97*(3), 262–280.

Szeri, F., Niaziorimi, F., Donnelly, S., Fariha, N., Tertyshnaia, M., Patel, D., Lundkvist, S., & van de Wetering, K. (2022). The mineralization regulator ANKH mediates cellular efflux of ATP, not pyrophosphate. *Journal of Bone and Mineral Research, 37*(5), 1024–1031. https://doi.org/10.1002/jbmr.4528

Termine, J. D., & Eanes, E. D. (1972). Comparative chemistry of amorphous and apatitic calcium phosphate preparations. *Calcified Tissue Research, 10*(1), 171–197. https://doi.org/10.1007/bf02012548

Termine, J. D., & Posner, A. S. (1966). Infra-red determination of the percentage of crystallinity in apatitic calcium phosphates. *Nature, 211*(5046), 268–270.

The HYP Consortium. (1995). A gene (PEX) with homologies to endopeptidases is mutated in patients with X-linked hypophosphatemic rickets. *Nature Genetics, 11*(2), 130–136.

Toroian, D., Lim, J. E., & Price, P. A. (2007). The size exclusion characteristics of type I collagen: implications for the role of noncollagenous bone constituents in mineralization. *Journal of Biological Chemistry, 282*(31), 22437–22447.

Urist, M. R. (1965). Bone: Formation by autoinduction. *Science, 150*(3698), 893–899. https://doi.org/10.1126/science.150.3698.893

van Driel, M., & van Leeuwen, J. P. T. M. (2017). Vitamin D endocrinology of bone mineralization. *Molecular and Cellular Endocrinology, 453*, 46–51. https://doi.org/10.1016/j.mce.2017.06.008

Wagermaier, W., Gupta, H. S., Gourrier, A., Burghammer, M., Roschger, P., & Fratzl, P. (2006). Spiral twisting of fiber orientation inside bone lamellae. *Biointerphases, 1*(1), 1–5.

Weiner, S., & Traub, W. (1986). Organization of hydroxyapatite crystals within collagen fibrils. *FEBS Letters, 206*(2), 262–266. https://doi.org/10.1016/0014-5793(86)80993-0

Weiner, S., & Wagner, H. D. (1998). The material bone: Structure-mechanical function relations. *Annual Review of Materials Research, 28*(1), 271–298.

West, J. B. (2014). Carl Wilhelm Scheele, the discoverer of oxygen, and a very productive chemist. *American Journal of Physiology - Lung Cellular and Molecular Physiology, 307*(11), Article L811. https://doi.org/10.1152/ajplung.00223.2014

White, K. E., Jonsson, K. B., Carn, G., Hampson, G., Spector, T. D., Mannstadt, M., Lorenz-Depiereux, B., Miyauchi, A., Yang, I. M., Ljunggren, O., Meitinger, T., Strom, T. M., Jüppner, H., & Econs, M. J. (2001). The autosomal dominant hypophosphatemic rickets (ADHR) gene is a secreted polypeptide overexpressed by tumors that cause phosphate wasting. *Journal of Clinical Endocrinology and Metabolism, 86*(2), 497–500. https://doi.org/10.1210/jcem.86.2.7408

Whyte, M. P. (2020). Hypophosphatasia: nature's window on alkaline phosphatase function in humans. In Bilezikian, John P., Martin, T. John, Clemens, Thomas L., & Rosen, Clifford J. (Eds.), In *Principles of Bone Biology* (4th ed., Vol 2.) (pp. 1569–1599). Academic Press (Elsevier).

Xu, Z., Yang, Y., Zhao, W., Wang, Z., Landis, W. J., Cui, Q., & Sahai, N. (2015). Molecular mechanisms for intrafibrillar collagen mineralization in skeletal tissues. *Biomaterials, 39*, 59–66. https://doi.org/10.1016/j.biomaterials.2014.10.048

Xue, Y., Karaplis, A. C., Hendy, G. N., Goltzman, D., & Miao, D. (2005). Genetic models show that parathyroid hormone and 1,25-dihydroxyvitamin D3 play distinct and synergistic roles in postnatal mineral ion homeostasis and skeletal development. *Human Molecular Genetics, 14*(11), 1515–1528. https://doi.org/10.1093/hmg/ddi160

Yadav, M. C., Bottini, M., Cory, E., Bhattacharya, K., Kuss, P., Narisawa, S., Sah, R. L., Beck, L., Fadeel, B., Farquharson, C., & Millán, J. L. (2016). Skeletal mineralization deficits and impaired biogenesis and function of chondrocyte-derived matrix vesicles in Phospho1-/- and Phospho1/Pit1 double-knockout mice. *Journal of Bone and Mineral Research, 31*(6), 1275–1286. https://doi.org/10.1002/jbmr.2790

Yadav, M. C., Simão, A. M. S., Narisawa, S., Huesa, C., McKee, M. D., Farquharson, C., & Millán, J. L. (2011). Loss of skeletal mineralization by the simultaneous ablation of PHOSPHO1 and alkaline phosphatase function: A unified model of the mechanisms of initiation of skeletal calcification. *Journal of Bone and Mineral Research, 26*(2), 286–297. https://doi.org/10.1002/jbmr.195

Yang, X., Matsuda, K., Bialek, P., Jacquot, S., Masuoka, H. C., Schinke, T., Li, L., Brancorsini, S., Sassone-Corsi, P., Townes, T. M., Hanauer, A., & Karsenty, G. (2004). ATF4 is a substrate of RSK2 and an essential regulator of osteoblast biology: Implication for Coffin-Lowry syndrome. *Cell, 117*(3), 387–398. https://doi.org/10.1016/S0092-8674(04)00344-7

Zhang, J., Lin, T., Qi, H., Zhao, Q., Liu, Y., & Zhang, Y. (2019). Dual function of magnesium in bone biomineralization. *Advanced Healthcare Materials, 8*(21). https://doi.org/10.1002/adhm.201901030

CHAPTER

Bone cells, roles and functions

3

Niovi Dollas[1], Oran Kennedy[2], Robert Majeska[1] and Mitchell Schaffler[1]

[1]The City College of New York, Department of Biomedical Engineering, New York, NY, United States; [2]Royal College of Surgeons in Ireland, Department of Anatomy and Regenerative Medicine, Dublin, Ireland

1. Introduction

Every tissue's properties are ultimately defined by its constituent cells. This also holds for bone, even though its extracellular matrix (ECM) comprises the vast majority of tissue mass and is a major determinant of both its mechanical and physiological characteristics. Bone contains four principal cell types: osteoblasts, osteoclasts, osteocytes, and bone lining cells. Osteoblasts are bone forming cells that secrete the organic matrix and control its initial mineralization. Osteoclasts are bone resorbing cells that break down the tissue as directed. Osteocytes and lining cells are retired osteoblasts and are both "permanent" residents of bone. Osteocytes are embedded within bone matrix while lining cells cover most bone surfaces. Osteocytes comprise the majority of adult bone cellular content. Osteocytes have multiple functions that include orchestrating the formation and activity of both osteoblasts and osteoclasts, mechanosensation, regulation of mineral exchange between bone and the body, regulation of tissue material properties, and generation of endocrine signals that act on other tissues to facilitate systemic coordination of mineral homeostasis and bioenergetics. Finally, bone lining cells act as an interface between bone and its surrounding tissues, coordinate with underlying osteocytes via multiple signaling pathways and, under appropriate stimuli, can be re-activated into osteoblasts. In this chapter we will discuss the origin of these cells, their morphology, their biological activities, and how each of these features is controlled. This chapter aims to serve as a basic primer on these topics, while directing readers to more detailed reviews where necessary, as well as providing clinically relevant insights where applicable.

2. Osteoblasts

Osteoblasts are the cells that produce the organic extracellular matrix of bone (i.e., osteoid) and facilitate the deposition of mineral within it (Fig. 3.1). Their cellular structure and molecular biology reflect this synthetic specialization. Osteoblasts

Biomechanics and Mechanobiology of Bone. https://doi.org/10.1016/B978-0-443-26739-0.00014-4

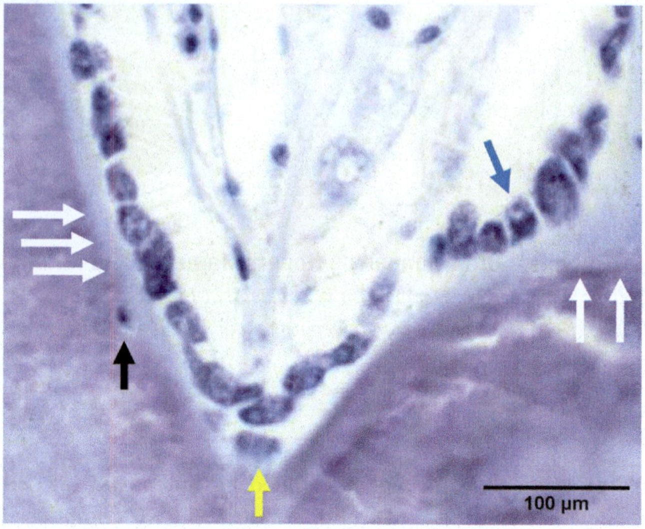

FIGURE 3.1 Photomicrograph of osteoblasts depositing bone.

Beneath the cuboidal-shaped cells (*blue arrow*) is a seam of recently deposited osteoid (pale staining region shown by *white arrows*), while darker staining matrix is more mature bone tissue. The yellow arrow points to an osteoblast in the process of becoming embedded to eventually become an osteocyte; the black arrow indicates a fully embedded osteocyte. Scale: 100 μm.

are cuboidal-shaped cells located, by definition, at the bone surfaces onto which they deposit new matrix. As a result, osteoblasts are polarized with respect to the bone surface—their synthetic/secretory organelles (rough endoplasmic reticulum and Golgi apparatus) are located near the bone, while the nucleus and other organelles are shifted toward the other end of the cell. As discussed in Section 2.2, osteoblasts belong to the mesenchymal developmental lineage that gives rise to all connective tissues.

2.1 Function

The major function of osteoblasts is to synthesize and secrete bone matrix; therefore, their major product is type I collagen, the majority of the organic bone matrix. Osteoblasts synthesize tropocollagen—the basic structural unit of collagen which is composed of three discrete protein chains (normally two $\alpha1$ and one $\alpha2$ chain). Tropocollagen molecules are secreted and self-assemble into collagen fibrils and fibers outside of the cells (See Chapter 1.2 for detailed discussion.). Even though osteoblasts do not directly regulate extracellular fibril assembly, they exert great influence over their structure and stability in several ways: i) they modify collagen amino acids (principally by hydroxylation of lysines and prolines), ii) they produce the enzyme lysyl oxidase that (extracellularly) catalyzes formation of cross-links between adjacent collagen chains that creates collagen fibril stability, and iii) they produce several minor collagens (Type V, as well as Type III during fracture healing) plus an array of proteoglycans and noncollagenous proteins (NCPs) that in part determine fibril width and overall organization (Table 3.1) The NCPs, while comprising only a small fraction of the organic matrix (~10%) also play critical roles in overall matrix organization and mineralization as well as cell adhesion (Morgan et al., 2015).

In addition to their role in facilitating the initial mineralization of bone, osteoblasts continue to exert indirect control over bone mineral by virtue of the composition and organization of osteoid. Collagen fibers, for example, impose the principal spatial constraints upon overall growth of the mineral phase, while NCPs like osteopontin (OPN) and osteocalcin (OCN), which can interact with both mineral ions and mineral crystals, can influence crystal size, their distribution in matrix and their ability to exchange ions at bone surfaces. These and other matrix molecules produced by osteocytes, including dentin matrix protein (DMP-1) and matrix extracellular phosphoglycoprotein (MEPE), are particularly important in regulating ion exchange between bone and tissue fluids. These interactions will be discussed further in the osteocyte section (see Section 5.1.4, Regulation of tissue material properties).

Under optimal and healthy conditions, osteoblasts lay down collagen matrix in organized sheets (also known as lamellae), which have a thickness of ~3–7 μm. In these sheets, collagen fibrils are oriented in parallel, resulting in a tissue known as lamellar bone. In contrast, when osteoblasts produce matrix at a high rate (e.g., in fracture healing or periods of extremely rapid growth) they deposit collagen fibrils in a more random isotropic manner, and the resulting tissue is known as woven bone. Woven bone also exhibits different proportions of non-collagenous matrix proteins compared to lamellar bone (Nanci, 1999). Across species, particularly in those that experience rapid growth, other types of bone exhibit organizational patterns lying somewhere between lamellar and woven bone, suggesting a continuum of matrix organization that is related to growth demands (Enlow, 1975).

Mechanical loading appears to play a role in influencing the direction in which osteoblasts lay down collagen during bone formation. In general, fiber direction tends to align with the principal loading axis in regions habitually under tension, while a cross-ply lamination at an angle of about 45 degrees tends to result in regions under compression. However, a variety of collagen orientations can be found in other bone areas, suggesting a more complex relationship between loading direction and fiber organization in bone. Bone formed under conditions of disuse shows a less distinctly lamellar organization (though not woven) than adjacent pre-existing bone that was formed under normal loading conditions, indicating further that loading influences prevailing collagen orientation. (Unpublished) Whether this is due to mechanical influences on osteoblasts or a direct effect of prevailing strain on matrix self-assembly remains unresolved.

In addition to forming the organic matrix of bone, osteoblasts regulate the deposition of mineral within it. Mechanistically, mineralization is essentially a two-step process that entails i) nucleation, in which soluble ions aggregate into regularly

Table 3.1 Bone non-collagenous proteins expressed by osteoblasts.

γ-Carboxyglutamic Acid-Containing Proteins (Gla Proteins)	
Osteocalcin (OCN, Bone Gla protein)	Regulates mineral maturation and bone formation Potentially involved in systemic bioenergetics Expressed in mature osteoblasts and osteocytes
Matrix Gla protein (MGP)	Mineralization activity unclear
Periostin	Interacts with cell-bound integrins Regulates collagen fibrillogenesis
Glycoproteins	
Osteonectin (ONN)	Promotes osteoblast differentiation and survival Regulates bone formation and remodeling Regulates collagen fibril assembly Binds to collagen, hydroxyapatite, vitronectin
Thrombospondins	Promote osteoblast differentiation and survival Regulate bone formation and remodeling
Fibronectin (FN)	Promotes osteoblast differentiation and survival Binds integrins May be involved in mineralization
Small integrin-binding ligand, N-linked glycoproteins (SIBLINGs)	
Osteopontin (OPN)	Early marker of matrix mineralization Binds calcium ions and bone mineral Mediates cell adhesion via integrins Expressed in osteoblasts and osteocytes
Bone sialoprotein (BSP)	Binds integrins Promote bone formation and mineralization (hydroxyapatite nucleator) May impair osteoclast activity
Dentin matrix protein-1 (DMP-1)	Regulates mineralization Expressed in osteoblasts to some extent, and constitutively in osteocytes
Matrix extracellular phosphoglycoprotein (MEPE)	Regulates mineralization Potential negative regulator of osteoblast activity Involved in systemic phosphate regulation Expressed in osteoblasts and osteocytes
Proteoglycans	
Heparan sulfate	Involved in cell-cell interactions Expressed in osteoblasts and osteoclasts
Small leucine-rich proteoglycans (SLRPs)	
Decorin (DCN) Biglycan (BGN) Fibromodulin	Regulate collagen fibrillogenesis Inhibit cell attachment to fibronectin Bind TGF-β

structured crystal nuclei and ii) appositional growth of crystals, as ions are added in an organized fashion to crystal surfaces. In bone mineral, a further process occurs as mineral crystals rearrange (mature) into highly stable hydroxyapatite-like bone mineral. The concentrations of calcium and phosphate in bone extracellular fluid are sufficient to support the growth of existing crystals, but osteoblasts likely facilitate the initial step in mineralization by contributing heterogeneous mineral nucleation sites on organic molecules or providing sequestered spaces (*e.g.*, within extracellular matrix vesicles) that allow nuclei to form (Hasegawa, 2018). However, the most widely known role for osteoblasts in promoting bone mineralization is through the enzyme alkaline phosphatase (ALP). ALP is highly expressed on osteoblast cell surfaces. ALP promotes mineralization by removing phosphate groups from molecules like pyrophosphate (PPi) and ATP, organic phosphate molecules produced by normal cellular metabolism that are inhibitors of mineralization. The importance of ALP to mineralization has long been appreciated. Loss-of-function mutations in the *Alpl*-gene encoding the form of ALP expressed in osteoblasts results in the genetic disease hypophosphatasia—lethal in its most severe form—which is characterized by deficient mineralization and, notably, elevated PPi levels (Wennberg et al., 2000). (Note: mouse gene nomenclature is used in this chapter; for details the authors direct readers to Mouse Genome Informatics, and for human equivalents HUGO Gene Nomenclature.)

It is interesting—and perhaps at first surprising—that while osteoblasts are necessary for proper bone mineralization, the appearance of readily detectable mineral (the "mineralization front") occurs separated in space and time from osteoblast activity at the surface of bone. This gap between matrix formation and mineralization may reflect the time needed for ALP to degrade mineralization inhibitors. In addition, mineral accumulation in the bone matrix is biphasic: once mineralization begins, the first phase (primary mineralization) occurs rapidly, with bone mineral content increasing to ~65%-70% of full bone mineral content over a period of days to weeks. Mineral deposition then slows in the secondary phase that lasts for months or longer. The decline in mineral apposition rate may reflect a reduction in permeability to calcium (Ca^{2+}) and phosphate (PO_4^{3-}) ions as mineral crystals fill the inter-fibrillar spaces.

2.2 Osteoblast origin, differentiation and fate

Osteoblasts originate from the lineage of pluripotent mesenchymal stem cells (MSCs) that also give rise to other connective tissue cell types including chondrocytes, adipocytes, tenocytes, myocytes and fibroblasts (Fig. 3.2). Osteoblast progenitors are found within the periosteum, marrow cavity, and perivascular connective tissue. Additionally, evidence indicates hypertrophic chondrocytes in the growth plate transdifferentiate into osteoblasts more commonly than originally thought, contributing to the osteogenic cell pool and to trabecular bone formation (Zhou et al., 2014).

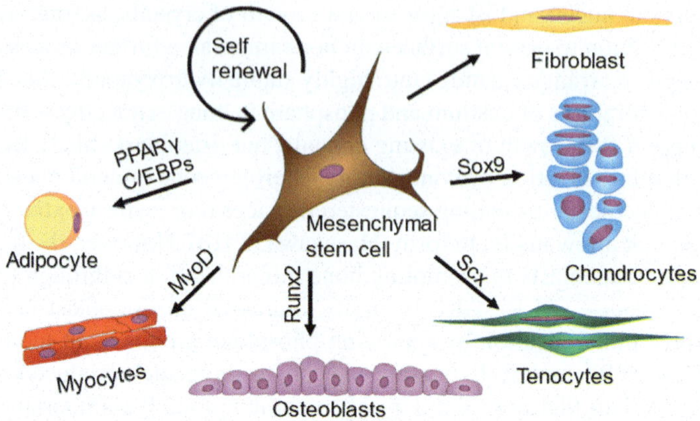

FIGURE 3.2 Schematic summary of mesenchymal stem cell differentiation.

Mesenchymal stem cells (MSCs) give rise to all connective tissues, with commitment to osteoblast, myocyte, tenocyte, chondrocyte, or adipocyte differentiation each controlled by the transcriptional regulators shown.

The formation of osteoblasts from progenitor cells involves essentially four steps: i) commitment of MSCs to osteo-chondroprogenitors, ii) further commitment to Osterix-expressing osteoprogenitors, iii) proliferation of osteoprogenitor cells, and iv) maturation of osteoprogenitors into functional osteoblasts. Osteoblast differentiation is normally triggered by morphogenic signaling molecules, most notably members of the transforming growth factor-β (TGF-β) superfamily that includes the bone morphogenetic proteins (BMPs), particularly BMP-2 and BMP-7. Importantly, commitment of a cell to the osteoblastic differentiation pathway is defined by expression of the Runx2 transcription factor (runt related transcription factor 2). Runx2 is indispensable for osteoblast differentiation, which was demonstrated most clearly by genetic evidence: deletion of the *Runx2* gene in mice resulted in failure of osteoblasts (and consequently bone tissue) to form, leaving animals with only a cartilage skeleton and thus unable to survive postnatally (Komori et al., 1997). For a detailed review on the signaling and transcriptional regulation of osteoblastogenesis, the authors direct readers to the review by Zhu et al. (2024).

Following commitment to osteoblastic differentiation, progenitor cells undergo sequential phenotypic changes characterized by expression of distinct markers (Fig. 3.3). The early stage of osteoblast differentiation is marked by the expression of ALP, while markers for more mature osteoblasts include expression of collagen I, OPN, bone sialoprotein (BSP), and OCN. Expression of ALP and collagen I (gene for α1 chain: *Col1a1*) essentially cease as osteoblasts further differentiate into osteocytes or lining cells, whereas other markers like OCN continue to be expressed.

The identification of stage-specific molecular markers of osteoblast differentiation has led to a range of techniques for understanding the differentiation process and to map the fates of osteoblasts and related lineages. Examples of these markers

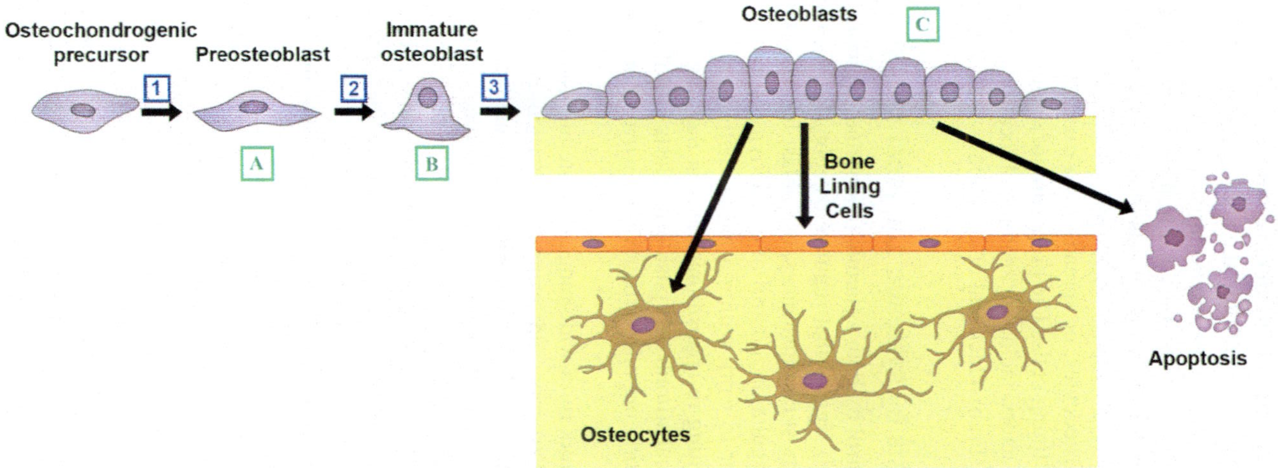

FIGURE 3.3 Schematic summary of osteoblast generation and fate.

Numbered boxes over arrows show critical signaling molecules and transcriptional factors in each major differentiation step: [1] Runx2, SATB2, BMPs, Wnts, FGF1/2; [2] Runx2, β-catenin, Osterix, BMPs, Wnts, FGF1/2, PTH, TGF-β; [3] Runx2, Osterix, ATF4, CREB, SATB2, Wnts, FGF1/2, PTH. Lettered boxes show expression of key osteoblast markers during each step of the differentiation process: [A] ALP (low), Col1a1 (low); [B] ALP, Col1a1, BSP; [C] ALP, Col1a1, OPN, OCN, ONN, BSP. Once they complete their bone formation activity, a portion of osteoblasts become osteocytes, others become bone lining cells, and the remainder undergo apoptosis.

include *Sp7*/Osterix for early osteoblasts, *Col1a1* for matrix producing osteoblasts, and DMP-1 and sclerostin for osteocytes. Promoters for these genes have been used to drive expression of fluorescent proteins to identify and track cells at each stage of differentiation in vivo and in vitro. In addition, the application of Cre-Lox technology to these genes has permitted stage-specific targeting for a host of purposes including gain- and loss-of-function mutations and transgene insertions.

Osteoblasts are active for relatively short periods (days to weeks) and ultimately face one of three fates: 60%–85% undergo apoptosis, while the remainder either become buried in matrix as osteocytes, or spread out to overlie the completed (i.e., quiescent) bone surface as lining cells. The proportion of osteoblasts that follow each fate depends on several factors, including anatomical location, type of bone being formed, age, and species (Franz-Odendaal et al., 2006).

2.3 Control of osteoblasts by systemic and local factors

Osteoblasts are subject to control by both systemic (endocrine, hormonal) and local (autocrine/paracrine, cytokine) signals that regulate osteoblast differentiation, cell activity, and lifespan. Under most circumstances, osteoblast formation and lifespan are the principal targets for regulation; control of cell activity is more constrained. Established osteoblasts appear to work within a narrow activity range to produce "normal" lamellar bone, resulting in appositional rates in the range of 0.5—2 μm/day. Biochemical or mechanical stimuli that push osteoblasts past their normal matrix synthesis rates lead to poorly organized matrix (i.e., woven bone), as discussed in Section 2.1.

Osteoblasts express receptors for a wide range of systemic hormones including parathyroid hormone (PTH), estrogens, androgens, glucocorticoids and β-adrenergic hormones. Two of the most-studied classes of endocrine regulators are the sex hormones and PTH. For an overview on the effects of sex hormones on bone cells, the authors recommend the review by Almeida et al. (2017). PTH produces complex and sometimes apparently contradictory effects on bone: intermittent PTH promotes osteoblastogenesis and decreases osteoblast apoptosis, while continuous PTH administration, well known to induce catabolic effects on bone, increases RANKL (receptor for activation of nuclear factor κB ligand) expression in osteoblasts and osteocytes, which triggers osteoclastic activation (Silva & Bilezikian, 2015).

3. Osteoclasts

Osteoclasts are the effector cells that resorb bone by secreting acid to dissolve its mineral and degradative enzymes to break down its organic matrix constituents. Unlike osteoblasts, they are part of the hematopoietic (blood-forming) cell family and are derived by differentiation of precursor cells within the monocyte-macrophage lineage—the same lineage that gives rise to dedicated phagocytic cells responsible for breakdown and removal of damaged or non-useful tissues throughout the body.

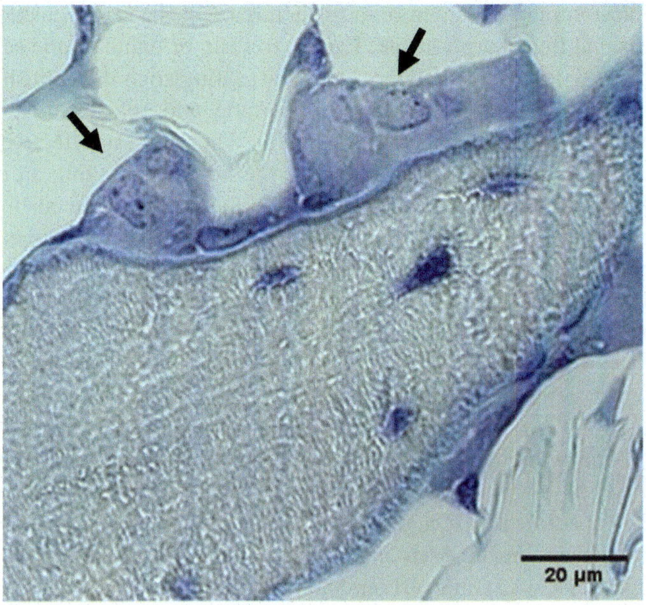

FIGURE 3.4 Photomicrograph of osteoclasts on bone surface.

Osteoclasts (large multinucleated cells, black arrows) resorbing bone on a trabecular surface. Scale: 20 μm.

Osteoclasts are large cells containing multiple nuclei, typically 2–25, and are formed by fusion of monocytic precursor cells (discussed further in Section 3.2). Osteoclast activity has been reported to correlate with size and nuclear number (Piper et al., 1992) (Fig. 3.4). Their size and multinuclearity, plus the presence of the enzyme tartrate-resistant acid phosphatase (TRAP), which is detectable by histological stains, make osteoclasts readily identifiable by light microscopy, both in vivo and in cell cultures. Osteoclasts are found on bone surfaces at resorption sites identified by the scalloped appearance of the bone surface (Howship's lacunae), and at the apex of resorption tunnels in the forming osteons of cortical bone.

3.1 Function

Resorption begins with the attachment of osteoclasts to the bone surface. Osteoclasts attach via $\alpha_V\beta_3$ integrin adhesion receptors to bone matrix proteins like OPN that contain Arg-Gly-Asp (RGD) integrin recognition sequences. The individual attachment sites soon organize into a closed ring-like structure that encloses and effectively isolates a discrete region of the bone surface. Binding of each integrin receptor to its ligand in bone also triggers a signaling cascade inside the cell mediated by c-Src, a non-receptor tyrosine kinase (NRTK). Activation of c-Src leads to formation of an F-actin ring linking the intracellular portions of the integrin receptors. This complex transmembrane structure creates a stable gasket or "sealing zone" (SZ) around the resorption site. Failure to form or maintain the actin ring prevents osteoclasts from resorbing bone. Once the sealing zone is established, intracellular lysosomes fuse with the plasma membrane and release their degradative enzymes into the resorption space (exocytosis). This process effectively turns the space between the osteoclast membrane and the bone matrix into an extracellular lysosome. Fusion of lysosomal membranes at the cell surface also results in a massive expansion of the plasma membrane within the sealing zone; this highly infolded membrane is termed the "ruffled border."

The ruffled border membrane is highly enriched in membrane-bound proton (H^+) pumps (vacuolar ATPases, vATPases) similar to those responsible for acidifying the interior of lysosomes. These enzymes continue to pump H^+ ions into the extracellular compartment inside the sealing zone, leading to a pH roughly between 3.5 and 4.5. The acidified fluid in the resorption pit dissolves the predominantly basic bone mineral and also provides a suitable environment for the major proteolytic enzyme involved in bone resorption, Cathepsin K (CatK), which functions optimally at acidic pH. In addition, since H^+ pumping by osteoclasts generates a net negative charge in the osteoclast cytoplasm, a bicarbonate/chloride (HCO_3^-/Cl^-) exchanger (SLC4A2) and carbonic anhydrase two act maintain electroneutrality. Impairment of either acid secretion or HCO_3^-/Cl^- exchange has been linked to some forms of osteopetrosis—a condition in which impaired bone resorption produces bone that is "stone-like." Defects in mineral dissolution are particularly severe as they also prevent the subsequent breakdown of organic matrix components.

Osteoclasts produce a variety of enzymes that contribute to degradation of organic matrix molecules (Fig. 3.5); principal among these are tartrate resistant acid phosphatases (TRAP), CatK, and matrix metalloproteinases MMP-9 and MMP-14. Importantly, MMP-1 (the archetypal collagenase) is not produced by osteoclasts. The predominant proteolytic enzyme responsible for matrix removal is CatK, a cysteine protease. Cathepsins are a family of lysosomal enzymes responsible for the normal breakdown and turnover of proteins. As noted, they are most active in the acidic intra-lysosomal environment. TRAP, the most widely used marker for osteoclast activity, dephosphorylates non-collagenous matrix proteins, like bone sialoprotein and osteopontin. TRAP has also been shown to generate reactive oxygen species (ROS) to further assist with matrix degradation (Halleen et al., 1999). Osteoclasts also utilize the matrix metalloproteinases MMP-9 (secreted) and MMP-14 (membrane bound; also known as MT1-MMP). It appears the MMPs and CatK are somewhat redundant, as CatK deficient mice are still able to undertake some bone resorption, and single knockouts of either MMP-9 or MMP-14 exhibit normal bone phenotypes, while a double knockout of MMP-9 and MMP-14 resulted in an osteopetrotic phenotype (Li et al., 2006; Zhu et al., 2020).

Following the dissolution of matrix, the ions released from bone mineral (mainly Ca^{2+} and PO_4^{3-}) and the organic products of enzymatic digestion are then removed from the resorption pit and released ultimately to the "bulk" extracellular fluid of bone. Some of the degradation products may escape at the sealing zone interface

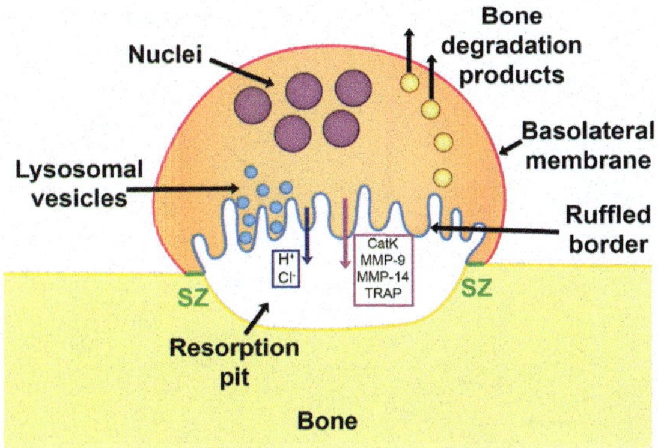

FIGURE 3.5 Schematic representation of osteoclast structure and function.

Resorbing osteoclast with multiple nuclei, ruffled border, sealing zone (SZ) and resorption pit. Protons (H^+) generated by carbonic anhydrase in the cytoplasm are secreted into the resorption pit via the vATPase proton pumps; chloride ions (Cl^-) are also secreted into the pit. Bicarbonate/chloride exchangers at the basolateral region of the cell (not shown) maintain cytoplasmic electroneutrality. Bone removal occurs through dissolution of the mineral by acid and enzymatic degradation of the matrix proteins by secreted cathepsin K (CatK), matrix metalloproteinases (MMPs), and tartrate-resistant acid phosphatase (TRAP). Bone matrix degradation products are taken up by the osteoclast by endocytosis, then secreted into the circulation from the basolateral aspect of the cell.

as osteoclasts reposition themselves; however, most transport to the extracellular fluid occurs through a combination of movement directly through the cytoplasm (for minerals and cations, which are pumped out across the basolateral membrane of the osteoclast) and vesicular transport (*e.g.*, for peptide fragments) followed by exocytosis at basolateral surfaces (i.e., the surfaces of the osteoclast positioned away from the bone).

3.2 Control of osteoclasts by systemic and local factors

Osteoclasts, like osteoblasts, are subject to control by both systemic and local signals that regulate their activation, recruitment, activity and lifespan.

Osteoclasts, as previously noted, are derived from cells of the monocyte-macrophage lineage (Fig. 3.6). Some 3 decades ago, Udagawa et al. found that coculture of monocytes with osteoblast lineage cells induced them to form TRAP-positive, multinucleated cells in vitro, providing experimental support identifying the source of critical signals to make osteoclasts (Udagawa et al., 1990). Subsequently, RANK (receptor for activation of nuclear factor κB) was found on osteoclast precursors and shown to be an obligate activator of osteoclast differentiation. Under the influence of RANKL and macrophage colony stimulating factor (M-CSF), monocyte precursors proliferate and differentiate into preosteoclasts, which then fuse and undergo maturation to become a functional (mature) osteoclast.

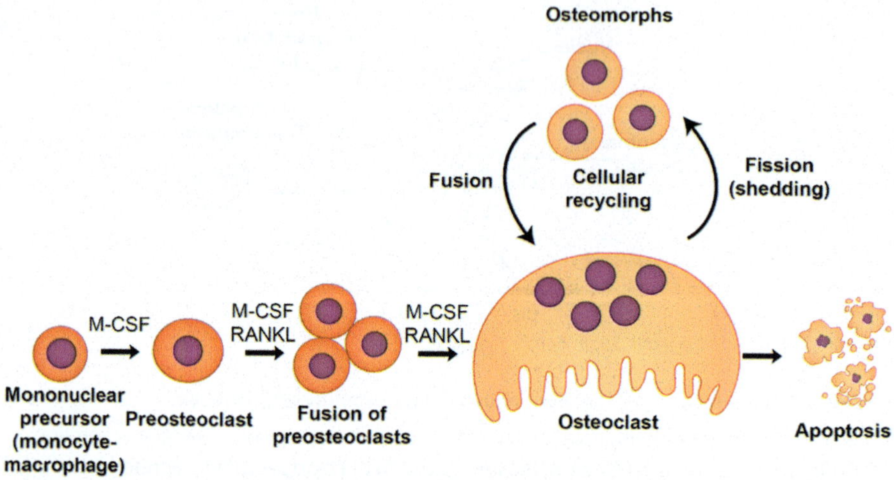

FIGURE 3.6 Schematic summary of osteoclast generation and fate.

Osteoclast formation is governed by RANKL produced by osteocytes and osteoblasts, and M-CSF from marrow stromal cells. M-CSF stimulates proliferation of monocyte precursors while RANKL is essential for their fusion and differentiation into mature osteoclasts. Mature osteoclasts are capable of shedding mononuclear cells termed "osteomorphs"; osteomorphs can subsequently undergo fusion with nearby osteoclasts.

While osteoblast lineage cells produce the RANKL necessary for osteoclastogenesis, they also produce a competing factor, osteoprotegerin (OPG), that allows for finer tuning of the signals that initiate resorption. OPG is a neutralizing decoy receptor that binds to RANKL and blocks its activity. Thus, the strength of a signal initiating osteoclastogenesis depends on the RANKL/OPG ratio. As a result, either the administration of excess OPG or an antibody to RANKL will prevent osteoclast formation (More on this in the Clinical correlate, Section 7.2.).

Although RANKL is essential for osteoclast differentiation, a second factor (M-CSF) is required to promote the proliferation of osteoclast precursors in order to achieve effective osteoclastogenesis. M-CSF can be produced by a range of cells, including bone marrow stromal cells, osteoblasts, and osteocytes. Mice with mutations in M-CSF display an osteopetrotic phenotype, in which bone is overly dense, brittle, and exhibits reduced marrow cavities (Lieschke et al., 1994).

Additionally, many systemic and local signals exert direct effects on extant osteoclast activity and lifespan. These include sex steroids, glucocorticoids, adrenergic hormones, and inflammatory cytokines such as tumor necrosis factors (TNFs, of which RANKL is a member) and interleukins (Boyle et al., 2003). Estrogen is of clinical interest, particularly as osteoclasts express estrogen receptors and its absence may prolong osteoclast lifespan. The precise molecular mechanism by which estrogen acts on osteoclasts remains an open field of investigation (Veis & O'Brien, 2023).

3.3 Nuclear recycling

It has long been appreciated that osteoclasts can apparently acquire and possibly shed nuclei over the course of their lifespan (Jaworski et al., 1981). In addition, a long-held presumption has been that osteoclasts undergo apoptosis when they complete their resorptive activity. In recent breakthrough studies, however, McDonald et al. showed that the life cycle of osteoclasts is more complex (McDonald et al., 2021). Osteoclasts do not simply shed nuclei; instead, they release a distinct mononuclear cell type termed an osteomorph. Furthermore, osteoclasts and osteomorphs engage in a dynamic fission-fusion process, in which osteomorphs are able to transfer between osteoclasts. This process of nuclear/cellular recycling is likely energetically conservative and allows for the expansion of osteoclast lifespan. When an OPG fraction is used to inhibit osteoclast activity, osteomorphs were found to accumulate as a pool of committed precursors (McDonald et al., 2021). This accumulation may also help account for the unexpectedly rapid recruitment of osteoclasts after discontinuation of the anti-RANKL antibody denosumab—a possible contributor to the "rebound effect" (or overshoot) discussed in the Clinical correlate (Section 7.2).

4. Bone modeling and remodeling

Bone undergoes two distinct formation/resorption activities throughout life: modeling and remodeling. In modeling, teams of osteoblasts or osteoclasts operate at the same time but on different bone surfaces, resulting in a net gain or loss of tissue and a change in overall bone shape. Modeling occurs mainly during skeletal growth and development, though a small amount occurs throughout life as exemplified by radial (periosteal) expansion of long bone diaphyses. By contrast, remodeling (also referred to as bone turnover) is a sequential process in which osteoclasts resorb a small, discrete packet of bone tissue, after which osteoblasts fill in the resorption cavity. Moreover, resorption and formation in remodeling are coupled. While a resorption-formation packet is active, it is referred to as a BMU or Bone Multicellular Unit. The complete remodeling unit has been referred to as a Bone Structural Unit (BSU), but these packets are more commonly referred to as secondary osteons or Haversian systems in cortical bone, and hemi-osteons on trabecular or endocortical surfaces. Bone remodeling is the predominant resorption/formation process occurring in the adult skeleton and is functionally analogous to turnover throughout the body which replaces cells and tissues that have reached the limits of their functional lifespan.

During bone remodeling, a short interval between resorption and formation results in an anatomic space between osteoclasts and osteoblasts, referred to as a "reversal zone." Mononuclear cells are observed in the reversal zone, and these cells have been suggested to provide a "fine finish" to the roughened bone surface left by osteoclasts, as well as to lay down a distinct layer of matrix (the cement line) onto which osteoblasts deposit osteoid (Andersen et al., 2013). While the origin of these reversal cells has been a subject of discussion, current consensus indicates that they belong to the osteoblast lineage. Osteoblastic cells have the capability both to modify bone surfaces (e.g., by proteases like MMP-1) and to produce the organic matrix molecules that comprise the cement line (Andersen et al., 2013).

5. Osteocytes

Osteocytes make up 90%–95% of the cells in mature bone. They are derived from the 10%–30% of osteoblasts that become buried in newly deposited osteoid. As osteoblasts transition to osteocytes, they undergo extensive structural and functional changes. Their cell bodies diminish in size, their secretory apparatus becomes greatly reduced, and they elaborate large numbers of dendritic processes using mechanisms similar to those that form dendrites in neurons (Zhang et al., 2006). The morphologic and molecular alterations that occur during the osteoblast-osteocyte transition are detailed in Dallas & Bonewald (2010). Once formed, osteocytes may live for years or decades in long-lived animals like humans and can persist throughout the entire lifespan of shorter-lived animals like mice and rats (Bonewald, 2011). This differs fundamentally from osteoclasts and osteoblasts that are formed only on demand where needed and generally function from days to weeks.

Osteocytes reside in a unique environment, encased within the mineralized matrix of bone. The spaces occupied by their cell bodies and processes are termed, respectively, lacunae and canaliculi, together comprising the lacuno-canalicular system (LCS). The small fluid layer separating osteocytes from their surrounding matrix imposes a distinctive set of constraints that dominate most aspects of osteocyte physiology.

Osteocyte cell bodies are separated from each other by tens of microns, but their dendritic processes are in contact with those of their neighbors, creating a linked cellular network (Fig. 3.7). Estimates of the number of primary processes emanating from a typical osteocyte are in the range of 40–100, prior to subsequent branching (Beno et al., 2006). These processes appear to connect not only nearest neighbors but also cells at greater distances. Connectomic analyses indicate that an average osteocyte could have approximately 1100 connections and touch an estimated 50 neighboring cells (Buenzli & Sims, 2015; Creecy et al., 2021). Scaled up to the entire human skeleton, this translates to approximately 42 billion osteocytes with some 23 trillion connections. The proximity of each osteocyte and its processes to the bone matrix provides a vast cell surface area (~215 m^2) over which osteocytes can regulate mineral exchange with the bone (Buenzli & Sims, 2015). For comparison, the lung alveolar surface area for gas exchange is on the order of ~120 m^2.

Contact points between dendritic processes of osteocytes—and also between osteocytes and neighboring cells on bone surfaces—are characterized by the presence of Connexin 43 (Cx43) based gap junctions (Fig. 3.8E). Gap junctions create an

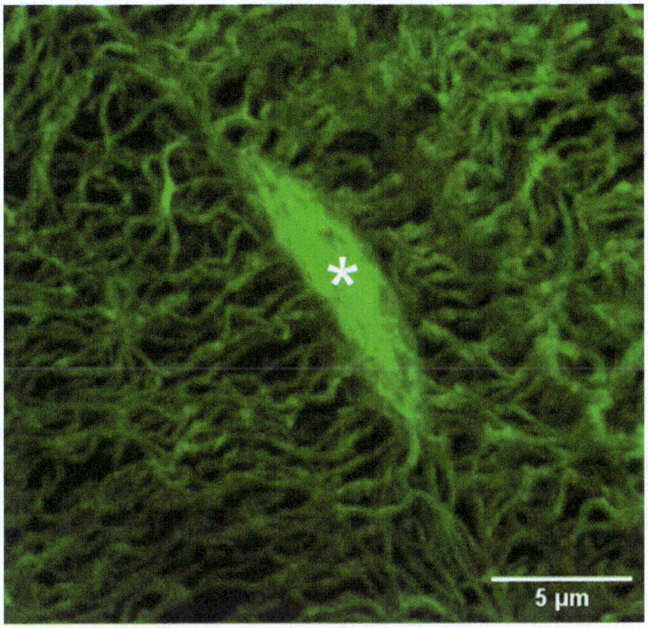

FIGURE 3.7 Photomicrograph of an osteocyte in situ.

3D super-resolution microscopy image (Z-axis depth = 3 μm), showing an osteocyte (asterisk) within its lacuna, and the extensive array of dendritic processes radiating outwards from the cell body. Scale: 5 μm.

electrochemical continuum of cells throughout the bone that allow signaling via direct transcellular movement of ions and other small molecules (<1 kDa). Osteocytes also rely on paracrine signaling via movement of molecules (*e.g.*, ATP, PGE2, cytokines) through the extracellular fluid spaces of the LCS.

The LCS fluid space is essential for movement of all metabolites, wastes, and signaling molecules among osteocytes and with cells at bone surfaces; however, the fluid space around osteocytes within the LCS is small. The distance between osteocyte cell bodies and their lacunar wall is ~0.5–1 μm, while the radial distance between the dendrites and the canalicular walls is in the range of ~100 nm. Moreover, movement through the LCS is further constrained by macromolecules, most notably proteoglycans, that can restrict the flow of both fluid and solutes. Movement of small solutes (below ~10 kDa molecular weight) through the LCS can occur by diffusion; however, movement of larger molecules, up to approximately 70 kDa molecular weight, requires convective fluid flow caused by mechanical loading (Wang et al., 2004). It is unclear whether molecules larger than this convective limit (e.g., antibodies) are able to penetrate through the LCS under normal conditions.

5.1 Osteocyte functions

The last few decades have led to an increasing understanding that osteocytes play central roles in virtually all physiological activities in bone (Schaffler et al., 2014). They regulate both the formation and activity of osteoblasts and osteoclasts during modeling and remodeling, sense mechanical loads on bone, regulate ion exchange between bone and extracellular fluids, regulate tissue material properties, and generate endocrine signals that communicate with non-skeletal tissues to modulate mineral homeostasis and systemic bioenergetics.

5.1.1 Orchestration of osteoblast and osteoclast activities during bone modeling and remodeling

5.1.1.1 Osteoblast formation and activity

Osteocytes exert direct control on osteoblasts through sclerostin (the product of the *Sost* gene). Sclerostin is expressed constitutively by osteocytes and inhibits osteoblast formation by interfering with the Wnt signaling pathway that stimulates transcription of pro-osteoblastogenic genes. Sclerostin inhibits Wnt activity by competitively binding to the co-receptors for low density lipoproteins 5/6 (LRP5/6) and Frizzled (Fzd) receptor complex. This leads to the intracellular degradation of the β-catenin transcription factor, which is then unable to activate target genes that lead to osteoblastogenesis and increased bone formation. Additionally, sclerostin antagonizes the activity of several BMPs.

Conversely, down-regulation of sclerostin removes these inhibitions and allows osteoblast formation. Sclerostin deficits in humans resulting from mutations in the *Sost* gene result in rare bone disorders characterized by high bone mass like Van Buchem's syndrome and sclerosteosis, and knockout of the *Sost* gene in mice similarly leads to atypically high bone mass (Li et al., 2008). Together, these findings show that the physiological role of sclerostin is to serve as a "molecular brake" that must be continuously applied in order to suppress new bone formation. Production of sclerostin has been shown to be modulated by a wide range of hormonal regulators of bone formation, including PTH, PTH-related peptide (PTHrP), and sex hormones (see the review by Drake and Khosla for details) (Drake & Khosla, 2017). These observations provide the basis for sclerostin-neutralizing antibodies as anabolic therapy to treat low bone mass (see Clinical correlate, Section 7.1).

5.1.1.2 Osteoclast formation and activity

In bone resorption, osteoclast formation is triggered by RANKL and antagonized by OPG (see Osteoclasts, Section 3.2), which acts as a decoy receptor for RANKL. Consequently, the ratio of RANKL to OPG is now considered the principal determinant of whether resorption occurs at a particular site. While osteoblast lineage cells were long known as the source of both RANKL and OPG, seminal studies by Xiong et al. and Kennedy et al. over the last decade established that osteocytes are the principal cell responsible for production of these pro-osteoclastogenic signals in bone (Kennedy et al., 2012; Xiong et al., 2011). Xiong et al. showed that if RANKL is developmentally absent from osteocytes (via gene deletion in mice), the resulting bone is severely osteopetrotic, i.e., completely lacking in resorption and thus having no marrow cavity, among other features (Xiong et al., 2011). Kennedy et al. discovered that upregulation of osteocyte RANKL expression near fatigue microdamage sites in bone is responsible for the activation and recruitment of osteoclasts to remove and remodel damage foci in adult mice (Kennedy et al., 2012). Furthermore, studies have shown that it is osteocyte apoptosis caused by a range of physiological challenges (e.g., fatigue microdamage, estrogen loss, disuse, glucocorticoid use) which is the trigger for surviving neighbor osteocytes to turn on RANKL production (Cabahug-Zuckerman et al., 2016; Emerton et al., 2010; Kennedy et al., 2014; O'Brien et al., 2004). More recent studies show osteocyte RANKL expression can also be triggered by extrinsic factors (i.e., in the absence of osteocyte apoptosis), such as in calcium deficiency and secondary hyperparathyroidism (Xiong et al., 2014). Recent data indicate that TNFα (tumor necrosis factor α), the master inflammatory mediator of the body and a potent driver of pathological bone resorption in autoimmune diseases, also triggers osteoclast formation by upregulating osteocyte RANKL expression (Marahleh et al., 2019).

Finally, there are circumstances where osteocyte apoptosis occurs but is insufficient to trigger a resorption response. In such circumstances, dying osteocytes will degenerate in situ in their lacunae, resulting in small regions of osteocyte-depleted ("dead") bone. The accumulation of such sites is commonly observed in bone with aging (Vashishth et al., 2000). Finally, if dying osteocytes are not removed, cellular debris in lacunae can either degrade completely, leaving empty lacunae, or some of the debris can mineralize, leading to infilled lacunae (i.e., micropetrosis) (Busse et al., 2010).

5.1.2 Mechanosensation

Osteocytes are the primary mechanosensory cells in bone. Theoretical and experimental studies have established that fluid movement in the LCS resulting from loading of bone is the predominant mechanical stimulus which osteocytes are subjected to. However, the nature of fluid movement interactions with osteocyte are complex. Key insights into the system stemmed from theoretical work by Weinbaum, Cowin, and Zeng; they modeled the fluid shear stress on osteocyte cell processes during mechanical loading and showed it to be similar in magnitude to that experienced by endothelial cells on arterial walls (Weinbaum et al., 1994). This arises from the flow rate through the narrow annular space between processes and canalicular walls. In contrast, the space between the cell body and lacunar wall is an order of magnitude larger, so the fluid shear on osteocyte cell bodies would be negligible.

The concept that osteocyte dendritic processes rather than cell bodies are the sites of greatest osteocyte mechanosensitivity was confirmed experimentally by Wu et al. using the MLO-Y4 osteocytic cell line (Wu et al., 2011). Using a Stokesian fluid stimulus probe to puff fluid at piconewton-level forces onto discrete regions of MLO-Y4 osteocytes in vitro, they found the force ($\leq$10 pN) needed to evoke an electrical response when applied to the process was 10 times smaller than that required for the cell body (Wu et al., 2011). Further studies revealed that the tiny fluid stimulus on the osteocyte dendrite triggered a localized cytoplasmic Ca^{2+} wave in the process that propagated to the cell body and subsequently spread to neighboring osteocytes (Wu et al., 2013).

The in situ environment for osteocytes sitting in the LCS is clearly more complex than in cell culture, as the cells reside within three-dimensional spaces and have direct connections to the surrounding bone matrix. These connections interact with the mechanically driven fluid movement around the cell and participate in mechanotransduction. You et al. found that loading-induced fluid flow in the canalicular space would deflect the tethering elements (consisting of long proteoglycans) that run from the cell processes membrane to canalicular walls, and this would result in high localized strains (i.e., strain amplification) at the membrane attachment points (You et al., 2001, 2004) (Fig. 3.8A). Thompson et al. identified the proteoglycan perlecan as the main constituent of these tethers (Thompson et al., 2011). McNamara and co-workers demonstrated that osteocyte processes also make direct contact with the canalicular walls at discrete sites where there are small hillocks or protrusions from the walls (McNamara et al., 2009). Wang et al. showed that these direct, rigid contacts could produce even greater local strain amplification during fluid flow than the tethers (Wang et al., 2008). Intriguingly, these process membrane-to-bone contact sites were revealed to be sites of $\alpha_v\beta_3$ integrin attachments. However, osteocyte processes lack the cytoplasmic components (i.e., vinculin, paxillin, talin, focal adhesion kinase) characteristic of classic integrin-based signal transduction, indicating that these $\alpha_v\beta_3$ foci do not function as conventional integrin transduction sites (Cabahug-Zuckerman et al., 2018). Rather, these $\alpha_v\beta_3$ attachment sites on osteocyte processes were found to anchor a novel osteocyte mechanosome complex, as will be detailed below. Importantly, Thi et al. showed that a tiny, localized fluid stimulus applied to osteocyte process attachment sites in vitro triggered Ca^{2+} signaling, but the same fluid stimulus applied elsewhere in the process had no effect (Thi et al., 2013).

5.1.2.1 The osteocyte mechanosome

The initiation of intracellular Ca^{2+} signaling following specific mechanical stimulation of osteocyte process attachment sites suggested that key mechanically activated channels are located in proximity to the $\alpha_v\beta_3$ integrin foci, i.e., in the localized membrane strain amplification regions. In their in vitro studies, Thi et al. found that the mechanically-induced Ca^{2+} signaling in osteocytes processes involved ATP release, which subsequent studies revealed involves the pannexin-1 (Panx-1)—P2X purinoreceptor 7 (P2X7R) purinergic signaling complex (Seref-Ferlengez et al., 2019; Thi et al., 2013). Cabahug-Zuckerman et al. confirmed the presence of Panx-1 and P2X7R, concentrated around these integrin foci in situ (Cabahug-Zuckerman et al., 2018) (Fig. 3.8B). Panx-1 opens in response to mechanical stimuli to release ATP, which when released from cells can act as a signaling molecule. ATP in turn binds to P2X7R and the complex then functions as a Ca^{2+} channel. Seref-Ferlengez et al. (2019) found that when Panx-1 was knocked out in a mouse model,

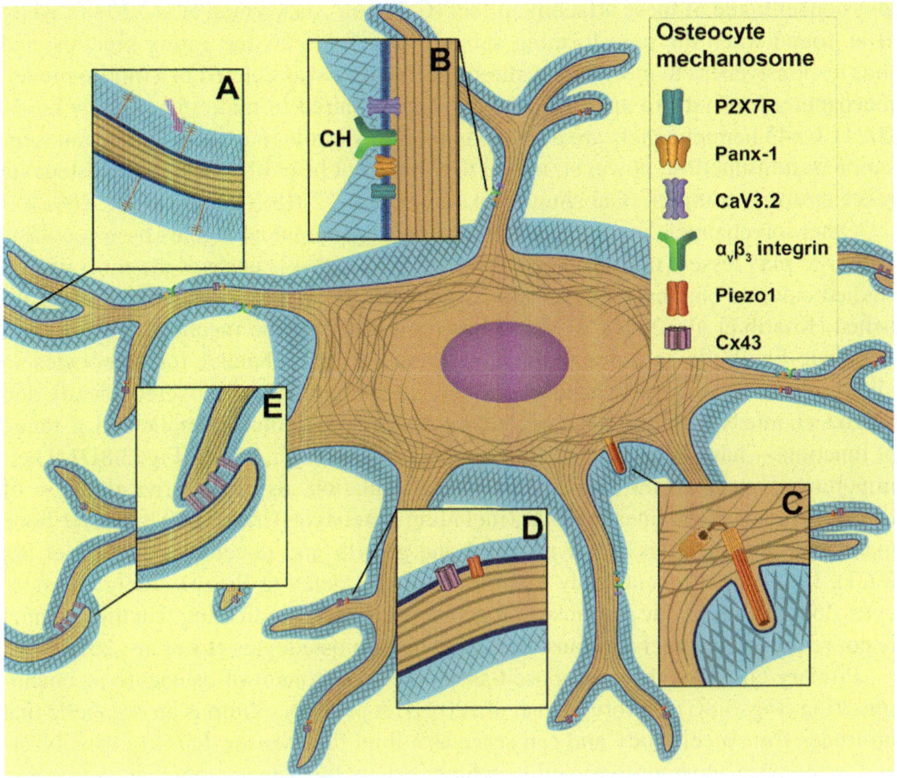

FIGURE 3.8 Schematic representation of an osteocyte in the LCS and key osteocyte mechanosensing/mechanotransducing elements.

Fluid and pericellular matrix molecules within the LCS are represented in blue with cross-hatching, and bone matrix is shown in yellow. A—shows tethering elements (arrow) running from the osteocyte process membrane to the bony canalicular wall; actin cytoskeleton within dendritic processes is organized in parallel fibers, in contrast to the meshwork of cortical actin in the osteocyte cell body. B—illustrates the osteocyte mechanosome found where collagen hillocks (CH) project from canalicular walls and attach to cell process membrane via αvβ3 integrins; key mechanically activated channels (Panx-1, P2X7R, CaV3.2) are located around the integrin attachment site, which is a region of membrane strain amplification during loading. C—shows a primary cilium projecting into the pericellular space of the lacuna. D—shows a Connexin-43 (Cx43) hemichannel and Piezo1 channel; Cx43 hemichannels are not located at the mechanosomes, while Piezo1 is depicted in association with Cx43 hemichannels based on their functional association per Zeng et al. (2022). E—shows dendrites from neighboring osteocytes communicating via gap junctions.

Wnt/β-catenin signaling was impaired in response to mechanical loading (Seref-Ferlengez et al., 2019). CaV3.2, a voltage sensitive calcium channel (VSCC), is also concentrated at these attachment foci (Cabahug-Zuckerman et al., 2018). Kelly et al. found when the pore-forming subunit (which affects the gating kinetics, and thus responsiveness to mechanical stimuli) of VSCCs was deleted in a mouse model, mechanoresponsivity to anabolic loading was impaired in male mice (Kelly et al., 2024). Cx43 hemichannels are also thought to play a role in osteocyte mechanosensation or transduction; however, in situ they have not been observed at the osteocyte mechanosome complex (Cabahug-Zuckerman et al., 2018; Cherian et al., 2005).

Other ion channels involved in the sensing of mechanical signals by osteoblasts appear to play lesser (if any) roles in osteocytes. TRPV4 is perhaps the most widely studied calcium channel in bone and its effect on bone development is well established (Khatib et al., 2023). While it has been shown to be mechanoresponsive in osteocyte-like cells in culture, the importance of this channel for osteocytes in vivo remains ill defined (Zhang et al., 2021). The recently discovered Piezo1 and Piezo2 channels—ubiquitous, mechanosensitive ion channels implicated a range of functions—have been identified in osteocytes (Qin et al., 2021) (Fig. 3.8D). Their importance has been the source of much speculation, as constitutive deletion of Piezo1 in osteoblast lineage cells (including osteocytes) results in reduced bone mass, pointing to its essential role in bone growth and development (Qin et al., 2021). However, a recent study demonstrated that deleting Piezo1 solely in osteocytes does not alter bone response to mechanical loading, indicating that this channel is not required for mechanotransduction in mature osteocytes (Li et al., 2025).

Primary cilia have been proposed as another component of osteocyte mechanosensation (Fig. 3.8C) (Verbruggen et al., 2023). A primary cilium is an organelle that protrudes from a cell body and can serve as a fluid flow sensor. In vitro, osteoblasts and osteocytes have primary cilia, which led to the suggestion that these cilia mediate responses to fluid flow, however, they do so for authentic osteocytes in situ remains unresolved (Verbruggen et al., 2023). The diameter of a primary cilium is fixed by its microtubule nanostructure and thus primary cilia need to be many microns in length in order to have the aspect ratio to be deflected by fluid movement, particularly in low flow situations like that in osteocyte lacunae. However, the pericellular space between osteocyte cell bodies and lacunar walls is small, on the order of 1 μm or less, and the short primary cilia able to fit into this space would lack a sufficient moment arm to be deflected in the low fluid flow. It has been speculated that short primary cilia in osteocyte lacunae might have integrins on their surface that interact with the hyaluronic acid in the lacunar space, allowing a different type of mechanotransduction, but this remains unconfirmed (Verbruggen et al., 2023). Bones from conditional knockout mice in which primary cilia were deleted from osteoblasts and osteocytes exhibited normal skeletal phenotypes; some attenuation of the bone formation response to experimental high-force mechanical loading was observed in these mice, though the extent to which this reflects changes in osteoblast or osteocyte function remains unclear (Temiyasathit et al., 2012).

5.1.3 Regulation of mineral exchange

Among the most critical osteocyte functions is regulating systemic mineral homeostasis. Mineral exchange between bone surfaces and the surrounding milieu occurs constantly, varying under normal circumstances with dietary intake and excretion. Direct osteocyte regulation of this exchange is facilitated by the proximity of osteocyte processes to bone surfaces throughout the vast LCS network. In addition, osteocytes can remove (and potentially replace) the matrix in their perilacunocanalicular environment, a process originally termed "osteocytic osteolysis" by Bélanger and more currently referred to as "peri-lacunocanalicular remodeling" (PLR) (Bélanger, 1969). Observations of enlarged osteocytic lacunae under pathological conditions led to the suggestion that this phenomenon might be the cause of bone loss in osteoporosis; however, this notion was abandoned after demonstrations that osteoclastic resorption drives osteoporotic bone loss (Parfitt, 1977). More recently, however, with a greater understanding of osteocyte physiology, PLR has reemerged as a focus of investigation.

The calcium phosphate mineral on the boundaries of the LCS is more easily dissolved (or labile) than the hydroxyapatite-like mineral comprising the majority of bone inorganic material, thus lending itself to mobilization under periods of calcium or phosphate demand (e.g., lactation). Osteocytes can facilitate mineral dissolution by acidifying their microenvironment using proton pumps. The pH level around osteocytes is less acidic than that in resorption pits underneath osteoclasts, but is sufficient to rapidly mobilize calcium and phosphate from the smaller, more labile crystals on the LCS wall. Osteocytes can also express a variety of proteolytic enzymes, including the matrix metalloproteinase MMP-13 (a collagenase), as well as TRAP and CatK (the same protease used by osteoclasts to resorb bone) (Qing et al., 2012; Tang et al., 2012). Physiological PLR, such as seen with lactation, is reversible, with osteocyte lacunae and canaliculi returning to their original size following cessation of the challenge (Kaya et al., 2017; Qing et al., 2012). However, it remains unclear whether osteocytes produce new bone matrix during the recovery, or if pre-existing matrix on the perilacunar boundary is retained and simply remineralizes. This subject has been reviewed by Heveran and Boerckel (2023)

5.1.4 Regulation of tissue material properties

The mechanical function of bones is determined both by their gross morphology and their material properties. In this case, material properties entail structural elements (e.g., trabecular thickness and spacing in cancellous bone, porosity of cortical bone), as well as the composition and biomechanics of the matrix level. "Indirect" osteocytic involvement in the regulation of these via the orchestration of remodeling is well established, however, "direct" interactions between the osteocyte and the bone matrix are increasingly being recognized.

While osteoblasts initiate mineral deposition, subsequent events in the mineralization process are controlled largely by osteocytes. Osteocytes produce a large amount of the NCPs found in bone, including DMP-1, OPN, MEPE, phosphate-regulating neutral endopeptidase on chromosome X (PHEX), and OCN. Of these,

DMP-1, OPN, and MEPE belong to the small integrin-binding ligand, N-linked glycoprotein (SIBLING) family, which express an acidic serine- and aspartate-rich motif (ASARM) that is cleaved by PHEX. The ability of NCPs of bone matrix both to bind free ions and to adhere to mineral surfaces suggests that they may play critical roles in defining spatial relationships between crystals during mineral growth, as well as between crystals and organic molecules like collagen fibers.

In addition to their modulatory role in the deposition of mineral, emerging evidence points to osteocytic involvement in the direct maintenance of matrix mechanical properties. Qing and Bonewald highlighted the involvement of PLR in skeletal function and brought the concept back into mainstream discussion (Qing & Bonewald, 2009). Dimensional increases with PLR correspond to a ~10%-15% increase in lacunar and canalicular diameter, and the LCS volume fraction represents approximately 2% of skeletal volume (Buenzli & Sims, 2015; Kaya et al., 2017; Qing et al., 2012). Kaya et al., using a murine lactation model, showed a decrease in tissue modulus related to the volumetric increases observed in the LCS during calcium stress (Kaya et al., 2017). This effect was reversible, with the cessation of lactation leading to a restoration of LCS volume to baseline values and tissue elastic modulus following suit. Mice with genetically-induced deficiencies in PLR, such as systemic MMP-13 (a critical enzyme involved in PLR) or osteocytic TGF-β receptor (a target for the regulation of PLR) knockouts, exhibit decreased fracture toughness (Dole et al., 2017; Tang et al., 2012). While the PLR deficiencies in these models are present in growth and development, the decline in tissue mechanical properties has been shown to worsen further with aging in the osteocyte TGF-β receptor knockout model, indicating an ongoing osteocytic role in the maintenance of the bone matrix (Schurman et al., 2024).

Osteocyte loss has consequences. Interstitial regions in human bone, which are older and frequently osteocyte-depleted, have long been appreciated to be more highly mineralized than adjacent and newer osteons. These regions also have higher stiffness and reduced resistance to crack propagation (Mullins et al., 2009; Rho et al., 2002). These observations have led to the widespread assumption that osteocyte loss and increased mineralization in these sites are causally related, and that this could be linked to fragility observed in aging bone. However, the extent to which osteocytes are able to actually maintain or influence matrix mineralization in adult bone remains poorly understood.

5.1.5 Endocrine function

Bone is becoming increasingly appreciated as an endocrine organ. Hormones produced by bone cells have been shown to enter the circulation and regulate other tissues in the body, including the kidney, gut, adipose tissue, and even the brain. Many of these hormones are produced by osteocytes, consistent with their overwhelming abundance in the adult skeleton. Of particular interest, a role for bone in the regulation of systemic energy metabolism is emerging.

5.1.5.1 Bone-gut-kidney axis for phosphate homeostasis

One of the first instances in which the skeletal system was identified as an endocrine organ involves the control of systemic phosphate homeostasis through osteocyte secretion of Fibroblast Growth Factor 23 (FGF23). FGF23 suppresses phosphate reabsorption in the proximal and distal renal tubules as well as renal synthesis of the vitamin D hormone, 1α,25-dihydroxyvitamin D_3 [$1,25(OH)_2D_3$], leading to decreased blood phosphate. FGF23 activity requires the simultaneous expression of αKlotho coreceptors, in order to bind to FGF receptors with high affinity. FGF23 deletion leads to hyperphosphatemia and elevated $1,25(OH)_2D_3$, while FGF23 excess is associated with hypophosphatemia, dysfunctional vitamin D metabolism, and osteomalacia/rickets (Martin et al., 2012). Importantly, when osteocytes are systemically ablated in a mouse model, FGF23 is depleted and phosphate metabolism is dysregulated throughout the body, leading to extreme morbidity and even death (Fujii et al., 2017).

5.1.5.2 Bioenergetic regulation with other tissues

A number of recent studies have implicated bone as a source of signals that can alter energy metabolism by several mechanisms. Such a role, while perhaps not immediately apparent, is consistent with the fact that forming and maintaining bone tissue are energy-intensive processes, and so coordination of skeletal metabolism with the systemic machinery that generates and stores metabolic energy would be beneficial.

An additional and complex endocrine role for osteocytes is centered on osteocalcin. OCN was identified as the most abundant noncollagenous protein in bone, and binds Ca^{2+} due to gamma L-carboxylated glutamic acid residues (Gla). However, OCN exists in both gamma-carboxylated and uncarboxylated (unOCN) forms; only the carboxylated form has high mineral affinity and is retained in bone (Zhou et al., 2021). When Ducy et al. showed elevated blood glucose in OCN-deficient mice, this indicated a potential systemic role for OCN; subsequent work has shown that unOCN induces insulin production in the pancreatic islets and improves insulin resistance in a variety of cell types including adipocytes, myocytes, and vascular endothelial cells (Ducy et al., 1996; Zhou et al., 2021).

Additionally, emerging evidence points to osteoblast-derived lipocalin-2 (LCN2) acting as an appetite suppressant by binding to the melanocortin-4 receptor (MC4R) in the hypothalamus; it has further been indicated to induce insulin secretion and improve glucose tolerance and insulin sensitivity (Mosialou et al., 2017). Given the relative abundance of osteocytes and osteoblasts in adult bone, and that mice deficient in osteocytic PTEN (phosphatase and tensin homolog deleted from chromosome 10, a LCN2 regulator) exhibit altered bone LCN2, it is reasonable to speculate osteocytes are also involved in this process (Kinoshita et al., 2025). The importance of this hormone, and how it interplays with muscle, energetic, and hepatic metabolism are active areas of research.

Additionally, sclerostin has been implicated in fat metabolism via its role in the Wnt signaling pathway; the LRP co-receptors for Wnts in osteoblast-lineage cells are in the same family as those found in adipocytes (Riddle, 2023). Homozygous sclerostin knockout mice (*Sost*$^{-/-}$) show reductions in visceral and subcutaneous fat, as well as enhanced glucose sensitivity (Kim et al., 2017). The activity of sclerostin on adipocytes resembles that in bone; mice with an adipose LRP4 deficiency show a similar metabolic phenotype to that of *Sost*$^{-/-}$ mice (Kim et al., 2019). Furthermore, a systemic effect on energy regulation via the peroxisome proliferator-activated receptors (PPAR) alpha and gamma appears to function in a sclerostin-independent manner; these actions are likely due to both the quantity of osteocytes and the magnitude of their metabolic needs (Lecka-Czernik et al., 2024).

6. Bone lining cells

The final cell type in bone is the bone lining cell (BLC). Formed by the differentiation of the final osteoblasts on bone surfaces, BLCs are flattened, contain few organelles, and form a boundary layer between bone and its adjacent tissues (Miller et al., 1989). These cells are difficult to discern visually, being extremely flat, and pose a number of challenges in their study, so relatively little is known about them compared to the other bone cell types.

The function of BLCs remains incompletely understood. As they form the boundary interface between the bone and the rest of the body, it can be expected these cells communicate with marrow cavity cells, periosteum, and vasculature. They are also part of the gap junction-linked osteocyte network, and as such likely have a role in integrating signals coming to and from osteocytes (Wein, 2017). In addition, they have functions related to bone formation and resorption. Lining cells can be activated rapidly to become active osteoblasts; this activation can occur in response to PTH or mechanical loading (Chow et al., 1998; Kim et al., 2012).

Another likely role of BLCs is to degrade the thin layer of organic material (the lamina limitans) that separates the BLCs from mineralized matrix before resorption takes place. Nakamura et al. localized MMP-13 within the Golgi apparatus of BLCs, and multiple observations have associated this collagenase activity with osteoclastic capacity to attach to bone (Nakamura et al., 2004; Paiva & Granjeiro, 2017). Given the capacity of osteocytes to turn over matrix on their peri-lacunocanalicular boundaries, it is highly likely that BLCs have similar mechanisms for the maintenance of this seam. Indeed, these cells have been implicated as a potential population that provides the mononuclear cells found behind osteoclasts at resorption and "prepare" the newly resorbed surface of bone for subsequent osteoblastic matrix deposition (Everts et al., 2002).

7. Clinical correlate

Improvements in our understanding of bone cell biology have led to many novel therapeutics aimed at skeletal diseases. These include small molecules targeted at specific bone cell functions and biologics—molecules derived from naturally occurring biological macromolecules. Biologics include monoclonal antibodies and modified peptide hormones.

7.1 Anabolic therapies

Analogs of parathyroid hormone and parathyroid hormone-related protein: Teriparatide is the 1–34 amino acid sequence of PTH and largely mimics the complex activities of native PTH 1–84 on bone. When present continuously, it promotes osteoclast activity, but when administered intermittently it is anabolic, enhancing both osteoblast recruitment and lifespan (Silva & Bilezikian, 2015). Abaloparatide, a synthetic PTHrP analog, is a selective PTH receptor type 1 (PTH1R) agonist that stimulates bone formation, similar to intermittent PTH.

Anti-sclerostin antibody: Romosozumab is an anti-sclerostin antibody that binds to and neutralizes sclerostin. This blocks the "off switch" suppressing osteoblastogenesis and allows bone formation to proceed.

7.2 Anti-catabolic (anti-resorptive) therapies

Bisphosphonates: Bisphosphonates are the gold standard for drugs used to prevent bone loss in osteoporosis by inhibiting osteoclasts. The fascinating history of bisphosphonate drug development has been thoroughly described in the review by Russell (2011). Bisphosphonates have two important characteristics mediated by different regions of the compound. First, they bind to bone mineral, and so are targeted to bone; second, they inhibit osteoclast function. The first generation of bisphosphonate drugs entered clinical use in the late 1970s; the second-generation bisphosphonate alendronate was FDA approved in 1995, and a host of others have been introduced subsequently. Alendronate is still one of the most commonly prescribed bisphosphonate drugs.

Given their propensity to bind strongly to mineral, and a lack of cellular clearance mechanisms, bisphosphonates have exceptionally long half-lives. This persistence in biological activity and high skeletal affinity makes them efficient for long-term use, and conjugating bisphosphonates with other therapeutic agents (e.g., cancer treatments, antibiotics) has been highlighted as a potential drug delivery mechanism targeting bone (Sun et al., 2021). On the other hand, long-term bisphosphonate treatment leads to an accumulation of drug in the skeleton such that resorption does not initiate rapidly following the discontinuation of treatment.

Anti-RANKL antibody: The anti-RANKL antibody denosumab binds to circulating RANKL, preventing it from accessing RANK receptors on pre-osteoclasts and thus inhibiting their differentiation. Denosumab was approved for clinical usage in 2010. Unlike bisphosphonates, denosumab is quickly cleared once treatment stops. Furthermore, denosumab exhibits a unique "rebound effect" in which overactive resorption occurs following discontinuation, leading to aggressive bone loss and increased fracture risk (Kim et al., 2022). The precise mechanisms responsible for this rebound (or overshoot) effect remain to be elucidated but may include compensatory overexpression of RANKL during the course of treatment, as well as rapid reactivation of accumulated osteomorphs (Kim et al., 2024; McDonald et al., 2021).

A common feature of all antiresorptives (including both bisphosphonates and denosumab) is that they suppress not only the overactive bone resorption leading to pathologic bone loss, but also the normal turnover of skeletal sites that have acquired damage. This leads to the accumulation of microcracks in bone and consequently the potential for increased fracture risk with long-term use. Indeed, atypical (brittle-appearing) fractures have been observed after 5–10 years of antiresorptive use in a small minority of patients (Collins et al., 2024). Accordingly, there has been wide discussion of the utility of a "drug holiday", i.e., treatment cessation to allow bone remodeling to resume and potentially restore bone material properties. How such a holiday would restore bone material properties is unclear, and currently there is little capability to predict which patients will experience atypical fractures, necessitating further investigation.

7.3 Less common therapies

Cathepsin K inhibitors: CatK is the principal lysosomal enzyme responsible for osteoclastic matrix degradation and thus was an obvious pharmaceutical target for the inhibition of resorption. However, all the enzyme inhibitors that have been tried to date in clinical trials have exhibited off-target effects on other cathepsins; cathepsins are the most abundant of the lysosomal proteases, and inhibition impairs their functions in intracellular degradation of proteins, energy metabolism, immune responses, as well as various extracellular activities, all critical lysosomal functions throughout the body (Drake et al., 2017).

Hormone replacement/estrogen therapy: Prior to our current understanding of the cellular mechanisms underlying post-menopausal bone loss, the first line of pharmaceutical intervention lay in delaying the onset of menopause via hormone replacement. The onset of bone loss following estrogen depletion has long been appreciated and investigated. Estrogen replacement regimens saw a sharp decline in utilization following the Women's Health Initiative trial (1991–2002) due to concerns of increased breast cancers. However, current hormone replacement regimes have been shown to be safe and effective at preventing, but not treating, osteoporosis without the previously identified side effects, though concerns linger due to historical data (Levin et al., 2018). Maintaining hormone levels prevents the onset of estrogen withdrawal, however, cessation of therapy causes the process to resume, and concomitant bone loss is observed (McClung, 2025).

Selective estrogen receptor modulators (SERMs): SERMs are a class of molecules which have estrogen-like activity (agonists) in certain tissues (e.g., bone) while antagonizing estrogen activity at other sites (*e.g.*, breast tissue). In bone, they mimic estrogen, providing a protective effect while avoiding certain concerns associated with estrogen treatment, such as increased risk of estrogen receptor positive breast cancers. The most common SERM in clinical use is raloxifene. In addition to estrogen/SERM activity on cells, Gallant et al. observed raloxifene alters the transfer of mechanical load between collagen and mineral in the bone matrix, suggesting reduction in fragility outcomes may in part be due to direct activity of estrogen on the bone matrix, independent of cellular mechanisms (Gallant et al., 2014).

Strontium ranelate: Approved for use in Europe in 2004 (but not in the United States), strontium ranelate is considered a dual-action bone agent which functions two-fold in promoting osteoblast formation and survival, as well as increasing osteoclast apoptosis. A direct effect of strontium ranelate on bone mineral has been suggested, though data indicate it is only internalized in newly deposited matrix (Li et al., 2010). While strontium ranelate moderately improved bone mass and reduced instances of fractures, it was pulled from the European market due to concerns regarding myocardial infarction.

References

Almeida, M., Laurent, M. R., Dubois, V., Claessens, F., O'Brien, C. A., Bouillon, R., Vanderschueren, D., & Manolagas, S. C. (2017). Estrogens and androgens in skeletal physiology and pathophysiology. *Physiological Reviews, 97*(1), 135–187. https://doi.org/10.1152/physrev.00033.2015

Andersen, T. L., Abdelgawad, M. E., Kristensen, H. B., Hauge, E. M., Rolighed, L., Bollerslev, J., Kjærsgaard-Andersen, P., & Delaisse, J. M. (2013). Understanding coupling between bone resorption and formation: Are reversal cells the missing link? *American Journal Of Pathology, 183*(1), 235–246. https://doi.org/10.1016/j.ajpath.2013.03.006

Bélanger, L. F. (1969). Osteocytic osteolysis. *Calcified Tissue Research, 4*(1), 1–12. https://doi.org/10.1007/BF02279101

Beno, T., Yoon, Y. J., Cowin, S. C., & Fritton, S. P. (2006). Estimation of bone permeability using accurate microstructural measurements. *Journal of Biomechanics, 39*(13), 2378–2387. https://doi.org/10.1016/j.jbiomech.2005.08.005

Bonewald, L. F. (2011). The amazing osteocyte. *Journal of Bone and Mineral Research, 26*(2), 229–238. https://doi.org/10.1002/jbmr.320

Boyle, W. J., Scott Simonet, W., & Lacey, D. L. (2003). Osteoclast differentiation and activation. *Nature, 423*(6937), 337–342. https://doi.org/10.1038/nature01658

Buenzli, P. R., & Sims, N. A. (2015). Quantifying the osteocyte network in the human skeleton. *Bone, 75*, 144–150. https://doi.org/10.1016/j.bone.2015.02.016

Busse, B., Djonic, D., Milovanovic, P., Hahn, M., Püschel, K., Ritchie, R. O., Djuric, M., & Amling, M. (2010). Decrease in the osteocyte lacunar density accompanied by hypermineralized lacunar occlusion reveals failure and delay of remodeling in aged human bone. *Aging Cell, 9*(6), 1065–1075. https://doi.org/10.1111/j.1474-9726.2010.00633.x

Cabahug-Zuckerman, P., Frikha-Benayed, D., Majeska, R. J., Tuthill, A., Yakar, S., Judex, S., & Schaffler, M. B. (2016). Osteocyte apoptosis caused by hindlimb unloading is required to trigger osteocyte RANKL production and subsequent resorption of cortical and trabecular bone in mice femurs. *Journal of Bone and Mineral Research, 31*(7), 1356–1365. https://doi.org/10.1002/jbmr.2807

Cabahug-Zuckerman, P., Stout, R. F., Majeska, R. J., Thi, M. M., Spray, D. C., Weinbaum, S., & Schaffler, M. B. (2018). Potential role for a specialized β3 integrin-based structure on osteocyte processes in bone mechanosensation. *Journal of Orthopaedic Research, 36*(2), 642–652. https://doi.org/10.1002/jor.23792

Cherian, P. P., Siller-Jackson, A. J., Gu, S., Wang, X., Bonewald, L. F., Sprague, E., & Jiang, J. X. (2005). Mechanical strain opens connexin 43 hemichannels in osteocytes: A novel mechanism for the release of prostaglandin. *Molecular Biology of the Cell, 16* (7), 3100–3106. https://doi.org/10.1091/mbc.E04-10-0912

Chow, J. W. M., Wilson, A. J., Chambers, T. J., & Fox, S. W. (1998). Mechanical loading stimulates bone formation by reactivation of bone lining cells in 13-Week-Old rats. *Journal of Bone and Mineral Research, 13*(11), 1760–1767. https://doi.org/10.1359/jbmr.1998.13.11.1760

Collins, L., Ronan, A., Hutcheon, E., Ebeling, P. R., Grill, V., & Nguyen, H. H. (2024). Atypical fractures at non-classical sites associated with anti-resorptive therapy: A systematic review. *Journal of Bone and Mineral Research, 39*(12), 1722–1734. https://doi.org/10.1093/jbmr/zjae159

Creecy, A., Damrath, J. G., & Wallace, J. M. (2021). Control of bone matrix properties by osteocytes. *Frontiers in Endocrinology, 11*. https://doi.org/10.3389/fendo.2020.578477

Dallas, S. L., & Bonewald, L. F. (2010/01). Dynamics of the transition from osteoblast to osteocyte. *Annals of the New York Academy of Sciences, 1192*, 437–443. https://doi.org/10.1111/j.1749-6632.2009.05246.x

Dole, N. S., Mazur, C. M., Acevedo, C., Lopez, J. P., Monteiro, D. A., Fowler, T. W., Gludovatz, B., Walsh, F., Regan, J. N., Messina, S., Evans, D. S., Lang, T. F., Zhang, B., Ritchie, R. O., Mohammad, K. S., & Alliston, T. (2017). Osteocyte-intrinsic TGF-β signaling regulates bone quality through perilacunar/canalicular remodeling. *Cell Reports, 21*(9), 2585–2596. https://doi.org/10.1016/j.celrep.2017.10.115

Drake, M. T., Clarke, B. L., Oursler, M. J., & Khosla, S. (2017). Cathepsin K inhibitors for osteoporosis: Biology, potential clinical utility, and lessons learned. *Endocrine Reviews, 38*(4), 325–350. https://doi.org/10.1210/er.2015-1114

Drake, M. T., & Khosla, S. (2017). Hormonal and systemic regulation of sclerostin. *Bone, 96*, 8–17. https://doi.org/10.1016/j.bone.2016.12.004

Ducy, P., Desbois, C., Boyce, B., Pinero, G., Story, B., Dunstan, C., Smith, E., Bonadio, J., Goldstein, S., Gundberg, C., Bradley, A., & Karsenty, G. (1996). Increased bone formation in osteocalcin-deficient mice. *Nature, 382*(6590), 448–452. https://doi.org/10.1038/382448a0

Emerton, K. B., Hu, B., Woo, A. A., Sinofsky, A., Hernandez, C., Majeska, R. J., Jepsen, K. J., & Schaffler, M. B. (2010). Osteocyte apoptosis and control of bone resorption following ovariectomy in mice. *Bone, 46*(3), 577–583. https://doi.org/10.1016/j.bone.2009.11.006

Enlow, D. H. (1975). *Handbook of facial growth*.

Everts, V., Delaissé, J. M., Korper, W., Jansen, D. C., Tigchelaar-Gutter, W., Saftig, P., & Beertsen, W. (2002). The bone lining cell: Its role in cleaning howship's lacunae and initiating bone formation. *Journal of Bone and Mineral Research, 17*(1), 77–90. https://doi.org/10.1359/jbmr.2002.17.1.77

Franz-Odendaal, T. A., Hall, B. K., & Witten, P. E. (2006). Buried alive: How osteoblasts become osteocytes. *Developmental Dynamics, 235*(1), 176–190. https://doi.org/10.1002/dvdy.20603

Fujii, O., Tatsumi, S., Ogata, M., Arakaki, T., Sakaguchi, H., Nomura, K., Miyagawa, A., Ikuta, K., Hanazaki, A., Kaneko, I., Segawa, H., & Miyamoto, K.-ichi (2017). Effect of osteocyte-ablation on inorganic phosphate metabolism: Analysis of bone–kidney–gut axis. *Frontiers in Endocrinology, 8*. https://doi.org/10.3389/fendo.2017.00359

Gallant, M. A., Brown, D. M., Hammond, M., Wallace, J. M., Du, J., Deymier-Black, A. C., Almer, J. D., Stock, S. R., Allen, M. R., & Burr, D. B. (2014). Bone cell-independent benefits of raloxifene on the skeleton: A novel mechanism for improving bone material properties. *Bone, 61*, 191–200. https://doi.org/10.1016/j.bone.2014.01.009

Halleen, J. M., Räisänen, S., Salo, J. J., Reddy, S. V., Roodman, G. D., Hentunen, T. A., Lehenkari, P. P., Kaija, H., Vihko, P., & Väänänen, H. K. (1999). Intracellular fragmentation of bone resorption products by reactive oxygen species generated by osteoclastic tartrate-resistant acid phosphatase. *Journal of Biological Chemistry, 274*(33), 22907–22910. https://doi.org/10.1074/jbc.274.33.22907

Hasegawa, T. (2018). Ultrastructure and biological function of matrix vesicles in bone mineralization. *Histochemistry and Cell Biology, 149*(4), 289–304. https://doi.org/10.1007/s00418-018-1646-0

Heveran, C. M., & Boerckel, J. D. (2023). Osteocyte remodeling of the lacunar-canalicular system: What's in a name? *Current Osteoporosis Reports, 21*(1), 11–20. https://doi.org/10.1007/s11914-022-00766-3

Jaworski, Z. F. G., Duck, B., & Sekaly, G. (1981). Kinetics of osteoclasts and their nuclei in evolving secondary Haversian systems. *Journal of Anatomy, 133*(3), 397–405.

Kaya, S., Basta-Pljakic, J., Seref-Ferlengez, Z., Majeska, R. J., Cardoso, L., Bromage, T. G., Zhang, Q., Flach, C. R., Mendelsohn, R., Yakar, S., Fritton, S. P., & Schaffler, M. B. (2017). Lactation-induced changes in the volume of osteocyte lacunar-canalicular space alter mechanical properties in cortical bone tissue. *Journal of Bone and Mineral Research, 32*(4), 688–697. https://doi.org/10.1002/jbmr.3044

Kelly, M. M., Sharma, K., Wright, C. S., Yi, X., Reyes Fernandez, P. C., Gegg, A. T., Gorrell, T. A., Noonan, M. L., Baghdady, A., Sieger, J. A., Dolphin, A. C., Warden, S. J., Deosthale, P., Plotkin, L. I., Sankar, U., Hum, J. M., Robling, A. G., Farach-Carson, M. C., & Thompson, W. R. (2024). Loss of the auxiliary α2δ1 voltage-sensitive calcium channel subunit impairs bone formation and anabolic responses to mechanical loading. *JBMR Plus, 8*(2). https://doi.org/10.1093/jbmrpl/ziad008

Kennedy, O. D., Herman, B. C., Laudier, D. M., Majeska, R. J., Sun, H. B., & Schaffler, M. B. (2012). Activation of resorption in fatigue-loaded bone involves both apoptosis and active pro-osteoclastogenic signaling by distinct osteocyte populations. *Bone, 50*(5), 1115–1122. https://doi.org/10.1016/j.bone.2012.01.025

Kennedy, O. D., Laudier, D. M., Majeska, R. J., Sun, H. B., & Schaffler, M. B. (2014). Osteocyte apoptosis is required for production of osteoclastogenic signals following bone fatigue in vivo. *Bone, 64*, 132–137. https://doi.org/10.1016/j.bone.2014.03.049

Khatib, N. S., Monsen, J., Ahmed, S., Huang, Y., Hoey, D. A., & Nowlan, N. C. (2023). Mechanoregulatory role of TRPV4 in prenatal skeletal development. *Science Advances, 9*(4). https://doi.org/10.1126/sciadv.ade2155

Kim, S. P., Da, H., Li, Z., Kushwaha, P., Beil, C., Mei, L., Xiong, W. C., Wolfgang, M. J., Clemens, T. L., & Riddle, R. C. (2019). Lrp4 expression by adipocytes and osteoblasts differentially impacts sclerostin's endocrine effects on body composition and glucose metabolism. *Journal of Biological Chemistry, 294*(17), 6899–6911. https://doi.org/10.1074/jbc.RA118.006769

Kim, S. P., Frey, J. L., Li, Z., Kushwaha, P., Zoch, M. L., Tomlinson, R. E., Da, H., Aja, S., Noh, H. L., Kim, J. K., Hussain, M. A., Thorek, D. L. J., Wolfgang, M. J., & Riddle, R. C. (2017). Sclerostin influences body composition by regulating catabolic and anabolic metabolism in adipocytes. *Proceedings of the National Academy of Sciences of the United States of America, 114*(52), Article E11238. https://doi.org/10.1073/pnas.1707876115

Kim, A. S., Girgis, C. M., & McDonald, M. M. (2022). Osteoclast recycling and the rebound phenomenon following denosumab discontinuation. *Current Osteoporosis Reports, 20*(6), 505–515. https://doi.org/10.1007/s11914-022-00756-5

Kim, S. W., Pajevic, P. D., Selig, M., Barry, K. J., Yang, J. Y., Shin, C. S., Baek, W. Y., Kim, J. E., & Kronenberg, H. M. (2012). Intermittent parathyroid hormone administration converts quiescent lining cells to active osteoblasts. *Journal of Bone and Mineral Research, 27*(10), 2075–2084. https://doi.org/10.1002/jbmr.1665

Kim, A. S., Taylor, V. E., Castro-Martinez, A., Dhakal, S., Zamerli, A., Mohanty, S., Xiao, Y., Simic, M. K., Wen, J., Chai, R., Croucher, P. I., Center, J. R., Girgis, C. M., & McDonald, M. M. (2024). Temporal patterns of osteoclast formation and activity following withdrawal of RANKL inhibition. *Journal of Bone and Mineral Research, 39*(4), 484–497. https://doi.org/10.1093/jbmr/zjae023

Kinoshita, S., Onuma, S., Yamazaki, N., Shibukawa, Y., Ozono, K., Michigami, T., & Kawai, M. (2025). Lack of PTEN in osteocytes increases Lipocalin-2 level and confers resistance to high-fat diet–induced obesity in mice. *Endocrinology, 166*(3). https://doi.org/10.1210/endocr/bqaf026

Komori, T., Yagi, H., Nomura, S., Yamaguchi, A., Sasaki, K., Deguchi, K., Shimizu, Y., Bronson, R. T., Gao, Y. H., Inada, M., Sato, M., Okamoto, R., Kitamura, Y., Yoshiki, S., & Kishimoto, T. (1997). Targeted disruption of Cbfa1 results in a complete lack of bone formation owing to maturational arrest of osteoblasts. *Cell, 89*(5), 755–764. https://doi.org/10.1016/S0092-8674(00)80258-5

Lecka-Czernik, B., Khan, M. P., Letson, J., Baroi, S., & Chougule, A. (2024). Regulatory effect of osteocytes on extramedullary and bone marrow adipose tissue development and function. *Current Osteoporosis Reports, 22*(3), 301–307. https://doi.org/10.1007/s11914-024-00871-5

Levin, V. A., Jiang, X., & Kagan, R. (2018). Estrogen therapy for osteoporosis in the modern era. *Osteoporosis International, 29*(5), 1049–1055. https://doi.org/10.1007/s00198-018-4414-z

Li, C. Y., Jepsen, K. J., Majeska, R. J., Zhang, J., Ni, R., Gelb, B. D., & Schaffler, M. B. (2006). Mice lacking cathepsin K maintain bone remodeling but develop bone fragility despite high bone mass. *Journal of Bone and Mineral Research, 21*(6), 865–875. https://doi.org/10.1359/jbmr.060313

Li, X., Ominsky, M. S., Niu, Q. T., Sun, N., Daugherty, B., D'Agostin, D., Kurahara, C., Gao, Y., Cao, J., Gong, J., Asuncion, F., Barrero, M., Warmington, K., Dwyer, D., Stolina, M., Morony, S., Sarosi, I., Kostenuik, P. J., Lacey, D. L., … Paszty, C. (2008). Targeted deletion of the sclerostin gene in mice results in increased bone formation and bone strength. *Journal of Bone and Mineral Research, 23*(6), 860–869. https://doi.org/10.1359/jbmr.080216

Li, C., Paris, O., Siegel, S., Roschger, P., Paschalis, E. P., Klaushofer, K., & Fratzl, P. (2010). Strontium is incorporated into mineral crystals only in newly formed bone during strontium ranelate treatment. *Journal of Bone and Mineral Research, 25*(5), 968–975. https://doi.org/10.1359/jbmr.091038Germany

Li, X., Zhang, C., Vail, C. E., Sherrill, J. T., & Xiong, J. (2025). Piezo1 expression in mature osteocytes is dispensable for the skeletal response to mechanical loading. *Bone, 190*. https://doi.org/10.1016/j.bone.2024.117276

Lieschke, G. J., Grail, D., Hodgson, G., Metcalf, D., Stanley, E., Cheers, C., Fowler, K. J., Basu, S., Zhan, Y. F., & Dunn, A. R. (1994). Mice lacking granulocyte colony-stimulating factor have chronic neutropenia, granulocyte and macrophage progenitor cell deficiency, and impaired neutrophil mobilization. *Blood, 84*(6), 1737–1746. https://doi.org/10.1182/blood.v84.6.1737.1737

Marahleh, A., Kitaura, H., Ohori, F., Kishikawa, A., Ogawa, S., Shen, W. R., Qi, J., Noguchi, T., Nara, Y., & Mizoguchi, I. (2019). TNF-α directly enhances osteocyte RANKL expression and promotes osteoclast formation. *Frontiers in Immunology, 10*. https://doi.org/10.3389/fimmu.2019.02925

Martin, A., David, V., & Quarles, L. D. (2012). Regulation and function of the FGF23/klotho endocrine pathways. *Physiological Reviews, 92*(1), 131–155. https://doi.org/10.1152/physrev.00002.2011

McClung, M. R. (2025). Sequential and long-term therapy for osteoporosis. *Current Osteoporosis Reports, 23*(1). https://doi.org/10.1007/s11914-025-00909-2

McDonald, M. M., Khoo, W. H., Ng, P. Y., Xiao, Y., Zamerli, J., Thatcher, P., Kyaw, W., Pathmanandavel, K., Grootveld, A. K., Moran, I., Butt, D., Nguyen, A., Warren, S., Biro, M., Butterfield, N. C., Guilfoyle, S. E., Komla-Ebri, D., Dack, M. R. G., Dewhurst, H. F., … Phan, T. G. (2021). Osteoclasts recycle via osteomorphs during RANKL-Stimulated bone resorption. *Cell, 184*(5), 1330. https://doi.org/10.1016/j.cell.2021.02.002

McNamara, L. M., Majeska, R. J., Weinbaum, S., Friedrich, V., & Schaffler, M. B. (2009). Attachment of osteocyte cell processes to the bone matrix. *The Anatomical Record, 292*(3), 355–363. https://doi.org/10.1002/ar.20869

Miller, S. C., De Saint-Georges, L., Bowman, B. M., & Jee, W. S. S. (1989). Bone lining cells: Structure and function. *Scanning Microscopy, 3*(3), 953–961.

Morgan, S., Poundarik, A. A., & Vashishth, D. (2015). Do non-collagenous proteins affect skeletal mechanical properties? *Calcified Tissue International, 97*(3), 281–291. https://doi.org/10.1007/s00223-015-0016-3

Mosialou, I., Shikhel, S., Liu, J. M., Maurizi, A., Luo, N., He, Z., Huang, Y., Zong, H., Friedman, R. A., Barasch, J., Lanzano, P., Deng, L., Leibel, R. L., Rubin, M., Nicholas, T., Chung, W., Zeltser, L. M., Williams, K. W., Pessin, J. E., & Kousteni, S. (2017). MC4R-dependent suppression of appetite by bone-derived lipocalin 2. *Nature, 543*(7645), 385–390. https://doi.org/10.1038/nature21697

Mullins, L. P., Sassi, V., McHugh, P. E., & Bruzzi, M. S. (2009). Differences in the crack resistance of interstitial, osteonal and trabecular bone tissue. *Annals of Biomedical Engineering, 37*(12), 2574–2582. https://doi.org/10.1007/s10439-009-9797-8

Nakamura, H., Sato, G., Hirata, A., & Yamamoto, T. (2004). Immunolocalization of matrix metalloproteinase-13 on bone surface under osteoclasts in rat tibia. *Bone, 34*(1), 48–56. https://doi.org/10.1016/j.bone.2003.09.001

Nanci, A. (1999). Content and distribution of noncollagenous matrix proteins in bone and cementum: Relationship to speed of formation and collagen packing density. *Journal of Structural Biology, 126*(3), 256–269. https://doi.org/10.1006/jsbi.1999.4137

O'Brien, C. A., Jia, D., Plotkin, L. I., Bellido, T., Powers, C. C., Stewart, S. A., Manolagas, S. C., & Weinstein, R. S. (2004). Glucocorticoids act directly on osteoblasts and osteocytes to induce their apoptosis and reduce bone formation and strength. *Endocrinology, 145*(4), 1835–1841. https://doi.org/10.1210/en.2003-0990

Paiva, K. B. S., & Granjeiro, J. M. (2017). Matrix metalloproteinases in bone resorption, remodeling, and repair. *Progress in Molecular Biology and Translational Science, 148*, 203–303. https://doi.org/10.1016/bs.pmbts.2017.05.001

Parfitt, A. M. (1977). The cellular basis of bone turnover and bone loss: A rebuttal of the osteocytic resorption-bone flow theory. *Clinical Orthopaedics and Related Research*, (127), 236–247.

Piper, K., Boyde, A., & Jones, S. J. (1992). The relationship between the number of nuclei of an osteoclast and its resorptive capability in vitro. *Anatomy and Embryology, 186*(4), 291–299. https://doi.org/10.1007/BF00185977

Qin, L., He, T., Chen, S., Yang, D., Yi, W., Cao, H., & Xiao, G. (2021). Roles of mechanosensitive channel Piezo1/2 proteins in skeleton and other tissues. *Bone Research, 9*(1). https://doi.org/10.1038/s41413-021-00168-8

Qing, H., Ardeshirpour, L., Divieti Pajevic, P., Dusevich, V., Jähn, K., Kato, S., Wysolmerski, J., & Bonewald, L. F. (2012). Demonstration of osteocytic perilacunar/canalicular remodeling in mice during lactation. *Journal of Bone and Mineral Research, 27*(5), 1018–1029. https://doi.org/10.1002/jbmr.1567

Qing, H., & Bonewald, L. F. (2009). Osteocyte remodeling of the perilacunar and pericanalicular matrix. *International Journal of Oral Science, 1*(2), 59–65. https://doi.org/10.4248/ijos.09019

Rho, J. Y., Zioupos, P., Currey, J. D., & Pharr, G. M. (2002). Microstructural elasticity and regional heterogeneity in human femoral bone of various ages examined by nanoindentation. *Journal of Biomechanics, 35*(2), 189–198. https://doi.org/10.1016/s0021-9290(01)00199-3

Riddle, R. C. (2023). Endocrine functions of sclerostin. *Current Opinion in Endocrine and Metabolic Research, 28*, Article 100433. https://doi.org/10.1016/j.coemr.2022.100433

Russell, R. G. G. (2011). Bisphosphonates: The first 40 years. *Bone, 49*(1), 2–19. https://doi.org/10.1016/j.bone.2011.04.022

Schaffler, M. B., Cheung, W. Y., Majeska, R., & Kennedy, O. (2014). Osteocytes: Master orchestrators of bone. *Calcified Tissue International, 94*(1), 5–24. https://doi.org/10.1007/s00223-013-9790-y

Schurman, C. A., Kaya, S., Dole, N., Luna, N. M. M., Castillo, N., Potter, R., Rose, J. P., Bons, J., King, C. D., Burton, J. B., Schilling, B., Melov, S., Tang, S., Schaible, E., & Alliston, T. (2024). Aging impairs the osteocytic regulation of collagen integrity and bone quality. *Bone Research, 12*(1). https://doi.org/10.1038/s41413-023-00303-7

Seref-Ferlengez, Z., Urban-Maldonado, M., Sun, H. B., Schaffler, M. B., Suadicani, S. O., & Thi, M. M. (2019). Role of pannexin 1 channels in load-induced skeletal response. *Annals of the New York Academy of Sciences, 1442*(1), 79–90. https://doi.org/10.1111/nyas.13914

Silva, B. C., & Bilezikian, J. P. (2015). Parathyroid hormone: Anabolic and catabolic actions on the skeleton. *Current Opinion in Pharmacology, 22*, 41–50. https://doi.org/10.1016/j.coph.2015.03.005

Sun, S., Tao, J., Sedghizadeh, P. P., Cherian, P., Junka, A. F., Sodagar, E., Xing, L., Boeckman, R. K., Srinivasan, V., Yao, Z., Boyce, B. F., Lipe, B., Neighbors, J. D., Russell, R. G. G., McKenna, C. E., & Ebetino, F. H. (2021). Bisphosphonates for delivering drugs to bone. *British Journal of Pharmacology, 178*(9), 2008–2025. https://doi.org/10.1111/bph.15251

Tang, S. Y., Herber, R. P., Ho, S. P., & Alliston, T. (2012). Matrix metalloproteinase-13 is required for osteocytic perilacunar remodeling and maintains bone fracture resistance.

Journal of Bone and Mineral Research, 27(9), 1936–1950. https://doi.org/10.1002/jbmr.1646

Temiyasathit, S., Tang, W. J., Leucht, P., Anderson, C. T., Monica, S. D., Castillo, A. B., Helms, J. A., Stearns, T., & Jacobs, C. R. (2012). Mechanosensing by the primary cilium: Deletion of Kif3a reduces bone formation due to loading. *PLoS One, 7*(3). https://doi.org/10.1371/journal.pone.0033368

Thi, M. M., Suadicani, S. O., Schaffler, M. B., Weinbaum, S., & Spray, D. C. (2013). Mechanosensory responses of osteocytes to physiological forces occur along processes and not cell body and require αvβ3 integrin. *Proceedings of the National Academy of Sciences of the United States of America, 110*(52), 21012–21017. https://doi.org/10.1073/pnas.1321210110

Thompson, W. R., Modla, S., Grindel, B. J., Czymmek, K. J., Kirn-Safran, C. B., Wang, L., Duncan, R. L., & Farach-Carson, M. C. (2011). Perlecan/Hspg2 deficiency alters the pericellular space of the lacunocanalicular system surrounding osteocytic processes in cortical bone. *Journal of Bone and Mineral Research, 26*(3), 618–629. https://doi.org/10.1002/jbmr.236

Udagawa, N., Takahashi, N., Akatsu, T., Tanaka, H., Sasaki, T., Nishihara, T., Koga, T., Martin, T. J., & Suda, T. (1990). Origin of osteoclasts: Mature monocytes and macrophages are capable of differentiating into osteoclasts under a suitable microenvironment prepared by bone marrow-derived stromal cells. *Proceedings of the National Academy of Sciences, 87*(18), 7260–7264. https://doi.org/10.1073/pnas.87.18.7260

Vashishth, D., Verborgt, O., Divine, G., Schaffler, M. B., & Fyhrie, D. P. (2000). Decline in osteocyte lacunar density in human cortical bone is associated with accumulation of microcracks with age. *Bone, 26*(4), 375–380. https://doi.org/10.1016/s8756-3282(00)00236-2

Veis, D. J., & O'Brien, C. A. (2023). Osteoclasts, master sculptors of bone. *Annual Review of Pathology: Mechanisms of Disease, 18*, 257–281. https://doi.org/10.1146/annurev-pathmechdis-031521-040919

Verbruggen, S. W., Sittichokechaiwut, A., & Reilly, G. C. (2023). Osteocytes and primary cilia. *Current Osteoporosis Reports, 21*(6), 719–730. https://doi.org/10.1007/s11914-023-00819-1

Wang, L., Ciani, C., Doty, S. B., & Fritton, S. P. (2004). Delineating bone's interstitial fluid pathway in vivo. *Bone, 34*(3), 499–509. https://doi.org/10.1016/j.bone.2003.11.022

Wang, Y., McNamara, L. M., Schaffler, M. B., & Weinbaum, S. (2008/12). Strain amplification and integrin based signaling in osteocytes. *Journal of Musculoskeletal and Neuronal Interactions, 8*(4), 332–334.

Wein, M. N. (2017). Bone lining cells: Normal physiology and role in response to anabolic osteoporosis treatments. *Current Molecular Biology Reports, 3*(2), 79–84. https://doi.org/10.1007/s40610-017-0062-x

Weinbaum, S., Cowin, S. C., & Zeng, Y. (1994). A model for the excitation of osteocytes by mechanical loading-induced bone fluid shear stresses. *Journal of Biomechanics, 27*(3), 339–360. https://doi.org/10.1016/0021-9290(94)90010-8

Wennberg, C., Hessle, L., Lundberg, P., Mauro, S., Narisawa, S., Lerner, U. H., & Millán, J. L. (2000). Functional characterization of osteoblasts and osteoclasts from alkaline phosphatase knockout mice. *Journal of Bone and Mineral Research, 15*(10), 1879–1888. https://doi.org/10.1359/jbmr.2000.15.10.1879

Wu, D., Ganatos, P., Spray, D. C., & Weinbaum, S. (2011). On the electrophysiological response of bone cells using a Stokesian fluid stimulus probe for delivery of quantifiable localized picoNewton level forces. *Journal of Biomechanics, 44*(9), 1702–1708. https://doi.org/10.1016/j.jbiomech.2011.03.034

Wu, D., Schaffler, M. B., Weinbaum, S., & Spray, D. C. (2013). Matrix-dependent adhesion mediates network responses to physiological stimulation of the osteocyte cell process. *Proceedings of the National Academy of Sciences of the United States of America, 110* (29), 12096–12101. https://doi.org/10.1073/pnas.1310003110

Xiong, J., Onal, M., Jilka, R. L., Weinstein, R. S., Manolagas, S. C., & O'Brien, C. A. (2011). Matrix-embedded cells control osteoclast formation. *Nature Medicine, 17*(10), 1235–1241. https://doi.org/10.1038/nm.2448

Xiong, J., Piemontese, M., Thostenson, J. D., Weinstein, R. S., Manolagas, S. C., & O'Brien, C. A. (2014). Osteocyte-derived RANKL is a critical mediator of the increased bone resorption caused by dietary calcium deficiency. *Bone, 66*, 146–154. https://doi.org/10.1016/j.bone.2014.06.006

You, L., Cowin, S. C., Schaffler, M. B., & Weinbaum, S. (2001). A model for strain amplification in the actin cytoskeleton of osteocytes due to fluid drag on pericellular matrix. *Journal of Biomechanics, 34*(11), 1375–1386. https://doi.org/10.1016/S0021-9290(01)00107-5

You, L. D., Weinbaum, S., Cowin, S. C., & Schaffler, M. B. (2004). Ultrastructure of the osteocyte process and its pericellular matrix. *The Anatomical Record Part A Discoveries in Molecular, Cellular, and Evolutionary Biology, 278*(2), 505–513. https://doi.org/10.1002/ar.a.20050

Zeng, Y., Riquelme, M. A., Hua, R., Zhang, J., Acosta, F. M., Gu, S., & Jiang, J. X. (2022). Mechanosensitive piezo1 calcium channel activates connexin 43 hemichannels through PI3K signaling pathway in bone. *Cell & Bioscience, 12*(1). https://doi.org/10.1186/s13578-022-00929-w

Zhang, K., Barragan-Adjemian, C., Ye, L., Kotha, S., Dallas, M., Lu, Y., Zhao, S., Harris, M., Harris, S. E., Feng, J. Q., & Bonewald, L. F. (2006). E11/gp38 selective expression in osteocytes: Regulation by mechanical strain and role in dendrite elongation. *Molecular and Cellular Biology, 26*(12), 4539–4552. https://doi.org/10.1128/MCB.02120-05

Zhang, K., Liu, X., Wang, L., Liu, Z., Yi, Q., Geng, B., Chen, X., Yu, D., & Xia, Y. Y. (2021). The mechanosensory and mechanotransductive processes mediated by ion channels and the impact on bone metabolism: A systematic review. *Archives of Biochemistry and Biophysics, 711*. https://doi.org/10.1016/j.abb.2021.109020

Zhou, R., Guo, Q., Xiao, Y., Guo, Q., Huang, Y., Li, C., & Luo, X. (2021). Endocrine role of bone in the regulation of energy metabolism. *Bone Research, 9*(1). https://doi.org/10.1038/s41413-021-00142-4

Zhou, X., von der Mark, K., Henry, S., Norton, W., Adams, H., & de Crombrugghe, B. (2014). Chondrocytes transdifferentiate into osteoblasts in endochondral bone during development, postnatal growth and fracture healing in mice. *PLoS Genetics, 10*(12). https://doi.org/10.1371/journal.pgen.1004820

Zhu, S., Chen, W., Masson, A., & Li, Y. P. (2024). Cell signaling and transcriptional regulation of osteoblast lineage commitment, differentiation, bone formation, and homeostasis. *Cell Discovery, 10*(1). https://doi.org/10.1038/s41421-024-00689-6

Zhu, L., Tang, Y., Li, X. Y., Keller, E. T., Yang, J., Cho, J. S., Feinberg, T. Y., & Weiss, S. J. (2020). Osteoclast-mediated bone resorption is controlled by a compensatory network of secreted and membrane-tethered metalloproteinases. *Science Translational Medicine, 12* (529). https://doi.org/10.1126/scitranslmed.aaw6143

CHAPTER

Mechanosensing and mechanotransduction in bone cells

4

Halima Kerdjoudj[1] and Jessica Schiavi-Tritz[2]
[1]*Université de Reims Champagne Ardenne, UR BIOS, Reims, France;* [2]*University of Lorraine, CNRS, LRGP, Vandoeuvre-les-Nancy, France*

Abbreviations

ALP	Alkaline Phosphatase
ATP	Adenosin TriPhosphate
BM-MSC	Bone Marrow Mesenchymal Stromal Cells
BMD	Bone Mineral Density
cAMP	cyclic Adenosine Monophosphate
Cox-2	Cyclooxygenase 2
Cx43	Connexin-43
Dkk1	Dickkopf-related protein 1
ECM	ExtraCellular Matrix
FA	Focal Adhesion
FSS	Fluid Shear Stress
Hh	Hedgehog
HSC	Hematopoietic Stem Cells
LCN	Lacuno-Canalicular Network
LINC	Linker of Nucleoskeleton and Cytoskeleton
NO	Nitric Oxide
Obl	Osteoblast
Ocl	Osteoclast
Ocy	Osteocyte
OPG	Osteoprotegerin
PGE-2	Prostaglandin E2
RANKL	Receptor Activator of NF-κB Ligand
TRPV4	Transient Receptor Potential Vanilloid 4

1. Introduction

Bone is a mechanosensitive and hierarchically organized tissue in which daily mechanical stimulations—among others—play a key role in maintaining homeostasis (i.e., the balance between the resorption of old or damaged tissue, and the formation of new functional bone (Fig. 4.1)).

Biomechanics and Mechanobiology of Bone. https://doi.org/10.1016/B978-0-443-26739-0.00023-5

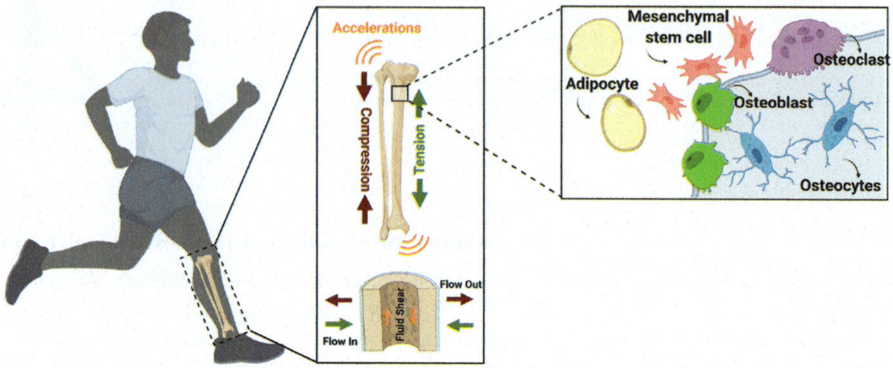

FIGURE 4.1 Bone mechanostimulation.

Mechanical forces on bone, providing a basis of mechanoregulation of bone tissue at the cellular level (Birks & Uzer, 2021).

-From Birks et al.

During embryogenesis, bone formation begins with the condensation of mesenchymal stromal cells, which prefigures each future skeletal elements. These processes have been described in vitro, where progenitors form three-dimensional structures commonly referred to as bone-*like* nodules. These nodules arise from colony-forming units—osteoprogenitor derived from Bone Marrow Mesenchymal Stromal Cells (BM-MSCs), which proliferate, condense, and then differentiate into Osteoblasts (Obl). Obl commitment involves a wide variety of cellular morphologies, including fibroblastic, polygonal, cuboidal, and stellar shapes, which characterize BM-MSCs, preosteoblasts, Obl, and Osteocytes (Ocy), respectively (Mechiche Alami et al., 2016). Osteoclasts (Ocl), the primary bone-resorbing cells, are highly specialized, and multinucleated, and of hematopoietic origin (bone marrow hematopoietic stromal cells). They are characterized by distinct features such as a ruffled border, which creates a localized microenvironment (i.e., acidic) to facilitate the matrix demineralization and its enzymatic degradation. Regular increases in loadings (i.e., for professional athletes) enhance Obl/Ocy-driven bone formation, whereas decreasing in physical activity (i.e., during bed rest or space flight) accelerates Ocl-driven bone resorption (Sims and Martin, 2014). In healthy individuals, diffuse damages and microcracks—caused by mechanical loading within the mineralized Extracellular Matrix (ECM)—serve as triggers for bone remodeling; a continuous cellular process primarily regulated by bone-resorbing Ocl and bone-forming Obl (Kennedy et al., 2012; Sims and Martin, 2014). A small amount of old or damaged bone is resorbed and subsequently replaced by immature osteoid, which is later reorganized and mineralized to restore the bone's structural and mechanical integrity. Bone remodeling renews an estimated 5%–10% of the bone matrix annually, ensuring the maintenance of the tissue's mechanical integrity.

Mechanical forces are thus essential for bone remodeling, and adaptations are dependent on the signal intensity and frequency. Long-term high-intensity exercises strengthen bone while long exposure to microgravity can lead to a monthly loss of 1%–1.5% of volumetric Bone Mineral Density (BMD) in the weight-bearing bones due to increased Ocl and decreased Obl activities. Loss of mechanical stimulation or diseases as type 2 diabetes can weaken the bone structure and change the matrix composition (Britton et al., 2024; Monahan et al., 2023), causing an increased risk of fracture. The mechanosensation of cells is made through several structures including membrane proteins, Linker of Nucleoskeleton and Cytoskeleton (LINC) proteins (Birks & Uzer, 2021), and primary cilia on the membrane of cells. Mechanical stimulations are transmitted by the ECM or mechanical stress (i.e., shear) from the cells surface to the cytoskeleton up to the nuclei with the nucleoskeleton (Fig. 4.2). Initially, Ocl were excluded from the category of bone mechanosensitive cells, but recent studies have shown that their activity is also mechanodriven (Bratengeier et al., 2020; Li et al., 2018; Xu, 2023). All these cellular processes are mediated by biochemical cues known as mechanotransduction pathways, which begin with the sensing of mechanical stimuli by external cellular structures. These stimuli are then translated into intracellular biochemical signals, ultimately leading to nuclei

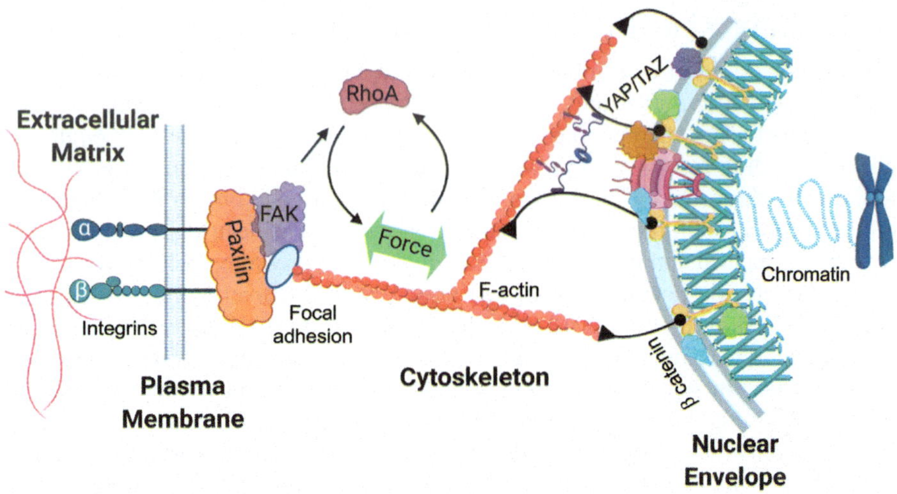

FIGURE 4.2 Mechanotransduction from the ECM to the nucleoskeleton.

Extracellular forces activate integrins and focal adhesion signaling to increase F-actin polymerization and cytoskeleton reorganization. LINC complexes are mechanically coupled to the cytoskeleton and regulate the access of important mechanical transducers such as β-catenin and YAP/TAZ. Inside the nucleus, G-actin is assembled into linear and branched networks to regulate chromatin dynamics and gene access (Birks & Uzer, 2021).

- From Birks et al.

remodeling, chromatin reorganization, and gene expression alteration, allowing the cells to adapt to new conditions.

The daily mechanical loading results in Fluid flow-induced Shear Stress (FSS), bending forces, and hydrostatic pressure which are primarily sensed by Ocy within the Lacuno-Canalicular Network (LCN). Therefore, the activation of Ocy precedes the differentiation of Ocl during bone remodeling. In addition to FSS, dynamic intramedullary pressurization results in hydrostatic pressure and fluid flow within the marrow cavity, on the endosteal surface and in the LCN, potentially exposing BM-MSCs, Obl, and Ocy to shear stress. These stimuli play essential roles from the embryonic stage—driving the shape of the developing bones (Verbruggen et al., 2018)—to adulthood, where they regulate bone tissue remodeling. Notably, the cellular sensitivity to these signals can be altered with aging and in pathological conditions such as osteoporosis (Verbruggen and McNamara, 2018).

In this chapter, we aim to highlight the key elements involved in cellular changes in response to the mechanical environment. After describing the main mechanosensors and mechanotransduction mechanisms, we will discuss the pathways involved in both healthy and pathological conditions, as well as potential future perspectives for bone mechanobiology.

2. Mechanosensitive structures in bone cells

Bone is composed of cell, mineral content, and an organic matrix, accounting for approximately 10%, 60%, and 30% of its composition, respectively. This distribution is essential for maintaining the mechanical properties of bone. Although mechanical loading is known to provide an anabolic stimulus for bone, it is difficult to correlate the behavior of cells to specific forces in bone due to the complexity of the mechanical forces. In vitro culture systems have been designed to apply several mechanical stimuli, including FSS, hydrostatic pressure, ECM strain, topography, and electric potentials (Arnsdorf et al., 2009; Freeman et al., 2017; Haugh et al., 2015; Morrell et al., 2018). Mechanotransduction is the conversion of these mechanical forces into biochemical signals that are integrated into a functional cell response. In bone, mechanotransduction occurs in four distinct phases: (i) mechanocoupling, where the mechanical loading causes deformation of bone cells; (ii) biochemical coupling, where the deformation of membrane is converted into intracellular signaling pathways; (iii) the transmission of signal, where the biochemical signal is transmitted from the sensor to the effector; and (iv) the effector cell response, where cells deposit new bone ECM or adapt the expression of genes (Ramaswamy et al., 2015; Wang et al., 2022). Mechanical forces directly stretch integrin binding sites at the cell surface-structure present on all eukaryotic cells. Although in bone, mechanical stimuli affect the three principal bone cells (Obl, Ocl, and Ocy), Ocy are the most abundant part, with almost 90% of the bony cells and are disseminated uniformly throughout the cortical and trabecular bone. In addition, they display several dendrites—cytoplasmic extensions—that increase the membrane surface area and enhance their ability to detect small shifts in movement and fluid, similarly to their cell bodies and cilia. Within the LCN, Ocy are enveloped by the glycocalyx and an interstitial fluid, which exert FSS on their membrane (Vaughan et al., 2015; Verbruggen et al., 2012). Ocy-ablated mice are resistant to unloading-induced bone loss, providing strong evidence for the role of Ocy in mechanotransduction. Whereas Ocy are responsive to FSS, deformation of Obl and osteoprogenitor membranes occurs in the presence of stretch. Furthermore, the ablation of Ocy increased the Ocl activity in mice by upregulating the expression of Receptor Activator of Nuclear factor- κB Ligand (RANKL) (Tatsumi et al., 2007). Conformational changes in the ECM, induced by mechanical stress, can modify the integrin structures and subsequently activate several intracellular secondary messenger pathways. These signaling cascades lead to changes in gene regulation that influence ECM synthesis and degradation, as well as cell proliferation and phenotype (Rammal et al., 2017). Another mechanism of mechanotransduction involves the deformation of gap junctions formed by connexin hemichannels and ion channels. Upon activation, these channels initiate secondary messenger signaling through pathways similar to those triggered by integrin activation, enabling the communication with adjacent cells. Additionally, mechanochemical transduction occurs through the primary cilium, a microtubule structure that protrudes from the cell membrane such as a cellular antenna, making it an optimal candidate for sensing extracellular mechanical loading. Moreover, the application of mechanical forces can activate hormone receptors and the release of growth factor—even in the absence of ligand binding—highlighting the sensitivity of bone cells to their mechanical environment.

2.1 ECM-integrin-cytoskeleton-nucleoskeleton "outside-in"

The ECM is a scaffolding that provides mechanical support and drives biological signaling in cells and has been recognized as a key aspect of the microenvironment that regulates cell behavior and phenotype. The ECM is composed of mineralized collagens, fibronectin, elastin, laminin, glycosaminoglycan, and glycoproteins, providing a specific topological, mechanical, and biochemical environment for the cells. As cells push and pull on the ECM during several biological processes, the ECM initially resists to these actions as governed by its stiffness, ranging from 100s of MPa in bone. In vitro, it was demonstrated that the undifferentiated BM-MSCs can commit and switch from adipogenesis to osteogenesis simply by changing the matrix stiffness (Engler et al., 2006). Furthermore, Ocy on stiff ECM pull harder than those on soft one (Klein-Nulend et al., 2012; McNamara et al., 2009). During bone formation, proliferating osteoprogenitors have the capacity to form multilayered 3D structures in which each cell interacts specifically with both nearby cells and its own ECM. The ECM thus plays a key regulatory role in maintaining proliferation and initiating signaling cascades through cell–ECM and cell–cell interactions, thereby mediating changes in cell shapes and regulating the expression of Obl-related genes. Furthermore, changes in Obl morphology are associated with

ECM-integrin-Focal Adhesion (FA) sites -cytoskeleton axis (Fig. 4.2) (Yourek et al., 2007). The extracellular domains of integrins interact with ECM while the cytoplasmic domains are involved in the assembly of signaling complexes (i.e., number of docking proteins, including talin and vinculin) with cytoplasmic F-actin bundles. Force loading at the FA causes a conformational change in talin proteins, revealing additional binding domains for vinculin. This last one binds and facilitates the localization of additional integrins, and the presence of the vinculin tail domain is necessary to propagate force from the FA to the F-actin bundles (Rammal et al., 2017). Other proteins such as zyxin and actinin also act to stabilize F-actin through other proteins including paxillin. The unique combinations of integrin subunits α and β determine the type of ECM molecules recognized by cells (Fig. 4.2), which is the first step of ultimate cell fate determination, and therefore play an important ECM organic composition and in matrix mineralization (Hong et al., 2010; Yourek et al., 2007). In parallel, proteins containing RGD sequences, such as fibronectin and osteopontin, are up-regulated earlier than other ECM proteins (such as collagen and osteocalcin), confirming their role in mediating cell attachment through integrins. The display of an RGD sequence in alginate hydrogels provides a strategy to control the phenotype of participating chondrocytes, Obl and MSCs (Schipani et al., 2009). Recently, Kindlin-2, an FA protein, has been identified as key in maintaining skeletal responses to mechanical stimulation by inhibiting Sclerostin expression during Ocy mechanotransduction (Qin et al., 2021).

The β_1 integrin is the major type I collagen-binding subchain (Lin et al., 2017). The inhibition of β_1 in Ocy induced a reduction in bone formation by Obl and an increase in adipogenesis differentiation of the BM-MSCs, signature of an altered bone homeostasis (Qin et al., 2022). As another example, the integrin $\alpha_V\beta_3$binds specifically to the RGD domains of proteins with a specific affinity for fibronectin, vitronectin, bone sialoprotein and osteopontin receptor, and seems to play an important role in the activation of Ocl during osteoporosis. Murphy et al. demonstrated that administrating to osteoporotic postmenopausal women an antagonist of the integrin $\alpha_v\beta_3$ (L-000845704) for 12 months, increased the BMD in the lumbar spine but had no effect on total BMD (Murphy et al., 2005). Another antagonist, tested on ovariectomized rats, showed an increase in 12.8% of the BMD compared with the SHAM rats (Hutchinson et al., 2003). The temporal distribution pattern of α_5/β_1 in bone-*like* nodules showed few differences among proliferation, remodeling and mineralization stages. But interfering $\alpha_5\beta_1$-ECM interactions suppress the osteogenic potential of progenitors (Moursi et al., 1997). The expression of $\alpha_3\beta_1$ and $\alpha_8\beta_1$ was maintained at a lower level after day 10 of culture, suggesting a possible downregulation of these receptors over time, and their disturbance blocked the upregulation of bone markers as Alkaline Phosphatase (ALP) and osteocalcin in Obl. However, cells retain the capacity to continue the osteogenic potential once the alteration is suppressed (Moursi et al., 1997). $\alpha_2\beta_1$-collagen binding requires the maintenance of type I collagen fibrillar conformation and the transcription of osteocalcin, suggesting its involvement at the onset of mineralization. The inhibition of β_1 or $\alpha_v\beta_3$ in MG-63 (i.e., osteoblast cell line from an osteosarcoma) upregulated the expression of osteogenic markers in response to FSS (Lee et al., 2008). Blockage of the integrin $\alpha_v\beta_3$ in MLO-Y4 (i.e., osteocyte-like cell line) impaired their chemosensitivity to FSS, disturbing Cyclo-oxygenase 2 (Cox-2) expression and Prostaglandin E2 (PGE-2) release (Haugh et al., 2015).

2.2 Gap junctions and hemichannels

Located at the membrane of bone cells, ion channels are sensitive to FSS and stretch. Cell–cell interactions and functional gap junctions were highlighted through the identification of Connexin-43 (Cx43) protein hexagonal-shaped monomers, one of the most abundant subtypes expressed in bone cells (Riquelme et al., 2020). Gap junctions facilitate therefore signal transduction between cells, allowing small molecules and electrical current to passively diffuse between adjacent cells in response to extracellular stimuli (Batra et al., 2012). During the osteoprogenitor differentiation, the morphologic changes initiate with an increase in the formation of gap junctions through microtubule actions, followed by the rapid assembly of preexisting gap junction protein on the membrane into an organized structure (Chiba et al., 1993). Disturbance of gap junctions prevents calcium transport, Obl maturation and ECM mineralization (Schiller et al., 2001). In terms of relaying mechanical cues to effector cells, Ocy communicate with Obl through gap junctions formed by Cx43 (Yellowley et al., 2000). Ocy are also involved in the regulation of Ocl differentiation and activation by increasing the Osteoprotegerin (OPG)/RANKL ratio (You et al., 2008; Zhao et al., 2002). RANKL stimulates Ocl formation by binding to its receptor RANK on the surface of Ocl precursors, while OPG, a decoy receptor for RANKL, inhibits osteoclastogenesis by competitively binding RANKL. Cx43 contributes to mechanical signaling in Ocy and Obl by producing calcium oscillations, providing evidence that calcium may use gap junctions to relay the mechanical signal detected by Ocy deep in the Obl on the bone surface (Yellowley et al., 2000). Paracrine communication between bone cells through indirect transmission forms a pivotal pathway in the generation of new bone. Ocy transmit mechanical signals from the external environment to Obl and Ocl by releasing molecules including ATP, cyclic Adenosine Monophosphate (cAMP), PGE-2, Nitric Oxide (NO) and growth factors into the extracellular environment. In this context, it was shown that MLO-Y4 in response to FSS releases PGE-2 through opening of Cx43 following its interaction with $\alpha_5\beta_1$ integrins. FSS also increased the expression of IgF-I in Ocy, increasing therefore the expression of type I collagen and osteocalcin in progenitors 48 hours later (Lean et al., 1995).

In addition to gap junctions, cadherins implement communication with nearby cells. In mature Obl, N-cadherin and cadherin-11 are predominantly expressed while their expression is low in undifferentiated MSCs (Marie et al., 2014; Mbalaviele et al., 2006). Cadherins also mediate the transmission of mechanical stimuli and the differentiation of MSC into Obl (Griffin et al., 2017).

2.3 Primary cilia

Primary cilia are immotile microtubule-based structures located at the surface of most mammalian cells. They originate at the basal body and extend into the extracellular compartment space (Hoey et al., 2012a, 2012b; Malone et al., 2007; Verbruggen et al., 2023). The primary cilium interacts with several cellular components including integrin-based adhesions and the cytoskeleton. Recent studies suggest that primary cilia act as mechanosensors in adult bone homeostasis (de Lima Perini et al., 2023; Moore and Jacobs, 2018; Verbruggen et al., 2023). It was shown that the embryonic cilia play a role in the earliest cellular events establishing left-right axis determination in mammals (Marszalek et al., 1999). A growing number of human pleiotropic syndromes including alterations in limb patterning, endochondral bone formation, craniofacial development and dentition are associated with ciliary or basal body dysfunction (Haycraft and Serra, 2008).

FSS along the Ocy body and primary cilia, induce bone adaptation through the regulation of Ocl (Geoghegan et al., 2021; You et al., 2008), Obl (Taylor et al., 2007) and osteoprogenitors (Hoey et al., 2011). However, it seems that enhancing human primary Obl cilia length decreases their mechanosensitivity, as it is observed in the idiopathic scoliosis bone cells (McNamara et al., 2009). Due to the specific localization of ion channels and receptors within this domain, the cilium is known to possess both mechanosensory and chemosensory functions in many cell types including bone cells (Hoey et al., 2012a, 2012b; Verbruggen et al., 2023). As bone cells respond to FSS with an increase in intracellular calcium (Chen et al., 2000), it was suggested that this response is cilia-dependent, as reported in kidney and liver. Some studies surprisingly suggested the absence of correlation between ciliated osteoblasts and Ca^{2+} flux under FSS (Kwon et al., 2010; Malone et al., 2007). The FSS induced cilia-dependent responses with a decrease in cAMP second messenger, which is necessary for the increase in Cox-2 expression and PGE2 (Kwon et al., 2010; Malone et al., 2007). Other studies highlight the presence of mechanosensitive calcium channel TRPV4, a member of the transient receptor potential channel, in MSCs, Obl, and Ocy cilia (Everaerts et al., 2010), demonstrating that primary cilia and TRPV4 channels collaborate to sense FSS, promoting osteogenesis and inhibiting adipogenesis (Corrigan et al., 2018). Ocy responding to FSS with cilia-dependent increases the Cox-2 expression and PGE2 release (Malone et al., 2007). Furthermore, an increase in OPG/RANKL ratio by FSS was abolished by cilia removal through specific inhibitor (Xiao et al., 2006).

Major evidence supporting a role for polycystin and primary cilia in bone comes from studies on mutant mouse model. Indeed, mice carrying a mutation in the gene encoding polycystin-1—a membrane protein that regulates calcium-permeable cation channels and intracellular calcium homeostasis—exhibit reduced BMD, trabecular bone volume, and cortical thickness. These skeletal defects are accompanied by decreased expression of Runx2, Osterix, osteocalcin, and a reduced OPG/RANKL ratio.

2.4 Channels

Mechanosensitive calcium channels, represented by the Piezo calcium channel protein family (Piezo 1 and 2), switch from closed to open in response to FSS, resulting in extracellular calcium ion influx and a series of biochemical reactions. The structures and functions of Piezo1 and Piezo2 proteins are highly similar, showing a three-bladed propeller-like homotrimer with two modules: a central ion-conduction pore module and a peripheral mechanotransduction module (Jiang et al., 2021; Zhao et al., 2019). By sensing the hydrostatic pressure, Piezo1 promotes the BM-MSC osteogenesis and inhibits adipogenesis. The decrease in Piezo1 channels is one of the causes of osteoporosis (Sun et al., 2019). FSS upregulates the expression of Piezo1 and induces the expression of Runx2 in preosteoblasts (Song et al., 2020). The microgravity simulation significantly reduced the expression of Piezo1 and ALP, whereas FSS upregulated their expression (Sun et al., 2019).

Studies of ion channels also provide evidence of the mechanosensitive capabilities of Ocl (Bratengeier et al., 2020; Guo et al., 2010; Li et al., 2018; McAllister et al., 2000; Suzuki et al., 2008). STIM1 and TRPV4 are Ca^{2+} channels that are highly expressed in the early and late stages, respectively, of Ocl differentiation. The blockage of these channels reduces the FSS-induced Ca^{2+}, indicating that STIM1 and TRPV4 may act as mechanosensitive ion channels for Ocl (Li et al., 2018; Ma et al., 2020).

As a mechanically gated ion channel, Piezo1 detects cell membrane tension changes, triggers Ca^{2+} influx, activates downstream signaling pathways (e.g., YAP/TAZ), and regulates stem cell (Baghdadi et al., 2024; Wang et al., 2024). Notably, its agonist Yoda1 enhances stem cell sensitivity to matrix stiffness (Wijerathne et al., 2022). The activation of Piezo1 with Yoda1, thus, maintains the cellular tension by upregulating the cytoskeleton-related gene expression and increases bone regeneration by upregulating osteogenic protein expression (Yang et al., 2023).

3. Mechanotransduction in bone cells

Bone homeostasis is continuously regulated by bone cells and influenced by mechanical stimuli (shear, hydrostatic pressure or strain) that are translated into biochemical cues altering the balance between bone formation and resorption. With 175,000 km of length and an approximate surface of 215 m^2 (Buenzli & Sims, 2015), the LCN in human bones might be the main place where mechanostimulation activates bone cells with various intensities (Schurman et al., 2021; Verbruggen et al., 2012). Several mechanotransduction phenomena and signaling pathways are implicated in osteogenesis with the differentiation of osteoprogenitors and the maturation of bone cells themselves, with several mechanosensitive structures i.e., cytoskeleton, integrins and channels (Fig. 4.3). Several biochemical reactions are taking place by implicating various pathways namely Wnt/β catenin and RhoA/ROCK pathways that depend on the influx of calcium ions and NO, and

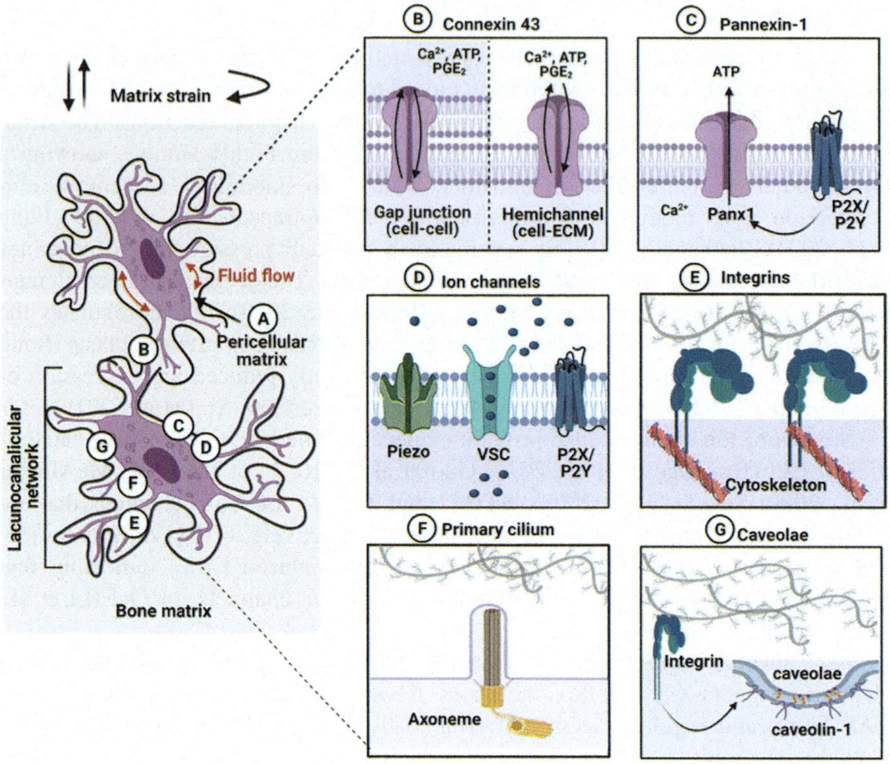

FIGURE 4.3 Osteocytes within the lacuno-canalicular network and their mechanosensors.

Osteocytes within the lacuno-canalicular network express mechanosensors, which can be activated by various mechanical stimuli as fluid flow in the pericellular matrix and matrix strain (e.g., compressive, tensile, and torsional loading). (A) Osteocytes are surrounded by the pericellular matrix, between the cell and the walls of lacunae and canaliculi, which acts as a tether for osteocytes to transduce the mechanical stimulation. Several structures are found on the surface of the cells: (B) Gap junctions as connexin 43 (Cx43), (C) pannexin-1 (Panx1), (D) mechanosensing ion channels such as Piezo, voltage-sensitive calcium channel (VSC), and purinergic receptor (P2X/P2Y), (E) integrins, (F) primary cilium, and (G) Caveolin-1 (Choi et al., 2022).

- Modified from Choi et al. Creative Commons Attribution License (CC BY).

Hippo. Other pathways have been extensively described in former reviews on bone cells including MSCs, and in healthy and diseased conditions (García-Aznar et al., 2021; Huang et al., 2013; Jacobs et al., 2010; Naqvi and McNamara, 2020; Uda et al., 2017; Verbruggen and McNamara, 2018; Xu, 2023).Young et al. showed that disruption of FAK signaling in Obl attenuated the load-induced activation of several key osteogenic outcomes including ERK phosphorylation, c-fos, Cox-2, PGE_2 release and osteopontin expression (Young et al., 2009).

3.1 Wnt/β catenin pathway

The Wnt/β catenin signaling is known to be implicated in bone formation, growth, development and repair (Duan and Bonewald, 2016), but as well in osteoarticular diseases. Through mechanical stimulation on osteoprogenitors, Obl or Ocy, the pathway is activated, then upregulates the expression of bone remodeling markers as Runx2 or Cox-2 (Robinson et al., 2006), but also OPG, which inhibits the differentiation of Ocl (Bolamperti et al., 2022). In addition, adhesion junction protein seems to be implicated in the signaling with an increase in the cytoplasmic β-catenin linked to N-cadherin, which increased bone marker expression (Arnsdorf et al., 2009). While in a static state, β-catenin mediates the connection between cadherins on the cell membrane and the cytoplasmic cytoskeleton. In response to FSS, β-catenin is released from cadherins, leading to its accumulation in the cytoplasm. Then, it translocates into the nucleus, where it triggers osteogenic gene expression and promotes osteogenesis (Arnsdorf et al., 2009; Stewart et al., 2020).

Antagonists such as sclerostin (SOST), a negative regulator of Obl through Wnt/β-catenin signaling pathway, reduce bone formation. Those downregulations are also downstream from mechanical stimuli sensed by Ocy as it increases their SOST expression and fine-tunes bone homeostasis. Dkk1 (Dickkopf) is another inhibitor of the Wnt signaling implicated in bone remodeling. Obl under compression increases the expression of Dkk1 along with the ratio RANKL/OPG and TNFα (Wu et al., 2023). On the contrary, Obl subjected to tensile forces expressed significantly less Dkk1 while showing an upregulation of bone formation markers Runx2, ALP and Osterix, and Wnt5a (Wu et al., 2023).

3.2 RhoA/ROCKII

RhoA signaling is involved in mechanotransduction through the regulation of cytoskeletal stabilization, which provides the shape of cells and is an essential mechanosensor for bone cells. This is probably due to their specific shape within the LCN for Ocy, or close to a hard tissue for Obl and Ocl. FSS applied in vitro on BM-MSC and Obl upregulate osteogenic markers depending on RhoA/ROCKII pathway and the actin cytoskeleton (Arnsdorf et al., 2009). In Obl, FSS induce an increase in the anisotropy of the cytoskeleton and an increase in osteogenic markers (PGE2), and a downregulation of bone resorption factors (Allison & McNamara, 2019; Simfia et al., 2020a, 2020b). In addition, in an estrogen withdrawal model, which mimics postmenopausal osteoporosis, the hypersensitivity of estrogen withdrawn Obl and Ocy was suppressed when ROCK was inhibited (Allison & McNamara, 2019; Geoghegan et al., 2021; Simfia et al., 2020a, 2020b). Taken together, these reports indicate the role of the cytoskeleton and RhoA/ROCKII pathway in bone resorption diseases such as osteoporosis. In addition, the extracellular osmolarity affects Obl migration by regulating the cytoskeletal dynamics, FA formation and cell polarity through the alteration of the RhoA/ROCKII pathway (Li et al., 2025). Specifically, TRPV4 expression was increased in hypotonic conditions with a rapid increase in cellular calcium concentration, peaking at approximately 20 seconds before gradually declining, and an upregulation of Rho proteins, which correlates with changes in Obl migration capacities (Li et al., 2025).

3.3 Hedgehog signaling

Hedgehog (Hh) signaling pathway is implicated in the osteogenesis of BM-MSC and specific gene expression is mechanically regulated and required for loading-induced osteogenic differentiation (Johnson et al., 2021). Interestingly, it has been established that RANKL production is mediated by Hh signaling in Obl and Ocy (Verbruggen et al., 2023), a key signaling pathway associated with the primary cilium. In an in vitro model of postmenopausal osteoporosis, it has been shown that the withdrawal of estrogen on Ocy leads to a disruption of focal adhesion, followed by an alteration of the actin contractility, an elongation of primary cilium and an activation of Hh upregulating the osteoclastogenesis (Geoghegan et al., 2021; Hoey et al., 2012a, 2012b).

3.4 Calcium signaling and NO

Mechanical loading on Ocy and Obl leads to several downstream signaling and the production of secondary mediators that induce bone formation (Chen et al., 2000; You et al., 2001). Calcium has an important role in the homeostasis of bone ECM and the regulation of bone cells. The mechanosensitive channel Piezo1 mediates preferentially Ca^{2+} current, is permeable for monovalent ions (K^+, Na^+ and Cs^+) and divalent ions (Ba^{2+}, Ca^{2+}, Mg^{2+} and Mn^{2+}) (Gnanasambandam et al., 2015), and is essential for bone remodeling (Sun et al., 2019). Ocy respond to only 10 minutes of FSS with a rapid and transient increase in $[Ca^{2+}]_i$ oscillations through mechanosensitive ion channels, along with an increase in PGE2 release (Ajubi et al., 1999). On the contrary, the blockage of Ca^{2+} release from internal stores significantly inhibits FSS-induced PGE_2 expression. Ocl are also regulated by mechanical stimuli and calcium flux, and the mechanosensitive channel anoctamin 1 regulates their responses to mechanical stress by increasing intracellular chloride ion concentration and inhibiting calcium signaling (Sun et al., 2023). NO is a crucial ubiquitous messenger molecule that is required for intercellular communication between Ocy and Obl after a mechanical stimulation. FSS-mediated NO release by Obl and Ocy is also associated with the actin cytoskeleton and microtubules, respectively, but oppositely regulated by the PGE2 release (McGarry et al., 2005).

3.5 Hippo

YAP (Yes-Associated transcriptional regulator) and TAZ (transcriptional coactivator with PDZ-binding motif) are the nuclear transducers of the Hippo pathway. Their activation is key for the osteogenic differentiation of BM-MSC mediated by external mechanical stimuli, showing its implication in osteogenesis, where a soft environment induces adipogenesis while a stiffer matrix an osteogenic differentiation (Dupont et al., 2011; Park et al., 2019). Mechanical stimuli such as ultrasound, activate the pathway with an increase in YAP expression, which is downstream from changes in F-actin, myosin and RhoA/ROCK signaling in BM-MSC (Xue et al., 2017). Besides BM-MSC, the Hippo pathway is implicated in changes in Obl capacity to sense mechanical stimuli (Wang et al., 2020) and in macrophages polarity (Wang et al., 2023), but it is also involved in the activation of the periosteal expansion after a fracture (Nijsure et al., 2024). YAP/TAZ is activated or inhibited differently depending on the stage of osteogenesis (Xiong et al., 2018). In fact, the deletion of YAP/TAZ in Obl progenitors increases Obl formation, along with an increase in Wnt signaling and Runx2 activity. In contrast, a deletion of YAP/TAZ in mature Obl and Ocy decreases ALP expression in an ex vivo culture of bone (Collins et al., 2024). Yoda1 is an agonist of Piezo1 and can decrease the mechanical threshold for its activation (Lacroix et al., 2018; Syeda et al., 2015). Indeed, when Yoda1 was upregulated in BM-MSC, it promoted osteogenic markers and upregulated the expression of the Piezo1/RhoA/ROCK1/YAP1 signaling pathways (Yang et al., 2023). Other proteins are activating YAP/TAZ as Pinch1 and Pinch2, which are mechanosensitive channels that have been studied for their implication in bone homeostasis as they regulate the activation of integrins and cell–ECM adhesion. In Ocy and mature Obl, Pinch1 is implicated in bone formation. Its inhibition induced a decrease in the integrin β_3 and an increase in apoptosis markers in Ocy, caspase 3 and 8 and an inhibition of the YAP/TAZ pathway (Wang et al., 2019).

3.6 EphrinB2/EphB4

EphrinB2 is a transmembrane ligand that interacts with its receptor, EphB4, to provide a checkpoint through which Obl must pass to reach full maturity, and is implicated in the regulation of bone mineralization (Takyar et al., 2013). Indeed, in adult bones from mice, Ephrin2 prevents Obl apoptosis to maintain their differentiation and bone strength (Tonna et al., 2014). An Obl-specific EphrinB2 deletion altered bone stiffness by impairing Obl differentiation and a twofold delay in the initiation of bone mineralization (Tonna et al., 2014). Recently, EphrinB2 has been identified as one pathway through which bone mineralization rate is controlled by Ocy. Mice with EphrinB2 depleted Ocy had more brittle bones due to secondary mineralization that occurred after Ocy autophagy (Vrahnas et al., 2019). Those bone homeostasis regulators are controlled by mechanobiological stimuli as Low-intensity Pulsed Ultrasound, which was applied on BM-MSC and the mandible of mice, enhancing therefore bone formation through the activation of EphrinB2/EphB4 (Zhou et al., 2023).

4. Future perspectives

It is well recognized that bone cells communicate and exchange information through direct cell-to-cell contact (adhesion, connexin interactions) or through soluble mediators (paracrine signaling). However, several studies have demonstrated that bone cells may also communicate through extracellular vesicles (EVs), which are produced by cells within the endosomal reticulum or by budding of the membrane, called the exosomes and microvesicles, respectively (Welsh et al., 2024). They enclose secretion from the cell cytoplasm including proteins, lipids, RNAs and glycoproteins. In recent years, EVs have received increasing attention as mediators of intercellular communication due to their capacity to transfer bioactive molecular content (cargo) to recipient cells, triggering a variety of biological responses in several tissues but as well to the microbiome. In vitro, EVs were firstly identified as Matrix Vesicles (MVs) within bone-*like* nodules before the ECM mineralization (Boonrungsiman et al., 2012). Early mineralization events were therefore found to localize around these MVs, providing evidence that they actively participate in the initiation of ECM mineralization. Indeed, they are enriched in essential components enzymes, annexin-calcium binding proteins, and phospholipids, which initiate mineral nucleation.

Few studies have aligned EVs concept with current definitions and guidelines established by the International Society for Extracellular Vesicles (Shapiro et al., 2015; Welsh et al., 2024). It has also been demonstrated that FSS-stimulated Ocy enhance the release of EVs, which in turn regulate Ocl maturation (Chen et al., 2024; Morrell et al., 2018). An emerging area of research focuses on optimizing EVs production processes, particularly to increase the yield of production to obtain clinical doses for therapeutic applications. EVs produced by Obl, Ocy and BM-MSCs have been tested successfully in vitro for their osteogenic effects (Man et al., 2022, 2023; Shen et al., 2023). Mechanical preconditioning of cells has been proposed as an effective strategy to enhance EV production and improve their functional efficacy.

All mechanobiological knowledge is key in exploring new therapeutic approaches by integrating mechanical stimuli at an appropriate phase of bone regeneration to treat fractures (Huang et al., 2013; Jacho and Yildirim-Ayan, 2024; Seo and Mooney, 2022), or in developing new biomedicine with the production of new drugs that will target the pathways or mechanosensors. In fact, the Romosozumab, a humanized sclerostin antibody, has been tested successfully to counteract the effects of postmenopausal osteoporosis on Obl or at a macroscale in several in vitro (Allison et al., 2020) and animal models (Li et al., 2009; Naqvi et al., 2025), and clinical trials (Cosman et al., 2016; Gatti et al., 2012; McClung et al., 2014). In addition the implication of the circadian rhythm on bone adaptation might have an impact on the mechanosensitiveness of bone cells as it is influencing some of the mechanotransduction pathways (Bouchard et al., 2022). Altogether, this would feed the health recommendation that performing exercise at any time of life is beneficial for preventing several diseases or improving symptoms. In addition, those research lead to scientific evidences on the intensities and at which frequency exercises are actively helping.

There is an increasing potential for further horizons with the combination of mechanobiology from biological and engineering perspectives that will inform and model computationally, at a molecular to macroscale, the effect of mechanical stimuli on bone cells and on improving EV bioprocesses (García-Aznar et al., 2021; Giorgi et al., 2016). Those studies will help in developing in vitro and ex vivo models of healthy bone and diseased states to create small bone substitutes, but primarily to test new biomedicine.

5. Conclusion

Despite the technologies available, Frost was the first to raise the hypothesis that mechanical stimuli are essential for the homeostasis of bone within a specific range of magnitude and outside which the tissue adapts to its mechanical microenvironment (Frost, 1994). In the last 2/3 decades, equipment and new knowledge have allowed scientists to confirm his hypothesis and to deeply identify several mechanosensors and mechanotransduction pathways. In addition, the input of computational modeling is leading to further understanding of the stress, strain and pressure felt by cells and to model mechanobiological pathways, while it is expected that Artificial Intelligence will help in predicting them or discovering new routes. The scientific advances will lead to new mechanotherapeutic biomedicine and strengthen the philosophy of a "healthy mind in a healthy body."

References

Ajubi, N. E., Klein-Nulend, J., Alblas, M. J., Burger, E. H., & Nijweide, P. J. (1999). Signal transduction pathways involved in fluid flow-induced PGE 2 production by cultured osteocytes. *American Journal of Physiology-Endocrinology and Metabolism, 276*(1), Article E171. https://doi.org/10.1152/ajpendo.1999.276.1.e171

Allison, H., Holdsworth, G., & McNamara, L. M. (2020). Scl-Ab reverts pro-osteoclastogenic signalling and resorption in estrogen deficient osteocytes. *BMC Molecular and Cell Biology, 21*(1). https://doi.org/10.1186/s12860-020-00322-w

Allison, H., & McNamara, L. M. (2019). Inhibition of osteoclastogenesis by mechanically stimulated osteoblasts is attenuated during estrogen deficiency. *American Journal of Physiology-Cell Physiology, 317*(5), Article C969. https://doi.org/10.1152/ajpcell.00168.2019

Arnsdorf, E. J., Tummala, P., Jacobs, C. R., & Bergmann, A. (2009). Non-Canonical wnt signaling and N-Cadherin related β-Catenin signaling play a role in mechanically induced osteogenic cell fate. *PLoS One, 4*(4), Article e5388. https://doi.org/10.1371/journal.pone.0005388

Arnsdorf, E. J., Tummala, P., Kwon, R. Y., & Jacobs, C. R. (2009). Mechanically induced osteogenic differentiation - The role of RhoA, ROCKII and cytoskeletal dynamics. *Journal of Cell Science, 122*(4), 546–553. https://doi.org/10.1242/jcs.036293

Baghdadi, M. B., Houtekamer, R. M., Perrin, L., Rao-Bhatia, A., Whelen, M., Decker, L., Bergert, M., Pérez-Gonzàlez, C., Bouras, R., Gropplero, G., Loe, A. K. H., Afkhami-Poostchi, A., Chen, X., Huang, X., Descroix, S., Wrana, J. L., Diz-Muñoz, A.,

Gloerich, M., Ayyaz, A., Vignjevic, D. M., & Kim, T. H. (2024). PIEZO-dependent mechanosensing is essential for intestinal stem cell fate decision and maintenance. *Science, 386*(6725). https://doi.org/10.1126/science.adj7615

Batra, N., Kar, R., & Jiang, J. X. (2012). Gap junctions and hemichannels in signal transmission, function and development of bone. *Biochimica et Biophysica Acta - Biomembranes, 1818*(8), 1909–1918. https://doi.org/10.1016/j.bbamem.2011.09.018

Birks, S., & Uzer, G. (2021). At the nuclear envelope of bone mechanobiology. *Bone, 151*, Article 116023. https://doi.org/10.1016/j.bone.2021.116023

Bolamperti, S., Villa, I., & Rubinacci, A. (2022). Bone remodeling: An operational process ensuring survival and bone mechanical competence. *Bone Research, 10*(1). https://doi.org/10.1038/s41413-022-00219-8

Boonrungsiman, S., Gentleman, E., Carzaniga, R., Evans, N. D., McComb, D. W., Porter, A. E., & Stevens, M. M. (2012). The role of intracellular calcium phosphate in osteoblast-mediated bone apatite formation. *Proceedings of the National Academy of Sciences, 109*(35), 14170–14175. https://doi.org/10.1073/pnas.1208916109

Bouchard, A. L., Dsouza, C., Julien, C., Rummler, M., Gaumond, M.-H., Cermakian, N., & Willie, B. M. (2022). Bone adaptation to mechanical loading in mice is affected by circadian rhythms. *Bone, 154*, Article 116218. https://doi.org/10.1016/j.bone.2021.116218

Bratengeier, C., Liszka, A., Hoffman, J., Bakker, A. D., & Fahlgren, A. (2020). High shear stress amplitude in combination with prolonged stimulus duration determine induction of osteoclast formation by hematopoietic progenitor cells. *Federation of American Societies for Experimental Biology Journal, 34*(3), 3755–3772. https://doi.org/10.1096/fj.201901458R

Britton, M., Monahan, G. E., Murphy, C. G., Kearns, S. R., Devitt, A. T., Okwieka, A., Jaisson, S., Van Gulick, L., Beljebbar, A., Kerdjoudj, H., Schiavi, J., & Vaughan, T. J. (2024). An investigation of composition, morphology, mechanical properties, and microdamage accumulation of human type 2 diabetic bone. *Bone, 187*, Article 117190. https://doi.org/10.1016/j.bone.2024.117190

Buenzli, P. R., & Sims, N. A. (2015). Quantifying the osteocyte network in the human skeleton. *Bone, 75*, 144–150. https://doi.org/10.1016/j.bone.2015.02.016

Chen, N. X., Ryder, K. D., Pavalko, F. M., Turner, C. H., Burr, D. B., Qiu, J., & Duncan, R. L. (2000). Ca2+ regulates fluid shear-induced cytoskeletal reorganization and gene expression in osteoblasts. *American Journal of Physiology - Cell Physiology, 278*(5), Article C989. https://doi.org/10.1152/ajpcell.2000.278.5.c989

Chen, Y., Zhao, R., Yang, L., & Guo, X. E. (2024). The roles of extracellular vesicles released by mechanically stimulated osteocytes in regulating osteoblast and osteoclast functions. *Mechanobiology in Medicine, 2*(2), Article 100065. https://doi.org/10.1016/j.mbm.2024.100065

Chiba, H., Sawada, N., Oyamada, M., Kojima, T., Nomura, S., Ishii, S., & Mori, M. (1993). Relationship between the expression of the gap junction protein and osteoblast phenotype in a human osteoblastic cell line during cell proliferation. *Cell Structure and Function, 18*(6), 419–426. https://doi.org/10.1247/csf.18.419

Choi, J. U. A., Kijas, A. W., Lauko, J., & Rowan, A. E. (2022). The mechanosensory role of osteocytes and implications for bone health and disease States. *Frontiers in Cell and Developmental Biology, 9*. https://doi.org/10.3389/fcell.2021.770143

Collins, J. M., Lang, A., Parisi, C., Moharrer, Y., Nijsure, M. P., (Thomas, J. H., Kim, Ahmed, S., Szeto, G. L., Qin, L., Gottardi, R., Dyment, N. A., Nowlan, N. C., & Boerckel, J. D. (2024). YAP and TAZ couple osteoblast precursor mobilization to angiogenesis and mechanoregulation in murine bone development. *Developmental Cell, 59*(2), 211. https://doi.org/10.1016/j.devcel.2023.11.029

Corrigan, M. A., Johnson, G. P., Stavenschi, E., Riffault, M., Labour, M. N., & Hoey, D. A. (2018). TRPV4-mediates oscillatory fluid shear mechanotransduction in mesenchymal stem cells in part via the primary cilium. *Scientific Reports, 8*(1). https://doi.org/10.1038/s41598-018-22174-3

Cosman, F., Crittenden, D. B., Adachi, J. D., Binkley, N., Czerwinski, E., Ferrari, S., Hofbauer, L. C., Lau, E., Michael Lewiecki, E., Miyauchi, A., Zerbini, C. A. F., Milmont, C. E., Chen, L., Maddox, J., Meisner, P. D., Libanati, C., & Grauer, A. (2016). Romosozumab treatment in postmenopausal women with osteoporosis. *Journal of Medicine, 375*(16), 1532–1543. https://doi.org/10.1056/NEJMoa1607948

de Lima Perini, M. M., Pugh, J. N., Scott, E. M., Bhula, K., Chirgwin, A., Reul, O. N., Berbari, N. F., & Li, J. (2023). Primary cilia in osteoblasts and osteocytes are required for skeletal development and mechanotransduction. *bioRxiv.* https://doi.org/10.1101/2023.12.15.570609

Duan, P., & Bonewald, L. F. (2016). The role of the wnt/β-catenin signaling pathway in formation and maintenance of bone and teeth. *The International Journal of Biochemistry & Cell Biology, 77*(Pt A), 23–29. https://doi.org/10.1016/j.biocel.2016.05.015

Dupont, S., Morsut, L., Aragona, M., Enzo, E., Giulitti, S., Cordenonsi, M., Zanconato, F., Le Digabel, J., Forcato, M., Bicciato, S., Elvassore, N., & Piccolo, S. (2011). Role of YAP/TAZ in mechanotransduction. *Nature, 474*(7350), 179–183. https://doi.org/10.1038/nature10137

Engler, A. J., Sen, S., Sweeney, H. L., & Discher, D. E. (2006). Matrix elasticity directs stem cell lineage specification. *Cell, 126*(4), 677–689. https://doi.org/10.1016/j.cell.2006.06.044

Everaerts, W., Nilius, B., & Owsianik, G. (2010). The vanilloid transient receptor potential channel TRPV4: From structure to disease. *Progress in Biophysics and Molecular Biology, 103*(1), 2–17. https://doi.org/10.1016/j.pbiomolbio.2009.10.002

Freeman, F. E., Schiavi, J., Brennan, M. A., Owens, P., Layrolle, P., & Mcnamara, L. M. (2017). Mimicking the biochemical and mechanical extracellular environment of the endochondral ossification process to enhance the in vitro mineralization potential of human mesenchymal stem cells. *Tissue Engineering Part A, 23*(23–24), 1466–1478. https://doi.org/10.1089/ten.tea.2017.0052

Frost, H. M. (1994). Wolff's Law and bone's structural adaptations to mechanical usage: An overview for clinicians. *The Angle Orthodontist, 64*(3), 175–188.

García-Aznar, J. M., Nasello, G., Hervas-Raluy, S., Pérez, M.Á., & Gómez-Benito, M. J. (2021). Multiscale modeling of bone tissue mechanobiology. *Bone, 151.* https://doi.org/10.1016/j.bone.2021.116032

Gatti, D., Viapiana, O., Fracassi, E., Idolazzi, L., Dartizio, C., Rosaria Povino, M., Adami, S., & Rossini, M. (2012). Sclerostin and DKK1 in postmenopausal osteoporosis treated with denosumab. *Journal of Bone and Mineral Research, 27*(11), 2259–2263. https://doi.org/10.1002/jbmr.1681

Geoghegan, I. P., McNamara, L. M., & Hoey, D. A. (2021). Estrogen withdrawal alters cytoskeletal and primary ciliary dynamics resulting in increased Hedgehog and osteoclastogenic paracrine signalling in osteocytes. *Scientific Reports, 11*(1). https://doi.org/10.1038/s41598-021-88633-6

Giorgi, M., Verbruggen, S. W., & Lacroix, D. (2016). In silico bone mechanobiology: Modeling a multifaceted biological system. *WIREs Systems Biology and Medicine, 8*(6), 485–505. https://doi.org/10.1002/wsbm.1356

Gnanasambandam, R., Bae, C., Gottlieb, P. A., Sachs, F., & Barnes, S. (2015). Ionic selectivity and permeation properties of human PIEZO1 channels. *PLoS One, 10*(5), Article e0125503. https://doi.org/10.1371/journal.pone.0125503

Griffin, F. E., Schiavi, J., McDevitt, T. C., McGarry, J. P., & McNamara, L. M. (2017). The role of adhesion junctions in the biomechanical behaviour and osteogenic differentiation of 3D mesenchymal stem cell spheroids. *Journal of Biomechanics, 59*, 71–79. https://doi.org/10.1016/j.jbiomech.2017.05.014

Guo, D., Zhang, Q., Li, J., Liang, X., Chen, Y., Wang, H., & Meng, B. (2010). Fluid shear stress changes cell morphology and regulates the expression of ATP6V1A and TCIRG1 mRNA in rat osteoclasts. *Molecular Medicine Reports, 3*(1), 173–178. https://doi.org/10.3892/mmr-00000236China

Haugh, M. G., Vaughan, T. J., & McNamara, L. M. (2015). The role of integrin αVβ3 in osteocyte mechanotransduction. *Journal of the Mechanical Behavior of Biomedical Materials, 42*, 67–75. https://doi.org/10.1016/j.jmbbm.2014.11.001

Haycraft, C. J., & Serra, R. (2008). Chapter 11 Cilia involvement in patterning and maintenance of the skeleton. *Current Topics in Developmental Biology, 85*, 303–332. https://doi.org/10.1016/S0070-2153(08)00811-9

Hoey, D. A., Chen, J. C., & Jacobs, C. R. (2012a). The primary cilium as a novel extracellular sensor in bone. *Frontiers in Endocrinology, 3*. https://doi.org/10.3389/fendo.2012.00075

Hoey, D. A., Kelly, D. J., & Jacobs, C. R. (2011). A role for the primary cilium in paracrine signaling between mechanically stimulated osteocytes and mesenchymal stem cells. *Biochemical and Biophysical Research Communications, 412*(1), 182–187. https://doi.org/10.1016/j.bbrc.2011.07.072

Hoey, D. A., Tormey, S., Ramcharan, S., O'Brien, F. J., & Jacobs, C. R. (2012b). Primary cilia-mediated mechanotransduction in human mesenchymal stem cells. *Stem Cells, 30*(11), 2561–2570. https://doi.org/10.1002/stem.1235

Hong, D., Chen, H.-X., Yu, H.-Q., Liang, Y., Wang, C., Lian, Q.-Q., Deng, H.-T., & Ge, R.-S. (2010). Morphological and proteomic analysis of early stage of osteoblast differentiation in osteoblastic progenitor cells. *Experimental Cell Research, 316*(14), 2291–2300. https://doi.org/10.1016/j.yexcr.2010.05.011

Huang, C., Holfeld, J., Schaden, W., Orgill, D., & Ogawa, R. (2013). Mechanotherapy: Revisiting physical therapy and recruiting mechanobiology for a new era in medicine. *Trends in Molecular Medicine, 19*(9), 555–564. https://doi.org/10.1016/j.molmed.2013.05.005

Hutchinson, J. H., Halczenko, W., Brashear, K. M., Breslin, M. J., Coleman, P. J., Duong, L. T., Fernandez-Metzler, C., Gentile, M. A., Fisher, J. E., Hartman, G. D., Huff, J. R., Kimmel, D. B., Leu, C. T., Meissner, R. S., Merkle, K., Nagy, R., Pennypacker, B., Perkins, J. J., Prueksaritanont, T., … Duggan, M. E. (2003). Nonpeptide αvβ3 antagonists. 8. In vitro and in vivo evaluation of a potent αvβ3 antagonist for the prevention and treatment of osteoporosis. *Journal of Medicinal Chemistry, 46*(22), 4790–4798. https://doi.org/10.1021/jm030306r

Jacho, D., & Yildirim-Ayan, E. (2024). Mechanome-guided strategies in regenerative rehabilitation. *Current Opinion in Biomedical Engineering, 29*, Article 100516. https://doi.org/10.1016/j.cobme.2023.100516

Jacobs, C. R., Temiyasathit, S., & Castillo, A. B. (2010). Osteocyte mechanobiology and pericellular mechanics. *Annual Review of Biomedical Engineering, 12*, 369–400. https://doi.org/10.1146/annurev-bioeng-070909-105302

Jiang, Y., Yang, X., Jiang, J., & Xiao, B. (2021). Structural designs and mechanogating mechanisms of the mechanosensitive Piezo channels. *Trends in Biochemical Sciences, 46*(6), 472–488. https://doi.org/10.1016/j.tibs.2021.01.008

Johnson, G. P., Fair, S., & Hoey, D. A. (2021). Primary cilium-mediated MSC mechanotransduction is dependent on Gpr161 regulation of hedgehog signalling. *Bone, 145*. https://doi.org/10.1016/j.bone.2021.115846

Kennedy, O. D., Herman, B. C., Laudier, D. M., Majeska, R. J., Sun, H. B., & Schaffler, M. B. (2012). Activation of resorption in fatigue-loaded bone involves both apoptosis and active pro-osteoclastogenic signaling by distinct osteocyte populations. *Bone, 50*(5), 1115–1122. https://doi.org/10.1016/j.bone.2012.01.025

Klein-Nulend, J., Bacabac, R. G., & Bakker, A. D. (2012). Mechanical loading and how it affects bone cells: The role of the osteocyte cytoskeleton in maintaining our skeleton. *European Cells and Materials, 24*, 278–291. https://doi.org/10.22203/eCM.v024a20

Kwon, R. Y., Temiyasathit, S., Tummala, P., Quah, C. C., & Jacobs, C. R. (2010). Primary cilium-dependent mechanosensing is mediated by adenylyl cyclase 6 and cyclic AMP in bone cells. *Federation of American Societies for Experimental Biology Journal, 24*(8), 2859–2868. https://doi.org/10.1096/fj.09-148007

Lacroix, J. J., Botello-Smith, W. M., & Luo, Y. (2018). Probing the gating mechanism of the mechanosensitive channel Piezo1 with the small molecule Yoda. *Nature Communications, 9*(1). https://doi.org/10.1038/s41467-018-04405-3

Lean, J. M., Jagger, C. J., Chambers, T. J., & Chow, J. W. (1995). Increased insulin-like growth factor I mRNA expression in rat osteocytes in response to mechanical stimulation. *American Journal of Physiology-Endocrinology and Metabolism, 268*(2), Article E318. https://doi.org/10.1152/ajpendo.1995.268.2.e318

Lee, D. Y., Yeh, C. R., Chang, S. F., Lee, P. L., Chien, S., Cheng, C. K., & Chiu, J. J. (2008). Integrin-mediated expression of bone formation-related genes in osteoblast-like cells in response to fluid shear stress: Roles of extracellular matrix, Shc, and mitogen-activated protein kinase. *Journal of Bone and Mineral Research, 23*(7), 1140–1149. https://doi.org/10.1359/jbmr.080302

Li, P., Bian, X., Liu, C., Wang, S., Guo, M., Tao, Y., & Huo, B. (2018). STIM1 and TRPV4 regulate fluid flow-induced calcium oscillation at early and late stages of osteoclast differentiation. *Cell Calcium, 71*, 45–52. https://doi.org/10.1016/j.ceca.2017.12.001

Li, X., Ominsky, M. S., Warmington, K. S., Morony, S., Gong, J., Cao, J., Gao, Y., Shalhoub, V., Tipton, B., Haldankar, R., Chen, Q., Winters, A., Boone, T., Geng, Z., Niu, Q.-T., Ke, H. Z., Kostenuik, P. J., Scott Simonet, W., Lacey, D. L., & Paszty, C. (2009). Sclerostin antibody treatment increases bone formation, bone mass, and bone strength in a rat model of Postmenopausal osteoporosis. *Journal of Bone and Mineral Research, 24*(4), 578–588. https://doi.org/10.1359/jbmr.081206

Li, Y., Yang, Y., Wang, X., Li, L., & Zhou, M. (2025). Extracellular osmolarity regulates osteoblast migration through the TRPV4-Rho/ROCK signaling. *Communications Biology, 8*(1). https://doi.org/10.1038/s42003-025-07946-8

Lin, T. H., Yang, R. S., Tu, H. J., Liou, H. C., Lin, Y. M., Chuang, W. J., & Fu, W. M. (2017). Inhibition of osteoporosis by the αvβ3 integrin antagonist of rhodostomin variants. *European Journal of Pharmacology, 804*, 94–101. https://doi.org/10.1016/j.ejphar.2017.03.019

Ma, C., Geng, B., Zhang, X., Li, R., Yang, X., & Xia, Y. (2020). Fluid shear stress suppresses osteoclast differentiation in RAW264.7 cells through extracellular signal-regulated kinase 5 (ERK5) signaling pathway. *Medical Science Monitor, 26*. https://doi.org/10.12659/msm.918370

Malone, A. M. D., Anderson, C. T., Tummala, P., Kwon, R. Y., Johnston, T. R., Stearns, T., & Jacobs, C. R. (2007). Primary cilia mediate mechanosensing in bone cells by a calcium-independent mechanism. *Proceedings of the National Academy of Sciences, 104*(33), 13325–13330. https://doi.org/10.1073/pnas.0700636104

Man, K., Brunet, M. Y., Federici, A. S., Hoey, D. A., & Cox, S. C. (2022). An ECM-mimetic hydrogel to promote the therapeutic efficacy of osteoblast-derived extracellular vesicles for bone regeneration. *Frontiers in Bioengineering and Biotechnology, 10*. https://doi.org/10.3389/fbioe.2022.829969

Man, K., Brunet, M. Y., Lees, R., Peacock, B., & Cox, S. C. (2023). Epigenetic reprogramming via synergistic hypomethylation and hypoxia enhances the therapeutic efficacy of mesenchymal stem cell extracellular vesicles for bone repair. *International Journal of Molecular Sciences, 24*(8). https://doi.org/10.3390/ijms24087564

Marie, P. J., Haÿ, E., Modrowski, D., Revollo, L., Mbalaviele, G., & Civitelli, R. (2014). Cadherin-mediated cell-cell adhesion and signaling in the skeleton. *Calcified Tissue International, 94*(1), 46–54. https://doi.org/10.1007/s00223-013-9733-7

Marszalek, J. R., Ruiz-Lozano, P., Roberts, E., Chien, K. R., & Goldstein, L. S. B. (1999). Situs inversus and embryonic ciliary morphogenesis defects in mouse mutants lacking the KIF3A subunit of kinesin-II. *Proceedings of the National Academy of Sciences of the United States of America, 96*(9), 5043–5048. https://doi.org/10.1073/pnas.96.9.5043

Mbalaviele, G., Shin, C. S., & Civitelli, R. (2006). Perspective: Cell–cell adhesion and signaling through cadherins: Connecting bone cells in their microenvironment. *Journal of Bone and Mineral Research, 21*(12), 1821–1827. https://doi.org/10.1359/jbmr.060811

McAllister, T. N., Du, T., & Frangos, J. A. (2000). Fluid shear stress stimulates prostaglandin and nitric oxide release in bone marrow-derived preosteoclast-like cells. *Biochemical and Biophysical Research Communications, 270*(2), 643–648. https://doi.org/10.1006/bbrc.2000.2467

McClung, M. R., Grauer, A., Boonen, S., Bolognese, M. A., Brown, J. P., Diez-Perez, A., Langdahl, B. L., Reginster, J. Y., Zanchetta, J. R., Wasserman, S. M., Katz, L., Maddox, J., Yang, Y. C., Libanati, C., & Bone, H. G. (2014). Romosozumab in postmenopausal women with low bone mineral density. *New England Journal of Medicine, 370*(5), 412–420. https://doi.org/10.1056/NEJMoa1305224

McGarry, J. G., Klein-Nulend, J., & Prendergast, P. J. (2005). The effect of cytoskeletal disruption on pulsatile fluid flow-induced nitric oxide and prostaglandin E2 release in osteocytes and osteoblasts. *Biochemical and Biophysical Research Communications, 330*(1), 341–348. https://doi.org/10.1016/j.bbrc.2005.02.175

McNamara, L. M., Majeska, R. J., Weinbaum, S., Friedrich, V., & Schaffler, M. B. (2009). Attachment of osteocyte cell processes to the bone matrix. *The Anatomical Record, 292*(3), 355–363. https://doi.org/10.1002/ar.20869

Mechiche Alami, S., Gangloff, S. C., Laurent-Maquin, D., Wang, Y., & Kerdjoudj, H. (2016). Concise review: In vitro formation of bone - Like nodules sheds light on the application of stem cells for bone regeneration. *Stem Cells Translational Med, 5*(11), 1587–1593. https://doi.org/10.5966/sctm.2015-0413

Monahan, G. E., Schiavi-Tritz, J., Britton, M., & Vaughan, T. J. (2023). Longitudinal alterations in bone morphometry, mechanical integrity and composition in Type-2 diabetes in a Zucker diabetic fatty (ZDF) rat. *Bone, 170*. https://doi.org/10.1016/j.bone.2023.116672

Moore, E. R., & Jacobs, C. R. (2018). The primary cilium as a signaling nexus for growth plate function and subsequent skeletal development. *Journal of Orthopaedic Research, 36*(2), 533–545. https://doi.org/10.1002/jor.23732

Morrell, A. E., Brown, G. N., Robinson, S. T., Sattler, R. L., Baik, A. D., Zhen, G., Cao, X., Bonewald, L. F., Jin, W., Kam, L. C., & Guo, X. E. (2018). Mechanically induced Ca2+ oscillations in osteocytes release extracellular vesicles and enhance bone formation. *Bone Research, 6*(1). https://doi.org/10.1038/s41413-018-0007-x

Moursi, A. M., Globus, R. K., & Damsky, C. H. (1997). Interactions between integrin receptors and fibronectin are required for calvarial osteoblast differentiation in vitro. *Journal of Cell Science, 110*(18), 2187–2196. https://doi.org/10.1242/jcs.110.18.2187

Murphy, M. G., Cerchio, K., Stoch, S. A., Gottesdiener, K., Wu, M., & Recker, R. (2005). Effect of L-000845704, an αVβ3 integrin antagonist, on markers of bone turnover and bone mineral density in postmenopausal osteoporotic women. *The Journal of Clinical Endocrinology & Metabolism, 90*(4), 2022–2028. https://doi.org/10.1210/jc.2004-2126

Naqvi, S. M., Ali, W., Allison, H., O'Sullivan, L. M., Holdsworth, G., Panadero-Perez, J. A., Schiavi-Tritz, J., & McNamara, L. M. (2025). Early intermittent low dose sclerostin antibody treatment promotes surface bone formation and reduces bone loss, but also decreases osteocyte apoptosis and mechanotransduction in ovariectomized rats. *bioRxiv*. https://doi.org/10.1101/2024.10.02.616263

Naqvi, S. M., & McNamara, L. M. (2020). Stem cell mechanobiology and the role of biomaterials in governing mechanotransduction and matrix production for tissue regeneration. *Frontiers in Bioengineering and Biotechnology, 8*. https://doi.org/10.3389/fbioe.2020.597661

Nijsure, M. P., Tobin, B., Jones, D. L., Lang, A., Hallström, G., Baitner, M., Tanner, G. I., Moharrer, Y., Panebianco, C. J., Seidl, E. G., Dyment, N. A., Szeto, G. L., Wood, L., & Boerckel, J. D. (2024). YAP regulates periosteal expansion in fracture repair. *bioRxiv*. https://doi.org/10.1101/2024.12.23.630086

Park, J. S., Kim, M., Song, N. J., Kim, J. H., Seo, D., Lee, J. H., Jung, S. M., Lee, J. Y., Lee, J., Lee, Y. S., Park, K. W., & Park, S. H. (2019). A reciprocal role of the Smad4-Taz axis in osteogenesis and adipogenesis of mesenchymal stem cells. *Stem Cells, 37*(3), 368–381. https://doi.org/10.1002/stem.2949

Qin, L., Fu, X., Ma, J., Lin, M., Zhang, P., Wang, Y., Yan, Q., Tao, C., Wen, L., Tang, B., Chen, Di, Bai, X., Cao, H., & Xiao, G. (2021). Kindlin-2 mediates mechanotransduction in bone by regulating expression of Sclerostin in osteocytes. *Communications Biology, 4*(1). https://doi.org/10.1038/s42003-021-01950-4

Qin, L., He, T., Yang, D., Wang, Y., Li, Z., Yan, Q., Zhang, P., Chen, Z., Lin, S., Gao, H., Yao, Q., Xu, Z., Tang, B., Yi, W., & Xiao, G. (2022). Osteocyte β1 integrin loss causes low bone mass and impairs bone mechanotransduction in mice. *Journal of Orthopaedic Translation, 34*, 60–72. https://doi.org/10.1016/j.jot.2022.03.008

Ramaswamy, G., Bidez, M. W., & Misch, C. E. (2015). *Bone response to mechanical loads dental implant prosthetics*. United States: Elsevier Inc. https://doi.org/10.1016/B978-0-323-07845-0.00006-3

Rammal, H., Dubus, M., Aubert, L., Reffuveille, F., Laurent-Maquin, D., Terryn, C., Schaaf, P., Alem, H., Francius, G., Quilès, F., Gangloff, S. C., Boulmedais, F., & Kerdjoudj, H. (2017). Bioinspired nanofeatured substrates: Suitable environment for bone regeneration. *ACS Applied Materials & Interfaces, 9*(14), 12791–12801. https://doi.org/10.1021/acsami.7b01665

Rammal, H., Harmouch, C., Maerten, C., Gaucher, C., Boulmedais, F., Schaaf, P., Voegel, J. C., Maquin, D. L., Menu, P., & Kerdjoudj, H. (2017). Upregulation of endothelial gene markers in Wharton's jelly mesenchymal stem cells cultured on polyelectrolyte multilayers. *Journal of Biomedical Materials Research Part A, 105*(1), 292–300. https://doi.org/10.1002/jbm.a.35868

Riquelme, M. A., Cardenas, E. R., Xu, H., & Jiang, J. X. (2020). The role of connexin channels in the response of mechanical loading and unloading of bone. *International Journal of Molecular Sciences, 21*(3). https://doi.org/10.3390/ijms21031146

Robinson, J. A., Chatterjee-Kishore, M., Yaworsky, P. J., Cullen, D. M., Zhao, W., Li, C., Kharode, Y., Sauter, L., Babij, P., Brown, E. L., Hill, A. A., Akhter, M. P., Johnson, M. L., Recker, R. R., Komm, B. S., & Bex, F. J. (2006). Wnt/β-catenin signaling is a normal physiological response to mechanical loading in bone. *Journal of Biological Chemistry, 281*(42), 31720–31728. https://doi.org/10.1016/s0021-9258(19)84086-3

Schiller, P. C., D'Ippolito, G., Balkan, W., A Roos, B., & A Howard, G. (2001). Gap-junctional communication is required for the maturation process of osteoblastic cells in culture. *Bone, 28*(4), 362–369. https://doi.org/10.1016/s8756-3282(00)00458-0

Schipani, E., Maes, C., Carmeliet, G., & Semenza, G. L. (2009). Regulation of osteogenesis-angiogenesis coupling by HIFs and VEGF. *Journal of Bone and Mineral Research, 24*(8), 1347–1353. https://doi.org/10.1359/jbmr.090602

Schurman, C. A., Verbruggen, S. W., & Alliston, T. (2021). Disrupted osteocyte connectivity and pericellular fluid flow in bone with aging and defective TGF-β signaling. *Proceedings of the National Academy of Sciences, 118*(25). https://doi.org/10.1073/pnas.2023999118

Seo, B. R., & Mooney, D. J. (2022). Recent and future strategies of mechanotherapy for tissue regenerative rehabilitation. *ACS Biomaterials Science & Engineering, 8*(11), 4639–4642. https://doi.org/10.1021/acsbiomaterials.1c01477

Shapiro, I. M., Landis, W. J., & Risbud, M. V. (2015). Matrix vesicles: Are they anchored exosomes? *Bone, 79*, 29–36. https://doi.org/10.1016/j.bone.2015.05.013

Shen, N., Maggio, M., Woods, I., Lowry, M. C., Almasri, R., Gorgun, C., Eichholz, K. F., Stavenschi, E., Hokamp, K., Roche, F. M., O'Driscoll, L., & Hoey, D. A. (2023). Mechanically activated mesenchymal-derived bone cells drive vessel formation via an extracellular vesicle mediated mechanism. *Journal of Tissue Engineering, 14*. https://doi.org/10.1177/20417314231186918

Simfia, I., Schiavi, J., & McNamara, L. M. (2020a). Alterations in osteocyte mediated osteoclastogenesis during estrogen deficiency and under ROCK-II inhibition: An in vitro study using a novel postmenopausal multicellular niche model. *Experimental Cell Research, 392*(1), Article 112005. https://doi.org/10.1016/j.yexcr.2020.112005

Simfia, I., Schiavi, J., & McNamara, L. M. (2020b). ROCK-II inhibition suppresses impaired mechanobiological responses in early estrogen deficient osteoblasts. *Experimental Cell Research, 396*(1), Article 112264. https://doi.org/10.1016/j.yexcr.2020.112264

Sims, N. A., & Martin, T. J. (2014). Coupling the activities of bone formation and resorption: A multitude of signals within the basic multicellular unit. *BoneKEy Reports, 3*. https://doi.org/10.1038/bonekey.2013.215

Song, J., Liu, L., Lv, L., Hu, S., Tariq, A., Wang, W., & Dang, X. (2020). Fluid shear stress induces Runx-2 expression via upregulation of PIEZO1 in MC3T3-E1 cells. *Cell Biology International, 44*(7), 1491–1502. https://doi.org/10.1002/cbin.11344

Stewart, S., Darwood, A., Masouros, S., Higgins, C., & Ramasamy, A. (2020). Mechanotransduction in osteogenesis. *Bone & Joint Research, 9*(1), 1–14. https://doi.org/10.1302/2046-3758.91.bjr-2019-0043.r2

Sun, W., Chi, S., Li, Y., Ling, S., Tan, Y., Xu, Y., Jiang, F., Li, J., Liu, C., Zhong, G., Cao, D., Jin, X., Zhao, D., Gao, X., Liu, Z., Xiao, B., & Li, Y. (2019). The mechanosensitive Piezo1 channel is required for bone formation. *eLife, 8*. https://doi.org/10.7554/elife.47454

Sun, W., Li, Y., Li, J., Tan, Y., Yuan, X., Meng, H., Ye, J., Zhong, G., Jin, X. Y., Liu, Z., Du, R., Xing, W., Zhao, D., Song, J., Li, Y., Pan, J., Zhao, Y., Qi, L., Wang, A., … Li, Y. (2023). Mechanical stimulation controls osteoclast function through the regulation of Ca2+-activated cl− channel anoctamin 1. *Communications Biology, 6*(1). https://doi.org/10.1038/s42003-023-04806-1

Suzuki, N., Yoshimura, Y., Deyama, Y., Suzuki, K., & Kitagawa, Y. (2008). Mechanical stress directly suppresses osteoclast differentiation in RAW264.7 cells. *International Journal of Molecular Medicine, 21*(3), 291–296. https://doi.org/10.3892/ijmm.21.3.291

Syeda, R., Xu, J., Dubin, A. E., Bertrand, C., Mathur, J., Huynh, T., Matzen, J., Lao, J., Tully, D. C., Engels, I. H., Petrassi, H. M., Schumacher, A. M., Montal, M., Bandell, M., & Patapoutian, A. (2015). Chemical activation of the mechanotransduction channel Piezo1. *eLife, 4*(MAY). https://doi.org/10.7554/elife.07369

Takyar, F. M., Tonna, S., Ho, P. W. M., Crimeen-Irwin, B., Baker, E. K., Martin, T. J., & Sims, N. A. (2013). EphrinB2/EphB4 inhibition in the osteoblast lineage modifies the anabolic response to parathyroid hormone. *Journal of Bone and Mineral Research, 28*(4), 912–925. https://doi.org/10.1002/jbmr.1820

Tatsumi, S., Ishii, K., Amizuka, N., Li, M., Kobayashi, T., Kohno, K., Ito, M., Takeshita, S., & Ikeda, K. (2007). Targeted ablation of osteocytes induces osteoporosis with defective mechanotransduction. *Cell Metabolism, 5*(6), 464–475. https://doi.org/10.1016/j.cmet.2007.05.001

Taylor, A. F., Saunders, M. M., Shingle, D. L., Cimbala, J. M., Zhou, Z., & Donahue, H. J. (2007). Mechanically stimulated osteocytes regulate osteoblastic activity via gap junctions. *American Journal of Physiology-Cell Physiology, 292*(1), Article C545. https://doi.org/10.1152/ajpcell.00611.2005

Tonna, S., Takyar, F. M., Vrahnas, C., Crimeen-Irwin, B., Ho, P. W. M., Poulton, I. J., Brennan, H. J., McGregor, N. E., Allan, E. H., Nguyen, H., Forwood, M. R., Tatarczuch, L., Mackie, E. J., John Martin, T., & Sims, N. A. (2014). EphrinB2 signaling in osteoblasts promotes bone mineralization by preventing apoptosis. *The FASEB Journal, 28*(10), 4482–4496. https://doi.org/10.1096/fj.14-254300

Uda, Y., Azab, E., Sun, N., Shi, C., & Pajevic, P. D. (2017). Osteocyte mechanobiology. *Current Osteoporosis Reports, 15*(4), 318–325. https://doi.org/10.1007/s11914-017-0373-0

Vaughan, T. J., Mullen, C. A., Verbruggen, S. W., & McNamara, L. M. (2015). Bone cell mechanosensation of fluid flow stimulation: A fluid–structure interaction model characterising the role integrin attachments and primary cilia. *Biomechanics and Modeling in Mechanobiology, 14*(4), 703–718. https://doi.org/10.1007/s10237-014-0631-3

Verbruggen, S. W., Kainz, B., Shelmerdine, S. C., Arthurs, O. J., Hajnal, J. V., Rutherford, M. A., Phillips, A. T. M., Nowlan, N. C., European Society of Biomechanics, S. M., & Perren, A. (2018). Altered biomechanical stimulation of the developing hip joint in presence of hip dysplasia risk factors. *Journal of Biomechanics, 78*(2018), 1–9. https://doi.org/10.1016/j.jbiomech.2018.07.016

Verbruggen, S. W., & McNamara, L. M. (2018). *Bone mechanobiology in health and disease mechanobiology in health and disease*. United States: Elsevier. https://doi.org/10.1016/B978-0-12-812952-4.00006-4

Verbruggen, S. W., Sittichokechaiwut, A., & Reilly, G. C. (2023). Osteocytes and primary cilia. *Current Osteoporosis Reports, 21*(6), 719–730. https://doi.org/10.1007/s11914-023-00819-1

Verbruggen, S. W., Vaughan, T. J., & McNamara, L. M. (2012). Strain amplification in bone mechanobiology: A computational investigation of the in vivo mechanics of osteocytes. *Journal of the Royal Society Interface, 9*(75), 2735–2744. https://doi.org/10.1098/rsif.2012.0286

Vrahnas, C., Blank, M., Dite, T. A., Tatarczuch, L., Ansari, N., Crimeen-Irwin, B., Nguyen, H., Forwood, M. R., Hu, Y., Ikegame, M., Bambery, K. R., Petibois, C., Mackie, E. J., Tobin, M. J., Smyth, G. K., Oakhill, J. S., John Martin, T., & Sims, N. A. (2019). Increased autophagy in EphrinB2-deficient osteocytes is associated with elevated secondary mineralization and brittle bone. *Nature Communications, 10*(1). https://doi.org/10.1038/s41467-019-11373-9

Wang, H. J., Wang, Y., Mirjavadi, S. S., Andersen, T., Moldovan, L., Vatankhah, P., Russell, B., Jin, J., Zhou, Z., Li, Q., Cox, C. D., Su, Q. P., & Ju, L. A. (2024). Microscale geometrical modulation of PIEZO1 mediated mechanosensing through cytoskeletal redistribution. *Nature Communications, 15*(1). https://doi.org/10.1038/s41467-024-49833-6

Wang, Y., Yan, Q., Zhao, Y., Liu, X., Lin, S., Zhang, P., Ma, L., Lai, Y., Bai, X., Liu, C., Wu, C., Feng, J. Q., Chen, Di, Cao, H., & Xiao, G. (2019). Focal adhesion proteins Pinch1 and Pinch2 regulate bone homeostasis in mice. *JCI Insight, 4*(22). https://doi.org/10.1172/jci.insight.131692

Wang, L., You, X., Lotinun, S., Zhang, L., Wu, N., & Zou, W. (2020). Mechanical sensing protein PIEZO1 regulates bone homeostasis via osteoblast-osteoclast crosstalk. *Nature Communications, 11*(1). https://doi.org/10.1038/s41467-019-14146-6

Wang, L., You, X., Zhang, L., Zhang, C., & Zou, W. (2022). Mechanical regulation of bone remodeling. *Bone Research, 10*(1). https://doi.org/10.1038/s41413-022-00190-4

Wang, H., Yu, H., Huang, T., Wang, B., & Xiang, L. (2023). Hippo-YAP/TAZ signaling in osteogenesis and macrophage polarization: Therapeutic implications in bone defect repair. *Genes & Diseases, 10*(6), 2528–2539. https://doi.org/10.1016/j.gendis.2022.12.012

Welsh, J. A., Goberdhan, D. C. I., O'Driscoll, L., Buzas, E. I., Blenkiron, C., Bussolati, B., Cai, H., Di Vizio, D., Driedonks, T. A. P., Erdbrügger, U., Falcon-Perez, J. M., Fu, Q. L., Hill, A. F., Lenassi, M., Lim, S. K., Mahoney, M.ỹ. G., Mohanty, S., Möller, A., Nieuwland, R., … Zubair, H. (2024). Minimal information for studies of extracellular vesicles (MISEV2023): From basic to advanced approaches. *Journal of Extracellular Vesicles, 13*(2). https://doi.org/10.1002/jev2.12404

Wijerathne, T. D., Ozkan, A. D., & Lacroix, J. J. (2022). Yoda1's energetic footprint on Piezo1 channels and its modulation by voltage and temperature. *Proceedings of the National Academy of Sciences, 119*(29). https://doi.org/10.1073/pnas.2202269119

Wu, Y., Jing, Z., Deng, D., Yan, J., Liu, M., Li, L., Zuo, Y., Wu, W., Hu, Q., & Xie, Y. (2023). Dkk-1–TNF-α crosstalk regulates MC3T3E1 pre-osteoblast proliferation and differentiation under mechanical stress through the ERK signaling pathway. *Molecular and Cellular Biochemistry, 478*(10), 2191–2206. https://doi.org/10.1007/s11010-022-04645-4

Xiao, Z., Zhang, S., Mahlios, J., Zhou, G., Magenheimer, B. S., Guo, D., Dallas, S. L., Maser, R., Calvet, J. P., Bonewald, L., & Quarles, L. D. (2006). Cilia-like structures and Polycystin-1 in osteoblasts/osteocytes and associated abnormalities in skeletogenesis and Runx2 expression. *Journal of Biological Chemistry, 281*(41), 30884–30895. https://doi.org/10.1074/jbc.m604772200

Xiong, J., Almeida, M., & O'Brien, C. A. (2018). The YAP/TAZ transcriptional co-activators have opposing effects at different stages of osteoblast differentiation. *Bone, 112*, 1–9. https://doi.org/10.1016/j.bone.2018.04.001

Xu, J. (2023). Mechanotransduction in osteoclasts: Novel strategies of bone repairs. *Mechanobiology in Medicine, 1*(1), Article 100008. https://doi.org/10.1016/j.mbm.2023.100008

Xue, X., Hong, X., Li, Z., Deng, C. X., & Fu, J. (2017). Acoustic tweezing cytometry enhances osteogenesis of human mesenchymal stem cells through cytoskeletal contractility and YAP activation. *Biomaterials, 134*, 22–30. https://doi.org/10.1016/j.biomaterials.2017.04.039

Yang, J., Yuan, K., Zhang, T., Zhou, S., Li, W., Chen, Z., & Wang, Y. (2023). Accelerated bone reconstruction by the Yoda1 bilayer membrane via promotion of osteointegration and angiogenesis. *Advanced Healthcare Materials, 12*(18). https://doi.org/10.1002/adhm.202203105

Yellowley, C. E., Li, Z., Zhou, Z., Jacobs, C. R., & Donahue, H. J. (2000). Functional gap junctions between osteocytic and osteoblastic cells. *Journal of Bone and Mineral Research, 15*(2), 209–217. https://doi.org/10.1359/jbmr.2000.15.2.209

You, J., Reilly, G. C., Zhen, X., Yellowley, C. E., Chen, Q., Donahue, H. J., & Jacobs, C. R. (2001). Osteopontin gene regulation by oscillatory fluid flow via intracellular calcium mobilization and activation of mitogen-activated protein kinase in MC3T3–E1 osteoblasts. *Journal of Biological Chemistry, 276*(16), 13365–13371. https://doi.org/10.1074/jbc.m009846200

You, L., Temiyasathit, S., Lee, P., Kim, C. H., Tummala, P., Yao, W., Kingery, W., Malone, A. M., Kwon, R. Y., & Jacobs, C. R. (2008). Osteocytes as mechanosensors in the inhibition of bone resorption due to mechanical loading. *Bone, 42*(1), 172–179. https://doi.org/10.1016/j.bone.2007.09.047

Young, S. R. L., Gerard-O'Riley, R., Kim, J. B., & Pavalko, F. M. (2009). Focal adhesion kinase is important for fluid shear stress-induced mechanotransduction in osteoblasts. *Journal of Bone and Mineral Research, 24*(3), 411–424. https://doi.org/10.1359/jbmr.081102

Yourek, G., Hussain, M. A., & Mao, J. J. (2007). Cytoskeletal changes of mesenchymal stem cells during differentiation. *ASAIO Journal, 53*(2), 219–228. https://doi.org/10.1097/MAT.0b013e31802deb2d

Zhao, S., Kato, Y., Zhang, Y., Harris, S., Ahuja, S. S., & Bonewald, L. F. (2002). MLO-Y4 osteocyte-like cells support osteoclast formation and activation. *Journal of Bone and Mineral Research, 17*(11), 2068–2079. https://doi.org/10.1359/jbmr.2002.17.11.2068

Zhao, Q., Zhou, H., Li, X., & Xiao, B. (2019). The mechanosensitive Piezo1 channel: A three-bladed propeller-like structure and a lever-like mechanogating mechanism. *FEBS Journal, 286*(13), 2461–2470. https://doi.org/10.1111/febs.14711

Zhou, J., Zhu, Y., Ai, D., Zhou, M., Han, L., Fu, Y., & Song, J. (2023). Low-intensity pulsed ultrasound regulates osteoblast-osteoclast crosstalk via EphrinB2/EphB4 signaling for orthodontic alveolar bone remodeling. *Frontiers in Bioengineering and Biotechnology, 11*. https://doi.org/10.3389/fbioe.2023.1192720

CHAPTER

Bone quantity and quality

5

Quinton Wright[1] and Eve Donnelly[1,2]

[1]*Cornell University, Department of Materials Science and Engineering, Ithaca, NY, United States;* [2]*Hospital for Special Surgery, Skeletal Health and Orthopedic Research Program, New York, NY, United States*

1. Introduction

Bone plays a crucial role in structural support and protection in the human body. However, bone fractures present a significant healthcare burden (Keen, 2003) and are associated with high rates of mortality in older people (Haentjens et al., 2010; Sattui & Saag, 2014). Bone's resistance to mechanical failure depends on bone quantity and quality. Bone strength refers to the ability of a bone to withstand a single overload. Other types of failure are possible, including failure from a single overload in the presence of a flaw (fracture) or from cyclic loading (fatigue). In this chapter, bone quantity and quality will be discussed in relation to bone strength unless another failure mode is explicitly stated. Although clinical assessment of bone fragility is focused on bone quantity assessed by bone mineral density (BMD), this parameter only partially explains fracture risk (Kanis et al., 2009); the remainder reflects changes in bone quality. Bone quality refers to characteristics of bone other than BMD that influence mechanical performance, such as morphology, architecture, and material properties. This chapter will explore techniques for measuring bone quantity and quality and explain how these properties contribute to the structural integrity of whole bones.

Bone quantity refers to the bone mass, as assessed by BMD. Bone mineral density can be evaluated in vivo using clinical imaging techniques such as dual-energy X-ray absorptiometry (DXA), the radiological gold standard for clinical assessment of areal BMD (aBMD), and computed tomography (CT), which enables assessment of volumetric BMD (vBMD). Bone mineral density assessed using DXA and CT is strongly associated with bone strength and fracture risk (Adams et al., 2018; Bouxsein et al., 2019; Chalhoub et al., 2016). The utility of Magnetic resonance imaging (MRI) for opportunistic assessment of vBMD is currently being evaluated (Sollmann et al., 2021). In the lab, BMD can be assessed in tissue samples with ashing. Although bone quantity predicts fracture risk, it is not the only determinant of bone strength.

Bone quality refers to bone characteristics other than BMD that contribute to bone strength (Table 5.1). These factors include bone geometry (Bouxsein &

Biomechanics and Mechanobiology of Bone. https://doi.org/10.1016/B978-0-443-26739-0.00025-9

Karasik, 2006), microarchitecture (Chappard et al., 2008; Cooper et al., 2016), microdamage (Seref-Ferlengez et al., 2015), and tissue composition (Currey, 1984; Garnero, 2012, 2015; Morgan et al., 2015; Surowiec et al., 2022). The *Bone Quality* section of this chapter (Section 3) will explore bone quality features and the association between bone quality measurements and bone strength in more detail (Adams et al., 2018).

2. Bone quantity

Bone quantity is assessed clinically as BMD. Here we focus on noninvasive methods, particularly those in use in the clinic, and briefly discuss direct measurement from a tissue explant with ashing (Tammisalo & Kiminki, 1969). BMD is typically evaluated with radiological techniques that quantify the X-ray attenuation of bone mineral and is usually reported as aBMD or vBMD. Both aBMD and vBMD are associated with greater whole-bone stiffness and strength (Ebbesen et al., 1999; Perilli et al., 2012) and reduced clinical fracture risk (Adams et al., 2018; Bouxsein et al., 2019; Chalhoub et al., 2016). In the following sections, we discuss methods, outcomes, advantages and disadvantages, and recent advances or novel analyses on the horizon for key techniques.

2.1 Noninvasive methods for in vivo assessment of BMD in human subjects

Dual-energy X-ray absorptiometry is the gold-standard technique for clinical assessment of bone mineral density (reviewed in Blake and Fogelman (1997); Jain and Vokes (2017); Roux and Briot (2017)). This technique measures the transmission of X-rays at two energies: a higher energy X-ray mostly attenuated by bone and a lower energy X-ray attenuated by bone and soft tissue. The difference in attenuation of the two X-rays is used to assess the density of the skeletal tissue based on the greater attenuation by denser bones. Measurements of aBMD are typically reported as a T-score, which compares an individual's aBMD to a reference value for a young adult of the same sex. Standard anatomic sites include the proximal femur, lumbar spine, and distal radius (typical resolution ~1 mm^2).

Measuring BMD with DXA has many advantages: low radiation dose suitable for screening [9 μSv at the hip, equivalent to 1 day background radiation (Damilakis et al., 2010)], widespread availability, and ability to predict fracture risk (Bouxsein et al., 2019; Cefalu, 2004; Marshall et al., 1996). However, DXA also has disadvantages (reviewed in Bolotin (2007); Pors Nielsen (2000)), which include a lack of volumetric assessment of BMD, inability to distinguish the cortical and trabecular compartments, and limited information about local variations in BMD, which is associated with the incidence of fractures (Kaiser et al., 2020). Additionally, DXA may overestimate BMD compared to quantitative computed tomography (Kim et al., 2021; Milisic et al., 2020). Recent advancements in DXA include the

Table 5.1 Selected bone quality parameters, measurement techniques, and associated outcomes.

Bone quality feature	Common measurement techniques	Resolution (if applicable)	Outcomes	Direction of association with bone mechanical properties	References
Geometry and microarchitecture	CT	50 nm - 3000 μm	Bone size	↑ Stiffness, strength	Bouxsein and Karasik (2006)
			Shape parameters (femoral neck-shaft angle, etc.)	Varies by anatomic site	
			Trabecular microarchitecture: BV/TV, connectivity, and plate-like morphology	↑ Stiffness, strength	Bregoli et al. (2024); Fields and Keaveny (2012); Zhou et al. (2014)
			Cortical porosity	↓ Stiffness, strength	Schaffler and Burr (1988); Vilayphiou et al. (2016)
Microdamage	Optical microscopy with fluorochrome staining, microCT with heavy element staining, SEM	nm-μm scale	Microcrack density, damaged volume	↓ Stiffness, strength, toughness	Carter and Hayes (1977); Lambers et al. (2013); Seref-Ferlengez et al. (2015)
Mineral properties	Ashing		Bone mineral content	↑Stiffness, strength ↓ductility	Currey (1969); Currey et al. (1996)
	FTIR and Raman spectroscopy	1–10 μm	Mineral:matrix	↑Stiffness, strength	Donnelly et al. (2010)
			Crystallinity	–	
			Carbonate:phosphate	–	
			Acid:phosphate	–	

Continued

Table 5.1 Selected bone quality parameters, measurement techniques, and associated outcomes.—*cont'd*

Bone quality feature	Common measurement techniques	Resolution (if applicable)	Outcomes	Direction of association with bone mechanical properties	References
	qBEI	1–10 μm	BMDD	–	
	XRD		Bone crystal structure	–	
	SAXS/WAXS		Bone mineral alignment, orientation	↑ Stiffness and strength (in direction of microstructural orientation)	Gustafsson et al. (2018)
Collagen organization	PLM, SHG, and SAXS/WAXS	μm-scale (PLM and SHG)	Alignment, orientation	↑ Stiffness and strength (in direction of alignment)	Garnero (2015)
Collagen crosslinking	Biochemical assays, bioanalytical techniques, IR spectroscopy	1–10 μm (IR spectroscopy)	Mature:immature crosslinks	–	Garnero (2012); McNerny et al. (2015)
			Non-enzymatic crosslinks	↓Toughness	Poundarik et al. (2015)
Non-collagenous proteins	Immunochemical assays, bioanalytical techniques		Concentration of individual NCPs	Varies by NCP	Morgan et al. (2015)
Water	Dehydration and UTE-MRI	200–5000 μm	Bound water	↑Strength and toughness	(Burr, 2019; Manhard et al., 2016)
			Free water	↓ Stiffness, strength	Manhard et al. (2016); Nyman et al. (2008)

The symbol (↑) denotes direct association. The symbol (↓) denotes inverse association. The symbol (–) denotes a relationship that is neutral or unclear.

trabecular bone score (TBS) and three-dimensional (3D) shaper. The DXA TBS (Silva et al., 2014) is a texture analysis technique that estimates the heterogeneity of trabecular microarchitecture and predicts fragility fractures independent of BMD (Hans et al., 2017; Shevroja et al., 2021). The 3D-Shaper is a software tool that estimates a model of the 3D shape and density distribution of the proximal femur from two-dimensional (2D) DXA scans based on CT reference data. The 3D-Shaper provides geometry reconstructions with average point-to-point surface distances of 1 mm relative to CT scans, but the reconstruction of the 3D density distribution is limited (Dudle et al., 2023). Despite its widespread use, predictive value, and recent advancements, the limitations of DXA highlight the need for complementary BMD assessment techniques.

Computed tomography is used for clinical assessment of vBMD (reviewed in Adams (2013); Schreiber et al. (2014)). Clinical CT scanners utilize a revolving X-ray source to construct a 3D image of bone (Fig. 5.1a). Because X-rays are attenuated proportionally to bone density, CT scans produce a spatially resolved (isotropic voxel dimension: 250–3000 μm) measure of vBMD. In QCT, X-ray attenuation values measured in Hounsfield Units (HU) are compared to a calibration

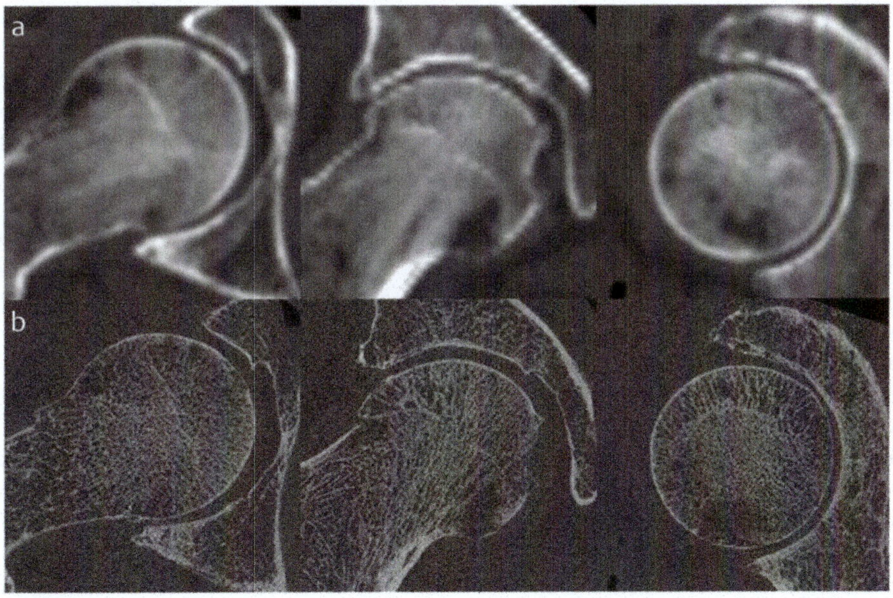

FIGURE 5.1

Clinical CT and HRpQCT images of cadaveric hip joint. Comparison of (a) clinical CT and (b) HRpQCT image of cadaveric hip joint in axial, coronal, and sagittal planes (left to right).

- From Turmezei, T. D., Treece, G. M., Gee, A. H., Houlden, R. & Poole, K. E. S. (2018). A new quantitative 3D approach to imaging of structural joint disease. Scientific Reports, 8(1). https://doi.org/10.1038/s41598-018-27486-. This figure is reprinted under Creative Commons license CC BY 4.0 (creativecommons.org/licenses/by/4.0/).

standard (phantom) to quantify vBMD (g/cm^3). Computed tomography scans taken to measure bone density are usually performed at the proximal femur and lumbar spine.

The main advantages of CT include the ability to capture 3D bone geometry and evaluate vBMD separately in cortical and cancellous bone. Clinical QCT is generally comparable to DXA for fracture risk assessment (Adams et al., 2018; Chalhoub et al., 2016; Löffler et al., 2019). The limitations of CT relative to DXA include longer scan times, greater cost, and a larger radiation dose [~4 mSv at the hip, equivalent to 1 year of background radiation (Damilakis et al., 2010; Huppertz et al., 2011)]. A recent advancement is the analysis of phantomless CT images for opportunistic osteoporosis screening in patient populations who do not routinely receive DXA scans (Deshpande et al., 2023). Other advancements in CT include high-resolution modalities, such as high-resolution peripheral quantitative computed tomography (HRpQCT; see Section 3.1.1) and micro-computed tomography (micro-CT; see Section 3.1.2). Compared to DXA, computed tomography offers 3D assessment of vBMD and geometry, making it a useful complementary tool for assessment of bone health.

Magnetic resonance imaging is increasingly evaluated for clinical assessment of BMD (reviewed in Majumdar and Genant (1995) and Sollmann et al. (2021)). In MRI, a magnetic field is applied to tissues, frequency pulses are applied, and relaxation processes of certain species within the tissue (usually hydrogen atoms) are measured. This modality provides good contrast between tissue types. Because the MR signal from bone is very low (Sollmann et al., 2021), most MRI-based BMD assessment techniques estimate trabecular bone volume from the bone marrow signal.

A key advantage of MRI-based measurement of BMD is that it does not rely on ionizing radiation and enables opportunistic assessment of bone fragility with a widely used clinical tool. However, measuring BMD from MRI presents several challenges: these techniques do not directly measure bone density and have only moderate diagnostic power for distinguishing between normal and osteoporotic BMD (Chen et al., 2023). Recent advancements in this field include the development of the vertebral bone quality score and microarchitecture imaging with ultrashort echo time magnetic resonance imaging (UTE-MRI). In clinical T1 MRI images (typical resolution approximately 1 mm^2) of the lumbar spine, the vertebral bone quality score is associated with DXA T-score (Bandirali et al., 2015) and fragility fracture risk (Ehresman et al., 2021; Huang et al., 2023). High-resolution UTE-MRI (typical voxel dimension 200–5000 μm) has achieved in-plane resolutions of 234 μm (slice thickness = 500 μm) in vivo at the hip (Kazakia et al., 2018). Comparisons of UTE-MRI and micro-CT have shown that UTE-MRI allows for visualization of some microarchitectural features in cadaveric specimens (Fig. 5.2). Although MRI cannot directly measure BMD, MRI-based opportunistic assessment of bone density shows promise for the early identification of patients who should be screened for osteoporosis.

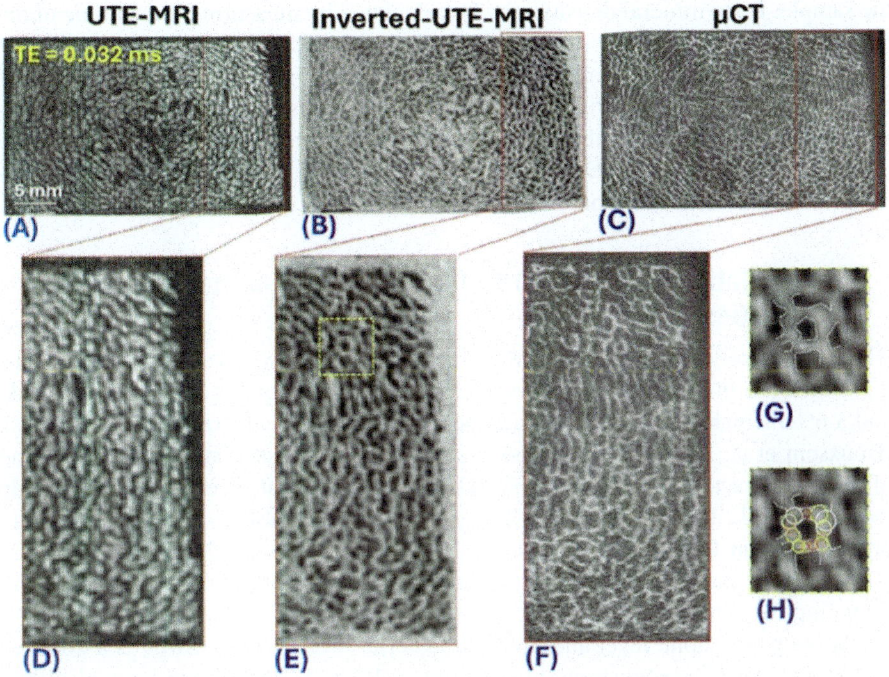

FIGURE 5.2

Distal tibia imaged with inverted ultrashort echo time MRI at the fat peak frequency versus micro-CT image at the same site (A) an excised cube from the distal tibia metaphysis imaged in the axial plane using 3D UTE sequence ($0.2 \times 0.2 \times 0.4$ mm^3 voxel size). (B) Signal inversion of 3D UTE image. (C) Isotropic μCT imaging ($0.05 \times 0.05 \times 0.05$ mm^3 voxel size) of trabeculae. (D–F) Corresponding zoomed regions for images (A–C). (G) Schematics of the trabecular bone boundaries. (H) The largest covering circles used to calculate the local bone thickness. The thin regions of bone are indicated with small covering circles in red, while the thick regions are indicated with larger covering circles in yellow and white.

- From Jerban, S., Moazamian, D., Mohammadi, H.S., Ma, Y., Jang, H., Namiranian, B., Shin, S.H., Alenezi, S., Shah, S.B., Chung, C.B., Chang, E.Y. & Du, J. (2024). More accurate trabecular bone imaging using UTE MRI at the resonance frequency of fat. Bone, 184. https://doi.org/10.1016/j.bone.2024.11709. Reprinted under Elsevier license 5998980249790.

2.2 Research methods for measuring BMD in tissue explants

Ashing is a laboratory technique for measuring vBMD. In ashing, bone is defatted in solvents, then heated to combust the organic components until only inorganic bone mineral remains. Heating 25–250°C removes water; heating 300–600°C combusts collagen and other organic components; further heating 550–900°C liberates some of the CO_3^{2-} in apatite as CO_2 gas (Greiner et al., 2019; Thotakura et al., 2022). The remaining "ash" is weighed to quantify the amount of bone mineral in the sample. Bone mineral density may be calculated by measuring the volume of the bone before ashing and dividing the ash weight by this volume. The advantage of ashing is that it is a direct measure of bone mineral content. The disadvantages of ashing are that it is a destructive technique, and errors may arise due to the loss of reactive compounds in the final product (Thotakura et al., 2022). Ashing is a direct measurement of BMD and has been used to validate radiological techniques for BMD assessment (Burghardt et al., 2008; Nagy et al., 2001; Schileo et al., 2008).

2.3 The relationship of BMD to fracture risk and the mechanical properties of bone

BMD is inversely associated with fracture risk. According to a recent meta-analysis, a 2% increase in hip aBMD is associated with a 16% decrease in hip fracture risk, and a 6% increase in hip aBMD is associated with a 40% decrease in fracture risk (Bouxsein et al., 2019). A 1 SD reduction in aBMD is associated with a doubling of vertebral fracture risk at the spine (Cefalu, 2004; Marshall et al., 1996) and hip (Marshall et al., 1996). Clinical QCT is generally comparable to or better at fracture prediction than DXA (Adams et al., 2018; Chalhoub et al., 2016; Löffler et al., 2019). These findings underscore the importance of clinical bone mineral density screening for predicting fracture risk.

At the whole-bone level and the bulk material level, greater BMD is associated with greater stiffness and strength (Ammann & Rizzoli, 2003; Ebbesen et al., 1999; Nobakhti & Shefelbine, 2018; Perilli et al., 2012). Bone is a composite material composed of type I collagen fibers reinforced with bone mineral crystals. The mineral component of bone provides stiffness and strength, whereas the fibrillar protein component provides ductility and post-yield toughness. Bone mineral density explains approximately 70% of the variation in human bone strength (Ammann & Rizzoli, 2003) and modulus (Nobakhti & Shefelbine, 2018) measured ex vivo. A power law relationship describes the connection between modulus and mineral density (modulus $\propto$ density$^{\alpha}$) in Nobakhti and Shefelbine (2018). Bone mineral density is a strong predictor of bulk mechanical properties and fracture risk.

2.4 Going beyond BMD

Additional factors beyond BMD affect bone strength. Age contributes to hip fracture risk independently of BMD; an 80-year-old woman with a T-score of −2 has double the fracture risk of a 60-year-old woman with the same BMD Fig. 5.3. These observations suggest that other factors beyond BMD contribute to age-related fragility. Bone quality refers to the skeletal factors other than BMD that contribute to bone strength. A large body of literature has identified some of these factors that affect bone fragility independently of bone density, including bone geometry (Bouxsein & Karasik, 2006), trabecular microarchitecture (Chappard et al., 2008), cortical porosity (Cooper et al., 2016), microdamage accumulation (Seref-Ferlengez et al., 2015), mineralization (Currey et al., 1996), collagen organization (Garnero,

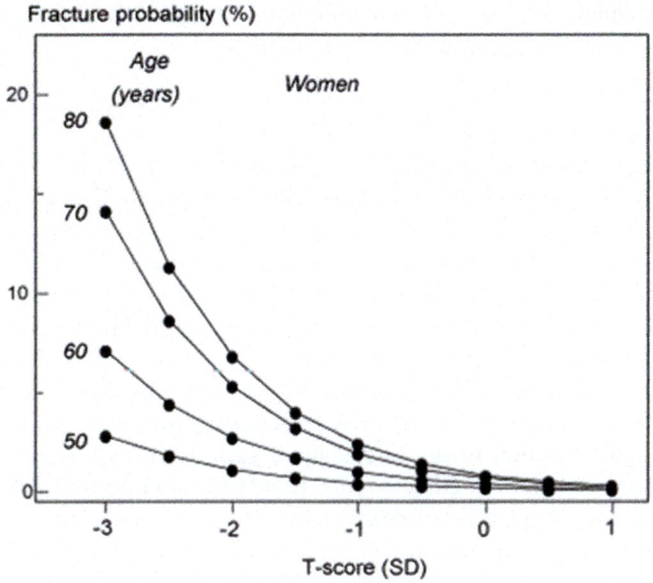

FIGURE 5.3

DXA T-score at the femoral neck versus hip fracture risk stratified by age. Older individuals are more likely to experience fractures than younger individuals with the same bone density.

- From Kanis, J.A., Johansson, H., Oden, A. & McCloskey, E.V. (2009). Assessment of fracture risk. European Journal of Radiology, 71(3), 392–397. https://doi.org/10.1016/j.ejrad.2008.04.06. Reprinted under Elsevier license 5998980249790.

2015), collagen post-translational modifications (Garnero, 2012), and hydration (Surowiec et al., 2022). Each of these will be explored further in Section 3 of this chapter.

3. Bone quality

Bone quality refers to bone characteristics other than BMD that affect bone strength. Bone quality encompasses the spatial distribution of bone material (macroscopic geometry, microarchitecture) and the properties of the bone material (tissue-level mechanical properties, composition, and microdamage). Another important factor that contributes to bone quality, bone remodeling, is reviewed in Bolamperti et al. (2022); Wang et al. (2022) and in this book in Chapters 1 and 3. In the last few decades, research interest in bone quality has grown tremendously; a PubMed search of "bone quality" yields 414 results in 1994, 1045 in 2004, 3045 in 2014, and 5606 in 2024. While some aspects of bone quality can be measured in the clinic, most are limited to the laboratory in practice. The growing research interest in bone quality highlights its importance in understanding bone strength and the need for greater clinical translation of current measurement tools.

3.1 Bone geometry and architecture

3.1.1 Measurement of geometry and architecture in human subjects

The strength of whole bones depends on both their macroscopic geometry and microscopic architecture, which includes trabecular organization and cortical porosity. Several imaging techniques can assess geometry or microarchitecture (see Sections 2.1 and 3.1.2). Some aspects of macroscopic geometry can be estimated from two-dimensional imaging techniques such as radiographs and DXA images. However, bone geometry and microarchitecture can only be characterized in detail with a 3D imaging modality, such as CT or MRI. Measuring small-scale features of bone structure, such as trabecular microarchitecture and cortical porosity, requires a high-resolution imaging technique, such as HRpQCT. Bone geometry can be captured with current clinical tools. Bone microarchitecture can be assessed noninvasively with HRpQCT, but this technique has not been adopted in the clinic. Micro-computed tomography and nano-computed tomography (nano-CT) provide detailed measurements of bone architecture but are currently limited to the laboratory due to high radiation dose and small maximum sample size restrictions in many scanners.

High-resolution peripheral quantitative computed tomography is a CT method used at peripheral sites such as the distal tibia and distal radius (reviewed in Mikolajewicz et al. (2020); Nishiyama and Shane (2013); van den Bergh et al. (2021)). High-resolution peripheral quantitative computed tomography scans achieve higher resolutions than clinical CT [isotropic voxel dimension: HRpQCT 60–90 μm; clinical CT 250–3000 μm (Fig. 5.1)], but not as high as micro-CT (1–100 μm) or nano-CT (300–700 nm; Table 5.2). Outcomes of HRpQCT include structural parameters such as bone volume/total volume, trabecular thickness, trabecular number, and cortical thickness. Microarchitecture outcomes correlate well between HRpQCT and micro-CT scans of the same sample, although the degree to which HRpQCT over/underestimates these parameters varies by study (Ohs et al., 2020). In addition, stiffness and strength of the radius and tibia can be estimated from HRpQCT-based finite element models, and these values correlate with stiffness and strength values measured in uniaxial compression testing of the same tissue samples (Schenk et al., 2022; Varga et al., 2011).

The advantages of HRpQCT are higher resolution lower radiation dose [3 μSv at the tibia, equivalent to half a day of background radiation (Burrows et al., 2010)] compared to DXA and clinical CT. The main limitation of HRpQCT is its restriction to peripheral sites due to size restrictions of scanning area, whereas most osteoporotic fractures occur at central sites. Other limitations include limited scan volume and lack of clinical availability. Novel HRpQCT analysis techniques include time-lapse HRpQCT, the application of 3D image registration to sequential HRpQCT scans taken over time, which allows for "virtual biopsies" (Nickolas et al., 2010)

Table 5.2 Comparison of CT techniques for vBMD assessment.

Technique	Typical isotropic voxel dimension	Application	Resolved microarchitectural features
CT	250–3000 μm	Human subjects	
HRpQCT	60–90 μm	Human subjects	Large trabeculae, cortical porosity
Micro-CT	1–100 μm	Rodents, tissue samples	Trabeculae, cortical porosity, osteocyte lacunae[a]
Nano-CT	300–700 nm	Tissue samples	Trabeculae, cortical porosity, osteocyte lacunae
Synchrotron nano-CT	<100 nm	Tissue samples	Trabeculae, cortical porosity, osteocyte lacunae, osteocyte cannaliculi

[a] *Only at the lowest voxel dimensions (~1 μm).*

and in vivo assessment of local bone turnover (Zhou et al., 2025). Although HRpQCT is currently limited to peripheral sites, it is a promising, noninvasive tool that provides insights into bone microarchitecture.

3.1.2 Measurement of geometry and architecture in small animals and tissue explants

Micro-CT and nano-CT are research QCT methods that image small objects (reviewed in Chappard et al. (2008); Peyrin et al. (2014); Ritman (2011)). Micro-CT is the gold standard for assessment of bone microarchitecture (Ohs et al., 2020) and is typically used to analyze bone explants ex vivo (e.g., human bone biopsies, rodent whole bones), or rodent skeletons in vivo, the former allowing for higher resolutions (1–100 μm) than the latter (3–100 μm). Outcomes include microarchitecture parameters: trabecular thickness, density, interconnectivity, and patterning, and cortical thickness, area, and porosity (Bouxsein et al., 2010). Nano computed tomography is based on similar principles as micro-CT and can achieve higher resolutions (300–700 nm isotropic voxel size). The resolution of nano-CT is an order of magnitude smaller than most cells, which enables the visualization of osteocyte lacunae. Even greater resolutions are possible in synchrotron nano-CT, which has been used to visualize the lacunar-canalicular network with 30–60 nm isotropic voxels (Hesse et al., 2015; Yu et al., 2020). The main advantage of micro-CT and nano-CT is high resolution. The limitations of these techniques are the high radiation doses required to achieve the highest resolutions, which may be a confounding factor for vivo rodent experiments, and the sample size restrictions of many scanners. Micro-CT and nano-CT offer the highest resolution for analyzing bone microarchitecture; however, their high radiation dose and compact scan volumes limit their clinical translatability.

3.1.3 The effects of geometry on the mechanical properties of bone

Bigger bones are typically stronger than smaller bones (reviewed in Bouxsein and Karasik (2006)). The effects of geometry on bone strength can be examined in a simplified geometry, such as the diaphysis of a long bone, which can be approximated with a hollow cylinder. If the mass and properties of bone material are held constant, then a bone with a larger diameter is more mechanically robust than one with a smaller diameter. A larger diameter dramatically increases the resistance to bending and torsion because distributing mass away from the center of the bone increases the area moment of inertia I, which characterizes the geometric resistance to bending ($I \propto \text{diameter}^4$) and the polar moment of inertia J, which characterizes the geometric resistance to torsion [$J \propto \text{diameter}^4$ (Turner & Burr, 1993)]. Bone size also plays a role in compressive strength. Increasing the cross-sectional area of the bone reduces the compressive stress on the bone (stress = force/cross-sectional area). These relationships are exemplified by findings of studies analyzing vertebral cross-sectional area with QCT and DXA, which showed that smaller vertebral cross-sections are associated with greater fracture risk (Bouxsein & Karasik, 2006). Larger bones are generally more resistant to compression, bending, and torsion.

Despite the mechanical benefits of large cross-sections, smaller bones are not inherently weaker than larger bones of the same mass. Bone strength can be achieved in a variety of phenotypes (reviewed in van der Meulen et al. (2001)). Across multiple species (rat, human) and ages (juvenile, adult), long bones with large, robust diaphyses had proportionally thinner cortices to minimize mass, and those with more slender diaphyses had proportionally thicker cortices to maximize stiffness (Jepsen et al., 2010; Pandey et al., 2009). These findings highlight the adaptability of bone structure and demonstrate that structural integrity can be maintained through diverse phenotypic strategies.

3.1.4 The effects of microarchitecture on the mechanical properties of bone

In cancellous bone, a robust trabecular network is necessary to maintain whole-bone structural integrity (reviewed in Chappard et al. (2008) and Fields and Keaveny (2012)). Cancellous bone comprises interconnected rods and plates that efficiently transfer loads to the dense cortical shell. Aspects of cancellous microarchitecture that contribute to bone stiffness and strength include volume fraction, plate-vs-rod morphology, and connectivity. The volume fraction of cancellous bone, assessed as bone volume to total volume (BV/TV), explains 60%–80% of the variation in cancellous bone strength (Fields & Keaveny, 2012). Trabecular plates and rods independently contribute to stiffness and strength (Liu et al., 2009; Zhou et al., 2014). In addition, as bone is lost during osteoporosis, stiffness disproportionately decreases as the number of trabeculae and connectivity of the trabecular network decreases (Bregoli et al., 2024). This is particularly true for rod-like trabeculae, which have the highest surface-to-volume ratio and consequently have the highest rate of remodeling. A more detailed discussion of the role of trabeculae in bone integrity can be found in other chapters of this text, including Chapter 7.

In cortical bone, porosity also influences bone stiffness and strength (reviewed in Cooper et al. (2016)). Cortical bone contains pores spanning many length scales, including osteocyte lacunae (~10 μm) and Haversian canals (~50 μm). Cortical porosity is inversely associated with bone stiffness (Schaffler & Burr, 1988) and strength (Vilayphiou et al., 2016) and is higher in individuals with fragility fractures than those without fractures (Bala et al., 2014; Patsch et al., 2013). Increased cortical porosity compromises bone stiffness and strength, underscoring its significant role in fracture susceptibility.

3.2 Bone material properties

3.2.1 Mechanical testing overview

The mechanical properties of bone material can be directly assessed with mechanical testing. This section provides a brief review of these techniques. For a more extensive discussion of bone mechanical properties, refer to Chapters 6–12.

3.2.2 In vivo mechanical testing of bone material

Some mechanical properties of bone tissue can be directly measured in the clinic using the Osteoprobe [ActiveLife Scientific, Santa Barbara, CA (Bridges et al., 2012)]. The Osteoprobe employs reference point indentation. In this technique, a needle-like probe tip applies a preset impact force to the bone. The indentation depth is measured by a displacement transducer in the device. The primary outcome of the Osteoprobe is the bone material strength index (BMSi), which measures the material's resistance to impact relative to a reference indentation distance into poly (methyl-methacrylate) plastic. In clinical cohorts, the BMSi was lower in patients with fragility fractures than those without fractures (Malgo et al., 2017; Rozental et al., 2018) and increased following osteoporosis treatment with bisphosphonates or denosumab (Schoeb et al., 2020). However, results of prospective studies examining BMSi to predict fractures are mixed (Jaiswal et al., 2023; Sosa & Eriksen, 2017). The key advantage of the Osteoprobe is that it is the only technique currently available for direct measurement of bone mechanical properties in vivo. Limitations include restriction to the anterior periosteal surface of the tibial diaphysis, which is not a site of fragility fractures, high inter-observer variability of measurements, and the lack of clarity on how BMSi relates to mechanical properties measured from traditional mechanical tests (Allen et al., 2015). The Osteoprobe has potential for directly measuring bone material integrity in the clinic.

3.2.3 Laboratory mechanical testing of bone material

The skeleton provides structural support and protection. Therefore, mechanical properties such as high strength, fatigue lifetime, and fracture toughness are essential to skeletal health (Hernandez & van der Meulen, 2017). Distinct failure mechanisms are more prevalent in certain types of clinical fractures than in others. For example, strength is most relevant to fractures that occur due to a single traumatic event, such as wrist fractures that occur during falls. Fatigue is most relevant to fractures that occur through damage accumulation over time, such as vertebral fractures, which often occur in the absence of a traumatic event (Schousboe, 2016). Mechanical testing in the clinic and laboratory provides direct insights into bone strength, with the type of test determining the measured properties and relevant failure modes.

Standard engineering tests that mechanically characterize solids may be applied to bone (reviewed in Fyhrie and Christiansen (2015) and Turner and Burr (1993)). The most common laboratory tests are single load to failure tests, which determine a bone's capacity to resist failure during a single overload event. Specimens include whole bones or bone explants machined into standard shapes. In these tests, an increasing load is slowly applied to the test specimen until failure occurs. Mechanical properties such as stiffness, strength, ductility (ability to deform), and toughness (energy to failure) can be obtained. Cortical bone is commonly tested in three-point bending (Fig. 5.4a), which approximates the in vivo bending forces experienced by diaphyseal cortical bone. Cancellous bone is commonly tested in uniaxial compression (Fig. 5.4b), which approximates the in vivo compressive forces exerted on epiphyseal cancellous bone.

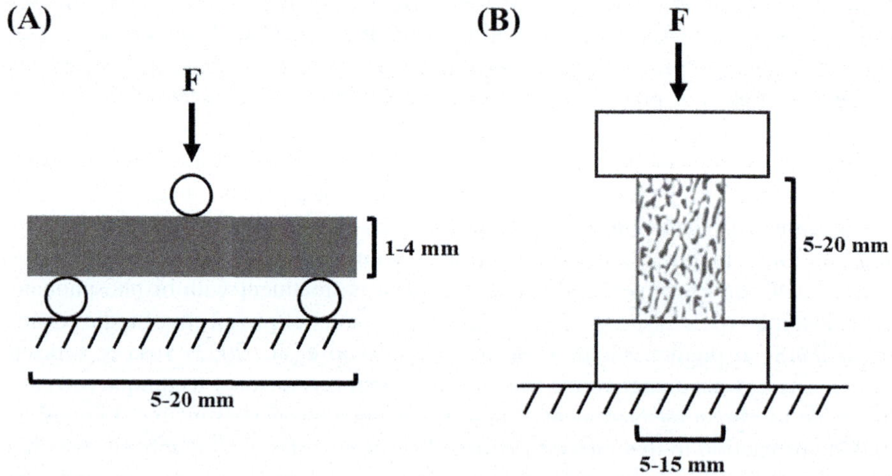

FIGURE 5.4

Diagrams of basic mechanical tests: (A) a three-point bending test, commonly done with cortical bone (B) a uniaxial compression test, commonly done with cancellous bone. Typical sample dimensions are indicated for each type of test. Sample sizes are typically limited by bone anatomy. In bending tests, human cortical bone samples typically have 1–4 mm thickness and depth, and 5–20 mm length (Zimmermann et al., 2016). Cancellous bone for compression tests are typically cylindrical cores with 5–15 mm diameters and 5–20 mm heights (Zhao et al., 2018). Occasionally, cancellous parallelipeds with similar dimensions (5–20 mm) are used for compression tests.

Other common tests include fatigue tests and fracture toughness tests. The purpose of a fatigue test is to determine how a bone resists failure under day-to-day loading in vivo. Fatigue tests may use the same setup as a single overload test. However, instead of applying a single large load, smaller sub-failure loads are applied for thousands to millions of cycles. Finally, the purpose of a fracture toughness test is to determine the bone's resistance to failure from a load in the presence of preexisting damage. In this test, a notch is machined into the specimen prior to application of a load. These mechanical tests provide direct insights into how bones respond to traumatic events, repetitive loading over time, and loading with preexisting damage.

3.2.4 Measurement of microdamage

Another important aspect of bone quality is microdamage accumulation (reviewed in Hoenig et al. (2022); Seref-Ferlengez et al. (2015)). Bone experiences cyclic loading during daily activities. As a result, damage accumulates in the form of linear microcracks (<100 μm) or clusters of diffuse damage (sub-lamellar cracking <3 μm). Microdamage is not directly assessed clinically because clinical imaging techniques cannot visualize microcracks. However, several bulk staining methods have been developed to visualize microdamage in tissue explants. Fluorochrome dyes such as basic fuchsin and calcein can label microcracks for fluorescence microscopy (Fig. 5.5), and X-ray contrast agents such as lead uranyl acetate and barium sulfide can label microdamage for scanning electron microscopy (SEM) and micro-CT (Poundarik & Vashishth, 2015). Microdamage can be visualized without a staining agent in nanometer-scale imaging modalities such as SEM and atomic force microscopy. Outcomes of microdamage imaging techniques include microcrack length, density, and damaged volume. Although many methods exist for assessing bone microdamage, clinical translation remains limited.

3.2.5 The effects of microdamage on the mechanical properties of bone

Microdamage has both beneficial and detrimental effects on bone mechanical properties. In bone and engineering composites, microdamage formation is a toughening mechanism; under loading, the formation of new interfaces can dissipate energy to prevent failure (Mayer, 2017). Additionally, microcrack interactions with osteocytes stimulate targeted bone remodeling (Hoenig et al., 2022; Seref-Ferlengez et al., 2015). However, microdamage formation that outpaces turnover compromises mechanical integrity; in both cortical and cancellous bone, microdamage formation is associated with decreased stiffness, strength, and toughness (Carter & Hayes, 1977; Lambers et al., 2013). Stress injuries may occur when microdamage formation outpaces its removal and replacement via turnover (Hoenig et al., 2022). As a person ages, microdamage accumulation becomes a greater concern, as the density of microcracks increases exponentially with age (Schaffler et al., 1995). While microdamage can dissipate energy and stimulate new bone formation, excessive accumulation with age undermines mechanical integrity and increases the risk of stress injuries.

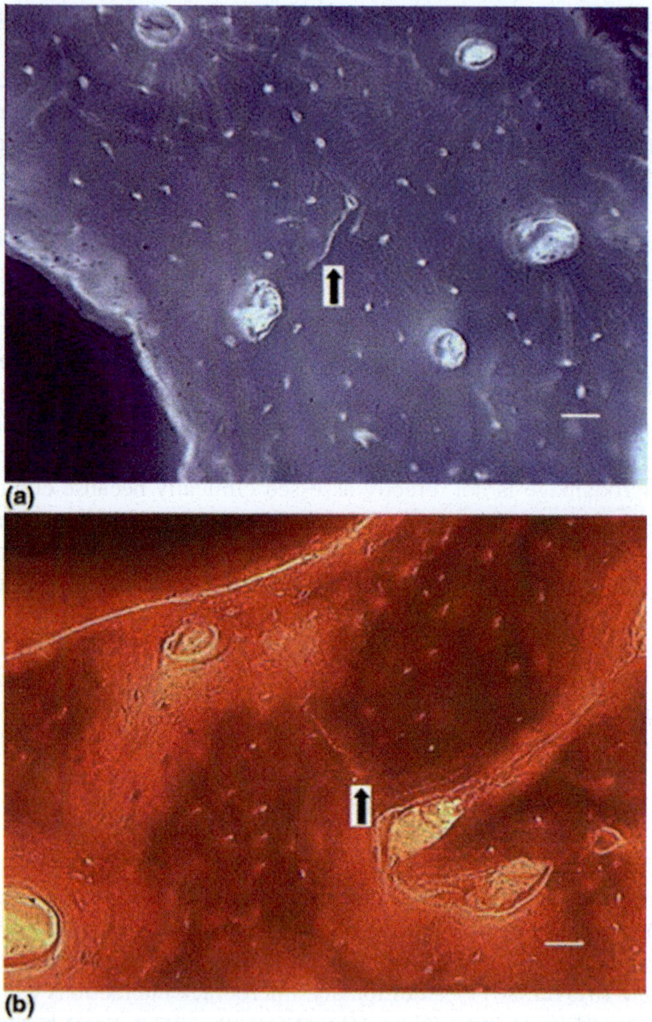

FIGURE 5.5

Fluorochrome-labeled microcracks in transverse section of human rib microcracks labeled with (A) calcein blue under UV epifluorescence ($\lambda = 365$ nm) and (B) xylenol orange under green epifluorescence ($\lambda = 546$ nm). Scale bar = 50 μm.

- From Lee, T.C., O'Brien, F.J. & Taylor, D. (2000). Nature of fatigue damage in bone. International Journal of Fatigue, 22(10), 847–853. https://doi.org/10.1016/S0142-1123(00)00054-2. Reprinted under Elsevier license 5998990694466.

3.3 Composition

Bone is a hierarchical composite material composed mainly of type I collagen, which provides ductility and post-yield toughness, and hydroxyapatite-like mineral, which provides stiffness and strength. Other mechanically relevant components of bone material include water and non-collagenous proteins. Non-collagenous proteins in the organic matrix of bone affect bone material quality through their roles in remodeling and matrix organization (reviewed in Morgan et al. (2015)), and possibly through a direct mechanical role (Wang et al., 2018). The following sections will focus on bone mineral, type I collagen, and water. The cellular component of bone directly contributes little to the mechanical properties of the tissue and will be excluded. Although compositional and organizational changes to bone mineral, collagen, and water alter tissue strength, clinical translation of bone composition measurements is limited.

3.3.1 Measurement of bone mineral properties

Bone mineral is composed primarily of poorly crystalline hydroxyapatite [$Ca_{10}(PO_4)_6(OH)_2$]. Although BMD is assessed clinically, tissue material-level mineral parameters are assessed only in the laboratory. Bone mineral may be characterized with research techniques such as ashing (see Section 2.2), quantitative backscattered electron imaging (qBEI), X-ray diffraction (XRD), Fourier transform infrared spectroscopy (FTIR), and Raman spectroscopy.

Quantitative backscattered electron imaging (qBEI) is an electron microscopy technique that utilizes differences in electron scattering between bone mineral and organic matrix to generate images of bone mineral content. The outcome of qBEI is a high-resolution (typically 1–10 μm isotropic pixel) image in which the regions with higher density appear brighter. A bone mineral density distribution (BMDD) can be obtained from each image, which can be analyzed quantitatively to identify metrics such as the distribution mean, full-width half-max, low concentration spread, and high concentration spread (Roschger et al., 1998). The advantages of qBEI are that it is a non-destructive technique that provides high-resolution, spatially resolved images of local tissue density. Limitations of qBEI include sample preparation to ensure flat, conductive surfaces, high inter-device variability (Hartmann et al., 2021), and the dependence of BMDD on acquisition time (Lukas et al., 2011). Quantitative backscattered electron imaging provides a high-resolution, quantitative analysis of bone mineral density, offering valuable insights into bone remodeling and structural properties.

X-ray diffraction probes bone mineral structure. X-ray diffractometers focus X-rays on a powder sample at varying angles. At certain angles, interactions between incident X-rays and the internal structure of the powder sample are evident as diffraction peaks. The outcome of XRD is the crystal structure of the powder sample. In addition, smaller crystals, defects, and strains in the crystal structure will cause XRD peaks to broaden; these effects cause the spectra of native bone tissue to be broader than synthetic hydroxyapatite (Peters et al., 2000). The advantage of XRD is that it is a rapid technique for probing the crystal structure of bone. The disadvantages of XRD are that it requires bone samples to be milled into a powder and does not provide spatially resolved data. X-ray diffraction is an effective technique for probing the crystal structure of bone.

X-ray scattering techniques, such as small-angle X-ray scattering (SAXS) and wide-angle X-ray scattering (WAXS), provide information about the size, shape, and orientation of bone mineral crystals (reviewed in Fratzl et al. (1996)). These techniques are based on the principle that X-rays interacting with inhomogeneities in a material are scattered, and the scattering angle is inversely associated with the size of the feature. In SAXS and WAXS, an incident X-ray beam strikes a sample, and the intensity and orientation of scattered radiation are measured by a 2D detector behind the sample. X-rays are scattered at small angles due to nm-scale features, which reflect properties such as the thickness and alignment of bone mineral particles. X-rays are scattered at wide angles by sub-nm features, which reflect properties such as crystallographic orientation. By rotating the sample around two axes and iterating the X-ray scattering experiment for each volume of interest, a 3D map of bone ultrastructure (Fig. 5.6) may be generated (Georgiadis et al., 2015). The

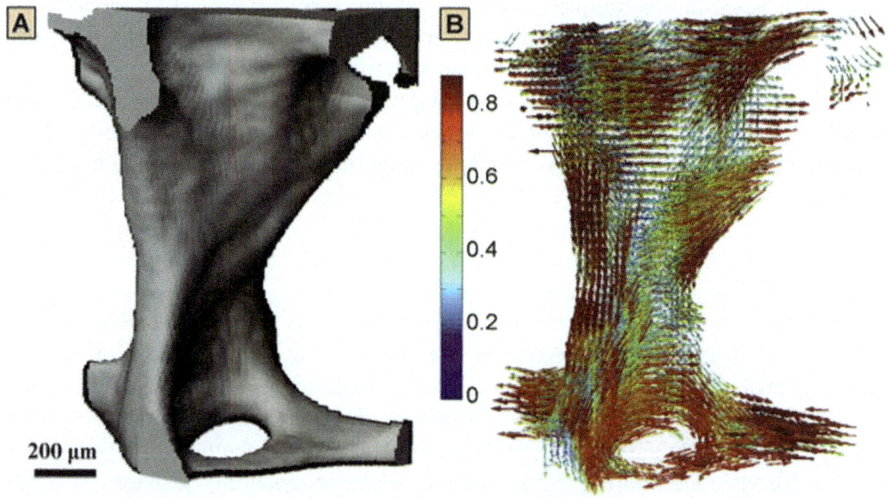

FIGURE 5.6

3D ultrastructure orientation of a trabecula from the T12 vertebra of a 50-year-old woman (a) μCT image at resolution 12.1 μm. (b) Reconstruction of 3D orientation of mineralized collagen fibrils for every bone sub-volume of the trabecula by vector field. The degree of orientation is encoded concurrently by the magnitude and the color of the vector, with longer and red vectors representing highly oriented ultrastructural sites, and smaller and blue vectors representing a sub-volume with no predominant ultrastructure orientation.

- From Georgiadis, M., Guizar-Sicairos, M., Zwahlen, A., Trüssel, A.J., Bunk, O., Müller, R. & Schneider, P. (2015). 3D scanning SAXS: A novel method for the assessment of bone ultrastructure orientation. Bone, 71, 42–52. https://doi.org/10.1016/j.bone.2014.10.002. Reprinted under Elsevier license 6011620895329.

advantages of SAXS and WAXS are that they offer insights into bone mineral size, shape, and orientation. The main disadvantages of SAXS and WAXS are high cost, complex data analysis, and impracticality for characterizing mineral over large volumes or numerous samples. Both SAXS and WAXS are powerful tools for probing nanoscale mineralization patterns in bone.

Fourier transform infrared spectroscopy and Raman spectroscopy measure the composition of bone by exciting molecular vibrations with infrared light (reviewed in Morris and Mandair (2011); Paschalis et al. (2017); Taylor and Donnelly (2020)). In FTIR spectroscopy, energy changes based on light absorption are measured. In Raman spectroscopy, energy changes in scattered light are measured. The outcome is a spectrum with peaks that correspond to characteristic energies for different kinds of chemical bonds (Fig. 5.7), a "molecular fingerprint" of the sample. Bonds with multiple vibrational modes have multiple peaks in the IR spectrum (e.g., amide I

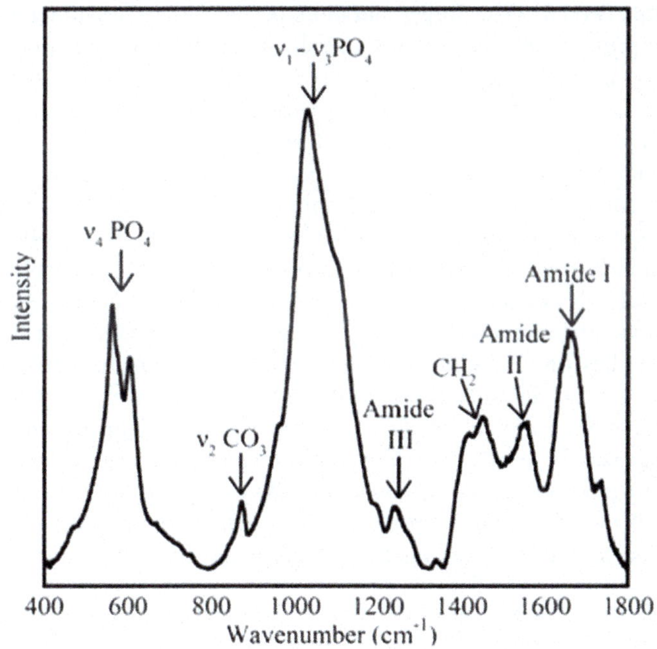

FIGURE 5.7

FTIR spectrum of cortical bone. Peaks correspond to characteristic energies for different kinds of chemical bonds. Peak intensity and area ratios are used to calculate parameters related to the mineral and matrix composition of bone.

- From Taylor, Erik A. & Donnelly, Eve. (2020). Raman and Fourier transform infrared imaging for characterization of bone material properties. Bone, 139. https://doi.org/10.1016/j.bone.2020.11549. Reprinted under Elsevier license 6022560855658.

and amide III). When a vibrational spectrometer is coupled to a microscope, FTIR and Raman spectroscopy can be used to map the composition of a material in a spatially resolved fashion. Bone mineral-related outcomes of FTIR and Raman spectroscopy are summarized in Table 5.3. Advantages of FTIR over Raman include a higher signal:noise ratio and faster scan time. However, Raman has finer spatial resolution than FTIR and does not require sample dehydration or a thin specimen. FTIR and Raman imaging provide unique insights through spatially resolved measurements of bone mineral composition.

3.3.2 The effects of mineral properties on the mechanical properties of bone

Bone mineral is mainly responsible for bone stiffness and strength (reviewed in Currey (1984)). Bone formation begins with initial deposition as unmineralized collagen (osteoid). During primary mineralization, amorphous calcium carbonate and calcium phosphate are rapidly deposited in the osteoid. As bone matures via secondary mineralization, crystals grow larger and more perfect. As tissue mineral content accrues during primary and secondary mineralization, stiffness and strength increase (Currey, 1969; Donnelly et al., 2010; Follet et al., 2004). However, the increase in stiffness and strength is achieved at the expense of ductility; bone with high values of mineral content becomes more brittle (Currey et al., 1996). This observation is further supported by studies in which SAXS and WAXS measurements are coupled with mechanical testing, which show that mineral strain contributes a small fraction of overall tissue strain (Gupta et al., 2006; Gustafsson et al., 2018). Mineral properties in addition to mineral content may also affect tissue mechanical behavior. In some studies, FTIR crystallinity is associated with nanoindentation modulus independent of mineralization (Bala et al., 2012; Gourion-Arsiquaud et al., 2009). However, other studies find no significant associations between mechanical properties and FTIR/Raman crystallinity (Wen et al., 2015). Observed relationships between carbonate:phosphate ratio and mechanical properties are highly variable, and there is no consensus on the effect of carbonate:phosphate ratio, as measured by FTIR or Raman, on the mechanical properties of bone (Taylor & Donnelly, 2020). Although tissue mineralization increases stiffness and strength, it reduces ductility. The relationship between other mineral parameters and bone strength is more variable.

3.3.3 Measurement of collagen composition and organization

Type I collagen comprises most of the organic components of bone. In bone, collagen forms hierarchical structures in which individual collagen molecules are arranged into triple helices, which are arranged in a staggered configuration into fibrils, which are then arranged into lamellae, and lamellae are further arranged into complex structures such as osteons (Garnero, 2015). Collagen undergoes several types of post-translational modification, including enzymatic crosslinking mediated by lysyl oxidase and non-enzymatic glycation reactions to form advanced glycation endproducts (AGEs). Enzymatic crosslinks are initially divalent (e.g., dihydroxylysinornorleucine, hydroxylysinonorleucine); with maturation, these

Table 5.3 Common FTIR and Raman parameters for measuring bone mineral content and composition.

Parameter	Measurement of	Common FTIR parameter	Common Raman parameter	FTIR Reference(s)	Raman Reference(s)
Mineral: Matrix ratio	Tissue mineralization	ν_1-ν_3 PO_4: Amide I peak area ratio	ν_2 PO_4: Amide III peak area ratio	Taylor et al. (2017)	Roschger, Gamsjaeger, et al. (2014); Taylor et al. (2017)
Carbonate: Phosphate ratio	CO_3 substitution in hydroxyapatite lattice	ν_2 CO_3: ν_1-ν_3 PO_4 peak area ratio	ν_1 CO_3: ν_1 PO_4 peak area ratio	Ou-Yang et al. (2001)	Awonusi et al. (2007)
Mineral maturity/ crystallinity (MMC)	Size and perfection of mineral crystals	1030: 1020 cm^{-1} intensity ratio	FWHM ν_1 PO_4	Pleshko et al. (1991); Termine and Posner (1966)	Kazanci et al. (2006)
Acid phosphate	HPO_4 substitution in hydroxyapatite structure	1127: 1096 cm^{-1} intensity ratio	No established parameter	Spevak et al. (2013)	

FWHM, *full width at half maximum.*

transform into trivalent crosslinks [e.g., pyridinoline, deoxypyridinoline (Garnero, 2012)]. Both the organization and post-translational modifications of collagen contribute to bone strength.

3.3.3.1 Methods of assessing collagen organization

Collagen organization can be measured with imaging techniques for assessing periodic fibrillar structures, including polarized light microscopy (PLM), second harmonic generation (SHG) microscopy, and X-ray scattering (reviewed in Georgiadis et al. (2016)).

Polarized light microscopy is an optical microscopy technique that leverages the interaction of polarized light with oriented collagen fibers to probe their orientation. In circular PLM, highly oriented fibers parallel to the focal plane produce the highest brightness (Fig. 5.8), and transverse fibers appear extinct. The principles of PLM with visible light can be applied to incident infrared light in FTIR (Bi et al., 2005) and Raman spectroscopy (Raghavan et al., 2010) to assess collagen structure. Advantages of PLM include relatively low cost, straightforward data analysis, and a long history of use for assessing structure-function relationships in bone (Ascenzi & Bonucci, 1967; Ascenzi et al., 2003). A downside of PLM is that image brightness depends on section thickness and collagen content. Additionally, there is a lack of standardization in the analysis of PLM, which can lead to incorrect interpretations of PLM outcomes (Georgiadis et al., 2016). Despite its limitations, polarized light microscopy remains an accessible technique for the assessment of collagen organization.

Second harmonic generation (SHG) microscopy is a nonlinear optical technique that detects pulsed laser scattering from oriented collagen molecules. Type I collagen, which has a noncentrosymmetric structure, contributes the majority of the SHG signal in bone because centrosymmetric structures (like hydroxyapatite) cannot emit SHG. The intensity of SHG signal reflects the concentration of oriented collagen molecules (Moreaux et al., 2000) enabling analyses of collagen orientation, alignment, and density via autocorrelation and texture analyses (Hao et al., 2018). SHG is a label-free method with high specificity for collagen and can be combined with confocal fluorescence microscopy. Limitations include complex data analysis and shallower propagation depth compared to X-ray scattering. Despite these challenges, SHG remains a useful optical technique for assessing collagen organization.

In addition to probing mineral properties (see Section 3.3.1), SAXS and WAXS can provide insights into collagen organization. These techniques have been used to probe parameters such as collagen fibril orientation (Georgiadis et al., 2015; Schaff et al., 2015) and d-spacing (Fratzl et al., 1996). A key advantage of SAXS/WAXS is that they can be performed in situ during a mechanical test to measure the change in collagen organization under load. For example, SAXS has been used to measure the collagen fibril contribution to overall tissue strain in cortical bone during a tensile test (Gupta et al., 2006; Zimmermann et al., 2011). Both SAXS and WAXS can probe nanoscale collagen organization in bone.

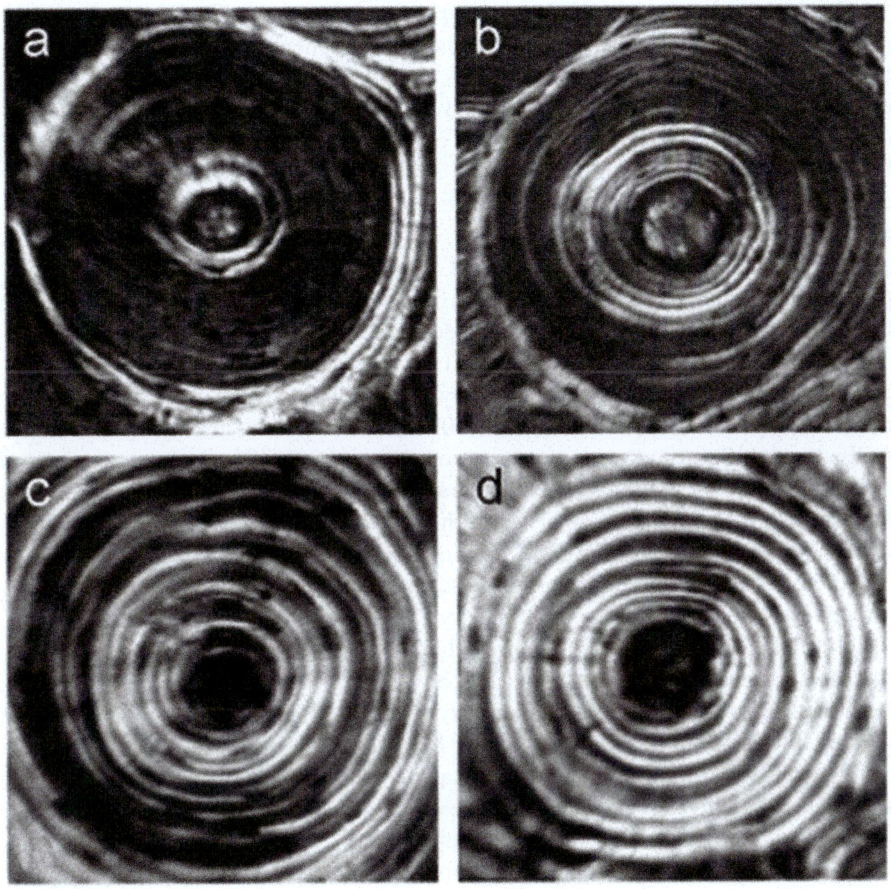

FIGURE 5.8

Human osteons as imaged with polarized light microscopy. In (a–d), bright regions correspond to transverse lamellae (aligned within the focal plane). Dark "extinct" regions correspond to longitudinal lamellae (aligned out of the focal plane).

- From Ascenzi, M.G., Di Comite, M., Mitov, P. & Michael Kabo, J. (2007). Hysteretic pinching of human secondary osteons subjected to torsion. Journal of Biomechanics, 40(12), 2619–2627. https://doi.org/10.1016/j.jbiomech.2007.01.014. Reprinted under Elsevier license 5999010817335.

3.3.3.2 Methods of assessing collagen post-translational modifications

Collagen crosslinks may be characterized in bone (reviewed in Vashishth (2009)), and in serum (reviewed in Perrone et al. (2020)). This section will focus on techniques characterizing enzymatic and non-enzymatic crosslinks and adducts in bone using bioanalytical techniques, biochemical assays, and vibrational spectroscopy.

High-performance liquid chromatography (HPLC) is a gold-standard bioanalytical technique for detection of enzymatic and nonenzymatic crosslinks and adducts. In HPLC, hydrolyzed bone samples are pumped through a stationary phase in a column. Components separate based on migration rates and are detected individually in the outflow. The main advantage of HPLC is that it can accurately quantify the concentration of individual chemical species in a sample, including immature enzymatic crosslinks, mature enzymatic crosslinks, and crosslinking and noncrosslinking AGEs. The disadvantages of HPLC compared to other crosslink characterization techniques are high cost and extensive sample preparation. High-performance liquid chromatography is a high-accuracy technique for quantifying concentrations of individual chemical species in bone.

Biochemical assays provide lower-cost options to quantify AGEs. Advanced glycation endproducts like carboxymethyl-lysine can be detected with commercially available ELISA kits (Vaidya et al., 2022). Alternatively, the subset of native AGEs that are fluorescent can be measured through an autofluorescence assay and then normalized by the collagen concentration measured from an absorbance assay that quantifies hydroxyproline (Vashishth, 2009). Compared to HPLC, this technique has the advantage of lower cost and complexity, but the disadvantages of lower specificity and capturing only the fluorescent subset of AGEs. Biochemical assays provide lower-cost options to quantify bone AGEs with reasonable accuracy.

Infrared spectroscopy techniques, including FTIR and Raman, assess the maturity of enzymatic collagen crosslinks and advanced glycation end-products (AGEs). Both methods determine collagen maturity by comparing intensity ratios associated with mature and immature crosslinks. FTIR spectroscopy quantifies glycation through the sugar:matrix ratio, while Raman measures AGEs using pentosidine and carboxy-methyl lysine integrated peak area ratios (Table 5.4). An advantage of these techniques relative to biochemical assays is that many compositional properties can be assessed simultaneously from a single spectrum (Fig. 5.7; Table 5.4), and their imaging modalities can be used to assess collagen properties in a spatially resolved fashion. A disadvantage of these techniques is that they do not directly probe crosslinking but rather detect the effects of crosslinking on the secondary structure of collagen reflected in the amide 1 band. Both FTIR and Raman provide valuable insights into collagen aging.

3.3.4 The effects of collagen composition and organization on the mechanical properties of bone

The post-yield properties of bone, such as ductility and post-yield toughness, depend on an intact collagen network (reviewed in Currey (2003); Garnero (2012, 2015)). Collagen crosslinks can act as sacrificial bonds, and collagen fibrils can stretch and slide past one another under loading, allowing for deformation without failure (Depalle et al., 2015). Inhibiting enzymatic crosslinking reduces bone toughness (McNerny et al., 2015; Paschalis et al., 2011). However, increases in nonenzymatic crosslinking reduce ductility and post-yield toughness (Poundarik et al., 2015). X-ray scattering studies have shown that collagen contributes to tissue strain more

Table 5.4 Common FTIR and Raman parameters for measuring collagen crosslinks in bone.

Parameter	Measurement of	Common FTIR parameter	Common Raman parameter	FTIR References(s)	Raman References (s)
Collagen maturity	Ratio of mature trivalent to immature divalent enzymatic collagen crosslinks	1660 cm^{-1}: 1690 cm^{-1} intensity ratio	1660 cm^{-1}: 1690 cm^{-1} intensity ratio	Paschalis et al. (2001, 2004)	Gamsjaeger et al. (2017)
Sugar: Matrix ratio	Sugar concentration in demineralized bone matrix	Sugars (900–1100 cm^{-1}): Amide I area ratio	No established parameter	Boskey and Pleshkocamacho (2007); Ibrahim et al. (2006)	
Pentosidine	Concentration of the AGE pentosidine in bone matrix	No established parameter	1495 cm^{-1}: CH_2 (1450 cm^{-1}) area ratio		Rubin et al. (2016)
Carboxy-methyl-lysine	Concentration of the AGE CML in bone matrix	No established parameter	1150 cm^{-1}: CH_2 (1450 cm^{-1}) area ratio		Rubin et al. (2016)

than bone mineral (Gupta et al., 2006), and that glycation reduces collagen's ability to confer plasticity to bone (Zimmermann et al., 2011). Disorders that impair collagen fibril organization, such as Padget's disease (Garnero, 2015) and rheumatoid arthritis (Ozasa et al., 2022) degrade bone strength. Collagen's role in maintaining the post-yield properties of bone is affected by its organization and crosslinking.

3.3.5 Bone water: Measurement and influence on mechanical properties

Water is present throughout bone and directly influences mechanical properties (reviewed in Burr (2019); Granke et al. (2015); Surowiec et al. (2022)). Water in bone can be directly incorporated into the bone mineral structure, tightly bound to collagen and NCPs, loosely bound at the collagen-mineral interface, or free to occupy pores within the bone. A simple measurement of bone water may be made by weighing a bone sample before and after dehydration with heat or solvents. Thermal dehydration can be used to study the water present in individual bone water compartments by controlling the dehydration temperature in a vacuum furnace. Free water is removed at 21–40°C, loosely bound water can be removed at 100°C, tightly bound water can be removed at 200°C, and structural water can only be removed at 340–540°C (Surowiec et al., 2022). A more complex measurement of bone water may be made with UTE-MRI, which can measure and distinguish bound versus free water in a spatially resolved fashion [typical voxel dimensions 200–5 mm (Chen et al., 2015; Du & Bydder, 2013; Jerban et al., 2019)]. Both bound and free water have significant mechanical effects on bone mechanical properties.

Bound and free water have opposite associations with bone stiffness, strength, and toughness. Bound water is associated with strength and toughness. Dehydrated bone becomes brittle (Granke et al., 2015), and increases in UTE-MRI measurements of bound water are directly associated with strength (Manhard et al., 2016). Bound water decreases with age and is thought to be one of the many factors contributing to increased fracture risk with aging (Burr, 2019). Free water typically resides in pores and is thus linked with cortical bone porosity (Rajapakse et al., 2015), which is associated with bone fragility (see Section 3.1.4). Greater free water is associated with losses in stiffness and strength (Manhard et al., 2016; Nyman et al., 2008). Bound water is directly associated with bone mechanical integrity and free water, likely due to its relationship with porosity, is inversely associated with bone mechanical integrity.

4. The links between bone quality and bone quantity

Processes within bone tissue, such as remodeling, interconnect bone quantity and bone quality [reviewed in Bolamperti et al. (2022); Wang et al. (2022)]. In osteoporosis, an imbalance in resorption and formation reduces BMD and causes bone fragility. Simultaneously, several bone quality parameters are negatively affected. Trabeculae thin and lose connectivity, which impairs the quality of the microarchitecture. A high rate of resorption also reduces the time for secondary mineralization to occur (Roschger, Misof, et al., 2014), which reduces the stiffness and strength of the bone tissue. In patients taking antiresorptive drugs for postmenopausal osteoporosis, BMD and tissue mineral content are increased (Cummings et al., 2002; Hochberg et al., 2002). Although increased mineralization improves stiffness and strength, there is a tradeoff with ductility and post-yield toughness. Furthermore, the decreased bone resorption allows additional time for bone material to accumulate non-enzymatic collagen crosslinks and microdamage, which may degrade bone integrity. When tissue samples are mechanically tested, non-enzymatic glycation is associated with reduced post-yield toughness (Poundarik et al., 2015) and microdamage is associated with reduced stiffness, strength, and toughness (Carter & Hayes, 1977; Lambers et al., 2013). Clinically, bisphosphonate use substantially decreases the incidence of "typical" single-overload hip fragility fractures, but long-term use may increase the incidence of rare atypical femoral fractures, which are thought to occur through a fatigue mechanism (Shane et al., 2014). Overall, changes to bone quantity and bone quality rarely occur in isolation. Therefore, a holistic understanding of the bone quantity, bone quality, and the processes that connect them is necessary for accurate assessment of bone fragility.

5. Conclusion

Bone quantity and bone quality both contribute to skeletal integrity. Clinical assessment of bone fragility is focused on bone quantity, and many bone quality parameters are only assessed in the laboratory. Developing better clinical tools to measure bone quality parameters would improve bone fragility assessment, leading to fewer fragility fractures. Bone fragility assessments and treatments should be implemented with the consideration that bone quantity and bone quality are linked.

References

Adams, J. E. (2013). Advances in bone imaging for osteoporosis. *Nature Reviews Endocrinology, 9*(1), 28–42. https://doi.org/10.1038/nrendo.2012.217

Adams, A. L., Fischer, H., Kopperdahl, D. L., Lee, D. C., Black, D. M., Bouxsein, M. L., Fatemi, S., Khosla, S., Orwoll, E. S., Siris, E. S., & Keaveny, T. M. (2018). Osteoporosis and hip fracture risk from routine computed tomography scans: The fracture, osteoporosis, and CT utilization study (FOCUS). *Journal of Bone and Mineral Research, 33*(7), 1291–1301. https://doi.org/10.1002/jbmr.3423

Allen, M. R., McNerny, E. M. B., Organ, J. M., & Wallace, J. M. (2015). True gold or pyrite: A review of reference point indentation for assessing bone mechanical properties in vivo. *Journal of Bone and Mineral Research, 30*(9), 1539–1550. https://doi.org/10.1002/jbmr.2603

Ammann, P., & Rizzoli, R. (2003). Bone strength and its determinants. *Osteoporosis International, 14*(S3), 13–18. https://doi.org/10.1007/s00198-002-1345-4

Ascenzi, M. G., Ascenzi, A., Benvenuti, A., Burghammer, M., Panzavolta, S., & Bigi, A. (2003). Structural differences between "dark" and "bright" isolated human osteonic lamellae. *Journal of Structural Biology, 141*(1), 22–33. https://doi.org/10.1016/S1047-8477(02)00578-6

Ascenzi, A., & Bonucci, E. (1967). The tensile properties of single osteons. *The Anatomical Record, 158*(4), 375–386. https://doi.org/10.1002/ar.1091580403

Awonusi, A., Morris, M. D., & Tecklenburg, M. M. J. (2007). Carbonate assignment and calibration in the Raman spectrum of apatite. *Calcified Tissue International, 81*(1), 46–52. https://doi.org/10.1007/s00223-007-9034-0

Bala, Y., Depalle, B., Farlay, D., Douillard, T., Meille, S., Follet, H., Chapurlat, R., Chevalier, J., & Boivin, G. (2012). Bone micromechanical properties are compromised during long-term alendronate therapy independently of mineralization. *Journal of Bone and Mineral Research, 27*(4), 825–834. https://doi.org/10.1002/jbmr.1501

Bala, Y., Zebaze, R., Ghasem-Zadeh, A., Atkinson, E. J., Iuliano, S., Peterson, J. M., Amin, S., Bjørnerem, A., Melton, L. J., Johansson, H., Kanis, J. A., Khosla, S., & Seeman, E. (2014). Cortical porosity identifies women with osteopenia at increased risk for forearm fractures. *Journal of Bone and Mineral Research, 29*(6), 1356–1362. https://doi.org/10.1002/jbmr.2167

Bandirali, M., Di Leo, G., Papini, G. D. E., Messina, C., Sconfienza, L. M., Ulivieri, F. M., & Sardanelli, F. (2015). A new diagnostic score to detect osteoporosis in patients undergoing lumbar spine MRI. *European Radiology, 25*(10), 2951–2959. https://doi.org/10.1007/s00330-015-3699-y

Bi, X., Li, G., Doty, S. B., & Camacho, N. P. (2005). A novel method for determination of collagen orientation in cartilage by fourier transform infrared imaging spectroscopy (FT-IRIS). *Osteoarthritis and Cartilage, 13*(12), 1050–1058. https://doi.org/10.1016/j.joca.2005.07.008

Blake, G. M., & Fogelman, I. (1997). Technical principles of dual energy x-ray absorptiometry. *Seminars in Nuclear Medicine, 27*(3), 210–228. https://doi.org/10.1016/S0001-2998(97)80025-6

Bolamperti, S., Villa, I., & Rubinacci, A. (2022). Bone remodeling: An operational process ensuring survival and bone mechanical competence. *Bone Research, 10*(1). https://doi.org/10.1038/s41413-022-00219-8

Bolotin, H. H. (2007). DXA in vivo BMD methodology: An erroneous and misleading research and clinical gauge of bone mineral status, bone fragility, and bone remodelling. *Bone, 41*(1), 138–154. https://doi.org/10.1016/j.bone.2007.02.022

Boskey, A., & Pleshkocamacho, N. (2007). FT-IR imaging of native and tissue-engineered bone and cartilage. *Biomaterials, 28*(15), 2465–2478. https://doi.org/10.1016/j.biomaterials.2006.11.043

Bouxsein, M. L., Boyd, S. K., Christiansen, B. A., Guldberg, R. E., Jepsen, K. J., & Müller, R. (2010). Guidelines for assessment of bone microstructure in rodents using micro-computed tomography. *Journal of Bone and Mineral Research, 25*(7), 1468–1486. https://doi.org/10.1002/jbmr.141

Bouxsein, M. L., Eastell, R., Lui, L. Y., Wu, L. A., de Papp, A. E., Grauer, A., Marin, F., Cauley, J. A., Bauer, D. C., & Black, D. M. (2019). Change in bone density and reduction in fracture risk: A meta-regression of published trials. *Journal of Bone and Mineral Research, 34*(4), 632–642. https://doi.org/10.1002/jbmr.3641

Bouxsein, M. L., & Karasik, D. (2006). Bone geometry and skeletal fragility. *Current Osteoporosis Reports, 4*(2), 49–56. https://doi.org/10.1007/s11914-006-0002-9

Bregoli, C., Biffi, C. A., Tuissi, A., & Buccino, F. (2024). Effect of trabecular architectures on the mechanical response in osteoporotic and healthy human bone. *Medical, & Biological Engineering & Computing, 62*(11), 3263–3281. https://doi.org/10.1007/s11517-024-03134-8

Bridges, D., Randall, C., & Hansma, P. K. (2012). A new device for performing reference point indentation without a reference probe. *Review of Scientific Instruments, 83*(4). https://doi.org/10.1063/1.3693085

Burghardt, A. J., Kazakia, G. J., Laib, A., & Majumdar, S. (2008). Quantitative assessment of bone tissue mineralization with polychromatic micro-computed tomography. *Calcified Tissue International, 83*(2), 129–138. https://doi.org/10.1007/s00223-008-9158-x

Burr, D. B. (2019). Changes in bone matrix properties with aging. *Bone, 120*, 85–93. https://doi.org/10.1016/j.bone.2018.10.010

Burrows, M., Liu, D., & McKay, H. (2010). High-resolution peripheral QCT imaging of bone micro-structure in adolescents. *Osteoporosis International, 21*(3), 515–520. https://doi.org/10.1007/s00198-009-0913-2

Carter, D. R., & Hayes, W. C. (1977). Compact bone fatigue damage—I. Residual strength and stiffness. *Journal of Biomechanics, 10*(5–6), 325–337. https://doi.org/10.1016/0021-9290(77)90005-7

Cefalu, C. A. (2004). Is bone mineral density predictive of fracture risk reduction? *Current Medical Research and Opinion, 20*(3), 341–349. https://doi.org/10.1185/030079903125003062

Chalhoub, D., Orwoll, E. S., Cawthon, P. M., Ensrud, K. E., Boudreau, R., Greenspan, S., Newman, A. B., Zmuda, J., Bauer, D., Cummings, S., & Cauley, J. A. (2016). Areal and volumetric bone mineral density and risk of multiple types of fracture in older men. *Bone, 92*, 100–106. https://doi.org/10.1016/j.bone.2016.08.014

Chappard, D., Baslé, M.-F., Legrand, E., & Audran, M. (2008). Trabecular bone microarchitecture: A review. *Morphologie, 92*(299), 162–170. https://doi.org/10.1016/j.morpho.2008.10.003

Chen, A., Feng, S., Lai, L., & Yan, C. (2023). A meta-analysis of the value of MRI-based VBQ scores for evaluating osteoporosis. *Bone Reports, 19*, Article 101711. https://doi.org/10.1016/j.bonr.2023.101711

Chen, J., Grogan, S. P., Shao, H., D'Lima, D., Bydder, G. M., Wu, Z., & Du, J. (2015). Evaluation of bound and pore water in cortical bone using ultrashort-TE MRI. *NMR in Biomedicine, 28*(12), 1754–1762. https://doi.org/10.1002/nbm.3436

Cooper, D. M. L., Kawalilak, C. E., Harrison, K., Johnston, B. D., & Johnston, J. D. (2016). Cortical bone porosity: What is It, why is it important, and how can we detect it? *Current Osteoporosis Reports, 14*(5), 187–198. https://doi.org/10.1007/s11914-016-0319-y

Cummings, S. R., Karpf, D. B., Harris, F., Genant, H. K., Ensrud, K., Lacroix, A. Z., & Black, D. M. (2002). Improvement in spine bone density and reduction in risk of vertebral fractures during treatment with antiresorptive drugs. *The American Journal of Medicine, 112*(4), 281–289. https://doi.org/10.1016/S0002-9343(01)01124-X

Currey, J. D. (1969). The mechanical consequences of variation in the mineral content of bone. *Journal of Biomechanics, 2*(1), 1–11. https://doi.org/10.1016/0021-9290(69)90036-0

Currey, J. D. (1984). Effects of differences in mineralization on the mechanical properties of bone. *Philosophical Transactions of the Royal Society of London - Series B: Biological Sciences, 304*(1121), 509–518. https://doi.org/10.1098/rstb.1984.0042

Currey, J. D. (2003). Role of collagen and other organics in the mechanical properties of bone. *Osteoporosis International, 14*, 29–36.

Currey, J. D., Brear, K., & Zioupos, P. (1996). The effects of ageing and changes in mineral content in degrading the toughness of human femora. *Journal of Biomechanics, 29*(2), 257–260. https://doi.org/10.1016/0021-9290(95)00048-8

Damilakis, J., Adams, J. E., Guglielmi, G., & Link, T. M. (2010). Radiation exposure in X-ray-based imaging techniques used in osteoporosis. *European Radiology, 20*(11), 2707–2714. https://doi.org/10.1007/s00330-010-1845-0

Depalle, B., Qin, Z., Shefelbine, S. J., & Buehler, M. J. (2015). Influence of cross-link structure, density and mechanical properties in the mesoscale deformation mechanisms of collagen fibrils. *Journal of the Mechanical Behavior of Biomedical Materials, 52*, 1–13. https://doi.org/10.1016/j.jmbbm.2014.07.008

Deshpande, N., Hadi, M. S., Lillard, J. C., Passias, P. G., Linzey, J. R., Saadeh, Y. S., LaBagnara, M., & Park, P. (2023). Alternatives to DEXA for the assessment of bone density: A systematic review of the literature and future recommendations. *Journal of Neurosurgery: Spine, 38*(4), 436–445. https://doi.org/10.3171/2022.11.SPINE22875

Donnelly, E., Chen, D. X., Boskey, A. L., Baker, S. P., & Van Der Meulen, M. C. H. (2010). Contribution of mineral to bone structural behavior and tissue mechanical properties. *Calcified Tissue International, 87*(5), 450–460. https://doi.org/10.1007/s00223-010-9404-x

Du, J., & Bydder, G. M. (2013). Qualitative and quantitative ultrashort-TE MRI of cortical bone. *NMR in Biomedicine, 26*(5), 489–506. https://doi.org/10.1002/nbm.2906

Dudle, A., Gugler, Y., Pretterklieber, M., Ferrari, S., Lippuner, K., & Zysset, P. (2023). 2D-3D reconstruction of the proximal femur from DXA scans: Evaluation of the 3D-Shaper software. *Frontiers in Bioengineering and Biotechnology, 11*.

Ebbesen, E. N., Thomsen, J. S., Beck-Nielsen, H., Nepper-Rasmussen, H. J., & Mosekilde, L. (1999). Lumbar vertebral body compressive strength evaluated by dual-energy X-ray absorptiometry, quantitative computed tomography, and ashing. *Bone, 25*(6), 713–724. https://doi.org/10.1016/s8756-3282(99)00216-1

Ehresman, J., Schilling, A., Yang, X., Pennington, Z., Ahmed, A. K., Cottrill, E., Lubelski, D., Khan, M., Moseley, K. F., & Sciubba, D. M. (2021). Vertebral bone quality score predicts fragility fractures independently of bone mineral density. *The Spine Journal, 21*(1), 20–27. https://doi.org/10.1016/j.spinee.2020.05.540

Fields, A. J., & Keaveny, T. M. (2012). Trabecular architecture and vertebral fragility in osteoporosis. *Current Osteoporosis Reports, 10*(2), 132–140. https://doi.org/10.1007/s11914-012-0097-0

Follet, H., Boivin, G., Rumelhart, C., & Meunier, P. J. (2004). The degree of mineralization is a determinant of bone strength: A study on human calcanei. *Bone, 34*(5), 783–789. https://doi.org/10.1016/j.bone.2003.12.012

Fratzl, P., Schreiber, S., & Klaushofer, K. (1996). Informa healthcare Austria bone mineralization as studied by small-angle x-ray scattering. *Connective Tissue Research, 34*(4), 247–254. https://doi.org/10.3109/03008209609005268

Fyhrie, D. P., & Christiansen, B. A. (2015). Bone material properties and skeletal fragility. *Calcified Tissue International, 97*(3), 213–228. https://doi.org/10.1007/s00223-015-9997-1

Gamsjaeger, S., Robins, S. P., Tatakis, D. N., Klaushofer, K., & Paschalis, E. P. (2017). Identification of pyridinoline trivalent collagen cross-links by Raman microspectroscopy.

Calcified Tissue International, 100(6), 565–574. https://doi.org/10.1007/s00223-016-0232-5

Garnero, P. (2012). The contribution of collagen crosslinks to bone strength. *BoneKEy Reports, 1*, 182. https://doi.org/10.1038/bonekey.2012.182

Garnero, P. (2015). The role of collagen organization on the properties of bone. *Calcified Tissue International, 97*(3), 229–240. https://doi.org/10.1007/s00223-015-9996-2

Georgiadis, M., Guizar-Sicairos, M., Zwahlen, A., Trüssel, A. J., Bunk, O., Müller, R., & Schneider, P. (2015). 3D scanning SAXS: A novel method for the assessment of bone ultrastructure orientation. *Bone, 71*, 42–52. https://doi.org/10.1016/j.bone.2014.10.002

Georgiadis, M., Müller, R., & Schneider, P. (2016). Techniques to assess bone ultrastructure organization: Orientation and arrangement of mineralized collagen fibrils. *Journal of The Royal Society Interface, 13*(119), Article 20160088. https://doi.org/10.1098/rsif.2016.0088

Gourion-Arsiquaud, S., Burket, J. C., Havill, L. M., DiCarlo, E., Doty, S. B., Mendelsohn, R., Van Der Meulen, M. C. H., & Boskey, A. L. (2009). Spatial variation in osteonal bone properties relative to tissue and animal age. *Journal of Bone and Mineral Research, 24* (7), 1271–1281. https://doi.org/10.1359/jbmr.090201

Granke, M., Does, M. D., & Nyman, J. S. (2015). The role of water compartments in the material properties of cortical bone. *Calcified Tissue International, 97*(3), 292–307. https://doi.org/10.1007/s00223-015-9977-5

Greiner, M., Rodríguez-Navarro, A., Heinig, M. F., Mayer, K., Kocsis, B., Göhring, A., Toncala, A., Grupe, G., & Schmahl, W. W. (2019). Bone incineration: An experimental study on mineral structure, colour and crystalline state. *Journal of Archaeological Science: Reports, 25*, 507–518. https://doi.org/10.1016/j.jasrep.2019.05.009

Gupta, H. S., Seto, J., Wagermaier, W., Zaslansky, P., Boesecke, P., & Fratzl, P. (2006). Cooperative deformation of mineral and collagen in bone at the nanoscale. *Proceedings of the National Academy of Sciences, 103*(47), 17741–17746. https://doi.org/10.1073/pnas.0604237103

Gustafsson, A., Mathavan, N., Turunen, M. J., Engqvist, J., Khayyeri, H., Hall, S. A., & Isaksson, H. (2018). Linking multiscale deformation to microstructure in cortical bone using in situ loading, digital image correlation and synchrotron X-ray scattering. *Acta Biomaterialia, 69*, 323–331. https://doi.org/10.1016/j.actbio.2018.01.037

Haentjens, P., Magaziner, J., Colón-Emeric, C. S., Vanderschueren, D., Milisen, K., Velkeniers, B., & Boonen, S. (2010). Meta-analysis: Excess mortality after hip fracture among older women and men. *Annals of Internal Medicine, 152*(6), 380–390. https://doi.org/10.7326/0003-4819-152-6-201003160-00008

Hans, D., Šteňová, E., & Lamy, O. (2017). The trabecular bone score (TBS) complements DXA and the FRAX as a fracture risk assessment tool in routine clinical practice. *Current Osteoporosis Reports, 15*(6), 521–531. https://doi.org/10.1007/s11914-017-0410-z

Hao, J., Yao, W., Harris, W. B. R., Vink, J. Y., Myers, K. M., & Donnelly, E. (2018). Characterization of the collagen microstructural organization of human cervical tissue. *Reproduction, 156*(1), 71–79. https://doi.org/10.1530/REP-17-0763

Hartmann, M. A., Blouin, S., Misof, B. M., Fratzl-Zelman, N., Roschger, P., Berzlanovich, A., Gruber, G. M., Brugger, P. C., Zwerina, J., & Fratzl, P. (2021). Quantitative backscattered electron imaging of bone using a thermionic or a field emission electron source. *Calcified Tissue International, 109*(2), 190–202. https://doi.org/10.1007/s00223-021-00832-5

Hernandez, C. J., & van der Meulen, M. C. H. (2017). Understanding bone strength is not enough. *Journal of Bone and Mineral Research, 32*(6), 1157–1162. https://doi.org/10.1002/jbmr.3078

Hesse, B., Varga, P., Langer, M., Pacureanu, A., Schrof, S., Männicke, N., Suhonen, H., Maurer, P., Cloetens, P., Peyrin, F., & Raum, K. (2015). Canalicular network morphology is the major determinant of the spatial distribution of mass density in human bone tissue: Evidence by means of synchrotron radiation phase-contrast nano-CT. *Journal of Bone and Mineral Research, 30*(2), 346–356. https://doi.org/10.1002/jbmr.2324

Hochberg, M. C., Greenspan, S., Wasnich, R. D., Miller, P., Thompson, D. E., & Ross, P. D. (2002). Changes in bone density and turnover explain the reductions in incidence of non-vertebral fractures that occur during treatment with antiresorptive agents. *Journal of Clinical Endocrinology and Metabolism, 87*(4), 1586–1592. https://doi.org/10.1210/jcem.87.4.8415

Hoenig, T., Ackerman, K. E., Beck, B. R., Bouxsein, M. L., Burr, D. B., Hollander, K., Popp, K. L., Rolvien, T., Tenforde, A. S., & Warden, S. J. (2022). Bone stress injuries. *Nature Reviews Disease Primers, 8*(1). https://doi.org/10.1038/s41572-022-00352-y

Huang, W., Gong, Z., Wang, H., Zheng, C., Chen, Y., Xia, X., Ma, X., & Jiang, J. (2023). Use of MRI-based vertebral bone quality score (VBQ) of S1 body in bone mineral density assessment for patients with lumbar degenerative diseases. *European Spine Journal, 32*(5), 1553–1560. https://doi.org/10.1007/s00586-023-07643-2

Huppertz, A., Radmer, S., Asbach, P., Juran, R., Schwenke, C., Diederichs, G., Hamm, B., & Sparmann, M. (2011). Computed tomography for preoperative planning in minimal-invasive total hip arthroplasty: Radiation exposure and cost analysis. *European Journal of Radiology, 78*(3), 406–413. https://doi.org/10.1016/j.ejrad.2009.11.024

Ibrahim, M. A., Allam, M., El-Haes, H., Jalbout, A. F., & De Leon, A. (2006). Analysis of the structure and vibrational spectra of glucose and fructose. *Ecletica Quimica, 31*(3), 15–21. https://doi.org/10.26850/1678-4618eqj.v31.3.2006.p15-21

Jain, R. K., & Vokes, T. (2017). Dual-energy X-ray absorptiometry. *Journal of Clinical Densitometry, 20*(3), 291–303. https://doi.org/10.1016/j.jocd.2017.06.014

Jaiswal, R., Zoulakis, M., Axelsson, K. F., Darelid, A., Rudäng, R., Sundh, D., Litsne, H., Johansson, L., & Lorentzon, M. (2023). Increased bone material strength index is positively associated with the risk of incident osteoporotic fractures in older Swedish women. *Journal of Bone and Mineral Research, 38*(6), 860–868. https://doi.org/10.1002/jbmr.4816

Jepsen, K. J., Courtland, H. W., & Nadeau, J. H. (2010). Genetically determined phenotype covariation networks control bone strength. *Journal of Bone and Mineral Research, 25*(7), 1581–1593. https://doi.org/10.1002/jbmr.41

Jerban, S., Ma, Y., Wong, J. H., Nazaran, A., Searleman, A., Wan, L., Williams, J., Du, J., & Chang, E. Y. (2019). Ultrashort echo time magnetic resonance imaging (UTE-MRI) of cortical bone correlates well with histomorphometric assessment of bone microstructure. *Bone, 123*, 8–17. https://doi.org/10.1016/j.bone.2019.03.013

Kaiser, J., Allaire, B., Fein, P. M., Lu, D., Adams, A., Kiel, D. P., Jarraya, M., Guermazi, A., Demissie, S., Samelson, E. J., Bouxsein, M. L., & Morgan, E. F. (2020). Heterogeneity and spatial distribution of intravertebral trabecular bone mineral density in the lumbar spine is associated with prevalent vertebral fracture. *Journal of Bone and Mineral Research, 35*(4), 641–648. https://doi.org/10.1002/jbmr.3946

Kanis, J. A., Johansson, H., Oden, A., & McCloskey, E. V. (2009). Assessment of fracture risk. *European Journal of Radiology, 71*(3), 392–397. https://doi.org/10.1016/j.ejrad.2008.04.061

Kazakia, G. J., Carballido-Gamio, J., Lai, A., Nardo, L., Facchetti, L., Pasco, C., Zhang, C. A., Han, M., Parrott, A. H., Tien, P., & Krug, R. (2018). Trabecular bone microstructure is impaired in the proximal femur of human immunodeficiency virus-infected men with normal bone mineral density. *Quantitative Imaging in Medicine and Surgery, 8*(1), 5–13. https://doi.org/10.21037/qims.2017.10.10

Kazanci, M., Fratzl, P., Klaushofer, K., & Paschalis, E. P. (2006). Complementary information on in vitro conversion of amorphous (precursor) calcium phosphate to hydroxyapatite from raman microspectroscopy and wide-angle X-Ray scattering. *Calcified Tissue International, 79*(5), 354–359. https://doi.org/10.1007/s00223-006-0011-9

Keen, R. W. (2003). Burden of osteoporosis and fractures. *Current Osteoporosis Reports, 1*(2), 66–70. https://doi.org/10.1007/s11914-003-0011-x

Kim, K., Song, S. H., Kim, I.-J., & Jeon, Y. K. (2021). Is dual-energy absorptiometry accurate in the assessment of bone status of patients with chronic kidney disease? *Osteoporosis International, 32*(9), 1859–1868. https://doi.org/10.1007/s00198-020-05670-z

Lambers, F. M., Bouman, A. R., Rimnac, C. M., & Hernandez, C. J. (2013). Microdamage caused by fatigue loading in human cancellous bone: Relationship to reductions in bone biomechanical performance. *PLoS One, 8*(12). https://doi.org/10.1371/journal.pone.0083662

Liu, X. S., Bevill, G., Keaveny, T. M., Sajda, P., & Guo, X. E. (2009). Micromechanical analyses of vertebral trabecular bone based on individual trabeculae segmentation of plates and rods. *Journal of Biomechanics, 42*(3), 249–256. https://doi.org/10.1016/j.jbiomech.2008.10.035

Löffler, M. T., Jacob, A., Valentinitsch, A., Rienmüller, A., Zimmer, C., Ryang, Y. M., Baum, T., & Kirschke, J. S. (2019). Improved prediction of incident vertebral fractures using opportunistic QCT compared to DXA. *European Radiology, 29*(9), 4980–4989.

Lukas, C., Kollmannsberger, P., Ruffoni, D., Roschger, P., Fratzl, P., & Weinkamer, R. (2011). The heterogeneous mineral content of bone—using stochastic arguments and simulations to overcome experimental limitations. *Journal of Statistical Physics, 144*(2), 316–331. https://doi.org/10.1007/s10955-011-0209-8

Majumdar, S., & Genant, H. K. (1995). A review of the recent advances in magnetic resonance imaging in the assessment of osteoporosis. *Osteoporosis International, 5*(2), 79–92. https://doi.org/10.1007/bf01623308

Malgo, F., Hamdy, N. A. T., Papapoulos, S. E., & Appelman-Dijkstra, N. M. (2017). Bone material strength index as measured by impact microindentation is low in patients with fractures irrespective of fracture site. *Osteoporosis International, 28*(8), 2433–2437. https://doi.org/10.1007/s00198-017-4054-8

Manhard, M. K., Uppuganti, S., Granke, M., Gochberg, D. F., Nyman, J. S., & Does, M. D. (2016). MRI-derived bound and pore water concentrations as predictors of fracture resistance. *Bone, 87*, 1–10. https://doi.org/10.1016/j.bone.2016.03.007

Marshall, D., Johnell, O., & Wedel, H. (1996). Meta-analysis of how well measures of bone mineral density predict occurrence of osteoporotic fractures. *BMJ, 312*(7041), 1254–1259. https://doi.org/10.1136/bmj.312.7041.1254

Mayer, G. (2017). Mechanical energy dissipation in natural ceramic composites. *Journal of the Mechanical Behavior of Biomedical Materials, 76*, 21–29. https://doi.org/10.1016/j.jmbbm.2017.06.019

McNerny, E. M. B., Gong, B., Morris, M. D., & Kohn, D. H. (2015). Bone fracture toughness and strength correlate with collagen cross-link maturity in a dose-controlled lathyrism mouse model. *Journal of Bone and Mineral Research, 30*(3), 446–455. https://doi.org/10.1002/jbmr.2356

Mikolajewicz, N., Bishop, N., Burghardt, A. J., Folkestad, L., Hall, A., Kozloff, K. M., Lukey, P. T., Molloy-Bland, M., Morin, S. N., Offiah, A. C., Shapiro, J., van Rietbergen, B., Wager, K., Willie, B. M., Komarova, S. V., & Glorieux, F. H. (2020). HR-pQCT measures of bone microarchitecture predict fracture: Systematic review and meta-analysis. *Journal of Bone and Mineral Research, 35*(3), 446–459. https://doi.org/10.1002/jbmr.3901

Milisic, L., Vegar-Zubovic, S., Valjevac, A., & Hasanovic-Vučković, S. (2020). Bone mineral density assessment by DXA vs. QCT in postmenopausal females with central obesity. *Current Aging Science, 13*(2), 153–161. https://doi.org/10.2174/1874609812666190912155525

Moreaux, L., Sandre, O., & Mertz, J. (2000). Membrane imaging by second-harmonic generation microscopy. *Journal of the Optical Society of America B: Optical Physics, 17*(10), 1685–1694. https://doi.org/10.1364/JOSAB.17.001685

Morgan, S., Poundarik, A. A., & Vashishth, D. (2015). Do non-collagenous proteins affect skeletal mechanical properties? *Calcified Tissue International, 97*(3), 281–291. https://doi.org/10.1007/s00223-015-0016-3

Morris, M. D., & Mandair, G. S. (2011). Raman assessment of bone quality. *Clinical Orthopaedics and Related Research*, (8), 2160–2169. https://doi.org/10.1007/s11999-010-1692-y

Nagy, T. R., Prince, C. W., & Li, J. (2001). Validation of peripheral dual-energy X-ray absorptiometry for the measurement of bone mineral in intact and excised long bones of rats. *Journal of Bone and Mineral Research, 16*(9), 1682–1687. https://doi.org/10.1359/jbmr.2001.16.9.1682

Nickolas, T. L., Shirazian, S., & Shane, E. (2010). High-resolution computed tomography imaging: A virtual bone biopsy. *Kidney International, 77*(11), 1046. https://doi.org/10.1038/ki.2009.558

Nishiyama, K. K., & Shane, E. (2013). Clinical imaging of bone microarchitecture with HR-pQCT. *Current Osteoporosis Reports, 11*(2), 147–155. https://doi.org/10.1007/s11914-013-0142-7

Nobakhti, S., & Shefelbine, S. J. (2018). On the relation of bone mineral density and the elastic modulus in healthy and pathologic bone. *Current Osteoporosis Reports, 16*(4), 404–410. https://doi.org/10.1007/s11914-018-0449-5

Nyman, J. S., Ni, Q., Nicolella, D. P., & Wang, X. (2008). Measurements of mobile and bound water by nuclear magnetic resonance correlate with mechanical properties of bone. *Bone, 42*(1), 193–199. https://doi.org/10.1016/j.bone.2007.09.049

Ohs, N., Collins, C. J., & Atkins, P. R. (2020). Validation of HR-pQCT against micro-CT for morphometric and biomechanical analyses: A review. *Bone Reports, 13*. https://doi.org/10.1016/j.bonr.2020.100711

Ou-Yang, H., Paschalis, E. P., Mayo, W. E., Boskey, A. L., & Mendelsohn, R. (2001). Infrared microscopic imaging of bone: Spatial distribution of CO32. *Journal of Bone and Mineral Research, 16*(5), 893–900. https://doi.org/10.1359/jbmr.2001.16.5.893

Ozasa, R., Matsugaki, A., Ishimoto, T., Kamura, S., Yoshida, H., Magi, M., Matsumoto, Y., Sakuraba, K., Fujimura, K., Miyahara, H., & Nakano, T. (2022). Bone fragility via degradation of bone quality featured by collagen/apatite micro-arrangement in human rheumatic arthritis. *Bone, 155*, Article 116261. https://doi.org/10.1016/j.bone.2021.116261

Pandey, N., Bhola, S., Goldstone, A., Chen, F., Chrzanowski, J., Terranova, C. J., Ghillani, R., & Jepsen, K. J. (2009). Interindividual variation in functionally adapted trait sets is established during postnatal growth and predictable based on bone robustness. *Journal of Bone and Mineral Research, 24*(12), 1969–1980. https://doi.org/10.1359/jbmr.090525

Paschalis, E. P., Gamsjaeger, S., & Klaushofer, K. (2017). Vibrational spectroscopic techniques to assess bone quality. *Osteoporosis International, 28*(8), 2275–2291. https://doi.org/10.1007/s00198-017-4019-y

Paschalis, E. P., Shane, E., Lyritis, G., Skarantavos, G., Mendelsohn, R., & Boskey, A. L. (2004). Bone fragility and collagen cross-links. *Journal of Bone and Mineral Research, 19*(12), 2000–2004. https://doi.org/10.1359/JBMR.040820

Paschalis, E. P., Tatakis, D. N., Robins, S., Fratzl, P., Manjubala, I., Zoehrer, R., Gamsjaeger, S., Buchinger, B., Roschger, A., Phipps, R., Boskey, A. L., Dall'Ara, E., Varga, P., Zysset, P., Klaushofer, K., & Roschger, P. (2011). Lathyrism-induced alterations in collagen cross-links influence the mechanical properties of bone material without affecting the mineral. *Bone, 49*(6), 1232–1241. https://doi.org/10.1016/j.bone.2011.08.027

Paschalis, E. P., Verdelis, K., Doty, S. B., Boskey, A. L., Mendelsohn, R., & Yamauchi, M. (2001). Spectroscopic characterization of collagen cross-links in bone. *Journal of Bone and Mineral Research, 16*(10), 1821–1828. https://doi.org/10.1359/jbmr.2001.16.10.1821

Patsch, J. M., Burghardt, A. J., Yap, S. P., Baum, T., Schwartz, A. V., Joseph, G. B., & Link, T. M. (2013). Increased cortical porosity in type 2 diabetic postmenopausal women with fragility fractures. *Journal of Bone and Mineral Research, 28*(2), 313–324. https://doi.org/10.1002/jbmr.1763

Perilli, E., Briggs, A. M., Kantor, S., Codrington, J., Wark, J. D., Parkinson, I. H., & Fazzalari, N. L. (2012). Failure strength of human vertebrae: Prediction using bone mineral density measured by DXA and bone volume by micro-CT. *Bone, 50*(6), 1416–1425. https://doi.org/10.1016/j.bone.2012.03.002

Perrone, A., Giovino, A., Benny, J., & Martinelli, F. (2020). Advanced glycation end products (AGEs): Biochemistry, signaling, analytical methods, and epigenetic effects. *Oxidative Medicine and Cellular Longevity, 2020*, 1–18. https://doi.org/10.1155/2020/3818196

Peters, F., Schwarz, K., & Epple, M. (2000). The structure of bone studied with synchrotron X-ray diffraction, X-ray absorption spectroscopy and thermal analysis. *Thermochimica Acta, 361*(1–2), 131–138. https://doi.org/10.1016/s0040-6031(00)00554-2

Peyrin, F., Dong, P., Pacureanu, A., & Langer, M. (2014). Micro- and Nano-CT for the study of bone ultrastructure. *Current Osteoporosis Reports, 12*(4), 465–474. https://doi.org/10.1007/s11914-014-0233-0

Pleshko, N., Boskey, A., & Mendelsohn, R. (1991). Novel infrared spectroscopic method for the determination of crystallinity of hydroxyapatite minerals. *Biophysical Journal, 60*(4), 786–793. https://doi.org/10.1016/s0006-3495(91)82113-0

Pors Nielsen, S. (2000). The fallacy of BMD: A critical review of the diagnostic use of dual X-ray absorptiometry. *Clinical Rheumatology, 19*(3), 174–183. https://doi.org/10.1007/s100670050151

Poundarik, A. A., & Vashishth, D. (2015). Multiscale imaging of bone microdamage. *Connective Tissue Research, 56*(2), 87–98. https://doi.org/10.3109/03008207.2015.1008133

Poundarik, A. A., Wu, P. C., Evis, Z., Sroga, G. E., Ural, A., Rubin, M., & Vashishth, D. (2015). A direct role of collagen glycation in bone fracture. *Journal of the Mechanical Behavior of Biomedical Materials, 52*, 120–130. https://doi.org/10.1016/j.jmbbm.2015.08.012

Raghavan, M., Sahar, N. D., Wilson, R. H., Mycek, M.-A., Pleshko, N., Kohn, D. H., & Morris, M. D. (2010). Quantitative polarized Raman spectroscopy in highly turbid bone tissue. *Journal of Biomedical Optics, 15*(3), Article 037001. https://doi.org/10.1117/1.3426310

Rajapakse, C. S., Bashoor-Zadeh, M., Li, C., Sun, W., Wright, A. C., & Wehrli, F. W. (2015). Volumetric cortical bone porosity assessment with MR imaging: Validation and clinical feasibility. *Radiology, 276*(2), 526–535. https://doi.org/10.1148/radiol.15141850

Ritman, E. L. (2011). Current status of developments and applications of micro-CT. *Annual Review of Biomedical Engineering, 13*, 531–552. https://doi.org/10.1146/annurev-bioeng-071910-124717

Roschger, P., Fratzl, P., Eschberger, J., & Klaushofer, K. (1998). Validation of quantitative backscattered electron imaging for the measurement of mineral density distribution in human bone biopsies. *Bone, 23*(4), 319–326. https://doi.org/10.1016/s8756-3282(98)00112-4

Roschger, A., Gamsjaeger, S., Hofstetter, B., Masic, A., Blouin, S., Messmer, P., Berzlanovich, A., Paschalis, E. P., Roschger, P., Klaushofer, K., & Fratzl, P. (2014). Relationship between the v2PO4/amide III ratio assessed by Raman spectroscopy and the calcium content measured by quantitative backscattered electron microscopy in healthy human osteonal bone. *Journal of Biomedical Optics, 19*(6), Article 065002. https://doi.org/10.1117/1.jbo.19.6.065002

Roschger, P., Misof, B., Paschalis, E., Fratzl, P., & Klaushofer, K. (2014). Changes in the degree of mineralization with osteoporosis and its treatment. *Current Osteoporosis Reports, 12*(3), 338–350. https://doi.org/10.1007/s11914-014-0218-z

Roux, C., & Briot, K. (2017). Current role for bone absorptiometry. *Joint Bone Spine, 84*(1), 35–37. https://doi.org/10.1016/j.jbspin.2016.02.032

Rozental, T. D., Walley, K. C., Demissie, S., Caksa, S., Martinez-Betancourt, A., Parker, A. M., Tsai, J. N., Yu, E. W., & Bouxsein, M. L. (2018). Bone material strength index as measured by impact microindentation in postmenopausal women with distal radius and hip fractures. *Journal of Bone and Mineral Research, 33*(4), 621–626. https://doi.org/10.1002/jbmr.3338

Rubin, M. R., Paschalis, E. P., Poundarik, A., Sroga, G. E., McMahon, D. J., Gamsjaeger, S., Klaushofer, K., & Vashishth, D. (2016). Advanced glycation endproducts and bone material properties in type 1 diabetic mice. *PLoS One, 11*(5). https://doi.org/10.1371/journal.pone.0154700

Sattui, S. E., & Saag, K. G. (2014). Fracture mortality: Associations with epidemiology and osteoporosis treatment. *Nature Reviews Endocrinology, 10*(10), 592–602. https://doi.org/10.1038/nrendo.2014.125

Schaff, F., Bech, M., Zaslansky, P., Jud, C., Liebi, M., Guizar-Sicairos, M., & Pfeiffer, F. (2015). Six-dimensional real and reciprocal space small-angle X-ray scattering tomography. *Nature, 527*(7578), 353–356. https://doi.org/10.1038/nature16060

Schaffler, M. B., & Burr, D. B. (1988). Stiffness of compact bone: Effects of porosity and density. *Journal of Biomechanics, 21*(1), 13–16. https://doi.org/10.1016/0021-9290(88)90186-8

Schaffler, M. B., Choi, K., & Milgrom, C. (1995). Aging and matrix microdamage accumulation in human compact bone. *Bone, 17*(6), 521–525. https://doi.org/10.1016/8756-3282(95)00370-3

Schenk, D., Indermaur, M., Simon, M., Voumard, B., Varga, P., Pretterklieber, M., Lippuner, K., & Zysset, P. (2022). Unified validation of a refined second-generation HR-pQCT based homogenized finite element method to predict strength of the distal segments in radius and tibia. *Journal of the Mechanical Behavior of Biomedical Materials, 131*, Article 105235. https://doi.org/10.1016/j.jmbbm.2022.105235

Schileo, E., Dall'Ara, E., Taddei, F., Malandrino, A., Schotkamp, T., Baleani, M., & Viceconti, M. (2008). An accurate estimation of bone density improves the accuracy of subject-specific finite element models. *Journal of Biomechanics, 41*(11), 2483–2491. https://doi.org/10.1016/j.jbiomech.2008.05.017

Schoeb, M., Malgo, F., Peeters, J. J. M., Winter, E. M., Papapoulos, S. E., & Appelman-Dijkstra, N. M. (2020). Treatments of osteoporosis increase bone material strength index in patients with low bone mass. *Osteoporosis International, 31*(9), 1683–1690. https://doi.org/10.1007/s00198-020-05375-3

Schousboe, J. T. (2016). Epidemiology of vertebral fractures. *Journal of Clinical Densitometry, 19*(1), 8–22. https://doi.org/10.1016/j.jocd.2015.08.004

Schreiber, J. J., Anderson, P. A., & Hsu, W. K. (2014). Use of computed tomography for assessing bone mineral density. *Neurosurgical Focus, 37*(1). https://doi.org/10.3171/2014.5.FOCUS1483

Seref-Ferlengez, Z., Kennedy, O. D., & Schaffler, M. B. (2015). Bone microdamage, remodeling and bone fragility: How much damage is too much damage? *BoneKEy Reports, 4*. https://doi.org/10.1038/bonekey.2015.11

Shane, E., Burr, D., Abrahamsen, B., Adler, R. A., Brown, T. D., Cheung, A. M., Cosman, F., Curtis, J. R., Dell, R., Dempster, D. W., Ebeling, P. R., Einhorn, T. A., Genant, H. K., Geusens, P., Klaushofer, K., Lane, J. M., McKiernan, F., McKinney, R., Ng, A., … Whyte, M. P. (2014). Atypical subtrochanteric and diaphyseal femoral fractures: Second report of a task force of the American society for bone and mineral research. *Journal of Bone and Mineral Research, 29*(1), 1–23. https://doi.org/10.1002/jbmr.1998

Shevroja, E., Cafarelli, F. P., Guglielmi, G., & Hans, D. (2021). DXA parameters, trabecular bone score (TBS) and bone mineral density (BMD), in fracture risk prediction in endocrine-mediated secondary osteoporosis. *Endocrine, 74*(1), 20–28. https://doi.org/10.1007/s12020-021-02806-x

Silva, B. C., Leslie, W. D., Resch, H., Lamy, O., Lesnyak, O., Binkley, N., McCloskey, E. V., Kanis, J. A., & Bilezikian, J. P. (2014). Trabecular bone score: A noninvasive analytical method based upon the DXA image. *Journal of Bone and Mineral Research, 29*(3), 518–530. https://doi.org/10.1002/jbmr.2176

Sollmann, N., Löffler, M. T., Kronthaler, S., Böhm, C., Dieckmeyer, M., Ruschke, S., Kirschke, J. S., Carballido-Gamio, J., Karampinos, D. C., Krug, R., & Baum, T. (2021). MRI-based quantitative osteoporosis imaging at the spine and femur. *Journal of Magnetic Resonance Imaging, 54*(1), 12–35. https://doi.org/10.1002/jmri.27260

Sosa, D. D., & Eriksen, E. F. (2017). Reduced bone material strength is associated with increased risk and severity of osteoporotic fractures. An impact microindentation study. *Calcified Tissue International, 101*(1), 34–42. https://doi.org/10.1007/s00223-017-0256-5

Spevak, L., Flach, C. R., Hunter, T., Mendelsohn, R., & Boskey, A. (2013). Fourier transform infrared spectroscopic imaging parameters describing acid phosphate substitution in biologic hydroxyapatite. *Calcified Tissue International, 92*(5), 418–428. https://doi.org/10.1007/s00223-013-9695-9

Surowiec, R. K., Allen, M. R., & Wallace, J. M. (2022). Bone hydration: How we can evaluate it, what can it tell us, and is it an effective therapeutic target? *Bone Reports, 16*. https://doi.org/10.1016/j.bonr.2021.101161

Tammisalo, E. H., & Kiminki, A. (1969). Density and mineral content of human mandibles. *Acta Odontologica Scandinavica, 27*(4), 403–407. https://doi.org/10.3109/00016356909040419

Taylor, E. A., & Donnelly, E. (2020). Raman and Fourier transform infrared imaging for characterization of bone material properties. *Bone, 139*, Article 115490. https://doi.org/10.1016/j.bone.2020.115490

Taylor, E. A., Lloyd, A. A., Salazar-Lara, C., & Donnelly, E. (2017). Raman and fourier transform infrared (FT-IR) mineral to matrix ratios correlate with physical chemical properties of model compounds and native bone tissue. *Applied Spectroscopy, 71*(10), 2404–2410. https://doi.org/10.1177/0003702817709286

Termine, J. D., & Posner, A. S. (1966). Infra-red determination of the percentage of crystallinity in apatitic calcium phosphates. *Nature, 211*(5046), 268–270. https://doi.org/10.1038/211268a0

Thotakura, A., Quenneville, C., & Schwarcz, H. P. (2022). Ashing of bone: Errors due to loss of CO2 and their correction. *Journal of Bone and Mineral Metabolism, 40*(4), 594–601. https://doi.org/10.1007/s00774-022-01327-5

Turner, C. H., & Burr, D. B. (1993). Basic biomechanical measurements of bone: A tutorial. *Bone, 14*(4), 595–608. https://doi.org/10.1016/8756-3282(93)90081-k

Vaidya, R., Rezaee, T., Edwards, T., Bender, R., Vickneswaran, A., Chalivendra, V., & Karim, L. (2022). Accumulation of fluorescent advanced glycation end products and carboxymethyl-lysine in human cortical and trabecular bone. *Bone Reports, 17*, Article 101634. https://doi.org/10.1016/j.bonr.2022.101634

van den Bergh, J. P., Szulc, P., Cheung, A. M., Bouxsein, M., Engelke, K., & Chapurlat, R. (2021). The clinical application of high-resolution peripheral computed tomography (HR-pQCT) in adults: State of the art and future directions. *Osteoporosis International, 32*(8), 1465–1485. https://doi.org/10.1007/s00198-021-05999-z

van der Meulen, M. C. H., J Jepsen, K., & Mikić, B. (2001). Understanding bone strength: Size isn't everything. *Bone, 29*(2), 101–104. https://doi.org/10.1016/s8756-3282(01)00491-4

Varga, P., Dall'Ara, E., Pahr, D. H., Pretterklieber, M., & Zysset, P. K. (2011). Validation of an HR-pQCT-based homogenized finite element approach using mechanical testing of ultra-distal radius sections. *Biomechanics and Modeling in Mechanobiology, 10*(4), 431–444. https://doi.org/10.1007/s10237-010-0245-3

Vashishth, D. (2009). Advanced glycation end-products and bone fractures. *IBMS BoneKEy, 6*(8), 268–278. https://doi.org/10.1138/20090390

Vilayphiou, N., Boutroy, S., Sornay-Rendu, E., Van Rietbergen, B., & Chapurlat, R. (2016). Age-related changes in bone strength from HR-pQCT derived microarchitectural parameters with an emphasis on the role of cortical porosity. *Bone, 83*, 233–240. https://doi.org/10.1016/j.bone.2015.10.012

Wang, Z., Vashishth, D., & Picu, R. C. (2018). Bone toughening through stress-induced non-collagenous protein denaturation. *Biomechanics and Modeling in Mechanobiology, 17*(4), 1093–1106. https://doi.org/10.1007/s10237-018-1016-9

Wang, L., You, X., Zhang, L., Zhang, C., & Zou, W. (2022). Mechanical regulation of bone remodeling. *Bone Research, 10*(1). https://doi.org/10.1038/s41413-022-00190-4

Wen, X. X., Wang, F. Q., Xu, C., Wu, Z. X., Zhang, Y., Feng, Y. F., Yan, Y. B., & Lei, W. (2015). Time related changes of mineral and collagen and their roles in cortical bone mechanics of ovariectomized rabbits. *Public Library of Science, China PLoS ONE, 10*(6). https://doi.org/10.1371/journal.pone.0127973

Yu, B., Pacureanu, A., Olivier, C., Cloetens, P., & Peyrin, F. (2020). Assessment of the human bone lacuno-canalicular network at the nanoscale and impact of spatial resolution. *Scientific Reports, 10*(1). https://doi.org/10.1038/s41598-020-61269-8

Zhao, S., Arnold, M., Ma, S., Abel, R. L., Cobb, J. P., Hansen, U., & Boughton, O. (2018). Standardizing compression testing for measuring the stiffness of human bone. *Bone & Joint Research, 7*(8), 524–538. https://doi.org/10.1302/2046-3758.78.BJR-2018-0025.R1

Zhou, M., Sadoughi, S., Go, L., Ramil, G., Yu, I., Saeed, I., Fan, B., Wu, P.-H., Salusky, I. B., Nickolas, T. L., Ix, J. H., & Kazakia, G. J. (2025). Time-lapse HR-pQCT reliably assesses and monitors local bone turnover in patients with chronic kidney disease. *Journal of Bone and Mineral Research*. https://doi.org/10.1093/jbmr/zjaf006

Zhou, B., Sherry Liu, X., Wang, J., Lucas Lu, X., Fields, A. J., & Edward Guo, X. (2014). Dependence of mechanical properties of trabecular bone on plate-rod microstructure determined by individual trabecula segmentation (ITS). *Journal of Biomechanics, 47*(3), 702–708. https://doi.org/10.1016/j.jbiomech.2013.11.039

Zimmermann, E. A., Schaible, E., Bale, H., Barth, H. D., Tang, S. Y., Reichert, P., Busse, B., Alliston, T., Ager, J. W., & Ritchie, R. O. (2011). Age-related changes in the plasticity and toughness of human cortical bone at multiple length scales. *Proceedings of the National Academy of Sciences of the United States of America, 108*(35), 14416–14421. https://doi.org/10.1073/pnas.1107966108

Zimmermann, E. A., Schaible, E., Gludovatz, B., Schmidt, F. N., Riedel, C., Krause, M., Vettorazzi, E., Acevedo, C., Hahn, M., Püschel, K., Tang, S., Amling, M., Ritchie, R. O., & Busse, B. (2016). Intrinsic mechanical behavior of femoral cortical bone in young, osteoporotic and bisphosphonate-treated individuals in low- and high energy fracture conditions. *Scientific Reports, 6*(1), Article 21072. https://doi.org/10.1038/srep21072

PART

2

Mechanical and multiphysical properties of the bone and constitutive laws

CHAPTER

Bone poromechanics

6

Luis Cardoso

Department of Biomedical Engineering, City University of New York, New York, NY, United States

1. Composition, hierarchical organization, and fluid-saturated porosity of bone

Bone is a highly complex organ and tissue that consists of a mineralized bone matrix, bone cells, fluid, connective tissue, blood vessels, and nerves. The bone matrix is mainly composed of type I collagen, minerals, and noncollagenous proteins. The fluids in bone include both blood within blood vessels and interstitial fluid outside blood vessels, within the bone's pore spaces. Bone provides a framework for movement, offers mechanical support, protects organs, stores minerals and fat, and participates in hormone production and blood cell formation. It is a dynamic tissue capable of adding and removing bone mass through cell-driven processes such as modeling, remodeling, and fracture healing, as well as self-healing at the matrix level. Bone has a well-organized structure that integrates its solid and fluid constituents across multiple levels (Fig. 6.1), from the molecular to the anatomical scale (Reznikov et al., 2018; Rodriguez-Palomo et al., 2024), consisting of distinct, hierarchically nested, fluid-saturated porous networks (Cowin, 1999; Cowin et al., 2009). These fluid-filled porous structures and tissue composition enable bone to fulfill its biomechanical, metabolic, cellular, and hormonal functions.

The vascular porosity (VP) comprises pore spaces containing blood vessels and interstitial fluid in cortical bone, including both Haversian and Volkmann canals, which are about 50 μm in diameter (Cowin, 1999). The lacunar–canalicular porosity (LCP) contains osteocyte lacunae pores and canaliculi channels, measuring roughly 10–20 μm and 100–700 nm, respectively (Cardoso, Fritton, et al., 2013). The pericellular space contains the osteocyte's glycocalyx, proteoglycans, and interstitial fluid. Recently, a matrix nanoporosity (MNP) network intertwined with the LCP within the solid matrix was discovered (Tang, Landis, Blouin, et al., 2023; Tang, Landis, Raguin, et al., 2022). The smallest porosity, about 1 nm in size, is the collagen-apatite porosity (CAP), where water is bound to the mineral phase (Cowin, 1999; Wehrli & Fernández-Seara, 2005).

Biomechanics and Mechanobiology of Bone. https://doi.org/10.1016/B978-0-443-26739-0.00003-X

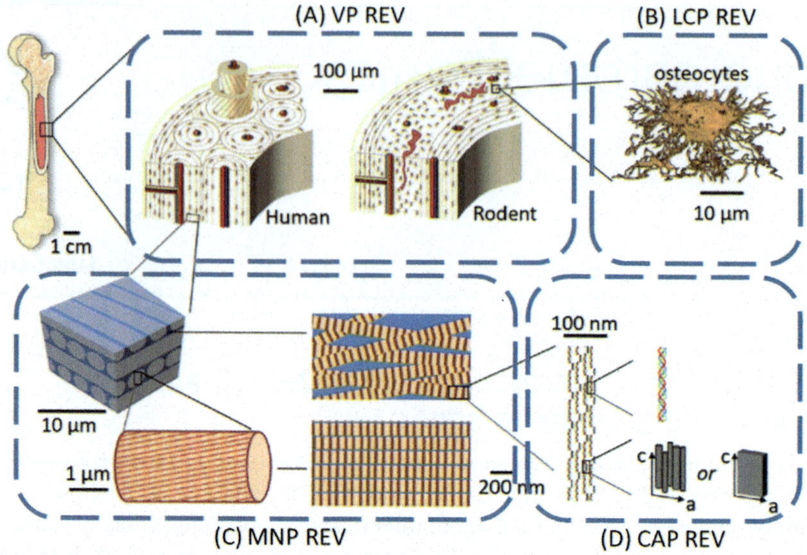

FIGURE 6.1

Nested fluid-saturated poroelastic systems at different scale levels in bone. (A) Osteons and vascular canals in cortical bone, (B) lacunar-canalicular system in cortical bone, (C) matrix nanoporosity system, (D) collagen-apatite porosity system.

- *Adapted from Rodriguez-Palomo, A., Østergaard, M., & Birkedal, H. (2024). Bone hierarchical structure: Heterogeneity and uniformity.* Advanced Functional Materials, 34*(35), 2307026.*

2. Bone poromechanics

Bone poroelasticity is the mathematical framework that integrates theories for elastic solids, viscous fluids, and fluid flow within the porous structure of bone. It consists of equations derived from a mix of conservation principles and constitutive laws. Poroelasticity provides a continuum description for materials whose macroscopic mechanical response results from interactions between a deformable solid skeleton and an interstitial fluid occupying the pore network. Applied to bone, this theory considers the biphasic nature of the tissue as a fluid-saturated porous medium and its hierarchical structure to explain the biomechanical functions of bone (Cowin, 1999, 2001). Poroelasticity enables us to study and better understand phenomena such as solute transport, convection-driven fluid flow, mechanotransduction, growth, adaptation, fluid flow across nested bone porosities, wave propagation, and more (Cowin & Doty, 2007; Cowin et al., 1991; Weinbaum et al., 1994). There are various approaches to developing the theory of poroelastic media, including the effective medium, micromechanics, mixture theory, homogenization theory, and variational energy formulation (Biot, 1941; Boer, 2000; Cheng, 2016; Coussy, 2003, 2010; Cowin, 2013; Cowin & Cardoso, 2012; Nemat-Nasser, 1993). In this chapter, we will follow the effective medium theory approach used by Biot (1941), in which fluid-saturated porous media are viewed as a single bulk material whose deformation results from both the stresses and the pore pressure acting on it (Cheng, 2016; Cowin, 1999, 2013). The core principle of Biot's poroelasticity theory is the averaging process of biphasic medium mechanics across two different length scales. At the larger length scale, the averaged, apparent mechanical behavior of an inhomogeneous, fluid-saturated porous medium arises from the behavior and interactions of its solid and fluid components, which are considered homogeneous at the smaller length scale. This allows an inhomogeneous medium at the larger scale to be treated as homogeneous through the averaging process. A homogeneous medium is understood as one with continuous derivatives within the considered length scale. These length scales are relative, and the stresses and strains are averaged within a Representative Elementary Volume (REV) at the relevant scale (e.g., vascular, lacunar-canalicular, nanoporosity, and crystal-apatite porosity levels; see Fig. 6.1). The pores are typically one to two orders of magnitude smaller than the REV, which is sufficiently small compared to the phenomenon's overall length scale so that the fluid-saturated porous medium can be treated mathematically as infinitesimal and homogeneous.

A key role of blood and interstitial fluid in bone is to carry nutrients, oxygen, calcium, and other molecules from the vascular system to the tissue cells, and to remove the cells' metabolic waste products (Cowin & Cardoso, 2015). Convection-driven interstitial fluid flow is crucial for solute transport in bone, and it is essential for the amplification of bone tissue loading at the osteocytes' processes within the lacunar-canalicular system (LCS) (Weinbaum et al., 1994; Zeng et al., 1994). In bone, the stiffness of the solid phase (e.g., mineralized matrix material) is 6–7 times greater than the fluid's bulk modulus (Cowin, 1999). Therefore, deformation of a poroelastic medium within an REV generally induces fluid motion within the pores (convection-driven fluid flow). Simultaneously, an increase in pore fluid pressure has only a minor effect on the deformation of the solid phase of the porous medium when the pores at the REV boundary are open (the drained condition). However, when the pores are closed, an increase in pore fluid pressure can oppose the free deformation of the solid porous structure and contribute to the overall stress in the matrix material (undrained condition) (Wang, Fritton, et al., 1999; Zhang et al., 1998). The pores in bone are fully connected within the REV, enabling fluid flow through them. The REV pore boundary condition can be either drained or undrained. The drained condition occurs when the pores at the REV boundary are open, allowing fluid to move freely out of the pore medium; in this case, we assume the pore fluid pressure is zero ($p = 0$). Conversely, the undrained condition applies to pores at the REV boundary that are closed, preventing fluid movement in or out, resulting in a nonzero pore fluid pressure when the porous medium deforms (Cheng, 2016; Cowin & Mehrabadi, 2007; Gailani & Cowin, 2008).

3. Constitutive equations

3.1 Notation

In this and the following sections, the equations of poroelasticity are presented for the case of material anisotropy, adapted mainly from the work of Cowin (1999, 2001, 2003, 2004, 2013) and Cheng (2016). The notation employed follows the convention of using italic letters for scalar quantities (e.g., x), bold lower case symbols for vectors (e.g., $\boldsymbol{x}$), bold capitals for second rank tensors (e.g., $\boldsymbol{X}$), and outline capital letters for fourth rank tensors (e.g., $\mathbb{X}$). The components of vectors and tensors can be expressed using indicial notation, following Einstein's convention of repeated indices summation. Also, a comma in the subscript followed by an index indicates partial differentiation with respect to the spatial coordinate (e.g., $u_{i,j} = \partial u_i / \partial x_j$). In tensor notation, we use a dot to indicate summation over one index (i.e., $\boldsymbol{Z} = \boldsymbol{X}\cdot\boldsymbol{Y}$, $Z_{ij} = X_{ik}Y_{kj}$), and a colon symbol to express the summation (contraction) over two indices (i.e., $\boldsymbol{Z} = \mathbb{X} : \boldsymbol{Y}$, $Z_{ij} = X_{ijkl}Y_{kl}$). The tensor product of two vectors, $\boldsymbol{x}$ and $\boldsymbol{y}$, is given by the outer product $\boldsymbol{x} \otimes \boldsymbol{y} = \boldsymbol{x}\boldsymbol{y}^T = \boldsymbol{Z}$, which results in a second rank tensor, $\boldsymbol{Z}$, and the tensor product of 2 second rank tensors, say $\boldsymbol{X}$ and $\boldsymbol{Y}$, is $\boldsymbol{X} \otimes \boldsymbol{Y} = \boldsymbol{X}\boldsymbol{Y}^T = \mathbb{Z}$, a fourth rank tensor, $\mathbb{Z}$.

3.2 Solid elastic materials

The reader is reminded that in an elastic solid material, the elasticity constitutive equation linearly relates the Cauchy stress tensor, $\boldsymbol{T} = \sigma_{ij}$, to the strain tensor, $\boldsymbol{E} = \varepsilon_{ij}$, by a fourth-rank tensor, $\mathbb{C} = C_{ijkl}$, the stiffness tensor representing the material constants (elastic moduli) of the material, as follows:

$$\boldsymbol{T} = \mathbb{C} : \boldsymbol{E} \tag{6.1}$$

and the inverse of Eq. (6.1) is the strain-stress constitutive relation for an elastic solid material, given by

$$\boldsymbol{E} = \mathbb{S} : \boldsymbol{T} \tag{6.2}$$

where $\mathbb{S}$ is known as the elastic compliance tensor, and $\mathbb{S} = \mathbb{C}^{-1}$. The relation between the solid displacement vector $\boldsymbol{u} = u_i$ and the total strain tensor, $\boldsymbol{E}$, is

$$\boldsymbol{E} = \frac{1}{2}\left([\nabla \otimes \boldsymbol{u}]^T + \nabla \otimes \boldsymbol{u}\right) \tag{6.3}$$

The stiffness tensor in Eq. (6.1) follows the symmetry relations $C_{ijkl} = C_{jikl}$ and $C_{ijkl} = C_{ijlk}$, because both stress and strain tensors are symmetric, $\sigma_{ij} = \sigma_{ji}$ and $\varepsilon_{ij} = \varepsilon_{ji}$. Additionally, $C_{ijkl} = C_{lkij}$ due to energy conservation in elastic materials, where the displacement gradient and the stress tensor are work conjugates of the strain energy density, U,

$$\sigma_{ij} = \frac{\partial U}{\partial \varepsilon_{ij}}, \rightarrow C_{ijkl} = \frac{\partial^2 U}{\partial \varepsilon_{ij}\varepsilon_{kl}} \tag{6.4}$$

3.3 The stress-strain-pore pressure constitutive relation

In Biot's theory, a fluid-saturated porous material is treated as an ordinary elastic material, recognizing that the porous medium deforms under an effective (total) stress acting on both the solid and fluid phases (Biot, 1941, 1955). Within the poroelastic framework, the stress transmitted through the material may be interpreted as the superposition of a matrix-borne component and a pressure-mediated fluid contribution, reflecting the shared load-bearing role of both phases (Fig. 6.2). Consequently, two constitutive equations, involving both solid and fluid phases, relate the stresses and strains in the porous medium via bulk material constants.

Let's first consider the hydrostatic pressure loading, which acts on both the inner and outer surfaces of the porous medium (Fig. 6.2A), and results in a uniform strain $E^{(p)}$ in the solid matrix material,

$$\boldsymbol{E}^{p} = -\mathbb{S}^{m} : \boldsymbol{I}p, \tag{6.5}$$

where $\boldsymbol{I} = \delta_{ij}$, is the second-rank identity tensor, and the fourth-rank tensor $\mathbb{S}^m$ represents the anisotropic elastic compliance tensor of the solid matrix material. At the analyzed length scale, the solid matrix material is often regarded as isotropic and pore-free (Cowin, 1999). Under these conditions, the $\mathbb{S}^m$ tensor becomes equal to the inverse of the effective bulk modulus of the solid, $1/K^m$.

The second loading condition involves a surface traction that deforms the porous frame and its pore space, thereby altering the pore fluid pressure (Fig. 6.2B). The

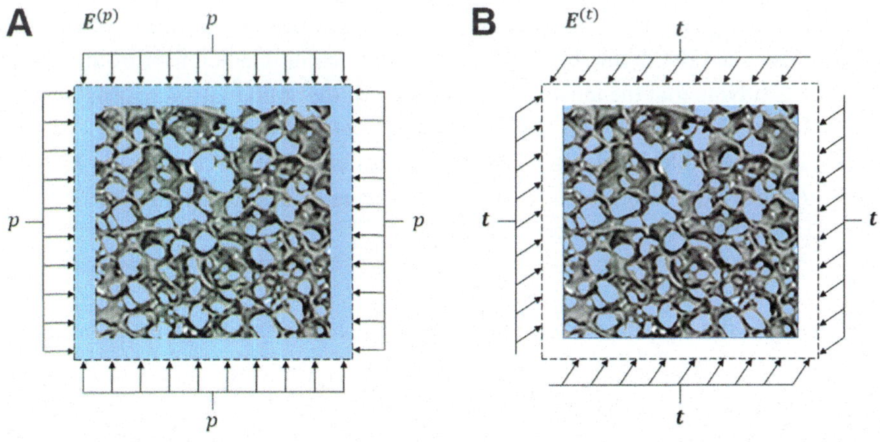

FIGURE 6.2

Loading on a fluid-saturated porous medium by superposition of a hydrostatic pressure, p, and a traction force, t. (A) Hydrostatic pressure, p, is produced by the fluid in both the inner and outer surfaces of the solid material. (B) Surface traction force, $\boldsymbol{t}$, acting on the fluid-saturated porous medium, producing deformation of the pore frame and a change in the pore fluid pressure.

strain from this loading condition results from both pressure, p, and the average stress, $\boldsymbol{T}$, acting on the porous frame:

$$\boldsymbol{E}^t = -\,\mathbb{S}^d : \boldsymbol{T} + \boldsymbol{I}p. \tag{6.6}$$

The total stresses (i.e., $\boldsymbol{T} + \boldsymbol{I}p$) applied to the porous structure and the resulting total strains $\boldsymbol{E}$ are connected by the drained anisotropic elastic compliance tensor of the porous frame, $\mathbb{S}^d$. Therefore, the total strain caused by both hydrostatic pressure loading and solid phase traction loading is

$$\boldsymbol{E} = \boldsymbol{E}^p + \boldsymbol{E}^t = \mathbb{S}^d : \boldsymbol{T} + \left(\mathbb{S}^d - \mathbb{S}^m\right) : \boldsymbol{I}p \tag{6.7}$$

In the previous equations, the fluid pressure term is negative, and the solid stress is positive due to the different sign conventions for stress and pressure. Eq. (6.7) was rearranged so that the average total strain $\boldsymbol{E}$ in a fluid-saturated porous medium is a function of the average stress $\boldsymbol{T}$, the pore fluid pressure p, the drained anisotropic compliance elastic constants of the porous medium, $\mathbb{S}^d$, and a symmetric second-rank tensor $\boldsymbol{A}$:

$$\begin{aligned} \boldsymbol{E} &= \mathbb{S}^d : \boldsymbol{T} + \mathbb{S}^d : \left(\mathbb{I} - \mathbb{C}^d : \mathbb{S}^m\right) : \boldsymbol{I}p = \mathbb{S}^d : \boldsymbol{T} + \mathbb{S}^d : \boldsymbol{A}p \\ &= \mathbb{S}^d : (\boldsymbol{T} + \boldsymbol{A}p) \end{aligned} \tag{6.8}$$

where $\mathbb{I} = I_{ijkl} = \delta_{ik}\delta_{jl}$, is the fourth rank identity tensor, and the tensor

$$\boldsymbol{A} = \left(\mathbb{I} - \mathbb{C}^d : \mathbb{S}^m\right) : \boldsymbol{I}, \tag{6.9}$$

is known as the *Biot effective stress coefficient tensor* (Carroll, 1979; Nur & Byerlee, 1971; Thompson & Willis, 1991). This Biot effective stress coefficient tensor $\boldsymbol{A}$ is related to the ratio of the effective anisotropic compliance tensor of the solid matrix material, $\mathbb{S}^m$, and the drained effective anisotropic compliance tensor of the porous medium, $\mathbb{S}^d$. Note that the tensor $\boldsymbol{A}$ is a property of the solid matrix material and the porous frame only, and it does not involve the fluid properties.

It is also possible to identify in Eq. (6.8) an effective stress, $\boldsymbol{T}^{eff} = \boldsymbol{T} + \boldsymbol{A}p$, which reduces Eq. (6.8) into $\boldsymbol{E} = \mathbb{S}^d : \boldsymbol{T}^{eff}$. Inversion of the relation in Eq. (6.8) leads to the stress-pore pressure-strain relation:

$$\boldsymbol{T} + \boldsymbol{A}p = \mathbb{C}^d : \boldsymbol{E}, \tag{6.10}$$

where $\mathbb{C}^d$, the drained anisotropic stiffness elastic constants tensor, is the inverse of $\mathbb{S}^d$.

3.4 The fluid content-stress-pore pressure constitutive relation

The work done on the poroelastic medium is given by $dU = \boldsymbol{T} : d\boldsymbol{E} + p d\zeta$, and it is a function of the total stress $\boldsymbol{T}$, pore pressure p, solid phase strain $\boldsymbol{E}$, and the variation in (dimensionless) fluid content ζ. The variation in fluid content is a scalar quantity equivalent to the change of fluid mass per unit volume in the porous medium. Such a change in fluid content may happen because of a variation in the fluid volume fraction (porosity), ϕ, or a change in the mass density of the fluid phase, ρ_f. Therefore, the variation in fluid content can be written as:

$$\zeta = \frac{\rho_f \phi - \phi_{f0} \phi_0}{\rho_{f0}} = \frac{\phi_f}{\phi_{f0}} \phi - \phi_0 \tag{6.11}$$

where ρ_{f0} and ϕ_0 are the initial values of the fluid mass density and porosity, respectively. Rewriting the constitutive Eq. (6.10) in terms of the variation in fluid content, ζ, instead of the pore pressure, p, and introducing a second-rank tensor $\boldsymbol{M}$,

$$\boldsymbol{T} = \mathbb{C}^u : \boldsymbol{E} - \boldsymbol{M}\zeta, \tag{6.12}$$

and the pore pressure is related to the strain tensor and the variation in fluid content, ζ, following

$$p = -\boldsymbol{M} : \boldsymbol{E} + M\zeta \tag{6.13}$$

where $\boldsymbol{M} = M\boldsymbol{A}$, and M is known as the *Biot modulus*, thus,

$$\boldsymbol{T} = \mathbb{C}^u : \boldsymbol{E} - M\boldsymbol{A}\zeta \tag{6.14}$$

$$p = -M\boldsymbol{A} : \boldsymbol{E} + M\zeta. \tag{6.15}$$

The inverse relation of Eq. (6.15) gives the variation in fluid content ζ in terms of the strain and the pore pressure,

$$\zeta = \boldsymbol{A} : \boldsymbol{E} + \Lambda p, \Lambda = M^{-1} \tag{6.16}$$

and the inverse of the Biot modulus, Λ, is also known as the *storage coefficient*, defined as the volume of exchanged fluid per unit volume of porous media, due to a change in pore pressure. Using Eq. (6.8) on Eq. (6.16), the variation in fluid content ζ expressed in terms of the stress $\boldsymbol{T}$ and the pore pressure, p, is

$$\begin{aligned} \zeta &= \boldsymbol{A} : \boldsymbol{E} + \Lambda p = \boldsymbol{A} : \mathbb{S}^d : \boldsymbol{T} + \left(\boldsymbol{A} : \mathbb{S}^d : \boldsymbol{A} + \Lambda\right) p \\ &= \boldsymbol{A} : \mathbb{S}^d : \boldsymbol{T} + S^d_{eff} p, \end{aligned} \tag{6.17}$$

where the effective drained compliance modulus, S^d_{eff}, was introduced as

$$S^d_{eff} = \boldsymbol{A} : \mathbb{S}^d : \boldsymbol{A} + \Lambda \tag{6.18}$$

The stress-strain-fluid content constitutive equation is obtained by substituting Eq. (6.15) into Eq. (6.10),

$$T = \mathbb{C}^d : \boldsymbol{E} - \boldsymbol{A}p = \mathbb{C}^d : E + M(\boldsymbol{A} \otimes \boldsymbol{A}) : E - M\boldsymbol{A}\zeta \tag{6.19}$$

where the term

$$\mathbb{C}^u = \mathbb{C}^d + M(\boldsymbol{A} \otimes \boldsymbol{A}) \tag{6.20}$$

relates the elastic constants of the drained to the undrained condition. Thus, Eq. (6.19) becomes

$$T = \mathbb{C}^u : E - MA\zeta, \tag{6.21}$$

and the stress-strain pore pressure relation is obtained by using Eq. (6.16) on Eq. (6.14):

$$\begin{aligned} T &= \mathbb{C}^u : E - MA\left(\frac{p}{M} + A : E\right) = \mathbb{C}^u - M(A \otimes A) : E - Ap \\ &= \mathbb{C}^d : E - Ap \end{aligned} \tag{6.22}$$

3.5 Skemptom tensor

The product $\mathbb{S}^d : A$ can be rewritten using an effective drained compliance modulus, S^d_{eff}, and a second rank tensor B, as:

$$\mathbb{S}^d : A = \frac{1}{3} S^d_{eff} B, \tag{6.23}$$

where the tensor B is known as the *Skempton compliance difference tensor* (Skempton, 1954):

$$B = \frac{3}{S^d_{eff}} \mathbb{S}^d : A = \frac{3}{S^d_{eff}} (\mathbb{S}^d - \mathbb{S}^m) : I \tag{6.24}$$

Substituting Eq. (6.24) into Eqs. (6.8) and (6.17),

$$E = \mathbb{S}^d : T + \mathbb{S}^d : Ap = \mathbb{S}^d : T + \frac{1}{3} S^d_{eff} Bp \tag{6.25}$$

$$\zeta = (\mathbb{S}^d : A) : T + S^d_{eff} p = \frac{1}{3} S^d_{eff} B : T + S^d_{eff} p \tag{6.26}$$

Then, Eq. (6.26) is used to eliminate the dependency on the pore pressure in Eq. (6.25), and write it as a function of the variation in fluid content:

$$E = \left(\mathbb{S}^d - \frac{1}{9} S^d_{eff} B \otimes B\right) : T + \frac{1}{3} B\zeta = \mathbb{S}^u : T + \frac{1}{3} B\zeta, \tag{6.27}$$

where

$$\mathbb{S}^u = \left(\mathbb{S}^d - \frac{1}{9} S^d_{eff} B \otimes B\right) = \mathbb{S}^d - \frac{1}{S^d_{eff}} \mathbb{S}^d : A \otimes \mathbb{S}^d : A \tag{6.28}$$

represents the relationship between undrained and drained elastic constants in the porous medium for the case of compressibility. Rearranging Eq. (6.27) and multiplying by C^u throughout,

$$T = \mathbb{C}^u : \left(E - \frac{1}{3} B\zeta\right). \tag{6.29}$$

Likewise, rearranging Eq. (6.25) and multiplying throughout by $\mathbb{C}^d$, we obtain

$$T = \mathbb{C}^d : E - \frac{1}{3} S^d_{eff} \mathbb{C}^d : Bp. \tag{6.30}$$

Finally, by comparing Eqs. (6.21) and (6.29), we identify

$$MA = \frac{1}{3}\mathbb{C}^u : \boldsymbol{B}, \tag{6.31}$$

and comparing the left-hand side of Eq. (6.21) with Eq. (6.30), we obtain

$$\boldsymbol{A} = \frac{1}{3}S^d_{eff}\mathbb{C}^d : \boldsymbol{B} = \frac{1}{3M}\mathbb{C}^d : \boldsymbol{B} \tag{6.32}$$

Thus, from Eq. (6.24),

$$\boldsymbol{B} = \left(\frac{3}{S^d_{eff}}\right)\mathbb{S}^d : \boldsymbol{A} = 3M\mathbb{S}^u : \boldsymbol{A} \tag{6.33}$$

and

$$S^d_{eff} = \left(\mathbb{C}^d : \boldsymbol{B}\right)^{-1} : \boldsymbol{A} = \frac{1}{M} + \boldsymbol{A} : \mathbb{S}^d : \boldsymbol{A} \tag{6.34}$$

Combining Eqs. (6.28) and (6.29),

$$M = \left(S^d_{eff} - \boldsymbol{A} : \mathbb{S}^d : \boldsymbol{A}\right)^{-1} = \frac{1}{S^d_{eff}} + \frac{1}{9}\boldsymbol{B} : \boldsymbol{B} : \mathbb{C}^u \tag{6.35}$$

3.6 Drained and undrained conditions

In the constitutive equations presented above (Eqs. 6.21 and 6.22), the stresses were linked to strains using either the drained or the undrained anisotropic elastic constant tensors, respectively. Constitutive Eq. (6.22), which involves the pore pressure, is associated with the drained tensor, while Eq. (6.21), which depends on the variation in fluid content, is related to the undrained elastic constant tensor. The undrained condition is understood as the case in which the fluid is unable to move into or out of the porous frame, either because the boundary pore condition prevents it (closed pores) or because the fluid does not have enough time to escape from open pores. Under this condition, there is no relative movement between the solid and fluid, and therefore, $\zeta = 0$, even though the solid frame and the fluid can deform. Consider the undrained condition being applied to Eq. (6.26),

$$\begin{aligned} \zeta = 0 &= \frac{1}{3}S^d_{eff}\boldsymbol{B} : \boldsymbol{T} + S^d_{eff}p, \\ \rightarrow p &= -\frac{1}{3}\boldsymbol{B} : \boldsymbol{T} \end{aligned} \tag{6.36}$$

which means that pore pressure changes may be produced by either normal or shear stresses acting on the porous structure, or by a combination of both. In contrast, the drained condition means the fluid can move in and out of the porous frame, as the pores are open and there is enough time for the fluid to do so. The consequence of the drained condition is that the pore fluid pressure is equivalent to zero, $p = 0$. It is thus easy to see that Eq. (6.21) under the undrained condition, $\zeta = 0$, becomes

$$\boldsymbol{T} = \mathbb{C}^u : \boldsymbol{E} - M\boldsymbol{A}\zeta = \mathbb{C}^u : \boldsymbol{E}, \tag{6.37}$$

and Eq. (6.22) under the drained condition, $p = 0$, is

$$T = \mathbb{C}^d : E - Ap = \mathbb{C}^d : E. \tag{6.38}$$

Both equations are similar to the constitutive equation of a solid elastic material. Still, the material constants have different meanings: they are either the undrained or drained anisotropic elastic constants of the porous medium. The difference between drained and undrained elastic constants is that in the drained case, the elastic constants refer to the mechanical response of the porous frame only. In contrast, in the undrained case, the elastic constants are a combination of contributions from the solid and fluid phases.

3.7 Effective moduli

The isotropic matrix elastic compliance tensor, $\mathbb{S}^m$, is estimated using effective medium theory, by contracting twice the tensor $\mathbb{S}^m$ with the identity matrix $\boldsymbol{I}$,

$$\boldsymbol{I} : \mathbb{S}^m : \boldsymbol{I} = \frac{3(1-2\nu)}{E} = \frac{1}{K^m} \tag{6.39}$$

which is equal to the reciprocal of the isotropic bulk modulus of the solid matrix material, K^m, the isotropic Young's modulus, E, and the Poisson's ratio, ν. Also, the effective drained compliance modulus, S^d_{eff}, is expressed as a function of effective moduli, i.e., K^d_{Reff}, K^m_{Reff}, bulk modulus, K^f, and porosity, ϕ, as

$$S^d_{eff} = \frac{1}{K^d_{Reff}} - \frac{1}{K^m_{Reff}} + \phi\left(\frac{1}{K^f} - \frac{1}{K^m_{Reff}}\right) \tag{6.40}$$

where K^d_{Reff} is the Reuss effective drained bulk modulus, K^m_{Reff}, the Reuss effective matrix bulk modulus, and K^f is the fluid bulk modulus. The effective bulk modulus of an anisotropic elastic compliance tensor falls between a lower and upper bound given by the Reuss, K_{Reff}, and Voigt, K_{Veff}, effective moduli:

$$(\boldsymbol{I} : \mathbb{S} : \boldsymbol{I})^{-1} = K_{Reff} \leq K_{eff} \leq K_{Veff} = (\boldsymbol{I} : \mathbb{S}^{-1} : \boldsymbol{I})/9. \tag{6.41}$$

4. Stress-strain-fabric tensor relations

Trabecular bone anisotropy can be characterized by a nondimensional measure of the mass orientation distribution (Whitehouse, 1974), and it can be expressed mathematically as a second rank, symmetric fabric tensor, $\boldsymbol{F}$ (Harrigan & Mann, 1984). The eigenvalues of the fabric tensor represent the magnitude of the preferential distribution of mass, and the eigenvalues correspond to the principal orientations of the mass (Fig. 6.3).

A general relation between the stress, strain, and fabric tensor was developed by considering the stress tensor $\boldsymbol{T}$ as an isotropic function of the stress tensor $\boldsymbol{E}$, and the fabric tensor, $\boldsymbol{F}$ (Cowin, 1985), leading to the expression:

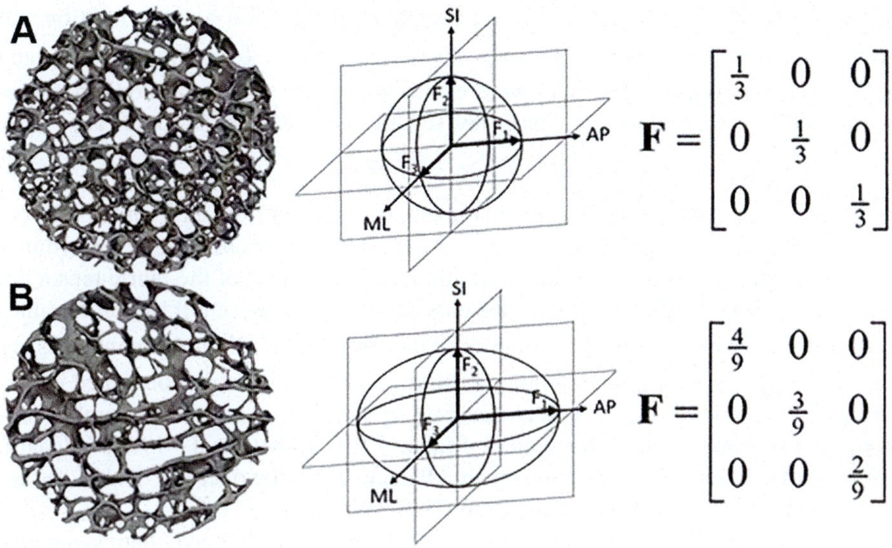

FIGURE 6.3

Trabecular bone microarchitecture, geometrical and mathematical representation of fabric tensor eigenvalues. (A) Bone sample with even distribution of solid phase (mineralized matrix material), represented by a sphere with equal eigenvalues ($F_1 = F_2 = F_3$). (B) Sample with a preferential direction of bone mass distribution in space along direction 1, corresponding to an ellipsoid, with unequal eigenvalues ($F_1 > F_2 > F_3$).

$$\begin{aligned} \boldsymbol{T} &= a_1 \boldsymbol{I} tr\boldsymbol{E} + a_2(\boldsymbol{F} tr\boldsymbol{E} + \boldsymbol{I} tr[\boldsymbol{E}\cdot\boldsymbol{F}]) + a_3(\boldsymbol{I} tr[\boldsymbol{F}\cdot\boldsymbol{E}\cdot\boldsymbol{F}] + tr\boldsymbol{E}\boldsymbol{F}^2) \\ &+ b_1 \boldsymbol{F} tr[\boldsymbol{F}\cdot\boldsymbol{E}] + b_2(\boldsymbol{F} tr[\boldsymbol{F}\cdot\boldsymbol{E}] + (tr[\boldsymbol{E}\cdot\boldsymbol{F}])\boldsymbol{F}^2) + b_3(\boldsymbol{F}^2 tr[\boldsymbol{F}\cdot\boldsymbol{E}\cdot\boldsymbol{F}]) \\ &+ 2c_1\boldsymbol{E} + 2c_2(\boldsymbol{F}\cdot\boldsymbol{E} + \boldsymbol{E}\cdot\boldsymbol{F}) + 2c_3(\boldsymbol{F}^2\cdot\boldsymbol{E} + \boldsymbol{E}\cdot\boldsymbol{F}^2), \end{aligned} \tag{6.42}$$

where the constants $a_1 \ldots a_3, b_1 \ldots b_3$, and $c_1 \ldots c_3$ are determined experimentally. The relationship 6.42 implies that the apparent level anisotropy of the porous medium is a consequence of the anisotropy of the porous frame. Based on Eq. (6.42), the relationship between the fourth rank, drained, anisotropic stiffness tensor, $\mathbb{C}^d$, the second-rank fabric tensor, $\boldsymbol{F}$, and nine parameters, $a_i^{cd}, b_i^{cd}, c_i^{cd}, i = 1 \ldots 3$, is (Cowin, 1985, 2004),

$$\begin{aligned} \mathbb{C}^d &= a_1^{cd}\boldsymbol{I}\otimes\boldsymbol{I} + a_2^{cd}(\boldsymbol{F}\otimes\boldsymbol{I} + \boldsymbol{I}\otimes\boldsymbol{F}) + a_3^{cd}\left(\boldsymbol{I}\otimes\boldsymbol{F}\cdot\boldsymbol{F} + (\boldsymbol{I}\otimes\boldsymbol{F}\cdot\boldsymbol{F})^T\right) \\ &+ b_1^{cd}\boldsymbol{F}\otimes\boldsymbol{F} + b_2^{cd}\left(\boldsymbol{F}\otimes\boldsymbol{F}\cdot\boldsymbol{F} + (\boldsymbol{F}\otimes\boldsymbol{F}\cdot\boldsymbol{F})^T\right) + b_3^{cd}\boldsymbol{F}\cdot\boldsymbol{F}\otimes\boldsymbol{F}\cdot\boldsymbol{F} \\ &+ c_1^{cd}\left[\boldsymbol{I}^T\otimes\boldsymbol{I}^T + (\boldsymbol{I}\otimes\boldsymbol{I})^T\right] + c_2^{cd}\left(\boldsymbol{F}\otimes\boldsymbol{I} + \boldsymbol{F}\otimes\boldsymbol{I}^T + \boldsymbol{F}^T\otimes\boldsymbol{I}^T + (\boldsymbol{F}^T\otimes\boldsymbol{I}^T)^T\right) \\ &+ c_3^{cd}\left(\left(\boldsymbol{F}\cdot\boldsymbol{F}\otimes\boldsymbol{I} + \boldsymbol{F}\cdot\boldsymbol{F}\otimes\boldsymbol{I}^T + (\boldsymbol{F}\cdot\boldsymbol{F})^T\right)\otimes\boldsymbol{I}^T + \left[(\boldsymbol{F}\cdot\boldsymbol{F})^T\otimes\boldsymbol{I}^T\right]^T\right), \end{aligned} \tag{6.43}$$

where $a_1^{cd}\ldots a_3^{cd}, b_1^{cd}\ldots b_3^{cd}, c_1^{cd}\ldots c_3^{cd}$ are functions of the solid volume fraction and invariants of the fabric tensor, $I_F = tr\boldsymbol{F}$, $II_F = \frac{1}{2}\left[(tr\boldsymbol{F})^2 - tr\boldsymbol{F}^2\right]$, and $III_F = \det(\boldsymbol{F})$. These parameters were evaluated against trabecular bone measurements, and curve-fitted against power law functions of the form (Kabel et al., 1999):

$$k_{ma} + k_{mb}(1-\phi)^n,\ m = 1\ldots9, n \in [0.5\ldots3], \tag{6.44}$$

in which ϕ is the porosity, n is a free parameter determined from experiments (Kabel et al., 1999). The eigenvalues of the second-rank fabric tensor determine the symmetry of the fourth-rank stiffness tensor. If all three eigenvalues of the fabric tensor are equal, the drained stiffness tensor represents an isotropic material; if two eigenvalues are equal, the symmetry of $\mathbb{C}^d$ is transverse isotropy, and if all three eigenvalues are different, $\mathbb{C}^d$ has orthotropic symmetry. Multiple studies have demonstrated experimentally that the fabric tensor is an accurate representation of the structural anisotropy of cancellous bone (Harrigan & Mann, 1984; Kabel et al., 1999; Odgaard, 1997; Odgaard et al., 1997; Souzanchi, Palacio-Mancheno et al., 2012; Turner et al., 1990; Van Rietbergen et al., 1996, 1998; Whitehouse, 1974).

The elasticity-fabric relationships (Eq. 6.43) were recast to make them symmetry invariant (Moesen et al., 2012), and an elegant approach based on power relationships of the fabric tensor was introduced (Zysset, 2003; Zysset & Curnier, 1995), which reduced the number of material constants from 18 to only 3 and 2 exponents. A detailed description of Zysset's approach can be found in the next Chapter of this book. In the current Chapter, we present the expressions of tensors used in poroelasticity, such as $\mathbb{C}^d$, $\boldsymbol{A}$, $\boldsymbol{A}\otimes\boldsymbol{A}$, $\boldsymbol{M}$, and $\mathbb{C}^u$, based on the general relations derived by Cowin; however, these relations can also be effectively derived using Zysset's fabric tensor approach.

The Biot effective stress coefficient tensor $\boldsymbol{A}$ is related to the fabric tensor $\boldsymbol{F}$ by

$$\boldsymbol{A} = \boldsymbol{I} - \frac{1}{3K^m}\left(a_0^{cd}\boldsymbol{I} + a_I^{cd}\boldsymbol{F} + a_{II}^{cd}\boldsymbol{F}\cdot\boldsymbol{F}\right) \tag{6.45}$$

where

$$\begin{aligned} a_0^{cd} &= 3a_1^{cd} + a_2^{cd} + a_3^{cd}(1-2II_F) + 2c_1^{cd}, \\ a_I^{cd} &= 3a_2^{cd} + b_1^{cd} + b_2^{cd}(1-2II_F) + 4c_2^{cd}, \\ a_{II}^{cd} &= 3a_3^{cd} + b_2^{cd} + b_3^{cd}(1-2II_F) + 4c_3^{cd}, \end{aligned} \tag{6.46}$$

and the open product of $\boldsymbol{A}$ with itself is,

$$\begin{aligned} \boldsymbol{A}\otimes\boldsymbol{A} &= \left[1 - \frac{a_0^{cd}}{3K^m}\right] \\ &\left[\left(1 - \frac{a_0^{cd}}{3K^m}\right)(\boldsymbol{I}\otimes\boldsymbol{I}) - \frac{a_I^{cd}}{3K^m}(\boldsymbol{F}\otimes\boldsymbol{I} + \boldsymbol{I}\otimes\boldsymbol{F}) - \frac{a_{II}^{cd}}{3K^m}\boldsymbol{I}\otimes\boldsymbol{F}\cdot\boldsymbol{F} + (\boldsymbol{I}\otimes\boldsymbol{F}\cdot\boldsymbol{F})^T\right] \\ &+ \left(\frac{a_I^{cd}}{3K^m}\right)^2\boldsymbol{F}\cdot\boldsymbol{F} + \left(\frac{a_I^{cd}a_{II}^{cd}}{3K^m}\right)^2\left(\boldsymbol{F}\otimes\boldsymbol{F}\cdot\boldsymbol{F} + (\boldsymbol{F}\otimes\boldsymbol{F}\cdot\boldsymbol{F})^T\right) + \left(\frac{a_{II}^{cd}}{3K^m}\right)^2(\boldsymbol{F}\cdot\boldsymbol{F}\otimes\boldsymbol{F}\cdot\boldsymbol{F}) \end{aligned} \tag{6.47}$$

The Biot's tensor $\boldsymbol{M}$ in terms of the fabric tensor is given by,

$$\boldsymbol{M} = M\boldsymbol{I} - \frac{M}{3K^m}\left(a_0^{cd}\boldsymbol{I} + a_I^{cd}\boldsymbol{F} + a_{II}^{cd}\boldsymbol{F}\cdot\boldsymbol{F}\right), \tag{6.48}$$

and the undrained anisotropic stiffness tensor, which will be used later for the theory of wave propagation, is written as a function of the fabric tensor as,

$$\begin{aligned}
\mathbb{C}^u = &\left[a_1^{cd} + M\left(\frac{3K^m - a_0^d}{3K^m}\right)^2\right](\boldsymbol{I}\otimes\boldsymbol{I}) + \left[a_2^{cd} - M\left(\frac{(3K^m - a_0^d)a_I^d}{3K^m}\right)^2\right](\boldsymbol{F}\otimes\boldsymbol{I} + \boldsymbol{I}\otimes\boldsymbol{F}) \\
&+ \left[a_3^{cd} - M\frac{(3K^m - a_0^{cd})a_{II}^{cd}}{(3K^m)^2}\right]\left(\boldsymbol{I}\otimes\boldsymbol{F}\cdot\boldsymbol{F} + (\boldsymbol{I}\otimes\boldsymbol{F}\cdot\boldsymbol{F})^T\right) + \left[b_1^{cd} + M\left(\frac{a_I^{cd}}{3K^m}\right)^2\right]\boldsymbol{F}\otimes\boldsymbol{F} \\
&+ \left[b_2^{cd} + M\frac{a_I^{cd}a_{II}^{cd}}{(3K^m)^2}\right]\left(\boldsymbol{F}\otimes\boldsymbol{F}\cdot\boldsymbol{F} + (\boldsymbol{F}\otimes\boldsymbol{F}\cdot\boldsymbol{F})^T\right) + \left[b_3^{cd} + M\frac{a_{II}^{cd}}{(3K^m)^2}\right](\boldsymbol{F}\cdot\boldsymbol{F}\otimes\boldsymbol{F}\cdot\boldsymbol{F}) \\
&+ c_1^{cd}\left(\boldsymbol{I}^T\otimes\boldsymbol{I}^T + (\boldsymbol{I}\otimes\boldsymbol{I})^T\right) + c_2^{cd}\left(\boldsymbol{F}\otimes\boldsymbol{I} + \boldsymbol{F}\otimes\boldsymbol{I}^T + \boldsymbol{F}^T\otimes\boldsymbol{I}^T + (\boldsymbol{F}^T\otimes\boldsymbol{I}^T)^T\right) \\
&+ c_3^{cd}\left(\boldsymbol{F}\cdot\boldsymbol{F}\otimes\boldsymbol{I} + \boldsymbol{F}\cdot\boldsymbol{F}\otimes\boldsymbol{I}^T + (\boldsymbol{F}\cdot\boldsymbol{F})^T\otimes\boldsymbol{I}^T + \left[(\boldsymbol{F}\cdot\boldsymbol{F})^T\otimes\boldsymbol{I}^T\right]^T\right)
\end{aligned} \tag{6.49}$$

5. Governing equations

5.1 Darcy's law

The relative motion between the pore fluid and the solid matrix is commonly modeled using a permeability-dependent relation that links pressure gradients to seepage velocity, known as Darcy's law. Darcy's law relates the fluid mass flow rate, $\boldsymbol{q}$, to the gradient of the pore pressure, ∇p, by the hydraulic conductivity, $\boldsymbol{H}$, a second-rank tensor for anisotropic porous media,

$$\boldsymbol{q} = -\frac{1}{\rho_f g}\boldsymbol{H}\cdot\nabla p, \boldsymbol{H} = \boldsymbol{H}^T \tag{6.50}$$

where ρ_f is the fluid density, and g is the gravitational acceleration. The hydraulic conductivity represents the ability of fluid to move through the porous medium. The fluid mass flow rate is a vector quantity defined as the volume of fluid moving through a porous medium, per unit area and time, and the gradient of the pore pressure, ∇p. Also, the hydraulic conductivity is related to the intrinsic permeability,

$$\boldsymbol{H} = \frac{\rho_f g}{\mu}\boldsymbol{K}, \tag{6.51}$$

where $\boldsymbol{K}$ is a second rank symmetric tensor known as the intrinsic permeability, dependent only on the geometry of the pores, with dimensions of length squared

$[L^2]$, and μ is the fluid viscosity. The fluid mass flow rate, $\boldsymbol{q}$, is proportional to the fluid velocity, $\boldsymbol{v}$, and the change in fluid mass density, ρ_f/ρ_{f0}, within the pore space, ϕ,

$$\boldsymbol{q}=\left(\frac{\phi\rho_f}{\rho_{f0}}\right)\boldsymbol{v}=-\frac{1}{\mu}\boldsymbol{K}\cdot\nabla p(x,t),\boldsymbol{K}=\boldsymbol{K}^T. \tag{6.52}$$

The permeability tensor can be anisotropic and described as a function of the fabric tensor (Cowin & Cardoso, 2011; Kreipke & Niebur, 2017)

$$\boldsymbol{K}=r_1\boldsymbol{I}+r_2\boldsymbol{F}+r_3\boldsymbol{F}^2, \tag{6.53}$$

where r_1, r_2, and r_3 are proportionality constants determined experimentally.

5.2 Conservation of mass and momentum

The conservation of mass is described by the continuity equation:

$$\frac{\partial\rho}{\partial t}+\nabla\cdot\rho\boldsymbol{v}=0, \tag{6.54}$$

and it is modified to the case of a fluid-saturated porous medium by replacing ρ by $\phi\rho_f$ and dividing through by ρ_{f0},

$$\frac{1}{\rho_{f0}}\frac{\partial\phi\rho_f}{\partial t}+\frac{1}{\rho_{f0}}\nabla\cdot\left(\phi\rho_f\boldsymbol{v}\right)=0. \tag{6.55}$$

The variation in fluid content, Eq. (6.11), is modified by using the time derivative as,

$$\frac{\partial\zeta}{\partial t}=\frac{1}{\rho_{f0}}\frac{\partial\phi\rho_f}{\partial t}, \tag{6.56}$$

and replacing Eq. (6.56) into Eq. (6.55), and using Eq. (6.52), the rate of change of fluid content is

$$\frac{\partial\zeta}{\partial t}+\nabla\cdot\boldsymbol{q}=0, \tag{6.57}$$

where the sum of the rate of change of fluid content and the divergence of the fluid flux in the porous medium is zero. In the case of incompressibility, $\rho_f=\rho_{f0}$, and Eq. (6.55) becomes

$$\frac{\partial\phi}{\partial t}+\nabla\cdot(\phi\boldsymbol{v})=0, \tag{6.58}$$

where the rate of change of the porosity is equal to the negative of the gradient of the fluid velocity in pore space.

6. Diffusion governing equations

The diffusion governing equation can be obtained for the variation in fluid content or for the pressure field. Combining Darcy's law (Eq. 6.52) with the expression (Eq. 6.57) for the conservation of mass, we obtain

$$\frac{\partial \zeta}{\partial t} - \frac{1}{\mu}\boldsymbol{K}: \nabla^2 p = 0, \tag{6.59}$$

which is given as a function of both the variation in fluid content and the pressure. We can eliminate the pore pressure by using (Eq. 6.16) to obtain the diffusion equation for the variation in fluid content,

$$\frac{\partial \zeta}{\partial t} - \frac{1}{\mu \Lambda}\boldsymbol{K}: \nabla^2 \zeta = -\frac{1}{\mu \Lambda}\boldsymbol{K}: \nabla^2 [\boldsymbol{A}: \boldsymbol{E}], \tag{6.60}$$

Here, the rate of change of the fluid content ζ is a consequence of either the fluid flux or volume changes resulting from deformation of the porous medium. An alternative form of this equation is obtained by replacing the strain on the right-hand side of Eq. (6.15) using Eq. (6.16), where the right-hand side of the diffusion equation is given in terms of the effective stress, $\boldsymbol{T} + \boldsymbol{A}p$:

$$\frac{\partial \zeta}{\partial t} - \frac{1}{\mu \Lambda}\boldsymbol{K}: \nabla^2 \zeta = -\frac{1}{\mu \Lambda}\boldsymbol{A}: \mathbb{S}^d : \boldsymbol{K} \cdot \nabla^2 [\boldsymbol{T} + \boldsymbol{A}p], \tag{6.61}$$

On the other hand, a diffusion equation for the pressure field, p, is obtained when the variation in fluid content, ζ, is eliminated from Eq. (6.60) by using Eq. (6.16),

$$\frac{\partial p}{\partial t} - \frac{1}{\mu \Lambda}\boldsymbol{K}: \nabla^2 p = -\frac{1}{\Lambda}\boldsymbol{A}: \frac{\partial \boldsymbol{E}}{\partial t} \tag{6.62}$$

This last diffusion equation can be written in terms of the stress by replacing the strain on the right-hand side with the expression of stress in Eq. (6.8) and using Eq. (6.18), thus

$$\begin{aligned} \frac{\partial p}{\partial t} - \frac{1}{\mu S^d_{eff}}\boldsymbol{K}: \nabla^2 p &= -\frac{1}{S^d_{eff}}\boldsymbol{A}: \mathbb{S}^d : \frac{\partial \boldsymbol{T}}{\partial t} \\ &= \frac{1}{S^d_{eff}}\left[\boldsymbol{I}: (\mathbb{S}^m - \mathbb{S}^d) : \frac{\partial \boldsymbol{T}}{\partial t}\right], \end{aligned} \tag{6.63}$$

7. Convection-driven fluid flow and mechanotransduction

Fluid flow within the vascular canals and the lacunar-canalicular system in bone is essential to osteocytes' role in mechanosensation and in controlling bone resorption and formation. It has long been known that bone adapts its mass according to the level of mechanical loading applied to the tissue. However, the tissue strain levels in bone are one to two orders of magnitude lower than the wall shear stresses needed to stimulate osteocytes in vitro. Such a paradox was addressed by building a detailed poroelasticity model of the lacunar-canalicular system and demonstrating that the time-varying mechanical loads applied to bone result in interstitial fluid flow within the lacunar-canalicular system (Weinbaum et al., 1994), in turn producing a tissue strain amplification at the osteocyte processes that is key for bone mechanosensation (You et al., 2001). The macroscopic–level loads applied along the z-direction of an isotropic bone tissue sample are related to the interstitial fluid flow in the lacunar-canalicular system (LCS) by using poroelasticity theory (equivalent to Eq. 6.63) as:

$$c\nabla^2\left(tr(\boldsymbol{T})+\frac{3}{B}p\right)=\frac{\partial}{\partial t}\left(tr(\boldsymbol{T})+\frac{3}{B}p\right), \tag{6.64}$$

where $tr(\boldsymbol{T})$, the trace of the stress tensor, is the sum of the normal stresses, and the compatibility condition $\nabla^2 tr(\boldsymbol{T}) = 0$. The coefficient B is the isotropic Skempton coefficient (Eq. 6.24) and c, the pore pressure diffusion coefficient,

$$c=\frac{k}{\mu S_{eff}^d}, \tag{6.65}$$

which depend on the dynamic viscosity of the fluid phase, the effective drained compliance modulus, S_{eff}^d, and the permeability of the LCS (Eq. 6.19 in (Weinbaum et al., 1994) gives c in terms of shear modulus and Poisson ratios). Weinbaum et al. defined three permeability constants associated with different length scales. The local permeability, k_p, is due to the presence of glycosaminoglycans (GAGs) fibers within the canaliculi without considering the boundaries of flow, and the effective permeability, $k_{p,eff}$, is due to the dimensions of the canaliculi and cell processes. These two local permeability parameters were used to estimate the LCS's overall permeability, k. This is equivalent to the LCS permeability determined by Darcy's law obtained when considering the fluid flow within the annulus representing the space between the cell process and the canaliculus wall, corresponding to

$$k=\frac{2\pi n a^4 q^3}{\gamma^3 L^2}\left\{A_1\left[I_1\left(\frac{\gamma}{q}\right)-qI_1(\gamma)\right]+B_1\left[qK_1(\gamma)-K_1\left(\frac{\gamma}{q}\right)\right]+\gamma\frac{(q^2-1)}{2q}\right\}, \tag{6.66}$$

and the shear stress acting at the surface of the cell process was calculated from the gradient of the fluid flow velocity as,

$$\tau(a)=\mu\frac{\partial u}{\partial r}(a)=\frac{b}{\gamma}\frac{p}{y}\left[A_1 I_1\left(\frac{\gamma}{q}\right)-B_1 K_1\left(\frac{\gamma}{q}\right)\right] \tag{6.67}$$

where parameters A_1 and B_1, are related to modified Bessel functions of zeroth order and the first (e.g., I_0) and third kind (e.g., K_0), $\gamma = b/\sqrt{k_p}$, and $q = b/a$, the ratio of the radius of the canaliculus, b, and the radius of the osteocyte process, a. After Eq. (6.64) was rendered nondimensional, and a sinusoidal stress load was specified in the z-direction, the fluid flow relaxation time of the vascular porosity system and the lacunar-canalicular system can be determined, as well as the shear stress produced at the osteocyte body and processes.

Microarchitecture details of the LCS, osteonal canal, and cement lines were added to the model and then used to investigate bone fluid relaxation under oscillatory four-point bending (Wang, Fritton, et al., 1999). It was shown that bone fluid in the LCS relaxes primarily through the vascular porosity and that a second relaxation behavior is associated with the thickness of the porous bone specimen at relatively lower loading frequencies. Also, the fluid pressure gradients and shear stresses near the osteonal canals are significantly amplified at higher loading frequencies (Wang, Fritton, et al., 1999).

Later, the whole-bone-level poroelasticity approach was combined with a cellular-level solute transport model and used to analyze solute transport within the LCS using fluorescence recovery after photobleaching (FRAP) (Zhou et al., 2008). The poroelasticity model for a hollow tube representing the midshaft of a mouse tibia cyclically loaded was given in cylindrical coordinates by

$$\frac{\partial^2 p}{\partial r^2}+\frac{1}{r}\frac{\partial p}{\partial r}+\frac{1}{r^2}\frac{\partial^2 p}{\partial \theta^2}-\frac{1}{c}\frac{\partial p}{\partial t}=\frac{B}{3c}\left[\frac{1}{A}+\frac{rL}{I}\sin\theta\right]\frac{df(t)}{dt} \tag{6.68}$$

where the loading function $f(t)$ was expanded into a Fourier series, and the transport of solutes in the LCS during FRAP is modeled using a modified diffusion-convection equation

$$\frac{\partial C}{\partial t}+\left(1-\sigma_f\right)u\frac{\partial C}{\partial x}=D\frac{\partial^2 C}{\partial x^2}, \tag{6.69}$$

in which C is the concentration of the solute tracer, σ_f is the tracer reflection coefficient, u is the convective fluid flow velocity, and D is the tracer diffusion coefficient. The recovery rate of the tracer after FRAP was exponential, and it was significant when the solute stroke displacement exceeded the distance between the source and sink (i.e., the canalicular length). Mechanical loading significantly increased convective tracer transport relative to diffusion for medium-sized tracers (Fig. 6.4), but the effect was much less for small and large tracers (Zhou et al., 2008). This research group quantified the convection of sodium fluorescein (a small tracer, 376Da) in the LCS of mice tibiae for an intermittent cyclic loading (Price et al., 2011), and parvalbumin (a larger tracer, 12.3 kDa), comparable in size to signaling proteins secreted by osteocytes (Wang, Zhou, et al., 2013). It was found that mechanical loading resulted in about 15% and 35% increases in tracer transport over diffusion-based transport at 2 and 0.5Hz loading, respectively. Overall, these experimental-poroelasticity studies support the hypothesis that osteocytes are mechanosensors that respond to mechanical loading-induced fluid flow to control bone resorption and formation (Weinbaum et al., 1994).

8. Quasi-static poroelasticity—Stress relaxation

The stress-relaxation of interstitial fluid within a single osteon has been investigated using poroelasticity theory under the assumption of quasi-static equilibrium. The

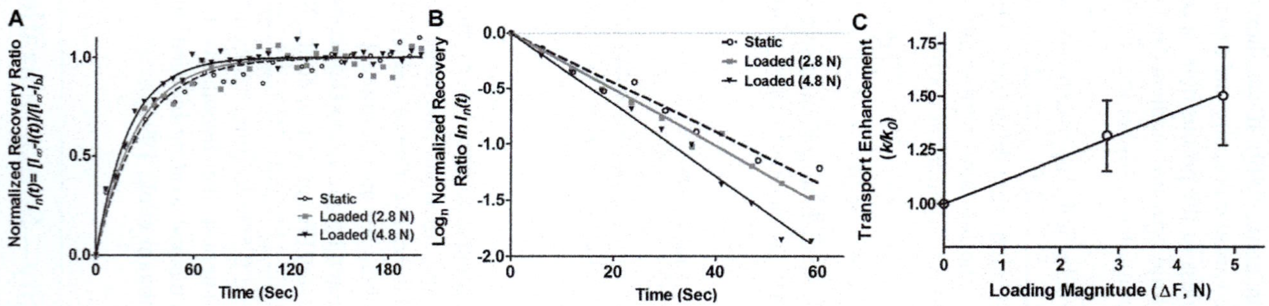

FIGURE 6.4

Effects of loading magnitude on the transport of parvalbumin in the bone LCS. (A) Fluorescence recovery ratio of a lacuna subjected to FRAP under varying peak loading conditions (0, 2.8 N, 4.8 N). (B) Increasing load magnitude increased the solute transport rate (k) during the FRAP experiment. (C) Mean transport enhancement (k/k_0) ($n = 12$ lacunae) increased linearly with increasing loading magnitude.

- Reproduced from Wang, B., Zhou, X., et al. (2013). Quantifying load-induced solute transport and solute-matrix interaction within the osteocyte lacunar-canalicular system. Journal of Bone and Mineral Research, 28*(5), 1075–1086 under CC BY 4.0 Open License.*

deformation of a fluid-saturated porous medium is time-dependent, leading to creep-like behavior when subjected to a constant load. Conversely, if the porous medium is under constant deformation, it will exhibit time-dependent stress relaxation. Importantly, no acceleration is produced by loading the osteon.

The equations of poroelasticity were adapted to the case of an unconfined annular cylindrical disk under ramp loading (Cowin & Mehrabadi, 2007; Gailani & Cowin, 2008) to estimate lacunar-canalicular permeability in isolated osteons from human samples by curve-fitting experimental stress-relaxation data to the model. The rationale for using isolated osteons was that such measurements were considered representative of the LCS system alone, without influence from vascular porosity (Cowin et al., 2009; Gailani, Benalla, et al., 2009; Gailani & Cowin, 2008). The boundary condition assumed that flow was allowed across the Haversian canal only, and the following expression for the pressure was derived:

$$\frac{p(t)}{\pi\left(b^2-c_0^2\right)\left(C_{rr}^d-C_{r\theta}^d\right)}=\frac{\left((M+F)a^2-\frac{N}{2}\right)\epsilon_0 t_0}{D+a^2(2C-D)}-\frac{b^2\epsilon_0}{c\alpha_n^4}\sum_{n=0}^{\infty}\frac{\left[\frac{2a^2F-N}{a(1-a^2)}+MR(\alpha_n)\right]\left(1-e^{\alpha_n^2 ct_0)/b^2}\right)e^{-\alpha_n^2 ct/b^2}}{\left[\frac{aC-3D}{a\alpha_n^2}+\frac{R(\alpha_n)(C+D)}{\alpha_n^2}-CZ(\alpha_n)-\frac{DV(\alpha_n)}{\alpha_n^2}\right]} \tag{6.70}$$

for $t \geq t_0$, where b and c refer to the inner and outer radii, $a = c_0/b$, parameters C, D, F, M, N, Y, J, and elastic constants C_{rr}^d and $C_{r\theta}^d$ are given in Tables 1 and 2 in Gailani, Benalla, et al. (2009). The pore fluid pressure diffusivity (referred to as m in Gailani's work) is

$$c=\frac{K_{rr}C_{rr}^d}{\mu J} \tag{6.71}$$

Where K_{rr} is the permeability in the radial direction, and μ is the viscosity of water. The functions $R(\alpha_n)$, $V(\alpha_n)$, and $Z(\alpha_n)$ depend on Bessel functions of the first and second kind, and α_n are the roots (solutions) of the Laplace transform of $p(t)$. The estimated LCS permeability value obtained using this approach was in the range of $O\left(10^{-24}m^2\right)$.

9. Quasi-static poroelasticity—Nested porosities

The equations of fluid-saturated porous media have also been applied to the case of two hierarchically nested pore spaces in bone, considering compressible constituents and small deformations (Cowin et al., 2009; Gailani, Benalla, et al., 2009; Gailani & Cowin, 2011). The porosity associated with the lacunar-canalicular system (LCS) within the bone matrix material is interconnected and hierarchically nested with the vascular porosity or the intertrabecular space, in cortical and trabecular bone, respectively (Cowin et al., 2009). The LCS is considered nested within the VS because the fluid within the LCS can only drain into and receive fluid from the VS. In Gailani and Cowin (2011), the two porosity levels considered were vascular porosity (VP) and lacunar-canalicular porosity (LCP). The pore fluid pressure in the VP and the LCP is deemed to differ significantly when bone tissue is loaded. The pressure in the VP can rapidly rise and fall, allowing for rapid fluid movement; however, the pressure in the LCP may take much longer to dissipate due to the LCS's much lower permeability. Therefore, the VS and LCS have different characteristic relaxation times.

The hierarchical, nested porosities model was applied to a cylindrical section of bone tissue representing an osteon and including the VP and the LCP (Gailani & Cowin, 2011). Under the action of ramp loading, $\varepsilon(t) = t\dot{\varepsilon}_0$, the coupling of the two porosity levels and their mutual influence is:

$$p^L(\lambda,t) = p^V(\beta,t) + \sum_{n=0}^{\infty}\left[C_n R_0(\lambda k_n)\frac{b^2 N^L}{k_n^2 c^L}\left(e^{-k_n^2 c^L t/b^2} - 1\right)\dot{\varepsilon}_0\right] - \left(M^L \sum_{n=0}^{\infty} C_n R_0(\lambda k_n) e^{-k_n^2 c^L t/b^2} \int_0^{\infty} e^{k_n^2 c^L \tau/b^2}\frac{\partial p^V(\tau)}{\partial \tau}d\tau\right) \tag{6.72}$$

where p^L is the pore pressure in the LCS, p^V is the pore pressure in the VS; λ and β are nondimensional radii of the osteon and the VP, respectively. Expressions for N^L, C_n, R_n, k_n, and M^L are given in (Gailani & Cowin, 2011), and they depend on the LCS permeability, fluid dynamic viscosity, elastic constants of the tissue matrix materials, and Bessel functions of the first kind.

The fluid exchange, γ, between the VP and the LCS is expressed by

$$\gamma^V(t) = h\gamma^L(t), \tag{6.73}$$

where h is a nondimensional ratio of the LCS to VS area. The conservation of mass, from Eq. (6.55), is

$$\frac{1}{\rho_{f0}}\frac{\partial \phi^V \rho_f}{\partial t} + \frac{1}{\rho_{f0}}\nabla\cdot\left(\phi^V \rho_f v\right) = h\frac{\phi^L \gamma^L}{\rho_{f0}} \tag{6.74}$$

and the expression for the volume flow rate in the LCS, γ^L,

$$\gamma^L(\lambda,t) = \frac{\rho_{f0}}{a_{cl}\mu b}K_{rr}\left[\sum_{n=0}^{\infty} C_n \frac{\partial R_0 \lambda k_n}{\partial \lambda}\frac{b^2 N^L}{k_n^2 c^L}\left(e^{-k_n^2 c^L t/b^2} - 1\right)\dot{\varepsilon}_0\right] - \frac{\rho_{f0}}{a_{cl}\mu b} K_{rr}\left[M^L \sum_{n=0}^{\infty} C_n \frac{\partial R_0 \lambda k_n}{\partial \lambda}\left[A_\gamma(t) + B_\gamma(\beta,t) - C_\gamma(\beta,t) + D_\gamma(\beta,t)\right]\right] \tag{6.75}$$

where A_γ, B_γ, C_γ, and D_γ are reported in (Gailani & Cowin, 2011). Overall, this study showed that the VP pore pressure has a small influence on the LCP pore pressure, of about 3%, and that the absence of VP will produce a 6% change in the fluid exchange between these two domains, supporting the hypothesis that the LCP pressure and flow are independent of the VP pressure. Also, the time scale of the response in the VS and LCS is very different (Zhang et al., 1998) and practically independent, with the fluid flow relaxation time inversely proportional to the LCS permeability. The authors reported that the total drainage time is 2.2s when the LCS permeability is 10–20 m^2 and 6.9s when it is assumed to be in the 10–24 m^2 range (Zhang et al., 1998).

10. Dynamic poroelasticity—Low frequency range

The mechanical response of poroelastic materials varies with excitation frequency as the relative influence of elastic deformation, inertia, and viscous dissipation evolves across loading regimes. A frequency-dependent, poroelasticity-based model of a fluid-saturated porous osteon was developed by Benalla et al. (2012) and tested under axial cyclic strains at low frequencies (1–100 Hz) using osteons extracted from human cortical bone samples (Benalla, Palacio-Mancheno et al., 2014). The model and test samples were developed to exclude the influence of vascular pressure on the proposed measurements. Briefly, the model considers a small segment of an isolated osteon to be deformed using axial cyclic loading in the form of an applied strain as

$$\varepsilon(t) = \varepsilon_0 e^{i\omega t}, \tag{6.76}$$

where ω and ε_0 are the frequency and strain amplitude applied to the sample, respectively. The relationship between stress and strain under harmonic loading (equation above), where $\widetilde{C}(\omega)$ is a frequency-dependent, complex-valued modulus, is

$$\widetilde{\sigma}(t) = \widetilde{C}(\omega)\varepsilon(t), \tag{6.77}$$

and such a complex modulus can be expressed as

$$\widetilde{C}(\omega) = C_0 e^{-i\delta(\omega)}, C_0 = \sigma_0/\varepsilon_0, \tag{6.78}$$

using Euler's formula, it can be rewritten as

$$\widetilde{C}(\omega) = C_0 e^{-i\delta(\omega)} = C_0 \cos[\delta(\omega)] + i \sin[\delta(\omega)] = C'(\omega) + iC''(\omega), \tag{6.79}$$

where $C'(\omega)$ is the storage modulus and $C''(\omega)$ is the loss modulus of the loaded material. From the equation above, the expression of the loss tangent is

$$\tan[\delta(\omega)] = C''(\omega)/C'(\omega) = \sin[\delta(\omega)]/\cos[\delta(\omega)]. \tag{6.80}$$

The loss tangent represents the phase delay between the sinusoidal stress and strain waveforms in the osteon. This complex modulus can be expressed in terms of the $\tan[\delta(\omega)]$,

$$\widetilde{C}(\omega) = \frac{C_0}{\sqrt{\tan^2[\delta(\omega)] + 1}}(1 + i\tan[\delta(\omega)]), \tag{6.81}$$

and in terms of the LCS permeability (K_{rr}) in the radial direction of the osteon and elastic constants $\left(C_{zz}^d, c_1 \ldots c_5\right)$ relations that can be found in Benalla et al. (2012), as

$$\widetilde{C}(\omega) = C_{zz}^d + \frac{c_1 - c_2 K_{rr}}{c_3 K_{rr} + c_4} - \frac{c_5}{K_{rr}}, \tag{6.82}$$

or expressing K_{rr} in terms of $\tan[\delta(\omega)]$

$$K_{rr}^2 \left(C_{zz}^d c_3 - c_2 - \widetilde{C}(\omega) c_3 \right) + K_{rr} \left(C_{zz}^d c_4 - \widetilde{C}(\omega) c_4 - c_3 c_5 \right) - c_4 c_5 = 0. \tag{6.83}$$

The lacunar–canalicular permeability was determined from curve fitting of the analytical model (Benalla et al., 2012) against experimental data of $\tan[\delta(\omega)]$. It was found that the LCS intrinsic permeability is on the order of $10^{-22} m^2$ for human osteonal bone, and the dynamic permeability has a linear dependence on loading frequency, decreasing at a rate of $2 \times 10^{-24} m^2/Hz$ from 1 to 100 Hz (Fig. 6.5). This result also indicates that the tissue matrix material behavior becomes purely elastic as the frequency increases, changing from a complex elastic modulus into a real-valued modulus, with a phase delay, $\tan[\delta(\omega)]$, that tends asymptotically to zero. In such a case, it is assumed that the fluid has no time to move in and out of the pore space, and the fluid-saturated porous medium behaves as if in an undrained condition, as in the wave propagation described next.

11. Dynamic poroelasticity—High frequency range: Wave propagation

Depending on the characteristic loading time scale, poroelastic behavior may be analyzed either by neglecting inertial contributions (quasi-static condition) or by explicitly incorporating the inertia associated with the motion of both the solid frame

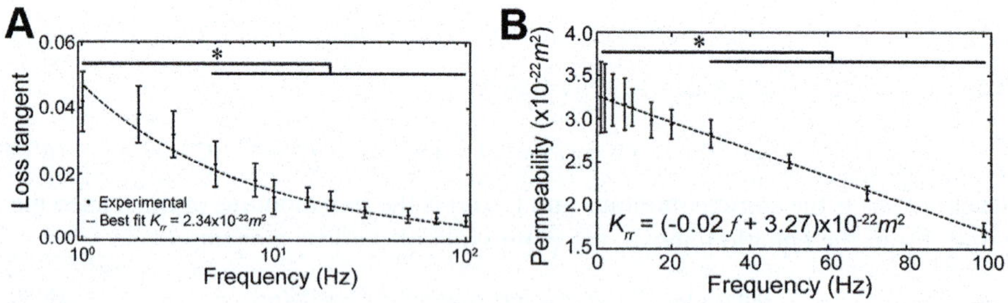

FIGURE 6.5

Loss tangent and LCS permeability of human osteons. (A) Curve fitting of experimental data with the poroelasticity model, resulting in an average LCS permeability, $K_{rr} = (2.34 \pm 0.33) \times 10^{-22} m^2$ ($*p < 0.05$, $n = 60$ osteons); (B) LCS permeability versus frequency, $Krr = (3.27 = 0.02f) \times 10^{-22} m^2$ $K_{rr} = (3.27 - 0.02f) \times 10^{-22} m^2$ ($*p < 0.05, n = 60$ osteons) obtained from the inverse poroelastic model.

- *Reproduced with permission from Benalla, M., Palacio-Mancheno, P. E., et al. (2014). Dynamic permeability of the lacunar-canalicular system in human cortical bone.* Biomechanics and Modeling in Mechanobiology, 13*(4), 801–812.*

and the pore fluid. The mechanical behavior of fluid-saturated porous media changes significantly when the strain rate of a ramp test or the cycling loading frequency becomes high enough that the fluid has no time to move in/out of the pores. Under such conditions, inertial effects become significant and must be accounted for in the study of this material's dynamic behavior. Biot introduced inertial effects into the constitutive equations of poroelasticity and formulated two coupled partial differential equations that characterize wave propagation in porous media. Here, we follow the original development of the poroelastic constitutive equations presented by Biot (1941, 1955, 1956a,b, 1962a,b), and subsequently further developed by others (Cardoso & Cowin, 2011; Cowin & Cardoso, 2011; Sharma, 2005, 2008).

11.1 Wave equation in poroelastic media

The kinematic variable used to represent the displacement vector of the solid phase material is $\boldsymbol{u}$, and for the fluid phase, it is $\boldsymbol{U}$, as in the early works of Biot (1956a, 1956b). Later, Biot replaced the use of $\boldsymbol{U}$ by a more convenient variable, $\boldsymbol{w}$, which corresponds to the relative motion between solid and fluid phases (Biot, 1962a, 1962b),

$$\boldsymbol{w}=\phi(\boldsymbol{U}-\boldsymbol{u}), \tag{6.84}$$

The variation in fluid content ζ is defined as the change of fluid volume per unit volume of the porous material, and in terms of $\boldsymbol{w}$ is written as:

$$\zeta= -\nabla\cdot\boldsymbol{w}, \tag{6.85}$$

Using the strain-displacement relations 3 and 87 into Eqs. (6.12) and (6.13) results in,

$$\boldsymbol{T}=\mathbb{C}^{u}\frac{\partial \boldsymbol{u}}{\partial \boldsymbol{x}}+\boldsymbol{M}\frac{\partial w}{\partial x} \tag{6.86}$$

$$p= -\boldsymbol{M}\frac{\partial \boldsymbol{u}}{\partial \boldsymbol{x}}+M\frac{\partial w}{\partial x}. \tag{6.87}$$

The balance of momentum in the form of the dynamical stress equations of motion in three dimensions is applied to both the solid and fluid fractions of the porous medium, as a function of $\ddot{\boldsymbol{u}}$ and $\ddot{\boldsymbol{U}}$,

$$\nabla\cdot\boldsymbol{T} = (1-\phi)\rho_s\ddot{\boldsymbol{u}} + \phi\rho_f\ddot{\boldsymbol{U}} \tag{6.88}$$

or, in terms of $\ddot{\boldsymbol{w}}$, by using Eq. (6.84),

$$\nabla\cdot\boldsymbol{T}=\left[(1-\phi)\rho_s+\phi\rho_f\right]\ddot{\boldsymbol{u}}+\rho_f\ddot{\boldsymbol{w}}=\rho\ddot{\boldsymbol{u}}+\rho_f\ddot{\boldsymbol{w}}, \tag{6.89}$$

where ρ_s and ρ_f are the mass densities of the solid and fluid, respectively, and the total mass density, ρ, is

$$\rho=(1-\phi)\rho_s+\phi\rho_f. \tag{6.90}$$

The expression of the balance of momentum of the fluid phase in the pore space may be written as $\rho_f \ddot{U}$ or in the form $\rho_f(\ddot{\boldsymbol{u}} + \ddot{\boldsymbol{w}}/\phi)$ by use of Eq. (6.84). However, Biot (1962a,b) introduced a scaling factor, J, the micro–macro velocity average tensor, in front of $\ddot{\boldsymbol{w}}$,

$$\rho_f \ddot{U} = \rho_f(\ddot{\boldsymbol{u}} + \ddot{\boldsymbol{w}}/\phi) = \rho_f(\ddot{\boldsymbol{u}} + \boldsymbol{J} \cdot \ddot{\boldsymbol{w}}). \tag{6.91}$$

Such micro–macro velocity average tensor $\boldsymbol{J}$ incorporates the factor $1/\phi$ in it, and is a density distribution function that relates the relative micro-solid-fluid velocity to its bulk volume average, $\boldsymbol{w}$ (Souzanchi et al., 2013). The use of Eq. (6.91) results in a balance of linear momentum for the fluid phase,

$$-\nabla p = \rho_f(\ddot{\boldsymbol{u}} + \boldsymbol{J} \cdot \ddot{\boldsymbol{w}}) \tag{6.92}$$

The viscous drag force of the fluid moving over the solid surface in the pores is incorporated into Eq. (6.92) by adding a viscous resistive force $\mu \boldsymbol{R} \cdot \dot{\boldsymbol{w}}$ that represents the effect of the fluid-solid viscous interaction. The flow-resistivity tensor, $\boldsymbol{R}$, is the inverse of the intrinsic permeability tensor $\boldsymbol{K}$ described in Eqs. (6.51) and (6.53),

$$\boldsymbol{R} = \boldsymbol{K}^{-1}. \tag{6.93}$$

The viscous resistive force opposes the pressure gradient, so it is subtracted from the left-hand side of Eq. (6.92), and it is then moved to the right-hand side of the equation, thus,

$$-\nabla p = \rho_f(\ddot{\boldsymbol{u}} + \boldsymbol{J} \cdot \ddot{\boldsymbol{w}}) + \mu \boldsymbol{R} \cdot \dot{\boldsymbol{w}}. \tag{6.94}$$

Now, Eq. (6.86) for the solid stress tensor and the expression 87 for the pore fluid pressure are used in the conservation of momentum for the solid phase, Eq. (6.89), and the conservation of momentum for the fluid phase, Eq. (6.94), respectively:

$$\mathbb{C}^u \frac{\partial^2 \boldsymbol{u}}{\partial \boldsymbol{x}^2} + \boldsymbol{M} \frac{\partial^2 \boldsymbol{w}}{\partial \boldsymbol{x}^2} = \rho \boldsymbol{I} \cdot \ddot{\boldsymbol{u}} + \rho_f \boldsymbol{I} \cdot \ddot{\boldsymbol{w}}, \tag{6.95}$$

$$\boldsymbol{M} \frac{\partial^2 \boldsymbol{u}}{\partial \boldsymbol{x}^2} + M \frac{\partial^2 \boldsymbol{w}}{\partial \boldsymbol{x}^2} = \rho_f \boldsymbol{I} \cdot \ddot{\boldsymbol{u}} + \rho_f \boldsymbol{J} \cdot \ddot{\boldsymbol{w}} + \mu \boldsymbol{R} \cdot \dot{\boldsymbol{w}}, \tag{6.96}$$

Eqs. (6.95) and (6.96) represent two coupled partial differential equations of the displacement fields $\boldsymbol{u}$ and $\boldsymbol{w}$. Second-rank tensors in Eqs. (6.95) and (6.96), such as $\boldsymbol{R}, \boldsymbol{K}, \boldsymbol{M}$, and $\boldsymbol{J}$, depend on the anisotropy of the porous frame and, thus, on the fabric tensor, as described by (Cowin & Cardoso, 2011). The second-rank, anisotropic, dynamic permeability tensor, $\boldsymbol{K}$, is complex valued and represents dissipation phenomena due to viscous losses at low frequencies of wave propagation (Cowin & Cardoso, 2011; Johnson et al., 1987):

$$\boldsymbol{K}(\omega) = K_0 \left[1 - 2 \frac{J_1(d\chi)}{d\chi J_0(d\chi)} \right] (K_1 \boldsymbol{I} + K_2 \boldsymbol{F} + K_3 \boldsymbol{F} \cdot \boldsymbol{F}), \tag{6.97}$$

where $\boldsymbol{K}$ depends on the average permeability K_0, $\boldsymbol{F}$, and Bessel functions. In here, J_1 and J_0 are the first-order and zeroth-order Bessel functions of the first kind, and

d corresponds to the average characteristic pore dimension. The inverse of the viscous skin depth, characterized by the parameter χ, is a function of the angular frequency ω, the mass density of the fluid phase ρ_f, and the dynamic viscosity of the fluid μ,

$$\chi = \sqrt{\frac{i\omega\rho_f}{\mu}}. \tag{6.98}$$

11.2 Plane wave solution

The propagation of a harmonic plane wave is expressed along a direction of propagation, denoted by $\boldsymbol{n} = (n_1, n_2, n_3)^T$, which is a unit normal to the wave front, and $\boldsymbol{a}$ or $\boldsymbol{b}$, the directions of particle displacement, or polarizations, of the wave fronts associated with $\boldsymbol{u}$ and $\boldsymbol{w}$, respectively. These two harmonic plane waves are represented by

$$\boldsymbol{u}(\boldsymbol{x}, t) = \boldsymbol{a}e^{-i(\widetilde{\boldsymbol{k}}\cdot\boldsymbol{x}-\omega t)}, \boldsymbol{w}(\boldsymbol{x}, t) = \boldsymbol{b}e^{-i(\widetilde{\boldsymbol{k}}\cdot\boldsymbol{x}-\omega t)}, \tag{6.99}$$

where $\boldsymbol{x}$ is the position vector, ω is the angular frequency, t is time, and $\widetilde{\boldsymbol{k}}$ is the complex wave vector,

$$\widetilde{\boldsymbol{k}} = (\boldsymbol{k} + i\boldsymbol{\alpha}) = (\widetilde{k}_1, \widetilde{k}_2, \widetilde{k}_3), \tag{6.100}$$

in which $\boldsymbol{k}$ is the real-valued wave vector, indicating the direction of wave propagation, and $\boldsymbol{\alpha}$ is the attenuation vector, indicating the direction of wave attenuation (Carcione, 2001; Cardoso & Cowin, 2011; Cowin & Cardoso, 2011; Sharma, 2005, 2008). Substituting the relations 99 for the plane waves into the field Eqs. (6.95) and (6.96) lead to the poroelastic Christoffel equation (Biot, 1962a, 1962b; Cowin & Cardoso, 2011; Sharma, 2005, 2008),

$$\begin{bmatrix} \boldsymbol{Q} - \rho\widetilde{z}\boldsymbol{I} & \boldsymbol{C} - \rho_f\widetilde{z}\boldsymbol{I} \\ \boldsymbol{C}^T - \rho_f\widetilde{z}\boldsymbol{I} & \boldsymbol{H} - \rho_f\widetilde{z}\widetilde{\boldsymbol{S}} \end{bmatrix} \begin{bmatrix} \boldsymbol{a} \\ \boldsymbol{b} \end{bmatrix} = \begin{bmatrix} \boldsymbol{0} \\ \boldsymbol{0} \end{bmatrix} \tag{6.101}$$

where the following notation has been introduced,

$$\begin{gathered} \boldsymbol{Q} = \left[\mathbb{C}^d + M(\mathbf{A} \otimes \mathbf{A})\right] : \boldsymbol{N} = \mathbb{C}^u : \boldsymbol{N}, \boldsymbol{C} = M\boldsymbol{N}, \boldsymbol{H} = M\boldsymbol{N}, \\ \widetilde{\mathbf{S}} = \left[\boldsymbol{J} + i\mu/\left(\omega\rho_f\right)\boldsymbol{R}\right], \boldsymbol{N} = n \otimes n, \end{gathered} \tag{6.102}$$

$\boldsymbol{Q}$ is the acoustic tensor, and $\boldsymbol{C}$ is the interaction of displacement fields $\boldsymbol{u}$ and $\boldsymbol{w}$. The detailed algebraic structure of the tensors $\boldsymbol{Q}$, $\boldsymbol{C}$, $\boldsymbol{H}$, and $\widetilde{\mathbf{S}}$ describing the wave propagation in the general direction $\boldsymbol{n}$ is recorded elsewhere (Cardoso & Cowin, 2011, 2012; Cowin & Cardoso, 2011). Eq. (6.101) represents an eigenvalue problem; the complex variable $\widetilde{z}$ corresponds to the eigenvalues, and the vectors $\boldsymbol{a}$ and $\boldsymbol{b}$ represent the eigenvectors. The determinant of Eq. (6.101),

$$\begin{vmatrix} \boldsymbol{Q} - \rho\tilde{z}\boldsymbol{I} & \boldsymbol{C} - \rho_f\tilde{z}\boldsymbol{I} \\ \boldsymbol{C}^T - \rho_f\tilde{z}\boldsymbol{I} & \boldsymbol{H} - \rho_f\tilde{z}\tilde{\boldsymbol{S}} \end{vmatrix} = 0 \tag{6.103}$$

has a characteristic equation that is a sixth-order polynomial in $\tilde{z}$, given by

$$c_6\tilde{z}^6 + c_5\tilde{z}^5 + c_4\tilde{z}^4 + c_3\tilde{z}^3 + c_2\tilde{z}^2 + c_1\tilde{z} + c_0 = 0, \tag{6.104}$$

where the square root of the complex-valued eigenvalues $\tilde{z}$ represents the frequency to complex wave vector ratio, or the velocity of the wave modes generated in a fluid-saturated porous medium (Cardoso & Cowin, 2011, 2012; Cowin & Cardoso, 2011),

$$\sqrt{\tilde{z}_i} = \omega/\tilde{k}_i = \tilde{v}_i, i = 1..6 \tag{6.105}$$

These are the six possible solutions of the two coupled wave Eq. (6.101). The dynamic response of poroelastic media admits two compressional modes and four shear waves. However, two out of the four shear waves are evanescent, do not propagate, and it is thus generally considered that only four wave modes can be observed in porous media: two longitudinal and two transverse waves. The two compressional waves propagate at different velocities, and Biot termed them as *fast* and *slow* waves. The fast wave is predominantly a matrix-controlled wave, and the second mode is strongly influenced by relative fluid–solid motion and viscous dissipation. The complex wave vector in Eq. (6.105) is decomposed as

$$\tilde{k}_i = \frac{\omega}{\tilde{v}_i} = \frac{\omega\left(\mathrm{Re}\tilde{v}_i\right)}{\left(\mathrm{Re}\tilde{v}_i\right)^2 + \left(\mathrm{Im}\tilde{v}_i\right)^2} - j\frac{\omega\left(\mathrm{Im}\tilde{v}_i\right)}{\left(\mathrm{Re}\tilde{v}_i\right)^2 + \left(\mathrm{Im}\tilde{v}_i\right)^2}, i = 1..4, \tag{6.106}$$

from which the real-valued phase velocity and attenuation are obtained

$$v_i = \frac{\omega}{\mathrm{Re}\tilde{k}_i} = \frac{\left(\mathrm{Re}\tilde{v}_i\right)^2 + \left(\mathrm{Im}\tilde{v}_i\right)^2}{\left(\mathrm{Re}\tilde{v}_i\right)}, \alpha_i = \frac{\omega\left(\mathrm{Im}\tilde{v}_i\right)}{\left(\mathrm{Re}\tilde{v}_i\right)^2 + \left(\mathrm{Im}\tilde{v}_i\right)^2}, \tag{6.107}$$

The phase velocity corresponds to the propagation velocity of a point of constant phase, $\mathrm{Re}\tilde{\boldsymbol{k}}\cdot\boldsymbol{x} - \omega t = constant$, thus $v_i = \omega/\mathrm{Re}\tilde{k}_i)$. Each phase velocity v_i is related to the common angular frequency ω and the real part of the corresponding complex wave vector $\mathrm{Re}\tilde{k}_i$.

11.3 Propagation of poroelastic waves in bone

Fast and slow waves have been observed in porous media, including trabecular bone (Cardoso, Teboul, et al., 2003; Hosokawa & Otani, 1997). The fast and slow wave velocities predicted by the poroelastic ultrasound (PEUS) wave propagation model were compared with experimental measurements of fast and slow wave velocities in human and bovine trabecular bone (Cardoso, Teboul, et al., 2003). Wave velocity measurements were taken along three orthogonal directions, perpendicular to the face of each cube-shaped sample ($n = 74$). The averaged velocities had a high correlation coefficient with the porosity, but directional measurements of velocity correlated best with the direction-dependent elastic constants from the PEUS model's predictions. The slow wave experiences much higher dissipation than the fast wave and may not propagate below a critical frequency (Biot, 1956a, 1956b, 1962a). Above this critical frequency, the theory accounts for viscous effects (absorption phenomena), making both waves difficult to observe past the viscous frequency, which is 10^4 times the critical frequency (Cardoso et al., 2008).

This anisotropic poroelastic ultrasound (PEUS) wave propagation approach was used to examine changes in the fabric tensor, bone mineral density (BMD), and apparent elastic constants (EC) in trabecular bone of an animal model experiencing bone loss from mechanical disuse and treated with antiresorptive therapy (Cardoso & Schaffler, 2015). Mechanical immobilization decreased trabecular thickness (Tb.Th.) and bone volume fraction (BV/TV). However, risedronate combined with immobilization resulted in less reduction in Tb.Th. and BV/TV, indicating that antiresorptive therapy slows bone loss but does not fully prevent it. Significantly, the elastic anisotropy of trabecular bone changed due to disuse and was restored by therapy. BMD accounted for only 58% of velocity variability; in contrast, the poroelastic model predicted 81% of velocity variability and also explained 91% of apparent ECs and anisotropy patterns (Fig. 6.6). The study also showed that PEUS wave propagation can detect changes in apparent ECs and elastic anisotropy after immobilization and antiresorptive treatment. This method offers an effective, noninvasive way to assess the structure–function relationship in immobilized and risedronate-treated anisotropic bone (Cardoso & Schaffler, 2015).

12. Conclusion

The poroelasticity theoretical framework integrates theories of elastic solids, viscous fluids, and fluid flow within porous media. At the core of bone poroelasticity, this theory describes bone's biphasic character as a fluid-saturated porous medium and its hierarchical structure, which together describe its biomechanical functions. Poroelasticity has been successfully applied in many studies of bone biomechanics, solute transport, mechanotransduction, the relationship between composition and structure and function, dynamic elasticity, permeability, and ultrasound wave propagation in fluid-saturated bone tissues at various scales. Specifically, quasi-static poroelasticity has been used to analyze solute transport and convection-driven fluid flow mechanotransduction in the lacunar-canalicular system. Both quasi-static and dynamic poroelasticity have also been employed to study stress-relaxation in single osteons, fluid flow between vascular and lacunar-canalicular porosities, and to estimate the intrinsic permeability of the LCS, which is otherwise very difficult to measure directly. Relationships among mechanical function, composition, and microarchitecture of the porous medium have been derived based on the fabric tensor of a porous frame structure. Additionally, poroelasticity-based ultrasound

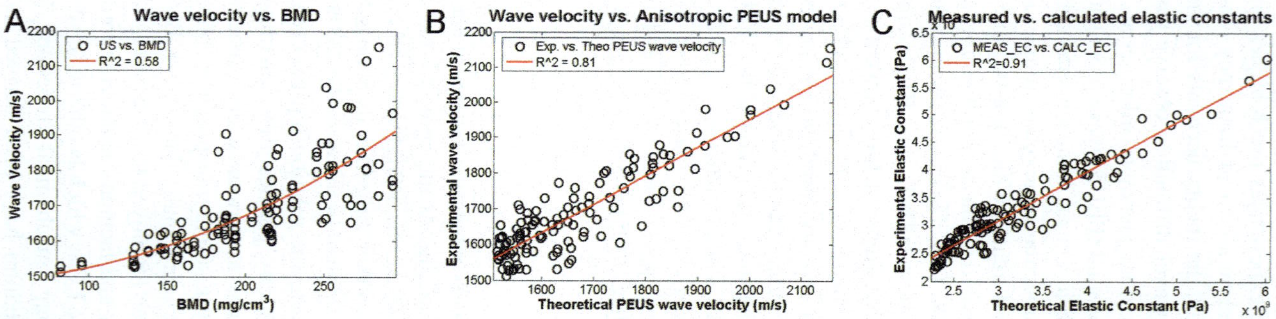

FIGURE 6.6

Ultrasound wave propagation in cancellous bone undergoing disuse and antiresorptive therapy. (A) Ultrasonic wave velocities versus vBMD ($R^2 = 0.58$), (B) experimental versus theoretical PEUS wave velocities ($R^2 = 0.81$), and (C) experimental apparent elastic constants versus theoretical elastic constants derived from anisotropic poroelastic model ($R^2 = 0.91$).

- Reproduced with permission from Cardoso, L., & Schaffler, M. B. (2015). Changes of elastic constants and anisotropy patterns in trabecular bone during disuse-induced bone loss assessed by poroelastic ultrasound. Journal of Biomechanical Engineering, 137, *1.*

theory has effectively predicted ultrasound wave velocities in anisotropic human trabecular bone, offering a potential noninvasive method to assess bone loss and changes in bone anisotropy. This summary of poroelasticity applications is not exhaustive but highlights many of the fields where this theoretical approach has made significant advances. Further development of poroelasticity applications in the bone field could enhance our understanding of bone biomechanics and mechanotransduction.

Acknowledgments

This work has been supported by NIH grants 1R16GM145474 and 1S10OD030324.

References

Benalla, M., Cardoso, L., & Cowin, S. C. (2012). Analytical basis for the determination of the lacunar-canalicular permeability of bone using cyclic loading. *Biomechanics and Modeling in Mechanobiology, 11*(6), 767–780.

Benalla, M., Palacio-Mancheno, P. E., et al. (2014). Dynamic permeability of the lacunar-canalicular system in human cortical bone. *Biomechanics and Modeling in Mechanobiology, 13*(4), 801–812.

Biot, M. A. (1941). General theory of three-dimensional consolidation. *Journal of Applied Physics, 12*(2), 155–164.

Biot, M. A. (1955). Theory of elasticity and consolidation for a porous anisotropic solid. *Journal of Applied Physics, 26*(2), 182–185.

Biot, M. A. (1956a). Theory of propagation of elastic waves in a fluid-saturated porous solid. I. Low-frequency range. *Journal of the Acoustical Society of America, 28*(2), 168–178.

Biot, M. A. (1956b). Theory of propagation of elastic waves in a fluid-saturated porous solid. II. Higher frequency range. *Journal of the Acoustical Society of America, 28*(2), 179–191.

Biot, M. A. (1962a). Generalized theory of Acoustic propagation in porous dissipative media. *Journal of the Acoustical Society of America, 34*, 1254–1264.

Biot, M. A. (1962b). Mechanics of deformation and acoustic propagation in porous media. *Journal of Applied Physics, 33*(4), 1482–1498.

Boer, R. de (2000). *Theory of porous media: Highlights in historical development and current state*. Berlin: Springer.

Carcione, J. (2001). *Wave fields in real media: Wave propagation in anisotropic, anelastic and porous media* (Vol 31). Elsevier Ltd.

Cardoso, L., & Cowin, S. C. (2011). Fabric dependence of quasi-waves in anisotropic porous media. *Journal of the Acoustical Society of America, 129*(5), 3302–3316.

Cardoso, L., & Cowin, S. C. (2012). Role of structural anisotropy of biological tissues in poroelastic wave propagation. *Mechanics of Materials, 44*, 174–188.

Cardoso, L., Fritton, S. P., et al. (2013). Advances in assessment of bone porosity, permeability and interstitial fluid flow. *Journal of Biomechanics, 46*(2), 253–265.

Cardoso, L., Meunier, A., & Oddou, C. (2008). In vitro acoustic wave propagation in human and bovine cancellous bone as predicted by Biot's theory. *Journal of Mechanics in Medicine and Biology, 08*(02), 183–201.

Cardoso, L., & Schaffler, M. B. (2015). Changes of elastic constants and anisotropy patterns in trabecular bone during disuse-induced bone loss assessed by poroelastic ultrasound. *Journal of Biomechanical Engineering, 137*, 1.

Cardoso, L., Teboul, F., et al. (2003). In vitro acoustic waves propagation in human and bovine cancellous bone. *Journal of Bone and Mineral Research, 18*(10), 1803–1812.

Carroll, M. M. (1979). An effective stress law for anisotropic elastic deformation. *Journal of Geophysical Research: Solid Earth, 84*(B13), 7510–7512.

Cheng, A. H.-D. (2016). *Poroelasticity* (Vol 27). Springer.

Coussy, O. (2003). *Poromechanics* (pp. 1–298). Wyley.

Coussy, O. (2010). *Mechanics and physics of porous solids* (pp. 1–281). Wyley.

Cowin, S. C. (1985). The relationship between the elasticity tensor and the fabric tensor. *Mechanics of Materials, 4*(2), 137–147.

Cowin, S. C. (1999). Bone poroelasticity. *Journal of Biomechanics, 32*(3), 217–238.

Cowin, S. C. (2001). *Bone mechanics handbook, second edition* (pp. 1–980). CRC Press.

Cowin, S. C. (2003). A recasting of anisotropic poroelasticity in matrices of tensor components. *Transport in Porous Media, 50*(1), 35–56.

Cowin, S. C. (2004). Anisotropic poroelasticity: Fabric tensor formulation. *Mechanics of Materials, 36*(8), 665–677.

Cowin, S. C. (2013). *Continuum mechanics of anisotropic materials*. New York, NY: Springer.

Cowin, S. C., & Cardoso, L. (2011). Fabric dependence of wave propagation in anisotropic porous media. *Biomechanics and Modeling in Mechanobiology, 10*(1), 39–65.

Cowin, S. C., & Cardoso, L. (2012). Mixture theory-based poroelasticity as a model of interstitial tissue growth. *Mechanics of Materials, 44*, 47–57.

Cowin, S. C., & Cardoso, L. (2015). Blood and interstitial flow in the hierarchical pore space architecture of bone tissue. *Journal of Biomechanics, 48*(5), 842–854.

Cowin, S. C., & Doty, S. (2007). Tissue mechanics. In *Tissue mechanics* (Vol. 1, p. 682). Springer.

Cowin, S. C., Gailani, G., & Benalla, M. (2009). Hierarchical poroelasticity: Movement of interstitial fluid between porosity levels in bones. *Philosophical transactions. Series A, Mathematical, physical, and engineering sciences, 367*(1902), 3401–3444.

Cowin, S. C., & Mehrabadi, M. M. (2007). Compressible and incompressible constituents in anisotropic poroelasticity: The problem of unconfined compression of a disk. *Journal of the Mechanics and Physics of Solids, 55*(1), 161–193.

Cowin, S. C., Moss-Salentijn, L., & Moss, M. L. (1991). Candidates for the mechanosensory system in bone. *Journal of Biomechanical Engineering, 113*(2), 191–197.

Gailani, G., Benalla, M., et al. (2009). Experimental determination of the permeability in the lacunar-canalicular porosity of bone. *Journal of Biomechanical Engineering, 131*(10), Article 101007.

Gailani, G. B., & Cowin, S. C. (2008). The unconfined compression of a poroelastic annular cylindrical disk. *Mechanics of Materials, 40*(6), 507–523.

Gailani, G., & Cowin, S. (2011). Ramp loading in Russian doll poroelasticity. *Journal of the Mechanics and Physics of Solids, 59*(1), 103–120.

Harrigan, T. P., & Mann, R. W. (1984). Characterization of microstructural anisotropy in orthotropic materials using a second rank tensor. *Journal of Materials Science, 19*(3), 761–767.

Hosokawa, A., & Otani, T. (1997). Ultrasonic wave propagation in bovine cancellous bone. *Journal of the Acoustical Society of America, 101*(1), 558–562. Hosokawa, A Otani, T 1997/1/1.

Johnson, D. L., Koplik, J., & Dashen, R. (1987). Theory of dynamic permeability and tortuosity in fluid-saturated porous media. *Journal of Fluid Mechanics, 176*, 379–402.

Kabel, J., et al. (1999). Constitutive relationships of fabric, density, and elastic properties in cancellous bone architecture. *Bone, 25*(4), 481–486.

Kreipke, T. C., & Niebur, G. L (2017). Anisotropic permeability of trabecular bone and its relationship to fabric and architecture: A computational study. *Annals of Biomedical Engineering, 45*, 1543–1554. https://doi.org/10.1007/s10439-017-1805-9

Moesen, M., Cardoso, L., & Cowin, S. C. (2012). A symmetry invariant formulation of the relationship between the elasticity tensor and the fabric tensor. *Mechanics of Materials, 54*, 70–83.

Nemat-Nasser, S. (1993). *Micromechanics: Overall properties of heterogeneous materials. North-Holland series in applied mathematics and mechanics*. Amsterdam; New York: North-Holland.

Nur, A., & Byerlee, J. D. (1971). An exact effective stress law for elastic deformation of rock with fluids. *Journal of Geophysical Research, 76*(26), 6414–6419.

Odgaard, A. (1997). Three-dimensional methods for quantification of cancellous bone architecture. *Bone, 20*(4), 315–328.

Odgaard, A., et al. (1997). Fabric and elastic principal directions of cancellous bone are closely related. *Journal of Biomechanics, 30*(5), 487–495.

Price, C., et al. (2011). Real-time measurement of solute transport within the lacunar-canalicular system of mechanically loaded bone: Direct evidence for load-induced fluid flow. *Journal of Bone and Mineral Research, 26*(2), 277–285. Price, Christopher Zhou, Xiaozhou Li, Wen Wang, Liyun 2010/8/18.

Reznikov, N., et al. (2018). Fractal-like hierarchical organization of bone begins at the nanoscale. *Science, 360*, 6388.

Rodriguez-Palomo, A., Østergaard, M., & Birkedal, H. (2024). Bone hierarchical structure: Heterogeneity and uniformity. *Advanced Functional Materials, 34*(35), Article 2307026.

Sharma, M. D. (2005). Propagation of inhomogeneous plane waves in dissipative anisotropic poroelastic solids. *Geophysical Journal International, 163*(3), 981–990.

Sharma, M. D. (2008). Propagation of harmonic plane waves in a general anisotropic porous solid. *Geophysical Journal International, 172*(3), 982–994.

Skempton, A. W. (1954). The pore-pressure coefficients A and B. *Géotechnique, 4*, 143–147.

Souzanchi, M. F., Cardoso, L., & Cowin, S. C. (2013). Tortuosity and the averaging of microvelocity fields in poroelasticity. *Journal of Applied Mechanics, 80*(2), 0209061–209065.

Souzanchi, M. F., Palacio-Mancheno, P., et al. (2012). Microarchitecture and bone quality in the human calcaneus: Local variations of fabric anisotropy. *Journal of Bone and Mineral Research, 27*(12), 2562–2572.

Tang, T., Landis, W., Blouin, S., et al. (2023). Subcanalicular nanochannel volume is inversely correlated with calcium content in human cortical bone. *Journal of Bone and Mineral Research, 38*(2), 313–325.

Tang, T., Landis, W., Raguin, E., et al. (2022). A 3D network of nanochannels for possible ion and molecule transit in mineralizing bone and cartilage. *Advanced NanoBiomed Research, 2*(8), Article 2100162.

Thompson, M., & Willis, J. R. (1991). A reformation of the equations of anisotropic poroelasticity. *Journal of Applied Mechanics, 58*(3), 612–616.

Turner, C. H., et al. (1990). The fabric dependence of the orthotropic elastic constants of cancellous bone. *Journal of Biomechanics, 23*(6), 549–561.

Van Rietbergen, B., et al. (1996). Direct mechanics assessment of elastic symmetries and properties of trabecular bone architecture. *Journal of Biomechanics, 29*(12), 1653–1657.

Van Rietbergen, B., et al. (1998). Relationships between bone morphology and bone elastic properties can be accurately quantified using high-resolution computer reconstructions. *Journal of Orthopaedic Research, 16*(1), 23–28.

Wang, L., Fritton, S. P., et al. (1999). Fluid pressure relaxation depends upon osteonal microstructure: Modeling an oscillatory bending experiment. *Journal of Biomechanics, 32*(7), 663–672.

Wang, B., Zhou, X., et al. (2013). Quantifying load-induced solute transport and solute-matrix interaction within the osteocyte lacunar-canalicular system. *Journal of Bone and Mineral Research, 28*(5), 1075–1086.

Wehrli, F. W., & Fernández-Seara, M. A. (2005). Nuclear magnetic resonance studies of bone water. *Annals of Biomedical Engineering, 33*(1), 79–86.

Weinbaum, S., Cowin, S. C., & Zeng, Y. (1994). A model for the excitation of osteocytes by mechanical loading-induced bone fluid shear stresses. *Journal of Biomechanics, 27*(3), 339–360.

Whitehouse, W. J. (1974). The quantitative morphology of anisotropic trabecular bone. *Journal of Microscopy, 101*(2), 153–168.

You, L., et al. (2001). A model for strain amplification in the actin cytoskeleton of osteocytes due to fluid drag on pericellular matrix. *Journal of Biomechanics, 34*(11), 1375–1386.

Zeng, Y., Cowin, S. C., & Weinbaum, S. (1994). A fiber matrix model for fluid flow and streaming potentials in the canaliculi of an osteon. *Annals of Biomedical Engineering, 22*(3), 280–292.

Zhang, D., Weinbaum, S., & Cowin, S. C. (1998). Estimates of the peak pressures in bone pore water. *Journal of Biomechanical Engineering, 120*(6), 697–703.

Zhou, X., Novotny, J. E., & Wang, L. (2008). Modeling fluorescence recovery after photobleaching in loaded bone: Potential applications in measuring fluid and solute transport in the osteocytic lacunar-canalicular system. *Annals of Biomedical Engineering, 36*(12), 1961–1977. Zhou, Xiaozhou Novotny, John E Wang, Liyun 2008/9/24.

Zysset, P. K. (2003). A review of morphology–elasticity relationships in human trabecular bone: Theories and experiments. *Journal of Biomechanics, 36*(10), 1469–1485.

Zysset, P. K., & Curnier, A. (1995). An alternative model for anisotropic elasticity based on fabric tensors. *Mechanics of Materials, 21*(4), 243–250.

CHAPTER

Morphology-mechanical property relationships in human trabecular bone

7

Philippe K. Zysset

ARTORG Centre for Biomedical Engineering Research, University of Bern, Switzerland

1. Introduction

Bone is a remarkable hierarchical composite biomaterial whose extracellular matrix (ECM) is primarily made of mineral, type I collagen and water. Bone can be found in compact and porous, trabecular forms and is known to adapt to its mechanical environment. The structural optimization of the bone diaphysis and the complementary arrangement of compact and trabecular bone to minimize bone mass in the human skeleton was originally described in (Currey, 2002).

Trabecular bone is lighter and more compliant than compact bone, covering a broad range of mechanical properties with a volume fraction (BV/TV) or structural density (ρ) extending from 5% to 45%. This makes trabecular bone an attractive solution for multiaxial load transfer in epiphyses, sandwich design in flat bones and core stability in small bones.

The 1D mechanical properties of trabecular bone, such as Young's modulus or yield/ultimate stress, were shown to follow power functions of ρ (Fleps et al., 2020; Rice et al., 1988), but they are not isotropic and depend on trabecular orientation distribution. Whitehouse (1974) characterized trabecular orientation in 2D histological sections using the concept of mean intercept length (MIL). He showed that the orientation distribution of MIL follows an ellipse with the major axis along the main direction of the trabecule. His concept was generalized to an ellipsoid in 3D that is represented mathematically by a positive definite, second-order fabric tensor (Harrigan & Mann, 1984).

This 3D generalization led Cowin to formulate quadratic relationships between density, the fabric tensor and the elasticity tensor (Cowin, 1985) involving 18 material constants that were first verified with orthogonal ultrasound measurements (Turner et al., 1990) and then by uniaxial mechanical tests (Goulet et al., 1994; Snyder et al., 1990). Alternatively, power relationships were proposed by Zysset and Curnier (1995) that ensure positivity of the stiffness tensor a priori with a reduced set of three material constants and two exponents, and deliver similar predictions of experimental data (Zysset, 2003).

Biomechanics and Mechanobiology of Bone. https://doi.org/10.1016/B978-0-443-26739-0.00010-7

Along the same approach, Cowin proposed a relationship between density, the fabric tensor and a Tsai-Wu strength criterion (Cowin, 1986), which was later extended to a more general quadric strength criterion in Schwiedrzik et al. (2013).

The continuum assumption for trabecular bone underlying the above relationships was estimated to two to three trabecular spacing (Harrigan et al., 1988), but this appeared insufficient for mechanical properties that required RVE up to 6 mm side (Zysset, Goulet, & Hollister, 1998). On the other hand, larger volume elements are limited by human anatomy.

Since the early 90's, 3D μCT reconstructions (Kuhn et al., 1990) allow a detailed 3D characterization of trabecular architecture (Goulet et al., 1994). Trabecular anisotropy is computed by MIL or numerous other methods such as mean surface length (MSL) from segmented images (Hosseini, Maquer, & Zysset, 2017) or alternatively by the gradient structure tensor (GST) from gray level images (Tabor, 2011).

The ability to solve large linear equation systems on high-performance computers opened the micro finite element (μFE) analysis era (VanRietbergen, 2001). The fully anisotropic apparent stiffness tensor of trabecular bone volume elements of 4–6 mm side length could be computed by homogenization on parallel CPUs (VanRietbergen et al., 1995). It was found that the apparent stiffness tensor of trabecular bone follows approximately orthotropic symmetry, which does not contradict the observation of nonorthogonal arrangement of trabecule (Skedros & Baucom, 2007), as the material symmetry is understood in an average statistical sense.

It was also shown that the principal directions of the fabric tensor coincide with the orthotropic axes of symmetry of the apparent stiffness tensor (Odgaard et al., 1997) and fabric elasticity-relationships (Cowin, 1985; Zysset & Curnier, 1995) could be validated with larger experimental and computational datasets (Kabel et al., 1999; VanRietbergen et al., 1995, 1998; Zysset, 2003; Zysset, Goulet, & Hollister, 1998).

However, the apparent stiffness tensors depend rather heavily on the applied boundary conditions for homogenization (Pahr & Zysset, 2008), especially for low volume fraction and size due to the fading representativity of the volume element. In contrast to the original kinematic uniform boundary conditions (KUBC) that overestimate the apparent stiffness tensors obtained with gold standard in situ boundary conditions, periodicity-compatible mixed uniform boundary conditions (PMUBC) underestimate these tensors but with a lower error (Daszkiewicz et al., 2017).

The homogenization scheme using μFE was also successfully applied using different boundary conditions to identify a multiaxial yield criterion in stress space for trabecular bone cubes from the proximal femur (Panyasantisuk et al., 2016).

A broad statistical analysis demonstrated that BV/TV and fabric explain up to 97% of apparent trabecular bone elastic property variations computed by μFE (Maquer et al., 2015). A similar result was obtained for apparent yield stresses of trabecular bone (Musy et al., 2017). However, the relative contribution of fabric was higher for yield stress (23%) than for elastic constants (9%). Most importantly, further architectural indices such as trabecular thickness, spacing, number, or connectivity did not improve these relationships.

Following this historical review, the present chapter aims at recalling the key notions of density, fabric, and RVE, exposing the mathematical formulations of morphology-mechanical property relationships, presenting the available material parameters for different anatomical locations, discussing the limitations of the approach and making a few recommendations for homogenized FE (hFE) applications.

Notations. Hereafter, scalars are denoted by italic letters (e.g., time t or density ρ), vectors by bold face minuscules (e.g., position $\mathbf{x}$), second-order tensors by bold face majuscules (e.g., fabric tensor $\mathbf{M}$) and fourth-order tensors by outline majuscules (e.g., $\mathbb{I}$).

With these notations, let $\mathbf{E}$ denote the Green–Lagrange strain $\mathbf{E} = \frac{1}{2}\left(\mathbf{F}^T\mathbf{F} - \mathbf{I}\right)$ defined from the gradient $\mathbf{F} = \nabla_{\mathbf{x}}\mathbf{y}$ of the motion $\mathbf{y} : (\mathbf{x}, t) \leftarrow \mathbf{y}(\mathbf{x}, t)$ and $\mathbf{S}$ the conjugate second Piola-Kirchhoff stress defined through the material version of Cauchy's theorem $\mathbf{s} = \mathbf{Sn}$, where $\mathbf{s} = \mathbf{F}^{-1}\mathbf{p}$ is transformed by the inverse gradient from the nominal stress vector $\mathbf{p}$, and $\mathbf{n}$ is the original normal. $\mathbf{S}$ and $\mathbf{E}$ are conjugate or dual because their internal power is equal to the external power supplied to the solid: $\int_{\Omega} \mathbf{S} : \mathbf{E}\, dV = \int_{\partial\Omega} \mathbf{p} \cdot \mathbf{y}\, dA$ where Ω is the solid original form and $\mathbf{y}$ the particle velocity. Both $\mathbf{E}$ and $\mathbf{S}$ are symmetric and objective, i.e., insensitive to a change of reference frame, which simplifies the direct formulation of constitutive laws in their terms. These strain and stress measures degenerate into their small-strain counterparts ϵ and σ in small strains and in the absence of rotations.

The reader is referred to (Curnier, 2004) for complete treatments. Finally, the following tensor products will be used.

- vector dyadic $(\mathbf{a} \otimes \mathbf{b})\, \mathbf{x} = (\mathbf{x} \cdot \mathbf{b})\, \mathbf{a}$, $\forall \mathbf{x}$ $\left({}^i\mathbf{e} \otimes {}^i\mathbf{e}\right) \mathbf{x} = x_i\, {}^i\mathbf{e}$
- tensor dyadic $(\mathbf{A} \otimes \mathbf{B})\, \mathbf{X} = (\mathbf{X} : \mathbf{B})\, \mathbf{A}$, $\forall \mathbf{X}$ $(\mathbf{I} \otimes \mathbf{I})\, \mathbf{X} = (\mathrm{tr}\mathbf{X})\mathbf{I}$
- tensor product $\left(\mathbf{A} \underline{\otimes} \mathbf{B}\right) \mathbf{X} = \mathbf{A}\,\mathbf{X}\,\mathbf{B}^T$, $\forall \mathbf{X}$ $\left(\mathbf{I} \underline{\otimes} \mathbf{I}\right) \mathbf{X} = \mathbf{X}$
- transp. product $[\mathbf{A} \overline{\otimes} \mathbf{B}]\, \mathbf{X} = \mathbf{A}\,\mathbf{X}^T\,\mathbf{B}^T$, $\forall \mathbf{X}$ $(\mathbf{I} \overline{\otimes} \mathbf{I})\, \mathbf{X} = \mathbf{X}^T$
- symm. product $\left(\mathbf{A} \underline{\overline{\otimes}} \mathbf{B}\right) \mathbf{X} = \mathbf{A}\,\mathbf{X}\,\mathbf{B}^T$ $\forall \mathbf{X} = \mathbf{X}^T$ $\left(\mathbf{I} \underline{\overline{\otimes}} \mathbf{I}\right) \mathbf{X} = \mathbf{X}$

hence $\mathbf{A} \underline{\overline{\otimes}} \mathbf{B} = \frac{1}{2}\left(\mathbf{A} \underline{\otimes} \mathbf{B} + \mathbf{A} \overline{\otimes} \mathbf{B}\right)$. The summation convention on repeated indices is also used. The spectral decomposition of the material strain and stress is expressed by $\mathbf{E} = \varepsilon_i\left({}^i\mathbf{e} \otimes {}^i\mathbf{e}\right)$ and $\mathbf{S} = \sigma_i\left({}^i\mathbf{s} \otimes {}^i\mathbf{s}\right)$ where σ_i and ε_i are the stress and strain eigenvalues.

2. Density and fabric

In the physiological range, trabecular bone exhibits an anisotropic, linear, poroviscoelastic behavior (Cowin, 1999; Peruzzi et al., 2021) with identical elastic moduli in compression and tension (Zysset & Curnier, 1996). Before introducing density and fabric, it should be emphasized that in the upcoming sections, the composition, ultrastructure, elastic and yield properties of bone ECM will be assumed to be homogeneous, rate-independent, and independent of sex and age.

Homogeneity of mechanical properties can be justified from the small variability of composition, especially of the degree of mineralization (Roschger et al., 2003). The time-dependence of the elastic and yield behavior of trabecular bone associated with viscosity of the organic component and fluid flow within the lacuno-canalicular network is established (Johnson et al., 2010), and consists of an amplification of rate-independent elastic and yield properties with a similar power law of strain rate (Peruzzi et al., 2021). Accordingly, the role of density and fabric is expected to remain uncoupled from strain-rate effects of the ECM in most applications. The independence of properties with sex and age is consistent with a large body of literature for an overall healthy population, but may be challenged by changes in bound water content, mineral conformation, collagen cross-links, lamellar organization, lacunar distribution, bone structural unit (BSU) pattern and accumulation of microdamage in pathological conditions that will be ignored in this chapter.

2.1 Density

The primary morphological characteristic of trabecular bone is structural density ρ, also named bone volume fraction BV/TV (bone volume over total volume). This positive variable belongs to the interval $[0, 1]$, 0 reflecting the absence of bone and 1 fully compact bone tissue. In contrast to apparent density that was originally introduced by Carter and Hayes (1977), ρ is uncoupled from ECM composition but contains the same information when the ECM is assumed homogeneous.

Trabecular bone exhibits a volume fraction between 5 and 45% for a RVE of approximately $5 \times 5 \times 5$ mm (Daszkiewicz et al., 2017; Harrigan et al., 1988; Zysset, Goulet, & Hollister, 1998). Similarly to cellular solids, the functional dependence of 1D elastic modulus, yield stress and ultimate stress of trabecular bone is typically a power function of ρ with a positive exponent (Fleps et al., 2020; Gibson, 1985; Rice et al., 1988; Wili et al., 2017; Zysset et al., 1998b, 1999).

2.2 Fabric

Mean intercept length (MIL) and many other measures of structural anisotropy, e.g., star length distribution (SLD) (Odgaard et al., 1990), line fraction deviation (LDF) (Geraets et al., 1997), consists of orientation distribution functions (ODF) that can be described mathematically by a scalar function of a normal vector $\mathbf{n}$. Such ODFs can be decomposed into spherical harmonics characterized by fabric tensors of growing order (Kanatani, 1984) or into a simple ellipsoidal function characterized by a single, second-order, positive-definite fabric tensor (Harrigan & Mann, 1984). Other methods, such as the gradient structure tensor (GST) (Tabor & Rokita, 2007), and mean surface length (MSL) (Hosseini, Maquer, & Zysset, 2017) provide directly a second-order positive definite tensor for a RVE of approximately 5 × 5 × 5 mm.

Accordingly, a generic second-order fabric tensor is now introduced:

$$\mathbf{M} = m_i{}^i\mathbf{M} \otimes {}^i\mathbf{M} \quad {}^i\mathbf{M} = {}^i\mathbf{m} \otimes {}^i\mathbf{m} \quad i = 1, d \tag{1.1}$$

where m_i are the positive fabric eigenvalues and ${}^i\mathbf{m}$ the orthonormal, unit fabric eigenvectors with ${}^i\mathbf{m} \cdot {}^j\mathbf{m} = \delta_{ij}$ and d is the dimension of the considered geometrical space (2 or 3). The fabric tensor is usually normed with $\text{tr}\mathbf{M} = d$ in order to make it independent from ρ (Fig. 7.1). This normalization is supported by the absence of a substantial correlation between density and degree of anisotropy in trabecular bone (Fig. 7.2). An alternative normalization is $\det \mathbf{M} = 1$, but most of the available results are for $\text{tr}\mathbf{M} = d$.

Without loss of generality, the fabric eigenvalues may be labeled along their increasing values, but the labeling also applies to the corresponding eigenvectors or eigendirections:

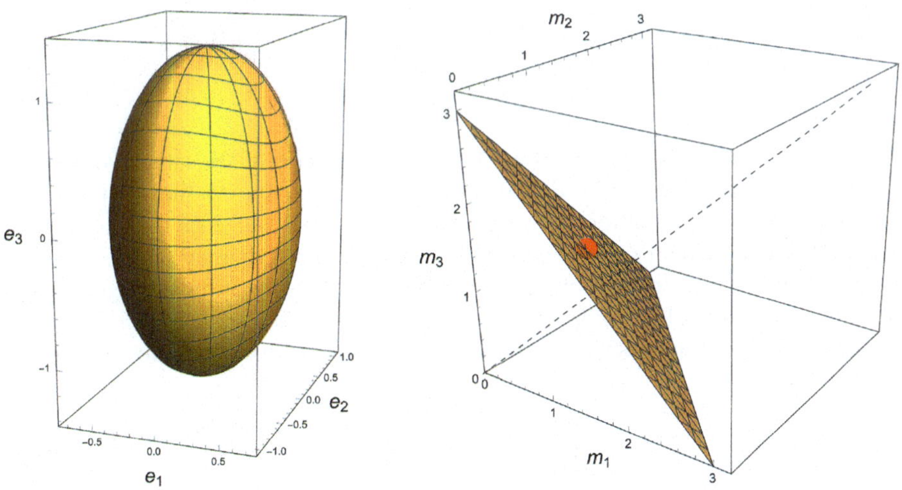

FIGURE 7.1

Left: Fabric tensor with eigenvalues representative of human trabecular bone. Right: normalization of the fabric tensor in the eigenvalue space. The surface representing the equation tr$M$ = 3 is a plane due to linearity. The identity is located at the intersection of the dashed line with the surface. A typical value of the fabric eigenvalues of trabecular bone is shown as a red dot.

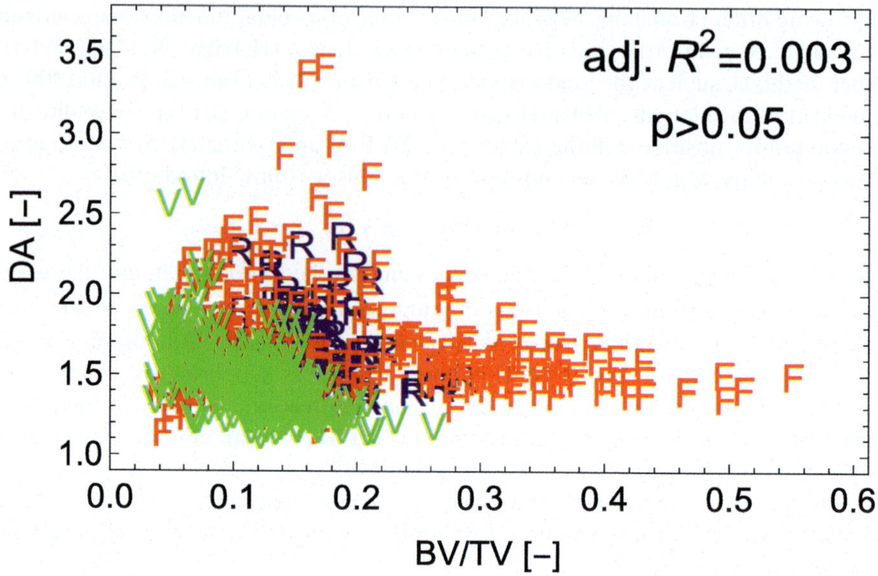

FIGURE 7.2

Independence of density and fabric for μCT data from trabecular bone of the proximal femur (F), radius (R) and vertebral body (V) (Gross et al., 2013).

$$0 \le m_1 \le m_2 \le m_3 \quad \sum_{i}^{d} m_i = d \tag{1.2}$$

The degree of anisotropy is often defined by the ratio of the largest versus the lowest eigenvalue $DA = m_d/m_1$ and is not affected by the scaling of the trace. Normalization makes the resulting tensor dimensionless, but does not account for the different sensitivities the various morphometric methods may have to capture structural anisotropy. In the frame of a second-order approximation, the eigenvectors of different **M** should coincide and the ranking of the eigenvalues as well. A practical and efficient way to relate fabric tensors obtained from different methods is the use of a power function (Larsson et al., 2014):

$$\mathbf{M} = \frac{d}{\mathrm{tr}\widetilde{\mathbf{M}}^n}\widetilde{\mathbf{M}}^n \quad m_i = \frac{d}{\mathrm{tr}\widetilde{\mathbf{M}}^n}\widetilde{m}_i^n \quad i = 1, d \tag{1.3}$$

The eigenvalues m_i and the degree of anisotropy DA follow the same power function. The normed fabric tensor **M** can therefore be used as a second-order approximation of structural anisotropy without consideration of the underlying method employed to quantify it. Nevertheless, a calibration is necessary to fit the single parameter n of the transformation (Table 7.1).

Table 7.1 Descriptive statistics of density, fabric eigenvalues and degree of anisotropy of human trabecular bone RVEs from three different anatomical locations that represent the sample basis for the material constants presented in this chapter. All variables are dimensionless.

Site/Var	N	ρ	m_1	m_2	m_3	DA	References
Femur A	126	0.266 ± 0.078	0.802 ± 0.050	0.944 ± 0.046	1.254 ± 0.071	1.575 ± 0.181	Panyasantisuk et al. (2016)
Femur B	264	0.187 ± 0.114	0.762 ± 0.088	0.992 ± 0.063	1.246 ± 0.072	1.673 ± 0.326	Gross et al. (2013)
Patella	200	0.29 ± 0.11	0.77 ± 0.07	0.99 ± 0.06	1.23 ± 0.09	1.62 ± 0.23	Latypova et al. (2016)
Radius	81	0.165 ± 0.039	0.761 ± 0.077	0.956 ± 0.065	1.283 ± 0.050	1.708 ± 0.233	Gross et al. (2013)
Tibia	603	0.282 ± 0.084	0.718 ± 0.062	0.926 ± 0.067	1.356 ± 0.071	1.907 ± 0.230	Simon et al. (2022)
Vertebra	356	0.113 ± 0.040	0.829 ± 0.068	0.968 ± 0.046	1.202 ± 0.062	1.466 ± 0.204	Gross et al. (2013)

Second-order fabric tensors can model a relatively broad range of isotropic, transverse isotropic, and orthotropic (with three orthogonal planes of symmetry) orientation distribution functions.

3. Fabric-mechanical property relationships

In this section, we expose the fabric-elasticity models and fabric-yield criteria describing the anisotropic elasticity and plastic/damage yield surface of trabecular bone at the continuum level.

3.1 Fabric-elasticity relationships

The basic, original fabric-based fourth-order stiffness tensor by Zysset & Curnier (1995) proposed for trabecular bone is expressed by

$$\mathbb{S}(\rho, \mathbf{M}) = f(\rho)\left(\lambda_0\, \mathbf{M} \otimes \mathbf{M} + 2\mu_0\, \mathbf{M} \underline{\overline{\otimes}} \mathbf{M}\right) \tag{1.4}$$

where ρ is bone volume fraction, f is a monotonic function of ρ, $\lambda_0 > 0$ and $\mu_0 > 0$ are strictly positive Lamé constants for the isotropic material when $\rho = 1$ and $\mathbf{M} = \mathbf{I}$. When two fabric eigenvalues degenerate, the model exhibits transverse isotropic (or transotropic) symmetry, while when all three eigenvalues are distinct, the model exhibits orthotropic symmetry with planes normal to the three fabric eigenvectors. The fourth order stiffness tensor can be visualized with two orientation distribution functions (He & Curnier, 1995) shown in Fig. 7.3. An important but somehow reasonable assumption of the model is the multiplicative effect of density and fabric, which implies that density scales but remains uncoupled from the anisotropic tensorial nature of stiffness.

The corresponding compliance tensor expressed with an elastic modulus ϵ_0 and a Poisson's ratio ν_0 is expressed by

$$\mathbb{E}(\rho, \mathbf{M}) = \frac{1}{f(\rho)}\left(-\frac{\nu_0}{\epsilon_0}\, \mathbf{M}^{-1} \otimes \mathbf{M}^{-1} + \frac{1+\nu_0}{\epsilon_0}\, \mathbf{M}^{-1} \underline{\overline{\otimes}} \mathbf{M}^{-1}\right) \tag{1.5}$$

where

$$\lambda_0 = \frac{\epsilon_0 \nu_0}{(1+\nu_0)(1-(d-1)\nu 0)} \qquad \mu_0 = \frac{\epsilon_0}{2(1+\nu_0)} \tag{1.6}$$

and $\mathbf{M}\mathbf{M}^{-1} = \mathbf{I}$.

In fact, the basic model corresponds to isotropic elasticity in which physical space is homogeneously stretched by the strictly positive definite tensor $\mathbf{M}$.

In the generalized fabric-elasticity relationship (Zysset, 2003) that better matches experimental and computational homogenization data for trabecular bone, the orthotropic, fourth order stiffness tensor $\mathbb{S}$ depends on the rank-one

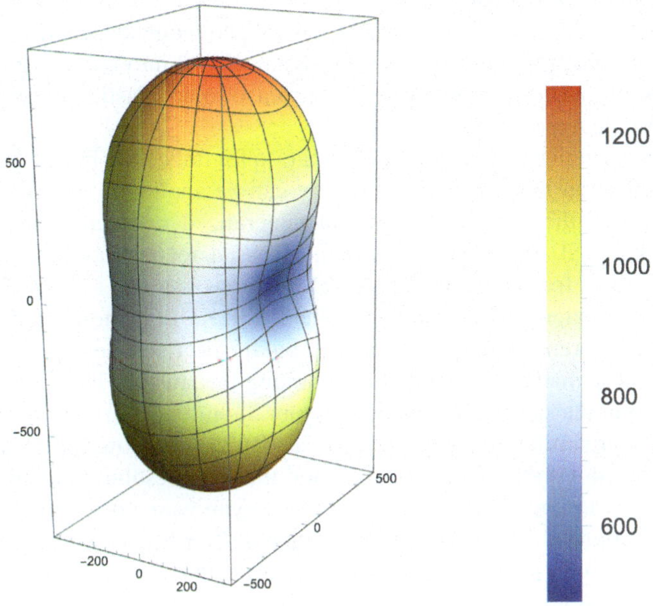

FIGURE 7.3

Visualization of the stiffness tensor (right) for the generic material constants with $\rho = 0.2$, $m_1 = 0.7$, $m_2 = 1.0$, $m_3 = 1.3$. The surface describes the elongation modulus and color the bulk modulus, both in MPa.

structural tensors $^i\mathbf{M}$ that result from the spectral decomposition of $\mathbf{M} = m_i{}^i\mathbf{M} = m_i{}^i\mathbf{m} \otimes {}^i\mathbf{m}$ rather than the whole fabric tensor $\mathbf{M}$ itself.

$$\mathbb{S}(\rho, \mathbf{M}) = f(\rho)\left(\sum_{i=1}^{d} (\lambda_0 + 2\mu_0) m_i^{2l}\,{}^i\mathbf{M} \otimes {}^i\mathbf{M} \right.$$
$$\left. + \sum_{\substack{i,j=1\\ i\neq j}}^{d} \lambda_0' m_i^l m_j^l\,{}^i\mathbf{M} \otimes {}^j\mathbf{M} + \sum_{\substack{i,j=1\\ i\neq j}}^{d} 2\mu_0 m_i^l m_j^l\,{}^i\mathbf{M}\,\underline{\overline{\otimes}}\,{}^j\mathbf{M}\right) \tag{1.7}$$

where d is the dimension, λ_0' is an additional elastic constant, $l \geq 0$ a positive exponent and the following conditions for strict positive-definiteness of the stiffness tensor

$$\lambda_0 + 2\lambda_0' + 2\mu_0 > 0 \quad \lambda_0 - \lambda_0' + 2\mu_0 > 0 \quad \mu_0 > 0 \tag{1.8}$$

In general,

$$f(\rho) = \rho^k \tag{1.9}$$

with k typically between 0 and 3.

In contrast to the basic model, the 3D generalized model degenerates into tetragonal symmetry when two eigenvalues are identical and into cubic symmetry when $\mathbf{M} = \mathbf{I}$. Moreover, the dependency of the anisotropic elastic properties can be adjusted with the exponent l.

Orthotropic symmetry is a fair approximation for trabecular bone (Zysset, 2003), while simplification into transverse isotropy can be applied in specific anatomical locations such as the vertebrae. A list of generalized material constants derived from computational homogenization of μCT-based reconstructions of trabecular bone architectures from several anatomical locations is given in Table 7.2.

The fabric-elasticity relationships obtained with KUBCs shown in Fig. 7.4 seem to diverge for different anatomical locations, but the μCT scanner, the original image resolution, and segmentation are different among the studies. The anatomical locations assessed in the same study, namely, femur, radius and vertebra by Gross et al., (2013) are rather similar, together with the second dataset from the proximal femur that was obtained with the same μCT scanner, the same resolution of 18 μm and the same image processing. The patella samples in cyan were collected with a similar original resolution of 18 μm, but with another μCT scanner and the images were then coarsened to 36 μm. The tibia samples in magenta were obtained from HR-pQCT reconstructions at 61 microns. Given the approximated voxel geometry, the distribution of the material is structurally not optimal and delivers lower microarchitectural stiffness for the same amount of material.

The Young and shear moduli obtained with alternative PMUBCs were found to be the closest to the ones computed with in situ boundary conditions (Daszkiewicz et al., 2017) and substantially lower than the ones derived with KUBC (Fig. 7.5).

3.2 Fabric-yield or fabric-strength relationships

Beyond linear elasticity, bone yields by microcracking, which produces a simultaneous accumulation of plastic strains and reduction in elastic property in stress-strain curves at the macroscopic scale (Zysset et al., 2016). A constitutive model was developed to account for this irreversible behavior in (Schwiedrzik & Zysset, 2013) and relies on an anisotropic quadric yield criterion that is nonsymmetric in tension and compression, proposed in (Schwiedrzik et al., 2013).

$$y(\mathbf{S};\rho,\mathbf{M}) = \sqrt{\mathbf{S}:\mathbb{F}\mathbf{S}} + \mathbf{F}:\mathbf{S} - 1 = 0 \tag{1.10}$$

where $\mathbf{F}$ and $\mathbb{F}$ are second and fourth-order tensors depending on volume fraction ρ, fabric $\mathbf{M}$ and material constants characterizing the yield surface.

The basic form of the fabric-based tensors can be expressed by

$$\mathbf{F}(\rho,\mathbf{M}) = \frac{1}{\hat{f}(\rho)}\frac{1}{2}\left(\frac{1}{\sigma_0^+} - \frac{1}{\sigma_0^-}\right)\mathbf{M}^{-2} = \frac{f_0}{\hat{f}(\rho)}\mathbf{M}^{-2} \tag{1.11}$$

Table 7.2 Elastic constants obtained by linear homogenization of N cubic μFE models of trabecular bone ($\epsilon_{tis} = 10$ GPa, $\nu_{tis} = 0.3$) from different anatomical locations using the fabric-based general orthotropy model with kinematic uniform boundary conditions (KUBC).

Variables	**N**	**BCs**	ϵ_0	ν_0	μ_0	k	l	**References**
Site/Units	–	–	**[MPa]**	[-]	**[MPa]**	[-]	[-]	–
Femur A	126	KUBC	7′704	0.2335	2′837	1.55	0.82	Panyasantisuk et al. (2025)
Femur A	126	PMUBC	9′759	0.2289	3′117	1.91	1.10	Panyasantisuk et al. (2025)
Femur B	264	KUBC	8′632	0.2340	3′115	1.60	0.99	Gross et al. (2013)
Patella	200	KUBC	10′603	0.2400	3′521	2.10	1.02	Latypova et al. (2016)
Radius	81	KUBC	10′041	0.2265	3′376	1.77	1.18	Gross et al. (2013)
Tibia	603	KUBC	11′627	0.1997	4′048	2.03	0.82	Simon et al. (2022)
Vertebra	356	KUBC	11′035	0.2232	3′463	1.71	1.12	Gross et al. (2013)

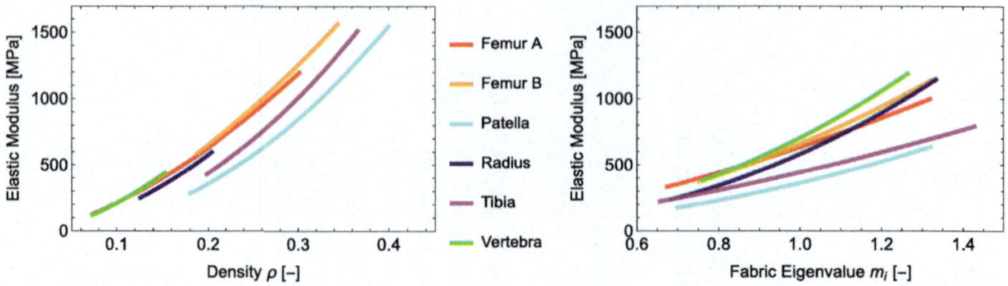

FIGURE 7.4

Influence of structural density for $m_i = 1$ (left) and influence of fabric eigenvalue for $\rho = 0.2$ (right) on elastic modulus computed for the typical density and fabric ranges ($\pm$ 1 standard deviation) of the different anatomical locations and boundary conditions.

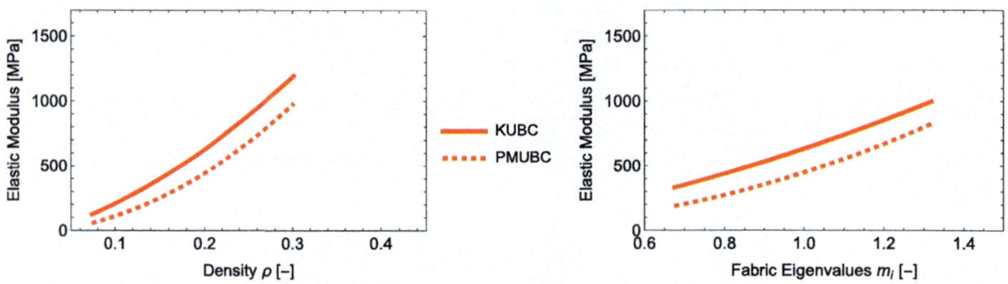

FIGURE 7.5

Effect of boundary conditions on the relationship between structural density for $m_i = 1$ (left), fabric eigenvalue for $\rho = 0.2$ (right) and elastic modulus for the relevant density and fabric ranges ($\pm$ 1 standard deviation).

$$\mathbb{F}(\rho, \mathbf{M}) = \frac{1}{\widehat{f}^2(\rho)} \frac{1}{4} \left(\frac{1}{\sigma_0^+} + \frac{1}{\sigma_0^-} \right)^2 (-\zeta_0 \, \mathbf{M}^{-2} \otimes \mathbf{M}^{-2}$$

$$+ (\zeta_0 + 1)\mathbf{M}^{-2} \,\overline{\underline{\otimes}}\, \mathbf{M}^{-2})$$

$$= \frac{F_0^2}{\widehat{f}^2(\rho)} \left(-\zeta_0 \, \mathbf{M}^{-2} \otimes \mathbf{M}^{-2} + (\zeta_0 + 1)\mathbf{M}^{-2} \,\overline{\underline{\otimes}}\, \mathbf{M}^{-2} \right) \tag{1.12}$$

where σ_0^-, σ_0^+ are the uniaxial yield stresses for $\mathbf{M} = \mathbf{I}$ in tension and compression, respectively, ζ_0 characterizes the shape of the surface, and

$$f_0 = \frac{1}{2} \left(\frac{1}{\sigma_0^+} - \frac{1}{\sigma_0^-} \right) \quad F_0 = \frac{1}{2} \left(\frac{1}{\sigma_0^+} + \frac{1}{\sigma_0^-} \right) \tag{1.13}$$

are alternative constants. The density function $\widehat{f}(\rho)$ may be slightly different from $f(\rho)$, the one used for elastic properties. In general, the ratio $h(\rho) = \widehat{f}(\rho)/f(\rho)$ reflecting yield strain is a constant or weakly decreasing function of ρ (Zysset et al., 1999).

The basic model degenerates into an isotropic model when the fabric tensor becomes identity, and the yield shear stress is related to the other material constants by

$$\tau_0 = \frac{1}{F_0}\sqrt{\frac{1}{2(1+\zeta_0)}} \tag{1.14}$$

The quadric yield criterion can also be expressed as a quadratic form that is shifted by a stress state **A**:

$$y(\mathbf{S};\rho,\mathbf{M}) = (\mathbf{S}-\mathbf{A}(\rho,\mathbf{M})) : \mathbb{A}(\rho,\mathbf{M})(\mathbf{S}-\mathbf{A}(\rho,\mathbf{M})) - 1 = 0 \tag{1.15}$$

where

$$\mathbb{A} = \frac{\mathbb{F}-\mathbf{F}\otimes\mathbf{F}}{1+\mathbf{A}:(\mathbb{F}-\mathbf{F}\otimes\mathbf{F})\mathbf{A}} \qquad \mathbf{A} = -(\mathbb{F}-\mathbf{F}\otimes\mathbf{F})^{-1}\mathbf{F} \tag{1.16}$$

Using the above expressions of **F** and $\mathbb{F}$, the tensor $\mathbb{A}$ is proportional to

$$\mathbb{F}-\mathbf{F}\otimes\mathbf{F} = \frac{1}{\hat{f}^2(\rho)}(-(F_0^2\zeta_0 + f_0^2)\mathbf{M}^{-2}\otimes\mathbf{M}^{-2} + F_0^2(\zeta_0+1)\mathbf{M}^{-2}\,\underline{\overline{\otimes}}\,\mathbf{M}^{-2} \tag{1.17}$$

Convexity of the yield criterion relies therefore on positive-definiteness of the fourth order tensor $\mathbb{A}$.

$$0 < \sigma_0^- \quad 0 < \sigma_0^+ - 1 \leq \zeta_0 \leq \frac{1}{2} \tag{1.18}$$

The yield surface degenerates into two parallel planes normal to the trisectrix for $\zeta_0 = -1$. The nondegenerated shape of quadratic forms is Tsai-Wu ellipsoids until an intermediate eigenvalue becomes zero

$$\zeta_{0,crit} = \frac{4\sigma_0^-\sigma_0^+ - (\sigma_0^+)^2 - (\sigma_0^-)^2}{(\sigma_0^+ + \sigma_0^-)^2}. \tag{1.19}$$

Beyond that value $\zeta_{0,crit}$, the yield surface takes the form of the positive leaf of a hyperboloid that degenerates into a Drucker-Prager cone for $\zeta_0 = \frac{1}{2}$ (Fig. 7.6).

The generalized form of these tensors are given by:

$$\mathbf{F}(\rho,\mathbf{M}) = \frac{1}{\hat{f}(\rho)}\sum_{i=1}^{3}\frac{1}{m_i^{2q}}\frac{1}{2}\left(\frac{1}{\sigma_0^+}-\frac{1}{\sigma_0^-}\right){}^i\mathbf{M} \tag{1.20}$$

$$\begin{aligned}\mathbb{F}(\rho,\mathbf{M}) = \frac{1}{\hat{f}^2(\rho)}\Bigg(&\sum_{i=1}^{3}\frac{1}{m_i^{4q}}\frac{1}{4}\left(\frac{1}{\sigma_0^+}+\frac{1}{\sigma_0^-}\right)^2 {}^i\mathbf{M}\otimes{}^i\mathbf{M}\\ &-\sum_{i,j=1;i\neq j}^{3}\frac{\zeta_0}{m_i^{2q}m_j^{2q}}\frac{1}{4}\left(\frac{1}{\sigma_0^+}+\frac{1}{\sigma_0^-}\right)^2 {}^i\mathbf{M}\otimes{}^j\mathbf{M}\\ &+\sum_{i,j=1;i\neq j}^{3}\frac{1}{2\tau_0^2 m_i^{2q}m_j^{2q}}{}^i\mathbf{M}\,\underline{\overline{\otimes}}\,{}^j\mathbf{M}\Bigg)\end{aligned} \tag{1.21}$$

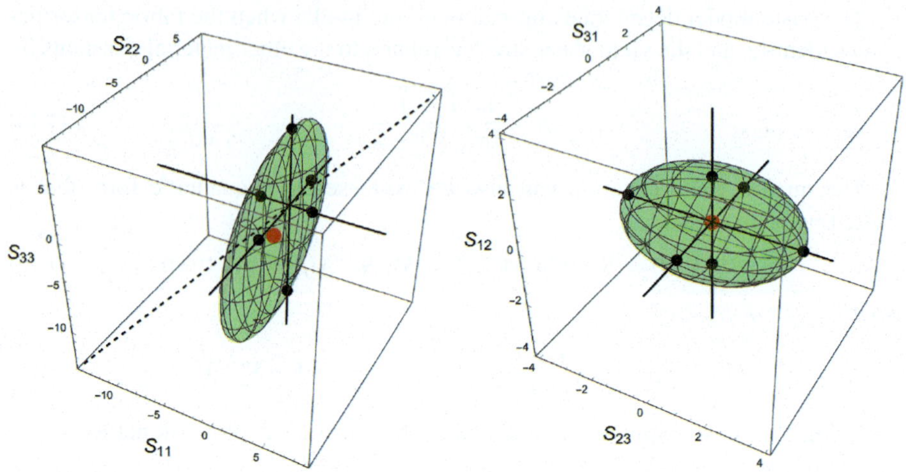

FIGURE 7.6

Visualization of the yield surface for the material constants in the uncoupled normal stress space and shear stress space with $\rho = 1.0$, $m_1 = 0.7$, $m_2 = 1.0$, $m_3 = 1.3$. The red point is the origin; the dashed line is the trisectrix corresponding to hydrostatic loading. The points represent the uniaxial yield and shear stresses along the three material directions. Units are MPa.

The material constants for the generalized yield criterion are provided in Table 7.3 for both KUBCs and PMUBCs. Dividing the uniaxial yield stress by the elastic modulus for $\rho = 0.2$ provides realistic yield strains of 0.0068–0.0069 in tension and 0.0087–0.0091 in compression that are somewhat higher than the ones selected for the trabecular ECM in validation experiments (0.0054 in tension and 0.0081in compression (Schwiedrzik et al., 2025). The ratio of compressive versus tensile yield stress of 1.27–1.32 for the respective BCs is lower than the ratio of the corresponding ECM yield stresses of 1.5. This can be understood by the fact that some tensile tissue yielding also occurs in compression of trabecular RVEs. The interaction coefficient ζ_0 which reflects plastic compressibility is slightly lower for PMUBC than KUBC. The exponent p describing the influence of ρ on yield stress is lower or similar to the exponent k describing the influence of ρ on elasticity, which suggests that yield properties scale in a weaker fashion than elastic properties and confirms the observed slight decrease of yield strain with volume fraction reported in previous studies, e.g. (Bayraktar et al., 2004; Rincon-Kohli & Zysset, 2009). Since the exponent q is also smaller than l, the influence of fabric seems also lower on the yield properties than on the elastic properties. This is consistent with the larger yield strains observed in secondary versus main fabric directions (Charlebois et al., 2010).

Table 7.3 Generalized quadric yield constants by nonlinear computational homogenization for femoral trabecular bone using KUBC and PMUBC from (Panyasantisuk et al., 2016). The tissue material properties were isotropic, and the yield properties were distinct in tension and compression: $\epsilon_{tis} = 10$ GPa, $\nu_{tis} = 0.3$, $\epsilon^{+}_{tis} = 0.0054$, $\epsilon^{-}_{tis} = 0.0081$, and $\zeta^{0}_{tis} = 0.49$.

Variables	N	BCs	σ_0^+	σ_0^-	ζ_0	τ_0	*p*	*q*
Site/Units			**[MPa]**	**[MPa]**	**[-]**	**[MPa]**	**[-]**	**[-]**
Femur A	126	KUBC	49.00	64.57	0.30	31.11	1.5	0.71
Femur A	126	PMUBC	57.69	73.10	0.28	29.61	1.82	0.98

The distinct relationships obtained with KUBC and PMUBC for yield stress are illustrated in Fig. 7.7 for tension and compression. The differences are substantial. In an exploratory study based on only three samples, PMUBCs provided much closer results with respect to embedded BCs than KUBC (Wili et al., 2017) and should be favored.

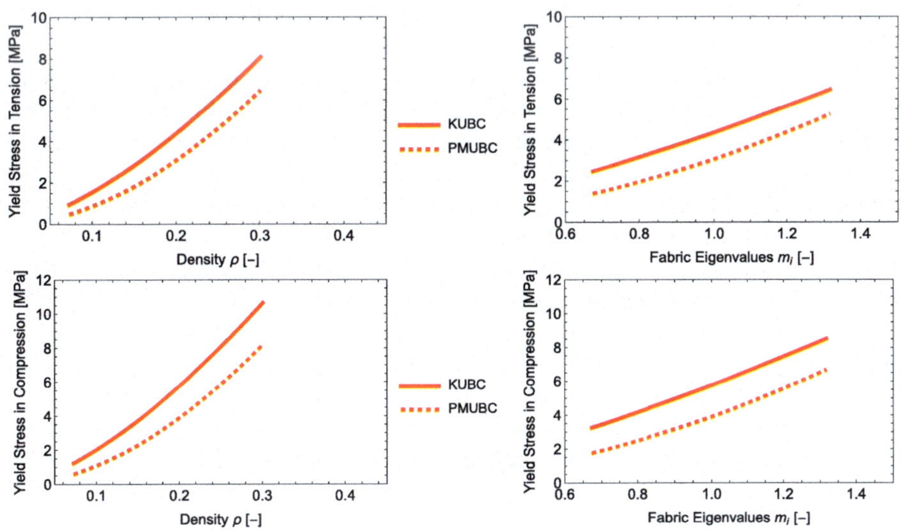

FIGURE 7.7

Effect of boundary conditions on the relationship between structural density for $m_i = 1$ (left), fabric eigenvalue for $\rho = 0.2$ (right), and yield stress in tension (top) and compression (bottom) for the relevant density and fabric ranges ($\pm$ 1 standard deviation).

4. Discussion

This chapter reviews the phenomenological morphology-mechanical property relationships developed for trabecular bone by the author and provides an update of the available material constants for the full 3D elastic and yield behavior.

Morphology is reduced to volume fraction and fabric. Despite the plethora of alternative morphological parameters for trabecular bone, including rod and plate decomposition or trabecular bone score (TBS), additional morphological parameters beyond volume fraction and fabric were not shown to improve the prediction of elastic and yield properties of trabecular bone (Maquer et al., 2015; Musy et al., 2017). Volume fraction is by far the dominant factor, and fabric contributes significantly to the predictions, especially for yield properties.

The basic models degenerate nicely into isotropic models when the fabric tensor becomes identity and are probably sufficient for first qualitative analyses, but the generalized models are more accurate in terms of shear properties.

The identified morphology-elastic property relationships appear to be similar across anatomical locations when the same scanner, image resolution and processing are used (Gross et al., 2013). This finding is consistent with a universal underlying topological optimization achieved by bone modeling and remodeling. Interestingly, no differences in morphology-elastic property relationships could be evidenced in osteoporosis (Homminga et al., 2003), hypoparathyroidism (Gross, Pahr, Mueller, et al., 2012) and osteogenesis imperfecta (Simon et al., 2022) when compared to controls, which suggests that the above topological optimization is not affected by these diseases. To my knowledge, such results are not available for morphology-yield property relationships.

Boundary conditions play an important role in the application of the homogenization method to compute apparent elastic or yield properties. Most data available were computed with KUBCs, which have the advantage that the VE can be selected in an arbitrary coordinate system. For PMUBCs, the VE must be rotated along the principal axes of the orthotropic symmetry, i.e., of the fabric tensor (Pahr & Zysset, 2008). KUBCs provide an upper bound estimate, and PMUBCs a closer lower bound estimate of the results calculated with in situ or embedded BCs. Accordingly, PMUBC is the first choice to select material parameters.

It should be noted that the homogenization method fails for highly heterogeneous VE, typically for very low volume fractions (<5%), as the faces of the cubic VE become poorly connected or not connected at all. Such samples are excluded in the calculation of material constants using a homogeneity index (Simon et al., 2022; Panyasantisuk et al., 2025). Nevertheless, extrapolation of morphology-mechanical property relationships to such low values of volume fraction in hFE analyses are not problematic as the related properties are negligible.

A few experiments attempted to identify morphology-mechanical property relationships for trabecular bone (Rincon-Kohli & Zysset, 2009), but the material constants were not reported here as the boundary conditions are poorly defined and introduce numerous artifacts (Keaveny et al., 1997; Odgaard & Linde, 1991).

A complete set of elastic and yield material parameters is provided for the proximal femur (A), but may also be applied to other anatomical locations. The material constants provided in this review refer to accurate volume fractions and MIL fabric tensors measured on high-resolution ($\leq 18\mu m$) μCT images. When using coarser resolution images or different measures of fabric, calibration equations transforming into the original BV/TV and **M** measures become necessary, see e.g, (Varga & Zysset, 2009).

Again, mechanical properties of bone ECM were assumed to be homogenous in this review, but pathological changes would obviously be reflected at the continuum level. Previous work suggests that the degree of mineralization of the ECM can be easily accounted for in elastic properties by a linear correction (Indermaur et al., 2025), but that heterogeneity of the mineral distribution has a minor effect on trabecular bone elasticity (Gross, Pahr, Peyrin, & Zysset, 2012).

Taken together, morphology-elastic properties and morphology-yield properties inform constitutive models of trabecular bone for homogenised finite element (hFE) analysis that account for anisotropic elasticity and the simultaneous flow of plastic strains and damage associated with micro-cracking beyond a certain anisotropic yield criterion (Schwiedrzik & Zysset, 2013).

For hFE analyses of whole bones or bone parts, the inclusion of fabric in constitutive models was shown to be particularly relevant for quantitative analyses and when multiple load cases along different directions are considered (Hazrati Marangalou et al., 2012; Panyasantisuk et al., 2018; Schenk et al., 2025).

In general, the extraction of a fabric tensor from a bone image requires at least a resolution of approximately 60 microns, which is delivered by HR-pQCT used clinically at the distal radius and tibia. Nevertheless, the distribution of the fabric tensor in a given bone can also be assumed (Chevalier et al., 2009) or extracted from generalized statistical shape models (Panyasantisuk et al., 2018; Taghizadeh et al., 2017; Chandran et al., 2025).

For qualitative hFE analyses of bone or bone-implant systems, isotropic constitutive models solely based on BV/TV ($\mathbf{M} = \mathbf{I}$) are often sufficient. However, comparative FE analyses for clinical purposes (Keaveny et al., 2020; Zysset et al., 2015) require a very careful calibration of BV/TV (Varga & Zysset, 2009; Dall'Ara et al., 2011; Hosseini, Dunki, et al., 2017).

Acknowledgments

I would like to acknowledge all the individuals who contributed to the original development and evolution of this subject: Alain Curnier, Vimal Chandran, Mathieu Charlebois, Yan Chevalier, Karol Daszkiewicz, Scott Hollister, Hadi Hosseini, David Larsson, Adeliya Latypova, Dieter Pahr, Thomas Gross, Ghislain Maquer, Sarah Musy, Jarunan Panyasantisuk, Liliana Rincón-Kohli, Denis Schenk, Jakob Schwiedrzik, Mathieu Simon, Elham Taghizadeh, Peter Varga and Patrik Wili.

References

Bayraktar, H., Morgan, E., Niebur, G., Morris, G., Wong, E., & Keaveny, T. (2004). Comparison of the elastic and yield properties of human femoral trabecular bone and cortical bone tissue. *Journal of Biomechanics, 37*, 27–35.

Carter, D., & Hayes, W. (1977). The compressive behaviour of bone as a two phase porous structure. *Journal of Bone and Joint Surgery American Volume, 59-A*, 954–962.

Chandran, V., Reyes, M., & Zysset, P. (2025). A novel registration-based methodology for prediction of trabecular bone fabric from clinical qct: A comprehensive analysis. *PLoS One, 12*(11).

Charlebois, M., Pretterklieber, M., & Zysset, P. K. (2010). The role of fabric in the large strain compressive behavior of human trabecular bone. *Journal of Biomechanical Engineering, 132*(12), Article 121006.

Chevalier, Y., Pahr, D., & Zysset, P. K. (2009). The role of cortical shell and trabecular fabric in finite element analysis of the human vertebral body. *Journal of Biomechanical Engineering, 131*(11), Article 111003.

Cowin, S. (1985). The relationship between the elasticity tensor and the fabric tensor. *Mechanics of Materials, 4*, 137–147.

Cowin, S. (1986). Fabric dependence of an anisotropic strength criterion. *Mechanics of Materials, 5*, 251.

Cowin, S. (1999). Bone poroelasticity. *Journal of Biomechanics, 32*, 217–238.

Curnier, A. (2004). *Mécanique des Solides Déformables*. Lausanne: Presses Polytechniques et Universitaires Romandes.

Currey, J. D. (2002). *Bones: Structure and mechanics*. Princeton: Princeton University Press.

Dall'Ara, E., Varga, P., Pahr, D., & Zysset, P. (2011). A calibration methodology of qct bmd for human vertebral body with registered micro-ct images. *Medical Physics, 38*(5), 2602–2608.

Daszkiewicz, K., Maquer, G., & Zysset, P. K. (2017). The effective elastic properties of human trabecular bone May be approximated using micro-finite element analyses of embedded volume elements. *Biomechanics and Modeling in Mechanobiology, 16*(3), 731–742.

Fleps, I., Bahaloo, H., Zysset, P. K., Ferguson, S. J., Palsson, H., & Helgason, B. (2020). Empirical relationships between bone density and ultimate strength: A literature review. *Journal of the Mechanical Behavior of Biomedical Materials, 110*, Article 103866.

Geraets, W., vanDerStelt, P., Lips, P., Elders, P., vanGinkel, F., & Burger, E. (1997). Orientation of the trabecular pattern of the distal radius around the menopause. *Journal of Biomechanics, 30*(4), 363–370.

Gibson, L. (1985). The mechanical behaviour of cancellous bone. *Journal of Biomechanics, 18*(5), 317–328.

Goulet, R., Goldstein, S., Ciarelli, M., Kuhn, J., Brown, M., & Feldkamp, L. (1994). The relationship between the structural and orthogonal compressive properties of trabecular bone. *Journal of Biomechanics, 27*(4), 375–389.

Gross, T., Pahr, D., Mueller, R., Rubin, M. R., Bilezikian, J. P., & Zysset, P. (2012). Morphology-elasticiy relationships in iliac crest biopsies are not affected by hypoparathyroidism. In J. F. Paulo Fernandez, & M. Silva (Eds.), *18th congress of the european society of biomechanics* (Vol 45). Lisbon: J Biomech, Article s253. suppl 1.

Gross, T., Pahr, D. H., Peyrin, F., & Zysset, P. K. (2012). Mineral heterogeneity has a minor influence on the apparent elastic properties of human cancellous bone: A srmuct-based finite element study. *Computer Methods in Biomechanics and Biomedical Engineering, 15*(11), 1137–1144.

Gross, T., Pahr, D. H., & Zysset, P. K. (2013). Morphology-elasticity relationships using decreasing fabric information of human trabecular bone from three major anatomical locations. *Biomechanics and Modeling in Mechanobiology, 12*(4), 793–800.

Harrigan, T., Jasty, M., Mann, R., & Harris, W. (1988). Limitations of the continuum assumption in cancellous bone. *Journal of Biomechanics, 21*(4), 269–275.

Harrigan, T., & Mann, R. (1984). Characterization of microstructural anisotropy in orthotropic materials using a second rank tensor. *Journal of Materials Science, 19*, 761–767.

Hazrati Marangalou, J., Ito, K., & van Rietbergen, B. (2012). A new approach to determine the accuracy of morphology-elasticity relationships in continuum fe analyses of human proximal femur. *Journal of Biomechanics, 45*(16), 2884–2892.

He, Q. C., & Curnier, A. (1995). A more fundamental approach to damaged elastic stress-strain relations. *International Journal of Solids and Structures, 32*(10), 1433–1457.

Homminga, J., McCreadie, B. R., Weinans, H., & Huiskes, R. (2003). The dependence of the elastic properties of osteoporotic cancellous bone on volume fraction and fabric. *Journal of Biomechanics, 36*(10), 1461–1467.

Hosseini, H. S., Dunki, A., Fabech, J., Stauber, M., Vilayphiou, N., Pahr, D., Pretterklieber, M., Wandel, J., Rietbergen, B. V., & Zysset, P. K. (2017). Fast estimation of colles' fracture load of the distal section of the radius by homogenized finite element analysis based on hr-pqct. *Bone, 97*, 65–75.

Hosseini, H. S., Maquer, G., & Zysset, P. K. (2017). muct-based trabecular anisotropy can be reproducibly computed from hr-pqct scans using the triangulated bone surface. *Bone, 97*, 114–120.

Indermaur, M., Casari, D., Kochetkova, T., Peruzzi, C., Zimmermann, E., Rauch, F., Willie, B., Michler, J., Schwiedrzik, J., & Zysset, P. (2025). Compressive strength of iliac bone ecm is not reduced in osteogenesis imperfecta and increases with mineralization. *Journal of Bone and Mineral Research.*

Johnson, T. P., Socrate, S., & Boyce, M. C. (2010). A viscoelastic, viscoplastic model of cortical bone valid at low and high strain rates. *Acta Biomaterialia, 6*(10), 4073–4080.

Kabel, J., VanRietbergen, B., Odgaard, A., & Huiskes, R. (1999). Constitutive relationships of fabric, density and elastic properties in cancellous bone architecture. *Bone, 25*(4), 481–486.

Kanatani, K. (1984). Distribution of directional data and fabric tensors. *International Journal of Engineering Science, 22*(2), 149–164.

Keaveny, T. M., Clarke, B. L., Cosman, F., Orwoll, E. S., Siris, E. S., Khosla, S., & Bouxsein, M. L. (2020). Biomechanical computed tomography analysis (bct) for clinical assessment of osteoporosis. *Osteoporosis International, 31*(6), 1025–1048.

Keaveny, T., Pinilla, T., Crawford, R., Kopperdahl, D., & Lou, A. (1997). Systematic and random errors in compression testing of trabecular bone. *Journal of Orthopaedic Research, 15*, 101–110.

Kuhn, J., Goldstein, S., Feldkamp, L., Goulet, R., & Jesion, G. (1990). Evaluation of a micro-computed tomography system to study trabecular bone structure. *Journal of Orthopaedic Research, 8*, 833–842.

Larsson, D., Luisier, B., Kersh, M. E., Dall'Ara, E., Zysset, P. K., Pandy, M. G., & Pahr, D. H. (2014). Assessment of transverse isotropy in clinical-level ct images of trabecular bone using the gradient structure tensor. *Annals of Biomedical Engineering, 42*(5), 950–959.

Latypova, A., Maquer, G., Elankumaran, K., Pahr, D., Zysset, P., Pioletti, D. P., & Terrier, A. (2016). Identification of elastic properties of human patellae using micro-finite element analysis. *Journal of Biomechanics, 49*(13), 3111–3115.

Maquer, G., Musy, S. N., Wandel, J., Gross, T., & Zysset, P. K. (2015). Bone volume fraction and fabric anisotropy are better determinants of trabecular bone stiffness than other morphological variables. *Journal of Bone and Mineral Research, 30*(6), 1000–1008.

Musy, S. N., Maquer, G., Panyasantisuk, J., Wandel, J., & Zysset, P. K. (2017). Not only stiffness, but also yield strength of the trabecular structure determined by non-linear microfe is best predicted by bone volume fraction and fabric tensor. *Journal of the Mechanical Behavior of Biomedical Materials, 65*, 808–813.

Odgaard, A., Jensen, E., & Gundersen, H. (1990). Estimation of structural anisotropy based on volume orientation. A new concept. *Journal of Microscopy, 159*, 335–342.

Odgaard, A., Kabel, J., vanRietbergen, B., Dalstra, M., & Huiskes, R. (1997). Fabric and elastic principal directions of cancellous bone are closely related. *Journal of Biomechanics, 30*(5), 487–495.

Odgaard, A., & Linde, F. (1991). The underestimation of young's modulus in compressive testing of cancellous bone specimens. *Journal of Biomechanics, 24*(8), 691–698.

Pahr, D. H., & Zysset, P. K. (2008). Influence of boundary conditions on computed apparent elastic properties of cancellous bone. *Biomechanics and Modeling in Mechanobiology, 7* (6), 463–476.

Panyasantisuk, J., Dall'Ara, E., Pretterklieber, M., Pahr, D. H., & Zysset, P. K. (2018). Mapping anisotropy improves qct-based finite element estimation of hip strength in pooled stance and side-fall load configurations. *Medical Engineering & Physics, 59*, 36–42.

Panyasantisuk, J., Pahr, D. H., Gross, T., & Zysset, P. K. (2025). Comparison of mixed and kinematic uniform boundary conditions in homogenized elasticity of femoral trabecular bone using microfinite element analyses. *Journal of Biomechanical Engineering, 137*(1).

Panyasantisuk, J., Pahr, D. H., & Zysset, P. K. (2016). Effect of boundary conditions on yield properties of human femoral trabecular bone. *Biomechanics and Modeling in Mechanobiology, 15*(5), 1043–1053.

Peruzzi, C., Ramachandramoorthy, R., Groetsch, A., Casari, D., Grönquist, P., Rüggeberg, M., Michler, J., & Schwiedrzik, J. (2021). Microscale compressive behavior of hydrated lamellar bone at high strain rates. *Acta Biomaterialia, 131*, 403–414.

Rice, J. C., Cowin, S. C., & Bowman, J. A. (1988). On the dependence of the elasticity and strength of cancellous bone on apparent density. *Journal of Biomechanics, 21*(2), 155–168.

Rincon-Kohli, L., & Zysset, P. K. (2009). Multi-axial mechanical properties of human trabecular bone. *Biomechanics and Modeling in Mechanobiology, 8*(3), 195–208.

Roschger, P., Gupta, H. S., Berzlanovich, A., Ittner, G., Dempster, D. W., Fratzl, P., Cosman, F., Parisien, M., Lindsay, R., Nieves, J. W., & Klaushofer, K. (2003). Constant mineralization density distribution in cancellous human bone. *Bone, 32*, 316–323.

Schenk, D., Indermaur, M., Simon, M., Voumard, B., Varga, P., Pretterklieber, M., Lippuner, K., & Zysset, P. (2025). Unified validation of a refined second-generation hr-pqct based homogenized finite element method to predict strength of the distal segments in radius and tibia. *Journal of the Mechanical Behavior of Biomedical Materials, 131.*

Schwiedrzik, J., Gross, T., Bina, M., Pretterklieber, M., Zysset, P., & Pahr, D. (2025). Experimental validation of a nonlinear fe model based on cohesive-frictional plasticity for trabecular bone. *International Journal for Numerical Methods in Biomedical Engineering, 32*(4).

Schwiedrzik, J. J., Wolfram, U., & Zysset, P. K. (2013). A generalized anisotropic quadric yield criterion and its application to bone tissue at multiple length scales. *Biomechanics and Modeling in Mechanobiology, 12*(6), 1155–1168.

Schwiedrzik, J. J., & Zysset, P. K. (2013). An anisotropic elastic-viscoplastic damage model for bone tissue. *Biomechanics and Modeling in Mechanobiology, 12*(2), 201–213.

Simon, M., Indermaur, M., Schenk, D., Hosseinitabatabaei, S., Willie, B. M., & Zysset, P. (2022). Fabric-elasticity relationships of tibial trabecular bone are similar in osteogenesis imperfecta and healthy individuals. *Bone, 155*, Article 116282.

Skedros, J. G., & Baucom, S. L. (2007). Mathematical analysis of trabecular 'trajectories' in apparent trajectorial structures: The unfortunate historical emphasis on the human proximal femur. *Journal of Theoretical Biology, 244*(1), 15–45.

Snyder, B. D., & Hayes, W. C. (1990). Multiaxial structure-property relations in trabecular bone. In V. Mow, A. Ratcliffe, & S.-Y. Woo (Eds.), *Biomechanics of diarthrodial joints* (pp. 31–59). New York: Springer Verlag.

Tabor, Z. (2011). A novel method of estimating structure model index from gray-level images. *Medical Engineering & Physics, 33*(2), 218–225.

Tabor, Z., & Rokita, E. (2007). Quantifying anisotropy of trabecular bone from gray-level images. *Bone, 40*(4), 966–972.

Taghizadeh, E., Chandran, V., Reyes, M., Zysset, P., & Buchler, P. (2017). Statistical analysis of the inter-individual variations of the bone shape, volume fraction and fabric and their correlations in the proximal femur. *Bone, 103*, 252–261.

Turner, C., Cowin, S., Rho, J., Ashman, R., & Rice, J. (1990). The fabric dependance of the orthotropic elastic constants of cancellous bone. *Journal of Biomechanics, 23*, 549–561.

VanRietbergen, B. (2001). Micro-fe analyses of bone: State of the art. *Advances in Experimental Medicine and Biology, 496*, 21–30.

VanRietbergen, B., Odgaard, A., Kabel, J., & Huiskes, R. (1998). Relationship between bone morphology and bone elastic properties can be accurately quantified using high-resolution computer reconstructions. *Journal of Orthopaedic Research, 16*, 23–28.

VanRietbergen, B., Weinans, H., Huiskes, R., & Odgaard, A. (1995). A new method to determine trabecular bone elastic properties and loading using micromechanical finite-element models. *Journal of Biomechanics, 28*(1), 69–81.

Varga, P., & Zysset, P. K. (2009). Assessment of volume fraction and fabric in the distal radius using hr-pqct. *Bone, 45*(5), 909–917.

Whitehouse, W. (1974). The quantitative morphology of anisotropic trabecular bone. *Journal of Microscopy, 101*, 153–168.

Wili, P., Maquer, G., Panyasantisuk, J., & Zysset, P. K. (2017). Estimation of the effective yield properties of human trabecular bone using nonlinear micro-finite element analyses. *Biomechanics and Modeling in Mechanobiology, 16*(6), 1925–1936.

Zysset, P. K. (2003). A review of morphology-elasticity relationships in human trabecular bone: Theories and experiments. *Journal of Biomechanics, 36*(10), 1469–1485.

Zysset, P. K., & Curnier, A. (1995). An alternative model for anisotropic elasticity based on fabric tensors. *Mechanics of Materials, 21*(4), 243–250.

Zysset, P. K., & Curnier, A. (1996). A 3d damage model for trabecular bone based on fabric tensors. *Journal of Biomechanics, 29*(12), 1549–1558.

Zysset, P. K., Goulet, R. W., & Hollister, S. J. (1998). A global relationship between trabecular bone morphology and homogenized elastic properties. *Journal of Biomechanical Engineering, 120*(5), 640–646.

Zysset, P. K., Ominsky, M. S., & Goldstein, S. A. (1998). A novel 3d microstructural model for trabecular bone: I. The relationship between fabric and elasticity. *Computer Methods in Biomechanics and Biomedical Engineering, 1*(4), 321–331.

Zysset, P. K., Ominsky, M. S., & Goldstein, S. A. (1999). A novel 3d microstructural model for trabecular bone: Ii. The relationship between fabric and the yield surface. *Computer Methods in Biomechanics and Biomedical Engineering, 2*(1), 1–11.

Zysset, P., Qin, L., Lang, T., Khosla, S., Leslie, W. D., Shepherd, J. A., Schousboe, J. T., & Engelke, K. (2015). Clinical use of quantitative computed tomography-based finite element analysis of the hip and spine in the management of osteoporosis in adults: The 2015 iscd official positions-part ii. *Journal of Clinical Densitometry, 18*(3), 359–392.

Zysset, P. K., Schwiedrzik, J., & Wolfram, U. (2016). European society of biomechanics s.m. perren award 2016: A statistical damage model for bone tissue based on distinct compressive and tensile cracks. *Journal of Biomechanics, 49*(15), 3616–3625.

CHAPTER

Experimental bone mechanics

8

Piyush Uniyal and G. Harry van Lenthe

KU Leuven, Department of Mechanical Engineering, Leuven, Belgium

Abbreviations

BMD	Bone Mineral Density
BV/TV	Bone Volume/Tissue Volume, or Bone Volume Fraction
CT	Computed Tomography
DXA	Dual-Energy X-ray Absorptiometry
FE	Finite element
QCT	Quantitative Computed Tomography
RPI	Reference Point Indentation

1. Introduction

Bone, a specialized connective tissue, serves as a load-bearing framework essential for human movement, protects vital internal organs, acts as a reservoir for minerals, and facilitates blood cell production. Its structure adapts dynamically to physiological loading conditions. Primary load-bearing bones, such as the femur and tibia in the lower extremities, are subjected to significant mechanical stresses during daily physiological activities. These bones are susceptible to fractures, especially when mechanical competence is compromised due to aging and diseases (Morgan et al., 2018). Bone fractures are prevalent in the elderly population and pose severe health risks, diminishing quality of life and increasing mortality rates. Accurate fracture risk prediction is essential for preventative strategies, which requires a thorough understanding of bone's mechanical behavior and integrity. Traditionally, bone mineral density (BMD) assessed through dual-energy X-ray absorptiometry (DXA) has been used as a fracture risk predictor (Cummings et al., 2002). However, the limitations in DXA's predictive accuracy (Ammann & Rizzoli, 2003) have led researchers to explore alternative markers and biomechanical methods to improve bone fracture risk prediction.

Bone fracture is a complex biomechanical event, resulting when bone is unable to withstand deformation under applied loads. Understanding bone's structural and mechanical properties at various hierarchical levels provides insight into its overall fracture resistance (Hunt & Donnelly, 2016). Over the past several decades, a range of experimental methods has been developed, each tailored to specific hierarchical

Biomechanics and Mechanobiology of Bone. https://doi.org/10.1016/B978-0-443-26739-0.00008-9

levels or length scales, with distinct benefits and limitations (Bailey & Vashishth, 2018). Technological advancements have facilitated new approaches to examining bone's mechanical behavior across these scales, from microstructural to macroscopic assessments.

This chapter provides an overview of various mechanical testing methodologies for assessing bone integrity, with a particular emphasis on 1–10 mm length scale mechanical properties. Additionally, this chapter discusses time-dependent mechanical behavior, reflecting bone's viscoelastic properties, and highlights advances in understanding fracture mechanics across different loading conditions and structural complexities.

2. Hierarchical structure of bone

Structurally, bone exhibits a hierarchical organization with progressively complex building blocks at defined length scales (Reznikov et al., 2014). At the largest scale, the *whole bone* represents the macrostructure, responsible for transferring joint and muscle loads. This scale typically spans from 1 cm to several decimeters.

At the *mesostructure* level, bone tissue is classified into two primary types based on porosity: trabecular and cortical bone. The trabecular bone is approximately 60%–95% porous whereas cortical bone is relatively dense and consists of 5%–15% porosity. The trabecular bone consists of an intricate network of trabecule, which facilitates efficient load distribution and metabolic activity. It is predominantly found in the joints and vertebral bodies. Cortical bone, characterized by a much lower porosity and provides structural rigidity and is primarily located in the outer shafts of long bones; it is also part of the surface layer of all bones. The *mesostructures* of bone typically range in size from 500 μm to 10 mm.

At the microstructural level, the fundamental unit of cortical bone is the osteon. Osteons are cylindrical structures composed of concentric lamellae of calcified matrix surrounding a central Haversian canal, which contains blood vessels and nerves. In contrast, the primary structural unit of trabecular bone is the trabecula, which forms a complex, three-dimensional lattice-like network. Microstructural features typically range in size from 5 to 500 μm.

At the *nanostructure* level, bone matrix is composed of collagen fibers, which are organized from arrays of mineralized collagen fibrils. Each fibril is made up of self-assembled type I collagen molecules, approximately 1.5 nm in diameter and 300 nm in length.

At the smallest scale, each tropocollagen molecule is composed of various amino acids. The compact alignment of collagen chains is facilitated by glycine, with the helix stabilized by numerous hydrogen bonds. The enzymatic collagen crosslinking not only stabilizes collagen I molecules but also promotes their self-assembly into fibrils. This hierarchical organization enables bone to combine toughness and strength, adapting to its functional and metabolic demands.

3. Experimental mechanical characterization of bone

Clinical bone fractures occur from the inability of bone to endure applied loads or deformations, making these events inherently mechanical in nature. Various mechanical characterization techniques have evolved over the years to investigate bone fracture behavior. The selection of a specific mechanical test depends on the hierarchical level at which mechanical properties need to be evaluated Fig. 8.1. At the macroscale, tests focus on whole bone properties, at the mesoscale, apparent tissue properties are assessed, and at the micro- and nanoscale, the material properties of bone tissue are explored. During these tests, controlled forces (or displacements) are applied to bone samples, and their mechanical responses are measured in terms of deformation (or force). Depending on the experimental design, the applied loading may or may not reflect physiologically relevant loading conditions. The primary mechanical properties quantified include stiffness, strength, toughness, fracture toughness and fatigue resistance. Each property offers unique insights into bone's mechanical competence and its capacity to perform under physiological or pathological conditions. These measurements are essential for a comprehensive

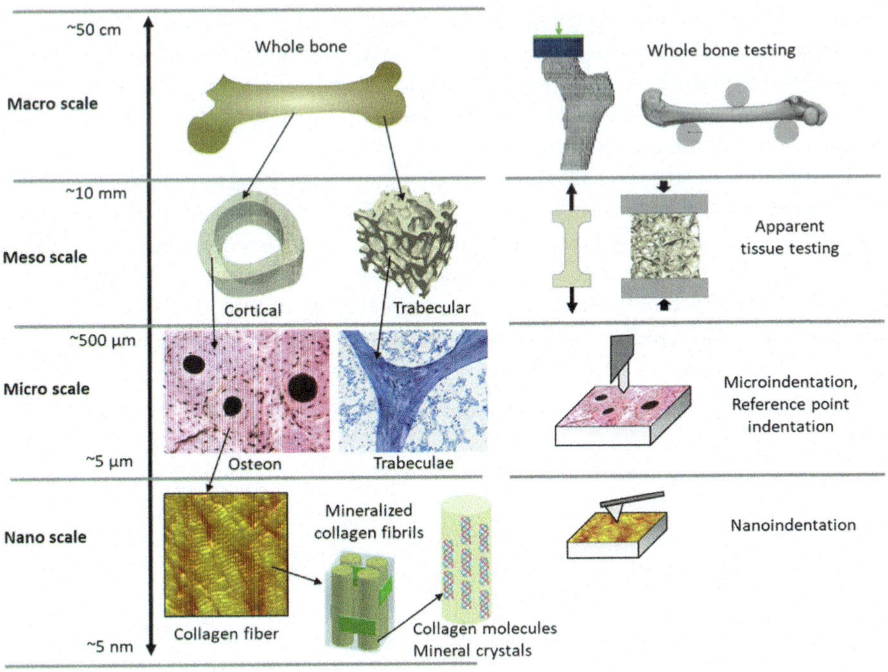

FIGURE 8.1

Hierarchical structure of bone and corresponding mechanical characterization techniques.

understanding of bone integrity and the factors contributing to its resistance or susceptibility to fractures.

3.1 Whole-bone mechanical tests

Whole-bone mechanical tests are conducted to evaluate the overall mechanical response of bone at the macroscale level. This response reflects contributions from various bone characteristics, including bone mineral density (BMD), bone geometry, and tissue material properties. Whole bone tests are inherently destructive and are typically performed ex vivo using animal or cadaveric bone specimens. These tests are commonly conducted in compression, bending, or torsional modes under either monotonic or cyclic load to failure. Long bones and vertebrae are most frequently used for such tests, with compression and bending being the predominant loading modes. In these experiments, the entire bone is subjected to mechanical loads up to failure and load-displacement data is collected at appropriate time intervals, which are subsequently used to plot the load-displacement curve. From this curve, key mechanical properties such as the maximum load, stiffness, and work-to-failure can be calculated Fig. 8.2.

Mechanical loading is applied in various configurations, often designed to replicate physiologically relevant loading conditions. Numerous studies have been performed on human femur bones subjected to such loading conditions (Keyak et al., 2005; Sas et al., 2022). During these tests, the distal end of the femur is typically embedded in PMMA or bone cement, while the proximal end is loaded using a loading cup. The resulting force-displacement curve is then analyzed to understand the bone's mechanical response.

For preclinical studies, rodent models are extensively used to evaluate the effects of therapies and treatments on bone mechanical behavior. Whole bone testing in rodent models is widely used because extracting standardized test specimens can be challenging due to the small size of rodent bones. As a result, bending tests are commonly employed to assess the mechanical behavior of rodent bones. These tests are generally conducted using either a 3-point or 4-point bending configuration. In

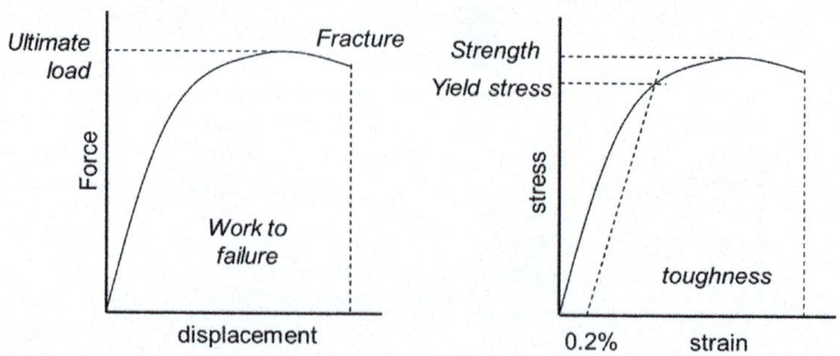

FIGURE 8.2

Typical force-displacement and stress-strain curve obtained from uniaxial testing.

these tests, the bone is placed on two rollers, and force is applied through either one or two rollers from the top for 3-point and 4-point bending, respectively. Force-displacement data is collected during these tests and used to investigate the mechanical properties of the bone.

Although whole bone mechanical tests are widely accepted and provide valuable insights into the mechanical behavior of bone, they come with several inherent limitations. These tests capture the mechanical response resulting from a combination of bone geometry and material properties. To isolate and derive material properties, theories of solid mechanics are typically applied. However, these theories are based on assumptions such as uniform cross-sections and homogeneous material composition. Since bone does not satisfy these assumptions due to its complex geometry and heterogeneous nature, the resulting material property estimations are inherently biased (van Lenthe et al., 2008). This may explain why the material properties estimated from whole-bone tests often show poor correlation with those obtained from direct assessment at smaller scales. Therefore, direct assessment of material properties at smaller scales is crucial for more accurate characterization. Another limitation is that the contributions of structural properties and material properties cannot be distinguished in whole bone testing, necessitating additional techniques to separate the contributions of these factors. A combination of image-based analysis and experimental mechanical testing can help achieve this (see Section 4).

3.2 Apparent-level testing (uniaxial tests on standard specimens)

Whole bone mechanical tests provide an overall assessment of bone behavior but cannot isolate the individual contributions of material properties and structural changes. To address this, mechanical tests are conducted on uniformly machined bone specimens, typically ranging from millimeters to centimeters in width and length, allowing for the evaluation of material properties at the apparent level. This approach accounts for features at lower hierarchical levels in the measured mechanical properties. Consequently, the results at this level of structural hierarchy are termed apparent-level properties, reflecting the integrated response of bone material. These tests are typically conducted on human or large animal bones, offering insights relevant to understanding load-bearing capacity and mechanical performance in physiological and clinical contexts.

Uniaxial testing can be classified into tension and compression tests. Tensile tests are predominantly conducted on cortical bone, with limited applications on trabecular bone. Specimens, typically in the form of a dog bone or dumb bell shape are machined. There are no standardized dimensions specified in the literature, and specimens with varying dimensions are commonly reported. During testing, the specimen is fixed between rigid grips, with force or displacement applied at one end while the opposite end remains stationary. Force is measured using a load cell, and displacement is monitored with an extensometer mounted within the gauge length. The resulting force-displacement curve is then normalized by the cross-sectional area and gauge length, allowing for the derivation of stress-strain behavior.

From the stress-strain curve, material properties such as Young's modulus (elastic modulus), yield strain, ultimate strain, and strain energy can be calculated Fig. 8.2. The stress-strain response of cortical bone under uniaxial testing typically exhibits an initial linear region followed by a non-linear region. The yield point is commonly estimated using the 0.2% strain offset method. The elastic modulus is determined through linear regression of the linear portion of the curve, while the highest point on the stress-strain curve is ultimate tensile strength of the bone.

Compared to tensile testing, compression tests are relatively simple to conduct and are commonly used for trabecular as well as cortical bone. Compression tests are typically performed on cylindrical or rectangular/cubic specimens. Cylindrical specimens are extracted using a core drill or trephine, while rectangular specimens are prepared with a diamond saw. Due to the susceptibility of trabecular bone cylindrical cores to fracture at the ends, they are often fitted with customized metallic end caps. Although specimens with varying dimensions are reported in the literature, a length-to-diameter (L/D) ratio of approximately 1.5 is generally recommended. Mechanical properties are derived in a similar way as in tensile testing. However, because trabecular bone is porous, its overall mechanical response is influenced by its microarchitecture. Consequently, the mechanical response obtained from compression tests on trabecular bone does not provide a direct quantification of material properties; instead, these properties are referred to as apparent mechanical properties.

3.3 Tissue-level mechanical testing

Indentation tests are widely utilized to evaluate the mechanical properties of bone at the tissue level. In these tests, a rigid indenter tip is pressed into the bone surface in a controlled manner, allowing for the measurement of mechanical properties based on the applied force, indentation depth, or indentation area. These tests capture material properties at the micro-to nanoscale, providing measurements that are independent of structural features, such as bone porosity, and focusing instead on the intrinsic characteristics of the bone tissue. In cyclic Reference Point Indentation (BioDent), a stiff metallic probe indents the bone matrix in subsequent steps and force versus indentation depth is recorded while for impact-micro indentation (OsteoProbe) only indentation depth is recorded. For the BioDent device, 90 degrees cono-spherical probe with 375-μm diameter, 2.5-μm tip radius is used. The OsteoProbe device have probe 375-μm tip diameter, <10-μm tip radius. Both devices create indent in the bone matrix which involves elastic, plastic, and damage processes. The details of these devices are provided in the following sections.

3.3.1 Cyclic Reference Point Indentation (BioDent)

The cRPI tests are performed using BioDent device (Active Life Scientific, Santa Barbara, CA). From the resulting load-displacement curve, several output variables are calculated, such as total indentation depth, indentation depth increase, creep indentation distance, loading and unloading slopes, and energy dissipated; these properties reflect the bone's strength, creep and fatigue responses. These properties are calculated as averages from the third-to-last loading cycle. The test is performed exclusively in force control mode, with the user selecting a maximum force between 2 and 10 N. This technique also allows customization of various parameters, including preconditioning cycles and the number of loading cycles. For accurate measurements, the device probe must first be calibrated on a standard PMMA block. All tips must be discarded after 150 indentation tests, as the surface of the tip wears over time. The measured mechanical properties primarily represent the material properties of the bone tissue, with minimal contribution from features such as microscopic porosity. However, the relationship between cRPI-measured properties and meso- and macroscale bone mechanical properties remains unclear. Some studies, such as those by Gallant et al. (2013), suggest a correlation between cRPI outcomes and traditional mechanical properties. In contrast, other research indicates that cRPI alone may be insufficient to determine conventional mechanical properties (Carriero et al., 2014; Krege et al., 2016). Therefore, further investigation is necessary to better understand the correlation between these properties.

3.3.2 Impact reference point indentation (OsteoProbe)

The OsteoProbe (Active Life Scientific, Santa Barbara, CA) is an impact-based microindentation device specifically developed for clinical use to assess in vivo mechanical properties of bone (Diez-Perez et al., 2016). The in-vivo indentation procedure consists of a preloading and loading phase. First, a 10 N linear preload is applied to penetrate soft tissue and reach the bone surface. Following this, a single impact load of 40 N is applied, driving the probe into the bone. A displacement transducer records the indentation distance during this impact phase. Finally, the probe is unloaded from the bone surface. The primary outcome, known as the Bone Material Strength index (BMSi) is determined by calculating the ratio of the indentation distance into the bone relative to that into a PMMA reference phantom, and then normalizing this value by multiplying it by 100 (Ly et al., 2020). This normalized value provides a quantifiable measure of bone material properties suitable for in vivo clinical applications. Yet, similar to cyclic RPI, the relationship to meso- and macroscale mechanical properties remains unclear.

3.4 Nanoscale mechanical testing

Nanoindentation is primarily used to characterize the mechanical properties of bone at the lamellar or interlamellar regions, offering insight into nanoscale mechanical integrity that arises from collagen fibrils and mineral crystals. This technique is one of the most effective methods to understand bone's nanoscale mechanical behavior. Nanoindentation tests utilize depth-sensing transducers to assess local material properties within the bone tissue.

For nanoindentation, bone samples are typically embedded in resin when working with rodent bones or trabecular bone. In contrast, embedding is generally not necessary for larger bones from humans or large animals. In all cases, samples must be ground and polished to remove surface artifacts prior to testing. Indentations can then be performed at multiple locations on the polished surface, either under force or displacement control. During testing, continuous measurement of tip displacement produces a load-displacement curve. Quasi-static tests are commonly used to determine elastic modulus and hardness, while dynamic tests are used to evaluate viscoelastic properties such as storage and loss moduli. Various tip geometries - such as Berkovich, spherical, and cube corner are selected based on the material being tested and specific properties to be measured. The reduced elastic modulus is calculated using the projected area of indentation and the contact depth (Oliver & Pharr, 1992). Hardness, which quantifies a material's resistance to localized plastic deformation, is determined by dividing the maximum applied load by the contact area of the indenter. It is important to consider the effect of viscoelastic deformation and thermal drift on measured material properties. A limitation of nanoindentation is the need for meticulous sample preparation, especially surface polishing, to ensure accurate and reliable data. A comprehensive overview of the nanoindentation technique as applied to mineralized hard tissues can be found in previous literature (Lewis & Nyman, 2008; Nyman et al., 2016).

4. Computational mechanical characterization of bone

The Finite Element Method (FEM) is a powerful numerical modeling technique widely used to simulate the mechanics of complex structures, making it well-suited for studying bone mechanics. FEM relies on dividing a structure into a finite number of elements, each governed by mathematical equations that capture the physics of the problem. This method has been applied to understand bone's mechanical behavior for nearly 5 decades, with most finite element (FE) models tailored to the macro-scale (whole bone) and mesoscale (cortical or trabecular bone). Finite element (FE) modeling incorporates both the geometry and material properties of bone, enabling the decoupling of geometric effects from the overall mechanical response to isolate and assess the intrinsic material properties of bone tissue.

FE models are typically constructed from computed tomography (CT) data, providing geometrically accurate representations of bone at the macro and mesoscale. At the macroscale, FE models have been developed that represent the bone's overall geometry, while treating the internal bone microstructure as a continuum with solid elements that are assigned apparent-level mechanical properties of cortical and trabecular bone obtained at the mesoscale. These models help assess the bone's load-bearing capacity in a holistic view. In contrast, mesoscale models focus on the detailed microstructure of cortical and trabecular bone, emphasizing the effects of porosity on mechanical behavior. The mechanical behavior at this level is governed by the mechanical properties derived from nanoscale tissue, also known as tissue-level properties. With advancements in the resolution of desktop micro-CT and the emergence of synchrotron radiation micro-CT, finite element modeling is increasingly including features at the nanoscale level such as osteocyte lacunae

(Hemmatian et al., 2021), enabling more detailed and precise simulations of bone behavior across hierarchical scales.

Finite element (FE) analysis has been widely used to examine the mechanical behavior of bones at various anatomical sites, with a particular focus on locations such as the vertebrae, radius, and femur at the macro scale, and trabecular bone at the meso scale (van Rietbergen & Ito, 2015). In macro-scale FE models, the spine and hip are typically scanned at around 0.5 mm resolution, where the grayscale values are calibrated to bone density. This approach includes both mineralized and marrow voxels in the mesh, an averaging method known as homogenized-FE, which uses apparent material properties of the bone-marrow mixture. When CT scans are conducted at resolutions below approximately 40 μm, the resulting models are referred to as micro-FE. Micro-FE modeling is being applied to peripheral sites such as the distal tibia and distal radius using high-resolution peripheral quantitative CT (HR-pQCT), cone-beam CT (Mys et al., 2021), and recently photon-counting CT (Quintiens, 2024).

Mesh generation is a critical aspect of FE modeling in bone analysis. In this context, voxel-based meshing is often used to directly convert CT image voxels into an FE mesh. While voxel meshes are computationally efficient, they present the drawback of irregular boundaries. A smoother boundary representation can be achieved using marching cube algorithm, which generates tetrahedral mesh elements. Alternatively, hexahedral or tetrahedral meshing offers smoother boundaries but requires more extensive processing efforts to create anatomically accurate and computationally efficient models (Viceconti et al., 1998).

Quantifying mechanical properties at the macro (or meso) level is only possible when the geometry as well as the mechanical properties at the meso (or nano) level is known. Stiffness or elastic modulus can be calculated using linear FEA, while properties such as yield point and failure load/stress require non-linear FEA for accurate estimation. This requires a precise selection of material properties, constitutive behavior, and boundary conditions. Furthermore, validation against experimental data is considered mandatory. Once validated, FEA offers distinct advantages over physical testing, such as eliminating end artifacts and off-axis measurement issues. Furthermore, FEA allows the simulation of various loading scenarios on a given bone microstructure.

To predict bone strength and identify potential failure zones, failure criteria must be included in the description of the constitutive model. Common failure criteria include von Mises stress (Keyak et al., 2005) and maximum strain (MacNeil & Boyd, 2008; Niebur et al., 2000; Sadoughi et al., 2020). The Pistoia criterion (Pistoia et al., 2002) is frequently applied to predict fragility fracture risk using finite element models based on quantitative computed tomography (QCT) or high-resolution peripheral QCT (HR-pQCT). According to this criterion, bone failure initiates when a specified percentage of elements (~2%) reach the tissue yield strain (0.007). This failure criterion is based on linear FEA and has been validated primarily for the distal radius. However, its applicability to other anatomical sites and under different loading conditions remains uncertain.

FEA in conjunction with mechanical testing can provide additional details such as tissue modulus of the bone (Niebur et al., 2000; van Lenthe et al., 2008). In this approach, macroscale properties are first determined via mechanical testing, while microstructure is captured through micro-CT imaging. After that virtual testing can then be conducted using FEA by implementing boundary conditions identical to those in physical experiments and assuming an initial tissue modulus. The true tissue modulus can subsequently be deduced by tuning the tissue modulus such that the observed behavior matches the simulated behavior. This approach is similarly applicable to whole bone failure behavior.

Mechanical properties of trabecular bone obtained through experimental testing often face limitations, such as end artifacts of the samples (Keaveny et al., 1997). However, these challenges can be mitigated through micro-CT-based finite element (micro-FE) analysis, which has been extensively used to model and analyze the mechanical properties of trabecular bone.

5. Mechanical properties of bone

Experimental mechanical tests are considered the gold standard to evaluate the mechanical competence of bone. An overview of these tests has been provided in the previous section. In this section the mechanical properties of bone obtained through these tests will be discussed in the detail.

5.1 Mechanical behavior of cortical bone

Cortical bone, the dense outer layer of most bones, is particularly well-adapted in load-bearing long bones like the femur and tibia. Osteons, the basic structural units of cortical bone, are aligned longitudinally, creating a preferred orientation. This alignment, combined with the presence of Haversian canals, results in anisotropic mechanical properties. Specifically, cortical bone exhibits higher elastic modulus and strength in the longitudinal direction compared to the circumferential and radial directions (Dong et al., 2012). As the difference between radial and circumferential properties is relatively small, cortical bone is often considered transversely isotropic, meaning its mechanical properties are uniform in the radial and circumferential directions but differ along the longitudinal axis. An overview of the mechanical properties of cortical bone is provided in Table 8.1.

Cortical bone demonstrates an asymmetric mechanical response under different loading conditions Fig. 8.3. It exhibits higher strength in compression compared to tension, with the lowest strength observed under shear loading (Wolfram & Schwiedrzik, 2016). The stress-strain behavior of bone under tension and compression is similar within the elastic region but diverges significantly in the post-yield region. More specifically, under tensile loading, bone exhibits exponential hardening followed by linear hardening in the post-yield zone. In contrast, under compressive loading, bone undergoes quasi-brittle softening after the yield point. The distinct

Table 8.1 Mechanical properties of human cortical bone.

Study	Anatomical site	Loading mode	Age (years)	Specimen dimensions (mm) cuboid (B×W×H), cylinder (D×H)	Elastic modulus (GPa)	Yield strain (%)	Yield stress (MPa)	Ultimate strain (%)	Ultimate stress (MPa)
Reilly et al. (1974)	Femur	T C	20–86	2 × 6		–	–	2.74 ± 0.57 2.11 ± 0.33	128.3 ± 12.3 190.4 ± 15.6
Carter et al. (1981)	Femur	T	51–84	3 × 10	17.5 1.9	0.84 ± 0.05	121 ± 11	1.57 ± 0.38	140 ± 12
Dickenson et al. (1981)	Femur	T	67–87 69–91(OP)	2.4 × 11	15.68 ± 2.1 11.55 ± 4.1	–	80.8 ± 13.7 75.8 ± 28.5	–	117 ± 17.3 95.1 ± 33.8
Courtney et al. (1996)	Femur	T	26 ± 5 72 ± 6	3 × 8	15.69 ± 1.66 14.78 ± 1.5	0.37 ± 0.06 0.35 ± 0.04	55.3 ± 8.6 49.3 ± 5.5		81.6 ± 10.6 78.7 ± 7.0
Currey (2004)	Femur	T	35	1.8 × 1.8 × 13	16.7	0.72	122.3	2.92	165.7
Bayraktar et al. (2004)	Femur	T	54–85 (71.8 ± 8.8)	2.5 × 3.0 × 11	19.9 ± 1.8	0.73 ± 0.05	107.9 ± 12.3	–	–
Duchemin et al. (2008)	Femur	T C	54–101 (81.8 ± 12.7)	3 × 3 × 25 3 × 3 × 5	14.3 ± 4.0 11.8 ± 3.6	–	–	–	53.8 ± 20.3 106.4 ± 29.4
Ebacher and Wang (2009)	Femur	C-Long C-Trans C-Oblique	55,64,69	3 × 3 x 6	16 ± 2 8 ± 1 13.3 ± 0.3	1.07 ± 0.14 1.15 ± 0.07 0.98 ± 0.08	141 ± 5 79.0 ± 7.0 104 ± 10	1.3 ± 0.2 3.3 ± 0.80 1.7 ± 0.4	143 ± 6 91.0 ± 9.0 113 ± 11
Dong et al. (2012)	Femur	C-Long C-Trans C-Radial	51.5 ± 3.3	3 × 5	19.1 ± 6.35 5.65 ± 1.61 6.49 ± 3.22	0.91 ± 0.09 0.83 ± 0.42 0.84 ± 0.23	111 ± 18.6 41.8 ± 19.4 44.1 ± 21.1	–	124 ± 15 65.2 ± 13.8 63.1 ± 20.7
Tang et al. (2015)	Femur	S	62–79	5 × 3 × 20 with V-notch	4.0 ± 0.4	–	–	–	49.9 ± 6.2
Mirzaali et al. (2016)	Femur	T C	46–99	3 × 6.5	18.16 ± 1.88 18.97 ± 1.84	0.45 ± 0.05 0.66 ± 0.09	71.6 ± 10.2 115.1 ± 16.4	1.9 ± 0.60 1.3 ± 0.30	92.9 ± 10.1 153.6 ± 21.6
Lee et al. (1997)	Tibia	T	53–98 (81 ± 12)	3 × 8	11.3 ± 5.8	0.56 ± 0.07	70.6 ± 31.6	2.0 ± 0.85	77.0 ± 42.2
Leng et al. (2009)	Tibia	C	53 ± 2 83 ± 6	3 × 5	19.1 ± 1.03 20.4 ± 1.27	0.82 ± 0.034 0.71 ± 0.075	118 ± 9.49 107 ± 15.7		132 ± 8.62 122 ± 14.0
Dong et al. (2013)	Tibia	S	76–88 (83.7 ± 5.1)	Double notch shear specimen	4.66 ± 1.10	0.88 ± 0.18	35.7 ± 9.9	–	61.4 ± 6.3

C, *compression;* Long, *longitudinal;* OP, *osteoporosis;* S, *shear;* T, *tension;* Trans, *transverse.*

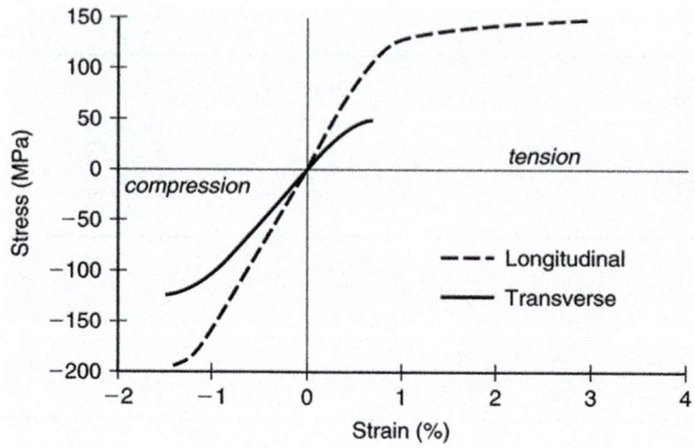

FIGURE 8.3

Asymmetric stress-strain behavior of bone under tension and compression modes.

post-yield behaviors of bone are attributed to different damage mechanisms under tensile and compressive loading conditions. Compressive loading in cortical bone is typically associated with the formation of microcracks, whereas tensile loading leads to the development of small, submicroscopic clusters of cracks, referred to as diffuse damage (Diab & Vashishth, 2005). These distinct damage morphologies contribute differently to the reduction in the bone's toughness. The ratio of tensile yield strain to compressive yield strain for cortical bone is approximately 0.68. Moreover, bone exhibits a more significant decrease in strength under compressive loading compared to tensile loading. While tensile testing maintains the bone's longitudinal alignment throughout the process, compressive loading can lead to instability and deviation from the loading axis after yielding. This deviation can initiate microcracks, which can further propagate and coalesce, resulting in non-uniform stress distribution within the sample. This disruption in stress distribution ultimately reduces the bone's ability to withstand load.

Cortical bone is a dense tissue with a porosity ranging from approximately 5%–15%. Cortical porosity negatively correlates with elastic modulus (Schaffler & Burr, 1988) and fracture toughness (Ural & Vashishth, 2007). The primary contributors to porosity include vascular canals, osteocyte lacunae, the canaliculi network, and small nanopores. This porosity is distributed throughout the bone matrix and is known to change during skeletal growth and in response to pathological conditions. These pores create stress concentrations within the bone matrix, serving as potential sites for microcrack initiation and propagation (Burr, 2019; Currey, 1962). Indeed, research has demonstrated that the local strain magnitudes near the porosities are significantly higher than the applied apparent strain, with experimental and finite element analyses reporting local strains up to 9 to 15 times greater than bulk tissue strain (Nicolella et al., 2006; Verbruggen et al., 2012). Hence, whereas in vivo organ-level strains rarely exceed the yield point, the localized amplification of strain near porosity can surpass this threshold, initiating microcracks. The analysis on the effects of stress concentrations is challenging because the size, orientation, and spatial distribution of bone porosities significantly influence the interaction between stress concentrations. It has been shown that a higher angle between the loading axis and orientation of vascular canals is more detrimental to bone integrity than aligned canals (Uniyal et al., 2021). Furthermore, the canal network also acts as a critical site for microcrack initiation and propagation (Voide et al., 2011). Collectively, these findings provide critical insights into how microdamage can develop under physiological loading conditions. These studies underscore the pivotal role of porosity distribution in the mechanical behavior of bone, highlighting its importance for understanding and assessing bone health in various pathological conditions.

5.2 Mechanical behavior of trabecular bone

Trabecular bone, also known as cancellous or spongy bone, is a highly porous structure. It is composed of a network of strut-like structures called trabecule, which are typically rod- or plate-shaped. The random arrangement of these trabecule imparts anisotropic mechanical properties at the apparent level. Due to the significant porosity in trabecular bone, its mechanical properties are largely influenced by the porosity. In addition to porosity, the microarchitectural arrangement of the trabecule also affects the apparent mechanical behavior. The microarchitecture of trabecular bone is commonly assessed using parameters such as bone volume fraction, trabecular thickness, trabecular number, trabecular spacing, degree of anisotropy, and structural model index; the treatment of these parameters is outside of scope for this chapter.

The apparent stress–strain behavior of trabecular bone typically includes an initial linear region followed by a non-linear region. The modulus is usually determined by fitting a linear or polynomial curve to the initial portion of the data, and the yield point is identified using the 0.2% offset method. Similar to cortical bone, trabecular bone exhibits asymmetric post-yield behavior under tension and compression. Under tensile loading, it shows non-linear stress-strain behavior after yielding, with progressive stiffness degradation leading to fracture. In contrast, under compressive loading, the specimen does not fracture; instead, after reaching peak stress, the stress–strain curve transitions into a plateau phase, followed by a densification phase at large strain. Although trabecular bone yields at around 0.7% strain in compression, it can sustain compressive strains of up to 50% while still maintaining a significant portion of its load-bearing capacity (Morgan et al., 2018). While trabecular bone demonstrates superior strength in compression compared to tension, the difference in strength between the two loading modes decreases as the bone volume fraction decreases. An overview of the mechanical properties of trabecular bone is provided in Table 8.2.

The mechanical properties of trabecular bone are primarily determined by the amount and arrangement of bone porosities. Microstructural parameters are typically obtained through micro-CT imaging, and these parameters are generally

Table 8.2 Mechanical properties of human trabecular bone.

Study	Anatomical Site	Loading mode	Age (years)	Specimen dimensions (mm) cuboid (B×W×H), cylinder (D×H)	Elastic modulus (GPa)	Yield strain (%)	Yield stress (MPa)	Ultimate strain (%)	Ultimate stress (MPa)
Martens et al. (1983)	Femur	C	27–77	8 × 8	0.900 ± 0.710				9.3 ± 4.5
Mosekilde et al. (1987)	Vertebra	C	15–87	7 × 5	0.067			7.4	2.45
Røhl et al. (1991)	Tibia	T C	42–76	9 × 9 × 20	0.487 ± 0.329 0.489 ± 0.331			1.55 ± 0.49 1.11 ± 0.63	2.54 ± 1.18 2.22 ± 1.42
Kopperdahl and Keaveny (1998)	Vertebra	T C	32–65	8 × 25	0.301 ± 0.100 0.291 ± 0.113	0.78 ± 0.04 0.84 ± 0.06	1.75 ± 0.65 1.92 ± 0.84	1.59 ± 0.33 1.45 ± 0.33	2.23 ± 0.76 2.23 ± 0.95
Morgan and Keaveny (2001)	Vertebra Proximal tibia Greater trochanter Femoral Neck	T C T C T C T C	20–90 (62 ± 17) 40–85 (69 ± 12) 49–101 (70 ± 13) 57–101 (74 ± 13)	8 × 25 6 × 8 6 × 8 6 × 8	0.349 ± 0.133 0.344 ± 0.148 1.068 ± 0.84 1.091 ± 0.634 0.597 ± 0.330 0.622 ± 0.302 2.7 ± 0.772 3.23 ± 0.936	0.70 ± 0.05 0.77 ± 0.06 0.65 ± 0.05 0.73 ± 0.06 0.61 ± 0.05 0.70 ± 0.05 0.61 ± 0.03 0.85 ± 0.10	1.72 ± 0.64 2.02 ± 0.92 4.50 ± 3.14 5.83 ± 3.42 2.44 ± 1.26 3.21 ± 1.83 10.93 ± 3.08 17.45 ± 6.15		
Kaneko et al. (2004)	Femur	C	45–88	15 × 15 × 15	0.774 ± 0.360		7.59 ± 3.29		7.89 ± 3.49
Rincón-Kohli and Zysset (2009)	Radius, Femur, Tibia	T C	30–89 (73.5 ± 16.8)	8 × 10	0.754 ± 0.454 0.597 ± 0.401	0.67 ± 0.1 1.52 ± 0.36	3.4 ± 1.9 8.97 ± 7.5	1.4 ± 0.0 2.3 ± 0.7	4.02 ± 2.3 10.16 ± 8.9
Wolfram et al. (2011)	Vertebra	T C	21–94	8 × 12	0.384 ± 0.155 0.383 ± 0.162		0. 72 0. 81		
Lambers et al. (2014)	Lumbar vertebra	T C	47–78 (70 ± 10)	8 × 15	0.276 0.231	0.73 0.75	1.48 1.19		
Hernandez et al. (2014)	Lumbar vertebra	C	64–92	8.3 × 19.90	0.221 ± 0.095	0.69 ± 0.08	1.1 ± 0.5	1.29 ± 0.31	1.29 ± 0.58
Wang et al. (2015)	Tibia, Femur	C	40–91	8.5 × 8	1.430 ± 1.457		8.16 ± 8.92		

C, compression; T, tension.

correlated with properties derived from mechanical testing. Various linear and power-law relationships have been proposed in the literature to model these correlations. Apparent density, which is the product of bone volume fraction (BV/TV) and bone tissue density, is one of the most important parameters for predicting the mechanical properties of trabecular bone. The general form of the relationship is expressed as $E = a.\left(\rho_{\text{app}}\right)^{b}$, where E is the elastic modulus, ρ_{app} is apparent density and a and b are constants. A comprehensive overview on the relationship has been presented by Helgason et al. (2008). A similar relationship between bone density and strength has been reviewed by Fleps et al. (2020). These relationships treat elastic modulus and strength as isotropic parameters. A more complex description is required to include the anisotropic behavior of trabecular bone Several studies have correlated the elastic properties of trabecular bone to the fabric tensor, which characterizes the anisotropy of trabecular bone. Cowin (1985) was the first to establish a relationship between the elasticity tensor and the fabric tensor. Building on this model, Turner et al. (1990) quantified the relationship specifically for trabecular bone. Later, Zysset and Curnier (1995) and Van Rietbergen et al. (1996) proposed alternative models to refine these correlations. Kabel et al. (1999) further demonstrated a reliable fit between the components of the fabric tensor and the elastic constants of trabecular bone.

6. The effect of strain rate

The mechanical behavior of bone is highly strain rate-dependent due to its viscoelastic nature. At low strain rates, bone exhibits more ductile characteristics, allowing for greater deformation before failure. In contrast, at high strain rates, as experienced during traumatic events, bone becomes stiffer and more brittle, with reduced deformation capacity but higher ultimate strength. In this section, the experimental techniques and rate dependent mechanical behavior is discussed.

6.1 Experimental techniques

Quasi-static mechanical testing of bone as described in Section 3 is typically performed using Universal Testing Machines (UTMs), where samples are held between grips and subjected to either tensile or compressive loads. UTMs are highly accurate for evaluating mechanical properties at low strain rates. However, as strain rates increase, measurement accuracy can be compromised. This is because the displacement sensor may take time to accelerate to the required speed, affecting the ability to maintain a constant deformation rate. If the specimen fails before reaching the desired strain rate, the test results are considered invalid. Therefore, special caution is needed when testing bone samples at higher strain rates.

For intermediate strain rate testing, techniques such as drop tests, impact loading with a propelled mass using a gas/pneumatic gun, or a wedge-type split Hopkinson pressure bar are used. For high strain rate testing, the split Hopkinson pressure bar (SHPB) is widely applied. In this method, the sample is placed between two co-axial bars, and a stress wave is generated by the impact of a striker bar at one end. Strain rate variation during these tests is a common issue, and several methods have been proposed to maintain a constant strain rate for accurate analysis.

6.2 Rate-dependent mechanical behavior of bone

Most studies on bone mechanics have focused on low strain rates. Yet, experimental tests have shown that bone exhibits time-dependent mechanical behavior, which becomes especially significant during high-impact and traumatic events. Many clinical fractures—such as those resulting from falls, vehicle accidents, sports injuries, or battlefield incidents—occur under high strain rate conditions, which are orders of magnitude higher than the strain rates experienced during daily physiological activities (Zioupos et al., 2020) Fig. 8.4. Under typical physiological conditions, bone is subjected to strain rates of about 0.001–0.01 s^{-1}. However, during traumatic events, strain rates can exceed the physiological range by 10-fold or more. Therefore, it is essential to understand the rate-dependent mechanical behavior of bone, especially for applications involving high loading rates. The properties of bone under these conditions are also critical inputs for computational models to predict fracture accurately and to inform the design of protective equipment, enhancing safety and injury prevention. The strain rate dependent mechanical properties of bone are summarized in Table 8.3.

Previous research has extensively examined the impact of strain rate on the mechanical properties of bone. While most studies have focused on low strain rates, relatively few have investigated bone's mechanical properties at high strain rates. In general, findings indicate that as strain rate increases, both elastic modulus and failure strength also increase, whereas failure strain decreases, leading bone to fracture in a more brittle manner (McElhaney, 1966). At low strain rates, bone fractures in a more ductile manner, and the energy absorbed, represented by the area under the stress-strain curve is higher. However, as strain rate increases, the fracture becomes more brittle, and the energy absorbed decreases, indicating a reduction in toughness.

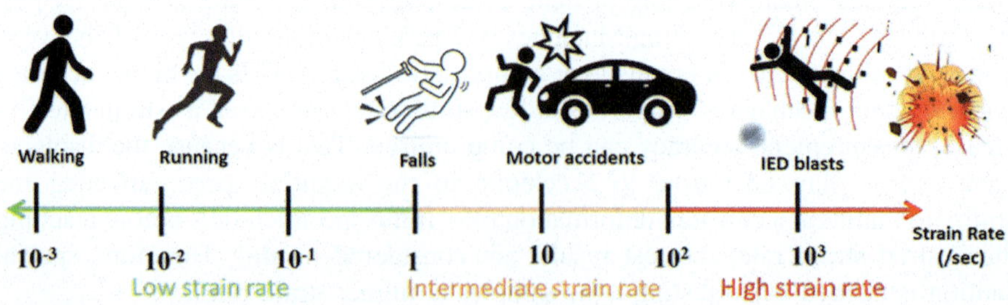

FIGURE 8.4

Strain rates observed by skeletal system during various events.

Table 8.3 Mechanical properties of bone at different strain rates.

Study	Species	Anatomical site	Age	Type of bone	Test condition	Specimen geometry, (mm), cuboid (B×W×H), cylinder (D×H)	Strain rate (s^{-1})	Modulus (GPa)	Max. strength (MPa)	Failure strain (%)
McElhaney (1966)	Human	Femur	24 Year	Cortical	Long	4.45 × 4.45 × 6.35	0.001	15.17	150.34	1.65
							0.01	17.24	179.31	1.75
							0.1	17.93	200	1.8
							1	22.07	220.69	1.78
							300	29.66	279.31	1.1
							1500	40.69	317.24	0.95
Carter and Hayes (1976)	Human	Tibia		Trabecular		10.3 × 5	0.001–10		$68 \cdot \rho_{app}^{2} \cdot \dot{\varepsilon}^{0.06}$	
Katsamanis and Raftopoulos (1990)	Human	Femur	45–55 Years	Cortical (dry)	Long	4 × 10.5 × 54	0.00002	16.2		
						4 × 10.5 × 216	100	19.9		
Linde et al. (1991)	Human	Tibia	65 Years	Trabecular		5.5 × 8.25	0.0001–10	$3101 \cdot \rho_{app}^{2} \cdot \dot{\varepsilon}^{0.042}$	$52.1 \cdot \rho_{app}^{2} \cdot \dot{\varepsilon}^{0.07}$	
Shim et al. (2005)	Human	Vertebra	40–79 Years	Trabecular		5 ×	0.001–1000	$242 \cdot \rho_{app}^{1.95}$	$5.09 \cdot \rho_{app}^{2.35} + 0.545 \cdot \dot{\varepsilon}^{0.45}$	
						5 × 8				
Weerasooriya et al. (2016)	Human	Femur	36–50 Years	Cortical (wet)	Long	3 × 3.25 × 4	0.001	9.15 ± 5.98	152.1 ± 21.6	3.06 ± 1.06
					Trans		1	10.2 ± 5.11	205.3 ± 16.5	3.83 ± 0.9
							1000	11.05 ± 3.46	319 ± 23.9	3.24 ± 0.79
							0.001	3.05 ± 1.14	86.60 ± 21.7	7.21 ± 3.61
							1	7.24 ± 3.8	146.98 ± 27.53	9.78 ± 4.79
							1000	8.30 ± 3.25	178.90 ± 26.02	3.17 ± 1.28
Zhai et al. (2020)	Human	Skull	70–74 Years	Cranial	Radial	7.2 × 4.7 × 7.9	0.01	0.40	24.8	11.2
					Trans	6.3 × 4.8 × 7.9	30	0.90	33.2	6.0
							625	1.67	37.8	3.7
							0.01	1.20	54.4	6.9
							30	1.62	76	6.3
							625	3.05	82.2	3.7
Uniyal et al. (2023)	Human	Tibia	73–86 Years	Cortical	Long	4.5 × 4.5 × 4.5	0.001	7.17	137.92	2.7
							0.01	7.76	154.15	3.1
							1000	10.76	317.40	4.0
McElhaney (1966)	Bovine	Femur	3 Year	Cortical	Long	4.45 × 4.45 × 6.35	0.001	18.62	175	1.88
							0.01	20	207	1.82
							0.1	24.14	231	1.75
							1	27.59	252	1.25
							300	33.11	283	1
							1500	42.08	366	0.9

Table 8.3 Mechanical properties of bone at different strain rates.—*cont'd*

Study	Species	Anatomical site	Age	Type of bone	Test condition	Specimen geometry, (mm), cuboid (B×W×H), cylinder (D×H)	Strain rate (s^{-1})	Modulus (GPa)	Max. strength (MPa)	Failure strain (%)
Currey (1975)	Bovine	Femur	–	Cortical	Tension	1.8 × 1.8 × 19	0.00013	21	91	0.40
							0.0013	24.1	137	0.55
							0.013	22.4	152	0.65
							0.062	24.2	185	0.65
							0.16	25.9	165	0.64
Wright and Hayes (1976)	Bovine	Femur	Adult	Cortical	Long (Tension)	3.18 × 54 (dog bone specimen)	0.00053	17.7 ± 0.8	99.2 ± 4.7	
							0.0053	19.9 ± 0.9	130.8 ± 8.2	
							0.053	23 ± 0.9	162.2 ± 6.1	
							0.174	22.3 ± 0.6	161.3 ± 8.9	
							14.3	27.5 ± 1	185.14 ± 12.2	
							59.2	23.8 ± 0.9	253.3 ± 14.6	
							237	40.4 ± 4	271.4 ± 8.4	
Fischer et al. (1986)	Bovine	Metatarsals	Mature	Cortical	Tension	3.18 × 3.18 × 31.8	0.001	27.5	113 ± 6.89	1.09 ± 0.2
							0.01	49.8	156 ± 14.7	1.70 ± 0.36
Pithioux et al. (2004)	Bovine	Femur	5–7	Cortical	Long (Tension)	7 × 3 × 60 (gauge area of dog bone specimen)	0.00014–0.014	11.3–17.5	105–193	0.93–1.8
							16.67	10–21.7	33–50	0.23–0.44
Adharapurapu et al. (2006)	Bovine	Femur	Mature	Cortical	Long	5 × 5 × 5	0.001		165	
					Trans		0.01		164	
							0.1		206	
							1		236	
							1000		458	
							0.001		143	
							0.01		141	
							0.1		155	
							1		195	
							1000		296	
Ferreira et al. (2006)	Bovine	Femur	–	Cortical	Long	Area 50–150 mm^2 x length	637–1070	9.9 ± 2.7	281 ± 42.4	3.26 ± 0.72
					Trans	7.2–10 mm	896–1209	6.8 ± 2.1	240 ± 66.4	3.73 ± 0.45
Bekker et al. (2015)	Bovine	Femur	–	Cortical	Long	5.99 × 5.95	0.0001		149.6 ± 13.7	3.98
							0.001		171.9 ± 18.5	4.51
							0.01		202.1 ± 14.9	5.1
							0.1		234.8 ± 15.5	6.2
							250		383.9 ± 21.3	2.4
							1000		413.9 ± 28.7	2.7
Prot et al. (2015)	Bovine	Femur	3 Years	Trabecular		41 × 14	1000	0.135 ± 0.022	11.8 ± 3.4	8.8 ± 1.3
							1500 (confined)	0.329 ± 0.057	20.6 ± 1.3	6.9 ± 1.6

Prot et al. (2016)	Bovine	Femur	4 Years	Trabecular		10.5 × 7.5	0.001	0.322 ± 0.216	8.3 ± 6.1	4.8 ± 1
							0.01	0.365 ± 0.250	10.2 ± 7.3	5.4 ± 0.9
							0.1	0.299 ± 0.192	9.4 ± 6.7	5.5 ± 0.7
							1	0.285 ± 0.171	12 ± 7.8	11 ± 4.8
							10	0.585 ± 0.386	14.6 ± 8.4	4.7 ± 1.6
							100	0.629 ± 0.479	13.4 ± 8.6	3.2 ± 0.8
							400	0.471 ± 0.333	12.5 ± 7.8	4.6 ± 1
							600	0.573 ± 0.313	15.3 ± 8.2	4.3 ± 1.1
Pagano et al. (2016)	Bovine	Femur (dry)	–	Cortical	Long	3.5 × 3.5 × 3.5	0.001		336 ± 6	
					Trans		2700		559 ± 23	
							0.001		244 ± 22	
							2400		350 ± 34	
Enns-Bray et al. (2018)	Bovine	Femur & Tibia	6 Years	Trabecular		8 × 16	30–168		$24.1 \cdot \rho_{app}^{1.07} \cdot \dot{\varepsilon}^{0.11} \cdot DA^{0.83}$	
Uniyal et al. (2022)	Bovine	Femur	18–20 Months	Cortical	Long	4.5 × 4.5 × 6	0.0001	6.98 ± 0.53	104.93 ± 11.16	1.9
							0.001	7.45 ± 1.17	120.32 ± 19.61	2.1
							0.01	6.67 ± 1.34	138.91 ± 21.34	3
							1000	15.02 ± 1.16	373.77 ± 27.05	2.8
Kulin et al. (2011)	Equine	Femur	0.5–28 Years	Cortical (wet)	Long	4 × 4 × 6.5	0.001	14.07	173	
					Trans		1	15.61	332	
							1000	33.95	528	
							0.001	9.65	119	
							1	10.91	202	
							1000	15.82	320	
Tüfekci et al. (2014)	Equine	Radius, Tibia, Metacarpal, Metatarsal	8 Years	Cortical	Long	6 × 5	0.0045	17.44 ± 1.67	199.01 ± 6.43	2.7
							727 (±117)	20.74 ± 4.13	239.27 ± 25.25	2.12
Shunmugasamy et al. (2010)	Rabbit	Femur diaphysis Femur epiphysis	120–180 Days	Cortical	Long	Femur dia. X10	225	14.38	187	
							309	20.18	215.9	
							497	29.95	242.8	
							609	29.97	214.4	
							775	17.76	180	
							274	12.41	100.6	
							353	13.68	99.2	
							546	10.17	99.8	
							616	8.67	86.4	
							722	8.99	82.4	

$\dot{\varepsilon}$, *strain rate;* ρ_{app}, *apparent density;* DA, *degree of anisotropy;* Long, *longitudinal;* Trans, *transverse.*

The effect of strain rate is especially pronounced in the post-yield region, with higher levels of microcracking observed at lower strain rates compared to higher rates (Zioupos et al., 2008). These findings suggest that at lower strain rates, bone has sufficient time to develop microcracks, which helps to dissipate energy and contribute to greater toughness. Consequently, bone toughness declines at high strain rates, where less microcracking occurs.

The viscous or inelastic behavior of bone arises from complex mechanisms at multiple length scales. Inelastic deformation in bone is attributed to various microscale processes, including sliding or slippage between mineral and collagen phases, inter- and intra-fibrillar sliding, the formation of sacrificial bonds, the opening of dilatational bands, and microcrack formation at larger scales (Fantner et al., 2005; Gupta et al., 2005; Poundarik et al., 2012). Previous study has shown that at higher strain rates, fibril strain increases relative to tissue strain compared to lower strain rates (Zimmermann et al., 2014). These findings suggest that the inelastic deformation mechanisms in bone are constrained under high strain rates, reducing the capacity for energy dissipation through these mechanisms and leading instead to direct fibrillar deformation. At lower strain rates, fibrils exhibit plasticity due to sliding mechanisms within and between fibrils. However, at high strain rates, these sliding mechanisms become "locked," resulting in limited plastic deformation but increased strength at the macroscopic level. Thus, at slow strain rates, bone can dissipate energy through these inelastic deformation processes, while at higher strain rates, the restricted sliding leads to a stiffer, stronger response with reduced plasticity.

The rate-dependent mechanical properties of bone primarily result from its water content, organic phase, and hierarchical structure. Water, present in both bound and unbound forms, is crucial for bone's ductility, as dry bone fractures in a brittle manner (Granke et al., 2015). Additionally, research on collagen denaturation has highlighted collagen's influence on bone's rate sensitivity and it was found that strain rate sensitivity of bone decreased with increasing collagen denaturation (Uniyal et al., 2022).

Bone's hierarchical structure—from molecular to macroscopic levels—also plays a significant role, providing plasticity and viscoelasticity through multiple interfaces across scales. Previous work Gautieri et al. (2012) has shown that collagen fibrils display greater viscosity than individual collagen molecule, suggesting that the hierarchical organization could affect bone's rate-dependent behavior, though this effect has not yet been fully explored. The rate sensitivity of bone may originate from submicroscopic levels, as strain rate sensitivity observed in micro-pillar compression tests aligns with that seen at the macroscopic level (Peruzzi et al., 2021). Furthermore, bones from different species with varying microstructures exhibit similar strain rate sensitivities, suggesting that this characteristic may be an intrinsic property of bone tissue.

7. Summary

This chapter summarizes various experimental techniques used to assess bone's mechanical integrity across different length scales, with a specific focus on characterizing mesoscale mechanical properties, including time-dependent behaviors that are especially relevant in traumatic events. Since most clinical fractures occur at high strain rates, understanding bone's mechanical response under these conditions is critical. Although many mechanical characterization techniques are inherently destructive and, thus, impractical for in-vivo clinical applications, they offer valuable insights into the fracture mechanisms as clinical fractures are fundamentally mechanical events. These techniques are also instrumental in assessing the impact of therapies and treatments on preclinical models. The choice of mechanical testing method depends significantly on the hierarchical scale at which structural changes are anticipated. Additionally, finite element analysis (FEA) can complement experimental testing by providing further insights into fracture processes, enhancing our understanding of bone integrity under various loading conditions.

References

Adharapurapu, R. R., Jiang, F., & Vecchio, K. S. (2006). Dynamic fracture of bovine bone. *Materials Science and Engineering: C, 26*(8), 1325–1332. https://doi.org/10.1016/j.msec.2005.08.008

Ammann, P., & Rizzoli, R. (2003). Bone strength and its determinants. *Osteoporosis International, 14*(S3), 13–18. https://doi.org/10.1007/s00198-002-1345-4

Bailey, S., & Vashishth, D. (2018). Mechanical characterization of bone: State of the art in experimental approaches—what types of experiments do people do and how does one interpret the results? *Current Osteoporosis Reports, 16*(4), 423–433. https://doi.org/10.1007/s11914-018-0454-8

Bayraktar, H. H., Morgan, E. F., Niebur, G. L., Morris, G. E., Wong, E. K., & Keaveny, T. M. (2004). Comparison of the elastic and yield properties of human femoral trabecular and cortical bone tissue. *Journal of Biomechanics, 37*(1), 27–35. https://doi.org/10.1016/S0021-9290(03)00257-4

Bekker, A., Cloete, T. J., Chinsamy-Turan, A., Nurick, G. N., & Kok, S. (2015). Constant strain rate compression of bovine cortical bone on the split-Hopkinson pressure bar. *Materials Science and Engineering: C, 46*, 443–449. https://doi.org/10.1016/j.msec.2014.10.071

Burr, D. B. (2019). Stress concentrations and bone microdamage: John Currey's contributions to understanding the initiation and arrest of cracks in bone. *Bone, 127*, 517–525. https://doi.org/10.1016/j.bone.2019.07.015

Carriero, A., Bruse, J. L., Oldknow, K. J., Luis Millán, J., Farquharson, C., & Shefelbine, S. J. (2014). Reference point indentation is not indicative of whole mouse bone measures of stress intensity fracture toughness. *Bone, 69*, 174–179. https://doi.org/10.1016/j.bone.2014.09.020

Carter, D. R., Caler, W. E., Spengler, D. M., & Frankel, V. H. (1981). Fatigue behavior of adult cortical bone: The influence of mean strain and strain range. *Acta Orthopaedica, 52*(5), 481–490. https://doi.org/10.3109/17453678108992136

Carter, D. R., & Hayes, W. C. (1976). Bone compressive strength: The influence of density and strain rate. *Science, 194*(4270), 1174–1176. https://doi.org/10.1126/science.996549

Courtney, A. C., Hayes, W. C., & Gibson, L. J. (1996). Age-related differences in post-yield damage in human cortical bone. Experiment and model. *Journal of Biomechanics, 29*(11), 1463–1471. https://doi.org/10.1016/0021-9290(96)84542-8

Cowin, S. C. (1985). The relationship between the elasticity tensor and the fabric tensor. *Mechanics of Materials, 4*(2), 137–147. https://doi.org/10.1016/0167-6636(85)90012-2

Cummings, S. R., Bates, D., & Black, D. M. (2002). Clinical use of bone densitometry: Scientific review. *JAMA, 288*(15), 1889–1897. https://doi.org/10.1001/jama.288.15.1889

Currey, J. D. (1962). Stress concentrations in bone. *Journal of Cell Science, S3–103*(61), 111–133. https://doi.org/10.1242/jcs.s3-103.61.111

Currey, J. D. (1975). The effects of strain rate, reconstruction and mineral content on some mechanical properties of bovine bone. *Journal of Biomechanics, 8*(1), 81–86. https://doi.org/10.1016/0021-9290(75)90046-9

Currey, J. D. (2004). Tensile yield in compact bone is determined by strain, post-yield behaviour by mineral content. *Journal of Biomechanics, 37*(4), 549–556. https://doi.org/10.1016/j.jbiomech.2003.08.008

Diab, T., & Vashishth, D. (2005). Effects of damage morphology on cortical bone fragility. *Bone, 37*(1), 96–102. https://doi.org/10.1016/j.bone.2005.03.014

Dickenson, R. P., Hutton, W. C., & Stott. (1981). The mechanical properties of bone in osteoporosis. *The Journal of Bone and Joint Surgery. British, 63-B*(2), 233–238. https://doi.org/10.1302/0301-620x.63b2.7217148

Diez-Perez, A., Bouxsein, M. L., Eriksen, E. F., Khosla, S., Nyman, J. S., Papapoulos, S., & Tang, S. Y. (2016). Technical note: Recommendations for a standard procedure to assess cortical bone at the tissue-level in vivo using impact microindentation. *Bone Reports, 5*, 181–185. https://doi.org/10.1016/j.bonr.2016.07.004

Dong, X. N., Acuna, R. L., Luo, Q., & Wang, X. (2012). Orientation dependence of progressive post-yield behavior of human cortical bone in compression. *Journal of Biomechanics, 45*(16), 2829–2834. https://doi.org/10.1016/j.jbiomech.2012.08.034

Dong, X. N., Luo, Q., & Wang, X. (2013). Progressive post-yield behavior of human cortical bone in shear. *Bone, 53*(1), 1–5. https://doi.org/10.1016/j.bone.2012.11.029

Duchemin, L., Bousson, V., Raossanaly, C., Bergot, C., Laredo, J. D., Skalli, W., & Mitton, D. (2008). Prediction of mechanical properties of cortical bone by quantitative computed tomography. *Medical Engineering & Physics, 30*(3), 321–328. https://doi.org/10.1016/j.medengphy.2007.04.008

Ebacher, V., & Wang, R. (2009). A unique microcracking process associated with the inelastic deformation of Haversian bone. *Advanced Functional Materials, 19*(1), 57–66. https://doi.org/10.1002/adfm.200801234Canada

Enns-Bray, W. S., Ferguson, S. J., & Helgason, B. (2018). Strain rate dependency of bovine trabecular bone under impact loading at sideways fall velocity. *Journal of Biomechanics, 75*, 46–52. https://doi.org/10.1016/j.jbiomech.2018.04.042

Fantner, G. E., Hassenkam, T., Kindt, J. H., Weaver, J. C., Birkedal, H., Pechenik, L., Cutroni, J. A., Cidade, G. A. G., Stucky, G. D., Morse, D. E., & Hansma, P. K. (2005). Sacrificial bonds and hidden length dissipate energy as mineralized fibrils separate during bone fracture. *Nature Materials, 4*(8), 612–616. https://doi.org/10.1038/nmat1428

Ferreira, F., Vaz, M. A., & Simões, J. A. (2006). Mechanical properties of bovine cortical bone at high strain rate. *Materials Characterization, 57*(2), 71–79. https://doi.org/10.1016/j.matchar.2005.11.023

Fischer, R. A., Arms, S. W., Pope, M. H., & Seligson, D. (1986). Analysis of the effect of using two different strain rates on the acoustic emission in bone. *Journal of Biomechanics, 19*(2), 119–127. https://doi.org/10.1016/0021-9290(86)90142-9

Fleps, I., Bahaloo, H., Zysset, P. K., Ferguson, S. J., Pálsson, H., & Helgason, B. (2020). Empirical relationships between bone density and ultimate strength: A literature review.

Journal of the Mechanical Behavior of Biomedical Materials, 110, Article 103866. https://doi.org/10.1016/j.jmbbm.2020.103866

Gallant, M. A., Brown, D. M., Organ, J. M., Allen, M. R., & Burr, D. B. (2013). Reference-point indentation correlates with bone toughness assessed using whole-bone traditional mechanical testing. *Bone, 53*(1), 301–305. https://doi.org/10.1016/j.bone.2012.12.015

Gautieri, A., Vesentini, S., Redaelli, A., & Buehler, M. J. (2012). Viscoelastic properties of model segments of collagen molecules. *Matrix Biology, 31*(2), 141–149. https://doi.org/10.1016/j.matbio.2011.11.005

Granke, M., Does, M. D., & Nyman, J. S. (2015). The role of water compartments in the material properties of cortical bone. *Calcified Tissue International, 97*(3), 292–307. https://doi.org/10.1007/s00223-015-9977-5

Gupta, H. S., Wagermaier, W., Zickler, G. A., Aroush, D. R. B., Funari, S. S., Roschger, P., Wagner, H. D., & Fratzl, P. (2005). Nanoscale deformation mechanisms in bone. *Nano Letters, 5*(10), 2108–2111. https://doi.org/10.1021/nl051584b

Helgason, B., Perilli, E., Schileo, E., Taddei, F., Brynjólfsson, S., & Viceconti, M. (2008). Mathematical relationships between bone density and mechanical properties: A literature review. *Clinical Biomechanics, 23*(2), 135–146. https://doi.org/10.1016/j.clinbiomech.2007.08.024

Hemmatian, H., Bakker, A. D., Klein-Nulend, J., & Harry van Lenthe, G. (2021). Alterations in osteocyte lacunar morphology affect local bone tissue strains. *Journal of the Mechanical Behavior of Biomedical Materials, 123*, Article 104730. https://doi.org/10.1016/j.jmbbm.2021.104730

Hernandez, C. J., Lambers, F. M., Widjaja, J., Chapa, C., & Rimnac, C. M. (2014). Quantitative relationships between microdamage and cancellous bone strength and stiffness. *Bone, 66*, 205–213. https://doi.org/10.1016/j.bone.2014.05.023

Hunt, H. B., & Donnelly, E. (2016). Bone quality assessment techniques: Geometric, compositional, and mechanical characterization from macroscale to nanoscale. *Clinical Reviews in Bone and Mineral Metabolism, 14*(3), 133–149. https://doi.org/10.1007/s12018-016-9222-4

Kabel, J., van Rietbergen, B., Odgaard, A., & Huiskes, R. (1999). Constitutive relationships of fabric, density, and elastic properties in cancellous bone architecture. *Bone, 25*(4), 481–486. https://doi.org/10.1016/s8756-3282(99)00190-8

Kaneko, T. S., Bell, J. S., Pejcic, M. R., Tehranzadeh, J., & Keyak, J. H. (2004). Mechanical properties, density and quantitative CT scan data of trabecular bone with and without metastases. *Journal of Biomechanics, 37*(4), 523–530. https://doi.org/10.1016/j.jbiomech.2003.08.010

Katsamanis, F., & Raftopoulos, D. D. (1990). Determination of mechanical properties of human femoral cortical bone by the hopkinson bar stress technique. *Journal of Biomechanics, 23*(11), 1173–1184. https://doi.org/10.1016/0021-9290(90)90010-Z

Keaveny, T. M., Pinilla, T. P., Crawford, R. P., Kopperdahl, D. L., & Lou, A. (1997). Systematic and random errors in compression testing of trabecular bone. *Journal of Orthopaedic Research, 15*(1), 101–110. https://doi.org/10.1002/jor.1100150115

Keyak, J. H., Kaneko, T. S., Tehranzadeh, J., & Skinner, H. B. (2005). Predicting proximal femoral strength using structural engineering models. *Clinical Orthopaedics and Related Research, 437*, 219–228. https://doi.org/10.1097/01.blo.0000164400.37905.22

Kopperdahl, D. L., & Keaveny, T. M. (1998). Yield strain behavior of trabecular bone. *Journal of Biomechanics, 31*(7), 601–608. https://doi.org/10.1016/S0021-9290(98)00057-8

Krege, J. B., Aref, M. W., McNerny, E., Wallace, J. M., Organ, J. M., & Allen, M. R. (2016). Reference point indentation is insufficient for detecting alterations in traditional mechanical properties of bone under common experimental conditions. *Bone, 87*, 97–101. https://doi.org/10.1016/j.bone.2016.04.002

Kulin, R. M., Jiang, F., & Vecchio, K. S. (2011). Effects of age and loading rate on equine cortical bone failure. *Journal of the Mechanical Behavior of Biomedical Materials, 4* (1), 57–75. https://doi.org/10.1016/j.jmbbm.2010.09.006

Lambers, F. M., Bouman, A. R., Tkachenko, E. V., Keaveny, T. M., & Hernandez, C. J. (2014). The effects of tensile-compressive loading mode and microarchitecture on microdamage in human vertebral cancellous bone. *Journal of Biomechanics, 47*(15), 3605–3612. https://doi.org/10.1016/j.jbiomech.2014.10.011

Lee, S. C., Coan, B. S., & Bouxsein, M. L. (1997). Tibial ultrasound velocity measured in situ predicts the material properties of tibial cortical bone. *Bone, 21*(1), 119–125. https://doi.org/10.1016/s8756-3282(97)00076-8

Leng, H., Dong, X. N., & Wang, X. (2009). Progressive post-yield behavior of human cortical bone in compression for middle-aged and elderly groups. *Journal of Biomechanics, 42*(4), 491–497. https://doi.org/10.1016/j.jbiomech.2008.11.016

Lewis, G., & Nyman, J. S. (2008). The use of nanoindentation for characterizing the properties of mineralized hard tissues: State-of-the art review. *Journal of biomedical materials research. Part B, Applied biomaterials, 87*(1), 286–301. https://doi.org/10.1002/jbm.b.31092

Linde, F., Nørgaard, P., Hvid, I., Odgaard, A., & Søballe, K. (1991). Mechanical properties of trabecular bone. Dependency on strain rate. *Journal of Biomechanics, 24*(9), 803–809. https://doi.org/10.1016/0021-9290(91)90305-7

Ly, F. S., Proctor, A., Hoffseth, K., Yang, H. T., & Hansma, P. K. (2020). Significant correlation of bone material strength index as measured by the OsteoProbe with Vickers and Rockwell hardness. *Review of Scientific Instruments, 91*(8). https://doi.org/10.1063/5.0006133

MacNeil, J. A., & Boyd, S. K. (2008). Bone strength at the distal radius can be estimated from high-resolution peripheral quantitative computed tomography and the finite element method. *Bone, 42*(6), 1203–1213. https://doi.org/10.1016/j.bone.2008.01.017

Martens, M., Van Audekercke, R., Delport, P., De Meester, P., & Mulier, J. C. (1983). The mechanical characteristics of cancellous bone at the upper femoral region. *Journal of Biomechanics, 16*(12), 971–983. https://doi.org/10.1016/0021-9290(83)90098-2

McElhaney, J. H. (1966). Dynamic response of bone and muscle tissue. *Journal of Applied Physiology, 21*(4), 1231–1236. https://doi.org/10.1152/jappl.1966.21.4.1231

Mirzaali, M. J., Schwiedrzik, J. J., Thaiwichai, S., Best, J. P., Michler, J., Zysset, P. K., & Wolfram, U. (2016). Mechanical properties of cortical bone and their relationships with age, gender, composition and microindentation properties in the elderly. *Bone, 93*, 196–211. https://doi.org/10.1016/j.bone.2015.11.018

Morgan, E. F., & Keaveny, T. M. (2001). Dependence of yield strain of human trabecular bone on anatomic site. *Journal of Biomechanics, 34*(5), 569–577. https://doi.org/10.1016/S0021-9290(01)00011-2

Morgan, E. F., Unnikrisnan, G. U., & Hussein, A. I. (2018). Bone mechanical properties in healthy and diseased States. *Annual Review of Biomedical Engineering, 20*, 119–143. https://doi.org/10.1146/annurev-bioeng-062117-121139

Mosekilde, L., Mosekilde, Le, & Danielsen, C. C. (1987). Biomechanical competence of vertebral trabecular bone in relation to ash density and age in normal individuals. *Bone, 8*(2), 79–85. https://doi.org/10.1016/8756-3282(87)90074-3

Mys, K., Varga, P., Stockmans, F., Gueorguiev, B., Neumann, V., Vanovermeire, O., Wyers, C. E., van den Bergh, J. P. W., & Harry van Lenthe, G. (2021). High-resolution cone-beam computed tomography is a fast and promising technique to quantify bone microstructure and mechanics of the distal radius. *Calcified Tissue International, 108* (3), 314–323. https://doi.org/10.1007/s00223-020-00773-5

Nicolella, D. P., Moravits, D. E., Gale, A. M., Bonewald, L. F., & Lankford, J. (2006). Osteocyte lacunae tissue strain in cortical bone. *Journal of Biomechanics, 39*(9), 1735–1743. https://doi.org/10.1016/j.jbiomech.2005.04.032

Niebur, G. L., Feldstein, M. J., Yuen, J. C., Chen, T. J., & Keaveny, T. M. (2000). High-resolution finite element models with tissue strength asymmetry accurately predict failure of trabecular bone. *Journal of Biomechanics, 33*(12), 1575–1583. https://doi.org/10.1016/S0021-9290(00)00149-4

Nyman, J. S., Granke, M., Singleton, R. C., & Pharr, G. M. (2016). Tissue-level mechanical properties of bone contributing to fracture risk. *Current Osteoporosis Reports, 14*(4), 138–150. https://doi.org/10.1007/s11914-016-0314-3

Oliver, W. C., & Pharr, G. M. (1992). An improved technique for determining hardness and elastic modulus using load and displacement sensing indentation experiments. *Journal of Materials Research, 7*(6), 1564–1583. https://doi.org/10.1557/jmr.1992.1564

Pagano, S. J., Hogan, J. D., & Lamberson, L. (2016). Bone and bone surrogate fragmentation under dynamic compression. *Journal of Dynamic Behavior of Materials, 2*(2), 234–245. https://doi.org/10.1007/s40870-016-0061-7

Peruzzi, C., Ramachandramoorthy, R., Groetsch, A., Casari, D., Grönquist, P., Rüggeberg, M., Michler, J., & Schwiedrzik, J. (2021). Microscale compressive behavior of hydrated lamellar bone at high strain rates. *Acta Biomaterialia, 131*, 403–414. https://doi.org/10.1016/j.actbio.2021.07.005

Pistoia, W., van Rietbergen, B., Lochmüller, E.-M., Lill, C. A., Eckstein, F., & Rüegsegger, P. (2002). Estimation of distal radius failure load with micro-finite element analysis models based on three-dimensional peripheral quantitative computed tomography images. *Bone, 30*(6), 842–848. https://doi.org/10.1016/s8756-3282(02)00736-6

Pithioux, M., Subit, D., & Chabrand, P. (2004). Comparison of compact bone failure under two different loading rates: Experimental and modelling approaches. *Medical Engineering & Physics, 26*(8), 647–653. https://doi.org/10.1016/j.medengphy.2004.05.002

Poundarik, A. A., Diab, T., Sroga, G. E., Ural, A., Boskey, A. L., Gundberg, C. M., & Vashishth, D. (2012). Dilatational band formation in bone. *Proceedings of the National Academy of Sciences, 109*(47), 19178–19183. https://doi.org/10.1073/pnas.1201513109

Prot, M., Cloete, T. J., Saletti, D., & Laporte, S. (2016). The behavior of cancellous bone from quasi-static to dynamic strain rates with emphasis on the intermediate regime. *Journal of Biomechanics, 49*(7), 1050–1057. https://doi.org/10.1016/j.jbiomech.2016.02.021

Prot, M., Saletti, D., Pattofatto, S., Bousson, V., & Laporte, S. (2015). Links between mechanical behavior of cancellous bone and its microstructural properties under dynamic loading. *Journal of Biomechanics, 48*(3), 498–503. https://doi.org/10.1016/j.jbiomech.2014.12.002

Quintiens, J. (2024). Photon-counting CT-based bone stiffness and strength agree well with values obtained from high-resolution peripheral quantitative CT. In *24th international workshop on quantitative musculoskeletal imaging (QMSKI) the Barossa valley.*

Røhl, L., Larsen, E., Linde, F., Odgaard, A., & Jørgensen, J. (1991). Tensile and compressive properties of cancellous bone. *Journal of Biomechanics, 24*(12), 1143–1149. https://doi.org/10.1016/0021-9290(91)90006-9

Reilly, D. T., Burstein, A. H., & Frankel, V. H. (1974). The elastic modulus for bone. *Journal of Biomechanics, 7*(3), 271. https://doi.org/10.1016/0021-9290(74)90018-9

Reznikov, N., Shahar, R., & Weiner, S. (2014). Bone hierarchical structure in three dimensions. *Acta Biomaterialia, 10*(9), 3815–3826. https://doi.org/10.1016/j.actbio.2014.05.024

Rincón-Kohli, L., & Zysset, P. K. (2009). Multi-axial mechanical properties of human trabecular bone. *Biomechanics and Modeling in Mechanobiology, 8*(3), 195–208. https://doi.org/10.1007/s10237-008-0128-z

Sadoughi, S., Scheidt, A.vom, Nawathe, S., Zhu, S., Moini, A., & Keaveny, T. M. (2020). Effect of variations in tissue-level ductility on human vertebral strength. *Bone, 137*, Article 115445. https://doi.org/10.1016/j.bone.2020.115445

Sas, A., Sermon, A., & van Lenthe, G. H. (2022). Experimental validation of a voxel-based finite element model simulating femoroplasty of lytic lesions in the proximal femur. *Scientific Reports, 12*(1). https://doi.org/10.1038/s41598-022-11667-x

Schaffler, M. B., & Burr, D. B. (1988). Stiffness of compact bone: Effects of porosity and density. *Journal of Biomechanics, 21*(1), 13–16. https://doi.org/10.1016/0021-9290(88)90186-8

Shim, V. P. W., Yang, L. M., Liu, J. F., & Lee, V. S. (2005). Characterisation of the dynamic compressive mechanical properties of cancellous bone from the human cervical spine. *International Journal of Impact Engineering, 32*(1–4), 525–540. https://doi.org/10.1016/j.ijimpeng.2005.03.006

Shunmugasamy, V. C., Gupta, N., & Coelho, P. G. (2010). High strain rate response of rabbit femur bones. *Journal of Biomechanics, 43*(15), 3044–3050. https://doi.org/10.1016/j.jbiomech.2010.06.034

Tüfekci, K., Kayacan, R., & Kurbanoğlu, C. (2014). Effects of gamma radiation sterilization and strain rate on compressive behavior of equine cortical bone. *Journal of the Mechanical Behavior of Biomedical Materials, 34*, 231–242. https://doi.org/10.1016/j.jmbbm.2014.02.004

Tang, T., Vincent, E., Cripton, P., Guy, P., McKay, H., & Wang, R. (2015). Shear deformation and fracture of human cortical bone. *Bone, 71*, 25–35. https://doi.org/10.1016/j.bone.2014.10.001

Turner, C. H., Cowin, S. C., Rho, J. Y., Ashman, R. B., & Rice, J. C. (1990). The fabric dependence of the orthotropic elastic constants of cancellous bone. *Journal of Biomechanics, 23*(6), 549–561. https://doi.org/10.1016/0021-9290(90)90048-8

Uniyal, P., Kaur, S., Dhiman, V., Kumar Bhadada, S., & Kumar, N. (2023). Effect of inelastic deformation on strain rate-dependent mechanical behaviour of human cortical bone. *Journal of Biomechanics, 161*, Article 111853. https://doi.org/10.1016/j.jbiomech.2023.111853

Uniyal, P., Sihota, P., & Kumar, N. (2022). Effect of organic matrix alteration on strain rate dependent mechanical behaviour of cortical bone. *Journal of the Mechanical Behavior of Biomedical Materials, 125*, Article 104910. https://doi.org/10.1016/j.jmbbm.2021.104910

Uniyal, P., Sihota, P., Tikoo, K., & Kumar, N. (2021). Anatomical variation in intracortical canal network microarchitecture and its influence on bone fracture risk. *Journal of the Mechanical Behavior of Biomedical Materials, 123*, Article 104770. https://doi.org/10.1016/j.jmbbm.2021.104770

Ural, A., & Vashishth, D. (2007). Effects of intracortical porosity on fracture toughness in aging human bone: A μCT-based cohesive finite element study. *Journal of Biomechanical Engineering, 129*(5), 625–631. https://doi.org/10.1115/1.2768377

van Lenthe, G. H., Voide, R., Boyd, S. K., & Müller, R. (2008). Tissue modulus calculated from beam theory is biased by bone size and geometry: Implications for the use of three-point bending tests to determine bone tissue modulus. *Bone, 43*(4), 717–723. https://doi.org/10.1016/j.bone.2008.06.008

van Rietbergen, B., & Ito, K. (2015). A survey of micro-finite element analysis for clinical assessment of bone strength: The first decade. *Journal of Biomechanics, 48*(5), 832–841. https://doi.org/10.1016/j.jbiomech.2014.12.024

Van Rietbergen, B., Odgaard, A., Kabel, J., & Huiskes, R. (1996). Direct mechanics assessment of elastic symmetries and properties of trabecular bone architecture. *Journal of Biomechanics, 29*(12), 1653–1657. https://doi.org/10.1016/S0021-9290(96)80021-2

Verbruggen, S. W., Vaughan, T. J., & McNamara, L. M. (2012). Strain amplification in bone mechanobiology: A computational investigation of the in vivo mechanics of osteocytes. *Journal of the Royal Society Interface, 9*(75), 2735–2744. https://doi.org/10.1098/rsif.2012.0286

Viceconti, M., Bellingeri, L., Cristofolini, L., & Toni, A. (1998). A comparative study on different methods of automatic mesh generation of human femurs. *Medical Engineering & Physics, 20*(1), 1–10. https://doi.org/10.1016/s1350-4533(97)00049-0

Voide, R., Schneider, P., Stauber, M., van Lenthe, G. H., Stampanoni, M., & Müller, R. (2011). The importance of murine cortical bone microstructure for microcrack initiation and propagation. *Bone, 49*(6), 1186–1193. https://doi.org/10.1016/j.bone.2011.08.011

Wang, J., Zhou, B., Sherry Liu, X., Fields, A. J., Sanyal, A., Shi, X., Adams, M., Keaveny, T. M., & Edward Guo, X. (2015). Trabecular plates and rods determine elastic modulus and yield strength of human trabecular bone. *Bone, 72*, 71–80. https://doi.org/10.1016/j.bone.2014.11.006

Weerasooriya, T., Sanborn, B., Allan Gunnarsson, C., & Foster, M. (2016). Orientation dependent compressive response of human femoral cortical bone as a function of strain rate. *Journal of Dynamic Behavior of Materials, 2*(1), 74–90. https://doi.org/10.1007/s40870-016-0048-4

Wolfram, U., & Schwiedrzik, J. (2016). Post-yield and failure properties of cortical bone. *Bone Reports, 5*. https://doi.org/10.1038/bonekey.2016.60

Wolfram, U., Wilke, H. J., & Zysset, P. K. (2011). Damage accumulation in vertebral trabecular bone depends on loading mode and direction. *Journal of Biomechanics, 44*(6), 1164–1169. https://doi.org/10.1016/j.jbiomech.2011.01.018

Wright, T. M., & Hayes, W. C. (1976). Tensile testing of bone over a wide range of strain rates: Effects of strain rate, microstructure and density. *Medical & Biological Engineering, 14*(6), 671–680. https://doi.org/10.1007/bf02477046

Zhai, X., Nauman, E. A., Moryl, D., Lycke, R., & Chen, W. W. (2020). The effects of loading-direction and strain-rate on the mechanical behaviors of human frontal skull bone. *Journal of the Mechanical Behavior of Biomedical Materials, 103*, Article 103597. https://doi.org/10.1016/j.jmbbm.2019.103597

Zimmermann, E. A., Gludovatz, B., Schaible, E., Busse, B., & Ritchie, R. O. (2014). Fracture resistance of human cortical bone across multiple length-scales at physiological strain rates. *Biomaterials, 35*(21), 5472–5481. https://doi.org/10.1016/j.biomaterials.2014.03.066

Zioupos, P., Hansen, U., & Currey, J. D. (2008). Microcracking damage and the fracture process in relation to strain rate in human cortical bone tensile failure. *Journal of Biomechanics, 41*(14), 2932–2939. https://doi.org/10.1016/j.jbiomech.2008.07.025

Zioupos, P., Kirchner, H. O. K., & Peterlik, H. (2020). Ageing bone fractures: The case of a ductile to brittle transition that shifts with age. *Bone, 131*. https://doi.org/10.1016/j.bone.2019.115176

Zysset, P. K., & Curnier, A. (1995). An alternative model for anisotropic elasticity based on fabric tensors. *Mechanics of Materials, 21*(4), 243–250. https://doi.org/10.1016/0167-6636(95)00018-6

CHAPTER

Role of water in the mechanical behavior of bone

9

Jeffry S. Nyman[1, 2, 3] and Emily D. Berestesky[2]

[1]*Vanderbilt University Medical Center, Department of Orthopaedic Surgery, Nashville, TN, United States;* [2]*Vanderbilt University, Department of Biomedical Engineering, Nashville, TN, United States;* [3]*United States Department of Veteran Affairs, Tennessee Valley Healthcare System, Nashville, TN, United States*

Abbreviations

AGEs	Advanced glycation end-products
AIR	Adiabatic inversion recovery
ATR	Attenuated total reflectance (FTIR spectroscopy)
BMD	Bone mineral density
BW	Bound water (tightly and loosely, unless specified)
C_{bw}	Bound water concentration
CCD	Charge-coupled device
Col I	Type 1 collagen
Col-BW	Collagen-related bound water
CPMG	Carr-Purcell-Meiboom-Gill
C_{pw}	Concentration of pore water
CT	Computed tomography
DAFP	Double adiabatic full passage
DXA	Dual energy X-ray absorptiometry
E	Young's/Elastic modulus
ECM	Extracellular matrix
F_F	Fitted fraction of fat
FID	Free induction decay
FIR	Far-infrared (spectroscopy)
F_L	Fitted fraction of pore water
F_{MAX}	Maximum load
F_S	Fitted fraction of bound water
FTIR	Fourier transform infrared (spectroscopy)
GAGs	Glycosaminoglycans
GIO	glucocorticoid-induced Osteoporosis
HA	Hemiarthroplasty
I_{min}	Minimum moment of inertia
IR	Infrared (spectroscopy)
J_f	J-integral

Biomechanics and Mechanobiology of Bone. https://doi.org/10.1016/B978-0-443-26739-0.00013-2

K_J	Crack initiation toughness
L_{bone}	Mole ^{1}H/L of bone
MFS	Maximum flexural strength
Min-BW	Mineral-related bound water
MIR	Mid-infrared (spectroscopy)
MRI	Magnetic resonance imaging
NaClO	Sodium hypochlorite
NF1	Neurofibromatosis type 1
NIR	Near-infrared (spectroscopy)
NMR	Nuclear magnetic resonance
OI	Osteogenesis Imperfecta
PI	Porosity index
PTH	Parathyroid hormone
PW	Pore water
PYD	Post-yield displacement
PYT	Post-yield toughness
RF	Radiofrequency
ROI	Region of interest
RS	Raman Spectroscopy
RTP	Relative to total protein
SEM	Scanning electron microscope
SENB	Single-edge notched beam
T	Toughness
T-BW	Tightly bound water
T_1	Longitudinal relaxation time constant
T_2	Transverse relaxation time constant
T_2*	Transverse relaxation time constant from free induction decay
TE	Echo time
THA	Total hip arthroplasty
Thz	Tetrahertz (spectroscopy)
TW	Total water
UTE	Ultrashort echo time (magnetic resonance imaging)
VEGF	Vascular endothelial growth factor
W_f	Work-to-fracture
δ	Stiffness
σ_b	Maximum bending stress
μCT	micro-computed tomography
$\Delta K_J/\Delta a$	Crack growth toughness final
α_n	Relative amplitude of the nth spectral peak of fat

1. Introduction

The fracture resistance of bone depends on the hierarchical organization of this organ. In the arrangement of helical collagen type I (Col I) and semi-crystalline, carbonated hydroxyapatite into mineralized collagen fibrils, water is distributed between pore spaces and the extracellular matrix (ECM) of bone. The importance of water to the mechanical behavior of bone is evident from the mechanical testing of wet and dry bone ex vivo. Compared to wet cortical bone, dry bone is stronger and harder but more brittle (Table 9.1). Likewise, dehydration of trabecular bone increases the apparent modulus (Lievers et al., 2010; Mazurkiewicz, 2018). The ability of bone to resist fracture depends on the integration of the hard mineral within the soft organic matrix as well as the water interactions with both the inorganic mineral and the proteins of the ECM.

Like all musculoskeletal tissues of the body, proper physiological function of bone depends on the presence of water. By volume, water occupies ~15%–~25% of bone (Nyman et al., 2006). In comparison, the organic matrix and mineral occupy ~30%–~35% and ~55%–~60% of the bone (Arnold, 1960; Mueller et al., 1966). The relative volume fractions of water, organic matrix, and mineral depend on the rate of bone remodeling and porosity. Thus, they vary throughout life. Water has several key roles in bone. In addition to facilitating the coordinated response of bone to changes in mechanical loading (Chapter 4.), it allows waste products to diffuse from the tissue to the vasculature and molecules to diffuse from the vasculature through the bone tissue (Dillaman et al., 1991). The present chapter primarily covers its role in hydrating the ECM with respect to mechanical behavior of cortical bone and its ability to be a surrogate of cortical porosity, a known determinant of the mechanical properties of bone. Few studies have explored water content in trabecular bone (Li et al., 2024; Querido et al., 2025; Rai et al., 2013) because of the difficulty in separating bone marrow and blood from the water that exists between trabecule. Presumably, the distribution of pore and bound water is similar to cortical bone, but future studies are needed to elucidate the contribution of pore water and bound water to the mechanical behavior of trabecular bone.

Beyond the knowledge that water is fundamental to the ability of bone to resist fracture, there is growing evidence that the in vivo assessment of water distribution in cortical bone can help predict an individual's fracture risk (Jerban et al., 2023; Nyman et al., 2023; Xiong et al., 2021). As discussed in previous reviews (Granke, Does, & Nyman, 2015; Manhard, Nyman, & Does, 2017; Surowiec et al., 2024), water in bone has a distinct relaxation profile when placed in a magnetic field and subjected to radiofrequency (RF) pulses. Thus, it can be quantified as pore water (slow relaxation) or bound water (fast relaxation) using magnetic resonance imaging (MRI), specifically ultra-short echo-time (UTE) MRI (Surowiec et al., 2024). Herein, we provide evidence that the water compartments of bone are important to its mechanical behavior. Specifically, we describe (i) the distribution of water within cortical bone, (ii) how pore water and bound water influence its mechanical properties, (iii) the different techniques for non-destructively measuring the two compartments of water in bone, (iv) the evidence that these measurements correlate with the mechanical properties of bone, and (v) what is known about the determinants of pore water and bound water.

Table 9.1 The effect of drying on the mechanical properties of bone.

Species of bone	Conditions	Mechanical test	Drying effect	References
Human cortical	Wet versus air-dried	Tension	Elastic modulus and ultimate tensile stress increased Ultimate tensile strain decreased	Evans (1973)
		Rockwell	Hardness increased	
Human cortical	Wet versus air-dried for 5 minutes to 1 h	Flexural (three-point bending or cantilever)	Ultimate bending stress increased after 30 min but significant after 1 h	Sedlin and Hirsch (1966)
Human cortical	Stored dry versus overnight rehydration	Tension	Elastic modulus and ultimate stress increased by 55% and 50%	Dempster and Liddicoat (1952)
		Compression	Elastic modulus and ultimate stress increased by 26% and 63%	
Multiple species cortical	Wet versus air-dried	Tension	Young's modulus increased by 7% after 1 h of drying	Smith and Walmsley (1959)
Mouse cortical	Air dehydration for 0.25, 0.5, 1.5, 2.0, 3.0, 4.0, or 48.0 h followed by saline rehydrated for 0.08, 0.17, 0.25, 0.5, 1.0, 3.0, 6.0, or 12.0 h versus stored in saline for less than 1 h at room temperature	Three-point bending	Increased stiffness and strength, and decreased ductility with prolonged dehydration Rehydration linearly restored mechanical properties except maximum load up to 3 h, after which it did not significantly change	Broz et al. (1993)
Bovine cortical	Wet versus air-dried for 25 days, then rehydrated for 3 h prior to testing	Three-point bending	Bending strength decreased by 4.5%, while Young's modulus and the work under load-deformation curve did not significantly differ	Currey (1988)
Human cortical	Wet versus vacuum oven-dried at 20°C for 30 minutes versus 21, 50, 70, or 110°C for 4 h	Three-point bending	Bone strength increased with a 5% water loss by weight (under 21°C for 4 h) but decreased with water loss greater than 9% (at $\geq$ 70°C) Bone toughness decreased at 21°C and above Stiffness linearly increased with water loss	Nyman et al. (2006)

Human cortical	Wet versus vacuum oven-dried at 70°C for 8 h	Cyclic stress-relaxation in tension and compression	Elastic modulus and ultimate strength increased, while failure strain and toughness decreased under dry conditions for both tensile and compressive testing	Samuel et al. (2016)
Human and sheep cortical	Bio-bath versus ambient air pressure versus water vapor pressure-SEM chamber versus air pressure-SEM chamber versus high-vacuum-SEM chamber	Fracture toughness testing	Crack initiation toughness was lower in SEM environments (similar across SEM conditions) than *ex situ* testing *Ex* situ methods had similar crack initiation toughness, but ambient air testing had greater crack growth resistance than bio-bath testing Findings follow similar trends across species	Shin et al. (2022)
Porcine cortical	Fresh versus 4% formalin versus vacuum dehydration	Static and dynamic compressive testing	Ultimate strain and stress increased under formalin fixation and vacuum dehydration Strain-rate sensitivity exponent was highest in fresh group, then formalin and vacuum dehydration groups	Qiu et al. (2023)
Human cancellous	Wet versus dried in alcohol for 6 h, then air dried for 24 h	Compressive testing	Young's modulus increased by 24%	Townsend et al. (1975)
Bovine cancellous	Wet versus air-dried for 3 days, then vacuum desiccator dried for 24 h	Tensile testing	Apparent modulus increased by 14 ± 14%, which considers the reduction of bone volume fraction of dehydrated samples	Lievers et al. (2010)
Porcine cancellous	Storage at-21°C versus in air versus buffered 10% formalin versus 96% alcohol versus 50% alcohol at room temperature for 14 weeks	Compressive testing (repeated every 2 weeks for 14 weeks)	Elastic modulus of open-air samples increased and stabilized by week 6, with a maximum modulus increase of 84.1% over 14 weeks No significant modulus changes under other conditions	Mazurkiewicz (2018)

2. Distribution of water

Estimates of the weight fractions of water, organic matrix, and mineral come from thermal gravimetric analysis in which the mass of water in bone is primarily lost by 150°C, the organic matrix combusts between 200 and 450°C releasing additional water (Mkukuma et al., 2005), and the mass of the final bone sample is the weight of mineral after 600°C (Ellingham et al., 2015; Lozano et al., 2003; Mkukuma et al., 2005). So-called structural water is removed from bone at 1300°C during mineral decomposition (Mkukuma et al., 2005).

The conversion of the mass of water loss to volume fraction of water is based on measuring the volume of bone by Archimedes' principle such that water volume per bone volume is wet mass minus dry mass divided by bone volume (Nyman et al., 2006). In the heating of bone during thermal gravimetric analysis, the volume of water residing in pores and the water bound to the matrix cannot be readily delineated.

2.1 Pore water

As the name implies, pore water is unbound or free water that resides in the pores of bone (Fig. 9.1). At the length-scale of 50–500 μm, the pores of cortical bone are the vascular channels and sites of active resorption by osteoclasts. Exchange of water between the vasculature and the tissue occurs in the capillaries and sinusoids. At the length-scale of 100 nm–1 μm, the pores of bone are the lacunar-canalicular system. The canaliculi are connected to the Haversian canals, in which blood vessels are aligned with the long axis of long bones, and to the Volkmann canals, in which blood vessels branch from the longitudinally oriented vessels and run perpendicular from the large artery in the medullary canal to the periosteum. Water flows over the transverse processes within the canalicular walls that interconnect osteocytes (Chapter 6.) and diffuse into the ECM of bone. Water can also occupy microdamage and diffuse (nano) damage that forms during activities of daily living.

In a tracer study of fluid transport within the canine tibial diaphysis using intravenous injections of radioisotopes (e.g., indium-111-labeled transferrin, carbon-14 in sucrose, tritiated water), the estimated volume of total water (radioactive water), fluid volume within the ECM ([^{14}C]sucrose), and interstitial space ([^{14}C]sucrose—^{111}In-labeled transferrin) was 0.245 ± 0.003 mL, 0.043 ± 0.003 mL, and 0.034 per ml of cortical bone (Morris et al., 1982). Within 3 hours of circulation, these tracers were present in bone demonstrating that water and solutes diffuse from circulation into the matrix of bone as well as into the space between blood vessels and canal walls.

When tracers of different molecular weight (procion red < microperoxidase) were injected into the tail vein of anaesthetized rats (60-day-old and 180-day-old), the fluorescent red dye was present throughout the vascular canals and lacunae of the tibial cortex after 10 and 30 minutes of circulation (Knothe Tate et al., 1998). The dye appeared inside and outside the blood vessels; however, within the lacunae, the dye surrounded the osteocytes. After 2 hour of circulation while the animals remained anaesthetized, the larger microperoxidase was present in two concentric

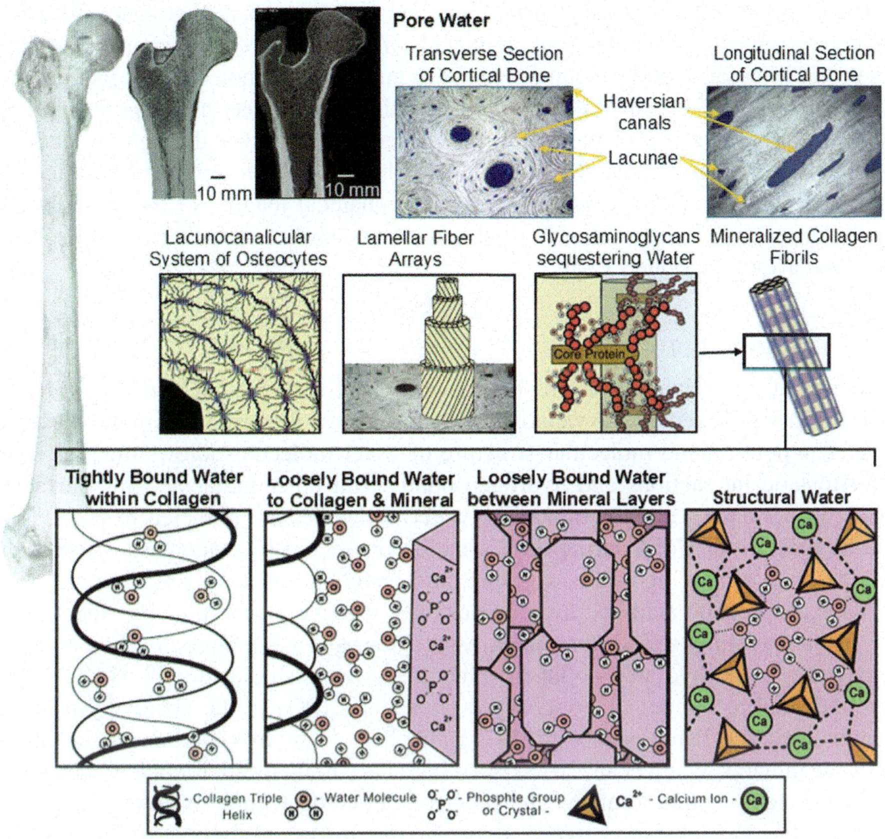

FIGURE 9.1 Overview of water distribution within cortical bone.

Within cortical bone, waters flows through the Haversian canals outside and inside the blood vessels and through canalicular channels around the transverse processes of osteocytes. The extracellular matrix of cortical and trabecular bone consists of lamellae of both mineralized collagen I fibrils and extrafibrillar mineral. Water is loosely and tightly bound to this matrix based on the bonding energy between H_2O molecules and the matrix. Bound water arises from (i) hydrogen bonding between water and hydrophilic amino acids of collagen I as well as negatively charged glycosaminoglycans that are attached to proteoglycans and (ii) electrostatic interactions with cations and anions on the surface of bone mineral. Loosely bound water also exists between apatite crystal layers of the extrafibrillar mineral. Water is also trapped within the crystal lattice of apatitic bone mineral and is called structural water because it is part of the molecular structure of bone mineral, which has similarity nano-sized platelets of hydroxyapatite but with multiple substitutions (e.g., CO_3 for OH or for PO_4,Mg or Zn for Ca).

bands—one near the periosteal surface and one near the endosteal surface, as well as in concentric rings around vascular channels. These dyes were not more prominent in the ECM because they bind proteins with a large enough molecular weight to limit their ability to diffuse into the matrix. When a tibia was loaded in four-point bending following the injection of the tracer, the presence of the dyes throughout lacunar-canalicular system was greater in the loaded bone than the contralateral unloaded bone (Tate et al., 2000). Presumably then, mechanical loading of the bone during activities of daily living enhances fluid transport and hence the movement of solutes and water into the bone ECM.

2.2 Bound water

Bound water is water that forms bonds with Col I and/or bone mineral, which is similar to nano-sized hydroxyapatite crystals, and is sequestered within the bone tissue. The polarity and molecular structure of water molecules allows them to form electrostatic interactions and hydrogen bonds with other molecules and surfaces. This occurs during biomineralization: water molecules are attracted to the hydrophilic residues of Col I during the formation of osteoid by osteoblasts (Robinson, 1975). Side groups of lysine, arginine, and histidine are positively charged; side groups of glutamic acid and aspartic acid are negatively charged; and side groups of hydroxyproline, serine, threonine, asparagine, and glutamine are polar but uncharged. As the hydrated collagen mineralizes, water molecules ($^{\delta+}$H–O$^{\delta-}$–H$^{\delta+}$) are attracted to the negatively-charged phosphates (PO_4^{3-}), the positively-charged calcium (Ca^{2+}) (Timmins & Wall, 1977), and the polar hydroxyl groups in the lattice of bone mineral (Wang et al., 2013). As mineral accumulates in osteoid, bound water is displaced, but some remains and retains its bonding with both protein and inorganic mineral (Robinson, 1979). Between the stacks of tightly packed apatitic plate-like layers (50–150 nm), water molecules are trapped within vacancies between and within the crystal lattice asstructural water. In these very small vacancies, water is bound to citrate anions bridging the calcium-phosphate platelets (Davies et al., 2014). A network of water molecules exists hydrating the collagen fibrils as bound water. The amount of bound water has been estimated to be ~45 mg of H_2O per gram of from dielectric measurements (Maeda & Fukada, 1982; Marino et al., 1967). Conductivity characterizes a material's ability to carry an electric current through the movement of ions, and bound water concentration directly correlated with conductivity, which directly correlated with toughness and ultimate strength of cortical bone (Unal et al., 2018).

Additionally, depending on where it exists spatially in relation to the fibrils of Col I and the strength of its bonding interactions with this protein, *matrix*-bound water can be further categorized as loosely or tightly bound (Kerch, 2018). Spatially, the closer the water molecules are to the charged collagen residues, the stronger their interactions are and the less their degrees of freedom there are. Tightly bound water exists within the triple helices of Col I, forming direct hydrogen bonds between two neighboring Col I peptide chains (ColIα1-ColIα1 or ColIα1-ColIα2), while loosely bound water forms topical bonds on the outside of mineralized collagen fibrils as well as between the collagen fibrils and apaptitic mineral crystals (Bella et al., 1995; Brodsky & Persikov, 2005). Combined, the loosely and tightly bound water provides stability to the collagen fibril through inter- and intra-chain water bridges. When heating wet bone from room temperature to 150°C, loosely bound water is lost at lower temperatures while tightly bound water is lost at higher temperatures. The removal of water during the heating of bone showed three distinct peaks, one at ~95°C, one at ~320°C, and one at ~1300°C (Mkukuma et al., 2005). The middle peak coincides with the peak of combustion (i.e., CO_2) and presumably is tightly bound water. The water peak at the highest temperatures occurs when inorganic mineral is melting and thought to be structural water. Thus, bound water exists on a continuum of bonding energy levels.

Recently, more attention has been given to glycosaminoglycans (GAGs) and their ability to sequester water in bone. One type of non-collagenous protein present within bone ECM is proteoglycans, which consist of a core protein that is covalently attached to negatively charged, linear polysaccharide chains of repeating disaccharide units known as GAGs (Iozzo & Schaefer, 2015). The bone ECM contains several types of GAGs—heparin sulfate, dermatan sulfate, and chondroitin sulfate—with chondroitin sulfate being the major GAG in the ECM of human (Wang et al., 2018) and mouse (Hua et al., 2020) bone. The negative charge of GAGs enables them to bind to collagen and mineral crystals, and most importantly water (Best et al., 2008; Hashimoto et al., 1995; Rees et al., 2001). Their ability to form hydrogen bonds with water molecules has been explored using molecular dynamics simulations (Samantray et al., 2021). GAG content has been found to correlate with the mechanical properties of bone (see Section 5). GAG concentration also appears to decrease with age, which is related to the age-related reduction in both bound water and toughness (Wang et al., 2018).

3. Non-destructive measurements of water

Pore water (i.e., free water) and bound water of any connective tissue can be quantified by first freezing the tissue to a constant temperature (e.g., −20°C) and then measuring the amount of energy absorbed as the tissue is heated at constant rate (i.e., calorimetry). Free water in the tissue sample freezes and melts at 0°C, while bound water freezes and melts at temperatures below −20°C. In effect, the mass of bound water is the total mass lost due to drying out the tissue sample minus the mass of free water, which is based on the measured amount of heat that melted ice and that 80 calories of heat melts 1 g of ice (Robinson, 1931). Such a method established the existence of bound water in musculoskeletal tissues and has been extended to differential scanning calorimetry for measuring free and bound water in muscle (Aktas et al., 1997). Over the past several years, NMR techniques have been implemented to quantify pore water and bound water in bone in a non-destructive manner.

3.1 ^{1}H nuclear magnetic resonance relaxometry

Since size and shape of the lacunar-canalicular system is smaller and distinct from the vascular channels of bone, the relaxation behavior of water protons (H–O–H) in bone subjected to ^{1}H NMR relaxometry is variable with proton signals relaxing quickly (time constant from 0.1 to 1 ms) and slowly (time constant from 10 to 600 ms). In effect, the internal surfaces of pores or local surface-to-volume ratios of different pore sizes in bone dictate both longitudinal (T_1) and transverse (T_2) relaxation behavior of pore water (Fantazzini et al., 2003). Using a low-field NMR technique (magnetic field strength, $B_0 = 0.047$ T), Ni and Wang showed how the spin-spin relaxation spectra of water in human cortical bone samples was directly related to the lacunar pores and Haversian canal pores as determined by histomorphometry (Wang & Ni, 2003). That is, area of lacunar pores and area of Haversian pores were related to the different areas of signal intensity as a function of T_2 relaxation time constants. In a subsequent study, Ni and colleagues showed how this technique could quantify fatigue-induced damage (creation of pores as cracks) in human cortical bone (Nicolella et al., 2011). The spin state of water protons (^{1}H) in bone relaxes faster in small pores like lacunae and slower in large pores like vascular channels when wet bone is subjected to RF pulses while in a magnetic field.

^{1}H NMR relaxometry works by subjecting a sample to alternating magnetic fields and collecting spectra of how the protons of a sample return to their original magnetization state through the process of relaxation, which is identified as a free induction decay (FID) signal. Since bound water is strongly associated with both the mineral and organic matrix of bone tissue, its FID signal decays more quickly than the FID of pore water (Granke, Does, & Nyman, 2015). Note that T_2 differs from T_2* in that T_2 is the decay of spin-spin interactions observed in the NMR signal without the effects of spatial variations in the magnetic field, while T_2* does consider magnetic field variations. Furthermore, the Carr-Purcell-Meiboom-Gill (CPMG) pulse sequence, which subjects the sample to a series of RF pulses at 180 degrees to cause the spine state of the protons to flip, stimulates repeated T_2 FID signals of the sample, extending the time for signal acquisition. This acquired data is fitted to multiple exponential functions to generate a "T_2 spectra" (Horch et al., 2010). For bone, solid protons (covalent bonds like C–H) have a faster rate of relaxation ($1/T_2$) than bound water, which in turn has a faster relaxation than pore water (Fig. 10.2). The peak signal from solid protons of the bone matrix occur at $T_2 \sim 100$ μs, while the peak signal from *matrix*-bound water occurs at $T_2 \sim 400$ μs (Granke, Does et al., 2015). Signals from pore water are broadly distributed ($T_2 > 1$ ms) because bone has pores of varying size (Fig. 9.2).

3.2 Magnetic resonance imaging

The acquired signals in ^{1}H NMR relaxometry underpin MRI. Details of MRI-derived measurements of water in bone can be found in several review articles (Manhard, Harkins, et al., 2017; Surowiec et al., 2022; Surowiec et al., 2024). In conventional MRI, bone is devoid of signal appearing black, but UTE imaging provides signal intensity contrast in cortical bone (Fig. 9.3A). There are effectively

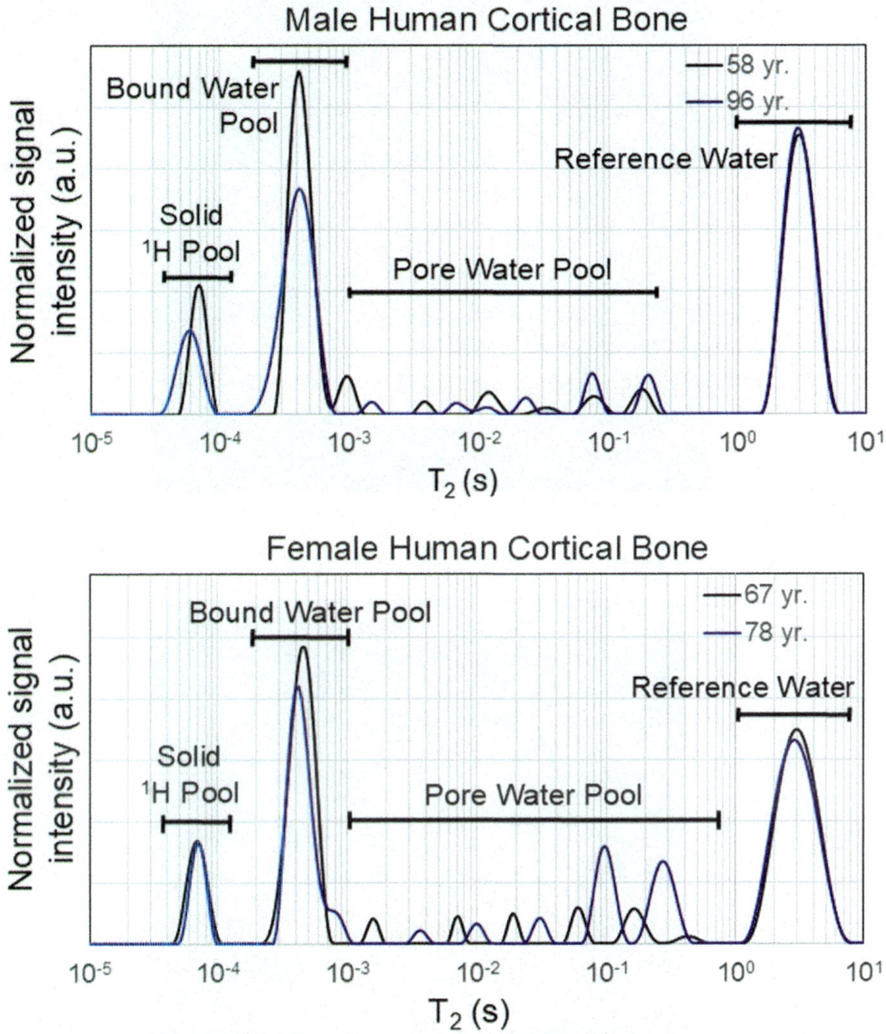

FIGURE 9.2 Representative T_2 spectra for hydrated human cortical.

The solid protons (^1H) of the bone tissue (e.g., –CH, –CH_2, –CH_3, –NH, –NH_2, OH) relax faster than water bound to the extracellular matrix of bone. There is a distribution of pore water signals in which water in small pores like lacuna relax faster than water in large pores like Haversian canals. A known volume of water is placed next to the bone during ^1H NMR relaxometry to convert the integrated areas to μl of H20 or mole of 1H. This reference volume of water is set to have a relaxation below pore water. T_2 is the spin-spin or transverse relaxation time constant of the proton signals. Signals were normalized to the area under the reference water band (T_2 from 1 to 10 s).

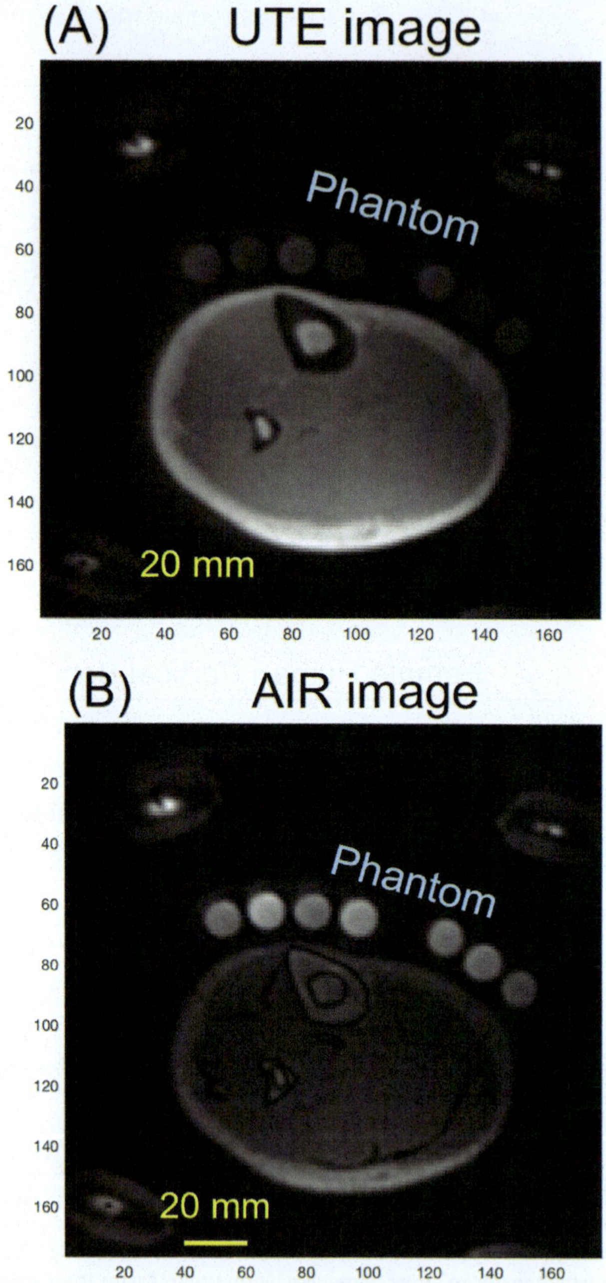

FIGURE 9.3 Images of bound and pore from UTE-MRI scans of human tibia.

In a clinical UTE scan, there is little signal intensity from bone (A). Using a pulse sequence called adiabatic inversion recovery (AIR), the signals from bone primarily pertain to bound water (B). Applying back-to-back adiabatic pulse sequences (double adiabatic full passage or DAFP) and no inversion recovery time delay, the signals from bone primarily pertain to pore water (C). By including a phantom of varying concentrations of deuterium (heavy water) relative to normal water doped with copper sulfate, the signal intensities can be converted to moles of proton (1H) per liter of bone such that color maps of bound water (D) and pore water (E) can be generated.

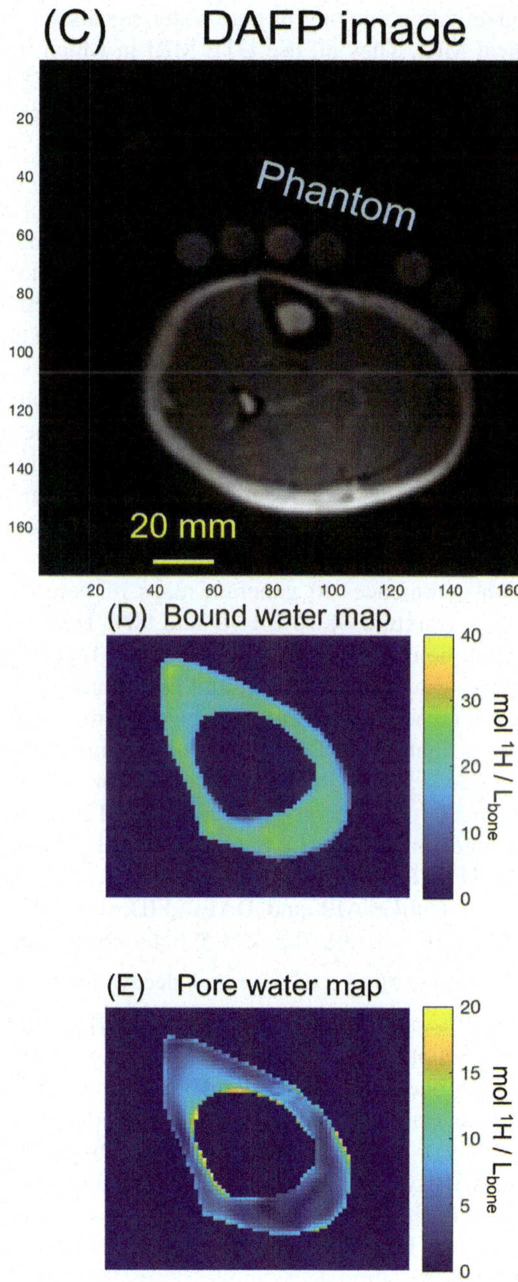

FIGURE 9.3 cont'd.

three strategies for quantitative imaging of pore water signals and bound water signals in bone by clinical MRI. They all use UTE-MRI in which the time delay between signal excitation and signal acquisition (echo time or TE) is shorter than conventional MRI in which bone appears black. In one strategy, specialized RF pulses are employed to either suppress signals from pore water, thereby acquiring signals from bound water, or suppress signals from bound water, thereby acquiring signals from pore water (Horch et al., 2012; Manhard et al., 2014, 2015; Manhard, Nyman et al., 2017). The adiabatic inversion recovery (AIR) method involves an adiabatic RF pulse to suppress the short T_2 of bound water (i.e., magnetization goes to zero) and inverts the long T_2 of pore water (i.e., magnetization goes from positive to negative). After an inversion delay period in which the pore water signals approach zero (i.e., magnetization is nulled) and the bound water signals approach a maximum (i.e., positive magnetization returns), UTE acquisition commences, thereby acquiring images of bound water levels (Fig. 9.3B). The double adiabatic full passage (DAFP) method involves two back-to-back adiabatic inversion pulses that saturates bound water signals (i.e., magnetization kept a zero) and maximizes the pore water signals (i.e., full magnetization goes from negative back to positive). There is no inversion delay period, so images of pore water signals are captured when UTE acquisition commences to generate maps of pore water (Fig. 9.3C). By including a phantom (mixture of 10% H_2O and 90% $D_2O + 80$ mM $CuSO_4$), the signal intensities can be converted to mole of ^1H per liter of bone to generate quantitative maps of bound water (Fig. 9.3D) and pore water (Fig. 9.3E).

In another strategy, UTE acquisition involves multiple images of short- and long-T^*_2 signals with varied weightings and then fitting the acquired signals to a bi- or tri-component relaxation decay model in which bound water and pore water are the short and long components, respectively (Bae et al., 2012; Biswas et al., 2012), and fat is the intermediate component (Lu et al., 2019). Like the adiabatic RF strategy described above, UTE-MRI of T^*_2 signals are acquired using a clinical 3 T magnetic resonance scanner. Unlike AIR and DAFP, FID signals are acquired after multiple echo times (e.g., TE = 0.008, 0.2, 2, 3.5, 6 ms, etc.). Then, a bicomponent analysis involving the fitting of two exponential decay functions ($e^{-t/T^*_{2\text{-short}}}$ and $e^{-t/T^*_{2\text{-long}}}$) to acquired T^*_2 signals at the different TEs (Bae et al., 2012). In this approach, the relative fractions of the short (i.e., bound water) and long (i.e., pore water) components equals one. In bovine cortical bone samples subjected to air drying, the short T^*_2 shifted from 85.2% (wet) to 94% (air dried for 30 min) and to 100% (air dried for 3 days), while the long T^*_2 shifted from 14.8% to 6.0% and then to 0%, respectively (Biswas et al., 2012). Ideally, bound water and pore water measurements would be independent of one another such that two different bones can have the same pore water level but different bound water levels. Fitting the decay signals to a tri-component relaxation decay model decouples bound water from pore water by adding a fat component, but there is still some dependence as indicated by the following models:

Bi-component model—$S^*(t) = F_S \cdot e^{-\frac{1}{T^*_{2-short}}} + F_L \cdot e^{-\frac{1}{T^*_{2-long}}} + \text{noise}\ [F_S + F_L = 1]$

Tri-component model—$S^*(t) = F_S \cdot e^{-\frac{1}{T^*_{2-short}}} + F_L \cdot e^{-\frac{1}{T^*_{2-long}}} + F_F \cdot e^{-\frac{1}{T^*_{2-fat}}} \cdot \sum_n \alpha_n e^{-i2\pi f_n t} + \text{noise}\ [F_S + F_L + F_F = 1]$ where F_S, F_L, and F_F are the fitted fractions of bound water, pore water, and fat, respectively; α_n is the relative amplitude of the nth spectral peak of fat; f_n is the corresponding multispectral peak frequency shift (Lu et al., 2019). Like the AIR technique, an adiabatic inversion recovery scheme can be applied to UTE acquisition of T^*_2 signals to suppress signals from pore water and fat, thereby maximizing short T^*_2 (bound) water (Carl et al., 2016; Chen et al., 2016). Notably, the quantification of bound water and pore water fractions using T^*_2 bi-component analysis depend on the field strength of the magnet such that the correlation between pore water and cortical porosity weakens as the field strength increases ($R^2 = 0.63$ at 1.5 T vs. $R^2 = 0.5$ at 3 T) (Seifert et al., 2015).

In the third strategy, a dual echo UTE acquisition of T^*_2 signals is processed to determine a signal ratio. With respect to the porosity index (PI), 2 MRI scans of cortical bone involve a short TE (e.g., 0.05 ms) and a long TE (e.g., 2 ms) and calculating the ratio (TE-long intensity/TE-short intensity) (Rajapakse et al., 2015). Since the signals are acquired as a function of time (i.e., FID) after the TE, TE-short acquisition includes proton signals from bound water and pore water in bone while TE-long acquisition includes proton signals primarily from pore water in bone. That is, waiting longer to acquire the signals ensures that fast relaxing proton signals, like bound water, are no longer present. By taking a ratio, there is no need for a phantom. The average ratio can be determined for the region of interest (ROI) like an entire cross-section of the tibia mid-diaphysis or can generate a map of the ratios (PI is 100 × ratio) within the ROI (Hong et al., 2019). Rajapakse and colleagues found that PI of cadaveric tibiae ($N = 16$) was directly correlated with cortical porosity ($R^2 = 0.79$), as derived from micro-computed tomography imaging of the bone and donor age ($R^2 = 0.64$) (Rajapakse et al., 2015) as well as inversely correlated with stiffness ($R^2 = 0.63$), as determined by compressive loading of segment of the tibia mid-diaphysis (Hong et al., 2019). Surowiec et al. extended this strategy to quantify a bound water index, which is the difference between TE-short intensity and TE-long intensity divided by the TE-short intensity (Jacobson et al., 2024).

3.3 High wavenumber Raman spectroscopy

A light-based technique, Raman spectroscopy (RS) is typically insensitive to water allowing it to quantify mineral-to-matrix ratio, type B carbonate substitution, and crystallinity of hydrated bone tissue in the so-called fingerprint region (Raman signals between 300 to 2000 cm^{-1}). In high wavenumber RS (2500 to 4000 cm^{-1}), water contributes to Raman signals between 3100 and 3800 cm^{-1} (Unal et al., 2014). An in-depth description of RS as applied to the analysis of bone can be found in a review by Unal et al. (2021). Briefly, Raman spectra are a measure of the number of photons undergoing a change in energy between the incident light (e.g., 785 nm laser) and the Raman-scattered photons. An RS instrument measures photon intensity as a function of the Raman shift (i.e., difference in wavelength of the incident photons and the Raman-scattered photons). The various vibration modes (e.g., OH stretching vibration) of H_2O causes a Raman band with a peak at ~3400 cm^{-1}.

Upon assembling a high wavenumber RS instrument, Unal et al. established the quantification of mineral-bound water and collagen-bound water of cortical bone samples (Unal et al., 2014). Using an 852 nm laser, a microscope objective with numerical aperture of 0.4, and spectrometer with a volume-phase holographic transmission grating, they acquired Raman spectra in the range between 2540 and 4540 cm^{-1} at a spectral resolution of 10 cm^{-1} from wet bovine cortical bone (Fig. 9.4) and as a function of (i) dehydration (air, oven at 40°C, and ethanol), (ii) demineralization, and (iii) hydrogen-deuterium exchange (i.e., D_2O replaces H_2O to varying degrees). Sub-peak intensities of the OH band at 3220, 3325, 3453, and 3584 cm^{-1} all decreased with thermal dehydration (Unal et al., 2014). These intensities also decreased upon immersing the bone samples in ethanol and D_2O. They attributed the sub-peak at 3584 cm^{-1} as water bound to the mineral phase of bone and the hydroxyl group within bone mineral because it was absent in demineralized

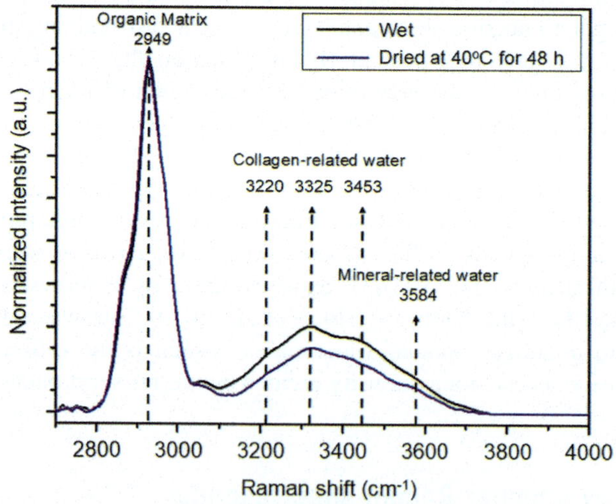

FIGURE 9.4 High wavenumber Raman spectra of cortical bone.

Dehydration by heating bovine bone at 40°C for 48 h decreased the level of bound water relative to the organic matrix. The spectra were normalized to the peak at 2949 cm^{-1}. The shoulder peak at 3584 cm^{-1} is related to bound water interacting with mineral, while the shoulder peak at 3453 cm^{-1} is related to bound water interacting with collagen. Dividing intensity height of each water peak by the intensity of CH stretching peak quantifies bound water in bone at the site of signal acquisition.

bone, while the sub-peaks at 3325 and 3453 cm^{-1} were attributed to collagen-bound water and either NH and OH (at 3325 cm^{-1}) chemical groups or hydroxyproline (3453 cm^{-1}) because these intensities were prominent in wet demineralized bone and weak in dry demineralized bone. The sub-peak at 3220 cm^{-1} was the most sensitive indicator of bound water in bone. The percent reduction of this sub-peak increased with time of deuterium exchange. RS-derived measurement of bound water is the intensity at 3220 cm^{-1} per the intensity at 2950 cm^{-1} where the latter peak arises from CH stretching of the proteins in bone.

Subsequent studies by Unal and colleagues found that fatigue testing (rotating beam) of human cortical bone increased RS-derived matrix-bound water (I_{3220}/I_{2950}), and gamma-irradiated sterilization at 25 kGy also increased I_{3220}/I_{2950} (Flanagan et al., 2017). One interpretation of these findings is that fatigue-induced denaturation of Col I and free radical scission of the Col I peptide backbone by gamma irradiation increases hydrogen bonding sites.

Commercial manufacturers sell Raman micro-spectroscopy instruments in which the diffraction grating (number of grooves per mm) disperses the Raman-scattered light at high wavenumbers onto a charge-coupled device (CCD) in the spectrograph. With fiber optic-probe based RS instruments, switching from 785 nm laser to a 680 nm laser without changing the diffraction grating causes the Raman shift (wavenumber range) to go from fingerprint (300 to 1800 cm^{-1}) to high wavenumber (2000 to 3800 cm^{-1}) (Masson et al., 2018). This technique requires a broad bandpass filter (625–800 nm) to allow the incident light at two different wavelengths to reach the sample and a long pass filter (800 nm) to block Raman signals at lower wavenumber (those below 900 cm^{-1} for 784 nm laser and below 2800 cm^{-1} for 680 nm laser) from reaching the CCD.

3.4 Infrared spectroscopy

Infrared (IR) spectroscopy uses the infrared region of light, which spans 12,500–10 cm^{-1} (or 800–1 mm), to discern sample composition and structure based on molecular vibrations (Ozaki, 2021). By subjecting a sample to a range of infrared light, the photons are absorbed when the wavelength frequency corresponds with the vibrational frequency of specific functional groups, while the rest of the light is transmitted through the sample and quantified by a detector. A sample's IR spectra consists of bond-specific absorbance bands as a function of wavenumber (Taylor & Donnelly, 2020). IR spectroscopy can be further sub-divided into mid-infrared (MIR, also known as FTIR), near-infrared (NIR), and far-infrared (FIR, also known as Thz) spectroscopy, each utilizing a portion of the infrared spectrum and detecting different types of energy absorptions. NIR spectroscopy discerns molecular structures by detecting combinations (two or more simultaneous vibrations) and overtones (light absorbed as electrons transition from their ground state to excited state) and is sensitive to bound water.

Water affects several absorption bands, making it difficult to study water compartments of hydrated bone using THz spectroscopy. Additionally, the penetration depth of light being absorbed requires 3–5 μm thick sections, meaning FTIR analysis requires thin, dehydrated bone sections or bone powder (Querido et al., 2021). Building off their work that identified the free (6890 cm^{-1}) and bound (5200 cm^{-1}) water in cartilage using NIR spectral imaging (Padalkar et al., 2013), Pleshko and colleagues developed NIR spectroscopy to investigate bound water in bone. Firstly, Rajapkse et al. validated the feasibility of NIR spectral imaging to detect bound water in bone by comparing NIR spectral imaging to UTE-MRI in the cortical bone of cadaveric tibiae (Rajapakse et al., 2017). NIR spectral peaks of 5152 and 7008 cm^{-1} are specific to water and significantly correlated with UTE-MRI-determined bound water content (Pearson's $r = +0.735$ and $r = +0.710$, respectively). After lyophilizing their bone samples, the group further identified bound water peaks as shifted versions of their pre-lyophilized water peaks (from 5152–~5168 cm^{-1} and 7008–~7040 cm^{-1}), which aligned with their previous findings of how free and bound water peaks in cartilage existed in relation to each other (Padalkar et al., 2013). Ailavajhala et al. addressed reproducibility issues by designing an environmental chamber to control humidity (Ailavajhala et al., 2019). Investigating the effect of relative humidity on NIR spectra, a new absorbance peak of 6560 cm^{-1} was observed and concluded to represent loosely bound water given its disappearance following lyophilization, while previously identified water peaks at 5184 cm^{-1} and 7008 cm^{-1} remained. The latter peaks represent both loosely and tightly bound water to either collagen and/or mineral (Ailavajhala et al., 2020). In NIR, partial dehydration is necessary to reveal peaks of tightly bound water.

NIR spectroscopy has translatable potential for clinical assessment of disease-related changes in bone matrix. Using the second metacarpal bone of cadaveric hands, Shanas et al. identified significant positive correlations between NIR-derived water absorbance bands and other bone characteristics such as dual-energy X-ray absorptiometry (DXA)-derived Z-scores (Pearson's $r = +0.71$) and relative cortical area (Pearson's $r = +0.89$) (Shanas et al., 2020). Interestingly, the peak intensity at 5300 cm^{-1} negatively correlated with participant age (Pearson's $r = -0.92$), further reaffirming the idea that bound water decreases with age. Combining both attenuated total reflectance (ATR)-FTIR and NIR spectroscopy, differences in water content within bone have been associated with the development and treatment of brittle bone disease in a mouse model of osteogenesis imperfecta (OI) as well (Shanas et al., 2022). Tightly bound water per protein and mineral-bound water were higher in OIM bone compared to wild-type bone. Bone samples from anti-resorptive-treated mice had lower amounts of loosely bound water, indicating that these treatments may help increase mineralization in OIM bones. Future NIR spectroscopy efforts are needed to develop in vivo imaging of water content within bone.

4. Role of pore water

A key role of pore water in the mechanical behavior of cortical bone is being a surrogate of porosity, a determinant of bone strength and toughness (McCalden et al., 1993).

Since the large pore space within trabecular bone is full of marrow fat, pore water within this type of bone is limited to the lacunar-canalicular system. Being free to flow through canals, pore water also confers hydraulic stiffening to bone.

4.1 Hydraulic stiffening

The computational techniques to characterize fluid flow in bone are covered elsewhere (Chapter 16.). Briefly, loading porous bone filled with a fluid like water causes pressure to build. The level of pressure depends on the forces acting on the bone, viscosity of the fluid (i.e., water plus various chemical compounds suspended in the water), and permeability of the pore network. When finite element models were generated from micro-computed tomography (μCT) scans of the femoral neck (male 61-year-old donor), hydraulic stiffening depended on the pore architecture and loading rate (Le Pense & Chen, 2017). Apparent modulus of bone when fluid is present in pores increases when porosity (or permeability) decreases and strain rate increases. However, the inclusion of a viscoelastic material behavior based on assumptions about permeability of cortical and trabecular bone did not drastically change the prediction of femoral neck strength (Haider et al., 2013). That is, the hydraulic stiffening or strengthening of bone due to changes in pore pressure under loading was negligible under a loading rate that mimics impact of bone during a fall to the ground. While it is reasonable that the flow of water in bone would confer some resistance to loading, the quality of the bone tissue itself likely dictates the energy required to fracture bone.

4.2 Pore water as surrogate of cortical porosity

Since water resides in all the vascular channels of bone, the quantity of pore water (PW) is a measure of cortical porosity. In their low-field NMR study of human cortical bone samples from 19 cadaveric femurs (16–89 years of age), Wang and Ni found a direct correlation (Pearson's $r = +0.85$) between PW and void area per total area (Wang & Ni, 2003). In their study using a CPMG spin echo method of acquiring T_2 signals, *lacunar pore water* (void area per total area = ~1.9%) had a T_2 peak between ~0.2 and ~8 ms and *Haversian pore water* (void area per total area = ~5–25%) had a T_2 peak between ~9 and ~200 ms. In another NMR study analyzing cortical bone samples from cadaveric tibia (6 donors between 27 and 83 years of age) at high field strength (9.4 T), pore water, which was total water minus bound water, strongly correlated (Pearson's $r = +0.95$) with μCT-derived cortical porosity (Ong et al., 2012). As determined by CPMG measurements of T_2 signals of human cortical bone (60 donors, 21–101 years of age) at 4.7 T and a known reference volume of water, the volume fraction of pore water directly correlated (Pearson's $r = +0.88$) with μCT-derived porosity (Granke, Makowski, et al., 2015).

With respect to imaging pore water by UTE-MRI (Section 3.2), multiple measurements of PW directly correlate with CT-derived measurements of cortical porosity (Chen et al., 2015; Jerban et al., 2020; Jerban, Ma, Li, et al., 2019; Jerban, Ma, Wong, et al., 2019; Li et al., 2014; Manhard et al., 2016; Seifert et al., 2015). In a study of discarded proximal femurs from hemi-arthroplasty (HA) cases to fix a femoral neck fracture (N = 35) and from total hip arthroplasty (THA) cases to treat arthritic pain due to osteoarthritis (N = 29), Porrelli et al. analyzed trabecular bone cores ex vivo using low-field NMR (0.47 T) (Porrelli et al., 2022). After rinsing in water, bleaching for 48 h, and autoclaving, pore water, as a unitless measurement of porosity, was significantly higher in HA than in THA.

In a clinical study involving UTE-MRI scans of the one-third radius, the concentration of pore water (C_{pw} in mol ^{1}H/L of bone, L_{bone}) was higher in adults with a recent fragility fracture than in age-matched and sex-matched adults without a history of osteoporosis (Nyman et al., 2023). In another cohort of patients, a similar UTE-MRI measurement of pore water (mol $^{1}H/L_{bone}$) in the tibia was significantly higher in post-menopausal women newly diagnosed with osteoporosis by dual energy X-ray absorptiometry (DXA) than in age-matched women without the disease (Nyman et al., 2023). Wehrli and colleagues also found higher C_{pw} of the tibia mid-diaphysis in post-menopausal women with T-scores less than −2.5 than in age-matched women without clinical osteoporosis (Jones et al., 2023). To date, there was one other UTE-MRI in vivo study linking pore water to osteoporosis. A measurement called PI of the tibia diaphysis was higher in women with DXA T-scores less than −2.5 than in women with T-scores greater than −2.5 but less than −1 as well as significantly higher than in women with T-scores above −1 (Jerban et al., 2023).

5. Role of bound water

Without bound water, bone is a strong but brittle material (Table 9.1). Likewise, bone is a tough material when Col I is properly organized into a triple helix, interconnected into collagen fibrils via enzymatic collagen crosslinking, and *hydrated* (Chapter 5.). The stability of the triple helix of the helical Col1α1 and Col1α2 peptide chains depends on hydrogen bonds within the helix but also hydrogen bonding with water. In strengthening the soft organic matrix, the nano-sized, bone mineral crystals displace collagen-bound water from the osteoid during mineralization, but they retain interactions with bound water via electronegative charge (PO_4^{3-}) (Timmins & Wall, 1977). As such, healthy bone is both strong and tough for its density.

5.1 Confers viscoelasticity

A viscoelastic material exhibits a combination of viscous and elastic behavior under time-dependent, stress-applied deformation. Bone is stiffer as the loading rate increases. Bone's viscoelastic behavior was originally attributed to bone being a two-phase material of mineral and matrix by Currey (1964), with the elastic and viscous behaviors of bone mainly attributed to its mineral phase (Burstein et al., 1975) and hydrated collagen (Lakes & Katz, 1979), respectively. Within the bone tissue, these three components interact such that the relative distribution of mineral, Col I, and water influence the degree of stress relaxation under a constant displacement, the degree of creep under a constant load, the phase shift between cyclic loading and displacement response, and the strain-rate dependency of modulus and yield stress. The network of collagen fibrils is both strengthened and constrained by the mineral crystals (Bonar & Glimcher, 1970; Kronick & Cooke, 1996), while the relative degree of bound water interacting with each phase (hard and soft) presumably affects viscoelastic characteristics (e.g., loss modulus during dynamic mechanical analysis and sensitivity of strength to loading rate) and viscoplastic properties (e.g., the degree of hysteresis during loading and unloading after the yield point and the amount of creep during fatigue). To date, empirical relationships between bound water and viscoelastic properties have not been established.

Dynamic Mechanical Analysis (DMA) is a non-destructive mechanical testing instrument that measures viscoelastic properties by sensitive time-dependent responses of materials to loading. Described in depth by Menard, the basis of DMA is the application of oscillating forces to a sample and recording how it responds to said applied forces (Menard & Menard, 2020). An important viscoelastic property derived from this technique is loss factor (Tan δ), which is a ratio of the amount of energy lost from the sample during oscillation per the amount of stored energy within the sample during oscillation. In effect, it is the viscous response (loss modulus, E'') per elastic response (storage modulus, E'). Yamashita et al. used DMA to explore the role of moisture content on the viscoelasticity of bone, finding that moisture content significantly affected the Tan δ values of cadaveric cortical bone specimens, where Tan δ values were higher in wetter specimens (Yamashita et al., 2001). Originally, the group attributed the relationship between Tan δ and viscoelasticity to reflect the hydration of Col I within the tissue, but later, the same group found that Tan δ was independent of collagen denaturation such that Tan δ—not E'—was higher in wet than in dry human cortical bone, regardless of whether the bone sample was subjected to 100 or 200°C for 1 hour to denature Col I (Yamashita et al., 2002). This was further supported by the molecular dynamics and modeling study of Milazzo et al. in which the capacity of mineralized fibrillar Col I to dissipate energy under sudden compressive loading (100, 500, and 100 m/s) depended on water content (Milazzo et al., 2021).

5.2 Confers plasticity

Plasticity is the ability of a material to withstand deformation before failure and after the onset of permanent damage (yielding) when subjected to increasing applied stress. A common mechanical property to quantify plasticity is post-yield toughness.

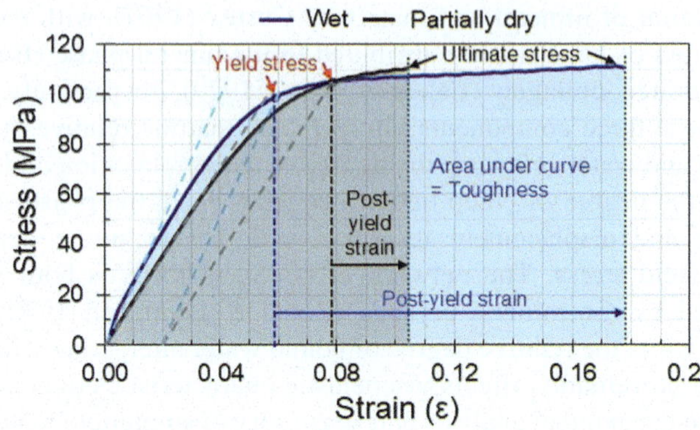

FIGURE 9.5 Stress versus strain curves from load-to-failure, tensile tests of wet and partially dehydrated human cortical bone.

Air drying the tensile specimen for several minutes (~3% loss in mass) decreased the toughness (area under the stress vs. strain curve) and post-yield strain compared to the wet specimen from the same donor. In this case, tensile strength (ultimate stress) and modulus were not higher for the partially dry versus wet bone, though dehydration generally increases both of these mechanical properties. The yield point was determined using the 0.2% offset rule.

Descriptions of mechanical tests of bone can be found in Chapter 8. The importance of water in supporting plasticity has been demonstrated in two ways—by thermally dehydrating bone through various methods of drying (Fig. 9.5) and by chemically dehydrating bone with the use of polar solvents. Both methods result in brittleness due to a loss of Col I ductility. Nyman et al. found that drying at room-temperature significantly decreased bone toughness through a loss of plasticity in cadaveric, cortical bone samples, emphasizing the importance of hydrated Col I in bone's post-yield deformation behavior (Nyman et al., 2006). Loss of water from air drying has been further shown to affect post-yield properties to a greater extent than elastic properties, where water loss of $<3\%$ by weight had no effect on the modulus or yield strength of cadaveric, cortical bone samples but did significantly decrease the post-yield strain (Nyman et al., 2013).

With the removal of water from the bone matrix, the interfaces between Col I and mineral become closer. As Col I dehydrates, it shrivels, pulling tightly on the mineral interface and pre-stressing the bone matrix, causing the bone to be stronger at the cost of being brittle. In addition to air-drying, vacuum-drying has also been shown to increase the elastic modulus and ultimate strength of bone, while decreasing failure strain and toughness (Table 9.1). Bone strength increased with a 5% water loss by weight at 21°C and then decreased with a water loss greater than 9% at 70°C, while modulus linearly increased with water loss (Nyman et al., 2006). Bone toughness, as measured by work-to-fracture, linearly decreased from fully hydrated to partially dehydrated at 21 and 50°C, but further decreased at 70°C (11% by weight) and 110°C (13% by weight). Bone specimens subjected to prolonged periods of dehydration exhibit increased ultimate bending stress and decreased ductility (Table 9.1). Shin et al. observed differences in crack mechanics of human and sheep cortical bone samples tested in SEM environments compared to in a bio-bath or ambient air pressure attributed to dehydration (Shin et al., 2022).

Additionally, some studies have examined the effects of rehydrating bone samples and have found conflicting evidence as to whether rehydrating samples fully recovers their mechanical properties. This is mainly dependent on whether the samples are exposed to long, dehydration periods on the scale of weeks verse shorter periods spanning hours (Currey, 1988; Lievers et al., 2010). When comparing human cortical specimens taken and subjected to tensile and compressive testing directly from cadavers to samples allowed to rehydrated overnight before testing, dried samples exhibited an increased elastic modulus and ultimate stress under tension (55% and 50%, respectively) and compression (26% and 63%, respectively) compared to rehydrated samples (Dempster & Liddicoat, 1952). When bovine cortical bone samples were air-dried for 25 days and rehydrated for 3 hours prior to three-point bending, the dehydrated-rehydrated samples could not fully recover, exhibiting a significantly decreased bending strength compared to immediately tested, wet samples (Currey, 1988). Considering all of this, it is important to maintain hydration during sample preparation and mechanical testing (ambient humidity can vary day-to-day).

During chemical dehydration, water is displaced out of the bone tissue by another polar solvent or miscible aqueous solution like ethanol. The solvent changes the strength of the hydrogen bonding with Col I from that of H_2O. The extent to which a non-aqueous solution can replace water depends on several chemical factors—its polarity, molecular size, and hydrogen bonding capability (Samuel et al., 2014). Organic solvents form fewer bridges between Col I chains due to their weaker hydrogen-bonding capability resulting in a greater number of direct Col I-to-Col I hydrogen-bonding. This combined with overall fewer hydrogen-bonded bridges in Col I results in less plastic bone tissue. For bone specimens stored in ethylene glycol, dimethylformamide, or carbon tetrachloride at 70°C for approximately 4 hour, specimens acquired a brittle behavior similar to that of air-dried bones with significantly lower post-yield deformation compared to rehydrated specimens (Samuel et al., 2014). Similarly, when human and bovine cortical bone samples were stored in ethyl alcohol-glycerine-phenol for 6 months and then tested in three-point bending, the samples displayed lower post-yield toughness compared to fresh-frozen controls (Stefan et al., 2010).

When human tibial cancellous bone was dried in alcohol for 6 hours, air-dried for 24 hours and then underwent compressive testing, dried samples exhibited a 24% increase in Young's modulus when compared to immediately tested wet specimens (Townsend et al., 1975). Mazurkiewicz found that the elastic modulus significantly increased for bone left drying in the air for 14 weeks, while the elastic modulus of specimens stored for 14 weeks at room temperature in buffered 10% formalin, in 96% alcohol, or in 50% alcohol did not significantly change (Mazurkiewicz, 2018).

Qiu et al. found that the strain-rate sensitivity exponent in dynamic compressive testing was highest in the fresh hydrated group, followed by the formalin group, and then vacuum-dehydration group (Qiu et al., 2023).

The interrelationship between plasticity, water content, and GAGs, which help sequester water within the bone tissue, has also begun to be explored. On the tissue-level, concentration of chondroitin sulfate directly confers nano-scratch toughness (Hua et al., 2020, 2025; Qingwen et al., 2021). Studies have further validated this relationship by dehydrating bone while leaving GAGs intact, showing a loss of toughness (T) (Wang et al., 2016) and increased elastic modulus (E) (Han et al., 2021). Removing GAGs through heating and enzymatic treatment also significantly decreased plastic energy dissipation of the extrafibrillar matrix, but not of the mineralized collagen fibrils, of hydrated, cadaveric cortical bone samples (Han et al., 2021).

5.3 Transfers forcers between mineral crystals and between mineral and collagen I

At the ultrastructural level of the hierarchical organization of bone (<100 nm), extrafibrillar mineral is dispersed around mineralized collagen fibrils (intrafibrillar mineral) (Bonar et al., 1985; Lees & Prostak, 1988; McNally et al., 2013; Schwarcz et al., 2014) forming a continuum (as reviewed in Shah, 2025). Water occupies any small void spaces (<10 nm) in both types of bone mineral. Applying a solid-state NMR technique to synthetic mineral and cortical bone powder (from human femur) as a function of dehydration (water content), Wilson et al. found the first evidence that water layers exist between the surfaces of mineral crystals and Col I molecules (Wilson et al., 2005). The presence of this so-called *interstitial* water was depleted after 3 h of air-drying, so presumably it is a type of loosely bound water. The signals from this water also exchanged magnetization with signals of free water (denoted bulk water) on the NMR timescale (6 ms), so *interstitial* water is likely not the strongly bound water stabilizing the triple helix of Col I. Based on oscillations of couplings in partially dehydrated bone (i.e., no bulk water), water protons existed at 2.30–2.55 Å (~0.243 nm) from PO_4 in the crystal lattice of bone mineral. It is possible such a highly ordered layer of water exists between extrafibrillar mineral crystals and/or between extrafibrillar and intrafibrillar mineral. In a similar solid-state NMR study of cortical bone (from goat fibula and filed to generates flakes) before and after dehydration, Dwivedi et al. found different relaxation behaviors of water and inferred that water molecules interact with both amorphous and crystalline mineral, with dehydration affecting interactions with apatitic mineral more so than amorphous mineral (Dwivedi et al., 2022). In effect, highly ordered water is not just bound between mineral and collagen but also present between neighboring mineral platelets. Presumably, *interstitial* water transfers local loads from mineral-to-mineral (amorphous or crystalline) and mineral-to-collagen via viscous mechanisms (load-rate dependent).

6. Correlations with mechanical properties

Using ^{1}H NMR relaxometry of human cortical bone, Does, Nyman, and colleagues found that multiple mechanical properties negatively and positively correlated with the amount of pore water (PW) and the amount of bone water (BW), respectively (Table 9.2). They also found that bound water concentration (C_{bw}) and pore water concentration (C_{pw}) positively and negatively, respectively, correlated with the bending strength of one-third radius when applying UTE-MRI to cadaveric forearms from older male and female donors (Table 9.2). Moreover, in a linear regression model, both C_{bw} (positive) and C_{pw} (negative) were significant predictors of the estimated material strength of cortical bone explaining 63.5% of the variance (Table 9.2). In a follow-up study by the same research group, a linear relationship between elastic modulus and these concentrations ($E = a_0 + a_1 \times C_{bw} + a_2 \times C_{pw}$) and a linear relationship between yield stress and these concentrations ($YS = b_0 + b_1 \times C_{bw} + b_2 \times C_{pw}$) could be incorporated in a linear elastic-perfectly plastic finite element analysis to better predict force versus displacement curves from three-point bending tests of cadaveric radii than FEAs in which E and YS was the same for all bones (Ketsiri et al., 2023). Du and colleagues also found that PW negatively and BW positively correlated with the bending strength of human cortical bone as determined by four-point bending tests (Table 9.2).

NIR spectroscopy and high wavenumber RS have identified correlations between bound water levels and the mechanical properties of cortical bone (Table 9.2). The benefit of IR spectroscopy is that hydrated, intact samples can be used to identify total water peaks and then the samples can be lyophilized to remove the pore water and loosely bound water, to isolate the more tightly bound mineral and matrix-bound water content in the specimens. The benefit of high wavenumber RS is the delineation of collagen-bound and mineral-bound water. Using an osteogenesis imperfecta (OI) mouse model, total water, both mineral and matrix-bound water, and mineral-bound water was found to negatively correlate with stiffness, maximum load, and post-yield displacement (Shanas et al., 2022). In a study of whole bone mechanics mimicking a sideways fall onto the hip, stiffness was negatively correlated to total water, and mineral and matrix-bound water of both hydrated, intact and lyophilized, cortical and trabecular bone samples (Querido et al., 2025). With regards to high wavenumber RS, collagen- and mineral-bound water positively correlated with toughness and bending strength, respectively, of bovine cortical bone (Table 9.2; see Section 3.3).

In the NIR analysis of cortical bone for wild-type and osteogenesis imperfecta (OI) mice, total water relative to total protein of hydrated intact bones and mineral-bound water relative to total protein inversely correlated with stiffness and post-yield displacement as determined by three-point bending the mouse humeri (Table 9.2). In the NIR analysis cadaveric bone, stiffness negatively correlated to total water and tightly bound water to matrix (Querido et al., 2025). Regression models predicted the stiffness of the proximal femur under loading conditions of a sideways fall when (Cross-validated $R^2 = 0.80$–0.92) partial least squares regression combined total and tightly bound water of cortical and trabecular bone (Querido et al., 2025).

Table 9.2 Significant correlations between water measurements and the mechanical properties of bone.

Species of bone (sample size[a])	Method of measurement (units)	Mechanical test (nominal dimensions)[b]	Correlation	References
Nuclear magnetic resonance relaxometry				
Human cortical (N = 18)	0.64 T NMR FID (integrated area per wet mass)	Three-point bending of beam (4.3 × 2.1 × 40.0 mm)	σ_b versus BW: r = +0.60, $P < .05$ σ_b versus PW: r = −0.46, $P < .1$ W_f/I_{min} versus BW: +0.64, $P < .05$	Nyman et al. (2008)
Human cortical (N = 40)	4.7 T NMR CPMG (mol $^1H/L_{bone}$)	Three-point bending of beam (5 × 2 × 40 mm)	σ_b versus BW: r = +0.82, $P < .05$ σ_b versus PW: r = −0.78, $P < .05$	Horch et al. (2011)
Human cortical (N = 62)	4.7 T NMR CPMG (water volume/bone volume as %)	R-curve of SENB (2.6 × 5.4 × 25 mm)	K_J versus BW: $r_S = +0.63$, $P < .05$ K_J versus PW: $r_S = -0.53$, $P < .05$ $\Delta K_J/\Delta a$ versus BW: $r_S = +0.35$, $P < .05$ J_f versus BW: $r_S = +0.45$, $P < .05$ J_f versus PW: $r_S = -0.34$, $P < .05$	Granke, Makowski, et al. (2015)
Magnetic resonance imaging				
Human cortical (N = 14)	3.0 T UTE-MRI FID (100% x short T2* fraction or total water intensity normalized to phantom intensity)	Four-point bending of beam (3.45 × 1.63 × 41.4 mm)	σ_b versus BW: r = −0.47, $P < .01$ σ_b versus TW: r = −0.50, $P < .001$	Bae et al. (2012)
Human radii (N = 40)	3.0 T UTE-MRI AIR and DAFP (mol $^1H/L_{bone}$)	Three-point bending of one-third radius	σ_b versus BW: r = +0.56, $P < .05$ σ_b versus PW: r = −0.68, $P < .05$	Manhard et al. (2016)
Human cortical (N = 37)	3.0 T UTE-MRI FID (% fraction of short, middle, or long T2%)	Four-point bending of beam (4 × 2 × 40 mm)	σ_b versus BW: r = +0.62, $P < .01$ σ_b versus Fat: r = −0.46, $P < .01$ σ_b versus PW: r = −0.57, $P < .01$	Jerban et al. (2020)
Human cortical (N = 18)	3.0 T UTE-MRI (dual-echo, PI as the ratio of long to short TE intensity)	Compressive testing	δ versus TW: r = −0.79, $P = .0028$ versus TW: $R^2 = 0.63$, $P = .002$	Hong et al. (2019)
Human cortical (N = 15)	3.0 T UTE-MRI (dual-echo, PI as the ratio of long to short TE intensity)	Whole bone testing under loading conditions for a sideways fall onto the hip	δ versus TW: r = −0.82, $P = .0014$δ versus TW: $R^2 = 0.67$, $P = .001$	Jones et al. (2021)

Infrared spectroscopy				
Mouse cortical ($N = 12–18$ per treatment group)	NIR (1800–1001 cm^{-1} and 2500–1801 cm^{-1}, 1 nm spectral resolution, hydrated and lyophilized samples)	Three-point bending of humeri (deltoid tuberosity in tension)	Hydrated, intact samples: (all RTP) δ versus δ versus TW (5250 cm^{-1}): $r = -0.20$, $P < .05$ F_{MAX} versus TW (5250 cm^{-1}): $r = -0.35$, $P < .05$ PYD versus TW (5250 cm^{-1}): $r = -0.27$, $P < .05$ PYD versus TW (7127 cm^{-1}): $r = -0.25$, $P < .05$ Lyophilized samples: (all RTP) δ versus Min- & Col-BWδ versus BW (5250 cm^{-1}): $r = -0.44$, $P < .05$ F_{MAX} versus Min- & Col-BW (5250 cm^{-1}): $r = -0.60$, $P < .05$ PYD versus Min-& Col-BW (5250 cm^{-1}): $r = -0.43$, $P < .05$ δ versus Min-BW (7050 cm^{-1}): $r = -0.40$, $P < .05$ F_{MAX} versus Min-BW (7050 cm^{-1}): $r = -0.60$, $P < .05$ PYD versus Min-BW (7050 cm^{-1}): $r = -0.51$, $P < .05$	Shanas et al. (2022)

Continued

Table 9.2 Significant correlations between water measurements and the mechanical properties of bone.—*cont'd*

Species of bone (sample size[a])	Method of measurement (units)	Mechanical test (nominal dimensions)[b]	Correlation	References
Human cortical (N = 12)	NIR (10,000–4000 cm^{-1}, 1 nm spectral resolution, hydrated and lyophilized samples)	Whole bone testing under loading conditions for a sideways fall onto the hip	Hydrated, intact samples: (all RTP) δ versus TW (5300 cm^{-1}): r = −0.63, $P = .02$ δ versus TW: r = −0.63, $P = .02$ δ versus TW: $R^2 = -0.40$, $P = .02$ Lyophilized samples: (all RTP) δ versus Min-& Col-BW (5300 cm^{-1}): r = −0.55, $P = .04$δ versus T-BW: $R^2 = 0.30$, $P = .04$	Querido et al. (2025)
Human trabecular (N = 12)			Hydrated, intact samples: (all RTP) δ versus TW (5300 cm^{-1}): r = −0.64, $P = .01$ δ versus TW: $R^2 = +0.44$, $P = .01$ Lyophilized samples: (all RTP) δ versus Min-& Col-BW (5300 cm^{-1}): r = −0.55, $P = .04$δ versus T-BW: $R^2 = +0.38$, $P = .02$	
Raman spectroscopy				
Bovine cortical (N = 28)	Raman (847 nm excitation, 1000–1400 nm range, 9 cm^{-1} resolution, intensity ratio of OH peaks (I_{OH}/I_{CH}) in range of 3100–3800 cm^{-1})	Three-point bending (3 × 1.1 × 30 mm)	T versus Col-BW (I_{3220}/I_{2949}): r = +0.9, $P < .001$ T versus Col-BW (I_{3325}/I_{2949}): r = +0.89, $P < .001$ PYT versus Col-BW (I_{3220}/I_{2949}): r = +0.81, $P < .001$ PYT versus Col-BW (I_{3325}/I_{2949}): r = +0.85, $P < .001$ E versus Min-BW (I_{3584}/I_{2949}): r = −0.88, $P < .001$ MFS versus Min-BW (I_{3584}/I_{2949}): r = +0.68, $P < .001$	Unal and Akkus (2015)

[a] *The sample size is the number of unique donors that provided bone samples.*
[b] *Width × Height × Length.*

7. Determinants of bound water

Little is known about the factors influencing the amount of bound water in the ECM of bone, let alone why bound water decreases with age (Granke, Does et al., 2015; Granke, Makowski, et al., 2015) (Fig. 9.6). As previously described (Section 2.2), the level of bound water in osteoid decreases as mineral accumulates. When cortical bone was deproteinated with sodium hypochlorite (NaClO, the active ingredient in bleach) for different periods of time, bound water decreased as the immersion time increased (i.e., linear relationship between bound water and the amount of organic matrix removed) while tissue mineral density remained constant (Nyman et al., 2019). In contrast, heat-treating human cortical bone caused BW to increase by 50% after the temperature exceeded 140°C. This coincided with denaturation (Nyman et al., 2019). Thus, bulk bound water requires organic matrix; degree of mineralization negatively modulates bulk bound water; uncoiled Col I can holds more water in a bound state than helical Col I; and glycation-mediated modifications to the organic matrix have subtle, complex effects on bound water, many of which have yet to be elucidated.

In a preclinical study involving MRI measurements of water, the selective estrogen receptor modulator called raloxifene was administered to skeletally mature female dogs (oral at 0.5 mg/kg of body weight/day) for 6 months. BW and PW in the tibia diaphysis were 14% lower and 20% higher, respectively, in raloxifene-treated animals than in saline-treated animals (Allen, Newman, et al., 2015; Allen, Territo, et al., 2015). In a similar study by the same research group, BW was higher while PW was lower in female dogs treated with raloxifene (+31% and −31%, respectively) or treated with raloxifene plus alendronate (+25% and −25, respectively)

	Intercept	Age (yr)	Sex (female)
Estimate	22.24	-0.04865	1.389
p-value	<0.0001	<0.0001	0.0028

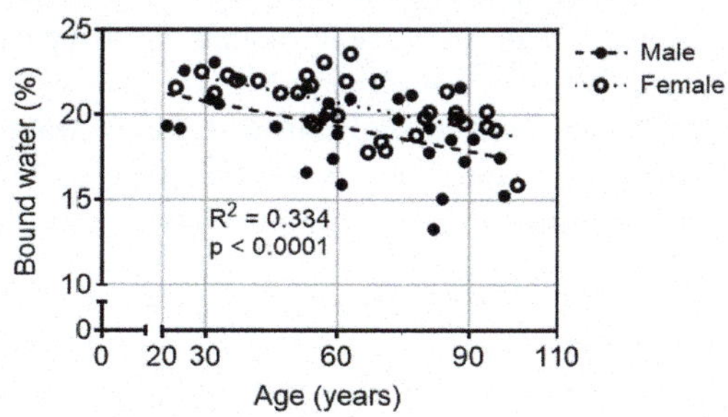

FIGURE 9.6 Age-related decrease in bound water.

The volume fraction of bound water was determined for 58 samples of cortical bone from female and male donors using ^{1}H NMR relaxometry. The level of bound water decreased with age, with bound water being higher in female donors than in male donors for a given age.

for 1 year compared to female dogs given the vehicle (Allen et al., 2017). There is evidence that raloxifene binds the bone matrix and improves bone toughness in cell-independent manner by sequestering water (Gallant et al., 2014). At present, there is no clinical evidence that osteoporosis medications like raloxifene and alendronate affect bound and pore water levels in human bone.

Several studies involving rodents indicate that aging, chronic kidney disease, glucocorticoid-induced osteoporosis (GIO), neurofibromatosis type 1 (NF1), and vascular disruption affect bound water. In both BALB/c mice and F344 rats, ex vivo BW measurements were lower in >20-mo animals than in 6-mo animals, while cortical tissue mineral density and AGEs were higher with age (Creecy et al., 2020; Uppuganti et al., 2016). When the *Nf1* gene was postnatally deleted in cells of the osteoblast lineage, BW (femur ex vivo) was lower in the conditional knockout mice than in the littermate controls, irrespective of sex, at 12-week and 20-week of age (Ahmed et al., 2022). Treating 10-week-old, male BALB/cJ mice with an anti-VEGF monoclonal antibody for 42 days decreased in vivo blood flow in the distal femur by 43% and decreased ex vivo BW in the whole femur by 14% when compared to age-, sex-, and strain-matched mice (Lane et al., 2019). Implanting 6-methylprednisolone pellets subcutaneously in 13-week-old male BALB/cJ mice for 60 and 120 days also decreased ex vivo BW compared to control mice (Dubrovsky et al., 2021). In a rat model of chronic kidney disease, ex vivo BW of the femur mid-diaphysis was lower at 35 weeks of age in the Cy/+ group with high bone turnover (high circulating levels of parathyroid hormone) than in the normal group (Allen, Newman, et al., 2015; Allen, Territo, et al., 2015). Overall, finding explanations for differences in BW in pre-clinical rodent studies is difficult because the manipulations tend have multiple effects on the bone tissue.

Strategies to increase the GAG content in bone ECM would presumably increase bound water and the overall fracture resistance of bone. In a proof-of-concept study, Jiang and colleagues intradermally administered chondroitin sulfate or vehicle for 4 weeks to wild-type mice and biglycan knock-out mice at 10-to-11-month of age and found that BW and nano-scratch toughness of the femur diaphysis was higher in chondroitin sulfate-treated than in vehicle-treated wild-type mice (Hua et al., 2020). This treatment effect was not observed in the knock-out mice presumably because this GAG could not bind to biglycan and attract water into the ECM. A similar administration of dermatan sulfate, another type of GAG, did not affect BW and nano-scratch toughness in wild-type mice (Hua et al., 2020).

8. Summary

Bone is full of water. Without water interacting with the organic matrix via hydrogen bonding and with the hard mineral phase via electrostatic bonding, bone is stiffer and stronger but less viscoelastic and less tough. In other words, water is essential to maintaining the resistance of bone to fracture. Water exists in two distinct states: (1) pore water flowing through vascular and canalicular channels and (2) bound water sequestered within the matrix at different bonding energy levels. Pore water is a surrogate of cortical porosity and can be non-destructively measured by proton nuclear magnetic resonance techniques including clinical MRI. It increases with age and negatively correlates with the fracture resistance of bone. Bound water can also be non-destructively measured by NMR and clinical MRI as well as by vibrational spectroscopy techniques. It decreases with age and positively correlates with the fracture resistance of bone. To date, the reason for the age-related decrease in bound water is not known, but several possibilities exists.

Acknowledgments

We thank Mustafa Unal, Ph.D., for contributing the graph in Figure 10.4 and Sasidhar Uppuganti, M.S., for contributing images and graphs in Figure 10.1, Figure 10.2, Figure 10.5, and Figure 10.6.

References

Ahmed, R., Uppuganti, S., Derasari, S., Meyer, J., Pennings, J. S., Elefteriou, F., & Nyman, J. S. (2022). Identifying bone matrix impairments in a mouse model of neurofibromatosis type 1 (NF1) by clinically translatable techniques. *Journal of Bone and Mineral Research, 37*(8), 1603–1621. https://doi.org/10.1002/jbmr.4633

Ailavajhala, R., Oswald, J., Rajapakse, C. S., & Pleshko, N. (2019). Environmentally-controlled near infrared spectroscopic imaging of bone water. *Scientific Reports, 9*(1). https://doi.org/10.1038/s41598-019-45897-3

Ailavajhala, R., Querido, W., Rajapakse, C. S., & Pleshko, N. (2020). Near infrared spectroscopic assessment of loosely and tightly bound cortical bone water. *Analyst, 145*(10), 3713–3724. https://doi.org/10.1039/c9an02491c

Aktas, N., Tülek, Y., & Gökalp, H. Y. (1997). Determination of freezable water content of beef semimembranous muscle DSC study. *Journal of Thermal Analysis, 48*(2), 259–266. https://doi.org/10.1007/bf01979269

Allen, M. R., McNerny, E., Aref, M., Organ, J. M., Newman, C. L., McGowan, B., Jang, T., Burr, D. B., Brown, D. M., Hammond, M., Territo, P. R., Lin, C., Persohn, S., Jiang, L., Riley, A. A., McCarthy, B. P., Hutchins, G. D., & Wallace, J. M. (2017). Effects of combination treatment with alendronate and raloxifene on skeletal properties in a beagle dog model. *PLoS One, 12*(8). https://doi.org/10.1371/journal.pone.0181750

Allen, M. R., Newman, C. L., Chen, N., Granke, M., Nyman, J. S., & Moe, S. M. (2015). Changes in skeletal collagen cross-links and matrix hydration in high- and low-turnover chronic kidney disease. *Osteoporosis International, 26*(3), 977–985. https://doi.org/10.1007/s00198-014-2978-9

Allen, M. R., Territo, P. R., Lin, C., Persohn, S., Jiang, L., Riley, A. A., McCarthy, B. P., Newman, C. L., Burr, D. B., & Hutchins, G. D. (2015). In vivo UTE-MRI reveals positive effects of raloxifene on skeletal-bound water in skeletally mature beagle dogs. *Journal of Bone and Mineral Research, 30*(8), 1441–1444. https://doi.org/10.1002/jbmr.2470

Arnold, J. S. (1960). Quantitation of mineralization of bone as an organ and tissue in osteoporosis. *Clinical Orthopaedics and Related Research, 17*.

Bae, W. C., Chen, P. C., Chung, C. B., Masuda, K., D'Lima, D., & Du, J. (2012). Quantitative ultrashort echo time (UTE) MRI of human cortical bone: Correlation with porosity and biomechanical properties. *Journal of Bone and Mineral Research, 27*(4), 848–857. https://doi.org/10.1002/jbmr.1535

Bella, J., Brodsky, B., & Berman, H. M. (1995). Hydration structure of a collagen peptide. *Structure, 3*(9), 893–906. https://doi.org/10.1016/S0969-2126(01)00224-6

Best, S. M., Duer, M. J., Reid, D. G., Wise, E. R., & Zou, S. (2008). Towards a model of the mineral-organic interface in bone: NMR of the structure of synthetic glycosaminoglycan- and polyaspartate-calcium phosphate composites. *Magnetic Resonance in Chemistry, 46* (4), 323–329. https://doi.org/10.1002/mrc.2168

Biswas, R., Bae, W., Diaz, E., Masuda, K., Chung, C. B., Bydder, G. M., & Du, J. (2012). Ultrashort echo time (UTE) imaging with bi-component analysis: Bound and free water evaluation of bovine cortical bone subject to sequential drying. *Bone, 50*(3), 749–755. https://doi.org/10.1016/j.bone.2011.11.029

Bonar, L. C., & Glimcher, M. J. (1970). Thermal denaturation of mineralized and demineralized bone collagens. *Journal of Ultrastructure Research, 32*(5–6), 545–557. https://doi.org/10.1016/S0022-5320(70)80027-2

Bonar, L. C., Lees, S., & Mook, H. A. (1985). Neutron diffraction studies of collagen in fully mineralized bone. *Journal of Molecular Biology, 181*(2), 265–270. https://doi.org/10.1016/0022-2836(85)90090-7

Brodsky, B., & Persikov, A. V. (2005). Molecular structure of the collagen triple helix. *Advances in Protein Chemistry, 70*, 301–339. https://doi.org/10.1016/S0065-3233(05)70009-7

Broz, J. J., Simske, S. J., Greenberg, A. R., & Luttges, M. W. (1993). Effects of rehydration state on the flexural properties of whole mouse long bones. *Journal of Biomechanical Engineering, 115*(4A), 447–449. https://doi.org/10.1115/1.2895510

Burstein, A. H., Zika, J. M., Heiple, K. G., & Klein, L. (1975). Contribution of collagen and mineral to the elastic-plastic properties of bone. *Journal of Bone and Joint Surgery, 57*(7), 956–961. https://doi.org/10.2106/00004623-197557070-00013

Carl, M., Bydder, G. M., & Du, J. (2016). UTE imaging with simultaneous water and fat signal suppression using a time-efficient multispoke inversion recovery pulse sequence. *Magnetic Resonance in Medicine, 76*(2), 577–582. https://doi.org/10.1002/mrm.25823

Chen, J., Carl, M., Ma, Y., Shao, H., Lu, X., Chen, B., Chang, E. Y., Wu, Z., & Du, J. (2016). Fast volumetric imaging of bound and pore water in cortical bone using three-dimensional ultrashort-TE (UTE) and inversion recovery UTE sequences. *NMR in Biomedicine, 29* (10), 1373–1380. https://doi.org/10.1002/nbm.3579

Chen, J., Grogan, S. P., Shao, H., D'Lima, D., Bydder, G. M., Wu, Z., & Du, J. (2015). Evaluation of bound and pore water in cortical bone using ultrashort-TE MRI. *NMR in Biomedicine, 28*(12), 1754–1762. https://doi.org/10.1002/nbm.3436

Creecy, A., Uppuganti, S., Girard, M. R., Schlunk, S. G., Amah, C., Granke, M., Unal, M., Does, M. D., & Nyman, J. S. (2020). The age-related decrease in material properties of BALB/c mouse long bones involves alterations to the extracellular matrix. *Bone, 130*. https://doi.org/10.1016/j.bone.2019.115126

Currey, J. D. (1964). Three analogies to explain the mechanical properties of bone. *Biorheology, 2*(1), 1–10. https://doi.org/10.3233/bir-1964-2101

Currey, J. D. (1988). The effects of drying and re-wetting on some mechanical properties of cortical bone. *Journal of Biomechanics, 21*(5), 439–441. https://doi.org/10.1016/0021-9290(88)90150-9

Davies, E., Müller, K. H., Wong, W. C., Pickard, C. J., Reid, D. G., Skepper, J. N., & Duer, M. J. (2014). Citrate bridges between mineral platelets in bone. *Proceedings of the National Academy of Sciences of the United States of America, 111*(14), Article E1354. https://doi.org/10.1073/pnas.1315080111

Dempster, W. T., & Liddicoat, R. T. (1952). Compact bone as a non-isotropic material. *American Journal of Anatomy, 91*(3), 331–362. https://doi.org/10.1002/aja.1000910302

Dillaman, R. M., Roer, R. D., & Gay, D. M. (1991). Fluid movement in bone: Theoretical and empirical. *Journal of Biomechanics, 24*(1), 163–177. https://doi.org/10.1016/0021-9290(91)90386-2

Dubrovsky, A. M., Nyman, J. S., Uppuganti, S., Chmiel, K. J., Kimmel, D. B., & Lane, N. E. (2021). Bone strength/bone mass discrepancy in glucocorticoid-treated adult mice. *JBMR Plus, 5*(3). https://doi.org/10.1002/jbm4.10443

Dwivedi, N., Dubey, R., Srivastava, S., & Sinha, N. (2022). Unraveling water-mediated 31 P relaxation in bone mineral. *ACS Omega, 7*(19), 16678–16688. https://doi.org/10.1021/acsomega.2c01133

Ellingham, S. T. D., Thompson, T. J. U., & Islam, M. (2015). Thermogravimetric analysis of property changes and weight loss in incinerated bone. *Palaeogeography, Palaeoclimatology, Palaeoecology, 438*, 239–244. https://doi.org/10.1016/j.palaeo.2015.08.009

Evans, F. G. (1973). Factors affecting the mechanical properties of bone. *Bulletin of the New York Academy of Medicine: Journal of Urban Health, 49*(9), 751–764.

Fantazzini, P., Brown, R. J. S., & Borgia, G. C. (2003). Bone tissue and porous media: Common features and differences studied by NMR relaxation. *Magnetic Resonance Imaging, 21*(3–4), 227–234. https://doi.org/10.1016/S0730-725X(03)00129-2

Flanagan, C. D., Unal, M., Akkus, O., & Rimnac, C. M. (2017). Raman spectral markers of collagen denaturation and hydration in human cortical bone tissue are affected by radiation sterilization and high cycle fatigue damage. *Journal of the Mechanical Behavior of Biomedical Materials, 75*, 314–321. https://doi.org/10.1016/j.jmbbm.2017.07.016

Gallant, M. A., Brown, D. M., Hammond, M., Wallace, J. M., Du, J., Deymier-Black, A. C., Almer, J. D., Stock, S. R., Allen, M. R., & Burr, D. B. (2014). Bone cell-independent benefits of raloxifene on the skeleton: A novel mechanism for improving bone material properties. *Bone, 61*, 191–200. https://doi.org/10.1016/j.bone.2014.01.009

Granke, M., Does, M. D., & Nyman, J. S. (2015). The role of water compartments in the material properties of cortical bone. *Calcified Tissue International, 97*(3), 292–307. https://doi.org/10.1007/s00223-015-9977-5

Granke, M., Makowski, A. J., Uppuganti, S., Does, M. D., & Nyman, J. S. (2015). Identifying novel clinical surrogates to assess human bone fracture toughness. *Journal of Bone and Mineral Research, 30*(7), 1290–1300. https://doi.org/10.1002/jbmr.2452

Haider, I. T., Speirs, A. D., & Frei, H. (2013). Effect of boundary conditions, impact loading and hydraulic stiffening on femoral fracture strength. *Journal of Biomechanics, 46*(13), 2115–2121. https://doi.org/10.1016/j.jbiomech.2013.07.004

Han, Y., Gomez, J., Hua, R., Xiao, P., Gao, W., Jiang, J. X., & Wang, X. (2021). Removal of glycosaminoglycans affects the in situ mechanical behavior of extrafibrillar matrix in bone. *Journal of the Mechanical Behavior of Biomedical Materials, 123*, Article 104766. https://doi.org/10.1016/j.jmbbm.2021.104766

Hashimoto, Y., Lester, G. E., Caterson, B., & Yamauchi, M. (1995). EDTA-insoluble, calcium-binding proteoglycan in bovine bone. *Calcified Tissue International, 56*(5), 398–402. https://doi.org/10.1007/bf00301609

Hong, A. L., Ispiryan, M., Padalkar, M. V., Jones, B. C., Batzdorf, A. S., Shetye, S. S., Pleshko, N., & Rajapakse, C. S. (2019). MRI-derived bone porosity index correlates to bone composition and mechanical stiffness. *Bone Reports, 11*. https://doi.org/10.1016/j.bonr.2019.100213

Horch, R. A., Gochberg, D. F., Nyman, J. S., & Does, M. D. (2011). Non-invasive predictors of human cortical bone mechanical properties: T2-discriminated 1H NMR compared with high resolution X-ray. *PLoS One, 6*(1). https://doi.org/10.1371/journal.pone.0016359

Horch, R. A., Gochberg, D. F., Nyman, J. S., & Does, M. D. (2012). Clinically compatible MRI strategies for discriminating bound and pore water in cortical bone. *Magnetic Resonance in Medicine, 68*(6), 1774–1784. https://doi.org/10.1002/mrm.24186

Horch, R. A., Nyman, J. S., Gochberg, D. F., Dortch, R. D., & Does, M. D. (2010). Characterization of1H NMR signal in human cortical bone for magnetic resonance imaging. *Magnetic Resonance in Medicine, 64*(3), 680–687. https://doi.org/10.1002/mrm.22459

Hua, R., Han, Y., Ni, Q., Fajardo, R. J., Iozzo, R. V., Ahmed, R., Nyman, J. S., Wang, X., & Jiang, J. X. (2025). Pivotal roles of biglycan and decorin in regulating bone mass, water retention, and bone toughness. *Bone Research, 13*(1). https://doi.org/10.1038/s41413-024-00380-2

Hua, R., Ni, Q., Eliason, T. D., Han, Y., Gu, S., Nicolella, D. P., Wang, X., & Jiang, J. X. (2020). Biglycan and chondroitin sulfate play pivotal roles in bone toughness via retaining bound water in bone mineral matrix. *Matrix Biology, 94*, 95–109. https://doi.org/10.1016/j.matbio.2020.09.002

Iozzo, R. V., & Schaefer, L. (2015). Proteoglycan form and function: A comprehensive nomenclature of proteoglycans. *Matrix Biology, 42*, 11–55. https://doi.org/10.1016/j.matbio.2015.02.003

Jacobson, A. M., Zhao, X., Sommer, S., Sadik, F., Warden, S. J., Newman, C., Siegmund, T., Allen, M. R., & Surowiec, R. K. (2024). A comprehensive set of ultrashort echo time magnetic resonance imaging biomarkers to assess cortical bone health: A feasibility study at clinical field strength. *Bone, 181*. https://doi.org/10.1016/j.bone.2024.117031

Jerban, S., Lu, X., Dorthe, E. W., Alenezi, S., Ma, Y., Kakos, L., Jang, H., Sah, R. L., Chang, E. Y., D'Lima, D., & Du, J. (2020). Correlations of cortical bone microstructural and mechanical properties with water proton fractions obtained from ultrashort echo time (UTE) MRI tricomponent T2* model. *NMR in Biomedicine, 33*(3). https://doi.org/10.1002/nbm.4233

Jerban, S., Ma, Y., Li, L., Jang, H., Wan, L., Guo, T., Searleman, A., Chang, E. Y., & Du, J. (2019b). Volumetric mapping of bound and pore water as well as collagen protons in cortical bone using 3D ultrashort echo time cones MR imaging techniques. *Bone, 127*, 120–128. https://doi.org/10.1016/j.bone.2019.05.038

Jerban, S., Ma, Y., Moazamian, D., Athertya, J., Dwek, S., Jang, H., Woods, G., Chung, C. B., Chang, E. Y., & Du, J. (2023). MRI-based porosity index (PI) and suppression ratio (SR) in the tibial cortex show significant differences between normal, osteopenic, and osteoporotic female subjects. *Frontiers in Endocrinology, 14*. https://doi.org/10.3389/fendo.2023.1148345

Jerban, S., Ma, Y., Wong, J. H., Nazaran, A., Searleman, A., Wan, L., Williams, J., Du, J., & Chang, E. Y. (2019a). Ultrashort echo time magnetic resonance imaging (UTE-MRI) of cortical bone correlates well with histomorphometric assessment of bone microstructure. *Bone, 123*, 8–17. https://doi.org/10.1016/j.bone.2019.03.013

Jones, B. C., Jia, S., Lee, H., Feng, A., Shetye, S. S., Batzdorf, A., Shapira, N., Noël, P. B., Pleshko, N., & Rajapakse, C. S. (2021). MRI-derived porosity index is associated with whole-bone stiffness and mineral density in human cadaveric femora. *Bone, 143*. https://doi.org/10.1016/j.bone.2020.115774

Jones, B. C., Lee, H., Cheng, C. C., al Mukaddam, M., Song, H. K., Snyder, P. J., Kamona, N., Rajapakse, C. S., & Wehrli, F. W. (2023). MRI quantification of cortical bone porosity, mineralization, and morphologic structure in postmenopausal osteoporosis. *Radiology, 307*(2). https://doi.org/10.1148/radiol.221810

Kerch, G. (2018). Distribution of tightly and loosely bound water in biological macromolecules and age-related diseases. *International Journal of Biological Macromolecules, 118*(Pt A), 1310–1318. https://doi.org/10.1016/j.ijbiomac.2018.06.187

Ketsiri, T., Uppuganti, S., Harkins, K. D., Gochberg, D. F., Nyman, J. S., & Mark, D. (2023). Does, finite element analysis of bone mechanical properties using MRI-derived bound and pore water concentration maps. *Computer Methods in Biomechanics and Biomedical Engineering, 26*(8), 905–916. https://doi.org/10.1080/10255842.2022.2098016

Knothe Tate, M. L., Niederer, P., & Knothe, U. (1998). In vivo tracer transport through the lacunocanalicular system of rat bone in an environment devoid of mechanical loading. *Bone, 22*(2), 107–117. https://doi.org/10.1016/s8756-3282(97)00234-2

Kronick, P. L., & Cooke, P. (1996). Thermal stabilization of collagen fibers by calcification. *Connective Tissue Research, 33*(4), 275–282. https://doi.org/10.3109/03008209609028885

Lakes, R. S., & Katz, J. L. (1979). Viscoelastic properties of wet cortical bone-II. Relaxation mechanisms. *Journal of Biomechanics, 12*(9), 679–687. https://doi.org/10.1016/0021-9290(79)90017-4

Lane, N. E., Nyman, J. S., Uppuganti, S., Chaudhari, A. J., Aguirre, J. I., Shidara, K., Liu, X. P., Yao, W., & Kimmel, D. B. (2019). Inhibition of vascular endothelial growth factor in young adult mice causes low bone blood flow and bone strength with no effect on bone mass in trabecular regions. *Bone Reports, 10*, Article 100210. https://doi.org/10.1016/j.bonr.2019.100210

Le Pense, S., & Chen, Y. (2017). Contribution of fluid in bone extravascular matrix to strain-rate dependent stiffening of bone tissue – A poroelastic study. *Journal of the Mechanical Behavior of Biomedical Materials, 65*, 90–101. https://doi.org/10.1016/j.jmbbm.2016.08.016

Lees, S., & Prostak, K. (1988). The locus of mineral crystallites in bone. *Connective Tissue Research, 18*(1), 41–54. https://doi.org/10.3109/03008208809019071

Li, Y., Liang, X., Liu, J., & Ma, Y. (2024). Assessment of osteoporosis at the lumbar spine using ultrashort echo time magnetization transfer (UTE-MT) MRI. *Journal of Magnetic Resonance Imaging: JMRI., 59*(4), 1285–1298. https://doi.org/10.1002/jmri.28910

Li, C., Seifert, A. C., Rad, H. S., Bhagat, Y. A., Rajapakse, C. S., Sun, W., Lam, S. C. B., & Wehrli, F. W. (2014). Cortical bone water concentration: Dependence of MR imaging measures on age and pore volume fraction. *Radiology, 272*(3), 796–806. https://doi.org/10.1148/radiol.14132585

Lievers, W. B., Poljsak, A. S., Waldman, S. D., & Pilkey, A. K. (2010). Effects of dehydration-induced structural and material changes on the apparent modulus of cancellous bone. *Medical Engineering & Physics, 32*(8), 921–925. https://doi.org/10.1016/j.medengphy.2010.06.001

Lozano, L. F., Peña-Rico, M. A., Heredia, A., Ocotlán-Flores, J., Gómez-Cortés, A., Velázquez, R., Belío, I. A., & Bucio, L. (2003). Thermal analysis study of human bone. *Journal of Materials Science, 38*(23), 4777–4782. https://doi.org/10.1023/A:1027483220584

Lu, X., Jerban, S., Wan, L., Ma, Y., Jang, H., Le, N., Yang, W., Chang, E. Y., & Du, J. (2019). Three-dimensional ultrashort echo time imaging with tricomponent analysis for human cortical bone. *Magnetic Resonance in Medicine, 82*(1), 348–355. https://doi.org/10.1002/mrm.27718

Maeda, H., & Fukada, E. (1982). Effect of water on piezoelectric, dielectric, and elastic properties of bone. *Biopolymers, 21*(10), 2055–2068. https://doi.org/10.1002/bip.360211010

Manhard, M. K., Harkins, K. D., Gochberg, D. F., Nyman, J. S., & Does, M. D. (2017). 30-Second bound and pore water concentration mapping of cortical bone using 2D UTE with optimized half-pulses. *Magnetic Resonance in Medicine, 77*(3), 945–950. https://doi.org/10.1002/mrm.26605

Manhard, M. K., Horch, R. A., Gochberg, D. F., Nyman, J. S., & Does, M. D. (2015). Vivo quantitative MR imaging of bound and pore water in cortical bone1. *Radiology, 277*(1), 221–229. https://doi.org/10.1148/radiol.2015140336

Manhard, M. K., Horch, R. A., Harkins, K. D., Gochberg, D. F., Nyman, J. S., & Does, M. D. (2014). Validation of quantitative bound- and pore-water imaging in cortical bone. *Magnetic Resonance in Medicine, 71*(6), 2166–2171. https://doi.org/10.1002/mrm.24870

Manhard, M. K., Nyman, J. S., & Does, M. D. (2017). Advances in imaging approaches to fracture risk evaluation. *Translational Research, 181*, 1–14. https://doi.org/10.1016/j.trsl.2016.09.006

Manhard, M. K., Uppuganti, S., Granke, M., Gochberg, D. F., Nyman, J. S., & Does, M. D. (2016). MRI-derived bound and pore water concentrations as predictors of fracture resistance. *Bone, 87*, 1–10. https://doi.org/10.1016/j.bone.2016.03.007

Marino, A. A., Becker, R. O., & Bachman, C. H. (1967). Dielectric determination of bound water of bone. *Physics in Medicine and Biology, 12*(3), 367–378. https://doi.org/10.1088/0031-9155/12/3/309

Masson, L. E., O'Brien, C. M., Pence, I. J., Herington, J. L., Reese, J., Van Leeuwen, T. G., & Mahadevan-Jansen, A. (2018). Dual excitation wavelength system for combined fingerprint and high wavenumber Raman spectroscopy. *Royal Society of Chemistry, United States Analyst, 143*(24), 6049–6060. https://doi.org/10.1039/c8an01989d

Mazurkiewicz, A. (2018). The effect of trabecular bone storage method on its elastic properties. *Acta of Bioengineering and Biomechanics, 20*(1), 21–27. https://doi.org/10.5277/ABB-00967-2017-03

McCalden, R. W., McGeough, J. A., Barker, M. B., & Court-Brown, C. M. (1993). Age-related changes in the tensile properties of cortical bone. The relative importance of changes in porosity, mineralization, and microstructure. *Journal of Bone and Joint Surgery, 75*(8), 1193–1205. https://doi.org/10.2106/00004623-199308000-00009

McNally, E., Nan, F., Botton, G. A., & Schwarcz, H. P. (2013). Scanning transmission electron microscopic tomography of cortical bone using Z-contrast imaging. *Micron, 49*, 46–53. https://doi.org/10.1016/j.micron.2013.03.002

Menard, K., & Menard, N. (2020). *Dynamic mechanical analysis*. https://doi.org/10.1201/9780429190308

Milazzo, M., David, A., Jung, G. S., Danti, S., & Buehler, M. J. (2021). Molecular origin of viscoelasticity in mineralized collagen fibrils. *Biomaterials Science, 9*(9), 3390–3400. https://doi.org/10.1039/d0bm02003f

Mkukuma, L. D., Imrie, C. T., Skakle, J. M. S., Hukins, D. W. L., & Aspden, R. M. (2005). Thermal stability and structure of cancellous bone mineral from the femoral head of patients with osteoarthritis or osteoporosis. *Annals of the Rheumatic Diseases, 64*(2), 222–225. https://doi.org/10.1136/ard.2004.021329

Morris, M. A., Lopez-Curto, J. A., Hughes, S. P. F., An, K. N., Bassingthwaighte, J. B., & Kelly, P. J. (1982). Fluid spaces in canine bone and marrow. *Microvascular Research, 23*(2), 188–200. https://doi.org/10.1016/0026-2862(82)90064-4

Mueller, K. H., Trias, A., & Ray, R. D. (1966). Bone density and composition: Age-related and pathological changes in water and mineral content. *JBJS, 48*(1).

Nicolella, D. P., Ni, Q., & Chan, K. S. (2011). Non-destructive characterization of microdamage in cortical bone using low field pulsed NMR. *Journal of the Mechanical Behavior of Biomedical Materials, 4*(3), 383–391. https://doi.org/10.1016/j.jmbbm.2010.11.007

Nyman, J. S., Gorochow, L. E., Adam Horch, R., Uppuganti, S., Zein-Sabatto, A., Manhard, M. K., & Does, M. D. (2013). Partial removal of pore and loosely bound water by low-energy drying decreases cortical bone toughness in young and old donors. *Journal of the Mechanical Behavior of Biomedical Materials, 22*, 136–145. https://doi.org/10.1016/j.jmbbm.2012.08.013

Nyman, J. S., Ketsiri, T., Louie, E. A., Harkins, K. D., Manhard, M. K., Gochberg, D. F., Lee, D. H., Desai, M. J., Maslow, J., Tanner, S. B., & Does, M. D. (2023). Toward the use of MRI measurements of bound and pore water in fracture risk assessment. *Bone, 176*. https://doi.org/10.1016/j.bone.2023.116863

Nyman, J. S., Ni, Q., Nicolella, D. P., & Wang, X. (2008). Measurements of mobile and bound water by nuclear magnetic resonance correlate with mechanical properties of bone. *Bone, 42*(1), 193–199. https://doi.org/10.1016/j.bone.2007.09.049

Nyman, J. S., Roy, A., Shen, X., Acuna, R. L., Tyler, J. H., & Wang, X. (2006). The influence of water removal on the strength and toughness of cortical bone. *Journal of Biomechanics, 39*(5), 931–938. https://doi.org/10.1016/j.jbiomech.2005.01.012

Nyman, J. S., Uppuganti, S., Unal, M., Leverant, C. J., Adabala, S., Granke, M., Voziyan, P., & Does, M. D. (2019). Manipulating the amount and structure of the organic matrix affects the water compartments of human cortical bone. *JBMR Plus, 3*(6). https://doi.org/10.1002/jbm4.10135

Ong, H. H., Wright, A. C., & Wehrli, F. W. (2012). Deuterium nuclear magnetic resonance unambiguously quantifies pore and collagen-bound water in cortical bone. *Journal of Bone and Mineral Research, 27*(12), 2573–2581. https://doi.org/10.1002/jbmr.1709

Ozaki, Y. (2021). Infrared Spectroscopy—Mid-infrared, near-infrared, and far-infrared/terahertz spectroscopy. *Analytical Sciences, 37*(9), 1193–1212. https://doi.org/10.2116/analsci.20r008

Padalkar, M. V., Spencer, R. G., & Pleshko, N. (2013). Near infrared spectroscopic evaluation of water in hyaline cartilage. *Annals of Biomedical Engineering, 41*(11), 2426–2436. https://doi.org/10.1007/s10439-013-0844-0

Porrelli, D., Abrami, M., Pelizzo, P., Formentin, C., Ratti, C., Turco, G., Grassi, M., Canton, G., Grassi, G., & Murena, L. (2022). Trabecular bone porosity and pore size distribution in osteoporotic patients – A low field nuclear magnetic resonance and microcomputed tomography investigation. *Journal of the Mechanical Behavior of Biomedical Materials, 125*, Article 104933. https://doi.org/10.1016/j.jmbbm.2021.104933

Qingwen, N., Rui, H., Holland, D., Tinajero, A., Yan, H., Jiang, J. X., & Xiaodu, W. (2021). Characterization of microstructural changes on Biglycan induced mice bone by low-field nuclear magnetic resonance. *Applied Physics, 4*(2), 58. https://doi.org/10.31058/j.ap.2021.42004

Qiu, J., Liao, Z., Xiang, H., Li, H., Yuan, D., Jiang, C., Xie, J., Qin, M., Li, K., & Zhao, H. (2023). Effects of different preservation on the mechanical properties of cortical bone under quasi-static and dynamic compression. *Frontiers in Bioengineering and Biotechnology, 11*. https://doi.org/10.3389/fbioe.2023.1082254

Querido, W., Kandel, S., & Pleshko, N. (2021). Applications of vibrational spectroscopy for analysis of connective tissues. *Molecules, 26*(4), 922. https://doi.org/10.3390/molecules26040922

Querido, W., Shanas, N., Radway, A. P., Jones, B. C., Ispiryan, M., Zhao, H., Hast, M. W., Rajapakse, C. S., & Pleshko, N. (2025). The multifactorial relationship between bone tissue water and stiffness at the proximal femur. *Calcified Tissue International, 116*(1). https://doi.org/10.1007/s00223-024-01327-9

Rai, R. K., Barbhuyan, T., Singh, C., Mittal, M., Khan, M. P., Sinha, N., & Chattopadhyay, N. (2013). Total water, phosphorus relaxation and inter-atomic organic to inorganic interface are new determinants of trabecular bone integrity. *PLoS One, 8*(12). https://doi.org/10.1371/journal.pone.0083478

Rajapakse, C. S., Bashoor-Zadeh, M., Li, C., Sun, W., Wright, A. C., & Wehrli, F. W. (2015). Volumetric cortical bone porosity assessment with MR imaging: Validation and clinical feasibility. *Radiology, 276*(2), 526–535. https://doi.org/10.1148/radiol.15141850

Rajapakse, C. S., Padalkar, M. V., Yang, H. J., Ispiryan, M., & Pleshko, N. (2017). Nondestructive NIR spectral imaging assessment of bone water: Comparison to MRI measurements. *Bone, 103*, 116–124. https://doi.org/10.1016/j.bone.2017.06.015

Rees, S. G., Hughes Wassell, D. T., Waddington, R. J., & Embery, G. (2001). Interaction of bone proteoglycans and proteoglycan components with hydroxyapatite. *Biochimica et Biophysica Acta - General Subjects, 1568*(2), 118–128. https://doi.org/10.1016/S0304-4165(01)00209-4

Robinson, W. (1931). Free and bound water determinations by the heat of fusion of ice method. *Journal of Biological Chemistry, 92*(3), 699–709. https://doi.org/10.1016/s0021-9258(17)32613-3

Robinson, R. A. (1975). Physicochemical structure of bone. *Clinical Orthopaedics and Related Research, 112*, 263–315. https://doi.org/10.1097/00003086-197510000-00034

Robinson, R. A. (1979). Bone tissue: Composition and function. *Johns Hopkins Medical Journal, 145*(1), 10–24.

Samantray, S., Olubiyi, O. O., & Strodel, B. (2021). The influences of sulphation, salt type, and salt concentration on the structural heterogeneity of glycosaminoglycans. *International Journal of Molecular Sciences, 22*(21), Article 11529. https://doi.org/10.3390/ijms222111529

Samuel, J., Park, J. S., Almer, J., & Wang, X. (2016). Effect of water on nanomechanics of bone is different between tension and compression. *Journal of the Mechanical Behavior of Biomedical Materials, 57*, 128–138. https://doi.org/10.1016/j.jmbbm.2015.12.001

Samuel, J., Sinha, D., Zhao, J. C. G., & Wang, X. (2014). Water residing in small ultrastructural spaces plays a critical role in the mechanical behavior of bone. *Bone, 59*, 199–206. https://doi.org/10.1016/j.bone.2013.11.018

Schwarcz, H. P., McNally, E. A., & Botton, G. A. (2014). Dark-field transmission electron microscopy of cortical bone reveals details of extrafibrillar crystals. *Journal of Structural Biology, 188*(3), 240–248. https://doi.org/10.1016/j.jsb.2014.10.005

Sedlin, E. D., & Hirsch, C. (1966). Factors affecting the determination of the physical properties of femoral cortical bone. *Acta Orthopaedica, 37*(1), 29–48. https://doi.org/10.3109/17453676608989401

Seifert, A. C., Wehrli, S. L., & Wehrli, F. W. (2015). Bi-component T2* analysis of bound and pore bone water fractions fails at high field strengths. *NMR in Biomedicine, 28*(7), 861–872. https://doi.org/10.1002/nbm.3305

Shah, F. A. (2025). Revisiting the physical and chemical nature of the mineral component of bone. *Acta Biomaterialia, 196*, 1–16. https://doi.org/10.1016/j.actbio.2025.01.055

Shanas, N., Querido, W., Dumont, A., Yonko, E., Carter, E., Ok, J., Karchner, J. P., Barbe, M. F., Ali, S., Patil, C., Raggio, C., & Pleshko, N. (2020). Clinical application of near infrared fiber optic spectroscopy for noninvasive bone assessment. *Journal of Biophotonics, 13*(4). https://doi.org/10.1002/jbio.201960172

Shanas, No'ad, Querido, W., Oswald, J., Jepsen, K., Carter, E., Raggio, C., & Pleshko, N. (2022). Infrared spectroscopy-determined bone compositional changes associated with anti-resorptive treatment of the oim/oim mouse model of osteogenesis imperfecta. *Applied Spectroscopy, 76*(4), 416–427. https://doi.org/10.1177/00037028211055477

Shin, M., Zhang, M., vom Scheidt, A., Pelletier, M. H., Walsh, W. R., Martens, P. J., Kruzic, J. J., Busse, B., & Gludovatz, B. (2022). Impact of test environment on the fracture resistance of cortical bone. *Journal of the Mechanical Behavior of Biomedical Materials, 129*. https://doi.org/10.1016/j.jmbbm.2022.105155

Smith, J. W., & Walmsley, R. (1959). Factors affecting the elasticity of bone. *Journal of Anatomy, 93*, 503–523.

Stefan, U., Michael, B., & Werner, S. (2010). Effects of three different preservation methods on the mechanical properties of human and bovine cortical bone. *Bone, 47*(6), 1048–1053. https://doi.org/10.1016/j.bone.2010.08.012

Surowiec, R. K., Allen, M. R., & Wallace, J. M. (2022). Bone hydration: How we can evaluate it, what can it tell us, and is it an effective therapeutic target? *Bone Reports, 16*. https://doi.org/10.1016/j.bonr.2021.101161

Surowiec, R. K., Does, M. D., & Nyman, J. S. (2024). In vivo assessment of bone quality without X-rays. *Current Osteoporosis Reports, 22*(1), 56–68. https://doi.org/10.1007/s11914-023-00856-w

Tate, M. L. K., Steck, R., Forwood, M. R., & Niederer, P. (2000). In vivo demonstration of load-induced fluid flow in the rat Tibia and its potential implications for processes associated with functional adaptation. *Journal of Experimental Biology, 203*(18), 2737–2745. https://doi.org/10.1242/jeb.203.18.2737

Taylor, E. A., & Donnelly, E. (2020). Raman and fourier transform infrared imaging for characterization of bone material properties. *Bone, 139*, Article 115490. https://doi.org/10.1016/j.bone.2020.115490

Timmins, P. A., & Wall, J. C. (1977). Bone water. *Calcified Tissue Research, 23*(1), 1–5. https://doi.org/10.1007/bf02012759

Townsend, P. R., Rose, R. M., & Radin, E. L. (1975). Buckling studies of single human trabeculae. *Journal of Biomechanics, 8*(3–4), 199–201.

Unal, M., Ahmed, R., Mahadevan-Jansen, A., & Nyman, J. S. (2021). Compositional assessment of bone by raman spectroscopy. *The Analyst, 146*(24), 7464–7490. https://doi.org/10.1039/d1an01560e

Unal, M., & Akkus, O. (2015). Raman spectral classification of mineral- and collagen-bound water's associations to elastic and post-yield mechanical properties of cortical bone. *Bone, 81*, 315–326. https://doi.org/10.1016/j.bone.2015.07.024

Unal, M., Cingoz, F., Bagcioglu, C., Sozer, Y., & Akkus, O. (2018). Interrelationships between electrical, mechanical and hydration properties of cortical bone. *Journal of the Mechanical Behavior of Biomedical Materials, 77*, 12–23. https://doi.org/10.1016/j.jmbbm.2017.08.033

Unal, M., Yang, S., & Akkus, O. (2014). Molecular spectroscopic identification of the water compartments in bone. *Bone, 67*, 228–236. https://doi.org/10.1016/j.bone.2014.07.021

Uppuganti, S., Granke, M., Makowski, A. J., Does, M. D., & Nyman, J. S. (2016). Age-related changes in the fracture resistance of male Fischer F344 rat bone. *Bone, 83*, 220–232. https://doi.org/10.1016/j.bone.2015.11.009

Wang, X., Hua, R., Ahsan, A., Ni, Q., Huang, Y., Gu, S., & Jiang, J. X. (2018). Age-related deterioration of bone toughness is related to diminishing amount of matrix glycosaminoglycans (GAGs). *JBMR Plus, 2*(3), 164–173. https://doi.org/10.1002/jbm4.10030

Wang, X., & Ni, Q. (2003). Determination of cortical bone porosity and pore size distribution using a low field pulsed NMR approach. *Journal of Orthopaedic Research, 21*(2), 312–319. https://doi.org/10.1016/s0736-0266(02)00157-2

Wang, Y., Von Euw, S., Fernandes, F. M., Cassaignon, S., Selmane, M., Laurent, G., Pehau-Arnaudet, G., Coelho, C., Bonhomme-Coury, L., Giraud-Guille, M. M., Babonneau, F., Azaïs, T., & Nassif, N. (2013). Water-mediated structuring of bone apatite. *Nature Materials, 12*(12), 1144–1153. https://doi.org/10.1038/nmat3787

Wang, X., Xu, H., Huang, Y., Gu, S., & Jiang, J. X. (2016). Coupling effect of water and proteoglycans on the in situ toughness of bone. *Journal of Bone and Mineral Research, 31*(5), 1026–1029. https://doi.org/10.1002/jbmr.2774

Wilson, E. E., Awonusi, A., Morris, M. D., Kohn, D. H., Tecklenburg, M. M. J., & Beck, L. W. (2005). Highly ordered interstitial water observed in bone by nuclear magnetic resonance. *Journal of Bone and Mineral Research, 20*(4), 625–634. https://doi.org/10.1359/JBMR.041217

Xiong, Y., He, T., Wang, Y., Liu, W. V., Hu, S., Zhang, Y., Wen, D., Hou, B., Li, Y., Zhang, P., Liu, J., He, F., & Li, X. (2021). CKD stages, bone metabolism markers, and cortical porosity index: Associations and mediation effects analysis. *Frontiers in Endocrinology, 12*. https://doi.org/10.3389/fendo.2021.775066

Yamashita, J., Furman, B. R., Rawls, H. R., Wang, X., & Agrawal, C. M. (2001). The use of dynamic mechanical analysis to assess the viscoelastic properties of human cortical bone. *Journal of Biomedical Materials Research, 58*(1), 47–53. https://doi.org/10.1002/1097-4636(2001)58:1<47::AID-JBM70>3.0.CO;2-U

Yamashita, J., Li, X., Furman, B. R., Ralph Rawls, H., Wang, X., & Mauli Agrawal, C. (2002). Collagen and bone viscoelasticity: A dynamic mechanical analysis. *Journal of Biomedical Materials Research, 63*(1), 31–36. https://doi.org/10.1002/jbm.10086

CHAPTER

Determinants of whole bone strength

10

Muyin Huang[1] and Marjolein van der Meulen[1, 2, 3]

[1]*Cornell University, Meinig School of Biomedical Engineering, Ithaca, NY, United States;* [2]*Cornell University, Sibley School of Mechanical & Aerospace Engineering, Ithaca, NY, United States;* [3]*Hospital for Special Surgery, New York, NY, United States*

1. Overview of bone mechanical competence

The musculoskeletal system is the loadbearing and locomotor structure for vertebrates. The skeleton bears high loads and is developed uniquely to fulfill its function, which includes supporting motion, bearing functional loads, and protecting our vital organs. Pathologies such as osteoporosis disrupt the ability of the skeleton, particularly sites with cancellous bone, to bear functional loads resulting in fracture and even death. Influences on this structural function include bone material quantity, architecture, and tissue properties. The influence of these factors individually and collectively determines the whole bone and tissue mechanical behavior. Understanding skeletal structure and function and using this knowledge to predict failure is of great clinical relevance to the etiology of skeletal pathologies and the design of therapies. Much research has been focused on osteoporosis and predicting fracture risk, but similar issues are present for skeletal diseases such as osteogenesis imperfecta and bone adaptation to implants.

Failure of a whole bone, in the form of fracture, can occur as the result of a single overload or fatigue. In fatigue, failure occurs as the result of damage accumulation following multiple sub-failure loads. Here, our focus will be on failure following a single high load. Whether a loaded structure will fail is determined by how the applied load compares to the structural strength, or maximum load, that the structure can bear. When the applied load exceeds the structural strength of a whole bone, fracture will occur. Maintaining bone strength is essential because fractures have profound effects on mortality. Hip fracture is associated with a threefold increase in mortality risk compared to the general population. Affected patients also exhibit a significantly elevated risk of premature death, often attributable to complications such as immobility, comorbid conditions, and extended hospitalization (Abrahamsen et al., 2009; Panula et al., 2011).

The load-bearing function of the skeleton depends on several intrinsic factors: total bone mass, bone geometry and tissue material properties. From a whole bone perspective, the material consists of two different tissues: cortical and cancellous bone. In addition to these intrinsic factors, bone strength is affected by a host of

Biomechanics and Mechanobiology of Bone. https://doi.org/10.1016/B978-0-443-26739-0.00017-X

extrinsic factors including sex, age, diet and disease. Together intrinsic and extrinsic factors determine whole bone strength.

In recent years, advances in biomechanics, imaging, and molecular biology have expanded our understanding of the determinants of bone strength. This chapter aims to provide a comprehensive overview of the state-of-the-art in bone mechanical competence, including intrinsic and extrinsic factors, methodologies for assessing bone strength, as well as mechanical behavior under stress and failure characteristics, culminating in a discussion of challenges and future trends in this critical field (Fig. 10.1).

1.1 Evaluating whole bone strength

Whole bone strength is an important indicator of bone mechanical competence and overall bone health. Mechanical testing of whole bones is the direct method to measure the load-bearing function of bone. In an ex vivo research setting, whole bone structural properties are measured by applying loads to intact bone samples. The applied loads can be compression, bending or torsion. For all loading modes, the outcomes depend on bone size and material properties. Key outcome measures include

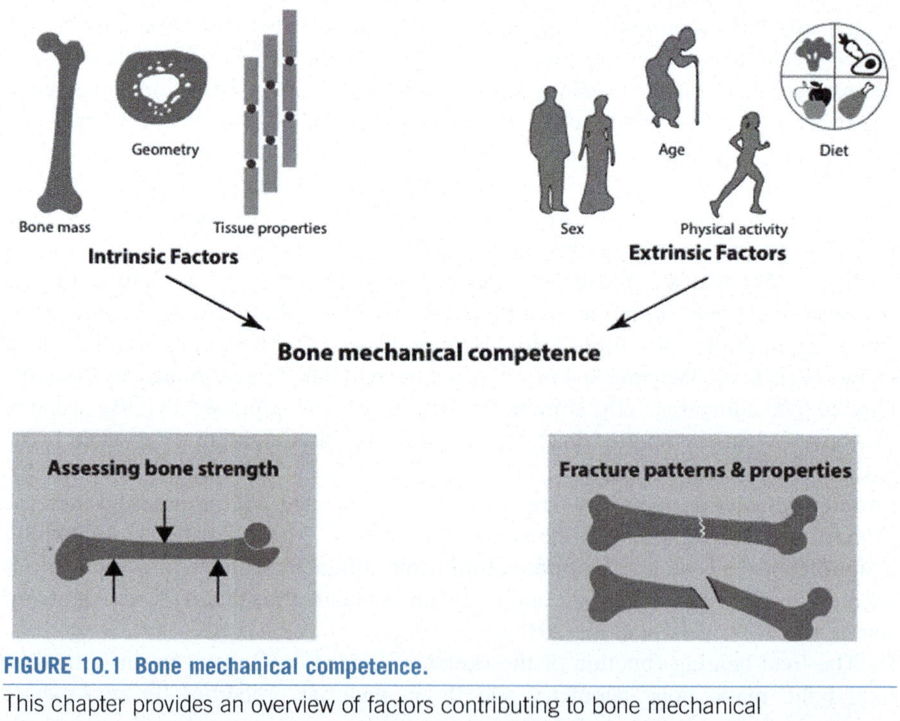

FIGURE 10.1 Bone mechanical competence.

This chapter provides an overview of factors contributing to bone mechanical competence, including intrinsic and extrinsic factors. Also discussed are key approaches for assessing bone strength and mechanical behavior and fracture characteristics.

the stiffness, the resistance of the whole bone to deformation for a given applied load, and structural strength, the amount of load required to fail the whole bone for a given applied loading mode.

Whole bone mechanical tests can be performed in axial loading, bending, and torsion (Fig. 10.2). The different testing methods are all performed by applying an increasing load to a bone sample with a load frame while recording the displacement and applied load. The differences lie in the loading mode: axial loading is performed in tension or compression on bone specimens that would physiologically experience axial loading, such as vertebrae. Relevant bending loading for assessing bone strength includes three-point and four-point bending. Bending tests are usually performed on long bone samples to measure bone strength of the cortical diaphysis. Torsional loading also is used for long bone specimens by applying a torsional moment about the long axis of the bone.

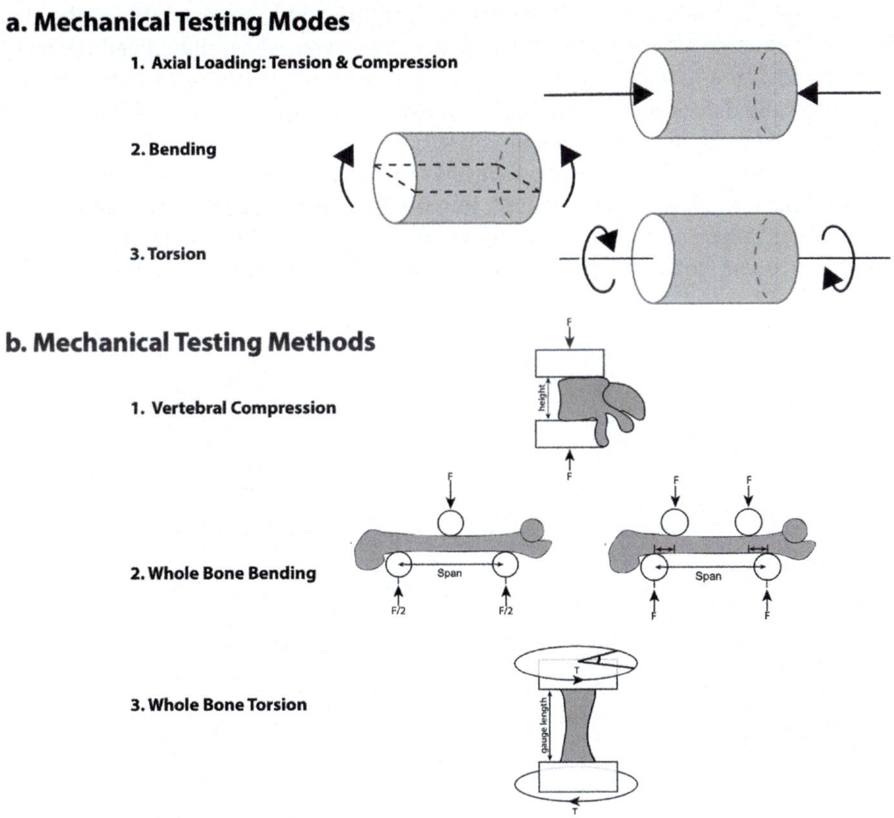

FIGURE 10.2 Mechanical testing

(a) Whole bone mechanical testing modes include axial loading, bending and torsion. (b) Examples of whole bone mechanical testing methods include vertebral compression, whole bone three- and four-point bending, and whole bone torsion.

The selection of a specific mechanical test for whole bone structural performance is influenced by factors such as the research question, physiological loading conditions, bone geometry, and ease of testing. Bones with prismatic cross sections and little curvature help minimize errors in strength and stiffness calculations. Curved bones introduce bending moments under compression, affect the stress distribution in bending, and increase shear stress in torsion. The distinct mechanical behavior of different testing modes influences the selection of testing methods. Torsion tests apply a constant moment along the test gauge length, thereby enabling failure at the weakest point along the entire length, and are unaffected by cross-sectional orientation, whereas bending tests are influenced by the direction of loading relative to the cross-sectional geometry. Four-point bending is preferable to three-point bending, because the maximum moment is constant between the two central loading points, and failure will occur at the weakest location within the central gauge section (Brodt et al., 1999). However, based on material test standards for conventional materials, such as those of ASTM International, whole bones are not sufficiently long for their cross-sectional width, introducing transverse shear into bending tests. Bending tests are more suitable for analyzing post-yield behavior, whereas torsion tests typically have linear load-displacement curves up to failure, without post-yield behavior (Burstein & Frankel, 1971).

Whole bone mechanical testing measures the applied load and displacement until the bone fails, a destructive procedure. Regardless of the specific loading mode chosen, all tests have similar features and outcome measures (Fig. 10.3). A monotonically increasing load is applied in compression, bending or torsion, and the

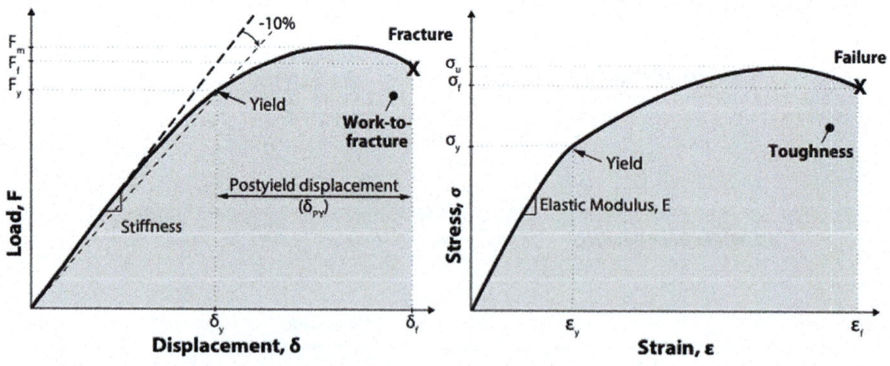

FIGURE 10.3 Load-displacement and stress-strain curves

(left) A load–displacement curve from whole-bone testing including the stiffness, maximum/ultimate load (F_m), failure load (F_f), yield load (F_y), yield displacement (δ_y), failure displacement (δ_f), and postyield displacement (δ_{py}). The area under the curve is the work to cause failure (work-to-failure). **(Right)** Stress-strain curve from tissue mechanical testing with strain (ε, a unitless measure of deformation), ultimate stress (σ_u), failure stress (σ_f), yield stress (σ_y), yield strain (ε_y), and failure strain (ε_f). Elastic modulus (E) is the slope of the linear region. Toughness is the area under the curve.

corresponding displacement is recorded. From the load (F) and displacement (δ) data, parameters such as structural or whole bone stiffness, and yield, ultimate or maximum, and failure loads (F_y, F_m, F_f), and yield, failure, and post-yield displacement (δ_y, δ_f, δ_{py}) can be either measured or calculated.

Whole bone stiffness describes the relationship between applied load and bone deformation, which is the slope of the linear region of the load-displacement curve. The stiffness is the amount of deformation upon application of a specific load and is the resistance of the whole bone to deformation for applied loads below the level of the yield load. The linear region before the yield point is referred to as the elastic region, during which the applied load does not permanently deform the bone, and the bone will return to its original shape when unloaded.

The point at which the curve changes from linear to non-linear behavior is the yield point. The load-displacement curve transitions from a linear region to a non-linear region (inelastic region), the stiffness decreases, and the curve flattens due to less increase in load. Loads greater than the yield load will subject the bone to permanent damage, thereby losing its ability to return to its original shape when the load is removed. The yield point is not directly measured, but rather calculated. One common method to determine yield is by plotting the regression line that represents a 10% loss in stiffness. By this method, the yield point is the intercept of the offset regression line with the load-displacement curve.

As the applied load increases beyond the yield load, further increases in load lead to more deformation in the bone tissue until failure. The applied load at which the bone fails is the fracture load, generally referred to as the whole bone strength. The displacement between yield (δ_y) and fracture (δ_f) is the post-yield displacement (PYD, $\delta_{py} = \delta_f - \delta_y$). The absence of PYD indicates failure without yielding, known as brittle behavior. A ductile failure has a large PYD and continues to deform after yielding without failing as the applied load increases. The area under the load-displacement curve is the work-to-fracture or energy-to-fracture, which represents the energy required to deform and fail the bone.

1.2 Evaluating bone tissue-level mechanical properties

When a whole bone is loaded, the material within the structure is subject to loads at the microscale. The ability of the material, either cortical or cancellous bone tissue, to bear these loads depends on the tissue-level mechanical properties. These tissue mechanical properties are independent of bone size or whole bone geometry. The tissue-level force is referred to stress (σ), the internal force exerted within a tissue per unit area, $\sigma = F/A$. The tissue-level deformation due to the applied force is the strain (ε), generally presented as the percentage change of the original length of the specimen, $\varepsilon = \delta/L$. Stress and strain can be tensile, compressive or shear. The stress-strain behavior is a fundamental characteristic of a material when loaded.

Similar to whole bone tests, stress and strain can be measured using small tissue samples of known geometry and homogenous material with loads applied in tension, compression or shear. These tests can be used to measure the tissue-level mechanical properties of either cortical or trabecular tissue (Fig. 10.3). The maximum stress is the ultimate stress (σ_u). The elastic modulus (E), or Young's modulus, is the tissue stiffness, relating the applied stress to the strain. Through a stress-strain curve, the elastic modulus is calculated as the slope of the linear region. Similar to a load-displacement curve, a non-linear or inelastic region starts beyond yield. Toughness, the tissue level-equivalent of work-to-fracture, is the energy required to reach tissue failure and is calculated as the area under the stress-strain curve. In vitro decalcification and cross-linking studies have demonstrated that bone mineral contributes mainly to the modulus of the bulk tissue, whereas collagen cross-links impact yield and plastic behavior (Burstein et al., 1975; Vashishth et al., 2001).

The stress-strain behavior of human cortical bone has been studied extensively. Characterizing bone tissue mechanical properties is challenging due to spatial heterogeneities and the small volumes that must be examined. Cortical bone samples can be machined from the middiaphysis and tested. The elastic modulus of human cortical bone is 17–20 GPa and is similar in tension and compression (Reilly et al., 1974). The ultimate strength is highest in compression ($\sigma_u = 193$ MPa), lower in tension ($\sigma_u = 133$ MPa) and lowest in shear ($\sigma_u = 68$ MPa) in cortical tissue (Reilly & Burstein, 1975). As will be explained below, these properties depend on bone mass and cortical microstructure. Cancellous mechanical properties are more difficult to generalize because bone mass and architecture vary substantially across different locations and even within a single anatomic location (Carter & Hayes, 1977; Gibson, 1985; Goldstein et al., 1983; Morgan et al., 2003). In rodent models, direct measurement of cortical bone tissue mechanical properties often is eschewed in favor of effective (or average) material properties estimated by measuring whole bone properties and geometry. These derived properties represent the average material properties for a cylinder of a constant cross-sectional geometry with the measured structural stiffness or strength. This approach can detect treatment-induced alterations in material on average but cannot localize variability and may not capture the magnitude and relative behavior of the true properties (Levenston et al., 1994; Silva et al., 2004). For example, mouse models of osteogenesis imperfecta had reduced whole bone properties when the femora were tested to failure, which suggested altered material-level mechanical properties when normalized for geometry (Camacho et al., 1999; Jepsen et al., 1996; Kozloff et al., 2004). However, when miniature beam specimens were directly tested for tissue flexural modulus and strength, the whole bone bending-derived material properties substantially underpredicted the microscopic mechanical properties by approximately 75% and were not representative of the relative relationships (Jepsen et al., 1997). Therefore, when possible, the mechanical behavior of the tissue should be measured directly and related to the material composition.

Nanoindentation is an alternate method to assess local tissue-level mechanical properties in compression (Donnelly et al., 2006; Zysset et al., 1999). In a nanoindentation experiment, the applied load and resulting displacement are measured as an indenter is pressed into and then withdrawn from a material's surface, following a specific load-time protocol. Then these measurements are processed to extract mechanical properties of the material including the hardness and indentation modulus. The resulting data also can be used to calculate dynamic mechanical properties such as storage and loss moduli. Because the indentation size can be extremely small, the local mechanical behavior is precisely evaluated at the microscale.

1.3 Intrinsic factors that contribute to bone strength

As mentioned above, the load-bearing function of skeleton is determined by several intrinsic factors: the total bone quantity or bone mass; the distribution of bone mass or whole bone geometry; and the intrinsic material properties of bone tissue.

1.3.1 Total bone mass

Bone mass is the total amount of tissue present. Total bone mass is an important factor that directly affects bone mechanical competence. Bone mass often is the first parameter evaluated when assessing fracture and osteoporosis risks. Clinically, bone mineral content (BMC, g) and bone mineral density (BMD, g/cm^2), measured with a dual-energy X-ray absorptiometry (DEXA) scan, are used as indicators of bone mass and fracture risk (Ulrich et al., 1999). DEXA scans usually are analyzed at the vertebrae and hips for cancellous bone mineral density, and at distal radius for cortical bone mineral density. Higher BMC and BMD generally indicate greater amount of mineral content within the bone, which accounts for 60%–90% of bone strength (Bouxsein, 2003). Clinically, bone density values are evaluated relative to data matched for age and gender (Z-scores) or compared to healthy young adults at their peak bone mass (T-scores). Despite not being measures of bone strength, clinically these scores are used to evaluate osteoporosis and fracture risk:

T-score $\geq = -1.0$ is classified as normal bone density.
$-1.0 >$ T-score > -2.5 indicates low bone density, also known as osteopenia.
T-score ≤ -2.5 signifies osteoporosis.

The risk of osteoporotic fracture increases as BMD declines. However, bone mineral density accounts for only ~60% of fracture risk in untreated patients, and conventional DXA criteria misdiagnose over 75% of men with prostate cancer on androgen deprivation therapy (Greenspan et al., 2013). Moreover, in the Rotterdam study analyzing the association between low BMD and fracture risk, researchers found that for individuals who experienced non-vertebral fractures, 18% of men and 13% of women had normal BMD and in those who had hip fractures, 3% of men and 5% of women had normal BMD. Having a T-score below −2.5 only accounted for 21% (men) and 44% (women) of all non-vertebral fractures, and 40% (men) and 64% (women) of all hip fractures (Schuit et al., 2004). Therefore, relying solely on BMD is inadequate for eliminating fracture risk.

In ex vivo and preclinical studies additional metrics of bone mass can be examined. Cancellous bone quantity can be assessed by bone volume fraction (bone volume per total volume, BV/TV), which can be determined from microcomputed tomography (microCT) scans, or bone apparent density, ρ_{app} (bone mass per total volume, BM/TV). BV/TV is the fraction of the volume occupied by cancellous bone within a ROI and is calculated as cancellous bone volume (BV) normalized to total volume of the sample (TV). In preclinical studies, evaluating cancellous bone mass helps to understand the ability of cancellous bone to withstand load in response to disease, pharmacological interventions, mechanical loading and other factors.

Information regarding bone geometry and tissue material properties helps to understand contributions and alterations in bone strength. Geometric properties include bone size and volume and architectural characteristics. Tissue properties mainly include tissue composition. Our ability to measure these parameters clinically is limited, but laboratory and preclinical studies contribute substantially to our knowledge. The sections below will explain geometric properties and tissue properties, respectively.

1.3.2 Whole bone geometric properties

Under loading, the geometric distribution of bone mass contributes to the resistance to fracture. These geometric properties encompass the morphological and architectural attributes of the skeleton. For long bones, the whole bone behavior effectively is measuring the cortical properties, for which the cortical geometry captures the bone tissue distribution. In the cortex, the effect of geometry can be modeled as a hollow circular section and can be understood based on the fundamentals of mechanics of materials if the material properties are homogeneous and invariant (van der Meulen et al., 2001). For conventional, nonbiological materials, these assumptions are reasonable and accurate. However, particularly cancellous tissue properties vary spatially and temporally (Boivin et al., 2000; Nuzzo et al., 2002; Paschalis et al., 1996, 1997a, 1997b). In addition, while these material assumptions are not valid for normal bone tissue, they may be even less applicable to disease states. Clinical conditions such as osteoporosis are known to affect not only the material quantity but also the tissue quality and distribution (Hasegawa et al., 1995; Paschalis et al., 1997b).

The relevant geometric measures depend on the loading mode. The cross-sectional area is the geometric resistance to compression. For bending and torsion loading, the section modulus (Z) is the geometric resistance of the structure to bending and torsion. The distribution of the bone tissue relative to the plane of bending and axis of twisting resists the load, and Z is determined by the area moment of inertia and polar moment of inertia, respectively. The moment of inertia is calculated in a section perpendicular to the plane of bending about the axis corresponding to the bending plane and varies between a maximum and minimum value depending on the section orientation relative to the load. The geometric resistance is least in the orientation for which the material is closest to the bending plane, corresponding to the minimum moment of inertia (I_{min}). The maximum moment of inertia (I_{max}) is the orientation for which the material is furthest from the bending plane and has the greatest resistance to bending. The torsional constant, K, is the geometric resistance to torsion along the longitudinal axis. For a circular cross section, the torsional constant is the polar moment of inertia, J, which is calculated as: $J = I_{min} + I_{max}$. However, predictions from simple geometric models can underestimate the maximum shear stresses (Levenston et al., 1994).

Cross-sectional geometric properties that contribute to the structural behavior of a long bone generally usually are evaluated at the mid-diaphysis of a long bone. In the laboratory, geometric properties can be measured by microCT scans. MicroCT scans provide high resolution cross-sectional geometry of the cortex and microstructural information for cancellous sites (Bouxsein et al., 2010; Jepsen et al., 2015). To understand the contribution of cortical geometry and periosteal mass to whole bone bending stiffness, we can compare a solid section to hollow cross sections with a similar outer diameter (D_p) or a similar area (Ct.Ar). A hollow section of similar periosteal diameter reduces bone mass by 25% but section modulus only 6%. When bone mass is matched between the solid and hollow sections, a 25% increase in periosteal diameter corresponds to a 70% increase in section modulus (Fig. 10.4).

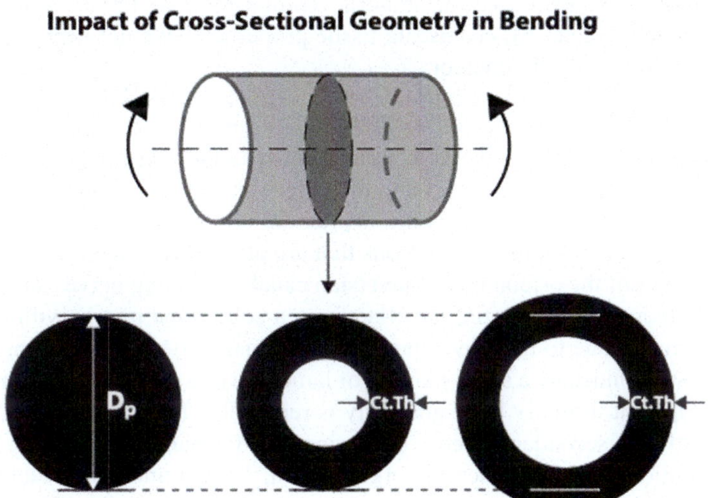

Periosteal Diameter (D_p)	**100%**	**100%**	125%
Cortical Area (A)	**100%**	75%	**100%**
Section Modulus (Z)	**100%**	94%	170%

FIGURE 10.4 Effect of cross-sectional geometry on cortical bone section modulus

Effect of cross-sectional geometry on cortical bone section modulus (Z), a measure of geometric resistance to bending and torsion. Compared to a solid cross section, a hollow section with the same periosteal diameter (D_p) but a smaller cortical area has slightly lower section modulus. For a hollow section with the cortical area matched to the solid section, the material is distributed farther from the neutral axis and thus resists bending more.

- *Adapted from Cole, J. H., & van der Meulen, M. C. H. (2011). Whole bone mechanics and bone quality.* Clinical Orthopedics and Related Research, 469*(8), 2139–2149. https://doi.org/10.1007/s11999-011-1784-3*

Material located further from the bending plane resists the load most, which means that changes in periosteal diameter have the greatest influence on Z.

1.3.3 Bone tissue material properties

From a materials science perspective, tissue composition and microstructure determine material mechanical properties. The mechanics of a whole bone arises from contributions of two primary tissue structures: dense cortical bone and lattice-like cancellous bone. Both tissue structures consist of an organic matrix containing 90% type-I collagen fibers and a mineral phase of calcium phosphate-based apatite. Bone cells (osteoblasts) deposit the organic matrix in layers, or lamellae, which then mineralize in situ. Distinctions between cortical and cancellous bone and their mechanical behavior primarily reflect differences in bone mass and microstructural organization and features. The material composition of these tissues is addressed elsewhere (Chapter 5). Nanoscale tissue properties have not been related to differences in mechanical behavior.

1.3.3.1 Tissue properties of cortical bone

Cortical bone, also known as compact bone, is the dense material that forms the tubular diaphysis of the cortex and the outer shell of cancellous bone sites. Cortical bone consists of a hierarchical microstructure of concentric lamellae organized into circular structures known as osteons that are aligned with the long axis of the bone. The center of the osteon is the Haversian canal containing nerves and blood vessels. The osteons and interstitial lamellae form a solid bone matrix with approximately 5% porosity due to the Haversian canals. Primary osteons, consisting of Haversian canal surrounded by a small number of lamellar rings, are formed during initial bone development. Primary bone gradually is remodeled to a more mature form of bone consisting of secondary osteons (Amprino, 1948; Chang & Liu, 2022; Enlow, 1962). In addition to microstructural organization, bone remodeling facilitates in vivo repair of damage (Raggatt & Partridge, 2010; Schaffler et al., 1995). The reversal of remodeling deposits a cement line at the outer surface of secondary osteons. Cement lines are less mineralized than bone tissue and serve to deflect crack growth (Burr et al., 1988; Schaffler et al., 1987).

Cortical bone is anisotropic due to the longitudinal alignment of osteons and the presence of embedded mineral crystals. In long bones, the mechanical properties of the cortex depend on the direction of load relative to the material orientation. When tissue samples from the cortical diaphysis are loaded, the highest modulus and ultimate strength are in the direction aligned with the longitudinal axis and lowest in both transverse directions perpendicular to the long axis (Fig. 10.5) (Reilly & Burstein, 1975). In addition, the mechanical behavior of the cortex depends on loading mode and loading rate. Cortical bone is stronger and stiffer in compression compared to tension, and under faster strain rates than slower ones (McElhaney, 1966; Reilly et al., 1974). Finally, porosity, which is the inverse of bone mass within a volume, is an important factor in the tissue properties of cortical bone (Schaffler & Burr, 1988).

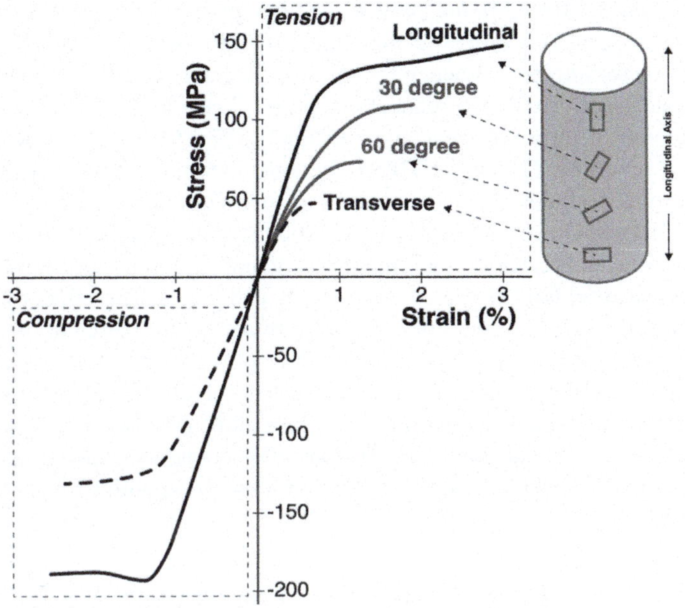

FIGURE 10.5 Anisotropic biomechanical behavior of cortical bone tissue.

The anisotropic microstructure of the diaphyses results in greater strength and stiffness when loaded in the longitudinal orientation than in either transverse direction. In addition, cortical bone is stronger when loaded in compression than in tension.

- *Adapted from Reilly, D. T., Burstein, A. H., & Frankel, V. H. (1974). The elastic modulus for bone.* Journal of Biomechanics, *7(3), 271. https://doi.org/10.1016/0021-9290(74)90018-9; Reilly, D. T., & Burstein, A. H. (1975). The elastic and ultimate properties of compact bone tissue.* Journal of Biomechanics, 8*(6), 393. https://doi.org/10.1016/0021-9290(75)90075-5.*

1.3.3.2 Tissue properties of cancellous bone

Cancellous bone, also known as trabecular bone or spongy bone, is the highly porous bone tissue found at the end of long bones such as the tibia and femur and in the center of irregular bones such as vertebrae. Cancellous bone does not contain osteons but rather consists of struts, known as trabecule, that consist of lamellae of bone tissue arranged as plates and rods. The lamellar tissue is similar to the material of cortical bone.

Bone mass is critical to cancellous bone mechanical performance. Site-to-site differences in apparent density can explain the majority of the variation in mechanical properties. Both elastic modulus and ultimate strength in compression depend on the apparent density, in the form of exponential or power law relationships (Carter & Hayes, 1977; Morgan et al., 2003). The exponential scaling results in disproportionately greater changes in the modulus or strength relative to small changes in bone mass. These relationships can explain 60%–80% of the variation in properties, based on data from small tissue samples and specific anatomic sites

(Keaveny & Hayes, 1993). In fact, within the tibial plateau of a single individual, the modulus varied over two orders of magnitude (Goldstein et al., 1983).

In addition to cancellous bone mass, microarchitecture plays an important role in the mechanical performance of cancellous bone. Cancellous architecture is generally considered in terms of trabecular thickness (Tb.Th) and separation (Tb.Sp), and trabecular number (Tb.N = 1/Tb.Sp). In computational models, trabecular number has a stronger influence on cancellous bone modulus and strength than trabecular thickness (Silva & Gibson, 1997). In these models, equivalent bone loss from fewer trabeculae caused a greater strength reduction than from thinner trabeculae. In addition, trabecular morphology can consist of rods and plates. The effect of trabecular form on cancellous bone strength has been examined with segmentation techniques that differentiate rods and plates (Liu et al., 2006). Accounting for trabecular plates and rods in finite element models predicts elastic modulus and yield strength of cancellous bone tissue better than BV/TV values alone (Wang et al., 2015). Trabecular plates have a more substantial role in cancellous bone mechanical integrity than rods, and a higher trabecular plate-to-rod ratio is associated with stronger bone tissue (Walker et al., 2014).

1.4 Extrinsic factors that determine bone strength

Bone strength is influenced by a variety of extrinsic factors that affect bone quantity, structure, composition, and cellular activity. Sex and age play significant roles, as hormonal differences between males and females, particularly estrogen levels, affect bone density. Aging leads to bone loss and structural deterioration due to reduced osteoblast activity and increased osteoclast resorption. Genetic background and race contribute to variations in peak bone mass and bone architecture, with some populations having higher mean bone mineral density than others. Additionally, diseases such as osteogenesis imperfecta, rheumatoid arthritis and chronic kidney disease alter bone remodeling, leading to weakened bone structure and compromised mechanical properties. Nutritional deficiencies, lifestyle factors such as physical activity, and environmental influences further modulate these biological mechanisms, ultimately determining bone strength and fracture risk.

Age is an important factor in bone strength. During growth and development, bone size increases, producing a more robust skeleton. Changes in tissue properties also contribute to increases in bone strength. The modulus of elasticity and bending strength of cortical bone is higher in adults than young children, reflecting increased mineral content (Currey & Butler, 1975). With advancing age, whole bone strength declines gradually and progressively due to reduced bone mass and altered geometry (Courtney et al., 1995). Cortical thinning, increased cortical porosity, trabecular thinning, and decreased trabecular connectivity are observed with aging and contribute to the overall reduced bone mechanical performance (Bala et al., 2015). These aging changes also contribute to weakened cortical and cancellous bone tissue (Burr, 1997; Burstein et al., 1976; Martin, 1993; Mosekilde et al., 1987). In cancellous bone, a transition in the trabecular microarchitecture from plate-like to rod-like trabecular bone is an important etiologic factor in age-and osteoporosis-related bone fragility (Wang et al., 2015). Sarcopenia due to aging reduces the loads applied to the skeleton and may result in a form of disuse-related bone loss.

Aging related changes in mass and architecture are the result of altered cell activities. The balance between bone formation and bone resorption depends on cell-to-cell communication, and osteocytes, embedded in the bone matrix, are responsible for transforming mechanosensory signals to biochemical signals and directing bone remodeling. The ability of osteocytes to sense mechanical stimuli is negatively affected by the shrinkage of the canalicular network with age (Hemmatian et al., 2017; Milovanovic et al., 2013; Tiede-Lewis & Dallas, 2019). With aging, the loss of connexin 43, a protein maintaining the canalicular network, triggered osteocyte cell death, empty lacunae, and recruitment of osteoclasts (Pignolo et al., 2021). Osteocyte lacunar density decreases with age in both men and women in cortical bone (Vashishth et al., 2000), and decreased osteocyte lacunar density is correlated with increased microdamage (Mori et al., 1997).

Bone size and strength differ by sex due to the effects of hormones and body size. Differences in bone size develop between boys and girls develop post-puberty and persist, contributing to differences with aging and fracture risk (Duan et al., 2003). In women, post-menopausal estrogen loss accelerates bone loss, increasing the risk of osteoporosis. The gradual decline in 17β-estradiol production leads to enhanced bone turnover, favoring bone resorption. The imbalance between bone formation and resorption leads to a rapid phase of bone loss and deterioration of bone microarchitecture for 5–10 years, followed by a prolonged period of gradual bone loss (Riggs & Melton, 1986). During perimenopause and menopause, bone metabolism also is affected by increased follicle stimulating hormone (FSH) and decreased inhibin (Klein et al., 1996, 2004; Welt et al., 1999). Increased FSH is correlated with increased bone resorption and decreased BMD. Decreased inhibin is associated with decreased BMD, bone volume, and deteriorated biomechanical properties, as well as decreased mineral apposition rate and osteoblastogenesis (Perrien et al., 2007). Age-induced bone loss differs in men and women. Women experience a greater decreased in trabecular number, whereas men lose cancellous bone primarily by trabecular thinning (Farr & Khosla, 2015).

Bone strength varies across different racial groups not only because of different body size, but also different bone microstructure. These differences are evident in fracture rates. Caucasian women have the highest rates of osteoporotic fractures, followed by Asian and Hispanic women, and Black women have the lowest rates. Individuals with smaller body sizes are at greater risk for osteoporosis due to having lower peak bone mass compared to those with larger body sizes. Interestingly, despite Asian populations having genetically lower bone mineral density, they experience fewer hip and forearm fractures compared to Caucasian individuals. Using the ITS technique, the trabecular bone of Chinese women consisted of more plates than rods, resulting in a higher plate-to-rod ratio and stronger bones (Walker et al., 2014).

These findings indicate that race influences bone strength through both bone mass and microarchitecture.

A variety of musculoskeletal conditions affect bone homeostasis and impair whole bone strength including osteoporosis, osteogenesis imperfecta and osteomalacia. As discussed earlier, osteoporosis is associated not only with changes in bone mass, but also cortical geometry and cancellous microarchitecture (Greenspan et al., 2013). Osteogenesis imperfecta (OI) or brittle bone disease, is a genetic disorder that negatively affects bone strength. The condition results from mutations in the genes responsible for producing collagen. The defective or insufficient collagen weakens the bone matrix, leading to brittleness, hyper-mineralized bone tissue, and a high susceptibility to fractures (Nijhuis et al., 2019). OI compromises the mechanical integrity of the skeleton at both cortical and cancellous sites. In OI, trabecular bone has reduced trabecular number and connectivity, along with decreased trabecular thickness and volumetric bone mass. Similarly, cortical bone of individuals with OI is characterized by reduced cortical thickness and increased osteocyte lacunae and vascular porosity.

Nutrition and diet play a crucial role in maintaining bone strength by supporting bone formation, remodeling, and overall skeletal health. Calcium is the primary mineral in bone, providing structural integrity. Other essential minerals include phosphorus, which works alongside calcium to form the bone matrix, and magnesium, which helps regulate bone metabolism. Vitamin D enhances calcium absorption in the intestines and promotes bone mineralization. Inadequate levels of vitamin D and calcium can lead to hypomineralization and rickets or osteomalacia. Additionally, vitamin K is necessary for producing osteocalcin, a protein essential for bone mineralization, and vitamin C supports collagen synthesis, a key component of bone structure. Protein intake is also vital, providing the building blocks for bone tissue and maintaining muscle mass, which in turn supports bone strength and provides necessary mechanical loading to the skeleton. A well-balanced diet rich in these nutrients is essential for building a healthy skeleton, maintaining bone mass, preventing bone loss and ensuring long-term skeletal health.

Physical activity and lifestyle choices are crucial to maintaining and improving bone strength. Weight-bearing activities, such as walking, running and resistance training, stimulate bone remodeling by promoting bone formation. Resistance training, in particular, enhances bone mass and improves muscle strength, reducing the risk of fractures (Watson et al., 2018). Bone microarchitecture and size are enhanced at both cancellous bone and cortical sites with exercise, and the cortical benefits during growth are maintained with age (Jones et al., 1977; Warden et al., 2014). Conversely, insufficient activity or immobilization reduces the loads applied to the skeleton, leading to bone loss (Schoutens et al., 1989). Additionally, lifestyle factors such as smoking and excessive alcohol consumption negatively impact bone health by interfering with calcium absorption and bone formation (Turner et al., 2021). Adequate UV exposure for vitamin D synthesis, a balanced diet rich in calcium, phosphorus, and other minerals, and maintaining a healthy body weight further contribute to bone strength. Adopting an active lifestyle with regular exercise and healthy habits is essential for preventing osteoporosis and ensuring long-term skeletal health.

1.5 Loading and bone fracture patterns

Preserving bone strength is critical, as fractures can significantly affect quality of life and overall health. Understanding the mechanisms of bone fracture in the context of bone mechanical properties can help prevent injuries, improve diagnosis, and optimize treatment (Chao et al., 1998, 2003). In particular, relating fractures to bone mechanical behavior can identify how and why a bone failed, allowing for better protective strategies, more accurate assessment of fracture risk, and informed decisions in surgical repair and rehabilitation. This knowledge also can guide the design of implants and biomaterials, and it sheds light on how diseases like osteoporosis affect bone strength. Ultimately, understanding failure mechanisms has significant implications for individual health outcomes and public health efforts aimed at reducing fracture-related mortality and disability.

Cortical bone is an anisotropic material due to its microstructure and is weaker when loaded in the transverse direction than in the longitudinal direction. Loading mode also affects the strength of cortical bone; bone tissue is stronger in compression than in tension in both transverse and longitudinal directions and weakest in shear. Three principal stress planes - maximum tensile stress plane, maximum compressive stress plane, and maximum shear stress - determine the fracture plane and help predict the location and failure mode.

The fracture patterns of long bones generally are predictable based on the mechanical properties of cortical bone (Fig. 10.6). In pure tension, a loading mode seldom encountered in vivo, a transverse fracture will occur. When loaded in pure compression, the cortex will fail along a shear plane, resulting in a transverse fracture surface. Bending of a whole bone induces tension and compression on opposite surfaces, and upon failure, the fracture surface will reflect the combination of the two loading modes. Because cortical bone is stronger in compression than in tension, fracture typically will initiate on the cortex loaded in tension. As the failure progresses and encounters tissue loaded in compression, the fracture surface will deflect along a shear plane. The interaction between the two loading modes results in the development of a fragment at the cortex loaded in compression. The magnitude of the fragment will increase if additional compression is present. In cases of torsional injury, a spiral fracture will occur along the plane of maximum shear. Shear stress along the spiral line can lead to small longitudinal cracks (Gitajn & Rodriguez, 2011).

Clinically, the rate of fracture healing is correlated to the amount of bone damage relative to soft tissue damage, as a competitive condition in the common hematoma might occur in fractures with extensive soft tissue damage. A fracture with limited soft tissue damage generally will heal faster than fracture with substantial multi-tissue trauma (Hulth, 1980). Fracture patterns may vary due to aging and bone metabolic diseases that alter tissue mechanical behavior. Understanding the fracture

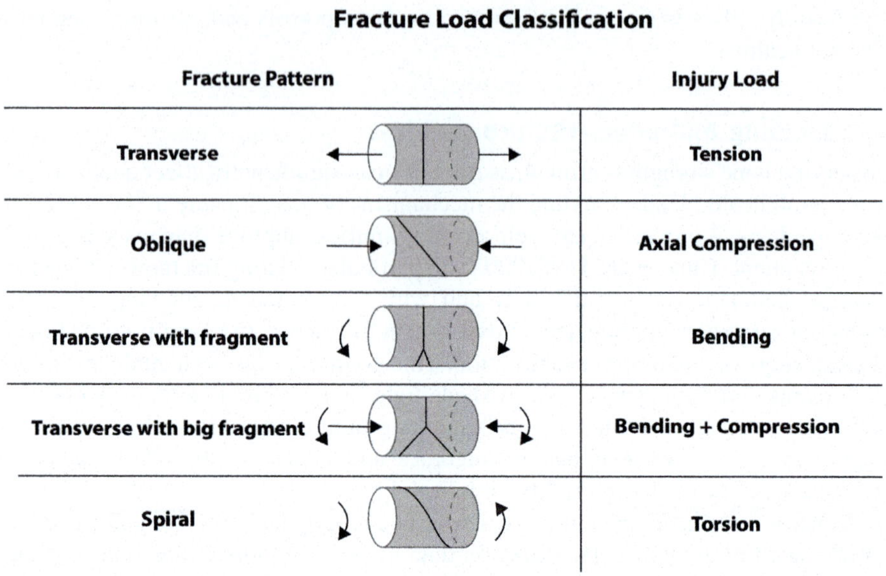

FIGURE 10.6 Fracture load classification including fracture pattern and the associated injury load.

Tensile loading will cause a transverse fracture, and compression an oblique fracture. Bending can cause a transverse fracture with a small fragment, but when combined with additional compression, bone will fracture transversely with a larger fragment. Torsion results in a spiral fracture.

pattern, tissue mechanics and applied loading, as well as patients' overall health condition, are critical factors for treatments and the design of fixation devices and surgical approaches.

1.6 Conclusion

In this chapter, we reviewed the intrinsic and extrinsic factors that influence bone mechanical competence, explored methods for measuring the mechanical behavior of whole bones and bone tissue, and examined common fracture patterns—laying the groundwork for understanding the structure and function of the human skeleton. Despite these insights, significant challenges remain in accurately evaluating bone strength and predicting fracture risk. One major obstacle is the heterogeneity of bone material, geometry and microstructure, both within an individual and across different individuals, that contribute to whole bone mechanical performance. Variations in geometry, density, and microstructure prevent uniform assessment of bone strength across populations or even within a single bone. For example, although DEXA is the current gold standard for clinically evaluating bone mass, bone microarchitecture is not accounted for when predicting fracture risk, and considerable overlap exists in BMD values between individuals who experience fractures and those who do not. Compounding this issue is the difficulty of modeling how bones respond to complex, multi-axial loading conditions that more closely represent real-life physiological scenarios. Together, these challenges limit the predictive power of biomechanical models and underscore the need for personalized, context-specific approaches to fracture risk assessment.

Fortunately, advances in imaging technology have provided new tools to assess skeletal health and predict fracture risk with greater accuracy. One example is high-resolution peripheral quantitative computed tomography (HR-pQCT), a non-invasive imaging technique capable of capturing volumetric bone mineral density (vBMD), a material property, and both trabecular and cortical bone microarchitecture. This level of detail allows for a more comprehensive assessment of bone quality beyond what is possible with DEXA, which only measures areal BMD and does not account for microstructure. HR-pQCT has potential for not only in identifying individuals at higher risk of fractures but also in tracking changes in bone health over time (Mikolajewicz et al., 2020).

In parallel, computational simulation methods such as finite element analysis (FEA) have continued to evolve, enabling the creation of more accurate and personalized bone models. These models incorporate HR-pQCT data to simulate how bone responds to various loading conditions, including those experienced during daily activities or traumatic events. The combination of FEA and HR-pQCT data, referred to as μFEA or biomechanical CT (BCT), captures tissue microarchitecture and spatial heterogeneity from HR-pQCT images, which facilitates the prediction of localized strain and stress distributions within the bone. These methods can identify potential sites of weakness and provide a case-by-case estimation of fracture risk.

Finite element analysis (FEA) based on QCT scans can predict fracture risk with greater accuracy than DXA-based BMD (Roberts et al., 2009). FEA models developed from QCT capture three-dimensional geometry 3D bone model that can be loaded in silico to estimate vertebral strength and understand factors that contribute to fracture. Based on computational models, in both men and women, the age-related reduction in femoral strength exceeds the decline suggested by BMD, and the prevalence of low femoral strength is greater than the prevalence of osteoporosis defined by the BMD T-score (Keaveny et al., 2010). Therefore, QCT-based FEA and similar computational approaches have the potential to advance the diagnosis of osteoporosis and fracture risk to early stages at which prevention may still be possible.

Leveraging existing CT scans has enabled retrospective studies to understand ethnic differences in bone strength (Hong et al., 2020), to monitor disease progression in patients with osteoporosis (Keaveny et al., 2014), and to validate the efficacy of osteoporosis treatment (Keaveny et al., 2024). The implementation of novel calibration techniques for CT has made these analyses possible (Lee et al., 2017). Future advances are expected to further improve the accuracy of these approaches. For example, capturing tissue-level material changes such as those present in diabetes could enhance the utility of these approaches (Hunt et al., 2019).

Together, novel imaging methods and computational simulations of bone mechanics offer a powerful framework for advancing bone health assessment and intervention planning (Karasik et al., 2017; Van Rietbergen, 2001). These methods will continue to be developed and improved in the future.

References

Abrahamsen, B., van Staa, T., Ariely, R., Olson, M., & Cooper, C. (2009). Excess mortality following hip fracture: A systematic epidemiological review. *Osteoporosis International, 20*(10), 1633–1650. https://doi.org/10.1007/s00198-009-0920-3

Amprino, R. (1948). A contribution to the functional meaning of the substitution of primary by secondary bone tissue. *Acta Anatomica, 5*(3), 291–300. https://doi.org/10.1159/000140331

Bala, Y., Zebaze, R., & Seeman, E. (2015). Role of cortical bone in bone fragility. *Current Opinion in Rheumatology, 27*(4), 406–413. https://doi.org/10.1097/bor.0000000000000183

Boivin, G. Y., M Chavassieux, P., C Santora, A., Yates, J., & Meunier, P. J. (2000). Alendronate increases bone strength by increasing the mean degree of mineralization of bone tissue in osteoporotic women. *Bone, 27*(5), 687–694. https://doi.org/10.1016/s8756-3282(00)00376-8

Bouxsein, M. L. (2003). Mechanisms of osteoporosis therapy: A bone strength perspective. *Clinical Cornerstone, 5*(2), S13. https://doi.org/10.1016/S1098-3597(03)90043-3

Bouxsein, M. L., Boyd, S. K., Christiansen, B. A., Guldberg, R. E., Jepsen, K. J., & Müller, R. (2010). Guidelines for assessment of bone microstructure in rodents using micro-computed tomography. *Journal of Bone and Mineral Research, 25*(7), 1468–1486. https://doi.org/10.1002/jbmr.141

Brodt, M. D., Ellis, C. B., & Silva, M. J. (1999). Growing C57B1/6 mice increase whole bone mechanical properties by increasing geometric and material properties. *Journal of Bone and Mineral Research, 14*(12), 2159–2166. https://doi.org/10.1359/jbmr.1999.14.12.2159

Burr, D. B. (1997). Muscle strength, bone mass, and age-related bone loss. *Journal of Bone and Mineral Research: The Official Journal of the American Society for Bone and Mineral Research, 12*(10), 1547–1551. https://doi.org/10.1359/jbmr.1997.12.10.1547

Burr, D. B., Schaffler, M. B., & Frederickson, R. G. (1988). Composition of the cement line and its possible mechanical role as a local interface in human compact bone. *Journal of Biomechanics, 21*(11), 939–945. https://doi.org/10.1016/0021-9290(88)90132-7

Burstein, A. H., & Frankel, V. H. (1971). A standard test for laboratory animal bone. *Journal of Biomechanics, 4*(2), 155–158. https://doi.org/10.1016/0021-9290(71)90026-1

Burstein, A. H., Reilly, D. T., & Martens, M. (1976). Aging of bone tissue: Mechanical properties. *Journal of Bone and Joint Surgery - Series A, 58*(1), 82–86. https://doi.org/10.2106/00004623-197658010-00015

Burstein, A. H., Zika, J. M., Heiple, K. G., & Klein, L. (1975). Contribution of collagen and mineral to the elastic-plastic properties of bone. *Journal of Bone and Joint Surgery, 57*(7), 956–961. https://doi.org/10.2106/00004623-197557070-00013

Camacho, N. P., Hou, L., Toledano, T. R., Alex Ilg, W., Brayton, C. F., Raggio, C. L., Root, L., & Boskey, A. L. (1999). The material basis for reduced mechanical properties in oim mice bones. *Journal of Bone and Mineral Research, 14*(2), 264–272. https://doi.org/10.1359/jbmr.1999.14.2.264

Carter, D. R., & Hayes, W. C. (1977). The compressive behavior of bone as a two-phase porous structure. *Journal of Bone and Joint Surgery - Series A, 59*(7), 954–962. https://doi.org/10.2106/00004623-197759070-00021

Chang, B., & Liu, X. (2022). Osteon: Structure, turnover, and regeneration. *Tissue Engineering Part B Reviews, 28*(2), 261–278. https://doi.org/10.1089/ten.teb.2020.0322

Chao, E. Y. S., Inoue, N., Elias, J. J., & Aro, H (1998). Enhancement of fracture healing by mechanical and surgical intervention. *Clinical Orthopaedics and Related Research, 355*, S163–S178S163. https://doi.org/10.1097/00003086-199810001-00018. https://journals.lww.com/clinorthop/pages/default.aspx

Chao, E. Y. S., Inoue, N., Ripamonti, U., & Fenwick, S. (2003). Biophysical stimulation of bone fracture repair, regeneration and remodelling. *European Cells and Materials, 6*, 72–85. https://doi.org/10.22203/eCM.v006a07

Courtney, A. C., Wachtel, E. F., Myers, E. R., & Hayes, W. C. (1995). Age-related reductions in the strength of the femur tested in a fall- loading configuration. *Journal of Bone and Joint Surgery, 77*(3), 387–395. https://doi.org/10.2106/00004623-199503000-00008

Currey, J. D., & Butler, G. (1975). The mechanical properties of bone tissue in children. *Journal of Bone and Joint Surgery, 57*(6), 810–814. https://doi.org/10.2106/00004623-197557060-00015

Donnelly, E., Baker, S. P., Boskey, A. L., & Van Der Meulen, M. C. H. (2006). Effects of surface roughness and maximum load on the mechanical properties of cancellous bone measured by nanoindentation. *Journal of Biomedical Materials Research Part A, 77*(2), 426–435. https://doi.org/10.1002/jbm.a.30633

Duan, Y., Beck, T. J., Wang, X. F., & Seeman, E. (2003). Structural and biomechanical basis of sexual dimorphism in femoral neck fragility has its origins in growth and aging. *Journal of Bone and Mineral Research, 18*(10), 1766–1774. https://doi.org/10.1359/jbmr.2003.18.10.1766

Enlow, D. H. (1962). Functions of the haversian system. *American Journal of Anatomy, 110* (3), 269–305. https://doi.org/10.1002/aja.1001100305

Farr, J. N., & Khosla, S. (2015). Skeletal changes through the lifespan - From growth to senescence. *Nature Reviews Endocrinology, 11*(9), 513–521. https://doi.org/10.1038/nrendo.2015.89

Gibson, L. J. (1985). The mechanical behaviour of cancellous bone. *Journal of Biomechanics, 18*(5), 317–328. https://doi.org/10.1016/0021-9290(85)90287-8

Gitajn, I. L., & Rodriguez, E. K. (2011). *Biomechanics of musculoskeletal injury.* InTech. https://doi.org/10.5772/20672

Goldstein, S. A., Wilson, D. L., Sonstegard, D. A., & Matthews, L. S. (1983). The mechanical properties of human tibial trabecular bone as a function of metaphyseal location. *Journal of Biomechanics, 16*(12), 965–969. https://doi.org/10.1016/0021-9290(83)90097-0

Greenspan, S. L., Wagner, J., Nelson, J. B., Perera, S., Britton, C., & Resnick, N. M. (2013). Vertebral fractures and trabecular microstructure in men with prostate cancer on androgen deprivation therapy. *Journal of Bone and Mineral Research, 28*(2), 325–332. https://doi.org/10.1002/jbmr.1771

Hasegawa, K., Turner, C. H., Recker, R. R., Wu, E., & Burr, D. B. (1995). Elastic properties of osteoporotic bone measured by scanning Acoustic microscopy. *Bone, 16*(1), 85–90. https://doi.org/10.1016/8756-3282(95)80016-j

Hemmatian, H., Bakker, A. D., Klein-Nulend, J., & van Lenthe, G. H. (2017). Aging, osteocytes, and mechanotransduction. *Current Osteoporosis Reports, 15*(5), 401–411. https://doi.org/10.1007/s11914-017-0402-z

Hong, N., Lee, D. C., Khosla, S., Keaveny, T. M., & Rhee, Y. (2020). Comparison of vertebral and femoral strength between white and Asian adults using finite element analysis of computed tomography scans. *Journal of Bone and Mineral Research, 35*(12), 2345–2354. https://doi.org/10.1002/jbmr.4149

Hulth, A. (1980). Fracture healing: A concept of competing healing factors. *Acta Orthopaedica, 51*(1–6), 5–8. https://doi.org/10.3109/17453678008990762

Hunt, H. B., Torres, A. M., Palomino, P. M., Marty, E., Saiyed, R., Cohn, M., Jo, J., Warner, S., Sroga, G. E., King, K. B., Lane, J. M., Vashishth, D., Hernandez, C. J., & Donnelly, E. (2019). Altered tissue composition, microarchitecture, and mechanical performance in cancellous bone from men with type 2 diabetes mellitus. *Journal of Bone and Mineral Research, 34*(7), 1191–1206. https://doi.org/10.1002/jbmr.3711

Jepsen, K. J., Goldstein, S. A., Kuhn, J. L., Schaffler, M. B., & Bonadio, J. (1996). Type-I collagen mutation compromises the post-yield behavior of Mov13 long bone. *Journal of Orthopaedic Research, 14*(3), 493–499. https://doi.org/10.1002/jor.1100140320

Jepsen, K. J., Schaffler, M. B., Kuhn, J. L., Goulet, R. W., Bonadio, J., & Goldstein, S. A. (1997). Type I collagen mutation alters the strength and fatigue behavior of Mov13 cortical tissue. *Journal of Biomechanics, 30*(11–12), 1141–1147. https://doi.org/10.1016/S0021-9290(97)00088-2

Jepsen, K. J., Silva, M. J., Vashishth, D., Guo, X. E., & Van Der Meulen, M. C. H. (2015). Establishing biomechanical mechanisms in mouse models: Practical guidelines for systematically evaluating phenotypic changes in the diaphyses of long bones. *Journal of Bone and Mineral Research, 30*(6), 951–966. https://doi.org/10.1002/jbmr.2539

Jones, H. H., Priest, J. D., Hayes, W. C., Tichenor, C. C., & Nagel, D. A. (1977). Humeral hypertrophy in response to exercise. *Journal of Bone and Joint Surgery, 59*(2), 204–208. https://doi.org/10.2106/00004623-197759020-00012

Karasik, D., Demissie, S., Lu, D., Broe, K. E., Boyd, S. K., Liu, C. T., Hsu, Y. H., Bouxsein, M. L., & Kiel, D. P. (2017). Bone strength estimated by micro-finite element analysis (μFEA) is heritable and shares genetic predisposition with areal BMD: The Framingham study. *Journal of Bone and Mineral Research, 32*(11), 2151–2156. https://doi.org/10.1002/jbmr.3200

Keaveny, T. M., Adams, A. L., Orwoll, E. S., Khosla, S., Siris, E. S., McClung, M. R., Bouxsein, M. L., Fatemi, S., Lee, D. C., & Kopperdahl, D. L. (2024). Osteoporosis treatment prevents hip fracture similarly in both sexes: The FOCUS observational study. *Journal of Bone and Mineral Research, 39*(10), 1424–1433. https://doi.org/10.1093/jbmr/zjae090

Keaveny, T. M., & Hayes, W. C. (1993). A 20-year perspective on the mechanical properties of trabecular bone. *Journal of Biomechanical Engineering, 115*(4), 534–542. https://doi.org/10.1115/1.2895536

Keaveny, T. M., Kopperdahl, D. L., Melton, L. J., Hoffmann, P. F., Amin, S., Riggs, B. L., & Khosla, S. (2010). Age-dependence of femoral strength in white women and men. *Journal of Bone and Mineral Research, 25*(5), 994–1001. https://doi.org/10.1002/jbmr.091033

Keaveny, T. M., McClung, M. R., Genant, H. K., Zanchetta, J. R., Kendler, D., Brown, J. P., Goemaere, S., Recknor, C., Brandi, M. L., Eastell, R., Kopperdahl, D. L., Engelke, K., Fuerst, T., Radcliffe, H. S., & Libanati, C. (2014). Femoral and vertebral strength improvements in postmenopausal women with osteoporosis treated with denosumab. *Journal of Bone and Mineral Research, 29*(1), 158–165. https://doi.org/10.1002/jbmr.2024

Klein, N. A., Houmard, B. S., Hansen, K. R., Woodruff, T. K., Sluss, P. M., Bremner, W. J., & Soules, M. R. (2004). Age-related analysis of inhibin A, inhibin B, and activin A relative to the intercycle monotropic follicle-stimulating hormone rise in normal ovulatory women. *Journal of Clinical Endocrinology and Metabolism, 89*(6), 2977–2981. https://doi.org/10.1210/jc.2003-031515

Klein, N. A., Illingworth, P. J., Groome, N. P., McNeilly, A. S., Battaglia, D. E., & Soules, M. R. (1996). Decreased inhibin B secretion is associated with the monotropic FSH rise in older, ovulatory women: A study of serum and follicular fluid levels of dimeric inhibin A and B in spontaneous menstrual cycles. *Journal of Clinical Endocrinology and Metabolism, 81*(7), 2742–2745. https://doi.org/10.1210/jc.81.7.2742

Kozloff, K. M., Carden, A., Bergwitz, C., Forlino, A., Uveges, T. E., Morris, M. D., Marini, J. C., & Goldstein, S. A. (2004). Brittle IV mouse model for osteogenesis imperfecta IV demonstrates postpubertal adaptations to improve whole bone strength. *Journal of Bone and Mineral Research, 19*(4), 614–622. https://doi.org/10.1359/JBMR.040111

Lee, D. C., Hoffmann, P. F., Kopperdahl, D. L., & Keaveny, T. M. (2017). Phantomless calibration of CT scans for measurement of BMD and bone strength—Inter-operator reanalysis precision. *Bone, 103*, 325–333. https://doi.org/10.1016/j.bone.2017.07.029

Levenston, M. E., Beaupré, G. S., & van der Meulen, M. C. H. (1994). Improved method for analysis of whole bone torsion tests. *Journal of Bone and Mineral Research, 9*(9), 1459–1465. https://doi.org/10.1002/jbmr.5650090919

Liu, X. S., Sajda, P., Saha, P. K., Wehrli, F. W., & Guo, X. E. (2006). Quantification of the roles of trabecular microarchitecture and trabecular type in determining the elastic modulus of human trabecular bone. *Journal of Bone and Mineral Research, 21*(10), 1608–1617. https://doi.org/10.1359/jbmr.060716

Martin, B. (1993). Aging and strength of bone as a structural material. *Calcified Tissue International, 53*(S1), S34. https://doi.org/10.1007/bf01673400

McElhaney, J. H. (1966). Dynamic response of bone and muscle tissue. *Journal of Applied Physiology, 21*(4), 1231–1236. https://doi.org/10.1152/jappl.1966.21.4.1231

Mikolajewicz, N., Bishop, N., Burghardt, A. J., Folkestad, L., Hall, A., Kozloff, K. M., Lukey, P. T., Molloy-Bland, M., Morin, S. N., Offiah, A. C., Shapiro, J., van Rietbergen, B., Wager, K., Willie, B. M., Komarova, S. V., & Glorieux, F. H. (2020). HR-pQCT measures of bone microarchitecture predict fracture: Systematic review and meta-analysis. *Journal of Bone and Mineral Research, 35*(3), 446–459. https://doi.org/10.1002/jbmr.3901

Milovanovic, P., Zimmermann, E. A., Hahn, M., Djonic, D., Püschel, K., Djuric, M., Amling, M., & Busse, B. (2013). Osteocytic canalicular networks: Morphological implications for altered mechanosensitivity. *ACS Nano, 7*(9), 7542–7551. https://doi.org/10.1021/nn401360u

Morgan, E. F., Bayraktar, H. H., & Keaveny, T. M. (2003). Trabecular bone modulus-density relationships depend on anatomic site. *Journal of Biomechanics, 36*(7), 897–904. https://doi.org/10.1016/S0021-9290(03)00071-X

Mori, S., Harruff, R., Ambrosius, W., & Burr, D. B. (1997). Trabecular bone volume and microdamage accumulation in the femoral heads of women with and without femoral neck fractures. *Bone, 21*(6), 521–526. https://doi.org/10.1016/s8756-3282(97)00200-7

Mosekilde, L., Mosekilde, Le, & Danielsen, C. C. (1987). Biomechanical competence of vertebral trabecular bone in relation to ash density and age in normal individuals. *Bone, 8*(2), 79–85. https://doi.org/10.1016/8756-3282(87)90074-3

Nijhuis, W. H., Eastwood, D. M., Allgrove, J., Hvid, I., Weinans, H. H., Bank, R. A., & Sakkers, R. J. (2019). Current concepts in osteogenesis imperfecta: Bone structure, biomechanics and medical management. *Journal of Children's Orthopaedics, 13*(1), 1–11. https://doi.org/10.1302/1863-2548.13.180190

Nuzzo, S., Lafage-Proust, M. H., Martin-Badosa, E., Boivin, G., Thomas, T., Alexandre, C., & Peyrin, F. (2002). Synchrotron radiation microtomography allows the analysis of three-dimensional microarchitecture and degree of mineralization of human iliac crest biopsy specimens: Effects of etidronate treatment. *Journal of Bone and Mineral Research, 17* (8), 1372–1382. https://doi.org/10.1359/jbmr.2002.17.8.1372

Panula, J., Pihlajamäki, H., Mattila, V. M., Jaatinen, P., Vahlberg, T., Aarnio, P., & Kivelä, S. L. (2011). Mortality and cause of death in hip fracture patients aged 65 or older - A population-based study. *BMC Musculoskeletal Disorders, 12*. https://doi.org/10.1186/1471-2474-12-105

Paschalis, E. P., Betts, F., DiCarlo, E., Mendelsohn, R., & Boskey, A. L. (1997a). FTIR microspectroscopic analysis of human iliac crest biopsies from untreated osteoporotic bone. *Calcified Tissue International, 61*(6), 487–492. https://doi.org/10.1007/s002239900372

Paschalis, E. P., Betts, F., DiCarlo, E., Mendelsohn, R., & Boskey, A. L. (1997b). FTIR microspectroscopic analysis of normal human cortical and trabecular bone. *Calcified Tissue International, 61*(6), 480–486. https://doi.org/10.1007/s002239900371

Paschalis, E. P., DiCarlo, E., Betts, F., Sherman, P., Mendelsohn, R., & Boskey, A. L. (1996). FTIR microspectroscopic analysis of human osteonal bone. *Calcified Tissue International, 59*(6), 480–487. https://doi.org/10.1007/bf00369214

Perrien, D. S., Akel, N. S., Edwards, P. K., Carver, A. A., Bendre, M. S., Swain, F. L., Skinner, R. A., Hogue, W. R., Nicks, K. M., Pierson, T. M., Suva, L. J., & Gaddy, D. (2007). Inhibin A is an endocrine stimulator of bone mass and strength. *Endocrinology, 148*(4), 1654–1665. https://doi.org/10.1210/en.2006-0848

Pignolo, R. J., Law, S. F., & Chandra, A. (2021). Bone aging, cellular senescence, and osteoporosis. *JBMR Plus, 5*(4). https://doi.org/10.1002/jbm4.10488

Raggatt, L. J., & Partridge, N. C. (2010). Cellular and molecular mechanisms of bone remodeling. *Journal of Biological Chemistry, 285*(33), 25103–25108. https://doi.org/10.1074/jbc.R109.041087Australia

Reilly, D. T., & Burstein, A. H. (1975). The elastic and ultimate properties of compact bone tissue. *Journal of Biomechanics, 8*(6), 393. https://doi.org/10.1016/0021-9290(75)90075-5

Reilly, D. T., Burstein, A. H., & Frankel, V. H. (1974). The elastic modulus for bone. *Journal of Biomechanics, 7*(3), 271. https://doi.org/10.1016/0021-9290(74)90018-9

Riggs, B. L., & Melton, L. J. (1986). Involutional osteoporosis. *New England Journal of Medicine, 314*(26), 1676–1686. https://doi.org/10.1056/NEJM198606263142605

Roberts, B. J., Kopperdahl, D., Thrall, E., Muller, J. A., Keaveny, T. M., & Bouxsein, M. L. (2009). Prediction of femoral strength in a sideways fall configuration using QCT-Based finite element analysis. *Bone, 44*, S72. https://doi.org/10.1016/j.bone.2009.01.158

Schaffler, M. B., & Burr, D. B. (1988). Stiffness of compact bone: Effects of porosity and density. *Journal of Biomechanics, 21*(1), 13–16. https://doi.org/10.1016/0021-9290(88)90186-8

Schaffler, M. B., Burr, D. B., & Frederickson, R. G. (1987). Morphology of the osteonal cement line in human bone. *The Anatomical Record, 217*(3), 223–228. https://doi.org/10.1002/ar.1092170302

Schaffler, M. B., Choi, K., & Milgrom, C. (1995). Aging and matrix microdamage accumulation in human compact bone. *Bone, 17*(6), 521–525. https://doi.org/10.1016/8756-3282(95)00370-3

Schoutens, A., Laurent, E., & Poortmans, J. R. (1989). Effects of inactivity and exercise on bone. *Sports Medicine, 7*(2), 71–81. https://doi.org/10.2165/00007256-198907020-00001

Schuit, S. C. E., van der Klift, M., M Weel, A. E. A., H de Laet, C. E. D., Burger, H., Seeman, E., Hofman, A., Uitterlinden, A. G., M van Leeuwen, J. P. T., & P Pols, H. A. (2004). Fracture incidence and association with bone mineral density in elderly men and women: The rotterdam study. *Bone, 34*(1), 195–202. https://doi.org/10.1016/j.bone.2003.10.001

Silva, M. J., Brodt, M. D., Fan, Z., & Rho, J. Y. (2004). Nanoindentation and whole-bone bending estimates of material properties in bones from the senescence accelerated mouse SAMP6. *Journal of Biomechanics, 37*(11), 1639–1646. https://doi.org/10.1016/j.jbiomech.2004.02.018

Silva, M. J., & Gibson, L. J. (1997). Modeling the mechanical behavior of vertebral trabecular bone: Effects of age-related changes in microstructure. *Bone, 21*(2), 191–199. https://doi.org/10.1016/s8756-3282(97)00100-2

Tiede-Lewis, L. A. M., & Dallas, S. L. (2019). Changes in the osteocyte lacunocanalicular network with aging. *Bone, 122*, 101–113. https://doi.org/10.1016/j.bone.2019.01.025

Turner, R. T., Sattgast, L. H., Jimenez, V. A., Grant, K. A., & Iwaniec, U. T. (2021). Making sense of the highly variable effects of alcohol on bone. *Clinical Reviews in Bone and Mineral Metabolism, 19*(1–4). https://doi.org/10.1007/s12018-021-09277-8

Ulrich, D., van Rietbergen, B., Laib, A., & Ruegsegger, P. (1999). The ability of three-dimensional structural indices to reflect mechanical aspects of trabecular bone. *Bone, 25*(1), 55–60. https://doi.org/10.1016/s8756-3282(99)00098-8

van der Meulen, M. C. H., J Jepsen, K., & Mikić, B. (2001). Understanding bone strength: Size isn't everything. *Bone, 29*(2), 101–104. https://doi.org/10.1016/s8756-3282(01)00491-4

Van Rietbergen, B. (2001). Micro-FE analyses of bone: State of the art. *Advances in Experimental Medicine and Biology, 496*, 21–30. https://doi.org/10.1007/978-1-4615-0651-5_3. http://www.springer.com/series/5584

Vashishth, D., Gibson, G. J., I Khoury, J., Schaffler, M. B., Kimura, J., & P Fyhrie, D. (2001). Influence of nonenzymatic glycation on biomechanical properties of cortical bone. *Bone, 28*(2), 195–201. https://doi.org/10.1016/s8756-3282(00)00434-8

Vashishth, D., Verborgt, O., Divine, G., Schaffler, M. B., & Fyhrie, D. P. (2000). Decline in osteocyte lacunar density in human cortical bone is associated with accumulation of microcracks with age. *Bone, 26*(4), 375–380. https://doi.org/10.1016/s8756-3282(00)00236-2

Walker, M. D., Shi, S., Russo, J. J., Liu, X. S., Zhou, B., Zhang, C., Liu, G., McMahon, D. J., Bilezikian, J. P., & Guo, X. E. (2014). A trabecular plate-like phenotype is overrepresented in Chinese-American versus Caucasian women. *Osteoporosis International, 25*(12), 2787–2795. https://doi.org/10.1007/s00198-014-2816-0

Wang, J., Kazakia, G. J., Zhou, B., Shi, X. T., & Guo, X. E. (2015). Distinct tissue mineral density in plate- and rod-like trabeculae of human trabecular bone. *Journal of Bone and Mineral Research, 30*(9), 1641–1650. https://doi.org/10.1002/jbmr.2498

Warden, S. J., Mantila Roosa, S. M., Kersh, M. E., Hurd, A. L., Fleisig, G. S., Pandy, M. G., & Fuchs, R. K. (2014). Physical activity when young provides lifelong benefits to cortical bone size and strength in men. *Proceedings of the National Academy of Sciences of the United States of America, 111*(14), 5337–5342. https://doi.org/10.1073/pnas.1321605111

Watson, S. L., Weeks, B. K., Weis, L. J., Harding, A. T., Horan, S. A., & Beck, B. R. (2018). High-intensity resistance and impact training improves bone mineral density and physical function in postmenopausal women with osteopenia and osteoporosis: The LIFTMOR randomized controlled trial. *Journal of Bone and Mineral Research, 33*(2), 211–220. https://doi.org/10.1002/jbmr.3284

Welt, C. K., McNicholl, D. J., Taylor, A. E., & Hall, J. E. (1999). Female reproductive aging is marked by decreased secretion of dimeric inhibin. *Journal of Clinical Endocrinology and Metabolism, 84*(1), 105–111. https://doi.org/10.1210/jc.84.1.105

Zysset, P. K., Edward Guo, X., Edward Hoffler, C., Moore, K. E., & Goldstein, S. A. (1999). Elastic modulus and hardness of cortical and trabecular bone lamellae measured by nanoindentation in the human femur. *Journal of Biomechanics, 32*(10), 1005–1012. https://doi.org/10.1016/S0021-9290(99)00111-6

CHAPTER

Bone fracture toughness 11

Alessandra Carriero[1] and Elizabeth A. Zimmermann[2]

[1]*The City College of New York, Department of Biomedical Engineering, New York, NY, United States;* [2]*McGill University, Faculty of Dental Medicine and Oral Health Sciences, Montreal, QC, Canada*

1. Introduction

Bone is a dynamic living tissue with exceptional mechanical properties and the ability to adapt to mechanical loads (Docaj & Carriero, 2024; Muñoz et al., 2021). Bone derives its strength and toughness from its composite nature, made primarily of mineral crystals and organic matrix (mainly collagen), organized in a complex hierarchical structure (Fig. 11.1). Because of its composite nature, bone gains properties from all its phase constituents (Muñoz et al., 2021; Ritchie, 2011). Understanding how bone develops its unique mechanical properties and preserves its function and adaptability is extremely important to understand healthy and compromised bone, develop treatments for aging and disease, and inform biomimetic materials. This chapter focuses on bone fracture toughness, that is bone resistance to fracture, which currently is the best descriptor of bone fragility. In this chapter, we define bone fracture toughness, ii) explore methods to quantify toughness in human and animal bone, iii) dive into the ability of bone to develop fracture toughness and shield growing cracks, and iv) explain how the hierarchical structure of bone contributes to its resistance to fracture.

2. How to measure bone fracture toughness in the lab?

2.1 Strength versus toughness versus fracture toughness for bone

Bone fracture toughness must not be confused with *bone strength* and *bone toughness*. Bone strength measures the ability of bone to resist deformation under loads. However, unlike the yield strength, the ultimate strength is not a material constant in bone (as in any other material with a complex hierarchical structure), but depends on the presence of preexisting defects within the tested samples (Ritchie et al., 2008). The bone toughness measured by the *work to fracture*, i.e., the area under the load-displacement curve, represents the energy dissipated during break. Bone strength and bone toughness are two parameters assessed from mechanical testing of unnotched samples, and can be tested in any sample shape and loading configuration

Biomechanics and Mechanobiology of Bone. https://doi.org/10.1016/B978-0-443-26739-0.00016-8

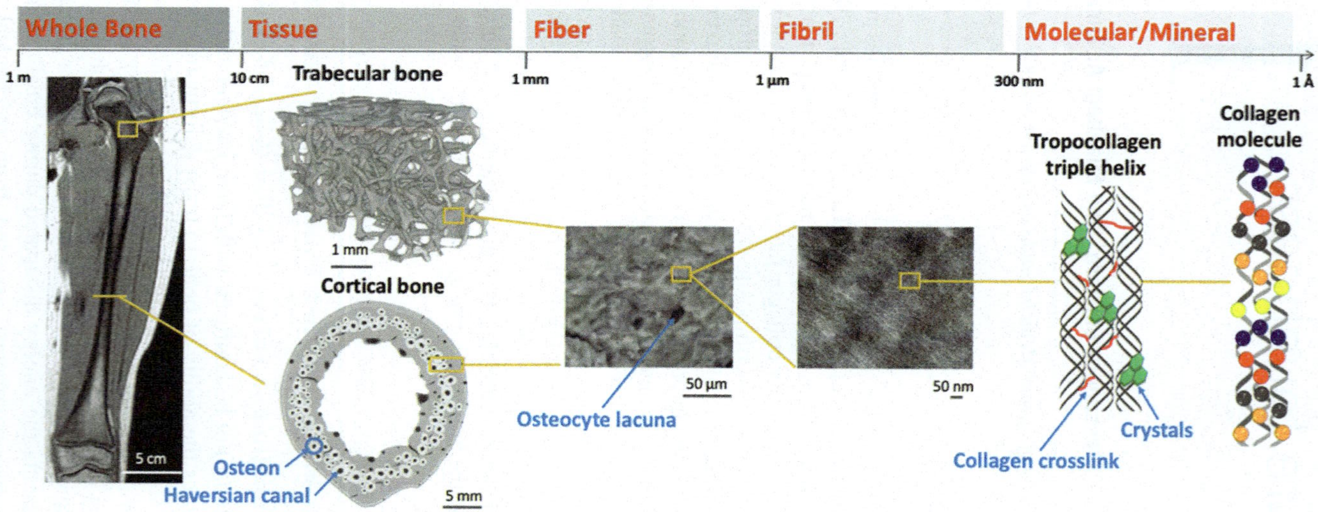

FIGURE 11.1 The hierarchical structure of human bone from macro- to nanoscale.

For an accurate description of bone matrix properties and the relevance of bone quality versus quantity refer to Docaj and Carriero (2024) and Muñoz et al. (2021), respectively.

-Adapted from Docaj, A., & Carriero, A. (2024). Bone health: Quality vs. quantity. Journal of the Pediatric Orthopedic Society of North America, 7. *https://doi.org/10.1016/j.jposna.2024.100054*

(compression, tension, bending, torsion) (Ritchie et al., 2008). However, the value of bone toughness tells us little about how bone actually fractures. To do that in mechanics we use what is called ***fracture toughness***, that is the driving force required to fracture bone in presence of a defect, i.e., a crack. Although Arnold Griffith (1920) is the official father of the fracture mechanics field, observations that materials with defects or flaws would have different ultimate strength were recorded already several centuries before by Leonardo da Vinci. Fracture mechanics investigates the influence of defects in materials and establishes the quantitative relationship between the defects dimensions, the applied load, and the material resistance (Anderson, 2005). Fracture mechanics tests are more sophisticated than strength and toughness tests, as they require the insertion of a sharp notch, a "worst-case" crack, from which the crack will start growing, to minimize the mechanical contribution of preexisting flaws and microcracks present in the bone sample (Bonfield, 1987). Fracture mechanics also requires the use of imaging techniques, such as an environmental scanning electron microscopy (ESEM), to follow the crack extension and obtain geometrical parameters of the fracture surface.

ASTM standards have been developed for fracture mechanics calculations (ASTM E1820-18, 2019; ASTM E399-24, 2024) that can be applied on bone in the presence of a worst-case flaw, an introduced defect such as a precrack or sharp notch (Bonfield, 1987). This notch is usually created either manually or with a notcher machine by using a razor blade irrigated with 0.05 μm diamond suspension to generate a notch root radius of less than 10 μm. Fracture mechanics tests can be conducted in any of these three different modes (Fig. 11.2A) or a combination of two or three modes:

- Mode I—opening mode: principal load applied normal to the crack surface pulls the crack open.
- Mode II—sliding mode: in-plane shear loading slides one crack face with respect to the other.
- Mode III—tearing mode: out-of-plane shear loading slides the crack faces in parallel and perpendicular to the crack front.

For such tests, the critical stress-intensity factor K_{IC} can be measured as it characterizes the local stress field at the crack tip (ASTM E1820-18). For example, for mode I, K_{IC} is expressed as:

$$K_{IC} = F\sigma(\pi a)^{1/2}$$

where F is a function of sample geometry, σ is the remote tensile stress perpendicular to the crack plane, and a is the crack length. K_C can also be calculated for other fracture modes, and, therefore, is a mode-, size-, and geometry-dependent quantity and serves as a measurement of the bone fracture toughness to initiate and progress the crack.

If significant plastic deformation is expected, LEFM should not be used as it assumes negligible plastic deformation. Indeed, to use LEFM stress intensity solutions, the plasticity should satisfy small-scale yielding and plane strain conditions

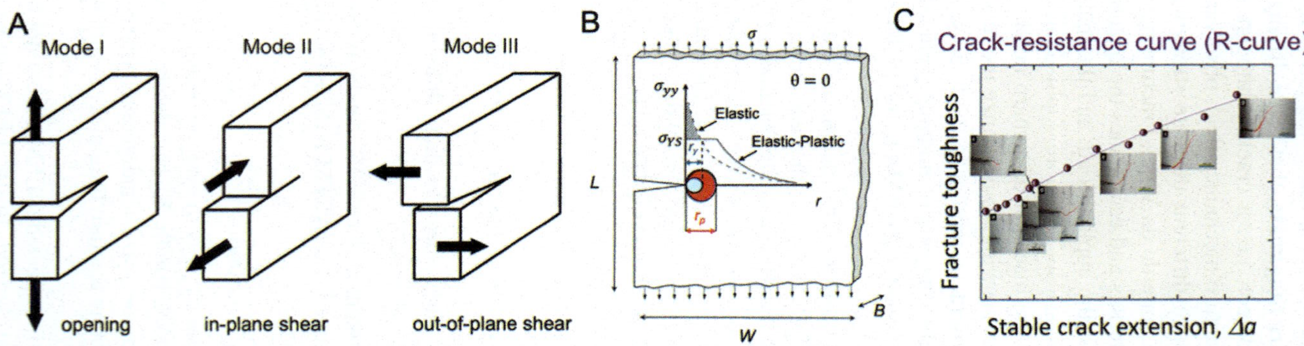

FIGURE 11.2 Fracture mechanics modes, plastic zone around a crack, and crack-resistance curves.

(A) The three modes of fracture mechanics. (B) Schematic of the plastic zones for an elastic material (light blue) and for an elastic-plastic material (red) around an ideal infinite sharp crack. (C) Schematic demonstrating how to build a crack-resistance curve (R-curve) of bone by plotting the instantaneous fracture toughness at each instance of the crack extension shown in the ESEM pictures and highlighted in red for visual purposes.

- Fig. 11.2A is adapted with permission from Zimmermann, E. A., Launey, M. E., & Ritchie, R. O. (2010). The significance of crack-resistance curves to the mixed-mode fracture toughness of human cortical bone. Biomaterials, 31*(20), 5297–5305. https://doi.org/10.1016/j.biomaterials.2010.03.056.*

(Fig. 11.2B), according to ASTM standards (ASTM E1820-18). For small-scale yielding, the plastic zone size should be an order of magnitude smaller than the crack length (a) and the uncracked ligament ($W-a$). For plane strain conditions to be satisfied, plastic zone size should be an order of magnitude smaller than the samples thickness (B), which ensures that the measured toughness is independent of geometry and sample thickness. The plastic zone size, r_p, for an infinitely sharp crack is approximate as:

$$r_p = \frac{1}{\pi}\left(\frac{K}{\sigma_{YS}}\right)^2$$

where K is the predicted toughness in terms of stress-intensity factor, and σ_{YS} is the tensile yield strength of the material. To satisfy small-scale yielding and plane strain conditions, ASTM standards (ASTM E1820-09, ASTM_E399-24) specify:

$$a, W - a, B > 2.5\left(\frac{K}{\sigma_{YS}}\right)^2$$

where a is the crack length, B is the thickness of the sample and $W - a$ is the uncracked ligament.

Bone fracture toughness can also be expressed in terms of energy release rate G, which is the increase in energy released with crack growth:

$$G = -\frac{d\Pi}{da}$$

where Π is the potential energy of an elastic body, defined as: $\Pi = U - F$ where U is the strain energy stored in the body and F is the work done by the external forces. G approach is a good choice for mixed-mode fracturing. For linear elastic materials loaded in mode I:

$$G = {K_I}^2/E$$

In presence of plastic deformation, non-linear-elastic J-integral fracture mechanics is used as representing the non-linear energy release rate per fracture area that is formed. The J-integral includes the contributions from plastic deformation and can be broken into the elastic, J_{el}, and plastic, J_{pl}:

$$J = J_{el} + J_{pl}$$

For linear elastic materials, G and J are equivalent. For linear elastic mode I loading,

$$J = {K_I}^2/E$$

Bone is an extraordinary material that develops fracture toughness during crack growth; thus, a single value of the toughness cannot completely describe its fracture behavior. Crack resistance curves, or R-curves, allow to study the contribution of crack growth to bone fracture toughness (Fig. 11.2C). R-curves can be built for any type of fracture toughness and sample geometry by considering the increments of the stable crack growth in bone and measuring the instantaneous fracture toughness at each progression of crack extension (Fig. 11.2C). As such, crack extension is monitored with the compliance method (not covered here) or collected using in situ microtesting rigs usually operated in an environmental scanning electron or light microscope (Carriero, Zimmermann, Shefelbine, et al., 2014). In the crack growth resistance curves, the K_R at $\Delta a \rightarrow 0$ is the crack-initiation toughness ($K_{\text{Initiation}}$), which represents only the intrinsic toughness of the bone before the crack even starts, and the slope of the R-curve is the crack-growth toughness (K_{Growth}), which represents the extrinsic toughness of bone developed during crack growth.

Once the bone is completely broken, the fracture surface of the bone can be imaged in a SEM in backscattering mode to determine the exact extent of the notch and of the stable crack (Fig. 11.3). In SEM images, the notch, the region of stable crack growth and the onset of instability all have distinguishable morphologies. The notch can be identified by the smooth flat appearance that contrasts the darker groove-like surface of the stable crack region, and the less rough instable crack region (Carriero, et al., 2014; Ritchie et al., 2008). Surface fractography is very informative—some examples will be presented later in this Chapter.

2.2 Fracture toughness measurements in human or large vertebrate bone

The fracture toughness in long bones of humans or larger vertebrates is most often measured under tensile mode I conditions at the crack tip, which in most materials generates the least resistance to crack extension (Olvera et al., 2012; Zimmermann et al., 2009, 2010). The most common geometries are the single edge bend, SE(B), specimen or the compact tension specimen (Fig. 11.3). In the SE(B), the notched sample is loaded in three-point bending with the middle pin aligned with the notch, to put the notch in tension (Fig. 11.3). For the SE(B), the geometry is often created from a piece of cortical bone using a hand saw and low speed saw followed by grinding and polishing, if needed. The compact tension specimen is also popular; it has two pin holes above and below the notch that are used to apply tensile loads to create tension at the notch (Fig. 11.3). As such, the compact tension specimen requires a milling machine.

2.2.1 Linear elastic fracture mechanics

Linear elastic fracture toughness can be used to investigate the toughness of large vertebrate bone samples. A critical value of the stress intensity factor, K_{IC}, or the crack growth resistance curve, K_R, can be measured. To perform the SE(B) test, ASTM E1820 provides the recommended test setup (ASTM E1820-18). The sample is notched such that the ratio of crack length, a, to sample width, W, is between $0.45 \leq a/W \leq 0.70$. The support span, S, for SE(B) should equal four times the width, W, and the thickness, B, should be approximately $0.5W$. Then, the stress intensity, K_i, can be calculated as follows:

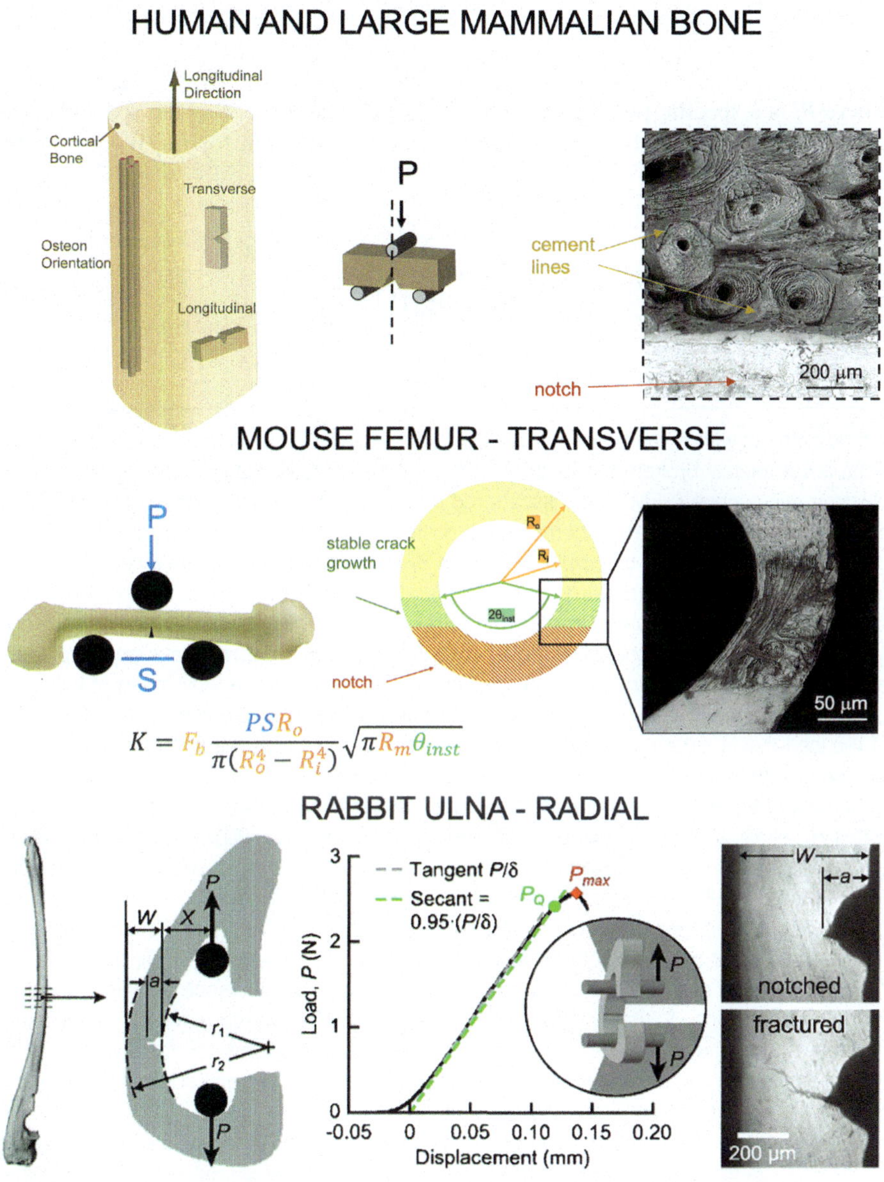

FIGURE 11.3 Fracture toughness measurements in large and small vertebrate bone.

Schematic of methodologies for calculating fracture toughness in mode I in human and large mammalian bones in any orientation, and in small animals, specifically, in mouse bone in the transverse direction and in rabbit ulna in the radial direction. (Top) Schematic of the human or large mammalian bone showing how to obtain transverse and longitudinal beams to be tested in three-point bending or compact tension samples that

$$K_i = \left[\frac{P_i S}{BW^{3/2}}\right] f(a_i / W)$$

where P_i is a specific point on the load-displacement curve, S is the span, B is the sample thickness, and W is the sample width. The solution for $f(a_i/W)$ for the SE (B) geometry is:

$$f\left(\frac{a_i}{W}\right) = \frac{3\left(\frac{a_i}{W}\right)^{1/2}\left[1.99 - \left(\frac{a_i}{W}\right)\left(1 - \frac{a_i}{W}\right)\left(2.15 - 3.93\left(\frac{a_i}{W}\right) + 2.7\left(\frac{a_i}{W}\right)^2\right)\right]}{2\left(1 + 2\frac{a_i}{W}\right)\left(1 - \frac{a_i}{W}\right)^{3/2}}$$

will be notched with a sharp single side notch and tested in tension. After the test when the bone is completely broken, the fracture surface of the bone is imaged with a scanning electron microscope (SEM) in backscattered mode to show the geometry of the notch and crack growth area. (Center) Schematic of a posteriorly notched mouse femur under three-point bending, where P is the applied load and S is the span, together with a notched mouse femur cross section after fracture has propagated though the bone. Bone fracture toughness can be calculated by imaging the fracture surface with an SEM in backscattered mode and taking measurements of the inner radius R_i, outer radius R_o, the half-crack angles of the notch θ_{notch}, and instantaneous crack growth θ_{inst}, as shown in the SEM insert of the crack growth. For the crack-initiation toughness, the load P is equal to P_c given by the intersection of the load-displacement curve with the 5% secant stiffness. For the crack-growth toughness, the load P is the load P_f at the instability point, that is the highest value of the load before a sudden drop. For more information, please refer to Carriero, Zimmermann, Paluszny, et al., 2014, Carriero, Zimmermann, Shefelbine, et al., 2014, Ritchie et al., 2008. (Bottom) MicroCT reconstruction of a rabbit ulna mid-diaphysis cross-section showing a gap cut into the cranial cortex to isolate the caudal cortex for tensile loading, and a notch in the endosteal caudal cortex of the bone. Two pins are inserted in the cross-section to apply tensile load, P, for mode I fracture mechanics. Geometrical properties of the bone are reported (internal radius, $r1$, external radius, $r2$) as well as the distance between the notch mouth and load line, X, that will be used to calculate the fracture toughness. For initiation toughness, the load is considered as the 5% secant stiffness. However, instability fracture toughness can be calculated by considering the extension of the crack and instability load to include the contribution of crack growth toughness.

- Figure 11.3 is made with some figures adapted with permission from Zimmermann, E. A., Gludovatz, B., Schaible, E., Busse, B., & Ritchie, R. O. (2014). Fracture resistance of human cortical bone across multiple length-scales at physiological strain rates. Biomaterials, 35*(21), 5472–5481. https://doi.org/10.1016/j.biomaterials.2014.03.066; Zimmermann, E. A., Launey M. E., & Ritchie R. O. (2010). The significance of crack-resistance curves to the mixed-mode fracture toughness of human cortical bone. Biomaterials 31(20): 5297–5305. https://doi.org/10.1016/j.biomaterials.2010.03.056; Hunckler M.D., Chu, E.D., Baumann, A. P., Curtis, T.E., Ravosa, M.J., Allen, M.R., & Roeder, R.K. (2017). The fracture toughness of small animal cortical bone measured using arc-shaped tension specimens: Effects of bisphosphonate and deproteinization treatments.* Bone, 105, *67–74. https://doi.org/10.1016/j.bone.2017.08.015.*

2.2.2 J-integral

A different approach is to use the J-integral, which is the non-linear strain energy release rate. The elastic, J_{el}, and plastic, J_{pl}, components given in ASTM E1820-18 for the SE(B) geometry are:

$$J_{el} = \frac{K_I^2}{E} \text{ and } J_{pl} = \frac{1.9A_{pl}}{Bb}$$

Where E is the Young's modulus, A_{pl} is the plastic area under the force-displacement curve, B is the sample thickness, and b is the uncracked ligament (equivalent to W-a). From the J-integral results, the stress intensity K_J can be back calculated using the formula for J-K equivalence for mode I failure:

$$K_J = (JE)^2$$

Where E is the Young's modulus of the material.

2.2.3 R-curves

To measure the R-curve using K or J, the crack growth must be monitored. The compliance method can be used or the crack length can be observed using a camera or microscope. Refer to ASTM E1820-18 for further explanation of the compliance method (ASTM_E1820-18). There is a valid range for the solution of the J-integral. The maximum crack extension valid for the J-integral is $\Delta a_{\max} = 0.25b$, where b is the uncracked ligament length ($b = W - a$). Additionally, the maximum J-integral capacity is either $J_{\max} = b\sigma_{YS}/10$ or $J_{\max} = B\sigma_{YS}/10$. However, it should be noted that these limits were derived for metallic materials.

2.2.4 Orientation

During mechanical testing of bone, the orientation of the microstructure with respect to the crack is critical. Notably, the fracture toughness of large vertebrate bone can be assessed in the longitudinal (in-plane), transverse, or radial (i.e., antiplane longitudinal) orientations (Fig. 11.3). Here, we will focus on the longitudinal and transverse orientations, where the initial crack is either parallel or perpendicular to the weak material interfaces at the microstructural scale. As such, these two orientations give rise to different toughening mechanisms, which will be described later, and to fracture toughness that are an order of magnitude difference (transverse ~25 MPa$\sqrt{\text{m}}$ and longitudinal ~2.5 MPa$\sqrt{\text{m}}$ (Koester et al., 2008)), thus demonstrating that bone is naturally optimized to resist transverse or spiral fractures happening in bending or torsion (twisting).

2.2.5 Modes of loading

Most fracture toughness tests are performed in mode I, where the crack tip is subject to tensile opening loads (Fig. 11.2A). Most materials are weakest in mode I (Zimmermann et al., 2009); therefore, it is the first choice to understand failure of materials. In reality, cracks are subject to mixed mode loading with combinations of mode I (tensile opening), mode II (in plane shear) or mode III (antiplane shear).

There are a number of different testing configurations to apply pure mode I, mode II or mode III as well as mixed-mode loading. Some notable geometries are the compact tension or the compact shear geometries used by (Brown et al., 2000) in human bone. The asymmetric three-point bending configuration (Olvera et al., 2012; Zimmermann et al., 2010) and the longitudinal crack in far field compression (Olvera et al., 2012) apply combinations of mode I and mode II.

2.3 Fracture toughness in small animal bone

Small animals, particularly mice, are widely used as models of human bone for the studies of bone mechanobiology, disease and therapies. Therefore, there has been a great interest in developing methods for the bone fracture toughness characterization of small animal bones, particularly for testing mode I fracture toughness. Here, we present the most widely used methods.

2.3.1 Transverse fracture toughness

In very small animals, such as mice, it is difficult to mill out perfect beam samples as in large vertebrate bone; thus, a full bone is used for the fracture mechanics testing in the transverse direction. In mice, femurs are ideal bones for the assessment of fracture toughness properties and a solution of LEFM for a circular thick-walled cylinder with a circumferential through-thickness crack has been used (Carriero, Zimmermann, Paluszny, et al., 2014; Carriero, Zimmermann, Shefelbine, et al., 2014; Ritchie et al., 2008). Mouse femurs have been tested in three-point bending with a micronotch at their mid-diaphysis (Carriero, Zimmermann, Paluszny, et al., 2014; Carriero, Zimmermann, Shefelbine, et al., 2014; Ritchie et al., 2008), mostly at their flat posterior side (Fig. 11.3). The micronotch is generated using either a notcher machine with a razor blade irrigated with a 0.05 μm diamond suspension, or with a slow-speed diamond saw to make a cut through the cortical wall that is then sharpened manually using a razor blade with a 0.05 μm diamond suspension. The notch resulting from this micronotching technique usually generates a root radius of ~5 μm, perfect to simulate the presence of a worst-case defect in the bone from which the notch would start. Fracture mechanics tests are conducted in wet conditions at 25–37°C, and performed in general accordance with ASTM standards (ASTM E1820-18). Previously, a displacement rate of 1–1.6 μm/s was used (Carriero, Zimmermann, Shefelbine, et al., 2014; Carriero, Zimmermann, Paluszny, et al., 2014; Carriero, J.L. Bruse, Oldknow, Millán, et al., 2014; Miller et al., 2017; Muñoz et al., 2026b). Other rates can also be used, but increasing the rate will reduce the mechanisms contributing to the fracture toughness.

In the case of a single notched, three-point bend mouse femur, the critical stress intensity factor K_c can be used to measure the bone fracture toughness, after getting geometry and stable crack growth information from SEM images of the bone fracture surface (Fig. 11.3). Particularly, the crack-initiation toughness and the fracture toughness at the instability (method that accounts for the contributions to the toughness of the stable crack growth as well as the initiation of cracking) (Carriero,

Zimmermann, Paluszny, et al., 2014; Carriero, Zimmermann, Shefelbine, et al., 2014; Muñoz et al., 2026b; Ritchie et al., 2008) can be measured with the following equation:

$$K_c = F_b \frac{PSR_o}{\pi(R_0^4 - R_i^4)} \sqrt{\pi R_m \theta}$$

where, F_b is a geometry factor, which depends on the bone cortex thickness, S is the span in three-point bending and R_o, R_i and R_m are the outer, inner, and mean radius of the cortex. P_c and θ_c change relative to the point on the load versus displacement curve where the toughness is assessed. Specifically,

$$F_b = \left(1 + \frac{t}{2R_m}\right)\left[A_b + B_b\left(\frac{\theta}{\pi}\right) + B_b\left(\frac{\theta}{\pi}\right) + C_b\left(\frac{\theta}{\pi}\right)^2 + D_b\left(\frac{\theta}{\pi}\right)^3 + E_b\left(\frac{\theta}{\pi}\right)^4\right]$$

$$A_b = 0.65133 - 0.5774\varepsilon - 0.3427\varepsilon^2 - 0.0681\varepsilon^3$$

$$B_b = 1.879 + 4.795\varepsilon + 2.343\varepsilon^2 - 0.6197\varepsilon^3$$

$$C_b = -9.779 - 38.14\varepsilon - 6.611\varepsilon^2 + 3.972\varepsilon^3$$

$$D_b = 34.56 + 129.9\varepsilon + 50.55\varepsilon^2 + 3.374\varepsilon^3$$

$$E_b = -30.82 - 147.6\varepsilon - 78.38\varepsilon^2 - 15.54\varepsilon^3$$

$$\varepsilon = log\left(\frac{t}{R_m}\right)$$

For the crack-initiation toughness, $P = P_c$ is the critical load at yielding (Fig. 11.3, given by the intersection of the loading curve with a line that has a 5% lower slope than the elastic deformation slope (5% secant construction)) and $\theta = \theta_c$ the notch angle (Ritchie et al., 2008). For the instability fracture toughness approach, which reliably and appropriately considers the contribution of both crack-initiation and propagation, $P = P_f$ given by the load at unstable fracture (instability, given by the value of the applied load right before a sudden drop - Fig. 11.3) and $\theta = \theta_{inst}$ is the instability angle (Fig. 11.3), given that the instability angle is smaller than 220 degrees (Carriero, Zimmermann, Paluszny, et al., 2014; Carriero, Zimmermann, Shefelbine, et al., 2014; Ritchie et al., 2008). The crack-initiation toughness and instability fracture toughness in terms of stress-intensity factor for mouse femurs are usually between 2–4 MPa$\sqrt{\text{m}}$ and 4.5–8 MPa$\sqrt{\text{m}}$, respectively (Ritchie et al., 2008; Carriero, Zimmermann, Shefelbine, et al., 2014; Carriero, Zimmermann, Paluszny, et al., 2014; Carriero, J.L. Bruse, Oldknow, Millán, et al., 2014; Miller et al., 2017; Muñoz et al., 2026b). As this is a LEFM fracture toughness measurement, it considers the bone as elastic with very limited plasticity when the fracture is happening in the sample according to ASTM Standards (ASTM E1820-18).

In the occurrence of plastic deformation, non-linear-elastic fracture mechanics is considered to provide a more accurate evaluation of toughness. A solution for J for a crack growth in a pipe was proposed (Ritchie et al., 2008), but it is only valid for R_m/

$t \geq 5$, where R_m is the mean cortical bone radius and t the mean wall thickness (Ritchie et al., 2008). In most small mammalian bones, such as mice, the ratio ranges from 1 to 4, invalidating the *J*-solution. Thus, the work of fracture W_f, has been used to account for the energy released during excessive plastic fracture toughness. The work of fracture is defined as the work per unit area of fracture. As such, it is determined from the area under the load-displacement curve divided by twice the area of the fracture surface (without counting the notch area) (Carriero, Zimmermann, Paluszny, et al., 2014; Carriero, Zimmermann, Shefelbine, et al., 2014):
where E_f is the energy of fracture (*i.e.*, the instantaneous area under the load-displacement curve) and A_{res} is the area of the bone surface resisting the crack propagation, *i.e.*, the total area of the bone minus the notch area.

2.3.2 Longitudinal fracture toughness

If the interest is to calculate fracture toughness in the longitudinal direction, in the bigger small animals, such as rabbits, cortical bone radial fracture toughness can be measured using ulna mid-diaphysis notched arc-shaped tension specimens (Fig. 11.3) (Hunckler et al., 2017). Samples of 1 mm thickness can be prepared by cutting the ulnae with a low-speed saw and notching them on the endosteal surface of their caudal side with a notch initiating a crack in the radial direction. In a previous study (Hunckler et al., 2017), the notch was created by using a low-speed diamond wire saw with a 0.3 mm diameter followed by a razor blade and 1 μm diamond suspension for a notch depth, *a*, of ~0.15 mm with a root radius of 10 μm. Samples were loaded to failure at a displacement rate of 0.02 mm/s, while hydrated in PBS at ambient temperature (Hunckler et al., 2017), and following guidelines for arc-shaped specimens outlined in ASTM E399-24 (2024). Because bone is easier to split than break, the radial fracture toughness for healthy rabbit bone is the radial orientation was very small, on the order of 1–1.1 $\text{MPa}\sqrt{\text{m}}$ (Hunckler et al., 2017), which is smaller than the toughness of longitudinal human humeral cortical bone (~1–2 $\text{MPa}\sqrt{\text{m}}$) measured with SE(B) beams (Koester et al., 2008). This method was used to only measure initiation toughness. If tests are conducted in an ESEM, or if fracture surface is imaged after complete break, it is possible to examine the stable crack extension and measure the contribution of crack-growth toughness to overall bone resistance to fracture.

2.3.3 R-curves

Similar to the human and large vertebrate bone samples, we can assess the contribution of crack growth to fracture toughness K_C by calculating the R-curve for the small animal bones. In mouse bone, R-curves can be measured for a notched femur undergoing three-point bending using an in situ microtesting rig in an ESEM in backscattered mode, for example (Carriero, Zimmermann, Paluszny, et al., 2014; Carriero, Zimmermann, Shefelbine, et al., 2014), so that the applied load and the length of the extending crack can be monitored. This allows for the fracture toughness to be calculated at each increment of the stable crack extension using an in situ loading stage in an ESEM (Carriero, Zimmermann, Paluszny, et al., 2014; Carriero,

Zimmermann, Shefelbine, et al., 2014). Bones for the R-curves are prepared in the same way as for the ex situ testing. However, when conducting the in situ testing, the periosteal surface of the mouse femurs needs to be polished to expose the osteocyte lacunae and the canals so they are visible within the ESEM (Carriero, Zimmermann, Paluszny, et al., 2014; Carriero, Zimmermann, Shefelbine, et al., 2014). Also before testing, it is recommended that the samples are immersed in ambient physiological solution for at least 12 h before testing (Carriero, Zimmermann, Paluszny, et al., 2014; Carriero, Zimmermann, Shefelbine, et al., 2014). R-curves for mouse bone can be calculated in terms of the crack-driving force, *i.e.*, the stress intensity factor K, and the energy absorbed, *i.e.*, the work to fracture W_f, as a function of the crack extension, Δa (Carriero, Zimmermann, Paluszny, et al., 2014; Carriero, Zimmermann, Shefelbine, et al., 2014). The instantaneous W_f is determined from the area under the load-displacement curve divided by twice the area of the fracture surface at the specific increment of the crack extension (Carriero, Zimmermann, Paluszny, et al., 2014; Carriero, Zimmermann, Shefelbine, et al., 2014). For the longitudinal fracture toughness R-curve, the value of the instantaneous crack extension, Δa, can be geometrically calculated along the circumference of the bone (Carriero, Zimmermann, Paluszny, et al., 2014; Carriero, Zimmermann, Shefelbine, et al., 2014).

These test methods to measure fracture toughness in small animal bones are precise, detect small differences within and from normal, and meet the ASTM requirements for fracture toughness (ASTM E1820-18). These methods have greatly improved the scientific understanding of bone fracture mechanics in relation to bone quality, and metabolic bone disease (Carriero, Zimmermann, Paluszny, et al., 2014; Carriero, Zimmermann, Shefelbine, et al., 2014; Hunckler et al., 2017; Ritchie et al., 2008). In particular, R-curves help to identify the dominant sources of toughness in bone and their contributions to the bone's resistance to the growth of cracks. In this way, researchers can comprehensively characterize and compare the fracture toughness of bones with a different phenotype, e.g., healthy and brittle bone, by looking at the extent of the stable crack, the slope and the value of the crack-initiation toughness.

3. Bone toughening mechanisms

Bone is a damage tolerant material. Indeed, bone absorbs mechanical energy through the creation of damage in the form of microcracks or diffuse damage (at smaller length-scales) (Boyce et al., 1998; Burr et al., 1998). Damage does not necessarily result in catastrophic failure and with time this damage can be repaired (Cardoso et al., 2009; Kennedy et al., 2014; Seref-Ferlengez et al., 2014). When a crack begins to propagate, physical mechanisms within the structure become activated and physically resist further growth of the crack (Zimmermann & Ritchie, 2015). Many of these toughening mechanisms were first observed in the field of ceramics, where manipulation or engineering of the microstructure induces resistance to the growth of cracks in naturally brittle ceramics.

In bone, toughening occurs in two forms: *i)* intrinsic plasticity mechanisms active ahead of the crack tip and *ii)* extrinsic crack tip shielding mechanisms activate in the crack wake (Fig. 11.4). As a sample with a notch is loaded in mode I, an elevated tensile stress occurs ahead of the notch. At the material level, elastic deformation occurs first, characterized by stretching of bonds. Then, if the stress ahead of the crack tip surpasses the yield stress, plastic deformation occurs. Plastic deformation dissipates energy and intrinsically toughens bone; these mechanisms usually occur at small length-scales. Some examples of intrinsic toughening in bone that will be discussed below include fibrillar sliding, sacrificial bonding, reorientation, inhomogeneous stretching. Extrinsic toughening mechanisms are activated once the crack begins to grow. These mechanisms interact with the crack to shield it from the full stress intensity. In bone, some examples are crack deflection, crack bridging, microcracks toughening, and splitting.

3.1 Intrinsic

3.1.1 Collagen deformation

Collagen is found in many mineralized and nonmineralized tissues within the body, such as skin, ligaments, cornea, nerves, bone and dentin. Within these tissues, collagen imparts toughness. As cortical bone is mechanically loaded, synchrotron small-angle X-ray scattering studies show that 59%–75% of the tissue level strain is transferred to the mineralized collagen fibril (Manrique et al., 2024). At first, the fibrils stretch elastically and there is a linear relationship between fibril strain and tissue strain. After elastic deformation, the fibrils strain either plateaus or becomes inhomogeneous indicating that deformation is taken up at different length-scales and not transferred to the fibril (Manrique et al., 2024; Zimmermann et al., 2011).

3.1.2 Fibrillar sliding and sacrificial bonding

During elastic deformation, there is a linear relationship between collagen fibril deformation and the tissue-level deformation. After this regime, it has been observed that the collagen fibril strain can plateau. This implies that the fibrils are maintaining a level of strain while the whole tissue continues to deform—thus the fibrils are sliding past one another (Gupta et al., 2005). This mechanism is consistent with the model of collagen fibrils stretching until yield of the material, after which point the interfibrillar matrix shears (Gupta et al., 2005). Indeed, there may be aspects of the interfibrillar or extracellular matrix that break and reform (i.e., sacrificial debonding) to allow plastic deformation (Fantner et al., 2005).

3.1.3 Inhomogeneous stretch

Another important but under explored mechanism is inhomogeneous stretching. Here, as the fibrils begin to deform, some fibrils may debond and unload; thus, these debonded fibrils return to their unstrained state, while others continue to deform. In the SAXS pattern, this is visible with broadening of the peak, where multiple

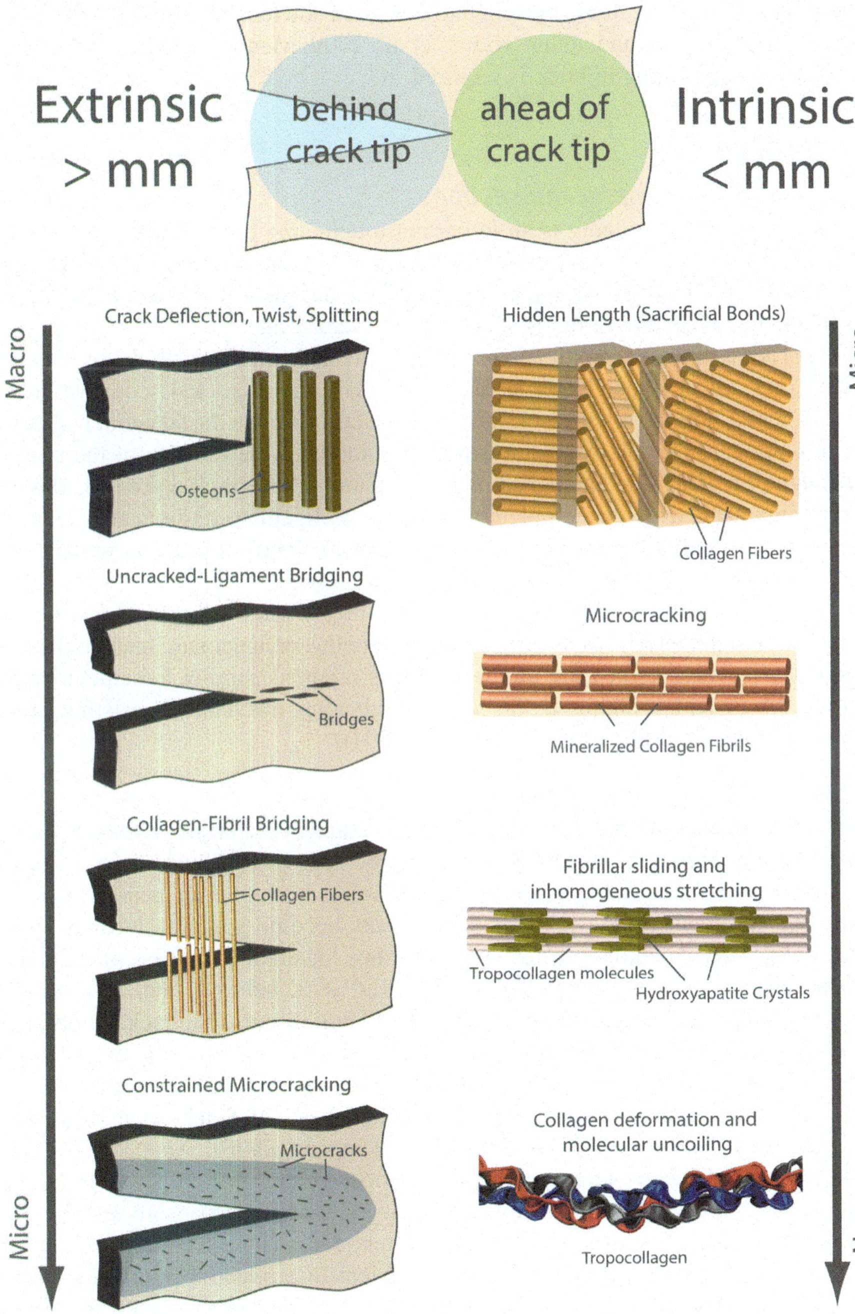

FIGURE 11.4 Intrinsic plasticity and extrinsic mechanisms developing in bone during fracture toughness.

In this figure, the inhomogeneous stretching is not illustrated.

- Adapted from Zimmermann, E. A., Launey, M. E., & Ritchie, R. O. (2010). The significance of crack-resistance curves to the mixed-mode fracture toughness of human cortical bone. Biomaterials, 31*(20), 5297–5305. https://doi.org/10.1016/j.biomaterials.2010.03.056*

subbands are resolvable that correspond to unloaded and loaded fibrils. This was first reported in soft tissues and antler (Krauss et al., 2009; Manrique et al., 2024); however, further exploration in bone is required.

3.2 Extrinsic

3.2.1 Bridging by uncracked sections

Uncracked-ligament bridging is an important toughening mechanism in bone. The term uncracked ligament comes from the ceramics literature and implies a piece of bone tissue uniting the material above and below the crack wake; it does not imply that ligament connective tissue is spanning the crack wake.

The process of uncracked ligament bridging involves microcracks forming ahead of the crack tip in bone (Nalla et al., 2004). The main crack is considered the "mother" crack and the microcracks ahead of the crack tip are the "daughter" cracks. The full amount of crack extension is thus considered to be the sum of the mother and daughter cracks. The uncracked material spanning the crack wake (i.e., the uncracked ligament, which is completely unrelated to ligament tissue) carries part of the load that would otherwise drive the crack forward. Thus, it is acting in the crack wake to shield the crack tip from the full driving force.

Uncracked-ligament bridging is commonly observed in bone. Most often microcracks occur at the interfaces in bone between lamellae or at cement lines. In the longitudinal orientation (when the mother crack is oriented parallel to cement lines), uncracked-ligament bridging is the main toughening mechanism resisting crack extension (Koester et al., 2008; Nalla et al., 2004).

3.2.2 Microcrack toughening

Microcrack toughening has been described in ceramics materials (Evans & Faber, 1984; Shum & Hutchinson, 1990), where microcracks open ahead of the crack tip at flaws or interfaces (such as grain boundaries in ceramics or cement lines in bone). As the crack progresses, the microcracks become spatially located in the crack wake. Microcracking creates voids and thus lowers the stiffness of the material, which could increase toughness. However, it is believed that microcracks mainly create a toughening effect when their dilatation puts the crack tip or crack wake in compression, thereby shielding the tip from the full opening stress (Evans & Faber, 1984; Nalla et al., 2004; Shum & Hutchinson, 1990). This force acting against the crack opening force increases the toughness with crack extension (Evans & Faber, 1984; Nalla et al., 2004; Shum & Hutchinson, 1990).

It has been debated whether this toughening mechanism is at work in human bone. Bone is a damage tolerant material that microcracks to absorb energy. The microcracks have been observed to occur after loading (Vashishth et al., 1997), but the level of microcracking remains insufficient to produce a significant toughening effect (Nalla et al., 2004). Both crack bridging and microcracks are visible in bone during stable crack growth. Nalla et al. (2004) measured the experimental compliance of bone and dentin during fracture toughness testing and compared it with the theoretical compliance in the absence of bridges and microcracks. They found that the compliance of the sample is lower than the theoretical compliance with no toughening mechanisms in the wake of the crack (Nalla et al., 2004); therefore, crack bridging is the dominant mechanism. Furthermore, they estimated that microcrack volume fraction on the order of 60%–90% would be needed to increase toughness by 1.5–3 MPa$\sqrt{m}$, which is extremely high (Nalla et al., 2004).

3.2.3 Crack deflection and twist

If a crack is loaded in opening mode I, such that tensile forces are applied ahead of the crack tip, a homogeneous isotropic material would exhibit crack extension in a plane perpendicular to the tensile forces. Crack deflection out of this mode I plane occurs then the crack has mixed mode at the crack tip, with combinations of tension (mode I) and shear (mode II) depending on the angle of crack deflection.

In bone and ceramic materials, crack deflection occurs because the material is not homogeneous: they contain interfaces that the cracks follow. In human bone, crack deflections mostly follow the cement lines (Fig. 11.3). Thus, in the transverse orientation, when the crack is perpendicular to the cement lines, the crack deflection toughening mechanism is most potent and causes greater increases in toughness than uncracked ligament bridging in the longitudinal orientation (Koester et al., 2008). Deflections are observable on the surface during mechanical testing. However, with 3D imaging using synchrotron microcomputed tomography or by observing the fracture surface, it is apparent that the deflections have a 3D nature (Fig. 11.5). The crack surface is not flat in purely mode I but has a 3D topography that depends on the microstructural features encountered by the extending crack. This crack path phenomenon is called crack twist (Fig. 11.5). Crack deflections and crack twists have also been observed in healthy mouse bone, possibly forming at the lamellar interphases (Fig. 11.5).

3.3 Factors that influence bone toughening mechanisms arising from multiscale structure

3.3.1 Enzymatic cross-links

Enzymatic cross-links, enzyme-catalyzed covalent bonds between molecules, stabilize collagen molecules and fibrils in bone, and, therefore, are important enhancer of crack-initiation toughness. Cross-links affect the plasticity of bone originating at the nanoscale level through the intermolecular fibrillar sliding. Perturbed enzymatic cross-linking amount and/or quality decreases bone fracture toughness in diseases, such as osteogenesis imperfecta (Carriero, Zimmermann, Paluszny, et al., 2014), where collagen can exhibit several defects as a consequence of its primary structure changes, altered post-translational modifications, abnormal intracellular folding and transport, or reduced ability to secrete/incorporate molecules into the matrix. Increased cross-linking also embrittles the material because it stiffens the collagen fibrils, reducing their ability to dissipate energy, leading to molecular fractures rather than plastic deformations (Buehler, 2007).

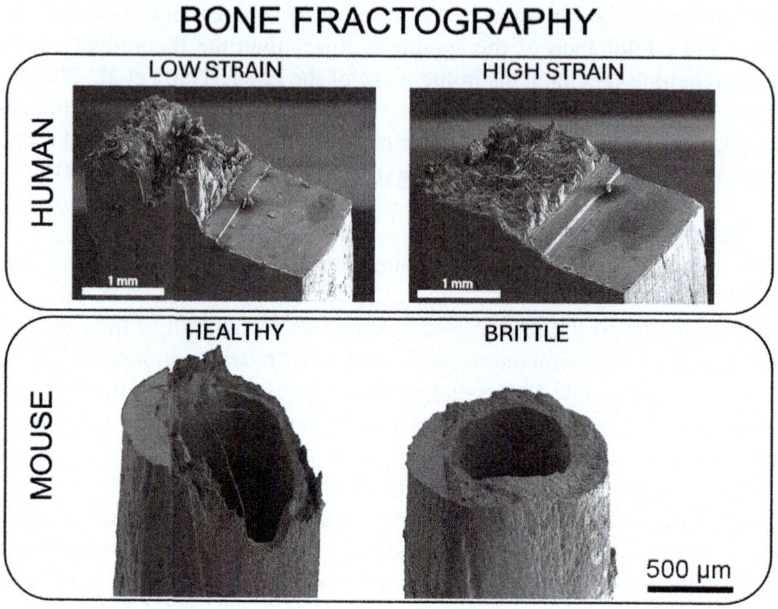

FIGURE 11.5 Bone fractography of human and mouse bone.

(Top) Bone fractography of healthy human bone in the transverse direction at low and high strain. Healthy bone fracture surface show many deflections and twists when tested in the transverse direction. These extrinsic toughening mechanisms are reduced if the bone splits, or if it breaks at high strains, showing a less rough fracture surface than when broken in the transverse direction, and with low strain, respectively. (Bottom) Bone fractography of healthy and brittle mouse femurs. Similarly to healthy human bones, healthy mouse bone shows many deflections and twists in the fracture surface. In brittle bones, these toughening mechanisms are absent or not tough as in normal bone, thus showing no resistance to the crack growth (flat R-curve in Fig. 11.6) and a flat fracture surface.

- Adapted with permission from Carriero, A., Zimmermann, E.A., Paluszny, A., Tang, S.Y., Bale, H., Busse, B., Alliston, T., Kazakia, G., Ritchie, R. O., & Shefelbine, S. J. (2014). How tough is brittle bone? Investigating osteogenesis imperfecta in mouse bone. Journal of Bone and Mineral Research, 29*(6), 1392–1401. https://doi.org/10.1002/jbmr.2172; Zimmermann, E. A., Gludovatz, B., Schaible, E., Busse, B., & Ritchie, R.O. (2014). Fracture resistance of human cortical bone across multiple length-scales at physiological strain rates.* Biomaterials, 35*(21), 5472–5481. https://doi.org/10.1016/j.biomaterials.2014.03.066*

3.3.2 Advanced glycation end products (AGEs)

Advanced glycation end products (AGEs) result from non-enzymatic glycation (NEG) of collagen, and are spontaneous enzyme-independent formation of chemical bonds between a sugar and a molecule. Their effect on bone fracture toughness has been studied by in vitro glycation of bone samples. In vitro glycation with ribose have shown conflicting results with one study finding lower K_{IC} values in ribose-treated compared with non-glycated bone (0.6M D-Ribose for 10 days) (Merlo et al., 2020), while two other studies did not observe significant differences in overall toughness between groups after in vitro glycation (0.1M D-Ribose for 4 weeks (Unal et al., 2023) or 0.6M D-Ribose for 7, 14, 21 days (Lin et al., 2025)). In R-curves, in vitro glycation (1.5M D-Ribose for 7 days) had no effect on crack-initiation toughness but reduced its crack-growth toughness by 35% by creating lower microcrack density, larger crack segments and smaller bridges (Poundarik et al., 2015). Glycated organic matrix reduce the ability of bone to sustain deformation and dissipate energy, which may result in long microcracks that coalesce in extensive crack length (Poundarik et al., 2015). However, the process of ribosylation in vitro may not fully replicate the degradation in collagen happening with AGEs accumulation in humans (Willett et al., 2022); thus, in vitro formation of AGEs through ribose incubation may not be a good model for biological AGE accumulation in humans.

In a diabetic mouse model, it was found that glycation widens intermolecular space, hampering fibril packing, and depresses lysyl oxidase-driven cross-linking as glucose carbonyls (AGEs precursors) and telopeptide aldehydes (enzymatic cross-link precursors) target the same triple-helical sites (Hudson et al., 2018). In aging and disease, excessive cross-linking of collagen fibrils due to AGEs glycation, oxidative stress or low bone turnover, reduced fracture toughness (Allen et al., 2015; Carriero, Zimmermann, Paluszny, et al., 2014; Thomas et al., 2018; Zimmermann et al., 2015).

3.3.3 Hydroxyapatite

Hydroxyapatite significantly stiffens the mineralized fibrils, reaching the peak at 2 nm thickness (Qin et al., 2012). Thicker minerals embrittle the fibril structure (Qin et al., 2012), while smaller mineral particles (as in diseases such as osteogenesis imperfecta, type 2 diabetes mellitus, and glucocorticoid-induced osteoporosis) make it softer (Carriero, Zimmermann, Paluszny, et al., 2014; Shahar & Weiner, 2018; Xi et al., 2018). Increased mineral content, particularly in the overlap regions, increases the stiffness and resistance to deformation of the bone tissue (Wang & Ural, 2019), which may embrittle the bone tissue and reduce bone resistance to fracture.

3.3.4 Non-collagenous proteins

Molecular interactions between mineralized collagen fibrils and extra-fibrillar mineral aggregates involve the action of non-collagenous proteins (NCPs), particularly, osteopontin and osteocalcin sitting at the interfibrillar interface. Osteopontin and osteocalcin count for 2%–3% of bone weight; however, they can achieve large non-linear deformations and great ductility (Morsali et al., 2019), thus contributing by more than 30% to the bone fracture toughness by a synergistic deformation mechanism of the two NCPs in bone (Morsali et al., 2019; Poundarik et al., 2012; Thurner et al., 2010). While osteocalcin is rigid and highly affine with mineral, osteopontin breaks and (re)forms sacrificial bonds, detaching from one osteocalcin and rolling toward neighboring osteocalcins in shear direction (Morsali et al., 2019; Poundarik et al., 2012; Tavakol & Vaughan, 2020). Osteocalcin function can be affected by collagen glycation, losing much of its energy dissipation ability (Tavakol & Vaughan, 2020). Osteopontin phosphorylation adds additional negative charge and hydrophilicity that potentially eases its adhesion to mineral (Sroga & Vashishth, 2018), while its binding to free calcium ions may prevent unwanted adsorption to other components and facilitate its deposition on mineral surface (Kläning et al., 2014). The absence of NCPs triggers alterations in mineralization degree, fibrillogenesis, bone formation, and microdamage accumulation (Depalle et al., 2021; Hua et al., 2020; Morgan et al., 2015). For example, removal of decorin and/or biglycan generates highly disorganized collagen fibrils (Kram et al., 2020). Instead, administration of chondroitin sulfate (CS) improves bound water amount and bone toughness (Hua et al., 2020).

3.3.5 Water

Structural water contained within hydroxyapatite is proposed to play a role in bone mineralization by facilitating the assembly of hydroxyapatite platelets (Von Euw et al., 2018). Tightly bound water within the tropocollagen molecules contributes to the conformational stability of the collagen triple helix structure by forming hydrogen bonds with the hydroxyl groups of hydroxyproline (Granke et al., 2015; Unal & Akkus, 2015). Intrafibrillar water-mediated hydrogen bridges between hydroxyapatite and tropocollagen restrain sliding between these molecules, aid in transferring load, and improve energy dissipation within the fibril (Nair et al., 2014). The removal of intrafibrillar water significantly reduces the bone matrix's capacity to absorb energy by one ninth of that in hydrated conditions (Maghsoudi-Ganjeh et al., 2020). Intrafibrillar loosely bound water also contributes to bone ductility acting as a plasticizer at the interface between mineral and collagen (Granke et al., 2015; Maghsoudi-Ganjeh et al., 2020). Perturbation of collagen triple helices affinity to bind with each other in favor of water bonds, observed in presence of deamidation of asparagine and glutamine residues, result in thicker fibrils and low bone fracture toughness (Creecy et al., 2021). With collagen maturation, interfibrillar cross-linking predominates over intrafibrillar cross-linking, and loosely bound water within fibrils is replaced by mineral (Creecy et al., 2020; Wang et al., 2018), reducing bone plasticity (Fielder & Nair, 2019; Maghsoudi-Ganjeh et al., 2020). When hydration was removed with high vacuum in human bone samples a 50% reduction of fracture toughness at initiation and propagation was found (Shin et al., 2022).

3.3.6 Bone porosity

Pores in bone extend from the nanoscale to macroscale. The porous nanostructure, osteocyte lacunar-canalicular network, and blood vessels in bone all represent discontinuities and imperfections at different length-scales, from which the crack will initiate and propagate (Muñoz et al., 2021). Intracortical vascular porosity density within the bone matrix inversely relates to bone fracture toughness (Granke et al., 2016; Ural & Vashishth, 2007; Yeni et al., 1997). Also, vascular canals architecture is important for bone fracture resistance as it affects the extent of the stable crack growth (Carriero, Zimmermann, Paluszny, et al., 2014). Indeed, many branched canals closely connected, disrupt the continuity of the bone matrix material increasing stress concentrations from which the crack will start (Carriero, Doube, Vogt, Busse, et al., 2014; Rodriguez-Florez et al., 2017) and progress (Carriero, Zimmermann, Paluszny, et al., 2014; Rodriguez-Florez et al., 2017), with a 10% longer crack extension than in presence of a single big canal with same overall porosity (Carriero, Zimmermann, Paluszny, et al., 2014).

Similar to vascular porosity, cellular and subcellular porosities generate strain concentrations in bone at smaller length scales promoting the initiation and growth of micro- and nano-cracks. Specifically, daily physiological activities (Muñoz et al., 2026a) drive convective-fluid flow in the lacunar canalicular systems in the order of nanometer per second (Muñoza et al., 2025a) so that osteocyte-lacunar morphology and canaliculi arrangement affect bone fragility: sperical lacunae, and canaliculi oriented perperndicular to the osteocyte major axis generating high local strains in bone that favor crack formation and progress (Muñoz et al., 2025b). When the bone is loaded, high strain concentrated at the canaliculi within the osteons create intralamellar circumferential microcracks associated with the development of shear bands (Ebacher et al., 2012), which form during inelastic deformations and control the cracking at the mineralized collagen fibril bundles level.

3.3.7 Lamellae

At the tissue level, bone is considered as a continuous material with periodically inhomogeneous layers: the lamellae. Lamellae exhibit variable mechanical properties depending on the orientation of their fibers (Razi et al., 2020) and disorganized zones lacking a consistent distribution of fibers, the woven bone (Grandfield et al., 2018; Reznikov et al., 2014). It has been proposed that the interfaces between the lamellae act as delamination barriers for the crack growth, generating deflections and twists that can double the bone fracture toughness during crack growth (Carriero, Zimmermann, Paluszny, et al., 2014; Carriero, Zimmermann, Shefelbine, et al., 2014; Peterlik et al., 2006). Accordingly, a large osteon with many lamellae should increase bone fracture resistance to crack propagation. Conversely, woven bone should have very low-to-no crack growth fracture toughness. However, these need yet to be demonstrated.

3.3.8 Cement lines

Osteons increase bone toughness (Granke et al., 2016; Yeni et al., 1997) because of the presence of cement lines, which creates interfaces within bone where microcracks form and crack deflects (Koester et al., 2008). Crack deflections can increase toughness 4–6 fold. Osteon orientation makes bone five times tougher to break than to split (Ritchie, 2010): a crack needs to propagate across the cement line to break the bone creating crack deflections, but can travel between cement lines to split it creating crack bridging (Koester et al., 2008). Crack-growth toughness is indeed one order of magnitude smaller when splitting the bone rather than breaking it.

As said above, isotropic materials are often weaker in mode I (tensile opening loading). Bone is not isotropic as it contains lamellae and cement lines oriented longitudinally that make bone actually tougher in mode I than combined mode I-II in the transverse orientation (Olvera et al., 2012; Zimmermann et al., 2009; Zimmermann et al., 2010). The toughness of bone is a result of a competition between driving force and the orientation of the microstructure; when the driving force is oriented parallel to the microstructure, the toughness is low (in the case of mode II in the transverse orientation or mode I in longitudinal orientation). Then, as the orientations of the driving force and the microstructure diverge, the toughness increases (Olvera et al., 2012; Zimmermann et al., 2009; Zimmermann et al., 2010).

3.3.9 Tissue-level mineral distribution and total mineral content

Fracture toughness may be proportional to mineral content; especially when comparing tissues with widely different mineral contents (Currey et al., 2004, 2019). For instance, antler, dentin, and narwhal tusk have lower mineral content and higher toughness than bone, even at different strain rates. Currey et al. (Currey et al., 2004, 2019) demonstrated a negative power-law relationship between energy to break a notched sample and calcium content (using bovine bone, tiger bone, narwal dentine, and antler), suggesting that lower mineralization imparts damage tolerance and notch sensitivity to the tissue. However, within human bone, the role of mineralization is less clear. In a cohort of 62 human bone samples from the ages of 21–101 years, Granke et al. (2016) only found a weak correlation between heterogeneity in mineralization (using quantitative backscatter electron imaging) and fracture toughness. Their best predictor of fracture toughness was a combination of age, vascular porosity, and osteonal area. Similarly, in a cohort of 26 human femurs from individuals from 50 to 95 years, Gardegaront et al. (2017) found no significant correlation between fracture toughness and degree of bone mineralization (measured with an X-ray scanner). Instead, Yeni et al. (Yeni et al., 1998; Yeni & Norman, 2000) found a relationship between bone fracture toughness (femurs and tibiae) and apparent density, with no relation to mineral or organic content. These results are different from the results of Ural and Vashishth (2007), who found that porosity and material properties of bone are both important for fracture toughness. In the same cohort as Granke et al. (2016), mineral and collagen parameters measured from Raman spectroscopy weakly correlated with bone fracture toughness (Makowski et al., 2017). However, using principal component analysis or general linear models, it was found that age, volumetric bone mineral density, and aspects of the Raman spectroscopy signal explained up to 48%–50% of the variance in fracture toughness (Makowski et al., 2017; Unal et al., 2019).

Fig. 11.6 summarizes the influence of the bone components and structure on the extrinsic and intrinsic toughening mechanics and on the R-curve fracture toughness.

3.3.10 Strain rate

The effect of strain rate on human bone has been investigated comparing quasistatic loading rate (~0.0001 s^{-1}) with high strain rates corresponding to a fall (0.01–0.1

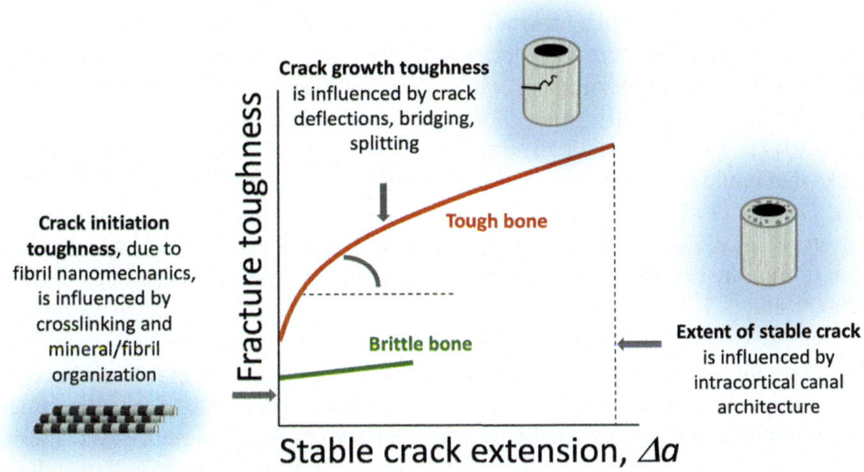

FIGURE 11.6 Schematic of the fracture toughness R-curves of healthy (tough) and fragile (brittle) bone.

This schematic shows the influence of the intrinsic and extrinsic toughening mechanisms arising from different levels of the bone hierarchical structure on the crack-initiation toughness ($\Delta a \rightarrow 0$), the crack-growth toughness (slope of the R-curve) and the extent of the stable crack. Crack-initiation toughness is high in tough bone where plasticity is developed at the nanoscale level to shield from the crack initiation thanks to the intrinsic toughening mechanisms. Changes in the brittle bones composition and structure at the nanoscale level impairs the fibril nanomechanics, affecting the initiation toughness. Crack-growth toughness is large in tough bone as the toughness can be 2–6-times its initial value thanks to deflections, splitting and bridging at the end of the stable crack growth, while it is almost null (flat R-curve) in brittle bones. These mechanisms result in extremely rough fracture surfaces in healthy bone and their absence in brittle bone results in a flat fracture surface (Fig. 11.5) high crack-growth toughness (high slope of R-curve). Indeed, in brittle bones, it requires far lower driving force or energy to propagate the crack generating smooth fracture surfaces with little-to-no crack deflection (Fig. 11.5). The stable crack extension is always affected by bone porosity, both amount and architecture. Usually brittle bones have more porosity and/or branched canals than healthy bone, which facilitative the quick growth of the crack.

- Adapted with permission from Carriero, A., Zimmermann, E.A., Paluszny, A., Tang, S. Y., Bale, H., Busse, B., Alliston, T., Kazakia, G., Ritchie, R. O., & Shefelbine, S. J. (2014). How tough is brittle bone? Investigating osteogenesis imperfecta in mouse bone. Journal of Bone and Mineral Research, *29(6), 1392–1401. https://doi.org/10.1002/jbmr.2172*

s^{-1}). Unal et al. (2023) found significantly lower $J_{plastic}$ in samples loaded to high strain rate versus low strain rate. In fracture toughness samples performed on bone from the distal tibia, Gauthier et al. (2015) found that plastic deformation at low strain rates was roughly 60% greater with only elastic behavior at strain rates corresponding to a fall. In a follow-up study on paired femoral neck as well as the diaphysis of the femur, tibia, and radius from 32 donors, Gauthier et al. (2017) found significantly lower fracture toughness at higher strain rates with no difference in toughness between skeletal sites; whereas, at quasistatic strain rates, the bone at the radius has significantly greater toughness than the other skeletal sites (Gauthier et al., 2017). The quasi-statically loaded samples had a greater volume of microcracks that could contribute toward dissipating energy (Gauthier et al., 2019). Similarly, Dapaah et al. (2022) observed a 42% smaller microdamage process zone at high strain rates. Zioupos et al. (2020) performed fracture mechanics tests measuring work to fracture on chevron notched samples at seven different strain rates between 0.00009 and 3.5 s^{-1}. A strain rate around 0.01s^{-1} (corresponding to walking in humans) gives the maximum work to fracture. This maximum work to fracture occurs at smaller strain rates for older bone samples suggesting that bone becomes embrittled with age (Zioupos et al., 2020). Zimmermann et al. (2014) performed tensile testing of bone samples combined with synchrotron small angle X-ray scattering finding that at high strain rates toughness decreases and similarly the collagen deformation locks up reducing the amount of fibril scale plastic deformation. As the bone loses its ability to activate plasticity mechanisms, the relative modulus and strain energy release rate between the different components of the microstructure change to favor cracks penetrating the interfaces rather than deflecting (Ming-Yuan & Hutchinson, 1989; Zimmermann et al., 2014) (Fig. 11.5).

4. Conclusions

Bone is tough and develops its high resistance to fracture thanks to its composite nature organized in a complex hierarchical structure. In this chapter, we have presented the definition of bone fracture toughness and the methods used for assessing fracture resistance in human and animal bones. We have shown how fracture mechanics uses very sensitive tools that have allowed researchers to discover how bone fractures and characterize mechanisms of toughness in bone at its multiscale, and characterize the effect of each component. These methodologies are being applied in models of aging and disease to determine how changes in bone composition and/or hierarchical structure result in bone fragility. A comprehensive knowledge on bone fracture toughness can inform new therapy approaches for bone fragility and man-made materials with strategies to achieve high resistance to fracture.

Acknowledgments

This work was supported by the National Science Foundation (CBET 1829310), the National Institute of Health (R21HD106607, R03AR084261), the Human Frontier Research Program (RGP0023/2021) for AC, and the Natural Sciences and Engineering Research Council of Canada (NSERC RGPIN-2021-03251) to E.A.Z.

References

Allen, M. R., Newman, C. L., Chen, N., Granke, M., Nyman, J. S., & Moe, S. M. (2015). Changes in skeletal collagen cross-links and matrix hydration in high- and low-turnover chronic kidney disease. *Osteoporosis International, 26*(3), 977–985. https://doi.org/10.1007/s00198-014-2978-9

Anderson, T. L. (2005). *Fracture mechanics*. Taylor and Francis Group.

ASTM E399-24. (2024). *Standard test method for linear-elastic plane-strain fracture toughness of metallic materials*. Advancing Standards Transforming Materials.

ASTM E1820-18. (2019). *Standard test method for measurement of fracture toughness*. Advancing Standards Transforming Materials.

Bonfield, W. (1987). Advances in the fracture mechanics of cortical bone. *Journal of Biomechanics, 20*(11–12), 1071–1081. https://doi.org/10.1016/0021-9290(87)90025-x

Boyce, T. M., Fyhrie, D. P., Glotkowski, M. C., Radin, E. L., & Schaffler, M. B. (1998). Damage type and strain mode associations in human compact bone bending fatigue. *Journal of Orthopaedic Research, 16*(3), 322–329. https://doi.org/10.1002/jor.1100160308

Brown, C. U., Yeni, Y. N., & Norman, T. L. (2000). Fracture toughness is dependent on bone location - A study of the femoral neck, femoral shaft, and the tibial shaft. *Journal of Biomedical Materials Research, 49*(3), 380–389. https://doi.org/10.1002/(SICI)1097-4636(20000305)49:3<380::AID-JBM11>3.0.CO;2-W

Buehler, M. J. (2007). Molecular nanomechanics of nascent bone: Fibrillar toughening by mineralization. *Nanotechnology, 18*(29), Article 295102. https://doi.org/10.1088/0957-4484/18/29/295102

Burr, D. B., Turner, C. H., Naick, P., Forwood, M. R., Ambrosius, W., Sayeed Hasan, M., & Pidaparti, R. (1998). Does microdamage accumulation affect the mechanical properties of bone? *Journal of Biomechanics, 31*(4), 337–345. https://doi.org/10.1016/S0021-9290(98)00016-5

Cardoso, L., Herman, B. C., Verborgt, O., Laudier, D., Majeska, R. J., & Schaffler, M. B. (2009). Osteocyte apoptosis controls activation of intracortical resorption in response to bone fatigue. *Journal of Bone and Mineral Research, 24*(4), 597–605. https://doi.org/10.1359/jbmr.081210

Carriero, A., Bruse, J. L., Oldknow, K. J., Millán, J. L., Farquharson, C., & Shefelbine, S. J. (2014). Reference point indentation is not indicative of whole mouse bone measures of stress intensity fracture toughness. *Bone, 69*, 174–179. https://doi.org/10.1016/j.bone.2014.09.020

Carriero, A., Doube, M., Vogt, M., Busse, B., Zustin, J., Levchuk, A., Schneider, P., Müller, R., & Shefelbine, S. J. (2014). Altered lacunar and vascular porosity in osteogenesis imperfecta mouse bone as revealed by synchrotron tomography contributes to bone fragility. *Bone, 61*, 116–124. https://doi.org/10.1016/j.bone.2013.12.020

Carriero, A., Zimmermann, E. A., Paluszny, A., Tang, S. Y., Bale, H., Busse, B., Alliston, T., Kazakia, G., Ritchie, R. O., & Shefelbine, S. J. (2014). How tough is brittle bone? Investigating osteogenesis imperfecta in mouse bone. *Journal of Bone and Mineral Research, 29*(6), 1392–1401. https://doi.org/10.1002/jbmr.2172

Carriero, A., Zimmermann, E. A., Shefelbine, S. J., & Ritchie, R. O. (2014). A methodology for the investigation of toughness and crack propagation in mouse bone. *Journal of the Mechanical Behavior of Biomedical Materials, 39*, 38–47. https://doi.org/10.1016/j.jmbbm.2014.06.017

Creecy, A., Brown, K. L., Rose, K. L., Voziyan, P., & Nyman, J. S. (2021). Post-translational modifications in collagen type I of bone in a mouse model of aging. *Bone, 143*. https://doi.org/10.1016/j.bone.2020.115763

Creecy, A., Uppuganti, S., Girard, M. R., Schlunk, S. G., Amah, C., Granke, M., Unal, M., Does, M. D., & Nyman, J. S. (2020). The age-related decrease in material properties of BALB/c mouse long bones involves alterations to the extracellular matrix. *Bone, 130*. https://doi.org/10.1016/j.bone.2019.115126

Currey, J. D., Brear, K., & Zioupos, P. (2004). Notch sensitivity of Mammalian mineralized tissues in impact. *Proceedings of the Royal Society B: Biological Sciences, 271*(1538), 517–522. https://doi.org/10.1098/rspb.2003.2634

Currey, J. D., Brear, K., & Zioupos, P. (2019). Strain rate dependence of work of fracture tests on bone and similar tissues: Reflections on testing methods and mineral content effects. *Bone, 128*, Article 115038. https://doi.org/10.1016/j.bone.2019.115038

Dapaah, D., Martel, D. R., Laing, A. C., & Willett, T. L. (2022). The impact of fall-related loading rate on the formation of micro-damage in human cortical bone fracture. *Journal of Biomechanics, 142*. https://doi.org/10.1016/j.jbiomech.2022.111254

Depalle, B., McGilvery, C. M., Nobakhti, S., Aldegaither, N., Shefelbine, S. J., & Porter, A. E. (2021). Osteopontin regulates type I collagen fibril formation in bone tissue. *Acta Biomaterialia, 120*, 194–202. https://doi.org/10.1016/j.actbio.2020.04.040

Docaj, A., & Carriero, A. (2024). Bone health: Quality versus quantity. *Journal of the Pediatric Orthopaedic Society of North America, 7*, Article 100054. https://doi.org/10.1016/j.jposna.2024.100054

Ebacher, V., Guy, P., Oxland, T. R., & Wang, R. (2012). Sub-lamellar microcracking and roles of canaliculi in human cortical bone. *Acta Biomaterialia, 8*(3), 1093–1100. https://doi.org/10.1016/j.actbio.2011.11.013

Evans, A. G., & Faber, K. T. (1984). Crack-growth resistance of microcracking brittle materials. *Journal of the American Ceramic Society, 67*(4), 255–260. https://doi.org/10.1111/j.1151-2916.1984.tb18842.x

Fantner, G. E., Hassenkam, T., Kindt, J. H., Weaver, J. C., Birkedal, H., Pechenik, L., Cutroni, J. A., Cidade, G. A. G., Stucky, G. D., Morse, D. E., & Hansma, P. K. (2005). Sacrificial bonds and hidden length dissipate energy as mineralized fibrils separate during bone fracture. *Nature Materials, 4*(8), 612–616. https://doi.org/10.1038/nmat1428

Fielder, M., & Nair, A. K. (2019). Effects of hydration and mineralization on the deformation mechanisms of collagen fibrils in bone at the nanoscale. *Biomechanics and Modeling in Mechanobiology, 18*(1), 57–68. https://doi.org/10.1007/s10237-018-1067-y

Gardegaront, M., Gauthier, R., Farlay, D., Rizzo, S., Mitton, D., & Follet, H. (2017). Influence of the degree of mineralization of the cortical bone on toughness. *Computer Methods in Biomechanics and Biomedical Engineering, 20*, 87–88. https://doi.org/10.1080/10255842.2017.1382874

Gauthier, R., Follet, H., Langer, M., Meille, S., Chevalier, J., Rongiéras, F., Peyrin, F., & Mitton, D. (2017). Strain rate influence on human cortical bone toughness: A comparative study of four paired anatomical sites. *Journal of the Mechanical Behavior of Biomedical Materials, 71*, 223–230. https://doi.org/10.1016/j.jmbbm.2017.03.015

Gauthier, R., Follet, H., & Mitton, D. (2015). Effect of strain rate on the toughness of human tibial cortical bone. *Computer Methods in Biomechanics and Biomedical Engineering, 18* (Suppl. 1), 1942–1943. https://doi.org/10.1080/10255842.2015.1070590

Gauthier, R., Langer, M., Follet, H., Olivier, C., Gouttenoire, P. J., Helfen, L., Rongiéras, F., Mitton, D., & Peyrin, F. (2019). Influence of loading condition and anatomical location on human cortical bone linear micro-cracks. *Journal of Biomechanics, 85*, 59–66. https://doi.org/10.1016/j.jbiomech.2019.01.008

Grandfield, K., Vuong, V., & Schwarcz, H. P. (2018). Ultrastructure of bone: Hierarchical features from nanometer to micrometer scale revealed in focused ion beam sections in the TEM. *Calcified Tissue International, 103*(6), 606–616. https://doi.org/10.1007/s00223-018-0454-9

Granke, M., Does, M. D., & Nyman, J. S. (2015). The role of water compartments in the material properties of cortical bone. *Calcified Tissue International, 97*(3), 292–307. https://doi.org/10.1007/s00223-015-9977-5

Granke, M., Makowski, A. J., Uppuganti, S., & Nyman, J. S. (2016). Prevalent role of porosity and osteonal area over mineralization heterogeneity in the fracture toughness of human cortical bone. *Journal of Biomechanics, 49*(13), 2748–2755. https://doi.org/10.1016/j.jbiomech.2016.06.009

Gupta, H. S., Wagermaier, W., Zickler, G. A., Aroush, D. R. B., Funari, S. S., Roschger, P., Wagner, H. D., & Fratzl, P. (2005). Nanoscale deformation mechanisms in bone. *Nano Letters, 5*(10), 2108–2111. https://doi.org/10.1021/nl051584b

Hua, R., Ni, Q., Eliason, T. D., Han, Y., Gu, S., Nicolella, D. P., Wang, X., & Jiang, J. X. (2020). Biglycan and chondroitin sulfate play pivotal roles in bone toughness via retaining bound water in bone mineral matrix. *Matrix Biology, 94*, 95–109. https://doi.org/10.1016/j.matbio.2020.09.002

Hudson, D. M., Archer, M., King, K. B., & Eyre, D. R. (2018). Glycation of type I collagen selectively targets the same helical domain lysine sites as lysyl oxidase-mediated cross-linking. *Journal of Biological Chemistry, 293*(40), 15620–15627. https://doi.org/10.1074/jbc.ra118.004829

Hunckler, M. D., Chu, E. D., Baumann, A. P., Curtis, T. E., Ravosa, M. J., Allen, M. R., & Roeder, R. K. (2017). The fracture toughness of small animal cortical bone measured using arc-shaped tension specimens: Effects of bisphosphonate and deproteinization treatments. *Bone, 105*, 67–74. https://doi.org/10.1016/j.bone.2017.08.015

Kennedy, O. D., Laudier, D. M., Majeska, R. J., Sun, H. B., & Schaffler, M. B. (2014). Osteocyte apoptosis is required for production of osteoclastogenic signals following bone fatigue in vivo. *Bone, 64*, 132–137. https://doi.org/10.1016/j.bone.2014.03.049

Kläning, E., Christensen, B., Sørensen, E. S., Vorup-Jensen, T., & Jensen, J. K. (2014). Osteopontin binds multiple calcium ions with high affinity and independently of phosphorylation status. *Bone, 66*, 90–95. https://doi.org/10.1016/j.bone.2014.05.020

Koester, K. J., Ager, J. W., & Ritchie, R. O. (2008). The true toughness of human cortical bone measured with realistically short cracks. *Nature Materials, 7*(8), 672–677. https://doi.org/10.1038/nmat2221

Kram, V., Shainer, R., Jani, P., Meester, J. A. N., Loeys, B., & Young, M. F. (2020). Biglycan in the skeleton. *Journal of Histochemistry and Cytochemistry, 68*(11), 747–762. https://doi.org/10.1369/0022155420937371

Krauss, S., Fratzl, P., Seto, J., Currey, J. D., Estevez, J. A., Funari, S. S., & Gupta, H. S. (2009). Inhomogeneous fibril stretching in antler starts after macroscopic yielding: Indication for a nanoscale toughening mechanism. *Bone, 44*(6), 1105–1110. https://doi.org/10.1016/j.bone.2009.02.009

Lin, M. C., Sihota, P., Kolibová, S. D., Fiedler, I. A. K., Krug, J., Wölfel, E. M., Moritz, M., Riedner, M., Ondruschka, B., Citak, M., Klebig, F., von Brackel, F. N., Qwamizadeh, M., Jähn-Rickert, K., & Busse, B. (2025). Fracture characteristics of human cortical bone influenced by the duration of in vitro glycation. *JBMR Plus, 9*(2). https://doi.org/10.1093/jbmrpl/ziae151

Maghsoudi-Ganjeh, M., Wang, X., & Zeng, X. (2020). Computational investigation of the effect of water on the nanomechanical behavior of bone. *Journal of the Mechanical Behavior of Biomedical Materials, 101*. https://doi.org/10.1016/j.jmbbm.2019.103454

Makowski, A. J., Granke, M., Ayala, O. D., Uppuganti, S., Mahadevan-Jansen, A., & Nyman, J. S. (2017). Applying full spectrum analysis to a raman spectroscopic assessment of fracture toughness of human cortical bone. *Applied Spectroscopy, 71*(10), 2385–2394. https://doi.org/10.1177/0003702817718149

Manrique, L., Moussa, M. S., Khan, M. T., Tahboub, K., Ritchie, R. O., Asgari, M., & Zimmermann, E. A. (2024). Deformation of collagen-based tissues investigated using a systematic review and meta-analysis of synchrotron X-ray scattering studies. *Cell Reports Physical Science, 5*(10). https://doi.org/10.1016/j.xcrp.2024.102212

Merlo, K., Aaronson, J., Vaidya, R., Rezaee, T., Chalivendra, V., & Karim, L. (2020). In Vitro-induced high sugar environments deteriorate human cortical bone elastic modulus and fracture toughness. *Journal of Orthopaedic Research, 38*(5), 972–983. https://doi.org/10.1002/jor.24543

Miller, B., Spevak, L., Lukashova, L., Javaheri, B., Pitsillides, A. A., Boskey, A., Bou-Gharios, G., & Carriero, A. (2017). Altered bone mechanics, architecture and composition in the skeleton of TIMP-3-Deficient mice. *Calcified Tissue International, 100*(6), 631–640. https://doi.org/10.1007/s00223-017-0248-5

Ming-Yuan, H., & Hutchinson, J. W. (1989). Crack deflection at an interface between dissimilar elastic materials. *International Journal of Solids and Structures, 25*(9), 1053–1067. https://doi.org/10.1016/0020-7683(89)90021-8

Morgan, S., Poundarik, A. A., & Vashishth, D. (2015). Do non-collagenous proteins affect skeletal mechanical properties? *Calcified Tissue International, 97*(3), 281–291. https://doi.org/10.1007/s00223-015-0016-3

Morsali, R., Dai, Z., Wang, Y., Qian, D., & Minary-Jolandan, M. (2019). Deformation mechanisms of "two-Part" natural adhesive in bone interfibrillar nano-interfaces. *ACS Biomaterials Science & Engineering, 5*(11), 5916–5924. https://doi.org/10.1021/acsbiomaterials.9b00588

Muñoz, A., De Paolis, A., Cardoso, L., & Carriero, A. (2025b). Osteocyte-lacuna shape and canaliculi architecture dictate fluid flow around osteocyte, and strain of cell and bone matrix: Implications for cell mechanobiology and bone fragility. *Bone, 117613*. https://doi.org/10.1016/j.bone.2025.117613

Muñoz, A., De Paolis, A., Cardoso, L., & Carriero, A. (2026a). Multiaxial physiological loading generates complex fluid flow and strain patterns in the osteocyte lacunar-canalicular system. *Bone, 117765*. https://doi.org/10.1016/j.bone.2026.117765.

Muñoz, A., Docaj, A., Ugarteburu, M., & Carriero, A. (2021). Poor bone matrix quality: What can be done about it? *Current Osteoporosis Reports, 19*(5), 510–531. https://doi.org/10.1007/s11914-021-00696-6

Muñoz, A., Silvan, A., & Carriero, A. (2026b). Bone fragility persists in long-term bisphosphonate-treated oim mice despite sex-specific changes in bone quality. *Acta Biomateriala, S1742-7061(26)00274-6*. https://doi.org/10.1016/j.actbio.2026.04.054. Epub ahead of print. PMID: 42061485.

Muñoza, A., De Paolis, A., Cardoso, L., & Carriero, A. (2025a). Mechanical deformations of bone generate interstitial fluid flow at nanoscale velocities around osteocytes. *Frontiers in*

Bioengineering and Biotechnology, 13, 1639788. https://doi.org/10.3389/fbioe.2025.1639788

Nair, A. K., Gautieri, A., & Buehler, M. J. (2014). Role of intrafibrillar collagen mineralization in defining the compressive properties of nascent bone. *Biomacromolecules, 15*(7), 2494–2500. https://doi.org/10.1021/bm5003416

Nalla, R. K., Kruzic, J. J., & Ritchie, R. O. (2004). On the origin of the toughness of mineralized tissue: Microcracking or crack bridging? *Bone, 34*(5), 790–798. https://doi.org/10.1016/j.bone.2004.02.001

Olvera, D., Zimmermann, E. A., & Ritchie, R. O. (2012). Mixed-mode toughness of human cortical bone containing a longitudinal crack in far-field compression. *Bone, 50*(1), 331–336. https://doi.org/10.1016/j.bone.2011.11.004

Peterlik, H., Roschger, P., Klaushofer, K., & Fratzl, P. (2006). From brittle to ductile fracture of bone. *Nature Materials, 5*(1), 52–55. https://doi.org/10.1038/nmat1545

Poundarik, A. A., Diab, T., Sroga, G. E., Ural, A., Boskey, A. L., Gundberg, C. M., & Vashishth, D. (2012). Dilatational band formation in bone. *Proceedings of the National Academy of Sciences of the United States of America, 109*(47), 19178–19183. https://doi.org/10.1073/pnas.1201513109UnitedStates

Poundarik, A. A., Wu, P. C., Evis, Z., Sroga, G. E., Ural, A., Rubin, M., & Vashishth, D. (2015). A direct role of collagen glycation in bone fracture. *Journal of the Mechanical Behavior of Biomedical Materials, 52*, 120–130. https://doi.org/10.1016/j.jmbbm.2015.08.012

Qin, Z., Gautieri, A., Nair, A. K., Inbar, H., & Buehler, M. J. (2012). Thickness of hydroxyapatite nanocrystal controls mechanical properties of the collagen-hydroxyapatite interface. *Langmuir, 28*(4), 1982–1992. https://doi.org/10.1021/la204052a

Razi, H., Predan, J., Fischer, F. D., Kolednik, O., & Fratzl, P. (2020). Damage tolerance of lamellar bone. *Bone, 130*. https://doi.org/10.1016/j.bone.2019.115102

Reznikov, N., Shahar, R., & Weiner, S. (2014). Three-dimensional structure of human lamellar bone: The presence of two different materials and new insights into the hierarchical organization. *Bone, 59*, 93–104. https://doi.org/10.1016/j.bone.2013.10.023

Ritchie, R. O. (2010). How does human bone resist fracture? *Annals of the New York Academy of Sciences, 1192*, 72–80. https://doi.org/10.1111/j.1749-6632.2009.05232.x. http://onlinelibrary.wiley.com/journal/10.1111/(ISSN)1749-6632

Ritchie, R. O. (2011). The conflicts between strength and toughness. *Nature Materials, 10*(11), 817–822. https://doi.org/10.1038/nmat3115

Ritchie, R. O., Koester, K. J., Ionova, S., Yao, W., Lane, N. E., & Ager, J. W. (2008). Measurement of the toughness of bone: A tutorial with special reference to small animal studies. *Bone, 43*(5), 798–812. https://doi.org/10.1016/j.bone.2008.04.027

Rodriguez-Florez, N., Carriero, A., & Shefelbine, S. J. (2017). The use of XFEM to assess the influence of intra-cortical porosity on crack propagation. *Computer Methods in Biomechanics and Biomedical Engineering, 20*(4), 385–392. https://doi.org/10.1080/10255842.2016.1235158

Seref-Ferlengez, Z., Basta-Pljakic, J., Kennedy, O. D., Philemon, C. J., & Schaffler, M. B. (2014). Structural and mechanical repair of diffuse damage in cortical bone in vivo. *Journal of Bone and Mineral Research, 29*(12), 2537–2544. https://doi.org/10.1002/jbmr.2309

Shahar, R., & Weiner, S. (2018). Open questions on the 3D structures of collagen containing vertebrate mineralized tissues: A perspective. *Journal of Structural Biology, 201*(3), 187–198. https://doi.org/10.1016/j.jsb.2017.11.008

Shin, M., Zhang, M., vom Scheidt, A., Pelletier, M. H., Walsh, W. R., Martens, P. J., Kruzic, J. J., Busse, B., & Gludovatz, B. (2022). Impact of test environment on the fracture resistance of cortical bone. *Journal of the Mechanical Behavior of Biomedical Materials, 129*. https://doi.org/10.1016/j.jmbbm.2022.105155

Shum, D. K. M., & Hutchinson, J. W. (1990). On toughening by microcracks. *Mechanics of Materials, 9*(2), 83–91. https://doi.org/10.1016/0167-6636(90)90032-B

Sroga, G. E., & Vashishth, D. (2018). Phosphorylation of extracellular bone matrix proteins and its contribution to bone fragility. *Journal of Bone and Mineral Research, 33*(12), 2214–2229. https://doi.org/10.1002/jbmr.3552

Tavakol, M., & Vaughan, T. J. (2020). The structural role of osteocalcin in bone biomechanics and its alteration in Type-2 diabetes. *Nature Research, Ireland Scientific Reports, 10*(1). https://doi.org/10.1038/s41598-020-73141-w

Thomas, C. J., Cleland, T. P., Sroga, G. E., & Vashishth, D. (2018). Accumulation of carboxymethyl-lysine (CML) in human cortical bone. *Bone, 110*, 128–133. https://doi.org/10.1016/j.bone.2018.01.028

Thurner, P. J., Chen, C. G., Ionova-Martin, S., Sun, L., Harman, A., Porter, A., Ager, J. W., Ritchie, R. O., & Alliston, T. (2010). Osteopontin deficiency increases bone fragility but preserves bone mass. *Bone, 46*(6), 1564–1573. https://doi.org/10.1016/j.bone.2010.02.014

Unal, M., & Akkus, O. (2015). Raman spectral classification of mineral- and collagen-bound water's associations to elastic and post-yield mechanical properties of cortical bone. *Bone, 81*, 315–326. https://doi.org/10.1016/j.bone.2015.07.024

Unal, M., Uppuganti, S., Dapaah, D. Y., Ahmed, R., Pennings, J. S., Willett, T. L., Voziyan, P., & Nyman, J. S. (2023). Effect of ribose incubation on physical, chemical, and mechanical properties of human cortical bone. *Journal of the Mechanical Behavior of Biomedical Materials, 140*. https://doi.org/10.1016/j.jmbbm.2023.105731

Unal, M., Uppuganti, S., Timur, S., Mahadevan-Jansen, A., Akkus, O., & Nyman, J. S. (2019). Assessing matrix quality by Raman spectroscopy helps predict fracture toughness of human cortical bone. *Scientific Reports, 9*(1). https://doi.org/10.1038/s41598-019-43542-7

Ural, A., & Vashishth, D. (2007). Effects of intracortical porosity on fracture toughness in aging human bone: A μCT-based cohesive finite element study. *Journal of Biomechanical Engineering, 129*(5), 625–631. https://doi.org/10.1115/1.2768377

Vashishth, D., Behiri, J. C., & Bonfield, W. (1997). Crack growth resistance in cortical bone: Concept of microcrack toughening. *Journal of Biomechanics, 30*(8), 763–769. https://doi.org/10.1016/s0021-9290(97)00029-8

Von Euw, S., Chan-Chang, T.-H.-C., Paquis, C., Haye, B., Pehau-Arnaudet, G., Babonneau, F., Azaïs, T., & Nassif, N. (2018). Organization of bone mineral: The role of mineral–water interactions. *Geosciences, 8*(12), 466. https://doi.org/10.3390/geosciences8120466

Wang, X., Hua, R., Ahsan, A., Ni, Q., Huang, Y., Gu, S., & Jiang, J. X. (2018). Age-related deterioration of bone toughness is related to diminishing amount of matrix glycosaminoglycans (GAGs). *JBMR Plus, 2*(3), 164–173. https://doi.org/10.1002/jbm4.10030

Wang, Y., & Ural, A. (2019). A finite element study evaluating the influence of mineralization distribution and content on the tensile mechanical response of mineralized collagen fibril networks. *Journal of the Mechanical Behavior of Biomedical Materials, 100*, Article 103361. https://doi.org/10.1016/j.jmbbm.2019.07.019

Willett, T. L., Voziyan, P., & Nyman, J. S. (2022). Causative or associative: A critical review of the role of advanced glycation end-products in bone fragility. *Bone, 163*. https://doi.org/10.1016/j.bone.2022.116485

Xi, L., De Falco, P., Barbieri, E., Karunaratne, A., Bentley, L., Esapa, C. T., Terrill, N. J., Brown, S. D. M., Cox, R. D., Davis, G. R., Pugno, N. M., Thakker, R. V., & Gupta, H. S. (2018). Bone matrix development in steroid-induced osteoporosis is associated with a consistently reduced fibrillar stiffness linked to altered bone mineral quality. *Acta Biomaterialia, 76*, 295–307. https://doi.org/10.1016/j.actbio.2018.05.053

Yeni, Y. N., Brown, C. U., & Norman, T. L. (1998). Influence of bone composition and apparent density on fracture toughness of the human femur and tibia. *Bone, 22*(1), 79–84. https://doi.org/10.1016/s8756-3282(97)00227-5

Yeni, Y. N., Brown, C. U., Wang, Z., & Norman, T. L. (1997). The influence of bone morphology on fracture toughness of the human femur and tibia. *Bone, 21*(5), 453–459. https://doi.org/10.1016/s8756-3282(97)00173-7

Yeni, Y. N., & Norman, T. L. (2000). Fracture toughness of human femoral neck: Effect of microstructure, composition, and age. *Bone, 26*(5), 499–504. https://doi.org/10.1016/s8756-3282(00)00258-1

Zimmermann, E. A., Gludovatz, B., Schaible, E., Busse, B., & Ritchie, R. O. (2014). Fracture resistance of human cortical bone across multiple length-scales at physiological strain rates. *Biomaterials, 35*(21), 5472–5481. https://doi.org/10.1016/j.biomaterials.2014.03.066

Zimmermann, E. A., Köhne, T., Bale, H. A., Panganiban, B., Gludovatz, B., Zustin, J., Hahn, M., Amling, M., Ritchie, R. O., & Busse, B. (2015). Modifications to nano- and microstructural quality and the effects on mechanical integrity in Paget's disease of bone. *Journal of Bone and Mineral Research, 30*(2), 264–273. https://doi.org/10.1002/jbmr.2340

Zimmermann, E. A., Launey, M. E., Barth, H. D., & Ritchie, R. O. (2009). Mixed-mode fracture of human cortical bone. *Biomaterials, 30*(29), 5877–5884. https://doi.org/10.1016/j.biomaterials.2009.06.017

Zimmermann, E. A., Launey, M. E., & Ritchie, R. O. (2010). The significance of crack-resistance curves to the mixed-mode fracture toughness of human cortical bone. *Biomaterials, 31*(20), 5297–5305. https://doi.org/10.1016/j.biomaterials.2010.03.056

Zimmermann, E. A., & Ritchie, R. O. (2015). Bone as a structural material. *Advanced Healthcare Materials, 4*(9), 1287–1304. https://doi.org/10.1002/adhm.201500070

Zimmermann, E. A., Schaible, E., Bale, H., Barth, H. D., Tang, S. Y., Reichert, P., Busse, B., Alliston, T., Ager, J. W., & Ritchie, R. O. (2011). Age-related changes in the plasticity and toughness of human cortical bone at multiple length scales. *Proceedings of the National Academy of Sciences of the United States of America, 108*(35), 14416–14421. https://doi.org/10.1073/pnas.1107966108

Zioupos, P., Kirchner, H. O. K., & Peterlik, H. (2020). Ageing bone fractures: The case of a ductile to brittle transition that shifts with age. *Bone, 131*. https://doi.org/10.1016/j.bone.2019.115176

CHAPTER

Bone damage and fatigue 12

Bowen Wang[1], Deepak Vashishth[1,2] and Mitchell Schaffler[3]

[1]*Rensselaer Polytechnic Institute - Icahn School of Medicine at Mount Sinai Center for Engineering and Precision Medicine, New York, NY, United States;* [2]*Rensselaer Polytechnic Institute, Department of Biomedical Engineering, New York, NY, United States;* [3]*The City College of New York, Department of Biomedical Engineering, New York, NY, United States*

1. How does physiological loading of bone lead to fractures?

When subjected to repeated fluctuating stress over a long period (defined as cyclic stress), fatigue occurs in the load-bearing material, leading to the initiation and propagation of cracks and overall a loss of structural integrity. Historically, fractures of engineered materials, such as metal and ceramics, have most often been attributed to fatigue. The cyclic stress that causes fatigue failure can be much lower than the ultimate stress or even the yield stress determined under static loading conditions (Reifsnider, 2012, p. 532). During the loading phases, microscopic damage within the composite material can be initiated at localized areas of stress concentration due to microscopic structural defects, geometric discontinuity, material inhomogeneity, etc. The initiated microcracks then continue to propagate as stress/strain cycles progress. Shorter cracks can be arrested, deflected, or accelerated depending on conditions at the crack tip and can eventually grow into long cracks of critical size, leading to catastrophic failure. In engineered materials, existing microdamage accumulates and cannot be reversed through unloading. However, various toughening mechanisms exist in structural materials to dissipate energy and resist the propagation of cracks, thereby extending their lifetime (Ye, 1989). The fatigue life, defined as the number of cycles to failure under a given cyclic stress, is usually used to characterize the material's resistance to fatigue failure. In addition, fracture toughness, independent of material strength, describes a material's ability to resist crack propagation under a preexisting flaw and can significantly influence its fatigue life.

Bone is a prime example of a quasi-brittle composite material subjected to cyclic loading (i.e., walking, running, etc.). It is well-documented that normal daily loading accumulates damage within bone tissue, as numerous structural defects and interfaces exist at the microscale and nanoscale levels. But bones in a healthy state are intricately designed to be both strong enough to provide structural support and resist inelastic deformation in day-to-day loading, and tough enough to avoid breaking.

Biomechanics and Mechanobiology of Bone. https://doi.org/10.1016/B978-0-443-26739-0.00007-7

The hierarchical composite structure in bone determines its mechanical fate (Fig. 12.1). From the molecular level, the organic matrix phase, consisting primarily of type I collagen, provides resilience and flexibility, while the mineral phase of bone contributes to its rigidity. The various forms of water (free, bound, and structural) also modulate the bone's ductility. The mineralized collagen fibrils arrange into lamellae and form the basic structural unit in cortical bone, an osteon. Similarly, in trabecular bone, the lamellae get assembled into plate- or rod-like structures called trabeculae (Cowin, 2009; Rho et al., 1998).

The hierarchical structure allows for multiple intrinsic and extrinsic toughening mechanisms. Intrinsic toughening mechanisms at nano- and submicron scales (e.g., microcracking, detailed in the sections below) develop plastic deformation at the crack tip, blunting the tip. Extrinsic toughening mechanisms at the microscale (e. g., crack deflection) shield the crack to impede growth. These toughening mechanisms dissipate energy at the crack tip, decelerate and eventually arrest the crack growth (Sabet et al., 2016; Zimmermann et al., 2015). Furthermore, bones can conduct targeted self-repair of microdamage at a specific rate (detailed in a section below), so we *usually* do not need to worry that repetitive daily loading will result in catastrophic fractures, as it might in engineered materials.

Nevertheless, bone fractures as a result of fatigue loading remain a significant clinical concern. As seen among athletes and military service members, stress fractures occur when excessive loading causes damage to accumulate faster than it can be repaired. The incidence of stress fractures accounts for up to 10% of orthopedic-related injuries and up to 20% of sports and exercise-related injuries (Abbott et al., 2020; Knapik et al., 2016). On the other hand, fragility fractures, or insufficiency fractures, occur among older individuals or those with metabolic bone diseases. In these cases, the energy required for fracture is below the level of a fall from standing height. A global study in 2019 showed that a total of 37 million fragility fractures occurred annually among individuals over 55 years old (Wu et al., 2021). The occurrence of a fragility fracture could be due to either diminished repair mechanisms, where microdamage cannot be sufficiently repaired, or losses of bone mass and/or matrix quality, allowing microdamage to propagate much more easily. Although clinically assessed reduced bone mineral density (BMD) confers an increase in fracture risk, it is insufficient to accurately predict the risk of fragility fracture (Unnanuntana et al., 2010). Furthermore, no in vivo imaging modality currently exists to detect microdamage in bone. Consequently, understanding how and why daily physical activity leads to fractures has been an active clinical research topic since the 1970s, with much still to be uncovered.

In this chapter, the following sections will break this topic down into multiple specific questions. We will seek to understand the essence of bone microdamage and its relationship to bone fragility, the mechanisms of microdamage repair, and how these two aspects can be altered with aging, diseases, and treatments.

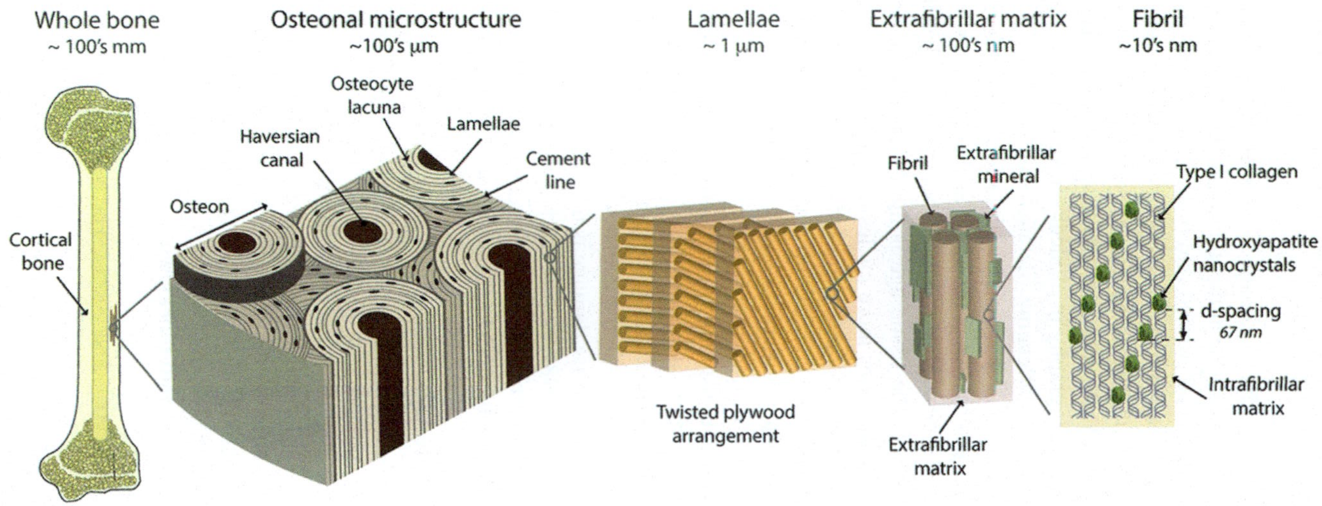

FIGURE 12.1

The hierarchical structure of cortical bone.

- Reprinted from Zimmermann, E.A., Schaible, E., Gludovatz, B., Schmidt, F.N., Riedel, C., Krause, M., Vettorazzi, E., Acevedo, C., Hahn, M., Puschel, K., Tang, S., Amling, M., Ritchie, R.O., & Busse, B. (2016). Intrinsic mechanical behavior of femoral cortical bone in young, osteoporotic and bisphosphonate-treated individuals in low-and high energy fracture conditions. Scientific Reports, 6. *https://doi.org/10.1038/srep21072*

2. Bone microdamage in vivo—What is it in multiple length scales?

As a result of habitual cyclic loading, damage occurs at multiple length scales (Fig. 12.2) and accumulates in bone tissue at the microscale before catastrophic failure. Microdamage in bone can be categorized into two distinct types: linear microcracks and diffuse damage. Viewed from the cross-sections of bone tissue, the length scale of linear microcracks is primarily above 50 μm, whereas the diffuse damage exists in length less than 10 μm (Vashishth, 2007). Due to their small sizes in nature, neither can be detected under radiography in clinical settings, and the visualization of microdamage ex vivo remains challenging (detailed below).

The formation of linear microcracks is typically linked to repetitive compression under physiological loading, resulting in shear stress at the interfaces between the lamellae. Individually formed linear microcracks appear as sharply defined lines in interstitial bone tissues, where the tissue age is higher compared with the surrounding osteonal tissue and undergoes more cycles of fatigue loading. The crack goes along the lamellar interfaces and gets deflected and arrested at the osteonal or interstitial cement lines (Boyce et al., 1998; Diab & Vashishth, 2005; Lee, 2003). If left unrepaired, the linear microcracks could continue to grow and become macrocracks that eventually lead to fractures.

In contrast, diffuse damage, consisting of many small cracks at microscopic scales woven into a cluster, is often associated with tensile loading and tends to appear around osteonal tissues. Diffuse damage occurs rapidly and more easily in response to cyclic loading compared with linear microcracks. The two microdamage mechanisms are independent, and diffuse damage is not considered the precursor of linear microcracks (Boyce et al., 1998; Diab & Vashishth, 2005; Vashishth et al., 2000). Additionally, a mesh of microcracks at intermediate lengths can also form, appearing to be cross-hatching patterns (Fazzalari et al., 1998; Vashishth et al., 2000).

The history of bone microdamage examination dates back to 1960, when Frost identified the natural existence of microcracks in bone using a transmitted light microscope with bone sections stained with 1% basic fuchsin (Frost, 1960). Thirty years later, this en bloc staining technique was validated by Burr and Stafford (1990), revised by Burr and Hooser (1995), and updated more recently by Dominguez and Agnew (2019). It remains the most popular approach to assess linear microcracks in 2D, which effectively distinguishes the microdamage that occurred in vivo from the damage created artificially during sample preparation.

Briefly, bone tissue is prepared into blocks with ~2 cm in length, defatted, and stored in 70% ethanol (EtOH). The dye—1% basic fuchsin—is prepared in a gradient of EtOH solution (80%, 90%, and 100%). The bone blocks are then stained with dye solutions containing progressively increasing EtOH concentrations under vacuum conditions, and embedded in methyl methacrylate (MMA) for long-term storage and to aid in sectioning. Bone tissue cured in MMA can then be sectioned

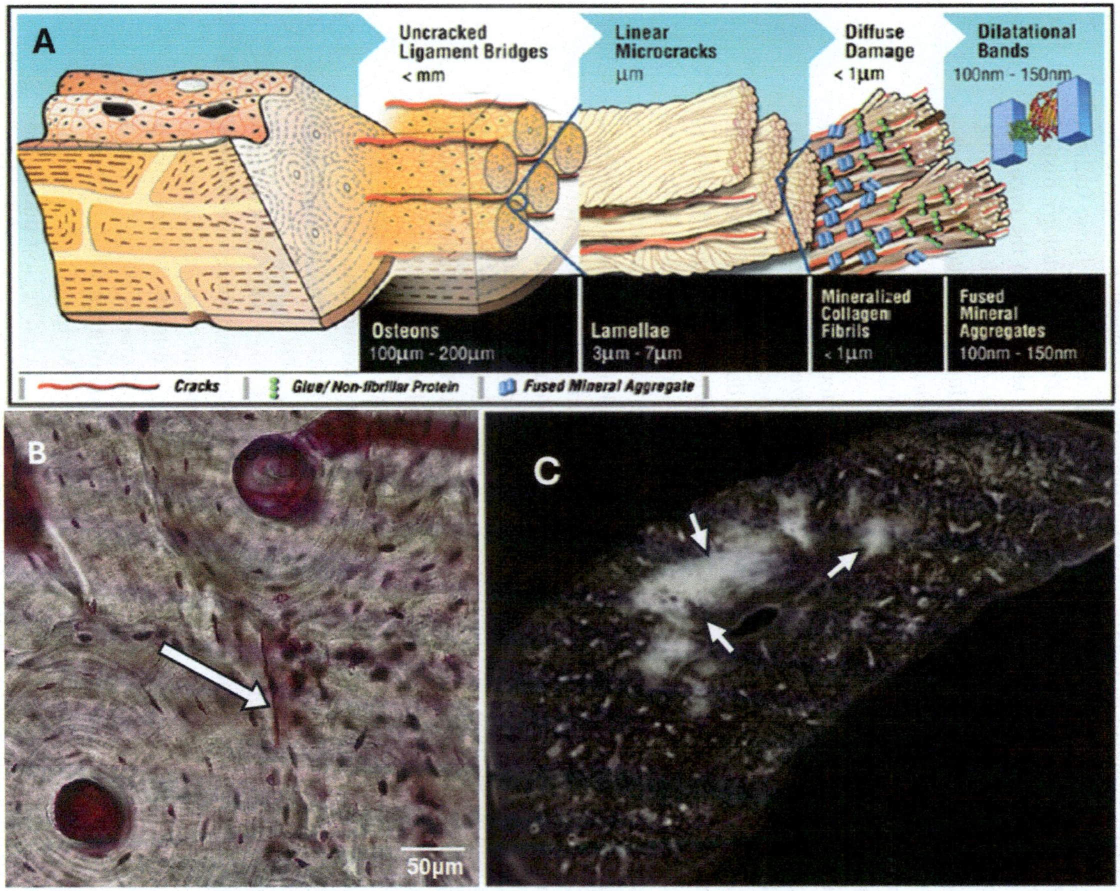

FIGURE 12.2

(A) Illustration of bone microdamage in multiple length scales, including dilatational bands, diffuse damage, linear microcracks, and uncracked ligament bridges. (B) Linear microdamage in human femoral cortical bone. The white arrows indicate basic fuchsin-stained linear microcracks. The black arrows show the arrest of microcracks at around cement lines. (C) Diffuse damage in the rat ulnae cortex. The white arrows indicate diffuse damage shown in fluorescence photomicrographs.

- *A. Reprinted from Poundarik, A. A., Diab, T., Sroga, G. E., Ural, A., Boskey, A. L., Gundberg, C. M., & Vashishth, D. (2012). Dilatational band formation in bone.* Proceedings of the National Academy of Sciences of the United States of America, 109*(47), 19178–19183. https://doi.org/10.1073/pnas.1201513109, B. Adpated from Diab, T. & Vashishth, D. (2007). Morphology, localization and accumulation of in vivo microdamage in human cortical bone.* Bone, 40*(3), 612–618. https://doi.org/10.1016/j.bone.2006.09.027, C. Adapted from Seref-Ferlengez, Z., Basta-Pljakic, J., Kennedy, O. D., Philemon, C. J., & Schaffler, M. B. (2014). Structural and mechanical repair of diffuse damage in cortical bone in vivo.* Journal of Bone and Mineral Research, 29*(12), 2537–2544. https://doi.org/10.1002/jbmr.2309*

and ground to 30–150 μm thickness. Newly generated microdamage during cutting and grinding does not absorb the dye; therefore, it would not be counted as in vivo microcracks. Finally, the sections can be mounted on glass slides for subsequent imaging under a transmission light microscope (Dominguez & Agnew, 2019).

Under similar principles, a double staining technique was developed, where toluidine blue (water soluble) is applied before mechanical testing and basic fuchsin (EtOH soluble) is applied following mechanical testing. This methodology distinguishes the microcracks generated during the mechanical testing from the preexisting in vivo microdamage (Poundarik & Vashishth, 2015). Along with the labeling techniques for bone histomorphometry, other in vivo fluorescent labeling techniques, such as using chelating agents (e.g., calcein blue), to bind to calcium ions around the microdamage sites, are also effective in assessing microdamage (O'Brien et al., 2005).

Alternatively, lead-uranyl acetate, a common staining technique for electron microscopy, can be utilized as a contrast enhancer to visualize bone microdamage. In 1995, a methodology for using scanning electron microscopy (SEM) with backscattered mode (BSE) to examine bone microdamage was developed. The technique was validated with the basic fuchsin staining, but the detected microcrack density is lower compared with the method described in reference (Schaffler et al., 1994). Due to its nanoscale resolution, SEM has also been widely employed to examine bone microdamage postfractures to understand how the morphology and density of microdamage mechanistically alter bone fracture properties at the tissue level. Branching, bridging, and deflection of fatigue microcracks have been visualized in multiple SEM studies and were identified both as causes of stress fractures and as energy dissipation mechanisms (see next section) (Nalla et al., 2003; Zhang et al., 2013). Under similar lead-uranyl acetate staining principles as in BSE, bone microdamage can be visualized in 3D using microcomputed tomography (μCT, Fig. 12.3). This technique allows for quantification of bone microdamage per bone volume (DV/BV). The damage surface per damage volume (DS/DV) could indicate the distribution of microdamage type, distinguishing whether it is more likely to consist of linear microcracks or diffuse damage (Tang & Vashishth, 2007).

In recent years, more research advances have been made to evaluate bone microdamage using various other imaging modalities, such as atomic force microscopy (AFM). With AFM, studies have discovered the formation and propagation mechanisms of microcracks, which lead to the separation of individual mineralized collagen fibrils. A nanoscale toughening mechanism, sacrificial bonding mediated by calcium, was identified, which acts as a "glue" to bind collagen fibrils together and prevent cracks from opening (Fantner et al., 2005). AFM imaging also established the presence of dilatational bands in the diffuse damaged region due to the deformation of noncollagenous protein (NCP) complexes, including osteocalcin (OC) and osteopontin (OPN), both crucial regulators of bone mineralization. The shearing and rupture of the dilatational bands could result in the formation of diffuse damage at a larger scale (Poundarik et al., 2012). More recently, a study utilized Optical Coherence Tomography (OCT) to render the 3D quantitative analysis of

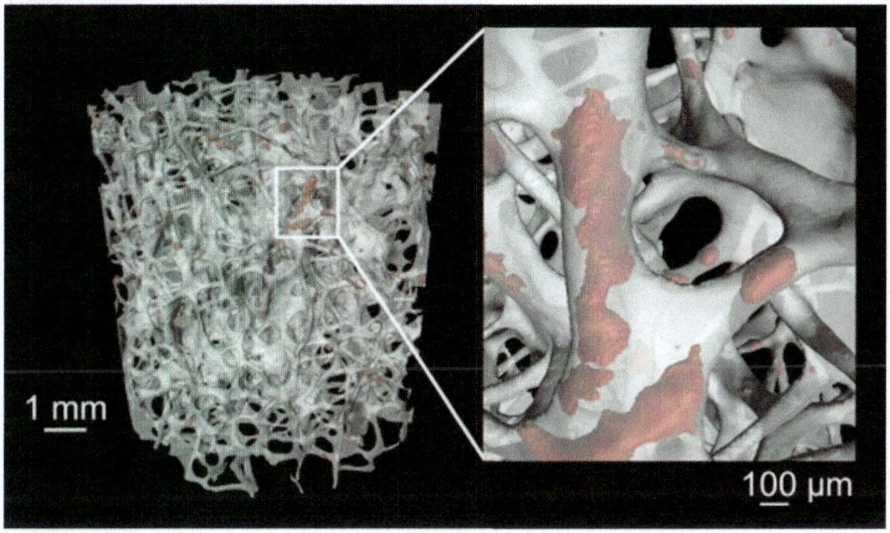

FIGURE 12.3

Visualization of bone microdamage in trabecular bone using microcomputed tomography with lead-uranyl acetate staining.

- Adapted from Lambers, F. M., Bouman, A. R., Rimnac, C. M., & Hernandez, C. J. (2013). Microdamage caused by fatigue loading in human cancellous bone: Relationship to reductions in bone biomechanical performance. PLoS ONE, 8*(12). https://doi.org/10.1371/journal.pone.0083662*

microcracks formed on cortical bone tissue caused by the insertion of orthodontic microimplants (Lakshmikantha et al., 2019).

3. How does microdamage affect bone fragility?

Intuitively, as seen in other engineered materials, the increased accumulation of microdamage within the bone structure weakens its mechanical properties, making it more prone to fracture. In general, it has been established that the accumulation of microdamage in bone exhibits a nonlinear relationship with cycles of fatigue loading (Fig. 12.4). The model can be simplified into three phases:

1. In the primary phase, damage initiates rapidly within interstitial tissue, corresponding to a significant drop in bone modulus;
2. Next, the aforementioned microstructural-level toughening mechanisms are activated in the secondary phase, where the initiated microdamage propagates more slowly, and the decline in mechanical integrity is decelerated;
3. Finally, in the tertiary phase, the excessive loading cycles accumulate a sufficient amount of damage that coalesces into a structural defect of higher length scales. At this point, osteons are debonded and pulled out. The mechanical

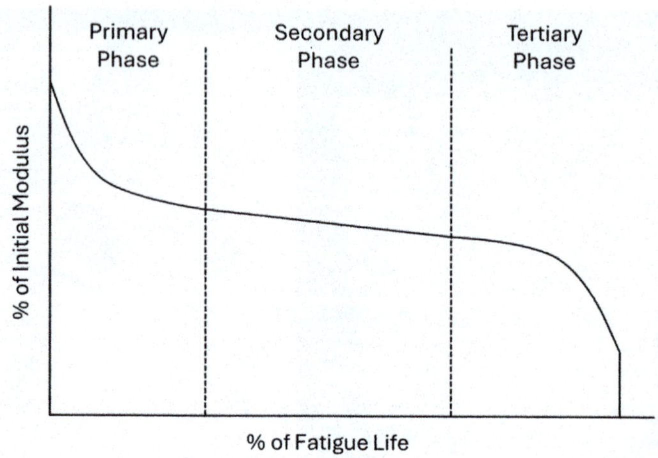

FIGURE 12.4

The nonlinear relationship between fatigue loading cycles and bone residual modulus.

properties of bone rapidly deteriorate and eventually lead to catastrophic fractures (Diab & Vashishth, 2005; Seref-Ferlengez et al., 2015; Varvani-Farahani & Najmi, 2010).

The impact of linear microcracks on bone residual properties was first tested in 1977, where microdamage was induced with fatigue loading of ex vivo bone tissue. When bone tissue was cyclically loaded to 50% of its fatigue life, a significant reduction in the residual tensile strength (~13% loss) was observed. Similarly, samples with a homogeneous microstructure showed a progressive loss of stiffness (~20% at the cycle to failure). Both the residual strength ratio and secant modulus ratio were found to have a significant negative relationship with fatigue life ratio (Carter & Hayes, 1977). This was further confirmed in a later study, where fatigue loading within physiological strain range created an increased number of microcracks and subsequently resulted in a decrease of stiffness in cortical bone (Schaffler et al., 1989). In terms of fracture toughness, which is more relevant than stiffness and strength to predict bone fractures, a study discovered that in human femora, tensile fracture toughness was negatively associated with in vivo preexisting microdamage parameters, such as microcrack density, microcrack surface density, and average microcrack length. However, the relationships were nonsignificant in human tibiae, possibly due to their inherently different fracture behavior (Norman et al., 1998).

Early studies primarily focused on the influence of linear microcracks on bone mechanical properties. A later work used a four-point bending setup to compare the spatial distribution of the two damage morphologies (linear damage vs. diffuse damage) and how they differently alter bone residual properties. During the primary phase of bending fatigue, twice more linear microcracks were found on the compressive side than on the tensile side, whereas the diffuse damage was four times higher on the tensile side than on the compressive side. Fractures initiated at the tensile side led to a greater loss of toughness. In contrast, during the tertiary phase, only linear microcracks were significantly increased on the compressive side of bending fatigue. A loss of toughness was only found when fracture was initiated on the compressive side at this stage (Diab & Vashishth, 2005).

Using a bending creep model, diffuse damage can be selectively formed in vivo to enable investigation of its contribution to bone fragility. Elastic modulus assessed from microindentation at the diffuse damage region showed a marked decrease compared with the nondamaged region. Whole bone stiffness was significantly reduced by 15% immediately following loading-induced diffuse damage formation. Both properties were restored to the baseline levels after 14 days with in vivo damage repair (Seref-Ferlengez et al., 2014).

A similar relationship has been identified between microdamage and mechanical properties in trabecular bone. Studies demonstrated that the preexisting microcrack density independently and significantly predicts vertebral trabecular bone strength. The negative relationship is best described with exponential fitting (Burr et al., 1997). Another study generated microdamage gradually in trabecular bone through in vitro fatigue loading, suspending at various stages before failure. The study showed a significant association between damage volume fraction and reductions in Young's Modulus. DV/BV was positively correlated with maximum strain, creep strain, cyclic strain, and energy dissipation. Moreover, it was discovered that most of the microdamage in trabecular bone was accumulated close to the end of the fatigue life (tertiary phase) (Lambers et al., 2013).

Paradoxically, microdamage plays a more complicated role in bone fracture properties. Similar to other quasi-brittle materials (Azmah Hanim et al., 2019, pp. 157–180; Li et al., 2018; Muju et al., 1998), the de novo formation of microcracks as a result of loading can act as a toughening mechanism to resist crack propagation. In a 1997 study, a significant increase in the amount of microcrack accumulation was observed during crack propagation. Correspondingly, the fracture toughness of bone tissue also increased as the crack extended, exhibiting a rising crack-growth-resistant curve (R-curve) behavior. Further examination of the fracture surface revealed that the microcracks were distributed behind the main crack tip (frontal process zone), and as the crack extended, around the main crack tip (process zone wake). The formation of these microcracks dissipated energy from the main crack and diverted the crack into a tortuous route until it was arrested (Vashishth et al., 1997). Thus, the toughness of bone can be determined by its ability to form de novo microcracks. The crack propagation approach (rising R-curve) is therefore a superior method to assess cortical bone fracture toughness.

The toughening mechanism of de novo diffuse damage was demonstrated through combined experimental and modeling approaches. When diffuse damage forms during loading, a large area of small interfaces is created to efficiently absorb energy (Seref-Ferlengez et al., 2015). Under in vitro fatigue loading, both the local region of diffuse damage and 50 μm within linear microcracks showed softened matrix. The formation of diffuse damage resulted in longer uncracked ligaments and was predicted to have much higher levels of propagation toughness. It was also predicted that the toughening mechanism of crack bridging between the collagen fibrils in diffuse damage contributed to more effective bone toughening compared with linear microcracks (Poundarik et al., 2012).

4. How does bone repair the microdamage?

As mentioned above, in engineered materials, microstructural defects initiated during repeated loading and unloading could progressively accumulate and propagate, and reach a critical crack size without reversal. Fortunately, bone constantly regenerates and replaces itself, and the regions of microdamage can be preferentially targeted and repaired through signaling mechanisms to prevent microcrack propagation.

It is now clear that the repair mechanisms for linear microcracks and diffuse damage are distinct, with the former more thoroughly understood. Back in the 1960s, Frost first proposed an osteonal remodeling mechanism to renew the damaged tissue, in the first study where linear microcracks in bone were visualized (Frost, 1960). A few years later, multiple studies demonstrated the significant association between the microdamage accumulation and the activated intracortical remodeling at the same damaged site (Burr et al., 1985; Mori & Burr, 1993). Mechanistic studies revealed that osteocyte apoptosis is the key modulator in this process. Osteocytes, acting as mechanosensors within lacunae, sense alterations in fluid shear stress due to mechanical stimuli and regulate bone homeostasis. At the sites where microcracks form, the osteocyte network is disrupted, inducing osteocyte apoptosis. This increases RANKL (receptor activator of nuclear factor-kB ligand) signaling from the surrounding viable osteocytes, which promotes the formation and activation of osteoclasts locally. The osteoclasts then break down the damaged bone tissue through intracortical resorption. Subsequently, osteoblast precursors are recruited to the site, and osteoblasts lay down the new matrix and form new bone at the damaged region (Fig. 12.5) (Cardoso et al., 2009; Kennedy et al., 2012, 2014; Verborgt et al., 2000, 2002).

In contrast, the repair of diffuse damage does not appear to be associated with intracortical remodeling. Although diffuse damage has the ability to heal autonomously, the study, which selectively induced in vivo diffuse damage, demonstrated that diffuse damage did not affect osteocytes in the same region and therefore does not involve the remodeling process (Seref-Ferlengez et al., 2014). The mechanisms of diffuse damage repair remain an active area of investigation. Independent of osteocyte activities, ex vivo remineralization of diffuse damaged bone showed that it can be repaired physiochemically through deposition of calcium and phosphorus (Mehraban Alvandi, 2020). Furthermore, the induction of diffuse damage leads to an extracellular Ca^{2+} efflux, suggesting the role of the voltage-gated Ca^{2+} channels (VGCC) in the anabolic response of osteoblasts (Sun et al., 2012). When VGCC is inhibited, the ability of osteoblasts to synthesize OC and mineralization at the

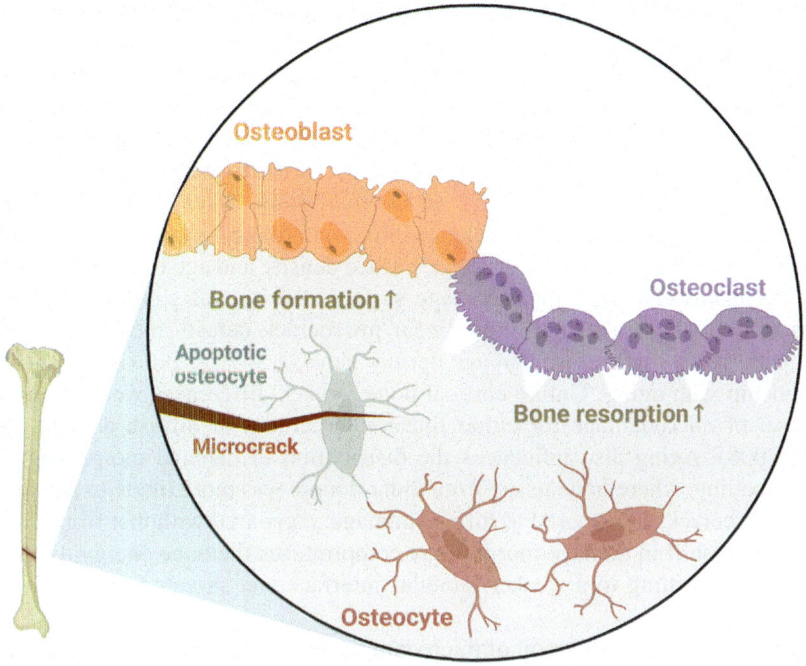

FIGURE 12.5

The repair mechanism of linear microcracks mediated by osteocytes. During fatigue loading, propagating microcracks induce osteocyte apoptosis, triggering responses of nearby osteocytes and activating the repair cycle.

- Created in BioRender. (2025). https://BioRender.com/ch57lmw.

diffuse damage region was correspondingly decreased (Jung & Akkus, 2020). The rupture of sacrificial bonds between the NCPs and mineralized fibrils, as noted earlier, may release calcium ions to facilitate the efflux. Furthermore, diffuse damage potentially exposes OC and OPN (inhibitors of mineral growth) to the action of proteases that break down OC and OPN and allow for mineral growth in the protein-depleted area (Poundarik et al., 2012).

5. How do the accumulation and repair of damage change under aging and disease conditions?

Understanding of microdamage contribution to bone mechanical integrity and its repair mechanisms raises an important clinical question: how do changes to microdamage explain the increase in bone fragility associated with aging and disease conditions? This can be addressed in two parts: 1) reduced ability to repair the existing microdamage, leading to increased damage accumulation; 2) altered matrix quality, compromising the toughening mechanism that restricts crack propagation.

In cortical bone, microcrack density increases significantly in individuals over 40 years old compared with younger individuals. The relationship between microcrack density and age can be described as a power relationship separately in males and females, with females showing a steeper increase in microcrack density with aging (Fig. 12.6) (Schaffler et al., 1995). In another study, when males and females were combined, significant linear relationships were found between microcrack density and age, and between microdamage surface density and age in femora. In tibiae, only the relationship with microdamage surface density was significant (Wang & Norman, 1997). In trabecular bone, linear microcrack density increases exponentially with age, while diffuse damage density shows a weaker but significant linear relationship with aging. Unlike cortical bone, no sex differences were observed in the level of microdamage for either linear microcracks or diffuse damage (Arlot et al., 2008). Aging also influences the distribution of damage morphology with in vivo loading, where bone tissue from older donors was more likely to accumulate linear microcracks as opposed to diffuse damage, compared with that from younger donors. This shift in damage morphology compromises the bone energy dissipation mechanisms, leading to a weaker lamellar interface and shorter fatigue life (Diab et al., 2006).

The increased accumulation of microdamage with aging could be explained by a decreased rate of intracortical remodeling to repair the microdamage. As discussed earlier, osteocytes play a critical role in microdamage repair. However, age-related

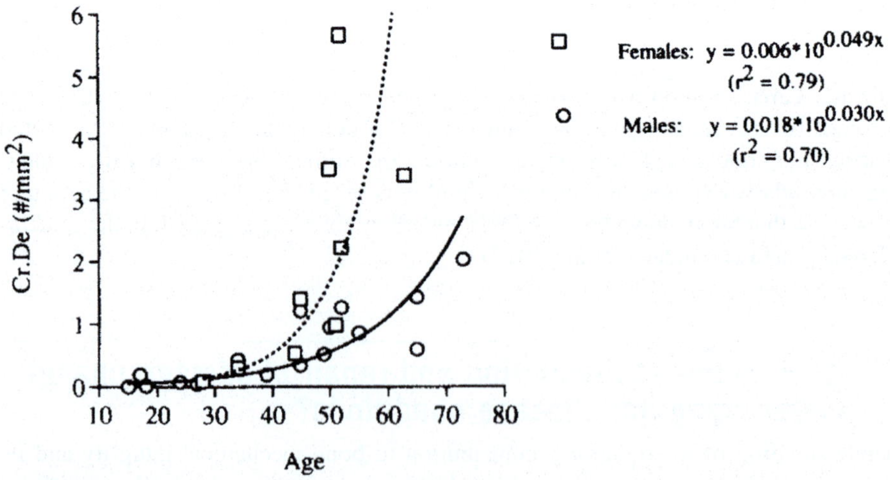

FIGURE 12.6

Microcrack density significantly increased with age for both females and males.

- Reprinted from Schaffler, M. B., Choi, K., & Milgrom, C. (1995). Aging and matrix microdamage accumulation in human compact bone. Bone, 17*(6), 521–525. https://doi.org/10.1016/8756-3282(95)00370-3*

changes, such as osteocyte apoptosis with subsequent micropetrosis (mineralized empty lacunae) (Milovanovic & Busse, 2020) and impaired osteocytic mechanosensitivity (Farr et al., 2017; Javaheri & Pitsillides, 2019), delay the detection of bone microdamage (Dragoun Kolibová et al., 2023). Evidence from in vivo studies on rats demonstrates that microcrack removal is significantly more efficient in younger rats compared with older ones following loading-induced microdamage (Waldorff et al., 2007). Similarly, increased osteocyte apoptosis has been seen in metabolic bone diseases in which higher levels of bone microdamage were observed, such as sex hormone deficiency-induced osteoporosis (Brennan et al., 2011; Kreipke et al., 2016) and type 1 diabetes (T1D) (Dragoun Kolibová et al., 2023; Liu et al., 2020; Mohsin et al., 2019). In contrast, despite that low bone turnover in type 2 diabetes (T2D) has been posited to promote microdamage accumulation (Rubin & Patsch, 2016), current experimental evidence indicates that the levels of microdamage are not significantly different in bone tissues between T2D and age-matched controls (Mohsin et al., 2019; Sacher et al., 2022).

In certain conditions, elevated accumulation of bone microdamage was observed, but the mechanisms are independent of remodeling rate. For instance, mutation of type I collagen in osteogenesis imperfecta negatively impacts collagen integrity. As a result, the bone tissues of osteogenesis imperfecta during loading have higher propensities to generate microcracks, and they propagate more easily (Davis et al., 2012). In end-stage osteoarthritis, loss of cartilage alters the stress concentration in subchondral bone, leading to a higher tendency for microdamage formation, while no change in remodeling and repair was observed (Shimamura et al., 2019).

This collective evidence of extensive damage accumulation and altered damage morphology could result in the degradation of bone residual properties, as mentioned above, therefore playing an important role in enhanced bone fragility in aging and metabolic bone diseases. On the other hand, in all the conditions above, the toughening mechanism through the formation of de novo microdamage during crack propagation is compromised, further exacerbating bone fragility. One of the factors that significantly impairs bone toughness in aging and metabolic bone diseases is the accumulation of advanced glycation and glycoxidation end-products (AGEs/AGOEs) in the bone matrix. AGEs/AGOEs, formed due to hyperglycemia and elevated oxidative stress, stiffen the collagen matrix and reduce the ductility, making bone tissue more susceptible to fragility fractures and insufficiency fractures (Wang & Vashishth, 2023). In vitro glycation studies to enhance the AGEs/AGOEs level in bone tissues demonstrated that the microcracks generated during crack propagation in glycated bones were significantly reduced, while the individual crack lengths and uncracked ligaments were longer. This indicates that the already initiated cracks in glycated bone tissue are more likely to propagate from easier pathways at a physiologically relevant energy level, without dissipating energy through the formation of nearby microcracks (Poundarik et al., 2015).

6. How do the accumulation and repair of damage change with antiosteoporotic medications?

Since the repair mechanisms of linear microcracks are directly linked to intracortical remodeling, antiosteoporotic medications, which act on altering bone remodeling, can significantly influence the accumulation and repair of microdamage in bone tissue. The most common types of antiosteoporotic medication are antiresorptive agents. These agents, including the first-line therapy bisphosphonates, increase bone mass and reduce the risk of fracture by slowing down the bone resorption process (Chen & Sambrook, 2011). However, due to this mechanism of action, the use of bisphosphonates (especially long-term) attenuates bone remodeling, significantly diminishing the activation frequency (a measure of new osteonal units formed) (Black et al., 2003; Drake et al., 2008), resulting in little to no repair of existing microcracks. The association between microcrack accumulation in bisphosphonate use has been demonstrated in multiple animal studies, with a dose-dependent relationship (Allen, Iwata, Phipps, et al., 2006; Mashiba et al., 2000, 2001; Tang et al., 2009). Notably, even at a clinically relevant low dosage, bisphosphonates still resulted in a significant 40% reduction in bone turnover and corresponded to a three-fold increase in microcrack surface density (Allen, Iwata, Phipps, et al., 2006). Moreover, other antiresorptive therapies, such as raloxifene, can also lead to increased microdamage in bone. A 20% decrease in bone turnover and a two-fold increase in microcrack surface density were found in raloxifene-treated groups, although neither of these parameters was statistically significant in the post hoc test of this study (Allen, Iwata, Sato, et al., 2006).

Evidence from clinical studies on the relationship between bisphosphonate use and microdamage accumulation is mixed. In one study of human iliac crest biopsies, crack density was significantly higher in patients treated with alendronates only after adjustments for potential confounders. The reduced femoral neck BMD and higher age were significantly associated with microdamage accumulation only in the alendronate treatment group (Stepan et al., 2007). In contrast, another study showed that microcrack frequency was not different between bisphosphonate-treated and controls, and the microcrack frequency was not dependent on the treatment duration of bisphosphonates. However, it is important to note that more than half of the biopsies in either group did not have observable microcracks (Chapurlat et al., 2007). It is also debatable whether the findings on iliac crest biopsies, a non-load-bearing site with less remodeling, can be extrapolated to other clinically relevant fracture sites, particularly since the microdamage accumulation associated with bisphosphonate use varies by sites (Mashiba et al., 2005).

Additionally, the use of bisphosphonates with associated attenuation in bone remodeling could also lead to an increase in AGEs/AGOEs accumulation, which in turn further suppresses bone formation and resorption (Wang & Vashishth, 2023). Higher doses of risedronate and alendronate both led to decreased turnover, elevated fluorescent AGEs, and reduced postyield work-to-fracture. A nonlinear relationship can describe the relationship between increased AGEs and decreased postyield work-to-fracture, suggesting that high doses of bisphosphonate can also contribute to embrittlement of bone matrix (Tang et al., 2009). The combined effects of microdamage accumulation and AGEs/AGOEs lead to a loss of bone toughness typically observed with bisphosphonate treatment (Allen & Burr, 2007; Komatsubara et al., 2003; Tang et al., 2009), and may explain that stress fracture incidences, such as atypical femoral fractures, are often related to long-term bisphosphonate usage (Black et al., 2020).

However, clinically, it is clear that established antiresorptive therapies effectively reduce bone fracture risk in patients with osteoporosis (Cedeno-Veloz et al., 2022; MacKey et al., 2011). When evaluating the effects of antiresorptive agents use and bone biomechanical properties for individuals with osteoporosis, it is important to consider the trade-off between the increase in bone strength with higher bone mass and the decrease in bone toughness with poorer matrix quality. While quasistatic testing is useful to determine these parameters, it does not highlight the role of microdamage accumulation on bone fragility under daily loading. A recent study sought to investigate whether the use of antiresorptive agents affects bone fatigue properties. Ovariectomized cynomolgus monkeys were treated for 12 months with alendronate or denosumab. Unexpectedly, all antiresorptive-treated groups showed significantly elevated fatigue life compared with the untreated group (Haider et al., 2023). This may suggest that, consistent with current clinical guidelines, the increase in bone density with antiresorptive therapies may dominate bone's resistance to fatigue fracture in osteoporosis. Nevertheless, the improved fatigue life with the use of antiresorptive agents may not be sufficient to reflect the loss of material properties, such as fracture toughness (Hernandez, 2023). The study is also unable to address the relationship between atypical femoral fracture occurrence and the use of antiresorptive agents.

In contrast, anabolic agents, the other type of antiosteoporosis medication, increase bone mass through acceleration of bone renewal. The most common anabolic agents are parathyroid hormone (PTH) analogs, which stimulate both formation and resorption of bone tissue, with a greater stimulation on bone formation when administered intermittently at lower doses (Kraenzlin & Meier, 2011). As a result, PTH therapy enhances the repair of microdamage by promoting the replacement of the damaged region with new bone. In a mouse study, PTH (1–34) significantly reduced the presence of both linear microcracks and diffuse damage after fatigue loading, compared with their loaded controls (Gardinier et al., 2018). In postmenopausal women previously treated with alendronate, 24 months of subsequent treatment with teriparatide significantly reduced crack density, crack surface density, and crack length in iliac crest biopsies (Dobnig et al., 2009). Another antiosteoporotic therapy, romosozumab, is a dual-action drug that simultaneously enhances bone formation and suppresses bone resorption and has shown great efficacy in increasing bone mass and reducing fracture risk (Liu et al., 2024). Currently, there is no evidence of romosozumab on bone microdamage and fatigue properties. Whether romosozumab alters favorably the repair mechanisms of bone microdamage should be considered as a crucial aspect of bone matrix quality in future studies.

7. Concluding remarks

In this chapter, we reviewed the current literature on how microdamage accumulates in bone under physiological fatigue loading and its role in bone fracture, and how bone responds to loading and damage to achieve self-repair. It is evident that the extent of microdamage significantly influences bone residual properties and toughening mechanisms. The imbalance between microdamage accumulation and repair, due to alterations in bone remodeling, may explain the increased risk of fragility fractures and insufficiency fractures. Consequently, microdamage should be considered as a critical bone quality component when estimating fracture risk. While the techniques in assessing microdamage in ex vivo tissues are well-established, challenges remain in applying clinical imaging modalities to evaluate in vivo microdamage. Alternative clinical approaches, such as blood-based and urine-based biomarkers as indicators of bone remodeling, can potentially serve as surrogate measures of microdamage and bone quality. Gaps of knowledge persist in the understanding of the microdamage repair mechanisms, particularly for diffuse damage. Additionally, greater clinical focus is needed to evaluate how aging, metabolic bone diseases, and the use of medication alter the matrix quality and damage repair mechanisms. Such insight can help more accurately predict, prevent, and manage the risk of fragility fracture associated with fatigue loading. Treatment strategies targeting the acceleration of bone damage repair may be more effective in bone health improvement.

AI Disclosure

During the preparation of this work, the author(s) used ChatGPT in order to check the grammar and edit text for clarity. After using this tool/service, the author(s) reviewed and edited the content as needed and take(s) full responsibility for the content of the publication.

References

Abbott, A., Bird, M. L., Wild, E., Brown, S. M., Stewart, G., & Mulcahey, M. K. (2020). Part I: Epidemiology and risk factors for stress fractures in female athletes. *The Physician and Sportsmedicine, 48*(1), 17–24. https://doi.org/10.1080/00913847.2019.1632158

Allen, M. R., & Burr, D. B. (2007). Three years of alendronate treatment results in similar levels of vertebral microdamage as after one year of treatment. *Journal of Bone and Mineral Research, 22*(11), 1759–1765. https://doi.org/10.1359/jbmr.070720

Allen, M. R., Iwata, K., Phipps, R., & Burr, D. B. (2006). Alterations in canine vertebral bone turnover, microdamage accumulation, and biomechanical properties following 1-year treatment with clinical treatment doses of risedronate or alendronate. *Bone, 39*(4), 872–879. https://doi.org/10.1016/j.bone.2006.04.028

Allen, M. R., Iwata, K., Sato, M., & Burr, D. B. (2006). Raloxifene enhances vertebral mechanical properties independent of bone density. *Bone, 39*(5), 1130–1135. https://doi.org/10.1016/j.bone.2006.05.007

Arlot, M. E., Burt-Pichat, B., Roux, J. P., Vashishth, D., Bouxsein, M. L., & Delmas, P. D. (2008). Microarchitecture influences microdamage accumulation in human vertebral trabecular bone. *Journal of Bone and Mineral Research, 23*(10), 1613–1618. https://doi.org/10.1359/jbmr.080517

Azmah Hanim, M. A., Brabazon, D., & Hashmi, M. S. J. (2019). *Cracks, microcracks, and fracture toughness of polymer composites*. Elsevier BV. https://doi.org/10.1016/b978-0-08-102293-1.00008-5

Black, D. M., Geiger, E. J., Eastell, R., Vittinghoff, E., Li, B. H., Ryan, D. S., Dell, R. M., & Adams, A. L. (2020). Atypical Femur fracture risk versus fragility fracture prevention with bisphosphonates. *New England Journal of Medicine, 383*(8), 743–753. https://doi.org/10.1056/NEJMoa1916525

Black, D. M., Greenspan, S. L., Ensrud, K. E., Palermo, L., McGowan, J. A., Lang, T. F., Garnero, P., Bouxsein, M. L., Bilezikian, J. P., & Rosen, C. J. (2003). The effects of parathyroid hormone and alendronate alone or in combination in postmenopausal osteoporosis. *New England Journal of Medicine, 349*(13), 1207–1215. https://doi.org/10.1056/NEJMoa031975

Boyce, T. M., Fyhrie, D. P., Glotkowski, M. C., Radin, E. L., & Schaffler, M. B. (1998). Damage type and strain mode associations in human compact bone bending fatigue. *Journal of Orthopaedic Research, 16*(3), 322–329. https://doi.org/10.1002/jor.1100160308

Brennan, O., Kennedy, O. D., Lee, T. C., Rackard, S. M., & Obrien, F. J. (2011). Effects of estrogen deficiency and bisphosphonate therapy on osteocyte viability and microdamage accumulation in an ovine model of osteoporosis. *Journal of Orthopaedic Research, 29*(3), 419–424. https://doi.org/10.1002/jor.21229

Burr, D. B., Forwood, M. R., Fyhrie, D. P., Martin, R. B., Schaffler, M. B., & Turner, C. H. (1997). Bone microdamage and skeletal fragility in osteoporotic and stress fractures. *Journal of Bone and Mineral Research, 12*(1), 6–15. https://doi.org/10.1359/jbmr.1997.12.1.6

Burr, D. B., & Hooser, M. (1995). Alterations to the en bloc basic fuchsin staining protocol for the demonstration of microdamage produced in vivo. *Bone, 17*(4), 431–433. https://doi.org/10.1016/s8756-3282(95)00241-3

Burr, D. B., Martin, R. B., Schaffler, M. B., & Radin, E. L. (1985). Bone remodeling in response to in vivo fatigue microdamage. *Journal of Biomechanics, 18*(3), 189–200. https://doi.org/10.1016/0021-9290(85)90204-0

Burr, D. B., & Stafford, T. (1990). Validity of the bulk-staining technique to separate artifactual from in vivo bone microdamage. *Clinical Orthopaedics and Related Research, 260*, 305–308. https://doi.org/10.1097/00003086-199011000-00047

Cardoso, L., Herman, B. C., Verborgt, O., Laudier, D., Majeska, R. J., & Schaffler, M. B. (2009). Osteocyte apoptosis controls activation of intracortical resorption in response to bone fatigue. *Journal of Bone and Mineral Research, 24*(4), 597–605. https://doi.org/10.1359/jbmr.081210

Carter, D. R., & Hayes, W. C. (1977). Compact bone fatigue damage-I. Residual strength and stiffness. *Journal of Biomechanics, 10*(5–6), 325–337. https://doi.org/10.1016/0021-9290(77)90005-7

Cedeno-Veloz, B. A., Erviti Lopez, J., Gutiérrez-Valencia, M., Leache Alegría, L., Saiz, L. C., Rodríguez García, A. M., Sánchez Latorre, M., Ramírez Vélez, R., Izquierdo, M., & Martínez-Velilla, N. (2022). Efficacy of antiresorptive treatment in osteoporotic older adults: A systematic review and meta-analysis of randomized clinical trials. *The Journal of Nutrition, Health & Aging, 26*(8), 778–785. https://doi.org/10.1007/s12603-022-1825-5

Chapurlat, R. D., Arlot, M., Burt-Pichat, B., Chavassieux, P., Roux, J. P., Portero-Muzy, N., & Delmas, P. D. (2007). Microcrack frequency and bone remodeling in postmenopausal osteoporotic women on long-term bisphosphonates: A bone biopsy study. *Journal of Bone and Mineral Research, 22*(10), 1502–1509. https://doi.org/10.1359/jbmr.070609

Chen, & Sambrook. (2011). Antiresorptive therapies for osteoporosis: A clinical overview. *Nature Reviews Endocrinology, 8*(2), 81–91.

Cowin, S. C. (2009). Chapter I: Integrated bone tissue physiology: Anatomy and physiology. *Bone Mechanics Handbook*.

Davis, M. S., Kovacic, B. L., Marini, J. C., Shih, A. J., & Kozloff, K. M. (2012). Increased susceptibility to microdamage in Brtl/+ mouse model for osteogenesis imperfecta. *Bone, 50*(3), 784–791. https://doi.org/10.1016/j.bone.2011.12.007

Diab, T., Condon, K. W., Burr, D. B., & Vashishth, D. (2006). Age-related change in the damage morphology of human cortical bone and its role in bone fragility. *Bone, 38*(3), 427–431. https://doi.org/10.1016/j.bone.2005.09.002

Diab, T., & Vashishth, D. (2005). Effects of damage morphology on cortical bone fragility. *Bone, 37*(1), 96–102. https://doi.org/10.1016/j.bone.2005.03.014

Dobnig, H., Stepan, J. J., Burr, D. B., Li, J., Michalská, D., Sipos, A., Petto, H., Fahrleitner-Pammer, A., & Pavo, I. (2009). Teriparatide reduces bone microdamage accumulation in postmenopausal women previously treated with alendronate. *Journal of Bone and Mineral Research, 24*(12), 1998–2006. https://doi.org/10.1359/jbmr.090527

Dominguez, V. M., & Agnew, A. M. (2019). Microdamage as a bone quality component: Practical guidelines for the two-dimensional analysis of linear microcracks in human cortical bone. *JBMR Plus, 3*(6). https://doi.org/10.1002/jbm4.10203

Dragoun Kolibová, S., Wölfel, E. M., Hemmatian, H., Milovanovic, P., Mushumba, H., Wulff, B., Neidhardt, M., Püschel, K., Failla, A. V., Vlug, A., Schlaefer, A., Ondruschka, B., Amling, M., Hofbauer, L. C., Rauner, M., Busse, B., & Jähn-Rickert, K. (2023). Osteocyte apoptosis and cellular micropetrosis signify skeletal aging in type 1 diabetes. *Acta Biomaterialia, 162*, 254–265. https://doi.org/10.1016/j.actbio.2023.02.037

Drake, M. T., Clarke, B. L., & Khosla, S. (2008). Bisphosphonates: Mechanism of action and role in clinical practice. *Mayo Clinic Proceedings, 83*(9), 1032–1045. https://doi.org/10.4065/83.9.1032

Fantner, G. E., Hassenkam, T., Kindt, J. H., Weaver, J. C., Birkedal, H., Pechenik, L., Cutroni, J. A., Cidade, G. A. G., Stucky, G. D., Morse, D. E., & Hansma, P. K. (2005). Sacrificial bonds and hidden length dissipate energy as mineralized fibrils separate during bone fracture. *Nature Materials, 4*(8), 612–616. https://doi.org/10.1038/nmat1428

Farr, J. N., Xu, M., Weivoda, M. M., Monroe, D. G., Fraser, D. G., Onken, J. L., Negley, B. A., Sfeir, J. G., Ogrodnik, M. B., Hachfeld, C. M., LeBrasseur, N. K., Drake, M. T., Pignolo, R. J., Pirtskhalava, T., Tchkonia, T., Oursler, M. J., Kirkland, J. L., &

Khosla, S. (2017). Targeting cellular senescence prevents age-related bone loss in mice. *Nature Medicine, 23*(9), 1072–1079. https://doi.org/10.1038/nm.4385

Fazzalari, N. L., Forwood, M. R., Smith, K., Manthey, B. A., & Herreen, P. (1998). Assessment of cancellous bone quality in severe osteoarthrosis: Bone mineral density, mechanics, and microdamage. *Bone, 22*(4), 381–388. https://doi.org/10.1016/s8756-3282(97)00298-6

Frost, H. (1960). Presence of microscopic cracks in vivo in bone. *Henry Ford Hospital Medical Journal, 8*(1), 25–35.

Gardinier, J. D., Al-Omaishi, S., Rostami, N., Morris, M. D., & Kohn, D. H. (2018). Examining the influence of PTH(1-34) on tissue strength and composition. *Bone, 117*, 130–137. https://doi.org/10.1016/j.bone.2018.09.019

Haider, I. T., Loundagin, L. L., Sawatsky, A., Kostenuik, P. J., Boyd, S. K., & Edwards, W. B. (2023). Twelve months of denosumab and/or alendronate is associated with improved bone fatigue life, microarchitecture, and density in ovariectomized cynomolgus monkeys. *Journal of Bone and Mineral Research, 38*(3), 403–413. https://doi.org/10.1002/jbmr.4758

Hernandez, C. J. (2023). Suppression of remodeling and bone fragility. *Journal of Bone and Mineral Research, 38*(3), 370–371. https://doi.org/10.1002/jbmr.4781

Javaheri, B., & Pitsillides, A. A. (2019). Aging and mechanoadaptive responsiveness of bone. *Current Osteoporosis Reports, 17*(6), 560–569. https://doi.org/10.1007/s11914-019-00553-7

Jung, H., & Akkus, O. (2020). Diffuse microdamage in bone activates anabolic response by osteoblasts via involvement of voltage-gated calcium channels. *Journal of Bone and Mineral Metabolism, 38*(2), 151–160. https://doi.org/10.1007/s00774-019-01042-8

Kennedy, O. D., Herman, B. C., Laudier, D. M., Majeska, R. J., Sun, H. B., & Schaffler, M. B. (2012). Activation of resorption in fatigue-loaded bone involves both apoptosis and active pro-osteoclastogenic signaling by distinct osteocyte populations. *Bone, 50*(5), 1115–1122. https://doi.org/10.1016/j.bone.2012.01.025

Kennedy, O. D., Laudier, D. M., Majeska, R. J., Sun, H. B., & Schaffler, M. B. (2014). Osteocyte apoptosis is required for production of osteoclastogenic signals following bone fatigue in vivo. *Bone, 64*, 132–137. https://doi.org/10.1016/j.bone.2014.03.049

Knapik, J. J., Reynolds, K., & Hoedebecke, K. L. (2016). Stress fractures: Etiology, epidemiology, diagnosis, treatment, and prevention. *Journal of Special Operations Medicine: A Peer Reviewed Journal for SOF Medical Professionals, 17*(2), 120–130.

Komatsubara, S., Mori, S., Mashiba, T., Ito, M., Li, J., Kaji, Y., Akiyama, T., Miyamoto, K., Cao, Y., Kawanishi, J., & Norimatsu, H. (2003). Long-term treatment of incadronate disodium accumulates microdamage but improves the trabecular bone microarchitecture in dog vertebra. *Journal of Bone and Mineral Research, 18*(3), 512–520. https://doi.org/10.1359/jbmr.2003.18.3.512

Kraenzlin, M. E., & Meier, C. (2011). Parathyroid hormone analogues in the treatment of osteoporosis. *Nature Reviews Endocrinology, 7*(11), 647–656. https://doi.org/10.1038/nrendo.2011.108

Kreipke, T. C., Garrison, J. G., Easley, J., Turner, A. S., & Niebur, G. L. (2016). The roles of architecture and estrogen depletion in microdamage risk in trabecular bone. *Journal of Biomechanics, 49*(14), 3223–3229. https://doi.org/10.1016/j.jbiomech.2016.08.009

Lakshmikantha, H. T., Ravichandran, N. K., Jeon, M., Kim, J., & Park, H. S. (2019). 3-Dimensional characterization of cortical bone microdamage following placement of orthodontic microimplants using optical coherence tomography. *Scientific Reports, 9*(1). https://doi.org/10.1038/s41598-019-39670-9

Lambers, F. M., Bouman, A. R., Rimnac, C. M., & Hernandez, C. J. (2013). Microdamage caused by fatigue loading in human cancellous bone: Relationship to reductions in bone biomechanical performance. *PLoS One, 8*(12). https://doi.org/10.1371/journal.pone.0083662

Lee, T. C. (2003). Microdamage in osteoporosis, bone quality and remodelling. *Journal of Anatomy, 203*(2), 159. https://doi.org/10.1046/j.1469-7580.2003.00217.x

Li, H. F., Wang, S. G., Zhang, P., Qu, R. T., & Zhang, Z. F. (2018). Crack propagation mechanisms of AISI 4340 steels with different strength and toughness. *Materials Science and Engineering: A, 729*, 130–140. https://doi.org/10.1016/j.msea.2018.05.056

Liu, L., Clifton-Bligh, R. J., Girgis, C. M., & Gild, M. L. (2024). Extending the therapeutic potential: Romosozumab in osteoporosis management. *Journal of the Endocrine Society, 8*(11). https://doi.org/10.1210/jendso/bvae160

Liu, X., Li, W., Cai, J., Yan, Z., Shao, X., Xie, K., Guo, X. E., Luo, E., & Jing, D. (2020). Spatiotemporal characterization of microdamage accumulation and its targeted remodeling mechanisms in diabetic fatigued bone. *Federation of American Societies for Experimental Biology Journal, 34*(2), 2579–2594. https://doi.org/10.1096/fj.201902011RR

MacKey, D. C., Black, D. M., Bauer, D. C., McCloskey, E. V., Eastell, R., Mesenbrink, P., Thompson, J. R., & Cummings, S. R. (2011). Effects of antiresorptive treatment on nonvertebral fracture outcomes. *Journal of Bone and Mineral Research, 26*(10), 2411–2418. https://doi.org/10.1002/jbmr.446

Mashiba, T., Hirano, T., Turner, C. H., Forwood, M. R., Johnston, C. C., & Burr, D. B. (2000). Suppressed bone turnover by bisphosphonates increases microdamage accumulation and reduces some biomechanical properties in dog rib. *Journal of Bone and Mineral Research, 15*(4), 613–620. https://doi.org/10.1359/jbmr.2000.15.4.613

Mashiba, T., Hui, S., Turner, C. H., Mori, S., Johnston, C. C., & Burr, D. B. (2005). Bone remodeling at the iliac crest can predict the changes in remodeling dynamics, microdamage accumulation, and mechanical properties in the lumbar vertebrae of dogs. *Calcified Tissue International, 77*(3), 180–185. https://doi.org/10.1007/s00223-005-1295-x

Mashiba, T., Turner, C. H., Hirano, T., Forwood, M. R., Johnston, C. C., & Burr, D. B. (2001). Effects of suppressed bone turnover by bisphosphonates on microdamage accumulation and biomechanical properties in clinically relevant skeletal sites in beagles. *Bone, 28*(5), 524–531. https://doi.org/10.1016/S8756-3282(01)00414-8

Mehraban Alvandi, L. (2020). Diffuse damage repair mechanism in bone. https://academicworks.cuny.edu/cc_etds_theses/950.

Milovanovic, P., & Busse, B. (2020). Phenomenon of osteocyte lacunar mineralization: Indicator of former osteocyte death and a novel marker of impaired bone quality? *Endocrine Connections, 9*(4), Article R70. https://doi.org/10.1530/ec-19-0531

Mohsin, S., Kaimala, S., AlTamimi, E. K. Y., Tariq, S., & Adeghate, E. (2019). In vivo labeling of bone microdamage in an animal model of type 1 diabetes mellitus. *Emirates Scientific Reports, 9*(1). https://doi.org/10.1038/s41598-019-53487-6

Mori, S., & Burr, D. B. (1993). Increased intracortical remodeling following fatigue damage. *Bone, 14*(2), 103–109. https://doi.org/10.1016/8756-3282(93)90235-3

Muju, S., Anderson, P. M., & Mendelsohn, D. A. (1998). Microcrack toughening in two-phase multilayered media. *Acta Materialia, 46*(15), 5385–5397. https://doi.org/10.1016/S1359-6454(98)00195-5

Nalla, R. K., Kinney, J. H., & Ritchie, R. O. (2003). Mechanistic fracture criteria for the failure of human cortical bone. *Nature Materials, 2*(3), 164–168. https://doi.org/10.1038/nmat832

Norman, T. L., Yeni, Y. N., Brown, C. U., & Wang, Z. (1998). Influence of microdamage on fracture toughness of the human femur and tibia. *Bone, 23*(3), 303–306. https://doi.org/10.1016/S8756-3282(98)00103-3

O'Brien, F. J., Taylor, D., & Lee, T. C. (2005). The effect of bone microstructure on the initiation and growth of microcracks. *Journal of Orthopaedic Research, 23*(2), 475–480. https://doi.org/10.1016/j.orthres.2004.08.005

Poundarik, A. A., Diab, T., Sroga, G. E., Ural, A., Boskey, A. L., Gundberg, C. M., & Vashishth, D. (2012). Dilatational band formation in bone. *Proceedings of the National Academy of Sciences of the United States of America, 109*(47), 19178–19183. https://doi.org/10.1073/pnas.1201513109

Poundarik, A. A., & Vashishth, D. (2015). Multiscale imaging of bone microdamage. *Connective Tissue Research, 56*(2), 87–98. https://doi.org/10.3109/03008207.2015.1008133

Poundarik, A. A., Wu, P. C., Evis, Z., Sroga, G. E., Ural, A., Rubin, M., & Vashishth, D. (2015). A direct role of collagen glycation in bone fracture. *Journal of the Mechanical Behavior of Biomedical Materials, 52*, 120–130. https://doi.org/10.1016/j.jmbbm.2015.08.012

Reifsnider, K. L. (2012). *Fatigue of composite materials.*

Rho, J. Y., Kuhn-Spearing, L., & Zioupos, P. (1998). Mechanical properties and the hierarchical structure of bone. *Medical Engineering & Physics, 20*(2), 92–102. https://doi.org/10.1016/S1350-4533(98)00007-1

Rubin, M. R., & Patsch, J. M. (2016). Assessment of bone turnover and bone quality in type 2 diabetic bone disease: Current concepts and future directions. *Bone Research, 4*. https://doi.org/10.1038/boneres.2016.1

Sabet, F. A., Najafi, A. R., Hamed, E., & Jasiuk, I. (2016). Modelling of bone fracture and strength at different length scales: A review. *Interface Focus, 6*(1), Article 20150055. https://doi.org/10.1098/rsfs.2015.0055

Sacher, S. E., Hunt, H. B., Lekkala, S., Lopez, K. A., Potts, J., Heilbronner, A. K., Stein, E. M., Hernandez, C. J., & Donnelly, E. (2022). Distributions of microdamage are altered between trabecular rods and plates in cancellous bone from men with type 2 diabetes mellitus. *Journal of Bone and Mineral Research, 37*(4), 740–752. https://doi.org/10.1002/jbmr.4509

Schaffler, M. B., Choi, K., & Milgrom, C. (1995). Aging and matrix microdamage accumulation in human compact bone. *Bone, 17*(6), 521–525. https://doi.org/10.1016/8756-3282(95)00370-3

Schaffler, M. B., Pitchford, W. C., Choi, K., & Riddle, J. M. (1994). Examination of compact bone microdamage using back-scattered electron microscopy. *Bone, 15*(5), 483–488. https://doi.org/10.1016/8756-3282(94)90271-2

Schaffler, M. B., Radin, E. L., & Burr, D. B. (1989). Mechanical and morphological effects of strain rate on fatigue of compact bone. *Bone, 10*(3), 207–214. https://doi.org/10.1016/8756-3282(89)90055-0

Seref-Ferlengez, Z., Basta-Pljakic, J., Kennedy, O. D., Philemon, C. J., & Schaffler, M. B. (2014). Structural and mechanical repair of diffuse damage in cortical bone in vivo. *Journal of Bone and Mineral Research, 29*(12), 2537–2544. https://doi.org/10.1002/jbmr.2309

Seref-Ferlengez, Z., Kennedy, O. D., & Schaffler, M. B. (2015). Bone microdamage, remodeling and bone fragility: How much damage is too much damage? *BoneKEy Reports, 4*. https://doi.org/10.1038/bonekey.2015.11

Shimamura, M., Iwata, K., Mashiba, T., Miki, T., & Yamamoto, T. (2019). Accumulation of microdamage in subchondral bone at the femoral head in patients with end-stage osteoarthritis of the hip. *Journal of Bone and Mineral Metabolism, 37*(5), 880–885. https://doi.org/10.1007/s00774-019-00988-z

Stepan, J. J., Burr, D. B., Pavo, I., Sipos, A., Michalska, D., Li, J., Fahrleitner-Pammer, A., Petto, H., Westmore, M., Michalsky, D., Sato, M., & Dobnig, H. (2007). Low bone mineral density is associated with bone microdamage accumulation in postmenopausal women with osteoporosis. *Republic Bone, 41*(3), 378–385. https://doi.org/10.1016/j.bone.2007.04.198

Sun, X., McLamore, E., Kishore, V., Fites, K., Slipchenko, M., Porterfield, D. M., & Akkus, O. (2012). Mechanical stretch induced calcium efflux from bone matrix stimulates osteoblasts. *Bone, 50*(3), 581–591. https://doi.org/10.1016/j.bone.2011.12.015

Tang, S. Y., Allen, M. R., Phipps, R., Burr, D. B., & Vashishth, D. (2009). Changes in non-enzymatic glycation and its association with altered mechanical properties following 1-year treatment with risedronate or alendronate. *Osteoporosis International, 20*(6), 887–894. https://doi.org/10.1007/s00198-008-0754-4

Tang, S. Y., & Vashishth, D. (2007). A non-invasive in vitro technique for the three-dimensional quantification of microdamage in trabecular bone. *Bone, 40*(5), 1259–1264. https://doi.org/10.1016/j.bone.2006.10.031

Unnanuntana, A., Gladnick, B. P., Donnelly, E., & Lane, J. M. (2010). The assessment of fracture risk. *Journal of Bone and Joint Surgery, 92*(3), 743–753. https://doi.org/10.2106/JBJS.I.00919

Varvani-Farahani, A., & Najmi, H. (2010). A damage assessment model for cadaveric cortical bone subjected to fatigue cycles. *International Journal of Fatigue, 32*(2), 420–427. https://doi.org/10.1016/j.ijfatigue.2009.08.002

Vashishth, D. (2007). Hierarchy of bone microdamage at multiple length scales. *International Journal of Fatigue, 29*(6), 1024–1033. https://doi.org/10.1016/j.ijfatigue.2006.09.010

Vashishth, D., Behiri, J. C., & Bonfield, W. (1997). Crack growth resistance in cortical bone: Concept of microcrack toughening. *Journal of Biomechanics, 30*(8), 763–769. https://doi.org/10.1016/S0021-9290(97)00029-8

Vashishth, D., Koontz, J., Qiu, S. J., Lundin-Cannon, D., Yeni, Y. N., Schaffler, M. B., & Fyhrie, D. P. (2000). In vivo diffuse damage in human vertebral trabecular bone. *Bone, 26*(2), 147–152. https://doi.org/10.1016/S8756-3282(99)00253-7

Verborgt, O., Gibson, G. J., & Schaffler, M. B. (2000). Loss of osteocyte integrity in association with microdamage and bone remodeling after fatigue in vivo. *Journal of Bone and Mineral Research, 15*(1), 60–67. https://doi.org/10.1359/jbmr.2000.15.1.60

Verborgt, O., Tatton, N. A., Majeska, R. J., & Schaffler, M. B. (2002). Spatial distribution of bax and Bcl-2 in osteocytes after bone fatigue: Complementary roles in bone remodeling regulation? *Journal of Bone and Mineral Research, 17*(5), 907–914. https://doi.org/10.1359/jbmr.2002.17.5.907

Waldorff, E. I., Goldstein, S. A., & McCreadie, B. R. (2007). Age-dependent microdamage removal following mechanically induced microdamage in trabecular bone in vivo. *Bone, 40*(2), 425–432. https://doi.org/10.1016/j.bone.2006.08.011

Wang, Z., & Norman, T. L. (1997). Microdamage of human cortical bone: Incidence and morphology in long bones. *Bone, 20*(4), 375–379. https://doi.org/10.1016/S8756-3282(97)00004-5

Wang, B., & Vashishth, D. (2023). Advanced glycation and glycoxidation end products in bone. *Bone, 176*, Article 116880. https://doi.org/10.1016/j.bone.2023.116880

Wu, A. M., Bisignano, C., James, S. L., Abady, G. G., Abedi, A., Abu-Gharbieh, E., Alhassan, R. K., Alipour, V., Arabloo, J., Asaad, M., Asmare, W. N., Awedew, A. F., Banach, M., Banerjee, S. K., Bijani, A., Birhanu, T. T. M., Bolla, S. R., Cámera, L. A., Chang, J. C., … Vos, T. (2021). Global, regional, and national burden of bone fractures in 204 countries and territories, 1990–2019: A systematic analysis from the global burden of disease study 2019. *The Lancet. Healthy Longevity, 2*(9), Article e580. https://doi.org/10.1016/S2666-7568(21)00172-0

Ye, L. (1989). On fatigue damage accumulation and material degradation in composite materials. *Composites Science and Technology, 36*(4), 339–350. https://doi.org/10.1016/0266-3538(89)90046-8

Zhang, W., Tekalur, S. A., Baumann, M., & McCabe, L. R. (2013). The effects of damage accumulation on the tensile strength and toughness of compact bovine bone. *Journal of Biomechanics, 46*(5), 964–972. https://doi.org/10.1016/j.jbiomech.2012.12.005

Zimmermann, E. A., Busse, B., & Ritchie, R. O. (2015). The fracture mechanics of human bone: Influence of disease and treatment. *BoneKEy Reports, 4*. https://doi.org/10.1038/bonekey.2015.112

PART

Biomechanics and mechanobiology modeling

3

CHAPTER

Bone adaptation models: From Wolff's law to modern views 13

Natalia M. Castoldi[1], **Peter Pivonka**[2] **and Vittorio Sansalone**[3]

[1]*Faculty of Mechanical Engineering, Technische Hochschule Nürnberg Georg Simon Ohm, Nuremberg, Germany;* [2]*School of Mechanical, Medical and Process Engineering, Queensland University of Technology, Brisbane, QLD, Australia;* [3]*Univ Paris Est Creteil, Univ Gustave Eiffel, CNRS, UMR 8208, MSME, Créteil, France*

1. Introduction

Bone is a remarkable living tissue that constantly repairs, renews, and adapts in response to biochemical and mechanical cues. This dynamic behavior arises from the coordinated processes of growth, modeling, and remodeling, which collectively maintain skeletal integrity across the lifespan. Capturing these processes through theoretical and computational models has become a central endeavor in biomechanics: models not only help unravel the mechanisms driving bone health and disease, but also provide predictive tools for assessing therapeutic strategies and clinical interventions.

The clinical significance of bone adaptation becomes evident when these processes are disrupted. Altered remodeling or modeling affects the skeleton at multiple structural levels. Osteoporosis, for instance, compromises both trabecular and cortical bone (Parfitt et al., 1983; Wehrli et al., 2001); osteomalacia and certain kidney diseases impair mineralization (Roschger et al., 1998; Zaloszyc et al., 2023); and abnormal growth patterns can lead to conditions such as scoliosis (Cheng et al., 2001; Smit, 2020). These diverse pathologies underscore the complexity of bone adaptation, which integrates chemical, biological, and mechanical interactions spanning scales from cells to the whole organ. Mathematical and mechanical models provide a way to formalize this complexity, offering frameworks to connect mechanisms across scales and to simulate both physiological and pathological scenarios.

Repair, renewal, and adaptation of bone can involve three distinct yet interconnected processes: growth, modeling, and remodeling (Mosekilde et al., 1995). *Growth* dominates from embryogenesis to early adulthood, establishing skeletal size and shape (Raimondi et al., 2012). *Modeling* reshapes bones through the uncoupled actions of osteoblasts and osteoclasts (Langdahl et al., 2016), typically yielding net bone gain during childhood and adolescence, though it may reemerge in adulthood under unusual mechanical or hormonal conditions (Castoldi et al., 2025; Sugiyama et al., 2012). *Remodeling*, in contrast, is a lifelong process carried

Biomechanics and Mechanobiology of Bone. https://doi.org/10.1016/B978-0-443-26739-0.00006-5

out by basic multicellular units (BMUs) conceptualized by Frost in the 1960s (Frost, 1969). It adjusts bone mass and architecture in accordance with prevailing mechanical demands, repairs microdamage, regulates mineral homeostasis, and supports hematopoiesis (Parfitt, 1994). Remodeling relies on tightly orchestrated cycles of resorption and formation, followed by mineralization, thereby controlling turnover rates and structural renewal (Parfitt, 1994, 2002). Because remodeling is lifelong and its impairment underlies several pathologies, it has been a natural target for mechanistic and numerical modeling since the earliest theoretical studies of bone adaptation. In this chapter, we concentrate on models of remodeling as the principal driver of lifelong adaptation, while directing the reader to (Chapter 14) for a more detailed treatment of growth and modeling processes.

At the cell level, bone adaptation is regulated by a wide range of systemic and local biochemical factors, including neuroendocrine, endocrine, and paracrine signals involving metabolites, cytokines, growth factors, hormones, vitamins, and minerals (Frost, 2003; Guadalupe-Grau et al., 2009; Lee & Lanyon, 2004; Rittweger, 2007; Rosen, 2008; Rubin et al., 2006; Scott et al., 2008; Skerry, 2006). Besides these regulators, mechanical loading is widely regarded as a primary driver of adaptation, not only directly but also by modulating the expression of many of these signaling molecules. Still, the exact nature of the mechanical stimulus sensed by bone cells remains debated, with models emphasizing strain magnitude and rate (Frost, 1987; Saxon & Lanyon, 2008), stress (Carter et al., 1989), or strain energy density (Huiskes et al., 2000). In all cases, adaptive changes occur only when mechanical cues exceed a threshold or fall outside a physiological window. Increasing bone mass or density typically requires mechanical loading that exceeds everyday levels. Both high-magnitude loading—capable of inducing controlled microdamage (Burr et al., 1997; Frost, 1990)—and high-frequency, low-magnitude loading (Rubin et al., 2001, 2002; Rubin & Lanyon, 1984) can promote bone formation, whereas mechanical disuse accelerates bone loss (Frost, 1987; Langdahl et al., 2016). Importantly, cell-scale mechanical cues ultimately originate from mechanical forces applied at the organ scale, highlighting the inherently multiscale nature of bone adaptation. Because experimental evidence points to multiple, possibly coexisting stimuli and mechanisms acting across cellular, tissue, and organ levels, mathematical and mechanical modeling has been indispensable for testing hypotheses, quantifying thresholds, and predicting long-term outcomes of mechanical loading.

This multifactorial regulation has motivated the development of mathematical models at multiple biological scales. Building upon the foundational ideas of Wolff (1870) and Roux (1881), researchers have sought to formalize how mechanical and biochemical factors shape bone structure at different scales. At the cellular scale, models capture signaling pathways and interactions among osteoblasts, osteoclasts, and osteocytes (Komarova et al., 2003; Lemaire et al., 2004; Maldonado et al., 2006; Pivonka et al., 2008). At the tissue and organ levels, phenomenological laws describe the adaptation of bone density and microarchitecture in response to mechanical stimuli (Beaupré et al., 1990a; Frost, 1987, 2001; Huiskes et al., 1987), while optimization approaches frame bone as an evolving optimal structure (Carter et al., 1987; Cowin, 1994; Fernandes et al., 1999; Geraldes et al., 2016). Bridging scales remains a central challenge as bone adaptation is inherently multiscale in both cause and effect: mechanical forces are applied at the organ level, trigger cell-level events, and produce tissue-level outcomes. To address this, researchers have developed agent-based simulations (Boaretti et al., 2023; Checa & Prendergast, 2009; Jaber et al., 2022), micro-mechanical frameworks linking tissue stresses to cellular stimuli (Pastrama et al., 2018; Scheiner et al., 2013), and increasingly, multiscale and multiphysics approaches coupling mechanics, biology, and transport phenomena. Together, these models aim not only to explain observed patterns of bone structure and adaptation, but also to provide predictive tools for future experimental and clinical applications.

This chapter reviews the evolution of bone adaptation models, from early phenomenological observations to state-of-the-art multiscale computational frameworks. We begin with the historical foundations established by Wolff, Roux, and pioneering experimental studies (Section 2), then examine the classical theories that shaped the field, including the influential models of Cowin, Carter, Frost, and Huiskes (Section 3). We next survey modern approaches, encompassing continuum mechanics-based models (Section 4.1), optimization frameworks (Section 4.2), microscale remodeling models (Section 4.3), and multiscale/multiphysics methods (Section 4.4). By tracing this progression, we highlight how bone adaptation modeling has evolved from descriptive laws into mechanistically grounded, predictive frameworks—an evolution that continues to open new avenues for research and clinical translation.

The classification adopted in this chapter was chosen to reflect overarching conceptual principles. Alternative categorizations are possible, for instance by focusing on the evolutionary nature of the models (Calvo-Gallego et al., 2022), and many models naturally span more than one family, highlighting the interconnectedness of ideas in this field.

2. Historical foundations

The close link between mechanics and stresses has been noted since the 19th century and the early observations of Culmann (1866); Meyer (1867); Wolff (1870) and Roux (1881). Georg Hermann von Meyer was a German professor of anatomy with a strong interest in the human skeleton. Benefiting from his interactions with the German civil engineering professor Carl Culmann, he got interested in the trabecular patterns emerging in the human proximal femur. Based on mechanical arguments, Culmann noticed that the trabecular struts appeared to be aligned along the expected principal stress directions in the femur. The result of this collaboration deeply influenced the work of Julius Wolff, a German surgeon convinced that Culmann and von Meyer's findings could not be a coincidence (Culmann, 1866; Meyer, 1867). In 1870, Wolff introduced a theory of bone remodeling, nowadays referred to as Wolff's law: "Every change in the form and function of bone or of their function alone is followed by certain definite changes in their internal architecture, and equally definite alterations in their external configuration, in accordance with mathematical laws" (Wolff, 1870, 2012). The theory posits that bone adapts its shape and microstructure in response to mechanical loads, and that trabeculae align along trajectories of principal stress (see Fig. 13.1).

Wolff's law played a key role in shaping early biomechanical theories of bone remodeling and guided research in the field for over a century. Nevertheless, Wolff's doctrine was the subject of considerable debate and strong criticism from some researchers, as reported by Roesler (1987). By the late 20th century, important limitations had become increasingly apparent. In particular, the original trajectorial interpretation of Wolff's law—suggesting that trabeculae align along two approximately orthogonal principal stress directions—does not fully capture the complexity of trabecular bone architecture. In practice, trabeculae often exhibit non-orthogonal arrangements, especially near joints, prompting several authors to question the adequacy of Wolff's law in this context. Cowin (1989) clarified the scale-dependent validity of Wolff's law by introducing the concept of a resolution length restriction, arguing that the law is more applicable at the tissue scale and may not accurately describe bone structure at smaller scales. Pidaparti and Turner (1997) suggested

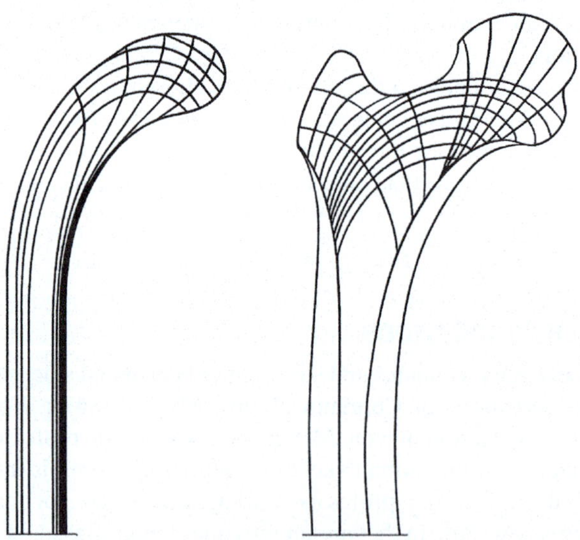

FIGURE 13.1

Cartoons depicting the stress trajectories in a curved Culmann crane (left) and the trabecular pattern in the proximal human femur (right), inspired from the original figure of Wolff (1870).

that multidirectional loading could explain the non-orthogonal alignment of trabeculae observed in regions such as the proximal femur.

Some authors have even questioned the very foundations of Wolff's vision, arguing that Wolff's law, as an explanation for bone architecture, may be fundamentally flawed. Roesler (1981) suggested that Wolff's conclusions were based on a misinterpretation of stress trajectories in Culmann's crane. Lee and Taylor (1999) pointed out that Wolff's view, rooted in static engineering, neglected the key dynamic aspect of bone remodeling—namely, the role of bone resorption. Building on computer simulations of bone remodeling as a regulated biological process, Huiskes (2000) argued that "the similarity between trabecular orientation and stress trajectories is circumstantial, not causal." All three—Roesler, Lee and Taylor, and Huiskes—emphasized that a more appropriate conceptual framework might be found in the work of Wilhelm Roux, a contemporary of Wolff.

Roux was a German zoologist interested in developmental mechanics. Ten years after Wolff, he published a book in which he introduced a broader mechanistic theory of "developmental mechanics," suggesting that organs adapt their internal structure to new functional demands through mechanistic principles (Roux, 1881). Applied to bone tissue, Roux's paradigm viewed bone remodeling as a self-organizing process locally driven by mechanical stimuli and regulated by bone cells. In contrast to Wolff's static interpretation of bone architecture, Roux emphasized the dynamic and regulatory nature of remodeling, laying the groundwork for modern theories of bone adaptation and mechanobiology. As highlighted by Roesler (1981), Roux's perspective captures the essence of bone functional adaptation as it is understood today. Accordingly, the concept of "functional adaptation to bone forces" is now often preferred to "Wolff's law" to describe the general idea that bone develops in response to mechanical loading (Bertram & Swartz, 1991; Cowin, 1998; Hammer, 2015; Lee & Taylor, 1999; Ruff et al., 2006). Some authors have even proposed renaming this generalized concept "Roux's law" to distinguish it from the interpretation originally proposed by Wolff (Cowin, 1998; Lee & Taylor, 1999).

While Wolff and Roux focused on how mature bone adapts its internal architecture through remodeling, other researchers of the same era investigated how mechanical forces influence bone growth during development, leading to the formulation of the Hueter–Volkmann law. The mechanical effect on growth was first noticed separately by Carl Hueter and Richard Von Volkmann in 1862 (Bartoníček & Naňka, 2024). Their discovery, now called the Hueter-Volkmann law, was validated by other studies (Stokes et al., 2006), and discloses that an increase in compression acting on growth plates decreases bone growth, where the opposite is also true (Mehlman et al., 1997; Simon, 1978; Villemure & Stokes, 2009; Willy et al., 2008). Note that, as first mentioned by the famous German pathologist Rudolf Virchow (Wolff & Virchow, 1875), growth without an increase in volume is not growth, and that growth should not be confused with a change in bone architecture, as described by Wolff and Roux's laws.

During this same period, pioneering investigations into the mechanical properties of bone tissue were initiated by Wertheim (1847). Wertheim's groundbreaking study represented perhaps the first systematic examination of tensile mechanical properties across human tissues, encompassing both soft and hard tissues. This foundational work on bone elasticity subsequently motivated Rauber's research, who developed specialized testing equipment to overcome technical challenges associated with specimen fixation (Rauber, 1876; Roesler, 1987). Rauber (1876) comprehensive investigations examined both compact and cancellous bone responses under various loading conditions, including tension, compression, shear, and torsion. Notably, Rauber appears to have been the first researcher to identify several fundamental characteristics now well-recognized in biomechanical experimentation: the significant effects of environmental humidity and temperature on bone strength and elastic properties, as well as the directional sensitivity of bone mechanical behavior (anisotropy). Building upon these early efforts, Messerer (1880) conducted the most extensive mechanical testing program of his era, examining fresh human whole bones from a collection of 500 specimens obtained from 90 cadavers. This comprehensive dataset remained the largest available for an entire century (Sherwood Mather, 1967). The contributions of Koch (1917) in the early 20th century also merit recognition for his extensive mechanical characterization of the femoral head. Drawing from contemporary research and his experimental findings, Koch proposed in 1917 a mathematical framework based on the principle that bone anatomy is optimized for maximum mechanical efficiency. After these early investigations, experimental research on bone mechanical properties remained relatively limited, with notable exceptions including studies by Bell et al. (1941), Dempster and Liddicoat (1952), and Sherwood Mather (1967). The field of bone mechanics research experienced a prolonged period of reduced activity until a resurgence of interest emerged between approximately 1950 and 1970, marked by renewed investigations (Evans, 1969; Roesler, 1987). This revival in bone mechanics research can be attributed to emerging technological demands related to terrestrial, aerial, and space transportation systems, as well as advancing methodologies in competitive athletics (Evans, 1969; Roesler, 1987). At the same time, rapid advances in imaging technologies facilitated discoveries such as Whitehouse's innovative research examining bone tissue anisotropy through low-magnification scanning electron microscopy (Whithehouse, 1974). This era also witnessed the emergence of the first experimental in vivo studies investigating the relationship between mechanical environment and bone remodeling processes. Animal studies (Goodship et al., 1979; Woo et al., 1981) and human research (Krølner et al., 1983) have shown that bone mass increases in response to enhanced exercise regimens, with dynamic loading proving more effective than static loading (O'Connor et al., 1982).

3. Classical theories (1970s & 1980s)

In the second half of the 20th century, renewed interest in experimental bone mechanics paved the way for several influential modeling studies of bone adaptation. These works aimed to explain bone mass density and trabecular architecture—often regarded as evidence of Wolff's law—in terms of mechanical stimuli. The early models, initially developed in the 1970 and 1980s, were formulated at different scales—from the microstructure to the tissue or organ level—and were all phenomenological in nature, prescribing empirical laws that link a chosen mechanical stimulus (such as strain, stress, or strain energy density) to a remodeling response. Although they do not explicitly account for the underlying cellular or molecular processes, they remain influential for their conceptual clarity, simplicity, and predictive capability. Four major conceptual frameworks stand out:

- **Adaptive Elasticity**: Continuum theory linking bone density and fabric orientation to local mechanical fields at equilibrium.
- **Stanford model**: Bone maintenance theory linking bone apparent density to an effective stimulus accounting for the loading history.
- **Mechanostat**: Regulatory model based on strain thresholds that control bone formation and resorption.
- **Adaptive Bone Remodeling**: Quantitative stimulus–response laws that extend or combine earlier frameworks for predictive modeling.

3.1 Adaptive elasticity

The theory of *Adaptive Elasticity* was developed by Stephen Cowin and co-workers in the late 1970s and early 1980s, using the conceptual framework of Truesdell's continuum thermodynamics.

In its earliest version, bone was described as a porous elastic solid, and bone remodeling emerged from mass deposition or resorption processes modifying the porosity of the solid (Cowin & Hegedus, 1976). In two companion papers, Cowin and coworkers explored the same theory under the small strain assumption (Cowin & Nachlinger, 1978; Hegedus & Cowin, 1976). The theory assumes that bone senses mechanical strain and that a specific range of strains defines a remodeling equilibrium (Cowin, 1984). Deviations from this equilibrium act as signals for the tissue to remodel, either at the surface or through internal remodeling.

The theory was further extended to account for *anisotropic remodeling*. A key concept in this respect was the introduction of the fabric tensor (Cowin, 1985), a second rank tensor which characterizes the average geometric arrangement of the porous material microstructure. Cowin (1986) studied the conditions under which there is no realignment, i.e., the *remodeling equilibrium*, showing that this is achieved when the principal axes of the stress, strain, and fabric tensors are aligned, laying Wolff's law on a sound mathematical ground. The remodeling equilibrium was further characterized in terms of minimization and maximization of the strain energy density (Cowin, 1995).

In the early 1990s, Cowin et al. (1992) further extended the theory describing the *temporal evolution* of both the apparent density and the orientation of the trabecular architecture. The central assumption is that bone tends toward an elastic state that minimizes the difference between the current strain and a "preferred" strain determined by habitual loading. Bone adaptation is thus described as a mechanical process in which the solid volume fraction and trabecular orientation—and therefore the elastic properties of bone—evolve in response to mechanical usage. The theory naturally accommodates anisotropic adaptation by evolving the orthotropic symmetry axes of the stiffness tensor toward the principal strain directions.

In the early 2000s, Cowin (2003) provided a comprehensive summary of the main results of adaptive elasticity theory, with applications to internal remodeling, surface adaptation, and architectural changes. The characterization of remodeling equilibrium developed by Cowin and coworkers remains a key contribution in the field and continues to serve as a reference for studying anisotropic remodeling. Notably, Martin et al. (2020) independently derived similar results, showing that Cowin's conditions for remodeling equilibrium apply only in specific cases.

3.2 Stanford model

One of the most influential early phenomenological frameworks was developed at Stanford University by Denis Carter and colleagues in the late 1980s. This line of work, sometimes referred to as the *Stanford Model* or the *Bone Maintenance Theory*, sought to connect daily loading histories to local changes in bone apparent density and architecture, under the guiding idea of a *bone maintenance* stimulus.

In an early form of the bone maintenance theory, Carter (1987) proposed that multiple cyclic loading events could be condensed into an *osteogenic index* based on a combination of octahedral shear stress and hydrostatic stress. This approach reflected the view that distinct components of the stress state contribute differently to the bone maintenance stimulus.

Carter et al. (1987) expanded the theory by formalizing the relationship between apparent density and an effective *daily stress stimulus* accounting for the loading history. The framework considered various formulations of the mechanical stimulus, including an effective stress approach, a fatigue damage approach, and a strain energy density approach (leading to the concept of *energy stress*). In addition, the theory introduced the concept of a homeostatic stimulus—a reference level of mechanical stimulus that maintains local bone mass, promoting neither formation nor resorption. Carter et al. (1989) further extended the model to account for both mass density change and trabecular orientation. They introduced a finite element (FE) implementation, in which apparent density was iteratively updated based on the daily loading history (energy stress approach), and trabecular orientation was linked to an equivalent cyclic normal stress.

Beaupré et al. (1990a) reformulated the theory in explicitly time-dependent terms. In this framework, bone modeling and remodeling were driven by the difference between the actual local tissue stimulus and an attractor state. This formulation unified internal and external remodeling within a single framework, with the apposition and resorption rates proportional to the stimulus error. Notably, the authors introduced two important concepts. First, they explicitly—albeit in a simplified way—accounted for multiple spatial scales by distinguishing stresses at the continuum and tissue levels. Second, they proposed the concept of available bone surface area as a modulating factor for remodeling activity, marking one of the earliest attempts to integrate bone cell activity into a mechanically based model of bone adaptation. In a companion study, Beaupré et al. (1990) applied the model in an FE analysis (FEA) using two simplified versions for the mechanical stimulus. They showed that incorporating a *dead zone*, closely resembling Frost's mechanostat (Frost, 1987), yielded realistic patterns of bone apparent density.

By integrating cyclic loading effects, tissue-level mechanics, attractor states, and remodeling kinetics in a unified framework, the Stanford Model remains a milestone in the field and opened the way to later continuum mechanics–based formulations.

3.3 Mechanostat

The *Mechanostat* theory, introduced by Harold Frost in the late 20th century (Frost, 1987), describes bone adaptation as a biologically regulated process aimed at maintaining mechanical strain within a physiological range. In this view, the skeleton is equipped with a "mechanostat," analogous to a thermostat, with defined strain thresholds that initiate either bone formation or resorption. Grounded in experimental evidence from animal studies, the model integrates mechanical, biological, and clinical perspectives into a unified conceptual framework.

Frost's original mechanostat model (Frost, 1987) and its subsequent developments (Frost, 1996, 2001, 2003) marked a pivotal advancement in bone and skeletal research. This theory proposed that the relationship between mechanical loading and bone adaptation is considerably more sophisticated than previously described by Wolff (1870), while remaining consistent with the earlier conceptual framework of Roux (1881). Its central innovation lies in the description of a biological feedback mechanism that continuously monitors mechanical usage and triggers adaptive responses to correct significant mismatches between bone mass and mechanical demands. This regulatory system operates analogously to a household thermostat: it activates ("turns ON") when a mechanical mismatch is detected and deactivates ("turns OFF") once equilibrium is restored—hence the term "mechanostat." Frost emphasized that sustained bone modeling and remodeling require persistent activation of biological processes that control the spatial, temporal, and quantitative aspects of bone formation and resorption.

The mechanostat framework is defined by discrete *mechanical setpoints*—specific strain thresholds that govern modeling and remodeling activity, as illustrated in Fig. 13.2. These thresholds delineate distinct operational windows: the minimum

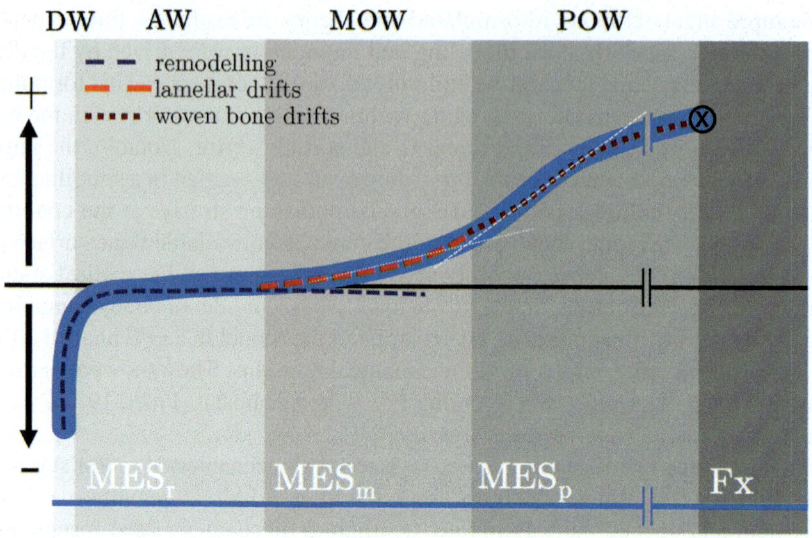

FIGURE 13.2

Frost's mechanostat model: Illustration of the mechanostat hypothesis, where the vertical axis represents the rate of bone loss or gain—typically expressed as (re)modeling velocity—and the horizontal axis represents bone strain. The following thresholds and operational windows can be identified: The minimum effective strain for remodeling (MES_r) defines the lower limit below which bone resorption occurs in the disuse window (DW); the minimum effective strain for modeling (MES_m) marks the strain level above which lamellar bone formation is triggered in the mild overload window (MOW); the strain range between MES_r and MES_m corresponds to the adapted window (AW), or "lazy zone," where neither formation nor resorption is stimulated. At higher strain magnitudes, the minimum effective strain for microdamage (MES_p) defines the threshold above which bone enters the pathologic overload window (POW), where woven bone is produced until the fracture strain (Fx) is reached. The white line represents the lamellar formation velocity modulus (FVM), i.e., the rate of (re)modeling velocity per unit strain.

- Illustration adapted from Frost's drawings in Frost, H. M. (1987). Bone "mass" and the "mechanostat": A proposal. The Anatomical Record, 219(1), 1–9, https://doi.org/10.1002/ar.1092190104; Frost, H. M. (2001). From Wolff's law to the Utah paradigm: Insights about bone physiology and its clinical applications. The Anatomical Record: An Official Publication of the American Association of Anatomists, 262(4), 398–419, https://doi.org/10.1002/ar.1049.

effective strain for remodeling (MES_r) marks the lower strain boundary below which bone resorption occurs, defining the disuse window (DW). Above this, the minimum effective strain for modeling (MES_m) is the threshold that triggers lamellar bone formation, corresponding to the mild overload window (MOW). The region between these two thresholds constitutes the adapted window (AW), or *lazy zone*, where strain levels are insufficient to stimulate either bone resorption or formation, thus maintaining bone mass equilibrium. At higher strain magnitudes, the minimum effective strain for microdamage (MES_p) marks the transition into the pathologic overload window (POW), where excessive loading induces woven bone formation until the fracture strain (Fx) is reached. The lamellar formation velocity modulus (FVM) quantifies the rate of modeling and remodeling per unit strain (Frost, 2001).

An individual's mechanical usage patterns regulate these mechano-biological feedback loops, continuously adjusting skeletal architecture to prevent mechanical overload and structural failure (Frost, 1996). These setpoint values can vary considerably even among healthy individuals, as postnatal load-bearing bones are designed to be strong enough to avoid spontaneous fractures during routine activities—-whether these activities represent chronically subnormal, normal, or supranormal loading conditions—while not necessarily protecting against traumatic injury (Frost, 1998). Furthermore, mechanostat parameters can be altered by non-mechanical factors such as disease, aging, nutritional status, and hormonal influences (Frost, 1987). These threshold shifts have important clinical implications, particularly in osteoporosis treatment. The capacity of therapeutic agents to modify thresholds and promote anabolic responses is a key mechanism of action (Jee & Tian, 2005). For example, parathyroid hormone (PTH) enhances bone cell mechanosensitivity by lowering MES_m, enabling bone formation at lower mechanical stimulus levels (Castoldi et al., 2025; Frost, 2003).

Over the past decades, Frost's mechanostat has become one of the most widely recognized and cited conceptual models of bone adaptation. Its simple yet powerful strain–threshold formulation has provided a coherent framework for interpreting bone adaptation and has served as the foundation for numerous bone remodeling models.

It should be noted that the mechanostat concept was originally developed for bone remodeling. Frost later proposed a complementary theory to describe bone modeling, i.e., the change of shape of cortical and trabecular bone (Frost, 1990). In this framework, bone modeling is governed by a drift operator, which determines whether a modeling drift occurs, and a drift rate function, which specifies the rate of this drift as a function of mechanical usage and biological parameters.

3.4 Adaptive bone remodeling

Adaptive bone remodeling theory was developed by Rik Huiskes and coworkers in the late 1980s (Huiskes et al., 1987). It extends earlier phenomenological approaches by introducing quantitative stimulus–response laws that predict the temporal evolution of bone density, structure, or mechanical properties in response to mechanical loading. These models build upon foundational ideas from the Stanford school, Cowin's theory of adaptive elasticity, and Frost's mechanostat, and are primarily intended for numerical implementation—particularly in FEA of bone adaptation under various loading scenarios.

The central idea is that local remodeling activity in bone tissue is governed by a mechanical stimulus—the Strain Energy Density (SED) (Huiskes et al., 1987)—compared to a reference stimulus associated with an adapted or equilibrium state.

When the difference between the actual and reference stimuli exceeds a certain threshold, a remodeling response is triggered to adjust local bone elasticity (internal remodeling) or bone shape (external remodeling). To reflect biological realism, the model includes a "lazy zone" around the equilibrium stimulus, in which no net remodeling occurs—thus echoing Frost's mechanostat.

This computational approach was thoroughly examined by Weinans et al. (1992), who analyzed the mass density patterns predicted by the algorithm—patterns that closely reproduced trabecular bone architecture—their seemingly chaotic nature, and their similarities to the Self-Optimization theory of Fyhrie and Carter (1986). The apparent paradox of a continuum model (for a discontinuous system) yielding discontinuous results led the authors to extend the concept by introducing a simple mechanobiological regulatory mechanism (Huiskes et al., 2000; Mullender et al., 1994; Mullender & Huiskes, 1995). In this framework, a spatial influence function modulates the mechanical stimulus for bone remodeling, thereby accounting for osteocyte-based mechanoregulation. This enhanced computational model successfully reproduced complex trabecular patterns as a function of specific loading conditions and initial configurations (Ruimerman et al., 2001). The key role of osteocytes was highlighted by contrasting bulk mechanosensation (related to osteocytes) with surface mechanosensation (related to lining cells) (Mullender & Huiskes, 1997).

Overall, the adaptive bone remodeling framework provides a powerful tool to simulate bone adaptation over time, under varying mechanical, anatomical, or pathological conditions. Huiskes' work brought several major contributions to the field. First, it underscored the limitations of Wolff's law and emphasized the potential of Roux's paradigm (Huiskes, 2000). Second, it marked a decisive step in shifting bone adaptation research toward a mechanobiology-driven science (van der Meulen & Huiskes, 2002). Finally, it opened the way for a variety of modeling and computational approaches, including micro-FEM, topology optimization, and discrete modeling.

4. Modern theories (1990s and later)

At the end of the 20th century, advances in experimental bone mechanics and 3D imaging technology (namely computed tomography (CT) and micro-computed tomography (μCT)) allowed more accurate experimental observations of bone adaptive behavior. Laying on the groundwork of Wolff and Roux and inspired by the classical theories described in the previous section, new mathematical and mechanical models were developed to formalize and quantify the processes underlying bone adaptation.

In this section, we categorize these models into four broad families:

- **Continuum Mechanics Approaches** are mostly conceived to describe bone remodeling at the tissue scale.

- **Optimization-Based Models** frame bone adaptation as a process of structural optimization.
- **Microstructure-scale Models** explicitly resolve the trabecular or cortical microarchitecture.
- **Multiscale and Multiphysics Models** couple macro-mechanical behavior with microstructural evolution and mechanobiology.

4.1 Continuum mechanics approaches

Several modern approaches to bone adaptation are grounded in continuum mechanics. Some models describe bone remodeling by introducing ad hoc evolution laws for material properties, such as mass density (Allena et al., 2024; Harrigan & Hamilton, 1992; Kuhl et al., 2003; Papastavrou et al., 2020), elastic properties (Geraldes et al., 2016; Miller et al., 2002; Watzky & Naili, 2004) or both (Pettermann et al., 1997). Other theories go further, treating bone remodeling as a damage-targeted process, or adopting the framework of continuum damage mechanics or enriched continuum mechanics and thermodynamics.

4.1.1 Damage-targeted remodeling laws

The central principle of damage-targeted remodeling models is that the primary stimulus for bone adaptation is the accumulation of microcracks in the bone matrix. Thus, remodeling is seen as a repair process designed to maintain bone health and prevent structural failure.

Martin (1992) proposed a mathematical model of bone remodeling grounded in the concept of damage-targeted remodeling introduced by Martin & Burr, 1982. Damage accumulation is assumed to depend on both loading intensity and frequency, and microcracks are considered to actively stimulate nearby bone remodeling, thereby promoting damage repair. Subsequently, Martin (1995) further explored the repair of fatigue damage in cortical bone, incorporating temporal aspects of internal remodeling and the effects of a periosteal response.

Prendergast and Taylor (1994) derived an integral mathematical law to predict bone adaptation, which naturally accounts for the stress history experienced by the tissue since formation. This formulation relies on the assumptions that bone at remodeling equilibrium contains a baseline level of damage (inter-constituent microcracks) and that the rate of damage repair is determined by the homeostatic stress. In a complementary study, Prendergast & Huiskes, 1996 have shown that the presence of damage significantly alters the local deformation around the osteocyte lacunae in cortical bone, thus potentially affecting osteocyte mechanoregulatory function. Furthermore, Taylor (1997) proposed a control system for bone remodeling driven by fatigue damage, demonstrating that a simple repair process could effectively regulate microcrack length without the need for complex sensing mechanisms.

More recently, McNamara and Prendergast (2007) showed that a mechanoregulatory system incorporating both strain and microdamage as remodeling stimuli—activated alternatively depending on a critical damage threshold—produces realistic results in describing the remodeling and repair of a single trabecula. Interestingly, their model proved more sensitive to surface stimuli than to bulk stimuli, in contrast with the findings of Mullender and Huiskes (1997).

4.1.2 Continuum damage mechanics

Another family of models is based on Continuum Damage Mechanics (CDM). CDM-based approaches provide a unified framework to describe microcrack accumulation, stiffness reduction, and the coupling between mechanical loading and remodeling. In these formulations, bone loss and repair are expressed through internal damage variables that evolve according to thermodynamically consistent laws.

A key contribution came from the work of Doblaré and Garcıa (2002). In their CDM formulation, damage was defined as a measure of void volume within the tissue, and an anisotropic *remodeling tensor*—analogous to a damage tensor—was introduced to simultaneously capture changes in apparent density and trabecular orientation (fabric tensor). The model was firmly grounded in the thermodynamics of irreversible processes, identifying a remodeling stimulus that is thermodynamically conjugated to the remodeling tensor. Importantly, while in classical CDM damage can only increase, the damage level encoded in the remodeling tensor could also decrease, thereby incorporating bone repair and formation alongside resorption.

This original framework laid the foundation for a series of García et al. 2001 generalized the model to describe both internal and external remodeling. García-Aznar et al. (2005) incorporated mineralization processes and biological activity, and Martínez-Reina et al. (2008) introduced a queue algorithm to account for the chronological age of bone patches.

Other works have been developed within the framework of CDM. For instance, Ramtani and Zidi (1999) proposed a damaged-bone remodeling theory that integrates mass transfer between the bone matrix and pores. Moreover, Hambli et al. (2013) developed an FE framework to simulate remodeling and fracture in the femoral head, which Hambli (2014) further enhanced to explicitly account for bone cell biology.

4.1.3 Enriched continuum mechanics & thermodynamics

A substantial body of work has addressed bone adaptation within the broad framework of enriched continuum mechanics and thermodynamics. These theories incorporate additional variables beyond those of the classical Cauchy continuum, representing microstructural features relevant to bone adaptation. (CDM can also be placed within this framework.) A central challenge lies in formulating thermodynamically consistent evolution laws for these variables, which requires adapting the classical principles of thermodynamics to properly account for the enriched continuum description.

Goda et al. (2016) developed a model of internal and external bone remodeling within the framework of irreversible thermodynamics, building on the formalism of Rodriguez et al. (1994). Internal remodeling, understood as changes in tissue structure, is related to the evolution of bulk density, with the remodeling stimulus identified as the Eshelby stress tensor. External remodeling, corresponding to changes in shape, is modeled by endowing the growing surface with a suitable surface potential. This formulation was applied to study the effects of mechanical loading at two scales: assemblies of trabeculae, and the homogenized effective continuum obtained through discrete homogenization. Using a similar two-scale strategy, Goda and Ganghoffer (2018) combined anisotropic internal remodeling with fatigue damage. In parallel, Ganghoffer et al. (2016) introduced a phase-field approach to describe the evolution of the bone-marrow interface, explicitly accounting for cell mechanobiology and tissue mineralization. This framework was later extended by Ganghoffer et al. (2019) with a second-gradient formulation, improving the description of strain concentration effects and the influence of microstructure on the effective response of bone.

DiCarlo and coworkers developed a model of material reorganization suitable for describing both growth (DiCarlo & Quiligotti, 2002) and remodeling (DiCarlo et al., 2006), formulated within the framework of generalized continuum mechanics (Capriz, 2013; Germain, 1973). In this approach, order parameters are introduced at the continuum scale to represent key features of bone microstructure. The model relies on generalized formulations of the virtual power principle and the dissipation inequality to derive evolution equations, which can be interpreted as growth and remodeling laws. Sansalone and coworkers later applied this framework to investigate rotary remodeling of trabecular bone tissue (Martin et al., 2020; Sansalone et al., 2011). They provided a detailed characterization of the conditions for stress-driven remodeling equilibrium. Interestingly, their model predicts—without any ad hoc assumptions—that remodeling equilibrium is attained when the principal axes of stress and strain are aligned. However, in contrast with Cowin (1986), alignment of the material principal axes occurs only under specific conditions. More recently, Sansalone et al. (2021) extended the model to incorporate bone turnover, explicitly accounting for bone resorption, apposition, and mineralization.

Other enriched-continuum models have been proposed to describe specific features of bone adaptation. George et al. (2018) developed a multiphysics model to describe bone density evolution, in which the remodeling stimulus includes mechanical, cellular, and chemical contributions. Giorgio et al. (2023) used the principle of virtual power to obtain a second-gradient model describing the evolution of orthotropic elastic coefficients of trabecular bone. Waffenschmidt et al. (2012) introduced the micro-sphere concept to describe the evolution of anisotropic bone material properties in the proximal tibia.

4.2 Optimization-based models

Optimization-based theories translate the descriptive nature of Wolff's law (bone follows load) into the concept of bone as an optimal structure. In most cases, structural optimization algorithms operate without explicit timescales, instead providing insights into how tissue structure may evolve toward potential steady states.

4.2.1 Optimality conditions in bone adaptation

Optimal material design has been extensively studied in the materials engineering community. For orthotropic materials, Rovati and Taliercio (1991) and later Banichuk (1996) investigated material optimization in relation to strain energy variations, demonstrating that extrema of strain energy correspond to alignment between material axes and principal strain directions. Similarly, Pedersen (1989) performed a two-dimensional study showing that strain energy density reaches a stationary value when the material axes align with the principal strain axes. These studies emphasized that the nature of energy optimization depends critically on the elastic moduli, particularly shear stiffness. Similar conclusions were reached independently by Cowin (1994, 1995); Martin et al. (2020). Furthermore, Cowin (1994, 1995) showed that, unlike materials with cubic symmetry, strain energy optimization in orthotropic materials depends on both material orientation and specific loading configurations.

The optimal character of bone remodeling was also acknowledged by other authors. Weinans et al. (1992) noted that their remodeling model could be interpreted as the solution of an optimization problem. Moreover, Harrigan and Hamilton (1992, 1994) highlighted the twofold nature of bone remodeling, which can be understood both as an adaptive response of bone tissue to its mechanical environment and as an optimization of tissue distribution within bones to sustain applied loads.

4.2.2 Mathematical optimization

A pioneering work explicitly formulating bone adaptation as a mathematical optimization process is the *self-optimization* theory developed by Fyhrie and Carter (1986). In this framework, bone is considered a self-optimizing material, where the relationship between trabecular bone apparent density, trabecular orientation, and stress is formulated as an optimization problem based either on stiffness (via the strain energy density principle) or on strength (via a failure criterion). This theory was later applied to predict the density distribution within the human femoral head (Fyhrie & Carter, 1990).

Alternatively, Jacobs et al. (1997) hypothesized that the remodeling response of bone is optimal and formulated a constrained optimization problem in which the objective function is the instantaneous dissipation rate, including a non-standard term to account for mass exchanges. The resulting optimality conditions led to the evolution law for both the full elasticity tensor and the mass density.

Another model was proposed by Lekszycki (1999), who treated the rate of change of Young's modulus as the design variable and introduced tailored objective functionals. In subsequent work, he extended the model to incorporate non-local mechanical stimuli arising from osteocyte communication (Lekszycki, 2002).

4.2.3 Topology optimization

Topology Optimization (TO), popularized by Bendsøe and Sigmund (2002), is a general form of structural optimization in engineering, aiming to determine the optimal topology of a system, i.e., the material distribution within a certain domain, given the boundary conditions and loads. Starting from an initial distribution of material, TO allows both the geometry of the boundaries and their number to vary—potentially creating holes—with the objective in most cases being to minimize compliance while satisfying constraints such as mechanical equilibrium. The approach can also modulate material properties to account for heterogeneity. The pioneering work of Bendsøe and Kikuchi (1988) on TO inspired a wealth of work on trabecular bone remodeling. Although developed independently, TO shares significant similarities with the Adaptive bone remodeling theory of Huiskes (Nutu, 2015).

Fernandes et al. (1999) developed a two-scale model of trabecular bone remodeling, where bone tissue was described as a cellular material and design variables were linked to its microstructure. The problem was formulated as a constrained global optimization, with a cost function combining structural stiffness across multiple load cases with the metabolic cost of maintaining bone tissue, and constraints related to the homogenized elastic properties. Applied to the proximal femur, the model successfully reproduced consistent density distributions and microstructural orientations.

Significant contributions have been developed by Nowak and coworkers in the early 2000s. Nowak (2006) proposed an algorithm to optimize the trabecular network in bone, based on the surface strain energy density (SSED)—nearly uniform in the optimal state—and the regulatory model of Huiskes et al. (2000). He also observed that the results of this model bear strong similarities to those obtained with topology optimization approaches. Nowak et al. (2018) further investigated this similarity, and Nowak et al. (2019, 2020) further developed a model of bone remodeling in the framework of shape optimization, whose solution features a constant SSED (that could be interpreted as a *lazy zone*), by including multiple load cases and surface curvature effects.

More details on current shape and topology optimization approaches can be found in (Chapter 15).

4.3 Microscale models

Microscale models aim to advance our understanding of the bone remodeling process at the cellular and tissue microstructural levels, providing a link between biological mechanisms and tissue-scale outcomes. A broad range of approaches has been developed. In the following sections, we review models specifically focused on the dynamics of basic multicellular units (Section 4.3.1), the adaptation of trabecular architecture (Section 4.3.2), and agent-based models (Section 4.3.3).

4.3.1 Models of BMU dynamics

Basic Multicellular Units (BMUs) represent the fundamental organizational entities of bone remodeling, coordinating the coupled activity of bone-resorbing osteoclasts and bone-forming osteoblasts. While the core function is the same, BMUs operate differently in cortical and trabecular bone. In cortical bone, BMUs act as a “cutting cone" that tunnels through existing tissue, whereas in trabecular bone they act on the surface, renewing bone in the form of pits or trenches. Modeling BMU dynamics offers a framework to connect cellular processes with tissue-level outcomes. Early approaches focused on reduced or discrete-event models, while more recent formulations account for spatial dynamics and mechanobiological feedback.

Martin (1984, 1991) developed a scalar model to study the evolution of cortical porosity. In this framework, BMUs are not treated as explicit geometric entities but rather provide the conceptual basis for a 0D description, assuming idealized cylindrical osteonal remodeling perpendicular to the cortical cross section. These models explored different remodeling patterns, including overlap feedback and random remodeling, in order to assess their influence on porosity evolution. The damage-based remodeling model of Martin (1992, 1995) (see Section 4.1.1) provided a foundation for further developments by Martin and coworkers. Expanding on these ideas, Hazelwood et al. (2001) proposed a remodeling algorithm for cortical bone in which BMUs are activated in response to either disuse or microdamage, and Nyman et al. (2004) extended this approach to include the effects of menopause and bisphosphonate treatments. Kimmel (1985) presented a discrete-event computer model to simulate the remodeling process in the adult trabecular skeleton. Building on this work, Heaney (1994) and Heaney et al. (1997) analyzed the temporal dynamics of remodeling and suggested that the apparent bone gain observed with some osteoporosis treatments may be attributable to a remodeling transient rather than to a true positive bone balance. Similarly, Thomsen et al. (1994) developed a stochastic model of vertebral trabecular bone remodeling to predict long-term changes in bone mass, trabecular thickness, and the occurrence of perforations.

These early discrete models typically considered only the transition from resorption to formation, neglecting the mineralization phase. More recently, Castoldi et al. (2024a) introduced a two-dimensional (2D) discrete model in both space and time that captures the entire remodeling process, from resorption to mineralization. This model investigates the effects of BMU activation frequency and mineralization kinetics on both the bone mineral density distribution (BMDD) (Ruffoni et al., 2007) and trabecular architecture. Applied to both healthy and osteoporotic bone, it provides new insights into how specific remodeling parameters contribute to overall bone quality.

Some BMU-oriented models are tightly related to the mechanobiological approaches discussed later in this chapter. For example, Buenzli et al. (2011, 2014) developed a one-dimensional model of cortical BMU dynamics extending the bone cell population model of Pivonka et al. (2008), and Ryser et al. (2009, 2010) proposed a two-dimensional model integrating the bone cell population model of Komarova et al. (2003) and Komarova (2005). In both cases, BMU dynamics are described by systems of nonlinear partial differential equations governing the spatio-temporal evolution of bone cells and key signaling molecules. In some cases, agent-based models are introduced to study BMU progression. Some examples (Buenzli et al., 2012; Van Oers et al., 2008) will be described in Section 4.3.3.

4.3.2 Models of trabecular remodeling

Numerous studies have investigated the emergence of trabecular patterns using regular lattice grids in two (2D) or three (3D) dimensions. Some models implemented stochastic processes based on phenomenological laws to simulate trabecular bone remodeling in the human vertebral body, either with mechanical feedback (Weinkamer et al., 2004) or without it (Langton et al., 1998).

Other models included mechanoregulatory mechanisms in the remodeling laws. The work of Huiskes et al. (2000) on 2D trabecular remodeling represents a key reference in this context. Adachi and coworkers simulated the remodeling of trabecular tissue in 2D (Tsubota et al., 2002) and of an individual trabecula in 3D (Adachi et al., 2001). Hartmann et al. (2011) described trabecular remodeling as a mechanoregulated stochastic process taking place at the bone surface.

The development of high-resolution imaging, particularly X-ray μCT, at the end of the 20th century enabled the study of remodeling in realistic trabecular geometries (Muller et al., 1994). Some studies used phenomenological remodeling laws, without mechanical feedback, to simulate microstructural changes (Mueller, 2005; Muller & Ruegsegger, 1996; Van Der Linden et al., 2001). Others combined μCT images with FEA to simulate the mechanical behavior of bone at the microscale (μFEA). Schulte et al. (2013); Christen et al. (2014); Marques et al. (2023) demonstrated a strong correlation between mechanical signals predicted by μFEA and experimentally observed microstructural remodeling. More recent studies have integrated remodeling models directly into μFEA, enabling the prediction of trabecular remodeling in real bone microstructures (Badilatti et al., 2016; Cheong et al., 2020; Marques et al., 2023).

Most of these models employ remodeling laws driven by mechanical stimuli related to strain energy density (Badilatti et al., 2016; Christen et al., 2014; Mullender & Huiskes, 1995; Ruimerman et al., 2003; Van Der Linden et al., 2001; Van Oers et al., 2008; van Oers et al., 2011), while others are based on strain (Schulte et al., 2013), stress (Tsubota et al., 2002), or interstitial fluid flow (Adachi et al., 2010; Kameo & Adachi, 2014).

4.3.3 Agent-based and lattice-based models

With advances in computing power, a broad class of computational approaches has emerged, including agent-based models (ABMs), in which agents represent individual cells, and lattice-based models (LBMs), in which each lattice site corresponds to an individual unit.

Some ABM formulations have been developed to capture remodeling dynamics at the microscale without explicit coupling with the macroscale. For instance, Ausk et al. (2006) presented an ABM simulating real-time signaling in osteocytic networks in response to mechanical stimuli. Van Oers et al. (2008) combined FEA with the cellular Potts model—a specialized form of ABM—to simulate both osteonal and hemi-osteonal remodeling. Buenzli et al. (2012) investigated bone resorption at the BMU level using a hybrid ABM/LBM approach to capture osteoclast dynamics. Paoletti et al. (2012) coupled shape calculus with ABM to describe remodeling dynamics. Recent work has extended ABMs by incorporating more detailed biological information. For example, Boaretti et al. (2023) introduced a micro-mechanical phenotyping ABM to simulate bone adaptation at the single-cell scale, accounting for mechanotransduction pathways in mice.

Other studies have combined FEA with ABMs/LBMs in multiscale frameworks to investigate how macroscopic loading influences cell activity at the microscale and, in turn, drives the evolution of macroscopic material properties. Significant contributions in this field have been made by Prendergast, Checa, Duda, and co-workers. For instance, Pérez and Prendergast (2007) proposed a mechanoregulatory model based on shear strain and fluid flow to study cell proliferation, migration, and differentiation around a bone implant (Lacroix & Prendergast, 2002). Checa and Prendergast extended the framework to simulate capillary network formation and tissue regeneration in the bone-implant gap (Checa & Prendergast, 2009) and within biodegradable scaffolds (Checa & Prendergast, 2010). Borgiani et al. (2021) integrated two ABMs with FEA to simulate the combined effects of cellular activity and drug treatment on bone healing in a critical-sized defect. Perier-Metz et al. (2020) and Jaber et al. (2022) explored different features of bone scaffold design to promote bone regeneration.

Inspired by these works, OReilly et al. (2016) incorporated Checa and Prendergast's revascularization algorithm into a broader ABM to study the effects of disrupted angiogenesis on bone regeneration. More recently, Kendall et al. (2023) proposed a micro-multiphysics ABM integrating tissue formation, differentiation, resorption, and revascularization, thereby providing a comprehensive framework for modeling bone healing.

4.4 Multiscale and multiphysics models

Most of the previously described bone adaptation models either do not account for the multiscale nature of bone and/or do not account for the underlying bone cellular activities that give rise to reshaping bone microarchitecture and modifying bone matrix composition.

The work of Lemaire et al. (2004) laid the foundation for incorporating biological aspects, i.e., major biochemical signaling pathways that couple cells of the osteoblastic and osteoclastic lineages in bone remodeling. Using this framework, bone diseases and drug interventions can be investigated. However, no mechanical aspects of bone were incorporated into that work. The link between bone cell biology and mechanics was first established by Pivonka et al. (2008), who formulated a bone cell population model at the scale of a representative volume element (RVE). Osteoclastic and osteoblastic activity in the RVE was linked to changes in bone volume fraction, BV/TV. This allowed tracking of changes in BV/TV due to bone diseases and evaluating the efficacy of drug treatments (Pivonka et al., 2008, 2010).

A further milestone in the development of mechanobiological models of bone remodeling was achieved by Scheiner et al. (2013), who included the first mechanobiological feedback into this type of modeling framework. Using a multiscale micromechanical model of cortical bone stiffness, the macroscopic loading could be downscaled to the strains engendered in the bone matrix that can be sensed by osteocytes. The latter stimulus (i.e., strain energy density in bone matrix) was used to formulate a mechanobiological feedback on both osteoblastic and osteoclastic cells (Scheiner et al., 2013). Note that this approach is fundamentally different from purely mechanics-based adaptation laws, as the mechanical stimulus is linked to biochemical regulations of bone cell activities. This approach was further refined to include osteocytes as an individual bone cell population into the bone cell population model (BCPM), together with Wnt signaling in osteoblasts, which regulates bone anabolic responses (Martin et al., 2019).

Mechanobiological stimuli used in bone remodeling and adaptation models have been frequently discussed in the research community. Frequently used mechanical stimuli include: strain energy density (on the tissue scale or bone matrix scale), hydrostatic pressure in the different pore spaces of bone (vascular and/or osteocyte lacunae), fluid velocity (osteocyte canalicular network), streaming potential, and piezoelectric stimulus, to name a few. Pastrama et al. (2018) used a micromechanics-based dual-porosity poroelastic model of cortical bone to compute the vascular and osteocyte lacunae pore pressures due to macroscopic mechanical loading. These pressures were used to formulate a mechanobiological feedback. It was concluded that the hydrostatic pressure in the lacunae pores under habitual loading (i.e., walking) is sufficiently high to give rise to bone cell responses. Subsequently, Estermann and Scheiner (2018) extended this work to analyze the fluid velocities that are induced in vascular and canalicular pores of cortical bone due to habitual loading. They found that while the fluid velocities in the vascular canals are sufficiently large to induce bone cell responses, the fluid velocity in canalicular pores is quite small. Computation of fluid velocity in the canaliculi network was pursued by many researchers over the years, indicating similar magnitudes to the ones computed by Estermann and Scheiner (Sansalone et al., 2013; Wang et al., 2013). These results suggest that fluid shear stress generated solely by pressure-driven flow in the canalicular network is unlikely to be sufficient to stimulate osteocyte processes. However, other coupled mechanisms may contribute to the effective shear stress experienced by osteocytes.

The works of T. Lemaire and coworkers suggest that coupled effects, including osmotic and electro-osmotic interstitial fluid flows, may prevail in determining the effective shear stress experienced by osteocytes in the lacuno-canalicular network (Kaiser et al., 2012; Lemaire et al., 2011; Sansalone et al., 2013) and, more generally, in transport phenomena related to bone remodeling (Lemaire et al., 2013, 2015). Bone fluid flow models will be described in (Chapter 16). Moreover, multiphysical aspects of bone remodeling and mechanostransduction will be thoroughly described in (Chapter 18).

Including mineralization kinetics of the bone matrix at the RVE scale into BCPMs was another critical step in advancing mechanobiological models of bone. This was accomplished by Martínez-Reina and Pivonka (2019). Incorporation of mineralization is critical for mechanistic predictions of antiresorptive drug therapies. This latter group of drugs inhibits osteoclastic bone resorption activity by inducing osteoclast apoptosis or impairing the functionality of osteoclastic resorption. Due to the coupling of bone formation and resorption, these types of drugs significantly reduce bone remodeling activity. Hence, antiresorptive drugs do not significantly change BV/TV. However, significant changes in bone mineral density (BMD) are observed experimentally. This increase in BMD is achieved through increased mineralization of the bone matrix (due to reduced bone turnover). Mechanobiological models that do not incorporate mineralization are not able to describe this phenomenon (Martínez-Reina & Pivonka, 2019; Scheiner et al., 2014).

A more recent step in the development of BCPMs has been the inclusion of bone matrix damage at the RVE scale due to dynamic habitual loading (Ruiz-Lozano et al., 2025). Using the framework of continuum damage mechanics, an isotropic damage variable was introduced in that work. Damage was assumed to induce osteocyte apoptosis and consequently affect bone remodeling. The aforementioned work goes beyond assessing BV/TV alone and also provides information on the damage induced in the bone matrix due to disease and/or treatment, which allows for the formulation of more complex optimization algorithms.

Chapter 4 provides an overview of the performance of in silico trials of osteoporosis treatments using a state-of-the-art mechanobiological pharmacokinetic-pharmacodynamic (PK/PD) model of bone remodeling.

5. Conclusions and perspectives

From Wolff's 19th-century formulation to the latest multiscale computational frameworks, bone adaptation models have progressively expanded in scope, rigor, and biological realism. Early phenomenological theories provided valuable intuition about mechanical regulation, while classical continuum and optimization approaches formalized structure–function relationships. More recently, microscale and multiscale models have begun to capture the cellular and molecular underpinnings of remodeling, pointing toward a unified view of bone as a living material shaped by both mechanical and biochemical forces.

While continuum models of bone remodeling are now well established, continuum models of bone modeling (external growth and shape change) remain comparatively less developed. Three-dimensional growth theories based on continuum mechanics have been widely applied to soft tissues (DiCarlo et al., 2009; Kuhl, 2014), but their application to hard tissues such as bone is still limited. Most bone growth models rely on phenomenological laws (Beaupré et al., 1990a; Cowin & Van Buskirk, 1979; Huiskes et al., 1987; Shefelbine & Carter, 2004; Stokes & Laible, 1990; van Der Meulen et al., 1993; Villemure et al., 2002), with only a few works adopting enriched continuum formulations explicitly designed to ensure thermodynamic consistency (Castoldi et al ., 2024b; García et al., 2001; Goda et al., 2016). Other approaches, not relying on 3D continuum mechanics, include Frost's drift theory of bone modeling (Frost, 1990), beam models of cortical surface evolution (Miller et al., 2024), and topology optimization strategies for predicting vertebral morphogenesis (Sakashita et al., 2021). These developments illustrate the diversity of possible approaches and underline that growth, modeling, and remodeling remain deeply interconnected processes.

Data-driven approaches are increasingly entering the field of bone adaptation. Recent systematic reviews highlight the strong potential of machine learning and deep learning techniques for automated segmentation, bone mineral density estimation, osteoporosis screening, and fracture risk prediction (Fleps & Morgan, 2022; Kalmet et al., 2020; Khalid et al., 2020; Smets et al., 2021; Zhang et al., 2025). Most applications in bone biomechanics focus on the organ scale, but also applications at the microscale are emerging, e.g., for the automatic distinction between cortical and trabecular compartments in μCT scans (Lagzouli et al., 2025). In the specific context of bone remodeling, only a few studies have applied data-driven models to complement or accelerate mechanistic simulations. Artificial neural networks (ANNs) have been integrated with FEA to simulate trabecular adaptation in a multiscale framework (Hambli et al., 2011) and to determine subject-specific forces from computer tomography images (Garijo et al., 2017). ANNs have also been trained to predict trabecular organization under varied geometries and load cases (Pais et al., 2023). Despite these advances, data-driven approaches face critical challenges, including limited longitudinal datasets, the difficulty of capturing causal biological mechanisms, and the need for model interpretability. For these reasons, they should not be seen as replacements for mechanistic theories but as complementary tools. They hold promise for calibrating, validating, and extending existing frameworks, while also generating new hypotheses to guide experiments.

Despite important advances, several challenges remain. Bridging scales—from molecular signaling pathways to tissue mechanics—remains a central obstacle. Equally important is the integration of increasingly complex biological mechanisms, including the interplay between bone and other organs, into tractable predictive models. Experimental validation and clinical translation also pose major hurdles: models must be tested against high-resolution imaging, in vivo data, and longitudinal studies to ensure their reliability in medical applications.

Looking ahead, the convergence of coarse-grained mechanistic modeling at the macroscale, fine-grained mechanobiological models at the microscale, and data-driven approaches—supported by high-performance computing—offers exciting opportunities. Hybrid models that integrate these approaches, together with advances in imaging and patient-specific modeling, hold promise for both fundamental research and personalized medicine. In this perspective, bone adaptation modeling is expected to evolve from a descriptive science into a predictive and translational discipline, capable of guiding experiments, informing therapeutic strategies, and ultimately improving patient outcomes.

References

Adachi, T., Kameo, Y., & Hojo, M. (2010). Trabecular bone remodelling simulation considering osteocytic response to fluid-induced shear stress. *Philosophical Transactions of the Royal Society of London,Series A: Mathematical, Physical and Engineering Sciences, 368* (1920), 2669–2682.

Adachi, T., Tsubota, K.-ichi, Tomita, Y., & Hollister, S. J. (04 2001). Trabecular surface remodeling simulation for cancellous bone using microstructural voxel finite element models. *Journal of Biomechanical Engineering, 123*(5), 403–409. https://doi.org/10.1115/1.1392315

Allena, R., Scerrato, D., Bersani, A. M., & Giorgio, I. (2024). Functional adaptation of bone mechanical properties using a diffusive stimulus originated by dynamic loads in bone remodelling. *Zeitschrift für Angewandte Mathematik und Physik, 75*(3), 85. https://doi.org/10.1007/s00033-024-02230-x

Ausk, B. J., Gross, T. S., & Srinivasan, S. (2006). An agent based model for real-time signaling induced in osteocytic networks by mechanical stimuli. *Journal of Biomechanics, 39*(14), 2638–2646.

Badilatti, S. D., Christen, P., Levchuk, A., Javad, H. M., Bert, van R., Parkinson, I., & Müller, R. (2016). Large-scale microstructural simulation of load-adaptive bone remodeling in whole human vertebrae. *Biomechanics and Modeling in Mechanobiology, 15*(1), 83–95. https://doi.org/10.1007/s10237-015-0715-8

Banichuk, N. V. (1996). Optimization of anisotropic properties for continuum bodies and structural elements using spectral methods of tensor analysis. *Journal of Structural Mechanics, 24*(1), 71–87.

Bartoníček, J., & Naňka, O. (2024). The true history of the Hueter-Volkmann law. *International Orthopaedics, 48*(10), 2755–2762.

Beaupré, G. S., Orr, T. E., & Carter, D. R. (1990). An approach for time-dependent bone modeling and remodeling-application: A preliminary remodeling simulation. *Journal of Orthopaedic Research, 8*(5), 662–670. ISSN 0736-0266.

Beaupré, G. S., Orr, T. E., & Carter, D. R. (1990a). An approach for time-dependent bone modeling and remodeling: Theoretical development. *Journal of Orthopaedic Research, 8*(5), 651–661. https://doi.org/10.1002/jor.1100080506

Bell, G. H., Cuthbertson, D. P., & Orr, J. (1941). Strength and size of bone in relation to calcium intake. *The Journal of Physiology, 100*(3), 299. https://doi.org/10.1113/jphysiol.1941.sp003944

Bendsøe, M. P., & Kikuchi, N. (1988). Generating optimal topologies in structural design using a homogenization method. *Computer Methods in Applied Mechanics and Engineering, 71*(2), 197–224. ISSN 0045-7825.

Bendsøe, M. P., & Sigmund, O. (2002). Topology optimization theory, methods and applications. In *Topology optimization.* Berlin/Heidelberg, Germany: Springer. ISBN 364207698X.

Bertram, J. E. A., & Swartz, S. M. (1991). The 'law of bone transformation': A case of crying Wolff? *Biological Reviews, 66*(3), 245–273.

Boaretti, D., Marques, F. C., Ledoux, C., Singh, A., Kendall, J. J., Wehrle, E., Kuhn, G. A., Bansod, Y. D., Schulte, F. A., & Müller, R. (2023). Trabecular bone remodeling in the aging mouse: A micro-multiphysics agent-based in silico model using single-cell mechanomics. *Frontiers in Bioengineering and Biotechnology, 11*, Article 1091294.

Borgiani, E., Duda, G. N., Willie, B. M., & Checa, S. (2021). Bone morphogenetic protein 2-induced cellular chemotaxis drives tissue patterning during critical-sized bone defect healing: An in silico study. *Biomechanics and Modeling in Mechanobiology, 20*(4), 1627–1644.

Buenzli, P. R., Jeon, J., Pivonka, P., Smith, D. W., & Cummings, P. T. (2012). Investigation of bone resorption within a cortical basic multicellular unit using a lattice-based computational model. *Bone, 50*(1), 378–389.

Buenzli, P. R., Pivonka, P., & Smith, D. W. (2011). Spatio-temporal structure of cell distribution in cortical bone multicellular units: A mathematical model. *Bone, 48*(4), 918–926. ISSN 8756-3282.

Buenzli, P. R., Pivonka, P., & Smith, D. W. (2014). Bone refilling in cortical basic multicellular units: Insights into tetracycline double labelling from a computational model. *Biomechanics and Modeling in Mechanobiology, 13*(1), 185–203. https://doi.org/10.1007/s10237-013-0495-y

Burr, D. B., Forwood, M. R., Fyhrie, D. P., Martin, R. B., Schaffler, M. B., & Turner, C. H. (1997). Bone microdamage and skeletal fragility in osteoporotic and stress fractures. *Journal of Bone and Mineral Research, 12*(1), 6–15.

Calvo-Gallego, J. L., Gutiérrez-Millán, F., Ojeda, J., Pérez, M.Á., & Martínez-Reina, J. (2022). The correlation between bone density and mechanical variables in bone remodelling models: Insights from a case study corresponding to the femur of a healthy adult. *Mathematics, 10*(18).

Capriz, G. (2013). *Continua with microstructure* (Vol 35). Springer Science & Business Media.

Carter, D. R. (1987). Mechanical loading history and skeletal biology. *Journal of Biomechanics, 20*(11), 1095–1109. ISSN 0021-9290.

Carter, D. R., Fyhrie, D. P., & Whalen, R. T. (1987). Trabecular bone density and loading history: Regulation of connective tissue biology by mechanical energy. *Journal of Biomechanics, 20*(8), 785–794. https://doi.org/10.1016/0021-9290(87)90058-3

Carter, D. R., Orr, T. E., & Fyhrie, D. P. (1989). Relationships between loading history and femoral cancellous bone architecture. *Journal of Biomechanics, 22*(3), 231–244.

Castoldi, N. M., Lagzouli, A., Pickering, E., Meakin, L., Cooper, D. M. L., Delisser, P., & Pivonka, P. (2025). Reverse engineering frost's mechanostat model in mouse tibia: Insights from combined pth and mechanical loading. *Bone*, Article 117491.

Castoldi, N. M., Pickering, E., Sansalone, V., Cooper, D., & Pivonka, P. (2024). Bone turnover and mineralisation kinetics control trabecular bmdd and apparent bone density: Insights from a discrete statistical bone remodelling model. *Biomechanics and Modeling in Mechanobiology, 23*(3), 893–909.

Castoldi, N. M., Pivonka, P., Teresi, L., & Sansalone, V. (2024b). Modelling and simulation of growth driven by mechanical and non-mechanical stimuli. *Journal of the Mechanics and Physics of Solids, 191*, Article 105769.

Checa, S., & Prendergast, P. J. (2009). A mechanobiological model for tissue differentiation that includes angiogenesis: A lattice-based modeling approach. *Annals of Biomedical Engineering, 37*, 129–145.

Checa, S., & Prendergast, P. J. (2010). Effect of cell seeding and mechanical loading on vascularization and tissue formation inside a scaffold: A mechano-biological model using a lattice approach to simulate cell activity. *Journal of Biomechanics, 43*(5), 961–968. ISSN 0021-9290.

Cheng, J. C. Y., Tang, S. P., Guo, X., Chan, C. W., & Qin, L. (2001). Osteopenia in adolescent idiopathic scoliosis: A histomorphometric study. *Spine, 26*(3), C1–C5.

Cheong, V. S., Marin, A. C., Lacroix, D., & Dall'Ara, E. (2020). A novel algorithm to predict bone changes in the mouse tibia properties under physiological conditions. *Biomechanics and Modeling in Mechanobiology, 19*, 985–1001. https://doi.org/10.1007/s10237-019-01266-7

Christen, P., Ito, K., Ellouz, R., Boutroy, S., Sornay-Rendu, E., Chapurlat, R. D., & Van Rietbergen, B. (2014). Bone remodelling in humans is load-driven but not lazy. *Nature Communications, 5*(1), 4855.

Cowin, S. C. (1984). Mechanical modeling of the stress adaptation process in bone. *Calcified Tissue International, 36*, S98–S103.

Cowin, S. C. (1985). The relationship between the elasticity tensor and the fabric tensor. *Mechanics of Materials, 4*(2), 137–147.

Cowin, S. C. (1986). Wolff's law of trabecular architecture at remodeling equilibrium. *Journal of Biomechanical Engineering, 108*(1), 83.

Cowin, S. C. (1989). A resolution restriction for Wolff's law of trabecular architecture. *Bulletin of the Hospital for Joint Diseases Orthopaedic Institute, 49*(2), 205–212.

Cowin, S. C. (1994). Optimization of the strain energy density in linear anisotropic elasticity. *Journal of Elasticity, 34*(1), 45–68. https://doi.org/10.1007/BF00042425

Cowin, S. C. (1995). On the minimization and maximization of the strain energy density in cortical bone tissue. *Journal of Biomechanics, 28*(4), 445–447.

Cowin, S. C. (1998). The false premise in Wolff's law. *Forma, 12*(3), 247–262.

Cowin, S. C. (2003). Adaptive elasticity: A review and critique of a bone tissue adaptation model. *Engineering Transactions, 51*(2–3), 113–193. https://doi.org/10.24423/engtrans.485.2003

Cowin, S. C., & Hegedus, D. H. (1976). Bone remodeling i: Theory of adaptive elasticity. *Journal of Elasticity, 6*(3), 313–326.

Cowin, S. C., & Nachlinger, R. R. (1978). Bone remodeling iii: Uniqueness and stability in adaptive elasticity theory. *Journal of Elasticity, 8*(3), 285–295. ISSN 0374-3535.

Cowin, S. C., Sadegh, A. M., & Luo, G. M. (1992). An evolutionary wolff's law for trabecular architecture. *Journal of Biomechanical Engineering, 114*(1), 129.

Cowin, S. C., & Van Buskirk, W. C. (1979). Surface bone remodeling induced by a medullary pin. *Journal of Biomechanics, 12*(4), 269–276. ISSN 0021-9290.

Culmann, C. (1866). Die graphische statik. Mayer and Zeller: Zurich, Switzerland.

Dempster, W. T., & Liddicoat, R. T. (1952). *Compact bone as a non-isotropic material.* Wiley Subscription Services, Inc., A Wiley Company.

DiCarlo, A., Naili, S., & Quiligotti, S. (2006). Sur le remodelage des tissus osseux anisotropes. *Comptes Rendus Mecanique, 334*(11), 651–661.

DiCarlo, A., & Quiligotti, S. (2002). Growth and balance. *Mechanics Research Communications, 29*(6), 449–456.

DiCarlo, A., Sansalone, V., Tatone, A., & Varano, V. (2009). Growth and remodeling of intracranial saccular aneurysms. In *Proceedings of the COMSOL conference*.

Doblaré, M., & Garcıa, J. M. (2002). Anisotropic bone remodelling model based on a continuum damage-repair theory. *Journal of Biomechanics, 35*(1), 1–17. https://doi.org/10.1016/S0021-9290(01)00178-6

Estermann, S.-J., & Scheiner, S. (2018). Multiscale modeling provides differentiated insights to fluid flow-driven stimulation of bone cellular activities. *Frontiers in Physics, 6*(76).

Evans, F. G. (1969). The mechanical properties of bone. *Artificial Limbs: A Review of Current Developments, 13*(1), 37–48.

Fernandes, P., Rodrigues, H., & Jacobs, C. (1999). A model of bone adaptation using a global optimisation criterion based on the trajectorial theory of wolff. *Computer Methods in Biomechanics and Biomedical Engineering, 2*(2), 125–138. https://doi.org/10.1080/10255849908907982

Fleps, I., & Morgan, E. F. (2022). A review of ct-based fracture risk assessment with finite element modeling and machine learning. *Current Osteoporosis Reports, 20*(5), 309–319. ISSN 1544-1873.

Frost, H. M. (1969). Tetracycline-based histological analysis of bone remodeling. *Calcified Tissue International, 3*, 211–237. https://doi.org/10.1007/BF02058664

Frost, H. M. (1987). Bone "mass" and the "mechanostat": A proposal. *The Anatomical Record, 219*(1), 1–9. https://doi.org/10.1002/ar.1092190104

Frost, H. M. (1990). Skeletal structural adaptations to mechanical usage (SATMU): 1. Redefining Wolff's law: The bone modeling problem. *The Anatomical Record, 226*(4), 403–413.

Frost, H. M. (1996). Perspectives: A proposed general model of the "mechanostat" (suggestions from a new skeletal-biologic paradigm). *The Anatomical Record: An Official Publication of the American Association of Anatomists, 244*(2), 139–147. https://doi.org/10.1002/(SICI)1097-0185(199602)244:2<139::AID-AR1>3.0.CO;2-X

Frost, H. M. (1998). *Osteoporoses: Their nature and therapeutic targets.*

Frost, H. M. (2001). From Wolff's law to the Utah paradigm: Insights about bone physiology and its clinical applications. *The Anatomical Record: An Official Publication of the American Association of Anatomists, 262*(4), 398–419. https://doi.org/10.1002/ar.1049

Frost, H. M. (2003). Bone's mechanostat: A 2003 update. *The Anatomical Record Part a: Discoveries in Molecular, Cellular, and Evolutionary Biology: An Official Publication of the American Association of Anatomists, 275*(2), 1081–1101. https://doi.org/10.1002/ar.a.10119

Fyhrie, D. P., & Carter, D. R. (1986). A unifying principle relating stress to trabecular bone morphology. *Journal of Orthopaedic Research, 4*(3), 304–317.

Fyhrie, D. P., & Carter, D. R. (1990). Femoral head apparent density distribution predicted from bone stresses. *Journal of Biomechanics, 23*(1), 1–10. ISSN 0021-9290.

Ganghoffer, J.-F., Rahouadj, R., Boisse, J., & Forest, S. (2016). Phase field approaches of bone remodeling based on tip. *Journal of Non-equilibrium Thermodynamics, 41*(1), 49–75.

Ganghoffer, J. F., Rahouadj, R., Boisse, J., & Schiavi, J. (2019). A phase field approach for bone remodeling based on a second-gradient model. *Mechanics Research Communications, 96*, 37–44.

García, J. M., Martínez, M. A., & Doblaré, M. (2001). An anisotropic internal-external bone adaptation model based on a combination of cao and continuum damage mechanics technologies. *Computer Methods in Biomechanics and Biomedical Engineering, 4*(4), 355–377. ISSN 1025-5842.

García-Aznar, J. M., Rueberg, T., & Doblaré, M. (2005). A bone remodelling model coupling microdamage growth and repair by 3d bmu-activity. *Biomechanics and Modeling in Mechanobiology, 4*(2–3), 147–167. ISSN 1617-7959.

Garijo, N., Verdonschot, N., Engelborghs, K., García-Aznar, J. M., & Pérez, M. A. (2017). Subject-specific musculoskeletal loading of the tibia: Computational load estimation. *Journal of the Mechanical Behavior of Biomedical Materials, 65*, 334–343. ISSN 1751-6161.

George, D., Allena, R., & Rémond, Y. (2018). A multiphysics stimulus for continuum mechanics bone remodeling. *Mathematics and mechanics of complex systems, 6*(4), 307–319. ISSN 2326-7186.

Geraldes, D. M., Modenese, L., & Phillips, A. T. M. (2016). Consideration of multiple load cases is critical in modelling orthotropic bone adaptation in the femur. *Biomechanics and Modeling in Mechanobiology, 15*(5), 1029–1042.

Germain, P. (1973). The method of virtual power in continuum mechanics. Part 2: Microstructure. *SIAM Journal on Applied Mathematics, 25*(3), 556–575.

Giorgio, I., Dell'Isola, F., Andreaus, U., & Misra, A. (2023). An orthotropic continuum model with substructure evolution for describing bone remodeling: An interpretation of the primary mechanism behind wolff's law. *Biomechanics and Modeling in Mechanobiology, 22*(6), 2135–2152. https://doi.org/10.1007/s10237-023-01755-w

Goda, I., & Ganghoffer, J.-F. (2018). Modeling of anisotropic remodeling of trabecular bone coupled to fracture. *Archive of Applied Mechanics, 88*(12), 2101–2121. ISSN 0939-1533.

Goda, I., Ganghoffer, J.-F., & Maurice, G. (2016). Combined bone internal and external remodeling based on eshelby stress. *International Journal of Solids and Structures, 94*, 138–157.

Goodship, A. E., Lanyon, L. E., & McFie, H. (1979). Functional adaptation of bone to increased stress. an experimental study. *JBJS, 61*(4), 539–546.

Guadalupe-Grau, A., Fuentes, T., Guerra, B., & Calbet, J. A. L. (2009). Exercise and bone mass in adults. *Sports Medicine, 39*, 439–468.

Hambli, R. (2014). Connecting mechanics and bone cell activities in the bone remodeling process: An integrated finite element modeling. *Frontiers in Bioengineering and Biotechnology, 2*, 6–18. ISSN 2296-4185.

Hambli, R., Katerchi, H., & Benhamou, C.-L. (2011). Multiscale methodology for bone remodelling simulation using coupled finite element and neural network computation. *Biomechanics and Modeling in Mechanobiology, 10*(1), 133–145. ISSN 1617-7959.

Hambli, R., Lespessailles, E., & Benhamou, C.-L. (2013). Integrated remodeling-to-fracture finite element model of human proximal femur behavior. *Journal of the Mechanical Behavior of Biomedical Materials, 17*, 89–106. ISSN 1751-6161.

Hammer, A. (2015). The paradox of Wolff's theories. *Irish Journal of Medical Science, 184*(1), 13–22.

Harrigan, T. P., & Hamilton, J. J. (1992). Optimality conditions for finite element simulation of adaptive bone remodeling. *International Journal of Solids and Structures, 29*(23), 2897–2906. ISSN 0020-7683.

Harrigan, T. P., & Hamilton, J. J. (1994). Bone remodeling and structural optimization. *Journal of Biomechanics, 27*(3), 323–328. ISSN 0021-9290.

Hartmann, M. A., Dunlop, J. W. C., Bréchet, Y. J. M., Fratzl, P., & Weinkamer, R. (2011). Trabecular bone remodelling simulated by a stochastic exchange of discrete bone packets from the surface. *Journal of the Mechanical Behavior of Biomedical Materials, 4*(6), 879–887. ISSN 1751-6161.

Hazelwood, S. J., Martin, R. B., Rashid, M. M., & Rodrigo, J. J. (2001). A mechanistic model for internal bone remodeling exhibits different dynamic responses in disuse and overload. *Journal of Biomechanics, 34*(3), 299–308. https://doi.org/10.1016/S0021-9290(00)00221-9

Heaney, R. P. (1994). The bone-remodeling transient: Implications for the interpretation of clinical studies of bone mass change. *Journal of Bone and Mineral Research, 9*(10), 1515–1523. https://doi.org/10.1002/jbmr.5650091003

Heaney, R. P., Yates, A. J., & Santora, A. C. (1997). Bisphosphonate effects and the bone remodeling transient. *Journal of Bone and Mineral Research, 12*(8), 1143–1151. https://doi.org/10.1359/jbmr.1997.12.8.1143

Hegedus, D. H., & Cowin, S. C. (1976). Bone remodeling-ii - small strain adaptive elasticity. *Journal of Elasticity, 6*(4), 337–352. ISSN 0374-3535.

Huiskes, R. (2000). If bone is the answer, then what is the question? *Journal of Anatomy, 197* (2), 145–156. ISSN 0021-8782.

Huiskes, R., Ruimerman, R., Harry Van Lenthe, G., & Janssen, J. D. (2000). Effects of mechanical forces on maintenance and adaptation of form in trabecular bone. *Nature, 405* (6787), 704–706. https://doi.org/10.1038/35015116

Huiskes, R., Weinans, H. H. J. G., Grootenboer, H. J., Dalstra, M., Fudala, B., & Slooff, T. J. (1987). Adaptive bone-remodeling theory applied to prosthetic-design analysis. *Journal of Biomechanics, 20*(11–12), 1135–1150. https://doi.org/10.1016/0021-9290(87)90030-3

Jaber, M., Poh, P. S. P., Duda, G. N., & Checa, S. (2022). Pcl strut-like scaffolds appear superior to gyroid in terms of bone regeneration within a long bone large defect: An in silico study. *Frontiers in Bioengineering and Biotechnology, 10*, Article 995266.

Jacobs, C. R., Simo, J. C., Beaupre, G. S., & Carter, D. R. (1997). Adaptive bone remodeling incorporating simultaneous density and anisotropy considerations. *Journal of Biomechanics, 30*(6), 603–613.

Jee, W. S. S., & Tian, X. Y. (2005). The benefit of combining non-mechanical agents with mechanical loading: A perspective based on the Utah paradigm of skeletal physiology. *Journal of Musculoskeletal and Neuronal Interactions, 5*(2), 110.

Kaiser, J., Lemaire, T., Naili, S., Sansalone, V., & Komarova, S. V. (2012). Do calcium fluxes within cortical bone affect osteocyte mechanosensitivity? *Journal of Theoretical Biology, 303*, 75–86. ISSN 0022-5193.

Kalmet, P. H. S., Sanduleanu, S., Primakov, S., Wu, G., Jochems, A., Refaee, T., Ibrahim, A., Hulst, L., Lambin, P., & Poeze, M. (01 2020). Deep learning in fracture detection: A narrative review. *Acta Orthopaedica, 91*, 1–6. https://doi.org/10.1080/17453674.2019.1711323

Kameo, Y., & Adachi, T. (2014). Modeling trabecular bone adaptation to local bending load regulated by mechanosensing osteocytes. *Acta Mechanica, 225*(10), 2833–2840. ISSN 0001-5970.

Kendall, J. J., Ledoux, C., Marques, F. C., Boaretti, D., Schulte, F. A., Morgan, E. F., & Müller, R. (2023). An in silico micro-multiphysics agent-based approach for simulating bone regeneration in a mouse femur defect model. *Frontiers in Bioengineering and Biotechnology, 11*, Article 1289127.

Khalid, H., Hussain, M., Mohammed, A., Al, G., Khalid, T., Khalid, K., Khan, M. A., Fatima, K., Masood, K., Almotiri, S. H., Farooq, M. S., & Ahmed, A. (2020). A comparative systematic literature review on knee bone reports from mri, x-rays and ct scans using deep learning and machine learning methodologies. *Diagnostics, 10*(8).

Kimmel, D. B. (1985). A computer simulation of the mature skeleton. *Bone, 6*(5), 369–372. ISSN 8756-3282.

Koch, J. C. (1917). The laws of bone architecture. *American Journal of Anatomy, 21*(2), 177–298.

Komarova, S. V. (2005). Mathematical model of paracrine interactions between osteoclasts and osteoblasts predicts anabolic action of parathyroid hormone on bone. *Endocrinology, 146*(8), 3589–3595. https://doi.org/10.1210/en.2004-1642

Komarova, S. V., Smith, R. J., Dixon, S. J., Sims, S. M., & Wahl, L. M. (2003). Mathematical model predicts a critical role for osteoclast autocrine regulation in the control of bone remodeling. *Bone, 33*(2), 206–215.

Krølner, B., Toft, B., Töndevold, E., et al. (1983). Physical exercise as prophylaxis against involutional vertebral bone loss: A controlled trial. *Clinical Science, 64*(5), 541–546.

Kuhl, E. (2014). Growing matter: A review of growth in living systems. *Journal of the Mechanical Behavior of Biomedical Materials, 29*, 529–543. https://doi.org/10.1016/j.jmbbm.2013.10.009

Kuhl, E., Menzel, A., & Steinmann, P. (2003). Computational modeling of growth - a critical review, a classification of concepts and two new consistent approaches. *Computational Mechanics, 32*(1–2), 71–88. ISSN 0178-7675.

Lacroix, D., & Prendergast, P. J. (2002). A mechano-regulation model for tissue differentiation during fracture healing: Analysis of gap size and loading. *Journal of Biomechanics, 35*(9), 1163–1171.

Lagzouli, A., Pivonka, P., Cooper, D. M. L., Sansalone, V., & Othmani, A. (2025). A robust deep learning approach for segmenting cortical and trabecular bone from 3d high resolution μct scans of mouse bone, 8656–18 *Scientific Reports, 15*(1). ISSN 2045-2322.

Langdahl, B., Ferrari, S., & Dempster, D. W. (2016). Bone modeling and remodeling: Potential as therapeutic targets for the treatment of osteoporosis. *Therapeutic Advances in Musculoskeletal Diseases, 8*(6), 225–235. https://doi.org/10.1177/1759720X16670154

Langton, C. M., Haire, T. J., Ganney, P. S., Dobson, C. A., & Fagan, M. J. (1998). Dynamic stochastic simulation of cancellous bone resorption. *Bone, 22*(4), 375–380.

Lee, K. C. L., & Lanyon, L. E. (2004). Mechanical loading influences bone mass through estrogen receptor *α*. *Exercise and Sport Sciences Reviews, 32*(2), 64–68.

Lee, T. C., & Taylor, D. (1999). Bone remodelling: Should we cry Wolff? *Irish Journal of Medical Science, 168*(2), 102–105. ISSN 0021-1265.

Lekszycki, T. (1999). Optimality conditions in modelling of bone adaptation phenomenon. *Journal of Theoretical and Applied Mechanics, 3*, 607–623.

Lekszycki, T. (2002). Modelling of bone adaptation based on an optimal response hypothesis. *Meccanica, 37*(4–5), 343–354. ISSN 0025-6455.

Lemaire, T., Capiez-Lernout, E., Kaiser, J., Naili, S., & Sansalone, V. (2011). What is the importance of multiphysical phenomena in bone remodelling signals expression? A multiscale perspective. *Journal of the Mechanical Behavior of Biomedical Materials, 4*(6), 909–920. ISSN 1751-6161.

Lemaire, T., Kaiser, J., Naili, S., & Sansalone, V. (2013). Textural versus electrostatic exclusion-enrichment effects in the effective chemical transport within the cortical bone: A numerical investigation. *International Journal for Numerical Methods in Biomedical Engineering, 29*(11), 1223–1242. ISSN 2040-7939.

Lemaire, T., Kaiser, J., Naili, S., & Sansalone, V. (2015). Three-scale multiphysics modeling of transport phenomena within cortical bone. *Mathematical Problems in Engineering, 2015*(1), Article 398970.

Lemaire, V., Tobin, F. L., Greller, L. D., Cho, C. R., & Suva, L. J. (2004). Modeling the interactions between osteoblast and osteoclast activities in bone remodeling. *Journal of Theoretical Biology, 229*(3), 293–309.

Maldonado, S., Borchers, S., Findeisen, R., & Allgower, F. (2006). Mathematical modeling and analysis of force induced bone growth. In *2006 international conference of the IEEE engineering in medicine and biology society* (pp. 3154–3157). IEEE. https://doi.org/10.1109/IEMBS.2006.260532

Marques, F. C., Boaretti, D., Walle, M., Scheuren, A. C., Schulte, F. A., & Müller, R. (2023). Mechanostat parameters estimated from time-lapsed in vivo micro-computed tomography data of mechanically driven bone adaptation are logarithmically dependent on loading frequency. *Frontiers in Bioengineering and Biotechnology, 11*, Article 1140673. ISSN 2296-4185.

Martínez-Reina, J., García-Aznar, J. M., Domínguez, J., & Doblaré, M. (2008). On the role of bone damage in calcium homeostasis. *Journal of Theoretical Biology, 254*(3), 704–712. https://doi.org/10.1016/j.jtbi.2008.06.007

Martínez-Reina, J., & Pivonka, P. (2019). Effects of long-term treatment of denosumab on bone mineral density: Insights from an in-silico model of bone mineralization. *Bone, 125*, 87–95. https://doi.org/10.1016/j.bone.2019.04.022

Martin, R. B. (1984). Porosity and specific surface of bone. *Critical Reviews in Biomedical Engineering, 10*(3), 179–222.

Martin, R. B. (1991). On the significance of remodeling space and activation rate changes in bone remodeling. *Bone, 12*(6), 391–400. https://doi.org/10.1016/8756-3282(91)90028-H

Martin, B. (1992). A theory of fatigue damage accumulation and repair in cortical bone. *Journal of Orthopaedic Research, 10*(6), 818–825. ISSN 0736-0266.

Martin, B. (1995). Mathematical model for repair of fatigue damage and stress fracture in osteonal bone. *Journal of Orthopaedic Research, 13*(3), 309–316. ISSN 0736-0266.

Martin, R. B., & Burr, D. B. (1982). A hypothetical mechanism for the stimulation of osteonal remodelling by fatigue damage. *Journal of Biomechanics, 15*(3), 137–139. ISSN 0021-9290.

Martin, M., Lemaire, T., Haiat, G., Pivonka, P., & Sansalone, V. (2020). Bone orthotropic remodeling as a thermodynamically-driven evolution. *Journal of Mechanics in Medicine and Biology, 20*(04), Article 1950084.

Martin, M., Sansalone, V., Cooper, D. M. L., Forwood, M. R., & Pivonka, P. (2019). Mechanobiological osteocyte feedback drives mechanostat regulation of bone in a multiscale computational model. *Biomechanics and Modeling in Mechanobiology, 18*, 1475–1496. https://doi.org/10.1007/s10237-019-01158-w

McNamara, L. M., & Prendergast, P. J. (2007). Bone remodelling algorithms incorporating both strain and microdamage stimuli. *Journal of Biomechanics, 40*(6), 1381–1391. ISSN 0021-9290.

Mehlman, C. T., Araghi, A., & Roy, D. R. (1997). Hyphenated history: The Hueter-Volkmann law. *American Journal of Orthopedics-Belle Mead, 26*, 798–800.

Messerer, O. (1880). Über Elasticität und Festigkeit der menschlichen Knochen. Cotta.

GH von Meyer. Die architectur der spongiosa, achiv fur anatomie. Physiologie und Wissenschaftliche Medizin, 34:615–625, 1867.

Miller, Z., Fuchs, M. B., & Arcan, M. (2002). Trabecular bone adaptation with an orthotropic material model. *Journal of Biomechanics, 35*(2), 247–256.

Miller, C. J., Pickering, E., Martelli, S., Dall'Ara, E., Delisser, P., & Pivonka, P. (2024). Cortical bone adaptation response is region specific, but not peak load dependent: Insights from μ ct image analysis and mechanostat simulations of the mouse tibia loading model. *Biomechanics and Modeling in Mechanobiology, 23*(1), 287–304. https://doi.org/10.1007/s10237-023-01775-6

Mosekilde, L., Thomsen, J. S., & Mosekilde, E. (1995). Dynamics of bone remodelling. In *Modelling the dynamics of biological systems: Nonlinear phenomena and pattern formation* (pp. 186–204). Springer.

Mueller, R. (2005). Long-term prediction of three-dimensional bone architecture in simulations of pre-, peri- and post-menopausal microstructural bone remodeling. *Osteoporosis International, 16*(S02), S25–S35. ISSN 0937-941X.

Mullender, M. G., & Huiskes, R. (1995). Proposal for the regulatory mechanism of wolff's law. *Journal of Orthopaedic Research, 13*(4), 503–512.

Mullender, M. G., & Huiskes, R. (1997). Osteocytes and bone lining cells: Which are the best candidates for mechano-sensors in cancellous bone? *Bone, 20*(6), 527–532. ISSN 8756-3282.

Mullender, M. G., Huiskes, R., & Weinans, H. (1994). A physiological approach to the simulation of bone remodeling as a self-organizational control process. *Journal of Biomechanics, 27*(11), 1389–1394. ISSN 0021-9290.

Muller, R., Hildebrand, T., & Ruegsegger, P. (1994). Non-invasive bone biopsy: A new method to analyse and display the three-dimensional structure of trabecular bone. *Physics in Medicine and Biology, 39*(1), 145–164. ISSN 0031-9155.

Muller, R., & Ruegsegger, P. (1996). Analysis of mechanical properties of cancellous bone under conditions of simulated bone atrophy. *Journal of Biomechanics, 29*(8), 1053–1060. ISSN 0021-9290.

Nowak, M. (2006). Structural optimization system based on trabecular bone surface adaptation. *Structural and Multidisciplinary Optimization, 32*(3), 241–249. ISSN 1615-147X.

Nowak, M., Sokołowski, J., & Żochowski, A. (2018). Justification of a certain algorithm for shape optimization in 3d elasticity. *Structural and Multidisciplinary Optimization, 57*(2), 721–734. ISSN 1615-147X.

Nowak, M., Sokołowski, J., & Żochowski, A. (2020). Biomimetic approach to compliance optimization and multiple load cases. *Journal of Optimization Theory and Applications, 184*(1), 210–225. ISSN 0022-3239.

Nowak, M., Sokolowski, J., & Zochowski, A. (07 2019). New aspects of the trabecular bone remodeling regulatory model resulting from the shape optimization studies. *Proceedings of the Institution of Mechanical Engineers - Part H: Journal of Engineering in Medicine, 234*, Article 095441191986267. https://doi.org/10.1177/0954411919862678

Nutu, E. (2015). Interpretation of parameters in strain energy density bone adaptation equation when applied to topology optimization of inert structures. *Mechanika, 21*(6), 443–449. ISSN 1392-1207.

Nyman, J. S., Yeh, O. C., Hazelwood, S. J., & Martin, R. B. (2004). A theoretical analysis of long-term bisphosphonate effects on trabecular bone volume and microdamage. *Bone, 35* (1), 296–305. https://doi.org/10.1016/j.bone.2004.03.007

O'Connor, J. A., Lanyon, L. E., & MacFie, H. (1982). The influence of strain rate on adaptive bone remodelling. *Journal of Biomechanics, 15*(10), 767–781.

OReilly, A., Hankenson, K. D., & Kelly, D. J. (2016). A computational model to explore the role of angiogenic impairment on endochondral ossification during fracture healing. *Biomechanics and Modeling in Mechanobiology, 15*, 1279–1294.

Pais, A., Lino Alves, J., & Belinha, J. (2023). Predicting trabecular arrangement in the proximal femur: An artificial neural network approach for varied geometries and load cases. *Journal of Biomechanics, 161*, Article 111860. ISSN 0021-9290.

Paoletti, N., Lio, P., Merelli, E., & Viceconti, M. (2012). Multilevel computational modeling and quantitative analysis of bone remodeling. *IEEE/ACM Transactions on Computational Biology and Bioinformatics, 9*(5), 1366–1378.

Papastavrou, A., Schmidt, I., Deng, K., & Steinmann, P. (2020). On age-dependent bone remodeling. *Journal of Biomechanics, 103*, Article 109701. ISSN 0021-9290.

Parfitt, A. M. (1994). Osteonal and hemi-osteonal remodeling: The spatial and temporal framework for signal traffic in adult human bone. *Journal of Cellular Biochemistry, 55* (3), 273–286. https://doi.org/10.1002/jcb.240550303

Parfitt, A. M. (2002). Targeted and nontargeted bone remodeling: Relationship to basic multicellular unit origination and progression. *Bone, 30*(1), 5–7. https://doi.org/10.1016/S8756-3282(01)00642-1

Parfitt, A. M., Mathews, C. H., Villanueva, A. R., Kleerekoper, M., Frame, B., Rao, D. S., et al. (1983). Relationships between surface, volume, and thickness of iliac trabecular bone in aging and in osteoporosis. implications for the microanatomic and cellular mechanisms of bone loss. *The Journal of Clinical Investigation, 72*(4), 1396–1409.

Pastrama, M.-I., Scheiner, S., Pivonka, P., & Hellmich, C. (2018). A mathematical multiscale model of bone remodeling, accounting for pore space-specific mechanosensation. *Bone, 107*, 208–221.

Pedersen, P. (1989). On optimal orientation of orthotropic materials. *Structural Optimization, 1*, 101–106.

Perier-Metz, C., Duda, G. N., & Checa, S. (2020). Mechano-biological computer model of scaffold-supported bone regeneration: Effect of bone graft and scaffold structure on large bone defect tissue patterning. *Frontiers in Bioengineering and Biotechnology, 8*, Article 585799. ISSN 2296-4185.

Pettermann, H. E., Reiter, T. J., & Rammerstorfer, F. G. (1997). Computational simulation of internal bone remodeling. *Archives of Computational Methods in Engineering, 4*, 295–323.

Pérez, M. A., & Prendergast, P. J. (2007). Random-walk models of cell dispersal included in mechanobiological simulations of tissue differentiation. *Journal of Biomechanics, 40*(10), 2244–2253.

Pidaparti, R. M. V., & Turner, C. H. (1997). Cancellous bone architecture: Advantages of nonorthogonal trabecular alignment under multidirectional joint loading. *Journal of Biomechanics, 30*(9), 979–983. ISSN 0021-9290.

Pivonka, P., Zimak, J., Smith, D. W., Gardiner, B. S., Dunstan, C. R., Sims, N. A., Martin, T. J., & Mundy, G. R. (2008). Model structure and control of bone remodeling: A theoretical study. *Bone, 43*(2), 249–263.

Pivonka, P., Zimak, J., Smith, D. W., Gardiner, B. S., Dunstan, C. R., Sims, N. A., Martin, T. J., & Mundy, G. R. (2010). Theoretical investigation of the role of the rank–rankl–opg system in bone remodeling. *Journal of Theoretical Biology, 262*(2), 306–316.

Prendergast, P. J., & Huiskes, R. (1996). Microdamage and osteocyte-lacuna strain in bone: A microstructural finite element analysis. *Journal of Biomechanical Engineering, 118*(2), 240–246. https://doi.org/10.1115/1.2795966

Prendergast, P. J., & Taylor, D. (1994). Prediction of bone adaptation using damage accumulation. *Journal of Biomechanics, 27*(8), 1067–1076. ISSN 0021-9290.

Raimondi, A. J., Choux, M., & Rocco, C. Di (2012). *The pediatric spine I: Development and the dysraphic state*. Springer Science & Business Media.

Ramtani, S., & Zidi, M. (1999). Damaged-bone remodeling theory: Thermodynamical approach. *Mechanics Research Communications, 26*(6), 701–708. https://doi.org/10.1016/S0093-6413(99)00081-6

Rauber, A.A. (1876). Elasticität und Festigkeit der Knochen: anatomisch-physiologische Studie. W. Engelmann.

Rittweger, J. (2007). What is new in neuro-musculoskeletal interactions: Mechanotransduction, microdamage and repair? *Journal of Musculoskeletal and Neuronal Interactions, 7* (2), 191.

Rodriguez, E. K., Hoger, A., & McCulloch, A. D. (1994). Stress-dependent finite growth in soft elastic tissues. *Journal of Biomechanics, 27*(4), 455–467. ISSN 0021-9290.

Roesler, H. (1981). Some historical remarks on the theory of cancellous bone structure (Wolff's law). *Mechanical properties of bone, 15*, 27–42.

Roesler, H. (1987). The history of some fundamental concepts in bone biomechanics. *Journal of Biomechanics, 20*(11–12), 1025–1034.

Roschger, P., Fratzl, P., Eschberger, J., & Klaushofer, K. (1998). Validation of quantitative backscattered electron imaging for the measurement of mineral density distribution in human bone biopsies. *Bone, 23*(4), 319–326. https://doi.org/10.1016/S8756-3282(98)00112-4

Rosen, C. J. (2008). . Bone remodeling, energy metabolism, and the molecular clock. *Cell Metabolism, 7*(1), 7–10.

W. Roux (1881). Der Kampf der Teile im Organismus.

Rovati, M., & Taliercio, A. (1991). Optimal orientation of the symmetry axes of orthotropic 3-d materials. In *Engineering optimization in design processes: Proceedings of the international conference Karlsruhe nuclear research center, Germany September 3–4, 1990* (pp. 127–134). Springer.

Rubin, C. T., & Lanyon, L. E. (1984). Regulation of bone formation by applied dynamic loads. *JBJS, 66*(3), 397–402.

Rubin, J., Rubin, C., & Jacobs, C. R. (2006). Molecular pathways mediating mechanical signaling in bone. *Gene, 367*, 1–16.

Rubin, C., Simon Turner, A., Bain, S., Craig, M., & McLeod, K. (2001). Low mechanical signals strengthen long bones. *Nature, 412*(6847), 603–604.

Rubin, C., Turner, A. S., Müller, R., Mittra, E., McLeod, K., Lin, W., & Qin, Y.-X. (2002). Quantity and quality of trabecular bone in the femur are enhanced by a strongly anabolic, noninvasive mechanical intervention. *Journal of Bone and Mineral Research, 17*(2), 349–357.

Ruff, C., Holt, B., & Trinkaus, E. (2006). Who's afraid of the big bad Wolff?: "Wolff's law" and bone functional adaptation. *American Journal of Physical Anthropology: The Official Publication of the American Association of Physical Anthropologists, 129*(4), 484–498.

Ruffoni, D., Fratzl, P., Roschger, P., Klaushofer, K., & Weinkamer, R. (2007). The bone mineralization density distribution as a fingerprint of the mineralization process. *Bone, 40*(5), 1308–1319. https://doi.org/10.1016/j.bone.2007.01.012

Ruimerman, R., Huiskes, R., Van Lenthe, G. H., & Janssen, J. D. (2001). A computer-simulation model relating bone-cell metabolism to mechanical adaptation of trabecular architecture. *Computer Methods in Biomechanics and Biomedical Engineering, 4*(5), 433–448. ISSN 1025-5842.

Ruimerman, R., Van Rietbergen, B., Hilbers, P., & Huiskes, R. (2003). A 3-dimensional computer model to simulate trabecular bone metabolism. *Biorheology, 40*(1–3), 315–320.

Ruiz-Lozano, R., Calvo-Gallego, J. L., Pivonka, P., & Martínez-Reina, J. (2025). Optimisation of romosozumab plus denosumab sequential treatments against postmenopausal osteoporosis. insights from in silico simulations. *Biomechanics and Modeling in Mechanobiology*, 1–22.

Ryser, M. D., Komarova, S. V., & Nigam, N. (2010). The cellular dynamics of bone remodeling: A mathematical model. *SIAM Journal on Applied Mathematics, 70*(6), 1899–1921. ISSN 0036-1399.

Ryser, M. D., Nigam, N., & Komarova, S. V. (2009). Mathematical modeling of spatio-temporal dynamics of a single bone multicellular unit. *Journal of Bone and Mineral Research, 24*(5), 860–870. ISSN 0884-0431.

Sakashita, M., Yamasaki, S., Yaji, K., Kawamoto, A., & Kondo, S. (2021). Three-dimensional topology optimization model to simulate the external shapes of bone. *PLoS Computational Biology, 17*(6), Article e1009043. ISSN 1553-7358.

Sansalone, V., Kaiser, J., Naili, S., & Lemaire, T. (2013). Interstitial fluid flow within bone canaliculi and electro-chemo-mechanical features of the canalicular milieu: A multi-parametric sensitivity analysis. *Biomechanics and Modeling in Mechanobiology, 12*, 533–553.

Sansalone, V., Martin, M., Haïat, G., Pivonka, P., & Lemaire, T. (2021). A new model of bone remodeling and turnover set up in the framework of generalized continuum mechanics. *Mathematics and Mechanics of Solids, 26*(9), 1376–1393. https://doi.org/10.1177/108128652098469

Sansalone, V., Naili, S., & Di Carlo, A. (2011). On the rotary remodelling equilibrium of bone. *Computer Methods in Biomechanics and Biomedical Engineering, 14*(Suppl. 1), 203–204.

Saxon, L. K., & Lanyon, L. E. (2008). Assessment of the in vivo adaptive response to mechanical loading. *Osteoporosis: Methods and Protocols*, 307–322.

Scheiner, S., Pivonka, P., & Hellmich, C. (2013). Coupling systems biology with multiscale mechanics, for computer simulations of bone remodeling. *Computer Methods in Applied Mechanics and Engineering, 254*, 181–196. https://doi.org/10.1016/j.cma.2012.10.015

Scheiner, S., Pivonka, P., Smith, D. W., Dunstan, C. R., & Hellmich, C. (2014). Mathematical modeling of postmenopausal osteoporosis and its treatment by the anti-catabolic drug denosumab. *International Journal for Numerical Methods in Biomedical Engineering, 30*(1), 1–27. https://doi.org/10.1002/cnm.2584

Schulte, F. A., Ruffoni, D., Lambers, F. M., Christen, D., Webster, D. J., Kuhn, G., & Müller, R. (2013). Local mechanical stimuli regulate bone formation and resorption in mice at the tissue level. *PLoS One, 8*(4), Article e62172.

Scott, A., Khan, K. M., Duronio, V., & Hart, D. A. (2008). Mechanotransduction in human bone: In vitro cellular physiology that underpins bone changes with exercise. *Sports Medicine, 38*, 139–160.

Shefelbine, S. J., & Carter, D. R. (2004). Mechanobiological predictions of growth front morphology in developmental hip dysplasia. *Journal of Orthopaedic Research, 22*(2), 346–352. https://doi.org/10.1016/S8756-3282(02)00700-7

Sherwood Mather, B. (1967). A method of studying the mechanical properties of long bones. *Journal of Surgical Research, 7*(5), 226–230. https://doi.org/10.1016/0022-4804(67)90056-X

Simon, M. R. (1978). The effect of dynamic loading on the growth of epiphyseal cartilage in the rat. *Cells Tissues Organs, 102*(2), 176–183.

Skerry, T. M. (2006). One mechanostat or many? Modifications of the site-specific response of bone to mechanical loading by nature and nurture. *Journal of Musculoskeletal and Neuronal Interactions, 6*(2), 122.

Smets, J., Shevroja, E., Hügle, T., Leslie, W., & Hans, D. (03 2021). Machine learning solutions for osteoporosis—a review. *Journal of Bone and Mineral Research, 36*. https://doi.org/10.1002/jbmr.4292

Smit, T. H. (2020). Adolescent idiopathic scoliosis: The mechanobiology of differential growth. *JOR Spine, 3*(4), Article e1115.

Stokes, I. A. F., Aronsson, D. D., Dimock, A. N., Cortright, V., & Beck, S. (2006). Endochondral growth in growth plates of three species at two anatomical locations modulated by mechanical compression and tension. *Journal of Orthopaedic Research, 24*(6), 1327–1334. https://doi.org/10.1002/jor.20189

Stokes, I. A. F., & Laible, J. P. (1990). Three-dimensional osseo-ligamentous model of the thorax representing initiation of scoliosis by asymmetric growth. *Journal of Biomechanics, 23*(6), 589–595. https://doi.org/10.1016/0021-9290(90)90051-4

Sugiyama, T., Meakin, L. B., Browne, W. J., Galea, G. L., Price, J. S., & Lanyon, L. E. (2012). Bones' adaptive response to mechanical loading is essentially linear between the low strains associated with disuse and the high strains associated with the lamellar/woven bone transition. *Journal of Bone and Mineral Research, 27*(8), 1784–1793. https://doi.org/10.1002/jbmr.1599

Taylor, D. (1997). Bone maintenance and remodeling: A control system based on fatigue damage. *Journal of Orthopaedic Research, 15*(4), 601–606. ISSN 0736-0266.

Thomsen, J. S., Mosekilde, L., Boyce, R. W., & Mosekilde, E. (1994). Stochastic simulation of vertebral trabecular bone remodeling. *Bone, 15*(6), 655–666. https://doi.org/10.1016/8756-3282(94)90314-X

Tsubota, K.-ichi, Adachi, T., & Tomita, Y. (2002). Functional adaptation of cancellous bone in human proximal femur predicted by trabecular surface remodeling simulation toward uniform stress state. *Journal of Biomechanics, 35*(12), 1541–1551.

Van Der Linden, J. C., Verhaar, J. A. N., & Weinans, H. (2001). A three-dimensional simulation of age-related remodeling in trabecular bone. *Journal of Bone and Mineral Research, 16*(4), 688–696.

van Der Meulen, M. C. H., Beaupré, G. S., & Carter, D. R. (1993). Mechanobiologic influences in long bone cross-sectional growth. *Bone, 14*(4), 635–642. ISSN 8756-3282.

van der Meulen, M. C. H., & Huiskes, R. (2002). Why mechanobiology?: A survey article. *Journal of Biomechanics, 35*(4), 401–414. ISSN 0021-9290.

Van Oers, R. F. M., Ruimerman, R., Tanck, E., Hilbers, P. A. J., & Huiskes, R. (2008). A unified theory for osteonal and hemi-osteonal remodeling. *Bone, 42*(2), 250–259.

van Oers, R. F. M., van Rietbergen, B., Ito, K., Hilbers, P. A. J., & Huiskes, R. (2011). A sclerostin-based theory for strain-induced bone formation. *Biomechanics and Modeling in Mechanobiology, 10*, 663–670.

Villemure, I., Aubina, C. E., Dansereau, J., & Labelle, H. (2002). Simulation of progressive deformities in adolescent idiopathic scoliosis using a biomechanical model integrating vertebral growth modulation. *Journal of Biomechanical Engineering, 124*(6), 784–790. https://doi.org/10.1115/1.1516198

Villemure, I., & Stokes, I. A. F. (2009). Growth plate mechanics and mechanobiology. a survey of present understanding. *Journal of Biomechanics, 42*(12), 1793–1803.

Waffenschmidt, T., Menzel, A., & Kuhl, E. (2012). Anisotropic density growth of bone—a computational micro-sphere approach. *International Journal of Solids and Structures, 49*(14), 1928–1946.

Wang, B., Zhou, X., Price, C., Li, W., Pan, J., & Wang, L. (2013). Quantifying load-induced solute transport and solute-matrix interaction within the osteocyte lacunar-canalicular system. *Journal of Bone and Mineral Research, 28*(5), 1075–1086.

Watzky, A., & Naili, S. (2004). Orthotropic bone remodeling: Case of plane stresses. *Mechanics Research Communications, 31*(5), 617–625.

Wehrli, F. W., Gomberg, B. R., Saha, P. K., Song, H. K., Hwang, S. N., & Snyder, P. J. (2001). Digital topological analysis of in vivo magnetic resonance microimages of trabecular bone reveals structural implications of osteoporosis. *Journal of Bone and Mineral Research, 16* (8), 1520–1531. https://doi.org/10.1359/jbmr.2001.16.8.1520

Weinans, H., Huiskes, R., & Grootenboer, H. J. (1992). The behavior of adaptive bone-remodeling simulation models. *Journal of Biomechanics, 25*(12), 1425–1441. https://doi.org/10.1016/0021-9290(92)90056-7

Weinkamer, R., Hartmann, M. A., Brechet, Y., & Fratzl, P. (2004). Stochastic lattice model for bone remodeling and aging. *Physical Review Letters, 93*(22), Article 228102. https://doi.org/10.1103/PhysRevLett.93.228102

Wertheim, G. (1847). Mémoire sur l'élasticité et la cohésion des principaux tissus du corps humain, Vol. 21.

Whitehouse, W. J. (1974). The quantitative morphology of anisotropic trabecular bone. *Journal of Microscopy, 101*(2), 153–168.

Willy, C., Schneider, P., Engelhardt, M., Hargens, A. R., & Mubarak, S. J. (2008). Richard von volkmann: surgeon and renaissance man. *Clinical Orthopaedics and Related Research, 466*, 500–506.

Wolff, J. (1870). Ueber die innere architectur der knochen und ihre bedeutung für die frage vom knochenwachsthum. Archiv für Pathologische Anatomie und Physiologie und für Klinische Medicin, 50(3), 389–450. doi: 10.1007/BF01944490.

Wolff, J. (2012). *The law of bone remodelling*. Springer Science & Business Media.

Wolff, J., & Virchow, R. (1875). Einige bemerkungen zum gegenwärtigen stand der knochenwachsthumsfrage. Archiv für Pathologische Anatomie und Physiologie und für Klinische Medicin, 64(1):140–144.

Woo, S.-L., Kuei, S. C., Amiel, D., Gomez, M. A., Hayes, W. C., White, F. C., & Akeson, W. H. (1981). The effect of prolonged physical training on the properties of long bone: A study of Wolff's law. *JBJS, 63*(5), 780–787.

Zaloszyc, A., Bernardor, J., Bacchetta, J., Laverny, G., & Schmitt, C. P. (2023). Mouse models of mineral bone disorders associated with chronic kidney disease. *International Journal of Molecular Sciences, 24*(6), 5325.

Zhang, Y., Ma, M., Huang, X., Liu, J., Tian, C., Duan, Z., Fu, H., Huang, L., & Geng, B. (2025). Machine learning is changing osteoporosis detection: An integrative review. *Osteoporosis International, 36*(8), 1313–1326. ISSN 0937-941X.

CHAPTER

Biomechanical modeling of bone: Topology optimization and microstructural analysis

14

Ibrahim Goda and Jean-François Ganghoffer

Universite de Lorraine, LEM3 - CNRS - Arts et Métiers Paris Tech, Nancy, France

1. Introduction

Bone is a complex hierarchical material that exhibits remarkable mechanical adaptability, enabling it to withstand various physiological loading conditions while maintaining an optimal balance between strength and weight. Its trabecular microstructure plays a crucial role in determining its mechanical behavior, as demonstrated by Goda et al. (2014) and Goda and Ganghoffer (2015) through their 3D micropolar analysis of vertebral trabecular bone, which revealed how microscale architecture influences macroscale mechanical properties. These findings align with earlier work by Gibson (2005) on cellular solid mechanics and micro-CT analyses by Liu et al. (2009). Understanding this structure is essential for applications in orthopedics, biomedical engineering, and implant design. The integration of computational modeling, particularly through homogenization techniques, topology optimization, and microstructural analysis, will provide valuable insights into bone mechanics.

As a hierarchical and heterogeneous material, bone exhibits a complex microstructure optimized through evolutionary adaptation for mechanical efficiency. Its mechanical behavior, particularly load-bearing capacity and failure mechanisms, has been extensively studied using homogenization techniques to derive effective macroscopic properties from its intricate trabecular architecture. Homogenization provides a crucial link between microscale structural features and macroscale mechanical responses, enabling accurate and computationally efficient models of bone behavior. Goda et al. (2014) demonstrated the utility of these techniques in biomechanics, facilitating the translation of microscale trabecular properties into macroscale predictive models. These approaches, validated against experimental measurements by Morgan et al. (2003), Morgan and Keaveny (2001), Wolfram et al. (2010), contribute to advancements in patient-specific implant designs, biomimetic materials, and improved strategies for bone adaptation and repair.

One particularly effective approach for capturing the anisotropic and heterogeneous nature of trabecular bone is lattice homogenization. For instance, Goda and Ganghoffer (2015) developed a 3D elastic couple-stress model of vertebral

Biomechanics and Mechanobiology of Bone. https://doi.org/10.1016/B978-0-443-26739-0.00018-1

trabecular bone, demonstrating how microstructural characteristics influence effective mechanical properties. These techniques successfully capture the anisotropic nature of trabecular bone by considering both classical and higher-order continuum mechanics. Recent studies, such as the asymptotic homogenization of tesseract lattice structures by Mohammadi et al. (2024), further reinforce the significance of homogenization techniques in modeling complex architectures. These approaches continue to enhance our understanding of hierarchical materials and their mechanical performance, with direct implications for biomedical applications and advanced material design.

Beyond mechanical characterization, optimizing bone's internal architecture is crucial for designing biomimetic materials and medical implants (Noman et al., 2025; Wu et al., 2021). Homogenization-based relaxation methods have been employed to determine optimal bone architectures, ensuring that structural configurations align with mechanical efficiency criteria (e.g. (Goda et al., 2016)). By considering bone as a porous material and incorporating isotropic material design, these methods enable the prediction of its optimal structural adaptation under physiological conditions (Allaire & Pantz, 2006). Such approaches highlight the efficiency of bone's natural remodeling process and provide a foundation for developing bioinspired structures.

Additionally, the mechanical performance of trabecular bone under complex loading scenarios, such as twisting and bending, is an essential consideration for structural integrity. Goda et al. (2016) constructed effective plastic yield surfaces using 3D microstructural models that allowed assessment of bone failure mechanisms under multi-axial stress states. These models build upon earlier yield criteria proposed by Fenech and Keaveny (1999) and offer crucial insights into the failure behavior of vertebral trabecular bone. Such studies support the development of patient-specific treatments and align with clinical observations by Kopperdahl and Keaveny (1998), contributing to improved diagnostic and therapeutic strategies.

This chapter presents a comprehensive framework linking bone microarchitecture to effective mechanical behavior and failure prediction. Beginning with classical homogenization approaches in Cauchy elasticity, we examine how microstructure governs effective properties. We then introduce structural optimization strategies, including SIMP and isotropic material design, which predict biologically optimal architectures. Finally, we explore advanced failure analysis through yield surface modeling under complex loading, with implications for biomedical engineering and patient-specific applications.

2. Homogenization of trabecular bone microstructure within cauchy elasticity framework

Homogenization has emerged as a powerful method for deriving macroscopic material properties from detailed microstructural features. It is widely applied in material science, especially in the analysis of cellular solids like bone, to provide a continuous description of their macroscopic behavior based on microscopic properties (Feng et al., 2021; Goda & Ganghoffer, 2015; Gibson, 2005; Liu et al., 2009). This method allows for simplifications and approximations that capture the essence of material behavior without directly simulating every microscopic detail.

In the context of Cauchy elasticity, the primary focus is on the classical elastic behavior, where stress and strain are related through conventional elasticity constants, such as Young's modulus, shear modulus, and Poisson's ratio. The key goal of homogenization here is to compute these effective properties at the macroscopic scale, considering the periodicity and structure of the material at the microscopic scale. This approach is especially relevant for complex materials like vertebral trabecular bone, where the micro-architecture significantly influences mechanical performance. By analyzing the microstructure of bone and applying the principles of classical elasticity, we can derive the effective elastic properties that describe the overall mechanical behavior of the material.

2.1 The homogenization framework

Trabecular bone is a highly porous, anisotropic material with a hierarchical architecture that plays a pivotal role in load distribution, energy dissipation, and mechanical adaptation within the vertebral skeleton. At the microscale, it comprises a three-dimensional network of rod- and plate-like trabecule that form a lattice-like structure (Fig. 14.1). These trabecule typically range in thickness from 80 to 300 μm, with the overall porosity reaching 50%–90% depending on anatomical location and age (Chen et al., 2013; Dagan et al., 2004; Syahrom et al., 2013). The architectural organization, including trabecular thickness, number, spacing, and connectivity, directly influences the mechanical performance of bone, particularly under complex loadings. High-resolution imaging studies using micro-computed tomography (micro-CT) and synchrotron tomography have reinforced the importance of microstructural anisotropy in determining failure behavior and have highlighted age- and disease-related microarchitectural deterioration as key factors in fragility fractures (Kadir et al., 2010; Syahrom et al., 2013).

Understanding how this microarchitecture influences the mechanical behavior of trabecular bone requires a transition from image-based morphological analysis to effective continuum modeling. To this end, various theoretical frameworks have been employed to link the microscale architecture with macroscale mechanical properties. The mechanical behavior of trabecular bone has traditionally been modeled using classical Cauchy elasticity, which assumes a locally homogeneous continuum and does not account for microstructural length scales.

To better capture the role of microarchitecture in the overall mechanical response, this subsection presents an asymptotic homogenization approach aimed at deriving the effective macroscopic properties of vertebral trabecular bone, modeled as a periodic 3D hexagonal lattice. While alternative frameworks such as micropolar (Cosserat) theories have been developed to include rotational degrees of freedom and couple stresses (Ganghoffer & Goda, 2018; Goda et al., 2012), we

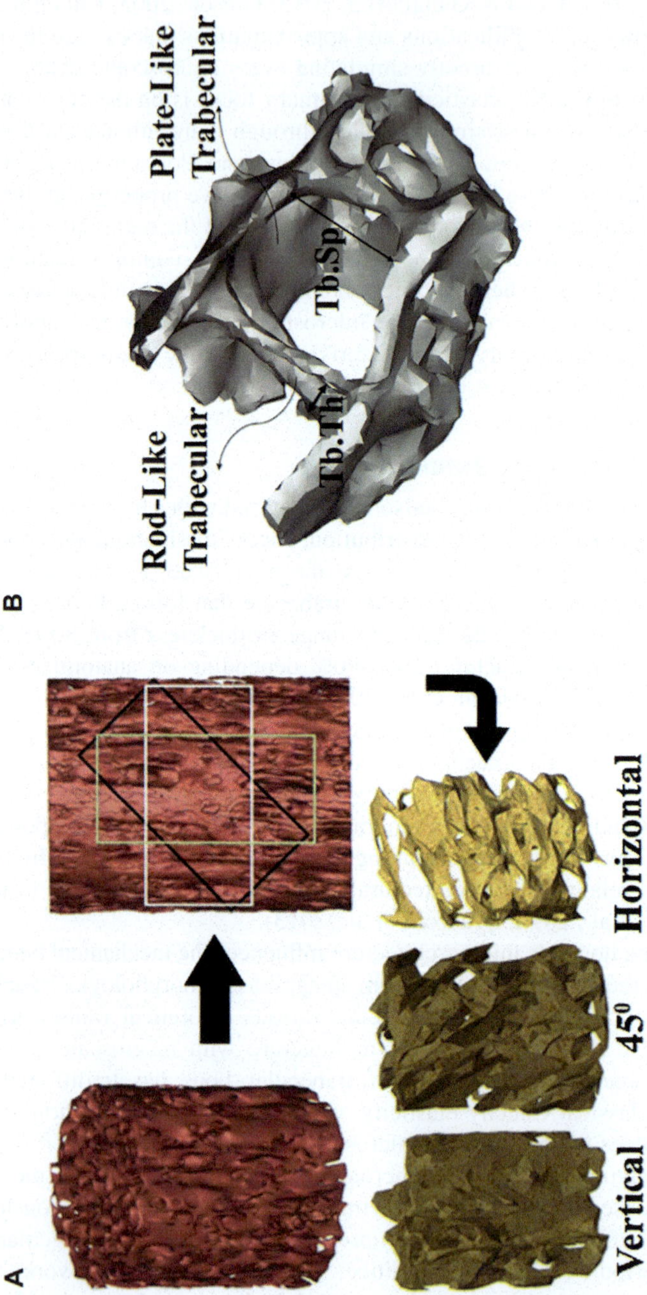

FIGURE 14.1

(a) Micro-CT reconstruction of trabecular bone showing three selected regions for morphological analysis, (b) detailed visualization of trabecular bone microarchitecture, highlighting key morphological parameters such as trabecular thickness and spacing.

- Adapted from Syahrom, A., Abdul Kadir, M. R., Abdullah, J., Öchsner, A. (2013). Permeability studies of artificial and natural cancellous bone structures. Malaysia Medical Engineering & Physics, *35(6), 792–799. doi: 10.1016/j.medengphy.2012.08.011.*

show that a suitably formulated Cauchy continuum model can still effectively predict the anisotropic elastic behavior when appropriate microstructural kinematics are incorporated in the analysis.

The homogenization process begins by considering the microstructure of the material, typically represented by a periodic unit cell. The material is assumed to have repeating geometric features that are small compared to the overall structure's size. The homogenization technique aims to derive effective continuum properties from the behavior of the microstructure. To simplify the complex analysis, we start by analyzing a representative volume element (RVE) of the material, which is a small sample that encapsulates the material's periodicity. In Cauchy elasticity, the stress and strain within the material are governed by classical continuum mechanics, where the stress tensor is related to the strain tensor through an isotropic or anisotropic stiffness matrix. The periodicity of the material's microstructure is incorporated into the problem, and the objective is to calculate the effective moduli that govern the macroscopic behavior of the material.

The detailed step-by-step homogenization algorithm, which translates microstructural mechanics into macroscopic elastic properties using an asymptotic expansion approach, is provided in Appendix A. This algorithm focuses on the linear elastic regime and derives the effective elastic moduli of trabecular bone modeled as a periodic 3D lattice structure. It forms the basis for computing directional stiffnesses such as Young's moduli, shear moduli, and Poisson's ratios in the subsequent sections.

2.2 Effective classical elastic moduli of trabecular bone

The classical elastic constants of vertebral trabecular bone, derived from the homogenization of a geometrically idealized 3D hexagonal lattice, provide critical insights into its anisotropic mechanical behavior. This analysis focuses exclusively on the effective classical elastic moduli, namely, the Young's moduli E_x^*, E_y^*, E_z^*, shear moduli G_{xy}^*, G_{xz}^*, G_{yz}^*, and Poisson's ratios ν_{xy}^* and ν_{yx}^* which are obtained from the homogenized stiffness tensor. These properties are strongly influenced by the underlying lattice geometry and the relative density of the trabecular architecture. The hexagonal unit cell geometry, characterized by an inclination angle θ, in-plane strut lengths (L, h) and out-of-plane strut length (L_v), see Fig. 14.2, serves as a simplified but representative model of vertebral trabecular bone. This geometric idealization is supported by microstructural image, which highlights the anisotropic nature of the trabecular network Fig. 14.2(a).

The variation of effective elastic properties with respect to the inclination angle θ (10°–80 degrees) is shown in Fig. 14.3(a–d). A 3D hexagonal unit cell is used to simulate the anisotropic trabecular structure, with geometric parameters $L = 1$ mm, $h = 0.5$ mm, $L_v = 1$ mm, and strut diameter $d = 0.25$, assuming identical circular cross-sections for all trabecule. The results, normalized by bulk material values, highlight the directional dependence of the mechanical response of the bone architecture.

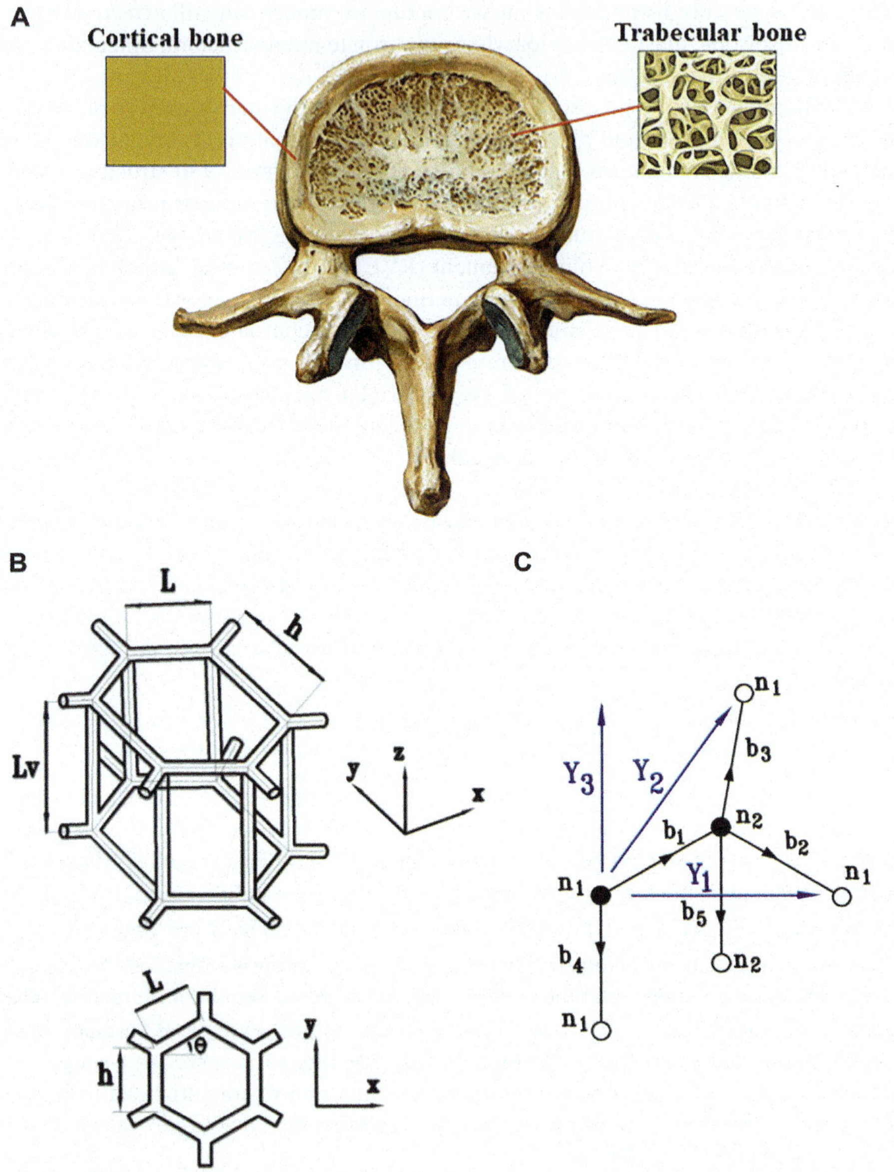

FIGURE 14.2

Geometric representation and periodic modeling of vertebral trabecular bone using a 3D hexagonal lattice framework. (a) Morphology of a typical vertebra, (b) 3D hexagonal unit cell with geometric parameters: Cell angle θ, in-plane strut lengths L, h, and vertical strut length Lv, (c) periodic lattice formed by repeating the unit cell along vectors Y1, Y2, Y3.

- (A) Based on Thompson, J.C. (2010). Biomechanical modeling of bone: topology optimization and microstrctural analysis. In: Netter's Concise orthopaedic Anatomy, second Ed., Elsevier.

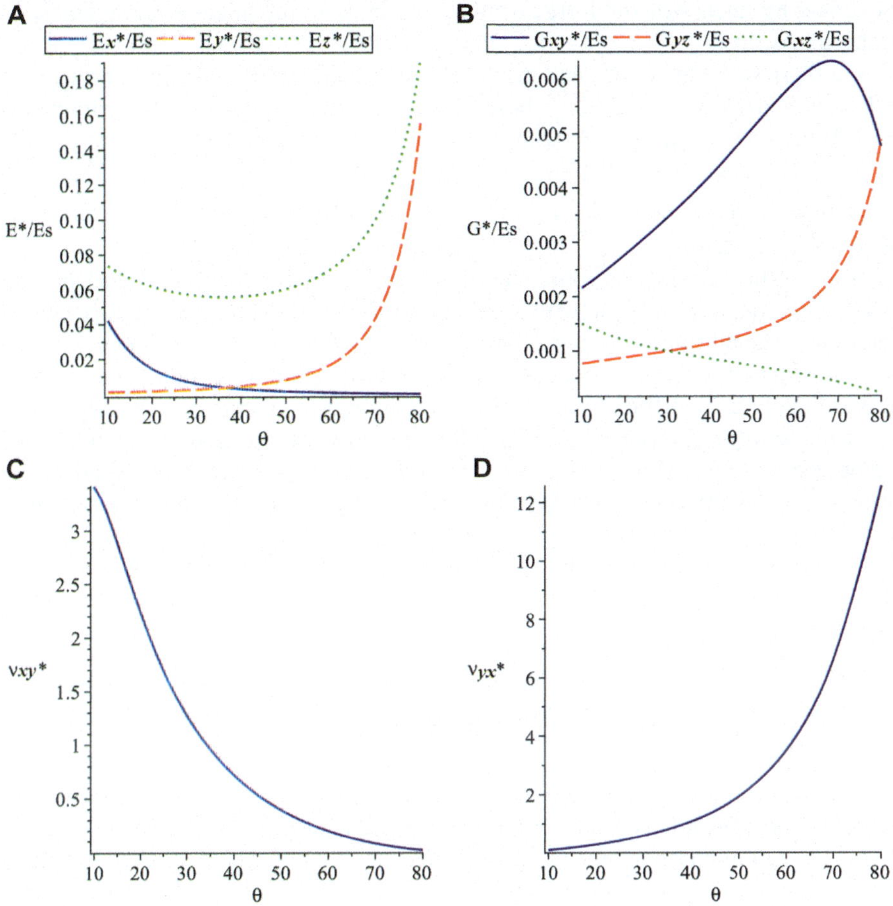

FIGURE 14.3

Influence of the geometrical angle θ on the effective elastic properties of vertebral trabecular bone. (a) Young's moduli, (b) shear moduli, (c) and (d) in-plane Poisson's ratios.

The in-plane Young's modulus E_x^* peaks at low angles (~10 degrees), while E_y^* reaches its maximum at high angles (~80 degrees), indicating a directional shift in stiffness driven by unit cell geometry. The out-of-plane modulus E_z^* shows a non-monotonic trend, with a minimum near 40 degrees. Similarly, the in-plane shear modulus G_{xy}^* increases with θ up to ~70 degrees, then drops sharply. The out-of-plane shear moduli G_{yz}^* and G_{xz}^* follow trends similar to E_y^* and E_x^*. Poisson's ratios also vary with θ: ν_{xy}^* generally decreases as θ increases, while ν_{yx}^* increases, further reflecting the directional dependency of the elastic response. These observations support the conclusion that trabecular bone mechanical anisotropy is strongly governed by microstructural organization, particularly the angular arrangement of trabecular struts.

Furthermore, Fig. 14.4(a and b) present the scaling of effective properties with relative density ρ^*/ρ_s, based on the geometric parameters $L = 1\text{mm}, L_v = 0.5\text{mm}, h = 0.5\text{mm}, \theta = 30°$. Elastic and shear moduli (E_x^*; E_y^*, G_{ij}^*) follow a power-law relationship, while E_z^* scales linearly. Neglecting transverse shear overestimates moduli E_x^*, E_y^* and G_{ij}^*, thus emphasizing the role of shear effects in accurate modeling.

Despite the absence of comprehensive experimental validation, model predictions align reasonably well with reported values in the literature. Experimental data for vertebral trabecular bone are limited and often inconsistent due to varying sample orientations, geometry, and measurement techniques. Most studies only report longitudinal stiffness and overlook the orthotropic nature of trabecular bone.

To assess the relevance of our predictions, Fig. 14.5(a) and (b) compare the computed effective Young's and shear moduli with a wide range of experimental values from the literature. The predicted values align well with trends observed in the data, despite some scatter, which is attributed to differences in apparent density, trabecular geometry, and specimen orientation. These results support the validity of the homogenized lattice model in capturing the anisotropic mechanical behavior of vertebral trabecular bone.

3. Topology optimization approaches for bone structure design

This section explores three complementary approaches for optimizing bone microstructure: the Solid Isotropic Material with Penalization (SIMP) method, a homogenization-based approach, and the Isotropic Material Design (IMD) method. Each approach provides a unique framework for predicting the optimal bone architecture under physiological loading conditions. The SIMP method focuses on penalizing intermediate densities to achieve a binary (solid-void) material distribution. In contrast, the homogenization-based approach models bone as a composite material, incorporating microstructural details to capture its inherent anisotropy. Finally, the IMD method optimizes the full stiffness tensor while maintaining isotropic properties, offering an alternative to traditional density-based methods. By comparing these approaches, we aim to highlight their respective strengths and limitations in replicating the structural organization of bone. The following subsections provide a detailed explanation of each method and its application to the femoral bone.

3.1 SIMP approach

The SIMP (Solid Isotropic Material with Penalization) method is commonly used as a topology optimization algorithm. It operates on the principle that the material of the optimized structure exists within the design domain Ω but is not initially optimally distributed. At the start of the optimization process, the material density θ

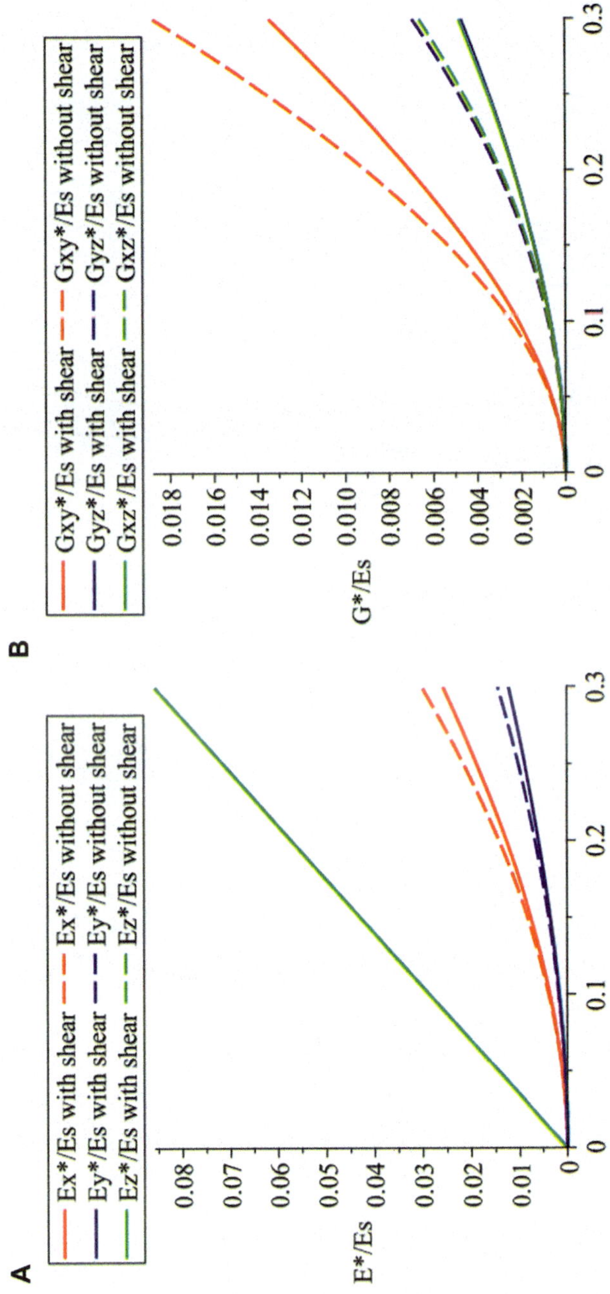

FIGURE 14.4

Effective elastic and shear moduli versus relative density, showing the effect of transverse shear.

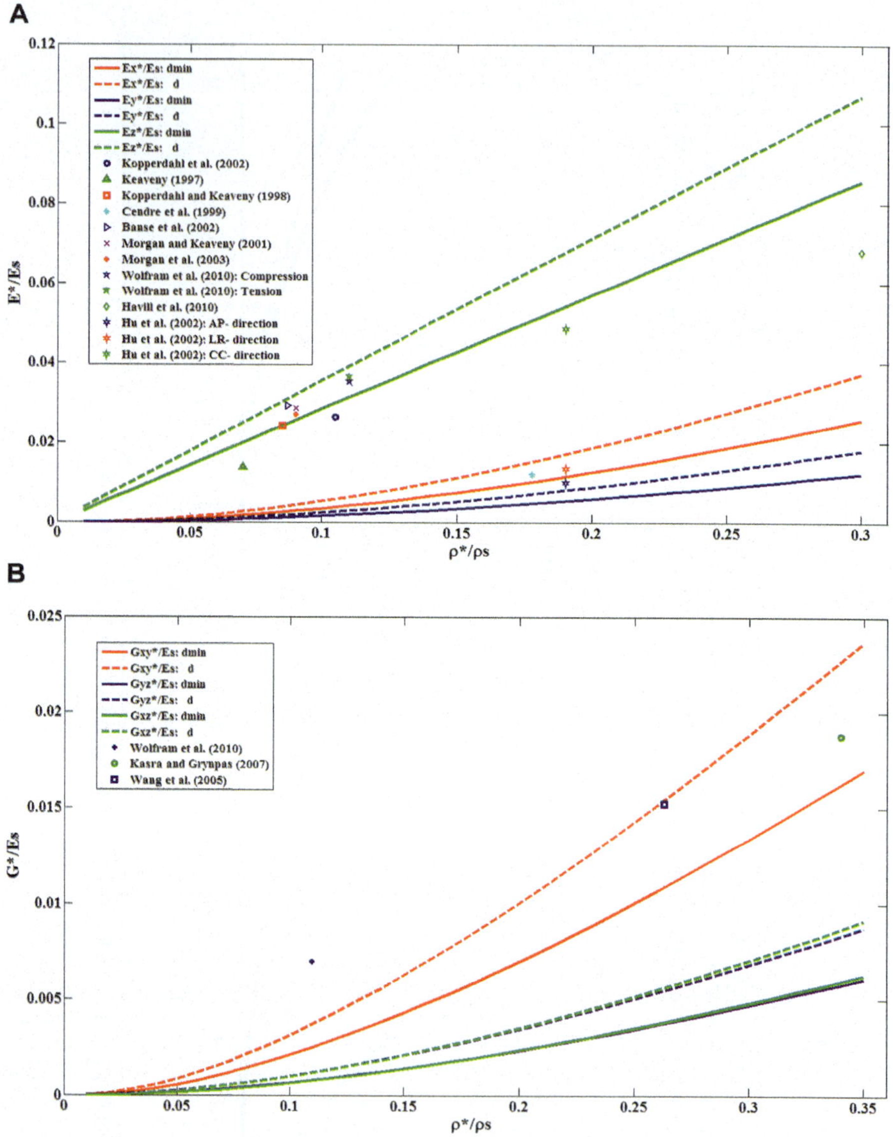

FIGURE 14.5

(a) Normalized effective Young's moduli, (b) shear moduli of vertebral trabecular bone versus relative density, compared with experimental data from literature.

is uniformly distributed across the domain. Throughout the optimization, the material condenses in areas where it is needed and is removed from regions with low values of the design variable θ, thereby achieving the desired structural performance. The method is presently used in the context of isotropic linear elasticity.

In the SIMP based topology optimization method, the dependence of material properties with design variables is expressed in terms of the material density using a power law approach, which means the intermediate values are suppressed by penalizing the bulk densities. The SIMP scheme is actually to replace the integer variables with continuous variables, and then introduce some form of penalty that steers the solution to discrete 0/1 values. Once the original problem is relaxed in this way, regions with intermediate densities between 0 and 1 may occur in the optimum topologies. To ensure an easy interpretation of distinct 0/1 optimal topologies, penalty schemes are often employed to suppress the intermediate densities.

Structural optimization relies on a specific choice of the objective functions, like minimization of stress, strain, weight, stiffness. A classical and widespread objective function is the compliance of the structural system under consideration; minimization of compliance (equivalent to maximization of stiffness) is equivalent to the internal energy minimization as proposed originally in Bendsøe and Kikuchi (1988).

The structural optimization problem is formulated here as minimization of the compliance (the work done by the exterior forces) in the framework of linear elasticity, so equivalently one aims at achieving the highest global stiffness, under the constraint of a fixed overall volume. The SIMP based topology optimization approach amounts to consider the following state equations

$$\begin{cases} -\text{div}\,\sigma = \rho g \text{ in } \Omega \\ \sigma = \theta(x)^p C\varepsilon(u) \text{in } \Omega, \text{ with } \varepsilon(u) = \frac{1}{2}(\nabla u + \nabla^t u), \\ \sigma.n = P\text{on } \partial\Omega \end{cases} \tag{14.1}$$

with $\theta(x)$ the density function taking values between 0 and 1. In fact, the elasticity tensor $\theta(x)C$ is replaced by $\theta(x)^p C$ for the penalization process. The Lamé constants λ, μ are related to the stiffness tensor C_{ijkl} by

$$C_{ijkl} = \lambda\delta_{ij}\delta_{kl} + \mu\left(\delta_{ik}\delta_{jl} + \delta_{il}\delta_{jk}\right), \tag{14.2}$$

where δ is the Kronecker delta function.

The objective is to minimize the compliance over all possible density fields satisfying the fixed volume condition. The Lagrange multiplier method is used and the min-max problem thus obtained is rearranged to the max-min problem. The nested minimization problem assumes the form, see Allaire (2002).

$$\min_{0\le\theta(x)\le 1}\left(S(\theta) + l\int_\Omega \theta(x)dx\right), \text{Lagrange multiplier} l \ge 0, \tag{14.3}$$

in which the compliance is defined as $S(\theta) = f(u(\theta))$, where $f(.)$ is the linear form expressed as the virtual work of the load, i.e.,

$$f(v) = \int_{\partial\Omega} P.v + \int_\Omega \rho g.v, \tag{14.4}$$

or, equivalently

$$S(\theta) = \int_{\Omega} (\theta(x)^p C)^{-1} \sigma(\theta).\sigma(\theta), \tag{14.5}$$

The argument θ stresses that the solution depends on the distribution of the density function. By using the principle of complementary energy and switching the two minimizations, we get the following equivalent formulation

$$\min_{0 \le \theta(x) \le 1} \min_{\substack{div\sigma=0 \text{ in}\Omega \\ \sigma.n=P\text{on}\partial\Omega}} \int_{\Omega} (\theta(x)^p C)^{-1} \sigma.\sigma dx + l \int_{\Omega} \theta(x) dx, \tag{14.6}$$

This problem is implemented numerically by using the alternate directions strategy which is an iterative method: for a given density θ, we compute the stress tensor σ solution of the linear elasticity equilibrium equation; then we update the design variable θ by the following optimality condition

$$\theta(x) = \begin{cases} 1 \text{ if} pC^{-1}\sigma.\sigma \ge l \\ \sqrt[p+1]{pl^{-1}C^{-1}\sigma.\sigma} \text{ if} pC^{-1}\sigma.\sigma \le l \end{cases}, \tag{14.7}$$

The algorithm of alternate directions method to solve problem (14.6) consists of the following steps: 1. Initialize the shape θ_0 2. Iterate $n \ge 1$ until convergence:—given a shape $\theta_{n\text{-}1}$, we compute the stress σ_n by solving elasticity problem (14.1)—given a stress tensor σ_n, we update the new material density θ_n with the explicit optimality formula in terms of σ_n. The numerical penalization parameter p is taken equal to 3. By this way, the intermediate densities are penalized by the exponent p which results in a solid-and-void topolog which is very desirable in structural topology optimization.

Let Ω be the 2D region representing part of the human femur, enclosed by boundary $\partial\Omega$. The boundary is divided into Γ_1, where the femur is fixed (u = v = 0), and Γ_2, where loads P1 = 2317 N and P2 = 703 N are applied at specific angles (Fig. 14.6a). These loads, representing the stance phase of gait for an 80 kg person, are uniformly distributed along their application lines. The shape of the human femur is described by B-splines, and it is then meshed with the open-source FE software Freefem++. We assume that the trabecular bone tissue obeys isotropic linear elasticity law with Young's modulus of 15 GPa representative of the bone tissue, and Poisson's ratio of 0.3. This approach requires the optimized structure to consist of both solid material regions and void regions. The optimization problem is formulated as a material distribution issue, where material is either present ($\theta = 1$) or absent ($\theta = 0$), with intermediate values being penalized during the optimization process. In practice, the design variables represent the relative material densities, with values ranging from 0 to 1. A lower bound of $\theta_{\min}$, i.e., 0.001 is usually imposed to avoid stiffness matrix singularities when solving for equilibrium across the entire domain. The results of the optimal bone design are shown in Fig. 14.6b, in which zones of $\theta = 1$ or close to unity indicate where bone material shall be present to make the design as stiff as possible to resist internal stresses; conversely, bone

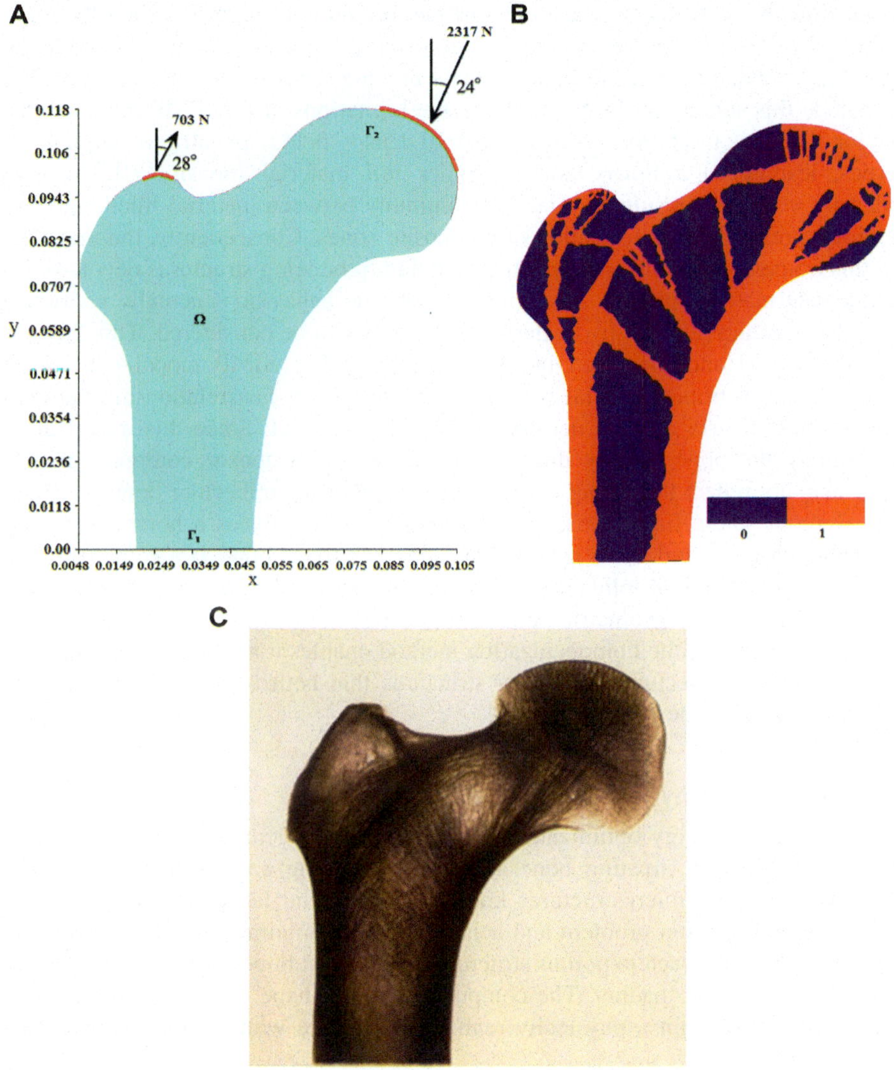

FIGURE 14.6

(a) Initial geometry and load, (b) optimum solution for the internal architecture of the femoral head using SIMP, (c) Roentgenogram of the femoral internal structure.

density could be removed within zones with low values of θ (close to zero). From a physiological standpoint, this design is consistent with the bone remodeling process that constantly occurs at the trabecular level.

In order to validate the optimal internal structures found with the developed SIMP approach, a comparison to the real internal structures in Fig. 14.6c shows that topology optimization has recreated the real macroscopic features of the internal bone structure. The results make obvious the formation of high bone density zones with unity highlighting the two cortical tissue layers along the diaphysis and in the femoral head and neck, connected by dense arching struts within the epiphysis. We conclude this subsection by highlighting the limitations of the SIMP method. One major drawback is its reliance on an artificial density field to penalize material properties, leading to a binary solution-either full material (density = 1) or void (density = 0). This results in a lack of continuity between material-filled and void regions, preventing the formation of composite zones. Consequently, the materials in this framework are inherently self-penalizing, producing structures devoid of intermediate material phases. Assuming an isotropic behavior is actually restrictive and the structural anisotropy of the bone tissue has to be considered. This fact has been observed and measured in several models. The SIMP model consists in choosing an isotropic material with an empirical power law relationship between the stiffness matrix and the density of the bone material. Since this model gives essentially the bone density distribution, the elasticity tensor computed in this way may not reflect the strong existing anisotropy of the trabecular bone structure. To overcome these limitations, the next subsection introduces a homogenization-based topology optimization model. This approach treats bone as a composite material composed of both solid and void phases, accounting for its inherent anisotropy. By incorporating the geometric parameters of microstructures and their orientation as design variables, the homogenization method enables a more realistic optimization of bone architecture, producing structures that better reflect the mechanical behavior of real bone tissue.

3.2 Relaxation by homogenization approach

We present a topology optimization model using homogenization to predict bone density distribution, treating bone as a self-optimizing composite material. The method introduces microstructures known as sequential laminated composites to relax the optimization problem and achieve a global minimum of compliance. By extending this approach to porous structures, we predict bone's internal architecture under physiological loading. The computed optimal shape is then refined through penalization to obtain a physically realizable structure with distinct material and void regions.

3.2.1 Problem definition

This section provides a concise overview of topology optimization using homogenization, with further details available in (Allaire, 2002). The objective is to formulate a topology optimization problem that models bone adaptation to mechanical loads. Bone tissue is treated as a porous composite material within a fixed domain, characterized by spatially varying density and microstructure, and governed by an effective form of Hooke's law.

Let Ω be a bounded reference domain subjected to surface loads P on its boundary. The boundary of Ω is partitioned into three regions: Γ_u, Γ_t, and Γ_0. The displacement of the structure, denoted by u is constrained to be clamped on Γ_u, subjected to surface loads P on Γ_t, and remains free of any external force on Γ_0.The equilibrium equations governing the linear elasticity of the structure are then given by:

$$\begin{cases} -\mathrm{div}\sigma = \rho g \ \mathrm{in}\omega \\ \sigma.n = P \ \ \mathrm{onS}_t \\ u(\omega) = 0 \ \mathrm{onS}_u \\ \sigma.n = 0 \ \ \mathrm{on}\partial\omega(S_t \cup S_u) \end{cases}, \tag{14.8}$$

In Eq. (14.8), we define ω as a subset of the reference domain Ω, $\sigma = C : \varepsilon(u)$ as the Cauchy stress tensor, $\varepsilon(u) = \frac{(\nabla u + \nabla u^T)}{2}$ as the strain tensor, ρg as the body force vector, and n as the outward normal to the boundary. The elasticity tensor, denoted by $\boldsymbol{C}$, follows Hooke's law and is expressed for any symmetric matrix as:

$$C\xi = 2\mu\xi + \lambda(\mathrm{Tr}\xi)I, \tag{14.9}$$

where λ and μ represent the Lamé coefficients of the material. These coefficients are often related to the Young's modulus and Poisson's ratio through the expressions:

$$\lambda = \frac{E\nu}{(1+\nu)(1\text{-}2\nu)}; \mu = \frac{E}{2(1+\nu)}, \tag{14.10}$$

The structural flexibility is quantified by the compliance functional. To optimize the structure, we consider the compliance minimization problem, formulated as in Eq. (14.3):

$$\min_{\omega\subset\Omega} \int_{S_t} P.u(\omega)dx + l|\omega|, \tag{14.11}$$

The compliance minimization problem (14.11) may lack a conventional solution, as shown by Allaire (2002) and Allaire and Pantz (2006). Minimizing sequences often generate an increasing number of voids, preventing convergence to a classical shape and leading instead to a composite structure. To address this, the problem is relaxed by expanding the design space to include material-void mixtures, with homogenization theory used as a key tool in this process.

3.2.2 Homogenization relaxation method

The relaxed formulation of the optimization problem is derived using homogenization theory, which replaces the original 0–1 problem with a composite material formulation. Instead of directly determining the domain ω, we define composite domains characterized by two key functions:

- Material density $\theta(x) : \Omega \rightarrow (0, 1)$: A continuous variable ranging from 0 (void) to 1 (solid material).
- Homogenized elastic tensor $C^{Hom}(x)$: Describes the effective mechanical properties of the underlying microstructure.

We model bone as a composite material occupying a fixed domain Ω under external forces P applied at the boundary. The governing equations are given by:

$$\begin{cases} -\mathrm{div}\sigma = \rho g \ \mathrm{in}\Omega \\ \sigma = \mathrm{C}^{Hom} : \varepsilon(u)\mathrm{in}\Omega\,, \\ \sigma.n = P\mathrm{on}\partial\Omega \end{cases} \tag{14.12}$$

where $\mathbf{u}$ and $\boldsymbol{\sigma}$ represent the homogenized displacement and stress fields, respectively. The objective is to minimize compliance while maintaining a fixed volume constraint. This results in the following nested minimization problem:

$$minJ\left(\theta, \mathrm{C}^{Hom}\right) = S\left(\theta, \mathrm{C}^{Hom}\right) + l\int_{\Omega}\theta dx, \tag{14.13}$$

where minimization occurs over all pairs $\left(\theta, C^{Hom}\right)$, ensuring that C^{Hom} corresponds to the homogenized Hooke's law for a given material density. The compliance function can be expressed in terms of the unknown stress field as:

$$S\left(\theta, C^{Hom}\right) = \int_{\Omega} C^{Hom^{-1}}\sigma.\sigma dx, \tag{14.14}$$

The minimum is achieved by using specialized composites known as sequential laminates, formed through successive layering of void and material in $\boldsymbol{N}$ orthogonal directions with appropriate proportions. In 2D, optimality is reached in Eq. (14.14) for $\mathbf{N = 2}$, or by using a rank-2 sequential laminate with a homogenized elastic tensor, where the lamination directions are aligned with the eigenvectors of the stress tensor σ. The optimal lamination fractions are given by

$$m_1 = \frac{|\sigma_2|}{|\sigma_1| + |\sigma_2|}; m_2 = \frac{|\sigma_1|}{|\sigma_1| + |\sigma_2|}, \tag{14.15}$$

where σ_1 and σ_2 are the eigenvalues of the stress tensor σ:

$$\sigma_1, \sigma_2 = \frac{\sigma_{11} + \sigma_{22}}{2} \pm \sqrt{\left(\frac{\sigma_{11}\text{-}\sigma_{22}}{2}\right)^2 + \sigma_{12}{}^2}, \tag{14.16}$$

The homogenized Hooke's law for the optimal composite is expressed as

$$C^{Hom^{-1}} = C^{-1} + \frac{1\text{-}\theta}{\theta}\left(m_1 f_C^S(e_1) + m_2 f_C^S(e_2)\right)^{-1}, \tag{14.17}$$

where e_1 and e_2 the eigenvectors of σ and $f_C^S(e_i)$ is the fourth order tensor defined by its quadratic form:

$$f_C^S(e_i)\xi : \xi = C\xi : \xi\text{-}\mu^{-1}|C\xi|^2 + \frac{\mu + \lambda}{\mu(2\mu + \lambda)}((C\xi)e_i.e_i)^2, \tag{14.18}$$

Finally, the optimal density θ_{opt} is determined as

$$\theta_{opt} = min\left(\sqrt{\frac{k + \mu}{4\mu\kappa l}}(|\sigma_1| + |\sigma_2|), 1\right), \tag{14.19}$$

where $k = \lambda + \mu$ is the bulk modulus.

We present here an iterative algorithm for applying the homogenization method to the optimal design problem. The process follows an alternate minimization approach over the stress field σ and design parameters θ, C^{Hom}, ensuring convergence toward an optimized structure. The steps are as follows: **Initialization:** Define the initial material distribution by setting the design parameters θ_0, C_0^{Hom} **Iterative optimization:** Repeat until convergence: i. Solve the linear elasticity system to compute the stress tensor σ for the given design parameters. ii. Update the design parameters using the optimality conditions (14.17) and (14.19) to improve material distribution.

The homogenization method naturally produces structures with intermediate material densities, representing a relaxed formulation of the problem. However, practical applications often require a discrete distribution (solid or void). To achieve this, a post-processing step penalizes intermediate densities by modifying the update scheme, progressively steering density toward values of either 0 or 1. This ensures that the final optimized structure is physically realizable.

To evaluate the effectiveness of the proposed model, we apply it to the same loading conditions previously described for the proximal femur. The results illustrate the optimal bone density distribution under static loading during the single stance phase (Fig. 14.7). Distinct bone density regions emerge: high-density areas primarily along the diaphysis, likely representing cortical bone; intermediate-density regions in the epiphysis, corresponding to trabecular bone; and low-density areas, shown in blue, mainly within the medullary canal and at the extremities of the femoral head and greater trochanter. The predicted density patterns align well with the observed distribution of cortical and trabecular bone in the real proximal femur (see Fig. 14.6c).

The homogenization method produces an optimal material layout, which often includes large regions of composite materials with intermediate density. While this is mathematically effective, it is typically undesirable in practical applications where the goal is to obtain a real, physical shape with only two distinct densities: material or void. This is achieved through a post-processing step that penalizes intermediate density regions. Numerically, the optimal bone density is first calculated (Fig. 14.7a) and then projected onto a real shape using a penalization procedure that eliminates areas with intermediate densities (Fig. 14.7b). The penalization results in a classic optimal structure with clear regions of bone material and void, as shown in Fig. 14.7b.

The resulting high-density regions (shown in red) are distributed in two cortical layers along the diaphysis and in the femoral head and neck, connected by dense arches within the epiphysis. Low-density regions (in blue) are mainly located inside the medullary canal and in the epiphysis, including the femoral head and trochanter. This density distribution closely resembles that obtained using the SIMP method (Fig. 14.6b), though there are some differences, particularly in the femoral head, which appears less voided in the SIMP result.

3.3 IMD method

The isotropic material design (IMD) method is a recently developed approach, introduced by Czarnecki (2015) and further refined by Czarnecki and Wawruch (2015).

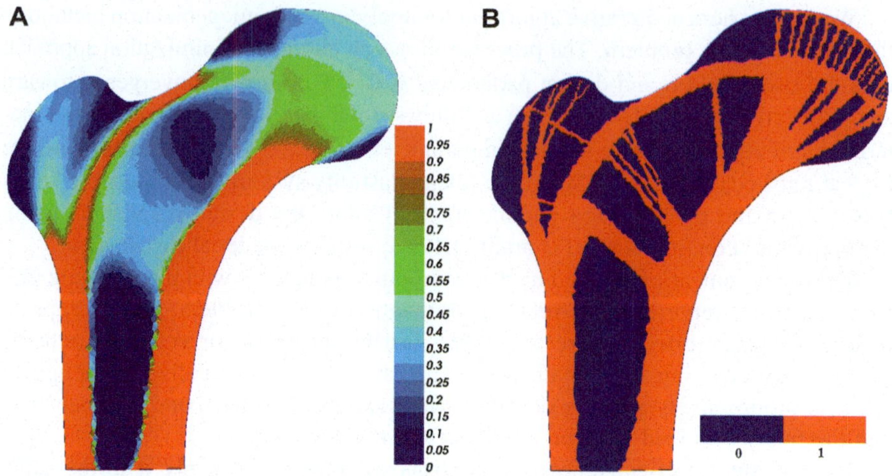

FIGURE 14.7

(a) Optimized femoral bone density distribution derived from the homogenization approach, illustrating regions of varying material density, (b) final internal bone architecture after penalization.

This method is as an isotropic counterpart to free material design (FMD), initially proposed by Bendsøe et al. (1994) and later extended across various optimization domains (see Czarnecki & Lewiński, 2014).

In FMD, the design variables are the spatially varying components of the Hooke tensor, defined over a given domain. The optimization objective is compliance minimization under a given loading condition. Compliance is chosen due to its unique self-adjoint property, which eliminates the need to solve an adjoint problem. To regulate the material cost, an isoperimetric constraint is introduced, expressed as the integral of the Hooke tensor's trace. This constraint, set as a fixed or upper-bounded constant V, ensures a rational optimization process.

Bendsøe et al. (1994) demonstrated that the optimal Hooke tensor is singular, meaning only one eigenvalue (Kelvin modulus) remains positive, while the remaining eigenvalues (two in 2D and five in 3D) vanish. This result implies that the optimal material is perfectly tailored to the given load, with stress and strain trajectories aligning along the directions where elastic moduli remain positive. Despite this degeneracy, the equilibrium problem remains solvable, as the governing variational equation remains compatible (Czarnecki & Lewiński, 2014).

To overcome the singularity issue in FMD, the IMD method imposes isotropy from the outset (Czarnecki, 2015; Czarnecki & Wawruch, 2015). In this approach, theHooke tensor is determined solely by two material parameters: bulk modulus k and shear modulus μ. The isoperimetric constraint remains applicable, with the trace of the Hooke tensor taking values $3k + 10\mu$ in 3D and $2k + 4\mu$ in 2D. The IMD method optimizes both moduli, leading to the emergence of subdomains where one modulus vanishes. This effect pushes Poisson's ratio toward either its minimum value (−1) or maximum limit (1/2 in 3D, 1 in 2D).

3.3.1 Principles of the IMD method

The IMD method is a specialized form of topology optimization, derived from the FMD method. It focuses on determining the optimal isotropic distribution of material properties, specifically the bulk modulus k and shear modulus μ, within a structure to achieve minimal compliance.

In both FMD and IMD, the minimal compliance S of an optimized structure is given by:

$$S=f(u)=Z^2/V \quad Z=min\left\{\int \|\tau\| \middle| \tau \text{being statically admissible}\right\}, \tag{14.20}$$

where $\mathbf{u}$ represents the displacement field of the optimal structure, $\tau=\left(\tau_{ij}\right)$ virtual stress fields, and the integral norm depends on the assumed material symmetry. The optimization problem (14.20) shares strong theoretical connections with the Monge-Kantorovich mass transport theory and is equivalent to formulations in shape optimization with total mass constraints (Bouchitté & Buttazzo, 2001).

A key feature of this method is that it simultaneously solves two coupled problems:

1 .Determining the optimal shape of the structure (by identifying the support of the minimizing measure).
2 .Optimizing the distribution of elastic moduli within that shape.

For the case of a single load variant, the original FMD formulation imposes the isoperimetric condition:

$$\int_{\Omega} \mathrm{tr}C dx = V, \tag{14.21}$$

where C is the Hooke tensor. In FMD, the norm in Eq. (14.20) is typically chosen as the Euclidean norm, defined by: ; $\|\tau\| = \left(\tau_{ij}\tau_{ij}\right)^{\frac{1}{2}}$.

Czarnecki (2015) and Czarnecki and Wawruch (2015) proposed an isotropic adaptation of FMD, known as IMD, where the material distribution is controlled through the spatial variation of bulk modulus$k$ and shear modulus μ. In 2D, the Hooke tensor takes the form:

$$C=2kJ+2\mu\Lambda, \tag{14.22}$$

$$J=\frac{1}{2}\overset{2}{I}\otimes\overset{2}{I}, \Lambda=\overset{4}{I}\text{-}J, \overset{2}{I}=\left(\delta_{ij}\right), \overset{4}{I}=\left(\frac{1}{2}\left(\delta_{ik}\delta_{jl}+\delta_{il}\delta_{kj}\right)\right), \tag{14.23}$$

where the eigenvalues of $\mathbf{C}$ are ($2k$, μ, μ), and its trace is given by:

$$\mathrm{tr}\,\mathbf{C}=2k+4\mu$$

The optimization problem (14.20) in IMD involves an integrand with linear growth in stress, expressed as:

$$||\tau|| = \alpha|\mathrm{tr}\tau| + \beta|dev\tau|, dev\tau = \tau\text{-}(1/2)\mathrm{tr}\tau, \alpha = \sqrt{2}/2, \beta = \sqrt{2}, \tag{14.24}$$

Assuming that $\tau*$ is the minimizer of Eq. (14.20), the optimal isotropic moduli can be obtained using the following expressions:

$$k^*(x) = \frac{\sqrt{2}}{4} V \frac{|tr\tau^*(x)|}{\int_\Omega ||\tau^*||dx}, \mu^* = \frac{1}{2} V \frac{|dev\tau^*(x)|}{\int_\Omega ||\tau^*||dx}, \tag{14.25}$$

Eq. (14.20) contains an integrand with linear growth, which leads to the emergence of subdomains where the minimizer vanishes. The effective domain of the minimizer defines the optimal structure, within which the optimal elastic moduli are determined using expressions Eq. (14.25). Once these moduli are obtained, the optimal Poisson ratio can also be computed. A notable feature of the solution is that, in some regions, the Poisson ratio approaches -1, while in others, it reaches 1.

To solve problem (14.20) with the Euclidean norm, Czarnecki and Lewiński (2012) developed a numerical method, later adapted for IMD by Czarnecki (2015). The method involves discretizing the set of admissible stress fields using finite element techniques with local shape functions defined over a given mesh. The equilibrium equations are then solved as a rectangular system using singular value decomposition (SVD), reducing the minimization problem to optimizing a set of free parameters. Since the optimization is unconstrained, standard numerical solvers can be employed.

It is important to emphasize that this numerical approach does not fall within the finite element method (FEM) framework, as it does not rely on constructing a global stiffness matrix. Traditional FEM assumes the existence of constitutive equations linking stresses and strains, which is not the case here due to the linear growth property of Eq. (14.24). The following example illustrates this method, where the optimal elastic moduli fields are computed using expressions (14.25).

3.3.2 Optimal material property distributions via IMD

This section presents the results of the optimal distribution of isotropic material properties within the femoral bone domain, as determined by the IMD method described previously. The analysis considers the loading conditions acting on the proximal femur, as outlined in Section 3.1 The results are visualized through plots depicting the spatial variations of the following material properties: Young's modulus (E), Bulk modulus (K), Shear modulus (G), Poisson's ratio (ν). These distributions are shown in Fig. 14.8.

A key characteristic of the IMD solution is the emergence of distinct bone subdomains where the optimal Poisson's ratio reaches its theoretical lower and upper bounds. In the two-dimensional case considered here, these bounds are -1 and 1, as shown in Fig. 14.8. The blue-colored regions correspond to areas where the

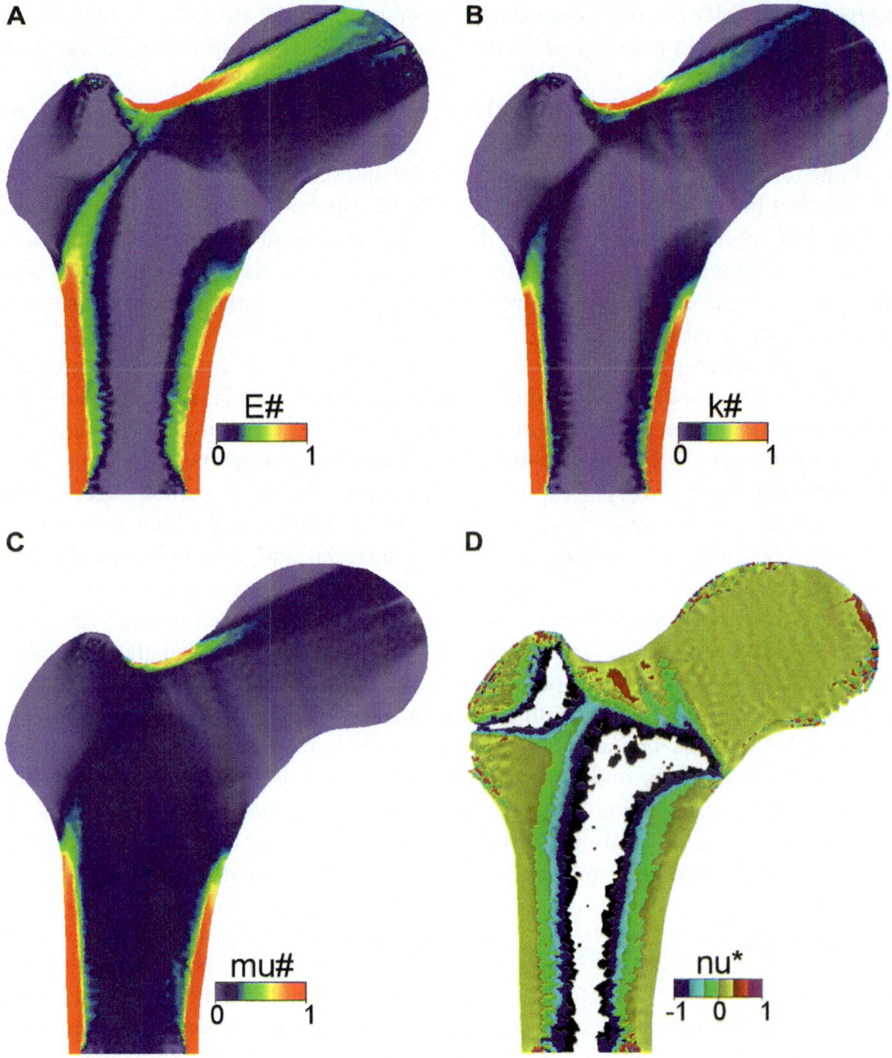

FIGURE 14.8

(a)–(c) Optimal normalized distributions of Young's modulus (E), bulk modulus (K), shear modulus (G), (d) Optimal Poisson's ratio in regions where the Young's modulus remains nonzero (i.e., all non-white regions).

Poisson ratio attains its lower bound (−1), whereas the purple-colored regions indicate areas where it reaches its upper bound (14.1). Furthermore, the IMD approach highlights the presence of voids or material-free regions within the optimized structure Fig. 14.8d. These voids correspond to areas where both the Young's modulus and bulk modulus vanish, as depicted in Fig. 14.8a–c. The presence of such regions suggests that IMD not only determines the optimal spatial distribution of material properties but also implicitly defines the structural shape by eliminating unnecessary material.

The isotropic version of the FMD approach in topology optimization differs fundamentally from methods like relaxation by homogenization. While homogenization-based approaches predict the internal density distribution within the femoral bone, IMD focuses on optimizing the spatial layout of the full stiffness tensor. Instead of defining a material density field that evolves during optimization, IMD determines the optimal distributions of isotropic elastic moduli (bulk and shear) resulting in a structure that efficiently supports the applied loads while maintaining isotropic properties.

This distinction is significant in the context of bone modeling. The homogenization method, by allowing intermediate material densities, provides a relaxed solution that mimics the trabecular and cortical regions of bone, reflecting biological adaptation. In contrast, IMD seeks an optimal arrangement of material stiffness without directly modeling porosity. Despite this difference, the IMD method successfully identifies the optimal shape and internal material distribution within the femoral bone by selectively eliminating unnecessary material and determining where the stiffness should be maximized. The results obtained using IMD highlight the emergence of distinct structural features, including regions where the Poisson's ratio reaches its extreme values of -1 and 1. These patterns suggest that the method is capable of capturing anisotropic-like behavior within an isotropic formulation, potentially providing insights into how bone material is optimally arranged under mechanical loads. A key advantage of IMD over homogenization-based methods is its ability to directly optimize the full stiffness tensor while avoiding the need for subsequent penalization steps to obtain a clear 0–1 material distribution. This makes IMD particularly useful for applications where explicit material property optimization is more relevant than simply identifying a density field.

3.4 Comparative analysis of the optimization approaches

The three topology optimization methods presented, namely SIMP, homogenization-based, and IMD, provide distinct perspectives on optimizing bone microstructure. While they share the common goal of predicting optimal bone architecture under mechanical loading, their underlying mathematical frameworks and the nature of their solutions differ significantly.

3.4.1 Comparison of optimization approaches

1 SIMP Method

- Uses a density-based penalization approach to enforce a binary material distribution (solid-void).
- Provides a clear, manufacturable structure but does not directly account for bone's anisotropy.

- The optimization result aligns well with observed trabecular and cortical bone patterns but requires post-processing to remove intermediate densities.

2 Homogenization-based method

- Models bone as a composite material with variable microstructures, capturing anisotropic mechanical behavior.
- Predicts a relaxed material distribution, allowing intermediate densities that reflect real trabecular bone heterogeneity.
- Requires an additional penalization step to convert the relaxed solution into a binary material layout.

3 IMD Method

- Focuses on optimizing the full stiffness tensor while maintaining isotropy, unlike the density-based SIMP and homogenization approaches.
- Does not explicitly model material density but instead optimizes bulk and shear moduli distributions.
- Naturally eliminates unnecessary material while determining regions where stiffness should be maximized, making it advantageous for cases where explicit material property optimization is crucial.

3.4.2 Key insights and practical considerations

- Biological realism: The homogenization-based approach most closely mimics the real trabecular-cortical transition observed in bone, as it allows intermediate material densities that reflect biological adaptation. The SIMP method, while effective in generating a manufacturable structure, does not inherently account for anisotropic behavior. The IMD method, despite assuming isotropy, still captures essential structural features by optimizing stiffness distribution.
- Computational complexity: SIMP is the simplest in terms of implementation and computational efficiency. The homogenization approach, due to its incorporation of microstructural details, requires more complex modeling. The IMD method, while computationally intensive, avoids the need for penalization, which simplifies post-processing.
- **Suitability for bone structure optimization**: If the goal is to reproduce realistic trabecular and cortical bone patterns, the homogenization method offers the best balance between structural accuracy and mechanical consistency. However, if manufacturability is a priority, SIMP provides a practical framework for obtaining clear solid-void structures. The IMD method, by optimizing the full stiffness tensor, provides a way to design materials with optimal mechanical properties without explicitly defining a density distribution.

4. Trabecular bone plasticity and failure analysis

Trabecular bone is a highly porous and anisotropic material that plays a fundamental role in load distribution and structural support within the vertebral column. Unlike cortical bone, which is denser and more homogeneous, trabecular bone consists of an intricate network of interconnected struts and plates, making its mechanical behavior highly dependent on microstructural features. Understanding the plastic yielding and failure behavior of trabecular bone under complex loading conditions, particularly twisting and bending, is essential for predicting fracture risk and improving orthopedic treatments.

Traditional studies on bone mechanics have primarily focused on uniaxial compression and tension, often neglecting the effects of multi-axial loading scenarios. However, in real-life activities such as bending, twisting, and lateral flexion of the spine, trabecular bone is subjected to combined stresses that influence its failure mechanisms. This section aims to construct effective plastic yield surfaces of vertebral trabecular bone under twisting and bending moments using a 3D microstructural model, ultimately identifying the critical conditions leading to failure.

4.1 3D microstructural model for predicting plastic yield and failure

To accurately capture the plastic yielding behavior of vertebral trabecular bone under torsion and bending, a 3D microstructural model is employed. This model represents trabecular bone as a periodic lattice of interconnected beam-like trabecule, mimicking its complex cellular architecture. The micropolar continuum approach is adopted to account for both translational and rotational degrees of freedom, ensuring a more realistic mechanical representation. The trabecular lattice is idealized as a double-prismatic hexagonal unit cell, with each trabecular element represented as a beam-like structure. The individual trabecule are modeled as Timoshenko beams, which account for: axial deformations (tension and compression), shear effects, flexural bending, and torsional twisting.

This beam-based representation allows for a more accurate description of the bone's response to bending and twisting loads, particularly in regions where trabecular elements experience significant shear and rotational effects. Unlike conventional methods that rely solely on continuum mechanics, this lattice-based approach retains the essential anisotropic and heterogeneous nature of trabecular bone.

Each node in the trabecular lattice possesses 6° of freedom, three translational and three rotational. This enhanced degree of freedom enables the model to:

- Capture micro-rotations and their effects on local stress distributions.
- Consider couple-stress interactions, which play a significant role in torsional and flexural deformations.
- Ensure both force equilibrium and moment equilibrium across the lattice structure.

By incorporating these rotational degrees of freedom, the model aligns with the micropolar continuum theory, where an infinitesimal material element can undergo independent microrotations without a corresponding macro-displacement. This is particularly important for trabecular bone, where the microstructural geometry leads to complex stress redistributions under torsion and bending loads.

To bridge the gap between the discrete microstructural lattice and a macroscopic continuum representation, an asymptotic homogenization approach is employed. This process involves:

- Converting the discrete nodal equilibrium equations into continuum-level couple-stress formulations.
- Expressing the effective material properties of trabecular bone as a function of its microstructural topology.
- Establishing the relationship between microscopic strain measures (such as bending curvature and torsional curvature) and macroscopic stress measures (such as force-stress and couple-stress tensors).

Through this homogenization process, the trabecular lattice is mapped onto an equivalent micropolar couple-stress medium, which can be analyzed using continuum mechanics. The couple-stress tensor is decomposed into its torsional and bending components, allowing for a clear distinction between pure twisting loads and flexural bending loads at the macroscopic level.

Ultimately, this 3D microstructural model forms the foundation for constructing the plastic yield surfaces of trabecular bone under twisting and bending moments. By incorporating the complex interplay between trabecular microarchitecture, material anisotropy, and rotational degrees of freedom, this framework enhances our ability to predict bone failure mechanisms and guide the design of biomechanically optimized implants and orthopedic treatments.

To determine the plastic yield surfaces of trabecular bone microstructures under combined bending and torsion moments, a computational framework based on the micropolar continuum theory is employed. This method derives effective couple-stress responses from a homogenized 3D beam-lattice representation of trabecular bone. The step-by-step computational algorithm, including coordinate transformations, microrotation formulations, moment vector construction, and failure criteria, is detailed in Appendix B. This framework enables the generation of plastic yield surfaces by mapping microscale behavior to macroscopic failure envelopes.

4.2 Significance and applications

The macroscopic collapse surfaces in couple stress space are formed by intersecting failure surfaces, each corresponding to a specific collapse mode. Here, we detail the calculations of these collapse surfaces for nine essential combinations of macroscopic couple stress loadings. The plastic yield surfaces represent the inner envelopes of these intersecting surfaces, defining the critical combinations of non-symmetrical micropolar couple stresses that induce failure via plastic collapse.

At the microscopic scale, trabecular bone consists of interconnected struts forming a three-dimensional porous structure. In the vertebral body, the trabecule are arranged in a relatively regular pattern (Fig. 14.9a), allowing idealization as a cellular solid. Accordingly, we model the rod-like architecture of vertebral bone as a three-dimensional lattice with the unit cell geometry depicted in (Fig. 14.9b). The entire structure is generated by repeating the reference unit cell (Fig. 14.9c) along three periodicity vectors defined in a Cartesian basis.

The architecture of the model, featuring thick vertical columns and thinner horizontal struts, is inspired by studies of samples taken from the central regions of vertebral bodies from healthy individuals aged 30–90 years. Four morphological indices characterize the unit cell: the horizontal trabecular thickness (D_h), vertical trabecular thickness (D_v), and the mean spacing between horizontal trabecule (L_h) and vertical trabecule (L_v). Structural variations of human trabecular bone with age, particularly changes in trabecular thickness and spacing, are well-documented by Mosekilde (1989), as summarized in the following relationships:

Horizontal trabecule thickness $(\mu m) : D_h = -1.03 \times \text{Age} + 189$, vertical trabecule thickness $(\mu m) : D_v = 0.14 \times \text{Age} + 208$, distance between horizontal trabecule (μm): $L_h = 13.74 \times \text{Age} + 288$, distance between vertical trabecule (μm) : $L_v = 6.74 \times \text{Age} + 456$.

Building upon the previous framework, we define plastic yielding for vertebral trabecular bone under multiaxial couple stresses. The micro-architectural parameters (L_v, L_h, D_h, D_y) are selected for a 40-year-old individual, with an oblique trabecular inclination angle $\theta = 60$ degrees yielding a relative density of ~12%. The plastic yield surface, a closed surface in couple stress space (m_{ij}, where i,j = x, y, z), is plotted for various subspaces (Fig. 14.10). These surfaces are normalized by the trabecular shear yield strength (τ_Y). Key observations include:

- The yield domains are closed, anisotropic, and consist of convex curved surfaces with planar facets.
- All collapse envelopes are symmetric under positive/negative torsion and bending.
- Positive torsion indicates clockwise rotation; negative torsion, counterclockwise.
- Positive bending induces concave-upward deflection; negative bending, concave-downward.

The plastic collapse strength of vertebral trabecular bone exhibits strong dependence on both age and relative density (ρ^*/ρ_s). To quantify these relationships, we analyze the normalized torsional and bending yield strengths (scaled by the trabecular shear strength τ_Y) across different ages and densities. Fig. 14.11a and b present these dependencies under nine distinct loading conditions: three pure twisting modes (Fig. 14.11a) and six bending modes (Fig. 14.11b), combining in-plane and out-of-plane cases). The results demonstrate two key trends: (i) yield strengths decrease progressively with age, and (ii) yield strengths increase monotonically with relative density. These dual dependencies reflect the microstructural degradation (trabecular thinning and increased spacing) and density variations that occur with aging.

The plastic yield surfaces derived from the couple-stress framework provide a comprehensive understanding of trabecular bone failure under combined torsional

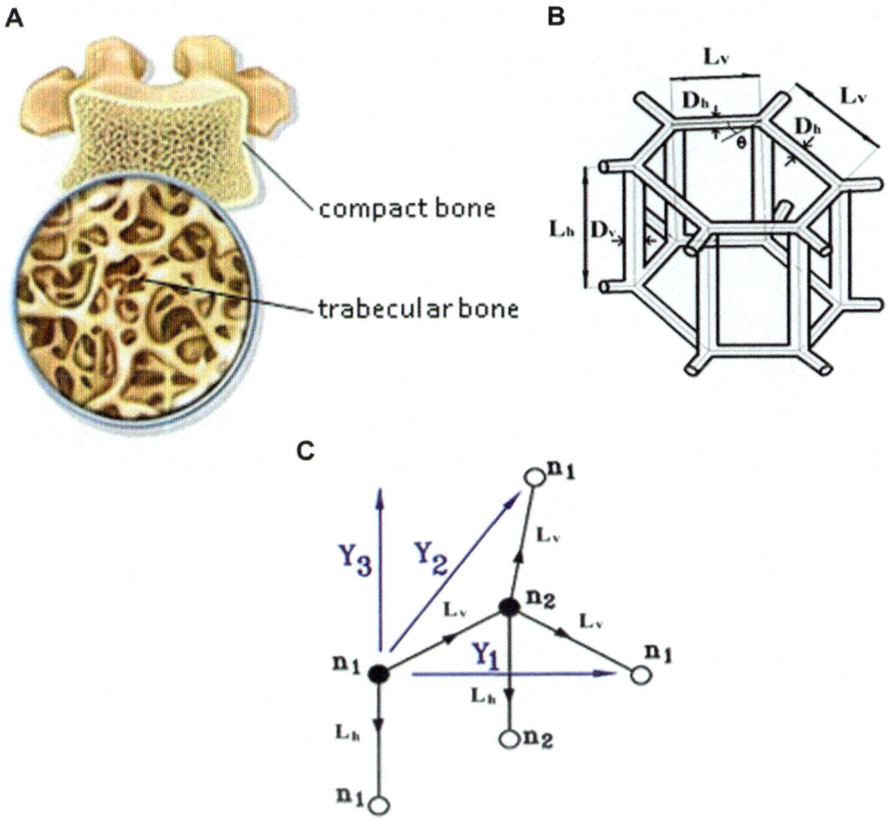

FIGURE 14.9

Microstructural geometry and architectural parameters of vertebral trabecular bone. (a) Cross-sectional view of human vertebral bone showing both trabecular and cortical bone regions, (b) 3D hexagonal unit cell representation featuring thick vertical trabecule and thinner horizontal struts, modeling human vertebral bone microstructure, (c) reference unit cell with periodicity vectors Y1, Y2, Y3.

and bending loads. These yield envelopes, shaped by microstructural geometry and material anisotropy, illustrate the critical stress conditions leading to plastic collapse. Notably, aging-induced trabecular thinning and increased porosity significantly reduce the bone's torsional and bending strengths, reinforcing the importance of density-dependent failure analysis.

5. Conclusions

This chapter bridges the microscale architecture and macroscale mechanical behavior of trabecular bone through an integrated computational framework

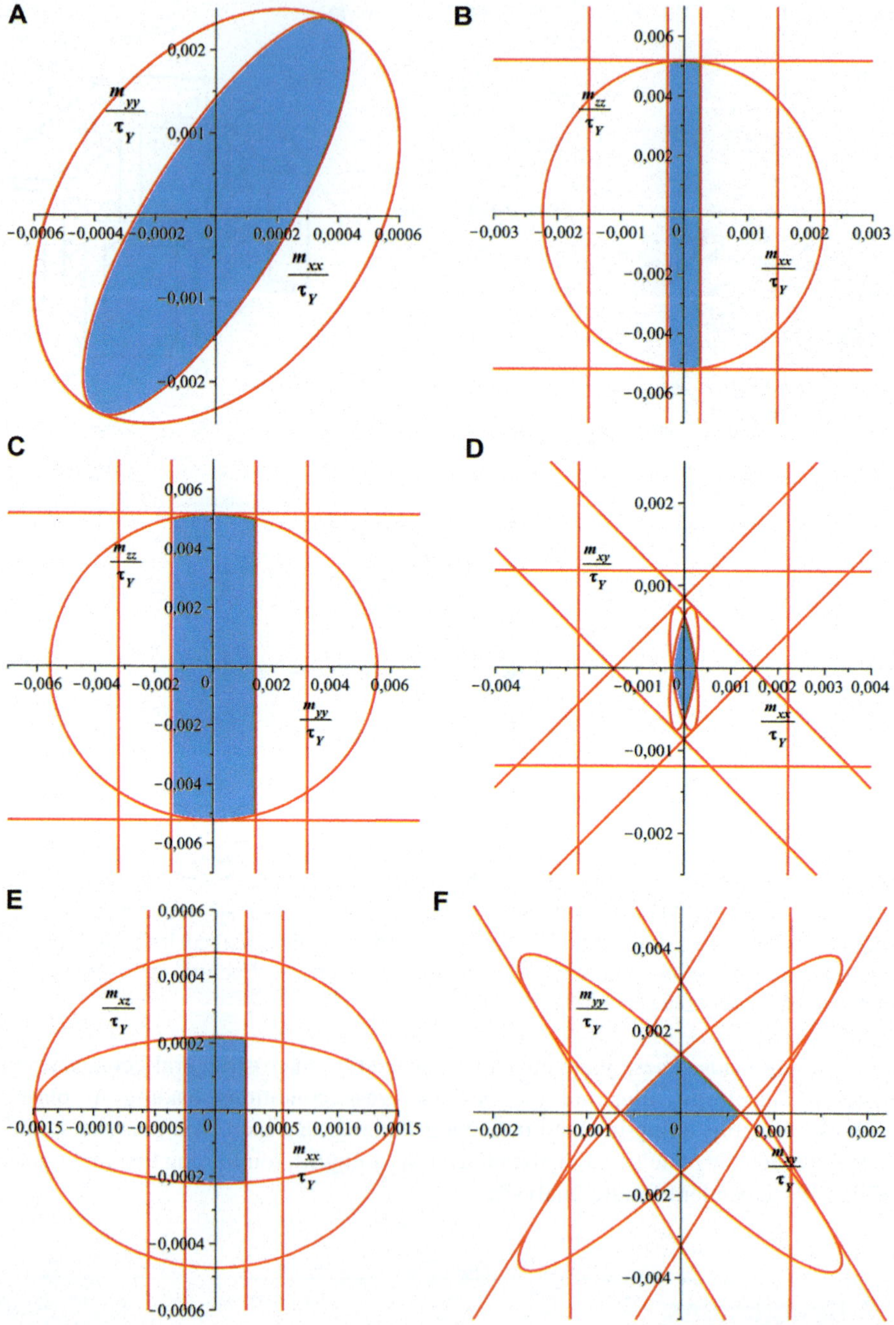
A
0,002
$\frac{m_{yy}}{\tau_Y}$
0,001
-0,0006 -0,0004 -0,0002 0 0,0002 0,0004 0,0006
$\frac{m_{xx}}{\tau_Y}$
-0,001
-0,002
B
0,006
$\frac{m_{zz}}{\tau_Y}$
0,004
0,002
-0,003 -0,002 -0,001 0 0,001 0,002 0,003
$\frac{m_{xx}}{\tau_Y}$
-0,002
-0,004
-0,006
C
0,006
$\frac{m_{zz}}{\tau_Y}$
0,004
0,002
-0,006 -0,004 -0,002 0 0,002 0,004 0,006
$\frac{m_{yy}}{\tau_Y}$
-0,002
-0,004
-0,006
D
0,002
$\frac{m_{xy}}{\tau_Y}$
0,001
-0,004 -0,001 0 0,001 0,002 0,003 0,004
$\frac{m_{xx}}{\tau_Y}$
-0,001
-0,002
E
0,0006
0,0004
$\frac{m_{xz}}{\tau_Y}$
0,0002
-0,0015 -0,0010 -0,0005 0 0,0005 0,0010 0,0015
$\frac{m_{xx}}{\tau_Y}$
-0,0002
-0,0004
-0,0006
F
0,004
$\frac{m_{yy}}{\tau_Y}$
0,002
-0,002 -0,001 0 0,001 0,002
$\frac{m_{xy}}{\tau_Y}$
-0,002
-0,004

FIGURE 14.10

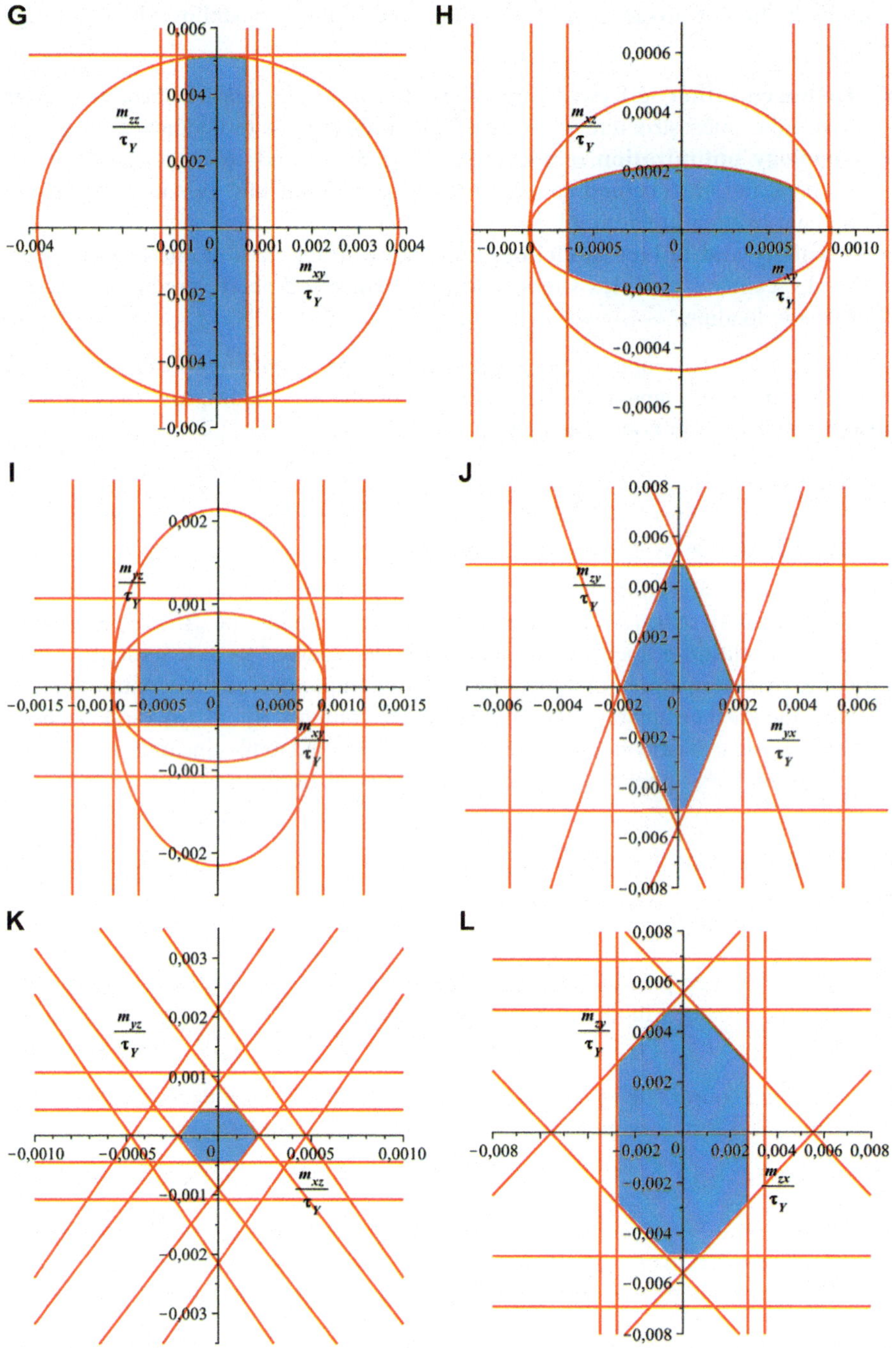

FIGURE 14.10 cont'd.

combining homogenization, optimization, and failure modeling. Key findings include:

- **Homogenization** effectively predicts macroscopic elastic behavior from trabecular microstructure, offering efficient models for biomedical design.
- **Topology optimization**, particularly using SIMP and IMD, reveals the mechanical rationale behind natural architecture of bone and aids in the development of biomimetic implants.
- **Plasticity and failure modeling**, using microstructural and micropolar frameworks, captures complex stress states and enhances fracture prediction under realistic loading.

Collectively, these approaches support the design of patient-specific orthopedic solutions, the development of bioinspired materials, and the improvement of diagnostic tools based on bone mechanics.

6. Appendix A

Homogenization algorithm for effective elastic properties of trabecular bone.

This appendix outlines the algorithm used to determine the effective (homogenized) linear elastic properties of trabecular bone modeled as a periodic 3D lattice structure. Building upon the theoretical foundation of Cauchy continuum mechanics, the method enables the translation of complex microstructural behavior into macroscopic elastic moduli. The analysis is confined to the linear elastic regime, excluding nonlinear phenomena such as plasticity or fracture.

6.1 Initialization

Define the geometry and mechanical properties of the periodic unit cell and its beams:

- Geometry: node positions x_i, beam connections $b \in B_R$, and beam lengths L^b
- Material: Young's modulus E^b, shear modulus G^b, cross-sectional area A, moments of inertia I_y, I_z

Define the coordinate transformation matrix $R_b \in R^{3\times3}$ for each beam b, which aligns the local beam frame to the global Cartesian coordinate system

Plastic yield surfaces in normalized couple-stress space for vertebral trabecular bone (age: 40 years, relative density: $\rho^*/\rho_s = 0.12$). Plots show biaxial interaction of: (a) *mxx* – *myy*, (b) *mxx* – *mzz*, (c) *myy* – *mzz*, (d) *mxx* – *mxy*, (e) *mxx* – *mxz*, (f) *mxy* – *myy*, (g) *mxy* – *mzz*, (h) *mxy* – *mxz*, (i) *mxy* – *myz*, (j) *myx* – *mzy*, (k) *mxz* – *myz*, (l) *mzx* – *mzy*. All stress components are normalized by the bulk material's shear yield strength (τ_Y).

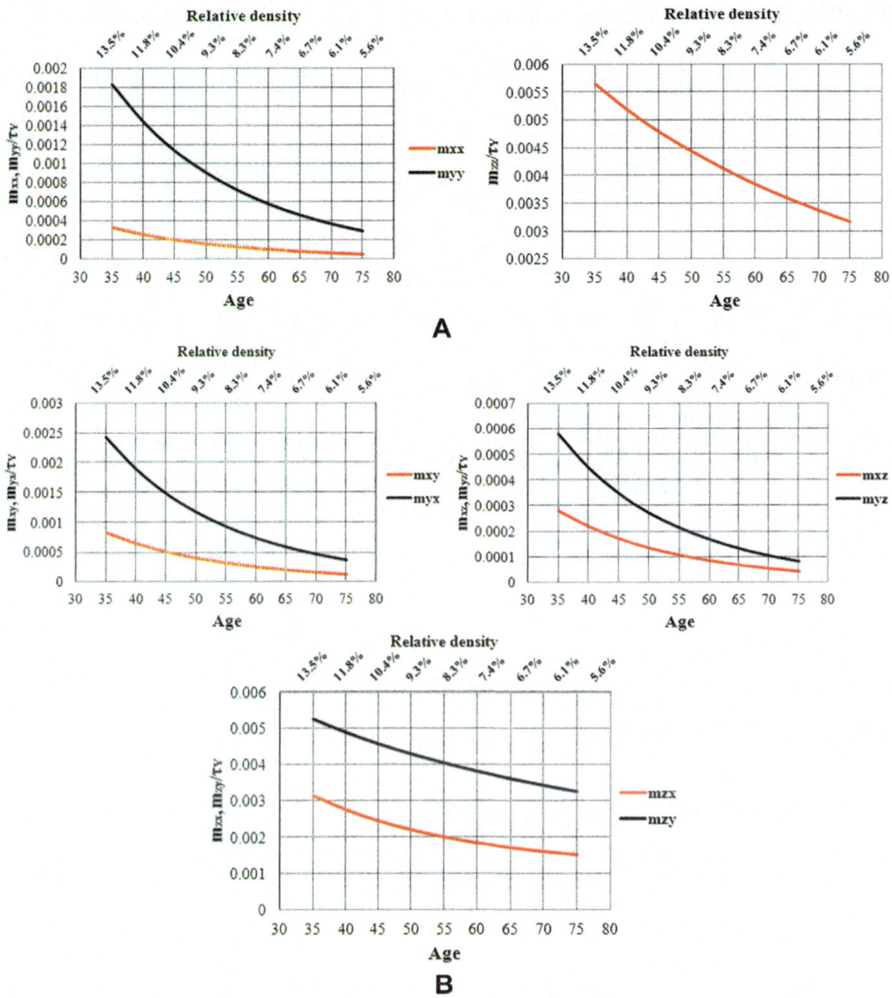

FIGURE 14.11

(a) Normalized torsional couple stresses versus bone age (years) and relative density (ρ^*/ρ_s), (b) bending couple stresses versus bone age (years) and relative density (ρ^*/ρ_s).

6.2 Beam-level kinematics and forces

For every beam b in the periodic unit cell:

(a) Identify its end nodes (origin and extremity) in the unit cell

(b) Assign its transverse shear correction factor $\Phi^b_{y,z} = 12E^b_s I^b_z \big/ G^b_s A^b k_s (L^b)^2$ to account for shear flexibility

(c) Define the relative displacement between the two ends of the beam using nodal displacements $\Delta U^b_1 = u^{E(b)}\text{-}u^{O(b)} + \frac{\partial u}{\partial \lambda^i}\delta^{ib}$

(d) Calculate forces and bending moments:

$$F_x^b = \frac{E_s^b A^b}{L^b}\left(e_x.\left(\Delta U_1^b\right)\right), F_y^b = \frac{12E_s^b I_z^b}{\left(L^b\right)^3\left(1+\Phi_y^b\right)}\left(e_y.\left(\Delta U_1^b\right) - \frac{L^b}{2}\left(e_z.\left(\phi^{O(b)}+\phi^{E(b)}\right)\right)\right),$$

$$F_z^b = \frac{12E_s^b I_y^b}{\left(L^b\right)^3\left(1+\Phi_z^b\right)}\left(e_z.\left(\Delta U_1^b\right) + \frac{L^b}{2}\left(e_y.\left(\phi^{O(b)}+\phi^{E(b)}\right)\right)\right),$$

$$M_y^{O(b)} = \frac{6E_s^b I_y^b}{\left(L^b\right)^2\left(1+\Phi_z^b\right)}\left(e_z.\left(\Delta U_1^b\right)\right) + \frac{E_s^b I_y^b}{L^b\left(1+\Phi_z^b\right)}\left(e_y.\left(\left(4+\Phi_z^b\right)\phi^{O(b)} + \left(2\text{-}\Phi_z^b\right)\phi^{E(b)}\right)\right),$$

$$M_y^{E(b)} = \frac{6E_s^b I_y^b}{\left(L^b\right)^2\left(1+\Phi_z^b\right)}\left(e_z.\left(\Delta U_1^b\right)\right) + \frac{E_s^b I_y^b}{L^b\left(1+\Phi_z^b\right)}\left(e_y.\left(\left(2\text{-}\Phi_z^b\right)\phi^{O(b)} + \left(4+\Phi_z^b\right)\phi^{E(b)}\right)\right),$$

$$M_z^{O(b)} = \frac{6E_s^b I_z^b}{\left(L^b\right)^2\left(1+\Phi_y^b\right)}\left(\text{-}e_y.\left(\Delta U_1^b\right)\right) + \frac{E_s^b I_z^b}{L^b\left(1+\Phi_y^b\right)}\left(e_z.\left(\left(4+\Phi_y^b\right)\right)\phi^{O(b)} + \left(2\text{-}\Phi_y^b\right)\right)\phi^{E(b)}\right)\right),$$

$$M_z^{E(b)} = \frac{6E_s^b I_z^b}{\left(L^b\right)^2\left(1+\Phi_y^b\right)}\left(\text{-}e_y.\left(\Delta U_1^b\right)\right) + \frac{E_s^b I_z^b}{L^b\left(1+\Phi_y^b\right)}\left(e_z.\left(\left(2\text{-}\Phi_y^b\right)\phi^{O(b)} + \left(4+\Phi_y^b\right)\phi^{E(b)}\right)\right)$$

with u^n and ϕ^n the displacements and rotations unknowns

6.3 Assembly of equilibrium equations

Construct the global equilibrium system by assembling the contributions of individual beams based on force and moment balance:

$$\sum_{b \in B_R} F^b.\left(v^{E(b)}\text{-}v^{O(b)}\right) = 0, \sum_{b \in B_R}\left(M^{O(b)}.w^{O(b)} + M^{E(b)}.w^{E(b)}\right) = 0$$

6.4 Solve for nodal displacements

Solve the system of equilibrium equations for the nodal displacement and rotation fields: u^n and ϕ^n

6.5 Compute macroscopic stress tensor

Sum over all beams in the unit cell to compute the homogenized stress tensor:

$$S^i = \sum_{b \in B_R}\left(F_{x1}^b e_x + F_{y1}^b e_y + F_{z1}^b e_z\right)\delta^{ib} \Rightarrow \sigma = \frac{1}{g} S^i \otimes \frac{\partial R}{\partial \lambda^i}$$

6.6 Homogenized stiffness tensor

Determine the effective stiffness matrix $[K]$ from the stress-strain relationship $[\sigma] = [K][\epsilon]$. The inverse of stiffness tensor gives the compliance tensor $[S] = [K]^{-1}$.

7. Appendix B

Computational algorithm for plastic yield surface prediction.

This appendix outlines the computational procedure used to determine the plastic yield surfaces of trabecular bone microstructures under combined bending and torsion. The approach is based on a micropolar continuum framework derived from a homogenized 3D beam-lattice model.

7.1 Coordinate transformation and curvilinear representation

- Define the coordinate transformation function R, such that:

$$R = R(\lambda_1, \lambda_2, \lambda_3)$$

  where λ_i are the curvilinear coordinates associated with the unit basis vectors.
- This transformation facilitates the mapping of the discrete lattice structure into an equivalent continuum representation, providing a smooth transition between the microstructure and the macroscopic continuum description.

7.2 Definition of key parameters for each beam in the elementary cell

For each beam b in the reference unit cell**,** the following parameters are defined:

(a) Node identification:
- Define the origin node ($O(b)$) and the end node ($E(b)$) of the beam.
- This establishes the beam connectivity and the local reference frame for further calculations.

(b) Non-dimensional shear factor:
- Introduce the **non-dimensional transverse shear correction factor** $\Phi^b_{y,z} = 12E^b_s I^b_z \Big/ G^b_s A^b k_s (L^b)^2$, which accounts for shear deformations in beams, particularly in cases where transverse shear effects are non-negligible.

(c) Displacement difference between beam extremities:
- Compute the difference in displacement between the extremity and origin node:

$$\Delta U^{b\varepsilon} = u^\varepsilon(E(b)) \text{-} u^\varepsilon(O(b)) = \varepsilon^2 \left(u^{E(b)}(\lambda^\varepsilon) \text{-} u^{O(b)}(\lambda^\varepsilon)\right)$$

where $u^{O(b)}$ and $u^{E(b)}$ are the displacements at the origin and end nodes, respectively.

$$F^b = \underbrace{\frac{A^b E_s}{L^b}(e_x.\Delta U^b)e_x}_{F_x^b} + \underbrace{\frac{12E_s I_z^b}{(L^b)^3(1+\Phi_y)}\left((e_y.\Delta U^b) - \frac{L^b}{2}\left(e_z.\left(\phi_1^{O_{R(b)}} + \phi_1^{E_{R(b)}} + \frac{\partial\phi}{\partial\lambda^i}\delta^{ib}\right)\right)\right)e_y}_{F_y^b}$$
$$+ \underbrace{\frac{12E_s I_y^b}{(L^b)^3(1+\Phi_z)}\left((e_z.\Delta U^b) + \frac{L^b}{2}\left(e_y.\left(\phi_1^{O_{R(b)}} + \phi_1^{E_{R(b)}} + \frac{\partial\phi}{\partial\lambda^i}\delta^{ib}\right)\right)\right)e_z}_{F_z^b}$$

where F_x^b, F_y^b, F_z^b are the axial and transverse force components.

(f) Moments at the beam extremities:

- Define the moments acting at the beam extremities:

$$M^{O(b)\varepsilon} = \left(M_x^{O(b)\varepsilon}e_x + M_y^{O(b)\varepsilon}e_y + M_z^{O(b)\varepsilon}e_z\right) = \frac{G_s J^{\varepsilon b}}{L^{\varepsilon b}}\varepsilon\left(e_x.\left(\phi_1^{O(b)} - \left(\frac{\partial\phi}{\partial\lambda^i}\delta^{ib} + \phi_1^{E(b)}\right)\right)\right)e_x$$
$$+ \left(\frac{6E_s I_y^{\varepsilon b}}{(L^{\varepsilon b})^2(1+\Phi_z^{\varepsilon})}(e_z.(\varepsilon^2\Delta U^b)) + \frac{E_s I_y^{\varepsilon b}}{L^{\varepsilon b}(1+\Phi_z^{\varepsilon})}\left(e_y.\left(\varepsilon\left((4+\Phi_z^{\varepsilon})\phi_1^{O(b)} + (2-\Phi_z^{\varepsilon})\phi_1^{E(b)}\right.\right.\right.\right.$$
$$\left.\left.\left.\left. + (2-\Phi_z^{\varepsilon})\frac{\partial\phi}{\partial\lambda^i}\delta^{ib}\right)\right)\right)\right)e_y + \left(\frac{6E_s I_z^{\varepsilon b}}{(L^{\varepsilon b})^2\left(1+\Phi_y^{\varepsilon}\right)}(-e_y.(\varepsilon^2\Delta U^b)) + \frac{E_s I_z^{\varepsilon b}}{L^{\varepsilon b}\left(1+\Phi_y^{\varepsilon}\right)}\right.$$
$$\left.\left(e_z.\left(\varepsilon\left(\left(4+\Phi_y^{\varepsilon}\right)\phi_1^{O(b)} + \left(2-\Phi_y^{\varepsilon}\right)\phi_1^{E(b)} + \left(2-\Phi_y^{\varepsilon}\right)\frac{\partial\phi}{\partial\lambda^i}\delta^{ib}\right)\right)\right)\right)e_z$$

$$M^{E(b)\varepsilon} = \left(M_x^{E(b)\varepsilon}e_x + M_y^{E(b)\varepsilon}e_y + M_z^{E(b)\varepsilon}e_z\right) = \frac{G_s J^{\varepsilon b}}{L^{\varepsilon b}}\varepsilon\left(e_x.\left(\phi_1^{E(b)} - \phi_1^{O(b)} + \frac{\partial\phi}{\partial\lambda^i}\delta^{ib}\right)\right)e_x$$
$$+ \left(\frac{6E_s I_y^{\varepsilon b}}{(L^{\varepsilon b})^2(1+\Phi_z^{\varepsilon})}(e_z.(\varepsilon^2\Delta U^b)) + \frac{E_s I_y^{\varepsilon b}}{L^{\varepsilon b}(1+\Phi_z^{\varepsilon})}\left(e_y.\left(\varepsilon\left((2-\Phi_z^{\varepsilon})\phi_1^{O(b)}\right.\right.\right.\right.$$
$$\left.\left.\left.\left. + (4+\Phi_z^{\varepsilon})\phi_1^{E(b)} + (4+\Phi_z^{\varepsilon})\frac{\partial\phi}{\partial\lambda^i}\delta^{ib}\right)\right)\right)\right)e_y + \left(\frac{6E_s I_z^{\varepsilon b}}{(L^{\varepsilon b})^2\left(1+\Phi_y^{\varepsilon}\right)}(-e_y.(\varepsilon^2\Delta U^b))\right.$$
$$\left. + \frac{E_s I_z^{\varepsilon b}}{L^{\varepsilon b}\left(1+\Phi_y^{\varepsilon}\right)}\left(e_z.\left(\varepsilon\left(\left(2-\Phi_y^{\varepsilon}\right)\phi_1^{O(b)} + \left(4+\Phi_y^{\varepsilon}\right)\phi_1^{E(b)} + \left(4+\Phi_y^{\varepsilon}\right)\frac{\partial\phi}{\partial\lambda^i}\delta^{ib}\right)\right)\right)\right)e_z$$

representing the bending and torsional moments at the origin and end nodes of each beam.

- These displacements act as the primary unknowns in solving the equilibrium equations.

(d) Nodal microrotations:

- Define the nodal microrotation variables: $\phi^{O(b)\varepsilon} = \varepsilon\phi_1^{O(b)}$, $\phi^{E(b)\varepsilon} = \varepsilon\left(\frac{\partial\phi}{\partial\lambda^i}\delta^{ib} + \phi_1^{E(b)}\right)$, with rotations unknowns $\phi_1^n = \left(\phi_x^n, \phi_y^n, \phi_z^n\right)$, where $\phi^{O(b)}$, $\phi^{E(b)}$ represent the rotations at the beam's origin and end node, respectively.

(e) Beam resultant force vector:

- Construct the resultant force vector acting along each beam:

7.3 Solving the equilibrium equations for kinematic unknowns

- Solve the systems of equilibrium equations for the unit cell:
- Rotation equilibrium equation:

$$\sum_{v^i \in Z^3}\sum_{b \in B_R}\left(M^{O(b)\varepsilon}.w^\varepsilon(O(b)) + M^{E(b)\varepsilon}.w^\varepsilon(E(b))\right) = 0$$

- Translation equilibrium equation:

$$\sum_{v^i \in Z^3}\sum_{b \in B_R} F^b.(v(O(b))\text{-}v(E(b))) = 0$$

- Express the solved displacement $u^n = \begin{pmatrix} u_x^n & u_y^n & u_z^n \end{pmatrix}$ and microrotation $\phi_1^n = \left(\phi_x^n, \phi_y^n, \phi_z^n\right)$ unknowns in terms of the micro-curvature tensor $[\chi]$

7.4 Construction of the twisting and bending moment vector

- Define a moment vector that gathers twisting and bending moments for all beams in the unit cell:

$$[M] = \begin{bmatrix} T^{O(1)} & T^{E(1)} & M_y^{O(1)} & M_y^{E(1)} & M_z^{O(1)} & M_z^{E(1)} & T^{O(2)} & \ldots \end{bmatrix}^T$$

where T is the twisting moment, M_y, M_z are the bending moments in the y and z directions.

7.5 Construction of the micromoment transformation matrix

- Build the matrix $[Q_m]$ that relates the micromoments of the unit cell $[M]$ to the homogenized micro-curvature tensor $[\chi]$: $[M] = [Q_m][\chi]$

This step is crucial for transitioning from the discrete beam-level moments to the continuum representation.

7.6 Construction of the couple stress vectors

- Compute the couple stress vectors for the homogenized continuum

$$\mu^i = \sum_{b \in B_R} \left(M_x^{E(b)} e_x + \frac{1}{2} \left(M_y^{E(b)} \text{-} M_y^{O(b)} \right) e_y + \frac{1}{2} \left(M_z^{E(b)} \text{-} M_z^{O(b)} \right) e_z \right) \delta^{ib}$$

7.7 Formation of the homogenized couple stress tensor

- Using the computed couple stress vectors, construct the homogenized couple stress tensor: $[m] = \frac{1}{g} \mu^i \otimes \frac{\partial R}{\partial \lambda^i}$
where each component represents the couple stresses associated with bending and torsion.

7.8 Constitutive equation for the homogenized couple-stress continuum

- Establish the constitutive equation for the homogenized continuum $[m] = [D][\chi]$:

$$[m] = \left[m_{xx}, m_{yy}, m_{zz}, m_{xy}, m_{yx}, m_{yz}, m_{zy}, m_{zx}, m_{xz} \right]^T$$
$$= [D] \left[\frac{\partial \phi_x}{\partial x}, \frac{\partial \phi_y}{\partial y}, \frac{\partial \phi_z}{\partial z}, \frac{\partial \phi_y}{\partial x}, \frac{\partial \phi_x}{\partial y}, \frac{\partial \phi_z}{\partial y}, \frac{\partial \phi_y}{\partial z}, \frac{\partial \phi_x}{\partial z}, \frac{\partial \phi_z}{\partial x} \right]^T$$

where D is the homogenized stiffness matrix of the couple-stress continuum.

$$\begin{bmatrix} D_{11} & D_{12} & D_{13} & 0 & 0 & 0 & 0 & 0 & 0 \\ D_{21} & D_{22} & D_{23} & 0 & 0 & 0 & 0 & 0 & 0 \\ D_{31} & D_{32} & D_{33} & 0 & 0 & 0 & 0 & 0 & 0 \\ 0 & 0 & 0 & D_{44} & D_{45} & 0 & 0 & 0 & 0 \\ 0 & 0 & 0 & D_{54} & D_{55} & 0 & 0 & 0 & 0 \\ 0 & 0 & 0 & 0 & 0 & D_{66} & D_{67} & 0 & 0 \\ 0 & 0 & 0 & 0 & 0 & D_{76} & D_{77} & 0 & 0 \\ 0 & 0 & 0 & 0 & 0 & 0 & 0 & D_{88} & D_{89} \\ 0 & 0 & 0 & 0 & 0 & 0 & 0 & D_{98} & D_{99} \end{bmatrix}$$

7.9 Construction of the compliance matrix

- Compute the compliance matrix, which is the inverse of the stiffness matrix:

$$[C] = [D]^{-1}$$

7.10 Relation between homogenized curvature and macroscopic couple stress

- Establish the relationship between the homogenized micro-curvature tensor and themacroscopic couple stress tensor: $[\chi] = [C][m]$

7.11 Relation between microscopic moment vector and macroscopic couple stress tensor

- Formulate the relation between the microscopic moment vector and the macroscopic couple stress tensor: $[M] = [Q_m][C][m]$

7.12 Normalization of the moment equation using shear yield strength

- Normalizing the equation from step 11 using the initial shear yield strength ensures that all torsion and bending couple stresses are expressed in a dimensionless form, which simplifies comparisons and makes the results independent of specific material units.

$$[M^N] = [M / \tau_Y] = \begin{bmatrix} T^{O(1)} \\ T^{E(1)} \\ M_y^{O(1)} \\ M_y^{E(1)} \\ M_z^{O(1)} \\ M_z^{E(1)} \\ T^{O(2)} \end{bmatrix}^N = [Q_m][C][m/\tau_Y] = [Q_m][C] \begin{bmatrix} m_{xx}/\tau_Y \\ m_{yy}/\tau_Y \\ m_{zz}/\tau_Y \\ m_{xy}/\tau_Y \\ m_{yx}/\tau_Y \\ m_{yz}/\tau_Y \\ m_{zy}/\tau_Y \\ m_{zx}/\tau_Y \\ m_{xz}/\tau_Y \end{bmatrix}$$

7.13 Failure criterion for plastic yield in torsion and bending

- Define the failure criterion for each beam b in the unit cell:

$$\left(\frac{M^b}{M_p}\right)^2 + \left(\frac{T^b}{T_p}\right)^2 = 1 \Rightarrow \left(\frac{6M^{bN}}{\sqrt{3}D_b^3}\right)^2 + \left(\frac{12T^{bN}}{\pi D_b^3}\right)^2 = 1$$

where M_p and T_p are the fully plastic bending and twisting moments of the beam cross-section and D_b is the beam diameter in the unit cell.
Final step: construction of the plastic yield Surface

- The plastic yield surface is obtained as the intersection of failure surfaces defined by the above criterion in the normalized couple-stress space.

References

Allaire, G. (2002). *Shape optimization by the homogenization method*. Springer-Verlag.

Allaire, G., & Pantz, O. (2006). Structural optimization with $\tt{FreeFem++}$. *Structural and Multidisciplinary Optimization, 32*(3), 173–181. https://doi.org/10.1007/s00158-006-0017-y

Bendsøe, M. P., Guedes, J. M., Haber, R. B., Pedersen, P., & Taylor, J. E. (1994). An analytical model to predict optimal material properties in the context of optimal structural design. *Journal of Applied Mechanics, Transactions ASME, 61*(4), 930–937. https://doi.org/10.1115/1.2901581

Bendsøe, M. P., & Kikuchi, N. (1988). Generating optimal topologies in structural design using a homogenization method. *Computer Methods in Applied Mechanics and Engineering, 71*(2), 197–224. https://doi.org/10.1016/0045-7825(88)90086-2

Bouchitté, G., & Buttazzo, G. (2001). Characterization of optimal shapes and masses through Monge-Kantorovich equation. *Journal of the European Mathematical Society, 3*(2), 139–168. https://doi.org/10.1007/s100970000027

Chen, H., Zhou, X., Fujita, H., Onozuka, M., & Kubo, K. Y. (2013). Age-related changes in trabecular and cortical bone microstructure. *International Journal of Endocrinology, 2013*. https://doi.org/10.1155/2013/213234

Czarnecki, S. (2015). Isotropic material design. *Computational Methods in Science and Technology, 21*(2), 49–64. https://doi.org/10.12921/cmst.2015.21.02.001

Czarnecki, S., & Lewiński, T. (2012). A stress-based formulation of the free material design problem with the trace constraint and single loading condition. *Bulletin of the Polish Academy of Sciences, Technical Sciences, 60*(2), 191–204. https://doi.org/10.2478/v10175-012-0026-5

Czarnecki, S., & Lewiński, T. (2014). The free material design in linear elasticity. *CISM International Centre for Mechanical Sciences, Courses and Lectures, 549*, 213–257. https://doi.org/10.1007/978-3-7091-1643-2_9

Czarnecki, S., & Wawruch, P. (2015). The emergence of auxetic material as a result of optimal isotropic design. *Physica Status Solidi (B), 252*(7), 1620–1630. https://doi.org/10.1002/pssb.201451733

Dagan, D., Be'ery, M., & Gefen, A. (2004). Single-trabecula building block for large-scale finite element models of cancellous bone. *Medical, & Biological Engineering & Computing, 42*(4), 549–556. https://doi.org/10.1007/bf02350998

Fenech, C. M., & Keaveny, T. M. (1999). A cellular solid criterion for predicting the axial-shear failure properties of bovine trabecular bone. *Journal of Biomechanical Engineering, 121*(4), 414–422. https://doi.org/10.1115/1.2798339

Feng, J., Liu, B., Lin, Z., & Fu, J. (2021). Isotropic octet-truss lattice structure design and anisotropy control strategies for implant application. *Materials & Design, 203*, Article 109595. https://doi.org/10.1016/j.matdes.2021.109595

Ganghoffer, J. F., & Goda, I. (2018). *Micropolar models of trabecular bone multiscale biomechanics* (pp. 263–316). France: Elsevier Inc.. https://doi.org/10.1016/B978-1-78548-208-3.50006-X

Gibson, L. J. (2005). Biomechanics of cellular solids. *Journal of Biomechanics, 38*(3), 377–399. https://doi.org/10.1016/j.jbiomech.2004.09.027

Goda, I., Assidi, M., Belouettar, S., & Ganghoffer, J. F. (2012). A micropolar anisotropic constitutive model of cancellous bone from discrete homogenization. *Journal of the*

Mechanical Behavior of Biomedical Materials, 16(1), 87–108. https://doi.org/10.1016/j.jmbbm.2012.07.012

Goda, I., Assidi, M., & Ganghoffer, J. F. (2014). A 3D elastic micropolar model of vertebral trabecular bone from lattice homogenization of the bone microstructure. *Biomechanics and Modeling in Mechanobiology, 13*(1), 53–83. https://doi.org/10.1007/s10237-013-0486-z

Goda, I., & Ganghoffer, J. F. (2015). Identification of couple-stress moduli of vertebral trabecular bone based on the 3D internal architectures. *Journal of the Mechanical Behavior of Biomedical Materials, 51*, 99–118. https://doi.org/10.1016/j.jmbbm.2015.06.036

Goda, I., Ganghoffer, J. F., Czarnecki, S., Wawruch, P., & Lewiński, T. (2016). Optimal internal architectures of femoral bone based on relaxation by homogenization and isotropic material design. *Mechanics Research Communications, 76*, 64–71. https://doi.org/10.1016/j.mechrescom.2016.06.007

Kadir, M. R. A., Syahrom, A., & Öchsner, A. (2010). Finite element analysis of idealised unit cell cancellous structure based on morphological indices of cancellous bone. *Medical, & Biological Engineering & Computing, 48*(5), 497–505. https://doi.org/10.1007/s11517-010-0593-2

Kopperdahl, D. L., & Keaveny, T. M. (1998). Yield strain behavior of trabecular bone. *Journal of Biomechanics, 31*(7), 601–608. https://doi.org/10.1016/S0021-9290(98)00057-8

Liu, X. S., Zhang, X. H., & Guo, X. E. (2009). Contributions of trabecular rods of various orientations in determining the elastic properties of human vertebral trabecular bone. *Bone, 45*(2), 158–163. https://doi.org/10.1016/j.bone.2009.04.201

Mohammadi, A., Sadighi, M., Hedayati, R., & Ganghoffer, J.-F. (2024). Asymptotic homogenization of tesseract lattice structures. *European Journal of Mechanics - A: Solids, 107*, Article 105368. https://doi.org/10.1016/j.euromechsol.2024.105368

Morgan, E. F., Bayraktar, H. H., & Keaveny, T. M. (2003). Trabecular bone modulus-density relationships depend on anatomic site. *Journal of Biomechanics, 36*(7), 897–904. https://doi.org/10.1016/S0021-9290(03)00071-X

Morgan, E. F., & Keaveny, T. M. (2001). Dependence of yield strain of human trabecular bone on anatomic site. *Journal of Biomechanics, 34*(5), 569–577. https://doi.org/10.1016/S0021-9290(01)00011-2

Noman, A. A., Shaari, M. S., Mehboob, H., & Azman, A. H. (2025). Recent advancements in additively manufactured hip implant design using topology optimization technique. *Results in Engineering, 25*. https://doi.org/10.1016/j.rineng.2025.103932

Syahrom, A., Abdul Kadir, M. R., Abdullah, J., & Öchsner, A. (2013). Permeability studies of artificial and natural cancellous bone structures. *Medical Engineering & Physics, 35*(6), 792–799. https://doi.org/10.1016/j.medengphy.2012.08.011

Wolfram, U., Wilke, H. J., & Zysset, P. K. (2010). Valid µ finite element models of vertebral trabecular bone can be obtained using tissue properties measured with nanoindentation under wet conditions. *Journal of Biomechanics, 43*(9), 1731–1737. https://doi.org/10.1016/j.jbiomech.2010.02.026

Wu, N., Li, S., Zhang, B., Wang, C., Chen, B., Han, Q., & Wang, J. (2021). The advances of topology optimization techniques in orthopedic implants: A review. *Medical, & Biological Engineering & Computing, 59*(9), 1673–1689. https://doi.org/10.1007/s11517-021-02361-7

CHAPTER

Shape and topology optimization approaches 15

Michał Nowak[1], Jan Sokołowski[2], Antoni Żochowski[2] and Jan Polak[1, 2]
[1]*Poznan University of Technology, Department of Virtual Engineering, Poznan, Poland;* [2]*Systems Research Institute of the Polish Academy of Sciences, Warszawa, Poland*

1. Introduction

Since there is an equivalence between the process of trabecular bone remodeling phenomenon and the process of stiffest design search, it can be assumed that both processes will lead to identical structures. This conclusion is the basis for the formulation of the paper's main objectives - the first two chapters are devoted to illustrating this thesis.

In mutual understanding the topology optimization involves finding the optimal distribution of materials in the design space. One method dominated the commercial software, namely the SIMP (Solid Isotropic Material with Penalization) method, a variant of a homogenization approach (Bendsoe & Kikuchi, 1988; Bendsoe & Sigmund, 2003; Deaton & Grandhi, 2014; Sigmund, 2001). The design of lightweight structures, allowing the actual weight of the structure to be minimized, is becoming increasingly important in the ever-expanding area of mechanical design. The problem is not just about using lightweight materials, but mostly about employing the right design methods. The main method used in the industry is structural optimization to design lightweight structures with high stiffness. Combined with additive manufacturing methods (Nowak & Polak, 2021; Nowak, 2022; Polak & Nowak, 2022, 2023), structural optimization is an excellent way to achieve solutions far superior to those possible with traditional design and manufacturing methods. The design process consists of two stages, the synthesis and analysis stage. Within the former, the most important decisions concerning the form of the designed structure are made. It is necessary to decide how to place the material in each design space so that the obtained structure has the highest stiffness and minimal weight. Topology optimization is a method that optimizes the material layout in each design space for a given set of boundary conditions and loads. As the topology optimization concerns the synthesis stage (the spatial shape of the structure at this stage is searched for and is not known), it differs from shape optimization and size optimization and so the tools and methods used to carry topology optimization out are specific and differ from other structural optimization methods. Among the various approaches to the topology optimization problem (Sigmund, 1999, 2001; Sigmund & Maute, 2013), the SIMP method is the most popular. This also applies to commercial software,

Biomechanics and Mechanobiology of Bone. https://doi.org/10.1016/B978-0-443-26739-0.00009-0

increasingly widely used in industrial practice (Allaire, 2002; Allaire et al., 2019; Gawe et al., 2017; Nowak & Boguszewski, 2021; Sigmund & Maute, 2013). **But there exists another approach** related to the simulation models of trabecular bone remodeling. The regulatory model of the trabecular bone remodeling phenomenon has been described in many papers and books (http://www.wccm2018.org/sites/default/files/WCCM2018-Abstracts-FINAL.pdf ISBN, 2018; Nowak, 2006). The first observation of the amazing behavior of the trabecular bone tissue was made by Wolff and published in 1892 (Wolff, 1892). The observation proposed by Julius Wolff—called the Wolff's law—can be described as a structural adaptation of the bone to the external forces. Since then, scientific research on the description and development of numerical models of the phenomenon of bone remodeling is still ongoing. As bone remodeling is extraordinarily complex, considering the dependence on external loads and various biological aspects, different numerical modeling approaches are considered for this phenomenon. One of the most interesting of them is the coupled model based on global stimulus of trabecular bone remodeling. The global stimulus approach allows to combine in one model varied factors related to the remodeling sequence. The factors are related both to mechanical stimulation and to other specific biological aspects of the remodeling process such as cellular migration and differentiation or nutriments supply (George et al., 2019; Kariem et al., 2015). Other models consider the relationship between mechanical and biological stimulation using the concept of mechanical and biological models coupling to describe the phenomenon of bone remodeling (Nowak et al., 2020a; http://www.wccm2018.org/sites/default/files/WCCM2018-Abstracts-FINAL.pdf ISBN, 2018; Lee et al., 2015; Marzban et al., 2015; Pivonka et al., 2013). From a medical point of view, it is also a principal issue to include in the models both bone tissue and the tissue reconstructed or reinforced by the addition of an artificial resorbable material (Kariem et al., 2015; Lekszycki et al., 2012). Regardless of the approach used, however, the modeling of structural adaptation to variable load becomes a central research topic. Our results show clearly that shape optimization principles and numerical procedures and simulation models of the trabecular bone remodeling phenomenon can be treated as the description of the solution of the same problem from the area of structural form modification with the goal in stored energy minimization. The interpretation of the results in the context of the trabecular bone adaptation are based on our previous work (Nowak et al., 2018a, 2018b, 2018d, 2020b, 2020c).

2. The stiffest design problem

To begin with, the problem of how to obtain the stiffest design and its setting could be presented as follows. To maximize the stiffness of a structure (the stiffest design goal), one needs to minimize the functional:

$$J(\Omega) = \int_{T1} t.u \, ds \tag{15.1}$$

The constraint is the given volume:

$$\int_{\Omega} dx - V_0 = 0 \tag{15.2}$$

The state equations are as follows:

$$\text{div}\quad \sigma(u) = 0 \quad \text{on} \quad \Omega,$$

$$\sigma(u).n = t \quad \text{on} \quad \Gamma_1,$$

$$\sigma(u).n = 0 \quad \text{on} \quad \Gamma_V,$$

$$u. = 0 \quad \text{on} \quad \Gamma_0 \tag{15.3}$$

Here in above, Ω represents domain of the elasticity system, u the displacement, Ω_0 a given volume, Γ_0 part of the boundary with Dirichlet condition, Γ_1 part of the boundary loaded by traction forces t, Γ_v part of the boundary subject to modification. The domain of the elasticity system and the featured parts are presented in (Fig. 15.1). By analyzing the picture above and construing the presented parts of the domain as the elements of the trabecular bone structure—even though a single trabecula size ranges from tens to hundreds of micrometers (Hamed & Jasiuk, 2012), one may imagine that the trabecular structure is the domain of the elasticity system. Subsequently, one can use their result from the area of shape optimization, which concerns Γ_V—a part of the boundary subject to modification. After defining the Lagrange function for the problem under considerations, where Ω_t is an image of Ω in transformation T_t,

$$L(\Omega_t,\lambda) = \int_{\Gamma_1} t.u\, ds + \lambda\left[\int_{\Omega_t} dx - V_0\right] \tag{15.4}$$

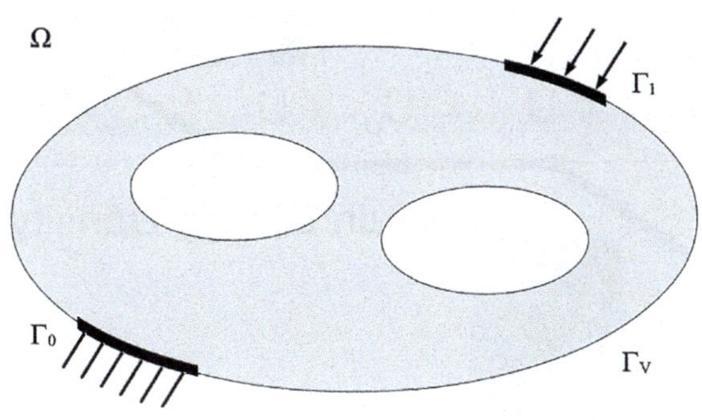

FIGURE 15.1

The domain of the elasticity system Ω, Γ_0 part of the boundary with Dirichlet condition, Γ_1 part of the boundary loaded by traction forces, Γ_v part of the boundary subject to modification.

The shape derivative (Sokołowski & Zolesio, 1992) of the state equation in the weak form gives the formula:

$$\int_{\Gamma_1} t.u'\, ds = \sigma(u) : \varepsilon(u)\ V.nds \tag{15.5}$$

Then the shape derivative of Lagrange function using speed method (Sokołowski & Zolesio, 1992) is taken.

$$[L(\Omega, \lambda)]' = \int_{\Gamma_1} t.u'\, ds + \lambda \int_{\Gamma_1} V.nds = 0 \tag{15.6}$$

and using the formula (Nowak & Polak, 2021) the derivative of Lagrange function gives.

$$\int_{\Gamma_V} [\lambda - \sigma(u) : \varepsilon(u)]\ V.nds = 0 \tag{15.7}$$

The derivative at the stationary point should vanish—for details see (Nowak et al., 2018d)—and it can be concluded that.

$$\sigma(u) : \varepsilon(u) = \lambda = \text{const} \tag{15.8}$$

It means that for the stiffest design, the strain energy density on the part of the boundary subject to modification Γ_V must be constant.

Going back to the numerical models of trabecular bone remodeling developed by Huiskes and others (2018), the regulatory model is based on the strain energy density U as the mechanical stimulation measure. The regulatory model presented in (Fig. 15.2) is described by the following equations:

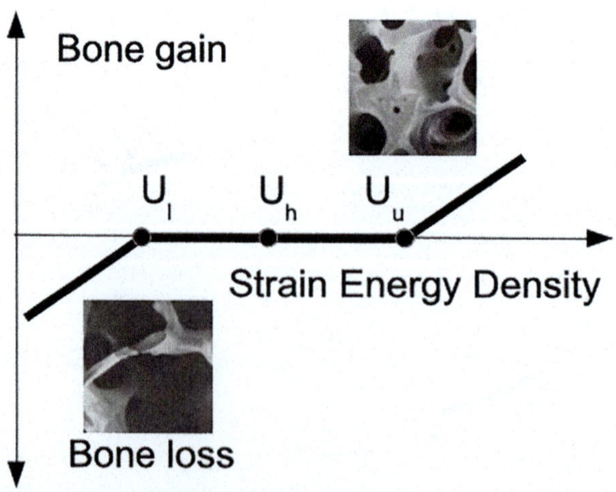

FIGURE 15.2

The regulatory model and the "lazy zone" concept.

$$dE/dt = C_e(U - (U_h + s)) \quad \text{for} \quad U > (U_h + s)$$
$$dE/dt = 0 \quad \text{for} \quad (U_h - s) \le U \le (U_h + s)$$
$$dE/dt = C_e(U - (U_h - s)) \quad \text{for} \quad U < (U_h + s) \qquad (15.9)$$

where U_h is the strain energy density value corresponding to homeostasis of bone loss and gain and it is a constant value, 2s is the size of the "lazy zone", C_e is a constant value, and E denotes Young's modulus of the tissue. The model assumes that the Young's modulus (as a local elastic modulus) of the tissue may vary depending on the level of the mechanical stimulation (strain energy density). The existence of homeostatic value of strain energy density surrounded by the "lazy zone" is assumed. In this way, the bone (trabecular tissue together with marrow) adapts its density value. The "lazy zone" concept corresponding to the clinical observations was originally proposed by Carter (1984), assuming that for a certain intermediate range of remodeling stimuli, no bone remodeling takes place. This concept was first used in a computational model by Huiskes. Now the analogy is clearly seen. The strain energy density value corresponding to homeostasis is just the value of the Lagrange multiplier λ strain and equation (Nowak et al., 2019, 2020b) will allow to formulate the postulate about the correspondence of trabecular bone remodeling phenomenon goal and stiffest design. To stay closer to the real trabecular remodeling phenomenon, one should look at the BMU (Basic Multicellular Unit) (van Oers et al., 2008), presented in (Fig. 15.3), i.e., the area where the remodeling process takes place. The BMU-s are on the trabecular tissue's surface to allow cells migration from the marrow. Accordingly, the remodeling process occurs on the surface rather than in the whole trabecular area. Subsequently, one can formulate **the postulate** regarding the modification of the trabecular bone remodeling regulatory model.

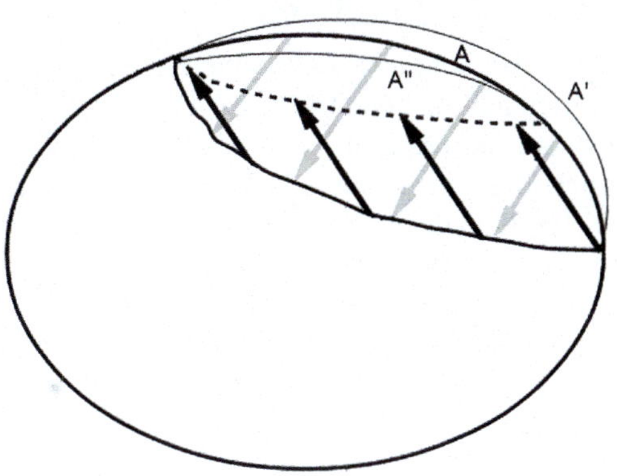

FIGURE 15.3

The cross-section of the single trabecula—a scheme of basic multicellular unit (BMU). The bright arrows depict bone loss while the dark arrows represent new tissue formation. A: The initial position of the trabecular tissue surface. A': a new position of the surface for a higher value of strain energy density than the "lazy zone" borders. A": a new position of the surface for a lower value of strain energy density value than "lazy zone" borders.

During the remodeling process, the trabecular bone tends to maximize the stiffness of a structure (i.e., to find the stiffest design) by the strain energy density equalization on the structural surface of the trabecular tissue. According to formula (Nowak et al., 2020b), the strain energy density on the part of the boundary subject to modification (in this case the trabecular bone tissue surrounded by the marrow) must be constant. The local change on the structural surface leads to global minimization of the strain energy for the whole structure. Importantly, the condition from formula (Nowak et al., 2020b) concerns the structural surface rather than the whole structural volume. However, the fulfillment of this surface condition is tantamount to reaching the goal in the form of minimizing the energy of the entire structure in the domain, i.e., minimization of the functional in formula (Nowak et al., 2024). In fact, assuming the possibility of moving the surface of the trabecular tissue in the virtual space, we can modify its position depending on the local value of strain energy density, as presented in Fig. 15.3. If the local value of the strain energy density (on the BMU surface) is higher than the "lazy zone" borders, the surface moves from position A to A'. If, in turn, the local value of the strain energy density (on the BMU surface) is lower than the "lazy zone" borders, the surface moves from position A to A". There is no need for the evolution in such a model, and the obtained structural configurations will reflect the real structure of the trabecular net.

Since there is an equivalence between the process of trabecular bone remodeling phenomenon and the process of stiffest design search, it can be assumed that both processes will lead to identical structures. This conclusion is the basis for the formulation of the paper's main objectives.

3. Pharmacokinetic models of osteoporosis treatment—New perspectives

Osteoporosis is a trabecular bone disease characterized by reduction of bone volume. These changes often lead to bone fractures. The development of new drags requires both the detailed description of the natural biological processes and theoretical studies concerned the possible influence on the bone gain process. Due to the increase of the elderly in the world, osteoporosis became a serious health issue for the whole community. Intensive studies on this issue have been conducted for many years (Baertschi et al., 2014; Bao et al., 2013; Calvo-Gallego et al., 2023; Cohen-Kfir et al., 2011; Courbon et al., 2017; Kariem et al., 2015; Koide et al., 2017; Martí et al., 2011; Martí et al., 2019; Morse et al., 2014; Perez-Campo et al., 2016; Pivonka et al., 2010, 2013; Scheiner et al., 2013; Sutherland et al., 2004; Winkler et al., 2003, 2005).

Pivonka and coworkers (Martí et al., 2019; Pivonka et al., 2010) proposed that the bone cell activities within the BMU are changing the bone volume fraction locally over time:

$$df_bm/dt = k_{form} \quad O_{b_a} - k_{res} \quad O_{c_a} \tag{15.10}$$

where it is assumed that osteoclasts resorb bone and osteoblasts form bone at constant rates, respectively k_{res} and k_{form} and Ob_a is number of active osteoblasts and Oc_a is number of active osteoclasts in the BMU. It is a universal way to describe the entire process. The bone volume fraction f_{bm} is then the result of the interrelation of the two quantities. Of course, the process is extremely complicated and here any number of factors affecting the actual bone turnover can be considered. Until now, the most effective way to effectively counteract osteoporosis was to use the RANKL-RANK-OPG pathway (Pivonka et al., 2010). This primarily involves the use of denosumab. Denosumab is a monoclonal human antibody directed against RANKL, a protein that is involved in the activation of osteoclasts, the so-called bone-setting cells. Denosumab binds to RANKL, thus preventing it from binding to the RANK receptor, resulting in no activation of osteoclast maturation. Denosumab inhibits the maturation and activity of osteoclasts, which prevents the resorption of bone tissue. The use of this approach leads, however, to atypical bone fractures. Atypical femur fractures have been particularly reported in patients receiving denosumab. Atypical femur fractures can occur in association with minor trauma to the subcondylar region and femoral shaft, or without association with trauma. These fractures have a characteristic radiographic picture. Such events occur in association with antiresorptive therapy. This can be explained based on the mechanostat theory in such a way that the very process of tissue replacement is interfered with, preventing the exchange of bone material. The question arises whether it is possible to find a better solution, allowing to maintain the mechanostat scheme with preservation of the sequence - removal-building of new tissue. It turns out that such a solution exists (Martí et al., 2023; Martin, 2020; Pivonka et al., 2013) and is related to the interaction of osteocytes, that is, an element of the system that controls the behavior of the tissue in context to the position and size of the lazy zone. The mechanism of action is based on romosozumab. Romosozumab is a humanized monoclonal antibody (IgG2) that binds sclerostin—a protein secreted by osteocytes that strongly inhibits the bone-forming activity of osteoblasts—and blocks its action, resulting in enhanced bone formation. In addition, romosozumab alters the expression of osteoclast formation mediators, thereby reducing bone resorption. The result of this dual action is a rapid increase in trabecular and cortical bone mass and improved bone structure and strength. What should be emphasized, in this approach, the action of romosumab is a much better solution, allowing to keep the modeling of the entire process close to the models relating to the mechanostat theory, that is, the regulation related directly to the mechanical stimulation of the trabecular bone structure. This opens the way to the construction of models in which the model of action and dosage of pharmaceuticals can be directly described based on the desired effects of mechanical stimulation of tissue structure. In this way it is possible to describe the use of drugs specific to the patient by optimization of the drug therapies as consequences of the desired signal of stimulation of the structure, not necessarily corresponding to the actual stimulation resulting from physical activity. Looking at the process in this way we can leave the BMU and treat the process as external in relation to the phenomena occurring in a single BMU. This, in turn, provides the opportunity and chance to model the remodeling process for the entire bone structure on a mechanostat basis.

Romosozumab binds sclerostin, a protein secreted by osteocytes. Osteocytes have reduced synthetic activity, but they are actively involved in the turnover of trabecular bone through mechanosensory mechanisms.

And this is the element of the trabecular bone remodeling process that can now be introduced into the research. The regulation by blocking the secretion can be interpreted as a change in the position of the homeostasis point, that is, the energy describing the mechanical stimulation and the definition of entire lazy zone, as shown in (Fig. 15.4).

The change in the position of the homeostasis point, U_hA versus U_hB and respectively the position and size of the entire lazy zone. This change in the position of lazy zone will now affect the rebuilding process not only through changes in the interior of each BMU, but looking from the point of view of the mechanostat the entire rebuilding process will have a different form. This interpretation of sclerostin quantity regulation will also preserve the earlier description of Huiskes' regulatory model. Now, depending on the effectiveness of the reduction of the amount of sclerostin Xs compared with the situation before the use of romosozumab Xs_0, a new position of the lazy zone can be given according to the formula:

$$U_hB = U_hA \cdot (X_s / X_{s_0}) \tag{15.11}$$

This is a completely fresh look at the role of the mechanostat and its description in the process of trabecular bone remodeling phenomenon. Consideration of its ability to affect the control function of osteocytes accounts for the significant **enhancement of pharmacokinetic models of osteoporosis treatment**. However, it is necessary to use the right numerical tools that will allow in silico analysis, and

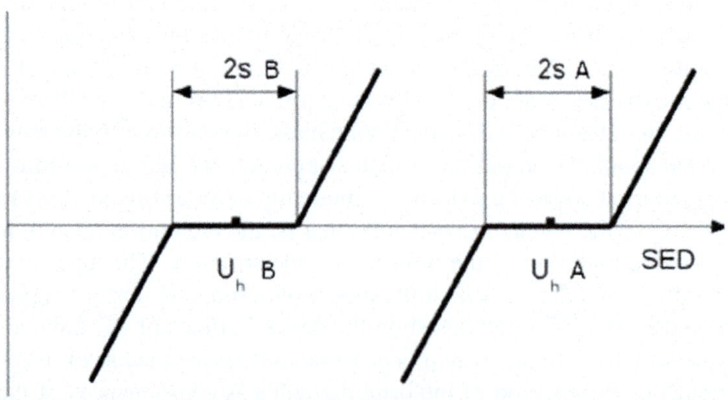

FIGURE 15.4

The regulation by blocking the secretion of sclerostin as a change in the position of the homeostasis point, U_hA versus U_hB and respectively the position and size of the entire lazy zone.

such tools are available to the team. This description does not exhaust the phenomena belonging to the research area. Studies show that sclerostin is a locally acting regulator of late-osteoblast/preosteocyte differentiation and regulates mineralization (Atkins et al., 2011). Considering this fact, extension and modification of the simulation algorithm is required. Similarly, the feedback observed in the bone formation process, which is coupled to resorption through suppression of sclerostin expression by osteoclasts (Koide et al., 2017).

4. The numerical tool enabling in silico research of pharmacokinetic models

To simulate trabecular bone structure, it is necessary to combine two areas: numerical simulation of deformation of the bone structure under load and structural evolution. The Finite Element Method will be used for this purpose. In terms of structural calculations, there is no alternative but to use the already existing FEM systems (Nowak., 2006; Nowak et al., 2018c, 2018d, 2022, 2024; Polak et al., 2022, pp. 459–471; Polak & Nowak, 2023). Tetrahedral, 10-node volume elements were used for computational purposes. The more challenging task is to plan how to reflect the evolution of the structure. Based on previous experience it was decided to build separate computational meshes for each simulation step. The idea of building a volumetric mesh based on data in the form of two-dimensional images containing the cross-section of tissue is presented schematically in Fig. 15.5.

This is a natural way of mesh generation since the visualization for the biological entities is based on the digital images e.g., Computer Tomography. Also, the evolution of the structure is based on the two-dimensional image's modification. According to remodeling scenario described by regulatory model, depending on the calculated value of the energy density the surface of the structure is modified by adding or removing material on its surface. The lazy zone, derived from the bone regulatory model, is calculated using material mechanic properties. In the biomimetic algorithm, if the strain energy density is less than the lower limit for the lazy zone,

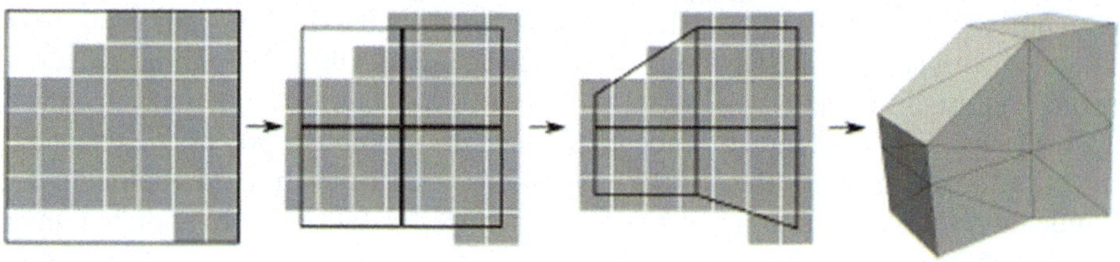

FIGURE 15.5

Data discretization and volumetric mesh building algorithms applied for sample images. Image is divided into squares, in this example size of three was selected. For each square, four points are selected and used for volumetric mesh building. The last image shows volumetric mesh created from two identical images.

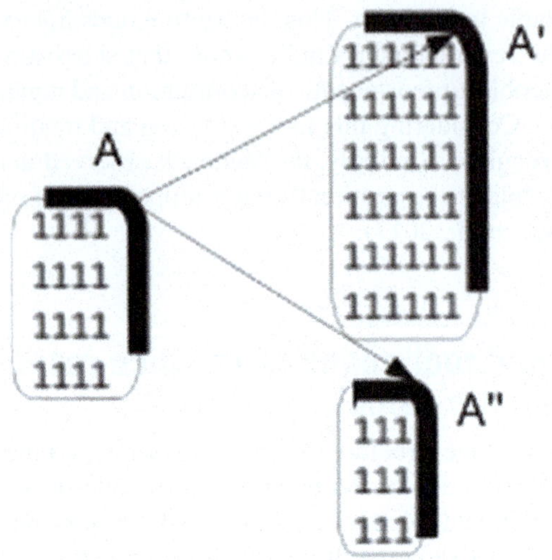

FIGURE 15.6

The modification of the structural surface position in the virtual space according to the strain energy distribution.

the small amount of material is removed from the surface. If the strain energy density is greater than the upper limit for the lazy zone, a small amount of material is added on the surface. The structural evolution is separated from the finite element method computations and modeled with use of 2-dimensional images. These images are converted to a binary bitmap and the evolution of the surface is modeled with it. If there is the need to add material on the surface, the additional layer of "1"'s, representing the material is placed instead of "0"'s (case A′ in Fig. 15.6). If there is the need to remove material from the surface, the additional layer of "0"'s, representing the void is placed instead of "1"'s (case A″ in Fig. 15.6). In this way the structural surface is moved in the virtual space, and it is possible to simulate structural evolution including separation or merging of the structural elements, without problems with finite element computations.

The system created (the system diagram is depicted in Fig. 15.7) has been designed with two main goals: high performance and flexibility for future enhancements and modifications.

Underlying phenomenon of trabecular bone remodeling is a process that happens over time hence the algorithm works in iterations, each run is full remodeling cycle. Single loop consists of a series of steps. The first step is data discretization to reduce the size of data used to build mesh. The scale factor is an input parameter for algorithms and allows choosing between fine and coarse meshes and consequent tradeoff between solution resolution and resources required to perform calculations. Each two-dimensional image (slice) is divided into squares of size depending on scale factor that are used as an input for Volumetric resulting tetragon approximates material

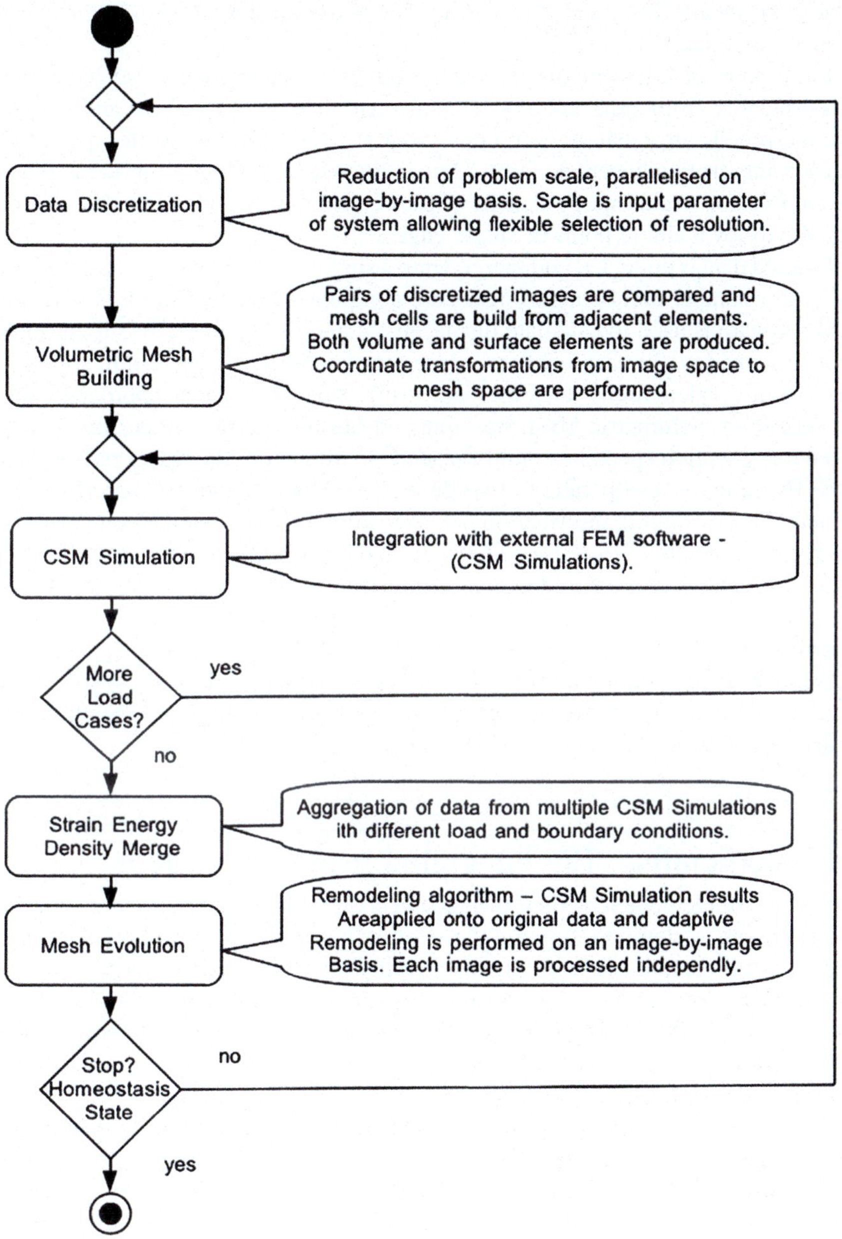

FIGURE 15.7

Activity diagram of system for efficient trabecular bone remodeling numerical simulation enabling multiple load case analysis.

inside each square. The second step is the Volumetric Mesh Building, which consists of three operations.

First, pairs of adjacent discretized images are converted into layers of mesh cells—squares from adjacent two-dimensional images are paired and for each pair, mesh cells are created, each consists of six tetrahedral elements spanned between points from adjacent squares. Data is filtered to remove mesh items that are not eligible for strain energy density (SED) calculation, mesh items that are connected only by a single point or single edge.

Filtered data is collected and merged into a single mesh—each cell and vertex get assigned a number and boundary faces are computed. Prepared mesh is saved, and FEM software is used to calculate displacement and SED values. To support multiple load scenarios, for each load case, separate simulation is executed in a loop. Results of all calculations are merged using the lazy zone approach. Slices Discretization, Volumetric Mesh Building and Mesh Evolution algorithms are optimized for execution speed in a multithreaded environment. Slices Discretization and Mesh Evolution steps operate on two-dimensional images (slices) directly without the need for exchanging information between workers hence parallelism is achieved by executing parallel processing of each item independently. Unlike two other algorithms, Volumetric Mesh Building was challenging to scale horizontally because the last phase—data collection—requires collecting all mesh data into the single data structure.

Described steps iteratively improve the initial structure, leading to the optimal solution, realizing the stiffest design through biomimetic approach. Sample result of such optimization is shown in (Fig. 15.8).

5. Superficialis—The integrated research platform

Algorithms implemented on their own do not create a ready to use research platform. The complete system, to support the needs of users, must additionally provide nonfunctional capabilities such as user-friendly interface, support for various input data formats and ability to adjust functionalities to facilitate research activities (Dongarra et al., 2020). Superficialis is a complete system created to enrich biomimetic algorithms with additional capabilities to create a research platform. Main capabilities of the system ae shown in Fig. 15.9.

The main requirement for the research system is to provide capabilities to explore new ideas and conduct in-silico research activities. The system provides this in two different ways: through modular design with Dependency Injection and through scripting. Modular design, where each part of the system consists of interface definition and one or more implementations. Those implementations can be selected through the Dependency Injection framework. For example, the system contains multiple remodeling algorithms developed during research. One adds/removes material when regulatory zone is exceeded always using the same amount, another one varies the amount of material depending on the magnitude of surface

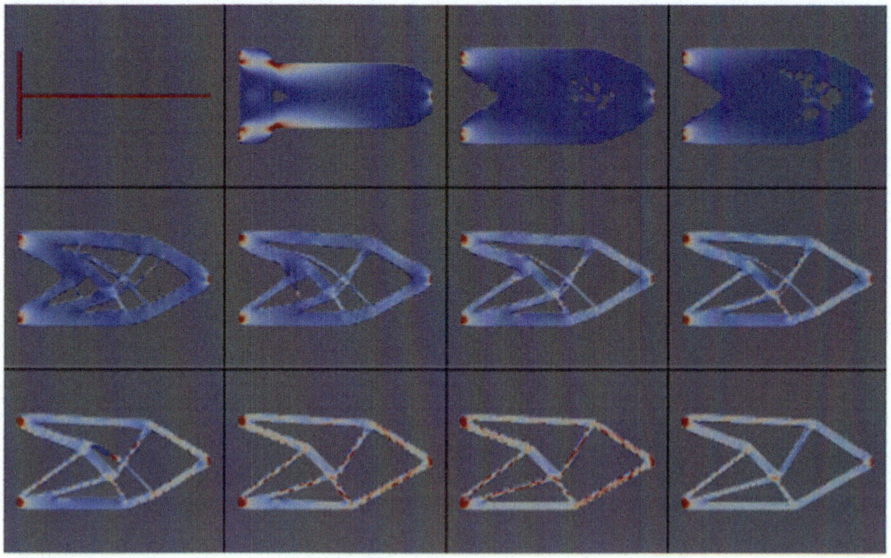

FIGURE 15.8

The example of structural optimization with use of the biomimetic method—the cantilever beam in bending, that all nodes of the one edge could be fixed (clamped wall) and the bending force is applied to the middle of the opposite edge.

stress. Consecutive tests of system can use different implementations to compare results without the need of making any source code changes. Additionally, the system supports python script that optionally can be used to calculate value of lazy zone for each simulation step. This allows for creation of custom heuristics that vary lazy zone over time, allow different values of lazy zone for different areas of simulation. It is possible to calculate lazy zone values based on surface stress distribution. This capability allows for convenient creation of custom heuristics as script is a part of optimization task that is submitted to the system, not the system itself. Thus, it is possible to simulate variable lazy zone or effects of changing the lazy zone (as shown in Fig. 15.4) on the result of the optimization. Sample scripted optimization is presented in (Fig. 15.10).

Research data comes in different formats—micro-CT scans are provided as sets of two-dimensional images; engineering problems are defined in CAD software. Superficialis system is able to read them both, as shown in (Fig. 15.11), and output simulation results in multiple formats. STL mesh format is the basic output format, which is de facto industrial standard for additional manufacturing hence it can be opened by most of available software. UCD format contains complete volumetric data with stress values and is used for visualization. CSV format contains result data as an unstructured cloud of points together with energy and displacement values and is convenient format for statistical analysis of the results. Available boundary conditions are supports, fixed displacements, forces and torques. Each can be

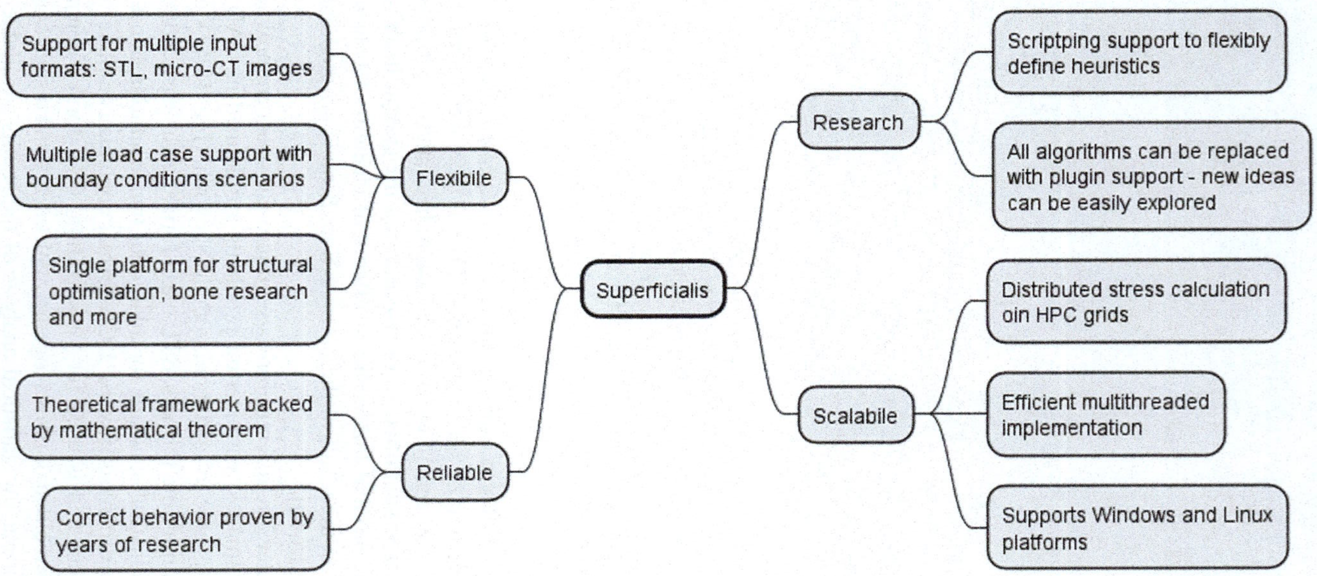

FIGURE 15.9

Main features of the superficialis platform. Biomimetic algorithms are enriched with additional capabilities to create a ready to use research platform.

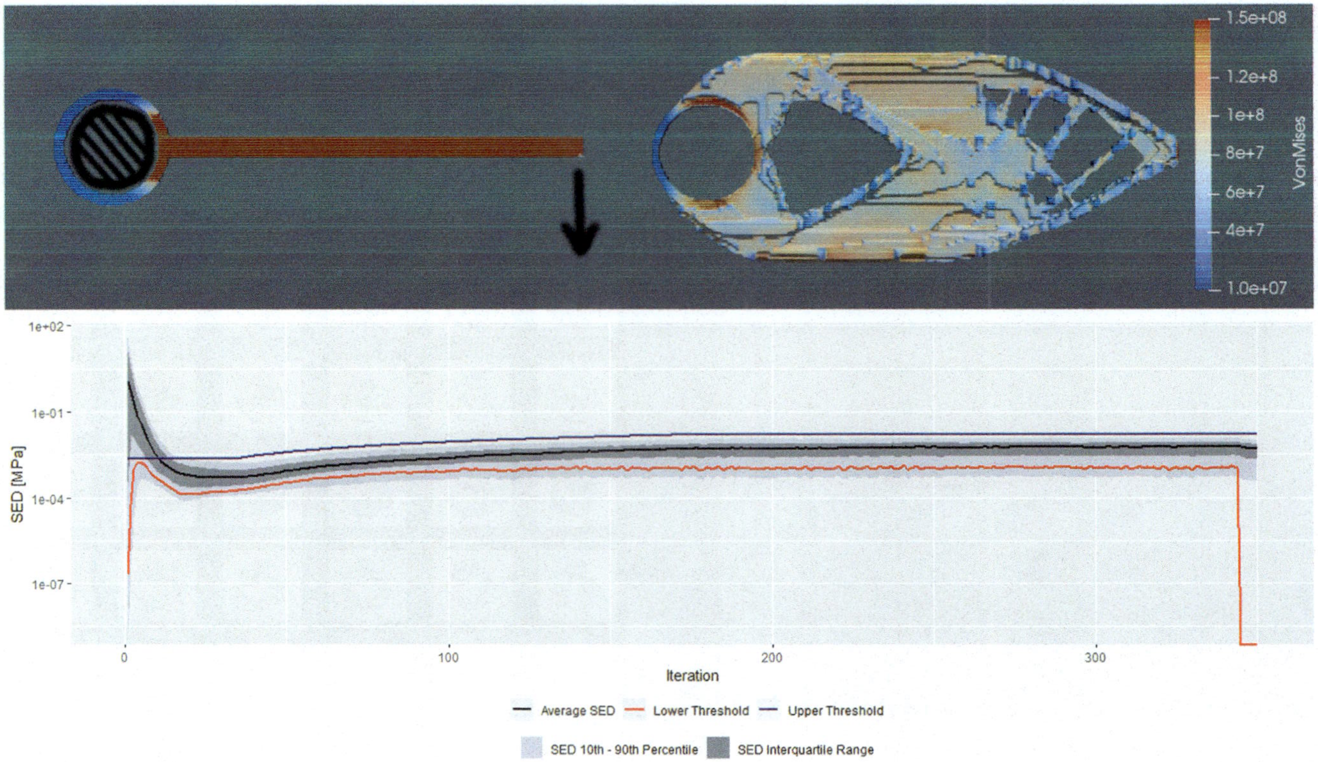

FIGURE 15.10

Example of optimization heuristic implemented with a custom script. Top left: input data—cantilever beam attacked to a cylindrical mount. Top right: The same element after structural optimization. Bottom: Plot of changes of surface energy density distribution and lazy zone thresholds during optimization. In the presented optimization upper threshold varies between fixed values, value of lower is dynamically selected based on the energy distribution.

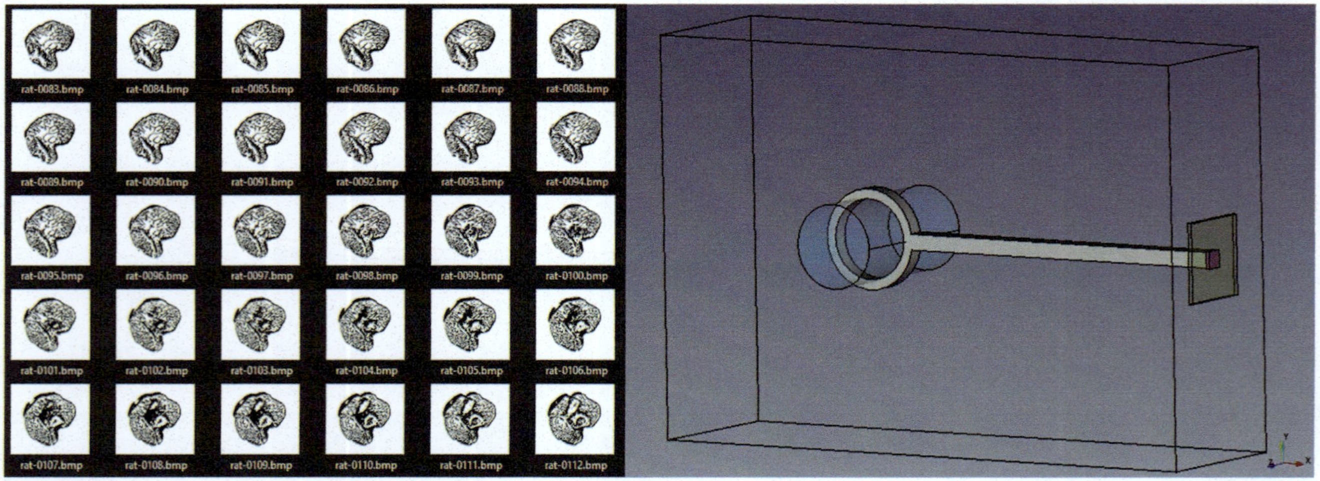

FIGURE 15.11

Two different formats of system input data. Left: Stack of micro-CT scan images. Right: Optimization task defined in a CAD system as a base element and support/stress areas. Both input formats are interchangeable and can be used depending on the available input data and research activities.

defined in various ways. The most basic is a fixed area defined by spatial coordinates. The more advanced selection method is custom formula, defined as predicate taking spatial coordinates as input and returning if point belongs to the boundary. Most flexible way of defining boundary conditions are STL files, defining volume inside which condition is active. Each boundary condition may consist of one or more STL volumes. Set of boundary conditions makes load scenario. A single optimization task may use multiple load scenarios; hence the system fully supports optimization with multiple loads. Sample definition of optimization with multiple load scenarios is shown in (Fig. 15.12) Flexibility of defining optimization task using heterogenous input data with different boundary condition makes possible to use the Superficialis platform for industrial structural optimization, research of biomimetic heuristics and modeling bone growth phenomenon, all in a single package.

The superficialis system supports both Windows and Linux platforms and through the use of Docker images it is easy to deploy it in the server environment. The core system algorithms work on a single machine with parallelism provided by multithreading. Stress simulation through Finite Elements Method software is the single most time-consuming operation, optimization of other system elements beyond single machine parallelism is considered not necessary. Stress simulation is realized through third party software. Currently, the system supports calculation FrontISTR, Elmer, and CalculiX, and FrontISTR is the recommended software. When used on a desktop PC the system is limited mostly by the amount of RAM available. The practical limit for a high-end desktop PC (with 128 GB of RAM), mostly imposed by the FEM software, was established as 13 million finite elements. The result of such computation is shown in (Fig. 15.13).

FrontISTR enables grid calculations using MPI technology and this capability is fully supported by the Superficialis system. Superficialis has been successfully run in the grid environment and tasks exceeding 50 million elements were successfully computed (see Fig. 15.14). When deployed in a grid environment computational times are no longer the limiting factor. For example, input data for detailed CT scan is hundreds of megabytes and input/output operations—the time required for writing intermediate files, distribute tasks over the grid and gather results back limits the system scalability.

6. Summary

In silico research has become a basic research method in many scientific fields, a prime example of which is genomics. Given the complexity and research challenges, progress in such fields is only possible by using this approach. Basic research, i.e., experimental or theoretical work undertaken primarily to acquire new knowledge about the foundations of phenomena and observable facts, finds the necessary support in the in silico approach when the complexity and breadth of the research area do not allow for the traditional method of acquiring new knowledge. The pharmacokinetics of new drugs, their interaction with other substances, their impact on

FIGURE 15.12

Example of multiple load case scenario, defined through the user interface of system. Nine STL files were used to define six defined boundary conditions—three supports and three forces. Two scenarios are defined by selecting boundary conditions that are active in each. New boundary conditions and scenarios can be easily added when required.

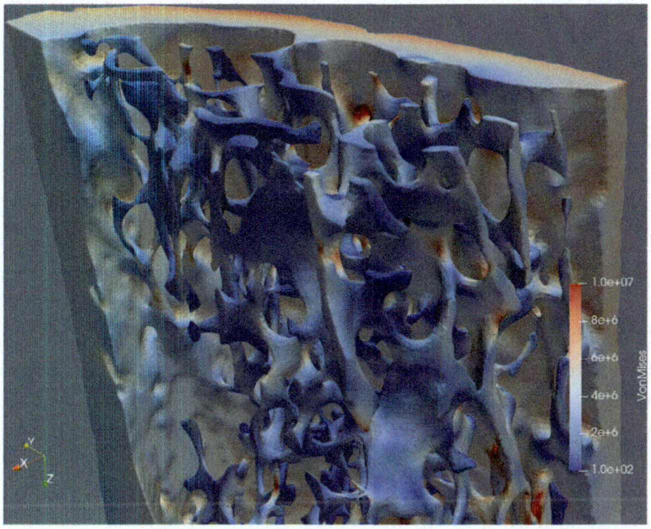

FIGURE 15.13

Example of superficialis system capabilities: Stress simulation performed for the micro-CT scan of rat femur bone. Finite mesh consists of 13 million elements and stress calculation was performed on a desktop PC. Results were smoothed for visualization purposes.

- Data from the MIAB project Waarsing, J.H., Day, J.S., & Weinans, H. (2004). An improved segmentation method for in vivo CT imaging. Journal of Bone and Mineral Research, *19 (10). Published online on July 12, 2004. 10.1359/JBMR.040705© 2004 American Society for Bone and Mineral Research, pp. 1640–1650, work supported by European Union GrantQLRT-1999-02024 (MIAB), 2004.*

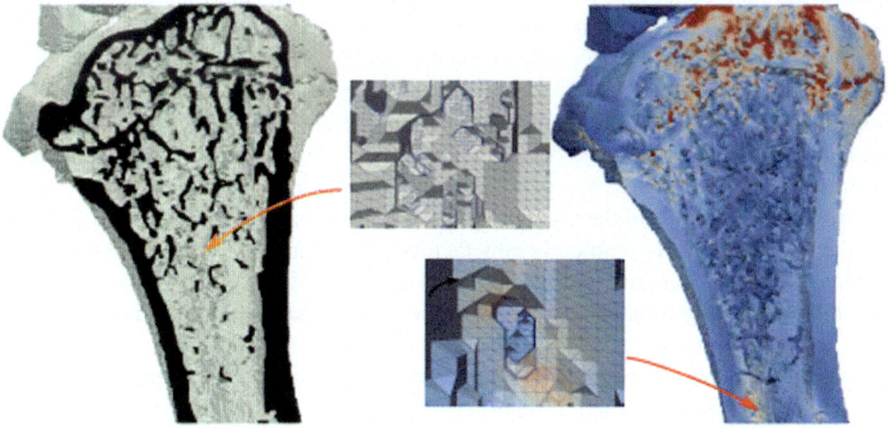

FIGURE 15.14

Left: Cross-section view of mesh with 54 million elements showing trabecula network of the bone (cross section plane in black). Insets show detailed structure of highlighted mesh fragments with edges of tetrahedral cells highlighted. Right: The same bone fragment after FEM simulation of shear stress applied to top part. Nonlinear coloring scale was used to increase contrast between trabeculae.

- Data from the MIAB project Waarsing, J.H., Day, J.S., & Weinans, H. (2004). An improved segmentation method for in vivo CT imaging. Journal of Bone and Mineral Research, *19 (10). Published online on July 12, 2004. 10.1359/JBMR.040705© 2004 American Society for Bone and Mineral Research, pp. 1640–1650, work supported by European Union GrantQLRT-1999-02024 (MIAB), 2004.*

individual organs of the body and their functioning, the impact of comorbidities, and finally the hope for personalized medicine, patient-specific details of therapy in the field of osteoporosis treatment also require such an approach. To consider such an impact, structural models are necessary, of large fragments, and preferably of whole bones, because only then can real conditions be simulated, or close to real ones, and only then can a proper picture of the course of the remodeling phenomenon be obtained. Such simulations are already possible. The team has an efficient computing environment (superficialis.com), a necessary element for starting in silico research. The key novelty proposed in the paper is a significant extension of the list of factors taken into account in the construction of simulation models. This primarily concerns the departure from focusing exclusively on phenomena occurring in a single BMU and taking into account the impact of the entire bone structure. The second important element is the exploitation of an important result, which indicates the identity of models such as the Huiske's remodeling model with the postulated lazy zone concept and structural optimization models developed based on theoretical studies. In the paper (Nowak et al., 2018d), the interpretation of the Lagrange multiplier in the Lagrange equation written for compliance minimization was pointed out, indicating that the Lagrange multiplier is the SED value postulated in the regulatory model, corresponding to homeostasis. In this approach, the regulatory model becomes a way to obtain an optimal structure in the sense of a stiffest design problem. With an appropriate computational environment, it is now possible to try to extend the regulatory model by including considerations regarding the change of the position of the lazy zone, and this is exactly what was presented in the paper. In the context of the effects of romosozumab on sclerostin and thus indirectly on the relationship between the effectiveness of the remodeling phenomenon and mechanical stimulation, this direction of research is particularly attractive. As for the computational side, the computational environment has already been tested and is ready to introduce modifications in the decision block. The software allows for modification of the remodeling scenario independent of the computational efficiency in the scope of structural analysis and the computational grids generation. Such conditions allow for the introduction of completely new factors into the model, concerning the whole organism, such as calcium and phosphorus balance. This way of building a simulation model is a way to develop ways to achieve an actual patient-specific approach. Another important aspect of the research possibilities described is space exploration. Problems with maintaining good bone condition in space travel conditions have become an important research direction for space agencies. The special role of sclerostin in regulating bone formation during long-term simulated microgravity has also been noted (Bloomfield, 2024). Including this and previously mentioned areas the use of the simulation platform should constitute a new quality in the design of algorithms describing pharmacokinetics in the osteoporosis treatment plan, which is reflected in several aspects:

- the research aspect—development of a new approach to the problem of describing the parameters of the regulatory model,

- the numerical aspect—the structural optimization process as a testing ground for developed algorithms,
- the technical aspect—significant improvement of the methods of designing drug strategies - novelty of the action in time planning and its integration with the regulatory model.

References

Allaire, G. (2002). *Shape optimization by the homogenization method.* Springer. https://doi.org/10.1007/978-1-4684-9286-6

Allaire, G., Cavallina, L., Miyake, N., Oka, T., & Yachimura, T. (2019). The homogenization method for topology optimization of structures: Old and new. *Interdisciplinary Information Sciences, 25*(2), 75–146. https://doi.org/10.4036/iis.2019.b.01

Atkins, G. J., Rowe, P. S., Lim, H. P., Welldon, K. J., Ormsby, R., Wijenayaka, A. R., Zelenchuk, L., Evdokiou, A., & Findlay, D. M. (2011). Sclerostin is a locally acting regulator of late-osteoblast/preosteocyte differentiation and regulates mineralization through a MEPE-ASARM-dependent mechanism. *Journal of Bone and Mineral Research, 26*(7), 1425–1436. https://doi.org/10.1002/jbmr.345

Baertschi, S., Baur, N., Lueders-Lefevre, V., Voshol, J., & Keller, H. (2014). Class I and IIa histone deacetylases have opposite effects on sclerostin gene regulation. *Journal of Biological Chemistry, 289*(36), 24995–25009. https://doi.org/10.1074/jbc.m114.564997

Bao, X., Liu, Y., Han, G., Zuo, Z., & Hu, M. (2013). The effect on proliferation and differentiation of cementoblast by using sclerostin as inhibitor. *International Journal of Molecular Sciences, 14*(10), 21140–21152. https://doi.org/10.3390/ijms141021140

Bendsøe, M. P., & Kikuchi, N. (1988). Generating optimal topologies in structural design using a homogenization method. *Computer Methods in Applied Mechanics and Engineering, 71*(2), 197–224. https://doi.org/10.1016/0045-7825(88)90086-2

Bendsoe, M., & Sigmund, O. (2003). *Topology optimization. Theory, methods and applications*. https://doi.org/10.1007/978-3-662-05086-6

Bloomfield, S. (2024). Sclerostin's role in regulating bone formation during long-term simulated microgravity and subsequent recovery. *NASA Human Research Roadmap*. Last Published: 10/31/24 https://humanresearchroadmap.nasa.gov/tasks/task.aspx?i=1907.

Calvo-Gallego, J. L., Manchado-Morales, P., Pivonka, P., & Martínez-Reina, J. (2023). Spatio-temporal simulations of bone remodelling using a bone cell population model based on cell availability. *Frontiers in Bioengineering and Biotechnology, 11*, 1060158. https://doi.org/10.3389/fbioe.2023.1060158

Carter. (1984). Mechanical loading histories and cortical bone remodelling. *Calcified Tissue International, 36*(Suppl. 1), 19–24. https://doi.org/10.1007/BF02406129.

Cohen-Kfir, E., Hanna, A., Levin, A., Abramowitz, E., Bajayo, A., Gurt, I., Zhong, L., D'Urso, A., Toiber, D., Mostoslavsky, R., & Dresner-Pollak, R. (2011). Sirt1 is a regulator of bone mass and a repressor of sost encoding for sclerostin, a bone formation inhibitor. *Endocrinology, 152*(12), 4514–4524. https://doi.org/10.1210/en.2011-1128

Courbon, G., Cleret, D., Linossier, M. T., Vico, L., & Marotte, H. (2017). Early subchondral bone loss at arthritis onset predicted late arthritis severity in a rat arthritis model. *Journal of Cellular Physiology, 232*(6), 1318–1325. https://doi.org/10.1002/jcp.25601

Deaton, J. D., & Grandhi, R. V. (2014). A survey of structural and multidisciplinary continuum topology optimization: Post 2000. *Structural and Multidisciplinary Optimization, 49*(1), 1–38. https://doi.org/10.1007/s00158-013-0956-z

Dongarra, J., Grigori, L., & Higham, N. J. (2020). Numerical algorithms for high-performance computational science. *Philosophical Transactions of the Royal Society A: Mathematical, Physical and Engineering Sciences, 378*(2166), Article 20190066. https://doi.org/10.1098/rsta.2019.0066

Gawel, D., Nowak, M., Hausa, H., & Roszak, R. (2017). New biomimetic approach to the aircraft wing structural design based on aeroelastic analysis. *Bulletin of the Polish Academy of Sciences, Technical Sciences, 65*(5), 741–750. https://doi.org/10.1515/bpasts-2017-0080

George, D., Allena, R., & Rémond, Y. (2019). Integrating molecular and cellular kinetics into a coupled continuum mechanobiological stimulus for bone reconstruction. *Continuum Mechanics and Thermodynamics, 31*(3), 725–740. https://doi.org/10.1007/s00161-018-0726-7

Hamed, E., & Jasiuk, I. (2012). Elastic modeling of bone at nanostructural level. *Materials Science and Engineering: R: Reports, 73*(3–4), 27–49. https://doi.org/10.1016/j.mser.2012.04.001

Huiskes, R., Rulmerman, R., Van Lenthe, G. H., & Janssen, J. D. (2000). Effects of mechanical forces on maintenance and adaptation of form in trabecular bone. *Nature, 405*(6787), 704–706. https://doi.org/10.1038/35015116

Kariem, H., Pastrama, M.-I., Roohani-Esfahani, S. I., Pivonka, P., Zreiqat, H., & Hellmich, C. (2015). Micro-poro-elasticity of baghdadite-based bone tissue engineering scaffolds: A unifying approach based on ultrasonics, nanoindentation, and homogenization theory. *Materials Science and Engineering: C, 46*, 553–564. https://doi.org/10.1016/j.msec.2014.10.072

Koide, M., Kobayashi, Y., Yamashita, T., Uehara, S., Nakamura, M., Yukihiro Hiraoka, B., Ozaki, Y., Iimura, T., Yasuda, H., Takahashi, N., & Udagawa, N. (2017). Bone formation is coupled to resorption via suppression of sclerostin expression by osteoclasts. *Journal of Bone and Mineral Research, 32*(10), 2074–2086. https://doi.org/10.1002/jbmr.3175

Lee, Y. H., Kim, Y., Kim, J. J., & Jang, I. G. (2015). Homeostasis-based aging model for trabecular changes and its correlation with age-matched bone mineral densities and radiographs. *European Journal of Radiology, 84*(11), 2261–2268. https://doi.org/10.1016/j.ejrad.2015.07.027

Lekszycki, T., & dell'Isola, F. (2012). A mixture model with evolving mass densities for describing synthesis and resorption phenomena in bones reconstructed with bioresorbable materials. *ZAMM Zeitschrift fur Angewandte Mathematik und Mechanik, 92*(6), 426–444. https://doi.org/10.1002/zamm.201100082

Martin, M., Sansalone, V., Cooper, D. M. L., Forwood, M. R., & Pivonka, P. (2020). Assessment of romosozumab efficacy in the treatment of postmenopausal osteoporosis: Results from a mechanistic PK-PD mechanostat model of bone remodeling. *Bone, 133*, Article 115223. https://doi.org/10.1016/j.bone.2020.115223

Martínez-Reina, J. (2023). *A mechanobiological bone remodelling model coupling bone physiology and systemic calcium and phosphorus homeostasis.*

Martínez-Reina, J., Domínguez, J., & García-Aznar, J. M. (2011). Effect of porosity and mineral content on the elastic constants of cortical bone: A multiscale approach. *Biomechanics and Modeling in Mechanobiology, 10*(3), 309–322. https://doi.org/10.1007/s10237-010-0236-4

Martínez-Reina, J., & Pivonka, P. (2019). Effects of long-term treatment of denosumab on bone mineral density: Insights from an in-silico model of bone mineralization. *Bone, 125*, 87–95. https://doi.org/10.1016/j.bone.2019.04.022

Marzban, A., Nayeb-Hashemi, H., & Vaziri, A. (2015). Numerical simulation of load-induced bone structural remodelling using stress-limit criterion. *Computer Methods in Biomechanics and Biomedical Engineering, 18*(3), 259–268. https://doi.org/10.1080/10255842.2013.792915

Metzger, C. E., Brezicha, J. E., Elizondo, J. P., Narayanan, S. A., Hogan, H. A., & Bloomfield, S. A. (2017). Differential responses of mechanosensitive osteocyte proteins in fore- and hindlimbs of hindlimb-unloaded rats. *Bone, 105*, 26–34. https://doi.org/10.1016/j.bone.2017.08.002. Epub 2017 Aug 3. PMID: 28782619.

Morse, A., McDonald, M. M., Kelly, N. H., Melville, K. M., Schindeler, A., Kramer, I., Kneissel, M., Ch van der Meulen, M., & Little, D. G. (2014). Mechanical load increases in bone formation via a sclerostin-independent pathway. *Journal of Bone and Mineral Research, 29*(11), 2456–2467. https://doi.org/10.1002/jbmr.2278

Nowak, M. (2006). Structural optimization system based on trabecular bone surface adaptation. *Structural and Multidisciplinary Optimization, 32*(3), 241–249. https://doi.org/10.1007/s00158-006-0027-9

Nowak, M. (2020). New aspects of the trabecular bone remodeling regulatory model—two postulates based on shape optimization studies. *Advanced Structured Materials., 132*, 97–105. https://doi.org/10.1007/978-3-030-50464-9_6

Nowak, M. (2022). *Biomimetyczna metoda optymalizacji strukturalnej jako element wzmacniający potencjał addytywnych metod wytwarzania, computer aided engineering: Nauka i przemysł/red.* Wrocław: Tadeusz Smolnicki.

Nowak, M., & Boguszewski, A. (2021). Topology optimization without volume constraint – The new paradigm for lightweight design. *Bulletin of the Polish Academy of Sciences, Technical Sciences, 69*(4). https://doi.org/10.24425/bpasts.2021.137732, 137732–137732.

Nowak, M., & Polak, J. (2021). Bioniamoto – The efficient numerical platform for shape and topology optimization (O107196 - FS10 - Optimization for solids and fluids – Oral). In *Abstract book ICTAM2021 – 25th International Congress of Theoretical and Applied Mechanics* (pp. 2905–2906).

Nowak, M., Polak, J., & Napierała, O. (2024). *The implementation of biomimetic structural optimization software in a supercomputing environment* (Vol 1146, pp. 139–153). Springer Science and Business Media LLC. https://doi.org/10.1007/978-3-031-73161-7_13

Nowak, M., Sokołowski, J., & Żochowski, A. (2018d). Justification of a certain algorithm for shape optimization in 3D elasticity. *Structural and Multidisciplinary Optimization, 57*(2), 721–734. https://doi.org/10.1007/s00158-017-1780-7

Nowak, M., Sokołowski, J., & Żochowski, A. (2020a). Biomimetic approach to compliance optimization and multiple load cases. *Journal of Optimization Theory and Applications, 184*(1), 210–225. https://doi.org/10.1007/s10957-019-01502-1

Nowak, M., Sokołowski, J., & Żochowski, A. (2020b). New aspects of the trabecular bone remodeling regulatory model resulting from the shape optimization studies. *Proceedings of the Institution of Mechanical Engineers - Part H: Journal of Engineering in Medicine, 234*(3), 282–288. https://doi.org/10.1177/0954411919862678

Nowak, M., Zochowski, A., & Sokolowski, J. (2018c). Benchmark problem for structural optimization with multiple load conditions in 3-D. In *41st Solid Mechanics Conference.*

Nowak, M., Zochowski, A., & Sokolowski, J. (2018b). The biomimetic method for shape optimization in 3D elasticity. In *13th World Congress on Computational Mechanics (WCCM XIII)/2nd Pan American Congress on Computational Mechanics (PANACM II), July 22-27, New York, NY, USA*.

Nowak, M., Zochowski, A., & Sokolowski, J. (2018a). Biomimetic topology optimization with multiple load cases. In *6th International Conference on Engineering Optimization*. Instituto Superior Técnico.

Pérez-Campo, F. M., Santurtún, A., García-Ibarbia, C., Pascual, M. A., Valero, C., Garcés, C., Sañudo, C., Zarrabeitia, M. T., & Riancho, J. A. (2016). Osterix and RUNX2 are transcriptional regulators of Sclerostin in human bone. *Calcified Tissue International, 99*(3), 302–309. https://doi.org/10.1007/s00223-016-0144-4

Pivonka, P., Buenzli, P. R., Scheiner, S., Hellmich, C., & Dunstan, C. R. (2013). The influence of bone surface availability in bone remodelling—A mathematical model including coupled geometrical and biomechanical regulations of bone cells. *Engineering Structures, 47*, 134–147. https://doi.org/10.1016/j.engstruct.2012.09.006

Pivonka, P., Zimak, J., Smith, D. W., Gardiner, B. S., Dunstan, C. R., Sims, N. A., John Martin, T., & Mundy, G. R. (2010). Theoretical investigation of the role of the RANK–RANKL–OPG system in bone remodeling. *Journal of Theoretical Biology, 262* (2), 306–316. https://doi.org/10.1016/j.jtbi.2009.09.021

Polak, J., & Nowak, M. (2022). *The efficient trabecular bone remodeling numerical tool enabling multiple load case simulation* (Vol 175, pp. 459–471). Springer Science and Business Media LLC. https://doi.org/10.1007/978-3-031-04548-6_22

Polak, J., & Nowak, M. (2023). From structural optimization results to parametric CAD modeling—automated, skeletonization-based truss recognition. *Applied Sciences, 13*(9), 5670. https://doi.org/10.3390/app13095670

Scheiner, S., Pivonka, P., & Hellmich, C. (2013). Coupling systems biology with multiscale mechanics, for computer simulations of bone remodeling. *Computer Methods in Applied Mechanics and Engineering, 254*, 181–196. https://doi.org/10.1016/j.cma.2012.10.015

Sigmund, O., (1999). On the optimality of bone microstructure. Springer Science and Business Media LLC. doi: 10.1007/0-306-46939-1_20.

Sigmund, O. (2001). A 99 line topology optimization code written in matlab. *Structural and Multidisciplinary Optimization, 21*(2), 120–127. https://doi.org/10.1007/s001580050176

Sigmund, O., & Maute, K. (2013). Topology optimization approaches: A comparative review. *Structural and Multidisciplinary Optimization, 48*(6), 1031–1055. https://doi.org/10.1007/s00158-013-0978-6

Sokołowski, J., & Zolésio, J. (1992). Introduction to shape optimization. Shape sensitivity analysis. In *Springer Series in Computational Mathematics, 16*. Springer-Verlag, Berlin, ISBN: 3-540-54177-2, 1992.

Sutherland, M. K., Geoghegan, J. C., Yu, C., Turcott, E., Skonier, J. E., Winkler, D. G., & Latham, J. A. (2004). Sclerostin promotes the apoptosis of human osteoblastic cells: A novel regulation of bone formation. *Bone, 35*(4), 828–835. https://doi.org/10.1016/j.bone.2004.05.023

van Oers, R. F. M., Ruimerman, R., Tanck, E., Hilbers, P. A. J., & Huiskes, R. (2008). A unified theory for osteonal and hemi-osteonal remodeling. *Bone, 42*(2), 250–259. https://doi.org/10.1016/j.bone.2007.10.009

Waarsing, J. H., Day, J. S., & Weinans, H. (2004). An improved segmentation method for in vivo μCT imaging. *Journal of Bone and Mineral Research, 19*(10), 1640–1650. https://doi.org/10.1359/jbmr.040705

Winkler, D. G., Kung Sutherland, M. S., Ojala, E., Turcott, E., Geoghegan, J. C., Shpektor, D., Skonier, J. E., Yu, C., & Latham, J. A. (2005). Sclerostin inhibition of Wnt-3a-induced C3H10T1/2 cell differentiation is indirect and mediated by bone morphogenetic proteins. *Journal of Biological Chemistry, 280*(4), 2498–2502. https://doi.org/10.1074/jbc.M400524200

Winkler, D. G., Sutherland, M. K., Geoghegan, J. C., Yu, C., Hayes, T., Skonier, J. E., Shpektor, D., Jonas, M., Kovacevich, B. R., Staehling-Hampton, K., Appleby, M., Brunkow, M. E., & Latham, J. A. (2003). Osteocyte control of bone formation via sclerostin, a novel BMP antagonist. *EMBO Journal, 22*(23), 6267–6276. https://doi.org/10.1093/emboj/cdg599

Wolff, J. Das Gesetz der Transformation der Knochen.

CHAPTER

Bone fluid flow models: Calculation of interstitial fluid flow around osteocytes

16

Eda Biricik and Susannah P. Fritton

The City College of New York, Department of Biomedical Engineering, New York, NY, United States

1. Introduction

Bone interstitial fluid plays a critical role in bone maintenance, converting whole bone-level mechanical loads to cellular-level signals and facilitating nutrient delivery, waste removal, and transport of signaling molecules throughout bone (Fritton & Weinbaum, 2009; Knothe Tate et al., 2000; Piekarski & Munro, 1977; Qin & Hu, 2014). Osteocytes are the primary mechanosensory cells of bone and are embedded within the bone matrix, connecting through a network of canals containing interstitial fluid (Bonewald, 2006; Burger et al., 1995; Cowin et al., 1991). Whole bone-level mechanical loading induces interstitial fluid flow around osteocytes, which is thought to generate key mechanical signals that initiate bone mechanotransduction (Klein-Nulend et al., 2013; Kufahl & Saha, 1990; Weinbaum et al., 1994). Quantifying bone interstitial fluid flow experimentally provides insights into molecular transport and mechanotransduction; however, the unique mineralized environment of osteocytes and the small dimensions of the interstitial fluid space make it difficult to make direct measurements. Thus, analytical and computational models are effective tools to estimate interstitial fluid flow throughout bone tissue (Cardoso et al., 2013). In this review, we examine how models across different length scales estimate mechanical signals experienced by osteocytes—interstitial fluid velocity, shear stress, hoop and axial strains, as well as how various loading modalities influence fluid movement. Additionally, we report a range of magnitudes for each mechanical signal across various model types to provide a comparative overview of modeling outcomes. Finally, we highlight how structural changes in bone porosity due to conditions such as aging and osteoporosis may alter interstitial fluid flow around osteocytes, which might impair bone function. We examine the modeling approaches and discuss their strengths and limitations, providing a summary of how models are contributing to our understanding of bone interstitial fluid flow.

Biomechanics and Mechanobiology of Bone. https://doi.org/10.1016/B978-0-443-26739-0.00005-3

2. Nested bone porosities, bone permeability, and the pathways of interstitial fluid flow

Understanding the characteristics and interconnectedness of the bone porosities is crucial for developing accurate models of interstitial fluid flow because features such as pore size and permeability strongly affect bone's mechanical properties and interstitial fluid movement. Both cortical and cancellous bone contain fluid-filled networks with different pore sizes that are nested and connected to the bone marrow cavity (Cowin, 1999). With whole bone-level deformations, fluid pressure builds up in the bone pores, and interstitial fluid flows in the direction of the fluid pressure gradient within and between the nested bone porosities. The permeability of each bone porosity depends on the characteristic pore size and geometry specific to that network (Cardoso et al., 2013). There are three levels of bone porosity, and each plays a distinct role in interstitial fluid dynamics, as described below.

2.1 The lacunar-canalicular porosity containing osteocytes and interstitial fluid

The lacunar-canalicular porosity houses the osteocyte network and includes lacunae—cavities occupied by the cell bodies—and canaliculi, the long, slender canals occupied by the osteocyte processes (Fig. 16.1). The diameter of canaliculi is ~100–600 nm, and the lacunar-canalicular porosity occupies approximately 1%–2% of the bone volume (Buenzli & Sims, 2015; Cardoso et al., 2013; Schneider

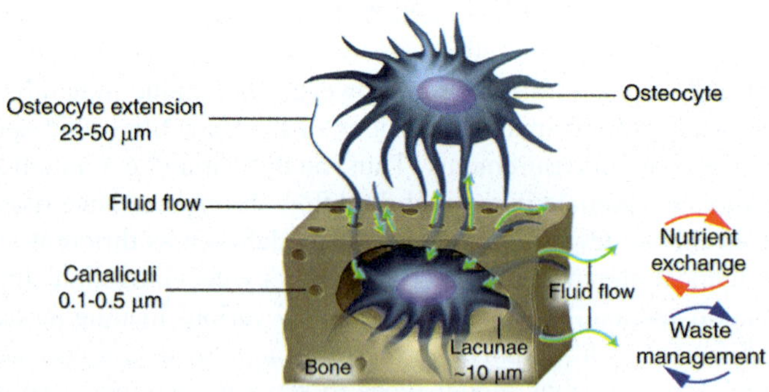

FIGURE 16.1

Osteocytes reside within the lacunae and canaliculi in the mineralized bone matrix. Interstitial fluid fills the gaps between the osteocyte and the bone extracellular matrix and flows within the lacunar-canalicular porosity after application of mechanical loading, facilitating nutrient exchange and removal of waste.

- From Wei, F., Flowerdew, K., Kinzel, M., Perotti, L.E., Asiatico, J., Omer, M., Hovell, C., Reumers, V. & Coathup, M.J. (2022). Changes in interstitial fluid flow, mass transport and the bone cell response in microgravity and normogravity. Bone Research, 10*(1), 65. https://doi.org/10.1038/s41413-022-00234-9*

et al., 2007). Interstitial fluid fills the gap between the walls of the lacunar-canalicular porosity and the osteocytes, and a fibrous matrix similar to endothelial glycocalyx coats the cells and fills the annulus (You et al., 2004). This pericellular matrix consists of negatively charged proteoglycans that were first theoretically estimated to have a spacing of ~7 nm (Weinbaum et al., 1994). Injectable tracer studies further supported this estimation by demonstrating that molecules smaller than 6 nm diameter can pass through the lacunar-canalicular porosity, whereas molecules larger than 12 nm cannot (Ciani et al., 2005; Knothe Tate et al., 1998; Tami et al., 2003; Wang et al., 2004). The proteoglycan of the pericellular matrix has been identified as heparan sulfate proteoglycan 2 (perlecan) with a fiber spacing between 10 and 13 nm in adult murine bone (Thompson et al., 2011; Wang, 2018; Wang et al., 2014). Transverse fibers of the pericellular matrix (tethering elements) position osteocyte processes centrally by connecting to the canalicular walls (Shapiro et al., 1995; You et al., 2001). In addition, periodical conical projections extending from the canalicular walls connect to the osteocyte processes via integrins, which generate strain concentrations on osteocyte processes during interstitial fluid flow (Cabahug-Zuckerman et al., 2018; McNamara et al., 2009; Thi et al., 2013; Wang et al., 2007). Strain amplification models have been proposed to explain the role of the osteocyte canalicular ultrastructure (the pericellular matrix, pericellular matrix tethering elements, and conical projections) in converting whole bone-level loading to cellular signals via interstitial fluid flow (see Section 3.2.1).

2.2 The vascular porosity containing bone vasculature and interstitial fluid

The vascular porosity in cortical bone consists of vascular canals (referred to as Haversian or osteonal and Volkmann canals in humans) that contain blood vessels, nerves, and interstitial fluid. This porosity has the largest pore size in cortical bone, with a diameter of ~20–100 μm (Cooper et al., 2003; Cowin & Cardoso, 2015; Roothaer et al., 2018), and occupies 1%–16% of bone volume depending on species, sex, age, and bone type (Andronowski et al., 2023; Cardoso et al., 2013). The vascular porosity connects to the lacunar-canalicular porosity at the vascular canal surfaces and to the bone marrow cavity at the endosteal surface, allowing interstitial fluid flow between these porosities and facilitating molecular exchange between blood and osteocytes (Cowin, 1999; Zhang et al., 1998) (Fig. 16.2). The vascular porosity does not sustain elevated fluid pressure due to its large pore size, allowing a rapid decay of fluid pressure (Cardoso et al., 2013). In addition, the interstitial fluid pressure in vascular canals cannot exceed the blood pressure in bone capillaries (40–60 mmHg) because higher pressure levels would collapse the blood vessels, leading to oxygen and nutrient deprivation (Cowin, 1999). Thus, the vascular porosity acts as a low-pressure reservoir for the lacunar-canalicular porosity, and interstitial fluid flows in and out during cyclic mechanical loading (Cowin & Cardoso, 2015).

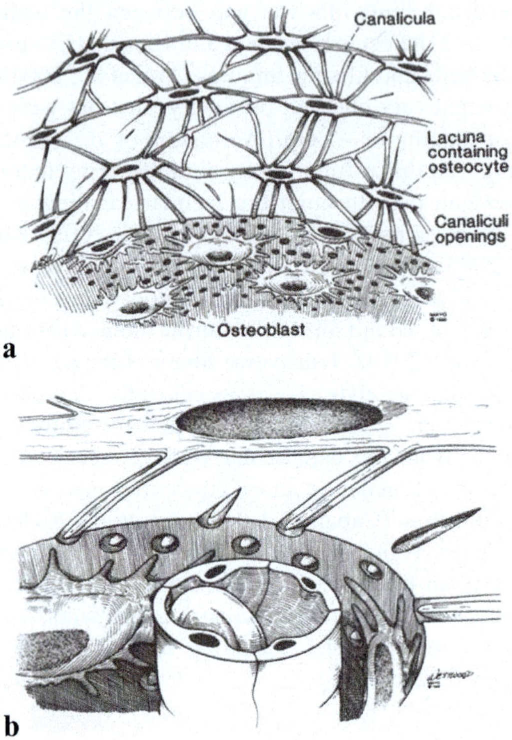

a

b

FIGURE 16.2

Nested porosities of bone. (a) Osteocytes reside in lacunae, while their processes extend through canaliculi. (b) The lacunar-canalicular porosity connects to the vascular porosity, which contains blood vessels and nerves, at vascular canal surfaces, facilitating interstitial fluid flow between the two porosities.

- From Kelly, P.J. (1983), Pathways of transport in bone. Comprehensive Physiology, 1983, 371–396. https://doi.org/10.1002/j.2040-4603.1983.tb00992.x

2.3 The collagen-apatite porosity in the mineralized matrix

The third and smallest bone porosity is the collagen-apatite porosity, which consists of the spaces between collagen fibrils and mineral apatite crystals and has been reported to have an approximate pore size of 5–10 nm (Cowin, 1999; Holmes et al., 1964). In vivo tracer studies have demonstrated minimal permeability to small molecules (1–2 nm) in this porosity, in contrast to the free (or pore) water found in the lacunar-canalicular and vascular porosities (Knothe Tate et al., 2000; Tanaka & Sakano, 1985; Wang et al., 2004). An NMR study in rabbit cortical bone showed that fluid could diffuse very slowly through the bone extracellular matrix (Fernández-Seara et al., 2002). Additionally, an interconnected network of nanochannels within the bone extracellular matrix in young mouse and older human bone has been identified with a diameter of ~10–50 nm, larger than previously estimated collagen-apatite pore size (Tang et al., 2022, 2023). These nanochannels are potentially an interconnected porosity of collagen-apatite that might facilitate small molecule transport (Tang et al., 2023). Additional studies are needed to clarify the composition and function of the nanochannels to further understand the permeability and interstitial fluid and solute movement in the collagen-apatite porosity.

2.4 Measurements of the permeabilities of the vascular and lacunar-canalicular porosities based on experimental studies

Accurately determining the permeability of the bone porosities is important when quantifying interstitial fluid flow throughout bone tissue. Permeability, a measure of the ability of fluid to flow through porous media, is governed by Darcy's law and relates the instantaneous flow rate to the fluid pressure gradient. Because vascular pores are two to three orders of magnitude larger than lacunar-canalicular pores, experimental measurements predominantly capture the permeability of vascular canals, and the permeability of the lacunar-canalicular porosity remains difficult to measure directly. Experimental measurements of permeability of the vascular porosity using Darcy's law have reported values ranging from 10^{-13} to 10^{-17} m^2 in canine and bovine bone (Li et al., 1987; Malachanne et al., 2008; Rouhana et al., 1981; Wen et al., 2010). To measure the lacunar-canalicular permeability, several experimental techniques, including single osteon stress relaxation, nanoindentation, and intramedullary pressure measurements, have been employed and combined with models. By fitting stress relaxation curves of single osteons of human, bovine, and canine femurs to poroelastic osteon models, the lacunar-canalicular permeability has been estimated to range from 10^{-22} to 10^{-25} m^2, several orders of magnitude smaller than the vascular permeability described above (Benalla et al., 2014; Gailani et al., 2009; Smit et al., 2002). Oyen estimated lacunar-canalicular permeability by fitting the creep displacement curve obtained from nanoindentation of equine bone to a poroelastic indentation model; a permeability of 10^{-24} m^2 was calculated (Oyen, 2008). Gardinier et al. fitted characteristic fluid pressure relaxation times obtained from intramedullary pressure measurements of canine metacarpal bones to an idealized poroelastic osteon model, calculating lacunar-canalicular permeability on the order of 10^{-23} m^2 (Gardinier et al., 2010). All these approaches have shown that estimations of lacunar-canalicular permeability based on experimental measurements range from 10^{-22} to 10^{-25} m^2; theoretical estimations of lacunar-canalicular permeability will be discussed in Section 3.2.

3. Analytical and computational models of interstitial fluid flow in bone

Analytical and computational models can overcome the limitations of experimental measurements to calculate interstitial fluid flow parameters. Models can capture the intricate structure of the bone porosities and quantify interstitial fluid flow at both cell and tissue levels. This paper reviews analytical and computational models at three length scales—osteocyte-level, osteon-level, and whole bone-level—to explore their capabilities and limitations. The following sections analyze these models in terms of key mechanical signals experienced by osteocytes—fluid velocity, shear stress, drag force, hoop and axial strains—that might initiate mechanotransduction. Although interstitial fluid flow is also crucial for solute transport in bone tissue, this review focuses on calculation of interstitial fluid velocity in the osteocyte lacunar-canalicular porosity, which is the most frequently reported fluid-flow parameter.

3.1 Models of interstitial fluid flow in the collagen-apatite porosity

The possibility of interstitial fluid flow in the collagen-apatite porosity has been considered by several models. Based on strain-generated potentials measured across mechanically loaded fluid-filled bone samples, early analytical models demonstrated that strain-generated potentials could result from the electrokinetic effect due to the flow of charged fluid on a charged surface within the collagen-apatite porosity (Pollack et al., 1984; Salzstein et al., 1987; Salzstein & Pollack, 1987). At the time, these strain-generated potentials were assumed to be the main stimuli sensed by bone cells (Starkebaum et al., 1979). However, tracer studies showed that the effective pore size of the collagen-apatite porosity was smaller than the pore size required to generate these potentials (Tanaka & Sakano, 1985). Additional analytical models then demonstrated that interstitial fluid flow within the lacunar-canalicular porosity could generate the strain-generated potentials (Cowin et al., 1995) (see Section 3.2.1). Mak et al. proposed a theoretical model that superposed collagen-apatite porosity with the other bone porosities, showing that, in addition to the interstitial fluid flow in the lacunar-canalicular porosity, fluid could flow within the collagen-apatite porosity as well (Mak et al., 1997). Interstitial fluid velocity within each porosity could influence the other depending on the permeabilities of these porosities (Mak & Zhang, 2001). Molecular dynamics simulations further supported the possibility of interstitial fluid flow in pores as small as 2–8 nm within the collagen-apatite porosity (Lemaire, Pham, Capiez-Lernout et al., 2015; Lemaire, Pham, de Leeuw, & Naili, 2015), even though in vivo tracer studies have not clearly demonstrated movement of nanometer-sized tracers in the collagen-apatite porosity (Wang et al., 2004). Current studies are underway to investigate the influence of potentially interconnected collagen-apatite nanochannels (Tang et al., 2022, 2023) that might contribute to small molecule transport in the mineralized matrix surrounding osteocytes.

3.2 Osteocyte-level analytical and computational models

Most osteocyte-level models focus on the effects of the geometry and ultrastructure of the lacunar-canalicular space on interstitial fluid flow, without considering flow in the collagen-apatite porosity. Since interstitial fluid flows within the gap between the osteocyte and the lacunar-canalicular walls, analytical and computational models have focused on investigating the effect of the lacunar-canalicular morphology on fluid flow and the mechanisms of mechanotransduction. Because the gap between the canaliculi and osteocyte processes is an order of magnitude smaller than the gap between lacunae and osteocyte cell bodies, the fluid velocity is higher within canaliculi and most studies focus on canalicular fluid flow.

3.2.1 The early development of interstitial fluid flow models and osteocyte mechanotransduction

Piekarski and Munro proposed that mechanical load-driven interstitial fluid flow in the lacunar-canalicular porosity was essential, as simple diffusion alone was insufficient to sustain osteocytes embedded throughout the bone matrix (Piekarski & Munro, 1977). Combining the nutritional pathway and strain-generated potential hypotheses, Weinbaum and colleagues proposed that mechanical load-driven interstitial fluid flow in the lacunar-canalicular porosity was the primary driver of bone mechanotransduction (Weinbaum et al., 1994). Using an idealized poroelastic analytical model of a periodic unit cell, they simulated the lacunar-canalicular porosity filled with pericellular matrix (fiber spacing ~7 nm) and demonstrated that whole-bone mechanical loading could generate sufficient fluid shear stress (0.8–3 Pa) to stimulate osteocytes (Weinbaum et al., 1994). In two complementary studies, the same group showed that when the lacunar-canalicular porosity contained charged fibers, interstitial fluid flow could generate strain-generated potentials consistent with experimental measurements by Starkebaum and colleagues (Cowin et al., 1995; Starkebaum et al., 1979; Zeng et al., 1994).

In addition to fluid shear stress, strains > ~10,000 microstrain (με) have been shown to stimulate bone cells in vitro (Klein-Nulend et al., 1995; Murray & Rushton, 1990; Smalt et al., 1997; You et al., 2000); however, physiological tissue strains are lower than the required strain levels for cell stimulation (Burr et al., 1996; Fritton et al., 2000; Lanyon et al., 1975). You et al. proposed a strain amplification theory, demonstrating that transverse tethering elements that connect osteocyte processes to the canalicular wall can amplify the strain experienced by osteocytes (You et al., 2001). Their model shows that a drag force acting on the pericellular matrix is generated by interstitial fluid flow through the canaliculi, exerting hoop strains on the osteocyte processes that can reach up to 100 times the tissue strains (You et al., 2001) (Table 16.1). Han et al. expanded this model by generating a refined 3D model, specifically examining the effect of the flexural rigidity of the tethering elements (Han et al., 2004) (Table 16.1).

While these strain amplification models showed possible mechanisms to enhance tissue-level strain via interstitial fluid flow at the cellular level, the initiation of

Table 16.1 Summary of modeling parameters and reported outcomes from models of strain amplification that enhance tissue-level strain via interstitial fluid flow at the cellular level.

	Modeling parameters								Outcomes		
Reference	Level of model	Model type	Model geometry	Applied load type	Applied strain magnitude (με)	Loading frequency (Hz)	Tissue-level lacunar-canalicular permeability (m^2)	Osteocyte pericellular permeability (m^2)	Hoop/radial strain on osteocyte process (με)	Axial strain on osteocyte process (με)	Tension on integrin attachment (pN)
You et al. (2001)	Osteocyte (canalicular)	Analytical poroelastic	Idealized	Cyclical compression	50	1	NR	NR	2900	–	–
						20			6800		
					1000	1			21,000		
						20			49,000		
Han et al. (2004)	Osteocyte (canalicular)	Analytical poroelastic	Idealized	Cyclical compression	50	1	NR	1.03×10^{-17}	~1000	–	–
						20			~2500		
					1000	1			~5000		
						20			~17,000		
Wang et al. (2007)	Osteocyte (canalicular)	Analytical poroelastic	Idealized	Cyclical compression	5	30	NR	1.03×10^{-17}	–	~15,000	–
					50	1			~500	~20,000	~1
						20			~1500	~35,000	~4
					1000	1			~3500	~60,000	~10
						20			~11,000	~120,000	~33

NR, *Not reported; parameter was not directly reported but was adopted from earlier research articles—refer to the original sources cited in the study.*

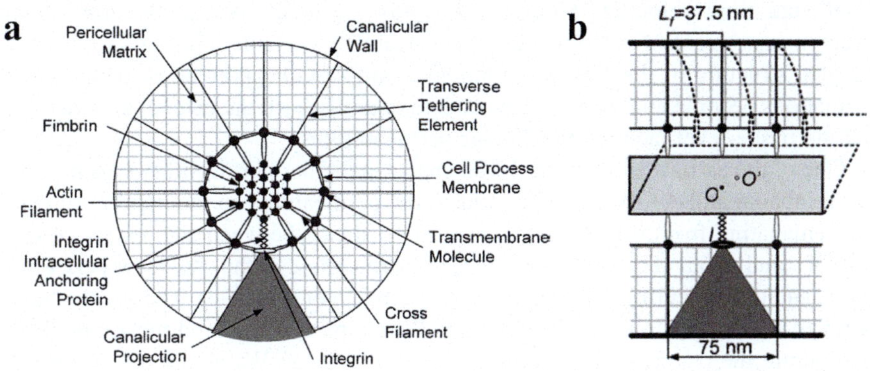

FIGURE 16.3

Strain amplification mechanism proposed by Wang et al. (2007): (a) Transverse cross section of an actin-filled osteocyte process surrounded by a pericellular matrix and interstitial fluid. The osteocyte process is connected to the canalicular wall via transverse tethering elements and canalicular projections connected via integrins. (b) A longitudinal section of the osteocyte process showing that when interstitial fluid flow is generated due to a pressure gradient in the lacunar-canalicular porosity, the tethering elements and the osteocyte process are deformed (dashed lines), producing local axial strains in the vicinity of the integrin attachment sites calculated to be one to two orders of magnitude larger than the tissue-level strains.

- From Wang, Y., McNamara, L. M., Schaffler, M. B., & Weinbaum, S. (2007). A model for the role of integrins in flow induced mechanotransduction in osteocytes. Proceedings of the National Academy of Sciences, 104*(40), 15941–15946. https://doi.org/10.1073/pnas.0707246104*

intracellular signaling remained unclear due to the absence of mediator molecules in the tethering elements. Wang et al. proposed that integrin-based conical projections can generate localized axial and radial strain concentrations from the drag force on the tethering elements due to interstitial fluid flow (Wang et al., 2007) (Fig. 16.3). Their analytical model demonstrated that a tissue-level strain of 1000 με at 1 Hz could generate approximately 60,000 axial με at the focal attachment sites of osteocyte processes (Wang et al., 2007) (Table 16.1), a strain level that has been shown to initiate Ca^{2+} signaling in bone cells (You et al., 2000).

3.2.2 The anisotropic microstructure of the lacunar-canalicular porosity affects analytical permeability estimations

Since interstitial fluid flow has been shown to amplify macro-level mechanical signals at the osteocyte level, permeability is a critical parameter for bone fluid flow models. The lacunar-canalicular permeability depends on the geometry of the lacunar-canalicular network and the permeability of the osteocyte pericellular matrix that fills the pores. In poroelastic models, tissue-level lacunar-canalicular permeability is used to calculate tissue-level fluid velocities, while osteocyte pericellular permeability defines the flow resistance in the canaliculi and is used to calculate canalicular fluid velocities around osteocytes. As mentioned in Section 2.4, the direct measurement of osteocyte-level permeability is challenging due to the intermingled structure of the lacunar-canalicular and vascular porosities, so theoretical estimations become an effective approach to quantify both the osteocyte pericellular and the tissue-level lacunar-canalicular permeabilities.

One of the earliest estimations of lacunar-canalicular permeability was proposed by Weinbaum and colleagues, who related the permeability to the number of canaliculi emanating from osteocyte lacunae, the distance between two osteocytes, the radii of canaliculi and osteocyte processes, the thickness of the osteocyte pericellular fibers, and the pericellular fiber spacing (Weinbaum et al., 1994). These calculations assumed a homogeneous 3D distribution of these parameters, and most models that incorporate the osteocyte pericellular matrix utilize the Weinbaum et al. formulation, with reported lacunar-canalicular permeabilities ~10^{-20} m^2 (Wang et al., 1999; Zhang et al., 1998). To examine the effect of anisotropy, Beno et al. estimated the number of canaliculi in three orthogonal directions using published measurements from various species (Beno et al., 2006). Their results indicated that although permeability constants differed across directions, the difference was less than an order of magnitude, and the largest effect on permeability was observed by varying canalicular annulus dimensions. Lemaire and colleagues (Lemaire et al., 2012) extended this analysis by incorporating canalicular tortuosity into their permeability calculations, obtaining lacunar-canalicular permeability values within the same range reported in earlier studies (~10^{-22}—10^{-18} m^2) (Anderson et al., 2008; Beno et al., 2006). Lemaire and colleagues also highlighted the discrepancies between experimental and theoretical estimations of lacunar-canalicular permeability, noting that experimental estimations were several orders of magnitude lower (~10^{-22}—10^{-25} m^2), potentially due to the electro-osmotic effects from negative charges on bone pore surfaces (Lemaire et al., 2012). While theoretical and experimental studies provide various estimates of lacunar-canalicular permeability, computational models are necessary to simulate the effect of these wide-range permeability values on interstitial fluid flow and osteocyte mechanotransduction.

3.2.3 Computational osteocyte-level lacunar-canalicular models to assess interstitial fluid flow

Lacunar-canalicular models mimic the environment of an osteocyte with its processes and aim to calculate interstitial fluid velocities in a lacunar-canalicular unit. Interstitial fluid flow is analyzed using analytical poroelastic models, computational fluid dynamics or transport models, as well as fluid-structure interaction simulations. The computational fluid dynamics method focuses solely on fluid behavior, assuming solid boundaries (cellular walls or osteocytes) are rigid, whereas the fluid-structure interactions approach integrates fluid dynamics with structural mechanics, allowing for interactions between solid deformations and fluid motion. These

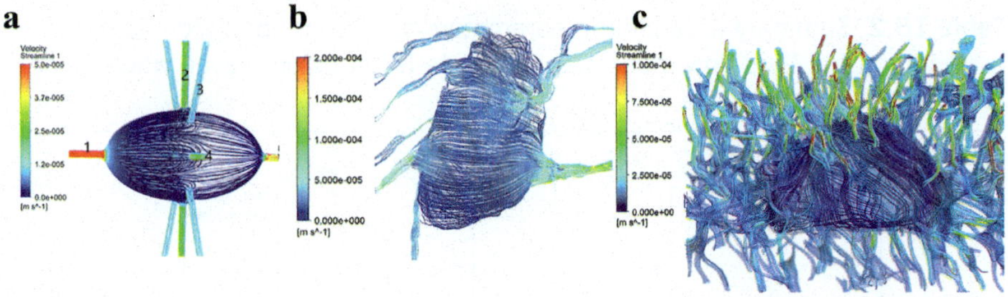

FIGURE 16.4

Osteocyte-level lacunar-canalicular models, ranging from idealized to anatomically realistic geometries, showing lacunar-canalicular fluid velocity streamlines, with maximum values on the order of 100 μm/s. (a) An idealized geometry with an ellipsoidal osteocyte lacuna and a few canaliculi; (b) and (c) lacunar-canalicular geometries obtained from FITC staining and confocal imaging, with b) showing an osteocyte from a rat tibia with only a few canaliculi and c) showing an osteocyte from a mouse femur with a denser canalicular network.

- Fig (a) and (c) From Niroobakhsh, M., Laughrey, L. E., Dallas, S.L., Johnson, M. L., & Ganesh, T. (2024). Computational modeling based on confocal imaging predicts changes in osteocyte and dendrite shear stress due to canalicular loss with aging. Biomechanics and Modeling in Mechanobiology, 23*(1), 129–143. https://doi.org/10.1007/s10237-023-01763-w, Fig (b) From Verbruggen, S.W., Vaughan, T.J., & McNamara, L.M. (2014). Fluid flow in the osteocyte mechanical environment: A fluid-structure interaction approach.* Biomechanics and Modeling in Mechanobiology, 13*(1), 85–97. https://doi.org/10.1007/s10237-013-0487-y*

osteocyte-level models can be either idealized, investigating specific morphological features separately, or physiologically realistic, assessing the total effect of morphology (Fig. 16.4, Table 16.2).

Idealized osteocyte lacunar-canalicular models usually consist of an ellipsoidal lacuna containing a cell body, cylindrical canaliculi occupied by osteocyte processes, and interstitial fluid flow driven by a fluid pressure gradient, mechanical loading, or both. These idealized models often neglect the osteocyte pericellular matrix. In a 2D analytical poroelastic model, Gururaja et al. demonstrated that mechanical loading elevates fluid pressure around lacunae and induces fluid flow in canaliculi (Gururaja et al., 2005). Anderson et al. used computational fluid dynamics to compare the differences in fluid pressure and shear stress profiles in canaliculi and around the osteocyte cell body (Anderson et al., 2005). A fluid pressure gradient of 300 Pa was introduced between arbitrarily assigned inlet and outlet canaliculi, and the shear stress around the cell body was calculated to be lower than in the canaliculi, with the highest shear stress observed at lacunar-canalicular junctions (Anderson et al., 2005). Fluid-structure interaction models have also used the same fluid pressure gradient of 300 Pa as well as 3000 με compression chosen to mimic

Table 16.2 Summary of modeling parameters and reported outcomes from bone interstitial fluid flow models at osteocyte-level, osteon-level, whole bone-level and multi-level.

	Modeling parameters								
	Reference	Level of model	Model type	Bone type	Model geometry	Applied load type	Maximum tissue-level strain or pressure magnitude or strain rate from loading	Loading frequency	Applied cellular-level fluid pressure
OSTEOCYTE-LEVEL	Weinbaum et al. (1994)	Lacunar-canalicular	Analytical poroelastic	NS	Idealized	Cyclical compression	Varied	1–100 Hz	N/A
	Anderson et al. (2005)	Lacunar-canalicular	Computational fluid dynamics	NS	Idealized	N/A	N/A	N/A	300 Pa
	Kwon and Frangos (2010)	Lacunar-canalicular	Computational transport model	Mouse femur	Idealized	Intra-medullary pressure	NR	~5 Hz	N/A
	Price et al. (2011)	Lacunar-canalicular	Computational transport model	Mouse tibia	Idealized	Cyclic end-compression	400 $\mu\varepsilon$	0.5 Hz	N/A
	Kwon et al. (2012)	Lacunar-canalicular	Computational transport model	Mouse femur	Idealized	Intra-medullary pressure	~9 kPa	~5 Hz	N/A
	Verbruggen et al. (2014)	Lacunar-canalicular	Fluid-structure interaction	NS Rat tibia	Idealized Image-based	Static compression	3000 $\mu\varepsilon$	N/A	300 Pa
	Niroobakhsh et al. (2024)	Lacunar-canalicular	Fluid-structure interaction	NS	Idealized	Static compression	3000 $\mu\varepsilon$	N/A	300 Pa
			Computational fluid dynamics	Mouse femur	Image-based	N/A	N/A	N/A	300 Pa
	Kamioka et al. (2012)	Canalicular	Computational fluid dynamics	Human femur	Image-based	N/A	N/A	N/A	1000 Pa/μm
	Yokoyama et al. (2021)	Canalicular	Fluid-structure interaction	Human femur	Image-based	N/A	N/A	N/A	1000 Pa/μm
	Kameo et al. (2022)	Canalicular	Fluid-structure interaction	NS	Idealized	Uniaxial cyclic loading	127 $\mu\varepsilon$	1 Hz	N/A
OSTEON-LEVEL	Wang et al. (2003)	Osteon	Analytical poroelastic	NS	Idealized	Sinusoidal axial loading	100 $\mu\varepsilon$ (locomotion)	1 Hz	N/A
							10 $\mu\varepsilon$ (posture)	20 Hz	
						Extra-vascular pressure	3.99 kPa (skeletal muscle contraction)	1 Hz	
							1.33 kPa (cardiac contraction)	2 Hz	
							0.27 kPa (venous stasis)	2 Hz	
	Rémond et al. (2008)	Osteon	Poroelastic FE	NS	Idealized	Sinusoidal longitudinal loading	1000 $\mu\varepsilon$	1 Hz 20 Hz	N/A
	Goulet et al. (2009)	Osteon	Poroelastic FE	NS	Idealized	Sinusoidal axial loading	~1000 $\mu\varepsilon$	1 Hz	N/A
								40 Hz	
	Nguyen et al. (2010)	Osteon	Poroelastic FE	NS	Idealized	Cyclical axial loading	0.001 s^{-1}	1 Hz	N/A
							0.003 s^{-1}		
							0.001 s^{-1}	10 Hz	
							0.003 s^{-1}		
	van Tol, Roschger, et al. (2020)	Osteon	Poroelasticity & Circuit theory	Human femur	Image-based	Dynamic axial loading	0.015 s^{-1}	NR	N/A
						N/A	N/A	N/A	13 kPa

					Outcomes			
Tissue-level lacunar-canalicular permeability (m^2)	Osteocyte pericellular matrix included?	Osteocyte pericellular permeability (m^2)	Permeable surfaces	Maximum tissue-level fluid velocity (μm/s)	Maximum canalicular fluid velocity (μm/s)	Fluid shear stress on osteocyte process (Pa)	Drag force on tethering elements (pN/μm)	Maximum principal strain on osteocyte process (με)
NR	✓	NR	N/A	–	–	0.8–3	–	–
N/A	×	N/A	N/A	–	–	~0.7 max	–	–
N/A	✓[a]	N/A	N/A	–	130	–	–	–
N/A	✓[a]	N/A	N/A	–	60	5	–	–
N/A	✓[a]	N/A	N/A	–	80	–	–	–
N/A	×	N/A	N/A	–	219[c] 298[c]	6.9 max[c] 11.6 max[c]	– –	– –
N/A	×	N/A	N/A	–	–	5.3 max	–	–
N/A		N/A	N/A	–	269	6.6 max	–	–
N/A	×	N/A	N/A	–	~2500	–	–	–
N/A	✓	7×10^{-18}	N/A	–	~15	–	–	~150,000
NR	✓	12×10^{-17} 6×10^{-17} 3×10^{-17}	N/A	–	~2.5 ~2 ~1.8	–	–	~18,000 ~23,000 ~30,000
NR	✓	NR	Vascular canal	–	–	1.6 max	–	–
						2.5 max		
						0.045 max		
						0.03 max		
						0.006 max		
~10^{-18}	b	N/A	Vascular canal	~0.16 ~3.3	–	–	–	–
1.05×10^{-20} [d] 1.05×10^{-19} [d] 1.05×10^{-18} [d] 1.05×10^{-20} [d] 1.05×10^{-19} [d] 1.05×10^{-18} [d]	b	N/A	Vascular canal	0.024 ~0.35 ~0.88 ~0.95	–	3.8 max 1.2 max 0.38 max ~57 max ~45 max ~15 max	–	–
10^{-18}	b	N/A	Vascular canal	0.11 0.34 0.067 0.45	–	–	–	–
~10^{-18}	✓	1.53×10^{-17}	Vascular canal	–	~100	–	–	–
					~10			

Continued

Table 16.2 Summary of modeling parameters and reported outcomes from bone interstitial fluid flow models at osteocyte-level, osteon-level, whole bone-level and multi-level.—*cont'd*

		Modeling parameters							
	Reference	**Level of model**	**Model type**	**Bone type**	**Model geometry**	**Applied load type**	**Maximum tissue-level strain or pressure magnitude or strain rate from loading**	**Loading frequency**	**Applied cellular-level fluid pressure**
WHOLE BONE -LEVEL	Fornells et al. (2007)	Whole bone	Dual porosity FE	Rat tibia	Image-based	Trapezoidal cantilever bending	1500 με	~1 Hz	N/A
	Goulet, Hamilton, et al. (2008)	Whole bone	Poroelastic FE	Human tibia	Image-based	Square-wave bending	1895 με	1 Hz	N/A
	Pereira and Shefelbine (2014)	Whole bone (Segment)	Poroelastic FE	Mouse tibia	Idealized	Haversine bending	2000 με	1–10, 15, 20, 30 Hz	N/A
	Pereira et al. (2015)	Whole bone	Poroelastic FE	Mouse tibia	Image-based	Trapezoidal axial loading	~5000 με	0.1 s load cycle with rest insertion	N/A
	Carriero et al. (2018)	Whole bone	Poroelastic FE	Mouse tibia	Image-based	Trapezoidal axial loading	~6000 με	0.1 s load cycle with rest insertion	N/A
	van Tol, Schemenz, et al. (2020)	Whole bone	Poroelasticity & Circuit theory	Mouse tibia	Image-based	Triangular bending	1200 με	4 Hz	N/A
	Meslier et al. (2022)	Whole bone	Poroelastic FE	Mouse tibia	Image-based	Triangular uniaxial compression	3500 με 2400 με	3.3 Hz 0.25 Hz 1 Hz	N/A
						Square-wave uniaxial compression	2400 με	1 Hz	
MULTI-LEVEL	Mi, Fritton, et al. (2005)	Whole bone	Analytical poroelastic	Turkey radius	Idealized	Trapezoidal bending	NR	NR	N/A
				Rooster tarsometatarsus	Image-based	Running			
	Zhou et al. (2008)	Multi-level	Analytical poroelastic	Mouse tibia	Idealized	Cyclical bending	800 με	0.5 Hz	N/A
	Fan et al. (2016)	Multi-level	Biphasic FE	Murine tibia	Image-based	Cyclical end-compression	~1840 με	0.5 Hz	N/A
	Sánchez et al. (2021)	Multi-level	Poroelastic FE	Turkey ulna Human femur	Image-based	Sinusoidal bending	NR	30 Hz	N/A
	Wang et al. (2022)	Multi-level	Poroelastic FE Computational fluid dynamics	Mouse tibia	Idealized osteocyte	Cyclical triangular axial loading	700 με	$0.01\ s^{-1}$	N/A
					Image-based osteocyte		700 με	$0.01\ s^{-1}$	

FE, Finite element; NS, *Not specified;* NR, *Not reported; parameter was not directly reported but was adopted from earlier research articles—refer to the original sources cited in the study.*
[a] *Canalicular fluid flow velocity was calculated based on experimental tracer concentration profiles; pericellular matrix was not modeled directly but the concentration profiles were affected by the osteocyte pericellular permeability.*
[b] *Osteocyte pericellular matrix was included in tissue-level lacunar-canalicular permeability.*
[c] *Reported model outcomes were the average of the group.*
[d] *Tissue-level canalicular permeability; lacunar permeability was reported separately:* $7.17 \times 10^{-17}\ m^2$.

					Outcomes			
Tissue-level lacunar-canalicular permeability (m^2)	**Osteocyte pericellular matrix included?**	**Osteocyte pericellular permeability (m^2)**	**Permeable surfaces**	**Maximum tissue-level fluid velocity (μm/s)**	**Maximum canalicular fluid velocity (μm/s)**	**Fluid shear stress on osteocyte process (Pa)**	**Drag force on tethering elements (pN/μm)**	**Maximum principal strain on osteocyte process (με)**
10^{-20}	b	N/A	Vascular canal, endosteal, periosteal	~0.01	–	–	–	–
10^{-19}	b	N/A	Vascular canal, endosteal	0.69	–	–	–	–
10^{-23}	b	N/A	Endosteal	0.0008	–	–	–	–
10^{-22}				0.008				
10^{-21}				0.08				
10^{-20}				0.68				
10^{-19}				2.9				
10^{-18}				4.0				
10^{-22}	b	N/A	Endosteal, periosteal	~0.15	–	–	–	–
10^{-22}	b	N/A	Endosteal, periosteal	~0.1	–	–	–	–
N/A	✓	NR	NR	–	~20	–	–	–
10^{-22}	b	N/A	Endosteal, periosteal	0.05 0.025 0.025 0.045	–	–	–	–
NR	✓	NR	Vascular network, endosteal	–	–	~1.1 max ~0.2 max	–	–
NR	✓	NR	Endosteal, periosteal	–	81	–	–	–
2.8×10^{-20}	✓	NR	Endosteal	~0.28	1.84	0.25	1.1	–
10^{-20} 1.7×10^{-20}	b	N/A	Vascular canal, endosteal	~0.1 ~0.075	–	–	–	–
10^{-17}	×	N/A	Endosteal	~0.13	~1300	~50 max	–	–
				~0.13	~5000	~175 max		

vigorous physical loading (Joukar et al., 2016; Niroobakhsh et al., 2024; Verbruggen et al., 2014). A limitation of these models is assigning canaliculi as separate inlets and outlets, as interstitial fluid could flow in or out of one canaliculus during a loading cycle. In addition, the exclusion of the pericellular matrix, which resists fluid flow and contributes to higher fluid shear stress, likely overestimates the canalicular fluid velocity (>200 μm/s) (Verbruggen et al., 2014) (Table 16.2). In addition to canalicular flow, stimulation by primary cilia on the osteocyte cell body has also been considered; however, unless the cilium was attached to the lacunar wall, the resulting strain remained under the stimulatory threshold (~10,000 με) due to the primary cilia's short length and the minimal deflection in the osteocyte lacuna (Vaughan et al., 2015).

Most idealized lacunar-canalicular models that do not include the osteocyte pericellular matrix predict canalicular fluid velocities of a few 100 μm/s, generating fluid shear stress in the range of 1–12 Pa (Fig. 16.4, Table 16.2). In contrast, Zhou and colleagues reported a lower canalicular fluid velocity (81 μm/s) when the pericellular permeability was included in a multi-level poroelastic analysis (Zhou et al., 2008) (Table 16.2). Interstitial fluid velocity estimates from in-situ fluorescence recovery after photobleaching (FRAP) experiments reported similar fluid velocity magnitudes, 60–130 μm/s (Kwon et al., 2012; Kwon & Frangos, 2010; Price et al., 2011) (Table 16.2). Unlike poroelastic analysis, these studies estimated the canalicular fluid velocities based on solute transport enhancements via mechanical loading or intramedullary pressurization. Additionally, Price and colleagues demonstrated that a canalicular fluid velocity of 60 μm/s was capable of producing shear stress of approximately 5 Pa on osteocyte processes (Price et al., 2011). While these studies estimated the interstitial fluid velocity based on experimental measurement of the lacunar-canalicular porosity, the authors noted that the model is highly sensitive to canalicular density and length, which might explain the higher shear stress magnitudes compared to the predictions (0.8–3 Pa) by Weinbaum and colleagues (Weinbaum et al., 1994) (Table 16.2). On the other hand, a multi-level poroelastic finite element model of the same FRAP experiment from Price et al. (2011) yielded a lower canalicular fluid velocity (~2 μm/s) when tissue-level velocities were scaled to canalicular velocities based on average lacunar-canalicular porosity (Fan et al., 2016). The authors noted that lower canalicular fluid velocities might be due to the higher lacunar-canalicular porosity (~15%) measured using confocal imaging.

Morphologically accurate 3D osteocyte lacunar-canalicular models have also been generated using the microstructure obtained from confocal microscopy images and applying a 300 Pa fluid pressure gradient based on earlier idealized models, as well as a compressive 3000 με to mimic vigorous mechanical loading (Niroobakhsh et al., 2024; Verbruggen et al., 2014). Similar to the idealized models, the osteocyte pericellular matrix was neglected in these models, leading to high canalicular fluid velocities, 298 μm/s (Verbruggen et al., 2014) and 269 μm/s (Niroobakhsh et al., 2024) (Table 16.2). Although both studies had similar boundary conditions, the difference in the fluid shear stress and velocity magnitudes could be due to loading direction differences in addition to the geometry of the lacunae.

Lacunar-canalicular models provide a broad view of interstitial fluid flow around osteocytes, capturing the mechanical environment osteocytes experience during loading. However, these osteocyte-level models often assume similar geometries for all canaliculi stemming from the same lacuna and overlook individual canalicular characteristics and their impact on interstitial fluid velocity.

3.2.4 Osteocyte-level canalicular models to assess the contribution of canalicular morphology to interstitial fluid flow

Canalicular models focus on how canalicular morphology affects interstitial fluid flow within a single canaliculus. Similar to lacunar-canalicular models, analytical poroelastic and computational fluid dynamics or fluid-structure interaction methods are used with either idealized or physiological geometries. These models either investigate the specific morphological features separately or are physiologically realistic and assess the total effect of morphology on interstitial fluid flow.

Idealized canaliculi models have examined the contribution of certain features of canalicular morphology, such as wall roughness, tortuosity, the density of tethering elements, and glycosaminoglycan content, to the interstitial fluid flow. Weinbaum et al. (1994) defined the interstitial fluid flow within the canalicular annulus as a strictly laminar "plug-like" flow with a low Reynolds number due to the distance between two adjacent lacunae and the low fluid velocity (Wang et al., 2000). The irregularities of the canalicular wall surface due to the orientation of collagen fibrils (i.e., wall roughness) could generate closed circulation regions but did not affect the plug flow (Wang et al., 2001). Barber and colleagues also investigated the contribution of canalicular wall roughness to fluid shear stress on osteocyte processes using an idealized canaliculi model and found that the shear stress was elevated only when the wall roughness decreased the canalicular annulus (Barber et al., 2024). Similarly, Kameo and colleagues investigated how canalicular curvature and osteocyte pericellular matrix composition, including glycosaminoglycan content and tethering elements, affected interstitial fluid flow (Kameo et al., 2022). Their analysis demonstrated that increased curvature enhanced both fluid velocity and the principal strains on osteocyte processes. In contrast, reduced glycosaminoglycan density (modeled as increased permeability) and fewer tethering elements had opposite effects: reduced glycosaminoglycan density increased canalicular fluid velocity but decreased osteocyte strains, while fewer tethering elements caused strain concentration points without affecting fluid velocity (Kameo et al., 2022) (Table 16.2).

In physiologically realistic computational canalicular models, high-resolution imaging techniques have been utilized to try to create morphologically accurate canalicular geometries to mimic physiological interstitial fluid flow. Anderson and Knothe Tate used transmission electron microscopy images to generate a canalicular geometry in 2D and showed that canalicular irregularities and projections increased interstitial fluid velocity and shear stress compared to idealized models (Anderson & Knothe Tate, 2008). Kamioka et al. used ultra-high voltage electron microscopy to reconstruct a 3D human canaliculus with collagen fibrils and calculated very high canalicular fluid velocities (2500 μm/s) with canalicular irregularities, likely due to the combination of the high fluid pressure gradient applied (1000 Pa/μm) as well as the exclusion of the pericellular matrix (Kamioka et al., 2012) (Table 16.2). Yokoyama et al. also utilized ultra-high voltage electron microscopy imaging to model the osteocyte process and canalicular details, including the tethering elements, and found that local strain concentration in the osteocyte process was induced by tethering elements with high tension (Yokoyama et al., 2021) (Fig. 16.5). With inclusion of the pericellular matrix in the calculation of the lacunar-canalicular permeability in this fluid-structure interaction model, the canalicular fluid velocity was much lower, ~15 μm/s (Yokoyama et al., 2021) (Table 16.2).

While osteocyte-level models are crucial for understanding localized interstitial fluid flow dynamics in canaliculi of a single osteocyte, they do not account for the influence of vascular porosity on interstitial fluid flow. Osteon-level models address this limitation by integrating the lacunar-canalicular network with the vascular pores and quantifying interstitial fluid velocity within an osteon.

3.3 Osteon-level models: Inclusion of vascular porosity

Osteon-level analytical and computational models have been used to simulate interstitial fluid flow within an osteon, a region of the lacunar-canalicular network around a vascular canal. Since vascular canals cannot maintain elevated interstitial fluid pressure, in models the interstitial fluid pressure within the osteonal canal is often set to zero. This assumption is a crucial feature of osteon-level models simulating interstitial fluid flow within bone.

3.3.1 Interstitial fluid within lacunar-canalicular porosity drains through vascular canals

Kufahl and Saha developed a mathematical model of the Haversian canal-lacunae-canaliculi system to investigate interstitial fluid flow in the lacunar-canalicular porosity under cyclical compressive stress (Kufahl & Saha, 1990). They showed that stress-generated fluid flow could reach four to five lacunae away from the vascular pore; however, the canaliculi were modeled as hollow cylindrical tubes instead of osteocyte-filled channels.

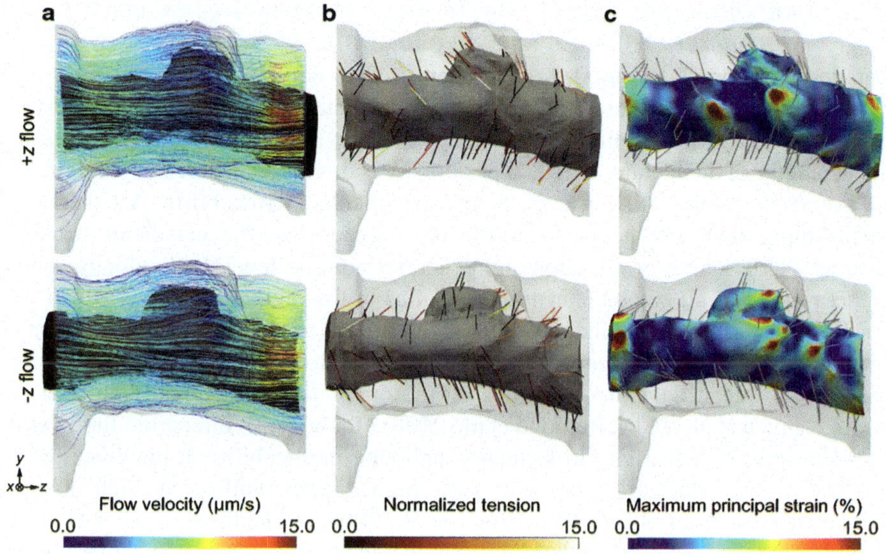

FIGURE 16.5

Image-based canalicular model showing the osteocyte process within the canaliculus, connected to the canalicular wall with randomly placed tethering elements. A fluid pressure gradient is applied in the +z direction (top row) and −z direction (bottom row). Interstitial fluid flow deflects the tethering elements, generating strain on the osteocyte process. (a) Canalicular fluid velocities; (b) tension in tethering elements, normalized by the median value (0.017 pN); (c) maximum principal strains induced on osteocyte process.

- *From Yokoyama, Y., Kameo, Y., Kamioka, H., & Adachi, T. (2021). High-resolution image-based simulation reveals membrane strain concentration on osteocyte processes caused by tethering elements.* Biomechanics and Modeling in Mechanobiology, 20*(6), 2353–2360. https://doi.org/10.1007/s10237-021-01511-y*

Wang et al. modeled osteonal bone as a poroelastic material with multiple connecting osteons, including osteonal canals (i.e., vascular porosity) and cement lines (Wang et al., 1999). Models prior to this study (Cowin et al., 1995; Weinbaum et al., 1994; Zeng et al., 1994) assumed that fluid pressure relaxes across the bone thickness; however, this homogenous approach could not explain the cusp-like shape in voltage profiles observed experimentally by Starkebaum and colleagues around the osteons (Starkebaum et al., 1979). Wang and colleagues proposed that low fluid pressure in the vascular porosity allows excess bone fluid to relax through these canals (Wang et al., 1999). Defining vascular canals as primary relaxation regions produced fluid pressure profiles comparable to voltage profiles reported by Starkebaum and colleagues (Starkebaum et al., 1979). Additionally, Wang et al. reported that pressure profiles are not affected by the permeability of osteon cement lines in the presence of vascular canals; however, higher loading frequency can produce increased pressure gradients across the bone (Wang et al., 1999).

3.3.2 Interstitial fluid flow is less sensitive to lacunar-canalicular permeability but more sensitive to strain rate

Osteon-level poroelastic models have been used to compute interstitial fluid flow while considering both vascular and lacunar-canalicular porosities. A single osteon is modeled as a hollow cylinder where the central cavity is the vascular porosity, and the geometry of the lacunar-canalicular porosity is often not included but assigned as an averaged isotropic tissue-level lacunar-canalicular permeability. A tissue-level permeability does not account for the heterogeneity in the canaliculi structure such as width, length or direction and generates a tissue-level fluid velocity, which is the average of all canalicular fluid velocities in the region.

Several studies have utilized osteon-level models to test the sensitivity of interstitial fluid flow to permeability and mechanical loading parameters, reporting tissue-level interstitial fluid velocities on the order of 0.01–1 μm/s (Goulet et al., 2009; Nguyen et al., 2010; Rémond et al., 2008) (Table 16.2). Interstitial fluid velocities showed low sensitivity to lacunar-canalicular permeability at low loading frequencies, but velocities increased with higher permeability at high loading frequency (Goulet et al., 2009; Rémond et al., 2008). In addition, Goulet et al. reported a 10-times increase in shear stress with decreased lacunar-canalicular permeability (from 10^{-18} to 10^{-20} m^2) (Goulet et al., 2009) (Table 16.2). These studies also analyzed the influence of mechanical loading magnitude, frequency, and rate, showing that tissue-level interstitial fluid velocity is highly dependent on strain rate, with higher strain rates producing higher velocities (Goulet et al., 2009; Nguyen et al., 2010; Rémond et al., 2008) (Table 16.2). Similarly, Wang et al. investigated the canalicular fluid flow in an osteon under different loading conditions including venous pressure from blood circulation and mechanical strains from physiological loading (Wang et al., 2003). The maximum shear stress generated by pulsatile extravascular pressure produced by skeletal muscle contraction, cardiac contraction, as well as venous stasis remained two orders of magnitude lower than that created by mechanical loading of the bone, showing that osteocyte stimulation via blood pressure is unlikely (Wang et al., 2003) (Table 16.2).

3.3.3 Interstitial fluid flow distribution depends on canalicular architecture

Most osteon-level studies model the lacunar-canalicular porosity as a poroelastic medium, neglecting the distribution and connectivity of the canaliculi. To include the connectivity, van Tol and colleagues developed a 3D hydraulic circuit network model based on confocal microscopy images of the lacunar-canalicular architecture (van Tol, Roschger, et al., 2020). Using Kirchhoff's circuit laws, interstitial fluid flow within the canaliculi was calculated for two osteon types: a standard osteon and an "osteon-in-osteon" (Fig. 16.6). The latter features a double-layer structure formed by a newly deposited osteon overlaying an older one, with limited connecting canaliculi between layers, creating longer interstitial fluid flow pathways to the osteonal canal. The highest canalicular fluid velocities occurred when canaliculi converged into narrow necks and were on the same order of magnitude as previously

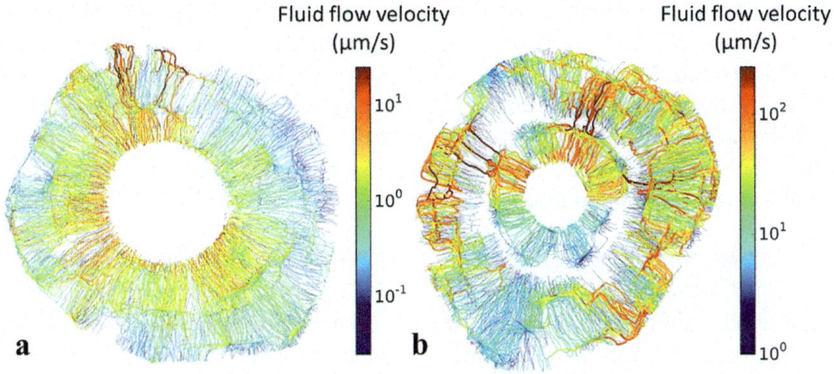

FIGURE 16.6

A circuit model of the lacunar-canalicular network of human osteons showing canalicular fluid velocities calculated after application of axial loading: (a) Ordinary osteon (b) osteon-in-osteon.

- From van Tol, A.F., Roschger, A., Repp, F., Chen, J., Roschger, P., Berzlanovich, A., Gruber, G.M., Fratzl, P., & Weinkamer, R. (2020). Network architecture strongly influences the fluid flow pattern through the lacunocanalicular network in human osteons. Biomechanics and Modeling in Mechanobiology, 19*(3), 823–840. https://doi.org/10.1007/s10237-019-01250-1*

described osteocyte-level models, ~100 μm/s (van Tol, Roschger, et al., 2020) (Table 16.2). Consistent with Nguyen and colleagues (Nguyen et al., 2010), van Tol et al. also reported increased strain-induced interstitial fluid velocity magnitudes when flow pathways were longer. Additionally, the inclusion of canalicular tortuosity, branching, and osteocyte pericellular permeability in the hydraulic circuit network provided canalicular fluid velocities in the osteon-level model; however, simplifications such as assuming constant canalicular dimensions limited the complexity of the model. This study demonstrated that the inclusion of the geometry of the lacunar-canalicular network generated localized elevated fluid velocities, which models using tissue-level permeability cannot capture (van Tol, Roschger, et al., 2020).

Osteon-level models provide valuable insights into the interstitial fluid flow within individual osteons. However, they still focus on relatively localized regions of bone and do not address how these interactions scale up to the entire bone. Whole bone-level models consider the global fluid flow within the bone, accounting for the mechanical loading and strain distribution across the entire structure.

3.4 Whole bone and multi-level models

Macroscale models investigate the interstitial fluid flow at larger scales, allowing the inclusion of various aspects of the whole bone, such as total geometry, vascular canals, and permeable surfaces. The non-symmetric shape and curvature of bones generate heterogeneous strain distributions that depend on the type of load applied.

Additionally, the inclusion of vascular canals, the definition of permeable surfaces, and the lacunar-canalicular permeability significantly affect the interstitial fluid flow. Whole bone-level poroelastic finite element models are frequently used to calculate averaged tissue-level interstitial fluid velocities, neglecting the ultrastructure of the lacunar-canalicular network. In contrast, multi-scale models couple the whole bone-scale dynamics to osteon- or osteocyte-level interstitial fluid flow, providing more detailed analyses.

3.4.1 Inclusion of vascular canals alters the interstitial fluid flow distribution

As described previously, interstitial fluid pressure within the lacunar-canalicular porosity relaxes through the bone vascular canals, which alters the interstitial fluid flow pattern. The vascular porosity was often neglected in early whole bone-level models due to challenges in imaging and segmentation; however, advancements in imaging techniques have enabled researchers to segment and include vascular pores in whole bone-level finite element models (see Sections 3.5.1 and 3.5.2).

Incorporating the influence of vascular porosity into the whole-bone level models has demonstrated global and local changes in interstitial fluid flow patterns in cortical bone. Fornells and colleagues employed a dual porosity method to investigate the effect of vascular porosity on interstitial fluid flow (Fornells et al., 2007). This method linked the interstitial fluid flow within the lacunar-canalicular porosity to the vascular porosity through a leakage parameter, calculating interstitial fluid velocities without explicitly including vascular pores in the model geometry. The dual porosity model was validated by replicating the osteonal model by Wang et al. (1999) (Section 3.3.1), showing good agreement with reported fluid pressure gradients. It was then applied to a rat tibia model to calculate interstitial fluid velocities, and incorporating vascular porosity reduced the magnitude of the tissue-level interstitial fluid velocity by approximately 50% (Fornells et al., 2007) (Table 16.2). However, since vascular canals were absent in the model geometry, the temporospatial velocity distribution could not be analyzed. Goulet and colleagues adopted a different approach, manually inserting vascular canals into the geometry of a human tibia cross-section (Goulet, Hamilton, et al., 2008). Their model predicted that when mechanically loaded, the tissue-level interstitial fluid velocity in the lacunar-canalicular porosity increased an order of magnitude to ~0.7 μm/s, particularly around the vascular pores, emphasizing the local effects of vascular canals (Fig. 16.7, Table 16.2). Yu et al. expanded this model by combining osteocyte-level, osteon-level and whole bone-level models in a multi-level study (Yu et al., 2025).

3.4.2 Interstitial fluid flow depends on loading frequency, strain rate, and waveform

Similar to what was shown with osteocyte and osteon-level models, interstitial fluid velocity within the lacunar-canalicular porosity also depends on loading parameters such as loading frequency, waveform, and strain rate. Pereira and Shefelbine used

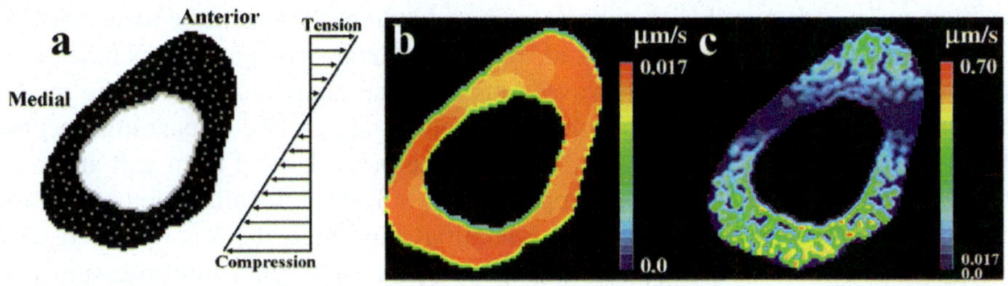

FIGURE 16.7

Cross-section of a human tibia showing the effect of vascular canals on interstitial fluid flow. (a) Segmented image showing permeable surfaces (i.e., vascular canals and endosteal surface) in gray and bone extracellular matrix in black. Bending was applied via a linear stress gradient along the longitudinal axis. Predicted tissue-level interstitial fluid velocities (µm/s) (b) without vascular canals and (c) with vascular canals. The magnitude of interstitial fluid velocity increased 40 times with the inclusion of vascular canals.

- From Goulet, G. C., Hamilton, N., Cooper, D., Coombe, D., Tran, D., Martinuzzi, R., & Zernicke, R. F. (2008). Influence of vascular porosity on fluid flow and nutrient transport in loaded cortical bone. Journal of Biomechanics, 41*(10), 2169–2175. https://doi.org/10.1016/j.jbiomech.2008.04.022*

poroelastic analysis to examine the relationship between loading frequency, lacunar-canalicular permeability, and the tissue-level interstitial fluid motion (displacement and velocity) (Pereira & Shefelbine, 2014). By applying bending at various frequencies to an idealized finite element model of a mouse tibia, they showed that fluid velocity increases with higher loading frequencies for lacunar-canalicular permeabilities in the range of ~10^{-21} to 10^{-18} m^2 (Pereira & Shefelbine, 2014) (Table 16.2). Conversely, they showed that tissue-level fluid velocity is not sensitive to loading frequency for lower permeabilities (~10^{-23} to 10^{-22} m^2) and highlighted that if the tissue permeability is low, interstitial fluid velocity requires a longer time to diminish to negligible levels after the load is removed and benefits from rest periods between loading cycles (Pereira & Shefelbine, 2014).

Higher loading magnitude or rate also increases the interstitial fluid velocity, as larger amounts of interstitial fluid are displaced in a short time, which results in higher fluid pressure gradients. Wang et al. investigated the individual and combined effects of strain parameters (e.g., magnitude, rate) on interstitial fluid velocity using a multi-level model (Wang et al., 2022). Various loading modalities were applied to simulate pore pressures in the entire tibia, which were subsequently applied to a single osteocyte model. While the tissue-level fluid velocity (~0.13 µm/s) is similar in magnitude to other studies, the high tissue-level permeability used along with the exclusion of the pericellular matrix led to an unrealistic canalicular fluid velocity (~1300 µm/s, Table 16.2); in addition, strain magnitude and rate were determined as primary factors influencing fluid velocity, though their combined effect was not directly additive (Wang et al., 2022). Another way to test the effect of loading magnitude and rate is to use different waveforms. By changing the shape of the loading curve, it is possible to achieve various loading rates with the same loading magnitude or vice versa. Kumar et al. reported a temporal-spatial dependency of interstitial fluid flow on loading waveforms using an analytical poroelastic model of murine tibia at the whole-bone level (Kumar et al., 2019). Expanding on this, Mertiya et al. incorporated realistic loading waveforms derived from gait analysis, simulating walking, running, and jumping in a whole-bone idealized femur and reported that the maximum tissue-level interstitial fluid velocity was observed with jumping since the load was applied instantaneously and higher fluid pressure was achieved; the maximum fluid displacement was achieved during running, which showed prolonged interstitial fluid flow (Mertiya et al., 2023). As most of the whole bone-level models mimic mechanical loading by applying loads at the ends of bone, Shrivas et al. included muscle forces, which were applied at various locations of the femur, to better replicate the mechanical environment experienced in vivo and demonstrated that higher fluid velocities can occur at low-strain sites (e.g., the neutral axis of bending) (Shrivas et al., 2023).

3.4.3 Models that use interstitial fluid flow to predict bone adaptation

As it has become more evident that interstitial fluid flow within bone is responsible for mechanotransduction (Fritton & Weinbaum, 2009; Wei et al., 2022) whole bone-level poroelastic models have focused on investigating the correlation between interstitial fluid flow and bone adaptation. Various flow parameters, including fluid velocity, fluid shear stress, fluid motion (displacement), and dissipation energy, have been tested as predictors for bone adaptation.

To analyze the contribution to bone adaptation of interstitial fluid flow through the lacunar-canalicular network, Mi et al. (Mi, Fritton, et al., 2005) calculated fluid shear stress in the lacunar-canalicular porosity of avian bone cross-sections based on in vivo experiments (Gross et al., 1997; Judex et al., 1997) (Table 16.2). The average shear stress showed no correlation with reported bone formation; instead, a neural network analysis demonstrated that bone formation was regulated by intercellular communication of osteocytes (Mi, Basu, et al., 2005). Pereira and colleagues used a weighted average of interstitial fluid velocity as a mechanical stimulus and estimated bone formation occurred when the stimuli exceeded a threshold on the endosteal and periosteal surfaces of murine tibia (Pereira et al., 2015). The same group also compared regions of bone formation to the distribution of interstitial fluid velocity in a whole mouse tibia, revealing an overlap between areas of high interstitial fluid flow and bone formation (Carriero et al., 2018). The applied loading in both studies produced very high-magnitude bone strains (Table 16.2). Interstitial fluid displacement has also been tested as a potential mechanical stimulus for bone adaptation; however, calculations of bone adaptation based only on surface nodes underestimated the contribution of the osteocyte network (Tiwari et al., 2018). The dissipation energy of interstitial fluid flow has also been used as an osteogenic stimulus to replicate endosteal bone formation in rat ulnae, demonstrating a non-linear frequency response of bone adaptation that aligns with experimental observations (Chennimalai Kumar et al., 2012). van Tol et al. included lacunar-canalicular architecture with circuit theory, predicting the weighted bone mechanoresponse from canalicular fluid velocity magnitudes (van Tol, Schemenz, et al., 2020) (Table 16.2). Predicted bone adaptation based on canalicular fluid flow showed a better correlation with experimentally measured bone response compared to predictions based on strain distributions calculated by finite element analysis (van Tol, Schemenz, et al., 2020). Because higher applied strain magnitudes often produce higher strain rates and higher fluid velocities, separating the contributions of strain and interstitial fluid flow remains difficult. To address this challenge, Meslier et al. designed four distinct loading profiles that dissociated interstitial fluid velocity from strain magnitude using poroelastic finite element models (Table 16.2); when applied to in vivo bone adaptation experiments, their results showed that fluid velocity was a more accurate predictor of bone adaptation than strain magnitude (Meslier et al., 2022).

Multi-level approaches have also been adopted to investigate the relationship between interstitial fluid flow and bone adaptation. Srinivasan and Gross calculated strains at several locations of avian ulnar cross-sections under bending and applied those strains to idealized osteocyte models (Srinivasan & Gross, 2000). The maximum interstitial fluid velocities were calculated at locations that experienced peak and minimal strains (i.e., the neutral axis, similar to Steck et al. (2003)), where Srinivasan et al. (2002) also reported bone formation in vivo. Sánchez and colleagues calculated macro-scale displacements and interstitial fluid pressures from a whole turkey ulna model, considering only the vascular porosity, and implemented them as boundary conditions in an osteon-level model that included the lacunar-canalicular porosity (Sánchez et al., 2021) (Table 16.2). This study found that osteons located in regions known for bone adaptation under sinusoidal bending (Qin et al., 1998) experienced higher magnitudes of tissue-level interstitial fluid velocity, further supporting the link between interstitial fluid flow and bone adaptation. In addition, their study showed that if the load on an osteon is reduced, in order to maintain the same interstitial fluid flow, the osteon diameter should be reduced (Sánchez et al., 2021). While this multi-level model provided a more detailed analysis, it did not capture fluid velocities at the canalicular level, which are the primary signals experienced by osteocytes.

3.5 Assessing interstitial fluid flow during aging, osteoporosis, and lactation

Conditions such as aging, osteoporosis, and lactation can alter bone structure and possibly affect interstitial fluid flow and the mechanosensitivity of osteocytes, which could contribute to skeletal fragility. The decline in bone mass and quality, which encompasses changes in vascular and lacunar-canalicular porosities as well as changes in material properties, has been frequently observed with aging (Cooper et al., 2007; Tiede-Lewis & Dallas, 2019; Zimmermann et al., 2011) as well as conditions that lead to osteoporosis (Brent et al., 2020; Gatti et al., 2019; Rolvien & Amling, 2022; Sharma et al., 2012; Westerlind et al., 1997). Analytical and computational models have been used to understand the effects of these physiological alterations on bone interstitial fluid flow.

3.5.1 Age-related increase in vascular and lacunar-canalicular porosity decreases interstitial fluid flow

Aging can change the vascular and lacunar-canalicular porosities of bone and alter the permeabilities of these networks, influencing interstitial fluid flow. Goulet et al. investigated the relationship between age-related changes in cortical bone and interstitial fluid velocity by modeling the 3D structure of a human femoral cortical bone sample reconstructed from micro-CT images, with the vascular porosity segmented (Goulet, Cooper, et al., 2008). Three different porosity values were generated with image thresholding, matching the published cortical canal porosities of various age groups (Bousson et al., 2001), and the vascular pores were modeled as low-pressure reservoirs. Calculations showed that a 3-fold increase in porosity decreased the maximum tissue-level interstitial fluid velocity by 40% in the aged model (Goulet, Cooper, et al., 2008) (Table 16.3). This study suggested that osteocytes experience a decrease in mechanical signals with aging; however, possible changes in the lacunar-canalicular porosity were neglected. Schurman and colleagues investigated the effects of age-related changes in the lacunar-canalicular porosity with osteocyte-level models; 3D models generated from confocal images showed aged mice had reduced interstitial fluid velocity due to the loss of the canalicular network (Schurman et al., 2021) (Table 16.3). Several factors, including osteocyte pericellular volume, and canalicular tortuosity and length, were considered to explain the reduction in fluid velocities due to age-related lacunar-canalicular degeneration, but only change in the pericellular volume showed a significant relation; doubling the width of the pericellular space increased shear stress approximately two times for all models (Schurman et al., 2021). Using a similar method, Niroobakhsh et al. demonstrated that the amount of canalicular surface area experiencing osteogenic fluid shear stress (higher than 0.8 Pa) was ~20 times greater in a young compared to an aged osteocyte model, and overall shear stresses on the osteocyte processes were reduced in aged bone (Niroobakhsh et al., 2024) (Table 16.3).

3.5.2 Osteoporosis-related changes in bone microstructure decrease the interstitial fluid velocity

Bone changes in osteoporotic conditions may also alter interstitial fluid flow and possibly the mechanical signals osteocytes receive, further disrupting bone homeostasis. Gatti et al. (2018) modeled the tibia of ovariectomized rats to mimic postmenopausal osteoporosis based on an in vivo study that observed increased vascular canal porosity and canalicular wall permeability after reduction of estrogen levels (Sharma et al., 2012). The whole-bone level poroelastic model of cortical bone was based on high-resolution micro-CT images of the proximal rat tibia, including the vascular porosity, and a cyclical 0.1% axial compression was applied, similar to the strain levels experienced during walking. The dominant effect of increased vascular porosity led to a net decrease in tissue-level interstitial fluid velocity around the vascular porosity (Table 16.3). This reduction in velocity occurred because the decrease in vascular separation prevented fluid pressure buildup; instead, interstitial fluid drained through the vascular porosity (Gatti et al., 2018) (Fig. 16.8). This redirection of interstitial fluid likely reduces the mechanical stimuli experienced by osteocytes, potentially impairing their role in bone remodeling and contributing to further bone loss.

A computational model of disuse osteoporosis, simulating the effects of reduced in vivo mechanical loading caused from Botox injection of rat hindlimb muscles (Gatti et al., 2019), demonstrated similar bone and fluid flow alterations as the ovariectomized rat model (Gatti et al., 2021). A significant decrease in tissue-level interstitial fluid velocity was observed around the enlarged vascular pores (Table 16.3); however, interstitial fluid flow around the lacunae remained unaffected. Additionally, a negative correlation between fluid velocity in lacunae and their distance from the vascular canals was demonstrated (Gatti et al., 2021).

3.5.3 Increased lacunar-canalicular porosity is compensated with increased pericellular permeability during lactation

Lactation can also produce bone microstructural changes, and osteocytes have been shown to degrade the surrounding bone matrix via perilacunar/canalicular remodeling, resulting in larger lacunae and canaliculi (Jähn et al., 2017; Kaya et al., 2017; Qing et al., 2012). Lai et al. investigated the effect of the sieving property of the pericellular matrix (i.e., permeability) on interstitial fluid flow during lactation (Lai et al., 2021). In addition to the increased lacunar-canalicular porosity, Lai et al. also considered the possibility of pericellular matrix degradation due to perilacunar/canalicular remodeling and used three different pericellular permeability values representing various fiber volume fractions (Lai et al., 2021) (Table 16.3). The fluid pressure gradient was calculated from a whole bone-level model using the lacunar-canalicular permeability and applied as a boundary condition to an osteocyte-level model. An increase in the tissue-level lacunar-canalicular permeability decreased canalicular fluid velocity and shear stress when osteocyte pericellular permeability remained the same. However, increasing the osteocyte pericellular permeability compensated for the reductions in fluid velocity, shear stress, and drag force, especially at higher loading frequences (2–4 Hz) (Table 16.3). Although the geometries were overly simplified and the pericellular fibrous volume fractions were hypothetical, this model showed that the pericellular and lacunar-canalicular permeabilities have intertwined effects on interstitial fluid flow and mechanotransduction signals. Similar to the effects of aging and osteoporosis, lactation alters bone microstructure, resulting in lower tissue-level and canalicular fluid velocities. However, degradation of the pericellular matrix may have the opposite effect, compensating for the losses due to these conditions.

Table 16.3 Summary of modeling parameters and reported outcomes from models of interstitial fluid flow in aging, osteoporosis, and lactation conditions.

	Modeling parameters								
	Reference	**Level of model**	**Model type**	**Bone type**	**Model geometry**	**Applied load type**	**Maximum tissue-level strain magnitude from loading**	**Loading frequency**	**Applied fluid pressure**
AGING	Goulet, Cooper, et al. (2008)	Whole bone (segment)	Poroelastic FE	Human femur	Image-based	Axial step compression	1109 $\mu\varepsilon$ 1169 $\mu\varepsilon$ 1247 $\mu\varepsilon$	N/A	N/A
	Schurman et al. (2021)	Osteocyte	Computational fluid dynamics	Mouse femur	Image-based	N/A	N/A	N/A	300 Pa
	Niroobakhsh et al. (2024)	Osteocyte	Computational fluid dynamics	Mouse femur	Image-based	N/A	N/A	N/A	300 Pa
OSTEOPOROSIS	Gatti et al. (2018)	Whole bone (segment)	Poroelastic FE	Rat tibia	Image-based	Haversine axial compression	1000 $\mu\varepsilon$	1 Hz	N/A
	Gatti et al. (2021)	Whole bone (segment)	Poroelastic FE	Rat tibia	Image-based	Haversine axial compression	1000 $\mu\varepsilon$	1 Hz	N/A
LACTATION	Lai et al. (2021)	Multi-level	Analytical poroelastic	Murine tibia	Idealized	Cyclical end-compression	320 $\mu\varepsilon$	0.5 Hz	N/A
								2 Hz	
								4 Hz	

FE, *Finite element*.
[a] *Osteocyte pericellular matrix was included in tissue-level lacunar-canalicular permeability.*
[b] *Reported model outcomes were the average of the group.*
[c] *Tissue-level canalicular permeability; lacunar permeabilities were reported separately: 3.83 × 10^{-17} m^2 for control and 3.63 × 10^{-17} m^2 for disuse groups.*
[d] *Mean canalicular fluid flow velocity.*

					Outcomes			
Conditions modeled	**Tissue-level lacunar-canalicular permeability (m^2)**	**Osteocyte pericellular matrix included?**	**Osteocyte pericellular permeability (m^2)**	**Permeable surfaces**	**Maximum tissue-level fluid velocity (μm/s)**	**Maximum canalicular fluid velocity (μm/s)**	**Fluid shear stress on osteocyte process (Pa)**	**Drag force on tethering elements (pN/μm)**
Young	10^{-19}	a	N/A	Vascular canal	2.3	–	–	–
Mid-aged					1.7			
Aged					1.4			
Young	N/A	×	N/A	N/A	–	526[b]	~2 max[b]	–
Aged						243[b]	~1 max[b]	
Young	N/A	×	N/A	N/A	–	269	0.57 avg	–
Aged						–	0.38 avg	
Sham	1.54×10^{-20}	a	N/A	Vascular canal, endosteal	~0.02	–	–	–
Ovariectomy	1.15×10^{-20}				~0.01			
Control	1.54×10^{-20}[c]	a	N/A	Vascular canal, endosteal	~0.01	–	–	–
Disuse					~0.0075			
Nulliparous	1×10^{-20}	✓	1.4×10^{-17}	Endosteal	–	44[d]	6.1	23
Lactating	1.2×10^{-20}		1.4×10^{-17}			38[d]	5.1	24
	1.8×10^{-20}		2.2×10^{-17}			42[d]	4.9	18
	3.2×10^{-20}		4.1×10^{-17}			45[d]	4.7	11
Nulliparous	1×10^{-20}		1.4×10^{-17}			71[d]	9.8	37
Lactating	1.2×10^{-20}		1.4×10^{-17}			69[d]	9.2	43
	1.8×10^{-20}		2.2×10^{-17}			99[d]	12	42
	3.2×10^{-20}		4.1×10^{-17}			142[d]	15	36
Nulliparous	1×10^{-20}		1.4×10^{-17}			73[d]	10	38
Lactating	1.2×10^{-20}		1.4×10^{-17}			73[d]	9.7	45
	1.8×10^{-20}		2.2×10^{-17}			113[d]	13	48
	3.2×10^{-20}		4.1×10^{-17}			189[d]	20	48

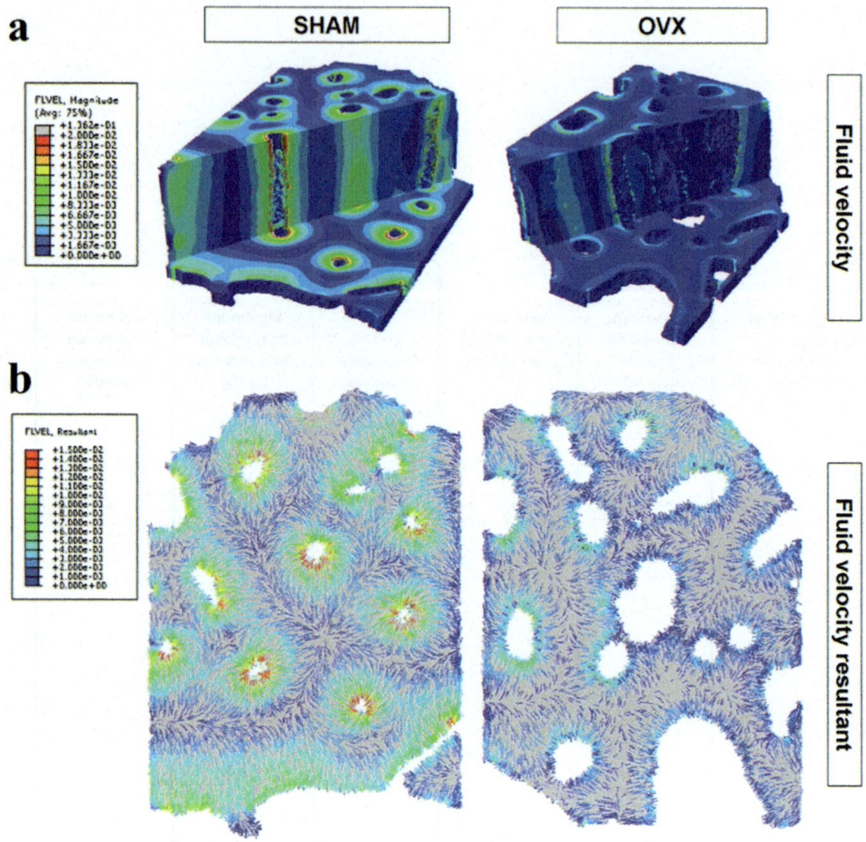

FIGURE 16.8

Whole bone-level modeling of interstitial fluid flow in an ovariectomized (OVX) rat tibia (right) and sham-operated (SHAM) control (left). (a) Axial and transverse sections showing the magnitudes of tissue-level interstitial fluid velocity (μm/s). (b) Cross-sections showing the interstitial fluid velocity vectors, where color represents the magnitude (μm/s) and vector orientation indicates the flow direction within the cortical bone.

- From Gatti, V., Azoulay, E. M., & Fritton, S. P. (2018). Microstructural changes associated with osteoporosis negatively affect loading-induced fluid flow around osteocytes in cortical bone. Journal of Biomechanics, 66, *127–136. https://doi.org/10.1016/j.jbiomech.2017.11.011*

4. Summary points

1. Mechanical loads are thought to be transmitted to osteocytes as fluid shear stress, drag forces, and hoop or axial strains generated by interstitial fluid flow within the lacunar-canalicular network. Because direct measurements of interstitial fluid flow are difficult in vivo, analytical and computational estimations are essential for predicting the mechanical signals received by osteocytes.

2. The contribution of vascular and lacunar-canalicular porosities to interstitial fluid flow is well-established; however, the role of collagen-apatite porosity is less clear. If the collagen-apatite porosity forms an interconnected pathway, it could contribute to fluid and solute movement within bone.
3. The reported range of tissue-level lacunar-canalicular permeability varies widely from 10^{-25} to 10^{-17} m^2; however, the most used tissue-level permeability values in models typically fall within 10^{-20} to 10^{-19} m^2, consistent with the estimates by Weinbaum et al. (1994) based on the lacunar-canalicular ultrastructure. In contrast, experimental estimates range from 10^{-25} to 10^{-22} m^2. Further investigation should be undertaken to resolve this discrepancy.
4. Interstitial fluid flow is highly dependent on loading parameters, such as strain magnitude, strain rate, and loading frequency. While selecting physiologically realistic loading parameters is crucial to accurately estimate interstitial fluid flow, sometimes relatively arbitrary values have been used based on the type of analysis being performed. For example, applying supra-physiological pressure gradients in osteocyte-level models or supra-physiological tissue-level strain magnitudes in whole bone models is not appropriate if the goal is to represent physiological conditions.
5. The lacunar-canalicular porosity has often been modeled as homogenous. However, the detailed structure of the lacunar-canalicular network, including its connectivity and branching patterns, can significantly influence canalicular fluid velocities. Incorporation of heterogenous lacunar-canalicular porosity into future models can enhance the accuracy of the calculated fluid flow parameters by capturing localized high fluid velocity regions.
6. Most osteocyte-level models report canalicular interstitial fluid velocities on the order of 1–100 μm/s, and even higher in some studies. This wide range is largely due to the omission of the osteocyte pericellular matrix in many models—since the pericellular matrix provides resistance to fluid flow, its absence results in an overestimation of canalicular fluid velocities. In addition, the choice of model parameters, particularly permeability values, also contributes to the wide range of canalicular velocities, with higher permeabilities producing higher velocities.
7. Tissue-level interstitial fluid velocities computed by osteon-level and whole-bone level models range from ~0.01 to 1 μm/s. The lower velocity magnitudes compared to canalicular fluid velocities are primarily due to poroelastic modeling, which does not explicitly represent individual canaliculi, and instead calculates a tissue-averaged fluid flow through the lacunar-canalicular porosity. In addition, the exclusion of vascular canals in whole-bone level models can lead to an underestimation of computed tissue-level interstitial fluid velocities, since the highest velocities are observed at vascular pore boundaries.

5. Conclusions

There has been a longstanding effort to enhance our understanding of interstitial fluid flow in bone across different length scales, offering insights into osteocyte mechanotransduction and bone adaptation. While analytical and computational models have been essential in analyzing interstitial fluid flow dynamics, they have simplifications and limitations. It is crucial to ensure that these models align with their intended purpose and scale, as not every model is appropriate for drawing the same conclusions. Significant variations in calculated magnitudes of interstitial fluid velocity emphasize the need for careful selection and clear acknowledgment of these limitations in line with study objectives. Future studies should consider coupling various levels of porosities and providing a more realistic representation of the lacunar-canalicular network based on high-resolution imaging when studying cellular-level stimulation. Additionally, validating model outcomes with experimental results is crucial for accurate and physiologically relevant modeling. Implementing these improvements will provide a more detailed analysis of the relationship between interstitial fluid flow and osteocyte mechanotransduction, contributing to a better understanding of mechanically induced bone adaptation.

References

Anderson, E. J., Kaliyamoorthy, S., Alexander, J. I. D., & Knothe Tate, M. L. (2005). Nano-microscale models of periosteocytic flow show differences in stresses imparted to cell body and processes. *Annals of Biomedical Engineering, 33*(1), 52–62. https://doi.org/10.1007/s10439-005-8962-y

Anderson, E. J., & Knothe Tate, M. L. (2008). Idealization of pericellular fluid space geometry and dimension results in a profound underprediction of nano-microscale stresses imparted by fluid drag on osteocytes. *Journal of Biomechanics, 41*(8), 1736–1746. https://doi.org/10.1016/j.jbiomech.2008.02.035

Anderson, E. J., Kreuzer, S. M., Small, O., & Knothe Tate, M. L. (2008). Pairing computational and scaled physical models to determine permeability as a measure of cellular communication in micro- and nano-scale pericellular spaces. *Microfluidics and Nanofluidics, 4*(3), 193–204. https://doi.org/10.1007/s10404-007-0156-5

Andronowski, J. M., Cole, M. E., Davis, R. A., Tubo, G. R., Taylor, J. T., & Cooper, D. M. L. (2023). A multimodal 3D imaging approach of pore networks in the human femur to assess age-associated vascular expansion and lacuno-canalicular reduction. *The Anatomical Record, 306*(3), 475–493. https://doi.org/10.1002/ar.25089

Barber, J., Mukhin, M., Maybruck, V., & Zhu, L. (2024). Modeling and simulation of osteocyte process–fluid interaction in a canaliculus. *Physics of Fluids, 36*(6), 061906. https://doi.org/10.1063/5.0208419

Benalla, M., Palacio-Mancheno, P. E., Fritton, S. P., Cardoso, L., & Cowin, S. C. (2014). Dynamic permeability of the lacunar-canalicular system in human cortical bone. *Biomechanics and Modeling in Mechanobiology, 13*(4), 801–812. https://doi.org/10.1007/s10237-013-0535-7

Beno, T., Yoon, Y. J., Cowin, S. C., & Fritton, S. P. (2006). Estimation of bone permeability using accurate microstructural measurements. *Journal of Biomechanics, 39*(13), 2378–2387. https://doi.org/10.1016/j.jbiomech.2005.08.005

Bonewald, L. F. (2006). Mechanosensation and transduction in osteocytes. *BoneKEy-Osteovision, 3*(10), 7–15. https://doi.org/10.1138/20060233

Bousson, V., Meunier, A., Bergot, C., Vicaut, E., Rocha, M. A., Morais, M. H., Laval-Jeantet, A. M., & Laredo, J. D. (2001). Distribution of intracortical porosity in human midfemoral cortex by age and gender. *Journal of Bone and Mineral Research, 16*(7), 1308–1317. https://doi.org/10.1359/jbmr.2001.16.7.1308

Brent, M. B., Thomsen, J. S., & Brüel, A. (2020). The efficacy of PTH and abaloparatide to counteract immobilization-induced osteopenia is in general similar. *Frontiers in Endocrinology, 11*. https://doi.org/10.3389/fendo.2020.588773

Buenzli, P. R., & Sims, N. A. (2015). Quantifying the osteocyte network in the human skeleton. *Bone, 75*, 144–150. https://doi.org/10.1016/j.bone.2015.02.016

Burger, E. H., Klein-Nulend, J., Van Der Plas, A., & Nijweide, P. J. (1995). Function of osteocytes in bone—their role in mechanotransduction. *The Journal of Nutrition, 125*(7), 2020S–2023S. https://doi.org/10.1093/jn/125.suppl_7.2020S

Burr, D. B., Milgrom, C., Fyhrie, D., Forwood, M., Nyska, M., Finestone, A., Hoshaw, S., Saiag, E., & Simkin, A. (1996). In vivo measurement of human tibial strains during vigorous activity. *Bone, 18*(5), 405–410. https://doi.org/10.1016/8756-3282(96)00028-2

Cabahug-Zuckerman, P., Stout, R. F., Majeska, R. J., Thi, M. M., Spray, D. C., Weinbaum, S., & Schaffler, M. B. (2018). Potential role for a specialized β3 integrin-based structure on osteocyte processes in bone mechanosensation. *Journal of Orthopaedic Research, 36*(2), 642–652. https://doi.org/10.1002/jor.23792

Cardoso, L., Fritton, S. P., Gailani, G., Benalla, M., & Cowin, S. C. (2013). Advances in assessment of bone porosity, permeability and interstitial fluid flow. *Journal of Biomechanics, 46*(2), 253–265. https://doi.org/10.1016/j.jbiomech.2012.10.025

Carriero, A., Pereira, A. F., Wilson, A. J., Castagno, S., Javaheri, B., Pitsillides, A. A., Marenzana, M., & Shefelbine, S. J. (2018). Spatial relationship between bone formation and mechanical stimulus within cortical bone: Combining 3D fluorochrome mapping and poroelastic finite element modelling. *Bone Reports, 8*, 72–80. https://doi.org/10.1016/j.bonr.2018.02.003

Chennimalai Kumar, N., Dantzig, J. A., & Jasiuk, I. M. (2012). Modeling of cortical bone adaptation in a rat ulna: Effect of frequency. *Bone, 50*(3), 792–797. https://doi.org/10.1016/j.bone.2011.12.008

Ciani, C., Doty, S. B., & Fritton, S. P. (2005). Mapping bone interstitial fluid movement: Displacement of ferritin tracer during histological processing. *Bone, 37*(3), 379–387. https://doi.org/10.1016/j.bone.2005.04.004

Cooper, D. M. L., Thomas, C. D. L., Clement, J. G., Turinsky, A. L., Sensen, C. W., & Hallgrímsson, B. (2007). Age-dependent change in the 3D structure of cortical porosity at the human femoral midshaft. *Bone, 40*(4), 957–965. https://doi.org/10.1016/j.bone.2006.11.011

Cooper, D. M. L., Turinsky, A. L., Sensen, C. W., & Hallgrímsson, B. (2003). Quantitative 3D analysis of the canal network in cortical bone by micro-computed tomography. *The Anatomical Record Part B: The New Anatomist, 274B*(1), 169–179. https://doi.org/10.1002/ar.b.10024

Cowin, S. C. (1999). Bone poroelasticity. *Journal of Biomechanics, 32*(3), 217–238. https://doi.org/10.1016/S0021-9290(98)00161-4

Cowin, S. C., & Cardoso, L. (2015). Blood and interstitial flow in the hierarchical pore space architecture of bone tissue. *Journal of Biomechanics, 48*(5), 842–854. https://doi.org/10.1016/j.jbiomech.2014.12.013

Cowin, S. C., Moss-Salentijn, L., & Moss, M. L. (1991). Candidates for the mechanosensory system in bone. *Journal of Biomechanical Engineering, 113*(2), 191–197. https://doi.org/10.1115/1.2891234

Cowin, S. C., Weinbaum, S., & Zeng, Y. (1995). A case for bone canaliculi as the anatomical site of strain generated potentials. *Journal of Biomechanics, 28*(11), 1281–1297. https://doi.org/10.1016/0021-9290(95)00058-p

Fan, L., Pei, S., Lucas Lu, X., & Wang, L. (2016). A multiscale 3D finite element analysis of fluid/solute transport in mechanically loaded bone. *Bone Research, 4*. https://doi.org/10.1038/boneres.2016.32

Fernández-Seara, M. A., Wehrli, S. L., & Wehrli, F. W. (2002). Diffusion of exchangeable water in cortical bone studied by nuclear magnetic resonance. *Biophysical Journal, 82*(1), 522–529. https://doi.org/10.1016/S0006-3495(02)75417-9

Fornells, P., García-Aznar, J. M., & Doblaré, M. (2007). A finite element dual porosity approach to model deformation-induced fluid flow in cortical bone. *Annals of Biomedical Engineering, 35*(10), 1687–1698. https://doi.org/10.1007/s10439-007-9351-5

Fritton, S. P., McLeod, K. J., & Rubin, C. T. (2000). Quantifying the strain history of bone: Spatial uniformity and self-similarity of low-magnitude strains. *Journal of Biomechanics, 33*(3), 317–325. https://doi.org/10.1016/S0021-9290(99)00210-9

Fritton, S. P., & Weinbaum, S. (2009). Fluid and solute transport in bone: Flow-induced mechanotransduction. *Annual Review of Fluid Mechanics, 41*, 347–374. https://doi.org/10.1146/annurev.fluid.010908.165136

Gailani, G., Benalla, M., Mahamud, R., Cowin, S. C., & Cardoso, L. (2009). Experimental determination of the permeability in the lacunar-canalicular porosity of bone. *Journal of Biomechanical Engineering, 131*(10). https://doi.org/10.1115/1.3200908

Gardinier, J. D., Townend, C. W., Jen, K. P., Wu, Q., Duncan, R. L., & Wang, L. (2010). In situ permeability measurement of the Mammalian lacunar-canalicular system. *Bone, 46*(4), 1075–1081. https://doi.org/10.1016/j.bone.2010.01.371

Gatti, V., Azoulay, E. M., & Fritton, S. P. (2018). Microstructural changes associated with osteoporosis negatively affect loading-induced fluid flow around osteocytes in cortical bone. *Journal of Biomechanics, 66*, 127–136. https://doi.org/10.1016/j.jbiomech.2017.11.011

Gatti, V., Gelbs, M. J., Guerra, R. B., Gerber, M. B., & Fritton, S. P. (2021). Interstitial fluid velocity is decreased around cortical bone vascular pores and depends on osteocyte position in a rat model of disuse osteoporosis. *Biomechanics and Modeling in Mechanobiology, 20*(3), 1135–1146. https://doi.org/10.1007/s10237-021-01438-4

Gatti, V., Ghobryal, B., Gelbs, M. J., Gerber, M. B., Doty, S. B., Cardoso, L., & Fritton, S. P. (2019). Botox-induced muscle paralysis alters intracortical porosity and osteocyte lacunar density in skeletally mature rats. *Journal of Orthopaedic Research, 37*(5), 1153–1163. https://doi.org/10.1002/jor.24276

Goulet, G. C., Coombe, D., Martinuzzi, R. J., & Zernicke, R. F. (2009). Poroelastic evaluation of fluid movement through the lacunocanalicular system. *Annals of Biomedical Engineering, 37*(7), 1390–1402. https://doi.org/10.1007/s10439-009-9706-1

Goulet, G. C., Cooper, D. M. L., Coombe, D., & Zernicke, R. F. (2008). Influence of cortical canal architecture on lacunocanalicular pore pressure and fluid flow. *Computer Methods in*

Biomechanics and Biomedical Engineering, 11(4), 379–387. https://doi.org/10.1080/10255840701814105

Goulet, G. C., Hamilton, N., Cooper, D., Coombe, D., Tran, D., Martinuzzi, R., & Zernicke, R. F. (2008). Influence of vascular porosity on fluid flow and nutrient transport in loaded cortical bone. *Journal of Biomechanics, 41*(10), 2169–2175. https://doi.org/10.1016/j.jbiomech.2008.04.022

Gross, T. S., Edwards, J. L., Mcleod, K. J., & Rubin, C. T. (1997). Strain gradients correlate with sites of periosteal bone formation. *Journal of Bone and Mineral Research, 12*(6), 982–988. https://doi.org/10.1359/jbmr.1997.12.6.982

Gururaja, S., Kim, H. J., Swan, C. C., Brand, R. A., & Lakes, R. S. (2005). Modeling deformation-induced fluid flow in cortical bone's canalicular-lacunar system. *Annals of Biomedical Engineering, 33*(1), 7–25. https://doi.org/10.1007/s10439-005-8959-6

Han, Y., Cowin, S. C., Schaffler, M. B., & Weinbaum, S. (2004). Mechanotransduction and strain amplification in osteocyte cell processes. *Proceedings of the National Academy of Sciences, 101*(47), 16689–16694. https://doi.org/10.1073/pnas.0407429101

Holmes, J. M., Davies, D. H., Meath, W. J., & Beebe, R. A. (1964). Gas adsorption and surface structure of bone mineral. *Biochemistry, 3*(12), 2019–2024. https://doi.org/10.1021/bi00900a042

Jähn, K., Kelkar, S., Zhao, H., Xie, Y., Tiede-Lewis, L. M., Dusevich, V., Dallas, S. L., & Bonewald, L. F. (2017). Osteocytes acidify their microenvironment in response to PTHrP in vitro and in lactating mice in vivo. *Journal of Bone and Mineral Research, 32*(8), 1761–1772. https://doi.org/10.1002/jbmr.3167

Joukar, A., Niroomand-Oscuii, H., & Ghalichi, F. (2016). Numerical simulation of osteocyte cell in response to directional mechanical loadings and mechanotransduction analysis: Considering lacunar-canalicular interstitial fluid flow. *Computer Methods and Programs in Biomedicine, 133*, 133–141. https://doi.org/10.1016/j.cmpb.2016.05.019

Judex, S., Gross, T. S., & Zernicke, R. F. (1997). Strain gradients correlate with sites of exercise-induced bone-forming surfaces in the adult skeleton. *Journal of Bone and Mineral Research, 12*(10), 1737–1745. https://doi.org/10.1359/jbmr.1997.12.10.1737

Kameo, Y., Ozasa, M., & Adachi, T. (2022). Computational framework for analyzing flow-induced strain on osteocyte as modulated by microenvironment. *Journal of the Mechanical Behavior of Biomedical Materials, 126*, Article 105027. https://doi.org/10.1016/j.jmbbm.2021.105027

Kamioka, H., Kameo, Y., Imai, Y., Bakker, A. D., Bacabac, R. G., Yamada, N., Takaoka, A., Yamashiro, T., Adachi, T., & Klein-Nulend, J. (2012). Microscale fluid flow analysis in a human osteocyte canaliculus using a realistic high-resolution image-based three-dimensional model. *Integrative Biology, 4*(10), 1198–1206. https://doi.org/10.1039/c2ib20092a

Kaya, S., Basta-Pljakic, J., Seref-Ferlengez, Z., Majeska, R. J., Cardoso, L., Bromage, T. G., Zhang, Q., Flach, C. R., Mendelsohn, R., Yakar, S., Fritton, S. P., & Schaffler, M. B. (2017). Lactation-induced changes in the volume of osteocyte lacunar-canalicular space alter mechanical properties in cortical bone tissue. *Journal of Bone and Mineral Research, 32*(4), 688–697. https://doi.org/10.1002/jbmr.3044

Klein-Nulend, J., Bakker, A. D., Bacabac, R. G., Vatsa, A., & Weinbaum, S. (2013). Mechanosensation and transduction in osteocytes. *Bone, 54*(2), 182–190. https://doi.org/10.1016/j.bone.2012.10.013

Klein-Nulend, J., van der Plas, A., Semeins, C. M., Ajubi, N. E., Erangos, J. A., Nijweide, P. J., & Burger, E. H. (1995). Sensitivity of osteocytes to biomechanical stress in vitro. *The FASEB Journal, 9*(5), 441–445. https://doi.org/10.1096/fasebj.9.5.7896017

Knothe Tate, M. L., Niederer, P., & Knothe, U. (1998). In vivo tracer transport through the lacunocanalicular system of rat bone in an environment devoid of mechanical loading. *Bone, 22*(2), 107–117. https://doi.org/10.1016/s8756-3282(97)00234-2

Knothe Tate, M. L., Steck, R., Forwood, M. R., & Niederer, P. (2000). In vivo demonstration of load-induced fluid flow in the rat tibia and its potential implications for processes associated with functional adaptation. *Journal of Experimental Biology, 203*(18), 2737–2745. https://doi.org/10.1242/jeb.203.18.2737

Kufahl, R. H., & Saha, S. (1990). A theoretical model for stress-generated fluid flow in the canaliculi-lacunae network in bone tissue. *Journal of Biomechanics, 23*(2), 171–180. https://doi.org/10.1016/0021-9290(90)90350-C

Kumar, R., Tiwari, A. K., Tripathi, D., Shrivas, N. V., & Nizam, F. (2019). Canalicular fluid flow induced by loading waveforms: A comparative analysis. *Journal of Theoretical Biology, 471*, 59–73. https://doi.org/10.1016/j.jtbi.2019.03.023

Kwon, R. Y., & Frangos, J. A. (2010). Quantification of lacunar-canalicular interstitial fluid flow through computational modeling of fluorescence recovery after photobleaching. *Cellular and Molecular Bioengineering, 3*(3), 296–306. https://doi.org/10.1007/s12195-010-0129-8

Kwon, R. Y., Meays, D. R., Meilan, A. S., Jones, J., Miramontes, R., Kardos, N., Yeh, J., & Frangos, J. A. (2012). Skeletal adaptation to intramedullary pressure-induced interstitial fluid flow is enhanced in mice subjected to targeted osteocyte ablation. *PLoS One, 7*(3), Article e33336. https://doi.org/10.1371/journal.pone.0033336

Lai, X., Chung, R., Li, Y., Liu, X. S., & Wang, L. (2021). Lactation alters fluid flow and solute transport in maternal skeleton: A multiscale modeling study on the effects of microstructural changes and loading frequency. *Bone, 151*, Article 116033. https://doi.org/10.1016/j.bone.2021.116033

Lanyon, L. E., Hampson, W. G. J., Goodship, A. E., & Shah, J. S. (1975). Bone deformation recorded in vivo from strain gauges attached to the human tibial shaft. *Acta Orthopaedica, 46*(2), 256–268. https://doi.org/10.3109/17453677508989216

Lemaire, T., Lemonnier, S., & Naili, S. (2012). On the paradoxical determinations of the lacuno-canalicular permeability of bone. *Biomechanics and Modeling in Mechanobiology, 11*(7), 933–946. https://doi.org/10.1007/s10237-011-0363-6

Lemaire, T., Pham, T. T., Capiez-Lernout, E., de Leeuw, N. H., & Naili, S. (2015). Water in hydroxyapatite nanopores: Possible implications for interstitial bone fluid flow. *Journal of Biomechanics, 48*(12), 3066–3071. https://doi.org/10.1016/j.jbiomech.2015.07.025

Lemaire, T., Pham, T. T., de Leeuw, N. H., & Naili, S. (2015). Bone water at the nanoscale: A molecular dynamics study. *Computer Methods in Biomechanics and Biomedical Engineering, 18*, 1982–1983. https://doi.org/10.1080/10255842.2015.1069586

Li, G., Bronk, J. T., An, K. N., & Kelly, P. J. (1987). Permeability of cortical bone of canine tibiae. *Microvascular Research, 34*(3), 302–310. https://doi.org/10.1016/0026-2862(87)90063-X

Mak, A. F. T., Huang, D. T., Zhang, J. D., & Tong, P. (1997). Deformation-induced hierarchical flows and drag forces in bone canaliculi and matrix microporosity. *Journal of Biomechanics, 30*(1), 11–18. https://doi.org/10.1016/S0021-9290(96)00121-2

Mak, A. F. T., & Zhang, J. D. (2001). Numerical simulation of streaming potentials due to deformation-induced hierarchical flows in cortical bone. *Journal of Biomechanical Engineering, 123*(1), 66–70. https://doi.org/10.1115/1.1336796

Malachanne, E., Dureisseix, D., Cañadas, P., & Jourdan, F. (2008). Experimental and numerical identification of cortical bone permeability. *Journal of Biomechanics, 41*(3), 721–725. https://doi.org/10.1016/j.jbiomech.2007.09.028

McNamara, L. M., Majeska, R. J., Weinbaum, S., Friedrich, V., & Schaffler, M. B. (2009). Attachment of osteocyte cell processes to the bone matrix. *The Anatomical Record, 292* (3), 355–363. https://doi.org/10.1002/ar.20869

Mertiya, A. S., Tiwari, A. K., Mishra, A., Main, R. P., Tripathi, D., & Tiwari, A. (2023). Computational modeling for osteogenic potential assessment of physical exercises based on loading-induced mechanobiological environments in cortical bone remodeling. *Biomechanics and Modeling in Mechanobiology, 22*(1), 281–295. https://doi.org/10.1007/s10237-022-01647-5

Meslier, Q. A., DiMauro, N., Somanchi, P., Nano, S., & Shefelbine, S. J. (2022). Manipulating load-induced fluid flow in vivo to promote bone adaptation. *Bone, 165*. https://doi.org/10.1016/j.bone.2022.116547

Mi, L. Y., Basu, M., Fritton, S. P., & Cowin, S. C. (2005). Analysis of avian bone response to mechanical loading, part two: Development of a computational connected cellular network to study bone intercellular communication. *Biomechanics and Modeling in Mechanobiology, 4*(2–3), 132–146. https://doi.org/10.1007/s10237-004-0066-3

Mi, L. Y., Fritton, S. P., Basu, M., & Cowin, S. C. (2005). Analysis of avian bone response to mechanical loading - Part one: Distribution of bone fluid shear stress induced by bending and axial loading. *Biomechanics and Modeling in Mechanobiology, 4*(2–3), 118–131. https://doi.org/10.1007/s10237-004-0065-4

Murray, D. W., & Rushton, N. (1990). The effect of strain on bone cell prostaglandin E2 release: A new experimental method. *Calcified Tissue International, 47*(1), 35–39. https://doi.org/10.1007/BF02555863

Nguyen, V. H., Lemaire, T., & Naili, S. (2010). Poroelastic behaviour of cortical bone under harmonic axial loading: A finite element study at the osteonal scale. *Medical Engineering & Physics, 32*(4), 384–390. https://doi.org/10.1016/j.medengphy.2010.02.001

Niroobakhsh, M., Laughrey, L. E., Dallas, S. L., Johnson, M. L., & Ganesh, T. (2024). Computational modeling based on confocal imaging predicts changes in osteocyte and dendrite shear stress due to canalicular loss with aging. *Biomechanics and Modeling in Mechanobiology, 23*(1), 129–143. https://doi.org/10.1007/s10237-023-01763-w

Oyen, M. L. (2008). Poroelastic nanoindentation responses of hydrated bone. *Journal of Materials Research, 23*(5), 1307–1314. https://doi.org/10.1557/jmr.2008.0156

Pereira, A. F., Javaheri, B., Pitsillides, A. A., & Shefelbine, S. J. (2015). Predicting cortical bone adaptation to axial loading in the mouse tibia. *Journal of the Royal Society Interface, 12*(110). https://doi.org/10.1098/rsif.2015.0590

Pereira, A. F., & Shefelbine, S. J. (2014). The influence of load repetition in bone mechanotransduction using poroelastic finite-element models: The impact of permeability. *Biomechanics and Modeling in Mechanobiology, 13*(1), 215–225. https://doi.org/10.1007/s10237-013-0498-8

Piekarski, K., & Munro, M. (1977). Transport mechanism operating between blood supply and osteocytes in long bones. *Nature, 269*(5623), 80–82. https://doi.org/10.1038/269080a0

Pollack, S. R., Petrov, N., Salzstein, R., Brankov, G., & Blagoeva, R. (1984). An anatomical model for streaming potentials in osteons. *Journal of Biomechanics, 17*(8), 627–636. https://doi.org/10.1016/0021-9290(84)90094-0

Price, C., Zhou, X., Li, W., & Wang, L. (2011). Real-time measurement of solute transport within the lacunar-canalicular system of mechanically loaded bone: Direct evidence for load-induced fluid flow. *Journal of Bone and Mineral Research, 26*(2), 277–285. https://doi.org/10.1002/jbmr.211

Qin, Y. X., & Hu, M. (2014). Mechanotransduction in musculoskeletal tissue regeneration: Effects of fluid flow, loading, and cellular-molecular pathways. *BioMed Research International, 2014*. https://doi.org/10.1155/2014/863421

Qin, Y. X., Rubin, C. T., & McLeod, K. J. (1998). Nonlinear dependence of loading intensity and cycle number in the maintenance of bone mass and morphology. *Journal of Orthopaedic Research, 16*(4), 482–489. https://doi.org/10.1002/jor.1100160414

Qing, H., Ardeshirpour, L., Divieti Pajevic, P., Dusevich, V., Jähn, K., Kato, S., Wysolmerski, J., & Bonewald, L. F. (2012). Demonstration of osteocytic perilacunar/canalicular remodeling in mice during lactation. *Journal of Bone and Mineral Research, 27*(5), 1018–1029. https://doi.org/10.1002/jbmr.1567

Rémond, A., Naïli, S., & Lemaire, T. (2008). Interstitial fluid flow in the osteon with spatial gradients of mechanical properties: A finite element study. *Biomechanics and Modeling in Mechanobiology, 7*(6), 487–495. https://doi.org/10.1007/s10237-007-0111-0

Rolvien, T., & Amling, M. (2022). Disuse osteoporosis: Clinical and mechanistic insights. *Calcified Tissue International, 110*(5), 592–604. https://doi.org/10.1007/s00223-021-00836-1

Roothaer, X., Delille, R., Morvan, H., Bennani, B., Markiewicz, E., & Fontaine, C. (2018). A three-dimensional geometric quantification of human cortical canals using an innovative method with micro-computed tomographic data. *Micron, 114*, 62–71. https://doi.org/10.1016/j.micron.2018.07.006

Rouhana, S. W., Johnson, M. W., Chakkalakal, D. R., Harper, R. A., & Katz, J. L. (1981). Permeability of compact bone. *Joint ASME-ASCE Conference Biomechanics Symposium, 43*, 169–172.

Sánchez, M. T., Pérez, M.Á., & García-Aznar, J. M. (2021). The role of fluid flow on bone mechanobiology: Mathematical modeling and simulation. *Computational Geosciences, 25*(2), 823–830. https://doi.org/10.1007/s10596-020-09945-6

Salzstein, R. A., & Pollack, S. R. (1987). Electromechanical potentials in cortical bone-II. Experimental analysis. *Journal of Biomechanics, 20*(3), 271–280. https://doi.org/10.1016/0021-9290(87)90294-6

Salzstein, R. A., Pollack, S. R., Mak, A. F. T., & Petrov, N. (1987). Electromechanical potentials in cortical bone-I. A continuum approach. *Journal of Biomechanics, 20*(3), 261–270. https://doi.org/10.1016/0021-9290(87)90293-4

Schneider, P., Stauber, M., Voide, R., Stampanoni, M., Donahue, L. R., & Müller, R. (2007). Ultrastructural properties in cortical bone vary greatly in two inbred strains of mice as assessed by synchrotron light based micro- and nano-CT. *Journal of Bone and Mineral Research, 22*(10), 1557–1570. https://doi.org/10.1359/jbmr.070703

Schurman, C. A., Verbruggen, S. W., & Alliston, T. (2021). Disrupted osteocyte connectivity and pericellular fluid flow in bone with aging and defective TGF-β signaling. *Proceedings of the National Academy of Sciences, 118*(25). https://doi.org/10.1073/pnas.2023999118

Shapiro, F., Cahill, C., Malatantis, G., & Nayak, R. C. (1995). Transmission electron microscopic demonstration of vimentin in rat osteoblast and osteocyte cell bodies and processes

using the immunogold technique. *The Anatomical Record, 241*(1), 39–48. https://doi.org/10.1002/ar.1092410107
Sharma, D., Ciani, C., Marin, P. A. R., Levy, J. D., Doty, S. B., & Fritton, S. P. (2012). Alterations in the osteocyte lacunar-canalicular microenvironment due to estrogen deficiency. *Bone, 51*(3), 488–497. https://doi.org/10.1016/j.bone.2012.05.014
Shrivas, N. V., Badhyal, S., Tiwari, A. K., Mishra, A., Tripathi, D., & Patil, S. (2023). Computation of physiological loading induced interstitial fluid motion in muscle standardized femur: Healthy vs. osteoporotic bone. *Computer Methods and Programs in Biomedicine, 237*, Article 107592. https://doi.org/10.1016/j.cmpb.2023.107592
Smalt, R., Mitchell, F. T., Howard, R. L., & Chambers, T. J. (1997). Induction of NO and prostaglandin E2 in osteoblasts by wall-shear stress but not mechanical strain. *American Journal of Physiology - Endocrinology and Metabolism, 273*(4), E751–E758. https://doi.org/10.1152/ajpendo.1997.273.4.e751
Smit, T. H., Huyghe, J. M., & Cowin, S. C. (2002). Estimation of the poroelastic parameters of cortical bone. *Journal of Biomechanics, 35*(6), 829–835. https://doi.org/10.1016/S0021-9290(02)00021-0
Srinivasan, S., & Gross, T. S. (2000). Canalicular fluid flow induced by bending of a long bone. *Medical Engineering & Physics, 22*(2), 127–133. https://doi.org/10.1016/S1350-4533(00)00021-7
Srinivasan, S., Weimer, D. A., Agans, S. C., Bain, S. D., & Gross, T. S. (2002). Low-magnitude mechanical loading becomes osteogenic when rest is inserted between each load cycle. *Journal of Bone and Mineral Research, 17*(9), 1613–1620. https://doi.org/10.1359/jbmr.2002.17.9.1613
Starkebaum, W., Pollack, S. R., & Korostoff, E. (1979). Midroelectrode studies of stress-generated potentials in four-point bending of bone. *Journal of Biomedical Materials Research, 13*(5), 729–751. https://doi.org/10.1002/jbm.820130506
Steck, R., Niederer, P., & Knothe Tate, M. L. (2003). A finite element analysis for the prediction of load-induced fluid flow and mechanochemical transduction in bone. *Journal of Theoretical Biology, 220*(2), 249–259. https://doi.org/10.1006/jtbi.2003.3163
Tami, A. E., Schaffler, M. B., & Knothe Tate, M. L. (2003). Probing the tissue to subcellular level structure underlying bone's molecular sieving function. *Biorheology, 40*(6), 577–590. https://doi.org/10.1177/0006355X2003040006003
Tanaka, T., & Sakano, A. (1985). Differences in permeability of microperoxidase and horseradish peroxidase into the alveolar bone of developing rats. *Journal of Dental Research, 64*(6), 870–876. https://doi.org/10.1177/00220345850640060201
Tang, T., Landis, W., Blouin, S., Bertinetti, L., Hartmann, M. A., Berzlanovich, A., Weinkamer, R., Wagermaier, W., & Fratzl, P. (2023). Subcanalicular nanochannel volume is inversely correlated with calcium content in human cortical bone. *Journal of Bone and Mineral Research, 38*(2), 313–325. https://doi.org/10.1002/jbmr.4753
Tang, T., Landis, W., Raguin, E., Werner, P., Bertinetti, L., Dean, M., Wagermaier, W., & Fratzl, P. (2022). A 3D network of nanochannels for possible ion and molecule transit in mineralizing bone and cartilage. *Advanced NanoBiomed Research, 2*(8). https://doi.org/10.1002/anbr.202100162
Thi, M. M., Suadicani, S. O., Schaffler, M. B., Weinbaum, S., & Spray, D. C. (2013). Mechanosensory responses of osteocytes to physiological forces occur along processes and not cell body and require αvβ3 integrin. *Proceedings of the National Academy of Sciences, 110*(52), 21012–21017. https://doi.org/10.1073/pnas.1321210110

Thompson, W. R., Modla, S., Grindel, B. J., Czymmek, K. J., Kirn-Safran, C. B., Wang, L., Duncan, R. L., & Farach-Carson, M. C. (2011). Perlecan/Hspg2 deficiency alters the pericellular space of the lacunocanalicular system surrounding osteocytic processes in cortical bone. *Journal of Bone and Mineral Research, 26*(3), 618–629. https://doi.org/10.1002/jbmr.236

Tiede-Lewis, L. M., & Dallas, S. L. (2019). Changes in the osteocyte lacunocanalicular network with aging. *Bone, 122*, 101–113. https://doi.org/10.1016/j.bone.2019.01.025

Tiwari, A. K., Kumar, R., Tripathi, D., & Badhyal, S. (2018). In silico modeling of bone adaptation to rest-inserted loading: Strain energy density versus fluid flow as stimulus. *Journal of Theoretical Biology, 446*, 110–127. https://doi.org/10.1016/j.jtbi.2018.03.009

van Tol, A. F., Roschger, A., Repp, F., Chen, J., Roschger, P., Berzlanovich, A., Gruber, G. M., Fratzl, P., & Weinkamer, R. (2020). Network architecture strongly influences the fluid flow pattern through the lacunocanalicular network in human osteons. *Biomechanics and Modeling in Mechanobiology, 19*(3), 823–840. https://doi.org/10.1007/s10237-019-01250-1

van Tol, A. F., Schemenz, V., Wagermaier, W., Roschger, A., Razi, H., Vitienes, I., Fratzl, P., Willie, B. M., & Weinkamer, R. (2020). The mechanoresponse of bone is closely related to the osteocyte lacunocanalicular network architecture. *Proceedings of the National Academy of Sciences, 117*(51), 32251–32259. https://doi.org/10.1073/pnas.2011504117

Vaughan, T. J., Mullen, C. A., Verbruggen, S. W., & McNamara, L. M. (2015). Bone cell mechanosensation of fluid flow stimulation: A fluid–structure interaction model characterising the role integrin attachments and primary cilia. *Biomechanics and Modeling in Mechanobiology, 14*(4), 703–718. https://doi.org/10.1007/s10237-014-0631-3

Verbruggen, S. W., Vaughan, T. J., & McNamara, L. M. (2014). Fluid flow in the osteocyte mechanical environment: A fluid-structure interaction approach. *Biomechanics and Modeling in Mechanobiology, 13*(1), 85–97. https://doi.org/10.1007/s10237-013-0487-y

Wang, B., Lai, X., Price, C., Thompson, W. R., Li, W., Quabili, T. R., Tseng, W. J., Liu, X. S., Zhang, H., Pan, J., Kirn-Safran, C. B., Farach-Carson, M. C., & Wang, L. (2014). Perlecan-containing pericellular matrix regulates solute transport and mechanosensing within the osteocyte lacunar-canalicular system. *Journal of Bone and Mineral Research, 29*(4), 878–891. https://doi.org/10.1002/jbmr.2105

Wang, H., Du, T., Li, R., Main, R. P., & Yang, H. (2022). Interactive effects of various loading parameters on the fluid dynamics within the lacunar-canalicular system for a single osteocyte. *Bone, 158*, Article 116367. https://doi.org/10.1016/j.bone.2022.116367

Wang, L. (2018). Solute transport in the bone lacunar-canalicular system (LCS). *Current Osteoporosis Reports, 16*(1), 32–41. https://doi.org/10.1007/s11914-018-0414-3

Wang, L., Ciani, C., Doty, S. B., & Fritton, S. P. (2004). Delineating bone's interstitial fluid pathway in vivo. *Bone, 34*(3), 499–509. https://doi.org/10.1016/j.bone.2003.11.022

Wang, L., Cowin, S. C., Weinbaum, S., & Fritton, S. P. (2000). Modeling tracer transport in an osteon under cyclic loading. *Annals of Biomedical Engineering, 28*(10), 1200–1209. https://doi.org/10.1114/1.1317531

Wang, L., Cowin, S. C., Weinbaum, S., & Fritton, S. P. (2001). In response to "mixing mechanisms and net solute transport in bone" by M. L. Knothe Tate. *Annals of Biomedical Engineering, 29*(9), 812–816. https://doi.org/10.1114/1.1397789

Wang, L., Fritton, S. P., Cowin, S. C., & Weinbaum, S. (1999). Fluid pressure relaxation depends upon osteonal microstructure: Modeling an oscillatory bending experiment. *Journal of Biomechanics, 32*(7), 663–672. https://doi.org/10.1016/S0021-9290(99)00059-7

Wang, L., Fritton, S. P., Weinbaum, S., & Cowin, S. C. (2003). On bone adaptation due to venous stasis. *Journal of Biomechanics, 36*(10), 1439–1451. https://doi.org/10.1016/S0021-9290(03)00241-0

Wang, Y., McNamara, L. M., Schaffler, M. B., & Weinbaum, S. (2007). A model for the role of integrins in flow induced mechanotransduction in osteocytes. *Proceedings of the National Academy of Sciences, 104*(40), 15941–15946. https://doi.org/10.1073/pnas.0707246104

Wei, F., Flowerdew, K., Kinzel, M., Perotti, L. E., Asiatico, J., Omer, M., Hovell, C., Reumers, V., & Coathup, M. J. (2022). Changes in interstitial fluid flow, mass transport and the bone cell response in microgravity and normogravity. *Bone Research, 10*, Article 65. https://doi.org/10.1038/s41413-022-00234-9

Weinbaum, S., Cowin, S. C., & Zeng, Y. (1994). A model for the excitation of osteocytes by mechanical loading-induced bone fluid shear stresses. *Journal of Biomechanics, 27*(3), 339–360. https://doi.org/10.1016/0021-9290(94)90010-8

Wen, D., Androjna, C., Vasanji, A., Belovich, J., & Midura, R. J. (2010). Lipids and collagen matrix restrict the hydraulic permeability within the porous compartment of adult cortical bone. *Annals of Biomedical Engineering, 38*(3), 558–569. https://doi.org/10.1007/s10439-009-9858-z

Westerlind, K. C., Wronski, T. J., Ritman, E. L., Luo, Z., An, K., Bell, N. H., & Turner, R. T. (1997). Estrogen regulates the rate of bone turnover but bone balance in ovariectomized rats is modulated by prevailing mechanical strain. *Proceedings of the National Academy of Sciences, 94*(8), 4199–4204. https://doi.org/10.1073/pnas.94.8.4199

Yokoyama, Y., Kameo, Y., Kamioka, H., & Adachi, T. (2021). High-resolution image-based simulation reveals membrane strain concentration on osteocyte processes caused by tethering elements. *Biomechanics and Modeling in Mechanobiology, 20*(6), 2353–2360. https://doi.org/10.1007/s10237-021-01511-y

You, J., Yellowley, C. E., Donahue, H. J., Zhang, Y., Chen, Q., & Jacobs, C. R. (2000). Substrate deformation levels associated with routine physical activity are less stimulatory to bone cells relative to loading-induced oscillatory fluid flow. *Journal of Biomechanical Engineering, 122*(4), 387–393. https://doi.org/10.1115/1.1287161

You, L., Cowin, S. C., Schaffler, M. B., & Weinbaum, S. (2001). A model for strain amplification in the actin cytoskeleton of osteocytes due to fluid drag on pericellular matrix. *Journal of Biomechanics, 34*(11), 1375–1386. https://doi.org/10.1016/S0021-9290(01)00107-5

You, L., Weinbaum, S., Cowin, S. C., & Schaffler, M. B. (2004). Ultrastructure of the osteocyte process and its pericellular matrix. *The Anatomical Record Part A Discoveries in Molecular, Cellular, and Evolutionary Biology, 278*(2), 505–513. https://doi.org/10.1002/ar.a.20050

Yu, W., Ou, R., Hou, Q., Li, C., Yang, X., Ma, Y., Wu, X., & Chen, W. (2025). Multiscale interstitial fluid computation modeling of cortical bone to characterize the hydromechanical stimulation of lacunar-canalicular network. *Bone, 193*, Article 117386. https://doi.org/10.1016/j.bone.2024.117386

Zeng, Y., Cowin, S. C., & Weinbaum, S. (1994). A Fiber matrix model for fluid flow and streaming potentials in the canaliculi of an osteon. *Annals of Biomedical Engineering, 22*(3), 280–292. https://doi.org/10.1007/BF02368235

Zhang, D., Weinbaum, S., & Cowin, S. C. (1998). Estimates of the peak pressures in bone pore water. *Journal of Biomechanical Engineering, 120*(6), 697–703. https://doi.org/10.1115/1.2834881

Zhou, X., Novotny, J. E., & Wang, L. (2008). Modeling fluorescence recovery after photobleaching in loaded bone: Potential applications in measuring fluid and solute transport in the osteocytic lacunar-canalicular system. *Annals of Biomedical Engineering, 36*(12), 1961–1977. https://doi.org/10.1007/s10439-008-9566-0

Zimmermann, E. A., Schaible, E., Bale, H., Barth, H. D., Tang, S. Y., Reichert, P., Busse, B., Alliston, T., Ager, J. W., & Ritchie, R. O. (2011). Age-related changes in the plasticity and toughness of human cortical bone at multiple length scales. *Proceedings of the National Academy of Sciences, 108*(35), 14416–14421. https://doi.org/10.1073/pnas.1107966108

CHAPTER

17

Simulation platform for in-silico trials of osteoporosis treatments—Biochemical signaling pathways: Mechanobiological bone cell population modeling framework

Peter Pivonka[1], **José Luis Calvo-Gallego**[2,3] **and Javier Martínez-Reina**[2,3]

[1]*School of Mechanical, Medical and Process Engineering, Queensland University of Technology, Brisbane, QLD, Australia;* [2]*Departmento de Ingeniería Mecánica y Fabricación, Universidad de Sevilla, Seville, Spain;* [3]*Instituto de Biomedicina de Sevilla (IBIS), Sevilla, Spain*

Abbreviations

Alend	Alendronate
ARO	Age-related osteoporosis
BCPM	Bone cell population model
BDG	Bone density gain
BDL	Bone density loss
BDV	Bone density variation
BMD	Bone mineral density
Dkk-1	Dickkopf-related protein 1
Dmab	Denosumab
FRI	Fracture risk index
IV	Intravenous
MLT	Mineralization lag time
NDuF	Number of days until failure
NO	Nitric oxide
OD	Oestrogen deficiency
OP	Osteoporosis
OPG	Osteoprotegerin
PD	Pharmacodynamic
PK	Pharmacokinetic
PMO	Postmenopausal osteoporosis
PTH	Parathyroid hormone
RANK	Receptor activator of nuclear factor $\kappa\beta$
RANKL	Receptor activator of nuclear factor $\kappa\beta$ ligand

Biomechanics and Mechanobiology of Bone. https://doi.org/10.1016/B978-0-443-26739-0.00019-3

RVE	Representative volume element
Scl	Sclerostin
TGF-β	Transforming growth factor beta
TSM	Time elapsed since menopause
TSSM	Time since skeletal maturity
VOI	Volume of interest

1. Introduction

Osteoporosis (OP) is a chronic progressive bone disease characterized by an increase in bone porosity and deterioration of bone matrix properties that results in a reduction of strength at the organ level and an increase of fracture risk (Lemaire et al., 2004). OP is a growing global health problem in society, placed alongside diseases such as breast cancer, cardiovascular diseases, and diabetes mellitus when considering prevalence, lifetime risk, and socioeconomic impact (Salari et al., 2021; Shen et al., 2022). A recent review by Yeh et al. (2023) reported that 22% of women aged 50 years or older in the European Union, the UK, and Switzerland had OP in 2019. Furthermore, based on a study analyzing the burden and management of fragility fractures in France, Germany, Italy, Spain, Sweden, and the UK, it was estimated that the total numbers of fractures for both men and women will increase dramatically in the coming years, from 2.7 million that occurred in 2017 to an expected 3.3 million in 2030 (Borgström et al., 2020), and the annual costs of fractures are expected to increase by 27%, from € 37.5 billion to € 47.4 billion (Borgström et al., 2020). Hip fractures constitute the most serious type of fracture, which is associated with ~ 27% mortality in men and ~ 20% mortality in women (Schnell et al., 2010). Less then 50% of hip fracture patients regain their previous level of function (Eastwood et al., 2002), and 20% of patients with hip fractures permanently lose independence and require long-term care (Landefeld, 2011). All these data highlight the importance of improving current treatments for the disease.

The most common forms of OP are age-related osteoporosis (ARO) and postmenopausal osteoporosis (PMO). The age-related increase in bone porosity is due to an imbalance in the bone remodeling process, with a predominance of resorption over formation. Some age- and environment-related factors may contribute to the onset and progression of OP, the most important being vitamin D and calcium deficiency, decreased physical activity, and age-related decrease in gonadal function (Compston, 2001; Garnero & Delmas, 2004; Riggs, 2003; Weitzmann & Pacifi, 2006). Other forms of OP include secondary OP, which can be linked to adverse effects of drug therapies (e.g., glucocorticoids and proton pump inhibitors), different types of cancer, endocrine disorders, eating disorders, immobilization, marrow-related disorders, disorders of the gastrointestinal or biliary tract, and kidney disease (Fitzpatrick, 2002).

The main objective in the treatment of OP is the prevention of future fracture events. This is managed at the clinical level by identifying individuals at high risk of OP, which is linked to risk factors such as age, sex, low body weight, and fracture history (Black & Rosen, 2016). Once identified, the current standard of care for patients with OP includes diet and lifestyle recommendations (e.g., calcium and vitamin D intake and moderate exercise) and pharmacologic therapy, depending on the level of fracture risk. Different drug treatments can be distinguished according to: mechanisms of action; route of administration (oral, intravenous (IV), or nasal); dosage regimens; degrees of efficacy and effectiveness (Black & Rosen, 2016; Close et al., 2006).

According to the terminology of Riggs and Pafitt (2005), drugs to treat OP can be divided into anticatabolic (previously also referred to as antiresorptive) and anabolic. The former, e.g., bisphosphonates (alendronate (Alend), risedronate, ibandronate, and zoledronic acid), denosumab (Dmab), and estrogen agonists/antagonists (raloxifene), act to reduce bone turnover and produce a moderate bone mass gain resulting from the increased mineralization of the existing bone matrix, with little impact on porosity. On the other hand, anabolic drugs, for example, synthetic parathyroid hormone (PTH) administered intermittently (teriparatide and abaloparatide) and sclerostin (Scl) antibody romosozumab (Rmab), strongly increase bone mass by acting on the osteoblastic lineage. PTH reduces the apoptosis rate of forming osteoblasts (Bellido et al., 2003), while Rmab augments the proliferation rate of osteoblast precursor cells, but it has also been attributed an anticatabolic role (Cosman et al., 2016) by suppressing the upregulation of RANKL caused by Scl (Wijenayaka et al., 2011).

The development of safe and effective treatments for OP is challenging and costly, namely the size and long duration of the clinical trials needed to evaluate the benefits of drugs, especially the reduction of fracture risk. This is due to the fact that fractures are low-incidence events that require large (e.g., a few thousand subjects) and prolonged trials (3 years for phase 3) to adequately quantify them. The case of combined treatments of several drugs presents even more difficulties due to the large number of combinations that can be considered.

An alternative to running clinical trials is to conduct in-silico trials based on mechanobiological pharmacokinetic–pharmacodynamic (PK-PD) models of OP treatments. These types of models are based on a detailed mechanobiological description of the bone remodeling process, including major signaling pathways and regulatory factors. Using in-silico trials avoids ethical issues with human data collection and costs involved and reduces animal experimentation. Furthermore, many different drug combinations can be investigated with the aim of identifying the most effective treatment combinations. Finally, this approach also allows for assessment of patient-specific treatment options, considering patient age, age at menopause, and weight, among other factors.

In this chapter, we give an overview of the comprehensive simulation platform for in-silico OP treatments developed by the authors over the last 15 years. The most sophisticated bone remodeling model to date was originally developed by Lemaire et al. (2004) and subsequently expanded by Pivonka et al. toward mechanobiological feedback (Pivonka et al., 2008, 2010; Scheiner et al., 2013). This type of model introduces the concentration of bone cells and biochemical factors that drive the bone remodeling process. It includes osteoblast precursor cells, active osteoblasts, and active osteoclasts, along with the RANK-RANKL-OPG signaling pathway and transforming growth factor beta (TGF-β), which couples the osteoblastic and osteoclastic cell lineages. PTH is used as a catabolic regulator of RANKL and OPG and consequently of bone resorption. Over the last decade, this model was continuously improved (Martínez-Reina and Pivonka, 2019; Ruiz-Lozano et al., 2025), together with adding new PK models of OP drugs, including Dmab, PTH, Rmab, and Alend (Lavaill et al., 2020; Martínez-Reina and Pivonka, 2019; Martínez-Reina et al., 2021a, 2021b, 2022; Ruiz-Lozano et al., 2025; Trichilo et al., 2019). This has resulted in the development of the most comprehensive platform of mechanobiological PK-PD models for OP treatments. With this platform, it is now feasible to perform in-silico trials of OP treatments and to investigate the efficacy of combined drug therapies.

In the following, we present the most important features of the latest version of this model. Section 2 provides a description of the bone cell population model (BCPM) for bone remodeling. Section 3 introduces a general framework for competitive ligand–receptor binding reactions that are used to formulate activator and repressor functions regulating bone cell behavior. Furthermore, the section discusses major signaling pathways (RANK-RANKL-OPG and Wnt), which are included in the model formulation. In Section 4, we describe the mechanobiological feedback that introduced Frost's mechanostat into the model formulation. The description of the composition of bone, including the degree of bone matrix mineralization and mechanical properties of bone defined on the scale of the representative volume element (RVE), is provided in Section 5. Effects of damage accumulation in the bone matrix are introduced in Section 6. Modeling of bone disease progression with focus on the effects of aging and PMO on bone turnover and balance is described in Section 7.

We then present simulation results of a variety of drug treatments and discuss outcomes in terms of improvements in drug efficacy (see Chapter 18).

2. Bone cell population model of bone remodeling

The independent variables of the model are the concentration of bone cells and biochemical factors in the volume of interest (VOI). In a biomechanical context, the VOI is considered an RVE, where continuum and/or micromechanical properties can be defined. Two different cell lineages (osteoblastic, Ob, and osteoclastic, Oc) are considered, in different stages of differentiation, from uncommitted stem cells $(\mathrm{Ob_u}, \mathrm{Oc_u})$ to precursors of osteoblasts $(\mathrm{Ob_p})$ and osteoclasts $(\mathrm{Oc_p})$, and finally active cells, i.e., forming osteoblasts $(\mathrm{Ob_a})$ and resorbing osteoclasts $(\mathrm{Oc_a})$. The concentration of osteocytes, Ot, is also included as an independent variable. The concentrations of uncommitted stem cells are considered much higher than those of the rest of the cells, and therefore, Ob_u and Oc_u are assumed constant. The balance equations for the rest of the cells in the VOI are formulated as in (Calvo-Gallego et al., 2022, 2023; Martínez-Reina et al., 2022):

$$\frac{dOb_p}{dt} = D_{Ob_u} \cdot Ob_u \cdot \pi_{act,Ob_u}^{TGF-\beta} - D_{Ob_p} \cdot Ob_p \cdot \pi_{rep,Ob_p}^{TGF-\beta} + P_{Ob_p} \cdot Ob_p \cdot \pi_{act,Ob_p}^{Wnt} \quad (17.1)$$

$$\frac{dOb_a}{dt} = D_{Ob_p} \cdot Ob_p \cdot \pi_{rep,Ob_p}^{TGF-\beta} - \left(D_{Ob_a} + A_{Ob_a}\right) \cdot Ob_a \quad (17.2)$$

$$\frac{dOc_p}{dt} = D_{Oc_u} \cdot Oc_u \cdot \pi_{act,Oc_u}^{RANK} - D_{Oc_p} \cdot Oc_p \cdot \pi_{act,Oc_p}^{RANK} \quad (17.3)$$

$$\frac{dOc_a}{dt} = D_{Oc_p} \cdot Oc_p \cdot \pi_{act,Oc_p}^{RANK} - A_{Oc_a} \cdot Oc_a \cdot \pi_{act,Oc_a}^{TGF-\beta} \quad (17.4)$$

$$\frac{dOt}{dt} = \eta \frac{df_{bm}}{dt} \quad (17.5)$$

where $D_{Ob_u}, D_{Oc_u}, D_{Ob_p}, D_{Oc_p}$ and D_{Ob_a} are the differentiation rates of uncommitted progenitor cells, precursor cells, and active osteoblasts, respectively. P_{Ob_p} denotes the proliferation rate of osteoblast precursor cells. A_{Ob_a} and A_{Oc_a} are, respectively, the apoptosis rates of active osteoblasts and osteoclasts. Differentiation of active osteoblasts, D_{Ob_a}, measures the rate of conversion of these cells into osteocytes. However, this rate is not needed to calculate the concentration of osteocytes, as Eq. (17.5) assumes that their population is proportional to the change in bone matrix volume fraction ($f_{bm} = BV/TV$), with proportionality constant η. Therefore, only the sum $\left(D_{Ob_a} + A_{Ob_a}\right)$, also called osteoblast clearance rate, is relevant for the model. Cell differentiation, proliferation, and apoptosis are regulated by several "activator" (π_{act}) and "repressor" (π_{rep}) functions, which model, respectively, the promotion or inhibition of the biological processes of differentiation, proliferation, or apoptosis. In these regulatory functions, the subscripts indicate the type of bone cell that is regulated by these functions, while the superscript indicates the regulatory molecule (e.g., RANKL, Wnt, TGF-β) responsible for the respective cell responses.

Bone cell activities within the VOI can change the bone volume fraction locally over time:

$$\frac{df_{bm}}{dt} = k_{form} Ob_a - k_{res} Oc_a \quad (17.6)$$

where k_{res} and k_{form} are the resorption and formation rates, respectively, which were assumed constant. The bone volume fraction f_{bm} can be linked to the apparent bone density, ρ_{app}, which also depends on the mineral content of the bone matrix (see Section 5). This mineral content was assumed constant until Martínez-Reina and Pivonka (2019) incorporated the mineralization process into the BCPM framework. By doing so, an alternative way of evaluating bone mineral density (BMD) changes is proposed in two steps: porosity through Eq. (17.6) and mineral content independently. A schematic representation of the BCPM is shown in Fig. 17.1,

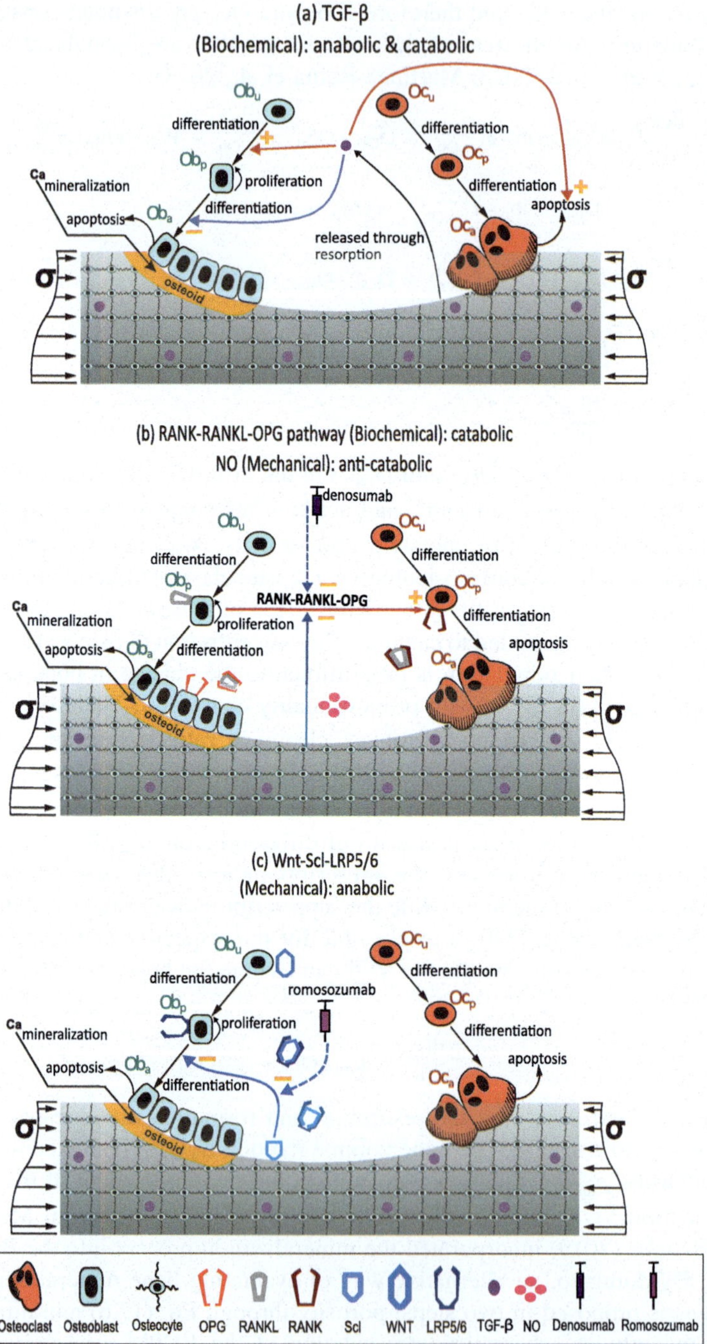

FIGURE 17.1

Schematic representation of the BCPM: bone cell differentiation stages are presented along with biochemical and biomechanical interactions.

which distinguishes the three main biochemical signaling pathways included in the model and described in detail in the next sections.

3. Biochemical signaling pathways

The processes of cell differentiation, proliferation, and apoptosis are governed by complex signaling pathways involving various biochemical factors. These signaling pathways have been modeled in a simplified way in different works (Lemaire et al., 2004; Pivonka et al., 2008) by adding the concentrations of those biochemical factors as state variables and modeling the binding reactions of ligands and receptors.

3.1 Competitive binding

Many biological processes are controlled by the binding of certain biochemical factors that act as receptors and ligands. In some cases, two or more ligands compete to bind to one receptor, or two (soluble) receptors compete for binding to one ligand. For the case of Wnt signaling, the ligands Wnt and Scl compete to bind to the receptor LRP5/6, which controls proliferation of osteoblast precursors. In the case of the RANK-RANKL-OPG pathway (see the next section for a detailed description), two receptors, i.e., RANK and OPG (i.e., decoy receptor), compete to bind to RANKL to regulate the differentiation of osteoclast precursors into mature active osteoclasts. This competitive binding can be modeled using the *law of mass action* (Lemaire et al., 2004).

Let us consider the binding of a given receptor R to its ligands A and B to form the complexes $A - R$ and $B - R$, respectively. Consider for each species $X = A, B, R$ a production term P_X and a degradation term D_X, together with the degradation of the complex D_{X-Y}. Let K^r_{X-Y} and K^f_{X-Y} be the reverse and forward binding reaction constants, respectively.

$$\begin{array}{ccccccc} \downarrow P_R & & \downarrow P_A & & & & \\ R & + & A & \underset{K^r_{A-R}}{\overset{K^f_{A-R}}{\rightleftharpoons}} & A - R & \overset{D_{A-R}}{\rightharpoonup} & \phi \\ \downarrow D_R & & \downarrow D_A & & & & \end{array} \tag{17.7}$$

$$\begin{array}{ccccccc} \downarrow P_R & & \downarrow P_B & & & & \\ R & + & B & \underset{K^r_{B-R}}{\overset{K^f_{B-R}}{\rightleftharpoons}} & B - R & \overset{D_{B-R}}{\rightharpoonup} & \phi \\ \downarrow D_R & & \downarrow D_B & & & & \end{array} \tag{17.8}$$

The law of mass action provides the following set of differential equations:

$$\frac{d[A-R]}{dt} = K^f_{A-R}[A][R] - K^r_{A-R}[A-R] - \widetilde{D}_{A-R}[A-R] \tag{17.9a}$$

$$\frac{d[B-R]}{dt} = K^f_{B-R}[B][R] - K^r_{B-R}[B-R] - \widetilde{D}_{B-R}[B-R] \tag{17.9b}$$

$$\frac{d[A]}{dt} = P_A - \widetilde{D}_A[A] + K^r_{A-R}[A-R] - K^f_{A-R}[A][R] \tag{17.9c}$$

$$\frac{d[B]}{dt} = P_B - \widetilde{D}_B[B] + K^r_{B-R}[B-R] - K^f_{B-R}[B][R] \tag{17.9d}$$

$$\frac{d[R]}{dt} = P_R - \widetilde{D}_R[R] + K^r_{A-R}[A-R] + K^r_{B-R}[B-R] - K^f_{A-R}[A][R] - K^f_{B-R}[B][R] \tag{17.9e}$$

where $[X]$ and $[X-Y]$ represent, respectively, the concentration of species X and complex $X-Y$. The degradation terms are assumed proportional to the concentration of the species, i.e., $D_X = \widetilde{D}_X[X]$, with $\widetilde{D}_X$ being the degradation rate.

It can be assumed that the binding reactions are much faster than the cell responses they lead to (Pivonka et al., 2008), and hence, a quasisteady state can be assumed, implying that the time derivatives of Eq. (17.9) are null. This condition in Eqs. (17.9a) and (17.9b) yields:

$$[A-R] = \frac{[A][R]}{K_{A-R}} \tag{17.10a}$$

$$[B-R] = \frac{[B][R]}{K_{B-R}} \tag{17.10b}$$

where:

$$K_{A-R} = \frac{K^r_{A-R} + \widetilde{D}_{A-R}}{K^f_{A-R}} \tag{17.11a}$$

$$K_{B-R} = \frac{K^r_{B-R} + \widetilde{D}_{B-R}}{K^f_{B-R}} \tag{17.11b}$$

The stationarity condition of Eqs. (17.9c)–(17.9e) yields:

$$[A] = \frac{P_A}{\widetilde{D}_A + \frac{\widetilde{D}_{A-R}}{K_{A-R}}[R]} \tag{17.12a}$$

$$[B] = \frac{P_B}{\widetilde{D}_B + \frac{\widetilde{D}_{B-R}}{K_{B-R}}[R]} \tag{17.12b}$$

$$[R] = \frac{P_R}{\widetilde{D}_R + \frac{\widetilde{D}_{A-R}}{K_{A-R}}[A] + \frac{\widetilde{D}_{B-R}}{K_{B-R}}[B]} \tag{17.12c}$$

If the degradation $\left(\widetilde{D}_X, \widetilde{D}_{X-Y}\right)$, production (P_X) and dissociation constants, (K_{X-Y}) are known, Eq. (17.12) constitutes a nonlinear system of three equations with three unknowns, namely [A], [B], [R].

The total concentration of receptors expressed at a particular cell type is the sum of the concentrations of free receptors and receptors bound to respective ligands. Using Eq. (17.10), this is given by:

$$[R]_{tot} = [R] + [A-R] + [B-R] = [R]\left(1 + \frac{[A]}{K_{A-R}} + \frac{[B]}{K_{B-R}}\right) \tag{17.13}$$

As described by Pivonka et al. (2012), the activation of a certain biological process regulated by the formation of the complex L − R is given by the ratio between the receptors R occupied by ligands L and the total number of ligands. In case of a single ligand, the latter expressions yield the first-order Hill activator and repressor functions:

$$\pi^{L}_{act,Y} = \frac{[R]}{[R] + K^{L-R}_{act,Y}} \tag{17.14}$$

Certain processes can be downregulated by the formation of the complex L-R. Thus, a repressor function can be defined as proportional to the complementary ratio, i.e., proportional to $\left(1 - \pi^{L}_{act,Y}\right)$. If a specific constant, $K^{L-R}_{rep,Y}$, is defined to describe such regulation, the repressor function can be written as follows:

$$\pi^{L}_{rep,Y} = \frac{K^{L-R}_{rep,Y}}{K^{L-R}_{rep,Y} + [R]} \tag{17.15}$$

3.2 RANK-RANKL-OPG signaling pathway

RANK-RANKL-OPG signaling pathway is key in regulating bone turnover and a good example of intercellular communication, as it explains how the osteoblast lineage regulates the differentiation and activity of osteoclasts (Boyce & Xing, 2008). It involves three major components: (i) RANK, expressed on the surface of Oc_u and Oc_p; (ii) RANKL, a polypeptide expressed on the surface of osteoblastic cells and also in soluble form after proteolytic cleavage; and (iii) OPG, a "decoy receptor" molecule released by osteoblastic cells (Simonet et al., 1997).

The procedure described in Section 17.3.1 can be used here, considering that R is now RANKL, which can bind to the receptors RANK (A) and OPG (B).

Differentiation of Oc_u and Oc_p requires the binding of RANKL to RANK, which can be inhibited by OPG, as this molecule can also bind to RANKL, thus preventing the formation of the RANKL-RANK complex. The upregulating role of this complex in the differentiation of Oc_u and Oc_p is modeled through π^{RANKL}_{act,Oc_u} and

π_{act,Oc_p}^{RANKL} in Eqs. (17.3) and (17.4), respectively. These Hill-type functions derive from Eq. (17.14):

$$\pi_{act,X}^{RANKL} = \frac{[RANKL]}{K_{act,X}^{RANKL} + [RANKL]} \text{ with } X = Oc_u, Oc_p \tag{17.16}$$

with $K_{act,X}^{RANKL}$ being constants corresponding to $X = Oc_u, Oc_p$. [•] denotes the concentrations of biochemical factors. In particular, [RANKL] is the concentration of free (unbound) RANKL and depends on the concentrations of RANK and OPG, along with the RANKL production rate. For this competitive binding system, Eq. (17.12c) can be expressed as:

$$[RANKL] = \frac{P_{RANKL}}{\tilde{D}_{RANKL} + \frac{\tilde{D}_{RANK\text{-}RANKL}}{K_{RANK\text{-}RANKL}}[RANK] + \frac{\tilde{D}_{OPG\text{-}RANKL}}{K_{OPG\text{-}RANKL}}[OPG]} \tag{17.17}$$

The concentration of OPG is given by:

$$[OPG] = \frac{\beta_{OPG,Ob_a}\pi_{rep,Ob_a}^{PTH} Ob_a\left(1 - \frac{[OPG]}{[OPG]_{max}}\right)}{\tilde{D}_{OPG} + \frac{\tilde{D}_{OPG-RANKL}[RANKL]}{K_{OPG-RANKL}}} \tag{17.18}$$

where β_{OPG,Ob_a} is the OPG production rate, $[OPG]_{max}$ is the maximum (or saturation) concentration of OPG, above which no more OPG is produced, π_{rep,Ob_a}^{PTH} is a Hill-type function that models the downregulating role of PTH on the production of OPG. The concentration of RANK is given by:

$$[RANK] = \frac{N_{Oc_p}^{RANK} Oc_p}{1 + \frac{[RANK]}{K_{RANK\text{-}RANKL}}} \tag{17.19}$$

where the constant $N_{Oc_p}^{RANK}$ measures the density of RANK receptors on the surface of osteoclast precursors. The RANKL production rate is modeled as:

$$P_{RANKL} = P_{Ob_p}^{RANKL} + P_{Ot}^{RANKL} \tag{17.20}$$

Following Nakashima et al. (2011) and Xiong et al. (2015), we assume that RANKL is expressed by osteocytes $\left(P_{Ot}^{RANKL}\right)$ and osteoblast precursors $\left(P_{Ob_p}^{RANKL}\right)$ and each production term is given by:

$$P_{Ob_p}^{RANKL} = \beta_{RANKL,Ob_p} \cdot \pi_{RANKL}^{OD} \cdot \pi_{act/rep,RANKL}^{PTH,NO} \cdot \left(1 - \frac{[RANKL]_{tot}}{[RANKL]_{max}}\right) \cdot Ob_p \tag{17.21}$$

$$P_{Ot}^{RANKL} = \beta_{RANKL,Ot} \cdot \pi_{RANKL}^{OD} \cdot \pi_{act,RANKL}^{dam} \cdot \pi_{act,RANKL}^{Scl} \cdot \left(1 - \frac{[RANKL]_{tot}}{[RANKL]_{max}}\right) \cdot Ot \quad (17.22)$$

where $[RANKL]_{max}$ is the maximum (or saturation) concentration of RANKL and $[RANKL]_{tot}$ is its total concentration (free and bound to its ligands), which can be calculated from Eq. (17.13). β_{RANKL,Ob_p} and $\beta_{RANKL,Ot}$ are the RANKL production rates of each cell and its inhibition by nitric oxide (NO) (Martin et al., 2019). $\pi_{act,RANKL}^{dam}$ is an activator function that accounts for the upregulation of RANKL expression by osteocytes due to the presence of microstructural damage (Martínez-Reina et al., 2021a). π_{RANKL}^{OD} is an activator function that models the increase in RANKL expression due to the oestrogen deficiency (OD) that occurs after menopause. OD has been reported to increase RANKL expression, decrease OPG production, and increase the responsiveness of osteoclasts to RANKL (Hofbauer & Schoppet, 2004). However, the three effects have similar consequences on the RANK-RANKL-OPG signaling pathway and can be modeled by either of them, as discussed by Ruiz-Lozano et al. (2024), who chose to simulate the increase in RANKL expression through the factor π_{RANKL}^{OD} in Eqs. (17.21) and (17.22), using a piecewise linear function of time:

$$\pi_{RANKL}^{OD}(t) = \begin{cases} 1 & \text{for } TSM < 0 \\ 1 + k_1 TSM & \text{for } 0 \le TSM < T^* \\ 1 + k_1\, T^* + k_2(TSM - T^*) & \text{for } T^* \le TSM \end{cases} \quad (17.23)$$

where $TSM = t - t_{meno}$ is the time elapsed since menopause and k_1, k_2 and T^* are constants. This factor is equal to 1, and consequently has no influence on the model, before menopause or in men.

As will be seen in Section 17.3.3, Scl has a major effect on the anabolic Wnt signaling pathway. However, Wijenayaka et al. (2011) showed that RANKL levels increase with high concentrations of Scl, indicating also an effect on the catabolic RANK-RANKL-OPG pathway, which is modeled through the sigmoidal function $\pi_{act,RANKL}^{Scl}$ in Eq. (17.22):

$$\pi_{act,RANKL}^{Scl} = 1 + (\alpha_{Scl} - 1)\frac{[Scl]^{\gamma_{Scl}}}{[Scl]^{\gamma_{Scl}} + \delta_{Scl}^{\gamma_{Scl}}} \quad (17.24)$$

with constants $\alpha_{Scl}, \gamma_{Scl}$, and δ_{Scl}.

3.3 Wnt/β-catenin signaling pathway

Extracellular Wnt ligands bind to Frizzled and lipoprotein receptor-related proteins (LRP5/6), triggering intracellular activation of β-catenin and promoting the proliferation of osteoblast precursors. Scl and Dickkopf-related protein 1 (Dkk-1) modulate the signaling pathway by interacting with LRP5/6 receptors and preventing the formation of the Frizzled–Wnt–LRP5/6 complex.

Scl has a higher affinity for LRP5/6 than Wnt (Bourhis et al., 2010). Therefore, enhanced Scl production, as occurs with increasing age (Ardawi et al., 2011), can lead to age-related bone loss (Ruiz-Lozano et al., 2024) that can potentiate the effects of PMO.

The procedure described in Section 17.3.1 can be used to model the competitive binding between the receptor LRP5/6 (which acts now as R) and the ligands Wht (A) and Scl (B). Activation of β-catenin is achieved if LRP5/6 binds to Wnt and is prevented if LRP5/6 binds to Scl. For that reason, the proliferation of osteoblast precursors is assumed to be proportional to a function $\pi_{\text{act,Ob}_\text{p}}^{\text{Wnt}}$ (see Eq. 17.1) defined as the ratio between LRP5/6-Wnt complexes and the total concentration of LRP5/6 receptors, $[\text{LRP5 }/6]_{\text{tot}}$, including those that bind to Scl.

$$\pi_{\text{act,Ob}_\text{p}}^{\text{Wnt}} = \frac{[\text{Wnt} - \text{LRP5}/6]}{[\text{LRP5}/6]_{\text{tot}}} = \frac{[\text{Wnt}][\text{LRP5}/6]}{\text{K}_{\text{Wnt}-\text{LRP5}/6}\cdot[\text{LRP5}/6]_{\text{tot}}} \tag{17.25}$$

where Eq. (17.10a) has been used. $[\text{LRP5}/6]_{\text{tot}}$ can be obtained from Eq. (17.13).

$$[\text{LRP5}/6]_{\text{tot}} = [\text{LRP5}/6]\left(1 + \frac{[\text{Wnt}]}{\text{K}_{\text{Wnt-LRP5/6}}} + \frac{[\text{Scl}]}{\text{K}_{\text{Scl-LRP5/6}}}\right) \tag{17.26}$$

The replacement of this equation in Eq. (17.25) finally yields the following.

$$\pi_{\text{act,Ob}_p}^{\text{Wnt}} = \frac{[\text{Wnt}]}{\text{K}_{\text{Wnt}-\text{LRP5}/6}\left(1 + \frac{[\text{Wnt}]}{\text{K}_{\text{Wnt}-\text{LRP5}/6}} + \frac{[\text{Scl}]}{\text{K}_{\text{Scl}-\text{LRP5}/6}}\right)} \tag{17.27}$$

The concentration of Wnt can be assumed constant based on two considerations: (1) the cells that produce Wnt, Ob_u, are much more numerous than other cells, and therefore their population is assumed constant in the model, i.e., not affected by their differentiation into Ob_p; and (2) the degradation rate of the Wnt-LRP5/6 complex is negligible (Martin et al., 2019).

The total concentration of LRP5/6 is given by:

$$[\text{LRP5}/6]_{\text{tot}} = \text{N}_{\text{Ob}_\text{p}}^{\text{LRP5}/6}\text{Ob}_\text{p} \tag{17.28}$$

where $\text{N}_{\text{Ob}_p}^{\text{LRP5}/6}$ is the number of LRP5/6 receptors per osteoblast precursor cell. The concentration of Scl can generally be written as:

$$[\text{Scl}] = \text{f}([\text{Wnt}], [\text{LRP5}/6], [\text{Ot}], \text{P}_{\text{Scl,aging}}, \Psi_{\text{bm}}, [\text{Rmab}]) \tag{17.29}$$

which is a function depending on the concentration of the other ligands (Wnt and LRP5/6), the cells that mainly produce Scl (osteocytes), the production of Scl by other cells considered in the aging process, ($\text{P}_{\text{Scl,aging}}$, see Section 17.7.2), the mechanical stimulus (Ψ_{bm}, see Section 17.4) and, in the case of drug treatments, the concentration of Rmab ([Rmab]), a drug that acts as an Scl antibody. The procedure for obtaining this function is not straightforward and has been omitted for the sake of brevity. The details can be consulted in (Martin et al., 2019, 2020).

3.4 Regulatory role of TGF-β

TGF-β is a cytokine stored in bone matrix and released by osteoclasts through bone resorption. Its concentration in bone marrow can be calculated as follows (Pivonka et al., 2008):

$$[\mathrm{TGF}-\beta] = \frac{\alpha_{\mathrm{TGF}-\beta}\, \mathrm{k_{res}}\, \mathrm{Oc_a}}{\tilde{\mathrm{D}}_{\mathrm{TGF}-\beta}} \tag{17.30}$$

where $\alpha_{\mathrm{TGF}\text{-}\beta}$ is the concentration of TGF-β in the bone matrix and $\tilde{\mathrm{D}}_{\mathrm{TGF}\text{-}\beta}$ is the TGF-β degradation rate. The concentration of TGF-β is used in the activator/repressor functions of Eqs. (17.1)–(17.3). These functions control the upregulation of the differentiation of $\mathrm{Ob_u}$ into $\mathrm{Ob_p}$, the upregulation of osteoclast apoptosis and the downregulation of the differentiation of $\mathrm{Ob_p}$ into $\mathrm{Ob_a}$:

$$\pi^{\mathrm{TGF}\text{-}\beta}_{\mathrm{act,Ob_u}} = \pi^{\mathrm{TGF}\text{-}\beta}_{\mathrm{act,Oc_a}} = \frac{[\mathrm{TGF}\text{-}\beta]}{\mathrm{K}^{\mathrm{TGF}\text{-}\beta}_{\mathrm{act}} + [\mathrm{TGF}\text{-}\beta]} \tag{17.31}$$

$$\pi^{\mathrm{TGF}\text{-}\beta}_{\mathrm{rep,Ob_p}} = \frac{\mathrm{K}^{\mathrm{TGF}\text{-}\beta}_{\mathrm{rep}}}{\mathrm{K}^{\mathrm{TGF}\text{-}\beta}_{\mathrm{rep}} + [\mathrm{TGF}\text{-}\beta]} \tag{17.32}$$

with $\mathrm{K}^{\mathrm{TGF}\text{-}\beta}_{\mathrm{act}}$ and $\mathrm{K}^{\mathrm{TGF}\text{-}\beta}_{\mathrm{rep}}$ the activation and repression constants, respectively. The functions $\pi^{\mathrm{TGF}\text{-}\beta}_{\mathrm{act,Ob_u}}$ and $\pi^{\mathrm{TGF}\text{-}\beta}_{\mathrm{act,Oc_a}}$ were assumed equal by Pivonka et al. (2008) but can be distinguished simply by defining a different constant $\mathrm{K}^{\mathrm{TGF}\text{-}\beta}_{\mathrm{act}}$ for each function.

4. Mechanobiological feedback—The mechanostat

Mechanical feedback is considered in the model through the functions: $\pi^{\mathrm{PTH,NO}}_{\mathrm{act/rep,RANKL}}$ (in Eq. 17.21), $\pi^{\mathrm{PTH}}_{\mathrm{rep,Ob_a}}$ (in Eq. 17.18), and $\pi^{\mathrm{Wnt}}_{\mathrm{act,Ob_p}}$ (in Eq. 17.1). The first two model the catabolic effect of disuse (or low mechanical stimulus) via PTH and NO, which control RANKL and OPG production. The latter controls the anabolic effect of overuse (high mechanical stimulus) through the proliferation of osteoblast precursor cells.

Martin et al. (2019) merged the antagonistic influence of PTH and NO on RANKL production, defining a coregulatory function that captured both effects.

$$\begin{aligned}\pi^{\mathrm{PTH,NO}}_{\mathrm{act/rep,RANKL}} = {} & \lambda_{\mathrm{s}}\left(\pi^{\mathrm{PTH}}_{\mathrm{act,RANKL}} + \pi^{\mathrm{NO}}_{\mathrm{rep,RANKL}}\right) + \\ & \lambda_{\mathrm{c}} \cdot \pi^{\mathrm{PTH}}_{\mathrm{act,RANKL}} \cdot \pi^{\mathrm{NO}}_{\mathrm{rep,RANKL}}\end{aligned} \tag{17.33}$$

The activator function that accounts for the effect of PTH is:

$$\pi_{\mathrm{act,RANKL}}^{\mathrm{PTH}} = \frac{[\mathrm{PTH}]}{[\mathrm{PTH}] + \mathrm{K}_{\mathrm{act}}^{\mathrm{PTH}}} \tag{17.34}$$

The repressor function that models the effect of NO is:

$$\pi_{\mathrm{rep,RANKL}}^{\mathrm{NO}} = \frac{\mathrm{K}_{\mathrm{rep}}^{\mathrm{NO}}}{[\mathrm{NO}] + \mathrm{K}_{\mathrm{rep}}^{\mathrm{NO}}} \tag{17.35}$$

The catabolic effect of PTH is two-fold, as it also suppresses the production of OPG by $\mathrm{Ob_a}$, an effect that is included in Eq. (17.18) through the function:

$$\pi_{\mathrm{rep,Ob_a}}^{\mathrm{PTH}} = \frac{\mathrm{K}_{\mathrm{rep}}^{\mathrm{PTH}}}{[\mathrm{PTH}] + \mathrm{K}_{\mathrm{rep}}^{\mathrm{PTH}}} \tag{17.36}$$

In these equations, λ_s, λ_c, $\mathrm{K}_{\mathrm{act}}^{\mathrm{PTH}}$, $\mathrm{K}_{\mathrm{rep}}^{\mathrm{NO}}$ and $\mathrm{K}_{\mathrm{rep}}^{\mathrm{PTH}}$ are constants. It is clear from the above that PTH has a catabolic effect while NO suppresses this catabolic effect, and therefore, it can be considered an anabolic factor. Similarly, Scl has an antianabolic effect as it inhibits $\mathrm{Ob_p}$ proliferation by interfering in the Wnt/β catenin pathway, which is modeled in Eq. (17.1) by means of $\pi_{\mathrm{act,Ob_p}}^{\mathrm{Wnt}}$

The production of Scl and NO by osteocytes is modeled by:

$$\mathrm{P_{Scl}} = \beta_{\mathrm{Scl,Ot}} \cdot \pi_{\mathrm{rep,Scl}}^{\Psi_{\mathrm{bm}}} \cdot \mathrm{Ot}\left(1 - \frac{[\mathrm{Scl}]}{[\mathrm{Scl}]_{\mathrm{max}}}\right) + \mathrm{P_{Scl,aging}} \tag{17.37}$$

$$\mathrm{P_{NO}} = \beta_{\mathrm{NO,Ot}} \cdot \pi_{\mathrm{act,NO}}^{\Psi_{\mathrm{bm}}} \cdot \mathrm{Ot}\left(1 - \frac{[\mathrm{NO}]}{[\mathrm{NO}]_{\mathrm{max}}}\right) \tag{17.38}$$

where $\mathrm{P_{Scl,aging}}$ is the production of Scl by cells other than osteocytes, which, following Ruiz-Lozano et al. (2024), is assumed to increase with age (see Section 17.7). The parameters $\beta_{\mathrm{Scl,Ot}}, \beta_{\mathrm{NO,Ot}}, [\mathrm{Scl}]_{\mathrm{max}}$ and $[\mathrm{NO}]_{\mathrm{max}}$ are, respectively, the Scl and NO production rates by osteocytes and their maximum concentrations.

Mechanical stimulus downregulates the production of Scl by osteocytes through the repressor function $\pi_{\mathrm{rep,Scl}}^{\Psi_{\mathrm{bm}}}$, and upregulates the production of NO through the activator function $\pi_{\mathrm{act,NO}}^{\Psi_{\mathrm{bm}}}$:

$$\pi_{\mathrm{act,NO}}^{\Psi_{\mathrm{bm}}} = \rho_{\mathrm{act}} + \frac{(\alpha_{\mathrm{act}} - \rho_{\mathrm{act}})\Psi_{\mathrm{bm}}^{\gamma_{\mathrm{act}}}}{\delta_{\mathrm{act}}^{\gamma_{\mathrm{act}}} + \Psi_{\mathrm{bm}}^{\gamma_{\mathrm{act}}}} \tag{17.39}$$

$$\pi_{\mathrm{rep,Scl}}^{\Psi_{\mathrm{bm}}} = \alpha_{\mathrm{rep}} - \frac{\left(\alpha_{\mathrm{rep}} - \rho_{\mathrm{rep}}\right)\Psi_{\mathrm{bm}}^{\gamma_{\mathrm{rep}}}}{\delta_{\mathrm{rep}}^{\gamma_{\mathrm{rep}}} + \Psi_{\mathrm{bm}}^{\gamma_{\mathrm{rep}}}} \tag{17.40}$$

where $\rho_{\sim}$ and $\alpha_{\sim}$ are, respectively, the minimum and maximum anticipated responses, the exponents $\gamma_{\sim}$ are sigmoidicity constants that modify the steepness of the response and $\delta_{\sim}$ is the value of the stimulus producing the half-maximal response. Ψ_{bm} is the variable used in the model to measure the mechanical stimulus.

The choice of this mechanical stimulus has been a matter of controversy in the literature. In previous versions of the BCPM, the authors used the strain energy density (Calvo-Gallego et al., 2022, 2023; Martínez-Reina and Pivonka, 2019; Martínez-Reina et al., 2021a, 2021b, 2022), thus following classical models (Beaupre et al., 1990; Huiskes et al., 1987). However, they later used the maximum principal strain (Ruiz-Lozano et al., 2024, 2025), following the conclusions of Martínez-Reina et al. (2024) and according to the mechanostat theory (Frost, 1987). This theory established that bone adapts its stiffness so that the strain it withstands remains within a certain range ("Adapted Window" in Frost's terminology). Martínez-Reina et al. (2024) analyzed through finite element modeling the strain and stress distributions that a number of habitual loads produce in a human femur and confirmed that stresses are positively correlated to bone density, but strains, and particularly the maximum principal strain, were almost uniformly distributed within a narrow range, as referred to above. The authors concluded that keeping strains within the Adapted Window could be the stimulus that drives bone remodeling, as suggested by the mechanostat theory. Furthermore, Martínez-Reina et al. (2024) confirmed that different habitual loads, such as walking, walking fast, jumping, and climbing up and down stairs, produced similar patterns of stresses and strains in the femur. Overloads could influence bone response as suggested by Carter and coworkers (Carter et al., 1987; Whalen et al., 1988), who proposed a mechanical stimulus for multiple loads based on the combination of the stimulus and the number of cycles of each individual load.

5. Composition and mechanical properties of bone

The composition of bone determines its mechanical properties to a large extent and must be considered to assess the changes in stiffness that control bone mechanoregulation. The mechanical properties of bone are defined in an RVE, based on continuum mechanics, thus assuming that it is possible to identify a suitable RVE size within which certain material characteristics can be assumed uniform, e.g., porosity, density, and Young's modulus. The RVE needs to be significantly larger than any material imperfection but much smaller than the structural length scale to account for the heterogeneities at this scale. Under such conditions, homogenization approaches can be applied to derive the stiffness tensor and strength properties of bone using information of lower-scale constituents (García-Rodríguez and Martínez-Reina, 2017; Kariem et al., 2015; Martínez-Reina et al., 2011).

A commonly made assumption is that bone at the tissue scale (i.e., using an RVE of typical length $l_{RVE} \approx 5 - 10$ mm) can be described as a two-phase porous material consisting of a solid phase (i.e., bone matrix, with volume V_{bm}) and a fluid phase (i.e., pores occupied by bone marrow, cells, and blood vessels, with volume V_p). In turn, the bone matrix is made up of an inorganic phase (mainly hydroxyapatite), an organic phase (mainly collagen), and water. If V_m, V_o, and V_w designate the volumes of each phase, the reference bone volume, V_{RVE}, can be decomposed as follows:

$$V_{RVE} = V_{bm} + V_p = V_m + V_o + V_w + V_p \tag{17.41}$$

The previously introduced bone matrix volume fraction ($f_{bm} = BV/TV$) and vascular volume fraction (or porosity, $f_{vas} = p$) are then defined as:

$$f_{bm} = \frac{V_{bm}}{V_{RVE}} \tag{17.42}$$

$$f_{vas} = \frac{V_p}{V_{RVE}} \tag{17.43}$$

such that $f_{bm} + f_{vas} = 1$. Volume fractions of each phase are used to measure the composition of the bone matrix throughout the mineralization process and are defined as:

$$v_i = \frac{V_i}{V_{bm}} \quad \text{with } i = o, w, m \tag{17.44}$$

such that

$$v_o + v_w + v_m = 1 \tag{17.45}$$

It is widely accepted that mineral accumulation in the bone matrix is produced by the replacement of the free water present in the matrix with mineral (Hernandez et al., 2000; Lees & Gobburu, 1987), so that v_o can be assumed constant, e.g., $v_o = 0.418$ (Currey, 2016, pp. 3–14) and then $\Delta v_m = -\Delta v_w$ during the mineralization process by virtue of Eq. (17.45). Mineral content is usually measured by the so-called ash fraction, which is defined as the ratio between the mass of mineral m_m (also called ash mass) and the dry mass (the sum of mineral and organic mass):

$$\alpha = \frac{m_m}{m_m + m_o} = \frac{\rho_m V_m}{\rho_m V_m + \rho_o V_o} \tag{17.46}$$

where the densities of the different phases are designated by ρ_i, again with $i = o, w, m$. Since v_o is assumed constant, v_m and α are univocally related by Eq. (17.46). Any of them can alternatively be used to specify the mineral content of bone matrix, which determines its density through:

$$\begin{aligned} \rho_{bm} &= \frac{m_{bm}}{V_{bm}} = \rho_m v_m + \rho_o v_o + \rho_w v_w \\ &= (\rho_m - 1)v_m + (\rho_o - 1)v_o + 1 \end{aligned} \tag{17.47}$$

where m_{bm} is the total mass of bone matrix. Eq. (17.45) and the value $\rho_w = 1\ g/cm^3$ have been used in Eq. (17.47) to show the relationship between the mineral content, v_m, and the density of the bone matrix, provided that v_o is constant. Finally, the apparent density of bone, ρ_{app}, is assessed taking into account the mass of the matter that fills the pores, typically the bone marrow (whose density can be assumed equal to that of water, i.e., $\rho_p = \rho_w = 1\ g/cm^3$). Thus:

$$\rho_{app} = \frac{m}{V_{RVE}} = f_{bm}\rho_{bm} + (1 - f_{bm})\rho_p \tag{17.48}$$

Bone apparent density is the most determinant factor of bone stiffness and strength, and many functional relationships can be found in the literature. For example, Jacobs (1994) provided the following correlation between Young's modulus and apparent density, distinguishing between trabecular and cortical bone and assuming both as isotropic materials:

$$E(MPa) = \begin{cases} 2014\,\rho_{app}^{2.5} & \text{if } \rho_{app} \leq 1.2\ \text{g/cm}^3 \\ 1763\,\rho_{app}^{3.2} & \text{if } \rho_{app} > 1.2\ \text{g/cm}^3 \end{cases} \tag{17.49}$$

In summary, bone stiffness can be estimated as a function of porosity (through f_{bm}) and mineral content (through ρ_{bm} or v_m).

5.1 Mineralization of bone matrix

The mineral content of the bone matrix influences the stiffness, strength, and fatigue properties of bone. Furthermore, mineralization is key to explain bone density gain (BDG) in anticatabolic treatments, and all this highlights the relevance of mineral content in the mechanical feedback and in the response to some pharmacological treatments.

Osteoid, the tissue laid de novo by osteoblasts, contains only water and organic phase comprising type I collagen (90% in volume) and a number of noncollagenous proteins (10%) (McLean & Urist, 1986). The inorganic phase is deposited later, after a certain mineralization lag time (MLT) of about 15 days (Eriksen et al., 1984). It consists of hydroxyapatite ($Ca_{10}(PO_4)_6(OH)_2$) with some substitutions, e.g., carbonate for the phosphate ion, which accounts for about 5% of the weight of bone ash (Eanes & Posner, 1970, pp. 1–26). After MLT, mineral is laid down very fast to reach from 60% (Parfitt and Recker, 1983) to 70% (Amprino & Engström, 1952) of the final mineral content in a few days during the so-called primary phase. The mineralization rate slows considerably as bone matrix becomes saturated with mineral during the secondary phase, and it can take years to reach the maximum mineral content (Parfitt and Recker, 1983). Thus, following Hernandez et al. (2000), v_m can be given as a function of time as:

$$v_m(t) = \begin{cases} 0 & \text{if } t \leq t_{mlt} \\ v_m^{prim}\dfrac{t - t_{mlt}}{t_{prim}} & \text{if } t_{mlt} < t \leq t_{prim} + t_{mlt} \\ v_m^{max} - (v_m^{max} - v_m^{prim})e^{-\kappa\cdot(t - t_{prim} - t_{mlt})} & \text{if } t_{prim} + t_{mlt} < t \end{cases} \tag{17.50}$$

where t_{mlt} and t_{prim} are, respectively, the length of the MLT and the primary phase; v_m^{prim} is the mineral content at the end of the primary phase, corresponding to α=0.45 (Hernandez et al., 2000); v_m^{max} corresponds to the maximum mineral

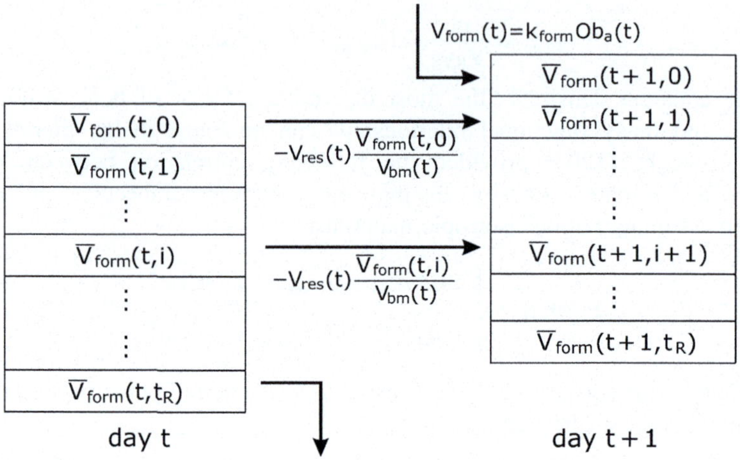

FIGURE 17.2

Mineralization algorithm: FIFO (first in–first out) queue algorithm used to update the distribution of tissue patches of different bone ages within the RVE.

content, 300 mg/g (Currey, 2004); and κ is a parameter measuring the rate of mineral deposition during the secondary phase.

The amount of mineral contained in an RVE depends on the age of the tissue through Eq. (17.50), but the RVE can be made up of tissue patches formed in the recent history, viz. of different ages. Moreover, the tissue within the RVE can be resorbed, which puts the mineral back into the blood flow. The amounts of tissue of different ages contained in the RVE are estimated using the algorithm depicted in Fig. 17.2 (Martínez-Reina et al., 2008). $\overline{V}_{form}(t, \tau)$ provides the bone volume formed τ days ago and still present (not yet resorbed) at time t. Knowing the distribution of tissue patches of different ages at day t (left column) and the volume formed $\left(V_{form}(t) = k_{form}\, Ob_a(t)\Delta t\right)$ and resorbed that day $\left(V_{res}(t) = k_{res}\, Oc_a(t)\Delta t\right)$,[1] the distribution at day $t+1$ (right column) can be estimated:

$$\overline{V}_{form}(t+1, i+1) = \overline{V}_{form}(t, i) - V_{res}(t)\,\frac{\overline{V}_{form}(t, i)}{V_{bm}(t)} \tag{17.51}$$

Finally, the mineral content of each patch is summed to estimate the average mineral content of the RVE at day $t+1$.

[1] The usual time step is Δt=1 day as specified in Fig. 17.2.

$$v_m(t+1) = \frac{\sum_{i=0}^{t_R} \overline{V}_{form}(t+1,i)\cdot v_m(i)}{V_{bm}(t+1)} \tag{17.52}$$

where the mineral contents of the patches, $v_m(i)$, are calculated through Eq. (17.50). t_R represents the residence time, e.g., the typical time the patch tissue remains within the bone before being resorbed. This residence time can be very large, but the queue can be truncated at a shorter time to reduce the computational cost. See Martínez-Reina et al. (2008) for more details.

6. Microstructural damage

Microstructural damage is another factor that determines bone stiffness, specifically by reducing it. Following the continuum damage mechanics approach (Lemaitre & Chaboche, 1994), damage is a variable linked to the local density of microcracks and to the loss of stiffness, such that $d \in [0, 1]$, with $d = 0$ corresponding to an undamaged bone and $d = 1$ to a fracture or local failure situation, such that:

$$E = E_0(1 - d) \tag{17.53}$$

where E is the current Young's modulus and E_0 is the intact or undamaged Young's modulus.

Damage can be accumulated in the bone matrix due to the mechanical loads of daily activity, through mechanisms similar to those involved in fatigue failure of engineering materials, i.e., nucleation, growth, and coalescence of microcracks. Osteoclasts resorb damaged tissue, while osteoblasts deposit initially intact osteoid. Therefore, bone turnover involves the repairing of microstructural damage, and a balance between repair (d_R) and accumulation due to fatigue loading (d_A) can be established to calculate the evolution of damage:

$$d = d_A - d_R \tag{17.54}$$

d_R is calculated by assuming that damage is uniformly distributed throughout the RVE and the amount of repaired damage is proportional to the damage present in the RVE and to the volume of tissue being resorbed, V_{res}, through the fraction that this volume represents within the bone matrix volume:

$$d_R = d\,\frac{V_{res}}{V_{bm}} = d\,\frac{k_{res}\cdot Oc_a}{f_{bm}} \tag{17.55}$$

The accumulation of damage by fatigue was evaluated following the procedure described in Martínez-Reina et al. (2009), using the differential equation provided by García-Aznar et al. (2005) for d_A under tensile stresses, which once integrated yields:

$$d_A = 1 - \left[\frac{1}{C_2}\ln\left(e^{C_2} - C_1\varepsilon_{max}^{\delta_f} N\right)\right]^{\frac{1}{\gamma_f}} \tag{17.56}$$

where N is the number of cycles applied, ε_{max} is the maximum principal strain[2] expressed in $\mu\varepsilon$ and C_1, C_2, γ_f and δ_f are constants fitted by García-Aznar et al. (2005) to reproduce the fatigue experiments of Pattin et al. (1996). These constants were made dependent on the mineral content to consider fragilization of the bone matrix with an excess mineral content. See Martínez-Reina et al. (2008) for more details.

Based on these equations, Martínez-Reina et al. (2021a) proposed a methodology to estimate the number of cycles needed to reach local failure (d = 1) at the RVE scale for a given overload.

1. A constant overload $\tilde{\sigma}$ is imposed at a given time instant t, which can induce local failure after a time period τ.
2. For the current instant t, the following variables are computed: $f_{bm}(t)$, $v_m(t)$, $Oc_a(t)$ and d(t). The first three variables are kept constant to calculate the evolution of damage from time t onward, i.e., $d(t+\tau)$.
3. Apparent density $\rho_{app}(t)$ is calculated using Eqs. (17.47) and (17.48) and kept constant, i.e., bone remodeling is not considered in the time period $[t, t+\tau]$.
4. The Young's modulus is calculated through Eq. (17.49), which is affected by damage with the factor (1−d) as in Eq. (17.53). The updated Young's modulus, $E(t+\tau)$, is used to calculate the updated strain, $\varepsilon_{max}(t+\tau)$ for the given overload $\tilde{\sigma}$.
5. Eq. (17.56) assumes that a constant ε_{max} is applied from the beginning of the fatigue test. This is not true in this procedure, but Eq. (17.56) can be used in an inverse manner, to calculate the number of cycles, $\tilde{N}$, needed to reach the current damage $d(t+\tau)$ with the current strain $\varepsilon_{max}(t+\tau)$, which can be worked out from the following equation.

$$d(t+\tau) = 1 - \left[\frac{1}{C_2}\ln\left(e^{C_2} - C_1(\varepsilon_{max}(t+\tau))^{\delta_f}\tilde{N}\right)\right]^{\frac{1}{\gamma_f}} \tag{17.57}$$

6. Then, if further N cycles are applied during the next day (the integration time step is fixed $\Delta\tau = 1$ day) and damage repair is taken into account through Eq. (17.55) for a constant population $Oc_a(t)$, damage can be updated by:

$$\begin{aligned} d(t+\tau+\Delta\tau) = {} & 1 - \left[\frac{1}{C_2}\ln\left(e^{C_2} - C_1\varepsilon_{max}^{\delta_f}\left(\tilde{N}+N\right)\right)\right]^{\frac{1}{\gamma_f}} \\ & - d(t+\tau)\frac{k_{res}\cdot Oc_a(t)}{f_{bm}(t)}\Delta\tau \end{aligned} \tag{17.58}$$

[2] In the damage model proposed by Martínez-Reina et al. (2009), cracks were assumed to grow normal to the maximum principal strain direction and only under tensile strains.

7. If the number of daily cycles, N, is assumed constant, this procedure can be used to estimate the number of days until failure for a given overload NDuF($\tilde{\sigma}$), by repeating steps 4–6 until d(t +NDuF).
8. Finally, the risk of local failure (at the RVE) is evaluated through the variable called FRI (fracture risk index), which is defined in terms of NDuF by:

$$\mathrm{FRI} = \frac{100}{\mathrm{NDuF}} \tag{17.59}$$

Note that FRI is not a probability index, though it can be interpreted as such for the extreme cases. Indeed, FRI = 100 if failure occurs immediately after the overload is applied, and FRI tends to 0 with an increasing NDuF, i.e., in case of no failure. In practical terms, the algorithm can be truncated at $\tau = 100$ days if no failure is observed and therefore FRI = 1.

7. Modeling bone diseases—osteoporosis

Bone diseases can readily be modeled with the above proposed framework by mechanistically introducing systems (i.e., bone remodeling) perturbations informed by clinical observations. The effect of aging on bone turnover was modeled by Ruiz-Lozano et al. (2024) through a decrease in the concentration of TGF-β within the bone matrix and an increase in the production of Scl.

7.1 Effect of aging: decreased TGF-β concentration

A decrease in TGF-β with age has been reported by several experimental studies (Nicolas et al., 1994; Okamoto et al., 2005; Pfeilschifter et al., 1998). The effect of aging is modeled as done by Lemaire et al. (2004), i.e., by reducing the TGF-β content in the bone matrix, $\alpha_{\mathrm{TGF-\beta}}$ (see Section 17.3.4). In contrast to Lemaire et al., who modeled an abrupt decrease of TGF-β concentration, in the present model, TGF-β is reduced continuously over time. Given that Riggs et al. (2004) reported a decrease in bone mass of vertebral trabecular bone starting at age 20, it can be assumed that the effects begin at that age. On the other hand, Pfeilschifter et al. (1998) detected a clear reduction in TGF-β content within the 30- to 50-year interval, but no differences were found after 50 years of age. Based on these findings, a bilinear function is used to simulate the decrease of TGF-β:

$$\alpha_{\mathrm{TGF}\text{-}\beta} = \begin{cases} \alpha_{\mathrm{TGF}\text{-}\beta,0} & \text{if } \mathrm{TSSM} < 0 \\ \alpha_{\mathrm{TGF}\text{-}\beta,0} - \mathrm{k_1 TSSM} & \text{if } 0 \le \mathrm{TSSM} < \mathrm{T_1} \\ \alpha_{\mathrm{TGF}\text{-}\beta,0} - \mathrm{k_1\, T_1} - \mathrm{k_2(TSSM - T_1)} & \text{if } \mathrm{T_1} \le \mathrm{TSSM} \end{cases} \tag{17.60}$$

where TSSM stands for time since skeletal maturity and is defined as $\mathrm{TSSM} = \mathrm{t} - \mathrm{t_{mat}}$, expressed in days, with $\mathrm{t_{mat}} = 20$ years, is the age at which bone loss begins, as stated before. The slopes $\mathrm{k_1}$ and $\mathrm{k_2}$ measure the rate of decay of the concentration of TGF-β in the bone matrix over time.

7.2 Effect of aging: Increase in the production of sclerostin

In their model, Martin et al. (2019) assumed that increased serum Scl concentration was an effect of menopause. However, several studies (Ardawi et al., 2011; Mödder et al., 2011) suggested that this must be an effect of aging, since increasing levels of Scl are already observed in premenopausal women (Ardawi et al., 2011) and also in men (Mödder et al., 2011). Roforth et al. (2014) suggested that aging could be associated with a reduced clearance of Scl from the circulation but also with an increased Scl production by nonskeletal sources. Indeed, circulating Scl may be derived not only from osteocytes (Ke et al., 2012), but also from several other sources, including myeloid cells (Kusu et al., 2003; Ota et al., 2013; Pederson et al., 2009), vascular smooth muscle cells (Zhu et al., 2011), and the kidney (Ke et al., 2012). Based on these studies, an increase in the external production of Scl, $P_{Scl,aging}$ (see Eq. 17.37) is considered in the model:

$$P_{Scl,aging} = \begin{cases} 0 & \text{if } TSSM < 0 \\ k_3 TSSM & \text{if } 0 \le TSSM < T_2 \\ k_3 T_2 + k_4 (TSSM - T_2) & \text{if } T_2 \le TSSM \end{cases} \tag{17.61}$$

where the slopes k_3 and k_4 measure the rate of increase in the external production of Scl, which may change after T_2 days elapsed since t_{mat}.

Another effect of aging that should not be forgotten and that greatly affects bone density is the decrease in physical activity with age. A lower mechanical stimulus also increases Scl levels and limits the proliferation of osteoblast precursors (Martin et al., 2019) and should be taken into account for a more realistic simulation of the aging process.

7.3 Simulations of primary osteoporosis

Primary OP is the most common form of the disease and includes PMO and ARO. Understanding the etiology of OP and, more specifically, discerning which factors are intrinsic to PMO and which are intrinsic to ARO, as we will highlight in this section, is key to the design of patient-specific treatments, as both can occur in postmenopausal women, but only some can be counteracted by medication.

BMD is commonly used to measure the severity of bone diseases and the efficacy of drug treatments, but it is convenient to define the variation with respect to a baseline. Thus, bone density variation (BDV) can be defined as:

$$BDV(\%) = \frac{\rho(t) - \rho(t_0)}{\rho(t_0)} \tag{17.62}$$

where $\rho(t)$ is the current bone apparent density, and $\rho(t_0)$ is the baseline value. More specifically, we will refer to BDG (bone density gain, if BDV is positive) to measure the efficacy of drugs, and to BDL (bone density loss, if BDV is negative) to measure the severity of bone diseases.

Fig. 17.3 summarizes the simulations of ARO and PMO performed by Ruiz-Lozano et al. (2024). BDL is measured from TSSM, which was assumed in that work to be equal to 20 years of age, and the results are compared with the regressions obtained by Riggs et al. (2004) (dashed lines). An accurate fit of these data was achieved for men (red lines), for whom only the effect of aging (ARO) was considered through Eqs. (17.60) and (17.61). For women, the fit with the experimental data was also quite good. In the latter simulations, both the effect of OD (PMO) and aging were considered through Eq. (17.23).

If only the effect of OD is considered in the simulation (black line), no BDL is seen until menopause, assumed to occur at 49 years of age. Then, an abrupt bone loss is seen during the first 5 years, followed by a stabilization of that bone loss. If only the effect of aging is considered, a continuous descent in bone density is obtained (red line). The simulation of the combined effect of aging and OD led to a very marked drop in bone mass right after menopause (blue line), which is much stronger than in the case where only OD was considered and does not stabilize. Furthermore, the effects of aging and menopause are coupled in the model to some extent, since the superposition of the BDL obtained with each separate effect (green line, which is the sum of the contributions of the red and black lines) underestimates the BDL when both effects are modeled jointly (blue line). In other words, the coupled effects of aging and OD manifest themselves in a more pronounced disease state.

Fig. 17.4 analyzes the influence of age at menopause on the evolution of BDL, by comparing the case analyzed before (49 years) with the cases of early and late onset

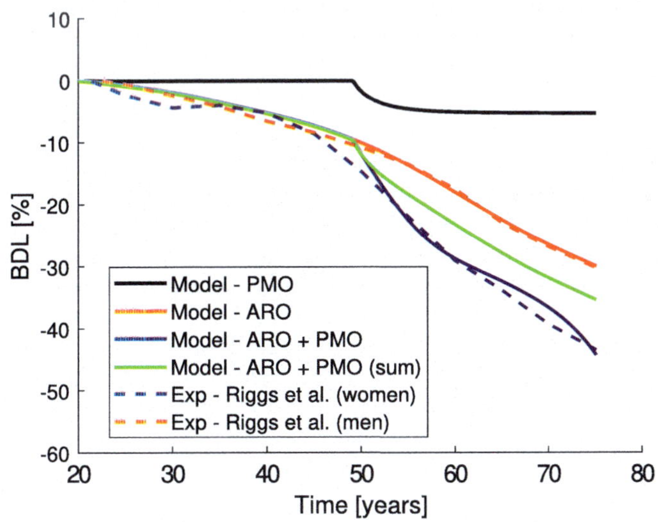

FIGURE 17.3

Simulations of bone diseases: Comparison of relative effects of age and menopause on BDL, measured from 20 years of age onwards (experimental data; Riggs et al., 2004). Effect of OD only (black line), effect of age only (red line) and combined effects of age and OD (blue line).

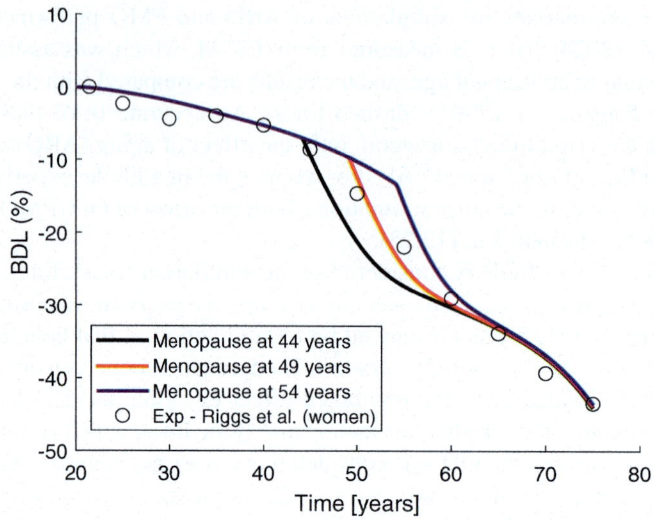

FIGURE 17.4

Simulations of bone diseases: Comparison of ARO + PMO simulations for different ages at menopause: 44 (black), 49 (red), and 54 (blue) years.

of menopause (44 and 54 years, respectively). The simulations predict a shift of the curves at the onset of menopause and a long-term convergence. Early menopause can be dangerous, as it significantly reduces average bone density from an earlier age and increases the risk of fracture for a greater part of the patient's life.

Acknowledgments

Dr. Calvo-Gallego and Dr. Martínez-Reina would like to gratefully acknowledge funding provided by MICIU/AEI/ 10.13039/501100011033 and by "ERDF/EU" through the research project entitled *Integración de un modelo de remodelación ósea y un modelo de homeostasis del Ca. Predicción de riesgo de fractura osteoporótica. Diseño de tratamientos de paciente-específico* (PID2024-161475OB-I00).

Dr. Pivonka would also like to gratefully acknowledge funding received through the Australian Research Council (ARC) Industrial Transformation Training Center for Joint Biomechanics (IC190100020) and Discovery Project (DP230101404).

References

Amprino, R., & Engström, A. (1952). Studies on X ray absorption and diffraction of bone tissue. *Acta Anatomica, 15*(1–2), 1–22. https://doi.org/10.1159/000140734

Ardawi, M. S. M., Al-Kadi, H. A., Rouzi, A. A., & Qari, M. H. (2011). Determinants of serum sclerostin in healthy pre- and postmenopausal women. *Journal of Bone and Mineral Research, 26*(12), 2812–2822. https://doi.org/10.1002/jbmr.479

Beaupre, G. S., Orr, T. E., & Carter, D. R. (1990). An approach for time-dependent bone modeling and remodeling – Theoretical development. *Journal of Orthopaedic Research, 8*(5), 651–661. https://doi.org/10.1002/jor.1100080506

Bellido, T., Ali, A. A., Plotkin, L. I., Fu, Q., Gubrij, I., et al. (2003). Proteasomal degradation of Runx2 shortens parathyroid hormone-induced anti-apoptotic signaling in osteoblasts. A putative explanation for why intermittent administration is needed for bone anabolism. *Journal of Biological Chemistry, 278*(50), 50259–50272. https://doi.org/10.1074/jbc.M307444200

Black, D. M., & Rosen, C. J. (2016). Postmenopausal osteoporosis. *The New England Journal of Medicine, 374*, 254–262. https://doi.org/10.1056/NEJMcp1513724

Borgström, F., Karlsson, L., Ortsäter, G., Norton, N., Halbout, P., et al. (2020). Fragility fractures in Europe: Burden, management and opportunities. *Archives of Osteoporosis, 15* (59). https://doi.org/10.1007/s11657-020-0706-y

Bourhis, E., Tam, C., Franke, Y., Bazan, J. F., Ernst, J., Hwang, J., et al. (2010). Reconstitution of a Frizzled8.Wnt3a.LRP6 signaling complex reveals multiple Wnt and Dkk1 binding sites on LRP6. *Journal of Biological Chemistry, 285*(12), 9172–9179. https://doi.org/10.1074/jbc.M109.092130

Boyce, B. F., & Xing, L. (2008). Functions of RANKL/RANK/OPG in bone modeling and remodeling. *Archives of Biochemistry and Biophysics, 473*(2), 139–146. https://doi.org/10.1016/j.abb.2008.03.018

Calvo-Gallego, J. L., Manchado-Morales, P., Pivonka, P., & Martínez-Reina, J. (2023). Spatio-temporal simulations of bone remodelling using a bone cell population model based on cell availability. *Frontiers in Bioengineering and Biotechnology, 11*. https://doi.org/10.3389/fbioe.2023.1060158. Article 1060158.

Calvo-Gallego, J. L., Pivonka, P., Ruiz-Lozano, R., & Martínez-Reina, J. (2022). Mechanistic PK-PD model of alendronate treatment of postmenopausal osteoporosis predicts bone site-specific response. *Frontiers in Bioengineering and Biotechnology, 10*, Article 940620. https://doi.org/10.3389/fbioe.2022.940620

Carter, D., Fyhrie, D., & Whalen, R. (1987). Trabecular bone density and loading history: Regulation of connective tissue biology by mechanical energy. *Journal of Biomechanics, 20*(8), 789–794. https://doi.org/10.1016/0021-9290(87)90058-3

Close, P., Neuprez, A., & Reginster, J. Y. (2006). Developments in the pharmacotherapeutic management of osteoporosis. *Expert Opinion on Pharmacotherapy, 7*(12), 1603–1615. https://doi.org/10.1517/14656566.7.12.1603

Compston, J. E. (2001). Sex steroids and bone. *Physiological Reviews, 81*(1), 419–447. https://doi.org/10.1152/physrev.2001.81.1.419

Cosman, F., Crittenden, D. B., Adachi, J. D., Binkley, N., Czerwinski, E., Ferrari, S., et al. (2016). Romosozumab treatment in postmenopausal women with osteoporosis. *The New England Journal of Medicine, 375*(16), 1532–1543. https://doi.org/10.1056/NEJMoa1607948

Currey, J. D. (2004). Tensile yield in compact bone is determined by strain, post-yield behaviour by mineral content. *Journal of Biomechanics, 37*(4), 549–556.

Currey, J. (2016). Cortical bone. In W. Murphy, J. Black, & G. Hastings (Eds.), *Handbook of Biomaterial Properties* (2nd ed.). Springer.

Eanes, E. D., & Posner, A. S. (1970). *Structure and chemistry of bone mineral*. Boston, MA: Springer US. https://doi.org/10.1007/978-1-4684-8485-4_1

Eastwood, E. A., Magaziner, J., Wang, J., Silberzweig, S. B., Hannan, E. L., et al. (2002). Patients with hip fracture: Subgroups and their outcomes. *Journal of the American Geriatrics Society, 50*(7), 1240–1249. https://doi.org/10.1046/j.1532-5415.2002.50311.x

Eriksen, E. F., Gundersen, H. J. G., Melsen, F., & Mosekilde, L. (1984). Reconstruction of the formative site in iliac trabecular bone in 20 normal individuals employing a kinetic model for matrix and mineral apposition. *Metabolic Bone Disease & Related Research, 5*(5), 243–252. https://doi.org/10.1016/0221-8747(84)90066-3

Fitzpatrick, L. A. (2002). Secondary causes of osteoporosis. *Mayo Clinic Proceedings, 77*(5), 453–468. https://doi.org/10.4065/77.5.453

Frost, H. M. (1987). The mechanostat: A proposed pathogenic mechanism of osteoporoses and the bone mass effects of mechanical and nonmechanical agents. *Bone & Mineral, 2* (2), 73–85.

García-Aznar, J., Rueberg, T., & Doblaré, M. (2005). A bone remodelling model coupling microdamage growth and repair by 3D BMU-activity. *Biomechanics and Modeling in Mechanobiology, 4*(2–3), 147–167. https://doi.org/10.1007/s10237-005-0067-x

García-Rodríguez, J., & Martínez-Reina, J. (2017). Elastic properties of woven bone: Effect of mineral content and collagen fibrils orientation. *Biomechanics and Modeling in Mechanobiology, 16*(1), 159–172. https://doi.org/10.1007/s10237-016-0808-z

Garnero, P., & Delmas, P. D. (2004). Contribution of bone mineral density and bone turnover markers to the estimation of risk of osteoporotic fracture in postmenopausal women. *Journal of Musculoskeletal and Neuronal Interactions, 4*(1).

Hernandez, C. J., Beaupré, G. S., & Carter, D. R. (2000). A model of mechanobiologic and metabolic influences on bone adaptation. *Journal of Rehabilitation Research and Development, 37*(2), 235–244.

Hofbauer, L. C., & Schoppet, M. (2004). Clinical implications of the osteoprotegerin/RANKL/RANK system for bone and vascular diseases. *JAMA, 292*(4), 490–495. https://doi.org/10.1001/jama.292.4.490

Huiskes, R., Weinans, H., Grootenboer, H. J., Dalstra, M., Fudala, B., & Sloof, T. J. (1987). Adaptive bone remodeling theory applied to prosthetic-design analysis. *Journal of Biomechanics, 20*(11–12), 1135–1150. https://doi.org/10.1016/0021-9290(87)90030-3

Jacobs, C. R. (1994). *Numerical simulation of bone adaptation to mechanical loading*. Ph.D. thesis. Stanford University.

Kariem, H., Pastrama, M. I., Roohani-Esfahani, S. I., Pivonka, P., Zreiqat, H., et al. (2015). Micro-poroelasticity of baghdadite-based bone tissue engineering scaffolds: A unifying approach based on ultrasonics, nanoindentation, and homogenization theory. *Materials Science and Engineering: C, 46*(1), 553–564. https://doi.org/10.1016/j.msec.2014.10.072

Ke, H. Z., Richards, W. G., Li, X., & Ominsky, M. S. (2012). Sclerostin and dickkopf-1 as therapeutic targets in bone diseases. *Endocrine Reviews, 33*(5), 747–783. https://doi.org/10.1210/er.2011-1060

Kusu, N., Laurikkala, J., Imanishi, M., Usui, H., Konishi, M., et al. (2003). Sclerostin is a novel secreted osteoclast-derived bone morphogenetic protein antagonist with unique ligand specificity. *Journal of Biological Chemistry, 278*(26), 24113–24117. https://doi.org/10.1074/jbc.M301716200

Landefeld, C. S. (2011). Goals of care for hip fracture: Promoting independence and reducing mortality. *Archives of Internal Medicine, 171*(20), 1837–1838. https://doi.org/10.1001/archinternmed.2011.534

Lavaill, M., Trichilo, S., Scheiner, S., Forwood, M. R., Cooper, D. M. L., & Pivonka, P. (2020). Study of the combined effects of PTH treatment and mechanical loading in post-menopausal osteoporosis using a new mechanistic PK-PD model. *Biomechanics and Modeling in Mechanobiology, 19*, 1765–1780. https://doi.org/10.1007/s10237-020-01307-6

Lees, S., & Gobburu, J. V. S. (1987). Considerations regarding the structure of the mammalian osteoid from viewpoint of the generalized packing model. *Connective Tissue Research, 16*, 281–303. https://doi.org/10.3109/03008208709005616

Lemaire, V., Tobin, F. L., Greller, L. D., Cho, C. R., & Suva, L. J. (2004). Modeling the interactions between osteoblast and osteoclast activities in bone remodeling. *Journal of Theoretical Biology, 229*(3), 293–309. https://doi.org/10.1016/j.jtbi.2004.03.023

Lemaitre, J., & Chaboche, J. L. (1994). *Mechanics of solid materials*. Cambridge, UK: Cambridge University Press.

Mödder, U. I., Hoey, K. A., Amin, S., McCready, L. K., Achenbach, S. J., et al. (2011). Relation of age, gender, and bone mass to circulating sclerostin levels in women and men. *Journal of Bone and Mineral Research, 26*(2), 373–379. https://doi.org/10.1002/jbmr.217

Martínez-Reina, J., Calvo-Gallego, J. L., Martin, M., & Pivonka, P. (2022). Assessment of strategies for safe drug discontinuation and transition of denosumab treatment in PMO—Insights from a mechanistic PK/PD model of bone turnover. *Frontiers in Bioengineering and Biotechnology, 10*, Article 886579. https://doi.org/10.3389/fbioe.2022.886579

Martínez-Reina, J., Calvo-Gallego, J. L., & Pivonka, P. (2021a). Combined effects of exercise and denosumab treatment on local failure in post-menopausal osteoporosis — Insights from bone remodelling simulations accounting for mineralisation and damage. *Frontiers in Bioengineering and Biotechnology, 9*, Article 635056. https://doi.org/10.3389/fbioe.2021.635056

Martínez-Reina, J., Calvo-Gallego, J. L., & Pivonka, P. (2021b). Are drug holidays a safe option in treatment of osteoporosis? — Insights from an in silico mechanistic PK-PD model of denosumab treatment of postmenopausal osteoporosis. *Journal of the Mechanical Behavior of Biomedical Materials, 113*, Article 104140. https://doi.org/10.1016/j.jmbbm.2020.104140

Martínez-Reina, J., Domínguez, J., & García-Aznar, J. M. (2011). Effect of porosity and mineral content on the elastic constants of cortical bone: A multiscale approach. *Biomechanics and Modeling in Mechanobiology, 10*(3), 309–322. https://doi.org/10.1007/s10237-010-0236-4

Martínez-Reina, J., García-Aznar, J. M., Domínguez, J., & Doblaré, M. (2008). On the role of bone damage in calcium homeostasis. *Journal of Theoretical Biology, 254*(3), 704–712. https://doi.org/10.1016/j.jtbi.2008.06.007

Martínez-Reina, J., García-Aznar, J., Domínguez, J., & Doblaré, M. (2009). A bone remodelling model including the directional activity of BMUs. *Biomechanics and Modeling in Mechanobiology, 8*(2), 111–127. https://doi.org/10.1007/s10237-008-0122-5

Martínez-Reina, J., Ojeda, J., Calvo-Gallego, J. L., Pivonka, P., & Martelli, S. (2024). Assessment of mechanical variables best describing bone remodelling responses based on their correlation with bone density. *Journal of the Mechanical Behavior of Biomedical Materials, 160*, Article 106773. https://doi.org/10.1016/j.jmbbm.2024.106773

Martínez-Reina, J., & Pivonka, P. (2019). Effects of long-term treatment of denosumab on bone mineral density: Insights from an in-silico model of bone mineralization. *Bone, 125*, 87–95. https://doi.org/10.1016/j.bone.2019.04.022

Martin, M., Sansalone, V., Cooper, D. M. L., Forwood, M. R., & Pivonka, P. (2019). Mechanobiological osteocyte feedback drives mechanostat regulation of bone in a multiscale computational model. *Biomechanics and Modeling in Mechanobiology, 18*(5), 1475–1496. https://doi.org/10.1007/s10237-019-01158-w

Martin, M., Sansalone, V., Cooper, D. M. L., Forwood, M. R., & Pivonka, P. (2020). Assessment of romosozumab efficacy in the treatment of postmenopausal osteoporosis: Results from a mechanistic PK-PD mechanostat model of bone remodeling. *Bone, 133*, Article 115223. https://doi.org/10.1016/j.bone.2020.115223

McLean, M. C., & Urist, M. R. (1986). *Bone: Fundamentals of the physiology of skeletal tissue*. Chicago: University of Chicago Press.

Nakashima, T., Hayashi, M., Fukunaga, T., Kurata, K., Oh-hora, M., et al. (2011). Evidence for osteocyte regulation of bone homeostasis through RANKL expression. *Nature Medicine, 17*(10), 1231–1234. https://doi.org/10.1038/nm.2452

Nicolas, V., Prewett, A., Bettica, P., Mohan, S., Finkelman, R. D., et al. (1994). Age-related decreases in insulin-like growth factor-I and transforming growth factor-beta in femoral cortical bone from both men and women: Implications for bone loss with aging. *Journal of Clinical Endocrinology & Metabolism, 78*(5), 1011–1016. https://doi.org/10.1210/jcem.78.5.8175953

Okamoto, Y., Gotoh, Y., Uemura, O., Tanaka, S., Ando, T., & Nishida, M. (2005). Age-dependent decrease in serum transforming growth factor (TGF)-beta 1 in healthy Japanese individuals; population study of serum TGF-beta 1 level in Japanese. *Disease Markers, 21*(2), 71–74. https://doi.org/10.1155/2005/381215

Ota, K., Quint, P., Ruan, M., Pederson, L., Westendorf, J. J., et al. (2013). Sclerostin is expressed in osteoclasts from aged mice and reduces osteoclast-mediated stimulation of mineralization. *Journal of Cellular Biochemistry, 114*(8), 1901–1907. https://doi.org/10.1002/jcb.24537

Parfitt, A. M. (1983). The physiologic and clinical significance of bone histomorphometric data. In R. R. Recker (Ed.), *Bone histomorphometry techniques and interpretation*. Boca Raton: CRC Press.

Pattin, C. A., Caler, D. R., & Carter, D. R. (1996). Cyclic mechanical property degradation during fatigue loading of cortical bone. *Journal of Biomechanics, 29*(1), 69–80. https://doi.org/10.1016/0021-9290(94)00156-1

Pederson, L., Ruan, M., Westendorf, J. J., Khosla, S., & Oursler, M. J. (2009). Regulation of bone formation by osteoclasts involves Wnt/BMP signaling and the chemokine sphingosine-1-phosphate. *Proceedings of the National Academy of Sciences of the United States of America, 105*(52), 20764–20769. https://doi.org/10.1073/pnas.0805133106

Pfeilschifter, J., Diel, I., Scheppach, B., Bretz, A., Krempien, R., et al. (1998). Concentration of transforming growth factor beta in human bone tissue: Relationship to age, menopause, bone turnover, and bone volume. *Journal of Bone and Mineral Research, 13*(4), 716–730. https://doi.org/10.1359/jbmr.1998.13.4.716

Pivonka, P., Buenzli, P., & Dunstan, C. R. (2012). *A systems approach to understanding bone cell interactions in health and disease*. USA: InTech.

Pivonka, P., Zimak, J., Smith, D. W., Gardiner, B. S., Dunstan, C. R., et al. (2008). Model structure and control of bone remodeling: A theoretical study. *Bone, 43*(2), 249–263. https://doi.org/10.1016/j.bone.2008.03.025

Pivonka, P., Zimak, J., Smith, D. W., Gardiner, B. S., Dunstan, C. R., et al. (2010). Theoretical investigation of the role of the RANK-RANKL-OPG system in bone remodeling. *Journal of Theoretical Biology, 262*(2), 306–316. https://doi.org/10.1016/j.jtbi.2009.09.021

Riggs, B. L. (2003). Role of the vitamin D-endocrine system in the pathophysiology of postmenopausal osteoporosis. *Journal of Cellular Biochemistry, 88*(2), 209–215. https://doi.org/10.1002/jcb.10345

Riggs, B. L., Melton, J. L., (III), Robb, R. A., Camp, J. J., Atkinson, E. J., Peterson, J. M., et al. (2004). Population-based study of age and sex differences in bone volumetric density, size, geometry, and structure at different skeletal sites. *Journal of Bone and Mineral Research, 19*(12), 1945–1954. https://doi.org/10.1359/JBMR.040916

Riggs, B. L., & Pafitt, A. M. (2005). Drugs used to treat osteoporosis: The critical need for a uniform nomenclature based on their action on bone remodeling. *Journal of Bone and Mineral Research, 20*(2), 177–184. https://doi.org/10.1359/JBMR.041114

Roforth, M. M., Fujita, K., McGregor, U. I., Kirmani, S., McCready, L. K., et al. (2014). Effects of age on bone mRNA levels of sclerostin and other genes relevant to bone metabolism in humans. *Bone, 59*, 1–6. https://doi.org/10.1016/j.bone.2013

Ruiz-Lozano, R., Calvo-Gallego, J. L., Pivonka, P., & Martínez-Reina, J. (2025). Optimisation of romosozumab plus denosumab sequential treatments against postmenopausal osteoporosis. Insights from in silico simulations. *Biomechanics and Modeling in Mechanobiology.* https://doi.org/10.1007/s10237-024-01900-z

Ruiz-Lozano, R., Calvo-Gallego, J. L., Pivonka, P., McDonald, M., & Martínez-Reina, J. (2024). An in-silico approach to elucidate the pathways leading to primary osteoporosis: Age-related vs. postmenopausal. *Biomechanics and Modeling in Mechanobiology, 23*(4), 1393–1409. https://doi.org/10.21203/rs.3.rs-3432139/v1

Salari, N., Ghasemi, H., Mohammadi, L., Behzadi, M. H., Rabieenia, E., et al. (2021). The global prevalence of osteoporosis in the world: A comprehensive systematic review and meta-analysis. *Journal of Orthopaedic Surgery and Research, 16*(609). https://doi.org/10.1186/s13018-021-027720

Scheiner, S., Pivonka, P., & Hellmich, C. (2013). Coupling systems biology with multiscale mechanics, for computer simulations of bone remodeling. *Computer Methods in Applied Mechanics and Engineering, 254*, 181–196. https://doi.org/10.1016/j.cma.2012.10.015

Schnell, S., Friedman, S. M., Mendelson, D. A., Bingham, K. W., & Kates, S. L. (2010). The 1-year mortality of patients treated in a hip fracture program for elders. *Geriatric Orthopaedic Surgery & Rehabilitation, 1*(1), 6–14. https://doi.org/10.1177/2151458510378105

Shen, Y., Huang, X., Wu, J., Lin, X., Zhou, X., et al. (2022). The global burden of osteoporosis, low bone mass, and its related fracture in 204 countries and territories, 1990-2019. *Frontiers in Endocrinology, 13*, Article 882241. https://doi.org/10.3389/fendo.2022.882241

Simonet, W. S., Lacey, D. L., Dunstan, C. R., Kelley, M., Chang, M. S., et al. (1997). Osteoprotegerin: A novel secreted protein involved in the regulation of bone density. *Cell, 89* (2), 309–319. https://doi.org/10.1016/s0092-8674(00)80209-3

Trichilo, S., Scheiner, S., Forwood, M., Cooper, D. M. L., & Pivonka, P. (2019). Computational model of the dual action of PTH — Application to a rat model of osteoporosis. *Journal of Theoretical Biology, 473*, 67–79. https://doi.org/10.1016/j.jtbi.2019.04.020

Weitzmann, M. N., & Pacifi, R. (2006). Estrogen deficiency and bone loss: An inflammatory tale. *The Journal of Clinical Investigation, 116*(5), 1186–1194. https://doi.org/10.1172/JCI28550

Whalen, R. T., Carter, D. R., & Steele, C. R. (1988). Influence of physical activity on the regulation of bone density. *Journal of Biomechanics, 21*, 825–837. https://doi.org/10.1016/0021-9290(88)90015-2

Wijenayaka, A. R., Kogawa, M., Lim, H. P., Bonewald, L. F., Findlay, D. M., et al. (2011). Sclerostin stimulates osteocyte support of osteoclast activity by a RANKL-dependent pathway. *PLoS One, 6*(10), Article e25900. https://doi.org/10.1371/journal.pone. 0025900

Xiong, J., Piemontese, M., Onal, M., et al. (2015). Osteocytes, not osteoblasts or lining cells, are the main source of the RANKL required for osteoclast formation in remodeling bone. *PLoS One, 10*(9), Article e0138189. https://doi.org/10.1371/journal.pone.0138189

Yeh, E. J., Gitlin, M., Sorio, F., & McCloskey, E. (2023). Estimating the future clinical and economic benefits of improving osteoporosis diagnosis and treatment among postmeno-pausal women across eight European countries. *Archives of Osteoporosis, 18*(68). https://doi.org/10.1007/s11657-023-01230-0

Zhu, D., Mackenzie, N. C., Millan, J. L., Farquharson, C., & MacRae, V. E. (2011). The appearance and modulation of osteocyte marker expression during calcification of vascular smooth muscle cells. *PLoS One, 6*(5), Article e19595. https://doi.org/10.1371/journal.pone.0019595

CHAPTER 18

Simulation platform for in-silico trials of osteoporosis treatments—PK-PD models: Towards optimizing single and multiple-drug treatments

José Luis Calvo-Gallego[1, 3], Peter Pivonka[2] and Javier Martínez-Reina[1, 3]

[1]*Departmento de Ingeniería Mecánica y Fabricación, Universidad de Sevilla, Seville, Spain;* [2]*School of Mechanical, Medical and Process Engineering, Queensland University of Technology, Brisbane, QLD, Australia;* [3]*Instituto de Biomedicina de Sevilla (IBIS), Sevilla, Spain*

List of abbreviations

Alend	Alendronate
BC	Bone compartment
BCPM	Bone cell population model
BDG	Bone density gain
BMD	Bone mineral density
BMU	Basic multicellular unit
BSAP	Bone-specific alkaline phosphatase
BW	Body weight
CC	Central compartment
Dmab	Denosumab
FRI	Fracture risk index
IV	Intravenous
NCT	Non-calcified tissues
NTx	N-terminal telopeptide of type 1 collagen
OP	Osteoporosis
OPG	Osteoprotegerin
PD	Pharmacodynamic
PK	Pharmacokinetic
PMO	Postmenopausal osteoporosis
RANK	Receptor activator of nuclear factor $\kappa\beta$
RANKL	Receptor activator of nuclear factor $\kappa\beta$ ligand
Rmab	Romosozumab
RVE	Representative volume element

Biomechanics and Mechanobiology of Bone. https://doi.org/10.1016/B978-0-443-26739-0.00004-1

SC	Subcutaneous
Scl	Sclerostin
TSM	Time elapsed since Menopause
WHO	World Health Organization

1. Introduction

In Chapter 17 we presented a bone cell population model (BCPM) capable of simulating the process of bone aging and the consequences of the oestrogen decline triggered by menopause. The joint effect of both processes can lead to OP, but they affect men and women differently. OP can nowadays be treated with a wide range of drugs with different mechanisms of action. In addition, research into the development of other drugs continues. It is therefore a challenge for the health system to find the optimal treatment for each patient, tailored to the specific etiology of their disease, e.g., distinguishing between the effect of aging and oestrogen level, among many other factors. However, current regulatory requirements significantly hinder both the development of new drug treatments and the comparison of their effectiveness against existing therapies. This makes it necessary to study an alternative to clinical trials and this is when in silico trials appear, if not to completely replace clinical trials, which are strictly necessary, then to guide and accelerate them as much as possible.

In this chapter, we will present in silico trials of three drugs against OP. Alend is a drug that infiltrates the bone matrix and is released during osteoclastic resorption. Dmab and Rmab are more recently developed drugs, the former being an anticatabolic drug and the latter mainly anabolic. The different mechanisms of action suggest the possibility of combining them in sequential treatments, as has been tested in some recent clinical trials of the latter two. Our results on Rmab + Dmab treatments indicate that a combination starting with Rmab is better compared to swapping this order. We also propose other promising combinations that have not been tested in clinical trials, but are easily tested in in silico studies. This demonstrates how the proposed simulation tool can help inform clinical usage of current drugs and development of new drugs by identifying the most promising scenarios to be tested clinically, optimizing the trial design, and identifying subgroups of the population that show benefit-risk profiles that are different from those of the average patient. Another important objective of our model is the development of a computational tool to design a patient-specific treatment with currently available drugs.

2. PK-PD models of drugs against osteoporosis

PK-PD models, usually used in combination, are fundamental tools in pharmacological research. PK models intend to explain the absorption, distribution, and elimination of a drug in the body, and constitute the first step to develop PD models, which analyze the effect of the drug on the body. In simpler terms, as Holford and Sheiner

(1982) define in their article, PK studies *what the body does to the drug*, while PD studies *what the drug does to the body.*

Due to the complexity of the human body, the kinetics of drug treatments is often simplified by using compartmental models (Shargel & Yu, 2016). A compartment is a hypothetical body region characterized by similar blood flow and drug affinity (Shargel & Yu, 2016). It is important to note that these compartments do not correspond to actual organs or anatomical structures. Within each compartment, the drug is assumed to be evenly distributed and thus its concentration is considered an average concentration (Shargel & Yu, 2016). Rate constants are used to represent the general processes of entry and exit of the drug from a given compartment. Linear differential equations are used to describe the processes involved (Shargel & Yu, 2016).

To validate a PK model, simulation results are compared with experimental data on drug concentration in blood serum after administration. If available, it can also be compared with elimination data, e.g., in urine. The validation of the PD model is usually based on bone density gain (BDG) and/or biomarker data (i.e., bone formation and resorption markers in blood or urine).

In the following, we provide an overview of PK-PD models for three of the most commonly used OP drugs: Alend, Dmab and Rmab. Bone disease simulations are discussed first based on single drug treatment effects, followed by combined drug treatments.

2.1 Alendronate

Bisphosphonates, a group of anti-catabolic treatments with a high affinity for the bone matrix (Fleisch et al., 1969; Francis et al., 1969; Jung et al., 1973), are the first-line and most commonly used treatment for postmenopausal women with OP (Brown & Josse, 2002). Among bisphosphonates, Alend is the most widely used (Cummings et al., 2020). Alend inhibits bone resorption through two distinct mechanisms: reducing the resorptive capacity of osteoclasts and inducing their apoptosis (Takagi et al., 2021; Yu et al., 2001, 2018). It is typically administered orally in 70 mg tablets once a week, although other dosing regimens are also prescribed.

Alend is distributed throughout the body via the blood plasma. It directly reaches plasma in the case of intravenous (IV) administration or through the gastrointestinal tract when administered orally. It does not appear to be metabolized in humans. Once in the plasma, Alend is quickly distributed into the noncalcified tissues of the body, followed by a redistribution in bone or renal elimination (Lin et al., 1991; Porras et al., 1999). In bone, its deposition is not uniform; it tends to concentrate in areas of high physiological activity where bone remodeling is most active (Sato et al., 1991). Once embedded in the bone matrix, Alend remains inactive until the resorption process removes the upper bone layers, allowing osteoclasts to access it. This mechanism explains the prolonged effect of Alend after administration (Porras et al., 1999). The way Alend is embedded into different bone compartments and how it is released during osteoclastic resorption poses a challenge for modeling its pharmacokinetics and pharmacodynamics, as we will show in this section.

2.1.1 Alendronate PK model

The most recent PK model for Alend, developed by Calvo-Gallego et al. (2022), is a three-compartment model. This fits plasma concentration and both short- and long-term urine excretion. In Fig. 18.1a, CC represents the central compartment (plasma), NCT the non-calcified tissues and BC the bone compartment, which in turn is subdivided into an active (A) and an inactive (I) part (see section below for justification). k_i and $k_{el,i}$ are the absorption and elimination rate constants respectively, with $i = CC, NCT, BC$ or urine. Urine excretion is represented by an elimination from the CC.

The differential equations governing the temporal evolution of the amount of Alend in the different compartments are:

$$\frac{dAle_{CC}}{dt} = F \cdot k_{CC} \cdot Ale_{Gut} + k_{el,BC} \cdot Ale_{BC} + k_{el,NCT} \cdot Ale_{NCT} - (k_{BC} + k_{NCT} + k_{el,urine})Ale_{CC} \quad (18.1)$$

$$\frac{dAle_{BC}}{dt} = k_{BC} \cdot Ale_{CC} - k_{el,BC} \cdot Ale_{BC} \quad (18.2)$$

$$\frac{dAle_{NCT}}{dt} = k_{NCT} \cdot Ale_{CC} - k_{el,NCT} \cdot Ale_{NCT} \quad (18.3)$$

$$\frac{dAle_{Gut}}{dt} = -k_{CC} \cdot Ale_{Gut} \quad (18.4)$$

$$\frac{dAle_{Urine}}{dt} = k_{el,urine} \cdot Ale_{CC} \quad (18.5)$$

where Ale_{CC}, Ale_{BC}, Ale_{NCT}, Ale_{Gut} and Ale_{Urine} are the amounts of Alend in the CC, BC, NCT, gut, and urine, respectively. F is the bioavailability. If an IV dose is administered, the term $F \cdot k_{CC} \cdot Ale_{Gut}$ in Eq. (18.1) must be replaced by the IV dose rate, while Eq. (18.4) is no longer needed. The concentration of Alend in the CC is $[Ale]_{CC} = \frac{Ale_{cc}}{V_c}$, where V_c is the volume of distribution of the CC. The concentration of Alend in the BC is assumed to be the one that influences the bone remodeling process.

Eqs. (18.1) to (18.5) are systemic (corresponding to the whole body); however, the BCPM is implemented locally, at a given representative volume element (RVE). To pass from the systemic to the local level, Calvo-Gallego et al. (2022) made the simplification that Alend was uniformly distributed throughout the bone matrix of the entire skeleton, V_{bm}^{skel}, assumed to be equal to $2.23 \cdot 10^{-3} m^3$ (adapted from Valentín (2002) for a 60 kg adult female). Then, the concentration at the RVE scale is:

$$[Ale]_{BC} = \frac{Ale_{BC}}{V_{bm}^{skel}} = \frac{Ale_{RVE}}{V_{bm}} \quad (18.6)$$

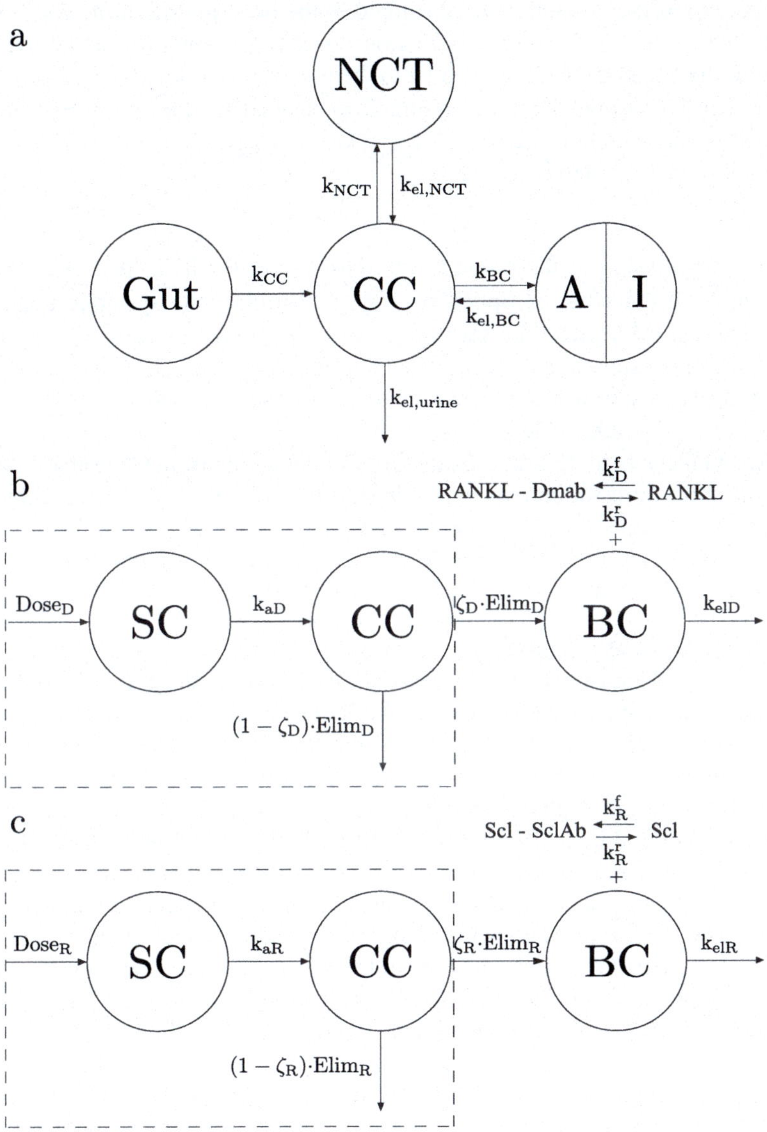

FIGURE 18.1

PK-PD models: (a) Alend, proposed by Calvo-Gallego et al. (2022). The BC is divided into active (A) and inactive (I) part; (b) Dmab, proposed by Martínez-Reina et al. (2022); and (c) Rmab, proposed by Ruiz-Lozano et al. (2025).

where Ale_{RVE} is the amount of Alend stored in the bone matrix of the RVE. Subsequently, it is necessary to switch back from local to systemic, to model the return flow from the BC to the CC. In contrast to the rest of k_i and $k_{el,i}$ (Fig. 18.1a), which were assumed constant, $k_{el,BC}$ is assumed to depend on the osteoclast-mediated release of Alend from bone matrix (Fisher et al., 2000; Yu et al., 2018) and therefore, on the PD model (explained later):

$$k_{el,BC} = k_{res} \cdot Oc_a \tag{18.7}$$

Thus, $k_{res} \cdot Oc_a \cdot [Ale]_{BC}$ is the Alend released locally, in the RVE, per unit volume of bone matrix. It must be multiplied by V_{bm}^{skel} to estimate the systemic quantity of Alend released. Moreover, the factor f_{SL} was introduced to consider that not all the RVEs in the skeleton have the same bone turnover rate. Thus, the second term on the right hand side of Eq. (18.1) (and therefore the last term of Eq. 18.2) would be equal to $f_{SL} \cdot k_{res} \cdot Oc_a \cdot [Ale]_{BC} \cdot V_{skel}$.

Calvo-Gallego et al. (2022) used the experimental data from Chae et al. (2014) to adjust the short-term serum concentration (see Fig. 18.2a), those obtained by Kang et al. (2006) to adjust the short-term urinary excretion (see Fig. 18.2b), and those obtained by Khan et al. (1997) to fit the long-term urinary excretion (see Fig. 18.2c), thus yielding the constants shown in Table 18.1.

2.1.2 Alendronate PD model

The amount of Alend in bone increases with each weekly dose. If the effect of the drug was proportional to that amount, such effect would be increasingly pronounced. But in clinical trials (Rizzoli et al., 2002), the increase in BMD is high during the first months of treatment and subsequently slows down. This behavior could be explained because a higher amount of Alend is deposited in trabecular bone compared to cortical bone (Fleisch, 2000; Khan et al., 1997; Lin et al., 1991; Porras et al., 1999), which explains the fast increase in BMD in the lumbar spine and the hip observed in clinical results. However, since trabecular bone is more scarce than cortical bone, in long treatments the amount of Alend that reaches the former is expected to be reduced over time in favor of the latter, due to the saturation of the tissue. This would explain the long-term stabilization of bone mass gain observed clinically.

In view of the clinical results, Calvo-Gallego et al. (2022) hypothesized that the BC can be divided in two parts. One part, termed active (A), is the closest one to the bone-marrow interface, where the retrieval of Alend is immediate through bone resorption. The second part, termed inactive (I), corresponds to the innermost regions of bone, where Alend is buried and, to some extent, inaccessible to bone resorption (Porras et al., 1999; Sato et al., 1991), as represented in Fig. 18.1a, where the subcompartment I is not involved in the exchange with the CC. The active subcompartment would predominate in trabecular bone and its proportion would decrease with porosity as the tissue becomes more compact (with a lower bone matrix - bone marrow interface). Calvo-Gallego et al. (2022) measured the compactness of the tissue by the specific surface, although expressed per volume of bone matrix

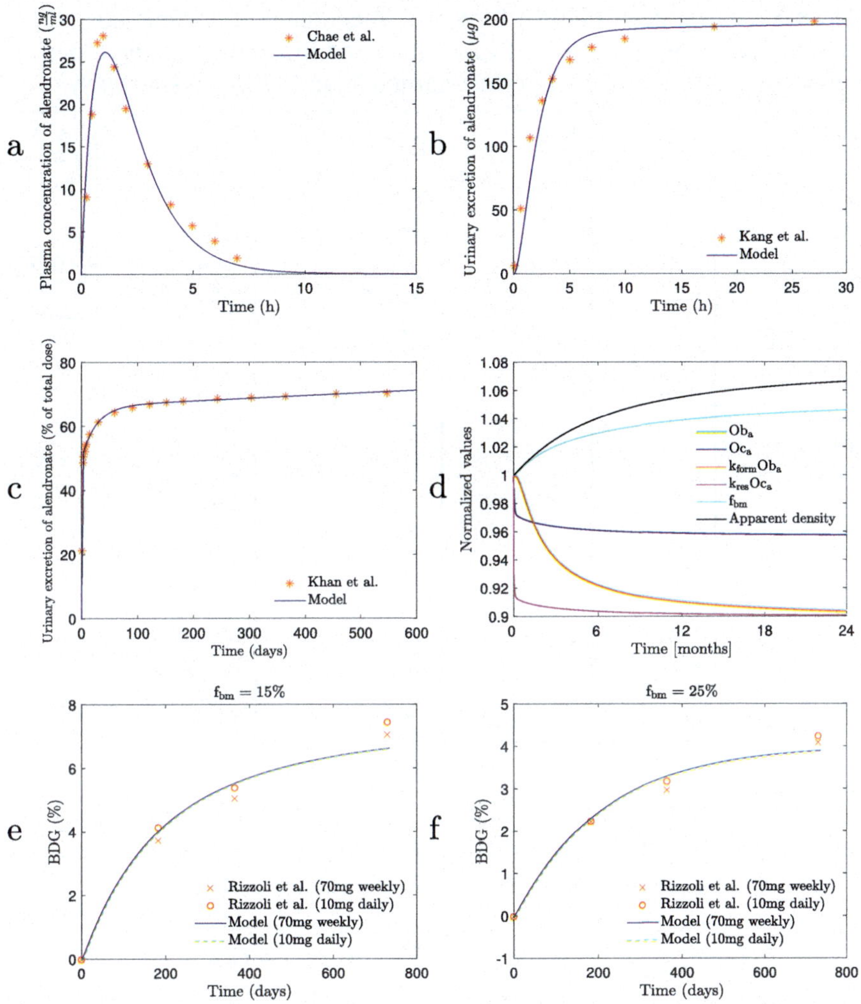

FIGURE 18.2

Alend PK-PD model results versus time: (a) Plasma concentration; (b) Short-term urinary excretion; (c) Long-term urinary excretion. (d) Osteoblasts and osteoclasts concentration, bone resorption and formation rates, apparent density and bone volume fraction f_{bm} at hip in a 70 mg weekly treatment (results normalized with values at the beginning of treatment). (e) and (f) comparison of BDG with clinical results (Rizzoli et al. (2002); (e) Hip ($f_{bm} = 15\%$) and (f) Lumbar vertebra, ($f_{bm} = 25\%$), for 70 mg weekly and 10 mg daily doses.

Table 18.1 Constants of the PK-PD models. Alendronate constants are taken from Calvo-Gallego et al. (2022). Dmab and Rmab were calibrated in Martínez-Reina et al. (2022) and Ruiz-Lozano et al. (2025), respectively.

Alendronate	k_{CC} 16.3 day^{-1}	$k_{el,urine}$ 14.8 day^{-1}	k_{NCT} 6.8 day^{-1}	$k_{el,NCT}$ 0.05 day^{-1}	k_{BC} 7.4 day^{-1}
	V_C 3.8 L	F 0.0054 –	Π_{act}^{Ale} 13.0 day mM^{-1}	Π_{rep}^{Ale} 14.3 day mM^{-1}	f 0.077 –
Denosumab	k_{aD} 0.17 day^{-1}	K_{MD} 27.96 nM	V_{cD}/F_D 77.90 $mL(kgBW)^{-1}$	V_{maxD} 18,177 ng $(kgBW)^{-1}$ day^{-1}	
	k_{elD} 0.0115 day^{-1}	ζ_D 0.4889 –	$\widetilde{D}_{Dmab_{BC}}$ 16.22 day^{-1}	$K_{Dmab-RANKL}$ 9.2 pM	
Romosozumab	k_{aR} 0.2 day^{-1}	K_{MR} 20.13 nM	V_{cR}/F_R 80.88 $mL(kgBW)^{-1}$	V_{maxR} 47,939 ng $(kgBW)^{-1} day^{-1}$	
	k_{elR} 0.1 day^{-1}	ζ_R 0.5571 –	$\widetilde{D}_{Rmab\ BC}$ 0.0118 day^{-1}	$K_{Rmab-Scl}$ $5 \cdot 10^5$ pM	

rather than per total volume, that is, $\frac{S_v}{f_{bm}}$, where Eq. (17.42) has been used. The amount of Alend contained in the active subcompartment is defined as:

$$Ale_{BC,act} = \begin{cases} Ale_{BC} \cdot \dfrac{S_v}{f_{bm}} \cdot \dfrac{f_{bm0}}{S_{v0}} & \text{if} \quad f_{bm} \geq f_{bm0} \\ Ale_{BC} & \text{if} \quad f_{bm} < f_{bm0} \end{cases} \tag{18.8}$$

where $f_{bm0} = 5\%$ is established as the minimum bone volume fraction below which all tissue can be considered superficial (keep in mind that Eq. (18.8) is not defined for $f_{bm} = 0$), and S_{v0} is the specific area corresponding to that reference value. The factor $\frac{S_v}{f_{bm}} \cdot \frac{f_{bm0}}{S_{v0}}$ is plotted against f_{bm} in Fig. 18.3.

The active subcompartment will become saturated relatively soon due to the low trabecular bone volume, whereas the total Alend in the BC will not, because of the large cortical bone volume capable to store high amounts of the drug. The active Alend was also saturated as a function of S_v/f_{bm}, being its maximum concentration:

$$[Ale]_{BC,act}^{max} = f \cdot \frac{S_v}{f_{bm}} \cdot \frac{f_{bm0}}{S_{v0}} \tag{18.9}$$

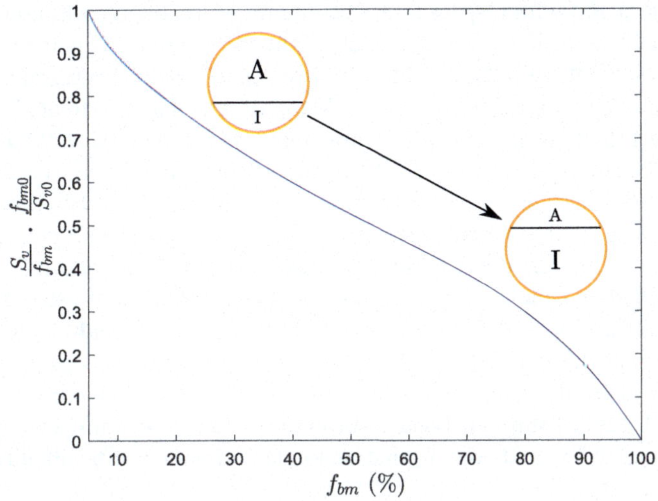

FIGURE 18.3

Alend phenomenological function that allows distinction between the active (A) and inactive (I) parts of the BC as function of bone volume fraction.

with f a constant to be fitted. Hence, as Alend enters the BC, a fraction is assigned to the active subcompartment, following Eq. (18.8), and the rest goes to the inactive subcompartment $(Ale_{BC,inact} = Ale_{BC} - Ale_{BC,act})$.

Alend has the following two effects. On the one hand, it disrupts the cytoskeleton of osteoclasts by inhibiting farnesyl pyrophosphate (FPP) synthase, leading to the disappearance of the clear zones and ruffled borders and limiting their resorbing capacity (Takagi et al., 2021; Yu et al., 2001), which is measured in the model by k_{res} (see Eq.17.6). Consequently, this parameter was modified as follows:

$$k_{res} = \frac{k_{res,nom}}{1 + \Pi_{rep}^{Ale} \cdot k_{res,nom} \cdot Oc_a \cdot [Ale]_{BC,act}} \tag{18.10}$$

where $k_{res,nom}$ is the nominal value and Π_{rep}^{Ale} is a constant quantifying the effect of Alend on osteoclast resorption capacity.

The second effect of Alend is the inhibition of FPP synthase in the mevalonate pathway and the reduction of protein prenylation, thus inducing apoptosis in osteoclasts (Sato et al., 1991; Takagi et al., 2021; Yu et al., 2001, 2018). To account for this effect, the apoptosis rate was increased in the model as follows:

$$A_{Oc_a} = A_{Oc_a,nom}\left(1 + \Pi_{act}^{Ale} \cdot k_{res} \cdot Oc_a \cdot [Ale]_{BC,act}\right) \tag{18.11}$$

where Π_{act}^{Ale} quantifies the effect of Alend on osteoclast apoptosis.

The fitted constants were: the PK constants, $k_{CC}, k_{NCT}, k_{BC}, k_{el,NCT}, k_{el,urine}, V_c$ and F; and the PD constants, $\Pi_{act}^{Ale}, \Pi_{rep}^{Ale}$ and f (see Table 18.1).

BDG, measured with respect to the beginning of treatment, was compared with clinical data taken from Rizzoli et al. (2002) who measured BDG in the lumbar spine ($f_{bm} = 15\%$) and the hip ($f_{bm} = 25\%$) in two sets of postmenopausal osteoporotic patients: i) treated with a weekly dose of 70 mg of Alend and ii) treated with a daily dose of 10 mg, both during 2 years. Postmenopausal osteoporosis (PMO) was simulated as explained in Section 17.7.3 of Chapter 17 and the time between the onset of menopause (49 years) and the beginning of treatment was set at 19 years, according to the mean age of the patients (Rizzoli et al., 2002). The BDG predicted by the model is plotted against the experimental results in Fig. 18.2e and f.

Fig. 18.2d shows the evolution of normalized osteoclast and osteoblast populations, bone resorption and formation rates, apparent density, and f_{bm} in the hip for a single weekly dose of 70 mg. The drug decreases both cell populations, thus reducing the bone turnover rate. This behavior was experimentally confirmed by Rizzoli et al. (2002) through bone resorption markers (N-terminal telopeptide of type 1 collagen, NTx) and bone formation markers (bone-specific alkaline phosphatase, BSAP).

Alend also decreases k_{res} and as a consequence, the formation rate is slightly higher than the resorption rate, and f_{bm} increases slightly. BDG is higher than the increase in f_{bm} due to mineralization of the existing tissue, similar to what occurs in other antiresorptive treatments that reduce bone turnover (Martínez-Reina & Pivonka, 2019).

This model has two main limitations. The first is the assumption that bone remodeling is a continuous process in time and space. In reality, it occurs in a cyclical manner at random bone sites, through the action of BMUs (Basic Multicellular Units), which follow a specific sequence (resorption, inversion, formation and quiescence). Therefore, the action of Alend would be restricted to the resorption phase and at those points where that specific process is occurring. The second limitation is the use of a systemic PK model in conjunction with a local PD model. The CC represents the serum of the entire body and the BC the whole skeleton. However, the PD model is applied locally, in a certain RVE at a specific bone site. Therefore assuming that the flux of Alend from BC to CC at a given bone site is representative of what occurs at the skeleton on average imply a strong simplification. A comprehensive PK-PD model would need to simulate the entire skeleton, which is generally disregarded for practical reasons.

2.2 Denosumab

Dmab is a fully human monoclonal antibody to receptor activator of nuclear factor $\kappa\beta$ (RANK) ligand (RANKL), which is a key regulator of bone resorption, as discussed earlier. Hence, the administration of its antibody, Dmab, at a dose of 60 mg every 6 months, reduces bone resorption and consequently bone turnover. This allows bone to become highly mineralized, which is the mechanism through which Dmab (and other anticatabolic drugs) increases bone mass, despite showing only minor effects on reducing bone porosity (Martínez-Reina & Pivonka, 2019).

2.2.1 Denosumab PK model

Martínez-Reina et al. (2022) proposed a two-compartment PK model for Dmab shown in Fig. 18.1b. A first-order rate process (k_{aD}) governs the absorption of the drug ($Dose_D$) from the subcutaneous (SC) injection site into the CC ($[Dmab]_{CC}$), being V_{cD}/F_D the volume of the CC adjusted for bioavailability.

$$\frac{d[Dmab]_{CC}}{dt} = \frac{Dose_D}{V_{cD}/F_D} k_{aD} e^{-k_{aD}t} - Elim_{CC,D} \tag{18.12}$$

To model the elimination from the CC $\left(Elim_{CC,D}\right)$ Marathe et al. (2011) used a non-linear Michaelis–Menten kinetics, described by a combination of a linear first-order process (k_{elD}) and a non-linear saturation process (V_{maxD}, K_{MD}):

$$Elim_{CC,D} = k_{elD}[Dmab]_{CC} + \frac{V_{maxD}}{\frac{V_{cD}}{F_D}} \frac{[Dmab]_{CC}}{K_{MD} + [Dmab]_{CC}} \tag{18.13}$$

in the model proposed by Martínez-Reina et al. (2022), this elimination from the CC was divided into two parts: a fraction excreted from the CC $(1 - \zeta_D)$, and another fraction that passes into the BC (ζ_D).

2.2.2 Denosumab PD model

Martínez-Reina et al. (2022) assumed that the fraction of the drug that enters the BC can be considered as $P_{Dmab_{BC}}$, i.e., the Dmab production rate in the competitive binding reactions between RANKL, osteoprotegerin (OPG) and Dmab:

$$P_{Dmab_{BC}} = \zeta_D Elim_{CC,D} \tag{18.14}$$

This production rate can be replaced in Eq. (17.12a) to yield the concentration of the new ligand (Dmab) of the RANK-RANKL-OPG-Dmab competitive binding:

$$[Dmab]_{BC} = \frac{P_{Dmab_{BC}}}{\tilde{D}_{Dmab_{BC}} + \frac{\tilde{D}_{RANKL-Dmab}}{K_{RANKL-Dmab}} \cdot [RANKL]} \tag{18.15}$$

where the constants $\tilde{D}_{Dmab_{BC}}$, $\tilde{D}_{Dmab\text{-}RANKL}$ and $K_{Dmab\text{-}RANKL}$ have the same interpretation as in Section 17.3.1 of Chapter 17. Eq. (17.17) is also modified to account for the new competitor, Dmab:

$$[RANKL] = P_{RANKL} \cdot \left[\tilde{D}_{RANKL} + \frac{\tilde{D}_{OPG-RANKL}}{K_{OPG-RANKL}} \cdot [OPG] + \frac{\tilde{D}_{RANK-RANKL}}{K_{RANK-RANKL}} \cdot [RANK] + \frac{\tilde{D}_{RANKL-Dmab}}{K_{RANKL-Dmab}} \cdot [Dmab]_{BC}\right]^{-1} \tag{18.16}$$

The Dmab PK-PD model was fitted by Martínez-Reina et al. (2022): the PK model using the clinical data given by Marathe et al. (2011) of Dmab plasma concentration, and the PD model using the clinical study by Bone et al. (2008). The adjustments are shown in Fig. 18.4a and b and the fitted constants in Table 18.1.

2.2.3 Simulations of denosumab treatments

The PK-PD model can be used to simulate different treatments and in different scenarios. Fig. 18.4c and d compares the effectiveness of World Health Organization (WHO)-approved Dmab treatment (60 mg of Dmab injected every 6 months, 60Q6) in patients with different ages at menopause (44, 49 and 54 years). The evolution of apparent density in a piece of trabecular bone corresponding to the femoral neck is represented over time. In the left graph, density is represented against the time elapsed since menopause (TSM) when the treatment commences 10 years after menopause, therefore, at different ages. It can be seen that BDG is similar for the three cases in absolute terms (around 0.025 g/cm^3 in 2 years). In the right graph,

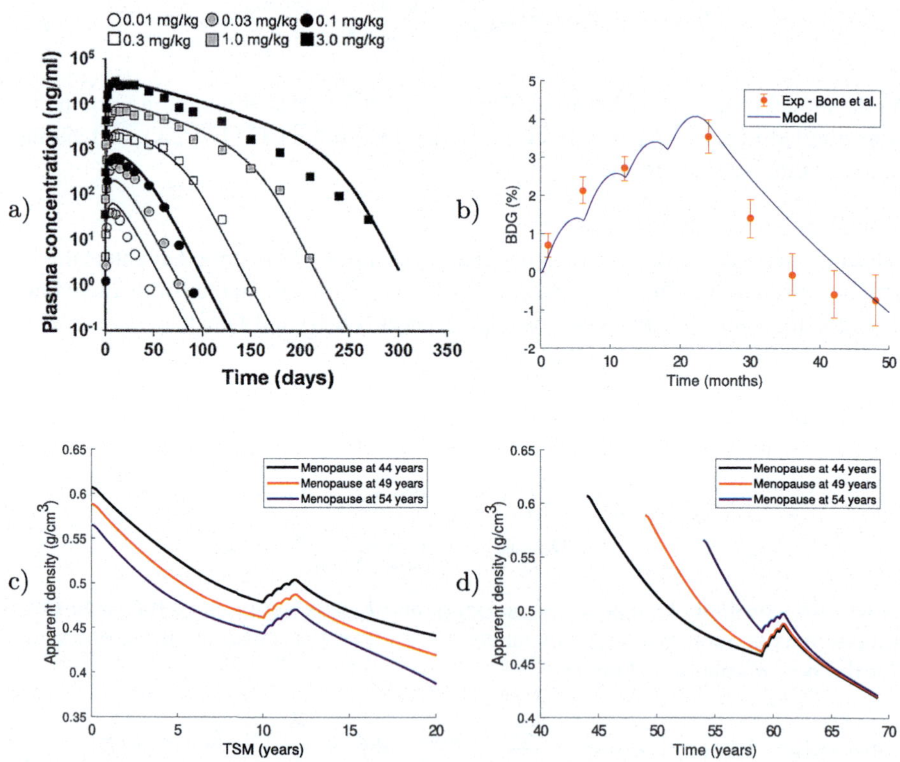

FIGURE 18.4

Dmab PK-PD model simulation results: (a) The constants of the PK model were fitted to reproduce the clinical data of Dmab concentration in plasma by Marathe et al. (2011). (b) The constants of the PD model were fitted to reproduce the clinical results of BDG from Bone et al. (2008). (c) and (d) Apparent density of trabecular bone of the femoral neck for women who went through menopause at different ages and were treated with Dmab (60Q6) for 2 years: (c) Treatment started 10 years after menopause; (d) Treatment started at the age of 59.

treatment begins at the age of 59 regardless of the age at menopause and BDG is similar for normal (49 years) or early menopause (44 years); respectively, 5.7% and 6.2% of BMD increase the beginning of the treatment. However, the drug is not as effective for women with late menopause (54 years), for whom treatment has been started only 5 years after menopause, probably too soon, and the BDG is only 3.6%.

2.3 Romosozumab

Rmab is a humanized monoclonal antibody targeting sclerostin (Scl), approved to treat postmenopausal women at high risk of fracture who cannot receive other OP treatments. As stated above, Scl binds to LRP5/6 to inhibit the Wnt/β-catenin pathway that promotes osteoblastogenesis. Therefore, by inhibiting Scl action, Rmab promotes osteoblast precursor proliferation, which subsequently leads to increased active osteoblast formation, responsible for the anabolic effect. In addition, some studies have confirmed that, in high concentrations, Scl may enhance osteoclastic activity by increasing RANKL levels (Wijenayaka et al., 2011). Hence, if the concentration of Scl is high, as it may occur in elderly people (Ardawi et al., 2011), an antibody to Scl might also have an anti-catabolic effect. For that reason, Rmab is considered to have a dual action: anabolic and anti-catabolic.

2.3.1 Romosozumab PK model

The PK-PD model of Rmab is based on the one proposed by Martin et al. (2020) and was modified to follow the same structure of the PK-PD model of Dmab (see Fig. 18.1c). In the original model, it was the concentration of Rmab in the CC that interfered with the Scl-LRP5/6-Wnt signaling pathway. The BC was added by Ruiz-Lozano et al. (2025) as the competitive binding Scl-LRP5/6-Wnt takes place on the surface of osteoblast precursors and only the fraction of Rmab reaching the BC can interfere with that signaling pathway. Therefore, all the equations are analogous to those of the PK-PD model of Dmab:

$$\frac{d[\mathrm{Rmab}]_{\mathrm{CC}}}{dt} = \frac{\mathrm{Dose}_{\mathrm{R}}}{V_{\mathrm{cR}}/F_{\mathrm{R}}} k_{\mathrm{aR}} e^{-k_{\mathrm{aR}} t} - \mathrm{Elim}_{\mathrm{CC,R}} \tag{18.17}$$

$$\mathrm{Elim}_{\mathrm{CC,R}} = k_{\mathrm{elR}} [\mathrm{Rmab}]_{\mathrm{CC}} + \frac{V_{\mathrm{maxR}}}{V_{\mathrm{cR}}/F_{\mathrm{R}}} \cdot \frac{[\mathrm{Rmab}]_{\mathrm{CC}}}{K_{\mathrm{MR}} + [\mathrm{Rmab}]_{\mathrm{CC}}} \tag{18.18}$$

where all the constants have the same interpretation as those of the Dmab PK-PD model (the subscript D has been changed for R). The term of elimination from the CC $(\mathrm{Elim}_{\mathrm{CC,R}})$ is divided into two parts: (1) the fraction of the drug that is degraded in the CC or excreted, $(1 - \zeta_{\mathrm{R}})$, and (2) the fraction that enters the BC, ζ_{R}.

2.3.2 Romosozumab PD model

As in the case of Dmab, the fraction of Rmab that enters the BC is identified with the production rate of Rmab in the competitive binding equations.

$$P_{Rmab_{BC}} = \zeta_R Elim_{CC,R} \tag{18.19}$$

This production rate can be replaced in Eq. (17.12a) to yield the concentration of Rmab in the BC.

$$[Rmab]_{BC} = \frac{P_{Rmab_{BC}}}{\widetilde{D}_{Rmab_{BC}} + \frac{\widetilde{D}_{Rmab-Scl}}{K_{Rmab-Scl}}[Scl]} \tag{18.20}$$

Similarly to Eq. (18.16), the concentration of Scl can be calculated using the competitive binding Eq. (17.12), with the new competitor, Rmab.

$$[Scl] = \frac{P_{Scl}}{\widetilde{D}_{Scl} + \frac{\widetilde{D}_{Scl-LRPS/6}}{K_{Scl-LRP5/6}}[LRP5/6] + \frac{\widetilde{D}_{Rmab-Scl}}{K_{Rmab-Scl}}[Rmab]_{BC}} \tag{18.21}$$

where the total production of Scl is divided into the production by osteocytes and by other sources considered in the aging process:

$$P_{Scl} = P_{Scl,Ot}(\Psi_{bm}, [Ot], [Scl]) + P_{Scl,ageing} \tag{18.22}$$

The dependence of $P_{Scl,Ot}$ on [Scl] makes it complex to work out an expression for this concentration, and has been omitted for the sake of brevity, but it can be consulted in Ruiz-Lozano et al. (2025).

The Rmab PK-PD model was fitted by Ruiz-Lozano et al. (2025): the PK model using the clinical data provided by Padhi et al. (2011) of Rmab plasma concentration, and the PD model using the clinical study by Langdahl et al. (2017) and McClung et al. (2018). The adjustments are shown in Fig. 18.5 and the fitted constants in Table 18.1.

2.4 Simulations of combined Denosumab and Romosozumab treatments

The combination of anabolic and anti-catabolic drugs has proven very efficient in the treatment of OP. The meta-analysis performed by Lou et al. (2019) showed that the combination of both types of drugs in sequential treatments was better than either as monotherapy to improve BMD of the spine and hip. Interestingly, the sequence of chosen treatments has a strong effect of overall BDGs. While the transition between two different anti-catabolic drugs produces only moderate BDGs, transitioning from an anti-catabolic to an anabolic drug improves efficacy, but BDG during the anabolic part of the sequence is lower than in untreated patients, at least for the transition between bisphosphonates to teriparatide (Cosman et al., 2017). Most interestingly, the transition from anabolic to anti-catabolic drugs further increased BMD, as the ARCH (Rmab to Alend) (Saag et al., 2017) and FRAME (Rmab to Dmab) (Cosman et al., 2018) studies demonstrated, and more notably than swapping the order of treatment (Langdahl, 2020).

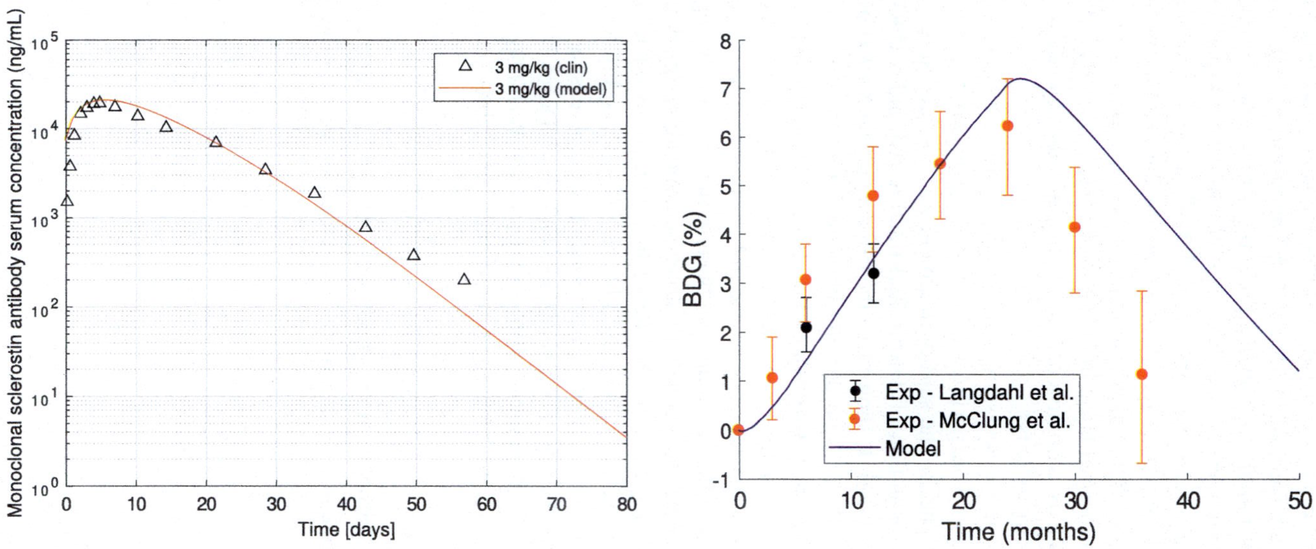

FIGURE 18.5

Rmab PK-PD model simulation results: (Left) the constants of the PK model were fitted to reproduce the clinical data of Rmab concentration in plasma by Padhi et al. (2011). (Right) the constants of the PD model were fitted to reproduce the clinical BDG results from Langdahl et al. (2017) and McClung et al. (2018).

Ruiz-Lozano et al. (2025) performed in silico trials of a series of treatments combining Dmab and Rmab by using the mechanobiological PK-PD models described in this chapter (see Table 18.2). WHO-approved dosages, i.e., 60 mg Dmab injected every 6 months (60Q6) and 210 mg Rmab injected every month (210Q1), were considered in every case. The sequence of drugs is specified through Dr_i and the length of each treatment phase through L_i (in years). In sequential treatments, the next phase begins after the normal lapse of the ongoing treatment. For example, in treatment *SeqDR* the first Rmab injection is administered 6 months after the last Dmab injection (normal lapse in Dmab treatments) and in treatment *SeqRD* the first Dmab injection is administered 1 month after the last Rmab injection (normal lapse in Rmab treatments).

The onset of menopause was assumed at age $t_{meno} = 49$ years to calculate TSM in Eq. (17.23). Simulations were performed for trabecular bone of $f_{bm} = 30\%$, representative for the femoral neck. The start of treatment was assumed at $t_0 = 59$ years. Fig. 18.6 compares the evolution of density and fracture risk index (FRI) (see Eq. 17.59) after the first injection. It shows that the efficacy of the *SeqRD* treatment is higher than that of *SeqDR* and the order of administration of the drug is important, as reported in different studies (Cosman et al., 2017, 2018; Langdahl, 2020; Saag et al., 2017). *ConcR + D* is the treatment that achieves the largest and fastest density increase and consequently the fastest decrease in FRI, followed by *SeqRDR*.

ConcR + D is the most efficient treatment, but also the one with the highest dose. This is an important fact and must be taken into account in the design of treatments, since, obviously, the higher the dose, the higher the reduction of FRI, but also the higher the probability of the appearance of adverse side effects. For this reason, it is interesting to assess to what extent FRI is reduced with increasing dose. This is analyzed in Fig. 18.7a for the *SeqRD* treatment in which doses $Dose_R$ (injected every month) and $Dose_D$ (every 6 months), different from the WHO-approved doses analyzed so far, have been tried. This figure shows the average FRI during treatment

Table 18.2 Summary of the analyzed treatments. Dr_i and L_i stand for the type of drug (D for Dmab and R for Rmab) and length (in years) of sequential phase *i*. In all cases, the WHO-approved dosages were used: 60Q6 for D and 210Q1 for R.

Name	Type	Dr_1	L_1	Dr_2	L_2	Dr_3	L_3
SeqRD		R	1	D	1	–	–
SeqDR	Sequential	D	1	R	1	–	–
SeqRDR		R	2	D	1	R	1
SeqDRD		D	2	R	1	D	1
Conc R + D	Concurrent	R + D	2	–	–	–	–

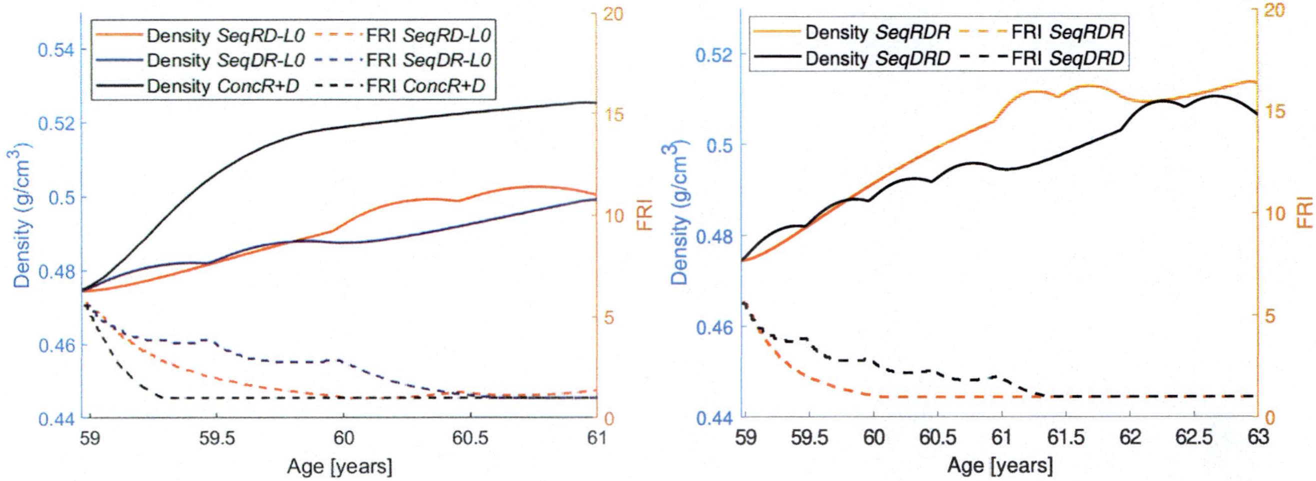

FIGURE 18.6

Comparison of the temporal evolution of apparent density (solid lines) and FRI (dashed lines) during treatments: *SeqRD* (left, red), *SeqDR* (left, blue), *ConcR* + *D* (left, black), *SeqRDR* (right, red) and *SeqDRD* (right, black). FRI is evaluated for an overload equal to 4 times the homeostatic stress. The start of treatment was set at age 59 and $t_{meno} = 49$ years.

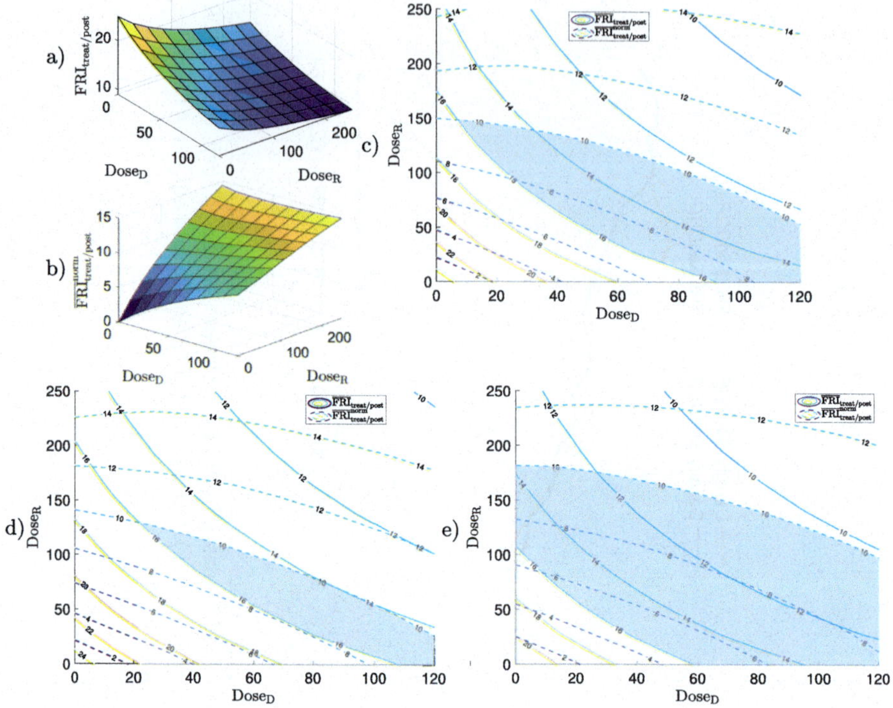

FIGURE 18.7

(a) FRI averaged through treatment and 5 years post-treatment, $\overline{\mathrm{FRI}}_{\mathrm{treat/post}}$; (b) FRI normalized with doses, $\overline{\mathrm{FRI}}^{\mathrm{norm}}_{\mathrm{treat/post}}$ (see Eq. 18.23); (c) Contour plot of both variables ($\overline{\mathrm{FRI}}_{\mathrm{treat/post}}$ in solid lines and $\overline{\mathrm{FRI}}^{\mathrm{norm}}_{\mathrm{treat/post}}$ in dashed lines), plotted against doses of Rmab and Dmab in mg. For those three figures, $t_{meno} = 49$ and BW $= 60$ kg. (d) and (e) influence of age at menopause and patient's body weight on contour plot of $\overline{\mathrm{FRI}}_{\mathrm{treat/post}}$ (solid lines) and $\overline{\mathrm{FRI}}^{\mathrm{norm}}_{\mathrm{treat/post}}$ (dashed lines). (d) $t_{meno} = 54$ and BW $= 60$ kg. (e) $t_{meno} = 49$ and BW $= 70$ kg. For all the figures, the start of treatment was set at age 59. The shaded area represents the region ($\mathrm{Dose}_D = \mathrm{Dose}_R$) producing $\overline{\mathrm{FRI}}_{\mathrm{treat/post}} < 16$ and $\overline{\mathrm{FRI}}^{\mathrm{norm}}_{\mathrm{treat/post}} < 10$.

and 5 years post-treatment $\left(\overline{\mathrm{FRI}}_{\mathrm{treat/post}}\right)$. It is observed that for very high doses the drop in FRI is no longer as significant as for low doses. Since doses cannot be increased indefinitely, due to adverse effects of the drugs, they should be optimized to achieve the highest efficacy per unit mass of drug injected. The following variable has been defined to evaluate this efficacy:

$$\overline{FRI}^{norm}_{treat/post} = \frac{1}{2}\overline{FRI}_{treat/post}\left(\frac{Dose_R}{210} + \frac{Dose_D}{60}\right) \quad (18.23)$$

The relative effect of drug doses is considered through the factor in parentheses, which penalizes those treatments in which the reduction in fracture risk is only achieved with a high dose. This expression ensures that $\overline{FRI}^{norm}_{treat/post} = \overline{FRI}_{treat/post}$ for WHO-approved doses. Single-drug treatments are represented by Eq. (18.23) simply considering the other dose as null.

$\overline{FRI}^{norm}_{treat/post}$ is plotted against doses in Fig. 18.7b, where it can be noted that $\overline{FRI}^{norm}_{treat/post}$ equals zero in the non-treated case ($Dose_D = Dose_R = 0$), though in this case $\overline{FRI}_{treat/post}$ is very high (see Fig. 18.7a). For this reason, both variables $\overline{FRI}_{treat/post}$ and $\overline{FRI}^{norm}_{treat/post}$ must be jointly minimized. Indeed, an ideal treatment must minimize the average fracture risk, i.e., $\overline{FRI}_{treat/post}$, but it must also achieve the lowest FRI at the lowest possible dose, thus minimizing $\overline{FRI}^{norm}_{treat/post}$ and therefore the likelihood of adverse effects. Fig. 18.7c shows a contour plot of both variables against doses and can be used to determine the combination of doses that keep $\overline{FRI}_{treat/post}$ and $\overline{FRI}^{norm}_{treat/post}$ below selected limits. For instance, the shaded area in Fig. 18.7c is the region where $\overline{FRI}^{norm}_{treat/post} < 10$ and $\overline{FRI}_{treat/post} < 16$.

The influence of patient body weight (BW) can be analyzed varying V_{cR}/F_R and V_{cD}/F_D of PK-PD models (see Table 18.1 and Eqs. (18.12) and (18.17)), while the age at menopause (t_{meno}) determines TSM in Eq. (17.23). The contour plot of Fig. 18.7c was obtained for BW = 60 kg and $t_{meno} = 49$ years and can be compared with that of Fig. 18.7d, obtained for BW = 60 kg and $t_{meno} = 54$ years, and that of Fig. 18.7e, obtained for BW = 70 kg and $t_{meno} = 49$ years. Patients undergoing a late menopause need lower doses to reduce the average fracture risk $\left(\overline{FRI}_{treat/post} < 16\right)$, in accordance with other studies (Ruiz-Lozano et al., 2024). On the contrary, heavier patients need higher doses to reduce the fracture risk and the efficiency of the WHO-approved dose is slightly reduced, making more difficult to find an optimal treatment, i.e., the feasible region for given bounds of $\overline{FRI}_{treat/post}$ and $\overline{FRI}^{norm}_{treat/post}$ is clearly reduced. An important idea emerging from this result is the possibility of optimizing treatments with a general objective function that minimizes fracture risk but limiting the doses, which could be patient-specific depending on BW, age at menopause, the sensitivity to each drug, etc.

3. Conclusions and future directions

In silico modeling and simulation are powerful tools that complement traditional methods for gathering evidence including bench-top (in vitro) testing, and animal or clinical (in vivo) studies. Many world regulatory bodies, i.e., the Food and

Drug Administration (FDA, USA), European Medicines Agency (EMA, Europe), Therapeutic Goods Administration (TGA, Australia) recognize that the use of in silico models has the potential to revolutionize the future of healthcare and is among the recognized emerging health technologies.

In this chapter, we have summarized the state-of-the-art of in silico mechanobiological PK-PD models of OP treatments. We have particularly focused on including essential mechanistic processes such as mineralization, microdamage and mechanobiological feedback on the local RVE scale. Based on three most commonly used OP drugs, i.e., Alend, Dmab and Rmab we have shown how such mechanistic PK-PD models for OP can be used either alone or in combination to predict bone mineral density changes and changes in FRI over time, which provides important insights into most effective treatment options. We also showed how these models can be applied in a patient-specific manner by adjusting variables such as body weight, time since onset of menopause among other parameters. These mechanobiological PK-PD models are continuously advancing, but there are still some modeling challenges which we discuss below.

BCPMs are typically formulated on the scale of a selected RVE. To simulate an entire bone organ response, e.g., femoral neck and/or whole vertebral body, these mechanobiological models need to be applied to the whole bone organ scale which is computationally challenging. Furthermore, bone resorption and formation biomarkers are measured in blood and/or urine and represent system responses, i.e., consequently remodeling events in the entire skeleton need to be taken into account. Hence, the single RVE scale model needs to be extended to the whole skeletal scale. The latter contributions from single RVE to whole skeleton response can be achieved as outlined in the description of the PK-model for Alend which involved consideration of different BCs.

Furthermore, the use of physical exercise as a therapeutical tool against OP, alone or in combination with drugs, must be further investigated by mechanobiological PK-PD models for its potential use in cases where drugs have failed, are contraindicated or cannot be extended.

Finally, the BCPMs need to be coupled to system calcium and phosphate homeostasis models which are highly important for simulations of bone disease such as chronic kidney disease and hyperparathyroidism and to consider nutritional aspects such as calcium intake and vitamin D exposure.

Acknowledgments

Dr. Calvo-Gallego and Dr. Martínez-Reina would like to gratefully acknowledge funding provided by MICIU/AEI/10.13039/501100011033 and by "ERDF/EU" through the research project entitled *Integración de un modelo de remodelación ósea y un modelo de homeostasis del Ca. Predicción de riesgo de fractura osteoporótica. Diseño de tratamientos de paciente-específico* (PID2024-161475OB-I00).

Dr. Pivonka would also like to gratefully acknowledge funding received through the Australian Research Council (ARC) Industrial Transformation Training Centre for Joint Biomechanics (IC190100020) and Discovery Project (DP230101404).

References

Ardawi, M. S. M., Al-Kadi, H. A., Rouzi, A. A., & Qari, M. H. (2011). Determinants of serum sclerostin in healthy pre- and postmenopausal women. *Journal of Bone and Mineral Research, 26*(12), 2812–2822. https://doi.org/10.1002/jbmr.479

Bone, H. G., Bolognese, M. A., Yuen, C. K., Kendler, D. L., Wang, H., Liu, Y., et al. (2008). Effects of denosumab on bone mineral density and bone turnover in postmenopausal women. *Journal of Clinical Endocrinology & Metabolism, 93*, 2149–2157.

Brown, J. P., & Josse, R. G. (2002). Clinical practice guidelines for the diagnosis and management of osteoporosis in Canada. *Canadian Medical Association Journal, 167*(10), 1–34.

Calvo-Gallego, J. L., Pivonka, P., Ruiz-Lozano, R., & Martínez-Reina, J. (2022). Mechanistic PK-PD model of alendronate treatment of postmenopausal osteoporosis predicts bone site-specific response. *Frontiers in Bioengineering and Biotechnology, 10*, Article 940620. https://doi.org/10.3389/fbioe.2022.940620

Chae, J., Seo, J., Mahat, B., et al. (2014). A simple pharmacokinetic model of alendronate developed using plasma concentration and urine excretion data from healthy men. *Drug Development and Industrial Pharmacy, 40*(10), 1325–1329.

Cosman, F., Crittenden, D. B., Ferrari, S., Khan, A., Lane, N. E., et al. (2018). FRAME study: The foundation effect of building bone with 1 year of romosozumab leads to continued lower fracture risk after transition to denosumab. *Journal of Bone and Mineral Research, 33*(7), 1219–1226. https://doi.org/10.1002/jbmr.3427

Cosman, F., Nieves, J. W., & Dempster, D. W. (2017). Treatment sequence matters: Anabolic and antiresorptive therapy for osteoporosis. *Journal of Bone and Mineral Research, 32*(2), 198–202. https://doi.org/10.1002/jbmr.3051

Cummings, S. R., Santora, A. C., Black, D. M., & Russell, R. G. G. (2020). History of alendronate. *Bone, 137*(115411). https://doi.org/10.1016/j.bone.2020.115411

Fisher, J. E., Rodan, G. A., & Reszka, A. A. (2000). In vivo effects of bisphosphonates on the osteoclast mevalonate pathway. *Endocrinology, 141*(12), 4793–4796. https://doi.org/10.1210/endo.141.12.7921

Fleisch, H. (2000). *Bisphosphonates in bone disease-from the laboratory to the patient*. New York: Academic Press.

Fleisch, H., Graham, R., Russell, G., & Francis, M. D. (1969). Diphosphonates inhibit hydroxyapatite dissolution in vitro and bone resorption in tissue culture and in vivo. *Science, 19*(165), 1262–1264. https://doi.org/10.1126/science.165.3899.1262

Francis, M. D., Graham, R., Russell, G., & Fleisch, H. (1969). Diphosphonates inhibit formation of calcium phosphate crystals in vitro and pathological calcification in vivo. *Science, 19*(165), 1264–1266. https://doi.org/10.1126/science.165.3899.1264

Holford, N. H. G., & Sheiner, L. B. (1982). Kinetics of pharmacologic response. *Pharmacology & Therapeutics, 16*(1), 143–166. https://doi.org/10.1016/0163-7258(82)90051-1

Jung, A., Bisaz, S., & Fleisch, H. (1973). The binding of pyrophosphate and two diphosphonates by hydroxyapatite crystals. *Calcified Tissue Research, 11*, 269–280. https://doi.org/10.1007/BF02547227

Kang, H., Hwang, S., Park, J., Kim, C., et al. (2006). HPLC method validation and pharmacokinetic study of alendronate sodium in human urine with fluorescence detection. *Journal of Liquid Chromatography & Related Technologies, 29*(11), 1589–1600.

Khan, S. A., Kanis, J. A., Vasikaranand, S., et al. (1997). Elimination and biochemical responses to intravenous alendronate in postmenopausal osteoporosis. *Journal of Bone and Mineral Research, 12*(10), 1700–1707.

Langdahl, B. L. (2020). Treatment of postmenopausal osteoporosis with bone-forming and antiresorptive treatments: Combined and sequential approaches. *Bone, 139*, Article 115516. https://doi.org/10.1016/j.bone.2020.115516

Langdahl, B. L., Libanati, C., Crittenden, D. B., Bolognese, M. A., Brown, J. P., et al. (2017). Romosozumab (Sclerostin monoclonal antibody) versus teriparatide in postmenopausal women with osteoporosis transitioning from oral bisphosphonate therapy: A randomised, open-label, phase 3 trial. *Lancet, 390*(10102), 1585–1594. https://doi.org/10.1016/S0140-6736(17)31613-6

Lin, J. H., Duggan, D. E., Chen, I. W., & Ellsworth, R. L. (1991). Physiological disposition of alendronate, a potent anti-osteolytic bisphosphonate, in laboratory animals. *Drug Metabolism & Disposition, 19*(5), 926–932. https://doi.org/10.1016/s0090-9556(25)08972-x

Lou, S., Lv, H., Yin, P., Li, Z., Tang, P., & Wang, Y. (2019). Combination therapy with parathyroid hormone analogs and antiresorptive agents for osteoporosis: A systematic review and meta-analysis of randomized controlled trials. *Osteoporosis International, 30*(1), 59–70. https://doi.org/10.1007/s00198-018-4790-4

Marathe, D. D., Marathe, A., & Mager, D. E. (2011). Integrated model for denosumab and ibandronate pharmacodynamics in postmenopausal women. *Biopharmaceutics and Drug Disposition, 32*(8), 471–481. https://doi.org/10.1002/bdd.770

Martínez-Reina, J., Calvo-Gallego, J. L., Martin, M., & Pivonka, P. (2022). Assessment of strategies for safe drug discontinuation and transition of denosumab treatment in PMO-insights from a mechanistic PK/PD model of bone turnover. *Frontiers in Bioengineering and Biotechnology, 10*, Article 886579. https://doi.org/10.3389/fbioe.2022.886579

Martínez-Reina, J., & Pivonka, P. (2019). Effects of long-term treatment of denosumab on bone mineral density: Insights from an in-silico model of bone mineralization. *Bone, 125*, 87–95. https://doi.org/10.1016/j.bone.2019.04.022

Martin, M., Sansalone, V., Cooper, D. M. L., Forwood, M. R., & Pivonka, P. (2020). Assessment of romosozumab efficacy in the treatment of postmenopausal osteoporosis: Results from a mechanistic PK-PD mechanostat model of bone remodeling. *Bone, 133*, Article 115223. https://doi.org/10.1016/j.bone.2020.115223

McClung, M. R., Brown, J. P., Diez-Perez, A., Resch, H., Caminis, J., et al. (2018). Effects of 24 months of treatment with romosozumab followed by 12 months of denosumab or placebo in postmenopausal women with low bone mineral density: A randomized, double-blind, phase 2, parallel group study. *Journal of Bone and Mineral Research, 33*(8), 1397–1406. https://doi.org/10.1002/jbmr.3452

Padhi, D., Jang, G., Stouch, B., Fang, L., & Posvar, E. (2011). Single-dose, placebo-controlled, randomized study of AMG 785, a sclerostin monoclonal antibody. *Journal of Bone and Mineral Research, 26*(1), 19–26. https://doi.org/10.1002/jbmr.173

Porras, A. G., Holland, S. D., & Gertz, B. J. (1999). Pharmacokinetics of alendronate. *Clinical Pharmacokinetics, 36*(5), 315–328. https://doi.org/10.2165/00003088-199936050-00002

Rizzoli, R., Greenspan, S., Bone, G., Watts, T. S. N., et al. (2002). Two-year results of once-weekly administration of alendronate 70 mg for the treatment of postmenopausal osteoporosis. *Journal of Bone and Mineral Research, 17*, 1988–1996.

Ruiz-Lozano, R., Calvo-Gallego, J. L., Pivonka, P., & Martínez-Reina, J. (2025). Optimisation of romosozumab plus denosumab sequential treatments against postmenopausal osteoporosis. Insights from in silico simulations. *Biomechanics and Modeling in Mechanobiology.* https://doi.org/10.1007/s10237-024-01900-z

Ruiz-Lozano, R., Calvo-Gallego, J. L., Pivonka, P., McDonald, M., & Martínez-Reina, J. (2024). An in-silico approach to elucidate the pathways leading to primary osteoporosis: Age-related vs. postmenopausal. *Biomechanics and Modeling in Mechanobiology, 23*(4), 1393–1409. https://doi.org/10.21203/rs.3.rs-3432139/v1

Saag, K. G., Petersen, J., Brandi, M. L., Karaplis, A. C., Lorentzon, M., et al. (2017). Romosozumab or alendronate for fracture prevention in women with osteoporosis. *New England Journal of Medicine, 377*(15), 1417–1427. https://doi.org/10.1056/NEJMoa1708322

Sato, M., Grasser, W., Endo, N., et al. (1991). Bisphosphonate action. Alendronate localization in rat bone and effects on osteoclast ultrastructure. *Journal of Clinical Investigation, 88*(6), Article 20952105.

Shargel, L., & Yu, A. B. C. (2016). *Applied biopharmaceutics and pharmacokinetics.* McGraw-Hill.

Takagi, Y., Inoue, S., Fujikawa, K., et al. (2021). Effect of nitrogen-containing bisphosphonates on osteoclasts and osteoclastogenesis: An ultrastructural study. *Microscopy, 70*(3), 302–307.

Valentín, J. (2002). Basic anatomical and physiological data for use in radiological protection: Reference values. *Annals of the ICRP, 32*(3–4), 1–277. https://doi.org/10.1016/S0146-6453(03)00002-2

Wijenayaka, A. R., Kogawa, M., Lim, H. P., Bonewald, L. F., Findlay, D. M., et al. (2011). Sclerostin stimulates osteocyte support of osteoclast activity by a RANKL-Dependent pathway. *PLoS One, 6*(10), Article e25900. https://doi.org/10.1371/journal.pone.0025900

Yu, Z., Surface, L. E., Park, C. Y., et al. (2001). Inhibition of bone resorption by alendronate and risedronate does not require osteoclast apoptosis. *Bone, 29*(6), 553–559.

Yu, Z., Surface, L. E., Park, C. Y., et al. (2018). Identification of a transporter complex responsible for the cytosolic entry of nitrogen-containing bisphosphonates. *eLife, 7,* Article e36620.

PART

4

Bone diseases, fractures, and frontiers of bone biomechanics

CHAPTER

Pathology and management of bone fractures 19

Peter Augat[1,2]

[1]*Paracelsus Medical University, Institute of Biomechanics, Salzburg, Austria;*
[2]*Berufsgenossenschaftliche Unfallklinik Murnau, Institute of Biomechanics, Murnau, Germany*

List of abbreviations

BMP	Bone morphogenetic protein
CT	Computed Tomography
FGF	Fibroblast growth factors
IL	Interleukin
TGF	Transforming growth factor
VEGF	Vascular endothelial growth factor

1. Introduction

Bone fractures are a common and significant medical issue, affecting individuals of all ages and backgrounds. While most fractures are neither life-threatening nor associated with chronic disease, certain types—such as hip fractures—are associated with increased morbidity and mortality, particularly in older adults. In this population, the elevated risk is often linked to the presence of multiple comorbidities, which can complicate recovery and increase mortality. For a fracture to heal properly, the broken bone must be immobilized to some extent, either through surgical fixation or conservative management. With adequate immobilization fractures generally heal well, often restoring bone to its original condition (Einhorn, 1995) mainly due to the musculoskeletal system's inherent repair capacity. Complications in healing can lead to extended immobilization, reoperation, and reduced functional capabilities. Serious issues like non-unions or persistent infections can even result in limb amputations. The health-care costs associated with fracture care and rehabilitation are a major burden for individuals, families, societies, and health-care systems. Fracture healing complications can escalate medical expenses and lead to

Biomechanics and Mechanobiology of Bone. https://doi.org/10.1016/B978-0-443-26739-0.00022-3

significant economic losses due to the need for repeated surgical interventions and prolonged work absences.

Understanding the pathology of bone fractures is essential for effective diagnosis, treatment, and rehabilitation. Accurate diagnosis is supported by the use of appropriate classification systems, which are based on the radiographic appearance of the fracture and include an assessment of the associated soft tissue damage. Effective classification schemes offer a meaningful and reliable description of the fracture and, more importantly, guide treatment decisions for the specific fracture situation. The choice of treatment is influenced by numerous factors including the anatomic location of the fracture, its severity, the patient's overall health status, and importantly, the preference of the patient. Numerous factors influence the fracture repair process. While some, such as initial trauma and patient-specific risk factors, are unchangeable, others can be modified or influenced by treatment approaches. A thorough understanding of underlying repair mechanisms, along with knowledge of the biological and mechanical factors that affect healing, is essential for optimizing fracture treatment (Buckwalter et al., 1996). Finally, the importance of aftercare and rehabilitation in ensuring successful fracture treatment is often overlooked. Effective aftercare strategies play a pivotal role in restoring optimal function and preventing long-term complications. These strategies encompass monitoring the progress of healing, managing pain effectively, and promoting mobility.

This chapter offers a comprehensive exploration of bone fractures, encompassing their epidemiology, various treatment options, and the diverse factors that play roles in the healing process. It will also cover advanced treatment approaches for fractures and discuss how physiological and environmental factors influence the recovery process.

2. Epidemiology of fractures

2.1 Incidence and burden of fractures

Bone fractures are a significant global public health issue, with a steady increase in absolute numbers as a result of population growth and aging. It is estimated that around 187 million fractures occur globally each year with considerable differences in incidence rates across regions Fig. 19.1 (Wu et al., 2021). Incidence rates strongly depend on anatomic location, sex and age (Fig. 19.2). Across all age groups, the top three anatomical sites for new fractures are the lower leg (patella, tibia, fibula, ankle), the lower arm (radius, ulna), and the upper leg (hip and femur), with incidence rates of 4.4, 4.0, and 3.6 fractures per 1000 person-years, respectively (Wu et al., 2021). The single location with the highest fracture incidence rate has consistently been found to be the distal radius, followed by the proximal femur (Bergh et al., 2020; van Staa et al., 2001). Males, for all ages combined haver higher rates of fractures than females. While fracture rates in females remain low from birth until the age of 55 after which point they increase, fracture rates in males peak between 15

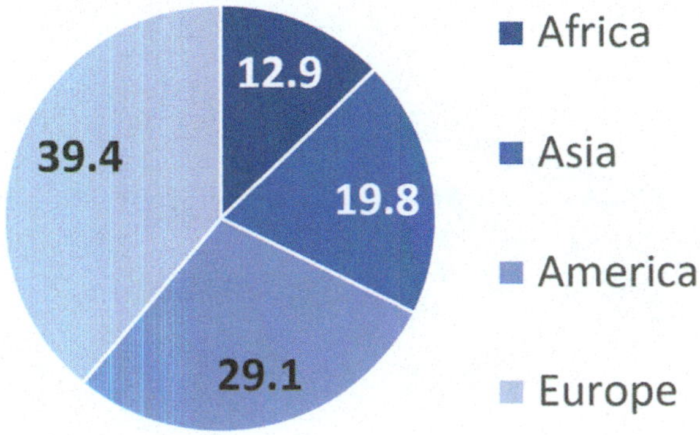

FIGURE 19.1 Global incidence of fractures

Global incidence of fractures across different world regions, presented as the number of fractures per 1000 persons per year (Wu et al., 2021).

and 40 years and increase later in life at the age of 70 years (Wu et al., 2021). This peak observed for fractures in males is likely due to the fact that in many locations around the world males work outside home in more dangerous jobs, and more hazardous sports activities and also are involved more in violence (James et al., 2020). The incidence rates dramatically increase in older age groups, predominantly in females being as high as 67 fractures per 1000 person-years in females of 80 years and older, highlighting the need for improved injury prevention, screening, and treatment of postmenopausal osteoporosis (Fuggle et al., 2024; McCloskey et al., 2024).

Fractures cause a considerable burden to individuals sustaining them, primarily due to a significant post-fracture disability and reduction in the quality of life. Key factors contributing to post-fracture disability include loss of muscle mass, strength and function (Zhao et al., 2013). Most of this muscle loss occurs within the first 3 months after a fracture due to reduced mobility and can amount to about 5% of total body lean mass (Fox et al., 2000). Persisting disabilities can further compound these issues, limiting a person's ability to perform daily activities and maintain independence. For elderly patients sustaining a hip fracture, less than half return to their pre-injury level of functional activity, and 20% lose their independence due to reduced functional capacities (Dyer et al., 2016).

Even if fractures heal successfully, they can lead to serious health complications. Fractures involving the joint are at a substantially increased risk of developing post traumatic osteoarthritis, characterized by degeneration of the articular cartilage and subchondral bone, often leading to pain, joint stiffness, and disability (Thomas et al., 2017). Approximately 12% of all cases of symptomatic osteoarthritis are related to previous joint injury or fracture (Brown et al., 2006) with the ankle and the knee being the most frequently affected joints. It is estimated that around one third of intra-

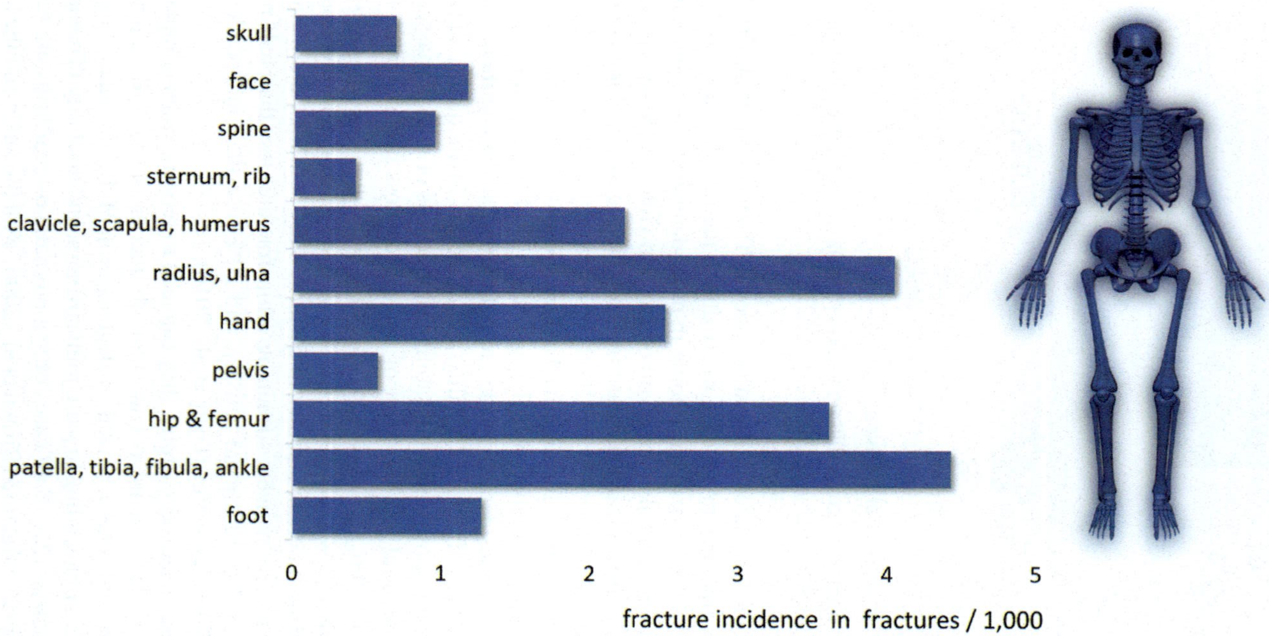

FIGURE 19.2 Local incidence of fractures

Incidence of fractures for various different anatomical regions, presented as the number of fractures per 1000 persons per year (Wu et al., 2021).

articular fractures of the knee will lead to post-traumatic osteoarthritis. For ankle fractures, the incidence of post-fracture osteoarthritis is even higher, reaching up to 70% in patients with additional risk factors such as age or obesity (Thomas et al., 2017).

2.2 Causes and risk factors for fractures

Bone is designed as a composite structure to withstand maximum loading with minimal weight (Currey, 2005). The mineral phase provides stiffness and strength, while the collagenous phase modulates the extreme brittleness of the mineral and increases toughness. This architecture enables optimized load transfer in daily life, incorporating a built-in safety factor for resilience (Huiskes et al., 2000). During everyday activities a bone deforms elastically and experiences strains of less than 0.2% (Lanyon et al., 1975), which are considered to be physiologic strain levels (Frost, 2003). As the applied load increases, the bone deforms plastically beyond its yield point, leading to the formation of microcracks. Microcracks are microscopic fissures (50–100 μm) within the bone matrix and are detected and repaired through the bone remodeling process (Seref-Ferlengez et al., 2015). Bone fractures can occur either when these microcracks accumulate beyond the bone's capacity for repair or when the single force exerted on the bone exceeds its strength. Bone failure typically begins when singular loading events cause strains exceeding 1%–2%, corresponding to ultimate stress levels of approximately 200 MPa for cortical bone (Morgan et al., 2018) and 1–10 MPa for trabecular bone, depending on factors such as age and bone density (Goldstein, 1987). Bone failure is induced by compressive loads resulting in compressive and shear stress in joints and vertebral bodies, or by bending moments resulting in shear and tensile stresses in long bones. The compressive strength of bone is primarily determined by its mineral content, which degrades as bone mineral is lost with aging or during menopause. The tensile strength and toughness of bone also depend on the quality and quantity of collagen fibrils, which deteriorate and change their composition with aging (Augat & Schorlemmer, 2006; Zioupos et al., 1999). As a result, bone strength generally decreases as we age, primarily due to a loss in bone mineral density (Osterhoff et al., 2016) but also due to cortical thinning, increased brittleness and decreased toughness (Augat & Schorlemmer, 2006).

In young and middle aged individuals, the leading causes of fractures are related to high or medium energy **trauma**. Primary causes include sports and recreational injuries, traffic accidents, falls, work-related injuries and acts of violence (Court-Brown et al., 1998). The fracture patterns resulting from traumatic events vary widely depending on the direction, magnitude, and nature of the force applied. High energy trauma or ballistic injuries often result in open wounds and extensive bone damage, frequently leading to bone defect or bone loss. Typical fracture locations in young and middle aged individuals include the foot (metacarpal bones), the lower leg, the hand but also the spine and the clavicle (Court-Brown & Caesar, 2006). Often long bone fractures in younger individuals are not isolated but are associated with other fractures or other musculoskeletal injuries, making these patients more challenging to adequately treat (Court-Brown et al., 1998).

Repetitive loading of otherwise healthy bone can lead to **stress or fatigue fractures** most frequently observed at the tibia, tarsal bones and metatarsals (Snyder et al., 2006) of athletes or military personnel. The stress fracture rate is highest for endurance activities with high impact, such as cross-country running, marathon running or military marches (Aicale et al., 2018). These overuse injuries are caused by the cumulative effect of repeated mechanical stress on bones causing accumulation of microdamage and overwhelming the bone's ability to repair itself. Increasing training volume and intensity elevates the risk for sustaining a stress fracture (Snyder et al., 2006). Women are more likely to suffer stress fractures than men, which might be related to anatomical predisposition, ovulatory disturbances and irregular steroid hormone levels triggered by high intensity activities (Saunier & Chapurlat, 2018).

A fracture occurring under normal physiological loading conditions or from minor impacts suggests that the bone is somehow weakened. Fractures that occur under normal mechanical loads due to a pre-existing weakening of the bone are called **insufficiency fractures**. This weakening can arise from a variety of factors such as osteoporosis, specific medical treatments, or various pathological disorders. Understanding the underlying causes of bone weakening is crucial for proper diagnosis and effective treatment of fractures. When an insufficiency fracture is localized at a focal neoplasm, which may be benign or malignant, it is classified as a **pathologic fracture**. Additionally, insufficiency fractures associated with metabolic bone diseases such as osteopetrosis, osteomyelitis or Paget's disease, are sometimes also referred to as pathologic fractures (Marshall et al., 2018). Pathological fractures often indicate advanced disease and require multidisciplinary management.

With advancing age, bones undergo significant changes in mass, architecture, and material properties, primarily due to **osteoporosis**. These changes result in a substantial reduction in mechanical competence. By the age of 80, the strength of the proximal femur is reduced by more than 50% compared to peak values in early adulthood (McCalden et al., 1997). Even more pronounced is the decline in spinal strength, with reductions of up to 70% reported over the course of a lifetime (Mosekilde et al., 1987). These alterations in bone properties significantly increase the risk of fractures, making low-impact incidents that are benign in younger individuals far more hazardous in the elderly. In addition to bone fragility, the mechanical loads acting on the skeleton change with age. Deterioration in muscle strength and coordination leads to a higher risk of falls, and older adults often have a diminished capacity to protect themselves during such events. The energy imparted during a fall frequently exceeds the threshold required to cause a fracture, particularly in the femur. Without adequate soft tissue cushioning, muscle contraction, or compensatory movement, the force generated during a fall will likely result in a hip fracture (Van Den Kroonenberg et al., 1996). For elderly individuals **fragility** fractures due to falls from standing height or stumbling often are an indicator of declining overall health. Not surprisingly, these fractures are associated with high rates of mortality and morbidity. Frail patients frequently experience peri- and post-operative complications resulting in prolonged stay in the hospital (Yan et al., 2022). Mortality rates following a fragility fracture of the hip have been reported to be 7% at one month and 30% at 1 year (Swift et al., 2016). In the first month, deaths are most commonly caused by acute conditions such as bacterial and aspiration pneumonia, followed by myocardial infarction. Over the 1 year period, mortality is primarily linked to chronic diseases, comorbidities and is often associated with hospital acquired delirium (Katsanos et al., 2023). Moreover, fractures in the elderly, particularly in individuals with osteoporosis, can also occur as a result of high-energy trauma (Augat et al., 2010). As activity levels increase in older adults due to lifestyle changes, the likelihood of accidents involving sufficient trauma to induce fractures rises. These fractures, however, may not always occur at typical osteoporotic sites but still require comparable clinical attention.

3. Types of bone fractures

Bone fractures are categorized based on their morphology and the mechanisms by which they occur (Fig. 19.3). Each fracture type presents distinct challenges and treatment considerations, making precise diagnosis critical for successful management and optimal patient outcomes (Love et al., 2011). Transverse fractures occur perpendicular to the long axis of the bone, often due to direct traumatic impact. Oblique fractures have an angled pattern or a diagonal fracture line and usually result from bending moments or from impact with larger objects. Wedge fractures result from bending moments and are characterized by a triangular or wedge-shaped defect on the compression side of the bone. They occur in long bones but also in the vertebral column and may contribute to spinal deformity, particularly in osteoporotic individuals. Spiral Fractures have a helical pattern and occur as a result of rotational forces or twisting moments. Comminuted fractures involve multiple bone fragments, typically resulting from high-energy trauma such as motor vehicle accidents. These

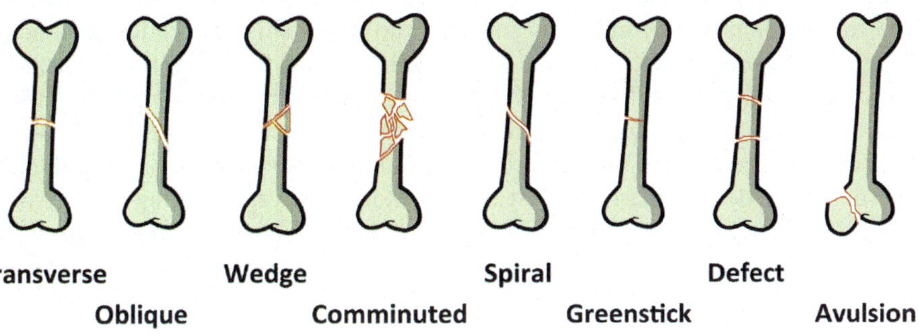

FIGURE 19.3 Fracture types

Types of bone fractures categorized by their morphological characteristics. These include transverse, oblique, wedge, comminuted, spiral, greenstick, defect and avulsion fractures.

fractures are often accompanied by significant soft tissue damage including open fracture scenarios. Greenstick fractures are incomplete fractures of a long bone commonly seen in children. Due to the higher elasticity of juvenile bone, these fractures present on only one side of the bone while the opposite side only bends and remains unfractured. Avulsion fractures occur when a fragment of bone is pulled away by a tendon or ligament under sudden force, typically during high-impact or abrupt movement but also due to muscular contraction. Stress fractures are small, often microscopic fractures caused by repetitive stress or overuse and are most commonly found in athletes and military personnel (Kaeding & Miller, 2013). Early detection is critical to prevent the fracture from progressing to a more significant break.

4. Regulatory mechanisms in fracture repair

4.1 Physiological response to fractures

Fracture repair is a regenerative process that restores the fractured bone to its original structure with the same biomechanical competence, rather than forming scar tissue. According to the 'diamond concept', five factors are required for this perfect interplay to achieve successful bone repair: mechanical stability, osteoconductive scaffolds, availability of growth factors, osteogenic cells around the fracture, and vascularity for perfusion (Giannoudis et al., 2008). Immediately after a fracture, the body triggers an acute inflammatory response, forming a fracture hematoma. This phase involves the release of inflammatory cytokines and the infiltration of various cells at the fracture site (Marsh & Li, 1999). Neutrophils, macrophages, and other immune cells infiltrate the area, removing debris and bacteria and also secrete growth factors that stimulate angiogenesis and the proliferation of osteoprogenitor cells. A healing cascade draws fibroblasts, endothelial cells, osteoblasts, and other repair cells to the fracture site, laying the groundwork for subsequent healing stages. Angiogenesis and connective tissue formation, lead to the development of a soft callus composed of collagen and fibrocartilage. This soft callus temporarily bridges the fracture gap, initiating the stabilization of the bone fragments. Over weeks to months, the soft callus is gradually replaced by a hard callus of woven bone through endochondral ossification. Osteoblasts synthesize new bone matrix, bridging the fracture gap and resulting in a radiologically visible periosteal callus. The final remodeling phase can last several months to years, during which woven bone is gradually resorbed and replaced by lamellar bone restoring the bone's pre-injury strength and structural integrity.

4.2 Biological factors

A range of biological factors impacts the different stages of fracture healing, with angiogenesis being among the most important for successful outcomes (Augat et al., 2005). Fractures disrupt blood circulation, resulting in acute necrosis and hypoxia of the surrounding bone and marrow. The restoration of local circulation, following trauma-induced interruption of blood supply, is essential for proper fracture repair. Multiple factors have been shown to influence endothelial cell proliferation and revascularization, including fibroblast growth factors (FGF), transforming growth factors (TGF-α and β), and vascular endothelial growth factor (VEGF). Age-related changes in angiogenesis result from alterations in the hemostatic cascade, growth factor expression, and endothelial cells. Conditions like diabetes impair vascularization and, consequently, fracture healing (Tanios et al., 2022).

Additional biochemical regulators during fracture repair include other growth factors, differentiation factors, hormones, cytokines, and matrix components (Dumic-Cule et al., 2018). Notable contributors include acidic and basic FGFs, insulin-like growth factors (IGF-1, IGF-2), platelet-derived growth factor (PDGF), transforming growth factor (TGF-β), and bone morphogenetic proteins (BMPs) (Simpson et al., 2006). These factors, synthesized by various cells during healing, work synergistically to regulate cell migration, proliferation, and differentiation. TGF-β, produced by osteoblasts and incorporated into the bone matrix, increases extracellular bone matrix production and enchondral bone formation. BMPs play crucial roles in chemotaxis, differentiation into cartilage, and subsequent replacement by bone. BMP-2 and BMP-7, the most studied BMPs, promote mesenchymal cell differentiation into bone- and cartilage-forming cells (Henssler et al., 2021).

Many of the molecular events in fracture healing are influenced by the mechanical environment, which is sensed by osteoblasts, osteocytes and other cells involved in the bone repair process. These cells respond to mechanical stimuli by expressing specific growth factors that drive and maintain bone formation. Studies demonstrate increased TGF-β1 levels in osteoblasts and serum during distraction osteogenesis. BMP expression is similarly influenced, with IGF-1, BMP-2, and BMP-4 being upregulated. Mechanical stimulation activates signaling pathways, such as the mitogen-activated protein kinase cascade, which regulates target gene transcription (Ma et al., 2023).

Fracture healing is also affected by various drugs (Pountos et al., 2008). Agents like cytostatics, corticosteroids, COX-2 inhibitors, certain antibiotics, and some thromboprophylactic medications can impair this process. Cytostatics suppress proliferation, corticosteroids reduce the initial inflammatory response, and COX-2 inhibitors hinder osseointegration, increasing the risk of implant failure. Conversely, certain drugs, including statins, beta-blockers, selective prostaglandin agonists, parathyroid hormone, and specific growth factors, have shown promise in accelerating fracture healing.

4.3 Mechanical factors

The process of bone healing is highly sensitive to the mechanical environment at the site of injury. Different mechanical conditions elicit distinct healing responses, with rigid fixation promoting direct (primary) healing, and flexible fixation leading to indirect (secondary) healing characterized by periosteal callus formation and endochondral ossification. Several mechanical parameters have been identified as critical to the fracture healing process, including fracture geometry, fracture type, and the magnitude, direction, and history of interfragmentary movement. These factors collectively determine local strain distributions, which serve as mechanobiological cues that regulate the repair process and cellular activity.

Fracture geometry is a key mechanical determinant of bone healing and includes both the fracture type and the size of the interfragmentary gap. Gap size, which reflects the quality of reduction, has a significant impact on healing, with larger gaps being associated with delayed or impaired osseous regeneration. Specifically, smaller gap sizes have been associated with faster and more effective healing, while larger gaps reduce periosteal callus formation and impede bone regeneration within the fracture site (Augat et al., 1998). For instance, fractures with a gap size of 2 mm or less exhibited over 50% bone and cartilage formation within the gap, achieving greater stability compared to fractures with larger gaps, which showed only approximately 20% bone and cartilage formation (Augat et al., 1998; Claes et al., 1997).

Interfragmentary movement, influenced by the mechanical loading at the fracture site and the fixation stability, plays a pivotal role in secondary fracture healing. A stiff fixation reduces interfragmentary movement, leading to limited callus stimulation. In contrast, a flexible fixation promotes callus formation, thereby enhancing the healing process. However, excessive instability can lead to nonunion, where the fracture fails to heal adequately (Claes et al., 1995; Kenwright & Goodship, 1989). The effect of interfragmentary movement on healing is modulated by the size of the fracture gap (Claes et al., 1998). For fractures with small gap sizes, increased interfragmentary movement results in a more substantial periosteal callus response. Conversely, in larger fracture gaps, excessive movement inhibits callus formation and delays fracture consolidation.

Callus formation begins with intramembranous ossification at the periosteum and endosteum, distant from the fracture gap, and later transitions to endochondral ossification as the callus expands toward the fracture. This expanding callus stabilizes the fracture by increasing the cross-sectional area of the bridging tissue (enhancing structural strength) and through tissue differentiation (improving material properties). Over time, the initial interfragmentary movement decreases as the stiffness of the callus increases. Ultimately, the hard callus bridges the fracture and reduces interfragmentary movement to a level that facilitates bone formation within the fracture gap. The rate of this reduction in movement is related to the initial movement, with larger interfragmentary displacements exhibiting a more rapid decline (Claes et al., 1998).

The direction of interfragmentary movement is another critical factor influencing fracture repair. Moderate axial interfragmentary movement has been shown to enhance healing by promoting periosteal callus formation and accelerating the overall healing process. In contrast, tensile or shear movements of comparable magnitude are less effective in promoting bone repair. Cyclic tensile strains can stimulate periosteal callus formation but do not lead to a significant enhancement in bone healing (Augat et al., 2001). Shear forces at the fracture site are particularly detrimental, leading to reduced periosteal callus formation, delayed bone formation within the fracture gap, and inferior mechanical stability compared to compressive forces (Augat et al., 2003). Despite this, shear movement has been observed to be compatible with successful healing in certain clinical contexts (Sarmiento et al., 1996). Moreover, experimental studies have demonstrated specific conditions under which shear movement can lead to superior healing outcomes compared to axial movement (BISHOP et al., 2022; Park et al., 1998). This suggests that the influence of shear versus axial motion may depend on variables such as timing, magnitude, and gap size, underscoring the complexity of mechanical influences on fracture healing.

5. Management of fractures

The management of fractures requires a systematic approach to stabilize the injury, minimize further damage, and restore the bone's function. The general strategy of fracture management is based on patient stabilization, fracture recognition, fracture reduction, fracture retention and rehabilitation.

The initial management of bone fractures is centered on **stabilizing** the patient and protecting the fracture site to prevent further injury which includes careful clinical assessment, immobilization and pain management. Immobilization is a priority, with the use of temporary splints, braces, or casts to stabilize the fracture and mitigate pain by preventing unnecessary movement. Properly immobilizing the fracture site reduces the risk of soft tissue injury and neurovascular compromise. Pain management is another cornerstone of initial fracture care. Effective analgesia not only improves patient comfort but also plays a vital role in patient compliance and cooperation during subsequent evaluation and treatment. The first essential step for developing an effective treatment plan is the accurate assessment of the patient for fracture **recognition**. After a thorough clinical evaluation imaging such as X-rays or CT scans are indispensable tools for visualizing the extent of the fracture location, pattern and displacement and also for the identification of associated injuries (Tucker, 2022). Definitive treatment of fractures is aimed at restoring anatomical alignment through fracture **reduction** and securing stable **retention** of the fracture to facilitate bone healing. Accurate fracture reduction and correct alignment are essential to obtain good functional outcome and an adequate healing response. Only with correct functional alignment the load transfer through the fracture and through the adjacent joints can be reconstituted.

5.1 Fracture fixation

The choice of treatment is influenced by the fracture type and location, as well as patient-specific factors such as age, comorbidities, activity level and patient expectations. For simple fractures with minimal displacement, where closed reduction can be achieved, conservative management using stabilization with a cast or brace can be appropriate (Sarmiento et al., 1995). This non-surgical treatment is particularly effective in pediatric patients, given their high bone healing potential. However, a significant drawback of conservative management is the risk of fracture displacement or malalignment in the course of healing. To mitigate these risks, regular and careful clinical follow-up and X-ray imaging is essential for early detection of any misalignment. Additionally, prolonged immobilization with a cast or brace may lead to muscle atrophy and joint stiffness (Drakos & Murphy, 2014).

Cases where fracture retention cannot be maintained by conservative management require fixation with osteosynthesis implants, which can be applied below (internally) or above the skin surface (externally). **Internal osteosynthesis** techniques include the use of metal plates, bone screws, rods, wires or intramedullary nails. When closed reduction of the fracture is achievable, these implants are inserted through surgical approaches that avoid direct access to the fracture site. Intramedullary nails are inserted through entry points in the bone ends, while minimally invasive plating utilizes small portals along the bone shaft to insert the plate and also the screws that secure the plate. Avoiding exposure of the fracture site preserves the local blood supply and protects the biologic factors naturally present in the fracture environment that are crucial for bone healing. This enhances the natural healing potential and minimizes soft tissue disruption, leading to improved outcomes and faster recovery.

Direct access to the fracture site becomes necessary when closed reduction methods are insufficient for achieving proper anatomical alignment. In particular fractures with significant displacement, comminution, or those involving joint surfaces often require surgical intervention with open access to the fracture site. This allows for precise manipulation and realignment of fracture fragments to restore anatomical alignment. Following the open reduction of the fracture internal fixation with osteosynthesis techniques is applied. This method, known as open reduction and internal fixation (**ORIF**) enables accurate reduction and alignment of the fracture while securing the fracture fragments in their correct anatomical positions (Augat & von Rüden, 2018; Wahnert and Gehweiler, 2017).

External fixation is primarily used in severe open fractures, cases with extensive soft tissue damage, or when immediate internal fixation is contraindicated. The procedure involves the insertion of pins or screws into the bone through the skin, which are then connected to an external frame that stabilizes the fracture (Bliven et al., 2019). This method allows for stabilization while leaving the soft tissues accessible for necessary interventions, such as wound care or grafting. External fixation is also valuable as a temporary measure in polytrauma patients, allowing for stabilization while other life-threatening injuries are managed before definitive surgical repair.

5.2 Fracture fixation techniques

Internal osteosynthesis techniques provide primary stability to fractured bones, enabling early mobilization. This is particularly beneficial because early load-bearing and weight-bearing have been shown to accelerate recovery following a fracture. By securely holding the fracture fragments in their correct anatomical positions, internal fixation reduces the need for prolonged immobilization, which is often associated with complications such as joint stiffness, muscle atrophy, and impaired function. These complications can hinder full recovery and prolong rehabilitation. Furthermore, by maintaining proper alignment and stability, internal fixation supports efficient bone healing and lowers the risk of malunion or delayed union.

Osteosynthesis techniques for internal fracture fixation encompass a variety of options, including screws, wires, plates, nails, and even joint replacements. These techniques are tailored to address the specific biomechanical demands and anatomical considerations of each fracture type. Two primary principles guide these techniques: **load-bearing osteosynthesis** and **load-sharing osteosynthesis** (Figure 19.4).

Load-bearing osteosynthesis relies on strong fixation systems that assume all functional loads at the fracture site. The resulting stability is primarily determined by the mechanical properties of the hardware employed, thereby allowing precise control over the degree of deformation. This approach is well-suited for fractures where the bone cannot effectively contribute to load sharing, such as cases with comminution, bone defects, or insufficient bone stock. By promoting controlled

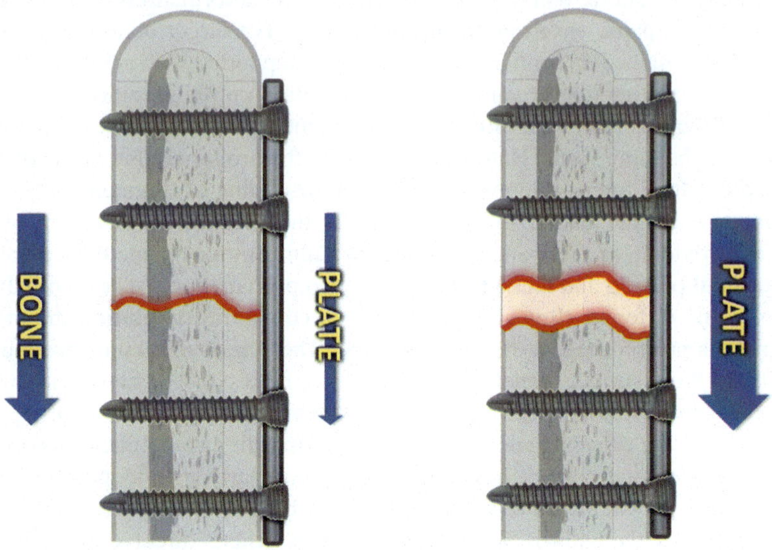

FIGURE 19.4 Load sharing versus load bearing osteosynthesis

The two fundamental principles of fracture fixation are load-sharing osteosynthesis (left) and load-bearing osteosynthesis (right). In load-sharing osteosynthesis, the bone and the fixation device share the mechanical load, whereas in load-bearing osteosynthesis, the fixation device bears the entire mechanical load.

interfragmentary motion of the fracture site, this technique facilitates a healing process, characterized by the formation of periosteal callus tissue. As the fracture site undergoes dynamic deformation, bone formation is typically preceded by the development of cartilaginous tissue at the site of injury. This sequential process is referred to as secondary fracture healing, represents the body's natural biological mechanism for restoring bone integrity.

In contrast, **load-sharing osteosynthesis** allows the bone to share the load with the fixation hardware. This principle operates through frictional contact and compression between the bone surfaces in combination with the mechanical support provided by the osteosynthesis implant. By combining load-bearing capacity of both the bone and the implant, this approach provides significant stability with minimal deformation at the fracture site. The minimal amount of deformation at the fracture site enables bone to directly bridge the gap and fill the fracture, a process commonly referred to as primary bone healing. These two principles form the foundation for selecting the most appropriate fixation method based on the desired balance between mechanical support and biological healing.

Screw osteosynthesis serves as the corner stone of load-sharing osteosynthesis, offering a versatile and effective method for stabilizing fractures. Using a lag screw or a lag screw technique, bone fragments are connected to each other and compressed as the screw threads engage with the bone opposite the fracture line. This compression creates the optimal conditions for primary fracture healing in both cortical and cancellous bone. To prevent rotational deformations, lag screws are often applied in pairs, ensuring enhanced stability. For larger bone fragments, the compression achieved through lag screws must be supported with additional plating. These plates can function in various roles depending on the specific fracture location. Neutralization plates are often employed at the fibula or the clavicle to manage torsional and bending loads. Buttressing plates at the proximal tibia or the distal femur take on shear forces to prevent the sliding or gliding of fragments.

Plating is another cornerstone technique in the surgical management of bone fractures, providing stability and promoting healing. Among the methods employed, **conventional plating** and **locked plating** are two primary approaches, with distinct principles and different clinical applications. In **conventional plating** the metallic bone plate is pressed against the bone surface. The screws that compress the plate against the bone surface generate the frictional forces that achieve the stability required for the fixation of the fracture. This technique requires contouring of the plate to match the bone's anatomy, ensuring optimal fit and stability. However, its reliance on frictional contact has limitations, particularly in osteoporotic bone or fractures with poor bone quality, where screw purchase is compromised. Conventional plating is well-suited for simple fractures where direct contact between bone fragments can be achieved and load-sharing principles can support the healing process. Specially designed gliding holes or pre-bending of the conventional plates enable the bone fragments to be compressed, thereby increasing the stability of the fixation. This interfragmentary compression also restores and helps to maintain anatomical alignment.

Locked plating, by contrast, represents an advanced fixation technique that provides angular stability of the screws through the mechanical interlock between screws and the plate (Augat & von Rüden, 2018). Screw heads are securely threaded or locked directly into the plate, forming a fixed-angle construct that ensures stability without relying on friction between the plate and the bone surface (Egol et al., 2004). As a result, locked plating reduces the need for precise contouring of the plate to fit the bone. Furthermore, because the plate does not rely on being tightly pressed against the bone surface, it helps preserve the bone's vascular supply, promoting better healing and reducing the risk of complications related to impaired periosteal blood flow and stress shielding. Locking plates are pre-contoured for specific locations and their application does not rely on perfect anatomic reduction because they function as a load bearing construct. This allows the plates to be inserted with a minimally invasive approach, an indirect reduction of the fracture, and sliding along the bone surface without opening the skin at the site of the fracture. The angular stability offered by the screws is especially beneficial for fractures around joints, where maintaining proper alignment and stability is critical. The locked screws can bear the loads transferred through the joint, ensuring consistent support and helping to maintain joint congruity during the healing process.

The primary distinction between conventional plating and locked plating lies in the mechanical environment created by each fixation method. In conventional plating, successful osteonal bridging requires fracture gaps to be less than 0.5 mm, with interfragmentary strain maintained below 2% (Perren, 1979). In contrast, locked plating is suitable for fracture treatment in cases where the bone fragments are not in direct contact. While larger fracture gaps may prolong the healing process, gaps exceeding 2 mm can still heal effectively under this approach (Augat et al., 1998). Unlike the absolute stability required in conventional plating, locked plating relies on relative stability. This allows for interfragmentary movement within the fracture gap, often exceeding 10% strain, facilitating the healing process (Augat & von Rüden, 2018; Tanveer et al., 2025).

Intramedullary nailing is a widely used technique for treating long bone fractures, particularly those involving the diaphysis (shaft) of the bone. This modestly invasive procedure involves the insertion of a metal rod, or nail, into the medullary canal through a distal or proximal entry point in the bone. The nail is stabilized by locking screws placed at both ends, securing its position. By avoiding direct manipulation of the fracture site, this technique preserves the periosteum, surrounding soft tissue and critical blood supply crucial for effective bone healing. The exceptional axial stability provided by intramedullary nailing enables most patients with lower extremity diaphyseal fractures to engage in immediate or early full weight-bearing activities. Consequently, this approach is regarded as the standard of care for treating the majority of long bone diaphyseal fractures (Wahnert and Gehweiler, 2017). The introduction of intramedullary reaming prior nail insertion has enabled the use of thicker nails, capable of accommodating larger and more robust locking screws. Innovations, such as multiple locking options, multiplanar locking configurations, and angular stable locking screws have enhanced the anchorage and stability of intramedullary nailing constructs, thereby increasing the versatility of intramedullary nails in fracture management (Kang et al., 2021). These advancements, particularly improved proximal and distal locking designs have expanded the application of intramedullary nailing to include the treatment of metaphyseal fractures of long bones. Another key feature of intramedullary nails is the option of so called dynamic locking, achieved by incorporating a slot rather than a hole in the nail to support the locking screw. This slot allows for either dynamization or compression of the fracture site. For dynamization, the locking screw slides within the slot under weight-bearing, thereby allowing the fracture to close. For compression, a specially designed insert integrated into the nail actively presses the locking screw along the slot, applying compressive forces across the fracture site. This ensures optimal alignment and increased mechanical stability (Högel et al., 2013). Compression of fracture ends is particularly beneficial for the treatment of transverse fractures, and has shown to result in faster bone healing and a reduction of healing complications (Gonschorek et al., 1998). Altogether, intramedullary nailing has become a versatile and reliable option for managing fractures across a wide range of anatomical locations and fracture patterns in long bones (Hierholzer et al., 2018).

5.3 Rehabilitation

Rehabilitation is a vital component of fracture management, ensuring the restoration of function, strength, and mobility post-fixation. A structured rehabilitation program includes physical therapy, occupational therapy, and weight-bearing exercises tailored to the individual needs of the patient. Physical therapy focuses on improving joint range of motion, muscle strength, and flexibility. Early mobilization, supervised by physical therapists, is essential to prevent stiffness and muscle atrophy. Rehabilitation following fractures, particularly in elderly and frail patients, should prioritize early postoperative weight-bearing. This approach significantly aids early restoration of function, minimizes complications, and improves quality of life (Fairhall et al., 2022). Restricting weight-bearing often stems from concerns about fixation failure or the risk for nonunion development, yet evidence increasingly highlights the benefits of early mobilization (Smeeing et al., 2015).

Weight-bearing plays a critical role in fracture healing, serving as both a catalyst and a potential risk factor. It stimulates bone repair by promoting callus formation and aligning mechanical stress with biological repair processes. However, excessive or premature weight-bearing can lead to complications, such as implant loosening, fragment dislocation, or fracture malalignment. Definitive weight-bearing guidelines are lacking for many extremity fractures, and this remains an area of active research (Dehghan et al., 2018). Progressive weight-bearing exercises are typically introduced gradually to stimulate bone remodeling and enhance fracture healing. Close monitoring of weight-bearing activities ensures they align with the fracture's healing stage, particularly for load-bearing bones, where proper mechanical stimulation is essential to accelerate recovery and optimize bone strength. However, immediate postoperative weight-bearing is increasingly recognized as a preferred practice for lower-limb fracture surgeries, except in rare cases where patient-specific factors necessitate temporary restrictions (Trompeter, 2020). To support effective rehabilitation, surgical planning must prioritize fixation techniques that allow for immediate mobilization. For instance, locked plating and dual-plating systems enhance stability, especially in osteoporotic or complex fractures. Clear communication with patients regarding weight-bearing expectations is crucial to avoid misinterpretation and ensure adherence to rehabilitation protocols. Finally, occupational therapy plays a complementary role by addressing the functional impact of fractures on daily activities. Interventions focus on improving the patient's ability to perform essential tasks such as self-care, mobility, and returning to work or hobbies (Simmel & Buehren, 2013). This aspect of rehabilitation is crucial for enhancing patient independence and promoting a successful transition back to normal life.

6. Complications in healing

Fracture healing complications arise from various pathological factors that impair the body's natural repair processes. Comorbidities like diabetes (Pscherer et al., 2017), rheumatoid arthritis (Claes et al., 2012), and vascular disease compromise healing due to poor blood supply, chronic inflammation, and weakened immune responses. Lifestyle factors such as smoking (Tümen et al., 2024) and excessive alcohol consumption exacerbate these issues, with nicotine reducing blood flow and osteoblast activity, while alcohol inhibits cell proliferation and decreases bone formation (Chakkalakal, 2005). Genetic factors also play a significant role, as conditions like osteogenesis imperfecta and variations in gene expression can affect bone quality and repair capacity. These pathological factors significantly contribute to, or even directly cause, many complications in the healing of bone fractures leading to delayed recovery and poor outcomes.

In osteoporotic fractures, the reduced bone density and weakened bone microarchitecture often lead to complications involving failure of internal fixation. Osteoporotic bone not only exhibits diminished mechanical strength and increased fracture susceptibility but also compromises the anchorage and stability of fixation implants (Goldhahn et al., 2008). This results in phenomena such as implant subsidence, hardware cut-through or cut-out, and eventual failure of fracture stabilization (Augat et al., 2005; von Rüden & Augat, 2016). Periprosthetic fractures, which occur around existing implants, are primarily caused by biomechanical stresses exceeding the capacity of osteoporotic bone, largely due to significant disparities in elasticity between metallic implants and the surrounding bone tissue (Hollensteiner et al., 2019). With an aging demographic and the rising prevalence of hip and knee arthroplasty procedures, the incidence of these fractures is increasing at an alarming rate. The management of osteoporotic and periprosthetic fractures necessitates the application of sophisticated surgical techniques and the implementation of advanced fixation systems specifically designed to address the challenges of reduced bone quality. These cases require a personalized approach, considering bone density, comorbidities, and functional needs to optimize outcomes and address their complexities.

6.1 Malunion, delayed union, non-union

Fracture healing is a complex process that can sometimes result in complications affecting the quality and timing of union. **Malunion** occurs when a bone heals in the wrong position, often due to improper stabilization. **Delayed union** refers to a slower-than-expected healing process in which the fracture ultimately achieves union despite the prolonged duration. **Non-union**, in contrast, occurs when a fracture fails to heal entirely, posing a significant clinical challenge that results in persistent pain and substantial limitations in movement and functionality.

Malunions occur when fractures heal in a clinically unacceptable position, resulting in deformities such as shortening, lengthening, abnormal rotation, or angulation (Patel et al., 2022). These complications are more common in conservatively treated fractures when casting is improperly or inadequately applied, or when patients fail to sufficiently immobilize the affected limb. Malunions may also arise after surgical fixation due to imperfect fracture reduction or alignment, as well as implant loosening or fragment dislocation. These malunion deformities disrupt the mechanical alignment of the limb, leading to chronic pain, altered gait, and an increased risk of post-traumatic arthritis. Additionally, the functional limitations caused by malunions can significantly impair patients' quality of life, often necessitating corrective procedures to restore proper alignment and functionality.

Delayed union refers to the failure of a fracture to consolidate within the expected time frame, which is influenced by the anatomical site, the nature of the fracture, and patient-specific factors such as age, comorbidities, and overall health status. Delayed union is characterized by ongoing biological activity at the fracture site, though the progression toward complete healing is slower than anticipated. This condition may arise due to inadequate stabilization, insufficient vascular supply, systemic factors such as malnutrition or smoking, or complications like infection. Radiographic evidence in delayed union typically reveals incomplete bridging of the fracture gap accompanied by callus formation, indicating potential for eventual recovery. Treatment approaches may involve dynamization or stabilization of fracture fixation, depending on the mechanical requirements of the fracture site. Enhancing local biological activity through biophysical modalities such as extracorporeal shockwave therapy or ultrasound stimulation can further support or reestablish the healing process (Aifantis et al., 2023). Pharmaceutical interventions such as bone morphogenetic proteins (BMPs) and teriparatide (PTH 1–34) have shown potential in enhancing fracture healing but have not demonstrated consistent clinical effectiveness across broader patient populations. Close monitoring and timely, targeted intervention remain crucial to prevent progression to non-union, which poses a more complex and persistent challenge to manage.

In contrast to delayed union, **non-union** occurs when the fracture fails to heal entirely within the expected timeframe, and biological healing activity has ceased.

Radiographic follow-ups of non-union typically demonstrate an absence of progressive callus formation or bridging bone across the fracture gap, signifying the lack of active healing. However, in plated fractures or fractures near joints, distinguishing between delayed union and non-union can be challenging (Hak et al., 2014). These fractures often heal with minimal or no periosteal callus formation, making it difficult to assess biological activity and determine the healing status accurately (Simpson & Tsang, 2018). The overall risk of non-union is approximately 2%, but certain fractures have a significantly higher risk, reaching up to 6% in some cases (Fig. 19.5). Fractures of the tibia and clavicle are among those with the highest rates of non-union. Younger and middle-aged adults experience the highest rates, while elderly individuals typically have the lowest (Fig. 19.6). Additionally, males are more frequently affected by non union than females (Mills et al., 2017).

The etiology of non-unions is multifaceted, with mechanical factors and infections being the two primary contributors. Mechanical factors account for approximately 50% of cases, while infections are responsible for about 20%–30% (Simpson & Tsang, 2018; Tsang et al., 2016). Mechanical factors often result from inadequate stabilization or excessive motion at the fracture site. Infections, particularly in open fractures, compromise the biological environment by inducing inflammation, causing tissue necrosis, and impairing vascularization.

Non-unions are classified into atrophic, characterized by minimal biological activity, and hypertrophic, defined by excessive callus formation without union. Atrophic non-unions commonly result from improperly reduced fractures with residual gaps, coupled with excessive rigidity that inhibits healing. They can also stem from biological inactivity caused by local factors such as avascular bone and

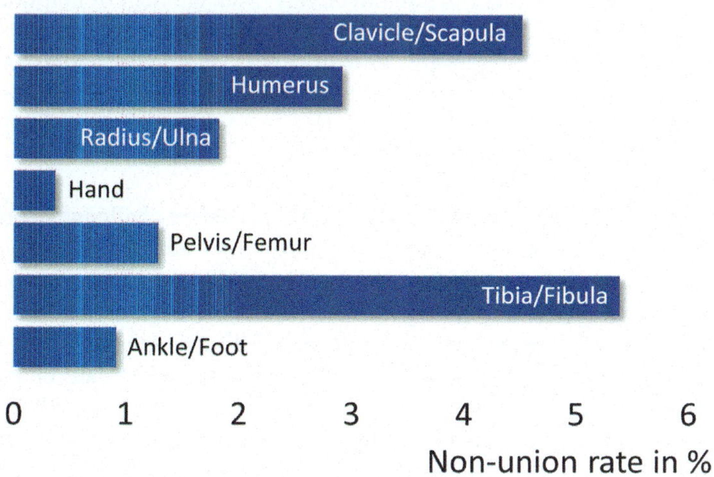

FIGURE 19.5 Non-union rates at different anatomical locations

The rate of non-union of fractures varies significantly depending on the anatomical location.

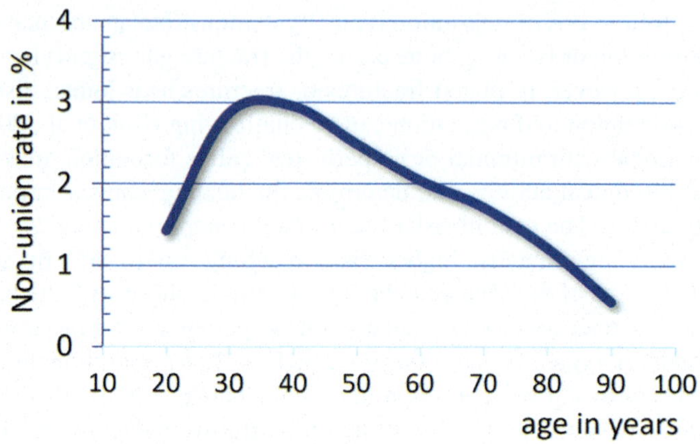

FIGURE 19.6 Non-union rate at different ages

The rate of fracture non-union varies with age, with the highest rates observed during the third and 4th decades of life and with a decrease in non-union rate with advancing age (Mills et al., 2017).

systemic host factors. Modifiable systemic factors, including chronic conditions like diabetes, hyperthyroidism, and vascular diseases, as well as the use of medications such as corticosteroids, bisphosphonates, and NSAIDs, can impair bone healing. Additionally, lifestyle factors such as alcohol and tobacco consumption should be addressed early to optimize healing potential. Hypertrophic non-unions typically result from inadequate mechanical fixation, which allows excessive movement at the fracture site. This excessive motion prevents the callus formation from progressing sufficiently to stabilize the fracture effectively. Management of non-union often requires a combination of surgical intervention, such as revision fixation, bone grafting, or the application of biologics like growth factors, along with addressing underlying systemic or local factors to reinitiate and sustain the healing process.

6.2 Infections

Fracture-related infections represent a critical and persistent challenge in orthopedic and trauma surgery, leading to extended hospitalizations, heightened morbidity, and substantially compromised patient outcomes. The incidence of fracture-related infections is influenced by a range of factors, with reported rates of less than 1% for closed fractures and increasing to as much as 30% for severely open fractures of the lower extremities (Walter et al., 2021). These infections are characterized by bacterial colonization and a chronic inflammatory response (Trenkwalder et al., 2024), which severely disrupt normal bone healing processes and may result in structural bone compromise. Management of fracture-related infections is highly complex and requires a multidisciplinary approach incorporating prolonged antibiotic regimens, advanced surgical techniques, and adherence to evidence-based guidelines (Walter et al., 2024). The primary surgical objectives are twofold: eliminating the infection and restoring both bone stability and structural integrity. This typically involves meticulous debridement of all necrotic and infected tissues, often combined with staged reconstructive procedures, which can include bone grafting and implant revision (von Rüden et al., 2024). The severity of the injury and the extent of the infection profoundly influence success rates, which in the most severe cases can be as low as 70%, underscoring the difficulty of achieving satisfactory outcomes. Furthermore, even after apparently successful treatment, many patients with long-bone fracture related infections report significantly diminished quality of life compared to normative populations, emphasizing the long-term consequences of these infections (Walter et al., 2021). Prevention, early identification, and timely intervention remain critical to reducing the incidence and impact of fracture-related infections and improving overall patient prognosis in fracture management.

7. Conclusion

Bone fractures are a common medical condition affecting individuals across all age groups, with nearly everyone at risk of experiencing a fracture during their lifetime. Fracture rates vary based on age and sex, with older adults, especially women, being particularly susceptible due to osteoporosis. While most fractures heal successfully with appropriate conservative or surgical treatment, certain fractures are prone to complications such as delayed union, malunion, or non-union. These complications often stem from comorbidities like diabetes, rheumatoid arthritis, or vascular diseases, which hinder the healing process. Lifestyle factors, including smoking and excessive alcohol consumption, further impair healing by reducing vascularization and bone formation. Additionally, mechanical issues such as inadequate stabilization or excessive movement at the fracture site exacerbate these challenges. Effective fracture treatment requires a thorough understanding and precise application of the two fundamental principles of fracture fixation: load-bearing and load-sharing osteosynthesis. Correct application of these principles supports primary healing or secondary healing through periosteal callus formation. Advances in surgical techniques, including minimally invasive procedures and enhanced fixation devices, have significantly improved patient outcomes. Comprehensive fracture management relies on accurate diagnosis, customized surgical strategies, and structured rehabilitation programs. Rehabilitation, particularly early weight-bearing and physical therapy, is essential to prevent complications like muscle atrophy and joint stiffness while providing the mechanical stimuli necessary for healing. Multidisciplinary care, involving surgeons, physiotherapists, and occupational therapists, ensures optimal recovery from fracture-related complications. Continued research into the biological and mechanical factors affecting bone repair is essential for optimizing treatment strategies, reducing the socioeconomic burden of fractures, and improving overall outcomes.

AI Disclosure

During the preparation of this work the author(s) used DEEPL for translation and to correct spelling and grammar. After using this tool/service, the author(s) reviewed and edited the content as needed and take(s) full responsibility for the content of the publication.

References

Aicale, R., Tarantino, D., & Maffulli, N. (2018). Overuse injuries in sport: A comprehensive overview. *Journal of Orthopaedic Surgery and Research, 13*(1). https://doi.org/10.1186/s13018-018-1017-5

Aifantis, I. D., Ampadiotaki, M.-M., Pallis, D., Tsivelekas, K. K., Papadakis, S. A., & Chronopoulos, E. (2023). Biophysical enhancement in fracture healing: A review of the literature. *Cureus, 15*(4). https://doi.org/10.7759/cureus.37704

Augat, P., Burger, J., Schorlemmer, S., Henke, T., Peraus, M., & Claes, L. (2003). Shear movement at the fracture site delays healing in a diaphyseal fracture model. *Journal of Orthopaedic Research, 21*(6), 1011–1017. https://doi.org/10.1016/s0736-0266(03)00098-6

Augat, P., Margevicius, K., Simon, J., Wolf, S., Suger, G., & Claes, L. (1998). Local tissue properties in bone healing: Influence of size and stability of the osteotomy gap. *Journal of Orthopaedic Research, 16*(4), 475–481. https://doi.org/10.1002/jor.1100160413

Augat, P., Merk, J., Wolf, S., & Claes, L. (2001). Mechanical stimulation by external application of cyclic tensile strains does not effectively enhance bone healing. *Journal of Orthopaedic Trauma, 15*(1), 54–60. https://doi.org/10.1097/00005131-200101000-00010

Augat, P., & Schorlemmer, S. (2006). The role of cortical bone and its microstructure in bone strength. *Age and Ageing, 35*(Suppl. 1_2), Article ii27. https://doi.org/10.1093/ageing/afl081

Augat, P., Ulrich, S., Liedert, A., & Claes, L. (2005). Mechanics and mechano-biology of fracture healing in normal and osteoporotic bone. *Osteoporosis International, 16*(S02), Article S36. https://doi.org/10.1007/s00198-004-1728-9

Augat, P., & von Rüden, C. (2018). Evolution of fracture treatment with bone plates. *Injury, 49*, S2. https://doi.org/10.1016/S0020-1383(18)30294-8

Augat, P., Weyand, D., Panzer, S., & Klier, T. (2010). Osteoporosis prevalence and fracture characteristics in elderly female patients with fractures. *Archives of Orthopaedic and Trauma Surgery, 130*(11), 1405–1410. https://doi.org/10.1007/s00402-010-1142-z

Bergh, C., Wennergren, D., Möller, M., Brisby, H., & Daniel Blank, R. (2020). Fracture incidence in adults in relation to age and gender: A study of 27,169 fractures in the Swedish fracture register in a well-defined catchment area. *PLoS One, 15*(12), Article e0244291. https://doi.org/10.1371/journal.pone.0244291

Bishop, N. E., Tami, I., Rhijn, Corveleijn, R., Schneider, E., & Ito, K. (2022). Effects of volumetric vs. shear deformation on tissue differentiation during secondary bone healing. *Translation Orthopaedic Research Society, 27.*

Bliven, E. K., Greinwald, M., Hackl, S., & Augat, P. (2019). External fixation of the lower extremities: Biomechanical perspective and recent innovations. *Injury, 50*, S10. https://doi.org/10.1016/j.injury.2019.03.041

Brown, T. D., Johnston, R. C., Saltzman, C. L., Marsh, J. L., & Buckwalter, J. A. (2006). Posttraumatic osteoarthritis: A first estimate of incidence, prevalence, and burden of disease.

Journal of Orthopaedic Trauma, 20(10), 739–748. https://doi.org/10.1097/01.bot.0000246468.80635.ef

Buckwalter, J. A., Einhorn, T. A., Bolander, M. E., & Cruess, R. L. (1996). *1996 Fractures in adults lippincott-raven healing of the musculoskeletal tissues.*

Chakkalakal, D. A. (2005). Alcohol-induced bone loss and deficient bone repair. *Alcoholism: Clinical and Experimental Research, 29*(12), 2077–2090. https://doi.org/10.1097/01.alc.0000192039.21305.55

Claes, L., Augat, P., Suger, G., & Wilke, H.-J. (1997). Influence of size and stability of the osteotomy gap on the success of fracture healing. *Journal of Orthopaedic Research, 15* (4), 577–584. https://doi.org/10.1002/jor.1100150414

Claes, L. E., Heigele, C. A., Neidlinger-Wilke, C., Kaspar, D., Seidl, W., Margevicius, K. J., & Augat, P. (1998). Effects of mechanical factors on the fracture healing process. *Clinical Orthopaedics and Related Research, 355.* https://doi.org/10.1097/00003086-199810001-00015

Claes, L., Recknagel, S., & Ignatius, A. (2012). Fracture healing under healthy and inflammatory conditions. *Nature Reviews Rheumatology, 8*(3), 133–143. https://doi.org/10.1038/nrrheum.2012.1

Claes, L. E., Wilke, H.-J., Augat, P., Rübenacker, S., & Margevicius, K. J. (1995). Effect of dynamization on gap healing of diaphyseal fractures under external fixation. *Clinical Biomechanics, 10*(5), 227–234. https://doi.org/10.1016/0268-0033(95)99799-8

Court-Brown, C. M., & Caesar, B. (2006). Epidemiology of adult fractures: A review. *Injury, 37*(8), 691–697. https://doi.org/10.1016/j.injury.2006.04.130

Court-Brown, C. M., Rimmer, S., Prakash, U., & McQueen, M. M. (1998). The epidemiology of open long bone fractures. *Injury, 29*(7), 529–534. https://doi.org/10.1016/s0020-1383(98)00125-9

Currey, J. D. (2005). Bone architecture and fracture. *Current Osteoporosis Reports, 3*(2), 52–56. https://doi.org/10.1007/s11914-005-0004-z

Dehghan, N., Mitchell, S. M., & Schemitsch, E. H. (2018). Rehabilitation after plate fixation of upper and lower extremity fractures. *Injury, 49*, S72. https://doi.org/10.1016/S0020-1383(18)30308-5

Drakos, M. C., & Murphy, C. I. (2014). Bracing versus casting in ankle fractures. *The Physician and Sportsmedicine, 42*(4), 60–70. https://doi.org/10.3810/psm.2014.11.2092

Dumic-Cule, I., Peric, M., Kucko, L., Grgurevic, L., Pecina, M., & Vukicevic, S. (2018). Bone morphogenetic proteins in fracture repair. *International Orthopaedics, 42*(11), 2619–2626. https://doi.org/10.1007/s00264-018-4153-y

Dyer, S. M., Crotty, M., Fairhall, N., Magaziner, J., Beaupre, L. A., Cameron, I. D., & Sherrington, C. (2016). A critical review of the long-term disability outcomes following hip fracture. *BMC Geriatrics, 16*(1). https://doi.org/10.1186/s12877-016-0332-0

Egol, K. A., Kubiak, E. N., Fulkerson, E., Kummer, F. J., & Koval, K. J. (2004). Biomechanics of locked plates and screws. *Journal of Orthopaedic Trauma, 18*(8), 488–493. https://doi.org/10.1097/00005131-200409000-00003

Einhorn, T. A. (1995). Enhancement of fracture-healing. *The Journal of Bone & Joint Surgery America, 77*(6), 940–956. https://doi.org/10.2106/00004623-199506000-00016

Fairhall, N. J., Dyer, S. M., Mak, J. C. S., Diong, J., Kwok, W. S., & Sherrington, C. (2022). Interventions for improving mobility after hip fracture surgery in adults. *Cochrane Database of Systematic Reviews, 2022*(9). https://doi.org/10.1002/14651858.CD001704.pub5

Fox, K. M., Magaziner, J., Hawkes, W. G., Yu-Yahiro, J., Hebel, J. R., Zimmerman, S. I., Holder, L., & Michael, R. (2000). Loss of bone density and lean body mass after hip fracture. *Osteoporosis International, 11*(1), 31–35. https://doi.org/10.1007/s001980050003

Frost, H. M. (2003). Bone's mechanostat: A 2003 update. *The Anatomical Record Part A Discoveries in Molecular, Cellular, and Evolutionary Biology, 275*(2), 1081–1101. https://doi.org/10.1002/ar.a.10119

Fuggle, N. R., Beaudart, C., Bruyère, O., Abrahamsen, B., Al-Daghri, N., Burlet, N., Chandran, M., Rosa, M. M., Cortet, B., Demonceau, C., Dere, W., Halbout, P., Hiligsmann, M., Kanis, J. A., Kaufman, J. M., Kurth, A., Lamy, O., Laslop, A., Maggi, S., … Harvey, N. C. (2024). Evidence-based guideline for the management of osteoporosis in men. *Nature Reviews Rheumatology, 20*(4), 241–251. https://doi.org/10.1038/s41584-024-01094-9

Giannoudis, P. V., Einhorn, T. A., Schmidmaier, G., & Marsh, D. (2008). The diamond concept–open questions. *Injury, 39*(S2), 5–8. https://doi.org/10.1016/S0020-1383(08)70010-X

Goldhahn, J., Suhm, N., Goldhahn, S., Blauth, M., & Hanson, B. (2008). Influence of osteoporosis on fracture fixation - A systematic literature review. *Osteoporosis International, 19*(6), 761–772. https://doi.org/10.1007/s00198-007-0515-9

Goldstein, S. A. (1987). The mechanical properties of trabecular bone: Dependence on anatomic location and function. *Journal of Biomechanics, 20*(11–12), 1055–1061. https://doi.org/10.1016/0021-9290(87)90023-6

Gonschorek, O., Hofmann, G. O., & Bühren, V. (1998). Interlocking compression nailing: A report on 402 applications. *Archives of Orthopaedic and Trauma Surgery, 117*(8), 430–437. https://doi.org/10.1007/s004020050287

Hak, D. J., Fitzpatrick, D., Bishop, J. A., Marsh, J. L., Tilp, S., Schnettler, R., Simpson, H., & Alt, V. (2014). Delayed union and nonunions: Epidemiology, clinical issues, and financial aspects. *Injury, 45*(2), S3. https://doi.org/10.1016/j.injury.2014.04.002

Henssler, L., Kerschbaum, M., Mukashevich, M. Z., Rupp, M., & Alt, V. (2021). Molecular enhancement of fracture healing - Is there a role for bone morphogenetic Protein-2, parathyroid hormone, statins, or sclerostin-antibodies? *Injury, 52*, S49. https://doi.org/10.1016/j.injury.2021.04.068

Hierholzer, C., Friederichs, J., Augat, P., Woltmann, A., Trapp, O., Bühren, V., & von Rüden, C. (2018). Entwicklung und Prinzipien der Verriegelungsmarknagelung. *Unfallchirurgie, 121*(3), 239–255. https://doi.org/10.1007/s00113-018-0461-3

Högel, F., Gerber, C., Bühren, V., & Augat, P. (2013). Reamed intramedullary nailing of diaphyseal tibial fractures: Comparison of compression and non-compression nailing. *European Journal of Trauma and Emergency Surgery, 39*(1), 73–77. https://doi.org/10.1007/s00068-012-0237-3

Hollensteiner, M., Sandriesser, S., Bliven, E., von Rüden, C., & Augat, P. (2019). Biomechanics of osteoporotic fracture fixation. *Current Osteoporosis Reports, 17*(6), 363–374. https://doi.org/10.1007/s11914-019-00535-9

Huiskes, R., Ruimerman, R., Harry van Lenthe, G., & Janssen, J. D. (2000). Effects of mechanical forces on maintenance and adaptation of form in trabecular bone. *Nature, 405*(6787), 704–706. https://doi.org/10.1038/35015116

James, S. L., Castle, C. D., Dingels, Z. V., Fox, J. T., Hamilton, E. B., Liu, Z., Roberts, N. L. S., Sylte, D. O., Bertolacci, G. J., Cunningham, M., Henry, N. J., Legrand, K. E., Abdelalim, A., Abdollahpour, I., Abdulkader, R. S., Abedi, A., Abegaz, K. H., Abosetugn, A. E., Abushouk, A. I., … Vos, T. (2020). Estimating global injuries morbidity and mortality: Methods and data used in the global burden of disease 2017 study. *Injury Prevention, 26*(1), Article I125. https://doi.org/10.1136/injuryprev-2019-043531

Kaeding, C. C., & Miller, T. (2013). The comprehensive description of stress fractures: A new classification system. *Journal of Bone and Joint Surgery, 95*(13), 1214–1220. https://doi.org/10.2106/JBJS.L.00890

Kang, N. W. W., Tan, W. P. J., Phua, Y. M. C., Min, A. T. G., Naidu, K., Umapathysivam, K., & Smitham, P. J. (2021). Intramedullary nail: The past, present and the future – A review exploring where the future May lead us. *Orthopedic Reviews, 13*(2). https://doi.org/10.52965/001C.25546

Katsanos, S., Sioutis, S., Reppas, L., Mitsiokapa, E., Tsatsaragkou, A., Mastrokalos, D., Koulalis, D., & Mavrogenis, A. F. (2023). What do hip fracture patients die from? *European Journal of Orthopaedic Surgery and Traumatology, 33*(4), 751–757. https://doi.org/10.1007/s00590-022-03250-x

Kenwright, J., & Goodship, A. E. (1989). Controlled mechanical stimulation in the treatment of tibial fractures. *Clinical Orthopaedics and Related Research, 241*, 36–47. https://doi.org/10.1097/00003086-198904000-00006

Lanyon, L. E., Hampson, W. G. J., Goodship, A. E., & Shah, J. S. (1975). Bone deformation recorded in vivo from strain gauges attached to the human tibial shaft. *Acta Orthopaedica, 46*(2), 256–268. https://doi.org/10.3109/17453677508989216

Love, J. C., Derrick, S. M., & Wiersema, J. M. (2011). *Skeletal atlas of child abuse*. Humana Press. https://doi.org/10.1007/978-1-61779-216-8

Ma, Q., Miri, Z., Haugen, H. J., Moghanian, A., & Loca, D. (2023). Significance of mechanical loading in bone fracture healing, bone regeneration, and vascularization. *Journal of Tissue Engineering, 14*. https://doi.org/10.1177/20417314231172573

Marsh, D. R., & Li, G. (1999). The biology of fracture healing: Optimising outcome. *British Medical Bulletin, 55*(4), 856–869. https://doi.org/10.1258/0007142991902673

Marshall, R. A., Mandell, J. C., Weaver, M. J., Ferrone, M., Sodickson, A., & Khurana, B. (2018). Imaging features and management of stress, atypical, and pathologic fractures. *RadioGraphics, 38*(7), 2173–2192. https://doi.org/10.1148/rg.2018180073

McCalden, R. W., McGeough, J. A., & Court-Brown, C. M. (1997). Age-related changes in the compressive strength of cancellous bone. The relative importance of changes in density and trabecular architecture. *Journal of Bone and Joint Surgery, 79*(3), 421–427. https://doi.org/10.2106/00004623-199703000-00016

McCloskey, E., Tan, A. T. H., & Schini, M. (2024). Update on fracture risk assessment in osteoporosis. *Current Opinion in Endocrinology Diabetes and Obesity, 31*(4), 141–148. https://doi.org/10.1097/MED.0000000000000871

Mills, L. A., Aitken, S. A., & Simpson, A. H. R. W. (2017). The risk of non-union per fracture: Current myths and revised figures from a population of over 4 million adults. *Acta Orthopaedica, 88*(4), 434–439. https://doi.org/10.1080/17453674.2017.1321351

Morgan, E. F., Unnikrisnan, G. U., & Hussein, A. I. (2018). Bone mechanical properties in healthy and diseased states. *Annual Review of Biomedical Engineering, 20*, 119–143. https://doi.org/10.1146/annurev-bioeng-062117-121139

Mosekilde, L., Mosekilde, Le, & Danielsen, C. C. (1987). Biomechanical competence of vertebral trabecular bone in relation to ash density and age in normal individuals. *Bone, 8*(2), 79–85. https://doi.org/10.1016/8756-3282(87)90074-3

Osterhoff, G., Morgan, E. F., Shefelbine, S. J., Karim, L., McNamara, L. M., & Augat, P. (2016). Bone mechanical properties and changes with osteoporosis. *Injury, 47*, S11. https://doi.org/10.1016/S0020-1383(16)47003-8

Park, S. H., O'Connor, K., Mckellop, H., & Sarmiento, A. (1998). The influence of active shear or compressive motion on fracture-healing. *Journal of Bone and Joint Surgery, 80*(6), 868–878. https://doi.org/10.2106/00004623-199806000-00011

Patel, I., Young, J., Washington, A., & Vaidya, R. (2022). Malunion of the Tibia: A systematic review. *Medicina, 58*(3), 389. https://doi.org/10.3390/medicina58030389

Perren, S. M. (1979). Physical and biological aspects of fracture healing with special reference to internal fixation. *Clinical Orthopaedics and Related Research, 138*, 175–196.

Pountos, I., Georgouli, T., Blokhuis, T. J., Pape, H. C., & Giannoudis, P. V. (2008). Pharmacological agents and impairment of fracture healing: What is the evidence? *Injury, 39*(4), 384–394. https://doi.org/10.1016/j.injury.2007.10.035

Pscherer, S., Sandmann, G. H., Ehnert, S., Nussler, A. K., Stöckle, U., & Freude, T. (2017). Delayed fracture healing in diabetics with distal radius fractures. *Acta Chirurgiae Orthopaedicae et Traumatologiae Cechoslovaca, 84*(1), 24–29. https://doi.org/10.55095/achot2017/003

Rüden, C.von, & Augat, P. (2016). Failure of fracture fixation in osteoporotic bone. *Injury, 47*, Article S3. https://doi.org/10.1016/s0020-1383(16)47002-6

Rüden, C.von, Wunder, J., Schirdewahn, C., Augat, P., & Hackl, S. (2024). Initial treatment of severe soft-tissue injuries in closed and open fractures to prevent fracture-related infection. *Injury, 55*, Article 111935. https://doi.org/10.1016/j.injury.2024.111935

Sarmiento, A., McKellop, H. A., Llinas, A., Park, S. H., Lu, B., Stetson, W., & Rao, R. (1996). Effect of loading and fracture motions on diaphyseal tibial fractures. *Journal of Orthopaedic Research, 14*(1), 80–84. https://doi.org/10.1002/jor.1100140114

Sarmiento, A., Sharpe, F. E., Ebramzadeh, E., Normand, P., & Shankwiler, J. (1995). Factors influencing the outcome of closed tibial fractures treated with functional bracing. *Clinical Orthopaedics and Related Research, 315*, 8–24. https://doi.org/10.1097/00003086-199506000-00003

Saunier, J., & Chapurlat, R. (2018). Stress fracture in athletes. *Joint Bone Spine, 85*(3), 307–310. https://doi.org/10.1016/j.jbspin.2017.04.013

Seref-Ferlengez, Z., Kennedy, O. D., & Schaffler, M. B. (2015). Bone microdamage, remodeling and bone fragility: How much damage is too much damage? *BoneKEy Reports, 4*. https://doi.org/10.1038/bonekey.2015.11

Simmel, S., & Buehren, V. (2013). Sequelae of severe injuries: Consequences for trauma rehabilitation. *Chirurgie, 84*(9), 764–770. https://doi.org/10.1007/s00104-013-2579-8

Simpson, A. H. R. W., Mills, L., & Noble, B. (2006). The role of growth factors and related agents in accelerating fracture healing. *Journal of Bone and Joint Surgery - Series B, 88*(6), 701–705. https://doi.org/10.1302/0301-620X.88B6.17524

Simpson, A. H. R. W., & Tsang, S. T. J. (2018). Non-union after plate fixation. *Injury, 49*, S78. https://doi.org/10.1016/S0020-1383(18)30309-7

Smeeing, D. P. J., Houwert, R. M., Briet, J. P., Kelder, J. C., Segers, M. J. M., Verleisdonk, E. J. M. M., Leenen, L. P. H., & Hietbrink, F. (2015). Weight-bearing and mobilization in the postoperative care of ankle fractures: A systematic review and meta-analysis of randomized controlled trials and cohort studies. *PLoS One, 10*(2). https://doi.org/10.1371/journal.pone.0118320

Snyder, R. A., Koester, M. C., & Dunn, W. R. (2006). Epidemiology of stress fractures. *Clinics in Sports Medicine, 25*(1), 37–52. https://doi.org/10.1016/j.csm.2005.08.005

Staa, T. P van, M Dennison, E., M Leufkens, H. G., & Cooper, C. (2001). Epidemiology of fractures in England and Wales. *Bone, 29*(6), 517–522. https://doi.org/10.1016/s8756-3282(01)00614-7

Swift, C., Ftouh, S., Langford, P., Chesser, T. S., & Johanssen, A. (2016). Interdisciplinary management of hip fracture. *Clinical Medicine, Journal of the Royal College of Physicians of London, 16*(6), 541–544. https://doi.org/10.7861/clinmedicine.16-6-541

Tanios, M., Brickman, B., Cage, E., Abbas, K., Smith, C., Atallah, M., Baroi, S., & Lecka-Czernik, B. (2022). Diabetes and impaired fracture healing: A narrative review of recent literature. *Current Osteoporosis Reports, 20*(5), 229–239. https://doi.org/10.1007/s11914-022-00740-z

Tanveer, M., Klein, K., von Rechenberg, B., Darwiche, S., & Dailey, H. L. (2025). Don't mind the gap: Reframing the Perren strain rule for fracture healing using insights from virtual mechanical testing. *Bone and Joint Research, 14*(1), 5–15. https://doi.org/10.1302/2046-3758.141.BJR-2024-0191.R2

Thomas, A. C., Hubbard-Turner, T., Wikstrom, E. A., & Palmieri-Smith, R. M. (2017). Epidemiology of posttraumatic osteoarthritis. *Journal of Athletic Training, 52*(6), 491–496. https://doi.org/10.4085/1062-6050-51.5.08

Trenkwalder, K., Hackl, S., Weisemann, F., & Augat, P. (2024). The value of current diagnostic techniques in the diagnosis of fracture-related infections: Serum markers, histology, and cultures. *Injury, 55*, Article 111862. https://doi.org/10.1016/j.injury.2024.111862

Trompeter, A. (2020). A call to arms: It's time to bear weight. *The Bone & Joint Journal, 102-B*(4), 403–406. https://doi.org/10.1302/0301-620x.102b4.bjj-2019-1482.r1

Tsang, S. T. J., Frantzias, J., Keating, J. F., Simpson, A. H. R. W., Mills, L. A., & Baren, J. P. (2016). Exchange nailing for nonunion of diaphyseal fractures of the tibia: Our results and an analysis of the risk factors for failure. *The Bone & Joint Journal, 98B*(4), 534–541. https://doi.org/10.1302/0301-620X.98B4.34870

Tucker, A. (2022). Management of common upper limb fractures in adults. *Surgery, 40*(3), 184–191. https://doi.org/10.1016/j.mpsur.2021.11.024

Tümen, L., Pollmann-Schweckhorst, L., Breinbauer, R., Hammour, M. M., Aspera-Werz, R. H., Blumenstock, G., Histing, T., Menger, M. M., Ehnert, S., & Nüssler, A. K. (2024). Smoking increases risk of complication after musculoskeletal surgery: Analysis of single immune parameter to predict complication risk. *EXCLI Journal, 23*, 967–990. https://doi.org/10.17179/excli2024-7306

Van Den Kroonenberg, A. J., Hayes, W. C., & McMahon, T. A. (1996). Hip impact velocities and body configurations for voluntary falls from standing height. *Journal of Biomechanics, 29*(6), 807–811. https://doi.org/10.1016/0021-9290(95)00134-4

Wahnert, D., & Gehweiler, D. (2017). Complications of intramedullary nailing-Evolution of treatment. *Injury, 48*(Suppl. 1), 59–63. https://doi.org/10.1016/j.injury.2017.04.032

Walter, N., Bärtl, S., Alt, V., & Rupp, M. (2024). Recent advancements and future directions in fracture related infections: A scoping review. *Injury, 55*, Article 111902. https://doi.org/10.1016/j.injury.2024.111902

Walter, N., Rupp, M., Hierl, K., Pfeifer, C., Kerschbaum, M., Hinterberger, T., & Alt, V. (2021). Long-term patient-related quality of life after fracture-related infections of the long bones. *Bone and Joint Research, 10*(5), 321–327. https://doi.org/10.1302/2046-3758.105.BJR-2020-0532

Wu, A. M., Bisignano, C., James, S. L., Abady, G. G., Abedi, A., Abu-Gharbieh, E., Alhassan, R. K., Alipour, V., Arabloo, J., Asaad, M., Asmare, W. N., Awedew, A. F., Banach, M., Banerjee, S. K., Bijani, A., Birhanu, T. T. M., Bolla, S. R., Cámera, L. A.,

Chang, J. C., … Vos, T. (2021). Global, regional, and national burden of bone fractures in 204 countries and territories, 1990–2019: A systematic analysis from the global burden of disease study 2019. *The Lancet. Healthy Longevity, 2*(9), Article e580. https://doi.org/10.1016/S2666-7568(21)00172-0

Yan, B., Sun, W., Wang, W., Wu, J., Wang, G., & Dou, Q. (2022). Prognostic significance of frailty in older patients with hip fracture: A systematic review and meta-analysis. *International Orthopaedics, 46*(12), 2939–2952. https://doi.org/10.1007/s00264-022-05605-9

Zhao, Y., Chen, S. Y., Lee, Y. C., & Wu, N. (2013). Clinical and economic characteristics of hip fracture patients with and without muscle atrophy/weakness in the United States. *Archives of Osteoporosis, 8*(1–2). https://doi.org/10.1007/s11657-013-0127-2

Zioupos, P., Currey, J. D., & Hamer, A. J. (1999). The role of collagen in the declining mechanical properties of aging human cortical bone. *Journal of Biomedical Materials Research, 45*(2), 108–116. https://doi.org/10.1002/(SICI)1097-4636(199905)45:2<108::AID-JBM5>3.3.CO;2-1

CHAPTER

Clinical advances and technical challenges of quantitative computed tomography

20

Matthias Walle, Bryn E. Matheson and Steven K. Boyd

University of Calgary, Department of Radiology, Department of Biomedical Engineering, Schulich School of Engineering, McCaig Institute for Bone and Joint Health, Calgary, AB, Canada

Abbreviations

3D	Three-Dimensional
AM	Autochthonous Muscles
BMAS	Bone Mineral Apparent Density Standard
BMD	Bone Mineral Density
CHA	Carbonated Hydroxyapatite
CT	Computed Tomography
DXA	Dual-Energy X-ray Absorptiometry
FDA	Food and Drug Administration
FE	Finite Element
Hex8	8-Node Hexahedral Element
HR-pQCT	High-Resolution Peripheral Quantitative Computed Tomography
HU	Hounsfield Unit
ICRU	International Commission on Radiation Units and Measurements
L1	First Lumbar Vertebra
NIST	National Institute of Standards and Technology
PCCT	Photon-Counting Computed Tomography
PMMA	Polymethyl Methacrylate
RCT	Randomized Controlled Trial
ROI	Region of Interest
Tet4	4-Node Tetrahedral Element
vBMD	Volumetric Bone Mineral Density

1. Introduction

Finite element (FE) has emerged as a key computational tool in musculoskeletal biomechanics, enabling the assessment of bone strength and fracture risk from clinical computed tomography (CT) scans (Verma et al., 2024). This numerical simulation technique translates CT-derived bone mineral density (BMD) and structural

Biomechanics and Mechanobiology of Bone. https://doi.org/10.1016/B978-0-443-26739-0.00015-6

information into meaningful biomechanical predictions, bridging the gap between medical imaging and patient-specific mechanical assessment. This chapter examines recent advances in CT-based FE analysis, focusing on its clinical applications, methodological developments, and impact on treatment monitoring and fracture risk assessment.

Over the past decade, CT-based FE analysis has transformed our understanding of osteoporosis treatments by providing insights that go beyond traditional density measurements. This advancement has been leveraged in evaluating newer therapeutics like romosozumab, which FE analysis has shown provides superior improvements in both trabecular and cortical bone strength compared to conventional treatments of bisphosphonates and teriparatide (Crack et al., 2024; Keaveny et al., 2017). Through FE modeling, researchers have also gained deeper insight into how antiresorptive medications work—pharmaceutical such as zoledronic acid and denosumab maintain bone stability primarily by preserving cortical integrity and reducing porosity (Crack et al., 2023; Zebaze et al., 2016). The FE technique has proven especially valuable in understanding complex treatment approaches, revealing how sequential teriparatide-denosumab therapy enhances lumbar spine stiffness (Agarwal et al., 2022), and demonstrating how different teriparatide dosing schedules achieve similar improvements in bone strength (Ganapathy et al., 2023). Importantly, FE analysis has uncovered critical site-specific treatment effects, showing that while pharmaceutical treatments like romosozumab and teriparatide effectively strengthen the spine and hip, their impact on other sites like the proximal tibia may be limited (Crack et al., 2024; Edwards et al., 2018). These mechanical insights continue to shape our approach to treating metabolic bone diseases.

Beyond treatment monitoring, FE analysis has proven valuable for clinical screening and risk assessment. Studies have validated its use in osteoporosis screening, showing that FE-derived vertebral strength thresholds predict reoperation risk in spinal fusion patients (Keaveny et al., 2023). The technique has revealed unexpected insights about bone fragility, such as the finding that weakness in type 2 diabetes stems from material rather than structural deficits (Burch et al., 1976; Kiyohara et al., 2015). FE analysis has also uncovered important sex differences in bone adaptation, showing women experience greater vertebral strength decline than men after hip fracture (Orwig et al., 2021). These findings highlight the role of FE analysis in individualized risk assessment, particularly in aging populations and those with metabolic bone disorders.

However, the clinical translation of FE modeling for bone assessment faces key challenges in standardization, model accuracy, and clinical integration (Engelke et al., 2016). Variability in density calibration methods affects the consistency of material property assignments (Bott et al., 2023), while differences in meshing strategies, element types, and failure criteria impact computational efficiency and predictive accuracy. While simplified models enable clinical feasibility, they may compromise mechanical accuracy. Validation against experimental and clinical data is essential to ensure reliability across different anatomical sites and patient populations. Additionally, segmentation quality and automation remain bottlenecks, requiring robust workflows to enhance reproducibility. For FE modeling to become a routine clinical tool, standardized protocols, efficient computational methods, and seamless integration into clinical workflows are necessary.

2. Overview of the chapter

This chapter examines the clinical translation of FE analysis for CT-based bone assessment, highlighting its potential to bridge the gap between imaging and biomechanical evaluation of bone strength. By evaluating methodological choices across the entire workflow—from image acquisition and density calibration to FE outputs (Fig. 20.1)—we aim to identify best practices that enhance clinical applicability.

A systematic literature search identified 21 relevant studies of the past decade, excluding those focused on high-resolution modalities like HR-pQCT or cone-beam CT (Table 20.1). We assessed study design (e.g., sample size, interventional or observational nature), imaging parameters (e.g., pixel size, slice thickness, anatomical site), and preprocessing steps (e.g., density calibration, segmentation methods). Additionally, we analyzed FE modeling choices, including material properties, element technology, boundary conditions, and strength assessment metrics (Table 20.2).

3. Density calibration for CT scans

To ensure accurate material properties for FE analysis, CT values must be calibrated to bone-equivalent densities. Of the 21 studies reviewed, 19 used density calibration phantoms, while two applied internal calibration methods (Table 20.2). These approaches correct for scanner-specific variations and allow for reliable biomechanical modeling. We further discuss the primary calibration techniques of phantom-based, asynchronous, and internal (phantomless) calibration below.

In CT, each pixel describes the property of a corresponding volume element, otherwise known as a voxel, in the patient. The gray scale value of a voxel is the volumetric average linear attenuation coefficient (μ) of the tissue in the voxel's location. For clinical applications, the gray scale values of the voxels in CT are calibrated to Hounsfield Units (HUs), defined as

$$HU(x,y,z) = 1000\frac{\mu_{x,y,z} - \mu_{\text{water}}}{\mu_{\text{water}}}, \tag{20.1}$$

where x, y, and z correspond to the specific 3D coordinates of the voxel (Bushberg et al., 2012).

Calibrating CT values to HUs enables the standardization and convenient expression of CT data, facilitating comparisons of data across clinical centers. However, despite calibration, HUs can vary between scanners from different manufacturers, between two scanners from the same manufacturer, and within the same scanner.

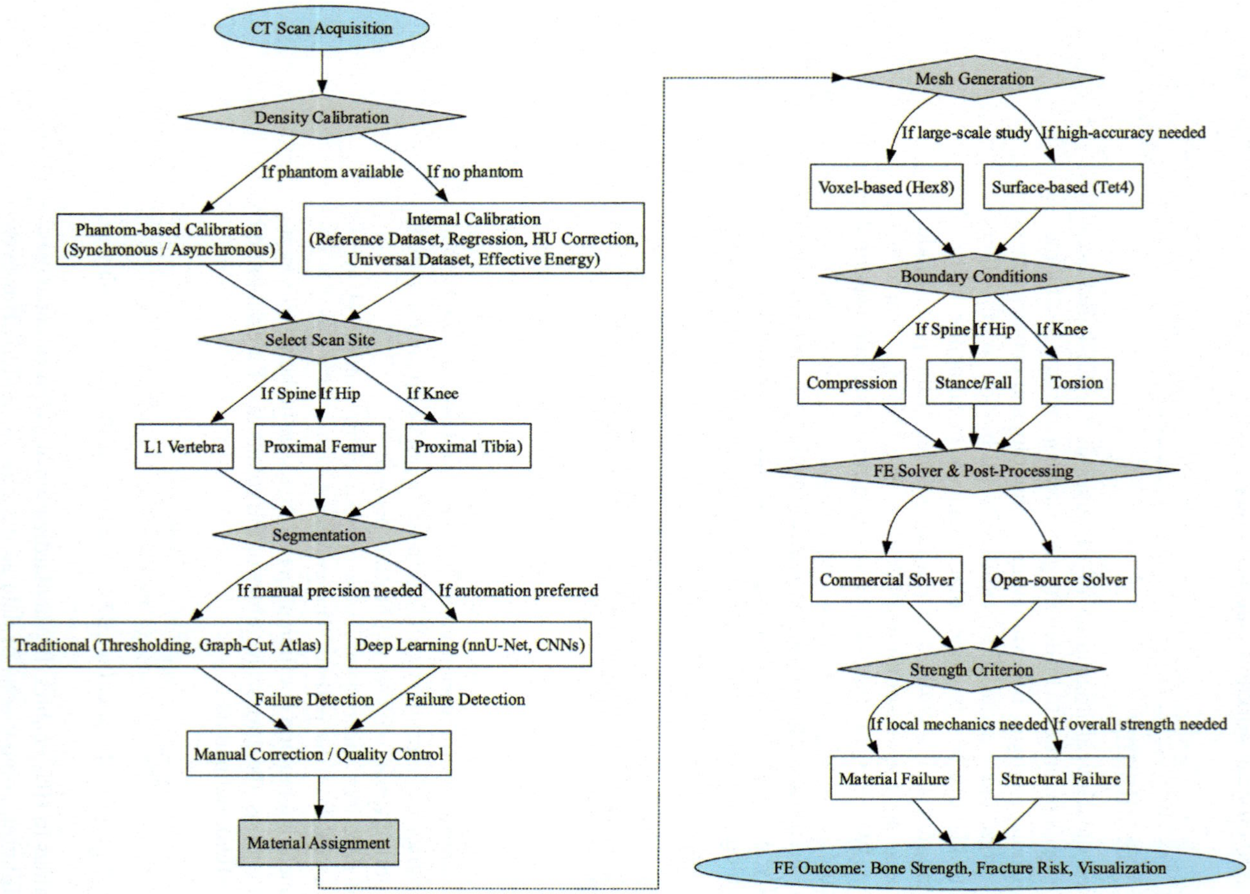

FIGURE 20.1

Schematic representation of a finite element analysis (FEA) workflow for bone biomechanics based on clinical CT imaging. First, CT scans are acquired and calibrated to map Hounsfield units (HU) to bone material properties. Next, the anatomical region of interest is selected, and bone structures are segmented using threshold-based, deep learning, or manual methods. After segmentation, material properties are assigned based on power-law density-modulus relationships. The finite element mesh is generated using either voxel-based 8-node hexahedral elements (Hex8) or surface-based 4-node tetrahedral elements (Tet4), followed by the application of boundary conditions specific to the anatomical site and loading scenario. The computational model is solved using commercial or open-source solvers, and results are analyzed to assess material or structural failure load. Finally, the output provides estimates of bone strength, fracture risk, and mechanical behavior under simulated loading conditions.

Table 20.1 Summary of search strategy.

Search component	Search terms
Databases	PubMed
Filters applied	Last 10 years
Initial results	99 articles identified
Screening process	Abstract screening for relevance to CT-based FE analysis
Exclusion criteria	Not CT-focused, case reports, small sample size (<10)
Final included studies	21 articles

These discrepancies arise from physical factors such as table height, X-ray spectrum, tube current, scan time, reconstruction kernel and filtration (Cann, 1987; Levi et al., 1982). To ensure accurate material properties and comparability across clinical sites in FE analysis of bone derived from clinical CT scans, the native in-scan HUs must be converted to their corresponding bone equivalent densities.

3.1 Phantom-based calibration

Phantom calibration is the current imaging gold standard for converting HUs to bone densities. Phantom calibration involves placing a calibration phantom with known density values of either solid CHA or liquid K_2HPO_4 under the patient in the scan field of view (Fig. 20.2A). The solid CHA or liquid K_2HPO_4 found in the phantom acts as the bone equivalent reference standard. The liquid K_2HPO_4 phantoms were the first calibration phantoms to be invented, and although K_2HPO_4 does not represent true bone material, it exhibits attenuation properties very similar to bone (Faulkner et al., 1993). Alternatively, solid CHA is one of the primary constituents of human bone tissue and has attenuation properties for X-rays that are closer to bone than K_2HPO_4. Both CHA and K_2HPO_4 calibration methods are valid for calibrating CT scans; however, the resulting vBMD values cannot be compared due to their differing bone equivalent units.

Phantom calibration requires sampling the in-scan HUs of the bone equivalent reference rods (from the calibration phantom) and comparing those values to the known reference values. A simple linear regression between the known values within the calibration phantom rods and the in-scan HUs can be created to convert the HUs to bone equivalent density (BMD). The calibration curve takes the form of

$$BMD = (\mathrm{HU} \cdot a) + b, \tag{20.2}$$

where a is the slope of the resulting calibration curve and b is the intercept. This calibration curve is applied to specific bone compartments in the image or to the entire image, creating a bone equivalent density image. The values of a and b depend on many factors, such as the composition of the X-ray spectrum emitted by the X-ray

Table 20.2 Summary of human studies using CT imaging and FE analysis, Covering study designs, imaging, modeling, outcomes, and tools.

Study	Study design						Imaging			Preprocess			Finite element modeling						
	Sample size	Inter-vent.	Rando-mized	Cont-rolled	Blin-ded	Pros-pective	Pixel size (mm)	Slice th. (mm)	Bone site	Density Calib.	Auto-matic segmet	Material symmetry	Yield criterion	Post-yield behavior	Element type	Edge length (mm)	Loading direction	Strength	Solver
Crack et al. (2024)	12					✓	0.4	1.0	Hip Knee	QRM	✓*	Orth	Hill	Lin	Hex8	1.5	Fall Torsion	d = 4% ε = 1.41%	Abaqus
Crack et al. (2023)	59	✓	✓	✓	✓	✓	0.4	1.0	Hip	QRM	✓*	Iso	vMises	pp	Hex8	1.5	Comp.	d = 4%	Abaqus
Ganapathy et al. (2023)	86	✓	✓	✓		✓	1.0	1.0	Spine Hip	K2HPO4	✓*	Asym	vMises	pp	Hex8	1 1.5	Comp. Fall	d = 2% d = 4%	VirtuOst
Keaveny et al. (2023)	469					✓	–	1.3	Spine	Internal	✓*	Asym	vMises	pp	Hex8	1	Comp.	d = 2%	VirtuOst
Agarwal et al. (2022)	41	✓	✓	✓		✓	0.9	3.0	Spine	Image		Trans	N.A.		Hex8	<3	Comp.	Stiffness	Abaqus
Doi et al. (2022)	157					✓	0.4	1.0	Spine Hip	BMAS		Asym	T: vMises, C: Drucker–Prager	Lin	Tet4	2 3	Comp. Stance/Fall	ε = 1%	MechFind
Brown et al (2021)	4093	✓	✓	✓	✓	✓	–	1.3	Spine	QRM	✓*	Asym	vMises	pp	Hex8	1	Comp.	d = 2%	VirtuOst
Ono et al. (2021)	39					✓	0.4	1.0	Spine Hip	BMAS		Asym	T: vMises, C: Drucker–Prager	Lin	Tet4	2 3	Comp. Stance/Fall	ε = 1%	MechFind
Orwig et al. (2021)	37			✓		✓	1.0	1.0	Spine	Internal	✓*	Asym	vMises	pp	Hex8	1	Comp.	d = 2%	VirtuOst
Haider et al (2019)	17	✓				✓	0.4	1.0	Knee	HA	✓*	Orth	Hill	Lin	Hex8	1.5	Torsion	ε = 1.41%	Abaqus
Edwards et al. (2018)	61	✓	✓	✓		✓	0.4	1.0	Knee	QRM	✓*	Orth	Hill	Lin	Hex8	1.5	Torsion	ε = 1.41%	Abaqus
Haider et al (2018)	101						0.4	1.0	Knee	QRM	✓*	Orth	Hill	Lin	Hex8	1.5	Comp./Tors.	Stiffness	Abaqus
Keaveny et al. (2017)	128	✓	✓	✓	✓	✓	–	1.3	Spine Hip	Mindw.	✓*	Asym	vMises	pp	Hex8	1 1.5	Comp. Fall	d = 2% d = 4%	VirtuOst
Snyder et al (2017)	211	✓		✓	✓	✓	–	1.0	Spine Hip	Mindw.	✓*	Asym	vMises	pp	Hex8	1 1.5	Comp. Fall	d = 2% d = 4%	VirtuOst
Bilezikian et al. (2016)	114	✓	✓	✓	✓	✓	–		Spine Hip	Mindw.	✓*	Asym	vMises	pp	Hex8	1 1.5	Comp. Fall	d = 2% d = 4%	VirtuOst
Burch et al. (2016)	98						1.0	1.0	Spine	Internal	✓*	Asym	vMises	pp	Hex8	1	Comp.	d = 2%	VirtuOst
Dall'Ara et al. (2016)	36						0.3	1.0	Hip	QRM	✓	Iso	Hill	Dmg	Hex8	3	Stance/Fall	F_u	Abaqus
Zebaze et al. (2016)	50	✓	✓	✓	✓	✓	–	0.7	Hip	Euro	✓*	Asym	von mises	PP	Hex8	1.5	Fall	d = 4%	VirtuOst

Graeff et al. (2015)	48	✓	✓	✓	✓	✓	0.6	3.0	Spine	Mindw.	✓*	Trans	N.A.		Tet4	2.5	Comp.	Stiffness	CalculiX
Kiyohara et al. (2015)	146						0.4	2.0	Spine	HA	N.A.	Asym	T: von Mises, C: Drucker–Prager	Lin	Tet4	2	Comp.	$\varepsilon = 1\%$	MechFind
Zysset, Qin, et al. (2015)	98					✓	0.7	1.0	Spine Hip	Euro	Automatic	Iso Trans	Eccentric elliptical	Dmg	Tet4	3	Comp. Stance/ Fall	F_u	ANSYS

with manual correction; Asym, *asymmetric;* BMAS, *bone mineral apparent density standard;* d, *deformation;* Dmg, *damage-based model;* Euro, *European calibration Standard;* F, *ultimate force;* HA, *hydroxyapatite;* Hex, *8-node hexahedral elements;* Hill, *hill yield criterion;* Internal, *internal calibration method;* Iso, *isotropic;* K2HPO4, *dipotassium hydrogen phosphate;* Lin, *linear plasticity;* Mindw., *mindways;* Orth, *orthotropic;* pp, *perfect plasticity;* QRM, *quality reference material;* Tet, *4-node tetrahedral elements;* Trans, *transversely isotropic;* vMises, *von mises yield criterion.*

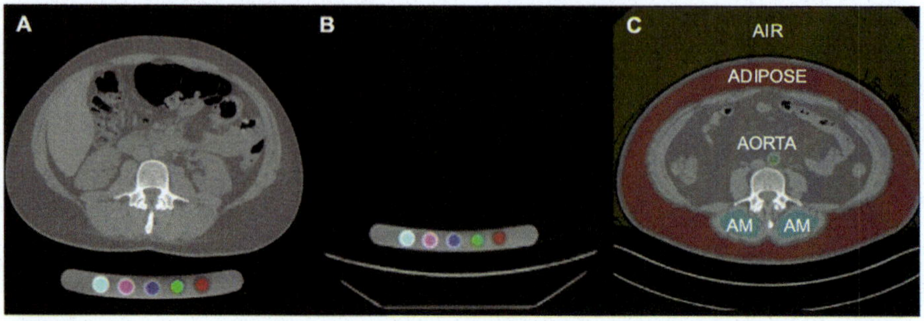

FIGURE 20.2

The three different methods for calibration. (A) Phantom calibration, with the calibration phantom placed underneath the patient. (B) Asynchronous calibration, with the calibration phantom scanned separately from the patient. (C) Internal calibration using different tissues regions within the patient. Tissues often include combinations of subcutaneous adipose tissue (ADIPOSE), blood from the aorta (AORTA), autochthonous muscles (AM) and cortical bone.

tube, tube current, reconstruction kernel or table height, and as a result are scanner and image protocol dependent.

An external calibration phantom in the scan field of view provides the ability to correct for scanner instabilities that affect both the patient and calibration phantom in the same manner (Engelke et al., 2015). However, there is evidence that including an external calibration phantom in the scan field of view may cause image artifacts such as beam hardening, scattering, partial volume averaging, and repositioning errors, which can degrade QCT precision (Boden et al., 1989; Cann, 1987). Further, when a phantom is located underneath the patient, the phantom will be subject to patient-specific attenuations depending on the patient's composition, size, and the positioning of the phantom relative to the patient (Boden et al., 1989). Due to practical constraints such as limited time and resources, calibration phantoms are rarely used in clinical settings and are typically reserved for research purposes.

3.2 Asynchronous calibration

To address the absence of calibration phantoms in clinical scans, an alternative approach known as asynchronous calibration is becoming increasingly common. Asynchronous calibration requires a scan of the calibration phantom separate from the patient, either before or after the patient scan, on the same scanner with the same imaging protocol and parameters (Fig. 20.2B). The same analysis of comparing the in-scan HUs of the bone equivalent rods to the known rod density values to determine a and b for Eq. (20.2) is performed, but instead uses asynchronous scan of the calibration phantom. Once the HU-to-BMD relationship has been derived (Eq. 20.2) it can be applied to the clinical scan of the patient to convert the image to a bone equivalent density image.

Asynchronous calibration, when compared to traditional in-scan phantom calibration, eliminates the risk of introducing image artifacts caused by the patient or the phantom and allows for the retrospective analysis of BMD. However, the same practical restraints of limited time and resources in a clinical setting make it impractical to take daily, or even monthly, asynchronous scans. As a result, asynchronous calibration must assume a very high degree of scanner stability to avoid precision errors (Engelke et al., 2015). Investigations into the influence of the time gap between patient scans and asynchronous calibration has yielded varying results. Dudle and colleagues reported moderate variability, noting that BMD measurement error doubled when calibrating with scans performed more than a year apart, whereas Engelke and colleagues observed an eight-fold increase in BMD measurement error with the same time gap (Dudle et al., 2024). These discrepancies may stem from hardware replacements, software upgrades, or X-ray source evolution over time (Dudle et al., 2024; Engelke et al., 2015). As such, to reduce the opportunity for precision errors in studies that employ asynchronous calibration, it is important to actively monitor the CT scanner stability.

3.3 Internal calibration

A phantomless alternative to both phantom and asynchronous calibration, known as internal calibration, replaces the phantom inserts of known density with in-scan regions of interest such as air, body tissues (adipose, skeletal muscle, cortical bone) or blood (Engelke et al., 2023). It is assumed that these in-scan referent tissues values are the same between individuals and have known density values, acting in a similar manner to the bone equivalent reference rods found in a calibration phantom. It is the task of the internal calibration method to determine what the a and b values are for Eq. (20.2). There are five main methods to performing internal calibration, which are:

1. **Reference Dataset Approach** (Adams et al., 2018; Ataei et al., 2022; Bartenschlager et al., 2022; Eggermont et al., 2019; Lee et al., 2017; Weaver et al., 2015). A reference dataset is created consisting of CT scans that have been calibrated to bone equivalent density using a calibration phantom, whether the phantom is synchronous or asynchronous. From this calibrated dataset, for each desired internal referent tissue (any tissue, though commonly some combination of air, adipose, blood, cortical bone, and skeletal muscle) bone density equivalents are determined by calculating the average density of each tissue across the entire reference dataset. To calibrate a new CT scan, the average reference tissue values (from the reference dataset) are compared to the in-scan HUs of the same reference tissue in the new CT scan, and a linear regression is created to determine a and b for Eq. (20.2). This creates a scan-specific conversion equation can be used to convert HUs to BMD.
2. **Multiple Linear Regression Approach** (Lee et al., 2019; Prado et al., 2021). This method also requires a reference dataset that has been calibrated with a

calibration phantom. The mean slope and intercept (a and b, respectively) and mean density for each tissue of interest (HU^R) from the calibration dataset is derived. Using HU^R of each tissue, along with a and b, the multiple linear regression equations

$$a = m_0 + m_1 HU_1^R + m_2 HU_2^R + \ldots + m_k HU_k^R \tag{20.3}$$

and

$$b = n_0 + n_1 HU_1^R + n_2 HU_2^R + \ldots + n_k HU_k^R \tag{20.4}$$

can be solved for the coefficients $m_0, m_1, \ldots, m_k$ and $n_0, n_1, \ldots, n_k$. To calibrate a new CT scan, the HU values for the same reference tissues can be substituted back into Eq. (20.3) and Eq. (20.4) to calculate a new slope (a^*) and intercept (b^*), where a^* and b^* are calculated by

$$a^* = m_0 + m_1 \mathrm{HU}_1^R + m_2 \mathrm{HU}_2^R + \ldots + m_k HU_K^R \tag{20.5}$$

and

$$b^* = n_0 + n_1 HU_1^R + n_2 HU_2^R + \ldots + n_k HU_K^R. \tag{20.6}$$

The CT image can now be calibrated by inserting the scan-specific a^* and b^* into Eq. (20.2).

3. **HU Number correction approach**(Boden et al., 1989; Gudmundsdottir et al., 1993). Instead of creating a new slope and intercept for individual images, the HU number correction approach adjusts the scan HUs so that the same slope and intercept can be used. First, the slope and intercept are determined by calibrating one or more scans in a reference dataset with a calibration phantom. In the same reference dataset, average HUs are derived for two internal reference tissues (HU_1^R and HU_2^R). The internal referent tissues are most commonly subcutaneous fat and paraspinal muscles. When calibrating a new scan that is not part of the reference dataset, an HU correction slope (s) and intercept (t) are derived by comparing the reference tissue HUs (HU_1^R and HU_2^R) to the uncalibrated tissue HUs (HU_1^N and HU_2^N) through equations

$$s = \frac{HU_1^R - HU_2^R}{HU_1^N - HU_2^N} \tag{20.7}$$

and

$$t = \left(HU_1^R - HU_1^N\right)\left(\frac{HU_1^R - HU_2^R}{HU_1^N - HU_2^N}\right). \tag{20.8}$$

The corrected HU value (HU_{corr}) for each subsequent CT scan to be calibrated is derived through the linear equation

$$HU_{corr} = HU \cdot s + t, \tag{20.9}$$

and this HU_{corr} can be substituted back into Eq. (20.2) with the reference dataset a and b values.

4. **Universal reference dataset approach** (Bartenschlager et al., 2023). While approaches one to three are all subject-specific, the universal reference dataset approach is a non-subject-specific approach. In this method, a reference dataset of calibrated images using a calibration phantom is required. A mean calibration equation of n CT scans is created by averaging all n image calibration slopes and intercepts to obtain a universal a and b for Eq. (20.2).
5. **Effective Energy Approach** (Michalski et al., 2020). While approaches one to four all require a reference dataset, the effective energy approach allows for subject-specific image calibration without requiring a reference dataset. Instead, this internal calibration method estimates the scan's effective energy level to determine reference mass attenuation coefficients, $\left(\frac{\mu}{\rho}\right)$, for each tissue used in the calibration (air, adipose, blood, cortical bone, and skeletal muscle) (Michalski et al., 2020). Reference mass attenuation values at increasing effective energy levels are obtained from the corresponding National Institute of Standards and Technology (NIST) datasets for the tissues of interest (White et al., 1989; Berger et al., 2010; Hubbell and Seltzer, 1996). The energy level is iteratively estimated by maximizing the coefficient of determination between the mean observed HUs from each tissue and the corresponding expected HUs at a given energy level based on the NIST data (White et al., 1989; Berger et al., 2010; Hubbell and Seltzer, 1996; Michalski et al., 2020). The mass attenuation coefficients at the effective energy that yielded the maximum coefficient of determination are used as the reference mass attenuation coefficients for each tissue. Using the five reference mass attenuation coefficients and the mean HUs from each tissue, two relationships are determined: (1) an HU-to-mass attenuation coefficient relationship $\left(\text{HU} - \frac{\mu}{\rho}\right)$, and (2) an HU-to-material density relationship $(\text{HU} - \rho_m)$. The HU-to-mass attenuation coefficient linear regression equation yields a voxel-specific mass attenuation coefficient, $\left(\frac{\mu}{\rho}\right)_{\text{voxel}}$, and the HU-to-material density linear regression equation yields a voxel-specific material density, ρ_{voxel}. The relationships are expressed as

$$\left(\frac{\mu}{\rho}\right)_{\text{voxel}} = a_{\left(\text{HU}-\frac{\mu}{\rho}\right)} + b_{\left(\text{HU}-\frac{\mu}{\rho}\right)} \tag{20.10}$$

and

$$\rho_{\text{voxel}} = a_{(\text{HU}-\rho_m)} + b_{(\text{HU}-\rho_m)}, \tag{20.11}$$

where a and b denote the resulting slope and intercept of the linear regression. The voxel-specific material density relationship is used to determine the total mass (m_{voxel}) of each voxel by multiplying by the voxel volume. The voxel-specific mass attenuation coefficient and the voxel-specific total mass are used in a two-component mass fraction model,

$$m_{\text{K}_2\text{HPO}_4} = m_{\text{voxel}} * \left(\frac{\left(\frac{\mu}{\rho}\right)_{\text{voxel}} - \left(\frac{\mu}{\rho}\right)_{\text{triglyceride}}}{\left(\frac{\mu}{\rho}\right)_{\text{K}_2\text{HPO}_4} - \left(\frac{\mu}{\rho}\right)_{\text{triglyceride}}} \right), \tag{20.12}$$

to determine the voxel-specific K_2HPO_4 mass ($m_{\text{K}_2\text{HPO}_4}$) for each voxel in the image, where both the mass attenuation of triglyceride, $\left(\frac{\mu}{\rho}\right)_{\text{triglyceride}}$, and the mass attenuation of bone, $\left(\frac{\mu}{\rho}\right)_{\text{K}_2\text{HPO}_4}$, are the reference NIST mass attenuation values at the determined effective energy level (Berger et al., 2010; Hubbell and Seltzer, 1996; White et al., 1989). The voxel-specific $m_{\text{K}_2\text{HPO}_4}$ can be used to determine the voxel specific density, $\rho_{\text{K}_2\text{HPO}_4}$, by dividing $m_{\text{K}_2\text{HPO}_4}$ by the voxel volume,

$$\rho_{\text{K}_2\text{HPO}_4} = \frac{m_{\text{K}_2\text{HPO}_4}}{\text{voxel volume}}. \tag{20.13}$$

The decision of which calibration method to use is situation dependent. If a calibration phantom is available, either phantom or asynchronous calibration should be used. However, if a calibration phantom is not available or the dataset comes from a myriad of sources (i.e., many different imaging centers as would be common in large clinical studies), the internal calibration methods should be used. If a reference dataset is available, any of the reference dataset internal calibration methods can be utilized (methods 1–4). It is important to note that the reference dataset should ideally be from the same scanner and taken with the same imaging protocol to minimize differences between scanners, scanner instabilities and variation caused by imaging protocol. The error in BMD estimates resulting from internal calibration methods can be estimated, and efforts such be made to reduce these sources of error whenever possible (Matheson et al., 2024). If an appropriate reference dataset is not available, the effective energy approach is the only method currently available that does not require a reference dataset.

4. Finite element workflow for model generation in clinical applications

After density calibration, finite element (FE) models can be generated to evaluate bone strength and mechanical properties. FE modeling leverages the density information to simulate mechanical behavior, offering a non-invasive means to assess bone strength and fragility.

4.1 Scan sites

The choice of anatomical site significantly influences both clinical utility and FE modeling approach. In our review of 21 studies, most examined the spine (14 studies) and hip (12 studies), with fewer investigating the knee (4 studies) (Table 20.2).

Spine (L1): The L1 vertebra serves as an ideal target for bone strength assessment due to several advantages. Its routine inclusion in both chest and abdominal CT scans enables wider potential opportunistic screening, while its location below the respiratory-sensitive cervical region provides stable imaging conditions. Studies have demonstrated successful FE analysis of L1 and adjacent vertebrae (T12-L3) (Brown et al., 2021; Keaveny et al., 2023). The L1's uniform geometry and consensus on axial loading as the primary loading mode enables consistent comparisons across studies, making it particularly suitable for clinical assessment. However, segmentation methods vary, as trabecular and cortical compartments are typically identified using either calibrated density values (Keaveny et al., 2008) or morphological distance-based approaches, which erode the periosteal segmentation to define the trabecular region (Haider et al., 2018). Additionally, some approaches only assess the vertebral body, while others retain the entire vertebra, including the spinal processes, introducing further variability in the analyzed region.

Hip (Proximal Femur): Preventing hip fractures is a major clinical priority, making the proximal femur a key site for FE analysis. Its complex trabecular and cortical architecture requires careful modeling, particularly regarding anisotropic material properties. Multiple studies (Ono et al., 2021; Doi et al., 2022; Dall'Ara et al., 2016; Zysset, Pahr, et al., 2015b) emphasize the importance of evaluating both stance and sideways fall loading conditions to comprehensively assess fracture risk.

Knee (Proximal Tibia): Although not a primary site for screening, FE analysis of the knee provides insights into bone strength, particularly under altered loading conditions. Some studies (Edwards et al., 2018; Haider et al., 2019) suggest that torsional loading is a principal failure mode, contributing to the spiral fracture patterns frequently observed in spinal cord injury patients (Martinez et al., 2002).

Other Sites: Emerging research explores the proximal humerus for FE analysis, particularly in assessing bone quality for shoulder biomechanics and surgical planning for stemless shoulder arthroplasty (Knowles et al., 2024). While fewer studies currently investigate this site, methodologies parallel those used for the femur.

4.2 Segmentation

Accurate delineation of bone structures from CT scans is essential for FE analysis. While our reviewed studies relied on traditional segmentation methods with manual corrections (Table 20.2), the field is advancing with deep learning approaches that promise improved accuracy and automation (Fig. 20.3).

4.2.1 Traditional segmentation approaches

Traditional segmentation begins with intensity-based approaches using thresholding and edge detection (Pham et al., 2000; Zhang, 1996). These methods classify bone using standardized ranges (700–1500 HU for cortical bone, 100–400 HU for trabecular bone), but struggle with partial volume effects and connected structures. While region-growing methods added spatial connectivity constraints (Kang et al., 2003; Wang et al., 2006), they remain sensitive to noise and seed point selection. Graph-cut methods represented a significant advance by incorporating smoothness constraints and anatomical priors (Besler et al., 2021; Krčah et al., 2011; Pauchard et al., 2016). Notable implementations include fully automatic femur segmentation achieving 81% accuracy across 197 scans (Krčah et al., 2011), and enhanced versions using Hessian filters that achieved Dice coefficients of 0.98 (Besler et al., 2021). Interactive approaches reduced segmentation time from 20 to 2–5 min while maintaining high accuracy (Pauchard et al., 2016). However, these methods still struggle with bones in jointed structures, where relative motion and close proximity make accurate separation and segmentation difficult. Atlas-based methods attempt to overcome these limitations through deformable registration with reference

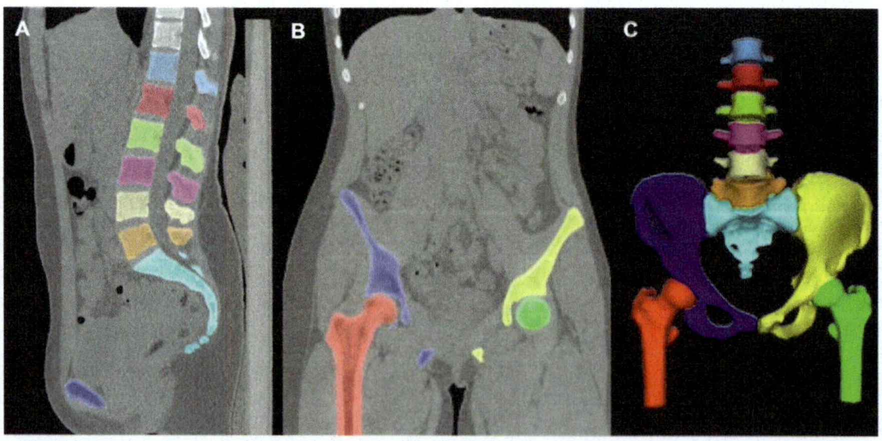

FIGURE 20.3

Example of an accurate automated deep learning-based segmentation of the lumbar spine, proximal femur, and pelvis in a patient with a sixth lumbar vertebra (L6). (A) Sagittal view, (B) coronal view, and (C) 3D rendering of the segmentation, with different colors representing distinct bone labels.

datasets (Besler et al., 2018). While showing promise for complex morphology, most clinical FE studies (Table 20.2) still require manual corrections for reliable segmentation.

4.2.2 Deep learning-based methods

Deep learning has transformed bone segmentation through models like nnU-Net, which automate the entire pipeline from preprocessing to post-processing (Isensee et al., 2021). Large-scale models such as TotalSegmentator now handle over 100 anatomical structures (Wasserthal et al., 2023), though challenges persist with pathological cases and low-dose scans (Krekiehn et al., 2023). A critical concern in clinical implementation is detecting segmentation failures, as errors compromise biomechanical assessments and treatment decisions. Recent advances in failure detection using ensemble-based uncertainty quantification and Dice score evaluation (Roy et al., 2019; Zenk et al., 2025) help identify unreliable segmentations, crucial for preventing silent failures (Fig. 20.4) in clinical applications. Standardization efforts through benchmark challenges like VerSe (Sekuboyina et al., 2021) are improving model performance across different scanners and populations. Combined with internal density calibration, these public datasets could serve as benchmarks for comparing FE modeling approaches, advancing the field toward reliable clinical implementation.

4.3 FE model generation

Generating a finite element mesh from segmented CT data requires addressing the inherent resolution anisotropy of clinical scans, where in-plane resolution (<1 mm) differs from slice thickness (1–4 mm). Most studies (15/16) handle this by resampling to isotropic voxels before mesh generation, though interpolation methods vary. The choice of meshing approach then depends on specific clinical requirements.

4.3.1 Meshing and element technology

An element is the fundamental unit of a finite element mesh, defining the geometry and resolution of the model. Clinical FE models predominantly use 8-node hexahedral (hex8) elements due to their direct compatibility with CT voxels and computational efficiency. These elements use linear shape functions, with displacement varying linearly along each direction. Our review shows hex8 elements (24 degrees of freedom) are most commonly used in clinical research (16/21 studies, Table 20.2), with standard element sizes of 1.0 mm for vertebrae and 1.5 mm for hips, though mesh convergence studies are lacking. 4-node tetrahedral (tet4) elements (12 degrees of freedom) serve as an alternative for complex geometries requiring adaptive meshing. While offering better geometric flexibility, they demand more careful mesh preparation, typically involving surface mesh generation followed by volume meshing. Studies have employed both open-source algorithms (Graeff et al., 2015; Labelle & Shewchuk, 2007) and commercial tools like Medtool (Zysset, Pahr, et al.,

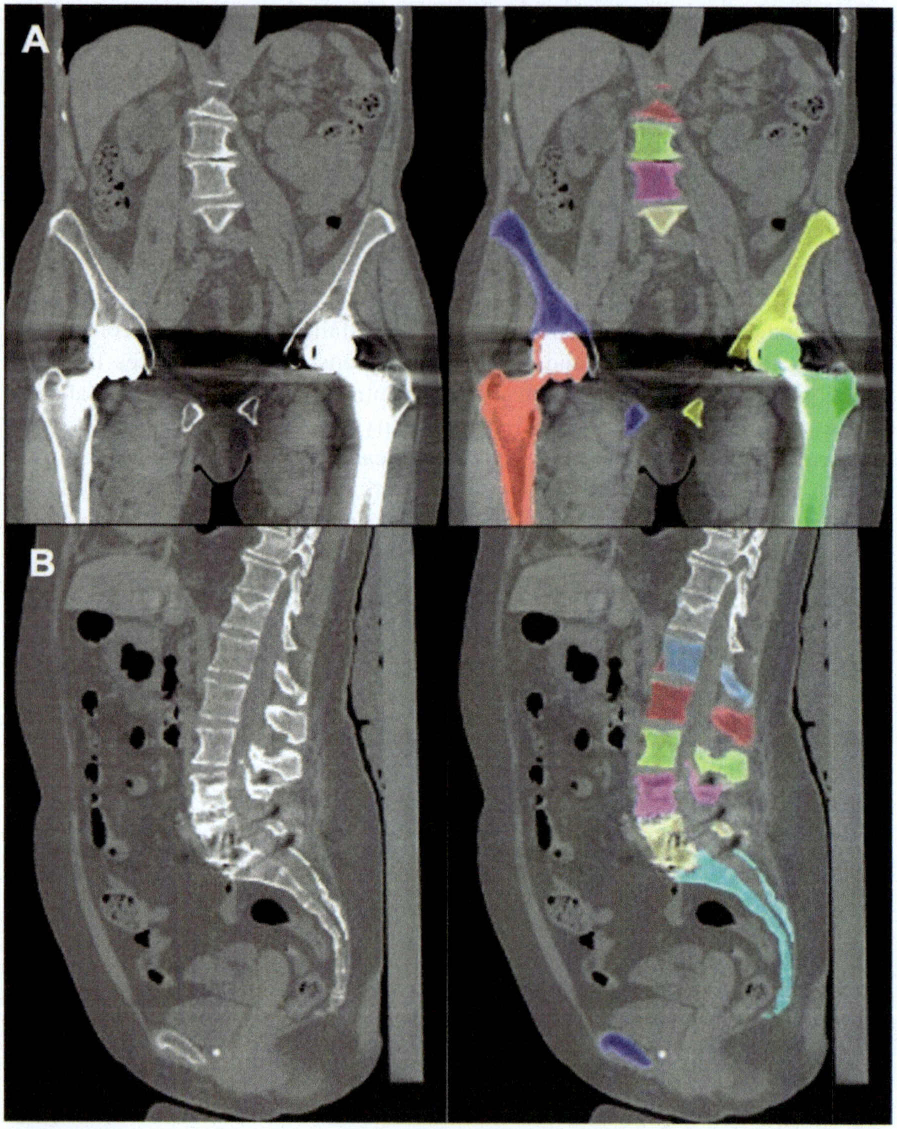

FIGURE 20.4

Segmentation errors caused by metal implants. Deep learning-based segmentations models can fail in the presence of metal artifacts from (A) spinal and (B) hip implants. These artifacts lead to significant segmentation inaccuracies, such as incorrect anatomical boundaries or missed structures. Proper evaluation and failure detection strategies are essential for mitigating such errors.

2015b) or Mechanical Finder (Doi et al., 2022; Kiyohara et al., 2015; Ono et al., 2021). Future research may benefit by evaluating higher-order elements (20-node hex, 10-node tet) against these linear elements, particularly for regions with high deformation gradients. However, the computational cost versus accuracy trade-off requires systematic investigation for different clinical applications.

4.3.2 Boundary conditions

Boundary conditions standardize how models interact with their environment across patient-specific analyses. For clinical studies, consistent application is achieved through model registration to reference anatomies, using techniques ranging from fast closest point to complex 3D alignments.

Each anatomical site requires specific loading conditions (Fig. 20.5). Spine models primarily use axial compression through the vertebral endplates, often incorporating simulated PMMA disks while excluding pedicles. Hip analysis employs two key configurations: stance loading (20 degrees femoral axis slant) (Bessho et al., 2007) and fall loading (10 degrees ground angle, 15 degrees internal rotation, 120 degrees neck-axis loading) (Greenspan et al., 1998). Both use simulated PMMA at loading surfaces, which ensures loads are distributed to the bone in a somewhat physiological fashion. Knee models focus on torsional loading due to observed spiral fracture patterns (Martinez et al., 2002) applying rotation to the proximal 2 cm

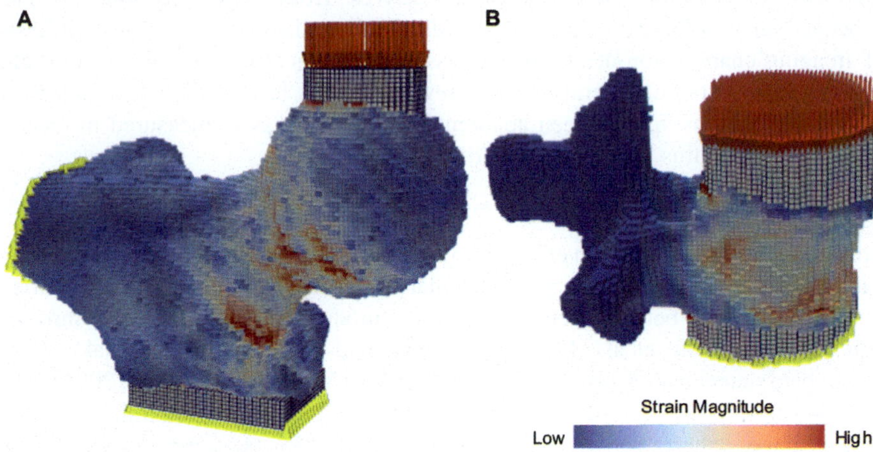

FIGURE 20.5

Representative finite element model solutions illustrating strain distribution in two loading configurations using 8-node hexahedral elements. (A) sideways fall configuration of the left proximal femur with polymethyl methacrylate (PMMA) disks ($E = 2500$ MPa, gray) added to the femoral head and greater trochanter. (B) compression of a lumbar vertebra with PMMA disks ($E = 2500$ MPa, gray) added to the superior and inferior endplates. Applied displacements are indicated with red arrows, and constrained nodes are marked with yellow triangles. Shades of red represent regions of high strain, while shades of blue indicate regions of low strain.

while constraining translation 13 cm distally (Edwards et al., 2013). PMMA properties in FE models require careful consideration. While experimental PMMA typically has a Young's modulus around 3 GPa, computational models use values ranging from 150 MPa (Zysset, Pahr, et al., 2015), 1.38 GPa (Dall'Ara et al., 2016) to 2.5 GPa (Edwards et al., 2013). Lower values may reduce computational artifacts (Zysset, Pahr, et al., 2015), but their influence on force transmission and stress distribution needs systematic evaluation.

4.3.3 Material properties

Understanding bone's mechanical behavior is fundamental to developing accurate FE models for clinical applications. This section examines the key material properties that govern bone's response to loading: Young's modulus, anisotropy, yield behavior, and post-yield characteristics.

4.3.4 Young's modulus and its relation to CT-derived density

Young's modulus (E) quantifies a bone's stiffness, defined as the ratio of stress (σ) to strain (ε) in the elastic region,

$$E = \frac{\sigma}{\varepsilon}. \tag{20.14}$$

Young's modulus plays a crucial role in predicting bone deformation under load. As a composite material, bone's stiffness derives from hydroxyapatite crystals embedded in a collagen matrix, with higher mineralization increasing stiffness. CT imaging enables non-destructive assessment of this structure, with Hounsfield Units relating to bone mineral density. Several studies have established empirical power-law relationships between apparent bone density (ρ_{app}, measured in g/cm^3) and Young's modulus (E) in the axial direction as

$$E_3 = A \times \rho_{\text{app}}^B, \tag{20.15}$$

(Edwards et al., 2013; Morgan et al., 2003) where A and B are regression coefficients that vary depending on bone type (trabecular or cortical) and anatomical location.

The relationship between density and elastic modulus has been validated through both ultrasonic (Rho et al., 1995) and mechanical testing (Morgan et al., 2003) though parameters vary by anatomical site. For knee models (Gray et al., 2008) validated a single power-law derived from trabecular bone for both trabecular and cortical bone, where cortical values remained within expected ranges (Crack et al., 2023; Edwards et al., 2013; Haider et al., 2018; Haider et al., 2019). Alternative approaches adjust trabecular properties to account for testing artifacts in cadaveric cores (Keaveny et al., 2008; Keaveny et al., 2017; Keaveny et al., 2023), suggesting traditional models may underestimate trabecular bone strength and that the mechanical difference between trabecular and cortical bone might be smaller than previously thought.

Given the variability in density-modulus relationships, standardized reporting protocols are essential to compare study outcomes between different approaches.

While some studies provide detailed documentation of density-to-modulus relationships (Edwards et al., 2013), others are more vague in reporting these details, making comparisons more challenging (Keaveny et al., 2008). Studies should always document their complete methodology, including CT-to-density conversion methods, power-law equations for different bone types, artifact corrections, validation approaches with error metrics, and material model assumptions. Such transparency is essential to allowing cross-study comparisons and advance patient-specific modeling.

4.3.5 Anisotropy in bone

Bone's mechanical properties vary directionally due to its hierarchical structure of oriented collagen fibers and mineral crystals. While high-resolution models can directly capture this microstructure, clinical CT-based models typically account for anisotropy through tension-compression asymmetry (Table 20.2). This distinction reflects bone's greater strength under compression than tension, as microcracks propagate more easily along collagen interfaces during tensile loading. The importance of this asymmetry varies by loading scenario—while less critical for primarily compressive loads in vertebrae, it becomes more important in hip fall simulations where both tensile and compressive stresses occur in different regions of the femoral neck.

To accurately capture these mechanical differences, finite element models often implement distinct yield criteria for compression and tension (Doi et al., 2022; Kiyohara et al., 2015; Ono et al., 2021). More sophisticated modeling approaches treat bone as an orthotropic material, incorporating three mutually perpendicular planes of symmetry (Crack et al., 2023; Crack et al., 2024; Edwards et al., 2013; Haider et al., 2018). This results in distinct elastic moduli (E_1, E_2, E_3), shear moduli (G_{12}, G_{23}, G_{31}), and Poisson's ratios (ν_{12}, ν_{23}, ν_{31}) along each principal direction. These properties can be related to the primary modulus (E_3) using the established relationships (Rho, 1996)

$$\begin{aligned} E_1 &= 0.574 \times E_3, E_2 = 0.577 \times E_3, \\ G_{12} &= 0.195 \times E_3, G_{23} = 0.265 \times E_3, \\ \nu_{12} &= 0.427, \nu_{23} = 0.234, \nu_{31} = 0.405. \end{aligned} \tag{20.16}$$

These relationships provide a framework for constructing finite element models that accurately capture the anisotropic nature of bone tissue; however, the accurate experimental determination of anisotropic properties remains a challenge.

4.3.6 Yield criterion

The yield point of a material defines the transition between elastic and plastic behavior. In bone mechanics, identifying this point is crucial for understanding how bone deforms under physiological and pathological conditions. Various yield criteria exist to define this transition, each tailored to different material assumptions.

A von Mises-like yield criterion (von MISES, 1913) is one of the most commonly used approaches when assuming isotropic behavior in bone (Keaveny et al., 2008; Keaveny et al., 2017; Keaveny et al., 2023). This criterion is based on the distortion energy theory, which states that yielding occurs when the equivalent stress exceeds a critical yield stress,

$$\sigma_{VM} = \sqrt{\frac{1}{2}\left[(\sigma_1 - \sigma_2)^2 + (\sigma_2 - \sigma_3)^2 + (\sigma_3 - \sigma_1)^2\right]}, \tag{20.17}$$

where σ_1, σ_2, σ_3 are the principal stresses. When the von Mises stress exceeds the material's yield stress, the local bone tissue undergoes plastic deformation.

Some models also differentiate between tensile and compressive yielding to account for bone's anisotropy. For example, certain approaches use von Mises in tension and the Drucker-Prager criterion (Drucker & Prager, 1952) in compression (Doi et al., 2022; Kiyohara et al., 2015; Ono et al., 2021). The Drucker-Prager model extends von Mises by incorporating a pressure-dependent term, which better represents bone's higher strength in compression than in tension. The Drucker-Prager yield criterion is defined as

$$\sigma_{DP} = \alpha I_1 + \sqrt{J_2} - k = 0, \tag{20.18}$$

where I_1 is the first invariant of the stress tensor, J_2 is the second invariant of the deviatoric stress tensor, and α and k are material constants related to cohesion and friction. This distinction may help accurately modeling bone strength, particularly in load-bearing bones subjected to complex stress states.

For orthotropic materials, Hill's yield criterion (Hill, 1948) is often used as an extension of von Mises, incorporating directional dependencies in the material's yield strength (Crack et al., 2023, 2024; Edwards et al., 2018; Edwards et al., 2013; Haider et al., 2018, 2019) through

$$F(\sigma_2 - \sigma_3)^2 + G(\sigma_3 - \sigma_1)^2 + H(\sigma_1 - \sigma_2)^2 + 2L\sigma_{23}^2 + 2M\sigma_{31}^2 + 2N\sigma_{12}^2 = 1, \tag{20.19}$$

where F, G, H, L, M, and N are material constants derived from experimental data. These constants are functions of the anisotropic yield stresses, defined as

$$\begin{aligned} F &= \frac{1}{2\sigma_{22}^2} + \frac{1}{2\sigma_{33}^2} - \frac{1}{\sigma_{22}\sigma_{33}}, \\ G &= \frac{1}{2\sigma_{33}^2} + \frac{1}{2\sigma_{11}^2} - \frac{1}{\sigma_{33}\sigma_{11}}, \\ H &= \frac{1}{2\sigma_{11}^2} + \frac{1}{2\sigma_{22}^2} - \frac{1}{\sigma_{11}\sigma_{22}}, \\ L &= \frac{1}{2\tau_{23}^2}, M = \frac{1}{2\tau_{31}^2}, N = \frac{1}{2\tau_{12}^2} \end{aligned} \tag{20.20}$$

and

$$\sigma_{DP} = \alpha I_1 + \sqrt{J_2} - k = 0.$$

In this formulation, σ_{11}, σ_{22}, σ_{33} represent the normal yield stresses in the principal material directions, while τ_{12}, τ_{23}, τ_{31} are the yield stresses in shear. The experimental determination of these parameters is challenging.

To estimate these yield stresses, it is often assumed that bone yields at a consistent strain threshold regardless of the direction. Based on experimental observations, yield strains are typically taken as 0.675% in normal loading and 1.215% in shear loading (Currey, 2004). The corresponding yield stresses are then computed by multiplying these yield strains by their respective elastic moduli: the normal moduli E_1,E_2,E_3 for normal stresses and the shear moduli G_{12},G_{23},G_{31} for shear stresses (Edwards et al., 2013).

Model complexity in clinical FE analysis depends on the specific application and required outcome. Tension-compression asymmetric models might offer a suitable balance for fracture risk assessment, capturing key mechanical differences while maintaining computational efficiency. More complex orthotropic models may be necessary for implant design where accurate stress distributions and bone-implant interactions are crucial, despite increased computational demands. However, while fully orthotropic or anisotropic models are often presented as a step forward, their parameters can be difficult to determine with meaningful accuracy, particularly in a heterogeneous material like bone. In large-scale screening scenarios, simpler isotropic models might prove practical, though potentially at the cost of post-yield prediction accuracy.

4.3.7 Post-yield behavior

Post-yield behavior describes how each element deforms after reaching its mechanical yield point, reflecting the material's complex response to loading. The mechanical behavior depends on various factors including loading rate, mineral content, and microstructural composition. The simplest modeling approach assumes perfect plasticity (pp), where stress remains constant after yielding. While this approach is widely used in computational models (Table 20.2), it oversimplifies bone's actual mechanical response. Bone's hierarchical structure means that its post-yield behavior is more nuanced than a perfectly plastic material.

A more advanced approach is linear plastic behavior, where stress increases proportionally with plastic strain, modeled by the equation

$$\sigma = \sigma_Y + H \cdot \varepsilon_p, \tag{20.21}$$

where σ_Y is the yield stress, H is the plastic modulus, and ε_p is the plastic strain. This approach captures strain hardening, where the material stiffens as plastic deformation progresses, better reflecting bone's ability to resist further deformation before failure.

Damage models further refine post-yield predictions by incorporating strain rate effects and progressive degradation (Dall'Ara et al., 2016; Zysset, Pahr, et al., 2015b). These models introduce a damage parameter, D(κ), which depends on accumulated plastic strain and reduces the stiffness matrix over time and is defined as

$$D(\kappa) = 1 - e^{-b\kappa}, \tag{20.22}$$

where b is a material-specific parameter, and κ represents the accumulated plastic strain. This approach effectively captures the progressive reduction in bone stiffness due to microdamage accumulation. This approach is particularly important for understanding fatigue and stress fracture mechanisms.

These constant or piecewise linear post-yield definitions allow for the modeling of nonlinear bone tissue behavior at the structural level, as individual element responses collectively shape the global force-displacement curve. In clinical finite element modeling, perfect plasticity may be useful for rapid screening but oversimplifies bone behavior. Linear plasticity may offer a better balance of accuracy and efficiency for fracture risk assessment and implant stability. Damage models provide the most realistic representation of progressive failure, making them ideal for patient-specific applications like osteoporosis monitoring and fatigue fracture prediction, though they require high computational resources. Ultimately, the choice of model should be guided by the required outcome measure, as even a simple linear model may be sufficient for certain predictive tasks like fracture risk assessment.

4.3.8 Bone strength and failure

Bone failure arises from both material and structural properties. Structural properties—force and displacement—depend on size and geometry, while material properties normalize these values to allow size-independent comparisons. Failure mechanisms differ between cortical and trabecular bone: cortical bone primarily fractures through crack propagation and brittle failure, whereas trabecular bone undergoes progressive microdamage accumulation and densification, requiring distinct strength definitions (Hernandez & van der Meulen, 2017).

Biomechanical models define yield strength using structural or material criteria. Structural definitions often use force-based thresholds, such as vertebral strength at 4% deformation and femoral neck strength at 2% deformation (Keaveny et al., 2008; Keaveny et al., 2017; Keaveny et al., 2023). Ultimate force, representing peak load before failure, is another widely used measure. In contrast, strain-based criteria define failure based on critical strain thresholds. One approach considers failure when 10% of surface elements exceed $\varepsilon = 1.41\%$ (Edwards et al., 2013; Rincón-Kohli & Zysset, 2009), while another defines failure at $\varepsilon = 1\%$ (10,000 microstrain) when the first element reaches this limit (Doi et al., 2022; Kiyohara et al., 2015; Ono et al., 2021). Strain-based methods better capture microdamage initiation and progression, particularly relevant for osteoporosis research. These strain-based failure criteria have also demonstrated strong correlations with experimental results even when applied to linear FE models, supporting their utility in computationally efficient strength estimation (Nishiyama et al., 2013).

Stiffness, indicating resistance to deformation, correlates with load-bearing capacity but does not fully describe post-yield behavior or energy absorption, both critical for fracture risk assessment. For clinical translation of CT-based finite element (FE) analysis, strength estimation must be clearly defined, as different criteria yield varying results. Whether based on ultimate force, deformation thresholds, or strain limits, standardizing or transparently reporting these measures is essential for study comparability, result interpretation, and clinical application.

4.3.9 Numerical solvers and FE solutions

The choice of finite element (FE) solver in CT-based bone analysis is influenced by regulatory approval, accessibility, and computational efficiency. Commercial solvers like VirtuOst and Mechanical Finder are widely used due to their streamlined workflows and regulatory endorsements. VirtuOst (O.N. Diagnostics, Berkeley, CA) is FDA-cleared as a Class II medical device, facilitating its adoption in large-scale studies. For instance (Keaveny et al., 2023), analyzed 469 participants, while (Brown et al., 2021) assessed 4093 individuals using VirtuOst for bone strength and fracture risk evaluation. These solvers optimize algorithms for specific loading conditions (e.g., vertebral compression, sideways falls), ensuring computational efficiency but limiting transparency and adaptability due to their proprietary nature.

General-purpose solvers such as Abaqus and ANSYS offer greater flexibility for defining complex material properties, boundary conditions, and loading scenarios but require custom workflows for large-scale studies. For example Edwards et al. (2018), used Abaqus in a knee biomechanics study with 61 participants, while Haider et al. (2019) applied it to torsional loading research in 17 subjects. Open-source alternatives like CalculiX provide cost-effective options, demonstrated in a study by Graeff et al. (2015) involving 48 participants. However, open-source tools demand more user expertise and development effort, limiting their clinical adoption. FAIM (version 9.0, Numerics88 Solutions Ltd., Canada), an open-source solver designed for CT-based FE analysis (MacNeil & Boyd, 2008), offers a well-documented and streamlined approach for bone strength estimation, reducing computational costs while maintaining accuracy (Nishiyama et al., 2013).

Selecting an FE solver for clinical bone studies requires balancing regulatory compliance, scalability, transparency, and modeling flexibility. Commercial solvers are ideal for high-throughput clinical applications but restrict customization. General-purpose solvers enable advanced modeling but pose challenges in large-scale analyses. Open-source options enhance transparency yet require technical expertise. Understanding these trade-offs is essential for advancing bone mechanics research and improving clinical decision-making.

5. Summary and conclusion

FE analysis is a powerful tool for assessing bone strength and fracture risk, yet its clinical adoption remains limited. Despite endorsement by the International Society for Clinical Densitometry (Zysset, Qin, et al., 2015), challenges persist in standardization, integration into clinical workflows, and validation across diverse populations. This chapter provides the technical foundation for understanding these challenges and highlights recent advancements that could help bridge the gap between research and clinical application. For FE analysis to be clinically useful, it must integrate seamlessly into workflows and provide reliable, actionable insights. FDA-approved tools like VirtuOst demonstrate feasibility and encourage randomized controlled trials (RCTs) to incorporate FE-derived bone strength (Table 20.2). However, greater transparency, benchmarking, and comparative studies are needed to ensure reliability and support broader regulatory approval.

A major challenge remains accurate density calibration, as clinical CT scans rarely include calibration phantoms. While phantom-based calibration is the gold standard, large-scale screening increasingly relies on internal calibration methods using in-scan tissue references (Bartenschlager et al., 2023). These methods require further validation across scanners and populations (Bott et al., 2023) to ensure accuracy. Segmentation remains a bottleneck in the modeling process. Most studies rely on traditional segmentation methods that require manual corrections, making the process time-consuming. Deep learning models offer automation but can fail silently, producing unreliable segmentations (Zenk et al., 2025). Robust quality control mechanisms, such as uncertainty quantification and failure detection, are essential for clinical implementation. Differences in modeling approaches also impact clinical utility. Simplified linear models are computationally efficient and may be sufficient to capture an outcome of interest, but may not fully capture bone's complex mechanical behavior. More advanced models incorporating anisotropic material properties and damage accumulation improve accuracy but increase computational demands. Balancing efficiency and accuracy remain a key challenge for future translational efforts.

Despite ongoing challenges, FE analysis has been increasingly proposed as a tool for clinical decision-making in fracture risk assessment. Yet, the optimal approach to integrating FE-derived measures with established clinical risk factors remains uncertain. Some studies have linked predefined bone strength thresholds to osteoporosis-equivalent risk classifications (Kopperdahl et al., 2014), but fracture risk depends not only on bone strength but also on external load demands. To address this, the load-to-strength ratio incorporates patient-specific loading conditions during falls. Importantly, variations in fall angles can alter fracture risk by up to 35% by shifting mechanical load distribution across the proximal femur, surpassing traditional risk factors such as age, sex, and weight (Awal & Faisal, 2024). Despite ongoing challenges, FE analysis has proven valuable for fracture risk assessment and treatment monitoring, outperforming traditional density-based metrics by providing a more comprehensive view of skeletal fragility (Keaveny et al., 2017). Integrating FE methods into Fracture Liaison Services could improve patient stratification and treatment planning.

While FE analysis is progressing toward clinical adoption, segmentation reliability, density calibration, and workflow integration remain key hurdles. Achieving greater transparency, standardized validation datasets, and robust method comparisons is essential for consistency and reproducibility. Since documentation standards differ across studies based on our review, we have developed a structured checklist to ensure consistency and facilitate reproducibility in clinical FE analysis (Table 20.3).

Table 20.3 Reviewer checklist for clinical finite element analysis.

☑	Category	Description
☐	**CT imaging parameters**	Reports pixel size, slice thickness, scan protocol, and CT scanner make and model to ensure reproducibility.
☐	**Density calibration**	Specifies calibration method (e.g., phantom-based, asynchronous, internal) and justifies its selection. Provides conversion equation details.
☐	**Segmentation method**	Clearly describes segmentation approach and parameters used (e.g., manual, thresholding, deep learning). Provides quality control measures and validation against reference segmentations.
☐	**Meshing strategy**	Reports element type (e.g., Hex8, Tet4), mesh size, degrees of freedom, and convergence criteria. Justifies chosen resolution for computational efficiency versus accuracy.
☐	**Material properties**	Specifies material model (e.g., isotropic, orthotropic, transversely isotropic) parameters. Provides density-modulus relationships and validation references.
☐	**Boundary conditions**	Defines anatomical site and loading conditions (e.g., axial compression, sideways fall, torsion). Reports assumptions and constraints.
☐	**Yield criteria**	Justifies choice of yield model (e.g., von mises, Drucker-Prager, Hill). Defines yield and failure threshold (e.g., stress, strain, energy).
☐	**Post-yield behavior**	Specifies post-yield model (e.g., perfect plasticity, linear plasticity, damage). Describes validation against experimental data.
☐	**Strength metrics**	Defines strength criteria (e.g., ultimate force, stiffness, deformation threshold, strain-based failure). Ensures comparability with prior studies.
☐	**Solver and computation**	Reports solver and computational resources used. Describes solver settings (e.g., implicit vs. explicit, convergence criteria).
☐	**Validation**	Provides experimental validation (e.g., cadaveric testing, mechanical testing). Compares FE results with clinical outcomes or previous studies.
☐	**Reproducibility & transparency**	Shares key parameters, model equations, and workflow steps. Indicates availability of scripts or code for reproducibility.
☐	**Clinical translation**	Discusses clinical relevance, potential integration into workflows, and limitations in real-world application.

Addressing these challenges will facilitate the transition of FE analysis from research to routine clinical practice, enhancing patient care and advancing precision medicine in bone health.

6. Future direction

Routine clinical CT is limited by its resolution, restricting its ability to capture detailed bone microstructure for advanced biomechanical analysis. However, emerging technologies like photon-counting CT (PCCT) offer higher spatial resolution and improved tissue contrast, narrowing the gap between clinical imaging and high-resolution peripheral quantitative CT (HR-pQCT) (Quintiens et al., 2024). These advancements could enhance fracture risk prediction and treatment monitoring, facilitating the transition of high-resolution micro-FE methods from research to clinical practice.

Higher-resolution imaging has enhanced treatment monitoring by providing deeper insights into bone metabolism and mechanical adaptation. HR-pQCT studies reveal that vitamin D supplementation may reduce bone density rather than improve it, underscoring the need for precise monitoring of treatment effects. In mechanical unloading conditions such as spaceflight, longitudinal imaging shows that bone loss involves both density reduction and microstructural deterioration, directly impacting FE-derived bone strength (Gabel et al., 2022a, 2022b). These findings demonstrate that bone mechanoregulation—the ability to respond to mechanical loading—can be quantified using FE-based approaches (Walle et al., 2021, 2023), offering a novel biomarker for monitoring conditions like osteoporosis (Christen et al., 2014) and diabetes (Walle et al., 2024).

In fracture risk prediction, combining FE modeling with machine learning-driven phenotyping approaches has significantly improved patient-specific risk assessment. HR-pQCT studies have identified distinct bone microarchitectural phenotypes that classify fracture risk beyond standard density-based methods (Bugbird et al., 2025; Whittier et al., 2022). Meanwhile, mFRAC, which integrates FEA-derived strength measures with predictive modeling, has outperformed conventional BMD-based tools in fracture prediction (Bugbird et al., 2024; Whittier et al., 2023). With clinical CT resolution improvements, these advancements in high-resolution imaging will drive the future of finite element modeling, integrating machine learning and leveraging imaging data to enable personalized fracture risk assessment, treatment monitoring, and disease progression tracking.

References

Adams, A. L., Fischer, H., Kopperdahl, D. L., Lee, D. C., Black, D. M., Bouxsein, M. L., Fatemi, S., Khosla, S., Orwoll, E. S., Siris, E. S., & Keaveny, T. M. (2018). Osteoporosis and hip fracture risk from routine computed tomography scans: The fracture, osteoporosis, and CT utilization study (FOCUS). *Journal of Bone and Mineral Research, 33*(7), 1291–1301. https://doi.org/10.1002/jbmr.3423

Agarwal, S., Shane, E., Lang, T., Shiau, S., Kamanda-Kosseh, M., Bucovsky, M., Lappe, J. M., Stubby, J., Recker, R. R., Hu, Y., Wang, Z., Guo, E., & Cohen, A. (2022). Spine volumetric BMD and strength in premenopausal idiopathic osteoporosis: Effect of teriparatide followed by denosumab. *Journal of Clinical Endocrinology and Metabolism, 107*(7), Article E2690. https://doi.org/10.1210/clinem/dgac232

Ataei, A., Eikhout, J., van Leeuwen, R. G. H., Tanck, E., Eggermont, F., & Deniz, C. M. (2022). The effect of variations in CT scan protocol on femoral finite element failure load assessment using phantomless calibration. *PLoS One, 17*(3), Article e0265524. https://doi.org/10.1371/journal.pone.0265524

Awal, R., & Faisal, T. (2024). QCT-based 3D finite element modeling to assess patient-specific hip fracture risk and risk factors. *Journal of the Mechanical Behavior of Biomedical Materials, 150*, Article 106299. https://doi.org/10.1016/j.jmbbm.2023.106299

Bartenschlager, S., Cavallaro, A., Pogarell, T., Chaudry, O., Uder, M., Khosla, S., Schett, G., & Engelke, K. (2023). Opportunistic screening with CT: Comparison of phantomless BMD calibration methods. *Journal of Bone and Mineral Research, 38*(11), 1689–1699. https://doi.org/10.1002/jbmr.4917

Bartenschlager, S., Dankerl, P., Chaudry, O., Uder, M., & Engelke, K. (2022). BMD accuracy errors specific to phantomless calibration of CT scans of the lumbar spine. *Bone, 157*, Article 116304. https://doi.org/10.1016/j.bone.2021.116304

Berger, M. J., Seltzer, S. M., Chang, J., Coursey, J. S., Sukumar, R., Zucker, D. S., & Olsen, K. (2010). *Photon cross sections database: NIST standard reference database*. XCOM.

Besler, B. A., Michalski, A. S., Forkert, N. D., & Boyd, S. K. (2018). Automatic full femur segmentation from computed tomography datasets using an atlas-based approach. In , *10734. Lecture notes in computer science (including subseries lecture notes in artificial intelligence and lecture notes in bioinformatics)* (pp. 120–132). https://doi.org/10.1007/978-3-319-74113-0_11

Besler, B. A., Michalski, A. S., Kuczynski, M. T., Abid, A., Forkert, N. D., & Boyd, S. K. (2021). Bone and joint enhancement filtering: Application to proximal femur segmentation from uncalibrated computed tomography datasets. *Medical Image Analysis, 67*. https://doi.org/10.1016/j.media.2020.101887

Bessho, M., Ohnishi, I., Matsuyama, J., Matsumoto, T., Imai, K., & Nakamura, K. (2007). Prediction of strength and strain of the proximal femur by a CT-based finite element method. *Journal of Biomechanics, 40*(8), 1745–1753. https://doi.org/10.1016/j.jbiomech.2006.08.003

Bilezikian, J. P., Watts, N. B., Usiskin, K., Polidori, D., Fung, A., Sullivan, D., & Rosenthal, N. (2016). Evaluation of bone mineral density and bone biomarkers in patients with type 2 diabetes treated with canagliflozin. *Journal of Clinical Endocrinology and Metabolism, 101*(1), 44–51. https://doi.org/10.1210/jc.2015-1860

Boden, S. D., Goodenough, D. J., Stockham, C. D., Jacobs, E., Dina, T., & Allman, R. M. (1989). Precise measurement of vertebral bone density using computed tomography without the use of an external reference phantom. *Journal of Digital Imaging, 2*(1), 31–38. https://doi.org/10.1007/BF03168013

Bott, K. N., Matheson, B. E., Smith, A. C. J., Tse, J. J., Boyd, S. K., & Manske, S. L. (2023). Addressing challenges of opportunistic computed tomography bone mineral density analysis. *Diagnostics, 13*(15). https://doi.org/10.3390/diagnostics13152572

Brown, J. P., Engelke, K., Keaveny, T. M., Chines, A., Chapurlat, R., Foldes, A. J., Nogues, X., Civitelli, R., De Villiers, T., Massari, F., Zerbini, C. A. F., Wang, Z., Oates, M. K., Recknor, C., & Libanati, C. (2021). Romosozumab improves lumbar spine bone mass and bone strength parameters relative to alendronate in postmenopausal women: Results from the active-controlled fracture study in postmenopausal women with osteoporosis at high risk (ARCH) trial. *Journal of Bone and Mineral Research, 36*(11), 2139–2152. https://doi.org/10.1002/jbmr.4409

Bugbird, A. R., Boisvert, N. M. J., Burt, L. A., & Boyd, S. K. (2025). Choose your mother wisely: The familial resemblance of bone adaptation. *Osteoporosis International, 36*(1), 141–149. https://doi.org/10.1007/s00198-024-07321-z

Bugbird, A. R., Whittier, D. E., & Boyd, S. K. (2024). Transferability of bone phenotyping and fracture risk assessment by μfRAC from first-generation high-resolution peripheral quantitative computed tomography to second-generation scan data. *Journal of Bone and Mineral Research, 39*(5), 571–579. https://doi.org/10.1093/jbmr/zjae039

Burch, S., Feldstein, M., Hoffmann, P. F., & Keaveny, T. M. (1976). Prevalence of poor bone quality in women undergoing spinal fusion using Biomechanical-CT analysis. *Spine, 41*.

Bushberg, J. T., Leidholdt, E. M., & Boone, J. M. (2012). *The essential physics of medical imaging - 3*.

Cann, C. E. (1987). Quantitative CT applications: Comparison of current scanners. *Radiology, 162*(1), 257–261. https://doi.org/10.1148/radiology.162.1.3786773

Christen, P., Ito, K., Ellouz, R., Boutroy, S., Sornay-Rendu, E., Chapurlat, R. D., & Van Rietbergen, B. (2014). Bone remodelling in humans is load-driven but not lazy. *Nature Communications, 5*. https://doi.org/10.1038/ncomms5855

Crack, L. E., Haider, I. T., Simonian, N., Barroso, J., Gabel, L., Schnitzer, T. J., & Edwards, W. B. (2023). Zoledronic acid after spinal cord injury mitigates losses in proximal femoral strength independent of ambulation ability. *Osteoporosis International, 34* (9), 1637–1645. https://doi.org/10.1007/s00198-023-06811-w

Crack, L. E., Simonian, N., Schnitzer, T. J., & Edwards, W. B. (2024). Monthly treatment with romosozumab for 1 year increases bone mineral at the hip, but not the knee, in women with chronic spinal cord injury. *JBMR Plus, 8*(7). https://doi.org/10.1093/jbmrpl/ziae077

Currey, J. D. (2004). Tensile yield in compact bone is determined by strain, post-yield behaviour by mineral content. *Journal of Biomechanics, 37*(4), 549–556. https://doi.org/10.1016/j.jbiomech.2003.08.008

Dall'Ara, E., Eastell, R., Viceconti, M., Pahr, D., & Yang, L. (2016). Experimental validation of DXA-based finite element models for prediction of femoral strength. *Journal of the Mechanical Behavior of Biomedical Materials, 63*, 17–25. https://doi.org/10.1016/j.jmbbm.2016.06.004

Doi, T., Ohashi, S., Ohtomo, N., Tozawa, K., Nakarai, H., Yoshida, Y., Ito, Y., Sakamoto, R., Nakajima, K., Nagata, K., Okamoto, N., Nakamoto, H., Kato, S., Taniguchi, Y., Matsubayashi, Y., Tanaka, S., & Oshima, Y. (2022). Evaluation of bone strength using finite-element analysis in patients with ossification of the posterior longitudinal ligament. *The Spine Journal, 22*(8), 1399–1407. https://doi.org/10.1016/j.spinee.2022.02.018

Drucker, D. C., & Prager, W. (1952). Soil mechanics and plastic analysis or limit design. *Quarterly of Applied Mathematics, 10*(2), 157–165. https://doi.org/10.1090/qam/48291

Dudle, A., Ith, M., Egli, R., Heverhagen, J., Gugler, Y., Wapp, C., Frauchiger, D. A., Lippuner, K., Jackowski, C., & Zysset, P. (2024). Asynchronous calibration of a CT scanner for bone mineral density estimation: Sources of error and correction. *JBMR Plus, 8*(9). https://doi.org/10.1093/jbmrpl/ziae096

Edwards, W. B., Schnitzer, T. J., & Troy, K. L. (2013). Torsional stiffness and strength of the proximal tibia are better predicted by finite element models than DXA or QCT. *Journal of Biomechanics, 46*(10), 1655–1662. https://doi.org/10.1016/j.jbiomech.2013.04.016

Edwards, W. B., Simonian, N., Haider, I. T., Anschel, A. S., Chen, D., Gordon, K. E., Gregory, E. K., Kim, K. H., Parachuri, R., Troy, K. L., & Schnitzer, T. J. (2018). Effects of teriparatide and vibration on bone mass and bone strength in people with bone loss and spinal cord injury: A randomized, controlled trial. *Journal of Bone and Mineral Research, 33*(10), 1729–1740. https://doi.org/10.1002/jbmr.3525

Eggermont, F., Verdonschot, N., van der Linden, Y., Tanck, E., & Angeles Pérez, M. (2019). Calibration with or without phantom for fracture risk prediction in cancer patients with femoral bone metastases using CT-based finite element models. *PLoS One, 14*(7), Article e0220564. https://doi.org/10.1371/journal.pone.0220564

Engelke, K., Chaudry, O., & Bartenschlager, S. (2023). Opportunistic screening techniques for analysis of CT scans. *Current Osteoporosis Reports, 21*(1), 65–76. https://doi.org/10.1007/s11914-022-00764-5

Engelke, K., Lang, T., Khosla, S., Qin, L., Zysset, P., Leslie, W. D., Shepherd, J. A., & Shousboe, J. T. (2015). Clinical use of quantitative computed tomography-based advanced techniques in the management of osteoporosis in adults: The 2015 ISCD official positions-part III. *Journal of Clinical Densitometry, 18*(3), 393–407. https://doi.org/10.1016/j.jocd.2015.06.010

Engelke, K., van Rietbergen, B., & Zysset, P. (2016). FEA to measure bone strength: A review. *Clinical Reviews in Bone and Mineral Metabolism, 14*(1), 26–37. https://doi.org/10.1007/s12018-015-9201-1

Faulkner, K. G., Glüer, C. C., Grampp, S., & Genant, H. K. (1993). Cross-calibration of liquid and solid QCT calibration standards: Corrections to the UCSF normative data. *Osteoporosis International, 3*(1), 36–42. https://doi.org/10.1007/bf01623175

Gabel, L., Liphardt, A. M., Hulme, P. A., Heer, M., Zwart, S. R., Sibonga, J. D., Smith, S. M., & Boyd, S. K. (2022a). Incomplete recovery of bone strength and trabecular microarchitecture at the distal tibia 1 year after return from long duration spaceflight. *Scientific Reports, 12*(2022).

Gabel, L., Liphardt, A. M., Hulme, P. A., Heer, M., Zwart, S. R., Sibonga, J. D., Smith, S. M., & Boyd, S. K. (2022b). Pre-flight exercise and bone metabolism predict unloading-induced bone loss due to spaceflight. *British Journal of Sports Medicine, 56*(2022), 196–203.

Ganapathy, A., Nieves, J. W., Keaveny, T. M., & Cosman, F. (2023). Effects of four-year cyclic versus two-year daily teriparatide treatment on volumetric bone density and bone strength in postmenopausal women with osteoporosis. *Bone, 167*. https://doi.org/10.1016/j.bone.2022.116618

Graeff, C., Campbell, G. M., Peña, J., Borggrefe, J., Padhi, D., Kaufman, A., Chang, S., Libanati, C., & Glüer, C. C. (2015). Administration of romosozumab improves vertebral trabecular and cortical bone as assessed with quantitative computed tomography and finite element analysis. *Bone, 81*, 364–369. https://doi.org/10.1016/j.bone.2015.07.036

Gray, H. A., Taddei, F., Zavatsky, A. B., Cristofolini, L., & Gill, H. S. (2008). Experimental validation of a finite element model of a human cadaveric tibia. *Journal of Biomechanical Engineering, 130*(3). https://doi.org/10.1115/1.2913335

Greenspan, S. L., Myers, E. R., Kiel, D. P., Parker, R. A., Hayes, W. C., & Resnick, N. M. (1998). Fall direction, bone mineral density, and function: Risk factors for hip fracture in frail nursing home elderly. *The American Journal of Medicine, 104*(6), 539–545. https://doi.org/10.1016/S0002-9343(98)00115-6

Gudmundsdottir, H., Jonsdottir, B., Kristinsson, S., Johannesson, A., Goodenough, D., & Sigurdsson, G. (1993). Vertebral bone density in icelandic women using quantitative computed tomography without an external reference phantom. *Osteoporosis International, 3*(2), 84–89. https://doi.org/10.1007/bf01623378

Haider, I. T., Lobos, S. M., Simonian, N., Schnitzer, T. J., & Edwards, W. B. (2018). Bone fragility after spinal cord injury: Reductions in stiffness and bone mineral at the distal femur and proximal tibia as a function of time. *Osteoporosis International, 29*(12), 2703–2715. https://doi.org/10.1007/s00198-018-4733-0

Haider, I. T., Simonian, N., Saini, A. S., Leung, F. M., Edwards, W. B., & Schnitzer, T. J. (2019). Open-label clinical trial of alendronate after teriparatide therapy in people with spinal cord injury and low bone mineral density. *Spinal Cord, 57*(10), 832–842. https://doi.org/10.1038/s41393-019-0303-3

Hernandez, C. J., & van der Meulen, M. C. H. (2017). Understanding bone strength is not enough. *Journal of Bone and Mineral Research, 32*(6), 1157–1162. https://doi.org/10.1002/jbmr.3078

Hill, R. (1948). A theory of the yielding and plastic flow of anisotropic metals. *Proceedings of the Royal Society of London. Series A. Mathematical and Physical Sciences, 193*(1033), 281–297. https://doi.org/10.1098/rspa.1948.0045

Hubbell, & Seltzer. (1996). *Tables of X-Ray Mass Attenuation Coefficients and Mass Energy-Absorption Coefficients 1 keV to 20 MeV for Elements Z = 1 to 92 and 48 Additional Substances of Dosimetric Interest.*

Isensee, F., Jaeger, P. F., Kohl, S. A. A., Petersen, J., & Maier-Hein, K. H. (2021). nnU-Net: A self-configuring method for deep learning-based biomedical image segmentation. *Nature Methods, 18*(2), 203–211.

Kang, Y., Engelke, K., & Kalender, W. A. (2003). A new accurate and precise 3-D segmentation method for skeletal structures in volumetric CT data. *IEEE Transactions on Medical Imaging, 22*(5), 586–598. https://doi.org/10.1109/TMI.2003.812265

Keaveny, T. M., Adams, A. L., Fischer, H., Brara, H. S., Burch, S., Guppy, K. H., & Kopperdahl, D. L. (2023). Increased risks of vertebral fracture and reoperation in primary spinal fusion patients who test positive for osteoporosis by biomechanical computed tomography analysis. *The Spine Journal, 23*(3), 412–424. https://doi.org/10.1016/j.spinee.2022.10.018

Keaveny, T. M., Crittenden, D. B., Bolognese, M. A., Genant, H. K., Engelke, K., Oliveri, B., Brown, J. P., Langdahl, B. L., Yan, C., Grauer, A., & Libanati, C. (2017). Greater gains in spine and hip strength for romosozumab compared with teriparatide in postmenopausal women with low bone mass. *Journal of Bone and Mineral Research, 32*(9), 1956–1962. https://doi.org/10.1002/jbmr.3176

Keaveny, T. M., Hoffmann, P. F., Singh, M., Palermo, L., Bilezikian, J. P., Greenspan, S. L., & Black, D. M. (2008). Femoral bone strength and its relation to cortical and trabecular changes after treatment with PTH, alendronate, and their combination as assessed by finite element analysis of quantitative CT scans. *Journal of Bone and Mineral Research, 23*(12), 1974–1982. https://doi.org/10.1359/jbmr.080805

Kiyohara, N., Yamamoto, M., Sugimoto, T., & Vashishth, D. (2015). Discordance between prevalent vertebral fracture and vertebral strength estimated by the finite element method based on quantitative computed tomography in patients with type 2 diabetes mellitus. *PLoS One, 10*(12), Article e0144496. https://doi.org/10.1371/journal.pone.0144496

Knowles, N. K., Quayyum, S., Ying, J., Stiles, C., & Beshay, D. (2024). Image-based finite element model stiffness and vBMD by single and dual energy CT reconstruction kernel. *Journal of Biomechanics, 177*, Article 112426. https://doi.org/10.1016/j.jbiomech.2024.112426

Kopperdahl, D. L., Aspelund, T., Hoffmann, P. F., Sigurdsson, S., Siggeirsdottir, K., Harris, T. B., Gudnason, V., & Keaveny, T. M. (2014). Assessment of incident spine and hip fractures in women and men using finite element analysis of CT scans. *Journal of Bone and Mineral Research, 29*(3), 570–580. https://doi.org/10.1002/jbmr.2069

Krčah, M., Székely, G., & Blanc, R. (2011). Switzerland fully automatic and fast segmentation of the femur bone from 3D-CT images with no shape prior. *Proceedings - International Symposium on Biomedical Imaging*, 2087–2090. https://doi.org/10.1109/ISBI.2011.5872823

Krekiehn, N. R., Yilmaz, E. B., Kruse, H. C., Meyer, C., & Glüer, C. C. (2023). Automated deep-learning-based vertebral body localization and instance segmentation for osteoporosis assessment using CT. *Informatik aktuell*, 171–176. https://doi.org/10.1007/978-3-658-41657-7_37. https://www.springer.com/series/2872

Labelle, F., & Shewchuk, J. R. (2007). Isosurface stuffing. *ACM Transactions on Graphics, 26* (3). https://doi.org/10.1145/1239451.1239508

Lee, D. C., Hoffmann, P. F., Kopperdahl, D. L., & Keaveny, T. M. (2017). Phantomless calibration of CT scans for measurement of BMD and bone strength—Inter-operator reanalysis precision. *Bone, 103*, 325–333. https://doi.org/10.1016/j.bone.2017.07.029

Lee, Y. H., Kim, J. J., & Jang, I. G. (2019). Patient-specific phantomless estimation of bone mineral density and its effects on finite element analysis results: A feasibility study. *Computational and Mathematical Methods in Medicine, 2019*. https://doi.org/10.1155/2019/4102410

Levi, C., Gray, J. E., McCullough, E. C., & Hattery, R. R. (1982). The unreliability of CT numbers as absolute values. *American Journal of Roentgenology, 139*(3), 443–447. https://doi.org/10.2214/ajr.139.3.443

MacNeil, J. A., & Boyd, S. K. (2008). Bone strength at the distal radius can be estimated from high-resolution peripheral quantitative computed tomography and the finite element method. *Bone, 42*(6), 1203–1213. https://doi.org/10.1016/j.bone.2008.01.017

Martinez, A., Cuenca, J., Herrera, A., & Domingo, J. (2002). Late lower extremity fractures in patients with paraplegia. *Injury, 33*(7), 583–586. https://doi.org/10.1016/s0020-1383(02)00163-8

Matheson, B. E., Neeteson, N. J., & Boyd, S. K. (2024). Establishing error bounds for internal calibration of quantitative computed tomography. *Medical Engineering & Physics, 124*. https://doi.org/10.1016/j.medengphy.2024.104109

Michalski, A. S., Besler, B. A., Michalak, G. J., & Boyd, S. K. (2020). CT-based internal density calibration for opportunistic skeletal assessment using abdominal CT scans. *Medical Engineering & Physics, 78*, 55–63. https://doi.org/10.1016/j.medengphy.2020.01.009

Morgan, E. F., Bayraktar, H. H., & Keaveny, T. M. (2003). Trabecular bone modulus-density relationships depend on anatomic site. *Journal of Biomechanics, 36*(7), 897–904. https://doi.org/10.1016/S0021-9290(03)00071-X

Nishiyama, K. K., Gilchrist, S., Guy, P., Cripton, P., & Boyd, S. K. (2013). Proximal femur bone strength estimated by a computationally fast finite element analysis in a sideways fall configuration. *Journal of Biomechanics, 46*(7), 1231–1236. https://doi.org/10.1016/j.jbiomech.2013.02.025

Ono, K., Ohashi, S., Oka, H., Kadono, Y., Yasui, T., Matsumoto, T., Omata, Y., & Tanaka, S. (2021). Evaluations of daily teriparatide using finite-element analysis over 12 months in rheumatoid arthritis patients. *Journal of Bone and Mineral Metabolism, 39*(2), 270–277. https://doi.org/10.1007/s00774-020-01146-6

Orwig, D. L., Kopperdahl, D., Keaveny, T., Magaziner, J., & Hochberg, M. (2021). Change in vertebral strength and bone mineral density in men and women over the year post-hip fracture: A subgroup analysis. *Archives of Osteoporosis, 16*(1). https://doi.org/10.1007/s11657-021-00907-8

Pauchard, Y., Fitze, T., Browarnik, D., Eskandari, A., Pauchard, I., Enns-Bray, W., Pálsson, H., Sigurdsson, S., Ferguson, S. J., Harris, T. B., Gudnason, V., & Helgason, B. (2016). Interactive graph-cut segmentation for fast creation of finite element models from clinical ct data for hip fracture prediction. *Computer Methods in Biomechanics and Biomedical Engineering, 19*(16), 1693–1703. https://doi.org/10.1080/10255842.2016.1181173

Pham, D. L., Xu, C., & Prince, J. L. (2000). Current methods in medical image segmentation. *Annual Review of Biomedical Engineering, 2*, 315–337. https://doi.org/10.1146/annurev.bioeng.2.1.315

Prado, M., Khosla, S., Chaput, C., & Giambini, H. (2021). Opportunistic application of phantom-less calibration methods for fracture risk prediction using QCT/FEA. *European Radiology, 31*(12), 9428–9435. https://doi.org/10.1007/s00330-021-08071-w

Quintiens, J., Coudyzer, W., Bevers, M., Vereecke, E., Van Den Bergh, J. P., Manske, S. L., & Van Lenthe, G. H. (2024). The quantification of bone mineral density using photon counting computed tomography and its implications for detecting bone remodeling. *Journal of Bone and Mineral Research, 39*(12), 1774–1782. https://doi.org/10.1093/jbmr/zjae163

Rho, J. Y. (1996). An ultrasonic method for measuring the elastic properties of human tibial cortical and cancellous bone. *Ultrasonics, 34*(8), 777–783. https://doi.org/10.1016/S0041-624X(96)00078-9

Rho, J. Y., Hobatho, M. C., & Ashman, R. B. (1995). Relations of mechanical properties to density and CT numbers in human bone. *Medical Engineering & Physics, 17*(5), 347–355. https://doi.org/10.1016/1350-4533(95)97314-f

Rincón-Kohli, L., & Zysset, P. K. (2009). Multi-axial mechanical properties of human trabecular bone. *Biomechanics and Modeling in Mechanobiology, 8*(3), 195–208. https://doi.org/10.1007/s10237-008-0128-z

Roy, A. G., Conjeti, S., Navab, N., & Wachinger, C. (2019). Bayesian QuickNAT: Model uncertainty in deep whole-brain segmentation for structure-wise quality control. *NeuroImage, 195*, 11–22. https://doi.org/10.1016/j.neuroimage.2019.03.042

Sekuboyina, A., Husseini, M. E., Bayat, A., Löffler, M., Liebl, H., Li, H., Tetteh, G., Kukačka, J., Payer, C., Štern, D., Urschler, M., Chen, M., Cheng, D., Lessmann, N., Hu, Y., Wang, T., Yang, D., Xu, D., Ambellan, F., … Kirschke, J. S. (2021). VERSE: A vertebrae labelling and segmentation benchmark for multi-detector CT images. *Medical Image Analysis, 73*. https://doi.org/10.1016/j.media.2021.102166

Snyder, P. J., Kopperdahl, D. L., Stephens-Shields, A. J., Ellenberg, S. S., Cauley, J. A., Ensrud, K. E., Lewis, C. E., Barrett-Connor, E., Schwartz, A. V., Lee, D. C., Bhasin, S., Cunningham, G. R., Gill, T. M., Matsumoto, A. M., Swerdloff, R. S., Basaria, S., Diem, S. J., Wan, C., Hou, X., … Keaveny, T. M. (2017). Effect of testosterone treatment on volumetric bone density and strength in older men with low testosterone a controlled clinical trial. *JAMA Internal Medicine, 177*(4), 471–479. https://doi.org/10.1001/jamainternmed.2016.9539

Verma, A., Jain, A., Sethy, S. S., Verma, V., Goyal, N., Vathulya, M., & Kandwal, P. (2024). Finite element analysis and its application in orthopaedics: A narrative review. *Journal of Clinical Orthopaedics and Trauma, 58*, Article 102803. https://doi.org/10.1016/j.jcot.2024.102803

von Mises. (1913). Mechanik der festen Körper im plastisch-deformablen Zustand. Nachrichten von der Gesellschaft der Wissenschaften zu Göttingen. *Mathematisch-Physikalische Klasse*, 582–592.

Walle, M., Duseja, A., Whittier, D. E., Vilaca, T., Paggiosi, M., Eastell, R., Müller, R., & Collins, C. J. (2024). Bone remodeling and responsiveness to mechanical stimuli in individuals with type 1 diabetes mellitus. *Journal of Bone and Mineral Research, 39*(2), 85–94. https://doi.org/10.1093/jbmr/zjad014

Walle, M., Marques, F. C., Ohs, N., Blauth, M., Müller, R., & Collins, C. J. (2021). Bone mechanoregulation allows subject-specific load estimation based on time-lapsed Micro-CT and HR-pQCT in vivo. *Frontiers in Bioengineering and Biotechnology, 9*. https://doi.org/10.3389/fbioe.2021.677985

Walle, M., Whittier, D. E., Schenk, D., Atkins, P. R., Blauth, M., Zysset, P., Lippuner, K., Müller, R., & Collins, C. J. (2023). Precision of bone mechanoregulation assessment in humans using longitudinal high-resolution peripheral quantitative computed tomography in vivo. *Bone, 172*. https://doi.org/10.1016/j.bone.2023.116780

Wang, L. I., Greenspan, M., & Ellis, R. (2006). Validation of bone segmentation and improved 3-D registration using contour coherency in CT data. *IEEE Transactions on Medical Imaging, 25*(3), 324–334. https://doi.org/10.1109/TMI.2005.863834

Wasserthal, J., Breit, H. C., Meyer, M. T., Pradella, M., Hinck, D., Sauter, A. W., Heye, T., Boll, D. T., Cyriac, J., Yang, S., Bach, M., & Segeroth, M. (2023). TotalSegmentator: Robust segmentation of 104 anatomic structures in CT images. *Radiology: Artificial Intelligence, 5*(5). https://doi.org/10.1148/ryai.230024

Weaver, A. A., Beavers, K. M., Hightower, R. C., Lynch, S. K., Miller, A. N., & Stitzel, J. D. (2015). Lumbar bone mineral density phantomless computed tomography measurements and correlation with age and fracture incidence. *Traffic Injury Prevention, 16*, Article S153. https://doi.org/10.1080/15389588.2015.1054029

White, J. B., Spokas, J. J., & Wilson, I. J. (1989). Tissue substitutes in radiation dosimetry and measurement. *International Commision on Radiation Units and Measurements (ICRU), 44*.

Whittier, D. E., Samelson, E. J., Hannan, M. T., Burt, L. A., Hanley, D. A., Biver, E., Szulc, P., Sornay-Rendu, E., Merle, B., Chapurlat, R., Lespessailles, E., Wong, A. K. O., Goltzman, D., Khosla, S., Ferrari, S., Bouxsein, M. L., Kiel, D. P., & Boyd, S. K. (2022). Bone microarchitecture phenotypes identified in older adults are associated with different levels of osteoporotic fracture risk. *Journal of Bone and Mineral Research, 37*(3), 428–439. https://doi.org/10.1002/jbmr.4494

Whittier, D. E., Samelson, E. J., Hannan, M. T., Burt, L. A., Hanley, D. A., Biver, E., Szulc, P., Sornay-Rendu, E., Merle, B., Chapurlat, R., Lespessailles, E., Wong, A. K. O., Goltzman, D., Khosla, S., Ferrari, S., Bouxsein, M. L., Kiel, D. P., & Boyd, S. K. (2023). A fracture risk assessment tool for high resolution peripheral quantitative computed tomography. *Journal of Bone and Mineral Research, 38*(9), 1234–1244. https://doi.org/10.1002/jbmr.4808

Zebaze, R., Libanati, C., McClung, M. R., Zanchetta, J. R., Kendler, D. L., Høiseth, A., Wang, A., Ghasem-Zadeh, A., & Seeman, E. (2016). Denosumab reduces cortical porosity of the proximal femoral shaft in postmenopausal women with osteoporosis. *Journal of Bone and Mineral Research, 31*(10), 1827–1834. https://doi.org/10.1002/jbmr.2855

Zenk, M., Zimmerer, D., Isensee, F., Traub, J., Norajitra, T., Jäger, P. F., & Maier-Hein, K. (2025). Comparative benchmarking of failure detection methods in medical image segmentation: Unveiling the role of confidence aggregation. *Medical Image Analysis, 101*, Article 103392. https://doi.org/10.1016/j.media.2024.103392

Zhang, Y. J. (1996). A survey on evaluation methods for image segmentation. *Pattern Recognition, 29*(8), 1335–1346. https://doi.org/10.1016/0031-3203(95)00169-7

Zysset, P., Pahr, D., Engelke, K., Genant, H. K., McClung, M. R., Kendler, D. L., Recknor, C., Kinzl, M., Schwiedrzik, J., Museyko, O., Wang, A., & Libanati, C. (2015). Comparison of proximal femur and vertebral body strength improvements in the FREEDOM trial using an alternative finite element methodology. *Bone, 81*, 122–130. https://doi.org/10.1016/j.bone.2015.06.025

Zysset, P., Qin, L., Lang, T., Khosla, S., Leslie, W. D., Shepherd, J. A., Schousboe, J. T., & Engelke, K. (2015a). Clinical use of quantitative computed tomography-based finite element analysis of the hip and spine in the management of osteoporosis in adults: The 2015 ISCD official positions-part II. *Journal of Clinical Densitometry, 18*(3), 359–392. https://doi.org/10.1016/j.jocd.2015.06.011

CHAPTER

21

Importance of reproducibility in bone fracture prediction using numerical models for clinical applications

David Mitton[1], Aurélie Levillain[1], Marc Gardegaront[1], Rémy Gauthier[2], Emile Saillard[3,4], Thomas Grenier[3], Jean-Baptiste Pialat[3,5,6], Cyrille Confavreux[4,5] and Hélène Follet[4]

[1]*Université Eiffel, Université Lyon 1, LBMC UMR_T9406, Lyon, France;* [2]*INSA Lyon, Université Lyon 1, CNRS, MATEIS, UMR5510, Villeurbanne, France;* [3]*INSA Lyon, Université Lyon 1, CNRS, INSERM, CREATIS, UMR 5220, UMR S1294, Villeurbanne, France;* [4]*Université Lyon 1, INSERM, LYOS UMR 1033, Lyon, France;* [5]*Hôpital Lyon Sud, Hospices Civils de Lyon, Centre Expert des Métastases Osseuses (CEMOS), Lyon, France;* [6]*Hôpital Lyon Sud, Hospices Civils de Lyon, Service de radiologie, Lyon, France*

1. Introduction

The increasing prevalence of skeletal pathologies, such as osteoporosis and bone metastases, have led to an increasing number of fractures.

Osteoporosis, is a skeletal disease in which the density and microarchitecture of bone are altered, leading to reduced mechanical strength and increased risk of fracture, particularly of the spine, wrist and hip. Osteoporosis is known as the "silent disease" because bone loss occurs without symptoms. Up to 70% of vertebral fractures remain undiagnosed (Cooper et al., 1992; Delmas et al., 2005). Lifetime fracture risk has been estimated at between 40% and 45% for women and 15%–27% for men. However, the risk for subsequent fractures after an initial fracture is the same in both men and women (Melton et al., 1992). Currently drugs are available for fracture prevention in the aging population, but many factors should be considered: e.g., prevention of osteoporosis in young individuals (to reach a high peak bone mass), the use of medical and nonmedical treatment options, and surgical techniques for fracture healing (Lems & Raterman, 2017).

Bone metastases result from the migration of cancer cells from a primary tumor to the bone through the bloodstream or lymphatic system. Bone metastases cause pathological fractures of the long bones, with the risk of impotence, and of the vertebrae, with the risk of spinal compression and spinal deformity. These are major possible complications impacting patient autonomy and mobility. Additionally,

Biomechanics and Mechanobiology of Bone. https://doi.org/10.1016/B978-0-443-26739-0.00024-7

depending on the cases, treatment can include decubitus and surgery, radiotherapy or interventional radiology. Such interventions can lead to suspension of oncological treatment, high costs and reduced patient survival (Decroisette et al., 2011; Yong et al., 2011). The clinician is faced with two situations: 1) patients with active cancer undergoing treatment and the challenge of not interrupting it; and 2) patients with cancer in remission but a persistent residual osteolytic lesion facing the "post-cancer" challenge of resuming physical activity. Given the number of cases and their consequences, assessing and preventing fracture risk of metastatic bone is a major challenge. Currently, fracture prediction tools used in clinical routine are of poor quality. For example, Mirels' score (Mirels, 1989) for long bones, has low sensitivity and specificity of around 0.5. The vertebral SINS score (Fisher et al., 2010) assesses stability and spinal cord neurological risk, but remains limited for prediction of fracture risk in intermediate cases (Confavreux et al., 2021).

3D numerical models based on quantitative computed tomography (QCT) by Faulkner et al. (1991) have been proposed in the literature to predict bone strength for osteoporosis (Keyak et al., 1997; Crawford, Cann, & Keaveny, 2003; Liebschner et al., 2003; Taddei et al., 2006; Keaveny et al., 2007; Duchemin, Mitton, et al., 2008; Dall'Ara et al., 2012; Viceconti et al., 2012; Zysset et al., 2013) or bone metastases (Benca et al., 2019; Cheal et al., 1993; Derikx et al., 2012; Falcinelli & Whyne, 2020; Johnson et al., 2020; Keyak et al., 2005; Sas, Tanck, et al., 2020; Spruijt et al., 2006; Stadelmann et al., 2020; Tanck et al., 2009; Yosibash et al., 2014). While promising, various steps limit their application in daily clinical practice. For example, a segmentation step is needed to build a 3D geometrical model of the bones. Both manual approaches (operator-dependent) or automatic methods (to avoid the influence of the operator) have been proposed. Providing patient-specific models from these clinical images requires their translation into quantitative data. Thus, different calibration methods with or without phantoms have been proposed (Ataei et al., 2022; Bartenschlager et al., 2023; Eggermont et al., 2019). Other parameters such as the relationships to convert density to mechanical properties or the constitutive law (Fleps et al., 2020; Knowles et al., 2016) and the failure criterion (see failure criteria paragraph below) could affect the bone fracture prediction. As no standard exists, the reproducibility of the proposed methods is essential for clinical applications. Model verification, validation and uncertainty quantification are critical to improve credibility (Oefner et al., 2021; Viceconti & Emili, 2024; Viceconti et al., 2005, 2021). This chapter reviews the various steps in model construction, model verification and model validation necessary for clinical applications, and specifically osteoporosis and bone metastases.

2. Model construction

The various steps in model construction are illustrated in Fig. 21.1 and detailed below.

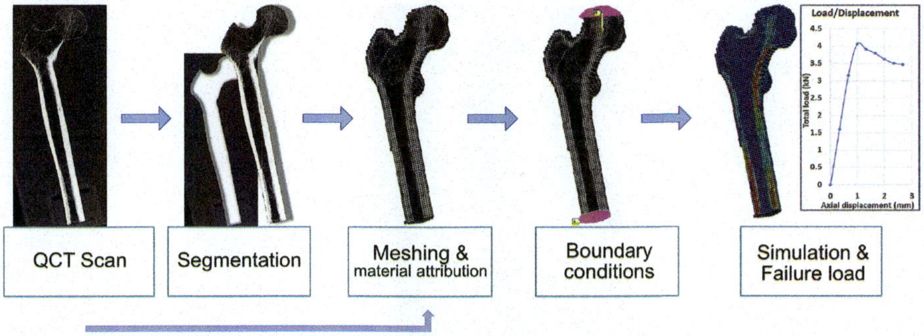

FIGURE 21.1

Various steps to build a subject-specific finite element model based on quantitative computed tomography (QCT): Example for the upper extremity of the femur.

- From Gardegaront, Marc, Sas, Amelie, Brizard, Denis, Levillain, Aurélie, Bermond, François, Confavreux, Cyrille B., Pialat, Jean-Baptiste, van Lenthe, G. Harry, Follet, Hélène & Mitton, David. (2024). Inter-laboratory reproduction and sensitivity study of a finite element model to quantify human femur failure load: Case of metastases. Journal of the Mechanical Behavior of Biomedical Materials, 158. https://doi.org/10.1016/j.jmbbm.2024.106676.

2.1 Segmentation

In the context of personalized bone fracture prediction, it is necessary to obtain a precise outline of the bones on the QCT image. This segmentation task can be done manually or automatically. Manual segmentations can be particularly tedious and time-consuming when drawn by expert clinicians and generally show poor reproducibility, with important operator variability. Semi-automatic (Pauchard et al., 2016) and automatic (Forsberg, 2015) methods are therefore better suited for this task. With the development of deep neural networks, specifically UNet based architectures (Isensee et al., 2021; Siddique et al., 2021), transformers (Dosovitskiy et al., 2021; Hatamizadeh et al., 2022), and foundation models (Ma et al., 2024), the performance of automatic algorithms for bone segmentation has greatly improved. For fracture prediction in patients with bone metastasis, obtaining accurate and automatic segmentation of the metastasis is an important challenge (Ataei et al., 2024; Rachmil, Artzi, et al., 2024). With the high diversity and variability of metastatic lesions, the task is difficult to achieve and methods have been proposed to tackle the data scarcity issue (Saillard, Gardegaront, et al., 2024).

Segmentation quality is generally evaluated with various metrics such as the Sørensen-Dice index, Hausdorff Distance, Mean Surface distance and many others (Maier-Hein et al., 2024).

The goal is to better understand the errors made by the algorithm. An example of distance errors on spine segmentation is shown in Fig. 21.2. This is especially important in the context of mechanical simulation on bone where the quantity of cortical

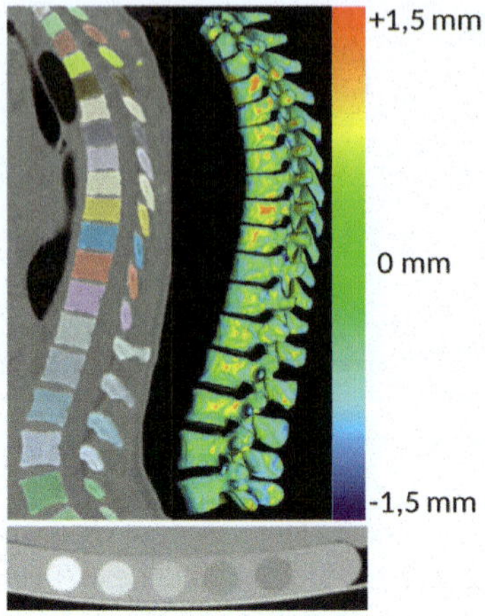

FIGURE 21.2

Quantitative computed tomography scan of spine on coronal view with associated automatic segmentation (left), distance map between automatic and manual segmentation (right) and acquisition phantom in axial view (bottom).

- From Saillard, E., Gardegaront, M., Levillain, A., Bermond, F., Mitton, D., Pialat, J.-B., Confavreux, C., Grenier, T. & Follet, H. (2024). Finite element models with automatic computed tomography bone segmentation for failure load computation. Scientific Reports, 14(1). https://doi.org/10.1038/s41598-024-66934-w.

bone plays an important role in bone resistance. To reduce systematic errors, dedicated post-processing can be considered.

Neural networks have to learn the segmentation task using manually annotated images with spatial, intensity and noise characteristics similar to the characteristic of the images on which the algorithm will later be used. Specific attention has to be paid to the acquisition parameters and pre-processing (resampling to median voxel size, filtering and intensity standardization) to result in smaller variability on contrast and image resolution. Domain adaptation methods (Guan & Liu, 2022) can be considered. This approach would allow improvements in segmentation performance by using already existing public databases of annotated CT images (Sekuboyina et al., 2021; Wasserthal et al., 2023). Sensitivity of the numerical models to input segmentation also has to be evaluated (Saillard, Gardegaront, et al., 2024) as the influence of the segmentation is not necessarily precisely determined by the different segmentation metrics.

2.2 Mesh

Segmented bone is meshed to build a Finite Element (FE) model. Two meshing approaches are commonly used: voxel-based (brick shaped elements) or geometry-based (tetrahedral or hexahedral elements) meshes (Sas, Tanck, et al., 2020) (Fig. 21.3). For the voxel-based approach, voxels from the scans are directly converted into elements. Therefore, this method is simple, fast, robust for mesh

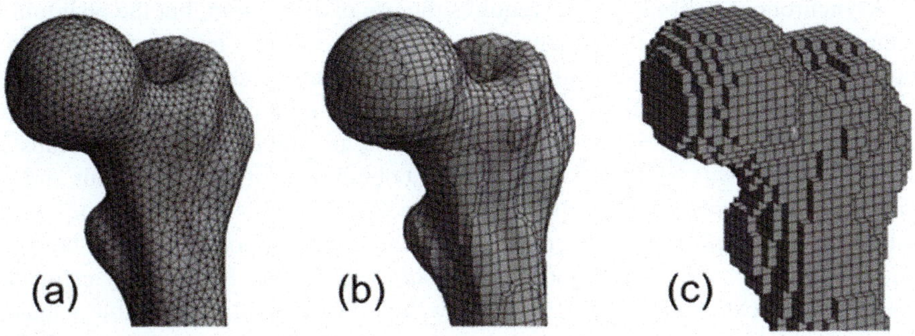

FIGURE 21.3

Different types of mesh based on the same femur geometry: Tetrahedral (a), hexahedral (b), voxel-based (c).

distortion (all elements have the same size), and can be fully automatized (Benca et al., 2019). However, inaccurate stress and strain distribution may be obtained at the surface due to the irregular mesh when the surface is curved. For the geometry-based approach, contours of the segmented bone surface are reconstructed, and the surface geometry is converted into a volumetric mesh (Sas, Tanck, et al., 2020). This geometry-based approach is now widely used and can be automatized, and both approaches give similar FE failure loads.

2.3 Density calibration

Bone mineral density (BMD, mg/cm^3), calculated from Hounsfield units (HU, with $-1000 =$ air, $0 =$ water, bone ~1000 HU), is extracted from the QCT scan and then converted into bone material properties that are integrated in the FE model. HU values depend on: (i) scanner settings and imaging parameters (electron density, beam-hardening, detector sensitivity, reconstruction algorithm, etc.), and (ii) patient-specific characteristics (body mass index, etc.) (Benca et al., 2020). Therefore, calibration of the QCT scan is required to convert HU to BMD (Lee et al., 2017). This calibration can be performed in three different ways.

The gold standard is done by using an external calibration phantom made of solid or liquid bone equivalent materials with known densities. The phantom is placed under the patient and is scanned simultaneously with the patient (in-scan calibration) (Fig. 21.2). Linear regression is then performed to obtain the relationship between HU and BMD. This method is the most widely used for creating FE models of bone fracture prediction, because it is patient-specific, accurate, and accounts for differences in imaging parameters (Ataei et al., 2022). However, these phantoms are expensive and in-scan calibration can be a logistical burden, making it difficult to use in clinical routine. Therefore, two other approaches have emerged: asynchronous calibration and phantomless calibration.

Asynchronous calibration also requires the use of a phantom, but the phantom is scanned separately from the patient. This method is easier to implement in clinical routine compared to in-scan calibration and allows the processing of CT-scans performed without a phantom (Brown et al., 2017). The relationship between HU and BMD is established using the calibration phantom and applied to patients' scans. A good correlation was found between the BMD obtained using asynchronous and in-scan calibration methods for the femoral neck ($r = 0.96$, standard error of the estimate $= 0.020$ g/cm^3) (Brown et al., 2017), as well as a good correlation between the FE failure loads obtained using both methods ($R^2 = 0.99$, estimated bias $= 16.4\%$) (Michalski et al., 2020). However, this method needs stable scanners with a standard imaging protocol, regular recalibration, and does not account for the variability in body mass index between patients, which influences HUs (Michalski et al., 2020).

Phantomless calibration is a patient-specific method based on the patient's own internal tissue. It consists of extracting the peak value of two or three materials of known density from an HU histogram, such as air, adipose tissue, muscle, blood, cortical or trabecular bone (Eggermont et al., 2019; Winsor et al., 2021). A good correlation was found between the FE failure loads obtained with the in-scan calibration and phantomless calibration using air, adipose tissue, and muscle ($R^2 = 0.94$, mean difference $= 20$ N) (Eggermont et al., 2019). However, similar to asynchronous calibration, this method is highly dependent on CT scan protocol, especially the peak kilovoltage, with a change from 120 kVp to 140 kVp leading to a 13.8% decrease in the failure load of femurs compared to phantom calibration (Ataei et al., 2022). To address this issue, Michalski et al. proposed to use mass attenuation coefficients to estimate the scan effective energy and then convert HU to equivalent BMD (Michalski et al., 2020).

2.4 Material properties attribution

Once the femur is meshed, each element is assigned material properties based on its density, using elasticity-density relationships (linear models), as well as yield strength-density and ultimate strength-density relationships (nonlinear models). These relationships are usually derived from ash density (ρ_{ash}), apparent density (ρ_{app}), QCT density (ρ_{QCT}), dry density (ρ_{dry}) or bone volume fraction (BV/TV) (Fleps et al., 2020).

Specific relationships exist for material attribution depending on: (i) site, such as vertebra (Keller, 1994; Kopperdahl et al., 2002; Morgan et al., 2003; Prado et al., 2021) and femur (Duchemin, Bousson, et al., 2008; Keller, 1994; Keyak et al., 1994; Morgan et al., 2003), (ii) type of bone (trabecular, cortical, pooled) (Fleps et al., 2020; Helgason et al., 2016), and (iii) bone pathology (osteoporosis, metastasis).

In FE models of the femur, elements with modulus values less than 0.01 MPa are commonly assigned a value of 0.01 MPa (Keyak, 2001).

2.5 Constitutive laws and failure criteria

Various constitutive laws have been used in the literature to model the stress-strain relationship for each element of the bone model. They can be divided into two categories: linear and non-linear constitutive laws. For linear models, each element is only assigned a Young's modulus value using a density-elasticity relationship. Because a linear model does not capture bone failure, a criterion needs to be defined to assess the failure load. Most commonly used criteria estimate that bone failure occurs when a critical volume of contiguous elements reaches a stress or strain value above a threshold value. The critical volume that is considered can be a fixed value or a percentage of elements. Several stress- and strain-based failure criteria exist, such as Coulomb-Mohr, Hoffmann, Hoffman strain analog, maximum principal strain, maximum shear strain (Johnson et al., 2020), Von Mises stress (Duchemin, Mitton, et al., 2008), and yield strain (Pistoia et al., 2002). For non-linear models, post-failure behavior is modeled, which can improve the precision of FE models (Keyak et al., 2005). For example, an elastic-perfectly plastic model has been used for the vertebra, with a constant stress after strain reaches a specific yield strain value (Kopperdahl et al., 2002). Failure was considered when the applied displacement on the vertebra reached a set value (Kopperdahl et al., 2002). In their FE model of femur, Keyak et al. added a strain softening phase followed by an indefinite perfectly plastic phase (Keyak et al., 2005). Each material parameter is defined from ash density. Failure load is then defined as the maximal force. Most models consider isotropic mechanical properties. However, more complex constitutive laws have been developed to capture bone anisotropy (Zysset & Rincón, 2006).

2.6 Boundary conditions

Boundary conditions are applied in the FE model to mimic the stress experienced by the bone in a specific position (e.g., standing position), during movement (e.g., walking, changing from sitting to standing position, etc.) or when falling. The direction and intensity of the in vivo applied forces can be obtained from studies using instrumented implants (Bergmann et al., 2001). These boundary conditions are simplified compared to in vivo loading, so that they can be easily controlled when validating the model experimentally (axial compression, torsion, etc.). The FE model can be created for the whole bone (vertebra, femur, tibia, radius, etc.), a portion of the bone (for example when the bone is not entirely scanned), or on bone segments (vertebral segments), resulting in various boundary conditions. Most FE models do not account for muscle or ligament forces, but approaches coupling musculoskeletal models with FE models have emerged to apply physiological loading (Altai et al., 2021), for example for osteoporotic femur (Viceconti et al., 2012).

3. Model verification

The accuracy of finite element models is fundamentally tied to rigorous verification and validation processes. Verification ensures that the mathematical formulation of the model and its implementation are correct, while validation assesses whether the model accurately reflects real-world behavior (Viceconti et al., 2021). In the context of bone biomechanics, these processes are critical, given the complexity of biological tissues, which exhibit inhomogeneous and anisotropic properties, complex geometries and inter individual variabilities linked to age, genetics or the environment. Finite element method verification involves a series of computational techniques designed to ensure that models are not only mathematically correct but also capable of converging toward accurate results. Sensitivity studies complement these efforts by testing how changes in model inputs, such as material properties or boundary conditions, influence the model's outputs (Saltelli et al., 2004). This two-step process—convergence through verification and robustness testing via sensitivity studies—is critical for ensuring that finite element methods used in biomechanics are both reliable and clinically applicable. In the following sections, we delve deeper into specific aspects of finite element method verification in biomechanics, focusing on convergence studies, sensitivity analyses, and the influence of different modeling parameters such as element type, material laws, and boundary conditions. By examining these aspects, we aim to highlight the importance of a comprehensive verification framework for reliable finite element method predictions in bone biomechanics.

3.1 Convergence study

Convergence studies are essential for verifying the accuracy and stability of finite element models, especially in personalized bone models that involve complex geometries and heterogeneous meshes. These studies determine whether refining the mesh—by increasing the number of elements or improving their quality—leads to stable, converged solutions. If further refinement causes minimal changes in outputs like stress, strain, resultant load, or displacement, the model is considered reliable.

In personalized bone modeling, computed tomography scans are used to segment the bone and define both the geometry and mechanical properties of each element. As the mesh is refined, not only does the element size change, but so do the mechanical parameters, since smaller elements capture different voxel data. Therefore, convergence studies must account for both geometric refinement and changes in material properties, ensuring the model is stable despite these variations.

While the h-version of the finite element method (refining mesh size) is commonly used, the p-version finite element method, which increases the polynomial order of the shape functions, has also been explored. Yosibash et al. (2010) demonstrated that for proximal femur models, the p-finite element method achieved rapid convergence with relatively coarse meshes, making it a powerful tool for modeling complex bone geometries without significantly increasing computational costs.

Conventional finite element methods in bone biomechanics typically use lower-order elements, like linear tetrahedral or hexahedral elements, where mesh refinement is concentrated in areas of high stress or strain gradients, such as joints (Klennert et al., 2017). However, incomplete convergence in these regions can lead to significant errors, as a coarse mesh may cause artificial stiffening and incorrect predictions of mechanical behavior. Henninger et al. reviewed verification and validation of biomechanical models (Henninger et al., 2010) and stressed the importance of mesh convergence studies, especially in cases where the geometry and material properties of bone are highly variable. Typically, convergence is considered satisfactory when the difference between successive refinements falls below a set threshold, often around 5% (Henninger et al., 2010).

Mesh quality is crucial for achieving convergence in finite element models. Distorted elements, such as poorly shaped tetrahedra, can lead to inaccurate results even when the mesh is refined. Therefore, maintaining high element quality is just as important for precise predictions as refining the mesh. Burkhart et al. (2013) highlighted that 95% of studies overlooked mesh quality, which can greatly affect the accuracy of simulations. Moreover, while 33% of studies performed mesh sensitivity analyses, only half of provided detailed results (Burkhart et al., 2013).

In conclusion, convergence studies and mesh quality are essential for ensuring that finite element method models in bone biomechanics produce stable and reliable results. These studies help avoid errors due to mesh sensitivity and are crucial for accurate predictions of bone mechanics, such as fracture risk.

3.2 Sensitivity studies

Sensitivity studies are a crucial to verifying the accuracy and robustness of FE models. These studies aim to evaluate how changes in model inputs—such as material properties, boundary conditions, and mesh characteristics—impact the simulation results. Understanding which factors have the most significant influence on model predictions is vital for ensuring that the simulation accurately represents the mechanical behavior of bones.

The key aspect of a sensitivity study is to correctly identify the input variables, their distribution—or at least their range of plausible values—and to choose the right method regarding the computational cost and information wanted as a result of the study. There are several commonly used sensitivity methods.

One-at-a-time method (Czitrom, 1999): Choosing a reference set of values that produces one simulation result, then changing one parameter and observing the difference between the reference simulation result and repeating this process for each parameter (Fig. 21.4a). The advantage of this method is a lower computational cost, however the interactions between parameters are not observed as the space of input parameters is lightly explored.

Morris method (Morris, 1991): Defining a trajectory composed of a reference set of values, then changing one parameter at a time by a small step, then repeating the trajectory process a number of time sufficient to explore the input space (Fig. 21.4b).

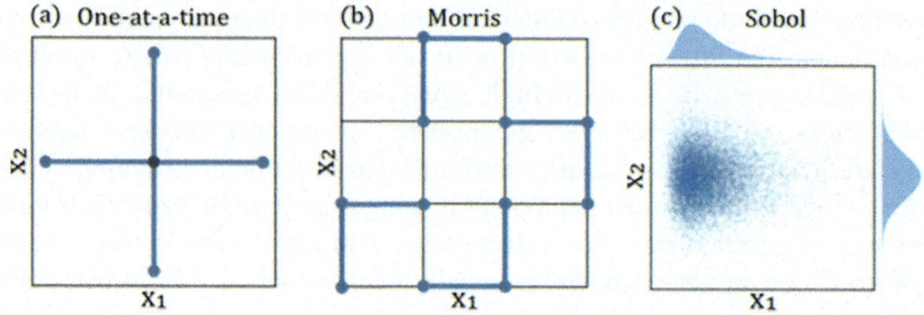

FIGURE 21.4

Illustration of a 2D input space exploration for (a) one-at-a-time method, (b) Morris method and (c) Sobol method.

This allows the measurement of the mean elementary effect of each parameter and the standard deviation of the elementary effect, which corresponds to the non-linear or interaction between parameters. The computational cost is relatively small, but the results are qualitative and depend on the range of each input parameter chosen to define the input space.

Sobol Analysis (Sobol′, 2001): Sampling an input set of values based on their probability distribution (Fig. 21.4c). This is a variance-based sensitivity analysis method that decomposes the output variance of a model into contributions from each input parameter and their interactions. It calculates Sobol indices, including the first-order index, which quantifies the direct effect of an individual parameter on the output, and the total-order index, which captures both the direct effects and interactions with other parameters. Higher-order indices measure the effects of interactions between two or more parameters. Sobol analysis provides quantitative and detailed insights into the influence of inputs, but it requires a large number of model evaluations, making the computational cost high.

Computed tomography-based FE models typically consist of four steps involving uncertainty: the definition of the mechanical parameters derived usually from the density of bone obtained from the conversion of the computed tomography data using a calibration method, the segmentation of the bone obtained from the computed tomography data, the orientation of the bone obtained from defining a local coordinate system, and the definition of the boundary conditions.

Global sensitivity analyses are not frequently used in bone models. However, uncertainty studies and local sensitivity give a hint about the behavior of bone models regarding the uncertainty of their input parameters.

Dozens of density to Young's modulus relationships exist (Eberle et al., 2013; Knowles et al., 2016). Eberle et al. (2013) investigated whether a single validated density-elasticity relationship could be applied to subject-specific FE models of human femurs. They found that the choice of the density-elasticity relationship significantly influenced the predicted mechanical responses. Sensitivity analysis showed that a small variation (2%) in the calibration curve led to a 4% change in model stiffness, indicating that the models are highly sensitive to the density-elasticity relationship. Additionally, the study showed there was no single equation universally optimal across all specimens.

An uncertainty quantification and global sensitivity study was performed on patient-specific FE models of femurs using Sobol indices (Wille et al., 2016). They focused on uncertainties related to bone density, Young's modulus, and loading conditions. Their results showed a 40% coefficient of variation in the principal strain response, with 60%–80% of this variance attributed to the relationship between bone density and Young's modulus.

Goodheart et al. (2015) investigated the fracture prediction accuracy of FE models under three loading scenarios: axial compression, level walking, and aggressive stair ascent. Their findings revealed fracture prediction rates of 80%, 80%, and 100%, respectively, suggesting consistent predictive capabilities across different boundary conditions.

Similarly, Amini et al. (2023) explored the sensitivity of femur model stiffness to variations in boundary conditions, material properties, and segmentation thresholds. The study concluded that boundary conditions related to embedding and loading were the most significant factors, contributing more than 5% to the variance in stiffness.

We conducted a global sensitivity analysis of a personalized femur model's fracture prediction capabilities, assessing factors such as density calibration, segmentation, orientation, and boundary conditions using the Morris method (Gardegaront et al., 2024). We found that all parameters were influential, except for rotation around the diaphyseal axis, contributed to more than 5% of the model's variance (Fig. 21.5a). In a separate model, the same sensitivity found that segmentation alone could influence fracture predictions by as much as 40% (Fig. 21.5b).

Similarly, Taddei et al. (2006) performed a sensitivity analysis on subject-specific femur models, investigating the effects of segmentation, density calibration, and material properties. Their study highlighted the dominant influence of

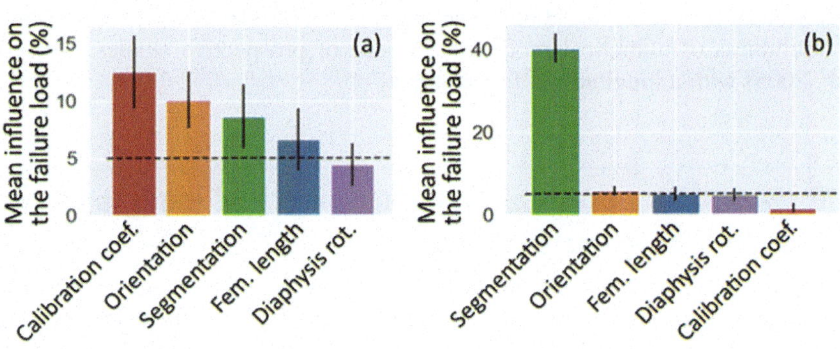

FIGURE 21.5

Results of a sensitive study using the Morris method on the simulated failure load for the Leuven FE model (a) (Gardegaront et al., 2024) and for the Lyon FE model (b).

segmentation and boundary conditions on prediction outcomes, although the results varied based on the individual subject analyzed.

Collectively, these studies underscore how accurate fracture prediction in patient-specific femur models relies heavily on proper characterization of material properties, precise boundary condition definition, and careful segmentation. Overall, comprehensive sensitivity analyses are crucial to improving the reliability of FE models.

4. Model validation ex vivo

The validation step is essential in bone model development. Model validations against a set of experimental data have been proposed. Table 21.1 and Table 21.2 summarize some of the studies assessing the prediction of the failure loads, respectively for femurs and vertebrae.

Even if the number of cases considered are important in terms of bone samples, validation performed in a specific laboratory could lead to overfitting. For instance, a model may not be as efficient compared to another experimental data set and in another laboratory.

Inter-laboratories comparisons could give more confidence in the simulation results (Allard et al., 2024; Kluess et al., 2019). As there is no standard, the models present different input characteristics as presented in the previous paragraph. Model assessment raises the influence of uncertainties for the model validation. Experimental uncertainties strongly influence model predictive ability (Amini et al., 2023). More recently, we showed the influence of the experimental data set in inter-laboratories comparisons (Gardegaront et al., 2024). These results (Fig. 21.6) highlight the importance of reproducibility evaluation, before considering clinical applications.

After ex vivo validation, the following steps should consider retrospective studies where the outcome of the patient is known to validate the prediction of the model (Viceconti et al., 2005). The best way to validate the clinical accuracy and usefulness of numerical models is the use of prospective studies (Viceconti et al., 2005) with a longitudinal follow-up.

5. Bone fracture prediction using numerical models in vivo

5.1 Case of osteoporosis

One major cause that needs reliable numerical models for bone fracture prediction is the osteoporosis affecting hundreds of millions of people across the world (Hernlund et al., 2013). This age-related pathology decreases bone mass and results in an increased risk of bone fracture. Osteoporosis could be defined based on the measurement of an areal Bone Mineral Density (aBMD, g/cm^2) through the 2D Dual X-ray

Table 21.1 Selection of some studies assessing the validation of the failure load prediction for the femur.

Articles	Elements types	Constitutive law	Failure criteria	Boundary conditions	Validation
Duchemin, Mitton, et al. (2008)	Hexahedral	Linear	Nb of continuous element over a stress threshold	Compression	40 femurs, $r^2 = 0.87$
Tanck et al. (2009)	Voxel-based	Nonlinear	Maximum load-displacement curve	Compression	5 pairs of femurs (half with artificial lesions), $r^2 = 0.92$
Yosibash et al. (2014)	Tetrahedral	Linear	One surface element over a strain threshold	Compression	7 pairs of femurs with metastases, $R^2 = 0.78$
Benca et al. (2019)	Voxel-based	Nonlinear	Maximum load-displacement curve	Compression	16 pairs of femurs, $r^2 = [0.77–0.98]$
Sas, Ohs, et al. (2020)	Voxel-based	Nonlinear	Maximum load-displacement curve	Compression	10 pairs of femurs with with artificial lesions, $r^2 = 0.93$
Johnson et al. (2020)	Voxel-based	Linear	Nb of continuous element over a strain threshold	Torsion	11 femurs with artificial lesions, $r^2 = 0.83$

Table 21.2 Selection of some studies assessing the validation of the failure load prediction for the vertebral body.

Articles	Elements types	Constitutive law	Failure criteria	Boundary conditions	Validation
Crawford, Rosenberg, and Keaveny (2003)	Voxel-based	Linear	Vertebral stiffness as a surrogate for vertebral strength	Compression	17 vertebrae, $r^2 = 0.92$
Buckley et al. (2007)	Voxel-based	Linearly elastic-perfectly plastic	Axial force at 3% apparent strain	Compression	81 vertebrae, $r^2 = 0.80$
Dall'Ara et al. (2010)	Voxel-based	Nonlinear	Maximum load-displacement curve	Anterior compression	37 vertebrae, $r^2 = 0.79$
Choisne et al. (2018)	Hexahedral	Linear	1 cm3 of contiguous element over 1.5% of strain	Anterior compression	28 vertebrae, $r^2 = 0.95$
Stadelmann et al. (2020)	Voxel-based	Nonlinear	Maximum load-displacement curve	Anterior compression	57 vertebrae with metastases, $r^2 = 0.78$

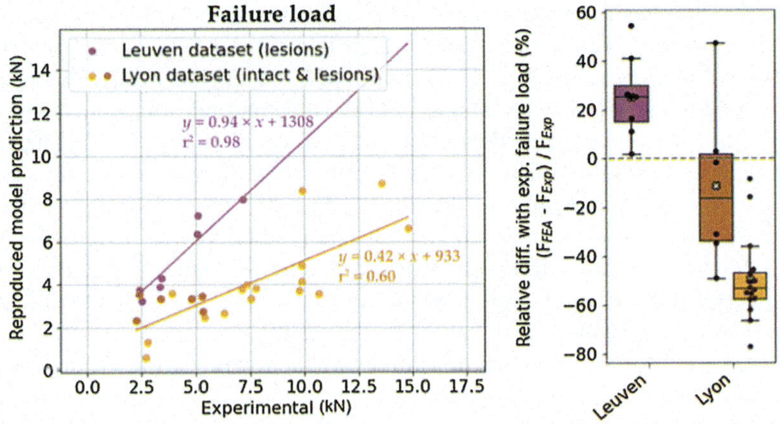

FIGURE 21.6

From left to right: Scatter plot and regression, box plot of the differences between reproduced model and experimental failure load on each dataset. In purple: Leuven's dataset; in orange: Lyon's dataset (femurs with created lesions); in yellow: Lyon's dataset (intact femurs) (Gardegaront et al., 2024).

- From Gardegaront, Marc, Sas, Amelie, Brizard, Denis, Levillain, Aurélie, Bermond, François, Confavreux, Cyrille B., Pialat, Jean-Baptiste, van Lenthe, G. Harry, Follet, Hélène & Mitton, David. (2024). Inter-laboratory reproduction and sensitivity study of a finite element model to quantify human femur failure load: Case of metastases. Journal of the Mechanical Behavior of Biomedical Materials, 158. https://doi.org/10.1016/j.jmbbm.2024.106676.

Absorptiometry (DXA) imaging method. With an aBMD under a defined threshold, osteoporosis is diagnosed for the patient. Some tools for the assessment of fracture risk have already been developed, such as the FRAX tool considering DXA results and some other clinical parameters (Kanis et al., 2008). It has been shown, however, that 50% of bone fractures occur in women with undetected osteoporosis based on DXA measurements (Siris et al., 2001). This highlights the need to introduce new criteria, in addition to aBMD, such as biomechanical criteria, in the prediction of osteoporotic fractures.

In that context, FE-based models have become a relevant tool to provide further insights regarding the risk of fracture and help for clinical decision. In 2018, Viceconti et al. highlighted that clinical QCT based FE-models showed better performance in predicting femoral strength compared to standard DXA-based aBMD (Viceconti et al., 2018). In 2021, Schileo and Taddei presented an additional study reporting better prediction using QCT-based FE-models for vertebral fracture compared to aBMD, but performance was lower than vBMD measured on 3D QCT scans (Schileo & Taddei, 2021). In 2022, Fleps et al., applied QCT-based FE analysis to predict hip fracture on a large cohort of 600 subjects. Again, the fracture group was better discriminated through the FE models compared to the standard aBMD parameter (Fleps et al., 2022). In this study, the authors also investigated the influence of the constitutive laws used to define elements mechanical properties. The elements properties were either linear, which is a similar behavior in tension and compression ($FEMs_{LIN}$, in Fig. 21.7), non-linear, signifying a different behavior in compression and tension ($FEMs_{NL}$, in Fig. 21.7), and non-linear with consideration of strain-rate properties dependency ($FEMs_{NLSR}$, in Fig. 21.7). The results showed that whatever the constitutive law considered, the prediction was always better using FE models compared to the aBMD (Fig. 21.7). By considering the non-linear behavior in tension and compression, the prediction was better than with a linear behavior. However, strain rate bone dependency did not improve the discrimination between the fracture and the control groups (Fig. 21.7). This highlights the determinant role of the model definition in the results.

Later, Taylor et al. showed that discriminating the fracture group from the control one on a cohort of 94 postmenopausal women using FE-models can be improved when used in combination with active shape and appearance modeling (AUROC = 0.72 and 0.86 for the aBMD and the FE-augmented model, respectively) (Taylor et al., 2021). They concluded that no further information can be extracted from image data and that a limit had been reached. Similarly, in 2024, Awal et al. showed that proximal femur fracture risk increased when considering different loading directions (Awal & Faisal, 2024), which was already highlighted by Viceconti et al. (2018).

This model based on the opportunistic use of routine clinical data can reduce the radiation dose applied to the patient and limit the number of tasks necessary by the medical team due to an additional CT scan. A major difficulty of building a model based on a previous CT scan performed for other indications lies in the lack of calibration at the time of the scanning procedure. To override this challenge,

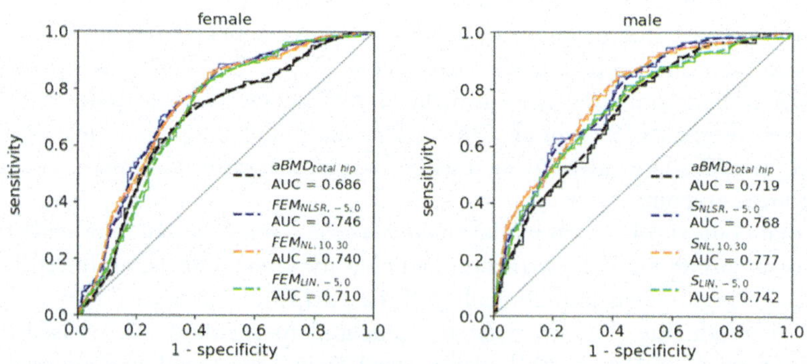

FIGURE 21.7

Classification performance using the receiver-operator characteristics curves (ROC curves) for the female and male populations.

- From Fleps, Ingmar, Pálsson, Halldór, Baker, Alexander, Enns-Bray, William, Bahaloo, Hassan, Danner, Michael, Singh, Navrag B., Taylor, William R., Sigurdsson, Sigurdur, Gudnason, Vilmundur, Ferguson, Stephen J. & Helgason, Benedikt. (2022). Finite element derived femoral strength is a better predictor of hip fracture risk than aBMD in the AGES Reykjavik study cohort. Bone, 154. https://doi.org/10.1016/j.bone.2021.116219.

asynchronous calibration methods have been applied, consisting of scanning a calibration phantom regularly and assessing scan stability, or internal calibration methods, based on the HU of the different tissues (fat, blood, etc.) composing the scanned anatomical region (Guenoun & Champsaur, 2023). In 2021, Prado et al. compared vertebral fracture strength prediction on 111 patients with the BMD estimated using a calibration phantom (Mindways Inc.) and using an internal calibration method based on the HU of the patient's fat, psoas muscle, as well as the air and the subject perimeters (Prado et al., 2021). They found that the fracture load obtained with the phantom and the phantomless method were highly correlated ($R^2 = 0.99$). Nevertheless, in this study, the authors did not assess the performance for such phantomless FE models to discriminate a fracture from a control group. In a previous study, made on a cohort of 1500 patients that underwent a hip fracture and 1300 control patients, Adams et al. found that a phantom-less conversion of HU into bone density was at least as effective as the standard DXA aBMD measurement to predict hip fracture strength (Adams et al., 2018).

Despite the promising results offered by FE-model based prediction of fracture strength compared to the standard DXA-based aBMD measurements, their implementation in routine clinical practices remains difficult. This is mainly due to the difficulty in providing true validations for these models and simulations in the field of biomechanics. To standardize the development of models to support clinical decision-making and promote their use in routine clinical practices, 10 rules of credible practice of modeling and simulation were established by an international committee (Erdemir et al., 2020). Despite the difficulty of implementing such model-based methods in clinics, VirtuOst software has been recently approved by the FDA as a class-II medical device (Keaveny et al., 2020). This method is based on the use of routine clinical CT scans to measure BMD and estimate bone strength through FE analysis. The estimated strength was highly correlated to the strength measured through standard biomechanical testing ($R^2 = 0.97$) and has shown better discrimination between fracture and control cases compared to the standard DXA method (Lee et al., 2016). In addition to bone strength, fall severity was considered in a multiscale model to predict the absolute risk of femoral fracture in a postmenopausal population (Bhattacharya et al., 2019).

5.2 Case of bone metastases

Bones are a common site of metastases in advanced cancers (De Angelis et al., 2024), with pain being predominant symptom. This pain increases morbidity and negatively impacts quality of life. Effective management of bone metastases requires a multidisciplinary approach, primary focused on alleviating pain and improving quality of life. Approximately 10% of all skeletal metastasis affect the long bones, with 70% affecting the axial skeleton (Coleman, 2006; Hage et al., 2000). Within the appendicular skeleton, the proximal femur and proximal humerus are the most frequently affected sites. Notably, almost 30% of patients with femoral metastases experiencing pathologic fractures, underscoring the clinical importance of timely diagnosis and intervention (Ashford et al., 2012; Bailey, Hackney, et al., 2020). The diagnosis and treatment of bone metastasis involve a variety of therapeutic modalities, specialists and caregivers. To optimize outcomes, diagnosis and treatment, a multidisciplinary team approach is essential, integrating expertise across disciplines to ensure comprehensive care (Confavreux et al., 2021; Shibata et al., 2016).

To adapt a simulation to a patient, pre-analysis in bone simulation need validation with ex vivo femurs (Derikx et al., 2012; Falcinelli & Whyne, 2020; Yosibash et al., 2020) or vertebrae (Alkalay et al., 2021, 2022; Bailey, Hackney, et al., 2020; Jackman et al., 2016; Soltani et al., 2024; Stadelmann et al., 2020). A multicenter prospective study was conducted to identify significant predictors of physicians' treatment plan for skeletal metastasis based on clinical fracture risk assessments and using the proposed CT-based Rigidity Analysis (Nazarian et al., 2015). One review established the state-of-the-art for the risk assessment of long bone metastases fractures, from simple radiologic scores to more complex multidimensional bone models, to define new decision-making tools (Nguyễ;n et al., 2021).

To date (2024), only a few studies have attempted to give a fracture risk prediction on patients.

Recently, Sas et al. provide an overview of the state-of-the-art in FE modeling for the treatment decision of metastatic bone lesions in the femur (Sas, Tanck, et al., 2020). They summarized the clinical and mechanical results of femoroplasty as a prophylactic treatment method and then reviewed current FE models for fracture risk assessment of metastatic femurs. They concluded that FE models could help to select appropriate candidates for whom femoroplasty provides a sufficient increase in strength and to further improve the mechanical benefit by optimizing the locations for cement augmentation.

Nijmegen developed the Bone Strength score (BOS) (Derikx et al., 2012; Eggermont et al., 2018, 2020; Tanck et al., 2009), which starts with prevention at a clinical level. The BOS score is based on a CT-based FE model that objectively calculates bone strength. In their pilot study, the added clinical value of the BOS score toward treatment-related decision making was assessed. In December 2019, the BOS score was implemented in four radiotherapy centers. The BOS scores and fracture risks of individual patients were calculated and returned to the physician to assist in treatment decisions. The authors concluded that the BOS score is an objective tool to assess fracture risk in femoral bone metastases and aids physicians and patients in making a more informed decision regarding the most appropriate treatment. In these studies, the knee joint center and centers of the condyles were needed to create the FE model, hence BOS scores of incompletely scanned femurs could not be calculated. Eggermont and colleagues then developed a statistical shape model to align finite element models of femurs with a removed anatomy, which finally gave relatively accurate BOS scores (Eggermont et al., 2024). Other strategies were also developed to obtain the fracture risk of metastatic long bones. For example, a retrospective study used CT-based finite element analyses (CTFEA) to identify a clear quantitative differentiating factor between patients who are at imminent risk of fracturing their femur and those who are not, and to identify the exact location of maximal weakness where the fracture is most likely to occur (Sternheim et al., 2018, 2020). The authors found that all the patients who sustained a pathological fracture of the femur had a strain fold ratio >1.48. And CTFEA were far better at accurately predicting the risk of fracture and its location compared with the Mirels' score. Other authors suggested combining several results, arguing that, for example, in the prediction of osteoporotic fractures, that the Fracture Risk Assessment Tool integrates clinical data with radiologic data (femoral-neck bone mineral density). Developing a similar tool for patients with metastatic disease and myeloma would be an important and realistic goal in orthopedics. Ideally, such a tool would incorporate a CT rigidity analysis, finite-element modeling, or positron emission tomography-CT for a biomechanical assessment and PATHFx for survival prediction (Damron, 2021; Damron & Mann, 2020). Following this, Lin and colleagues published a study on biomechanical computed tomography (BCT) which is a radiomic technique that measures bone mineral density (BMD) and bone strength from CT scans (Lin et al., 2024). They used a multi-center retrospective cohort study among patients with metastatic hormone-sensitive prostate cancer between 2013 and 2020 and who received CT abdomen/pelvis or PET/CT within 48 weeks before and during follow-up. They found BCT assessments were strongly correlated with dual-energy X-ray absorptiometry, predicted subsequent pathologic fracture, and identified additional men indicated for antiresorptive therapy (Lin et al., 2024).

Work on vertebrae is still in its early stages. After a review of the state-of-the-art in clinical and experimental approaches for prediction of pathologic fractures with bone metastases, Bailey and colleagues concluded that mechanics of solids (MOS) and finite element analysis (FEA) approaches offer the ability to directly compute the effects of metastasis-mediated damage to skeletal anatomy on the structural rigidity of the affected skeletal structure (Bailey, Hackney, et al., 2020; Stadelmann et al., 2020). They also suggested that, routinely collected as part of clinical care for metastatic patients, the use of bone turnover biomarkers is increasingly being advocated as prognostic indicators for the risk of skeletal events, and patient survival. Combined with imaging-based approaches, utilization of bone turnover markers values may provide improved prediction of the effect of disease trajectory on the risk of fracture, thus improving clinical management of patients with bone metastasis. Taken together, advancement in current approaches and the development of new methodologies will elucidate the differential effect of lesion type on bone quality, improving the prediction of pathologic fractures, and help guide treatment (Bailey, Stadelmann, et al., 2020). Alkalay et al. compared the loss of compressive strength and the ability of CT-derived bone lesion quality (classification of vertebral bone metastases [BM]) with a CT-measured volumetric bone mineral density (vBMD) for evaluating the strength and stiffness of cadaver vertebrae from donors with metastatic spinal disease (Alkalay et al., 2021, 2022). Soltani et al. evaluated the performance of specimen-specific material calibration versus global material calibration across osteosclerotic, osteolytic, and mixed lesion vertebrae that they derived using a machine learning approach (Soltani et al., 2024). Their results showed the FE damage-based simulations highlighted the differences in the pattern of spatial damage evolution between osteosclerotic and osteolytic vertebral bodies. They concluded that applying this FE approach may allow the prediction of the onset and anatomical location of vertebral failure, which is critical for developing image-based diagnostics of impending pathologic vertebral fractures (Soltani et al., 2024). Garcia-Andrés et al. developed a patient-specific methodology capable of immediately predicting the structural behavior and risk of failure of a vertebra with a spherical tumor of arbitrary characteristics (Garcia-Andrés et al., 2024). They claimed that this immediate evaluation of the results, together with the ease of use of the proposed methodology (including the creation of a personalized computational model from a CT scan of the patient's vertebra), makes this methodology suitable for its use in clinical practice. By running several personalized structural analyses of vertebrae with simulated tumors, using the Cartesian grid FEM in combination with the Sparse Subspace Learning technique, the authors generated a surrogate model of the vertebra. This model was able to predict the vertebra's behavior for different tumor growth scenarios. This could be useful as a clinical decision support tool (Garcia-Andrés et al., 2024). However, this study needs more clinical cases. In the same spirit, Sanhinova and colleagues developed spine lesion growth, using a novel method to automatically align longitudinal spine CTs and accurately assess lesion progression. They tested their approach on 37 vertebrae, from five patients, with baseline CTs and 3-, 6-, and 12-month follow-ups leading to 111 registrations. Their experiment showed accurate registration with an average Hausdorff distance of 0.65 mm and average Dice score of 0.92 (Sanhinova et al., 2024).

A review of state-of-the-art for risk assessment of long bone metastases fractures concluded that predicting long bone pathologic fractures is a multifactorial process. In this regard, machine learning holds significant potential to be of value by considering clinical survival prediction as well as clinical concerns and improve computed tomography rigidity analysis (CT-RA) and finite element analysis data (Nguyễ;n et al., 2021). The near future is also likely to see improvements in the automatic segmentation of metastatic bones (Yosibash et al., 2020, 2023, Rachmil, Artzi, et al., 2024; Rachmil, Myers, et al., 2024; Saillard, Gardegaront, et al., 2024; Saillard, Levillain, et al., 2024). Furthermore, there is a growing emphasis on incorporating model sensitivity and experimental variability into predictive frameworks to ensure robustness and reliability (Ataei et al., 2022, 2024; Gardegaront et al., 2024).

Acknowledgments

This work was partly funded by LabEx Primes (ANR-11-LABX-0063) and MSD Avenir Research Grant.

References

Adams, A. L., Fischer, H., Kopperdahl, D. L., Lee, D. C., Black, D. M., Bouxsein, M. L., Fatemi, S., Khosla, S., Orwoll, E. S., Siris, E. S., & Keaveny, T. M. (2018). Osteoporosis and hip fracture risk from routine computed tomography scans: The fracture, osteoporosis, and CT utilization study (FOCUS). *Journal of Bone and Mineral Research, 33*(7), 1291–1301. https://doi.org/10.1002/jbmr.3423

Alkalay, R. N., Adamson, R., Miropolsky, A., Davis, R. B., Groff, M. L., & Hackney, D. B. (2021). Large lytic defects produce kinematic instability and loss of compressive strength in human spines: An in vitro study. *Journal of Bone and Joint Surgery, 103*(10), 887–899. https://doi.org/10.2106/JBJS.19.00419

Alkalay, R. N., Groff, M. W., Stadelmann, M. A., Buck, F. M., Hoppe, S., Theumann, N., Mektar, U., Davis, R. B., & Hackney, D. B. (2022). Improved estimates of strength and stiffness in pathologic vertebrae with bone metastases using CT-derived bone density compared with radiographic bone lesion quality classification. *Journal of Neurosurgery: Spine, 36*(1), 113–124. https://doi.org/10.3171/2021.2.SPINE202027

Allard, V., Heidsieck, C., Bermond, F., Confavreux, C., Travert, C., Gajny, L., Skalli, W., Mitton, D., & Follet, H. (2024). Comparison of two models to predict vertebral failure loads on the same experimental dataset. *Lecture Notes in Computational Vision and Biomechanics, 39*, 319–329. https://doi.org/10.1007/978-3-031-55315-8_35

Altai, Z., Montefiori, E., van Veen, B., Paggiosi, M. A., McCloskey, E. V., Viceconti, M., Mazzà, C., Li, X., & Saul, K. (2021). Femoral neck strain prediction during level walking using a combined musculoskeletal and finite element model approach. *PLoS One, 16*(2), Article e0245121. https://doi.org/10.1371/journal.pone.0245121

Amini, M., Reisinger, A., Synek, A., Hirtler, L., & Pahr, D. (2023). The predictive ability of a QCT-FE model of the proximal femoral stiffness under multiple load cases is strongly influenced by experimental uncertainties. *Journal of the Mechanical Behavior of Biomedical Materials, 139*, Article 105664. https://doi.org/10.1016/j.jmbbm.2023.105664

Ashford, R. U., Benjamin, L., Pendlebury, S., & Stalley, P. D. (2012). The modern surgical and non-surgical management of appendicular skeletal metastases. *Orthopedie Traumatologie, 26*(3), 184–199. https://doi.org/10.1016/j.mporth.2012.02.002

Ataei, A., Eggermont, F., Verdonschot, N., Lessmann, N., & Tanck, E. (2024). The effect of deep learning-based lesion segmentation on failure load calculations of metastatic femurs using finite element analysis. *Bone, 179*, Article 116987. https://doi.org/10.1016/j.bone.2023.116987

Ataei, A., Eikhout, J., van Leeuwen, R. G. H., Tanck, E., Eggermont, F., & Deniz, C. M. (2022). The effect of variations in CT scan protocol on femoral finite element failure load assessment using phantomless calibration. *PLoS One, 17*(3), Article e0265524. https://doi.org/10.1371/journal.pone.0265524

Awal, R., & Faisal, T. (2024). QCT-based 3D finite element modeling to assess patient-specific hip fracture risk and risk factors. *Journal of the Mechanical Behavior of Biomedical Materials, 150*, Article 106299. https://doi.org/10.1016/j.jmbbm.2023.106299

Bailey, S., Hackney, D., Vashishth, D., & Alkalay, R. N. (2020). The effects of metastatic lesion on the structural determinants of bone: Current clinical and experimental approaches. *Bone, 138*, Article 115159. https://doi.org/10.1016/j.bone.2019.115159

Bailey, S., Stadelmann, M. A., Zysset, P. K., Vashishth, D., & Alkalay, R. N. (2020). Influence of metastatic bone lesion type and tumor origin on human vertebral bone architecture, matrix quality, and mechanical properties. *Journal of Bone and Mineral Research, 37*(5), 896–907. https://doi.org/10.1002/jbmr.4539

Bartenschlager, S., Cavallaro, A., Pogarell, T., Chaudry, O., Uder, M., Khosla, S., Schett, G., & Engelke, K. (2023). Opportunistic screening with CT: Comparison of phantomless BMD calibration methods. *Journal of Bone and Mineral Research, 38*(11), 1689–1699. https://doi.org/10.1002/jbmr.4917

Benca, E., Amini, M., & Pahr, D. H. (2020). Effect of CT imaging on the accuracy of the finite element modelling in bone. *European Radiology Experimental, 4*(1). https://doi.org/10.1186/s41747-020-00180-3

Benca, E., Synek, A., Amini, M., Kainberger, F., Hirtler, L., Windhager, R., Mayr, W., & Pahr, D. H. (2019). QCT-based finite element prediction of pathologic fractures in proximal femora with metastatic lesions. *Scientific Reports, 9*(1). https://doi.org/10.1038/s41598-019-46739-y

Bergmann, G., Deuretzbacher, G., Heller, M., Graichen, F., Rohlmann, A., Strauss, J., & Duda, G. N. (2001). Hip contact forces and gait patterns from routine activities. *Journal of Biomechanics, 34*(7), 859–871. https://doi.org/10.1016/s0021-9290(01)00040-9

Bhattacharya, P., Altai, Z., Qasim, M., & Viceconti, M. (2019). A multiscale model to predict current absolute risk of femoral fracture in a postmenopausal population. *Biomechanics and Modeling in Mechanobiology, 18*(2), 301–318. https://doi.org/10.1007/s10237-018-1081-0

Brown, J. K., Timm, W., Bodeen, G., Chason, A., Perry, M., Vernacchia, F., & DeJournett, R. (2017). Asynchronously calibrated quantitative bone densitometry. *Journal of Clinical Densitometry, 20*(2), 216–225. https://doi.org/10.1016/j.jocd.2015.11.001

Buckley, J. M., Loo, K., & Motherway, J. (2007). Comparison of quantitative computed tomography-based measures in predicting vertebral compressive strength. *Bone, 40*(3), 767–774. https://doi.org/10.1016/j.bone.2006.10.025

Burkhart, T. A., Andrews, D. M., & Dunning, C. E. (2013). Finite element modeling mesh quality, energy balance and validation methods: A review with recommendations associated with the modeling of bone tissue. *Journal of Biomechanics, 46*(9), 1477–1488. https://doi.org/10.1016/j.jbiomech.2013.03.022

Cheal, E. J., Hipp, J. A., & Hayes, W. C. (1993). Evaluation of finite element analysis for prediction of the strength reduction due to metastatic lesions in the femoral neck. *Journal of Biomechanics, 26*(3), 251–264. https://doi.org/10.1016/0021-9290(93)90363-j

Choisne, J., Valiadis, J.-M., Travert, C., Kolta, S., Roux, C., & Skalli, W. (2018). Vertebral strength prediction from Bi-Planar dual energy x-ray absorptiometry under anterior compressive force using a finite element model: An in vitro study. *Journal of the Mechanical Behavior of Biomedical Materials, 87*, 190–196. https://doi.org/10.1016/j.jmbbm.2018.07.026

Coleman, R. E. (2006). Clinical features of metastatic bone disease and risk of skeletal morbidity. *Clinical Cancer Research, 12*(20), Article 6243s. https://doi.org/10.1158/1078-0432.ccr-06-0931

Confavreux, C. B., Follet, H., Mitton, D., Pialat, J. B., & Clézardin, P. (2021). Fracture risk evaluation of bone metastases: A burning issue. *Cancers, 13*(22), 5711. https://doi.org/10.3390/cancers13225711

Cooper, C., Atkinson, E. J., MichaelO'Fallon, W., & Melton, J. L. D. (1992). Incidence of clinically diagnosed vertebral fractures: A population-based study in rochester, minnesota, 1985-1989. *Journal of Bone and Mineral Research, 7*(2), 221–227. https://doi.org/10.1002/jbmr.5650070214

Crawford, R. P., Cann, C. E., & Keaveny, T. M. (2003). Finite element models predict in vitro vertebral body compressive strength better than quantitative computed tomography. *Bone, 33*(4), 744–750. https://doi.org/10.1016/S8756-3282(03)00210-2

Crawford, R. P., Rosenberg, W. S., & Keaveny, T. M. (2003). Quantitative computed tomography-based finite element models of the human lumbar vertebral body: Effect of element size on stiffness, damage, and fracture strength predictions. *Journal of Biomechanical Engineering, 125*(4), 434–438. https://doi.org/10.1115/1.1589772

Czitrom, V. (1999). One-Factor-at-a-Time versus designed experiments. *The American Statistician, 53*(2), 126–131. https://doi.org/10.1080/00031305.1999.10474445

Dall'Ara, E., Pahr, D., Varga, P., Kainberger, F., & Zysset, P. (2012). QCT-based finite element models predict human vertebral strength in vitro significantly better than simulated DEXA. *Osteoporosis International, 23*(2), 563–572. https://doi.org/10.1007/s00198-011-1568-3

Dall'Ara, E., Schmidt, R., Pahr, D., Varga, P., Chevalier, Y., Patsch, J., Kainberger, F., & Zysset, P. (2010). A nonlinear finite element model validation study based on a novel experimental technique for inducing anterior wedge-shape fractures in human vertebral bodies in vitro. *Journal of Biomechanics, 43*(12), 2374–2380. https://doi.org/10.1016/j.jbiomech.2010.04.023

Damron, T. A. (2021). CORR insights®: Can a novel scoring system improve on the mirels score in predicting the fracture risk in patients with multiple myeloma? *Clinical Orthopaedics and Related Research, 479*(3), 531–533. https://doi.org/10.1097/CORR.0000000000001373

Damron, T. A., & Mann, K. A. (2020). Fracture risk assessment and clinical decision making for patients with metastatic bone disease. *Journal of Orthopaedic Research, 38*(6), 1175–1190. https://doi.org/10.1002/jor.24660

De Angelis, R., Demuru, E., Baili, P., Troussard, X., Katalinic, A., Lopez, M. D. C., Innos, K., Santaquilani, M., Blum, M., Ventura, L., Paapsi, K., Galasso, R., Guevara, M., Randi, G., Bettio, M., Botta, L., Guzzinati, S., Maso, L. D., Rossi, S., … Wyn Huws, D. (2024). Complete cancer prevalence in Europe in 2020 by disease duration and country (EUROCARE-6): A population-based study. *The Lancet Oncology, 25*(3), 293–307. https://doi.org/10.1016/s1470-2045(23)00646-0

Decroisette, C., Monnet, I., Berard, H., Quere, G., Le Caer, H., Bota, S., Audigier-Valette, C., Geriniere, L., Vernejoux, J.-M., & Chouaid, C. (2011). Epidemiology and treatment costs of bone metastases from lung cancer: A French prospective, observational, multicenter study (GFPC 0601). *Journal of Thoracic Oncology, 6*(3), 576–582. https://doi.org/10.1097/jto.0b013e318206a1e3

Delmas, P. D., Van Langerijt, L. D., Watts, N. B., Eastell, R., Genant, H., Grauer, A., & Cahall, D. L. (2005). Underdiagnosis of vertebral fractures is a worldwide problem: The IMPACT study. *Journal of Bone and Mineral Research, 20*(4), 557–563. https://doi.org/10.1359/JBMR.041214

Derikx, L. C., Van Aken, J. B., Janssen, D., Snyers, A., Van Der Linden, Y. M., Verdonschot, N., & Tanck, E. (2012). The assessment of the risk of fracture in femora with metastatic lesions: Comparing case-specific finite element analyses with predictions by clinical experts. *Journal of Bone and Joint Surgery - Series B, 94*(8), 1135–1142. https://doi.org/10.1302/0301-620x.94b8.28449

Dosovitskiy, A., Beyer, L., Kolesnikov, A., Weissenborn, D., Zhai, X., Unterthiner, T., Dehghani, M., Minderer, M., Heigold, G., Gelly, S., Uszkoreit, J., & Houlsby, N. (2021). An image is worth 16x16 words: Transformers for image recognition at scale.

In *ICLR 2021 - 9th international conference on learning representations international conference on learning representations.* https://iclr.cc/virtual/2021/papers.html?filter=titles.

Duchemin, L., Bousson, V., Raossanaly, C., Bergot, C., Laredo, J. D., Skalli, W., & Mitton, D. (2008). Prediction of mechanical properties of cortical bone by quantitative computed tomography. *Medical Engineering & Physics, 30*(3), 321–328. https://doi.org/10.1016/j.medengphy.2007.04.008

Duchemin, L., Mitton, D., Jolivet, E., Bousson, V., Laredo, J. D., & Skalli, W. (2008). An anatomical subject-specific FE-model for hip fracture load prediction. *Computer Methods in Biomechanics and Biomedical Engineering, 11*(2), 105–111. https://doi.org/10.1080/10255840701535965

Eberle, S., Göttlinger, M., & Augat, P. (2013). An investigation to determine if a single validated density-elasticity relationship can be used for subject specific finite element analyses of human long bones. *Medical Engineering & Physics, 35*(7), 875–883. https://doi.org/10.1016/j.medengphy.2012.08.022

Eggermont, F., Derikx, L. C., Verdonschot, N., van der Geest, I. C. M., de Jong, M. A. A., Snyers, A., van der Linden, Y. M., & Tanck, E. (2018). Can patient-specific finite element models better predict fractures in metastatic bone disease than experienced clinicians? *Bone & Joint Research, 7*(6), 430–439. https://doi.org/10.1302/2046-3758.76.bjr-2017-0325.r2

Eggermont, F., Mathijssen, E., Bakker, M., & Tanck, E. (2024). Using a statistical shape model to estimate the knee landmarks for aligning femurs for femoral finite element models. *Computer Methods and Programs in Biomedicine, 255*, Article 108324. https://doi.org/10.1016/j.cmpb.2024.108324

Eggermont, F., van der Wal, G., Westhoff, P., Laar, A., de Jong, M., Rozema, T., Kroon, H. M., Ayu, O., Derikx, L., Sander, D., Verdonschot, N., van der Linden, Y., & Tanck, E. (2020). Patient-specific finite element computer models improve fracture risk assessments in cancer patients with femoral bone metastases compared to clinical guidelines. *Bone, 130*, Article 115101. https://doi.org/10.1016/j.bone.2019.115101

Eggermont, F., Verdonschot, N., van der Linden, Y., Tanck, E., & Angeles Pérez, M. (2019). Calibration with or without phantom for fracture risk prediction in cancer patients with femoral bone metastases using CT-based finite element models. *PLoS One, 14*(7), Article e0220564. https://doi.org/10.1371/journal.pone.0220564

Erdemir, A., Mulugeta, L., Ku, J. P., Drach, A., Horner, M., Morrison, T. M., Peng, G. C. Y., Vadigepalli, R., Lytton, W. W., & Myers, J. G. (2020). Credible practice of modeling and simulation in healthcare: Ten rules from a multidisciplinary perspective. *Journal of Translational Medicine, 18*(1). https://doi.org/10.1186/s12967-020-02540-4

Falcinelli, C., & Whyne, C. (2020). Image-based finite-element modeling of the human femur. *Computer Methods in Biomechanics and Biomedical Engineering, 23*(14), 1138–1161. https://doi.org/10.1080/10255842.2020.1789863

Faulkner, K. G., Cann, C. E., & Hasegawa, B. H. (1991). Effect of bone distribution on vertebral strength: Assessment with patient-specific nonlinear finite element analysis. *Radiology, 179*(3), 669–674. https://doi.org/10.1148/radiology.179.3.2027972

Fisher, C. G., Dipaola, C. P., Ryken, T. C., Bilsky, M. H., Shaffrey, C. I., Berven, S. H., Harrop, J. S., Fehlings, M. G., Boriani, S., Chou, D., Schmidt, M. H., Polly, D. W., Biagini, R., Burch, S., Dekutoski, M. B., Ganju, A., Gerszten, P. C., Gokaslan, Z. L., Groff, M. W., … Fourney, D. R. (2010). A novel classification system for spinal instability in neoplastic disease: An evidence-based approach and expert consensus from the spine oncology study group. *Spine, 35*(22), Article E1221. https://doi.org/10.1097/BRS.0b013e3181e16ae2

Fleps, I., Bahaloo, H., Zysset, P. K., Ferguson, S. J., Pálsson, H., & Helgason, B. (2020). Empirical relationships between bone density and ultimate strength: A literature review. *Journal of the Mechanical Behavior of Biomedical Materials, 110*, Article 103866. https://doi.org/10.1016/j.jmbbm.2020.103866

Fleps, I., Pálsson, H., Baker, A., Enns-Bray, W., Bahaloo, H., Danner, M., Singh, N. B., Taylor, W. R., Sigurdsson, S., Gudnason, V., Ferguson, S. J., & Benedikt Helgason. (2022). Finite element derived femoral strength is a better predictor of hip fracture risk than aBMD in the AGES Reykjavik study cohort. *Bone, 154*, Article 116219. https://doi.org/10.1016/j.bone.2021.116219

Forsberg, D. (2015). *Atlas-based registration for accurate segmentation of thoracic and lumbar vertebrae in CT data. 20* (pp. 49–59). Springer Science and Business Media LLC. https://doi.org/10.1007/978-3-319-14148-0_5

Garcia-Andrés, X., Nadal, E., Arana, E., Gandía-Vañó, B., & José Ródenas, J. (2024). Methodology for the assessment of the risk of failure of metastatic vertebrae through ROM-based patient-specific simulations. *Computers & Structures, 296*, Article 107298. https://doi.org/10.1016/j.compstruc.2024.107298

Gardegaront, M., Sas, A., Brizard, D., Levillain, A., Bermond, F., Confavreux, C. B., Pialat, J.-B., Harry van Lenthe, G., Follet, H., & Mitton, D. (2024). Inter-laboratory reproduction and sensitivity study of a finite element model to quantify human femur failure load: Case of metastases. *Journal of the Mechanical Behavior of Biomedical Materials, 158*, Article 106676. https://doi.org/10.1016/j.jmbbm.2024.106676

Goodheart, J. R., Cleary, R. J., Damron, T. A., & Mann, K. A. (2015). Simulating activities of daily living with finite element analysis improves fracture prediction for patients with metastatic femoral lesions. *Journal of Orthopaedic Research, 33*(8), 1226–1234. https://doi.org/10.1002/jor.22887

Guan, H., & Liu, M. (2022). Domain adaptation for medical image analysis: A survey. *IEEE Transactions on Biomedical Engineering, 69*(3), 1173–1185. https://doi.org/10.1109/TBME.2021.3117407

Guenoun, D., & Champsaur, P. (2023). Opportunistic computed tomography screening for osteoporosis and fracture. *Seminars in Musculoskeletal Radiology, 27*(04), 451–456. https://doi.org/10.1055/s-0043-1771037

Hage, W. D., Aboulafia, A. J., & Aboulafia, D. M. (2000). Incidence, location, and diagnostic evaluation of metastatic bone disease. *Orthopedic Clinics of North America, 31*(4), 515–528. https://doi.org/10.1016/S0030-5898(05)70171-1

Hatamizadeh, A., Tang, Y., Nath, V., Yang, D., Myronenko, A., Landman, B., Roth, H. R., & Xu, D. (2022). UNETR: Transformers for 3D medical image segmentation. In *Proceedings - 2022 IEEE/CVF Winter Conference on Applications of Computer Vision, WACV 2022* (pp. 1748–1758). https://doi.org/10.1109/WACV51458.2022.00181. http://ieeexplore.ieee.org/xpl/mostRecentIssue.jsp?punumber=9706406

Helgason, B., Gilchrist, S., Ariza, O., Vogt, P., Enns-Bray, W., Widmer, R. P., Fitze, T., Pálsson, H., Pauchard, Y., Guy, P., Ferguson, S. J., & Cripton, P. A. (2016). The influence of the modulus–density relationship and the material mapping method on the simulated mechanical response of the proximal femur in side-ways fall loading configuration. *Medical Engineering & Physics, 38*(7), 679–689. https://doi.org/10.1016/j.medengphy.2016.03.006

Henninger, H. B., Reese, S. P., Anderson, A. E., & Weiss, J. A. (2010). Validation of computational models in biomechanics. *Proceedings of the Institution of Mechanical Engineers - Part H: Journal of Engineering in Medicine, 224*(7), 801–812. https://doi.org/10.1243/09544119jeim649

Hernlund, E., Svedbom, A., Ivergård, M., Compston, J., Cooper, C., Stenmark, J., McCloskey, E. V., Jönsson, B., & Kanis, J. A. (2013). Osteoporosis in the European union: Medical management, epidemiology and economic burden: A report prepared in collaboration with the international osteoporosis foundation (IOF) and the european Federation of pharmaceutical industry associations (EFPIA). *Archives of Osteoporosis, 8*(1–2). https://doi.org/10.1007/s11657-013-0136-1

Isensee, F., Jaeger, P. F., Kohl, S. A. A., Petersen, J., & Maier-Hein, K. H. (2021). nnU-Net: A self-configuring method for deep learning-based biomedical image segmentation. *Nature Methods, 18*(2), 203–211. https://doi.org/10.1038/s41592-020-01008-z

Jackman, T. M., DelMonaco, A. M., & Morgan, E. F. (2016). Accuracy of finite element analyses of CT scans in predictions of vertebral failure patterns under axial compression and anterior flexion. *Journal of Biomechanics, 49*(2), 267–275. https://doi.org/10.1016/j.jbiomech.2015.12.004

Johnson, J. E., Brouillette, M. J., Permeswaran, P. T., Miller, B. J., & Goetz, J. E. (2020). Simulated lesions representative of metastatic disease predict proximal femur failure strength more accurately than idealized lesions. *Journal of Biomechanics, 106*, Article 109825. https://doi.org/10.1016/j.jbiomech.2020.109825

Kanis, J. A., Johnell, O., Oden, A., Johansson, H., & McCloskey, E. (2008). FRAX™ and the assessment of fracture probability in men and women from the UK. *Osteoporosis International, 19*(4), 385–397. https://doi.org/10.1007/s00198-007-0543-5

Keaveny, T. M., Clarke, B. L., Cosman, F., Orwoll, E. S., Siris, E. S., Khosla, S., & Bouxsein, M. L. (2020). Biomechanical computed tomography analysis (BCT) for clinical assessment of osteoporosis. *Osteoporosis International, 31*(6), 1025–1048. https://doi.org/10.1007/s00198-020-05384-2

Keaveny, T. M., Donley, D. W., Hoffmann, P. F., Mitlak, B. H., Glass, E. V., & San Martin, J. A. (2007). Effects of teriparatide and alendronate on vertebral strength as assessed by finite element modeling of QCT scans in women with osteoporosis. *Journal of Bone and Mineral Research, 22*(1), 149–157. https://doi.org/10.1359/jbmr.061011

Keller, T. S. (1994). Predicting the compressive mechanical behavior of bone. *Journal of Biomechanics, 27*(9), 1159–1168. https://doi.org/10.1016/0021-9290(94)90056-6

Keyak, J. H. (2001). Improved prediction of proximal femoral fracture load using nonlinear finite element models. *Medical Engineering & Physics, 23*(3), 165–173. https://doi.org/10.1016/s1350-4533(01)00045-5

Keyak, J. H., Kaneko, T. S., Rossi, S. A., Pejcic, M. R., Tehranzadeh, J., & Skinner, H. B. (2005). Predicting the strength of femoral shafts with and without metastatic lesions. *Clinical Orthopaedics and Related Research, 439*, 161–170. https://doi.org/10.1097/01.blo.0000174736.50964.3b

Keyak, J. H., Lee, I. Y., & Skinner, H. B. (1994). Correlations between orthogonal mechanical properties and density of trabecular bone: Use of different densitometric measures. *Journal of Biomedical Materials Research, 28*(11), 1329–1336. https://doi.org/10.1002/jbm.820281111

Keyak, J. H., Rossi, S. A., Jones, K. A., & Skinner, H. B. (1997). Prediction of femoral fracture load using automated finite element modeling. *Journal of Biomechanics, 31*(2), 125–133. https://doi.org/10.1016/S0021-9290(97)00123-1

Klennert, B. J., Ellis, B. J., Maak, T. G., Kapron, A. L., & Weiss, J. A. (2017). The mechanics of focal chondral defects in the hip. *Journal of Biomechanics, 52*, 31–37. https://doi.org/10.1016/j.jbiomech.2016.11.056

Kluess, D., Soodmand, E., Lorenz, A., Pahr, D., Schwarze, M., Cichon, R., Varady, P. A., Herrmann, S., Buchmeier, B., Schröder, C., Lehner, S., & Kebbach, M. (2019). A round-robin finite element analysis of human femur mechanics between seven participating laboratories with experimental validation. *Computer Methods in Biomechanics and Biomedical Engineering, 22*(12), 1020–1031. https://doi.org/10.1080/10255842.2019.1615481

Knowles, N. K., Reeves, J. M., & Ferreira, L. M. (2016). Quantitative computed tomography (QCT) derived bone mineral density (BMD) in finite element studies: A review of the literature. *Journal of Experimental Orthopaedics, 3*(1). https://doi.org/10.1186/s40634-016-0072-2

Kopperdahl, D. L., Morgan, E. F., & Keaveny, T. M. (2002). Quantitative computed tomography estimates of the mechanical properties of human vertebral trabecular bone. *Journal of Orthopaedic Research, 20*(4), 801–805. https://doi.org/10.1016/S0736-0266(01)00185-1

Lee, D. C., Hoffmann, P. F., Kopperdahl, D. L., & Keaveny, T. M. (2017). Phantomless calibration of CT scans for measurement of BMD and bone strength—Inter-operator reanalysis precision. *Bone, 103*, 325–333. https://doi.org/10.1016/j.bone.2017.07.029

Lee, D. C., Varela, A., Kostenuik, P. J., Ominsky, M. S., & Keaveny, T. M. (2016). Finite element analysis of denosumab treatment effects on vertebral strength in ovariectomized cynomolgus monkeys. *Journal of Bone and Mineral Research, 31*(8), 1586–1595. https://doi.org/10.1002/jbmr.2830

Lems, W. F., & Raterman, H. G. (2017). Critical issues and current challenges in osteoporosis and fracture prevention. An overview of unmet needs. *Therapeutic Advances in Musculoskeletal Disease, 9*(12), 299–316. https://doi.org/10.1177/1759720X17732562

Liebschner, M. A. K., Kopperdahl, D. L., Rosenberg, W. S., & Keaveny, T. M. (2003). Finite element modeling of the human thoracolumbar spine. *Spine, 28*(6), 559–565. https://doi.org/10.1097/01.BRS.0000049923.27694.47

Lin, J. K., Hearn, C. M., Getzen, E., Long, Q., Lee, D. C., Keaveny, T. M., Jayadevappa, R., Robinson, K. W., Wong, Y. N., Maxwell, K. N., Narayan, V., Haas, N. B., Takvorian, S. U., Bikle, D. D., Chiang, J. M., Khan, A. N., Rajapakse, C. S., Morgans, A. K., & Parikh, R. B. (2024). Validation of biomechanical computed tomography for fracture risk classification in metastatic hormone-sensitive prostate cancer. *European Urology Oncology, 7*(4), 794–803. https://doi.org/10.1016/j.euo.2023.10.016

Ma, J., He, Y., Li, F., Han, L., You, C., & Wang, B. (2024). Segment anything in medical images. *Nature Communications, 15*(1). https://doi.org/10.1038/s41467-024-44824-z

Maier-Hein, L., Reinke, A., Godau, P., Tizabi, M. D., Buettner, F., Christodoulou, E., Glocker, B., Isensee, F., Kleesiek, J., Kozubek, M., Reyes, M., Riegler, M. A., Wiesenfarth, M., Emre Kavur, A., Sudre, C. H., Baumgartner, M., Eisenmann, M., Heckmann-Nötzel, D., Rädsch, T., … Jäger, P. F. (2024). Metrics reloaded: Recommendations for image analysis validation. *Nature Methods, 21*(2), 195–212. https://doi.org/10.1038/s41592-023-02151-z

Melton, L. J., Chrischilles, E. A., Cooper, C., Lane, A. W., & Riggs, B. L. (1992). Perspective how many women have osteoporosis? *Journal of Bone and Mineral Research, 7*(9), 1005–1010. https://doi.org/10.1002/jbmr.5650070902

Michalski, A. S., Besler, B. A., Michalak, G. J., & Boyd, S. K. (2020). CT-based internal density calibration for opportunistic skeletal assessment using abdominal CT scans. *Medical Engineering & Physics, 78*, 55–63. https://doi.org/10.1016/j.medengphy.2020.01.009

Mirels, H. (1989). Metastatic disease in long bones A proposed scoring system for diagnosing impending pathologic fractures. *Clinical Orthopaedics and Related Research, 249*, 256–264. https://doi.org/10.1097/00003086-198912000-00027

Morgan, E. F., Bayraktar, H. H., & Keaveny, T. M. (2003). Trabecular bone modulus-density relationships depend on anatomic site. *Journal of Biomechanics, 36*(7), 897–904. https://doi.org/10.1016/S0021-9290(03)00071-X

Morris, M. D. (1991). Factorial sampling plans for preliminary computational experiments. *Technometrics, 33*(2), 161–174. https://doi.org/10.1080/00401706.1991.10484804

Nazarian, A., Entezari, V., Zurakowski, D., Calderon, N., Hipp, J. A., Villa-Camacho, J. C., Lin, P. P., Cheung, F. H., Aboulafia, A. J., Turcotte, R., Anderson, M. E., Gebhardt, M. C., Cheng, E. Y., Terek, R. M., Yaszemski, M., Damron, T. A., & Snyder, B. D. (2015). Treatment planning and fracture prediction in patients with skeletal metastasis with CT-Based rigidity analysis. *Clinical Cancer Research, 21*(11), 2514–2519. https://doi.org/10.1158/1078-0432.ccr-14-2668

Nguyễn, M.-V., Carlier, C., Nich, C., Gouin, F., & Vincent, C. (2021). Fracture risk of long bone metastases: A review of current and new decision-making tools for prophylactic surgery. *Cancers, 13*(15), 3662. https://doi.org/10.3390/cancers13153662

Oefner, C., Herrmann, S., Kebbach, M., Lange, H.-E., Kluess, D., & Woiczinski, M. (2021). Reporting checklist for verification and validation of finite element analysis in orthopedic and trauma biomechanics. *Medical Engineering & Physics, 92*, 25–32. https://doi.org/10.1016/j.medengphy.2021.03.011

Pauchard, Y., Fitze, T., Browarnik, D., Eskandari, A., Pauchard, I., Enns-Bray, W., Pálsson, H., Sigurdsson, S., Ferguson, S. J., Harris, T. B., Gudnason, V., & Helgason, B. (2016). Interactive graph-cut segmentation for fast creation of finite element models from clinical ct data for hip fracture prediction. *Computer Methods in Biomechanics and Biomedical Engineering, 19*(16), 1693–1703. https://doi.org/10.1080/10255842.2016.1181173

Pistoia, W., van Rietbergen, B., Lochmüller, E.-M., Lill, C. A., Eckstein, F., & Rüegsegger, P. (2002). Estimation of distal radius failure load with micro-finite element analysis models based on three-dimensional peripheral quantitative computed tomography images. *Bone, 30*(6), 842–848. https://doi.org/10.1016/s8756-3282(02)00736-6

Prado, M., Rezaei, A., & Giambini, H. (2021). Density-dependent material and failure criteria equations highly affect the accuracy and precision of QCT/FEA-Based predictions of osteoporotic vertebral fracture properties. *Annals of Biomedical Engineering, 49*(2), 663–672. https://doi.org/10.1007/s10439-020-02595-w

Rachmil, O., Artzi, M., Iluz, M., Druckmann, I., Yosibash, Z., & Sternheim, A. (2024). Automatic segmentation of femoral tumors by nnU-net. *Clinical Biomechanics, 116*, Article 106265. https://doi.org/10.1016/j.clinbiomech.2024.106265

Rachmil, O., Myers, K., Merose, O., Sternheim, A., & Yosibash, Z. (2024). The influence of femoral lytic tumors segmentation on autonomous finite element analysis. *Clinical Biomechanics, 112*, Article 106192. https://doi.org/10.1016/j.clinbiomech.2024.106192

Saillard, E., Gardegaront, M., Levillain, A., Bermond, F., Mitton, D., Pialat, J.-B., Confavreux, C., Grenier, T., & Hélène, F. (2024). Finite element models with automatic computed tomography bone segmentation for failure load computation. *Scientific Reports, 14*(1). https://doi.org/10.1038/s41598-024-66934-w

Saillard, E., Levillain, A., Mitton, D., Pialat, J. B., Confavreux, C., Follet, H., & Grenier, T. (2024). *Enhanced segmentation of femoral bone metastasis in CT scans of patients using synthetic data generation with 3D diffusion models*. France: arXiv. https://doi.org/10.48550/arXiv.2409.11011

Saltelli, A., Tarantola, S., Campolongo, F., & Ratto, M. (2004). *Sensitivity analysis in practice: A guide to assessing scientific models*. Wiley. https://doi.org/10.1002/0470870958

Sanhinova, M., Haouchine, N., Pieper, S. D., Wells, W. M., Balboni, T. A., Spektor, A., Huynh, M. A., Guenette, J. P., Czajkowski, B., Caplan, S. D. P., Kang, H., Hackney, D. B., & Alkalay, R. N. (2024). Registration of longitudinal spine CTs for monitoring lesion growth. *Progress in Biomedical Optics and Imaging - Proceedings of SPIE, 12926*. https://doi.org/10.1117/12.3006621

Sas, A., Ohs, N., Tanck, E., & Harry van Lenthe, G. (2020). Nonlinear voxel-based finite element model for strength assessment of healthy and metastatic proximal femurs. *Bone Reports, 12*, Article 100263. https://doi.org/10.1016/j.bonr.2020.100263

Sas, A., Tanck, E., Sermon, A., & Harry van Lenthe, G. (2020). Finite element models for fracture prevention in patients with metastatic bone disease. A literature review. *Bone Reports, 12*, Article 100286. https://doi.org/10.1016/j.bonr.2020.100286

Schileo, E., & Taddei, F. (2021). Finite element assessment of bone fragility from clinical images. *Current Osteoporosis Reports, 19*(6), 688–698. https://doi.org/10.1007/s11914-021-00714-7

Sekuboyina, A., Husseini, M. E., Bayat, A., Löffler, M., Liebl, H., Li, H., Giles, T., Kukačka, J., Payer, C., Štern, D., Urschler, M., Chen, M., Cheng, D., Lessmann, N., Hu, Y., Wang, T., Yang, D., Xu, D., Ambellan, F., … Kirschke, J. S. (2021). VerSe: A vertebrae labelling and segmentation benchmark for multi-detector CT images. *Medical Image Analysis, 73*, Article 102166. https://doi.org/10.1016/j.media.2021.102166

Shibata, H., Kato, S., Sekine, I., Abe, K., Araki, N., Iguchi, H., Izumi, T., Inaba, Y., Osaka, I., Kato, S., Kawai, A., Kinuya, S., Kodaira, M., Kobayashi, E., Kobayashi, T., Sato, J., Shinohara, N., Takahashi, S., Takamatsu, Y., … Yoshida, M. (2016). Diagnosis and treatment of bone metastasis: Comprehensive guideline of the Japanese society of medical oncology, Japanese orthopedic association, Japanese urological association, and Japanese society for radiation oncology. *ESMO Open, 1*(2), Article e000037. https://doi.org/10.1136/esmoopen-2016-000037

Siddique, N., Paheding, S., Elkin, C. P., & Devabhaktuni, V. (2021). U-Net and its variants for medical image segmentation: A review of theory and applications. *IEEE Access, 9*, 82031–82057. https://doi.org/10.1109/access.2021.3086020

Siris, E. S., Miller, P. D., Barrett-Connor, E., Faulkner, K. G., Wehren, L. E., Abbott, T. A., Berger, M. L., Santora, A. C., & Sherwood, L. M. (2001). Identification and fracture outcomes of undiagnosed low bone mineral density in postmenopausal women: Results from the national osteoporosis risk assessment. *JAMA, 286*(22), 2815–2822. https://doi.org/10.1001/jama.286.22.2815

Sobol, I. M. (2001). Global sensitivity indices for nonlinear mathematical models and their monte Carlo estimates. *Mathematics and Computers in Simulation, 55*(1–3), 271–280. https://doi.org/10.1016/s0378-4754(00)00270-6

Soltani, Z., Xu, M., Radovitzky, R., Stadelmann, M. A., Hackney, D., & Alkalay, R. N. (2024). CT-based finite element simulating spatial bone damage accumulation predicts metastatic human vertebrae strength and stiffness. *Frontiers in Bioengineering and Biotechnology, 12*. https://doi.org/10.3389/fbioe.2024.1424553

Spruijt, S., Van Der Linden, J., Sander Dijkstra, P., Wiggers, T., Oudkerk, M., Snijders, C., Van Keulen, F., Verhaar, J., Weinans, H., & Swierstra, B. (2006). Prediction of torsional failure in 22 cadaver femora with and without simulated subtrochanteric metastatic defects: A CT scan-based finite element analysis. *Acta Orthopaedica, 77*(3), 474–481. https://doi.org/10.1080/17453670610046424

Stadelmann, M. A., Schenk, D. E., Maquer, G., Lenherr, C., Buck, F. M., Bosshardt, D. D., Hoppe, S., Theumann, N., Alkalay, R. N., & Zysset, P. K. (2020). Conventional finite element models estimate the strength of metastatic human vertebrae despite alterations of the bone's tissue and structure. *Bone, 141*, Article 115598. https://doi.org/10.1016/j.bone.2020.115598

Sternheim, A., Dadia, S., Gortzak, Y., Traub, F., Trabelsi, N., Snir, N., Gorfine, M., & Yosibash, Z. (2020). When and where do patients with bone metastases actually break their femurs? A CT-based finite element analysis. *The Bone & Joint Journal, 102*(5), 638–645. https://doi.org/10.1302/0301-620X.102B5.BJJ-2019-1328.R2

Sternheim, A., Giladi, O., Gortzak, Y., Drexler, M., Salai, M., Trabelsi, N., Milgrom, C., & Yosibash, Z. (2018). Pathological fracture risk assessment in patients with femoral metastases using CT-based finite element methods. A retrospective clinical study. *Bone, 110*, 215–220. https://doi.org/10.1016/j.bone.2018.02.011

Taddei, F., Martelli, S., Reggiani, B., Cristofolini, L., & Viceconti, M. (2006). Finite-element modeling of bones from CT data: Sensitivity to geometry and material uncertainties. *IEEE Transactions on Biomedical Engineering, 53*(11), 2194–2200. https://doi.org/10.1109/tbme.2006.879473

Tanck, E., van Aken, J. B., van der Linden, Y. M., Bart Schreuder, H. W., Binkowski, M., Huizenga, H., & Verdonschot, N. (2009). Pathological fracture prediction in patients with metastatic lesions can be improved with quantitative computed tomography based computer models. *Bone, 45*(4), 777–783. https://doi.org/10.1016/j.bone.2009.06.009

Taylor, M., Viceconti, M., Bhattacharya, P., & Li, X. (2021). Finite element analysis informed variable selection for femoral fracture risk prediction. *Journal of the Mechanical Behavior of Biomedical Materials, 118*, Article 104434. https://doi.org/10.1016/j.jmbbm.2021.104434

Viceconti, M., & Emili, L. (2024). *Toward good simulation practice: Best practices for the use of computational modelling and simulation in the regulatory process of biomedical products*. Springer Nature Switzerland.

Viceconti, M., Olsen, S., Nolte, L.-P., & Burton, K. (2005). Extracting clinically relevant data from finite element simulations. *Clinical Biomechanics, 20*(5), 451–454. https://doi.org/10.1016/j.clinbiomech.2005.01.010

Viceconti, M., Pappalardo, F., Rodriguez, B., Horner, M., Bischoff, J., & Tshinanu, F. M. (2021). In silico trials: Verification, validation and uncertainty quantification of predictive models used in the regulatory evaluation of biomedical products. *Methods, 185*, 120–127. https://doi.org/10.1016/j.ymeth.2020.01.011

Viceconti, M., Qasim, M., Bhattacharya, P., & Li, X. (2018). Are CT-Based finite element model predictions of femoral bone strengthening clinically useful? *Current Osteoporosis Reports, 16*(3), 216–223. https://doi.org/10.1007/s11914-018-0438-8

Viceconti, M., Taddei, F., Cristofolini, L., Martelli, S., Falcinelli, C., & Schileo, E. (2012). Are spontaneous fractures possible? An example of clinical application for personalised, multiscale neuro-musculo-skeletal modelling. *Journal of Biomechanics, 45*(3), 421–426. https://doi.org/10.1016/j.jbiomech.2011.11.048

Wasserthal, J., Breit, H.-C., Meyer, M. T., Pradella, M., Hinck, D., Sauter, A. W., Heye, T., Boll, D. T., Cyriac, J., Yang, S., Bach, M., & Segeroth, M. (2023). TotalSegmentator: Robust segmentation of 104 anatomic structures in CT images. *Radiology: Artificial Intelligence, 5*(5). https://doi.org/10.1148/ryai.230024

Wille, H., Ruess, M., Rank, E., & Yosibash, Z. (2016). Uncertainty quantification for personalized analyses of human proximal femurs. *Journal of Biomechanics, 49*(4), 520–527. https://doi.org/10.1016/j.jbiomech.2015.11.013

Winsor, C., Li, X., Qasim, M., Henak, C. R., Pickhardt, P. J., Ploeg, H., & Viceconti, M. (2021). Evaluation of patient tissue selection methods for deriving equivalent density calibration for femoral bone quantitative CT analyses. *Bone, 143*, Article 115759. https://doi.org/10.1016/j.bone.2020.115759

Yong, M., Jensen, A.Ø., Jacobsen, J. B., Nørgaard, M., Fryzek, J. P., & Toft Sørensen, H. (2011). Survival in breast cancer patients with bone metastases and skeletal-related events: A population-based cohort study in Denmark (1999–2007). *Breast Cancer Research and Treatment, 129*(2), 495–503. https://doi.org/10.1007/s10549-011-1475-5

Yosibash, Z., Katz, Y., Nir, T., & Sternheim, A. (2023). Femurs segmentation by machine learning from CT scans combined with autonomous finite elements in orthopedic and endocrinology applications. *Computers & Mathematics with Applications, 152*, 16–27. https://doi.org/10.1016/j.camwa.2023.09.044

Yosibash, Z., Mayo, R. P., Dahan, G., Trabelsi, N., Amir, G., & Milgrom, C. (2014). Predicting the stiffness and strength of human femurs with real metastatic tumors. *Bone, 69*, 180–190. https://doi.org/10.1016/j.bone.2014.09.022

Yosibash, Z., Myers, K., Trabelsi, N., & Sternheim, A. (2020). Autonomous FEs (AFE) - A stride toward personalized medicine. *Computers & Mathematics with Applications, 80* (11), 2417–2432. https://doi.org/10.1016/j.camwa.2020.03.012

Yosibash, Z., Tal, D., & Trabelsi, N. (2010). Predicting the yield of the proximal femur using high-order finite-element analysis with inhomogeneous orthotropic material properties. *Philosophical Transactions of the Royal Society A: Mathematical, Physical and Engineering Sciences, 368*(1920), 2707–2723. https://doi.org/10.1098/rsta.2010.0074

Zysset, P. K., Dall'Ara, E., Varga, P., & Pahr, D. H. (2013). Finite element analysis for prediction of bone strength. *Bone Reports, 2*. https://doi.org/10.1038/bonekey.2013.120

Zysset, P., & Rincón, L. (2006). *An alternative fabric-based yield and failure criterion for trabecular bone mechanics of biological tissue* (pp. 457–470). Berlin Heidelberg: Springer. https://doi.org/10.1007/3-540-31184-X_33

CHAPTER

Biomaterials for bone regeneration

22

Kishore Hazarika[1], Abhijit Vyas[1, 3], Subrata Bandhu Ghosh[1], Tanmay Gupta[2], Mohini Sain[2] and Sanchita Bandyopadhyah-Ghosh[1]

[1]*Manipal University Jaipur, Engineered Biomedical Materials Research and Innovation Centre, Department of Mechanical Engineering, Jaipur, Rajastan, India;* [2]*University of Toronto, Department of Mechanical and Industrial Engineering, Toronto, ON, Canada;* [3]*University Centre for Research and Development (UCRD), Chandigarh University, Mohali, Punjab, India*

1. Introduction

Bone is a hierarchical and complicated material with diverse structural arrangements at several scales of length that work in concert to perform diverse chemical, biological, and mechanical functions. The mechanical properties are influenced and multifaceted by aspects like location, water content, mineral, temperature, and disease (Evans, 1976; Morgan et al., 2018). At the macroscopic scale, these are categorized into two types known as trabecular and cortical bones. Cortical bone is stiffer and denser, while the trabecular bone is less stiff and more porous (Cook and Zioupos, 2009). The mechanical properties of the bone tissues, including trabecular and cortical bone, are determined by the architecture of the organic matrix consisting of collagen fiber and the inorganic bioreinforcement consisting of hydroxyapatite crystals (Bandyopadhyay-Ghosh, 2008; Gupta et al., 2025). The human skeleton, despite its complexity, has a remarkable ability to repair itself through a complicated procedure involving cellular regeneration. When the bone tissue is damaged, the body starts a healing cascade, engaging specialized cells like osteoclasts and osteoblasts (Albrektsson & Johansson, 2001). This natural healing relies on the underlying structure and mechanics of the bone itself. Moreover, in cases of extensive bone defects or chronic skeletal challenges, biomaterials can play a crucial role in bridging the gap between biological potential and clinical necessity (Perić Kačarević et al., 2020). These biomaterials, such as advanced ceramics, polymers, and bioactive glasses, can function as an intelligent scaffold to support bone regeneration (Mancuso et al., 2017). Advanced biomaterials are being established to mimic the natural environment of bone tissue, guiding cellular behavior, stimulating growth, and promoting integration (Baino et al., 2019). These materials can be designed to release targeted molecules and adapt to the surrounding conditions, representing a significant advancement in regenerative medicine (Rezwan et al., 2006). The focus is on exploring complex biomaterials and their mechanisms to reshape our comprehension of bone tissue regeneration.

Biomechanics and Mechanobiology of Bone. https://doi.org/10.1016/B978-0-443-26739-0.00001-6

2. Bone tissue grafts

Bone grafting is referred to as the surgical process that involves the transplantation of bone tissue to replace or repair missing or damaged bone (Govoni et al., 2021). Traditional treatments of bone grafting frequently entail the use of autografts, allografts, or substitutes for bone graft. These techniques are the traditional approach to grafting a bone and were effective in several cases (Fu, 2018). The techniques of bone grafting offer an alternative solution to improve the natural process of by acting as scaffold new bone regeneration. This can stimulate the growth of cellular bone and foster the regeneration of new bones (Pecci et al., 2020). Limitations of the natural process of healing include inadequate volume of bone, specific health issues, or elderly patients with impaired bone healing, and difficulty in stabilization and alignment of the damaged or fractured bone (Choi et al., 2020). Furthermore, traditional techniques of bone grafting might need additional surgeries, having the risk of graft rejection, comorbidity and general discomfort. To address these limitations, advanced methods of bone grafting were established. These techniques include using 3D printing technology to generate biomimetic bone grafts. This also entails shaping the living cells or tissues using biocompatible materials to integrate the natural shapes that promote formation of new bone. The advanced methods provide important advantages over several conventional techniques, like faster material manipulation, the capability to generate customized grafts for patients, and enhanced bone regeneration (Inzana et al., 2014; Zhang et al., 2016). Hence, the establishment of novel methods of bone grafting, like 3D printing of the biomimetic bone graft, has now become imperative to mitigate limitations of the natural process of healing and offer better solutions for bone defects and injuries (Moussa & Dym, 2020). The bone grafts are typically classified as artificial substitutes of bone grafts and conventional grafts of natural bones.

2.1 Natural bone grafts

The standard approach of bone tissue engineering includes natural bone grafts, like xenografts, allografts, and autografts, etc are classified based on the harvesting site (Laurencin & El-Amin, 2008; Orciani et al., 2017).

2.1.1 Autografts

Autografts are considered the gold standard for grafting of bone because of their high rate of success and low complication risks. They are acquired from healthy tissues of bone in the patient's body, typically from areas like the spinal, iliac crest, ribs, or fusion areas. These are osteoconductive, osteogenic, and osteoinductive, making them ideal for repairing bone. The trabecular porous bone is frequently harvested because of its rich composition of osteoblasts and osteocytes that fosters improved osteogenic response (Govoni et al., 2021; Zhang et al., 2021). The porosity of the trabecular area facilitated vascular growth, cell attachment, and proliferation, which are essential for the remodeling of bone. Autografts, thus provide a low risk of rejection and infection due to their histocompatibility, creating favourable conditions for bone regeneration. However, their use is limited by the amount of bone that can be harvested safely and the morbidity of potential donor sites (Bisht et al., 2021).

2.1.2 Allografts

Another form of bone grafting technique is the allografting method. In this method, the tissues that are implanted are mainly harvested from the cadaver of the donor patient (Wang & Yeung, 2017). Hence, do not need any additional surgery on the host patient for harvesting the tissues, preventing problems like morbidity of the donor site (Moussa & Dym, 2020; Wang & Yeung, 2017). However, the primary challenge that exists for this technique is lower osteogenic bioactivity. Storage and handling present another challenge requiring several steps before implantation, like lyophilization, cleansing, and sterilization (Grover et al., 2011). While, these steps prevent infection at the site of the host patient, they can negatively impact the overall effectiveness and strength of the allograft, which lowers its reliability (Block & Miller, 2011, p. 31).

2.1.3 Xenografts

Xenografts are obtained from animal sources, including bovine and porcine tissues. This technique is preferred in situations where there is a limited supply of bone grafts, to address the expensive nature of allografts and autografts (Correia et al., 2021; Moussa & Dym, 2020). There are both advantages and significant challenges of this technique, including a huge risk of disease transmission, viral infection, and chronic rejection, along with ethical concerns. As such, xenografts remain a limited option for transplantation (Govoni et al., 2021). In human trials, they are normally used for orthopedic application engaging maxillofacial sites. Hence, the use of this technique is restricted to repairing substantial fractures and loss of bone (Block & Miller, 2011, p. 31). The primary variables that impacts the features and their influence on the choices of the substitute of bone graft has been explored in Table 22.1 below.

Table 22.1 Bone graft substitutes based on their characteristics.

Variable	Autografts	Allografts	Xenografts
Osteoconduction	***	***	***
Osteoinduction	**	***	*
Osteointegration	**	***	**
Osteogenesis	–	***	–

****, Excellent; **, Average; *, Poor; –, None.*

2.2 Synthetic bone grafts (scaffolds)

Synthetic bone substitutes, often referred to as bone scaffolds, are the artificial materials used to support the fracture or damage site of bone tissues (Grunenwald et al., 2014). These scaffolds provide a promising, versatile alternative to the traditional bone grafting methods and can be engineered to mimic the natural structure of bone and its composition, towards generating a biomimetic environment that fosters new bone formation (Bergmann et al., 2010). Their functionalization and customizable design with bioactive molecules improve their osteoconductive and osteoinductive properties. Synthetic scaffolds can be consistently produced in large quantities, overcoming the scarcity of allogeneic and autologous grafts, potentially resulting in more accessible and cost-effective treatments. While engineered biomaterial candidates have shown promise, they demonstrate limitations in mechanical, physical, or chemical properties. Despite these limitations, synthetic bone scaffolds hold great potential as a therapeutic avenue in regenerative and orthopedic medicine (Garimella et al., 2023; Jang et al., 2014; Ranjan et al., 2020; Wu et al., 2023). Fig. 22.1 schematically highlights the strategies considered while designing and developing bone scaffolds for tissue repair and regeneration.

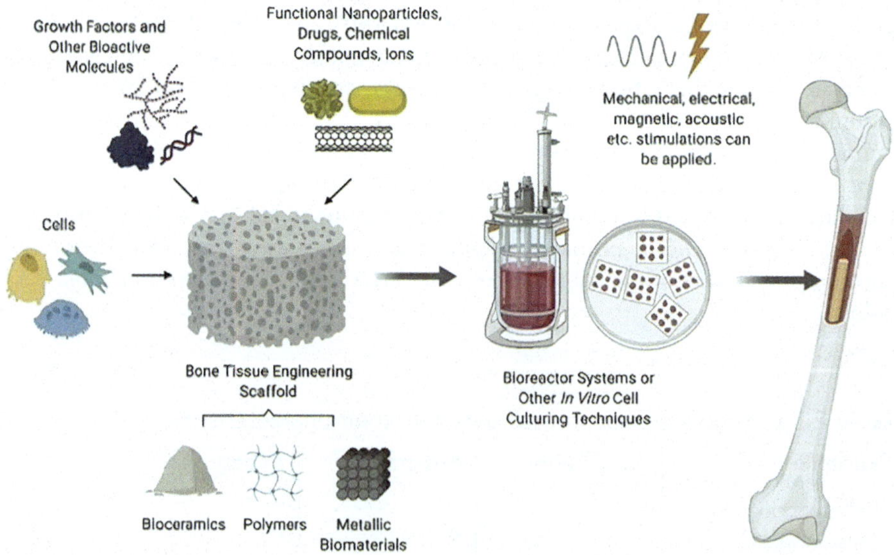

FIGURE 22.1

Bone tissue healing: Strategic design of scaffolds.

- From Jodati, H., Yılmaz, B. & Evis, Z. (2020). Ceramics International, 46(10, Part B), 15725–15739. https://doi.org/10.1016/j.ceramint.2020.03.192.

3. Biomaterials for bone tissue engineering (BTE)

Scaffolds are designed to replicate the intricate architecture of the natural bone, fostering blood vessel infiltration as well as providing mechanical strength. Ideally, the bone scaffold must be biocompatible, biomimetic, and osteoconductive with a structure that closely replicates the native bone tissue (Alknery et al., 2022; Boskey & Robey, 2018). They must also possess sufficient strength and permeability to enable delivery bioactive molecules and potentially incorporate sensors for real time monitoring. Biomaterials for regeneration of bone tissue are broadly categorized into composite materials, ceramics, metals, and polymers, each meeting the biocompatibility requirements (Prasad et al., 2017; Xia et al., 2013). The selection of material depends on the specific requirements of the bone scaffold, including mechanical strength, biocompatibility, and osteoconductivity.

3.1 Biometals

The concept of artificial bone scaffolds and implants to date has been widely explored with metallic materials for their fundamental advantage of having better mechanical strength. A range of metallic biomaterials, such as stainless steel, cobalt nickel chromium, titanium, tantalum, and magnesium-based alloys, have been explored as the choice of materials for bone tissue engineering (Garimella et al., 2019; Heissler et al., 1998; Jana et al., 2022; Shekhawat et al., 2021). For an implant, it is important to establish an interfacial bonding with the host bone during the period of healing. The presence of trace elements like Mg, Zn, Sr, Fe, Cu, Mn, Co, Li, Ag, etc, in the native bone has a significant impact on the bone growth and metabolism. A study reporting fabrication of metal reinforced polymer composite bone scaffolds with bioactive doping of the above elements established the role of these metallic ions on the hydrophilicity and thus improving interfacial bonding, imparting antibacterial activity, and improving mechanical properties along with the promotion of vascularization of the scaffolds. However, doped metals should not only exist on the surface of the composite scaffold but should be evenly dispersed throughout the whole composite scaffold, to ensure the smooth completion of the entire bone repair process (Li et al., 2022). Nevertheless, metallic materials consist of components that may be harmful or trigger allergies. Additionally, they are prone to releasing undesirable metallic ions in extremely harsh conditions caused by corrosion, wear, and friction within the physiological system (Habibovic & de Groot, 2007). This occurrence can potentially provoke inflammatory responses, resulting in gradual osteolysis of surrounding tissues and causing additional harm to local tissues. To address these challenges, researchers have developed metal-reinforced polymer composite bone scaffolds with bioactive doping of elements like magnesium, zinc, and strontium, which promote bone growth through lower immunogenecity, hydrophilicity, and antibacterial activity (Li et al., 2022). Despite these advancements, the need for less harmful alternatives to metallic implants still exists, as revision surgeries and complications can cause significant pain and distress to patients (Habibovic & de Groot, 2007).

3.2 Bioceramics

Bioceramics and bioglass have shown promising results as bone scaffolds and implants due to their bioresorbability, biocompatibility, and bioactivity (Brook & Hatton, 1998; Grover et al., 2011; Grunenwald et al., 2014; Vyas et al., 2021; Zhang et al., 2022). Various rigid biofillers, including bioglass, tricalcium phosphate cements, and synthetic hydroxyapatite, have been developed for bone rehabilitation and tissue growth (Banerjee et al., 2010; Goodridge et al., 2006; Kumar et al., 2018; Schmidt et al., 2018; Song et al., 2019). Bioactive glasses have gained attention for their ability to form a hydroxyapatite layer, promoting bonding to bone and exhibiting excellent osteoinductivity, osteoconductivity biocompatibility, making them suitable for applications in bone regeneration. Glass-ceramics, particularly chain silicate glass-ceramics, have emerged as a promising choice for bone tissue regenerative applications due to their tunable properties (Song et al., 2019). Despite their brittleness, glass-ceramics can be engineered to achieve desired mechanical properties by adjusting their precursor glass compositions (Kumawat et al., 2020). Chain silicates are referred to as the structural family of the silicate minerals, where silicate tetrahedra (SiO_4) are related together to form multiple or single chains. Interesting features like regulated anisotropic, bioactivity and mechanical behavior are imparted by this structural arrangement (Alizadeh & Marghussian, 2000; Bandyopadhyay-Ghosh et al., 2010). Chain silicates, such as diopside ($CaMgSi_2O_6$) and wollastonite ($CaSiO_3$), have shown excellent mechanical properties, bioactivity, and biocompatibility, making them suitable for load-bearing applications (Schmidt et al., 2018). Fluorcanasite glass-ceramic, a quadruple chain silicate, stands out for its exceptional fracture toughness, bioactivity, and ability to promote bone mineralization (Likitvanichkul & Lacourse, 1995). Its unique microstructure and mechanical properties make it an attractive option for bone tissue engineering applications (Mei et al., 2020). Fluorcanasite glass-ceramic (FC GC) is a type of chain silicate glass-ceramic that has been studied for its potential in bone tissue engineering (Fantini & Curto, 2018; Miao & Sun, 2010; Vyas et al., 2021; Wei et al., 2014). Fluorcanasite glass-ceramic acts as a biomimetic agent, mimicking the natural minerals present in bone tissue, and promotes bone mineralization by encouraging the development of fluorapatite and hydroxyapatite crystals (Bandyopadhyay-Ghosh et al., 2010; Fantini & Curto, 2018; Miao & Sun, 2010; Wei et al., 2014). Its high crystallinity and mechanical strength (>300 MPa), toughness of fracture (ʾ5 MPa $m^{1/2)}$, and reasonably excellent deflection of crack make it suitable for applications involving heavy loads (Vyas et al., 2021). Additionally, fluorcanasite glass-ceramic has volume nucleation characteristics, allowing it to be cast into intricate shapes (De Santis et al., 2018; Majumder et al., 2024; O'Brien, 2011; Williams et al., 2005). Fluorcanasite glass ceramics based biomaterials plays an important role in bone tissue regeneration because of their capability to foster angiogenesis, ensuring proper supply of blood to the new bone tissue formation (Bandyopadhyay-Ghosh et al., 2010). It also impacts cellular activities like migration, proliferation, and differentiation, making them necessary for the successful regeneration of bone. The diverse features of fluorcanasite glass ceramics generally render them indispensable in the establishment of biomaterials for the engineering of bone tissue, providing promising advancements in regenerative medicine.

3.3 Biopolymers

Polymeric materials are widely used in implantable devices due to their ease of fabrication, flexibility, and biocompatibility (Li et al., 2006). They can be combined with other materials to create composites with tailored properties, offering significant design flexibility. Polymers also exhibit excellent tensile strength (Li et al., 2019; Teo et al., 2016), ensuring they can securely enclose implants throughout their lifespan. The selection of polymeric materials for implantable devices focuses on creating a safe, durable, and functionally optimal environment that interacts harmoniously with the human body (Dutta Roy et al., 2003). Table 22.2 below shows a list of different polymeric biomaterials used in the cases of bone scaffolding and implantation. Based on the data from the US Food and Drug Administration (FDA), these polymeric biomaterials shown below are divided into categories.

From Table 22.2 and the respective data, as obvious as it looks, polymeric implants as biomaterials have major applications in the non-load bearing tissues of the body, with a limited range of materials for load-bearing (Seitz et al., 2005). This being an inherent limitation of the polymeric implants, only composite bone scaffolds consisting of a polymeric matrix have been found to influence the load-bearing bone tissue scaffolds.

3.4 Composites

In bone tissue engineering, it is desirable to create biomaterials that are comparable to native bone in terms of structure and composition. Bone is a naturally occurring biocomposite composed of an apatite-based mineral phase (approximately 50–70 wt %) that serves as a bioactive reinforcement, together with a collagenous organic matrix (approximately 20–40 wt%) (Vyas et al., 2022). Hence, to match the mechano-biological and overall physiological performance of native bone, biomimicry is generally considered the most reliable approach for biocomposite bone scaffold development. Biocomposite approach permits tuning the mechano-biological performance of bone scaffolds. A wide range of biodegradable and biocompatible materials has been explored extensively for the development of bone, including metals like zirconia, hydroxyapatite, ceramics, alloys, titanium, and magnesium (Li et al., 2017; Osman et al., 2022; Yang et al., 2010; Yang et al., 2020).

3.4.1 Biometal matrix composites

Conventional metallic scaffolds, encompassing chromium, cobalt, tantalum alloys, titanium, and stainless steel, present challenges for regenerative medicine. While appreciated for bio-inertia, these materials generally suffer from inherent limitations. The high density and low corrosion resistance contribute to substantial weight,

Table 22.2 Categorial medical implant materials and their applications.

Category	Applications	Materials used
Anesthesiology	Catheters	• Polyethylene • Polytetrafluorethylene • Polyamide
Cardiovascular	Pacemaker, implantable cardioverter/defibrillator, Left ventricular Assist device.	• Polypropylene • Polyethylene • Polytetrafluorethylene • Polyamide
Dental	Dentures, dental implants	• Polymethylmethacrylate
Orthopedic	Implants for bone fractures and injuries	• Polyethylene • Polyether ether ketone • Polyhydroxyalkanoates
Ear, nose, and throat	Strap implant Nasal implants for nose reconstruction	• Polydimethylsiloxane • Silicone • Parylene • Polyethylene
General and plastic surgery	Synthetic blood vessels Breast implants Cheek, jaw, and chin implants Lip implant. Titanium surgical implants Hip implant	• Polypropylene • Polyethylene terephthalate • Polytetrafluorethylene • Silicone • Polydimethylsiloxane

From Teo, A.J.T., Mishra, A., Park, I., Kim, Y.J., Park, W.T. & Yoon, Y.J. (2016). Polymeric Biomaterials for Medical Implants and Devices. ACS Biomaterials Science and Engineering, 2(4), 454–472. https://doi.org/10.1021/acsbiomaterials.5b00429.

impacting osteointegration and biomechanics, and potentially degradation. Metallic implants have limitations, including a high elastic modulus that can cause mechanical mismatch with native bone, leading to stress shielding and potential bone resorption. Furthermore, their insolubility makes them permanent implants, requiring revision surgeries if complications arise. To address these challenges while maintaining the reliability of metallic implants, metal matrix biocomposite bone scaffolds are being explored. These biocomposites can be tailored to achieve specific properties, such as strength and osteogenic response, by selecting suitable metals. One promising option is Magnesium (Mg) based alloys, which offer a range of benefits, including biocompatibility, degradability, and the ability to promote bone growth and regeneration. Magnesium (Mg) and its alloys are promising materials for next-generation bone scaffolds due to their biodegradability and lightweight properties. By incorporating bioactive molecules into Mg-based composites, researchers can develop advanced bone scaffolds that can withstand mechanical loads and promote biological integration (Garimella, Ashri, et al., 2022; Garimella et al., 2019, 2023; Garimella, Rathi, et al., 2022). This approach offers a potential solution for creating more effective bone scaffolds. Magnesium-based composite bone scaffolds have several potential benefits for bone tissue engineering, including improved biocompatibility, regulated biodegradation, and suitable mechanical performance (Su et al., 2020). These advantages can reduce scaffold failure and eliminate the need for additional surgery. For example, Xiong et al. incorporated hydroxyapatite (HAp) into magnesium composites at different weight percentages and reported enhanced mechanical properties, improved corrosion resistance compared with pure Mg, and tunable biological behavior depending on the HAp content (Xiong et al., 2016). Various studies have explored the use of different materials within Mg-based scaffolds, such as hydroxyapatite, β-tricalcium phosphate, and silicate-based bioactive glass. Khanra et al. found enhanced mechanical properties; however, with a shift from ductile to brittle behavior with the addition of HAp (Khanra et al., 2010). While these materials have shown promise, they also present challenges, including corrosion, pitting, and changes in mechanical behavior. Therefore, careful material selection and optimization are vital to balance key properties for successful applications of bone tissue engineering. Huang et al. noticed that β-tricalcium phosphate (β-TCP) in Mg-alloy (Mg–2Zn-0.5Ca) matrix improved grain refinement and micro-hardness; however, significant pitting and corrosion post-immersion in simulated body fluid was observed (Huang et al., 2015). Dutta et al. reported silicate bioactive glass (BG)-reinforced Mg matrix composites and found that the addition of BG improved microhardness, elastic modulus, and corrosion resistance, with the Mg-10 wt% BG composite exhibiting the highest corrosion resistance and good cytocompatibility (Dutta et al., 2019).

3.4.2 Bioceramic matrix composites

Bioactive ceramic matrices are combined with reinforcing phases to form mechanically strong structures with customized bioactivity for bone regeneration in ceramic matrix biocomposite scaffolds. Alumina, zirconia, or silicon carbide are high-strength ceramic fibers or whiskers that are used to reinforce bioactive glass matrices and hydroxyapatite (HAp) for enhanced fracture resistance (Sun et al., 2023). To combine mechanical strength and bioactivity, inert ceramic matrices (zirconia, alumina) have been mixed with bioactive glass/ceramic particles. Apatite-forming capabilities and customized mechanics are demonstrated by zirconia and akermanite bioceramics composites (Habraken et al., 2007; Sun et al., 2023). To improve qualities and offer nano topographical cues, carbon nanotubes and fibers have been combined with bioactive glasses, glass-ceramics, and reinforced calcium phosphates. The fracture resistance of HA and bioglass scaffolds has been enhanced using carbon nanotubes (Putlyaev et al., 2018; Shuai et al., 2014). The bioceramics used in these composites, such as bioactive glass and hydroxyapatite, play a crucial role due to their inherently high bioactivity, which allows them to form strong bonds with bone tissue and support new bone formation (Li et al., 2020). The combination with bioglass has shown high bioactivity, indicating strong potential for osteoconductive and osteoinductive reinforcement. Conventionally, biocomposite bone scaffold fabrication approaches incorporating hydroxyapatite (HAp), beta tricalcium phosphate (β-TCP) cement-based bioactive fillers have already been tried out (Sohn & Oh, 2019; Xing et al., 2018). However, these materials have demonstrated weak mechanical properties and are primarily used as filler materials at the fracture site (Li et al., 2019). In such scenarios, glass and glass-ceramics come to the rescue. The prime advantage of glass and glass-ceramics is their tunable mechanical and biological properties, making them suitable for applications where a multitude of performance characteristics are required. For a bone scaffold, not only are biological properties such as osteoconductivity, biocompatibility, cell viability, and proliferation essential, but appropriate mechanical properties for load bearing are also required (Mondal et al., 2020). Glass-ceramic, bioceramics, and bioglass find their suitability for such applications with their reinforcement capabilities. In case of glass-ceramics, the selectively tuned glass precursor compositions in glass-ceramics synthesis can be regulated to achieve desired crystallite presences and achieve favorable mechano-biological performance (Kanchanarat et al., 2005; Miller et al., 2002). For bone tissue and its repair, the necessary crystalline phases contributing to tissue regeneration are the apatite-based crystallites occurring naturally in the native bone in the form of hydroxyapatite (HAp). Similarly, the tailoring of glass-ceramics for tissue regenerative applications requires inducing apatite-based crystalline phases, specifically such as hydroxyapatite or fluorapatite, both of which have a significant contribution in regulating bioactivity, such as osteoconductivity as well as osteoinductivity (Bandyopadhyay-Ghosh et al., 2010). This can be obtained from oxides of silicate-based glass-ceramics. Also, bone scaffolds with load-bearing applications require strength, enduring reinforcement as well. Chain silicate glass-ceramics are one such class of materials which can be conducively tailored to achieve polycrystalline phases contributing to each performance characteristic, such as strength in terms of toughness, chemical stability, and essential biological characteristics for bone scaffolds. They notably promote osteogenesis by influencing the differentiation of mesenchymal stem cells into osteoblasts, which is vital for the formation of new bone tissue. Various approaches have been used to improve in vitro cell viability in silicate ceramic–polymer composite scaffolds, including enhancement of surface roughness, controlled release of bioactive ions, maintenance of a stable pH, apatite layer formation, improved three-dimensional architecture, enhanced hydrophilicity, increased availability of active sites for protein adsorption, and a high surface-area-to-volume ratio.

3.4.3 Biopolymer matrix composites

Polymer biocomposite bone scaffolds offer a promising solution for regenerating damaged or lost bone tissue. The polymer matrix biocomposite bone scaffolds are typically made from a combination of synthetic or natural polymers and other biocompatible materials such as ceramics or bioglass. They mimic the natural bone structure, providing crucial features for successful bone regeneration. Several polymeric matrix materials have been reported with use of biocompatible ceramics as bioactive reinforcements (Ito, 1991; Yunus Basha et al., 2015). Composite bone scaffolds are designed to revolutionize bone repair by combining enhanced mechanical strength, mimicking natural bone, with an improved cellular environment. This means they provide a suitable microenvironment for crucial nutrients like calcium and phosphate ions, or other beneficial components, to be delivered to cells, ultimately accelerating the healing process. On the polymeric matrices front, flexible and easily decomposable biopolymers, like polyglycolic acid, polyvinyl alcohol, polylactic acid, and polycaprolactone, are emerging as promising materials for scaffold matrices. These biocompatible and lightweight materials offer key advantages over traditional options, such as brittle bioceramics and rapidly degrading magnesium alloys. However, these biopolymeric materials, although biodegradable they lack the appropriate bioactivity and structural stability to form new bone tissue at the defect site (Costa et al., 2020; Dutta Roy et al., 2003; Elomaa et al., 2011; Guerra, Lara-Padilla et al., 2019; Musidang & Jiratumnukul, 2015). To tackle these inherent limitations and improve the scope for these materials in bone tissue repair and regeneration, bioactive glass, glass-ceramics, etc, have been found to be as suitable alternatives as osteoconductive tailored high-strength reinforcements (Kumar et al., 2018; Rottensteiner et al., 2014). The potential of these reinforcements lies in their tunable precursor compositions, which can be tuned to achieve near-human bone-like compositions and strength. These fillers possess additional characteristics such as bioactivity, hydrophilic osteoconductive surfaces, chemical stability, and potential biocompatibility for enhanced bone tissue healing (Lee et al., 2011; Wei & Ma, 2004; Yunus Basha et al., 2015; Zhang et al., 2007; Zhou & Lee, 2011). With regards to the strength enhancement, apatite-forming glass ceramics such as wollastonite, mullite, and fluorocanasite have demonstrated encouraging possibilities for mechanical functionality. Especially, the structural stability of fluorocanasite glass-ceramic and thereby its inherent fracture toughness and flexural strength has demonstrated a good potential (Kanchanarat et al., 2005; Kumawat et al., 2020). When it comes to the choice of a suitable biopolymer, several explored options are available to date, ranging from polypropylene fumarate to polycaprolactone as a biodegradable grade photopolymer in bone tissue regeneration (Fisher et al., 2003; Guerra, Lammel-Lindemann et al., 2019; Kumawat et al., 2019). The specific limitation for these biodegradable grade photopolymers lies in their load-bearing capabilities (Pongwisuthiruchte et al., 2022). In bone tissue engineering, fundamentally, the scaffold structure should be able to bear the load of the damaged bone at the site of implantation, with the polymeric matrix having load-bearing capabilities of its own. In this line, leaving aside biodegradability, several photopolymeric resins, which are typically biocompatible but not biodegradable, are employed in dental applications (Alifui-Segbaya, 2020). These biocompatible dental grade resins have sufficient strength from the load bearing perspective and have been least explored for the bone scaffolds, especially with critical-sized bone defect healing (Alifui-Segbaya et al., 2018; Ma et al., 2020; Vyas et al., 2024) (Shar).

4. Properties of biomaterials

The properties crucial to bone scaffold development primarily include biocompatibility and cytotoxicity. For enhanced osteointegration, characteristics such as,

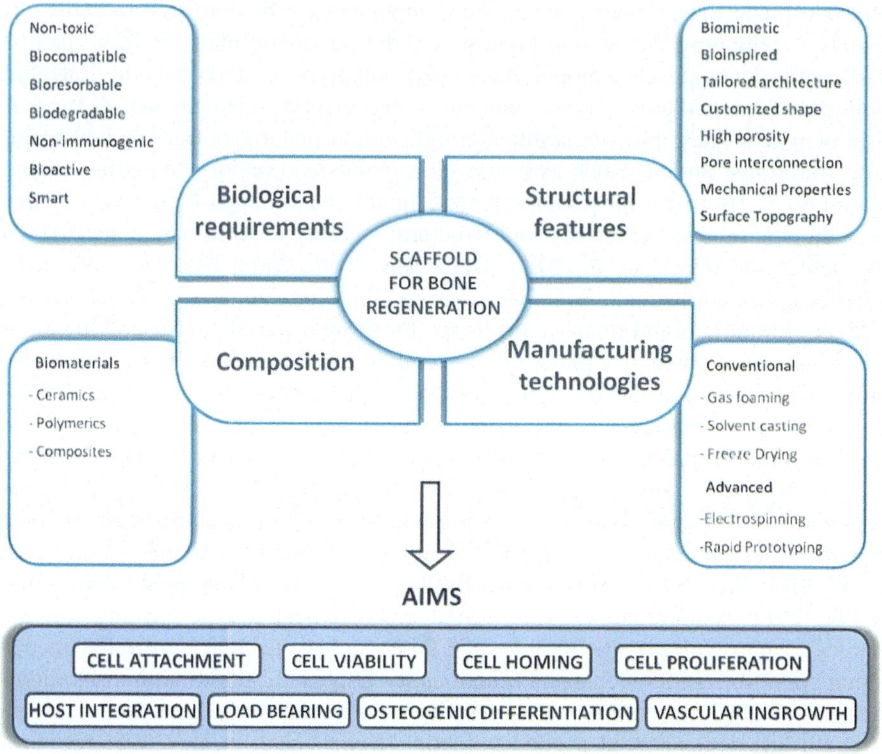

FIGURE 22.2

Bone scaffold characteristics for bone tissue repair and healing.

- From Roseti, Livia, Parisi, Valentina, Petretta, Mauro, Cavallo, Carola, Desando, Giovanna, Bartolotti, Isabella & Grigolo, Brunella. (2017). Scaffolds for Bone Tissue Engineering: State of the art and new perspectives. Materials Science and Engineering: C, 78, 1246–1262. https://doi.org/10.1016/j.msec.2017.05.017.

bioresorbability, osteoconductivity, and osteoinductivity have a vital function in the process of tissue development. Additionally, the biodegradation of the bone scaffold materials is an important consideration. These characteristics of the bone scaffolds are summarized in Fig. 22.2 and elaborated further in this section.

4.1 Biological properties

The effectiveness and clinical performance of bone scaffolds are significantly influenced by their biological characteristics. A crucial requirement is biocompatibility, or the capacity to interact with living tissues without causing adverse reactions (Schuster et al., 2007). For tissue integration to be effective, scaffolds need to be non-cytotoxic and promote cell adhesion, proliferation, and differentiation.

Additionally, bioresorbability and biodegradability enable synchronous replacement of scaffold with regenerated tissue (Taboas et al., 2003; Zhang et al., 2021). In addition, scaffolds ought to demonstrate bioactivity, which is the capacity to generate an apatite layer that is physiologically active and resembles bone, hence promoting osseointegration and bonding with native bone (Wang & Yeung, 2017).

4.1.1 Biocompatibility and cytocompatibility

Ensuring biocompatibility and cytocompatibility is a priority in bone tissue engineering to confirm the safety of a bone scaffold for human application. Biocompatibility, a crucial characteristic of a bone scaffold, indicates its ability to support cellular activities during the bone regeneration process, while avoiding toxic, inflammatory, or immune-related reactions when interacting with the body's physiological fluids (Li et al., 2015). Cytocompatibility test is a key component of biological evaluation and screening protocols, employing tissue cells in vitro to assess cell growth, reproduction, and morphological impacts induced by bone scaffolds or implant materials. In vitro cytocompatibility assays using living cell culture lines, can be supplemented with cell-free assays evaluating biomaterial's interaction with buffers, such as phosphate-buffered saline (PBS) or simulated bodily fluid (SBF), where the ion concentrations are like those in human blood plasma (Caughman et al., 1991; Kontonasaki et al., 2002).

4.1.2 Bioresorbability and biodegradability

As new research keeps evolving toward the development of biomaterials for bone tissue regeneration, each new class of material builds upon the limitations of previously researched materials. Key characteristic of synthetic bone scaffold materials is their biodegradability and bioresorbability. In its true form, the advantage of biodegradability and bioresorbability is that it eliminates the need for a second revision surgery post-implantation and, in tandem, matches up with the pace of bone tissue healing. The byproduct of biodegradation should be non-cytotoxic, and should be eliminated from the body within the clinically acceptable tenure (Vert et al., 1992). The mechanism of biodegradation significantly influences the rate of biodegradation. Typically, biodegradation mechanisms are categorized into enzymatic and hydrolytic biodegradation. Enzymatic degradation is carried out by specific enzymes that are secreted by the body's cells, that target specific sites within implant material, and cleave them. The process may be highly selective, depending on the types of enzymes that are present around the implant (Vert, 2009). Hydrolytic degradation, on the other hand, takes place when water molecules come into direct contact with the implant material, breaking it down over time. Both mechanisms contribute to implant resorption or disintegration, which has an impact on long-term functionality and the potential for removal. Understanding these degradation processes is critical for developing implants with optimal performance, biocompatibility, and long-term safety in the complex biological environment of the human body (Ronca et al., 2013).

4.1.3 Bioactivity

In terms of bone tissue engineering, bioactivity refers to the biological characteristics of bone scaffold biomaterials that enhance the osteoconductive and osteoinductive performances at the bone and scaffold interface. Biomaterials with superior bioactivity are the subject of recent research in bone tissue engineering. For instance, the existing bioactive glasses are being upgraded through the development of novel compositions, e.g., from silicate glasses, borate glasses, and the addition of trace elements (Fernandes et al., 2017). These upgrades may enhance bone and vascular tissue regeneration, including differentiation of osteoblasts from stem cells. Tuneable mechanical strength and biodegradation rate values make bioactive glasses even more attractive. Elastomeric composites are being investigated to match the elasticity of collagen, an important bone protein, with an eye toward materials that mimic both the strength and flexibility of bone proteins like collagen, paving the way for next-generation bone scaffolds that can promote bone regeneration effectively (Vert, 2009). Table 22.3 below showcases a list of bioactive materials with their characteristics, highlighting advantages and disadvantages.

4.2 Mechanical properties

A bone scaffold requires structural integrity to sustain loads in both the load-bearing and non-load-bearing regions. The inherent heterogeneity of bone mechanics, characterized by variations in fracture toughness, compressive strength, and Young's modulus across anatomical locations, necessitates the development of bone scaffolds that exhibit site-specific mechanical compatibility (Chen et al., 2012). Additionally, adequate mechanical strength is crucial to withstand surgical manipulation during implantation. Furthermore, the scaffold must retain sufficient structural integrity throughout the implantation period to sustain physiological loading. While advancements have been made in the development of biomaterials with superior mechanical properties, this often comes at the expense of porosity, a critical factor for cellular infiltration and differentiation (Xing et al., 2018). Unfortunately, numerous bone scaffolds demonstrating promising in vitro results have ultimately failed in vivo due to limitations in mechanical strength, along with vascularization, insufficient porosity for cellular activity, and inadequate nutrient delivery (Liu et al., 2021). Therefore, it is evident that the efficacy of a bone scaffold hinges on achieving an optimal balance between its mechanical characteristics and a porous architecture that facilitates cellular infiltration and vascularization.

4.3 Biomimetic properties

4.3.1 Osteoconduction and osteoinduction

Osteoconduction and osteoinduction represent complementary yet distinct mechanisms essential for effective bone tissue regeneration. Osteoconduction refers to a biomaterial's capacity to function as a passive structural scaffold that facilitates bone growth along its surface by providing a three-dimensional architecture with

Table 22.3 Advantages and limitations of bioactive bone scaffold biomaterials.

Bioactive materials	Advantages	Disadvantages
Calcium phosphates (e.g., HAp, β-TCP, and biphase CaP)	• Excellent biocompatibility • Supports cell viability. • High osteoconduction	• Brittle • Slow biodegradability in the crystalline state • Low strength in the amorphous state
Silicate bioactive glasses	• Excellent biocompatibility • Supports cell viability. • High osteoconduction • Tailorable mechanical properties and degradation • Vasculature • Rapid gene expression	• Low strength in the amorphous state
Bioactive glass-ceramics and bioglass	• Tunable mechano-biological performance • Excellent biocompatibility • Supports cell viability. • High osteoconduction	• Brittle • Toxicity risk due to the release of borate ions in case of bioactive glasses • Slow biodegradability in the crystalline state
Bulk biodegradable polymers (Poly (lactic acid)) Poly (glycolic acid) Poly (lactic-co-glycolic acid) Poly (propylene fumarate) Poly (polyol sebacate)	• Good biocompatibility • Biodegradable (with a wide range of degradation rates) • Bioresorbable • Good processability • Good ductility • Elasticity	• May cause inflammatory responses because of acidic byproducts of biodegradation. • Scaffold structural breakdowns due to accelerated biodegradation profiles
Composites (containing bioactive phases)	• Good biocompatibility • Supporting cell activity • Good osteoconductivity • Tailorable degradation rate • Improved mechanical reliability.	• Not as good as native bone matrix • There is a need to improve fabrication techniques.

From Chen, Qizhi, Zhu, Chenghao & Thouas, George A. (2012). Progress and challenges in biomaterials used for bone tissue engineering: bioactive glasses and elastomeric composites. Progress in Biomaterials, 1(1). https://doi.org/10.1186/2194-0517-1-2.

optimal porosity (typically 100–500 μm), interconnected pores, and surface properties conducive to cell attachment (Guerra, Lara-Padilla et al., 2019; Miao & Sun, 2010). This structural framework enables the migration and proliferation of osteoblasts and supports neovascularization, ultimately guiding the spatial organization of newly formed bone. In contrast, osteoinduction involves the active recruitment and stimulation of undifferentiated mesenchymal stem cells to differentiate into the osteoblastic lineage, initiating the new bone formation through biochemical signaling (Albrektsson & Johansson, 2001). This process is primarily mediated by growth factors such as bone morphogenetic proteins (BMPs), transforming growth factor-beta (TGF-β), and insulin-like growth factor (IGF), or through the controlled release of bioactive ions like silicon, strontium, or zinc from the biomaterial (Majumder et al., 2024; Williams et al., 2005). The most clinically effective bone regeneration strategies incorporate biomaterials that seamlessly integrate both osteoconductive and osteoinductive properties, creating a synergistic environment where the structural template guides the spatial organization of new bone while biochemical cues accelerate cellular differentiation and matrix production (Khan et al., 2005; Rottensteiner et al., 2014). This dual functionality allows these advanced biomaterials to potentially outperform traditional autografts by eliminating donor site morbidity while providing customizable properties that can be tailored to specific defect sites, patient populations, and clinical scenarios, representing a significant advancement in the treatment of challenging bone defects (Mahato et al., 2020; Parthasarathy, 2014).

4.3.2 Cell attachment, migration and differentiation

In tissue engineering, cell attachment, migration, and differentiation are fundamental processes that underpin the success of regenerative strategies. Cell adhesion refers to the initial attachment of cells to biomaterial surfaces, a crucial first step for tissue regeneration. The surface properties of biomaterials, such as surface chemistry, topography, and hydrophilicity, can significantly influence this process by promoting the binding of cells through molecular interactions with adhesion proteins like fibronectin and collagen (Carter et al., 1997; Han et al., 2020; Hazarika et al., 2025; Shuai et al., 2020). Once attached, cells must migrate to the injury site or throughout the scaffold to promote tissue formation. Migration is largely governed by chemical signals, such as growth factors, that direct cell movement toward areas of injury (chemotaxis), as well as by the physical structure of the material, which can guide movement through aligned microstructures or channels (Moore et al., 2004). Additionally, the mechanical properties of the biomaterial, such as its stiffness or flexibility, affect migration efficiency, with cells generally responding better to substrates that resemble the mechanical properties of the native tissue they are meant to regenerate (Fernandez-Yague et al., 2015). Once cells are positioned at the injury site, they undergo differentiation, a critical process in which stem or progenitor cells transform into specialized cell types like osteoblasts or chondrocytes, depending on the tissue being regenerated (Kanwar & Vijayavenkataraman, 2021; Visser et al., 2015). This process is regulated by biochemical cues from the biomaterial, such as growth factors embedded within the scaffold, and by the material's physical properties that can mimic the natural extracellular matrix. The stiffness or elasticity of the material can further influence which lineage the cells differentiate into; for example, stiffer materials may promote bone formation, while softer ones may favor neural or adipogenic differentiation (Jilka, 2003). Together, these interconnected processes of attachment, migration, and differentiation are essential for creating functional, integrated tissues. By understanding and optimizing how biomaterials influence these cellular behaviors, we can improve the regenerative capacity of implants and scaffolds, facilitating the successful repair of complex tissue defects and disorders (Wee et al., 2023).

5. Fabrication techniques for biomaterials

5.1 Conventional manufacturing

Conventional fabrication techniques for bone tissue engineering scaffolds have historically provided the foundation for clinical approaches to bone regeneration. Solvent casting/particulate leaching is one of the most widely employed methods, in which polymer solutions are cast over porogen particles (typically salt or sugar) that are subsequently leached out to create porous structures (Raeisdasteh Hokmabad et al., 2017). This relatively simple approach allows for control of porosity through porogen size and concentration, though it often results in limited pore interconnectivity and residual solvent concerns. In parallel, freeze-drying techniques exploit the formation of ice crystals within polymer solutions which, upon sublimation, leave a porous architecture with pore sizes dictated by freezing parameters (Gentile et al., 2012; Muthukumar et al., 2016). While offering good control over overall porosity, freeze-drying struggles to produce consistent pore morphologies or mechanical properties suitable for load-bearing applications (Raeisdasteh Hokmabad et al., 2017). Gas foaming represents another conventional approach, where pressurized gas (typically carbon dioxide) is used to create bubbles within polymer melts or solutions that form pores upon depressurization or solvent evaporation (Kim et al., 2012). This technique eliminates the need for organic solvents but often yields structures with limited interconnectivity and inconsistent pore distribution. Thermally induced phase separation similarly provides solvent-free processing by leveraging temperature changes to drive polymer-rich and polymer-poor phase separation, ultimately creating microporous structures after solvent removal (Nam & Park, 1999; Zeinali et al., 2021). While these techniques have enabled the creation of numerous clinically used bone graft substitutes, their application for complex defects remains constrained by processing limitations that typically result in random, poorly controlled internal architectures. The limitations of conventional fabrication techniques ultimately stem from their inability to precisely control scaffold microarchitecture and incorporate spatial heterogeneity. Most conventional methods produce scaffolds with randomly distributed pores that inadequately mimic the hierarchical structure of native bone tissue. These techniques generally cannot create precisely controlled patient-specific external geometries necessary for complex defects (Zadpoor, 2015). The inability to spatially control the distribution of bioactive factors or mechanical properties further limits their biological performance. Additionally, conventional methods often struggle to achieve reproducibility across batches, making quality control challenging. Perhaps most significantly, these approaches typically cannot produce scaffolds with vascular channels or defined nutrient transport pathways, which severely limits construct size due to diffusion constraints and results in poor core viability in larger implants (Bian et al., 2011). These collective limitations have driven the development of advanced manufacturing techniques that offer greater precision and architectural control for next-generation bone tissue engineering.

5.2 Additive manufacturing

Additive manufacturing techniques are promising tools for making scaffolds for bone tissue engineering. On the metallic side, titanium and its alloys like Ti–6Al–4V have been extensively investigated for SLS bone scaffolds due to their excellent biocompatibility and mechanical characteristics. Porous Ti–6Al–4V scaffolds via SLS with compression strengths up to 454 MPa demonstrated porosities ranging from 41% to 63%, making them suitable for load-bearing bone defects (Zheng et al., 2020). Other biocompatible metals like stainless steels, cobalt-chrome alloys, and shape memory alloys like nitinol have also been processed by SLS for bone scaffolds. Porous NiTi scaffolds with super elastic behavior, making them promising candidates for implant applications where flexibility and shape recovery are desired, have been reported (Elahinia et al., 2016). FDM offers several advantages in bone tissue engineering, including its capacity to produce scaffolds characterized by extensive porosity and interconnected pore structure (Buj-Corral et al., 2018). The movement of cells, the diffusion of nutrients, and the elimination of waste products are all made possible by these porous structures and are necessary for the effective regeneration of bone. With FDM, it is possible to replicate the complex architecture of actual bone tissue and manipulate the scaffold's porosity, pore size, and overall shape by varying process parameters like layer height, nozzle diameter, and printing speed. The possibility to include bioactive substances into the scaffold during the creation process, such as proteins, growth hormones, or medications, is another benefit of FDM. These bioactive compounds can provide localized and sustained administration to improve bone regeneration and encourage tissue integration. They can also be coated onto the scaffold's surface or encapsulated within the polymer matrix (Lian et al., 2021). Both DLP and SLA are light-based 3D printing techniques that utilize a photo-polymeric liquid resin as a raw material, which, upon irradiation with UV light (DLP) or laser (SLA), results in a completely cross-linked material with rigid structures (Bagheri & Jin, 2019). The major difference between the technologies, apart from the irradiation source, is the method of light projection. The DLP 3D printing operates on a mask projection technique wherein an entire part cross sections based on CAD slices is printed at once layer by layer, whereas in stereolithography a 2-photon polymerization (2 PP) approach is used where spot crosslinking mechanism takes places resulting in tracing the CAD slice point by point (Guerra, Lammel-Lindemann et al., 2019; Quan et al., 2020). Other advantages of DLP and SLA 3D printing techniques include the possibility of selection from a range of biocompatible and biodegradable photopolymers available. Selective photopolymers are shown below in Table 22.4. These materials can be tailored to exhibit desired mechanical properties, degradation rates, and biological responses, ensuring optimal integration with the surrounding tissue and facilitating the gradual replacement of the scaffold with newly formed bone (Lee and et al., 2018).

6. Challenges and future directions in biomaterials for bone tissue engineering

6.1 Achieving vascularization and innervation in engineered scaffolds

Bone is a highly innervated and vascularized connective tissue reliant on complex interactions between blood vessels, neuronal tissues, and bone cells for maintaining homeostasis and repair. The vascular network plays a pivotal role in ensuring adequate nutrient and oxygen transport, removal of metabolic waste, and supporting overall cellular activity (Chen et al., 2024). The vascular network is also crucial for the recruitment of hematopoietic cells necessary for bone repair (Clarke, 2008). The intricate influence of the nervous system on bone homeostasis, impacting processes such as bone formation, resorption, and healing, underscores the profound significance of incorporating strategies for innervation into engineered bone tissues (Orvalho et al., 2023). While vascular and neuronal ingrowth occurs passively over time in conventional bone grafts, significant efforts in tissue engineering focus on actively promoting these processes within biomaterial scaffolds. Targeted delivery of cytokines, such as growth factors like vascular endothelial growth factor (VEGF) and nerve growth factors (NGF), has been a popular strategy. Simultaneous delivery of angiogenic and neurotrophic factors to promote neurovascular coupling has also been reported. Similarly, co-culture of endothelial cells and neuronal cells along with osteogenic cells within or on scaffolds has been reported to enhance de-novo bone formation through enhanced crosstalk and in-situ cytokine activation (Tomlinson et al., 2016). Advanced fabrication techniques, such as multi-material deposition or printing, can further aid in enhancing the spatial distribution of different cell types or bioactive cues. Additionally, methods to create interconnected porosity and hollow channels within scaffolds have been investigated. For instance, the use of porogens or sacrificial materials (like gelatin or carbohydrate-based polymers) during fabrication can create voids upon removal (Malkani et al., 2025). These voids allow for the infiltration of blood vessels, allowing vascularization. Such

Table 22.4 Biopolymeric photocurable resins with their features.

Photo polymer	Tissue application	Advantages	Disadvantages	References
Poly (lactic-co-glycolic acid) (PLGA)	Bone, cartilage, muscle, Nerve, Skin	Biodegradable, tunable degradation rate, FDA *approved*	Can be brittle, potential inflammatory response	Lee et al. (2011)
Poly (ε-caprolactone) (PCL)	Bone, cartilage, Skin, vascular grafts	Biodegradable, good printability, tunable properties	Low mechanical strength, slow degradation rate	Elomaa et al. (2011)
Poly (ethylene glycol) (PEG) Hydrogels	Cartilage, Neural tissue, Skin	Highly tunable properties, good biocompatibility	Lower mechanical strength compared to some other options	Guerra, Lara-Padilla, et al. (2019)
Poly(γ-glutamic acid) (PGA) Hydrogels	Bone, cartilage, Neural tissue	Biodegradable, excellent cell adhesion	Can be brittle, pH sensitive, depending on formulation	Yang et al. (2020)
Polydopamine (PDA)	Neural tissue, Skin, vascular grafts	Biocompatible, promotes cell adhesion	Limited printability depending on formulation	Chiu et al. (2019)
Polypropylene fumarate (PPF)	Bone, cartilage, Neural tissue, Skeletal muscle	Biodegradable, tunable mechanical properties, and injectable	Lower printability compared to some other options	Walker et al. (2017)
PEGDA (Poly (ethylene glycol) diacrylate)	Cartilage, muscle, and neural tissue engineering	Highly tunable properties, good biocompatibility	Potentially limited default mechanical strength	Mondal et al. (2021)
Thiol-ene based polymers	Cartilage, neural tissue, skin, and blood vessels	Highly tunable properties, good biocompatibility	Limited commercially available options require specific curing mechanisms.	Zhu et al. (2019)

strategies have been associated with significantly enhanced in vivo neovascularization, thereby resulting in improved perfusion and tissue integration. Biomimicry through the incorporation of conductive biomaterials (like polydopamine, graphene, MXene and carbon nanotubes), as well as the application of external electrical stimulation, has been reported to promote neural differentiation and axonal guidance in BTE scaffolds (Zhang et al., 2023). Despite these advances, challenges persist in engineering hierarchical vascular and neuronal networks that mimic native tissue architecture, as given in Table 22.5 below. Uncontrolled or burst release of growth factors remains an ongoing challenge. Controlled delivery, through various means, is generally accepted as the way forward, but successful incorporation into anatomically relevant constructs needs further realization. Similarly, effective co-culture of distinct cell lines within a single scaffold remains difficult. Future research must focus on optimizing biomaterial compositions, refining fabrication techniques for complex multi-component structures, and integrating bioreactors with mechano- and electro-transduction capabilities that support dynamic vascular and neuronal growth.

6.2 Improving mechanical performance while maintaining bioactivity

The high strength and stiffness of bone tissue necessitate that BTE scaffolds exhibit sufficiently high mechanical properties (such as compressive strength ranging between 2–100 MPa, depending on the defect site). As highlighted in a recent review, a dichotomy often exists between the mechanical performance and the chemo-biological properties of biomaterials for BTE applications (Gupta et al., 2023). In essence, while many acellular biomaterials (such as certain thermoplastic polymers, metallic materials like zinc alloys, and ceramics like calcium silicates) exhibit high mechanical stiffness and strength, they may fail to fully replicate the biochemical niche of the host tissue, which has implications for cellular adhesion, scaffold integration, and new bone formation. In contrast, softer polymeric biomaterials like hydrogels (especially those formulated as bioinks or for delivering living cells, chemokines, and cytokines) can be tailored to closely mimic the biochemical niche of the host site but often suffer from poor mechanical and degradation stability. As such, balancing mechanical strength with bioactivity presents a major challenge for biomaterials in BTE. Strategies to address this concern include.

- Incorporation of bioactive fillers (such as calcium phosphates or bioactive glasses) within or at the surface of otherwise less bioactive but mechanically robust biomaterials like polycaprolactone, titanium alloys, and zirconia (Thangavel & Elsen Selvam, 2022).
- Strengthening of hydrogel scaffolds through the incorporation of reinforcing fillers like carbonaceous nanomaterials or bioceramics, which may also offer the advantage of improving osteogenic differentiation due to localized high stiffness (Cao et al., 2025).

Table 22.5 Comparative evaluation of the various vascularization strategies.

Vascularization strategy	Advantages	Disadvantages
Angiogenic growth factors	Technically simple	Optimization of the drug-release system is necessary
	Numerous angiogenic growth factors are commercially available	Time delay between implantation and achieving a sufficient blood supply to the implant
	A combination of angiogenic and osteogenic growth factors is possible	Short half-life of some angiogenic growth factors
Cell-based strategies	Prevascularization of vessel networks in the implant is possible by seeding endothelial cells	Limited survival of implanted cells
	A combination of endothelial cells and bone-forming cells is feasible	In a prospective clinical setting, autologous cells must be used
Biofabrication of vessel networks	Precise control of the spatial distribution of cells	Limited survival of implanted cells
	Combined printing of cells and drug release hydrogels is feasible	
	Combined printing of endothelial cells and bone-forming cells is feasible	In a prospective clinical setting, autologous cells must be used
	Acceleration of vessel formation by spatial alignment of endothelial cells	
Surgical strategies	Immediate blood supply can be achieved	Technically challenging
	Combination of immediate vascularization and naturally occurring vascular sprouting in the AV-loop model	A two-stage approach is necessary to create a vascularized implant
	Integration of growth factors and cells is feasible in the AV-loop model	

From Simunovic, Filip & Finkenzeller, Günter. (2021). Vascularization Strategies in Bone Tissue Engineering. Cells, 10(7). https://doi.org/10.3390/cells10071749.

- Implementing multi-material approaches, often enabled by advanced fabrication methods, to decouple material property requirements. For example, combining calcium phosphate cements with cell-laden hydrogels can enable the improvement of mechanical strength while maintaining biological advantages (Gupta et al., 2023, 2025, 2026). Challenges in developing high-strength or load-bearing scaffolds also stem from the need to balance porosity distribution and stiffness gradients with overall mechanical strength. Future research may benefit significantly from employing advanced computational modeling and machine learning to optimize scaffold architectures and develop material combinations aimed at achieving superior, integrated mechanical and biological performance.

6.3 Standardization and clinical translation

The successful clinical translation of biomaterials for bone tissue engineering (BTE) is contingent upon addressing challenges across regulatory approval, preclinical validation, personalization, standardization, manufacturing, and cost-effectiveness. Despite significant scientific advancements in biomaterials for BTE, the path to clinical application remains arduous. Very few scaffold-based BTE products have progressed beyond early-phase (Phase I/II) clinical trials in recent years (Le et al., 2023). A primary hurdle lies in navigating complex regulatory approval processes established by agencies such as the US Food and Drug Administration (FDA) and the European Medicines Agency (EMA). BTE products often straddle definitions between medical devices, biologics, and drugs. While simpler, acellular scaffolds may be classified as Class II or III medical devices, those incorporating cells, growth factors, or other bioactive agents are typically regulated as combination products under more stringent frameworks (Le et al., 2023). This dual classification necessitates integrated data packages that address both device performance (e.g., mechanical stability, degradation) and biological function (e.g., immunogenicity, pharmacodynamics, cell viability). It also requires evidence of the interaction and combined safety profile of all components. Further complicating this landscape is the lack of harmonized international standards. Although frameworks such as the ISO 10993 series guide the biocompatibility assessment of medical devices, these are often inadequate for evaluating novel bioactive or cell-laden scaffolds. Specific standards such as ISO 10993-6 (local effects after implantation) and ISO 10993-11 (systemic toxicity) are relevant, yet they fall short in covering the complexity of BTE constructs. ASTM standards, including F2150 for testing tissue-engineered medical products, offer additional guidance. However, their adoption and consistent application across jurisdictions remain limited (Simon et al., 2015; Babu, 2024; "ASTM F2150-19," 2019). Preclinical validation, especially using physiologically relevant animal models, is a critical requirement mandated by regulatory bodies before human trials. While small animal models (e.g., rodents) are useful for initial biocompatibility and efficacy screening, large animal models such as sheep, goats, and pigs are often necessary to replicate human bone physiology, defect sizes, and biomechanical loading conditions (Gray et al., 2022; Pearce et al., 2007). These models are particularly important for load-bearing applications. However, such studies are resource-intensive, time-consuming, and pose ethical concerns, making them a significant barrier to clinical translation. Rigorous study design and careful model selection are essential to generate translatable and convincing preclinical data. Standardization must also encompass material sourcing, scaffold fabrication, and characterization methodologies. Variability in raw material properties, processing techniques (e.g., 3D printing, electrospinning, freeze-drying, or molding), and scaffold testing protocols currently impairs reproducibility across studies. This hinders cross-study comparisons and generalization of findings. Topical reviews have also highlighted significant discrepancies in both in vitro and in vivo testing methodologies, including inconsistencies in mechanical testing environments (static vs. dynamic loading), choice of cell types (primary human cells vs. immortalized lines), and animal model selection (Burg et al., 2000; Khademhosseini & Langer, 2016; Unamuno et al., 2022). Establishing consensus on evaluation parameters such as scaffold biocompatibility, degradation kinetics, mechanical stability, and long-term immunogenicity is essential for reproducible preclinical assessment. Another critical consideration is patient-to-patient biological variability, which remains underrepresented in preclinical research. Variables such as age, metabolic health, bone density, and comorbid conditions (e.g., diabetes, osteoporosis, chronic inflammation) can significantly influence scaffold performance. These inter-individual differences introduce additional complexity to regulatory approval, manufacturing scalability, and cost-effectiveness, especially in settings with limited healthcare resources. Finally, economic considerations play a non-trivial role in clinical adoption. While biologically advanced or patient-specific scaffolds may demonstrate promising outcomes in experimental settings, they may not be scalable or affordable for widespread clinical use. Future strategies must incorporate principles of cost-benefit optimization and value-based design to ensure that biomaterials for BTE can be effectively integrated into existing healthcare systems (van Schalkwyk & Harris, 2018).

6.4 Emerging materials and strategies for bone engineering

The future of biomaterials for BTE increasingly involves strategies aimed at providing patient-specific, personalized treatment of bone defects. In situ fabrication approaches, in which scaffolds are formed or solidified directly at the defect site, offer the potential for immediate defect repair and reduced surgical complexity (Samandari et al., 2022). While conceptually attractive, applying these methods to bone repair is more challenging than for softer tissues (like skin grafts), owing partly to the need for sufficient penetration depth for curing mechanisms (e.g., light for photopolymerization) and achieving appropriate mechanical properties upon formation (Jain et al., 2024). Although stricter process and material property requirements need to be addressed, recent reports employing combinations of injectable materials and advanced techniques like two-photon lithography have shown promising results. Finally, "smart" biomaterials, such as stimuli-responsive hydrogels, are being designed to adapt to the physiological environment by altering their mechanical or chemical properties in response to intrinsic or extrinsic triggers (e.g., temperature, pH, enzymes) (Rodrigo-Navarro et al., 2021). For example, thermo-responsive and pH-sensitive biomaterials have been developed to dynamically modulate degradation rates and drug release profiles. Similarly, enabling mechano-transduction within scaffold implants through materials responsive to external triggers like magnetic or acoustic fields has been shown to potentially improve directional cell growth and enhance osteogenesis (Montoya et al., 2021). Future advancements in the development of novel and responsive biomaterials, dynamic and scalable scaffold fabrication techniques, and supportive regulatory frameworks will pave the way for the successful clinical application of modern biomaterials for BTE applications.

7. Conclusion

The development of effective biomaterials remains central to advancing bone tissue engineering (BTE), offering potential solutions for a spectrum of skeletal pathologies that exceed the body's innate repair capacity. This chapter reviewed key principles of bone biology and regeneration to contextualize the functional requirements for biomaterial-based scaffolds. The evolution from autografts and allografts to synthetic biomaterials has expanded therapeutic possibilities while introducing new design challenges. Metallic scaffolds provide the mechanical strength necessary for load-bearing applications but are often limited by issues such as bioinertness, ion leaching, and stress shielding. In contrast, bioceramics and bioactive glasses offer excellent biocompatibility and osteoconductivity, effectively mimicking mineral phase of bone. However, their inherent brittleness restricts their use in dynamic or high-load environments. Biopolymers afford versatility in processing and tunable degradation profiles but frequently lack the structural robustness and intrinsic bioactivity required for functional bone repair. As a result, there is increasing emphasis on composite biomaterials that integrate complementary material properties. Polymer-matrix composites reinforced with bioactive ceramics and biodegradable metal-matrix systems (e.g., magnesium alloys) are being explored to recapitulate the mechanical and biological complexity of native bone. These engineered systems aim to optimize the interplay between mechanical stability, degradation kinetics, and biological response. Engineering predictable and functional vascularization and innervation within scaffolds, particularly for large defects, remains a significant challenge crucial for long-term graft success. Resolving the persistent dichotomy between biomechanical competence and favorable bioactivity necessitates continued material science innovation and refined fabrication approaches. Furthermore, clinical adoption is impeded by inconsistencies in preclinical testing standards, complex regulatory pathways (especially for combination products), and challenges in achieving scalable, GMP-compliant manufacturing. Patient-specific biological variability adds another layer of complexity to scaffold design and performance validation. Future progress requires integrated, multidisciplinary strategies, incorporating advanced computational modeling, multi-material fabrication techniques, stimuli-responsive "smart" materials, and potentially in situ regenerative therapies. Addressing these interconnected challenges through rigorous, standardized investigation is essential to realize the full therapeutic potential of next-generation BTE scaffolds in restoring skeletal function.

References

Albrektsson, T., & Johansson, C. (2001). Osteoinduction, osteoconduction and osseointegration. *European Spine Journal, 10*, S96. https://doi.org/10.1007/s005860100282

Alifui-Segbaya, F. (2020). Biomedical photopolymers in 3D printing. *Rapid Prototyping Journal, 26*(2), 437–444. https://doi.org/10.1108/RPJ-10-2018-0268

Alifui-Segbaya, F., Bowman, J., White, A. R., Varma, S., Lieschke, G. J., & George, R. (2018). Toxicological assessment of additively manufactured methacrylates for medical devices in dentistry. *Acta Biomaterialia, 78*, 64–77. https://doi.org/10.1016/j.actbio.2018.08.007

Alizadeh, P., & Marghussian, V. K. (2000). The effect of compositional changes on the crystallization behaviour and mechanical properties of diopside–wollastonite glass-ceramics in the SiO 2 –CaO–MgO (Na 2 O) system. *Journal of the European Ceramic Society, 20*(6), 765–773. https://doi.org/10.1016/s0955-2219(99)00135-1

Alknery, Z., Ibrahim Sktani, Z. D., & Arab, A. (2022). Effect of cell geometry on the mechanical properties of 3D voronoi tessellation. *Journal of Functional Biomaterials, 13*(4), 302. https://doi.org/10.3390/jfb13040302

ASTM F2150-19. (2019). *Standard guide for characterization and testing of biomaterial scaffolds used in regenerative medicine and tissue-engineered medical products standard guide for characterization and testing of biomaterial scaffolds used in regenerative medicine and tissue-engineered medical products.*

Babu, S. (2024). *ISO 10993: Navigating biocompatibility testing for medical devices.*

Bagheri, A., & Jin, J. (2019). Photopolymerization in 3D printing. *ACS Applied Polymer Materials, 1*(4), 593–611. https://doi.org/10.1021/acsapm.8b00165

Baino, F., Minguella-Canela, J., Korkusuz, F., Korkusuz, P., Kankılıç, B., Montealegre, M., De los Santos-López, M., & Vitale-Brovarone, C. (2019). In vitro assessment of bioactive glass coatings on Alumina/Zirconia composite implants for potential use in prosthetic applications. *International Journal of Molecular Sciences, 20*(3), 722. https://doi.org/10.3390/ijms20030722

Bandyopadhyay-Ghosh. (2008). Bone as a collagen-hydroxyapatite composite and its repair. *Trends in Biomaterials and Artificial and LastNameLastName, 22*, 112–121.

Bandyopadhyay-Ghosh, S., Faria, P. E. P., Johnson, A., Felipucci, D. N. B., Reaney, I. M., Salata, L. A., Brook, I. M., & Hatton, P. V. (2010). Osteoconductivity of modified fluorcanasite glass-ceramics for bone tissue augmentation and repair. *Journal of Biomedical Materials Research - Part A, 94*(3), 760–768. https://doi.org/10.1002/jbm.a.32750

Banerjee, S. S., Tarafder, S., Davies, N. M., Bandyopadhyay, A., & Bose, S. (2010). Understanding the influence of MgO and SrO binary doping on the mechanical and biological properties of β-TCP ceramics. *Acta Biomaterialia, 6*(10), 4167–4174. https://doi.org/10.1016/j.actbio.2010.05.012

Bergmann, C., Lindner, M., Zhang, W., Koczur, K., Kirsten, A., Telle, R., & Fischer, H. (2010). 3D printing of bone substitute implants using calcium phosphate and bioactive glasses. *Journal of the European Ceramic Society, 30*(12), 2563–2567. https://doi.org/10.1016/j.jeurceramsoc.2010.04.037

Bian, W., Li, D., Lian, Q., Zhang, W., Zhu, L., Li, X., & Jin, Z. (2011). Design and fabrication of a novel porous implant with pre-set channels based on ceramic stereolithography for vascular implantation. *Biofabrication, 3*(3), Article 034103. https://doi.org/10.1088/1758-5082/3/3/034103

Bisht, B., Hope, A., Mukherjee, A., & Paul, M. K. (2021). Advances in the fabrication of scaffold and 3D printing of biomimetic bone graft. *Annals of Biomedical Engineering, 49*(4), 1128–1150. https://doi.org/10.1007/s10439-021-02752-9

Block, J., & Miller, L. E. (2011). *Perspectives on the clinical utility of allografts for bone regeneration within osseous defects: A narrative review.* Orthopedic Research and Reviews. https://doi.org/10.2147/ORR.S18397

Boskey, A. L., & Robey, P. G. (2018). *The composition of bone primer on the metabolic bone diseases and disorders of mineral metabolism* (pp. 84–92). United States: wiley. https://doi.org/10.1002/9781119266594.ch11

Brook, I. M., & Hatton, P. V. (1998). Glass-ionomers: Bioactive implant materials. *Biomaterials, 19*(6), 565–571. https://doi.org/10.1016/s0142-9612(98)00138-0

Buj-Corral, I., Bagheri, A., & Petit-Rojo, O. (2018). 3D printing of porous scaffolds with controlled porosity and pore size values. *Materials, 11*(9). https://doi.org/10.3390/ma11091532

Burg, K. J. L., Porter, S., & Kellam, J. F. (2000). Biomaterial developments for bone tissue engineering. *Biomaterials, 21*(23), 2347–2359. https://doi.org/10.1016/S0142-9612(00)00102-2

Cao, Y., Liu, C., Ye, W., Zhao, T., & Fu, F. (2025). Functional hydrogel interfaces for cartilage and bone regeneration. *Advanced Healthcare Materials, 14*(6). https://doi.org/10.1002/adhm.202403079

Carter, D. H., Sloan, P., Brook, I. M., & Hatton, P. V. (1997). Role of exchanged ions in the integration of ionomeric (glass polyalkenoate) bone substitutes. *Biomaterials, 18*(6), 459–466. https://doi.org/10.1016/s0142-9612(96)00150-0

Caughman, W. F., Caughman, G. B., Shiflett, R. A., Rueggeberg, F., & Schuster, G. S. (1991). Correlation of cytotoxicity, filler loading and curing time of dental composites. *Biomaterials, 12*(8), 737–740. https://doi.org/10.1016/0142-9612(91)90022-3

Chen, S., Zhou, X., Li, T., & He, C. (2024). Vascularization and innervation for bone tissue engineering. *Accounts of Materials Research, 5*(9), 1121–1133. https://doi.org/10.1021/accountsmr.4c00165

Chen, Q., Zhu, C., & Thouas, G. A. (2012). Progress and challenges in biomaterials used for bone tissue engineering: Bioactive glasses and elastomeric composites. *Progress in Biomaterials, 1*(1), 2. https://doi.org/10.1186/2194-0517-1-2

Chiu, Y.-C., Shen, Y.-F., Lee, A. K.-X., Lin, S.-H., Wu, Y.-C., & Chen, Yi-W. (2019). 3D printing of amino resin-based photosensitive materials on multi-parameter optimization design for vascular engineering applications. *Polymers, 11*(9), 1394. https://doi.org/10.3390/polym11091394

Choi, J.-W., Lee, J.-J., Bae, E.-B., & Huh, J.-B. (2020). Implant-supported fixed dental prosthesis with a microlocking implant prosthetic system: A clinical report. *The Journal of Prosthetic Dentistry, 123*(1), 15–19. https://doi.org/10.1016/j.prosdent.2018.11.021

Clarke, B. (2008). Normal bone anatomy and physiology. *Clinical Journal of the American Society of Nephrology, 3*(Suppl. ment_3), Article S131. https://doi.org/10.2215/cjn.04151206

Cook, R. B., & Zioupos, P. (2009). The fracture toughness of cancellous bone. *Journal of Biomechanics, 42*(13), 2054–2060. https://doi.org/10.1016/j.jbiomech.2009.06.001

Correia, F., Pozza, D. H., Gouveia, S., Felino, A. C., & Faria-Almeida, R. (2021). Advantages of porcine xenograft over autograft in sinus lift: A randomised clinical trial. *Materials, 14*(12), 3439. https://doi.org/10.3390/ma14123439

Costa, N. N., Lopes, L. D. F., Ferreira, D. F., de Prado, E. M. L., Severi, J. A., Resende, J. A., Careta, F. D. P., Ferreira, M. C. P., Carreira, L. G., de Souza, S. O. L., Cotrim, M. A. P., Boeing, T., de Andrade, S. F., Lambert Oréfice, R., & Villanova, J. C. O. (2020). Polymeric films containing pomegranate peel extract based on PVA/starch/PAA blends for use as wound dressing: In vitro analysis and physicochemical evaluation. *Materials Science and Engineering: C, 109*, Article 110643. https://doi.org/10.1016/j.msec.2020.110643

De Santis, R., Guarino, V., & Ambrosio, L. (2018). *Composite biomaterials for bone repair bone repair biomaterials: Regeneration and clinical applications* (2nd ed.). Italy: Elsevier. https://doi.org/10.1016/B978-0-08-102451-5.00010-X

Dutta, S., Devi, K. B., Gupta, S., Kundu, B., Krishna Balla, V., & Roy, M. (2019). Mechanical and in vitro degradation behavior of magnesium-bioactive glass composites prepared by SPS for biomedical applications. *Journal of Biomedical Materials Research Part B: Applied Biomaterials, 107*(2), 352–365. https://doi.org/10.1002/jbm.b.34127

Dutta Roy, T., Simon, J. L., Ricci, J. L., Rekow, E. D., Thompson, V. P., & Parsons, J. R. (2003). Performance of degradable composite bone repair products made via three-dimensional fabrication techniques. *Journal of Biomedical Materials Research - Part A, 66*(2), 283–291. https://doi.org/10.1002/jbm.a.10582

Elahinia, M., Moghaddam, N. S., & Andani, M. T. (2016). Fabrication of NiTi through additive manufacturing: A review. *Progress in Materials Science, 83*, 630–663. https://doi.org/10.1016/J.PMATSCI.2016.08.001

Elomaa, L., Teixeira, S., Hakala, R., Korhonen, H., Grijpma, D. W., & Seppälä, J. V. (2011). Preparation of poly(ε-caprolactone)-based tissue engineering scaffolds by stereolithography. *Acta Biomaterialia, 7*(11), 3850–3856. https://doi.org/10.1016/j.actbio.2011.06.039

Evans, F. G. (1976). Mechanical properties and histology of cortical bone from younger and older men. *The Anatomical Record, 185*(1), 1–11. https://doi.org/10.1002/ar.1091850102

Fantini, M., & Curto, M. (2018). Interactive design and manufacturing of a Voronoi-based biomimetic bone scaffold for morphological characterization. *International Journal on Interactive Design and Manufacturing, 12*(2), 585–596. https://doi.org/10.1007/s12008-017-0416-x

Fernandes, J. S., Gentile, P., Pires, R. A., Reis, R. L., & Hatton, P. V. (2017). Multifunctional bioactive glass and glass-ceramic biomaterials with antibacterial properties for repair and regeneration of bone tissue. *Acta Biomaterialia, 59*, 2–11. https://doi.org/10.1016/j.actbio.2017.06.046

Fernandez-Yague, M. A., Abbah, S. A., McNamara, L., Zeugolis, D. I., Pandit, A., & Biggs, M. J. (2015). Biomimetic approaches in bone tissue engineering: Integrating biological and physicomechanical strategies. *Advanced Drug Delivery Reviews, 84*, 1–29. https://doi.org/10.1016/j.addr.2014.09.005

Fisher, J. P., Timmer, M. D., Holland, T. A., Dean, D., Engel, P. S., & Mikos, A. G. (2003). Photoinitiated cross-linking of the biodegradable polyester poly(propylene fumarate). Part I. Determination of network structure. *Biomacromolecules, 4*(5), 1327–1334. https://doi.org/10.1021/bm030028d

Fu, Q. (2018). *Bioactive glass scaffolds for bone tissue engineering biomedical, therapeutic and clinical applications of bioactive glasses* (pp. 417–442). United States: Elsevier. https://doi.org/10.1016/B978-0-08-102196-5.00015-X

Garimella, A., Ashri, A., Kumar, A., Kumar, A., Ghosh, S. B., & Bandyopadhyay-Ghosh, S. (2022). Evaluation of microstructure and mechanical properties of magnesium alloy based porous bone substitutes. *Materials Today: Proceedings, 56*, 2708–2713. https://doi.org/10.1016/j.matpr.2021.09.392

Garimella, A., Awale, G., Parai, R., Ghosh, S. B., & Bandyopadhyay-Ghosh, S. (2019). An integrated approach to develop engineered metal composite bone scaffold with controlled degradation. *Materials Technology, 34*(14), 858–866. https://doi.org/10.1080/10667857.2019.1639004

Garimella, A., Ramya, M., Ghosh, S. B., Ghosh, S. B.-, & Agrawal, A. K. (2023). Bioactive fluorcanasite reinforced magnesium alloy-based porous bio-nanocomposite scaffolds with tunable mechanical properties. *Journal of Biomedical Materials Research Part B: Applied Biomaterials, 111*(2), 463–477. https://doi.org/10.1002/jbm.b.35166

Garimella, A., Rathi, D., Jangid, R., Ghosh, S. B., & Bandyopadhyay-Ghosh, S. (2022). Investigation of bio-degradation behaviour of porous magnesium alloy based bone scaffolds. *Materials Today: Proceedings, 50*, 2276–2279. https://doi.org/10.1016/j.matpr.2021.09.537

Gentile, P., Belmonte, M. M.-, Chiono, V., Ferretti, C., Baino, F., Turo, C. T.-, Brovarone, C. V.-, Pashkuleva, I., Reis, R. L., & Ciardelli, G. (2012). Bioactive glass/polymer composite scaffolds mimicking bone tissue. *Journal of Biomedical Materials Research Part A, 100A*(10), 2654–2667. https://doi.org/10.1002/jbm.a.34205

Goodridge, R. D., Dalgarno, K. W., & Wood, D. J. (2006). Indirect selective laser sintering of an apatite-mullite glass-ceramic for potential use in bone replacement applications. *Proceedings of the Institution of Mechanical Engineers - Part H: Journal of Engineering in Medicine, 220*(1), 57–68. https://doi.org/10.1243/095441105X69051

Govoni, M., Vivarelli, L., Mazzotta, A., Stagni, C., Maso, A., & Dallari, D. (2021). Commercial bone grafts claimed as an alternative to autografts: Current trends for clinical applications in orthopaedics. *Materials, 14*(12), 3290. https://doi.org/10.3390/ma14123290

Gray, M., Guido, S., & Kugadas, A. (2022). Editorial: The use of large animal models to improve pre-clinical translational research. *Frontiers in Veterinary Science, 9*. https://doi.org/10.3389/fvets.2022.1086912

Grover, V., Kapoor, A., Malhotra, R., & Sachdeva, S. (2011). Bone allografts: A review of safety and efficacy. *Indian Journal of Dental Research, 22*(3), 532–537. https://doi.org/10.4103/0970-9290.87084

Grunenwald, A., Keyser, C., Sautereau, A. M., Crubézy, E., Ludes, B., & Drouet, C. (2014). Revisiting carbonate quantification in apatite (bio)minerals: A validated FTIR methodology. *Journal of Archaeological Science, 49*(1), 134–141. https://doi.org/10.1016/j.jas.2014.05.004

Guerra, A. J., Lammel-Lindemann, J., Katko, A., Kleinfehn, A., Rodriguez, C. A., Catalani, L. H., Becker, M. L., Ciurana, J., & Dean, D. (2019). Optimization of photocrosslinkable resin components and 3D printing process parameters. *Acta Biomaterialia, 97*, 154–161. https://doi.org/10.1016/j.actbio.2019.07.045

Guerra, A. J., Lara-Padilla, H., Becker, M. L., Rodriguez, C. A., & Dean, D. (2019). Photopolymerizable resins for 3D-printing solid-cured tissue engineered implants. *Current Drug Targets, 20*(8), 823–838. https://doi.org/10.2174/1389450120666190114122815

Gupta, T., Ghosh, S. B., Bandyopadhyay-Ghosh, S., & Sain, M. (2023). Is it possible to 3D bioprint load-bearing bone implants? A critical review. *Biofabrication, 15*(4), Article 042003. https://doi.org/10.1088/1758-5090/acf6e1

Gupta, T., Jalali, A., Azarnia, F., et al. (2025). Bioprinting the bone. In N. Sultana, S. Bandyopadhyay-Ghosh, & C. Fhong Soon (Eds.), *Biomedical materials and biofabrication for regenerative medicine* (1st, pp. 76–117). Boca Raton: CRC Press. https://doi.org/10.1201/9781003454632

Gupta, T., Rath, P., Pakharenko, V., et al. (2026). Bioinspired thermoreversible bioink orchestrates focal adhesion-dependent osteogenesis. *Trends in Biotechnology, 44*(1), 239–265. https://doi.org/10.1016/j.tibtech.2025.09.007

Habibovic, P., & de Groot, K. (2007). Osteoinductive Biomaterials—Properties and relevance in bone repair. *Journal of Tissue Engineering and Regenerative Medicine, 1*(1), 25–32. https://doi.org/10.1002/term.5

Habraken, W. J. E. M., Wolke, J. G. C., & Jansen, J. A. (2007). Ceramic composites as matrices and scaffolds for drug delivery in tissue engineering. *Advanced Drug Delivery Reviews, 59*(4–5), 234–248. https://doi.org/10.1016/j.addr.2007.03.011

Han, S. B., Kim, J. K., Lee, G., & Kim, D. H. (2020). Mechanical properties of materials for stem cell differentiation. *Advanced Biosystems, 4*(11). https://doi.org/10.1002/adbi.202000247

Hazarika, K., Mahato, P. K., & Das, P. (2025). MAXene-enabled drug free Aloe Vera hydrogel expedites wound healing in New Zealand white rabbits. *Journal of Reinforced Plastics and Composites, 2025*. https://doi.org/10.1177/07316844251352934

Heissler, E., Fischer, F.-S., Boiouri, S., Lehrnann, T., Mathar, W., Gebhardt, A., Lanksch, W., & Bler, J. (1998). Custom-made cast titanium implants produced with CAD/CAM for the reconstruction of cranium defects. *International Journal of Oral and Maxillofacial Surgery, 27*(5), 334–338. https://doi.org/10.1016/s0901-5027(98)80060-x

Huang, Y., Liu, D., Anguilano, L., You, C., & Chen, M. (2015). Fabrication and characterization of a biodegradable Mg–2Zn–0.5Ca/1β-TCP composite. *Materials Science and Engineering: C, 54*, 120–132. https://doi.org/10.1016/j.msec.2015.05.035

Inzana, J. A., Olvera, D., Fuller, S. M., Kelly, J. P., Graeve, O. A., Schwarz, E. M., Kates, S. L., & Awad, H. A. (2014). 3D printing of composite calcium phosphate and collagen scaffolds for bone regeneration. *Biomaterials, 35*(13), 4026–4034. https://doi.org/10.1016/j.biomaterials.2014.01.064

Ito, M. (1991). In vitro properties of a chitosan-bonded hydroxyapatite bone-filling paste. *Biomaterials, 12*(1), 41–45. https://doi.org/10.1016/0142-9612(91)90130-3

Jain, P., Kathuria, H., Ramakrishna, S., Parab, S., Pandey, M. M., & Dubey, N. (2024). In situ bioprinting: Process, bioinks, and applications. *ACS Applied Bio Materials, 7*(12), 7987–8007. https://doi.org/10.1021/acsabm.3c01303

Jana, S., Datta, P., Das, H., Ranjan Ghosh, P., Kundu, B., & Kumar Nandi, S. (2022). Engineering vascularizing electrospun dermal grafts by integrating fish collagen and ion-doped bioactive glass. *ACS Biomaterials Science & Engineering, 8*(2), 734–752. https://doi.org/10.1021/acsbiomaterials.1c01098

Jang, D. W., Franco, R. A., Sarkar, S. K., & Lee, B. T. (2014). Fabrication of porous hydroxyapatite scaffolds as artificial bone preform and its biocompatibility evaluation. *ASAIO Journal, 60*(2), 216–223. https://doi.org/10.1097/MAT.0000000000000032

Jilka, R. L. (2003/09/01). Biology of the basic multicellular unit and the pathophysiology of osteoporosis. *Medical and Pediatric Oncology, 41*(3), 182–185. https://doi.org/10.1002/mpo.10334

Kanchanarat, N., Miller, C. A., Hatton, P. V., James, P. F., & Reaney, I. M. (2005). Early stages of crystallization in Canasite-Based glass ceramics. *Journal of the American Ceramic Society, 88*(11), 3198–3204. https://doi.org/10.1111/j.1551-2916.2005.00589.x

Kanwar, S., & Vijayavenkataraman, S. (2021). Design of 3D printed scaffolds for bone tissue engineering: A review. *Bioprinting, 24*, Article e00167. https://doi.org/10.1016/j.bprint.2021.e00167

Khademhosseini, A., & Langer, R. (2016). A decade of progress in tissue engineering. *Nature Protocols, 11*(10), 1775–1781. https://doi.org/10.1038/nprot.2016.123

Khan, S. N., Cammisa, F. P., Sandhu, H. S., Diwan, A. D., Girardi, F. P., & Lane, J. M. (2005). The biology of bone grafting. *Journal of the American Academy of Orthopaedic Surgeons, 13*(1), 77–86. https://doi.org/10.5435/00124635-200501000-00010

Khanra, A. K., Jung, H. C., Hong, K. S., & Shin, K. S. (2010). Comparative property study on extruded Mg-HAP and ZM61-HAP composites. *Materials Science and Engineering: A, 527*(23), 6283–6288. https://doi.org/10.1016/j.msea.2010.06.031

Kim, H. J., Park, I. K., Kim, J. H., Cho, C. S., & Kim, M. S. (2012). Gas foaming fabrication of porous biphasic calcium phosphate for bone regeneration. *Tissue Engineering and Regenerative Medicine, 9*(2), 63–68. https://doi.org/10.1007/s13770-012-0022-8

Kontonasaki, E., Zorba, T., Papadopoulou, L., Pavlidou, E., Chatzistavrou, X., Paraskevopoulos, K., & Koidis, P. (2002). Hydroxy carbonate apatite formation on particulate bioglass in vitro as a function of time. *Crystal Research and Technology, 37*(11), 1165–1171. https://doi.org/10.1002/1521-4079(200211)37:11<1165::aid-crat1165>3.0.co;2-r

Kumar, P., Dehiya, B. S., & Sindhu, A. (2018). Bioceramics for hard tissue engineering applications. *A Review Bioceramics, 13*(5), 2744–2752.

Kumawat, V. S., Ghosh, S. B., & Bandyopadhyay-Ghosh, S. (2019). Microporous biocomposite scaffolds with tunable degradation and interconnected microarchitecture-A synergistic integration of bioactive chain silicate glass-ceramic and poly(ε-caprolactone). *Polymer Degradation and Stability, 165*, 20–26. https://doi.org/10.1016/j.polymdegradstab.2019.04.017

Kumawat, V. S., Vyas, A., Bandyopadhyay-Ghosh, S., & Bandhu Ghosh, S. (2020). Selectively modified nanostructured fluorcanasite glass-ceramic with enhanced micromechanical properties. *Journal of Non-crystalline Solids, 547*, Article 120303. https://doi.org/10.1016/j.jnoncrysol.2020.120303

Laurencin, C. T., & El-Amin, S. F. (2008). Xenotransplantation in orthopaedic surgery. *Journal of the American Academy of Orthopaedic Surgeons, 16*(1), 4–8. https://doi.org/10.5435/00124635-200801000-00002

Le, H. T., Mahara, A., Nagasaki, T., & Yamaoka, T. (2023). Prevention of anastomotic stenosis for decellularized vascular grafts using rapamycin-loaded boronic acid-based hydrogels mimicking the perivascular tissue function. *Biomaterials Advances, 147*, Article 213324. https://doi.org/10.1016/j.bioadv.2023.213324

Lee, et al. (2018). *Materials*. https://doi.org/10.3390/ma11091711

Lee, J. W., Kang, K. S., Lee, S. H., Kim, J. Y., Lee, B. K., & Cho, D. W. (2011). Bone regeneration using a microstereolithography-produced customized poly(propylene fumarate)/diethyl fumarate photopolymer 3D scaffold incorporating BMP-2 loaded PLGA microspheres. *Biomaterials, 32*(3), 744–752. https://doi.org/10.1016/j.biomaterials.2010.09.035

Li, J., Chen, M., Wei, X., Hao, Y., & Wang, J. (2017). Evaluation of 3D-Printed polycaprolactone scaffolds coated with freeze-dried platelet-rich plasma for bone regeneration. *Materials, 10*(7), 831. https://doi.org/10.3390/ma10070831

Li, M., Mondrinos, M. J., Chen, X., Gandhi, M. R., Ko, F. K., & Lelkes, P. I. (2006). Co-electrospun poly(lactide-co-glycolide), gelatin, and elastin blends for tissue engineering scaffolds. *Journal of Biomedical Materials Research - Part A, 79*(4), 963–973. https://doi.org/10.1002/jbm.a.30833

Li, F., Shan, L., Liu, Y., Zhang, Z., & Li, Z. (2022). Current advances in the roles of doped bioactive metal in biodegradable polymer composite scaffolds for bone repair: A mini review. *Advanced Engineering Materials, 24*(8). https://doi.org/10.1002/adem.202101510

Li, Y., Zheng, L., Peng, S., Miao, J. T., Zhong, J., Wu, L., & Weng, Z. (2020). Structure-property relationship of stereolithography resins containing polysiloxane core-shell nanoparticles. *ACS Applied Materials and Interfaces, 12*(4), 4917–4926. https://doi.org/10.1021/acsami.9b20417

Li, Y., Zhong, J., Wu, L., Weng, Z., Zheng, L., Peng, S., & Zhang, X. (2019). High performance POSS filled nanocomposites prepared via UV-curing based on 3D stereolithography printing. *Composites Part A: Applied Science and Manufacturing, 117*, 276–286. https://doi.org/10.1016/j.compositesa.2018.11.024

Li, W., Zhou, J., & Xu, Y. (2015). Study of the in vitro cytotoxicity testing of medical devices. *Biomedical Reports, 3*(5), 617–620. https://doi.org/10.3892/br.2015.481

Lian, M., Sun, B., Han, Y., Yu, B., Xin, W., Xu, R., Ni, B., Jiang, W., Yongqiang, H., Zhang, X., Shen, Y., Qiao, Z., & Dai, K. (2021). A low-temperature-printed hierarchical porous sponge-like scaffold that promotes cell-material interaction and modulates paracrine activity of MSCs for vascularized bone regeneration. *Biomaterials, 274*, Article 120841. https://doi.org/10.1016/j.biomaterials.2021.120841

Likitvanichkul, S., & Lacourse, W. C. (1995). Effect of fluorine content on crystallization of canasite glass-ceramics. *Journal of Materials Science, 30*(24), 6151–6155. https://doi.org/10.1007/bf00369661

Liu, S., Mo, L., Bi, G., Chen, S., Yan, D., Yang, J., Jia, Y.-G., & Ren, L. (2021). DLP 3D printing porous β-tricalcium phosphate scaffold by the use of acrylate/ceramic composite slurry. *Ceramics International, 47*(15), 21108–21116. https://doi.org/10.1016/j.ceramint.2021.04.114

Ma, L., Feng, X., Liang, H., Wang, K., Song, Y., Tan, L., Wang, B., Luo, R., Liao, Z., Li, G., Liu, X., Wu, S., & Yang, C. (2020). A novel photothermally controlled multifunctional scaffold for clinical treatment of osteosarcoma and tissue regeneration. *Materials Today, 36*, 48–62. https://doi.org/10.1016/j.mattod.2019.12.005

Mahato, A., Sandy, Z., Bysakh, S., Hupa, L., Das, I., Bhattacharjee, P., Kundu, B., De, G., Nandi, S. K., Vallittu, P., Balla, V. K., & Bhattacharya, M. (2020). Development of nano-porous hydroxyapatite coated e-glass for potential bone-tissue engineering application: An in vitro approach. *Materials Science and Engineering: C, 111*, Article 110764. https://doi.org/10.1016/j.msec.2020.110764

Majumder, N., Seit, S., Sarovar Bhabesh, N., & Ghosh, S. (2024). An advanced bioconjugation strategy for covalent tethering of TGFβ3 with silk fibroin matrices and its implications in the chondrogenesis profile of human BMSCs and human chondrocytes: A paradigm shift in cartilage tissue engineering. *Advanced Healthcare Materials, 13*(10). https://doi.org/10.1002/adhm.202303513

Malkani, S., Prado, O., & Stevens, K. R. (2025). Sacrificial templating for accelerating clinical translation of engineered organs. *ACS Biomaterials Science & Engineering, 11*(1), 1–12. https://doi.org/10.1021/acsbiomaterials.4c01824

Mancuso, E., Alharbi, N., Bretcanu, O. A., Marshall, M., Birch, M. A., McCaskie, A. W., & Dalgarno, K. W. (2017). Three-dimensional printing of porous load-bearing bioceramic scaffolds. *Proceedings of the Institution of Mechanical Engineers - Part H: Journal of Engineering in Medicine, 231*(6), 575–585. https://doi.org/10.1177/0954411916682984

Mei, L., Yin, J., Xia, Y., Yao, D., Liang, H., Zuo, K., & Zeng, Y.-P. (2020). Preparation of high-strength β-CaSiO3 bioceramic with B2O3 and SiO2 sintering additives. *Ceramics International, 46*(16), 25970–25978. https://doi.org/10.1016/j.ceramint.2020.07.086

Miao, X., & Sun, D. (2010). Graded/Gradient porous biomaterials. *Materials, 3*(1), 26–47. https://doi.org/10.3390/ma3010026Australia

Miller, C. A., Kokubo, T., Reaney, I. M., Hatton, P. V., & James, P. F. (2002). Formation of apatite layers on modified canasite glass–ceramics in simulated body fluid. *Journal of Biomedical Materials Research, 59*(3), 473–480. https://doi.org/10.1002/jbm.10018

Mondal, S., Nguyen, T. P., Pham, Van H., Hoang, G., Manivasagan, P., Kim, M. H., Nam, S. Y., & Oh, J. (2020). Hydroxyapatite nano bioceramics optimized 3D printed poly lactic acid scaffold for bone tissue engineering application. *Ceramics International, 46*(3), 3443–3455. https://doi.org/10.1016/j.ceramint.2019.10.057

Mondal, D., Srinivasan, A., Comeau, P., Toh, Yi-C., & Willett, T. L. (2021). Acrylated epoxidized soybean oil/hydroxyapatite-based nanocomposite scaffolds prepared by additive manufacturing for bone tissue engineering. *Materials Science and Engineering: C, 118*, Article 111400. https://doi.org/10.1016/j.msec.2020.111400

Montoya, C., Du, Y., Gianforcaro, A. L., Orrego, S., Yang, M., & Lelkes, P. I. (2021). On the road to smart biomaterials for bone research: Definitions, concepts, advances, and outlook. *Bone Research, 9*(1). https://doi.org/10.1038/s41413-020-00131-z

Moore, M. J., Jabbari, E., Ritman, E. L., Lu, L., Currier, B. L., Windebank, A. J., & Yaszemski, M. J. (2004). Quantitative analysis of interconnectivity of porous biodegradable scaffolds with micro-computed tomography. *Journal of Biomedical Materials Research - Part A, 71*(2), 258–267. https://doi.org/10.1002/jbm.a.30138

Morgan, E. F., Unnikrisnan, G. U., & Hussein, A. I. (2018). Bone mechanical properties in healthy and diseased states. *Annual Review of Biomedical Engineering, 20*, 119–143. https://doi.org/10.1146/annurev-bioeng-062117-121139

Moussa, N. T., & Dym, H. (2020). Maxillofacial bone grafting materials. *Dental Clinics of North America, 64*(2), 473–490. https://doi.org/10.1016/j.cden.2019.12.011

Musidang, A., & Jiratumnukul, N. (2015). Preparation of poly(lactic acid) acrylate for UV-curable coating applications. *Key Engineering Materials, 659*, 570–574. https://doi.org/10.4028/www.scientific.net/KEM.659.570

Muthukumar, T., Aravinthan, A., Sharmila, J., Kim, N. S., & Kim, J.-H. (2016). Collagen/Chitosan porous bone tissue engineering composite scaffold incorporated with ginseng compound K. *Carbohydrate Polymers, 152*, 566–574. https://doi.org/10.1016/j.carbpol.2016.07.003

Nam, Y. S., & Park, T. G. (1999). Porous biodegradable polymeric scaffolds prepared by thermally induced phase separation. *Journal of Biomedical Materials Research, 47*(1), 8–17. https://doi.org/10.1002/(SICI)1097-4636(199910)47:1<8::AID-JBM2>3.0.CO;2-L

O'Brien, F. J. (2011). Biomaterials & scaffolds for tissue engineering. *Materials Today, 14*(3), 88–95. https://doi.org/10.1016/S1369-7021(11)70058-X

Orciani, M., Fini, M., Di Primio, R., & Mattioli-Belmonte, M. (2017). Biofabrication and bone tissue regeneration: Cell source, approaches, and challenges. *Frontiers in Bioengineering and Biotechnology, 5*. https://doi.org/10.3389/fbioe.2017.00017

Orvalho, J. M., Fernandes, J. C. H., Moraes Castilho, R., & Fernandes, G. V. O. (2023). The macrophage's role on bone remodeling and osteogenesis: A systematic review. *Clinical*

Reviews in Bone and Mineral Metabolism, 21(1–4), 1–13. https://doi.org/10.1007/s12018-023-09286-9

Osman, M. A., Virgilio, N., Rouabhia, M., & Mighri, F. (2022). Development and characterization of functional polylactic acid/chitosan porous scaffolds for bone tissue engineering. *Polymers, 14*(23), 5079. https://doi.org/10.3390/polym14235079

Parthasarathy, J. (2014). 3D modeling, custom implants and its future perspectives in craniofacial surgery. *Annals of Maxillofacial Surgery, 4*(1), 9. https://doi.org/10.4103/2231-0746.133065

Pearce, A. I., Richards, R. G., Milz, S., Schneider, E., & Pearce, S. G. (2007). Animal models for implant biomaterial research in bone: A review. *European Cells and Materials, 13*, 1–10. https://doi.org/10.22203/eCM.v013a01

Pecci, R., Baiguera, S., Ioppolo, P., Bedini, R., & Gaudio, C. Del (2020). 3D printed scaffolds with random microarchitecture for bone tissue engineering applications: Manufacturing and characterization. *Journal of the Mechanical Behavior of Biomedical Materials, 103*, Article 103583. https://doi.org/10.1016/j.jmbbm.2019.103583

Perić Kačarević, Ž., Rider, P., Alkildani, S., Retnasingh, S., Pejakić, M., Schnettler, R., Gosau, M., Smeets, R., Jung, O., & Barbeck, M. (2020). An introduction to bone tissue engineering. *The International Journal of Artificial Organs, 43*(2), 69–86. https://doi.org/10.1177/0391398819876286

Pongwisuthiruchte, A., Dubas, S. T., Aumnate, C., & Potiyaraj, P. (2022). Mechanically tunable resins based on acrylate-based resin for digital light processing (DLP) 3D printing. *Scientific Reports, 12*(1). https://doi.org/10.1038/s41598-022-24667-8

Prasad, A., Sankar, M. R., & Katiyar, V. (2017). State of art on solvent casting particulate leaching method for orthopedic ScaffoldsFabrication. *Materials Today: Proceedings, 4* (2), 898–907. https://doi.org/10.1016/j.matpr.2017.01.101

Putlyaev, V. I., Evdokimov, P. V., Filippov, Y. Y., Safronova, T. V., & Tikhonov, A. A. (2018). Investigation of highly concentrated calcium phosphate suspensions for forming bioceramic with complex architecture. *Federation Glass and Ceramics (English translation of Steklo i Keramika), 74*(9–10), 378–381. https://doi.org/10.1007/s10717-018-9998-4

Quan, H., Zhang, T., Xu, H., Luo, S., Nie, J., & Zhu, X. (2020). Photo-curing 3D printing technique and its challenges. *Bioactive Materials, 5*(1), 110–115. https://doi.org/10.1016/j.bioactmat.2019.12.003

Raeisdasteh Hokmabad, V., Davaran, S., Ali, R., & Salehi, R. (2017). Design and fabrication of porous biodegradable scaffolds: A strategy for tissue engineering. *Journal of Biomaterials Science, Polymer Edition, 28*(16), 1797–1825. https://doi.org/10.1080/09205063.2017.1354674

Ranjan, N., Singh, R., Ahuja, I. P. S., Kumar, R., Singh, J., Verma, A. K., & Leekha, A. (2020). On 3D printed scaffolds for orthopedic tissue engineering applications. *SN Applied Sciences, 2*(2). https://doi.org/10.1007/s42452-020-1936-8

Rezwan, K., Chen, Q. Z., & Blaker, J. J. (2006). Aldo roberto boccaccini, biodegradable and bioactive porous polymer/inorganic composite scaffolds for bone tissue engineering. *Biomaterials, 27*(18), 3413–3431. https://doi.org/10.1016/j.biomaterials.2006.01.039

Rodrigo-Navarro, A., Sankaran, S., Dalby, M. J., del Campo, A., & Salmeron-Sanchez, M. (2021). Engineered living biomaterials. *Nature Reviews Materials, 6*(12), 1175–1190. https://doi.org/10.1038/s41578-021-00350-8

Ronca, A., Ambrosio, L., & Grijpma, D. W. (2013). Preparation of designed poly(d,l-lactide)/nanosized hydroxyapatite composite structures by stereolithography. *Acta Biomaterialia, 9*(4), 5989–5996. https://doi.org/10.1016/j.actbio.2012.12.004

Rottensteiner, U., Sarker, B., Heusinger, D., Dafinova, D., Rath, S., Beier, J., Ulrich, K., Horch, R., Detsch, R., Boccaccini, A., & Arkudas, A. (2014). In vitro and in vivo biocompatibility of alginate dialdehyde/gelatin hydrogels with and without nanoscaled bioactive glass for bone tissue engineering applications. *Materials, 7*(3), 1957–1974. https://doi.org/10.3390/ma7031957

Samandari, M., Mostafavi, A., Jacob, Q., Memić, A., & Tamayol, A. (2022). In situ bioprinting: Intraoperative implementation of regenerative medicine. *Trends in Biotechnology, 40*(10), 1229–1247. https://doi.org/10.1016/j.tibtech.2022.03.009

Schmidt, J., Elsayed, H., Bernardo, E., & Colombo, P. (2018). Digital light processing of wollastonite-diopside glass-ceramic complex structures. *Journal of the European Ceramic Society, 38*(13), 4580–4584. https://doi.org/10.1016/j.jeurceramsoc.2018.06.004

Schuster, M., Turecek, C., Mateos, A., Stampfl, J., Liska, R., & Varga, F. (2007). Evaluation of biocompatible photopolymers II: Further reactive diluents. *Monatshefte für Chemie - Chemical Monthly, 138*(4), 261–268. https://doi.org/10.1007/s00706-007-0609-2

Seitz, H., Rieder, W., Irsen, S., Leukers, B., & Tille, C. (2005). Three-dimensional printing of porous ceramic scaffolds for bone tissue engineering. *Journal of Biomedical Materials Research Part B: Applied Biomaterials, 74B*(2), 782–788. https://doi.org/10.1002/jbm.b.30291

Shekhawat, D., Singh, A., Bhardwaj, A., & Patnaik, A. (2021). A short review on polymer, metal and ceramic based implant materials. *IOP Conference Series: Materials Science and Engineering, 1017*(1), Article 012038. https://doi.org/10.1088/1757-899x/1017/1/012038

Shuai, C., Liu, T., Gao, C., Feng, P., & Peng, S. (2014). Mechanical reinforcement of diopside bone scaffolds with carbon nanotubes. *International Journal of Molecular Sciences, 15*(10), 19319–19329. https://doi.org/10.3390/ijms151019319

Shuai, C., Yu, L., Feng, P., Gao, C., & Peng, S. (2020). Interfacial reinforcement in bioceramic/biopolymer composite bone scaffold: The role of coupling agent. *Colloids and Surfaces B: Biointerfaces, 193*, Article 111083. https://doi.org/10.1016/j.colsurfb.2020.111083

Simon, C. G., Yaszemski, M. J., Ratcliffe, A., Tomlins, P., Luginbuehl, R., & Tesk, J. A. (2015). ASTM international workshop on standards and measurements for tissue engineering scaffolds. *Journal of Biomedical Materials Research - Part B Applied Biomaterials, 103*(5), 949–959. https://doi.org/10.1002/jbm.b.33286

Sohn, H.-S., & Oh, J.-K. (2019). Review of bone graft and bone substitutes with an emphasis on fracture surgeries. *Biomaterials Research, 23*(1). https://doi.org/10.1186/s40824-019-0157-y

Song, S. Y., Park, M. S., Lee, D., Lee, J. W., & Sun Yun, J. (2019). Optimization and characterization of high-viscosity ZrO2 ceramic nanocomposite resins for supportless stereolithography. *Materials & Design, 180*, Article 107960. https://doi.org/10.1016/j.matdes.2019.107960

Su, J.l., Teng, J., Xu, Z.l., & Li, Y. (2020). Biodegradable magnesium-matrix composites: A review. *International Journal of Minerals, Metallurgy and Materials, 27*(6), 724–744. https://doi.org/10.1007/s12613-020-1987-2

Sun, J., Ye, D., Zou, J., Chen, X., Wang, Y., Yuan, J., Liang, H., Qu, H., Binner, J., & Bai, J. (2023). A review on additive manufacturing of ceramic matrix composites. *Journal of Materials Science & Technology, 138*, 1–16. https://doi.org/10.1016/j.jmst.2022.06.039

Taboas, J. M., Maddox, R. D., Krebsbach, P. H., & Hollister, S. J. (2003). Indirect solid free form fabrication of local and global porous, biomimetic and composite 3D polymer-ceramic scaffolds. *Biomaterials, 24*(1), 181–194. https://doi.org/10.1016/s0142-9612(02)00276-4

Teo, A. J. T., Mishra, A., Park, I., Kim, Y. J., Park, W. T., & Yoon, Y. J. (2016). Polymeric biomaterials for medical implants and devices. *ACS Biomaterials Science & Engineering, 2*(4), 454–472. https://doi.org/10.1021/acsbiomaterials.5b00429

Thangavel, M., & Elsen Selvam, R. (2022). Review of physical, mechanical, and biological characteristics of 3D-Printed bioceramic scaffolds for bone tissue engineering applications. *ACS Biomaterials Science & Engineering, 8*(12), 5060–5093. https://doi.org/10.1021/acsbiomaterials.2c00793

Tomlinson, R. E., Li, Z., Zhang, Q., Goh, B. C., Li, Z., Thorek, D. L. J., Rajbhandari, L., Brushart, T. M., Minichiello, L., Zhou, F., Venkatesan, A., & Clemens, T. L. (2016). NGF-TrkA signaling by sensory nerves coordinates the vascularization and ossification of developing endochondral bone. *Cell Reports, 16*(10), 2723–2735. https://doi.org/10.1016/j.celrep.2016.08.002

Unamuno, X., Gómez-Ambrosi, J., Becerril, S., Javier Álvarez-Cienfuegos, F., Ramírez, B., Rodríguez, A., Ezquerro, S., Valentí, V., Moncada, R., Mentxaka, A., Llorente, M., Silva, C., de los Reyes Elizalde, M., Catalán, V., & Frühbeck, G. (2022). Changes in mechanical properties of adipose tissue after bariatric surgery driven by extracellular matrix remodelling and neovascularization are associated with metabolic improvements. *Acta Biomaterialia, 141*, 264–279. https://doi.org/10.1016/j.actbio.2022.01.007

van Schalkwyk, M. C. I., & Harris, M. (2018). Translational health policy: Towards an integration of academia and policy. *Journal of the Royal Society of Medicine, 111*(1), 15–17. https://doi.org/10.1177/0141076817735692

Vert, M. (2009). Degradable and bioresorbable polymers in surgery and in pharmacology: Beliefs and facts. *Journal of Materials Science: Materials in Medicine, 20*(2), 437–446. https://doi.org/10.1007/s10856-008-3581-4

Vert, M., Li, S. M., Spenlehauer, G., & Guerin, P. (1992). Bioresorbability and biocompatibility of aliphatic polyesters. *Journal of Materials Science: Materials in Medicine, 3*(6), 432–446. https://doi.org/10.1007/bf00701240

Visser, J., Melchels, F. P. W., Jeon, J. E., van Bussel, E. M., Kimpton, L. S., Byrne, H. M., Dhert, W. J. A., Dalton, P. D., & Hutmacher, D. W. (2015). Jos malda, reinforcement of hydrogels using three-dimensionally printed microfibres. *Nature Communications, 6*(1). https://doi.org/10.1038/ncomms7933

Vyas, A., Ghosh, S. B., Ghosh, S. B.-, Agrawal, A. K., Khare, D., & Kumar Dubey, A. (2022). Digital light processing mediated 3D printing of biocomposite bone scaffolds: Physicochemical interactions and in-vitro biocompatibility. *Polymer Composites, 43*(5), 3175–3188. https://doi.org/10.1002/pc.26609

Vyas, A., Mondal, S., Shankar Kumawat, V., Ghosh, S. B., Mishra, D., Sen, J., Khare, D., Kumar Dubey, A., Nandi, S. K., & Bandyopadhyay-Ghosh, S. (2024). Biomineralized fluorocanasite-reinforced biocomposite scaffolds demonstrate expedited osteointegration of critical-sized bone defects. *Journal of Biomedical Materials Research Part B: Applied Biomaterials, 112*(1). https://doi.org/10.1002/jbm.b.35352

Vyas, A., Shankar Kumawat, V., Ghosh, S. B., & Bandyopadhyay-Ghosh, S. (2021). Microstructural analysis and bioactive response of selectively engineered glass-ceramics in simulated body fluid. *Materials Technology, 36*(8), 451–459. https://doi.org/10.1080/10667857.2020.1774208

Walker, J. M., Bodamer, E., Kleinfehn, A., Luo, Y., Becker, M., & Dean, D. (2017). Design and mechanical characterization of solid and highly porous 3D printed poly(propylene fumarate) scaffolds. *Progress in Additive Manufacturing, 2*(1–2), 99–108. https://doi.org/10.1007/s40964-017-0021-3

Wang, W., & Yeung, K. W. K. (2017). Bone grafts and biomaterials substitutes for bone defect repair: A review. *Bioactive Materials, 2*(4), 224–247. https://doi.org/10.1016/j.bioactmat.2017.05.007

Wee, C. Y., Lim, Q. R. T., Zhao, Y., Xu, X., Yang, Z., Wang, D., & Thian, E. S. (2023). Optimizing fabrication parameters via taguchi method for production of high yield hydroxyapatite microsphere scaffolds using drop-on-demand inkjet method. *Journal of Biomedical Materials Research - Part B Applied Biomaterials, 111*(11), 1938–1955. https://doi.org/10.1002/jbm.b.35297

Wei, L., Ke, J., Prasadam, I., Miron, R. J., Lin, S., Xiao, Y., Chang, J., Wu, C., & Zhang, Y. (2014). A comparative study of Sr-incorporated mesoporous bioactive glass scaffolds for regeneration of osteopenic bone defects. *Osteoporosis International, 25*(8), 2089–2096. https://doi.org/10.1007/s00198-014-2735-0

Wei, G., & Ma, P. X. (2004). Structure and properties of nano-hydroxyapatite/polymer composite scaffolds for bone tissue engineering. *Biomaterials, 25*(19), 4749–4757. https://doi.org/10.1016/j.biomaterials.2003.12.005

Williams, J. M., Adewunmi, A., Schek, R. M., Flanagan, C. L., Krebsbach, P. H., Feinberg, S. E., Hollister, S. J., & Das, S. (2005). Bone tissue engineering using polycaprolactone scaffolds fabricated via selective laser sintering. *Biomaterials, 26*(23), 4817–4827. https://doi.org/10.1016/j.biomaterials.2004.11.057

Wu, Y., Chen, R., Chen, X., Yang, Y., Qiao, J., & Liu, Y. (2023). Development of strong and tough β-TCP/PCL composite scaffolds with interconnected porosity by digital light processing and partial infiltration. *Materials, 16*(3), 947. https://doi.org/10.3390/ma16030947

Xia, Y., Zhou, P. Y., Cheng, X. S., Xie, Y., Liang, C., Li, C., & Xu, S. G. (2013). Selective laser sintering fabrication of nano-hydroxyapatite/poly-ε-caprolactone scaffolds for bone tissue engineering applications. *International Journal of Nanomedicine, 8*, 4197–4213. https://doi.org/10.2147/IJN.S50685China

Xing, Z., Liu, W., Chen, Y., & Li, W. (2018). Effect of plasticizer on the fabrication and properties of alumina ceramic by stereolithography-based additive manufacturing. *Ceramics International, 44*(16), 19939–19944. https://doi.org/10.1016/j.ceramint.2018.07.259

Xiong, G., Nie, Y., Ji, D., Li, J., Li, C., Li, W., Zhu, Y., Luo, H., & Wan, Y. (2016). Characterization of biomedical hydroxyapatite/magnesium composites prepared by powder metallurgy assisted with microwave sintering. *Current Applied Physics, 16*(8), 830–836. https://doi.org/10.1016/j.cap.2016.05.004

Yang, Z., Peng, H., Wang, W., & Liu, T. (2010). Crystallization behavior of poly(ε-caprolactone)/layered double hydroxide nanocomposites. *Journal of Applied Polymer Science, 116*(5), 2658–2667. https://doi.org/10.1002/app.31787

Yang, R., Wang, X., Liu, S., Zhang, W., Wang, P., Liu, X., Ren, Y., Tan, X., & Chi, B. (2020). Bioinspired poly (γ-glutamic acid) hydrogels for enhanced chondrogenesis of bone marrow-derived mesenchymal stem cells. *International Journal of Biological Macromolecules, 142*, 332–344. https://doi.org/10.1016/j.ijbiomac.2019.09.104

Yunus Basha, R., Sampath, S. K., & Doble, M. (2015). Design of biocomposite materials for bone tissue regeneration. *Materials Science and Engineering: C, 57*, 452–463. https://doi.org/10.1016/j.msec.2015.07.016

Zadpoor, A. A. (2015). Bone tissue regeneration: The role of scaffold geometry. *Royal Society of Chemistry, Netherlands Biomaterials Science, 3*(2), 231–245. https://doi.org/10.1039/c4bm00291a

Zeinali, R., del Valle, L. J., Torras, J., & Puiggalí, J. (2021). Recent progress on biodegradable tissue engineering scaffolds prepared by thermally-induced phase separation (TIPS). *International Journal of Molecular Sciences, 22*(7), 3504. https://doi.org/10.3390/ijms22073504

Zhang, F., Chang, J., Lu, J., Lin, K., & Ning, C. (2007). Bioinspired structure of bioceramics for bone regeneration in load-bearing sites. *Acta Biomaterialia, 3*(6), 896–904. https://doi.org/10.1016/j.actbio.2007.05.008

Zhang, H., Jiao, C., & Liu, Z. (2021). 3D-printed composite, calcium silicate ceramic doped with CaSO4·2H2O: Degradation performance and biocompatibility. *Journal of the Mechanical Behavior of Biomedical Materials, 121*. https://doi.org/10.1016/J.JMBBM.2021.104642

Zhang, H., Mao, X., Du, Z., Jiang, W., Han, X., Zhao, D., Han, D., & Li, Q. (2016). Three dimensional printed macroporous polylactic acid/hydroxyapatite composite scaffolds for promoting bone formation in a critical-size rat calvarial defect model. *Science and Technology of Advanced Materials, 17*(1), 136–148. https://doi.org/10.1080/14686996.2016.1145532

Zhang, X., Wang, T., Zhang, Z., Liu, H., Li, L., Wang, A., Ouyang, J., Xie, T., Zhang, L., Xue, J., & Tao, W. (2023). Electrical stimulation system based on electroactive biomaterials for bone tissue engineering. *Materials Today, 68*, 177–203. https://doi.org/10.1016/j.mattod.2023.06.011

Zhang, C., Yuan, Y., Zeng, Y., & Chen, J. (2022). DLP 3D printed silica-doped HAp ceramic scaffolds inspired by the trabecular bone structure. *Ceramics International, 48*(19), 27765–27773. https://doi.org/10.1016/j.ceramint.2022.06.077

Zheng, Y., Han, Q., Wang, J., Li, D., Song, Z., & Yu, J. (2020). Promotion of osseointegration between implant and bone interface by titanium alloy porous scaffolds prepared by 3D printing. *ACS Biomaterials Science & Engineering, 6*(9), 5181–5190. https://doi.org/10.1021/acsbiomaterials.0c00662

Zhou, H., & Lee, J. (2011). Nanoscale hydroxyapatite particles for bone tissue engineering. *Acta Biomaterialia, 7*(7), 2769–2781. https://doi.org/10.1016/j.actbio.2011.03.019

Zhu, Y., Shen, J., Yin, L., Wei, X., Chen, F., Zhong, M., Gu, Z., Xie, Y., Jin, W., Liu, Z., Chitrakar, C., & Chang, L. (2019). Directly photopatterning of polycaprolactone-derived photocured resin by UV-initiated thiol-ene "click" reaction: Enhanced mechanical property and excellent biocompatibility. *Chemical Engineering Journal, 366*, 112–122. https://doi.org/10.1016/j.cej.2019.02.045

CHAPTER

Additive manufacturing of scaffolds for bone tissue engineering

23

A. H. Rezwanuddin Ahmed and Sihong Wang

The City College of New York, Biomedical Engineering, New York, NY, United States

1. Introduction

Large segmental bone defects past a critical size require external intervention that involve implants, with the gold standard being autografts, followed by allografts. Both these intuitively natural sources have risky trade-offs. Having an autograft to replace a defect means removing the implanted material from a healthy tissue region, and this creates surgical trauma at the donor site and morbidity—coupled with the very limited resource availability as the graft comes from one's own body. Allografts taken from other donors can mildly address the low resource availability but increases the risk of immune rejection or disease transfer. Eliminating resource scarcity, reducing surgical trauma and risk of disease, maintaining excellent biocompatibility/bioactivity, and personalized customizability are all factors that coalesce to ideate the ideal bone substitute. The fast-developing additive manufacturing (AM) technology using biocompatible/bioactive materials fits these needs impeccably. AM has the added benefit of constructing shapes that would be untenable or imprecise using traditional manufacturing methods to create porous scaffolds, which for ceramics include gas foaming, fiber bonding, freeze drying, phase inversion (separation), and particulate leaching (Kim et al., 2020; Mota et al., 2015), while porous metals have traditionally been made by metal powder or fiber sintering, gas foaming, plasma spraying, combustion synthesis, space-holder technique, wire-mesh orientation, and replication (Bandyopadhyay et al., 2023). Current AM technologies can affordably construct porous scaffolds in polymers, ceramics, and metals with pores as small as 100um, readily accommodating the 100–350 μm pore size range that is notably optimal for osseointegration (Abbasi et al., 2020). With humble beginnings of 3D printing with single materials, the current frontier of AM explores composites and functionally graded scaffolds for bone tissue engineering (BTE).

The general process for 3D printing will require creating the 3D model, typically as an STL tile (standard tessellation language). This STL file is then taken into a software that has the appropriate software to generate the 2D cross-sections that are "sliced" out of the 3D form. The cross sections do not have to be equally spaced, as variable height printing can be done through most modern slicers, albeit still with

Biomechanics and Mechanobiology of Bone. https://doi.org/10.1016/B978-0-443-26739-0.00021-1

limited functionality. Currently well-known software for manipulating meshes and 3D models include Blender (2025), Meshmixer (2025), and 3D Slicer (Kikinis et al., no date; Fedorov et al., 2012)—which can help with refining models by cleaning up, segmenting the model, or adding features to the model directly to the STL mesh. Blender as a heavily used artist/CGI program has a rich and active support and development network, with no shortage of beginner to advanced tools—and specific ones for mesh adjustment and manipulation would benefit the designer of bone scaffolds. Similarly, Meshmixer and 3D Slicer have been actively used in 3D medical imaging, from simple segmentation to model generation through powerful mesh cleanup and manipulation functions, and are supported by an active community as well. To create the raw models, however, parametric CAD modules are recommended. Beginner-friendly software includes TinkerCAD and SketchUp, while more professional suites include SolidWorks and AutoCAD Fusion 360. Organic models can be easily created in Blender, Unity 3D, Unreal Engine, and ZBrush. As of the writing of this book chapter, Blender is open source and under continuous development. Unity 3D and Unreal Engine are free especially to students and hobbyists but require a small royalties fee when products created through their software generate over $100K-$1M. AutoCAD Fusion 360 is also free for educators/students and hobbyists, with only the most advanced functions requiring paid subscription. TinkerCAD and SketchUp are free but online-only, and are useful when a powerful computational workstation may be in limited availability for simple models.

2. Modes of additive manufacturing

Manufacturing models from a single material, multimaterial (MMAM, multimaterial additive manufacturing), or composites can generally be categorized under three broad technologies—(1) photo-sensitive or laser-based, (2) direct printing, and (3) nozzle-based/assisted with or without binders modalities (Fig. 23.1) (Germaini et al., 2022). The American Society for Testing and Materials' (ASTM) "Additive Manufacturing Technologies Committee F42" (Committee F42 on Additive Manufacturing Technologies | ASTM, no date; The 7 categories of Additive Manufacturing, 2025) has classified the processes themselves into seven distinct categories as per the ISO/ASTM 52900 standards: (1) Vat photopolymerization, (2) Material Jetting, (3) Binder Jetting, (4) Material Extrusion (extrusion-based), (5) Powder Bed Fusion, (6) Sheet Lamination, (7) Direct Energy Deposition (Pérez et al., 2020).

In terms of fabrication for bone implants, all seven categories are relevant—with Direct Energy Deposition (DED) being the least used as it is still undergoing extensive research (Ryu et al., 2021).

Typically, photopolymerization and extrusion-based additive manufacture has the lowest investment and can be set up within a relatively modest footprint, and this chapter details some procedural preparations that can enable the success of such prints in-house, based on our collective experience and shared experience

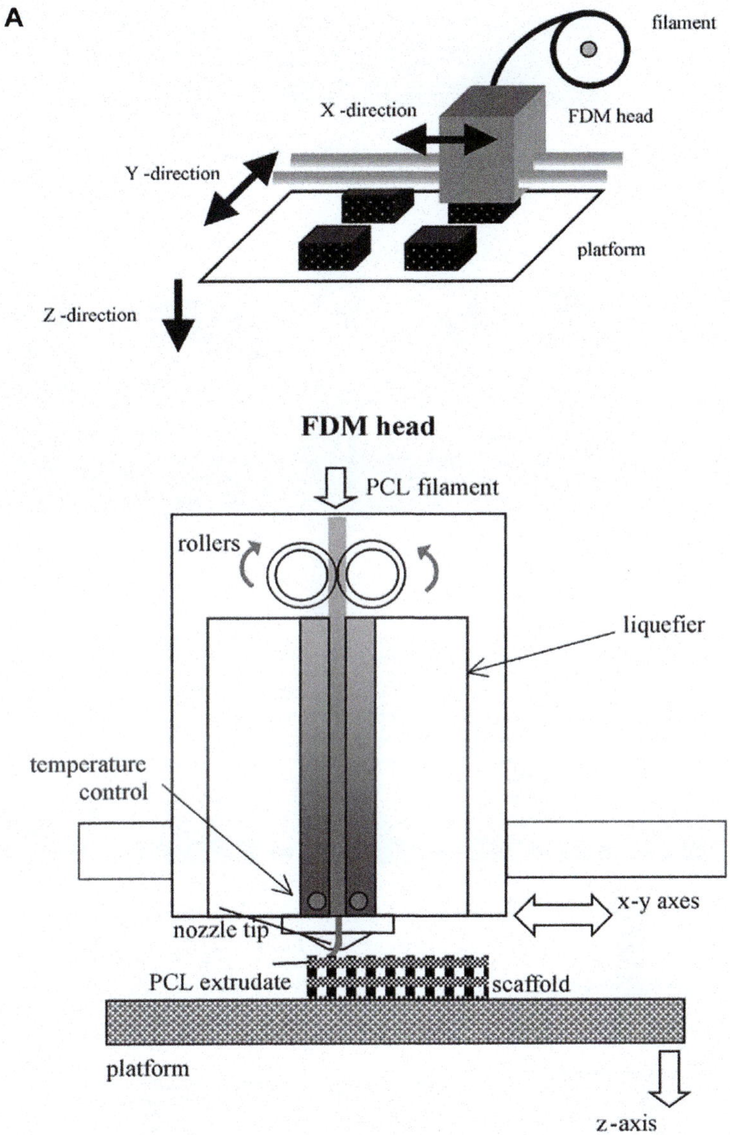

FIGURE 23.1 Keynote examples of additive manufacturing processes.

(A) Fused deposition manufacture (FDM), a basic and ubiquitous manufacturing technique. (B) Stereolithography (SLA), a form of laser- or light-cured additive process. (C) Multi-Jet Fusion (MJF) illustrating the steps of powder and binder deposition ending with UV curing at different stages. (D) Selective laser sintering (SLS) process—selective laser melting (SLM) is similar. (E) Laser-assisted bioprinting (LAB) through Laser-induced forward transfer (LIFT) (F) Working principle of laser-directed energy deposition (L-DED) for multimaterial deposition.

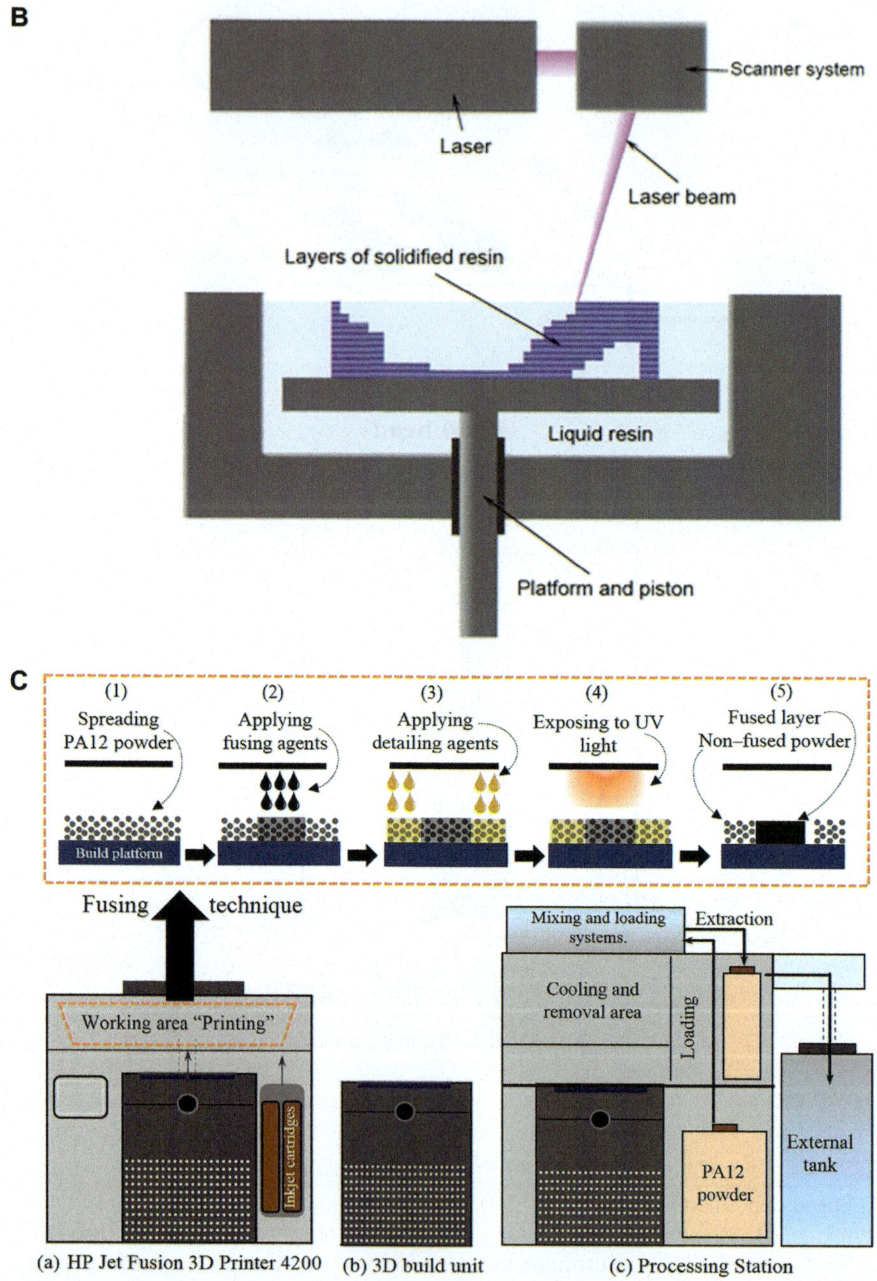

FIGURE 23.1 cont'd.

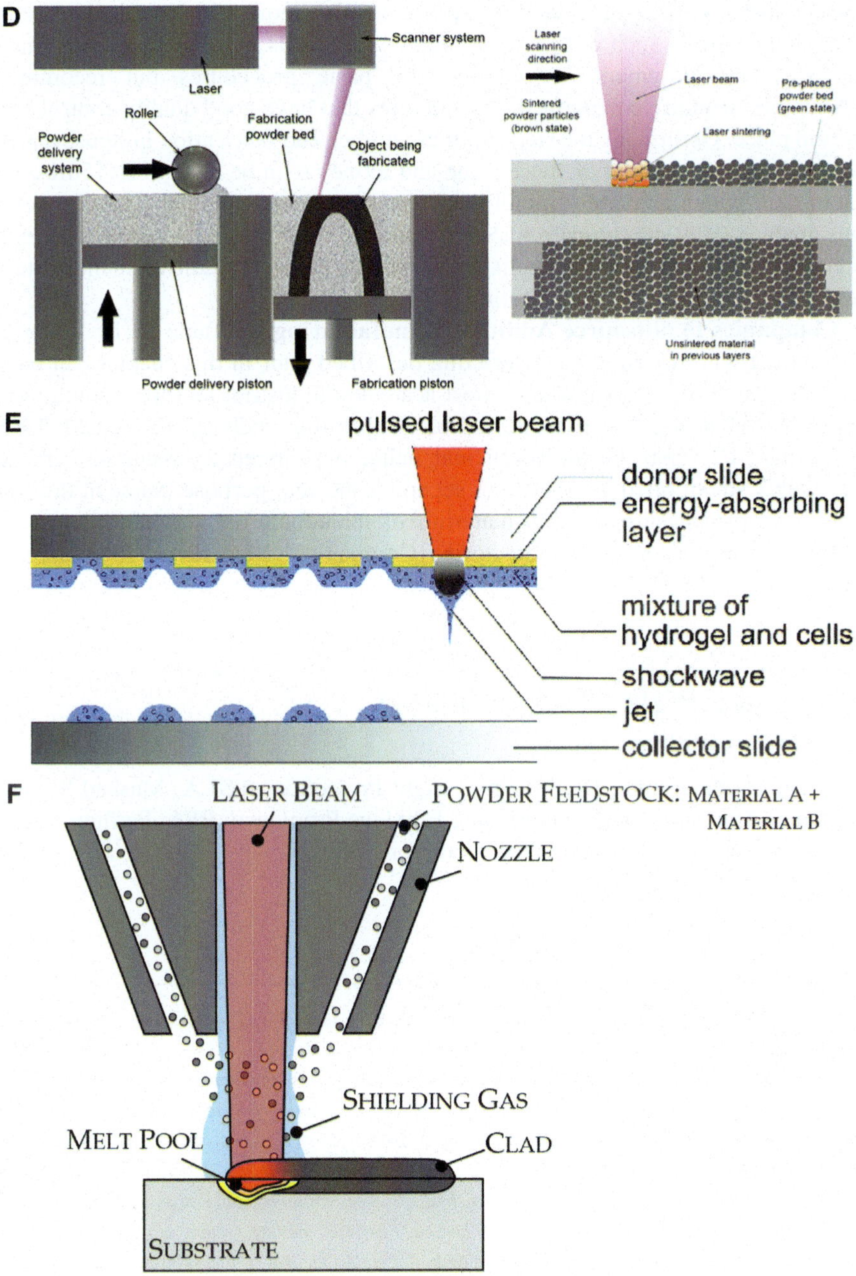

FIGURE 23.1 cont'd.

across makers. For the manufacturing processes that require highly regulated environmental control—such as direct laser melting of metals, selective laser sintering of metals/ceramics/polymers, we outline the basic procedures but recommend outsourcing to additive manufacturing facilities that have good quality control measures in place. Optimizing the steps for these higher-end fabrication in-house usually requires a big investment of space, time and money (can be as high as \$50K-\$70K just to start, with a moderate-to-industrial grade setup running up to \$500K+ before consumables (How to compare SLS 3D printer prices, 2025). Detailed sources on such procedures are usually published in whitepapers by additive manufacturers and in the *Additional Reading* section.

Companies to outsource Additive Manufacturing to (Table 23.1): While the photocuring and extrusion-based printing described later in this chapter can be set up at dedicated spaces in-house at most academic/industrial facilities, some of the more controlled and expensive forms of 3D printing such as SLS, SLM, EBM, DED, MJF, BJ—may be more efficient being outsourced to manufacturers that have highly streamlined process flows set up for the very purpose. Some of the companies that provide high-quality materials or manufacturing service currently includes (22 Best Additive Manufacturing Companies (2024 review) - Uptive, no date).

3. Photocuring/photocrosslinking

3.1 Definitions/acronyms

SLA: Stereolithography; **DLP**: Digital Light Projection; **mSLA**: Masked Stereolithography Apparatus; **FEP**: Fluorinated Ethylene Propylene; **PFA**: Perfluoroalkoxy Alkane; **nFEP**: "non" FEP, another name for PFA.

- (A) From Zein, I., Hutmacher, D. W., Tan, K. C. & Teoh, S. H. (2002). Fused deposition modeling of novel scaffold architectures for tissue engineering applications. Biomaterials, 23(4), 1169–1185. https://doi.org/10.1016/S0142-9612(01)00232-0. (B) From Materialgeeza. (n.d.). Stereolithograthy apparatus schematic. Retrieved October 18, 2025, from https://commons.wikimedia.org/wiki/File:Stereolithography_apparatus.jpg. (C) From Alomarah, A., Ruan, D., Masood, S. & Gao, Z. (2019). Compressive properties of a novel additively manufactured 3D auxetic structure. Smart Materials and Structures, 28(8), 085019. https://doi.org/10.1088/1361-665X/ab0dd6. (D) Image from Materialgeeza. (n.d.). SLS system schematic. Retrieved October 15, 2025, from https://commons.wikimedia.org/wiki/File:Selective_laser_melting_system_schematic.jpg. (E) From Gruene, Martin, Unger, Claudia, Koch, Lothar, Deiwick, Andrea & Chichkov, Boris. (2011). Dispensing pico to nanolitre of a natural hydrogel by laser-assisted bioprinting. BioMedical Engineering OnLine, 10(1). https://doi.org/10.1186/1475-925X-10-19. (F) Adapted from Ostolaza, M., Arrizubieta, J. I., Lamikiz, A. & Cortina, M. (2021). Functionally Graded AISI 316L and AISI H13 Manufactured by L-DED for Die and Mold Applications. Applied Sciences, 11(2), 771. https://doi.org/10.3390/app11020771.

Table 23.1 Companies and services for 3D additive manufacturing.

Company	Services	Location	Location and URL
Protolabs	Injection molding, CNC machining, sheet metal, SLA, SLS, FDM. Certifications and Compliance: ISO 9001: 2015, ISO 13485:2016, AS9100D, ITAR	Minnesota, USA	• https://www.protolabs.com/ • https://www.protolabs.com/services/3d-printing/
Aenium	DMLS, SLS, metrology, metallographic analysis, powder analysis, CAD optimization and more	Valladolid, Spain	https://aenium.es/
AP&C - a Colibrium additive company (GE subsidiary)	Aerospace and medical-grade metal powders, ISO 9001:2015 and ISO 13485:2016 certified	USA	https://www.advancedpowders.com/
Novus life sciences	Nano-hydroxyapatite crystal infused on nano-copolymers for bone repair - craniomaxillofacial (CMF) surgery, oral and maxillofacial (OMF) surgery, guided bone regeneration (GBR) implants, total knee or hip arthroplasty (TKA or THA).	Hong Kong	https://www.novusls.com/
Kuraray Poval	Specialty PVA binders for ceramic printing	Japan, also providing services in Europe, south American, and USA	https://www.kuraray-poval.com/
Restor3D	Personalized 3D printed implants for foot/ankle, knee, hip	North Carolina, USA	https://www.restor3d.com/
in3DTec	DMLS and SLM metal printing; FDM, SLA, SLS, MJF, and SLM ceramic printing	Shanghai, China; Hong Kong	https://www.in3dtec.com/direct-metal-laser-sintering-3d-printing/
Xometry	Printing services of thermoplastics, resin, and metals. ISO 9001: 2015, ISO 13485:2016, AS9100D and IATF 16949:2016 certified	Maryland, USA.	https://www.xometry.com/

Continued

Table 23.1 Companies and services for 3D additive manufacturing.—*cont'd*

Company	Services	Location	Location and URL
Stratasys direct (service name for statasys)	DLP, powder bed fusion, FDM, SLA, Polyjet (material jetting)	Minnesota, USA	https://www.stratasys.com/en/stratasysdirect/service-locations/minnesota/

The principal technologies that fall under this category include SLA, DLP, and mSLA, which are all traditionally known as *vat photopolymerization*, and are described herein. All three methods utilize a light source through laser, LED, or spot projection of UV light, that typically employ a 395 nm or 405 nm wavelength source to initiate photocuring of a resin into hardened plastic. Uncommon systems where the wavelengths can range from 400 to 680 nm or infrared two photon polymerization, depending on both resin curability and light source. By using a resin that can be cured by light, a CNC (computer numeric control) system can control a set thickness of resin to be cured, given a specific pattern to cure. By stacking the patterns layer by layer, the 3D object is constructed. This is essentially giving a third dimension to the traditional 2-dimensional photolithography done for MEMS/Bio-MEMS devices, where a photopolymer is used at the initial stage to cure a single pattern of predefined thickness to create a pattern, after which the pattern can be used directly or as a mold for casting soft elastomers such as PDMS (Polydimethylsiloxane).

The principle behind stereolithography was researched in the 1970s, Charles "Chuck" Hull's patent filed in 1984 details the process as it was conceived at the time (Hull, 1986). The process basically involves a **build plate** that moves down in an elevator fashion into the "vessel" (resin tank) as the top layers are cured into a desired pattern with a computer-controlled "synergistic stimulation," which is described as being an X-ray, a beam of light or UV light, high-energy particles, etc.—anything that can cure the resin. This is the *top-down* approach of rapid prototyping. The inverted form, also referred to as the *bottom-up* approach, is more common now, wherein the build plate is upside down in a resin vat with the light source coming through the bottom of the vat. In this conformation the build plate moves up after every resin curing step, away from the vat. This configuration was mentioned in Fig. 23.2 of Chuck Hull's patent, with the caveat that he made use of a heavy, immiscible liquid through which the light would pass, on top of which the resin would sit. Today's setup is much simpler as a transparent, flexible film can be used through which the curing light can pass. Without the innovation of flexible, nonadherent films, it would be difficult to peel off resin layers from the bottom of the vat as they cured if the bottom was rigid or very adhesive to the photopolymerized resin.

- **SLA**, or StereoLithography and entails the use of a vector scan instead of rasterized pixels, typically with a laser head that moves on an X–Y axis, or along a single axis of a "printhead" that resembles a scanner's wand-like scan head, with the head moving as a whole on a second axis, carrying the laser module with it. Formlabs calls their printhead as a "Light Processing Unit," as it not only carries the laser but also pushes against the flexible film to help peel the cured resin off the film (How does SLA Printing Work?, 2025).
- **DLP**, or Digital Light Projection/Processing, utilizes a micromirror array, which redirects light that is projected onto it to the bottom of the vat. This creates a raster image ("pixelated") of the pattern that needs to be cured. DLP systems last a lot longer than mSLA systems as they do not have to worry about the lifetime of the LCD screen burning out.
- **mSLA**, or Masked SLA, is inspired by modern LCD screens. The mechanism is an LED/LCD system, wherein a bed of LEDs emit the wavelength of light for curing the resin, collimated to make the rays parallel. Before contacting the resin, the light passes through an LCD screen, where pixels are turned off or on based on the pattern to create. This LED/LCD system allows for very high precision and resolution limited only by the LCD screens, with $1000 systems attaining resolutions of 16 × 20um pixel sizes. The downside is that LCD screens burn out after 200–2000 hours of use, needing replacement frequently, costing 20%–50% of the printer. LCD screens attenuate the curing light, requiring much longer exposure and build times compared with SLA or DLP systems where light enters through a clear glass window with minimal attenuation. Current generation LCD screens tend to be monochrome instead of RGB due to higher light transmittance and longer life. For example, resin manufacturer Liqcreate tabulates their monochrome LCD lifetime at ~2000 hours, but only 200–700 for RGB-LCDs. Using two test resins for curing a 100um layer, Liqcreate Premium and Tough-X, the exposure times for monochrome versus RGB ranges from a 2-fold (14 second mono vs. 30 second RGB) to an 8-fold (2 second mono vs. 16 second RGB) differential (Monochrome vs. RGB LCD screen in resin 3D-printing | Liqcreate, no date). Monochrome screens are the superior to speed up the printing process and reduce consumable cost of replacing the LCD.

3.2 Process

Regardless of the light source, the principle of vat photopolymerization is simple. "Inverted stereolithography" common today was commercially introduced in 2011 by Formlabs cofounders Max Lobovsky, David Cranor, and Natan Linder. The **printing process** for inverted stereolithography is shown in Fig. 23.1B, and described below (Darulis, 2022).

1. A **vat** (colloquially called a "tray") holds the resin in place, with the bottom of the vat being a transparent film (FEP, nFEP, or ACF films).

2. The **build plate** comes down onto the vat, gets immersed into the resin-filled vat, and is lowered until the gap to the transparent film at the bottom reaches a designated **layer height.** This is how thick a single layer will be and typically ranges 10–60 μm.
3. A patterned source of light is transmitted from the bottom of the vat through the film. The patterning mechanism varies across SLA, DLP, and mSLA, but this step creates a singular slice of 2D cured resin with the thickness of the layer height. Typically this is a "soft cure" at low exposure, and a "hard cure" at longer exposure times.
4. The build plate lifts up, and the flexible film peels off the cured resin. With the correct exposure settings and lifting speed of the build plate, the cured layer will adhere to the cured layers above it (or to the build plate) and will peel off the film.
5. The process continues until the full model is completed layer-by-layer.

Clean up and Post Processing: The prints at this stage need further cleanup to render them safe for use. The standard procedure is to clean all the parts by washing with isopropyl alcohol (IPA, rubbing alcohol) with a recommended IPA content >95%. Alcohol baths with stirrers are available from resin printer manufacturers. After a thorough IPA wash, **postcuring** with a flood lamp of appropriate wavelength is done for at least 30–120 minutes to completely cure the resin. The part is rotated during postcuring to allow generous exposure across the whole part. *UV Curing systems with a rotary platform and protective shields are commercially available from 3D printer manufacturers.For ceramic or metal printing, the part is called a **"green" part** and needs further drying, debinding, and sintering after the cleanup and post-processing to get a final, usable part.*

Underwater Post Curing. If the printed part will be used directly or needs coating for biomedical applications, the parts should be cured under water for several hours, ensuring adequate UV coverage everywhere as in the dry postcuring above. Curing under water alleviates problems that arise when casting a silicone mold around the resin, or to reduce cytotoxicity to meet biocompatibility standards (ISO 10993-5 and ISO 10993-10). For parts that remove photocured binder through sintering, this is not required. Photopolymer resins have sulfur residues that can leech through the surface into materials that come into contact with it. When casting silicones on a 3D resin print, there will be a region of 0.2–1 mm from the surface of the print where the silicone will never cure (called "*curing inhibition*"), attributed to "sulfur poisoning" of the silicone by the resin. Generally, this is a problem for platinum-cured silicones such as polydimethylsiloxane (PDMS), commonly used in BioMEMS devices. Tin-cured silicones do not suffer from this drawback. Prolonged periods of underwater curing can prevent curing inhibition of silicones, and is a recommended way of curing by resin manufacturers such as Siraya Tech. Other methods to prevent curing inhibition of silicones include spray-coating a modest layer of paint or varnish onto the resin surface where before silicone casting.

3.3 Technical information

The **film** at the bottom of the vat is typically made of FEP, nFEP, or ACF. The FEP/nFEP films are fluoropolymers—specifically, copolymers of monomers resembling PTFE (Teflon), with substitutions. *Like* PTFE, they prevent the cured resin from sticking, allowing the film to "peel off" when the build plate lifts up. They wear out after frequent and prolonged use (typically a few hundred to 1000 cycles of peeling). nFEP performs relatively better in peeling, earning it the moniker "FEP 2.0," and is relabeled PFA. The composition of ACF is proprietary (undisclosed), but is indicated to be another fluoropolymer—Ethylene tetrafluoroethylene (ETFE, DuPont brand name **Tefzel**), referred to as "Teflon film." The chemical, mechanical, thermal and electrical properties of all 4 "Teflon fluoroplastics"—PTFE, FEP, PFA (nFEP), and ETFE are available on Chemours' archived materials webpage (Fluoroplastic Comparison - Typical Properties, 2018). Once the film is expended through several hundred or 1000 peels, cured parts tend to adhere to the film more firmly than to the build plate side, causing the prints to detach from the build plate. This problem is remedied by replacing the with a new film or applying a thin coating (e.g., spray coating) of clear PTFE to restore the old film's peeling vigor.

Certain films, such as ACF or "matte films," come with bumpy or light-diffusing patterns on the film, which can imprint onto the model as (1) mildly noticeable textures on printed models, or (2) printed features appearing "blurred." If the smoothened, diffuse look is preferred, this is beneficial as ACF or patterned films can support faster peeling speed when the build plate lifts, translating to overall faster printing times. However, the diffusion of the light source at the pixel level requires adjusting exposure times, which could outweigh the gain in speed from faster peeling.

Elephant's foot issues can arise along the first few layers of the build. This is because a prolonged exposure time is required for the initial layers (usually 2–8) for strong adhesion to the build plate, which could be 10- to 40-fold of normal exposure times. This causes "bleeding" of light, making the bottom few layers "expand" out, creating features that resembles an elephant's foot. This is compensated for by geometric corrections in the CAD model to account for the expansion, or, by reducing the cure time (instead of 10-40X, use ~4-8X). Lifting speeds will need adjustments accordingly to ensure that the peeling is not harsh, and usually works better for clear resin since the light can weakly transmit through the current layer to the previously cured layers above.

Resin type is critical for photopolymerized prints. Print settings for embedded channels and pores within clear resins need carefully calibrated adjustments, since light can bleed through into previously cured layers, causing any uncured resin trapped within channels by surface tension to become cured (Fig. 23.2). Exposure times of curing is the first adjustment. For opaque and "high resolution" or "fast" resins, the problem is less severe, and where applicable, an opaque resin is recommended for parts with internal channels. However, the same opaqueness that prevents curing

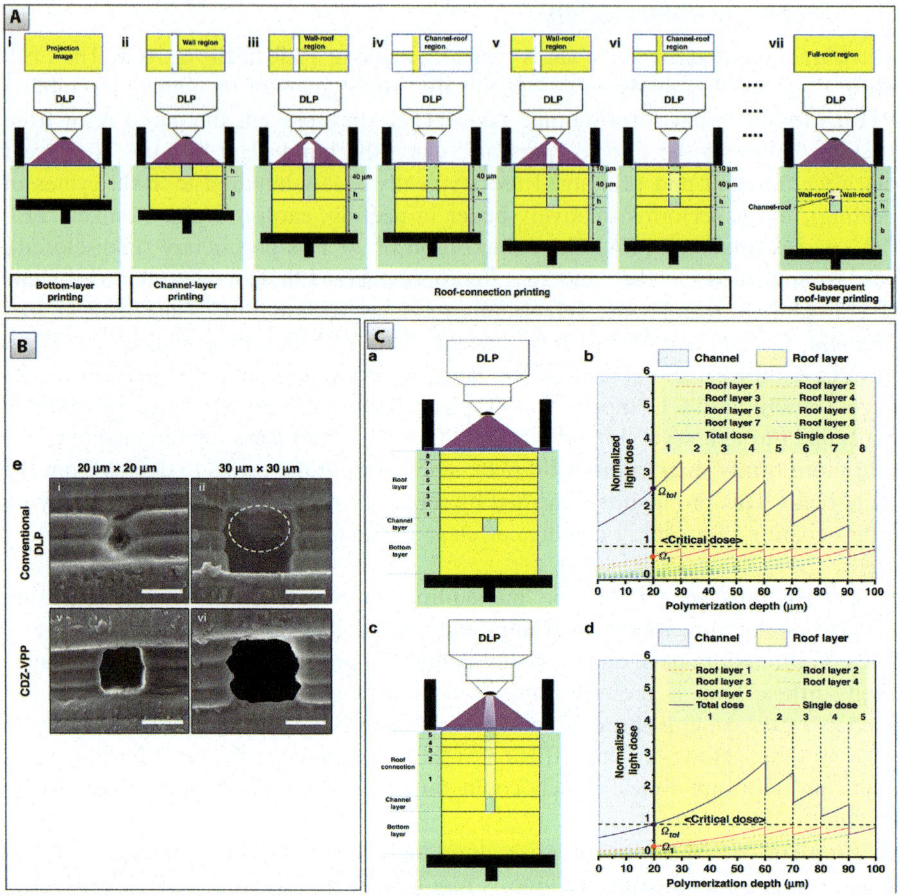

FIGURE 23.2 Adapting to technological challenges by process tuning.

(A) Zone adjusted light exposure for controlled and successful printing of embedded channels in 3D prints as a modification to the traditional DLP, showing a process that locally customizes exposure strength over different features, with lower exposures over channels. (B) Comparison of unadjusted default settings (conventional DLP) and zoning-controlled (CDZ-VPP) printing, showing closed channels from resin overexposure in the former and clear channels in the latter. (C) An exposé of the CDZ-VPP process tuning to the critical energy exposure along each layer, which is absent in default DLP processes.

- Figures adapted from Luo, Zhiming, Zhang, Haoyue, Chen, Runze, Li, Hanting, Cheng, Fang, Zhang, Lijun, Liu, Jia, Kong, Tiantian, Zhang, Yang & Wang, Huanan. (2023). Digital light processing 3D printing for microfluidic chips with enhanced resolution via dosing- and zoning-controlled vat photopolymerization. Microsystems & Nanoengineering, 9(1). https://doi.org/10.1038/s41378-023-00542-y.

through the resin also prevents postcuring or clear vision of residual resin during the printing process, so care must be taken to ensure uncured resin is fully removed. For biomedical applications, uncured resin can be highly cytotoxic.

Subpixel Printing. Resin printers will come with a nominal resolution typically listed as a "pixel" size for mSLA/DLP and a laser dot size for SLA. While the laser dot size cannot be made any smaller for SLA without hardware adjustments, a special technique allows "printing" a pixel smaller than the nominal pixel size for DLP and mSLA. If the pixel size for printing in the XY dimension is listed as 50umx50um, adjusting the exposure through antialiasing, blurring, and grayscale levels in the respective slicer software creates "gray" pixels instead of purely black (off) and white (on) pixels. When a gray pixel is created at edges near a solid pixel, it will form a cured voxel of resin whose volume is adjusted to the exposure intensity (grayscale)—e.g., a 50% exposure will create a voxel about half the size of a 100% exposure. This allows a "partial" voxel to be created, which by cohesive forces adhere to the bulk 3D print that is forming. There are instructive videos from the Autodesk Ember 3D printing team, using supersampling techniques for high-fidelity printing (High-Fidelity Printing Techniques by Richard Greene, 2016), and a video demo of grayscale slicing to allow subpixel printing (Subpixel resolution for DLP 3D printing, 2016). A clear example in the video can be found wherein a smooth drafted wall is built across the stretch of a single pixel (50um) with a height of 25 layers (25 steps within 50um).

Bottom Up versus **Top Down:** The methods above describe an "inverted stereolithograph" technique, where the build plate ("the bottom") moves up as the model is materialized layer by layer at the bottom, earning the moniker of a "bottom-up" approach. In a "top-down" approach, the build plate sits submerged inside a resin tank, and a light source cures the top layer of the resin. The build plate then moves *down* (the "top" moves down), getting immersed further into the resin tank. The top-down approach was the original technique but has mostly been replaced by the bottom-up approach due to more efficient resin use and reliable layer height control.

Support Your Model! The "inverted stereolithography" process has the object printing on the build plate upside down, hence the weight of the hanging object can become an issue mid-print if adequate supports are not printed throughout the model. Ensuring enough support on the printing model and hollowing out parts that need not be fully solid will ensure success for large-size prints (>800 cm^3).

Shells of Hollowed Models are Traps. hollowing out solid models into shells to reduce resin usage can cause liquid resin may become entrapped in the shells. Provide escape holes for the resin to run out of during the cleaning process. Trapped resin cures over time, which causes the model to crack due to the highly exothermic nature of resin curing.

4. Extrusion based printing: Fused deposition modeling/manufacture (FDM)

Fused Deposition modeling (trademarked as FDM), also nonproprietarily known as Fused Filament Fabrication (FFF), is the most basic gateway to 3D printing, as the simplest printers start at merely ~ USD $100 and goes up to $40K for commercially available "large-scale industrial" style printers. FDM technology is the poster-child for extrusion-based (EB) printing, as any thermoplastic material can be used directly or as a carrier to lay down layer-by-layer patterns on a build plate. Super-sized extrusion-based 3D printers recently been employed for 3D printing small homes by extruding cement out of the nozzle, with New Story and ICON teaming up to create affordable homes in Mexico for as cheap as $4000 (3D-Printed Houses: 12 Top Examples | Built In, no date; ICON and New Story/World's First 3D-printed community of homes, no date). The size, however, is far from what determines the price point. The concept behind FDM is simple—a material to be printed is melted (fused) through a nozzle and the melted material is drawn out into a desired shape layer-by-layer, ultimately forming a 3D volume. The type of material, nozzle, heat blocks, environment, build plate, and extrusion mechanisms all determine the type of appropriate printer required for printing specific materials.

The standard workflow in EB-type printing is to have a build plate as a floor, on top of which a three-axis controllable nozzle is run as it extrudes material. Several advanced improvements have been made over the traditional 3-axis movement of the nozzle, including 4,5, and 6-axis machines that allow the nozzle to move in free space over the topological 3D surface contours of a model in space instead of the plain 2D layer-by-layer approach. FDM is not the only extrusion-based additive manufacturing technique, as techniques that extrude viscous liquid (gels/bioinks) or from melting a large volume of pellets from a hopper/feeder are also readily available. Viscous bioinks do not necessarily require a heated component but simple extrusion control through mechanical means. However, filament-based printing has seen extremely prolific development, and their ubiquitous affordability and accessibility means the machines are serviceable for several years, with third-party replacement parts being easier to acquire for open source printers.

FDM printers have three components that must be compatible:

1. **Print Environment:** Standard starter materials that can be printed with low melting temperatures (~195–240°C) such as PLA, PETG, TPU can be printed without too much environmental control in an open-frame printer at room temperature. Very low temperature (~100–120°C) filaments such as polycaprolactone can be used to print objects that are remoldable in hot water after printing. PCL has been blended with PLA, PEG, and PET to increase the mechanical strength (BOC Sciences, no date), and is a common material used for surgical sutures and drug delivery scaffolds (Willerth, 2017). PEEK is a polymer that is popular in spinal and orthopedic implants, desirable due to its chemical stability, biocompatibility, and an elastic modulus of 3.6 GPa,

comparable with the normal human bone (McKeen, 2012). However, printing PEEK is not trivial. In addition to the very hot printing temperatures (365–440°C), it requires the print bed and environment to be maintained at high temperatures that are not achievable by hobbyist 3D printers, and demands closed-chamber industrial grade 3D printers. The typical bed temperature and environment ("build chamber temperature") has to be maintained at >133°C for the PEEK to be formed into a semicrystalline state, as colder temperatures lead to amorphous PEEK prints that have low or undesirable mechanical properties (Esmail, 2021). Any materials that require very high temperatures (>250°C) to print can suffer from heavy warping if previously printed layers cool down much faster than the layers deposited above it, so careful maintenance of the environment is key.

2. **Printing Nozzle:** A great guide on nozzle types is available at 3DJake's post, as they are also manufacturers/suppliers of several custom nozzles (3D Printer Nozzle Guide, 2025). Their information along with our experience-based recommendations are presented as below in Table 23.2. The nozzles themselves come in various sizes, ranging from 0.1 to 2 mm diameter nozzles. The earlier generations of 3D printers were designed to be used with 2.75 mm diameter filaments, but most printers nowadays default to 1.75 mm. Filament size can be changed in the slicer software.
3. **Filament Control During Printing:** The filament is generally pushed by an **extruder** into the hot end of the printer, which exudes the melting filament through the nozzle. Contrary to intuition, the nozzle does not actively extrude—FDM printers designate the **extruder** as strictly the geared apparatus that pulls and feeds the filament into the hot end. Various extruder types are available—direct, dual gear, full metal, and Bowden tube, to name the most common ones. Softer materials such as TPU (Shore hardness 60–90A) typically perform well with direct extrusion, as they otherwise get jammed inside a Bowden tube. For rapid printing, the filament may need to be retracted when hopping from one layer to the next, or else a stringy residue of filament can follow along with the nozzle as it moves to the next layer and Bowden tubes are beneficial when using retraction.

5. Powder bed fusion (PBF)

The standard layer-by-layer building procedure for powder bed fusion is shared across SLS, SLM, EBM:

1. The base material powder is deposited onto the build platform, typically with a rake or roller, to the thickness of the layer of the 2D cross section to be "printed."
2. The print chamber is brought to ambient conditions—an inert gas environment for SLS/SLM or a near-vacuum (0.0001 mbar) chamber for EBM, and

Table 23.2 Nozzle types and characteristics for fused-deposition modeling.

Nozzle type and description	Max printing temp	Advantages	Disadvantages
Brass Most common type of nozzle, can be found coated with nickel or chrome	300°C (typically <280°C)	• "Dirt" cheap amongst all nozzles • High thermal conductivity - allows quick and efficient melting of filament	• Soft metal, very low wear resistance - abrasive binder or filaments wear the nozzle out quickly
Coated brass or copper nozzles Nickel or Chrome coating is applied to the brass or copper nozzle	500°C	• Coating provides wear resistance and higher temperature tolerance • Same high thermal conductivity of brass or copper	• Not good for long term use with abrasive filaments • Hardened steel nozzles typically fare better for the same price
Stainless steel	500°C	• Safe to use for food-safe filaments • Improved wear resistance over standard brass	• Lower heat conductivity than brass/copper • Hardened steel is better for the same use cases
Hardened steel	500°C	• Compared against the few months of use of a brass nozzle, can be used for years due to high wear resistance • Suitable for abrasive filaments	• Lower thermal conductivity means that heat isn't transferred as quickly to the filament, calling for several steps of fine-tuning the extrusion rate • Suboptimal fine-tuning can lead to low print quality • The harder material means that the inner surface of the nozzle doesn't get smoothened out, also leading to low quality prints. • Higher cost per nozzle
Ruby nozzles/ Brass Ruby nozzles These are the cream of the crop when it comes to nozzles - they are brass nozzles	550°C	• The ruby tip prevents abrasives from wearing out the nozzle, as the tip is usually the highest	• Very expensive compared with the standard or hardened steel nozzles. Typically, one ruby nozzle can run up to

Table 23.2 Nozzle types and characteristics for fused-deposition modeling.—*cont'd*

Nozzle type and description	Max printing temp	Advantages	Disadvantages
(usually coated) seeded with a ruby at the tip, with a precision hole drilled through the ruby.		region of wear. The wear resistance • The high thermal conductivity of brass is retained. • The coated nozzle and ruby tip also ensures very high temperature tolerance.	10–30 times the price of the normal brass nozzle.

Data compiled from 3D Printer Nozzle Guide. (2025, April 19). Retrieved June 19, 2025 https://www.3djake.com/info/guide/3d-printer-nozzle-guide.

preheated base plate (build platform) of typically ~600°C for SLM and 600–1000°C for EBM. ***For polymers (SLS), preheating is done to just below the melting or glass transition temperature of the material*** (Selective Laser Sintering (SLS) 3D Printing Technology Overview, 2023).Preheating prevents severe temperature gradients between the sintering/melting spot and the ambient temperature of the model being printed, reducing warping and high internal stresses.

3. The energy source (laser or electron beam) traces out the 2D cross-section pattern.
4. The build plate drops by a layer height.
5. A fresh layer of powder is reapplied onto the build platform, and, step 3-5 is repeated as the 3D object is traced out layer-by-layer.
6. Parts must be left to cool before removal, which can take several hours (overnight).
7. Cleanup of residual powders and support structures are done after removal. Special cleaning hoods and/or sand-blasting equipment are required.

5.1 Selective laser sintering (SLS) and direct metal laser sintering (DMLS)

Selective laser Sintering typically employs polymer powders spread into a building bed/platform that is sintered by a guided laser (SLS vs. DMLS: What You Should Know About These 3D Printing Methods, no date; How to Compare SLS 3D Printer Prices, 2025). As the laser draws the pattern in 2D, the powder is sintered and becomes fused at the local spots. It could be seen as the dry analog of SLA, with a higher energy laser to almost melt the polymer. While ceramics can be used, the parts formed through sintering are of poor mechanical quality. The manufacture of bioglass ceramics and compatible ceramic structures from a calcium phosphate powder through SLS are still under investigation, and techniques to utilize binders that produce mechanically stronger parts after postprocess debinding and sintering are being actively investigated (Goodridge et al., 2006). In SLS, due to the build plate and chamber being heated near to the melting point or glass transition point of the polymer, the residual powder may not necessarily be reused as they may not remain in its pristine, granular form. SLS is ideal for polymers, ideally in an inert gas-filled chamber to prevent unintended oxidation and degradation of the polymer powder. Nitrogen is commonly used to fill the chamber similar to various nanofabrication work (Guide to Selective Laser Sintering (SLS) 3D Printing | Formlabs, no date).

With **Direct Metal Laser Sintering**, metal powder instead of polymer is used for creating metal parts.

***Terminology Alert*:** Various sources misrepresent DMLS and Selective Laser Melting (SLM) as one and the same, and that the trademarking of DMLS is tied to the misnomer of "sintering" versus "melting" (DMLS Trademark of EOS GmbH - Electro Optical Systems - Registration Number 4515227 - Serial Number 85592510: Justia Trademarks, no date; What are the main metal Additive Manufacturing processes?, 2016). While the same equipment could potentially be used for both processes, sintering and melting are inherently different processes, and SLM will typically produce mechanically superior parts to DMLS. *Sintering* is heating near the melting point of the powder at localized lasered spots, allowing the material to fuse through the glass transition temperature. *Melting* will turn the powder completely to liquid at the lasered spot. While the option for printing metals may default to SLM even if requesting a "DMLS" protocol, the reader should be aware that they are distinct and take precaution against using the terms interchangeably should the service provider actually execute the DMLS process if requested.

5.2 Selective laser melting (SLM)

SLM is like SLS/DMLS, except that the energy used is much higher such that the metal powder melts instead of being sintered. The process must be carried out in an inert gas environment, and for certain metals (such as magnesium) the generation of fumes can obstruct the laser, thus the printing process needs to be monitored and fine-tuned. SLM and DMLS can often be confused—"***Terminology Alert***" *above*.

5.3 Electron beam melting (EBM)

EBM is very similar to SLM, the difference being the use of an electron beam as an energy source to melt the metal powder as opposed to a high-powered laser. EBM demands extra care to ensure the printing environment must be a vacuum to prevent interference, leading to successful and faster prints. EBM is able to produce high-density parts with mechanical properties that are comparable with traditional casting or CNC machining methods.

EBM processes handle a very limited range of materials and require that the material for building parts is electrically conductive—these include titanium, chromium-cobalt alloys, select steel powders, and Inconel 718.

5.4 Binder jet (BJ) and Multi-Jet Fusion (MJF)

Binder jetting and Multi-Jet Fusion fall under the category of using binders on a powder bed with precision using an inkjet dropping method such as the everyday paper printer. Polymers, ceramics, and applicable metals are conducive to these techniques of manufacture.

In **Binder Jet**, a powder bed is used as with any of the other powder bed fusion methods above, however—on every layer of powder, a binding agent—typically an adhesive—is deposited through a printhead that binds the powder together and to the layer below it. HP has made big strides in proprietary advances in BJ and MJF technology, including a specialization in metal binder jetting ("HP Metal Jet").

In **Multi-jet fusion**, there are multiple "jets," with these jets being a prebinding instead of a complete binding agent (The Complete Guide to Multi Jet Fusion (MJF) 3D Printing, 2022). Similar to the BJ method, one printhead deposits a *fusion agent*, and a second head deposits a *detailing agent*. A heat source, such as a lamp, is passed over the substrate, fusing areas where the fusion agent is deposited. The detailing agent is also affected by the heat source, and modifies the fusion process, creating smoother curves, which in other AM methods would require several postprocessing steps (HP Multi Jet Fusion 3D Printing Technology, no date; What is Multi Jet Fusion? - Design Solutions, 2022).

6. Direct ink writing (robocasting)

Direct ink writing (DIW) is a close replica of FDM printing. It is another extrusion-based (EB) printing wherein a viscous slurry ("ink") is extruded out to form a 2D layer, which culminates into a 3D structure as in FDM. This method is typically useful for extruding out precursor ceramic structures that need to be sintered later. Ambient temperature control is critical in this step for the slurry to hold its shape before moving on to postprocessing. This technology allows precise control over ceramic part porosity. It is not unusual to select shear-thinning mixtures where applicable as to prevent nozzle clogging during extrusion. The slurries must also be well-mixed so that suspended particulates are not clumped in the nozzle tip, leading to inhomogeneous printing density or clogging.

7. Material jetting

Material Jetting can be seen as a variation of using inkjets—liquid droplets having photopolymerizable media are deposited onto a substrate, very much like the run-of-the-mill paper printer. Multiple nozzles can be used, which means various materials can be juxtaposed into the printheads—the most basic setup being the build material and support material. When one layer is deposited, a UV light cures the deposited ink, and the build plate moves down a layer, continuing the cycle until the object is built. Support materials are usually dissolvable in an appropriate solvent, leaving behind the build material. This mode of printing is also dubbed Drop on Demand (DOD) printing, as the droplets of ink are deposited where necessary. Common solvents for support material is a basic (sodium hydroxide) solution or simple waterjet. Unlike in vat polymerization, a postcure may not be necessary as the drops are cured fully on the go (Material Jetting | Additive Manufacturing Research Group | Loughborough University, 2025). 3D printing industry giants have developed proprietary technologies surrounding material jetting with biocompatible materials, such as Stratasys's PolyJet printers (PolyJet Technology for 3D Printing | Stratasys, no date).

8. Direct energy deposition (DED)

Direct Energy Deposition is an on-the-go format of creating 3D printed parts, and emulates FDM for printing directly in high-temperature melting materials such as metals. In this aspect, it resembles SLS, SLM, or EBM. However, in an FDM "style," instead of a large reserve of preformed material as in the powder bed fusion or vat photopolymerization, DED is used to dispense a preformed material powder or metal wire that is fed into a focused point under a moving head that supplies an intense burst of energy to melt the feedstock material. The mode of energy can be from a laser beam, an electron beam, or wire arc—although laser beams are the most commonly available format. The efficiency of DED comes from the fact that there is no extra powder handling and cleanup process afterward as in standard powder bed fusion methods. Fully functional metal alloy parts can be manufactured using this technology, and there are well-established protocols used by individual manufacturers for materials such as titanium alloys (Ti–6Al–4V), stainless steels, nickel alloys, copper alloys, and aluminum alloys. As was mentioned in the discussion of brass (copper alloy) and copper nozzle for FDM printing, these materials are excellent carriers of thermal energy, which makes it difficult to print. For bone implants, a subset of DED application includes the process of **Laser Cladding**, which can be used to coat a workpiece with bioactive materials for bone, such as hydroxyapatite coating or bioglass coating. Similar approaches that use liquid precursors to be cured by laser have also been studied to make bone implants less corrosive and more osteoconductive, and can be critical for biodegradable implants as in the case of magnesium alloys (Park et al., 2023).

9. Laser-assisted bioprint (LAB) via laser-induced forward transfer (LIFT)

This method is a unique way of "ejecting" or propelling a bioink drop, videographic details can be found on an NPTEL lecture series (Introduction to Laser-assisted Bioprinting, 2022) and by Chang and Sun (2023). In brief, there are three distinct parts to the LIFT setup: (1) a pulsed laser source, (2) a donor layer that donates the "bioink," and (3) a receiving substrate that receives the bioink ejected from the donor layer. The donor layer usually has three component layers—a top transparent layer (glass for near IR and visible light and quartz/fused silica for UV), an absorber layer comprising Ti, TiO_2, or gold, and a biomaterial layer, which is a viscous ink solution containing DNA, protein or cells in cell medium, glycerol, or fibrinogen. The pulsed laser will excite a spot on the donor layer, going through the transparent layer and onto the absorbing layer. The working theory is that the energy of the laser absorbed by the absorber layer will cause vapor to form, leading to cavitation and expelled jetting of the bioink from the biomaterial layer. By cascading the energy transfer from the laser to the biomaterial through the absorber layer, direct interaction between the laser and biological components within the bioink is avoided. The receiving substrate layer is placed underneath the donor layer to catch the ejected bioink, thus forming a desired pattern. The substrate itself may comprise biocompatible scaffolds, including natural/synthetic gels or 3D printed constructs. The donor or substrate layers could be on movable platforms (Chang & Sun, 2023). LIFT technology is still not easily accessible, and is not suitable to create full 3D constructs, however, they are a powerful technique to complement other 3D scaffold printing techniques to impregnate well-defined cellular components within the scaffolds themselves, in vivo or in situ. Directly transferring biologically active cells onto scaffolds that otherwise would be cell-free during implantation or could be superficially coated with cells permits osseointegration to levels conducive to repairing large defects that would otherwise not be healed just with scaffolds alone (Keriquel et al., 2017).

10. Printing materials

A rich amount of original publications and reviews have been published on materials used for additive manufacturing in bone tissue engineering (BTE), focusing on scaffolds with (Figs. 23.3 and 23.4) or without bioactivity (Fig. 23.5). Excellent reviews on the various biomaterials primarily categorized to polymers, ceramics, metals, bioinks, and composites can be found in Qu et al. (2019), Chen et al. (2020) and Pazhamannil and Alkhedher (2025).

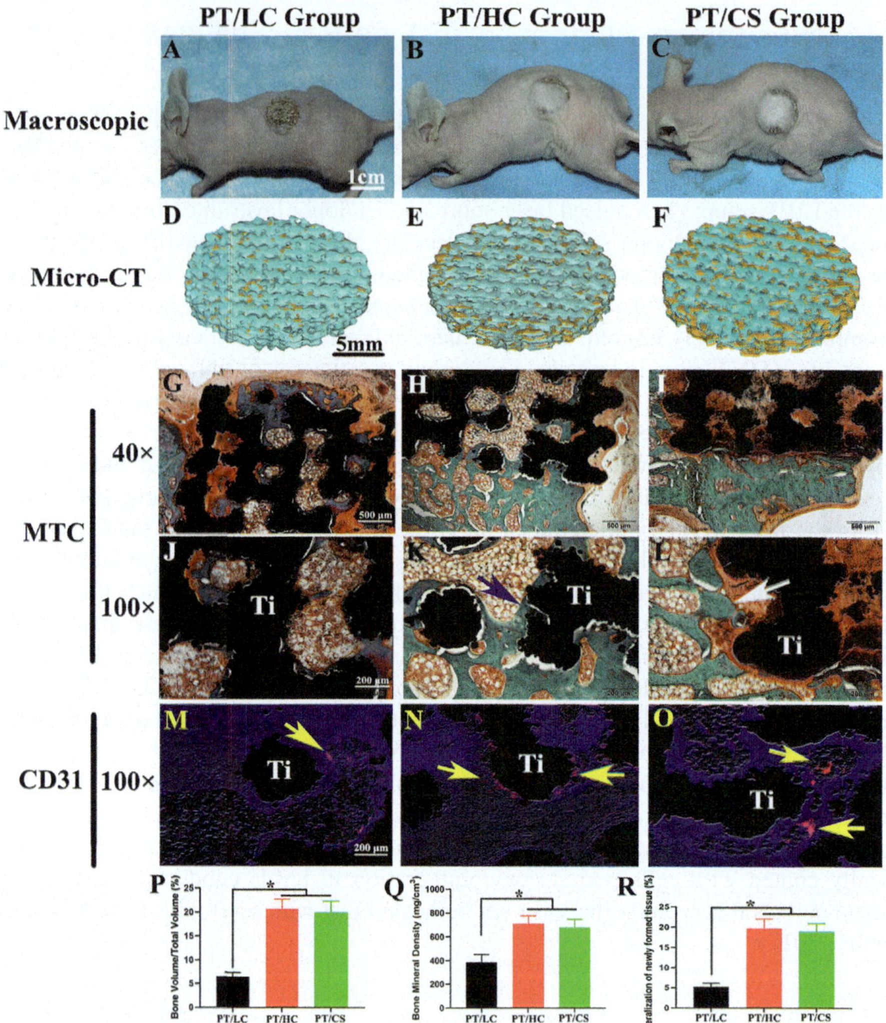

FIGURE 23.3 Evaluation effect of the combination of 3D-printed porous Ti–6Al–4V (PT) scaffolds and bone marrow mesenchymal stem cells (BMSCs) on the construction of biomimetic engineered bone at an ectopic site.

(A–C) Macroscopic images of the subcutaneous implantation sites at 12 weeks of implantation. (D–F) Micro-CT reconstruction of the bone regeneration of the complex of PT scaffolds with BMSCs. The newly formed bones were brownish yellow in color. (G–L) Masson's trichrome (MTC) staining of the PT scaffolds with BMSCs was performed to assess osteogenesis.PT/LC: PT scaffold + Low cell seeding; PT/HC: PT Scaffold + High cell seeding; PT/CS: PT scaffolds wrapped in a cell Sheet of BMSCs.

- From Wang, Z., Han, L., Zhou, Y., Cai, J., Sun, S., Ma, J., Wang, W., Li, X. & Ma, L. (2022). The combination of a 3D-Printed porous Ti–6Al–4V alloy scaffold and stem cell sheet technology for the construction of biomimetic engineered bone at an ectopic site. Materials Today Bio, 16, 100433. https://doi.org/10.1016/j.mtbio.2022.100433.

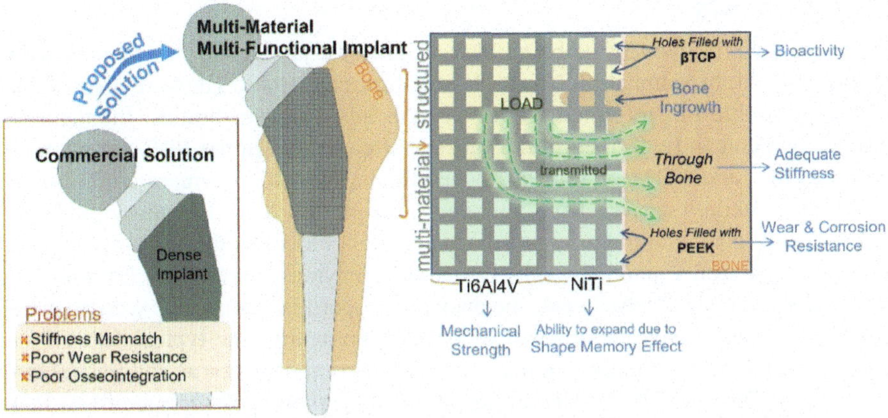

FIGURE 23.4 Design concept of multimaterial multifunctional hip implant solution.

Porosity (a multifunctional component) can be complementary to a multimaterial fabrication process in creating bioactive implants.

- From Costa, M. M., Lima, R., Alves, N., Silva, N. A., Gasik, M., Silva, F. S., Bartolomeu, F. & Miranda, G. (2022). Multi-material cellular structured orthopedic implants design: In vitro and bio-tribological performance. Journal of the Mechanical Behavior of Biomedical Materials, 131, 105246. https://doi.org/10.1016/j.jmbbm.2022.10524, licensed under CC-BY 4.0.

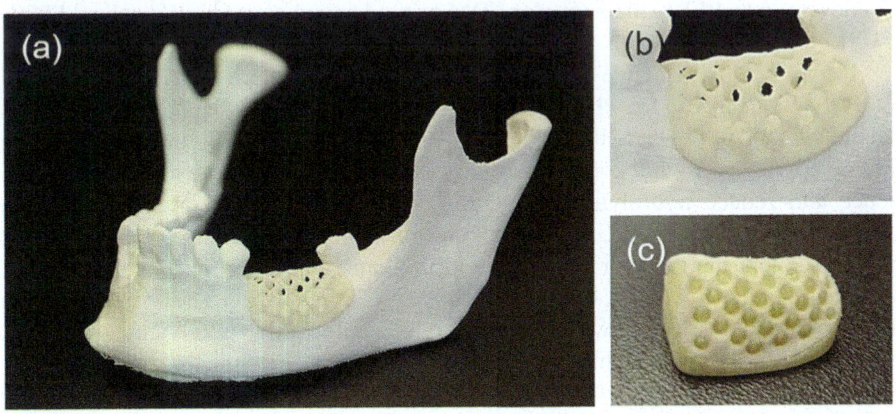

FIGURE 23.5 Three-dimensional (3D)-printed mesh (thickness 1.0 mm) for guided bone regeneration of the alveolar process.

(A) Visual assessment of the implant fit on 3D-printed model. (B) Magnified view of the open-porous alveolar mesh. (C) Mesh with support structure before postprocessing.

- From Maintz, M., Tourbier, C., de Wild, M., Cattin, P. C., Beyer, M., Seiler, D., Honigmann, P., Sharma, N. & Thieringer, F. M. (2024). Patient-specific implants made of 3D printed bioresorbable polymers at the point-of-care: material, technology, and scope of surgical application. 3D Printing in Medicine, 10(1), 13. https://doi.org/10.1186/s41205-024-00207-0 licensed under CC-BY 4.0.

10.1 Polymers

Polymers have the benefit of being easy to process, along with low cost, tunable biomechanical, biocompatibility and degradation properties. They are also usually readily available in preformed feedstock as beads/powders, liquids, filaments, or bioinks, which can be formed into a functional part through the PBF, vat photopolymerization, and FDM or robocasting methods. Natural polymers comprise the family of materials derived from collagen, gelatin, silk fibroin, chitosan, alginate, or hyaluronic acid. Natural polymers have excellent biocompatibility, with degradation rates between 2 and 24 weeks by natural enzymes, varying by cross-linking level. However, they have weak mechanical properties comparing with matched bone (e.g., compressive modulus of 0.005–90MPa against the typical 100 MPa of cortical bone), and are preferentially used within composites or as additives (Garot et al., 2021; Qu et al., 2019). The most prominent synthetic polymers are PLA, PCL, PLGA, PU (polyurethane), PVA (polyvinyl alcohol) and PEEK for bone tissue engineering in creating scaffolds and grafts. PCL is already well-known in its use as surgical sutures, and PEEK is readily used for lifetime use in dental, spinal, and orthopedic implants (DelMonico, 2022; The Role of Polyether Ether Ketone in the Medical Industry, no date). When using FDM, typically a higher melting-point polymer or composite may correspond to a higher elastic modulus or stronger mechanical properties, however, this can be affected by print orientation and loading directions—it is recommended to create small test prints for testing in the expected loading conditions before implementing a large, full-scale model.

Several new PEEK filaments for FDM printed spinal cages have received FDA's 510(k) clearance.

PLA is readily available and is usually the cheapest material ($20/Kg for a spool of PLA vs. $800/Kg for PEEK). There are several variants of PLA that often is labeled is "PLA+," where the "+" part is proprietary but usually involves a doping of polyethylene (or other polymers) to give the PLA material more mechanical resilience and reduce hydrolytic degradation of the filament that is exacerbated in warmer environments. PLGA, a copolymer of lactic (LA) and glycolic acid (GA), is an excellent material for bone scaffold implants due to its inherent mechanical and degradation tunability (Ge et al., 2009). A higher LA:GA ratio makes for a stiffer material with a slow degradation rate, while a higher GA content (lower LA:GA) allows the hydrophilic and easily degradable GA to impart a quicker degradation rate to the whole printed part (Factors Affecting PLGA Degradation: What You Need to Know, 2023). The LA:GA ratio can tune the degradation from a few weeks to several months (Miller et al., 1977). It is advised to always check with the manufacturer, other publications or whitepapers that enlist using a chosen material, before adapting for any medical use. There are "medical grade" implantable filaments that can be purchased, which meet different needs, from sutures to bulk implants (Implantable Filament for 3D printing – Medical grade, no date; ISO 13485 Medical 3D Printing Filaments, no date) and includes filaments, which include osseointegration-promoting minerals such as hydroxyapatite (HA), such as the PLGA/HA filament from Lattice Services (PLGA/HA filament résorbable et implantable, no date), which is blend of 90% PLGA and 10% HA. These blends can allow for better resorption after implantation. Another filament material seeded with bioactive (HA) nanoparticles in a PMMA base is OssFila's *Bonlecule* (Bonlecule Bioactive 3D Printer Filament, no date), which advertises modest printing temperatures of 220–240°C and certified for implant use (ISO 10993–23:2021, ISO 10993–10:2021, ISO 10993–5:2019 and REACH (EC1907/2006) certifications). However, the current recommendation seems to be for non-load-bearing applications, including cranioplasty implant, parietal bone, spinal cage, bone of rib cage, radius and ulna, carpal bones, metacarpal bones and phalanges. The use of integrated nano-HA/polymer filament maintains most of the mechanical strength of the base polymer with the added benefit toward osseointegration (Wei & Ma, 2004).

10.2 Metals

Metals are a classic material when it comes to the design of scaffolds/implants, and has a rich history in being manufactured through conventional methods before additive manufacturing. Some of the well-known metals for BTE include titanium-based alloys, cobalt chromium alloys, iron-manganese alloys, and magnesium alloys. Depending on the postprocessing and surface treatment, these implants can offer long-term stability and biomechanical compatibility. Tantalum has also been doped into titanium to produce alloys for BTE, wherein porous tantalum can enhance osseointegration (Bencharit et al., 2014) and reduce the high elastic modulus of titanium. CoCr has been known to be used in hip replacement or knee arthroplasty, and cobalt poisoning from wear of the implants—which is rare but induces debilitating neuropathy amongst other severely cytotoxic effects (Venkatraman et al., 2020)—have become a niche theme for medical shows (Payne, 2019). Ti and CoCr 3D printed implants for load bearing applications such as total talus replacement for avascular necrosis of the ankle have picked up traction following FDA clearance (FDA approves 3D-printed ankle implant for humanitarian use | MedTech Dive, no date; Total Talus Replacement – Paragon 28, no date). There is ongoing research on semidegradable materials such as Mg-infused Ti or CoCr alloys, as the biodegradable Mg could stimulate cell proliferation while it vacates space during its resorption. Ti6Al4V is a very common alloy used for biomedical prototypes, and extensive processes to match them to the mechanical properties of bone by constructing porous forms of the alloy is found extensively in literature. Niobium coating of Ti substrate has also improved bioactivity (Zhang et al., 2015). Iron (Fe), Iron-manganese (Fe–Mn), iron-manganese-calcium (Fe–Mn–Ca) were also explored as biodegradable metal implants, as Fe-based metals were the earliest 3D printed biodegradable metals used for bone regeneration (Chen et al., 2020). Hydroxyapatite (HA) coating of metal implants can also promote osseointegration, and the integration of β-tricalcium phosphate (β-TCP) is another lead to making implants more permissive to osteogenesis (Belgin Paul et al., 2024; Nandi et al., 2018; Schönegg et al., 2024).

FDM, PBF, and DED—are all ideal methods of metal manufacture, although PBF and DED are the most commonly used. Different 3D systems will use various proprietary methods and binder formulations optimized for their metal manufacture process. Typical binders can be polymer type (commonly polyvinyl alcohol, PVA), wax type, or acrylic binders (Hazizi, 2023). Once a part is printed, it is referred to as a **greenpart**, which is then put through a debinding process. The debinding process can involve chemical (gaseous) degradation, evaporation, or immersion into a debinding fluid (What is Debinding and Why is it Used?, no date). After debinding, the parts need to be sintered and annealed. One should expect a shrinkage of 15%–25% of the final part from the originally printed green part, as getting rid of the binder will create voids within the part that will be filled in during the sintering and annealing process. This expected shrinkage needs to be accounted for in the fabrication process as it will reduce not just the overall dimension of the object but also the pore sizes in porous 3D architecture.

10.3 Ceramics

Ceramics are a prime material for osteogenesis and as such is prolifically used for bone repair, and an everyday material for dental repair. A well-known inorganic mineral is hydroxyapatite (HA), a calcium phosphate ceramic is hydroxyapatite found naturally in the bones and teeth and used as a coating on bone implants for efficient osseointegration. Other bioactive ceramics include tricalcium phosphates (TCP), calcium sulfate, akermanite ($Ca_2MgSi_2O_7$), diopside ($CaMgSi_2O_6$), and bioactive glasses, while bioinert ceramics in common use are aluminum-based or zirconia (ZrO_2). Typical methods for ceramics additive manufacturing rely on Powder Bed Fusion, robocasting (i.e., Extrusion-Based methods), SLS, and SLA. The binders used for PBF, EB, and SLA determine the postprocessing steps needed for drying, debinding, sintering, and firing to produce the finished part. There also exist binders used that are not unlike those used in metal printing, including organic (PVA, starches, polyethylene glycols, waxes) and inorganic (silicates and bentonite) binders (Binders for Ceramic Bodies, no date). Companies such as Formlabs and Baikowski have extensive white papers and publications related to their line of ceramic materials (Ceramic 3D Printing Made Accessible: Introducing Alumina 4N Resin for Technical Ceramic Parts, 2023; "Ceramic 3D Printing Powders & Slurries," no date, p. 10; Innovation: white papers about advanced oxide solutions - Baikowski, no date). Rapidia touts their novel approach to using water-based binders in a manner such that overly complex or large parts can be manufactured in parts and simply "bonded" together after the parts are dipped into water, producing a delicate but manageable green part that is further processed to the final compound part (Iftikhar, 2019; Rapidia Metal 3D-water bonding process, 2019).

10.4 Gels/bioinks

Bioinks, aptly given this nomenclature as the innovation came about at the end of the 20th century by using the technology developed for regular inkjet printing to print "inks" that consisted of natural and synthetic materials—even live cells! Similar to inkjet printing, using thermal or piezoelectric triggers, the "ink" could be deposited layer by layer in picoliter amounts, and the rapid printing in this format would allow one to create a fully 3D construct. At the frontier of innovation of bioinks is the decellularized extracellular matrix (dECM), and various natural polymers including gelatin, methacrylated gelatin (Gel-MA), alginate, collagen, silk fibroin, starch, chitosan, hyaluronic acid, and various combinations of these polymers. Pluronic, a synthetic triblock copolymer of PEO-PPO-PEO, is also commonly used as a gel base for scaffolds with embedded structures (Guida et al., 2024; Zhou et al., 2024). The technology of using bioinks covers inkjet bioprinting (IB), extrusion-based printing (EBP), fused deposition modeling (FDM), and laser-assisted bioprinting (LAB). Any extrusion-based methods need to have the bioink retain enough viscosity to hold the shape once deposited onto a surface. On the other hand, LAB/LIFT methods can be employed as a complement to the standard scaffold methods to deposit low-viscosity inks containing active cells interspersed within a scaffold for improved osseointegration (Keriquel et al., 2017).

10.5 Composites

Single materials offer part of an ideal material, and just as rebars are used to reinforce concrete, the materials aforementioned—metals, polymers, ceramics, and biologics (bioinks) can be combined to create composites that bring the best of each constituent (Fig. 23.6). A metal implant can have a calcium phosphate (CaP) ceramic coating to enhance osteoinduction. A composite structure made of sacrificial elements could create a well-connected lumen structure within an otherwise difficult or randomly generated scaffold (Zhou et al., 2024). Functionally graded scaffolds (FGS) or gradiented interfaces can also be created where the material composition can change across the bulk of the manufactured part, matching expected tissue properties that naturally have gradients (such as the cancellous to cortical bone). Simple functional gradation uses the same material at varying porosities to create the physical gradient (Gao et al., 2022), but adding in capabilities of multimaterial additive manufacturing (MMAM) opens the door to endless combinations of metal, polymer, ceramic and biologics combinations to tune an implant to the structure and function of the tissue it is engineered to replace (Nazir et al., 2023). A novel application of one such method is highlighted in a coaxial-cable type functional gradient created for a bone-tendon interface in an in vivo rotator cuff tear (RCT) model, using bone bioink (bone dECM, hydroxyapatite) and tendon

A

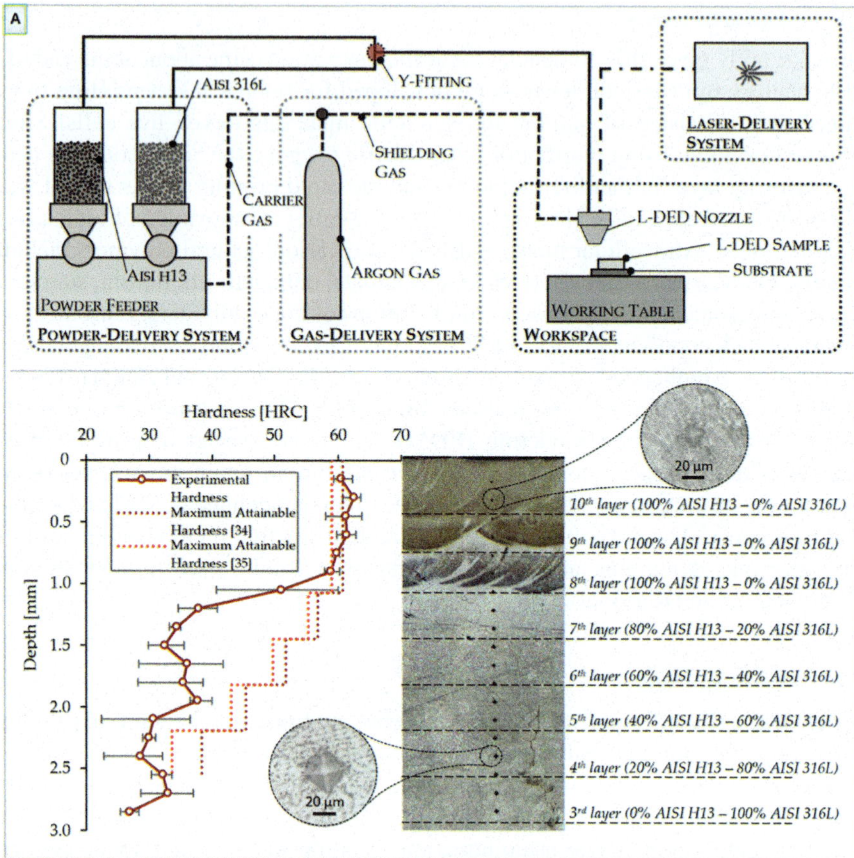

FIGURE 23.6 Multimaterials and composites in the additive manufacturing space.

Multimaterial manufacturing is at the frontier of additive manufacturing techniques, such that (A) functionally graded compositions of materials can be created by locally controlling the mixture ratio delivery of different material powders to be fused, (B) Adapting SLS processes to overcoat Nb on a previously formed Ti structure, showing the (a) top view (b) surface view, (c) cross section through the porous zone, and a (d) transverse section through the interface region. (C) A core–shell-based method of functionally gradiented additive manufacturing, created by altering the mixture composition between the core bioink (tendon dECM) and shell bioinks (bone dECM). (D) A marriage of traditional porous manufacturing method with additive manufacturing—additive manufacture of a custom part by fusing polymer pellets and salt allows the part to be porously redefined after printing and salt leaching.

- *(A) Adapted from Ostolaza, M., Arrizubieta, J. I., Lamikiz, A. & Cortina, M. (2021). Functionally Graded AISI 316L and AISI H13 Manufactured by L-DED for Die and Mold Applications. Applied Sciences, 11(2), 771. https://doi.org/10.3390/app11020771. (B) Adapted from Figs. 2 and 4 of Zhang, S., Cheng, X., Yao, Y., Wei, Y., Han, C., Shi, Y., Wei, Q. & Zhang, Z. (2015). Porous niobium coatings fabricated with selective laser melting on*

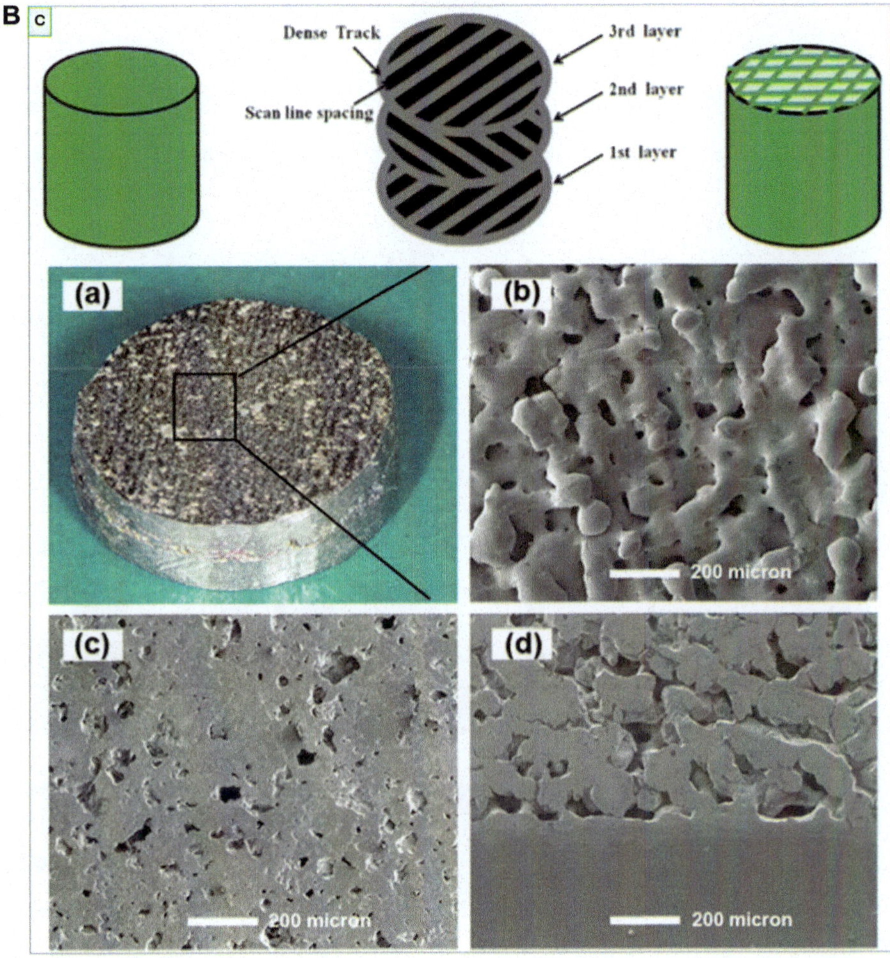

FIGURE 23.6 cont'd.

bioinks (tendon dECM, β-TCP), with both inks containing human adipose-derived stem cells (hASC, "human adipose stem cells"). The functionally graded MMAM construct showed vast improvements in mechanical properties and tissue integration, including elevated angiogenesis and ECM formation (Kim et al., 2025). Conventional paradigms for composites include (1) metal-based composites (Ti6Al4V with tantalum, Fe/HA, Ti6Al4V/HA, Ti/PVA, etc). (2) polymer composites (PCL/HA, PLGA/nHA, PLA/PDA/HA, PEEK/HA or PEEK/nHA, PEEK/Ag, PEG/PU,

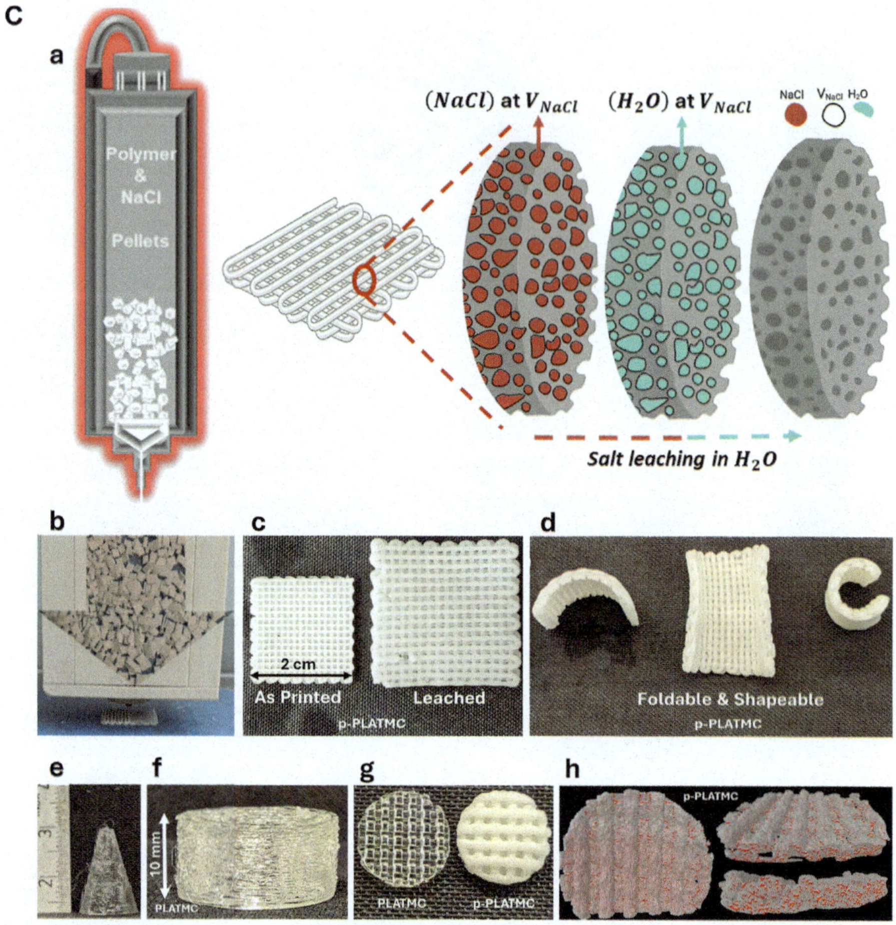

FIGURE 23.6 cont'd.

etc), (3) ceramic-based composites (HA/β-TCP, HA/PCL, β-TCP/PCL, α-TCP/collagen, etc), (4) functional composites and MMAM composites using the three major material categories (metals, polymers, ceramics) in combination with natural or synthetic biointegrative gels (Chen et al., 2020; Garot et al., 2021; Moreno Madrid et al., 2019; Qu et al., 2019).

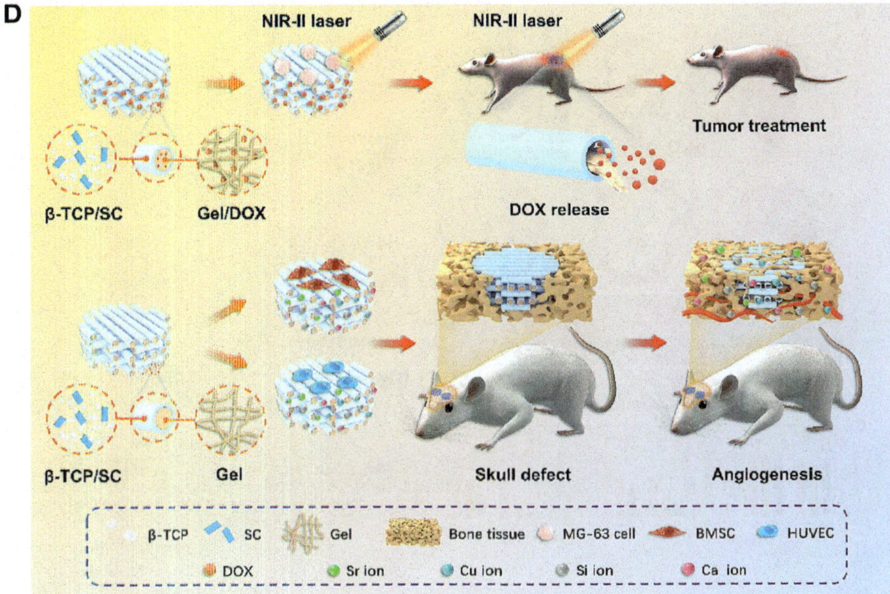

FIGURE 23.6 cont'd.

11. Clinical applications, hope and challenges

Due to limited and sometimes conflicting data on patients who received standard or patient-matched 3D printed orthopedic implants, in the eyes of the health insurance industry, the evidence of benefits of such implants to the overall health outcome and compared with conventional methods are insufficient for those with typical bone and joint anatomy. However, for those with bone or joint deformity, which can require "revision total hip arthroplasty with severe acetabular defects, reconstruction following orthopedic tumor resection, and spinal abnormalities," there is sufficient clinical data mostly from outside the United States supporting the advantages of patient-specific custom 3D printed implants (Three Dimensional Printed Orthopedic Implants | Providers | Blue Cross NC, no date; Wu et al., 2023). There has been substantial success in metal 3D implants, which rely on Ti6Al4V alloy, especially for spinal cages, pelvic prostheses, joint applications (e.g., feet and ankle joint, knee, hip, wrist and shoulder), and craniomaxillofacial implants. Excellent reviews of clinical trials of 3D printed implants can be found in Wu et al. (2023) and McAnena et al. (2025), showing that most clinical trials and longest follow-ups are from outside the United States. Dental and craniomaxillofacial applications of implants are usually a good precursor to testing bone replacement materials, as the load

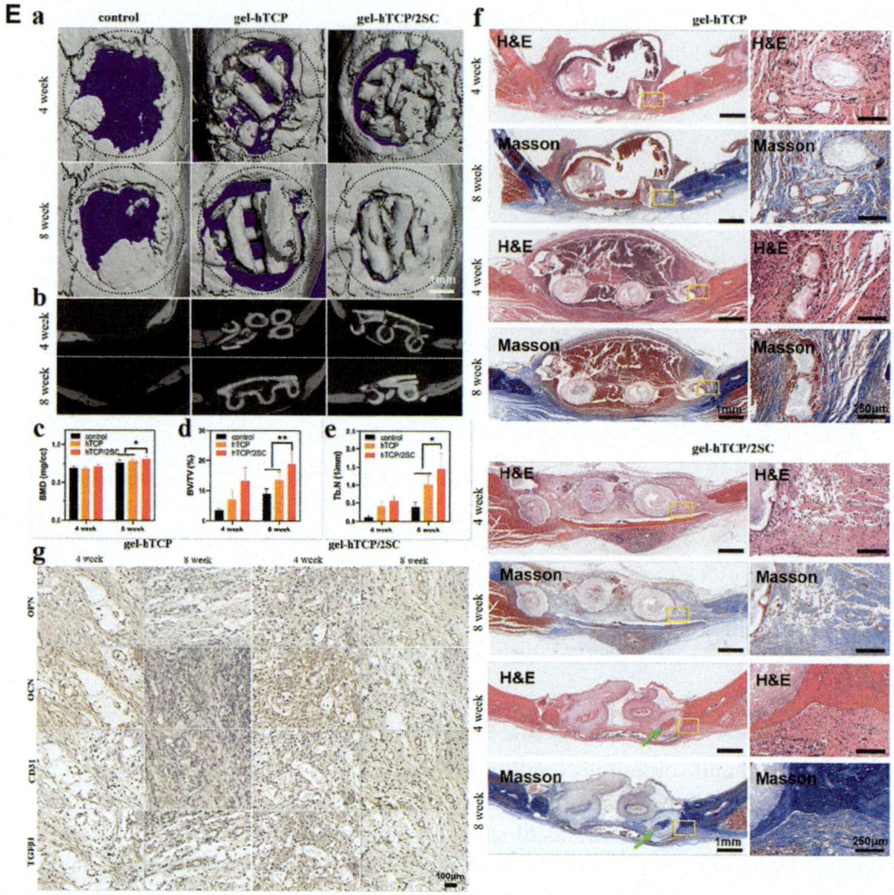

FIGURE 23.6 cont'd.

bearing is minimal. Relatively simple use of 3D printed scaffolds for alveolar augmentation can be an important first step toward fine-tuning osseointegration through "customized bone regeneration," and data is still being collected for familiar materials such as PCL (Aravena, 2025) and β-tricalcium phosphate (β-TCP) (Schönegg et al., 2024) These studies are also outsourced to outside the United States, in Chile and Switzerland, respectively.

While this chapter reflects on orthopedic implants, 3D printed biocompatible implants are useful in other soft tissue applications too—such as in the recently approved clinical trial (Ohye, 2025) for 3D printed tracheobronchial splints for infants suffering from tracheobronchomalacia, as a collaborative effort between Materialize and the University of Michigan (Materialise, 2024; Bioresorbable Implant 3D Printed by Materialise Enters Clinical Trial, no date; Clinical trial opens to study groundbreaking 3D printed device for babies with rare respiratory disease | Michigan Medicine, 2025) following the first successful case in 2012 (Groundbreaking 3D-Printed Splint Restores a Baby's Breathing, no date). The splints are manufactured through SLS using polycaprolactone, and the techniques described in this chapter are applicable to cases that are not limited to orthopedics alone.

Changing the landscape of healthcare to formulate customized treatment regimen is a big paradigm shift often facing severe roadblocks in the United States, which include regulatory complexities, limited expertise, high costs, limited access, and procuring evidence of benefit that justifies cost-effectiveness and reimbursement especially when targeting a small, select number of people—a "custom" part essentially will fit only one person. However, with technology becoming more accessible and advanced, certain manufacturing practices can be moved in-house, while more complex processes can be outsourced. The advantage of additive manufacturing over conventional methods is that there is no upfront investment into expensively machined molds that is not cost effective unless the molds are used for mass-production. The AM method creates every part uniquely, the "mold" is the digital blueprint file.

AM as a rapidly advancing field has a lot of room to improve, with the greatest challenge being biomimicry. It is not entirely possible yet by additive manufacturing to achieve fully vascularized or physiologically accurate tissue, although having strong osseointegrative properties within the scaffold material help immensely. Several of the techniques that are in active research for multimaterial printing can allow for discretely defined lumen and compartments down to the 100–300 μm

titanium substrates: Preparation, characterization, and cell behavior. Materials Science and Engineering C, 53, 50–59. https://doi.org/10.1016/j.msec.2015.04.005. (C) Adapted from Aydin, M. S., Nicolae, C.-V., Campodoni, E., Mohamed-Ahmed, S., Kadousaraei, M. J., Yassin, M. A., Gjerde, C., Sandri, M., Stancu, I.-C., Rashad, A. & Mustafa, K. (2025). Osteogenic Potential of 3D-Printed Porous Poly(lactide-co-trimethylene carbonate) Scaffolds Coated with Mg-Doped Hydroxyapatite. ACS Applied Materials & Interfaces, 17(21), 31411–31433. https://doi.org/10.1021/acsami.5c03945. (D) From Fig. 1 of Zhang, X., Wei, H., Dong, C., Wang, J., Zhang, T., Huang, L., Ni, D. & Luo, Y. (2023). 3D printed hydrogel/bioceramics core/shell scaffold with NIR-II triggered drug release for chemo-photothermal therapy of bone tumors and enhanced bone repair. Chemical Engineering Journal, 461, 141855. https://doi.org/10.1016/j.cej.2023.141855. (E) From Fig. 8 of Zhang, X., Wei, H., Dong, C., Wang, J., Zhang, T., Huang, L., Ni, D. & Luo, Y. (2023). 3D printed hydrogel/bioceramics core/shell scaffold with NIR-II triggered drug release for chemo-photothermal therapy of bone tumors and enhanced bone repair. Chemical Engineering Journal, 461, 141855. https://doi.org/10.1016/j.cej.2023.141855.

scale, and being able to integrate multiple materials along with bioinks, will boost bone tissue engineering to approach physiological integration and mimicry in the hopes of having readily available, personalized treatment for all patients in the future.

References

22 Best additive manufacturing companies (2024 review) - Uptive. https://uptivemfg.com/best-additive-manufacturing-companies/.

3D-printed houses: 12 top examples | built in. Available from: https://builtin.com/articles/3d-printed-house.

3D printer nozzle guide. (2025). https://www.3djake.com/info/guide/3d-printer-nozzle-guide.

Abbasi, N., Hamlet, S., Love, R. M., & Nguyen, N. T. (2020). Porous scaffolds for bone regeneration. *Journal of Science: Advanced Materials and Devices, 5*(1), 1–9. https://doi.org/10.1016/j.jsamd.2020.01.007

Aravena, P. C. (2025). *The objective of this study is to test the bone formation capacity of a 3D printed scaffold with a plastic material called polycaprolactone (PCL), compared to the use of synthetic bone grafts, for bone formation in the maxilla or mandible of elderly patients in the city of valdivia between the years 2023-2024.* NCT06773923 clinicaltrials.gov Effectiveness of Patient-specific 3D-printed Resorbable Scaffolds of Polycaprolactone, Beta-tricalcium Phosphate, and Metformin Used for Maxillary Bone Regeneration: Randomized Clinical Trial Effectiveness of 3D-printed Patient-specific Resorbable Scaffolds for Maxillary Bone Regeneration https://clinicaltrials.gov/study/NCT06773923.

Bandyopadhyay, A., Mitra, I., Avila, J. D., Upadhyayula, M., & Bose, S. (2023). Porous metal implants: Processing, properties, and challenges. *International Journal of Extreme Manufacturing, 5*(3), Article 032014. https://doi.org/10.1088/2631-7990/acdd35

Belgin Paul, D. L., et al. (2024). 3D-printed β-TCP/Ti6Al4V composite scaffolds for bone regeneration: Process parameter optimization and evaluation. *Annals of 3D Printed Medicine, 15*. https://doi.org/10.1016/j.stlm.2024.100168

Bencharit, S., Byrd, W. C., Altarawneh, S., Hosseini, B., Leong, A., Reside, G., Morelli, T., & Offenbacher, S. (2014). Development and applications of porous tantalum trabecular metal-enhanced titanium dental implants. In *Clinical implant dentistry and related research* (pp. 817–826). https://doi.org/10.1111/cid.12059

Binders for ceramic bodies. Available from: https://digitalfire.com/article/binders+for+ceramic+bodies.

Bioresorbable implant 3D printed by materialise enters clinical trial. https://www.materialise.com/en/news/press-releases/3d-printed-bioresorbable-implant-clinical-trial.

Blender. (2025). https://www.blender.org/.

BOC sciences. Available from: https://www.bocsci.com/polycaprolactone-pcl.html.

Bonlecule bioactive 3D printer filament. Available from: https://www.ossfila.com/bioactive-3d-printed-filament/.

Ceramic 3D printing powders & slurries. 10.

Ceramic 3D Printing Made Accessible: Introducing Alumina 4N resin for technical ceramic parts. (2023). Available from: https://formlabs.com/blog/ceramic-3d-printing-alumina-4n-resin/.

Chang, J., & Sun, X. (2023). Laser-induced forward transfer based laser bioprinting in biomedical applications. *Frontiers in Bioengineering and Biotechnology, 11*. https://doi.org/10.3389/fbioe.2023.1255782

Chen, Y., Li, W., Zhang, C., Wu, Z., & Liu, J. (2020). Recent developments of biomaterials for additive manufacturing of bone scaffolds. *Advanced Healthcare Materials, 9*(23). https://doi.org/10.1002/adhm.202000724

Clinical trial opens to study groundbreaking 3D printed device for babies with rare respiratory disease | Michigan Medicine Available from: https://www.michiganmedicine.org/news-release/clinical-trial-opens-study-groundbreaking-3d-printed-device-babies-rare-respiratory-disease.

Committee F42 on additive manufacturing technologies | ASTM https://www.astm.org/membership-participation/technical-committees/committee-f42.

Darulis, A. (2022). *Master resin 3D printing settings as a beginner! learn key adjustments for quality prints with safe tips. Start now! AmeraLabs the complete resin 3D printing settings guide for beginners*. https://ameralabs.com/blog/the-complete-resin-3d-printing-settings-guide-for-beginners/.

DelMonico, K. (2022). *FDA clears hydroxyapatite porous PEEK cervical implant*. Orthop. This Week. Available from: http://ryortho.com/2022/09/fda-clears-hydroxyapatite-porous-peek-cervical-implant/.

DMLS Trademark of EOS GmbH - Electro optical systems - Registration Number 4515227 - Serial Number 85592510: Justia Trademarks. Available from: https://trademarks.justia.com/855/92/dmls-85592510.html.

Esmail, I. (2021). *PEEK - Best 3D printing temperatures to maximize crystallinity*. Available from: https://www.aon3d.com/material-science/peek-3d-printing-temperatures/.

Factors affecting PLGA degradation: What you need to know. (2023). https://polylactide.com/plga-degradation/.

FDA approves 3D-printed ankle implant for humanitarian use | MedTech Dive. Available from: https://www.medtechdive.com/news/fda-approves-3d-printed-ankle-implant-for-humanitarian-use/595262/.

Fedorov, A., Beichel, R., Kalpathy-Cramer, J., Finet, J., Fillion-Robin, J. C., Pujol, S., Bauer, C., Jennings, D., Fennessy, F., Sonka, M., Buatti, J., Aylward, S., Miller, J. V., Pieper, S., & Kikinis, R. (2012). 3D slicer as an image computing platform for the quantitative imaging network. *Magnetic Resonance Imaging, 30*(9), 1323–1341. https://doi.org/10.1016/j.mri.2012.05.001

Fluoroplastic comparison - typical properties. (2018). Available from: https://web.archive.org/web/20181116163946/https://www.chemours.com/Teflon_Industrial/en_US/tech_info/techinfo_compare.html.

Gao, X., Zhao, Y., Wang, M., Liu, Z., & Liu, C. (2022). Parametric design of hip implant with gradient porous structure. *Frontiers in Bioengineering and Biotechnology, 10*. https://doi.org/10.3389/fbioe.2022.850184

Garot, C., Bettega, G., & Picart, C. (2021). Additive manufacturing of material scaffolds for bone regeneration: Toward application in the clinics. *Advanced Functional Materials, 31*(5). https://doi.org/10.1002/adfm.202006967

Ge, Z., Tian, X., Heng, B. C., Fan, V., Yeo, J. F., & Cao, T. (2009). Histological evaluation of osteogenesis of 3D-printed poly-lactic-co- glycolic acid (PLGA) scaffolds in a rabbit model. *Biomedical Materials, 4*(2). https://doi.org/10.1088/1748-6041/4/2/021001

Germaini, M.-M., Belhabib, S., Guessasma, S., Deterre, R., Corre, P., & Weiss, P. (2022). Additive manufacturing of biomaterials for bone tissue engineering – A critical review of the state of the art and new concepts. *Progress in Materials Science, 130*, Article 100963. https://doi.org/10.1016/j.pmatsci.2022.100963

Goodridge, R. D., Dalgarno, K. W., & Wood, D. J. (2006). Indirect selective laser sintering of an apatite-mullite glass-ceramic for potential use in bone replacement applications. *Proceedings of the Institution of Mechanical Engineers - Part H: Journal of Engineering in Medicine, 220*(1), 57–68. https://doi.org/10.1243/095441105X69051

Groundbreaking 3D-printed splint restores a baby's breathing. Available from: https://www.materialise.com/en/inspiration/cases/3d-printed-splint.

Guida, L., Cavallaro, M., & Levi, M. (2024). Advancements in high-resolution 3D bioprinting: Exploring technological trends, bioinks and achieved resolutions. *Bioprinting, 44*, Article e00376. https://doi.org/10.1016/j.bprint.2024.e00376

Guide to Selective Laser Sintering (SLS) 3D Printing | Formlabs. Available from: https://formlabs.com/blog/what-is-selective-laser-sintering/.

Hazizi, K. (2023). *Re: What are the binders used for cobalt-based alloy in binder jet 3D printing?*. Available from: https://www.researchgate.net/post/What_are_the_binders_used_for_cobalt-based_alloy_in_Binder_Jet_3D_Printing/64c22695962ec0835c094a32/citation/download.

High-fidelity printing techniques by Richard Greene. (2016). https://www.youtube.com/watch?v=PsK7An7ymYk.

How does SLA printing work? (2025) https://formlabs.com/blog/ultimate-guide-to-stereolithography-sla-3d-printing/?wvideo=67tcujl2sk.

How to compare SLS 3D printer prices. (2025). https://formlabs.com/blog/selective-laser-sintering-sls-3d-printer-price/.

HP Multi Jet Fusion 3D Printing Technology. Available from: https://www.hp.com/us-en/printers/3d-printers/products/multi-jet-technology.html.

Hull, C. W. (1986). *Apparatus for production of three-dimensional objects by stereolithography.*

ICON and New Story/World's First 3D-printed community of homes. Available from: https://www.iconbuild.com/media-gallery/icon-and-new-story-worlds-first-3d-printed-community-of-homes.

Iftikhar, U. (2019). *3D printing industry preview the rapidia two-step metal and ceramic 3D printer.* Available from: https://3dprintingindustry.com/news/3d-printing-industry-preview-the-rapidia-two-step-metal-and-ceramic-3d-printer-155420/.

Implantable filament for 3D printing – Medical grade. Available from: https://lattice-services.com/en/absorbable-implantable-filaments/.

Innovation: White papers about advanced oxide solutions - Baikowski. Available from: https://www.baikowski.com/en/white-paper-on-oxide-solutions-for-ceramics-and-polishing/.

27 Introduction, 2022 Introduction to laser-assisted bioprinting. (2022).

ISO 13485 Medical 3D printing filaments. Available from: https://www.novusls.com.

Keriquel, V., Oliveira, H., Rémy, M., Ziane, S., Delmond, S., Rousseau, B., Rey, S., Catros, S., Amédée, J., Guillemot, F., & Fricain, J. C. (2017). In situ printing of mesenchymal stromal cells, by laser-assisted bioprinting, for in vivo bone regeneration applications. *Scientific Reports, 7*(1). https://doi.org/10.1038/s41598-017-01914-x

Kikinis R. Pieper S. Fillion-Robin J.-C. Aucoin N. Aylward S. Fedorov A. Hata N. Johnson H. Kapur T. Fichtinger G. Lasso A. Pinter C. Miller J. Pujol S. Tokuda J. O'Donnell L. Enquobahrie A. Paniagua B. Slicer http://www.slicer.org.

Kim, J. W., Yang, B. E., Hong, S. J., Choi, H. G., Byeon, S. J., Lim, H. K., Chung, S. M., Lee, J. H., & Byun, S. H. (2020). Bone regeneration capability of 3D printed ceramic scaffolds. *International Journal of Molecular Sciences, 21*(14), 1–13. https://doi.org/10.3390/ijms21144837

Kim, W. J., Kwon, D. R., Lee, H., Lee, J. Y., Moon, Y. S., Lee, S. C., & Kim, G. H. (2025). 3D bioprinted multi-layered cell constructs with gradient core-shell interface for tendon-to-bone tissue regeneration. *Bioactive Materials, 43*, 471–490. https://doi.org/10.1016/j.bioactmat.2024.10.002

Material jetting | additive manufacturing research group | Loughborough University. (2025). https://www.lboro.ac.uk/research/amrg/about/the7categoriesofadditivemanufacturing/materialjetting/.

Materialise. (2024). *The airway splint is bioresorbable, meaning the child's body will absorb the splint over about five years. A multi-center trial to assess the safety and effectiveness of a bioresorbable tracheobronchial splint in pediatric subjects with clinically significant tracheobronchomalacia bioresorbable airway splint pivotal clinical trial.* https://clinicaltrials.gov/study/NCT06406452.

McAnena, A. P., McClennen, T., & Zheng, H. (2025). Patient-specific 3-Dimensional-Printed orthopedic implants and surgical devices are potential alternatives to conventional technology but require additional characterization. *CiOS Clinics in Orthopedic Surgery, 17* (1), 1–15. https://doi.org/10.4055/cios23294

McKeen, L. W. (2012). *High-temperature and high-performance polymers* (pp. 233–250). Elsevier BV. https://doi.org/10.1016/b978-1-4377-3469-0.10011-6

Meshmixer. (2025). https://apps.autodesk.com/FUSION/en/Detail/Index?id=4108920185261935100&appLang=en&os=Win64.

Miller, R. A., Brady, J. M., & Cutright, D. E. (1977). Degradation rates of oral resorbable implants (polylactates and polyglycolates): Rate modification with changes in PLA/PGA copolymer ratios. *Journal of Biomedical Materials Research, 11*(5), 711–719. https://doi.org/10.1002/jbm.820110507

Monochrome vs. RGB LCD screen in resin 3D-printing | Liqcreate. Available from: https://www.liqcreate.com/supportarticles/monochrome-rgb-lcd-screen-resin/.

Moreno Madrid, A. P., Vrech, S. M., Sanchez, M. A., & Rodriguez, A. P. (2019). Advances in additive manufacturing for bone tissue engineering scaffolds. *Materials Science and Engineering: C, 100*, 631–644. https://doi.org/10.1016/j.msec.2019.03.037

Mota, C., Puppi, D., Chiellini, F., & Chiellini, E. (2015). Additive manufacturing techniques for the production of tissue engineering constructs. *Journal of Tissue Engineering and Regenerative Medicine, 9*(3), 174–190. https://doi.org/10.1002/term.1635

Nandi, S. K., Fielding, G., Banerjee, D., Bandyopadhyay, A., & Bose, S. (2018). 3D-printed β-TCP bone tissue engineering scaffolds: Effects of chemistry on in vivo biological properties in a rabbit tibia model. *Journal of Materials Research, 33*(14), 1939–1947. https://doi.org/10.1557/jmr.2018.233

Nazir, A., Gokcekaya, O., Billah, K. Md M., Ertugrul, O., Jiang, J., Sun, J., & Hussain, S. (2023). Multi-material additive manufacturing: A systematic review of design, properties, applications, challenges, and 3D printing of materials and cellular metamaterials. *Materials & Design, 226*, Article 111661. https://doi.org/10.1016/j.matdes.2023.111661

Ohye, R. (2025). *A multi-center trial to assess the safety and effectiveness of a bioresorbable tracheobronchial splint in pediatric subjects with clinically significant Tracheobronchomalacia.*

Park, J., Um, S. H., Seo, Y., Lee, J., Kim, Y. C., Ok, M. R., Hwang, S. W., Sun, J. Y., Han, H. S., & Jeon, H. (2023). Improving hydroxyapatite coating ability on biodegradable metal through laser-induced hydrothermal coating in liquid precursor: Application in orthopedic implants. *Bioactive Materials, 25*, 796–806. https://doi.org/10.1016/j.bioactmat.2022.06.020

Payne, J. (2019). *New Amsterdam replacement*. https://www.youtube.com/watch?v=9IElQuOARYM.

Pazhamannil, R. V., & Alkhedher, M. (2025). Advances in additive manufacturing for bone tissue engineering: Materials, design strategies, and applications. *Biomedical Materials (Bristol), 20*(1). https://doi.org/10.1088/1748-605X/ad9dce

Pérez, M., Carou, D., Rubio, E. M., & Teti, R. (2020). Current advances in additive manufacturing. In *Procedia CIRP, 88* (pp. 439–444). Spain: Elsevier B.V.. https://doi.org/10.1016/j.procir.2020.05.076. http://www.sciencedirect.com/science/journal/22128271

PLGA/HA filament résorbable et implantable. Available from: https://lattice-services.com/filament-resorbables-et-implantables/filament-plga-hydroxyapatite/.

PolyJet technology for 3D printing | Stratasys. Available from: https://www.stratasys.com/en/guide-to-3d-printing/technologies-and-materials/polyjet-technology/.

Qu, H., Fu, H., Han, Z., & Sun, Y. (2019). Biomaterials for bone tissue engineering scaffolds: A review. *RSC Advances, 9*(45), 26252–26262. https://doi.org/10.1039/c9ra05214c

38 Rapidia Metal 3D-water, 2019 Rapidia metal 3D-water bonding process. (2019).

Ryu, D. J., Jung, A., Ban, H. Y., Kwak, T. Y., Shin, E. J., Gweon, B., Lim, D., & Wang, J. H. (2021). Enhanced osseointegration through direct energy deposition porous coating for cementless orthopedic implant fixation. *Scientific Reports, 11*(1). https://doi.org/10.1038/s41598-021-01739-9

Schönegg, D., Essig, H., Husain, A. A.-H., Weber, F. E., & Valdec, S. (2024). Patient-specific beta-tricalcium phosphate scaffold for customized alveolar ridge augmentation: A case report. *International journal of implant dentistry, 10*(1). https://doi.org/10.1186/s40729-024-00541-2

Selective laser sintering (SLS) 3D printing technology overview. (2023). https://xometry.pro/en-eu/articles/3d-printing-sls-overview/.

SLS vs. DMLS: What you should know about these 3D printing methods. Available from: https://www.xometry.com/resources/3d-printing/sls-vs-dmls-3d-printing/.

Sub-pixel resolution for DLP 3D printing. (2016). https://www.youtube.com/watch?v=5qTAmPrHLow.

The 7 categories of additive manufacturing | additive manufacturing research group | Loughborough University. (2025). https://www.lboro.ac.uk/research/amrg/about/the7categoriesofadditivemanufacturing/.

The complete guide to multi jet fusion (MJF) 3D printing. (2022a), Available from: https://www.3dnatives.com/en/the-complete-guide-to-multi-jet-fusion-mjf-3d-printing/.

The role of polyether ether ketone in the medical industry. Available from: https://www.thogus.com/blog/post/the-role-of-polyether-ether-ketone-in-the-medical-industry/.

Three dimensional printed orthopedic implants | Providers | Blue Cross NC. Available from: https://www.bluecrossnc.com/providers/policies-guidelines-codes/commercial/surgery/updates/three-dimensional-printed-orthopedic-implants.

Total Talus replacement – Paragon 28. Available from: https://paragon28.com/products/total-talus-replacement/.

Venkatraman, V., Wong, M. K., Shalita, C., Parente, B., & Lad, S. P. (2020). Cobalt-induced toxicity and spasticity secondary to hip arthroplasty: Case report and review of the literature. *Cureus, 12*(12). https://doi.org/10.7759/cureus.12368

Wei, G., & Ma, P. X. (2004). Structure and properties of nano-hydroxyapatite/polymer composite scaffolds for bone tissue engineering. *Biomaterials, 25*(19), 4749–4757. https://doi.org/10.1016/j.biomaterials.2003.12.005

What are the main metal additive manufacturing processes? (2016). Available from: https://www.metal-am.com/introduction-to-metal-additive-manufacturing-and-3d-printing/metal-3d-printing-additive-manufacturing-processes/.

What is debinding and why is it used? Available from: https://proto3000.com/resources/faqs/what-is-debinding-and-why-is-it-used/.

What is multi jet fusion? - Design solutions. (2022). Available from: https://designsolutionsmag.co.uk/what-is-multi-jet-fusion/.

Willerth, S. (2017). *Synthetic biomaterials for engineering neural tissue from stem cells* (pp. 127–158). Elsevier BV. https://doi.org/10.1016/b978-0-12-811385-1.00006-6

Wu, Y., Liu, J., Lin, K., Tian, J., Zhang, X., Hu, J., Huang, Y., Liu, F., Wang, H., & Wu, Z. (2023). An overview of 3D printed metal implants in orthopedic applications: Present and future perspectives. *Heliyon, 9*(7), Article e17718. https://doi.org/10.1016/j.heliyon.2023.e17718

Zhang, S., Cheng, X., Yao, Y., Wei, Y., Han, C., Shi, Y., Wei, Q., & Zhang, Z. (2015). Porous niobium coatings fabricated with selective laser melting on titanium substrates: Preparation, characterization, and cell behavior. *Materials Science and Engineering: C, 53*, 50–59. https://doi.org/10.1016/j.msec.2015.04.005

Zhou, S., Tirichenko, I. S., Zhang, X., Hong, Y., Payne, H., Withers, P. J., Bouville, F., & Saiz, E. (2024). Embedded 3D printing of microstructured multi-material composites. *Matter, 7*(2), 668–684. https://doi.org/10.1016/j.matt.2023.10.031

CHAPTER

Frontiers of bone biomechanics: overview of challenges

24

David Burr[1], Matthew Allen[2,3], Alexander Robling[1] and Joseph Wallace[4]

[1]*Indiana University School of Medicine, Department of Anatomy, Cell Biology, and Physiology, Indiana Center for Musculoskeletal Health, Indianapolis, IN, United States;* [2]*Indiana University School of Medicine, Department of Anatomy, Cell Biology, and Physiology, Indianapolis, IN, United States;* [3]*Richard L. Roudebush Veterans' Administration Medical Center, Indianapolis, IN, United States;* [4]*Purdue University, Weldon School of Biomedical Engineering, Indiana Center for Musculoskeletal Research, Indianapolis, IN, United States*

1. Introduction

Biomechanical testing has become an essential tool for evaluating the mechanical properties of bone following various interventions, including mechanical loading, pharmaceutical treatment, or genetic manipulations, among others. The techniques used are often relatively straightforward, and the results are easy to understand. The techniques themselves, and the engineering basis for them, have been described many times, along with descriptions of required equipment, parameters and calculations (e.g., see (Turner & Burr, 1993)). Tensile tests are among the most accurate for assessing material properties, but preparation requires machining to a specified shape, which may require larger specimens than other tests. Perhaps the most commonly used tests are 3- and 4-point bending, which can be applied to smaller specimens. Given the prevalence of rodent models for bone studies, tests that can be applied to smaller specimens are often the best choice for standard mechanical evaluation. However, the testing set-up necessary to ensure pure bending (e.g., using a length to width ratio of 20:1) is not possible in mouse bone, and even in larger specimens is often not used, making most data a combination of both shear and bending. Still, even with a shorter region between loading points in 4-point bending, that region is still under pure bending as opposed to 3 point bending where no region of pure bending exists. Compression tests can also be used for smaller specimens, and are routinely used for vertebral bodies, which are primarily loaded in compression in vivo in humans, but not in most animal models. Torsion (which creates shear but also tension and compression), or other modes of pure shear testing using machined test specimens (e.g., Iosipescu or Arcan test methods) are less commonly used but can be informative in certain situations (i.e., testing fracture callus integrity). All these techniques have their drawbacks, based on bone tissue's composite nature and complex hierarchical arrangement, the difficulty in creating a pure stress

Biomechanics and Mechanobiology of Bone. https://doi.org/10.1016/B978-0-443-26739-0.00020-X

state in the bone under loading (e.g., torsion also creates tensile stress, bending generates shear stress), and in some cases sensitivity to experimental errors, giving the tests a relatively high variability (10%–20% in some cases). They are also not representative of physiological loading; long bones, although loaded in bending in vivo, are not loaded in 3- or 4-point bending. The bending moment typically comes from eccentric axial loading on a curved structure.

Other approaches can be useful, and may be more physiological for some applications, but due to the difficulty of preparation or length of time required for testing, are used less frequently. This is particularly true for fatigue testing, which, if performed within physiological levels of stress/strain and frequency, can require days to test a single specimen. Fatigue testing also requires more sophisticated equipment capable of loading cyclically. Impact testing can be useful for those regions of the body, such as the femoral neck, that are more susceptible to fracture from falls than from fatigue. Again, these tests require a more complex experimental setup than simpler loading modalities. There is also much less data in the literature on these methods making it challenging to interpret and integrate results into the greater body of literature.

Although material properties can be estimated from some of these approaches, mechanical properties at the tissue-level are best determined using tests specifically designed for microlevel assessment. Micro- and nanoindentation are relatively simple ways to evaluate bone hardness and modulus at the material level but provide limited information about other tissue-level properties. Properties can also vary widely from one location in bone to a spot just microns distant from it, so nanomeasurements require the averaging of many measurements, which in itself can cause variability in the final assessment. Fracture mechanics is often used to probe tissue-level properties. Fracture mechanics tests the propensity of bone to initiate and/or to propagate a defect, such as a microcrack. Again, such tests usually require specialized preparation (e.g., a well-characterized and reproducible notch) and equipment, and the results of such tests can be more difficult to interpret. Most useful are dynamic fracture mechanics approaches that define *how crack growth processes accompanied by rapidly occurring changes in loading alter crack growth and crack/structure geometry.* This includes characterization of energy fields at the edges and tip of cracks as they grow, strain rate sensitivity and 3-dimensional effects on cracks. These include both intrinsic and extrinsic factors that can complicate interpretation (Ritchie, 2011).

While the above testing approaches are incredibly useful, none gives a complete or necessarily totally accurate assessment of the properties of bone, nor do they integrate the different hierarchical levels of bone to provide a more holistic view of bone's behavior under all conditions. They seldom (or never) reflect the different ways that bone is loaded in real life situations. This chapter will consider future needs in biomechanical analysis, some of which have not been used or only used sparingly. These include cell loading, which can provide information about the conditions under which cells are responsive or silenced; additional micro- and nanomeasurements that better characterize the dynamics of bone tissue and interaction between the constitutive elements of which bone is composed; multimodal loading, which may be a better reflection of the diversity of ways that bone is loaded in real life; and additional ways of mechanically stimulating bone in vivo. While all of the current methods used to assess bone mechanical properties have specialized utility, the goal of this chapter is to encourage the field to invest in exploring additional ways that bone's properties might be evaluated, with the ultimate goal of integrating measurements at different hierarchical levels to provide a more complete picture of bone properties and fracture risk.

2. Cell loading

Improving structural and material properties in the skeleton are key determinants of skeletal health and fracture resistance, but it is important to understand how the cells that control bone tissue properties can be manipulated to improve bone's mechanical competence. That is, how can the innate biology of skeletal cells be harnessed to improve the accumulation of tissue and fortify bone mechanical properties? One way to do this is through pharmacologic intervention, and there are many FDA-approved agents—small molecules and biologics—that reduce bone loss and increase bone formation. However, all these agents work by improving bone mass indiscriminately across the entire skeleton, i.e., in places where more bone is needed (hip, spine, and wrist) but also in locations where bone preservation/formation is not needed. An alternative approach is to harness or accentuate the antiresorptive and proformation effects of mechanical loading or exercise, which target improvements in skeletal properties only at locations that receive mechanical stimulation. Thus, mechanical loading presents a self-guided targeting strategy to improve mechanical properties where they are needed, while concurrently leaving nonstimulated sites unperturbed. These site-specific benefits to bone mass and strength are well-described in the clinical and preclinical literature. For example, subjects engaged in gymnastics over a 12-month period added significant amounts of bone mass to their hip and lumbar spine compared with controls and swimmers, but their whole-body BMD was not different from either control group, suggesting that the addition of new bone is restricted to vigorously loaded sites, but is not added globally (Taaffe et al., 1997). It is important to know how local skeletal cell populations are stimulated during exercise, so that we have a better understanding of the cellular and molecular biology associated with mechanical improvement of the skeleton. To tackle the problem experimentally, it is crucial to devise models to recapitulate the in vivo cell loading environment in both intact animal models and in vitro cell culture approaches.

2.1 Animal models for studying skeletal cell loading

Our ability to experimentally alter the loading environment experienced by skeletal cells began by using intact animal models. This work, in conjunction with measured and modeled strain environments within the bone, provided the basis for our understanding of what normal, subnormal, and supraphysiological loads look like in the mammalian skeleton. Early work in this field started with basic exercise models in which experimental animals were subjected to different compelled exercise such as treadmill running (Fig. 24.1A). While those studies were informative in establishing a relationship between mechanical deformation and osteogenic responses, voluntary exercise studies are notoriously variable in terms of the nature of the strain environment among different experimental animals let alone the systemic effects of exercise incurred on other organ systems (e.g., cardiovascular, neurological) that might influence skeletal responses. Later models took a more focused and controlled approach to deformation of various skeletal elements, using surgically implanted pins that could be loaded using standard materials testing machines (Fig. 24.1B). Subsequently, noninvasive (i.e., nonsurgical) exogenous loading models were developed that applied well-controlled loads through skin and other soft tissues to deform long bones in predictable ways (Fig. 24.1C). Those noninvasive exogenous loading models, in combination with elegant, cell selective genetically engineered mouse models, provided guidance on cell types within the bone environment that are primarily implicated for driving the response to skeletal loading. In short, multiple lines of evidence pointed to the osteocyte as the primary mechanosensory cell type in bone. That conclusion is based on many studies in mice where crucial mechanotransduction genes have been selectively knocked out in the osteocyte population, creating severe deficiencies in mechanical signaling and bone adaptation (Dallas et al., 2018).

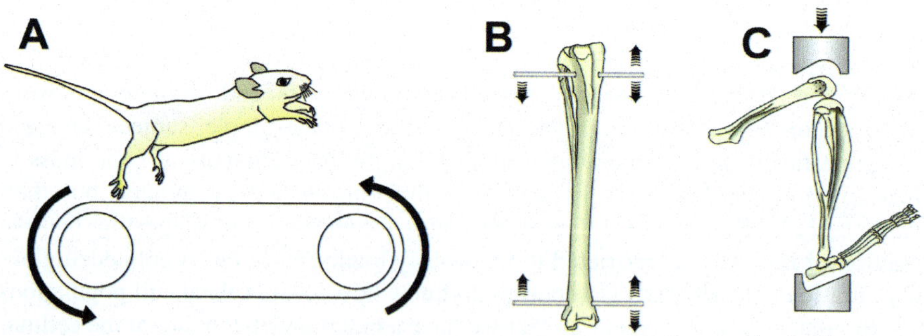

FIGURE 24.1

General categories of animal models that can be used to induce skeletal loading experimentally. (A) Exercise models, in this case treadmill running, are the most physiologically normal models, but they suffer from significant variation from animal to animal. (B) More precise skeletal loading can be achieved by surgically implanting pins at various regions of mammalian and avian skeletons, and those pins can be loaded with great control and precision using standard materials testing system. However, they are limited by the surgical procedures and potential regional acceleratory phenomena. (C) Noninvasive loading models have gained popularity in recent decades due to their lack of complications from surgery and natural bending environments induced by compression.

2.2 Translating the in vivo environment into in vitro cell loading models

Because of our current understanding that osteocytes are the primary sensor cell type in bone, it is of paramount concern to determine how we can replicate the parameters and conditions associated with mechanical stimulation of osteocytes in situ. Osteocytes are terminally differentiated osteoblasts that become encased in protein rich osteoid, which ultimately mineralizes and provides a mineral "shell" that surrounds the entire cell and all of its cell processes. The space between the osteocyte cell membrane and the mineral wall of the encasement is typically 80–100 nm, which provides enough space for extracellular fluid to bathe the cells, deliver nutrients, and transport waste to the venous system beyond the bone surfaces. These cells are suspended from the surrounding bony wall by a variety of structures, including integrins, proteoglycans, and glycolipids (Fig. 24.2).

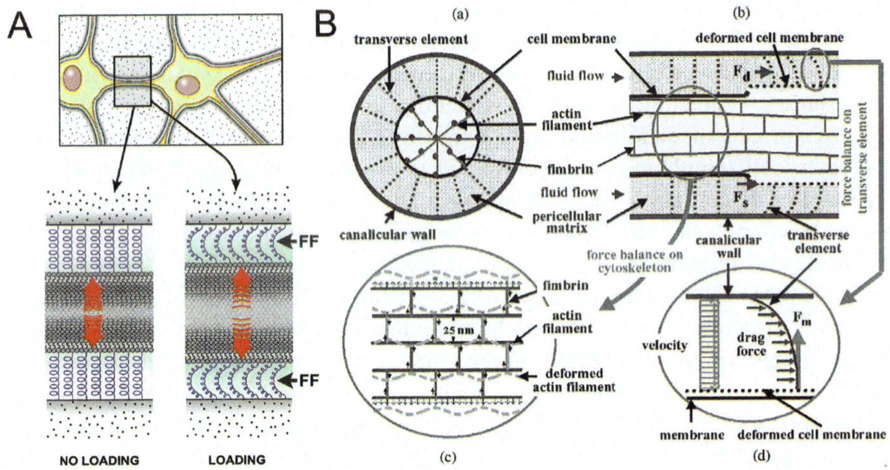

FIGURE 24.2

(A) Osteocytes exist in a unique environment within the bone tissue, surrounded by a mineralized cavity. The distance between the cell membrane and the cavity wall is roughly 100 nm, through which fluid moves. (B) Loading induces fluid movement in the extracellular space, which is postulated to deform tethering structures that connect the cell membranes to the canalicular wall, ultimately creating strains on the cell membrane that are amplified well beyond local bone matrix strains.

- Panel B is reproduced with permission from and Modified from You, L., Cowin, S.C., Schaffler, M.B. & Weinbaum, S. (2001). A model for strain amplification in the actin cytoskeleton of osteocytes due to fluid drag on pericellular matrix. Journal of Biomechanics, 34(11), 1375–1386. https://doi.org/10.1016/S0021-9290(01)00107-5.

Early attempts to model mechanical stimulation of bone cells in culture were focused on substrate deformation, applied to osteoblasts. At that time, it was thought that osteoblasts were the primary mechanosensory cell type in bone. In addition, these cells were easier to study experimentally (compared with osteocytes) as they are surface cells and are easily accessible for collection and culturing. Stretching the substrate beneath cultured osteoblasts can trigger a mechanotransduction response, but these cells require a great deal of substrate stretch, on the order of 1%–2% (10–20,000 με) to induce a response. In vivo, bone tissue will fracture well below that trigger point, at around 4000 με, so it is unlikely that osteoblasts in vivo would ever experience that much strain under normal, nontraumatic conditions. Given this disconnect, researchers turned to another stimulus that might explain the ability of bone to respond at strains much lower than those required to "turn on" osteoblasts. Here, the focus (appropriately) turned to the unique environment of the osteocyte, and the particular conditions associated with the fluid compartment around the cell. Computational modeling of the osteocyte in its surrounding mineralized matrix revealed a mechanism for normal strains to become amplified on the osteocyte cell membrane through fluid drag forces. Here, dynamic loading of the bone produces pressure gradients within the fluid compartment of the osteocyte network, which drives fluid movement from compressive areas to tensile areas during the loading cycle. Similar to fluid shear stresses that develop along the cell membranes of endothelial cells lining blood vessels, fluid movement across osteocyte cell membranes was postulated to be the driving mechanical stimulus for functional adaptation. Electron microscopic work conducted on long bone osteocytes in situ revealed the presence of very fine tethering structures that connect the osteocyte cell membrane to the canalicular wall. As these tethering structures traverse the extracellular area through which load-driven fluid moves, it was postulated that these structures serve to create hoop strains within the osteocyte cell process, thereby amplifying global strains on the bone matrix into much higher hoop strains on the osteocyte cell processes (Fig. 24.2). It is worthwhile to point out that subsequent work revealed that the osteocyte cell processes are much more sensitive to mechanical perturbation than are the cell bodies, suggesting that much of the mechanoreception occurs along osteocyte cell processes in the canaliculi. Hoop strains along the cell processes are likely to trigger the opening of mechanosensitive channels, which have manifold downstream consequences for the entire process of mechanotransduction.

Discussion about these tethering elements, and the disproportionate sensitivity of the osteocytes' cell processes, are relevant as we look to the future of in vitro experimentation for bone cell stimulation. The state of the field today is such that there is no standard cell culture model that reliably recapitulates complete entombment of the cells, with development of associated tethering structures, so that fluid forces can be applied in a way that models stimulated osteocytes in vivo. The parallel plate flow chamber has been used for decades to study fluid shear stimulation of a variety of cell types, including endothelial cells, osteoblasts, and osteocytes (Fig. 24.3). While that model represents an overall improvement compared with the substrate

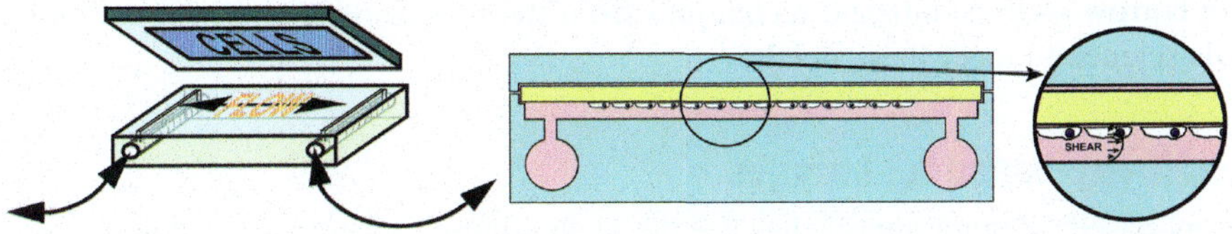

FIGURE 24.3

Idealized diagram of the parallel plate flow chamber, which is used to generate shear stress on cultured cell membranes. Flow can be unidirectional or oscillate back and forth across the cell bed. This approach has been useful but does not capture the three-dimensional environment in which osteocytes reside.

strain approach (at least in the context of bone cells), it fails to recapitulate the three-dimensional in vivo osteocyte environment because fluid passes along the entire apical membrane of cultured cells adhered to the support slide, without involvement of tethering structures along dendrites. Some approaches to remedying this problem have included the use of polymerizable scaffolds (collagen, alginate), in which cultured osteocytes can be embedded before polymerization, and can potentially mineralize the scaffold so that a more in vivo environment is achieved. Another area that has potential for development and widespread use is 3D bioprinting, where cell printers are capable of printing live osteocytes within coprinted bioinks of various composition. As methodological improvements in creating more "natural" in vitro environments for cultured osteocytes are made, we will undoubtedly gain a better understanding of exactly how osteocytes can be optimally stimulated, and in the process, reveal key signal transduction cascades that can be targeted to improve overall bone properties.

3. Nanoscale and microscale measurements in bone

As a tissue and structure, bone is somewhat unique in that it is composed of organic and inorganic building blocks. Being an elegant two-phase composite material allows bone to balance strength and toughness, for which there is usually a trade-off. In addition, organizing that material into a hierarchical structure allows for less material to meet the mechanical demands of the body. One downside of this arrangement is the difficulty in understanding how the various levels of that structural hierarchy contribute to the large-scale mechanical properties of a whole bone. Structural-level tests at the whole bone level are functionally important but dissecting out the impacts of tissue properties from size/shape/organization are hampered by the engineering principals underlying the tests. As the importance of tissue quality, physical properties of the bone material itself independent of quantity, is becoming more appreciated, developing methods of characterizing the microscopic (and even nanoscopic) properties of the tissue is critical. Fortunately, we can borrow well-established techniques from the worlds of engineering and materials science to meet these needs.

3.1 X-ray-based techniques

X-ray-based imaging approaches are the most common way to characterize the internal microstructure of a bone because of bone's ability to attenuate X-rays. Micro-computed tomography (μCT) has become widespread across preclinical studies, and high-resolution peripheral quantitative computed tomography (HR-pQCT) has allowed that expansion to continue into clinical research. While useful from a structural perspective, neither technique does much to expand our understanding of bone quality or function with the exception of measuring volumetric tissue mineral density, a property most often positively correlated with bone stiffness and strength (Turner, 2006). Beyond using the attenuation of X-rays to image bone, other X-ray-based techniques provide information on mineral crystal structure, composition, and physical properties of collagen and mineral at a much smaller length scale. Small-angle and wide-angle X-ray Scattering (SAXS and WAXS, respectively although WAXS is sometimes referred to as X-ray diffraction or XRD) are based on the elastic scattering of an X-ray beam as it passes through a sample. As their names imply, the difference between the techniques is related to the angle of scattering. SAXS generates data related to nm-range inhomogeneities at low angles of scatter (<6–7 degrees) but can also provide larger-scale information (up to hundreds of nanometers). This can include information about the shape and size of macromolecules or characteristic distances of partially ordered materials and as opposed to XRD, the sample does not have to be crystalline. In bone, the most frequently reported properties are related to collagen—either the D-spacing of collagen fibrils or the relative orientation of fibrils relative to the long axis of the sample (Puxkandl et al., 2002). In XRD or WAXS, atomic planes of a crystal cause an incident beam of X-rays to constructively interfere with one another as they scatter and leave the crystal, allowing for the determination of interatomic distances in the crystal lattice. By focusing on specific peaks within the diffractograms, one can further calculate the size and shape of crystalline regions in the sample (Boskey, 2003).

This description of SAXS and WAXS would imply that only structural quantitative data are possible using these techniques. However, a bit of creativity can allow these methods to give functional mechanical information. As an example, when a bone sample is loaded in tension, SAXS can be used to track collagen fibril strain (% change in D-spacing with loading), WAXS to track mineral strain (shifts in the position of the 002 c-axis plane of hydroxyapatite), and markers placed on the sample can be used to measure global strain (Gupta et al., 2006). Together, ratios of these strain measures can be used to determine the relative deformation of collagen and mineral when the sample is loaded.

3.2 Electron-based techniques

Electron microscopy can also be used to visualize bone at small length scales with high resolution, although the functional measures these techniques can provide are limited. Because of the wavelength of moving electrons, resolution on the order of hundreds of picometers is possible. Transmission electron microscopy (TEM) was the first form developed and requires samples that are extremely thin (<100 s nm) so as to be semitransparent to electrons. Imaging takes place in a vacuum, so biological samples are usually fixed, dehydrated, embedded, and then negatively stained to enhance contrast. Some newer techniques can avoid harsh processing by cutting samples at low temperature and then vitrifying in liquid ethane before imaging in cryoTEM (Quan & Sone, 2013). Several imaging modalities exist including bright field, electron energy loss, phase contrast, and diffraction contrast but little beyond crystal lattice spacing (similar to WAXS) and high-resolution images of tissue ultrastructure are possible.

TEM is not possible with thick samples but in scanning electron microscopy (SEM), the electron beam is raster-scanned across the surface of the sample. As electrons interact with the sample, energy is lost and can be converted into other forms of energy. The most common SEM mode collects low energy secondary electrons from loosely bound conduction bands close to the surface to produce high-resolution images (on the order of 1 nm) with a large depth of field, which gives the images a 3D appearance. Higher energy backscattered electrons come from deeper in the sample as incident electrons interact with the nucleus of atoms. The intensity of the resulting image is related to the atomic number of what the electrons interact with so in addition to topographic images similar to secondary electron imaging, backscatter SEM can provide information about tissue composition when proper standards are used for normalization (Hartmann et al., 2021). With enough incident energy, X-rays can be collected and used for energy-dispersive X-ray spectroscopy (EDS, EDX). X-ray energies are characteristic of the difference in energy between inner and outer shells, which allows elemental composition to be measured (Scimeca et al., 2018). Like TEM, dehydration and fixation are typically required since a high vacuum is needed for imaging. The sample surface must also be electrically conductive and grounded to prevent charge buildup so samples are usually coated with a thin layer of gold or carbon, which can obscure very small surface features. To bypass some of this processing, environmental SEM allows samples to be imaged in a low-pressure gaseous environment with high humidity, and surface coating is not required.

Although no direct mechanical measures can be made with EM, SEM can be used in conjunction with other techniques to derive functional information. In a fracture toughness test, a sharp notch is placed in a sample before loading the sample to failure by forcing the notch to open in tension. Ideally, the crack would be imaged during the test and crack propagation would be measured directly from these images. This is often not possible in small biological samples so in some tests, the fracture surface is imaged using SEM to extract information about how the crack propagated (Ritchie et al., 2008). Recent studies have been performed loading within an environmental SEM, which allows researchers to overcome some of these limitations by imaging small samples at high resolution in a moist environment as the sample is being loaded (Carriero et al., 2014).

3.3 Vibrational spectroscopy

Within materials, molecular bonds are constantly vibrating. The bonds can be excited by electromagnetic radiation and depending on the type of radiation (e. g., microwaves, infrared, etc.), the bonds can rotate, vibrate/stretch, or ionization can occur. By measuring the changes in energy associated with these bonds, vibrational spectroscopy can be used to create a compositional fingerprint of the sample. Two primary types of spectroscopy, Fourier Transform Infrared (FTIR) and Raman, are most often employed in bone research. FTIR uses IR radiation to alter bond vibrations and change their dipole moment. Depending on the type of molecular bone present, IR radiation is absorbed differently at a given frequency. By subjecting a sample to a large range of frequencies, one can determine the types of molecular bonds that are present. Raman spectroscopy uses visible light (in the form of a polarized monochromatic laser) to excite valance electrons and change their polarizability. These electrons shift the light characteristics of bonds present through inelastic scattering known as the Raman effect. By measuring how much the light is shifted, one can determine the types of molecular bonds that are present. Whether using FTIR or Raman, properties related to collagen, mineral, noncollagenous proteins, lipids, and water can be determined (Taylor & Donnelly, 2020). Rather than being absolute measures of any of these bone components, most outcome measures are relative concentrations of the various constituents in the form of ratios of either peak intensities, or peak areas. Neither Raman spectroscopy nor FTIR are better for the chemical characterization of bone, as both have strengths and weaknesses. Most FTIR systems work by passing radiation through the sample on its way to a detector. This requires the sample to be thin (for bone, perhaps 2–4 μm thick), which means the bone must be prepared for undecalcified histology by dehydrating, embedding, and sectioning. As with EM above, this processing can distort aspects of the sample one is trying to measure. The removal of all water is critical since water has a strong IR signal that can interfere with measurements. However, because of the short time needed for each measurement, spatial compositional maps are possible and at high enough resolution to capture local tissue heterogeneity. Raman can be performed on native samples surfaces with very little sample preparation. However, because of the relatively low signal to noise ratio, multiple spectra at each location in a sample need to be acquired and averaged, making imaging time longer than FTIR.

Regardless of which method is used, neither Raman nor FTIR have the ability to generate direct information related to the mechanical performance of bone tissue or structure. Despite this limitation, efforts have been made to use spectroscopic techniques to predict bulk mechanical properties with various levels of success.

4. Indentation-based assessment of nano- and micromechanical properties

The methods discussed to this point have a limitation in common—the inability to directly measure functional mechanical properties. In recent years, this limitation has been addressed with high-resolution mechanical probing using either instrumented nanoindentation (Managuli et al., 2023), or indentation based on atomic force microscopy (AFM) (Thurner, 2009). Either technique can be performed in force- or displacement-control, and both are adept at generating measures of true tissue stiffness (elastic modulus). AFM allows for characterization of nanoscale properties in bone with forces and deformations firmly below 100 nN and 100 nm, respectively, because the probes used for indentation have a radius of curvature less the 100 nm. In instrumented indentation, typical forces for indenting bone exceed 1 mN with indentation depths over 1 μm. Pyramidal Berkovich tips are often used but are sometimes replaced with spherical probes with a radius of up to 5 μm. These large forces and probes limit the "resolution" of the indentations to larger features and in practice should be considered microscale tissue measures. AFM can overcome some of these limitations allowing for characterization of nanoscale features including those of individual collagen fibrils (Wallace, 2012), or mapping tissue properties surrounding osteocytes (useful to study lacunar-canalicular remodeling following lactation, for instance (Vahidi et al., 2021). The most readily available and well-established property generated from indentation testing is elastic modulus. Using models established from the field of contact mechanics, force, indentation depth, and information related to the shape of the indentation probes are used to fit the loading (or preferably the unloading) curve. Beyond modulus, hardness is typically reported in instrumented indentation studies. This measure is related to stress (and like stress is calculated as force normalized by indentation area). Both modulus and hardness are considered preyield properties, which, while important, limit the extent of what we can interpret from the test. Efforts to extend to additional properties are ongoing (e.g., modeling creep behavior during a peak-load holding phase, area between the loading and unloading curve as a measure of energy dissipation, etc (Managuli et al., 2023). These properties have not been fully validated. Even less has been done in the area of Dynamic Mechanical Analysis (DMA) at the nanoscale. This technique typically superimposes an oscillating load on top of static load at increasing frequencies. By measuring the offset between the applied load and resulting displacement, information related to both the elastic and viscous aspects of bone can be ascertained. Given that bone is mildly viscoelastic, and the role water plays in bone mechanics is growing in interest, additional work in this area is needed.

5. Physiological ways to study mechanical properties ex vivo

Bone mechanical properties are an important outcome of studies aimed at revealing mechanisms of bone development/adaptation and skeletal disease, and the effects of interventions. One goal of any mechanical assessment is to provide insight into fracture risk. Despite the abundance of data in the literature, fractures are complicated events and controlled laboratory tests will always need to be carefully interpreted in their translation to the clinic and population health.

Most bone mechanical data in the current literature are derived from monotonic tests, in which loads or displacements are imparted to a bone at a controlled rate and in a single direction, and the resulting load versus displacement (or time) are recorded. Such tests are conducted on long bones in bending (3 or 4 point) or on vertebral bodies in compression. There are studies that utilize other testing methods, including impact, fatigue, torsion, and multiaxial loading (compression and torsion combined), but these are less frequently used. One key advantage of monotonic testing is that it can be experimentally reproduced over time and across laboratories making it a validated method to compare data. Monotonic tests are typically conducted at a controlled displacement rate (typically on the order of 1 mm/min) that permits both the organic and mineral phases to influence the test outcomes. Because operator inputs can also influence outcomes (most notably loading rate), the experimental conditions are generally standardized such that modest loading rates are used. The limitations of monotonic testing are well-known, yet often downplayed/ignored in data interpretation (including by the authors of this chapter!). The greatest limitation is that fractures are rarely due to a monotonic load. Other essential testing features, including having a uniform cross-section at the region of testing and assuring the contact points are equally loaded, are often difficult or impossible to achieve with bones.

Compression testing of vertebrae often involves the manipulation of vertebral structure (removing end plates, neural arches, pedicles, intervertebral disks), structures that are key to the overall mechanical integrity of the vertebral column. Although such manipulation can help standardize testing, the absence of important structures known to contribute to mechanical function and skeletal integrity, limits the translational value.

So, although there is, and always will be, value in quasi-static monotonic testing for bone evaluation, it seems reasonable to ask whether we can do "better?"

5.1 Impact testing

Fractures are usually caused by impacts.

Currey, 1962

Bone's composite nature makes it good at resisting both high loading-rate impacts and fatigue loading over a lifetime of exposure to such loading modalities

(Stock, 2015; Wagermaier et al., 2015). Impact fractures are common across the skeleton and occur with impact from objects, such as in a motor vehicle accident, or the ground when a person falls. They mainly affect the femoral neck, distal radius, or any of the long bones. Impact testing was a common mode of mechanical testing in the latter half of the last century. At that time, impact testing typically involved controlled drops of weighted hammers that impacted machined bone specimens (often bovine or human bone) in compression (A) or bending. Over time, test set-ups were adapted to focus on whole bone tests, modeling impacts from falls at clinically relevant sites such as the femoral neck (B). The historical literature (now 30–50 years old), by Currey, Hayes, Carter (and others) defined key features such as the strain-dependent nature of impact fracture (Courtney et al., 1994), age-associated loss of impact energy (Currey, 1979), and the higher strength of bone in impact versus static (monotonic) strength (Saha & Hayes, 1976) Fig. 24.4.

Impact testing provided much of the foundational understanding of how bone structure influences mechanical properties and is still used commonly today in material science/engineering to study bone as a material, and in studies investigating trauma from vehicular accidents. A PubMed search of impact studies over the past decade revealed only one biologically focused study, from 2021. The work was designed to study the effect of rest-insertion on mechanical properties by comparing in vivo impact and fatigue loading in rats (Yan et al., 2021). Fatigue loading in rodents produces two well-known responses, increased woven bone and microdamage. Previous work on this topic demonstrated partial recovery of mechanical properties (determined by monotonic bending tests) when rest days were allowed during in vivo loading. In contrast, when impact loading was used to assess mechanical properties, there was no sign of recovery with rest insertion. The authors concluded that the results of quasi-static bending tests, in which the addition of woven bone appears to provide a protective effect, do not apply to those from impact loading.

The reason that impact testing has fallen out of favor in the biological literature is unclear yet is an area important to address as exemplified by work from (Zioupos et al., 2008). Designed to assess the concept of strain-rate dependent properties, this work tested specimens at multiple strain rates and assessed the development of microdamage. The authors concluded that strain rate has a stronger effect on post-yield deformation than on the initiation of macroscopic yielding during the preyield phase of failure. This highlights how studying bone failure in relatively slow monotonic bending may not fully reveal how bone is damaged and ultimately fails during high-rate impact.

5.2 Load interaction

Interaction is potentially an important part for the unpredictability of in vivo bone fracture.

Fyhrie, 2015

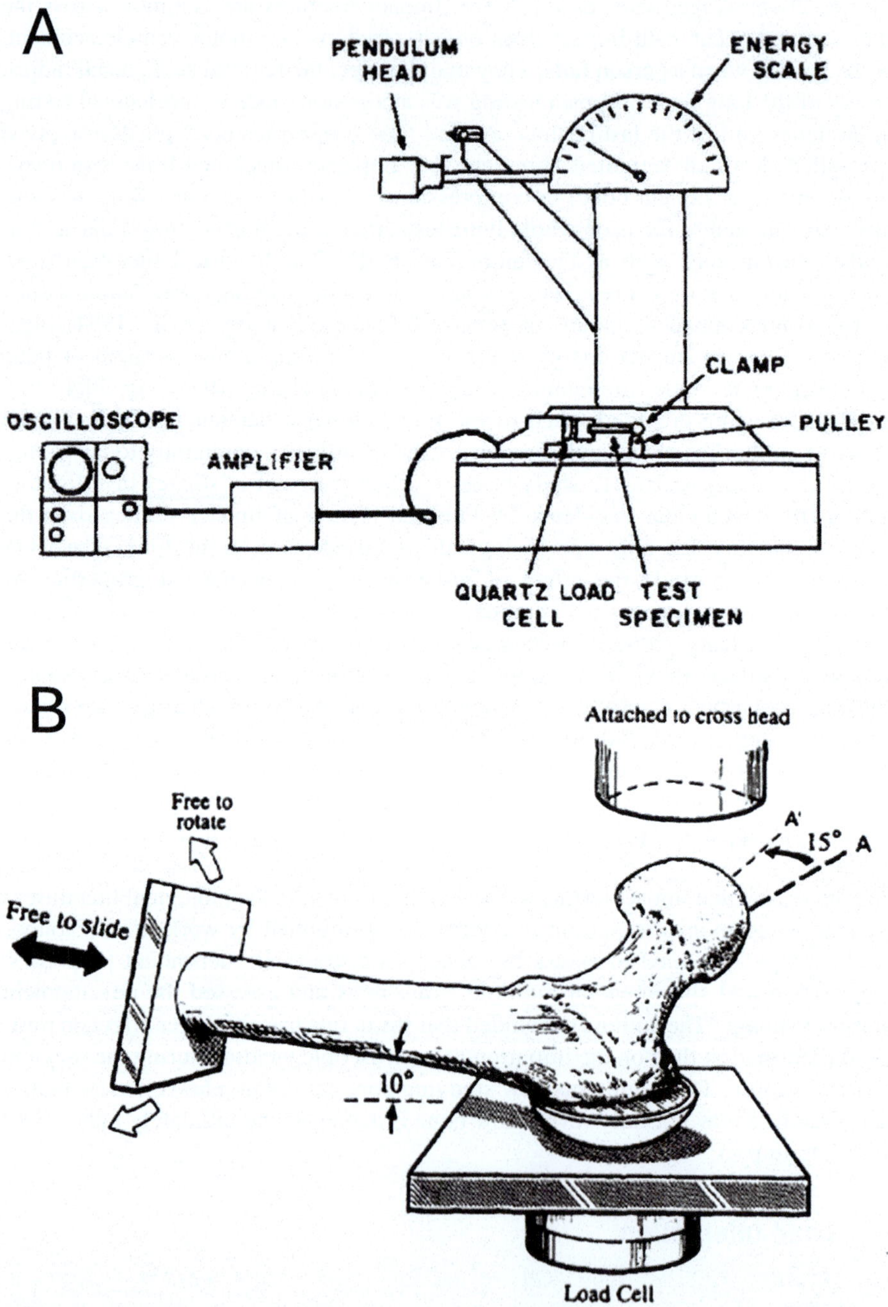

FIGURE 24.4 Impact testing of bone has evolved from controlled weight drop set-ups on machined specimens.

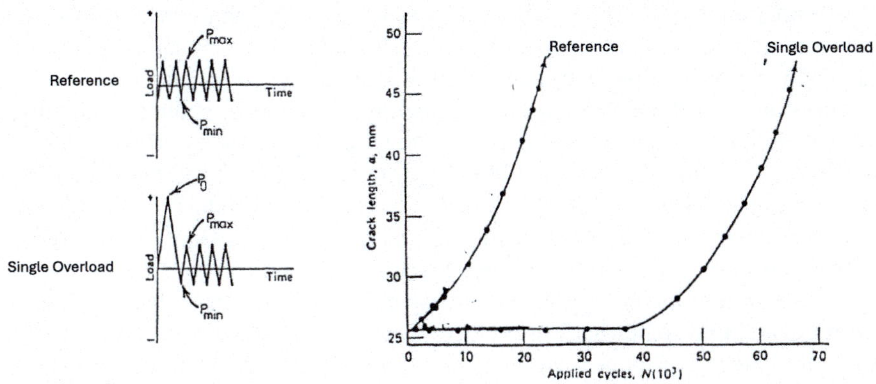

FIGURE 24.5

Mechanical interaction has complex effects on the mechanical properties of materials. These effects are illustrated, in 2024-T3 Al alloy, by the suppression of crack growth when a single overload precedes fatigue. Whether such interaction exists in bone is unknown.

- Adapted from Skorupa, M., (1998). Load interaction effects during fatigue crack growth under variable amplitude loading—a literature review. Part I: Empirical trends. Fatigue and Fracture of Engineering Materials and Structures. 8756-758X. 21 (8), 987–1006, https://doi.org/10.1046/j.1460-2695.1998.00083.x.

Composite materials can be modeled after bone—a field called biomimetics - designed with the goal of resisting both high impact and chronic fatigue loads. Composite materials (as well as metals) are subject to "loading interaction" where the insertion of an impact/overload event on top of a background of fatigue loading alters fatigue crack growth and residual mechanical properties (Skorupa, 1998, 1999). For example, a test sample subjected to a subcritical impact load, followed by a period of fatigue loading, displays considerably slower crack growth relative to fatigue loading alone (Fig. 24.5). Such tests illustrate the complex interaction between loading modalities and how studying a single modality likely presents an incomplete picture of true mechanical behavior. Whether interactions between these common loading modalities influence properties in bone is unknown (Fyhrie & Christiansen, 2015; Mirzaali et al., 2015), but given its composite nature and lifetime exposure to loading, it seems likely.

(A) to more sophisticated devices focused on clinically relevant sites such as the hip (B). Panels reproduced with permission from (A) (Saha & Hayes, 1976) and (B) (Courtney et al., 1994).

- (a), Saha, S. & Hayes, W.C. (1976). Tensile impact properties of human compact bone. Journal of Biomechanics, 9(4), 243–251. https://doi.org/10.1016/0021-9290(76)90010-5. (b). Courtney, A.C., Wachtel, E.F., Myers, E.R. & Hayes, W.C. (1994). Effects of loading rate on strength of the proximal femur. Calcified Tissue International, 55(1), 53–58. https://doi.org/10.1007/BF00310169.

Consider a femoral neck fracture that occurs in an aged individual due to a fall. The femoral neck of this individual would undoubtedly have contained microdamage, the result of both previous subcritical falls/impacts and fatigue loading due to weight-bearing. Although damage accumulation within a bone reduces both fatigue life and residual mechanical properties (Carter & Hayes, 1977), it's possible that some damage within the tissue could have protective effects when the bone is subjected to impact. Thus, events that modify microdamage production/removal may have more complex effects on bone mechanical properties than have been described using either monotonic, fatigue or impact testing in isolation.

Fatigue testing has been utilized extensively in both the early bone mechanics literature, and to a somewhat lesser extent in preclinical studies. In fatigue testing, bones are loaded at submaximal loads until failure, often defined as a 30% loss of stiffness. The underlying premise is that through low load/strains, microdamage is created and propagated until failure. A notable "challenge" for fatigue testing is that it can take a long time, depending on the bone and applied stresses. Human cortical bone specimens have been cycled at physiological strain and strain rate, though slightly higher frequencies (4 Hz) for one million cycles (almost 3 days) (Schaffler et al., 1989); bovine cortical bone has been cycled for over 20 million cycles (almost 19 days) (Schaffler et al., 1990). This can be impractical, especially for preclinical studies where overall experimental sizes can be 30–50 specimens. There is also the limitation that skeletal tissues will degrade over time even if the samples are hydrated, and in a temperature-controlled environment. One benefit of mechanical interaction studies is that they could assess the contributions of fatigue and impact using a shorter test duration.

Another advantage of mechanical interaction tests is that they would at least partially overcome limitations of fatigue experiments, which assess how internal flaws propagate. The skeletal elements of rodents, the most commonly used preclinical models, typically do not generate or accumulate in vivo microdamage unless it's imparted through intervention (such as loading). Without such "baseline" damage, fatigue loading interpretation can be challenging. Imparting damage through impact, followed by assessing the propagation of that damage with fatigue loading may be a more physiological way to mimic human bone failure mechanisms.

Some insight into mechanical interaction can be gained from experiments in which fatigue loading (to induce stiffness loss) is followed by impact loading. Reilly and Currey (2000) showed that damage formed in compression is highly detrimental to tensile mechanical properties suggesting that it is advantageous for bone to undergo only one mode of loading in vivo. Other similar experiments have been performed using fatigue loading followed by notched-specimen fracture mechanics testing (Yeni & Fyhrie, 2002). One key feature of these, however, is that fatigue-induced damage was a controlled process, and stopped at a certain point that was matched across specimens. This is different from a more "realistic" interaction experiment in which load and fatigue are imparted in a less controlled fashion and intermixed. For example, experiments that utilize a fatigue load for a given number of cycles, then impart an impact, followed by additional fatigue loading could more realistically assess the interaction of properties that form and propagate damage/failure under physiological loading conditions.

Together with the previous section suggesting expanding the use of impact mechanical testing, we further challenge our colleagues to take up the challenge of studying mechanical interactions in bone. Such interactions are simple to assess in theory, yet the combination of variables and their timing are numerous and likely would necessitate an organized collection of both biological and computational experiments. While this is not a trivial undertaking, it's possible that such a comprehensive assessment would reveal novel mechanical phenotypes that could uncover new material property interactions that could serve as interventional/therapeutic targets.

6. More physiological ways to study mechanical adaptation in vivo

Historically, many approaches have been used to evaluate the processes of skeletal adaptation to loading, and the features of the strain environment that exert some control over these processes. These include surgical approaches (ulnar osteotomy [canine] or insertion of pins through which bending or axial loads can be applied to either long bones or vertebrae [rabbit, turkey, rat]) or external loading approaches that involve 4-point bending [rat tibia], longitudinal loading from the ends of the long bones to create axial compression and bending moments [forelimb or tibia], or medio-lateral cantilever bending [mouse, rat] (Fig. 24.1, also see (Robling et al., 2019, pp. 203–233) for a review of these). Noninvasive, high- or low-frequency vibrational loading has been used in both animal models [e.g., sheep] and in humans (Gilsanz et al., 2006; Rajapakse et al., 2021; Spain et al., 2021). Using these models, a lot has been learned about how bone responds to specific aspects of the load/strain environment, allowing exploration of the roles individual aspects of the strain environment play (magnitude, frequency, rate, duration, periodicity between loads, strain distribution mode or polarization of the strain). However, the models do not satisfactorily reflect in vivo physiological loading conditions. Surgical approaches create responses unrelated to the mechanical loading environment—sometimes termed Regional Acceleratory Phenomena (RAP). These responses originate as local phenomena but can affect the contralateral limb or bones adjacent to the surgically manipulated bone as well. In some cases, these responses occur because of altered normal loading of the bones following surgery, but in other cases through processes of signaling (either between bones themselves, or between muscle and bones) that are not fully understood. For the most part, they involve loading in ways that are entirely nonphysiological. For instance, no bone in the body is physiologically loaded in 3- or 4-point bending in vivo. Therefore, the models provide a limited view of the adaptation of bone, or the mechanical properties of bone, as they occur during life.

The question arises: Are there methods of loading that can be used that are more physiological in that they replicate the types of loads experienced in vivo?

7. Multiaxial and multimodal loading

... a uniaxial loading mode... fails to capture the complexities of in vivo loading.
George et al., revision under review

Bone is rarely, if ever, loaded in a single direction. Although the primary loads on vertebrae are compressive, the curvature of the spine necessarily creates bending and shear stresses, and this is particularly true within the trabecular compartment. The directional nature of trabeculae in all areas of bone predispose those struts to multidirectional stresses. Understanding how this may affect bone adaptation is important not only to assessing fracture risk but can also be important to understand the mechanisms (or prevention of) adaptation at the junction between bone and a prosthesis. It is known that uniaxial loading models overestimate the fatigue life of bone (George & Vashishth, 2006) in part because multiaxial loading contributes to greater and more rapid loss of stiffness (Zysset & Curnier, 1996).

Mixed mode loading is more realistic of normal gait than uniaxial loading in that it incorporates a combination of axial and torsional loads (Burr et al., 1996; Vashishth et al., 2001). Torsion is known to reduce apparent fatigue life by 5–10 fold over uniaxial loading (Fatihhi et al., 2015; George & Vashishth, 2006). Therefore, it may provide a better representation of bone's properties and adaptive responses in vivo if methods of multiaxial loading are incorporated into experimental protocols. There is no doubt that complex loading would require sophisticated, perhaps novel, engineering approaches to analyze outcomes, and could confound interpretation of the results. However, physiology is complex, and complex loading could lead to a more valid comprehension of in vivo loading environments. This could also be used in other contexts, to study damage accumulation or age-related changes in bone under these complex loads.

These approaches could use some established databases to model multiaxial loading responses. Recently, George et al. (in revision) used in vivo human tibial strain gage data with four different failure criteria models to show that fatigue failure could occur within 5000–200,000 cycles of loading, depending on the activity. This is a much more reasonable estimate than prior studies based on ex vivo uniaxial loading (Schaffler et al., 1990). Further, they showed a 3.5-fold shorter fatigue life in older people (mean age 71 years, 35,000 loading cycles) compared with younger people (mean age 39 years, 125,000 loading cycles). In older people, this represents about 22 miles of running at a moderate pace. Of course, this estimate depends on many factors, including bone mineral density (BMD), bone material properties (in part determined by mineralization, collagen cross-linking, water content) and the physical condition of the individual, since muscle fatigue has been shown to affect bone strains in long-term physical activity (Clansey et al., 2012; Milgrom et al., 2007; Yoshikawa et al., 1994). Nevertheless, this is consistent with data showing that that there is a component of fracture risk that increases in older people independent of changes in BMD (Burr, 2019; Hui et al., 1988; Kanis et al., 2001).

New multiaxial methods, and perhaps new animal and computational models, will need to be developed to make these approaches feasible and interpretable. Models that include maximum shear strain rather than stress-based models may be a starting point as bone failure is dominated by shear-induced plastic strain (Taylor et al., 2003).

7.1 In vivo muscle stimulation

Complex interaction patterns [and pleiotropic genes that both regulate bone and muscle] integrate muscle and bone into such organismic interactions that still need to be unraveled in the sense of a systems biology network analysis.

Herrmann et al., 2020

Muscles exert greater forces on bone than does gravity, so it makes sense that in vivo loading to study mechanical adaptation should use muscle to provide the stimulus leading to adaptive remodeling. Using muscles to provide in vivo loading can also incorporate multiaxial and/or off-axis loading that is more physiologically relevant to everyday activities. Skeletal muscle contraction involves both longitudinal and lateral forces, unlike application of external loads in specific directions, and allows variation in magnitudes and velocities generated by concentric, eccentric, or isometric muscular contractions. It more accurately models location and direction specific loads (compressive, tensile and bending) that occur during physical activities, and allows the stimulation of several muscles to examine synergistic effects. It is well-known that there is a close link between muscle strength and bone mineral density and strength (and size during growth). This relationship could be exploited while also providing a more realistic loading scenario with which to study skeletal adaptation. Although more realistic than current mechanical loading approaches, one limitation is that electrical stimulation of muscle recruits large diameter fibers, whereas activity-related muscle activity first activates small muscle fibers. This could alter the response of bone to the load.

We now know there is significant muscle–bone cross talk (see (Goodman et al., 2015; Hart et al., 2017; Herrmann et al., 2020) for more detailed discussions of this). Therefore, when loading of bone is performed without muscle contraction, the skeletal response is not subject to cross talk influences and the local response may be different than when muscle loading causes release of factors that contribute to the bone response. Moreover, limbs that are not directly loaded can also be affected, likely because of soluble factors that are released either from muscle or bone upon loading and are distributed systemically (Zhao et al., 2021). This may be the explanation for the observation that bone not experimentally loaded and even at a distance from the loaded bone will sometimes also respond to the loading stimulus (Bab et al., 1985). This also means that loading bone through muscle might produce adaptive results that are not directly related to the mechanical stimulus, but which secondarily allow for other secretory factors that could affect both bone and muscle. Loading that is performed through muscle contraction may not have the same effects on gene transcription (or effects on pleiotropic genes that regulate both muscle and bone); cytoskeletal, nuclear and cell membrane changes; and changes in muscle and bone secretomes as external loading constructs. Loading through muscle stimulation could provide a more accurate picture of both bone and muscle response, and the interactions of bone and muscle. Physiological loading through direct muscle stimulation therefore provides a better approach to systems biology. Because both muscles and bones can sense load, and because we now know there is substantial cross talk between them outside of the mechanical stimulus itself, a clear picture of bone and muscle adaptation can only be achieved when they are loaded in synchrony.

Electrical stimulation of muscles to study bone mechanical adaptation avoids the invasiveness of some current in vivo loading models (e.g., that use pins, or osteotomies). Loading in this case can be performed by stimulation through noninvasive surface electrodes, or—even better—through needle electrodes (Berman et al., 2020; Marquez-Chin & Popovic, 2020). This also can be used to deliver stimulation to specific nerves rather than to whole muscle, which requires lower power and reduces the risk of tissue damage (Peckham & Knutson, 2005). This might avoid the complication that muscle stimulation can induce rapid muscle fatigue (Dolbow et al., 2014).

Although aimed at using functional electrical muscle stimulation for the treatment of spinal cord injuries Dorrian et al., 2023), provide an extensive analysis of the parameters for electrical stimulation, early and later functional outcomes, adverse events associated with stimulation, analysis methods, and potential mechanisms through which stimulation may work.

7.2 Impact loading

Most fractures of bone are not the result of the application of slowly applied load.

Currey, 1979

Most osteoporotic fractures occur either in the hip, wrist, or spine and these areas should be targeted for the development of appropriate in vivo tests that employ methods that recognize the causes and specific location of the fracture. It is important to utilize the appropriate mode of loading to evaluate fracture risk and adaptive remodeling at sites that are relevant to the information that is desired. This is nowhere more important than for the hip, a common site of age-related fracture in both women and men. The hip and radius are more susceptible to fractures due to impact loads from falling, especially in specific directions, than to preexisting damage or fatigue processes (Grisso et al., 1991; Parkkari et al., 1999) about 90% of hip fractures are associated with falling (Scott, Wagar and Elliott, 2010), although only a small fraction of falls result in fracture. While it has been suggested that quasi-static loading may provide reasonable estimates of results from impact loading

Table 24.1 Variation in different mechanical tests.

Test	Coefficient of variation
Elastic modulus, bending	9.4
Strength, bending	9.4
Elastic modulus, tension	10.5
Strength, tension	13.8
Work, bending	27.3
Work of fracture	27.0
Energy absorption, impact	38.1
Work, tension	38.9

From Currey, J.D., Pitchford, J.W. & Baxter, P.D. (2007). Variability of the mechanical properties of bone, and its evolutionary consequences. Journal of the Royal Society Interface, 4(12), 127–135. https://doi.org/10.1098/rsif.2006.0166.

(Jazinizadeh et al., 2020), impact testing at high strain rates has been shown to more greatly affect impact energy absorption, when quasi-static tests show little variation in test results (Currey et al., 1995). One limitation of impact studies on bone is that they tend to provide results that are more variable than those from quasi-static tests (Currey et al., 1995; Currey et al., 2007) (Table 24.1).

Regardless of this inherent variability, it is still better to simulate the actual type of loading that results in a fracture at a particular location when possible. Impacts from falls generate multiaxial and off-axis loading that contribute to hip and radial fractures. Therefore, the challenges in studying and fully understanding these fractures in vivo are dependent on specificities related to direction of load/impact (Abe et al., 2022; Augat, Bliven and Hackl, 2019; Levine et al., 2018), anatomical location (Abe et al., 2022; de Bakker et al., 2009), and variations in skeletal geometry (Levine et al., 2018). In turn, these conditions also are affected by age. It has been shown that hip fractures from falls generally begin in the superior cortex of the femoral neck (de Bakker et al., 2009; Verhulp et al., 2008), which is only about 20% as thick at age 85 as at age 25 (Poole et al., 2010). These changes increase stresses on the supero-lateral femoral neck where fractures initiate and increase the risk of fracture from falls (Bell et al., 1999; Duboeuf et al., 1997; Mayhew et al., 2005).

Beyond this, Currey (1979) showed that bone's impact energy absorption generally declines with age in cortical bone and this decline is not attributable to increased cortical porosity. Currey attributed this to changes in mineralization, but it is equally (or more) likely that it is a combination of mineralization, changes to collagen (e.g. greater accumulation of advanced glycation end products that can make bone less ductile), and loss of water, all of which occur with age and alter both elastic and plastic mechanical properties (Burr, 2019). As the Young's modulus of bone increases, the amount of energy that can be absorbed on impact declines (Currey, 2004; Currey et al., 2004; Currey et al., 1995).

Because of the complexities inherent in experimentally controlling fall direction, location of fracture initiation, and variations in bone geometry, the role of impact in hip fractures has often been estimated using finite element analyses (FEA) (Abe et al., 2018, 2022). This technique requires estimation of material properties, including heterogeneity and anisotropy inherent in both material and structural components of the hip. FEA provides an initial estimate of which factors may be more important, but the specific model must be validated through experimental studies. Experimental validation of FE models could be facilitated by the development of appropriate animal models, but this is complicated by the different geometry of the hip complex in animals compared with humans. Dogs and rabbits tend to have short femoral necks with thick cortices and may not be appropriate as models. Morphologically closest to humans are nonhuman primates, and even the smaller monkeys such as cynomolgus monkeys, have femoral neck structures, but not pelvic structures, similar to humans. One area of potential progress would be to identify/develop animal models that could be used to assess effects of impact forces on the hip in different directions, and at different impact rates. This could not only provide direct evidence of the types of impacts that place the femoral neck at greatest risk, but would also allow the collection of data that could verify properties used in FEA analyses of impact dynamics.

It is possible, however, to use certain types of experimental approaches in humans. For instance, de Bakker et al. (2009) used high speed video capture to simulate hip fractures. Levine et al. subjected young human females in their third decade to low-energy impact falls to estimate the magnitude and distribution of impact force and bending moments on the proximal femur. Yang et al. (2016) used high speed video to capture fall directions, methods of reducing impact such as hand impact, and balance in subjects in long-term care. Unfortunately, they were not able to measure the dynamics of the fall inherent in impact velocity, force, energy, and the distribution of pressure, force and energy. These factors could be incorporated experimentally in assessing falls using a younger cohort. Therefore, another area of potentially fruitful investigation is to investigate impact forces in vivo in humans, much like Levine et al. (2018) but using high speed video capture. As we develop a better idea of which directions older people often fall, and which directions are more likely to cause fracture, studies of forces specifically in these fall patterns could be performed using younger populations in which the risk of hip fracture is lower (or nonexistent, in an experimental setting). These data could certainly be applied to understanding hip fractures in the older population, and developing countermeasures to train people in fall patterns that would reduce falling, or to prevent fracture when a fall does occur.

7.3 The next big challenge: Integrating different hierarchical levels for a holistic view of bone's mechanical properties

This chapter has highlighted several new (and old) techniques that are used to characterize architectural, compositional, and mechanical properties of bone across hierarchical length scales. None of these are sufficient in isolation, and a complementary set of approaches is needed. This raises the important and unresolved question of how we relate any of these properties to whole bone mechanical integrity, which is of greatest importance in understanding how to prevent bone fracture. It also raises the question of how to integrate across these hierarchical levels of bone – a problem that has not yet been solved. This integration may be the next big challenge to fully understand bone's mechanical properties and its response to loading.

References

Abe, S., Kouhia, R., Nikander, R., Narra, N., Hyttinen, J., & Sievänen, H. (2022). Effect of fall direction on the lower hip fracture risk in athletes with different loading histories: A finite element modeling study in multiple sideways fall configurations. *Bone, 158*, Article 116351. https://doi.org/10.1016/j.bone.2022.116351

Abe, S., Narra, N., Nikander, R., Hyttinen, J., Kouhia, R., & Sievänen, H. (2018). Impact loading history modulates hip fracture load and location: A finite element simulation study of the proximal femur in female athletes. *Journal of Biomechanics, 76*, 136–143. https://doi.org/10.1016/j.jbiomech.2018.05.037

Augat, P., Bliven, E., & Simon, H. (2019). Biomechanics of Femoral Neck Fractures and Implications for Fixation. *Journal of Orthopaedic Trauma, 33*(1), Article S27. https://doi.org/10.1097/bot.0000000000001365

Bab, I., Gazit, D., Massarawa, A., & Sela, J. (1985). Removal of tibial marrow induces increased formation of bone and cartilage in rat mandibular condyle. *Calcified Tissue International, 37*(5), 551–555. https://doi.org/10.1007/bf02557840

Bell, K. L., Loveridge, N., Power, J., Garrahan, N., Stanton, M., Lunt, M., Meggitt, B. F., & Reeve, J. (1999). Structure of the femoral neck in hip fracture: Cortical bone loss in the inferoanterior to superoposterior axis. *Journal of Bone and Mineral Research, 14*(1), 111–119. https://doi.org/10.1359/jbmr.1999.14.1.111

Berman, A. G., Organ, J. M., Allen, M. R., & Wallace, J. M. (2020). Muscle contraction induces osteogenic levels of cortical bone strain despite muscle weakness in a mouse model of Osteogenesis Imperfecta. *Bone, 132*. https://doi.org/10.1016/j.bone.2019.115061

Boskey, A. (2003). Bone mineral crystal size. *Osteoporosis International, 14*(0), 16–21. https://doi.org/10.1007/s00198-003-1468-2

Burr, D. B. (2019). Changes in bone matrix properties with aging. *Bone, 120*, 85–93. https://doi.org/10.1016/j.bone.2018.10.010

Burr, D. B., Milgrom, C., Fyhrie, D., Forwood, M., Nyska, M., Finestone, A., Hoshaw, S., Saiag, E., & Simkin, A. (1996). In vivo measurement of human tibial strains during vigorous activity. *Bone, 18*(5), 405–410. https://doi.org/10.1016/8756-3282(96)00028-2

Carriero, A., Zimmermann, E. A., Paluszny, A., Tang, S. Y., Bale, H., Busse, B., Alliston, T., Kazakia, G., Ritchie, R. O., & Shefelbine, S. J. (2014). How tough is brittle bone? Investigating osteogenesis imperfecta in mouse bone. *Journal of Bone and Mineral Research, 29*(6), 1392–1401. https://doi.org/10.1002/jbmr.2172

Carter, D. R., & Hayes, W. C. (1977). Compact bone fatigue damage—I. Residual strength and stiffness. *Journal of Biomechanics, 10*(5–6), 325–337. https://doi.org/10.1016/0021-9290(77)90005-7

Clansey, A. C., Hanlon, M., Wallace, E. S., & Lake, M. J. (2012). Effects of fatigue on running mechanics associated with tibial stress fracture risk. *Medicine & Science in Sports & Exercise, 44*(10), 1917–1923. https://doi.org/10.1249/MSS.0b013e318259480d

Courtney, A. C., Wachtel, E. F., Myers, E. R., & Hayes, W. C. (1994). Effects of loading rate on strength of the proximal femur. *Calcified Tissue International, 55*(1), 53–58. https://doi.org/10.1007/BF00310169

Currey, J. D. (1962). Stress concentrations in bone. *Journal of Cell Science, S3–103*(61), 111–133. https://doi.org/10.1242/jcs.s3-103.61.111

Currey, J. D. (1979). Changes in the impact energy absorption of bone with age. *Journal of Biomechanics, 12*(6), 459–469. https://doi.org/10.1016/0021-9290(79)90031-9

Currey, J. (2004). Incompatible mechanical properties in compact bone. *Journal of Theoretical Biology, 231*(4), 569–580. https://doi.org/10.1016/j.jtbi.2004.07.013

Currey, J. D., Brear, K., & Zioupos, P. (2004). Notch sensitivity of mammalian mineralized tissues in impact. *Proceedings of the Royal Society B: Biological Sciences, 271*(1538), 517–522. https://doi.org/10.1098/rspb.2003.2634

Currey, J. D., Brear, K., Zioupos, P., & Reilly, G. C. (1995). Effect of formaldehyde fixation on some mechanical properties of bovine bone. *Biomaterials, 16*(16), 1267–1271. https://doi.org/10.1016/0142-9612(95)98135-2

Currey, J. D., Pitchford, J. W., & Baxter, P. D. (2007). Variability of the mechanical properties of bone, and its evolutionary consequences. *Journal of the Royal Society Interface, 4*(12), 127–135. https://doi.org/10.1098/rsif.2006.0166

Dallas, S. L., Xie, Y., Shiflett, L. A., & Ueki, Y. (2018). Mouse Cre models for the study of bone diseases. *Current Osteoporosis Reports, 16*(4), 466–477. https://doi.org/10.1007/s11914-018-0455-7

de Bakker, P. M., Manske, S. L., Ebacher, V., Oxland, T. R., Cripton, P. A., & Guy, P. (2009). During sideways falls proximal femur fractures initiate in the superolateral cortex: Evidence from high-speed video of simulated fractures. *Journal of Biomechanics, 42*(12), 1917–1925. https://doi.org/10.1016/j.jbiomech.2009.05.001

Dolbow, D. R., Holcomb, W. R., & Gorgey, A. S. (2014). Improving the efficiency of electrical stimulation activities after spinal cord injury. *Current Physical Medicine and Rehabilitation Reports, 2*(3), 169–175. https://doi.org/10.1007/s40141-014-0053-2

Duboeuf, F., Hans, D., Schott, A. M., Kotzki, P. O., Favier, F., Marcelli, C., Meunier, P. J., & Delmas, P. D. (1997). Different morphometric and densitometric parameters predict cervical and trochanteric hip fracture: The EPIDOS study. *Journal of Bone and Mineral Research, 12*(11), 1895–1902. https://doi.org/10.1359/jbmr.1997.12.11.1895

Dorrian, R. M., Berryman, C. F., Lauto, A., & Leonard, A. V. (2023). Electrical stimulation for the treatment of spinal cord injuries: A review of the cellular and molecular mechanisms that drive functional improvements. *Frontiers in Cellular Neuroscience, 17*. https://doi.org/10.3389/fncel.2023.1095259

Fatihhi, S. J., Harun, M. N., Abdul Kadir, M. R., Abdullah, J.r, Kamarul, T., Öchsner, A., & Syahrom, A. (2015). Uniaxial and multiaxial fatigue life prediction of the trabecular bone based on physiological loading: A comparative study. *Annals of Biomedical Engineering, 43*(10), 2487–2502. https://doi.org/10.1007/s10439-015-1305-8

Fyhrie, D. P., & Christiansen, B. A. (2015). Bone material properties and skeletal fragility. *Calcified Tissue International, 97*(3), 213–228. https://doi.org/10.1007/s00223-015-9997-1

George, WT, Debphadhaya, Stephen, SJ Multiaxial cyclic loading induces fatigue fractures of human bone. Bone.

George, W. T., & Vashishth, D. (2006). Susceptibility of aging human bone to mixed-mode fracture increases bone fragility. *Bone, 38*(1), 105–111. https://doi.org/10.1016/j.bone.2005.08.002

Gilsanz, V., Wren, T. A. L., Sanchez, M., Dorey, F., Judex, S., & Rubin, C. (2006). Low-Level, high-frequency mechanical signals enhance musculoskeletal development of young women with low BMD. *Journal of Bone and Mineral Research, 21*(9), 1464–1474. https://doi.org/10.1359/jbmr.060612

Goodman, C. A., Hornberger, T. A., & Robling, A. G. (2015). Bone and skeletal muscle: Key players in mechanotransduction and potential overlapping mechanisms. *Bone, 80*, 24–36. https://doi.org/10.1016/j.bone.2015.04.014

Grisso, J. A., Strom, B. L., Ghiu, G. Y., Maislin, G., O'brien, L. A., Kelsey, J. L., Hoffman, S., Kaplan, F., Junkin, D., Good, R., Ecker, M., DeLong, W., Hummer, C., Wohl, M., Stein, H., Anthony, J., Markman, W., Mooar, P., Cotler, J., & Habermann, E. (1991). Risk factors for falls as a cause of hip fracture in women. *New England Journal of Medicine, 324*(19), 1326–1331. https://doi.org/10.1056/NEJM199105093241905

Gupta, H. S., Seto, J., Wagermaier, W., Zaslansky, P., Boesecke, P., & Fratzl, P. (2006). Cooperative deformation of mineral and collagen in bone at the nanoscale. *Proceedings of the National Academy of Sciences of the United States of America, 103*(47), 17741–17746. https://doi.org/10.1073/pnas.0604237103

Hart, N. H., Nimphius, S., Rantalainen, T., Ireland, A., Siafarikas, A., & Newton, R. U. (2017). Mechanical basis of bone strength: Influence of bone material, bone structure and muscle action. *Australia Journal of Musculoskeletal Neuronal Interactions, 17*(3), 114–139.

Hartmann, M. A., Blouin, S., Misof, B. M., Fratzl-Zelman, N., Roschger, P., Berzlanovich, A., Gruber, G. M., Brugger, P. C., Zwerina, J., & Fratzl, P. (2021). Quantitative backscattered electron imaging of bone using a thermionic or a field emission electron source. *Calcified Tissue International, 109*(2), 190–202. https://doi.org/10.1007/s00223-021-00832-5

Herrmann, M., Engelke, K., Ebert, R., Müller-Deubert, S., Rudert, M., Ziouti, F., Jundt, F., Felsenberg, D., & Jakob, F. (2020). Interactions between muscle and bone—where physics meets biology. *Biomolecules, 10*(3), 432. https://doi.org/10.3390/biom10030432

Hui, S. L., Slemenda, C. W., & Johnston, C. C. (1988). Age and bone mass as predictors of fracture in a prospective study. *Journal of Clinical Investigation, 81*(6), 1804–1809. https://doi.org/10.1172/JCI113523

Jazinizadeh, F., Mohammadi, H., & Quenneville, C. E. (2020). Comparing the fracture limits of the proximal femur under impact and quasi-static conditions in simulation of a sideways fall. *Journal of the Mechanical Behavior of Biomedical Materials, 103*, Article 103593. https://doi.org/10.1016/j.jmbbm.2019.103593

Kanis, J. A., Johnell, O., Oden, A., De Laet, C., Dawson, A., & Jonsson, B. (2001). Ten year probabilities of osteoporotic fractures according to BMD and diagnostic thresholds. *Osteoporosis International, 12*(12), 989–995. https://doi.org/10.1007/s001980170006

Levine, I. C., Pretty, S. P., Nouri, P. K., Mourtzakis, M., & Laing, A. C. (2018). Pelvis and femur geometry: Relationships with impact characteristics during sideways falls on the hip. *Journal of Biomechanics, 80*, 72–78. https://doi.org/10.1016/j.jbiomech.2018.08.029

Managuli, V., Bothra, Y. S., Kumar, S. S., Gaur, P., & Chandracharya, P. L. (2023). Overview of mechanical characterization of bone using nanoindentation technique and its applications. *Engineered Science, 22*. https://doi.org/10.30919/es8d820

Marquez-Chin, C., & Popovic, M. R. (2020). Functional electrical stimulation therapy for restoration of motor function after spinal cord injury and stroke: A review. *BioMedical Engineering Online, 19*(1). https://doi.org/10.1186/s12938-020-00773-4

Mayhew, P. M., Thomas, C. D., Clement, J. G., Loveridge, N., Beck, T. J., Bonfield, W., Burgoyne, C. J., & Reeve, J. (2005). Relation between age, femoral neck cortical stability, and hip fracture risk. *Lancet, 366*(9480), 129–135. https://doi.org/10.1016/S0140-6736(05)66870-5

Milgrom, C., Radeva-Petrova, D. R., Finestone, A., Nyska, M., Mendelson, S., Benjuya, N., Simkin, A., & Burr, D. (2007). The effect of muscle fatigue on in vivo tibial strains. *Journal of Biomechanics, 40*(4), 845–850. https://doi.org/10.1016/j.jbiomech.2006.03.006

Mirzaali, M. J., Bürki, A., Schwiedrzik, J., Zysset, P. K., & Wolfram, U. (2015). Continuum damage interactions between tension and compression in osteonal bone. *Journal of the Mechanical Behavior of Biomedical Materials, 49*, 355–369. https://doi.org/10.1016/j.jmbbm.2015.05.007

Parkkari, J., Kannus, P., Palvanen, M., Natri, A., Vainio, J., Aho, H., Vuori, I., & Järvinen, M. (1999). Majority of hip fractures occur as a result of a fall and impact on the greater Trochanter of the Femur: A prospective controlled hip fracture study with 206 consecutive patients. *Calcified Tissue International, 65*(3), 183–187. https://doi.org/10.1007/s002239900679

Peckham, P. H., & Knutson, J. S. (2005). Functional electrical stimulation for neuromuscular applications. *Annual Review of Biomedical Engineering, 7*, 327–360. https://doi.org/10.1146/annurev.bioeng.6.040803.140103

Poole, K. E. S., Mayhew, P. M., Rose, C. M., Brown, J. K., Bearcroft, P. J., Loveridge, N., & Reeve, J. (2010). Changing structure of the femoral neck across the adult female lifespan. *Journal of Bone and Mineral Research, 25*(3), 482–491. https://doi.org/10.1359/jbmr.090734

Puxkandl, R., Zizak, I., Paris, O., Keckes, J., Tesch, W., Bernstorff, S., Purslow, P., & Fratzl, P. (2002). Viscoelastic properties of collagen: Synchrotron radiation investigations and structural model. *Philosophical Transactions of the Royal Society B: Biological Sciences, 357*(1418), 191–197. https://doi.org/10.1098/rstb.2001.1033

Quan, B. D., & Sone, E. D. (2013). Cryo-TEM analysis of collagen fibrillar structure. *Methods in Enzymology, 532*, 189–205. https://doi.org/10.1016/B978-0-12-416617-2.00009-6

Rajapakse, C. S., Johncola, A. J., Batzdorf, A. S., Jones, B. C., Al Mukaddam, M., Sexton, K., Shults, J., Leonard, M. B., Snyder, P. J., & Wehrli, F. W. (2021). Effect of low-intensity vibration on bone strength, microstructure, and adiposity in pre-osteoporotic postmenopausal women: A randomized placebo-controlled trial. *Journal of Bone and Mineral Research, 36*(4), 673–684. https://doi.org/10.1002/jbmr.4229

Reilly, G. C., & Currey, J. D. (2000). The effects of damage and microcracking on the impact strength of bone. *Journal of Biomechanics, 33*(3), 337–343. https://doi.org/10.1016/S0021-9290(99)00167-0

Ritchie, R. O. (2011). The conflicts between strength and toughness. *Nature Materials, 10*(11), 817–822. https://doi.org/10.1038/nmat3115

Ritchie, R. O., Koester, K. J., Ionova, S., Yao, W., Lane, N. E., & Ager, J. W. (2008). Measurement of the toughness of bone: A tutorial with special reference to small animal studies. *Bone, 43*(5), 798–812. https://doi.org/10.1016/j.bone.2008.04.027

Robling, A. G., Daly, R., Fuchs, R. K., & Burr, D. B. (2019). *Mechanical adaptation*. Elsevier BV. https://doi.org/10.1016/b978-0-12-813259-3.00011-7

Saha, S., & Hayes, W. C. (1976). Tensile impact properties of human compact bone. *Journal of Biomechanics, 9*(4), 243–251. https://doi.org/10.1016/0021-9290(76)90010-5

Schaffler, M. B., Radin, E. L., & Burr, D. B. (1989). Mechanical and morphological effects of strain rate on fatigue of compact bone. *Bone, 10*(3), 207–214. https://doi.org/10.1016/8756-3282(89)90055-0

Schaffler, M. B., Radin, E. L., & Burr, D. B. (1990). Long-term fatigue behavior of compact bone at low strain magnitude and rate. *Bone, 11*(5), 321–326. https://doi.org/10.1016/8756-3282(90)90087-f

Scimeca, M., Bischetti, S., Lamsira, H. K., Bonfiglio, R., & Bonanno, E. (2018). Energy dispersive X-ray (EDX) microanalysis: A powerful tool in biomedical research and diagnosis. *European Journal of Histochemistry, 62*(1), 89–99. https://doi.org/10.4081/ejh.2018.2841

Scott, L. W., & Elliott, S. (2010). Falls & related injuries among older Canadians: Fall-related hospitalizations & intervention initiatives. *Prepared on behalf of the Public Health Agency of Canada. Division of Aging and Seniors.*

Skorupa, M. (1998). Load interaction effects during fatigue crack growth under variable amplitude loading—a literature review. Part i: Empirical trends. *Fatigue & Fracture of Engineering Materials & Structures, 21*(8), 987–1006. https://doi.org/10.1046/j.1460-2695.1998.00083.x

Skorupa, M. (1999). Load interaction effects during fatigue crack growth under variable amplitude loading—a literature review. Part II: Qualitative interpretation. *Fatigue and Fracture of Engineering Materials and Structures, 22*(10), 905–926. https://doi.org/10.1046/j.1460-2695.1999.00158.x

Spain, L., Yang, L., Mark Wilkinson, J., & McCloskey, E. (2021). Transmission of whole body vibration – Comparison of three vibration platforms in healthy subjects. *Bone, 144*, Article 115802. https://doi.org/10.1016/j.bone.2020.115802

Stock, S. R. (2015). The mineral–collagen interface in bone. *Calcified Tissue International, 1*, 1–19.

Taaffe, D. R., Robinson, T. L., Snow, C. M., & Marcus, R. (1997). High-impact exercise promotes bone gain in well-trained female athletes. *Journal of Bone and Mineral Research, 12*(2), 255–260. https://doi.org/10.1359/jbmr.1997.12.2.255

Taylor, E. A., & Donnelly, E. (2020). Raman and Fourier transform infrared imaging for characterization of bone material properties. *Bone, 139*, Article 115490. https://doi.org/10.1016/j.bone.2020.115490

Taylor, D., O'Reilly, P., Vallet, L., & Lee, T. C. (2003). The fatigue strength of compact bone in torsion. *Journal of Biomechanics, 36*(8), 1103–1109. https://doi.org/10.1016/s0021-9290(03)00104-0

Thurner, P. J. (2009). Atomic force microscopy and indentation force measurement of bone. *WIREs Nanomedicine and Nanobiotechnology, 1*(6), 624–649. https://doi.org/10.1002/wnan.56

Turner, C. H. (2006). Bone strength: Current concepts. *Annals of the New York Academy of Sciences, 1068*(1), 429–446. https://doi.org/10.1196/annals.1346.039

Turner, C. H., & Burr, D. B. (1993). Basic biomechanical measurements of bone: A tutorial. *Bone, 14*(4), 595–608. https://doi.org/10.1016/8756-3282(93)90081-k

Vahidi, G., Rux, C., Sherk, V. D., & Heveran, C. M. (2021). Lacunar-canalicular bone remodeling: Impacts on bone quality and tools for assessment. *Bone, 143*, Article 115663. https://doi.org/10.1016/j.bone.2020.115663

Vashishth, D., Tanner, K. E., & Bonfield, W. (2001). Fatigue of cortical bone under combined axial-torsional loading. *Journal of Orthopaedic Research, 19*(3), 414–420. https://doi.org/10.1016/s0736-0266(00)00036-x

Verhulp, E., van Rietbergen, B., & Huiskes, R. (2008). Load distribution in the healthy and osteoporotic human proximal femur during a fall to the side. *Bone, 42*(1), 30–35. https://doi.org/10.1016/j.bone.2007.08.039

Wagermaier, W., Klaushofer, K., & Fratzl, P. (2015). Fragility of bone material controlled by internal interfaces. *Calcified Tissue International, 97*(3), 201–212. https://doi.org/10.1007/s00223-015-9978-4

Wallace, J. M. (2012). Applications of atomic force microscopy for the assessment of nanoscale morphological and mechanical properties of bone. *Bone, 50*(1), 420–427. https://doi.org/10.1016/j.bone.2011.11.008

Yan, C., Song, H., Pfister, J., Andersen, T. L., Warden, S. J., Bhargava, R., & Kersh, M. E. (2021). Effect of fatigue loading and rest on impact strength of rat ulna. *Journal of Biomechanics, 123*, Article 110449. https://doi.org/10.1016/j.jbiomech.2021.110449

Yang, Y., Mackey, D. C., Liu-Ambrose, T., Feldman, F., & Robinovitch, S. N. (2016). Risk factors for hip impact during real-life falls captured on video in long-term care. *Osteoporosis International, 27*(2), 537–547. https://doi.org/10.1007/s00198-015-3268-x

Yeni, Y. N., & Fyhrie, D. P. (2002). Fatigue damage-fracture mechanics interaction in cortical bone. *Bone, 30*(3), 509–514. https://doi.org/10.1016/s8756-3282(01)00696-2

Yoshikawa, T., Mori, S., Santiesteban, A. J., Sun, T. C., Hafstad, E., Chen, J., & Burr, D. B. (1994). The effects of muscle fatigue on bone strain. *Journal of Experimental Biology, 188*, 217–233.

You, L., Cowin, S. C., Schaffler, M. B., & Weinbaum, S. (2001). A model for strain amplification in the actin cytoskeleton of osteocytes due to fluid drag on pericellular matrix. *Journal of Biomechanics, 34*(11), 1375–1386. https://doi.org/10.1016/S0021-9290(01)00107-5

Zhao, W., Peng, Y., Hu, Y., Guo, X. E., Li, J., Cao, J., Pan, J., Feng, J. Q., Cardozo, C., Jarvis, J., Bauman, W. A., & Qin, W. (2021). Electrical stimulation of hindlimb skeletal muscle has beneficial effects on sublesional bone in a rat model of spinal cord injury. *Bone, 144*, Article 115825. https://doi.org/10.1016/j.bone.2020.115825

Zioupos, P., Hansen, U., & Currey, J. D. (2008). Microcracking damage and the fracture process in relation to strain rate in human cortical bone tensile failure. *Journal of Biomechanics, 41*(14), 2932–2939. https://doi.org/10.1016/j.jbiomech.2008.07.025

Zysset, P. K., & Curnier, A. (1996). A 3D damage model for trabecular bone based on fabric tensors. *Journal of Biomechanics, 29*(12), 1549–1558. https://doi.org/10.1016/s0021-9290(96)80006-6

Index

'*Note:* Page numbers followed by "f" indicate figures and "t" indicate tables.'

B

D

E

F

N

O

Q

R

Printed and bound by CPI Group (UK) Ltd, Croydon, CR0 4YY
14/09/2026
02216284-0012